T0207340

An Introduction to
Quantum
Field
Theory

An Introduction to
Quantum
Field
Theory

Michael E. Peskin
Stanford Linear Accelerator Center

Daniel V. Schroeder
Weber State University

THE ADVANCED BOOK PROGRAM

CRC Press
Taylor & Francis Group
Boca Raton London New York

CRC Press is an imprint of the
Taylor & Francis Group, an **informa** business

CRC Press
Taylor & Francis Group
6000 Broken Sound Parkway NW, Suite 300
Boca Raton, FL 33487-2742

First issued in paperback 2019

ISBN-13: 978-0-2015-0397-5 (hbk)
ISBN-13: 978-0-367-32056-0 (pbk)

Visit the Taylor & Francis Web site at
http://www.taylorandfrancis.com

and the CRC Press Web site at
http://www.crcpress.com

Figures 5.2 and 14.2 have been adapted, with permission, from figures in *Physical Review Letters*.

Table 13.1 has been adapted, with permission, from a set of tables in *Physical Review B*.

Table 17.1 and Figures 5.3, 5.7, 7.9, 17.13, 17.23, and 18.8 have been adapted, with permission, from tables/figures in *Physical Review D*.

Figure 17.5 has been adapted, with permission, from a table in *Zeitschrift für Physik C*.

Table 20.1 and Figures 17.22 and 20.5 have been reprinted, with permission, from *Physical Review D*.

In all cases, the specific source is cited in the caption of the figure or table.

Library of Congress Cataloging-in-Publication Data
Peskin, Michael Edward, 1951–
  Introduction to quantum field theory / Michael E. Peskin, Daniel
V. Schroeder.
    p. cm.
  Includes bibliographical references and index.
  ISBN-13 978-0-201-50397-5   ISBN-10 0-201-50397-2
  1. Quantum field theory.   2. Feynman diagrams.   3. Renormalization
(Physics)   4. Gauge fields (Physics)   I. Schroeder, Daniel V.
II. Title.
QC174.45.P465   1995
530. 1 43–dc20
                                                        89-27170
                                                        CIP

Cover art by Michael E. Peskin and Daniel V. Schroeder
Cover design by Lynne Reed
Typeset by Daniel V. Schroeder in 10 pt. Computer Modern using T$_E$Xtures for the Macintosh.

Contents

Part I: Feynman Diagrams and Quantum Electrodynamics

Part II: Renormalization

Part III: Non-Abelian Gauge Theories

Epilogue

*These sections may be omitted in a one-year course emphasizing the less formal aspects of elementary particle physics.

Preface

Quantum field theory is a set of ideas and tools that combines three of the major themes of modern physics: the quantum theory, the field concept, and the principle of relativity. Today, most working physicists need to know some quantum field theory, and many others are curious about it. The theory underlies modern elementary particle physics, and supplies essential tools to nuclear physics, atomic physics, condensed matter physics, and astrophysics. In addition, quantum field theory has led to new bridges between physics and mathematics.

One might think that a subject of such power and widespread application would be complex and difficult. In fact, the central concepts and techniques of quantum field theory are quite simple and intuitive. This is especially true of the many pictorial tools (Feynman diagrams, renormalization group flows, and spaces of symmetry transformations) that are routinely used by quantum field theorists. Admittedly, these tools take time to learn, and tying the subject together with rigorous proofs can become extremely technical. Nevertheless, we feel that the basic concepts and tools of quantum field theory can be made accessible to all physicists, not just an elite group of experts.

A number of earlier books have succeeded in making parts of this subject accessible to students. The best known of these is the two-volume text written in the 1960s by our Stanford colleagues Bjorken and Drell. In our opinion, their text contains an ideal mixture of abstract formalism, intuitive explanations, and practical calculations, all presented with great care and clarity. Since the 1960s, however, the subject of quantum field theory has developed enormously, both in its conceptual framework (the renormalization group, new types of symmetries) and in its areas of application (critical exponents in condensed matter systems, the standard model of elementary particle physics). It is long overdue that a textbook of quantum field theory should appear that provides a complete survey of the subject, including these newer developments, yet still offers the same accessibility and depth of treatment as Bjorken and Drell. We have written this book with that goal in mind.

An Outline of the Book

This textbook is composed of three major sections. The first is mainly concerned with the quantum theory of electromagnetism, which provided the first example of a quantum field theory with direct experimental applications. The third part of the book is mainly concerned with the particular quantum field theories that appear in the standard model of particle interactions. The second part of the book is a bridge between these two subjects; it is intended to introduce some of the very deep concepts of quantum field theory in a context that is as straightforward as possible.

Part I begins with the study of fields with linear equations of motion, that is, fields without interactions. Here we explore the combined implications of quantum mechanics and special relativity, and we learn how particles arise as the quantized excitations of fields. We then introduce interactions among these particles and develop a systematic method of accounting for their effects. After this introduction, we carry out explicit computations in the quantum theory of electromagnetism. These illustrate both the special features of the behavior of electrons and photons and some general aspects of the behavior of interacting quantum fields.

In several of the calculations in Part I, naive methods lead to infinite results. The appearance of infinities is a well-known feature of quantum field theory. At times, it has been offered as evidence for the inconsistency of quantum field theory (though a similar argument could be made against the classical electrodynamics of point particles). For a long time, it was thought sufficient to organize calculations in such a way that no infinities appear in quantities that can be compared directly to experiment. However, one of the major insights of the more recent developments is that these formal infinities actually contain important information that can be used to predict the qualitative behavior of a system. In Part II of the book, we develop this theory of infinities systematically. The development makes use of an analogy between quantum-mechanical and thermal fluctuations, which thus becomes a bridge between quantum field theory and statistical mechanics. At the end of Part II we discuss applications of quantum field theory to the theory of phase transitions in condensed matter systems.

Part III deals with the generalizations of quantum electrodynamics that have led to successful models of the forces between elementary particles. To derive these generalizations, we first analyze and generalize the fundamental symmetry of electrodynamics, then work out the consequences of quantizing a theory with this generalized symmetry. This analysis leads to intricate and quite nontrivial applications of the concepts introduced earlier. We conclude Part III with a presentation of the standard model of particle physics and a discussion of some of its experimental tests.

The Epilogue to the book discusses qualitatively the frontier areas of research in quantum field theory and gives references that can guide a student to the next level of study.

Where a choice of viewpoints is possible, we have generally chosen to explain ideas in language appropriate to the applications to elementary particle physics. This choice reflects our background and research interests. It also reflects our strongly held opinion, even in this age of intellectual relativism, that there is something special about unraveling the behavior of Nature at the deepest possible level. We are proud to take as our subject the structure of the fundamental interactions, and we hope to convey to the reader the grandeur and continuing vitality of this pursuit.

How to Use This Book

This book is an *introduction* to quantum field theory. By this we mean, first and foremost, that we assume no prior knowledge of the subject on the part of the reader. The level of this book should be appropriate for students taking their first course in quantum field theory, typically during the second year of graduate school at universities in the United States. We assume that the student has completed graduate-level courses in classical mechanics, classical electrodynamics, and quantum mechanics. In Part II we also assume some knowledge of statistical mechanics. It is not necessary to have mastered every topic covered in these courses, however. Crucially important prerequisites include the Lagrangian and Hamiltonian formulations of dynamics, the relativistic formulation of electromagnetism using tensor notation, the quantization of the harmonic oscillator using ladder operators, and the theory of scattering in nonrelativistic quantum mechanics. Mathematical prerequisites include an understanding of the rotation group as applied to the quantum mechanics of spin, and some facility with contour integration in the complex plane.

Despite being an "introduction", this book is rather lengthy. To some extent, this is due to the large number of explicit calculations and worked examples in the text. We must admit, however, that the total number of topics covered is also quite large. Even students specializing in elementary particle theory will find that their first research projects require only a part of this material, together with additional, specialized topics that must be gleaned from the research literature. Still, we feel that students who want to become experts in elementary particle theory, and to fully understand its unified view of the fundamental interactions, should master every topic in this book. Students whose main interest is in other fields of physics, or in particle experimentation, may opt for a much shorter "introduction", omitting several chapters.

The senior author of this book once did succeed in "covering" 90% of its content in a one-year lecture course at Stanford University. But this was a mistake; at such a pace, there is not enough time for students of average preparation to absorb the material. Our saner colleagues have found it more reasonable to cover about one Part of the book per *semester*. Thus, in planning a one-year course in quantum field theory, they have chosen either to reserve

Part III for study at a more advanced level or to select about half of the material from Parts II and III, leaving the rest for students to read on their own.

We have designed the book so that it can be followed from cover to cover, introducing all of the major ideas in the field systematically. Alternatively, one can follow an accelerated track that emphasizes the less formal applications to elementary particle physics and is sufficient to prepare students for experimental or phenomenological research in that field. Sections that can be omitted from this accelerated track are marked with an asterisk in the Table of Contents; none of the unmarked sections depend on this more advanced material. Among the unmarked sections, the order could also be varied somewhat: Chapter 10 does not depend on Chapters 8 and 9; Section 11.1 is not needed until just before Chapter 20; and Chapters 20 and 21 are independent of Chapter 17.

Those who wish to study some, but not all, of the more advanced sections should note the following table of dependencies:

Before reading . . .	one should read all of . . .
Chapter 13	Chapters 11, 12
Section 16.6	Chapter 11
Chapter 18	Sections 12.4, 12.5, 16.5
Chapter 19	Sections 9.6, 15.3
Section 19.5	Section 16.6
Section 20.3	Sections 19.1–19.4
Section 21.3	Chapter 11

Within each chapter, the sections marked with an asterisk should be read sequentially, except that Sections 16.5 and 16.6 do not depend on 16.4.

A student whose main interest is in statistical mechanics would want to read the book sequentially, confronting the deep formal issues of Part II but ignoring most of Part III, which is mainly of significance to high-energy phenomena. (However, the material in Chapters 15 and 19, and in Section 20.1, does have beautiful applications in condensed matter physics.)

We emphasize to all students the importance of working actively with the material while studying. It probably is not possible to understand any section of this book without carefully working out the intermediate steps of every derivation. In addition, the problems at the end of each chapter illustrate the general ideas and often apply them in nontrivial, realistic contexts. However, the most illustrative exercises in quantum field theory are too long for ordinary homework problems, being closer to the scale of small research projects. We have provided one of these lengthy problems, broken up into segments with hints and guidance, at the end of each of the three Parts of the book. The volume of time and paper that these problems require will be well invested.

At the beginning of each Part we have included a brief "Invitation" chapter, which previews some of the upcoming ideas and applications. Since these chapters are somewhat easier than the rest of the book, we urge all students to read them.

What This Book is Not

Although we hope that this book will provide a thorough grounding in quantum field theory, it is in no sense a complete education. A dedicated student of physics will want to supplement our treatment in many areas. We summarize the most important of these here.

First of all, this is a book about theoretical methods, not a review of observed phenomena. We do not review the crucial experiments that led to the standard model of elementary particle physics or discuss in detail the more recent experiments that have confirmed its predictions. Similarly, in the chapters that deal with applications to statistical mechanics, we do not discuss the beautiful and varied experiments on phase transitions that led to the confirmation of field theory models. We strongly encourage the student to read, in parallel with this text, a modern presentation of the experimental development of each of these fields.

Although we present the elementary aspects of quantum field theory in full detail, we state some of the more advanced results without proof. For example, it is known rigorously, to all orders in the standard expansion of quantum electrodynamics, that formal infinities can be removed from all experimental predictions. This result, known as *renormalizability*, has important consequences, which we explore in Part II. We do not present the general proof of renormalizability. However, we do demonstrate renormalizability explicitly in illustrative, low-order computations, we discuss intuitively the issues that arise in the complete proof, and we give references to a more complete demonstration. More generally, we have tried to motivate the most important results (usually through explicit examples) while omitting lengthy, purely technical derivations.

Any introductory survey must classify some topics as beyond its scope. Our philosophy has been to include what can be learned about quantum field theory by considering weakly interacting particles and fields, using series expansions in the strength of the interaction. It is amazing how much insight one can obtain in this way. However, this definition of our subject leaves out the theory of bound states, and also phenomena associated with nontrivial solutions to nonlinear field equations. We give a more complete listing of such advanced topics in the Epilogue.

Finally, we have not attempted in this book to give an accurate record of the history of quantum field theory. Students of physics do need to understand the history of physics, for a number of reasons. The most important is to acquire a precise understanding of the experimental basis of the subject. A second important reason is to gain an idea of how science progresses as a human endeavor, how ideas develop as small steps taken by individuals to become the major achievements of the community as a whole.*

In this book we have not addressed either of these needs. Rather, we have included only the kind of mythological history whose purpose is to motivate new ideas and assign names to them. A principle of physics usually has a name that has been assigned according to the community's consensus on who deserves credit for its development. Usually the real credit is only partial, and the true historical development is quite complex. But the clear assignment of names is essential if physicists are to communicate with one another.

Here is one example. In Section 17.5 we discuss a set of equations governing the structure of the proton, which are generally known as the Altarelli-Parisi equations. Our derivation uses a method due to Gribov and Lipatov (GL). The original results of GL were rederived in a more abstract language by Christ, Hasslacher, and Mueller (CHM). After the discovery of the correct fundamental theory of the strong interactions (QCD), Georgi, Politzer, Gross, and Wilczek (GPGW) used the technique of CHM to derive formal equations for the variation of the proton structure. Parisi gave the first of a number of independent derivations that converted these equations into a useful form. The combination of his work with that of GPGW gives the derivation of the equations that we present in Section 18.5. Dokhshitzer later obtained these equations more simply by direct application of the method of GL. Sometime later, but independently, Altarelli and Parisi obtained these equations again by the same route. These last authors also popularized the technique, explaining it very clearly, encouraging experimentalists to use the equations in interpreting their data, and prodding theorists to compute the systematic higher-order corrections to this picture. In Section 17.5 we have presented the shortest path to the end of this convoluted historical road and hung the name 'Altarelli-Parisi' on the final result.

There is a fourth reason for students to read the history of physics: Often the original breakthrough papers, though lacking a textbook's advantages of hindsight, are filled with marvelous personal insights. We strongly encourage students to go back to the original literature whenever possible and see what the creators of the field had in mind. We have tried to aid such students with references provided in footnotes. Though occasionally we refer to papers merely to give credit, most of the references are included because we feel the reader should not miss the special points of view that the authors put forward.

---

*The history of the development of quantum field theory and particle physics has recently been reviewed and debated in a series of conference volumes: *The Birth of Particle Physics*, L. M. Brown and L. Hoddeson, eds. (Cambridge University Press, 1983); *Pions to Quarks*, L. M. Brown, M. Dresden, and L. Hoddeson, eds. (Cambridge University Press, 1989); and *The Rise of the Standard Model*, L. M. Brown, M. Dresden, L. Hoddeson, and M. Riordan, eds. (Cambridge University Press, 1995). The early history of quantum electrodynamics is recounted in a fascinating book by Schweber (1994).

Acknowledgments

The writing of this book would not have been possible without the help and encouragement of our many teachers, colleagues, and friends. It is our great pleasure to thank these people.

Michael has been privileged to learn field theory from three of the subject's contemporary masters—Sidney Coleman, Steven Weinberg, and Kenneth Wilson. He is also indebted to many other teachers, including Sidney Drell, Michael Fisher, Kurt Gottfried, John Kogut, and Howard Georgi, and to numerous co-workers, in particular, Orlando Alvarez, John Preskill, and Edward Witten. His association with the laboratories of high-energy physics at Cornell and Stanford, and discussions with such people as Gary Feldman, Martin Perl, and Morris Swartz, have also shaped his viewpoint. To these people, and to many other people who have taught him points of physics over the years, Michael expresses his gratitude.

Dan has learned field theory from Savas Dimopoulos, Leonard Susskind, Richard Blankenbecler, and many other instructors and colleagues, to whom he extends thanks. For his broader education in physics he is indebted to many teachers, but especially to Thomas Moore. In addition, he is grateful to all the teachers and friends who have criticized his writing over the years.

For their help in the construction and writing of this book, many people are due more specific thanks. This book originated as graduate courses taught at Cornell and at Stanford, and we thank the students in these courses for their patience, criticism, and suggestions. During the years in which this book was written, we were supported by the Stanford Linear Accelerator Center and the Department of Energy, and Dan also by the Pew Charitable Trusts and the Departments of Physics at Pomona College, Grinnell College, and Weber State University. We are especially grateful to Sidney Drell for his support and encouragement. The completion of this book was greatly assisted by a sabbatical Michael spent at the University of California, Santa Cruz, and he thanks Michael Nauenberg and the physics faculty and students there for their gracious hospitality.

We have circulated preliminary drafts of this book widely and have been continually encouraged by the response to these drafts and the many suggestions and corrections they have elicited. Among the people who have helped us in this way are Curtis Callan, Edward Farhi, Jonathan Feng, Donald Finnell, Paul Frampton, Jeffrey Goldstone, Brian Hatfield, Barry Holstein, Gregory Kilcup, Donald Knuth, Barak Kol, Phillip Nelson, Matthias Neubert, Yossi Nir, Arvind Rajaraman, Pierre Ramond, Katsumi Tanaka, Xerxes Tata, Arkady Vainshtein, Rudolf von Bunau, Shimon Yankielowicz, and Eran Yehudai. For help with more specific technical problems or queries, we are grateful to Michael Barnett, Marty Breidenbach, Don Groom, Toichiro Kinoshita, Paul Langacker, Wu-Ki Tung, Peter Zerwas, and Jean Zinn-Justin. We thank our Stanford colleagues James Bjorken, Stanley Brodsky, Lance Dixon, Bryan Lynn, Helen Quinn, Leonard Susskind, and Marvin Weinstein, on whom we have tested many of the explanations given here. We are especially grateful to Michael Dine, Martin Einhorn, Howard Haber, and Renata Kallosh, and to their long-suffering students, who have used drafts of this book as textbooks over a period of several years and have brought us back important and illuminating criticism.

We pause to remember two friends of this book who did not see its completion—Donald Yennie, whose long career made him one of the heroes of Quantum Electrodynamics, and Brian Warr, whose brilliant promise in theoretical physics was cut short by AIDS.

We thank our editors at Addison-Wesley—Allan Wylde, Barbara Holland, Jack Repcheck, Heather Mimnaugh, and Jeffrey Robbins—for their encouragement of this project, and for not losing faith that it would be completed. We are grateful to Mona Zeftel and Lynne Reed for their advice on the formatting of this book, and to Jeffrey Towson and Cristine Jennings for their skillful preparation of the illustrations. The final typeset pages were produced at the WSU Printing Services Department, with much help from Allan Davis. Of course, the production of this book would have been impossible without the many individuals who have kept our computers and network links up and running.

We thank the many friends whose support we have relied on during the long course of this project. We are particularly grateful to our parents, Dorothy and Vernon Schroeder and Gerald and Pearl Peskin. Michael also thanks Valerie, Thomas, and Laura for their love and understanding.

Finally, we thank you, the reader, for your time and effort spent studying this book. Though we have tried to cleanse this text of conceptual and typographical errors, we apologize in advance for those that have slipped through. We will be glad to hear your comments and suggestions for further improvements in the presentation of quantum field theory.

Michael E. Peskin
Daniel V. Schroeder

Notations and Conventions

Units

We will work in "God-given" units, where

$$\hbar = c = 1.$$

In this system,

$$[\text{length}] = [\text{time}] = [\text{energy}]^{-1} = [\text{mass}]^{-1}.$$

The mass (m) of a particle is therefore equal to its rest energy (mc^2), and also to its inverse Compton wavelength $(mc/\hbar)$. For example,

$$m_{\text{electron}} = 9.109 \times 10^{-28}\,\text{g} = 0.511\,\text{MeV} = \left(3.862 \times 10^{-11}\,\text{cm}\right)^{-1}.$$

A selection of other useful numbers and conversion factors is given in the Appendix.

Relativity and Tensors

Our conventions for relativity follow Jackson (1975), Bjorken and Drell (1964, 1965), and nearly all recent field theory texts. We use the metric tensor

$$g_{\mu\nu} = g^{\mu\nu} = \begin{pmatrix} 1 & 0 & 0 & 0 \\ 0 & -1 & 0 & 0 \\ 0 & 0 & -1 & 0 \\ 0 & 0 & 0 & -1 \end{pmatrix},$$

with Greek indices running over 0, 1, 2, 3 or t, x, y, z. Roman indices—i, j, etc.—denote only the three spatial components. Repeated indices are summed in all cases. Four-vectors, like ordinary numbers, are denoted by light italic type; three-vectors are denoted by boldface type; unit three-vectors are denoted by a light italic label with a hat over it. For example,

$$x^\mu = (x^0, \mathbf{x}), \qquad x_\mu = g_{\mu\nu} x^\nu = (x^0, -\mathbf{x});$$

$$p \cdot x = g_{\mu\nu} p^\mu x^\nu = p^0 x^0 - \mathbf{p} \cdot \mathbf{x}.$$

A massive particle has

$$p^2 = p^\mu p_\mu = E^2 - |\mathbf{p}|^2 = m^2.$$

Note that the displacement vector x^μ is "naturally raised", while the derivative operator,

$$\partial_\mu = \frac{\partial}{\partial x^\mu} = \left(\frac{\partial}{\partial x^0}, \nabla\right),$$

is "naturally lowered".

We define the totally antisymmetric tensor $\epsilon^{\mu\nu\rho\sigma}$ so that

$$\epsilon^{0123} = +1.$$

Be careful, since this implies $\epsilon_{0123} = -1$ and $\epsilon^{1230} = -1$. (This convention agrees with Jackson but not with Bjorken and Drell.)

Quantum Mechanics

We will often work with the Schrödinger wavefunctions of single quantum-mechanical particles. We represent the energy and momentum operators acting on such wavefunctions following the usual conventions:

$$E = i\frac{\partial}{\partial x^0}, \qquad \mathbf{p} = -i\nabla.$$

These equations can be combined into

$$p^\mu = i\partial^\mu;$$

raising the index on ∂^μ conveniently accounts for the minus sign. The plane wave $e^{-ik\cdot x}$ has momentum k^μ, since

$$i\partial^\mu\left(e^{-ik\cdot x}\right) = k^\mu\, e^{-ik\cdot x}.$$

The notation 'h.c.' denotes the Hermitian conjugate.

Discussions of spin in quantum mechanics make use of the Pauli sigma matrices:

$$\sigma^1 = \begin{pmatrix} 0 & 1 \\ 1 & 0 \end{pmatrix}, \qquad \sigma^2 = \begin{pmatrix} 0 & -i \\ i & 0 \end{pmatrix}, \qquad \sigma^3 = \begin{pmatrix} 1 & 0 \\ 0 & -1 \end{pmatrix}.$$

Products of these matrices satisfy the identity

$$\sigma^i\sigma^j = \delta^{ij} + i\epsilon^{ijk}\sigma^k.$$

It is convenient to define the linear combinations $\sigma^\pm = \frac{1}{2}(\sigma^1 \pm i\sigma^2)$; then

$$\sigma^+ = \begin{pmatrix} 0 & 1 \\ 0 & 0 \end{pmatrix}, \qquad \sigma^- = \begin{pmatrix} 0 & 0 \\ 1 & 0 \end{pmatrix}.$$

Fourier Transforms and Distributions

We will often make use of the Heaviside step function $\theta(x)$ and the Dirac delta function $\delta(x)$, defined as follows:

$$\theta(x) = \begin{cases} 0 & x < 0, \\ 1 & x > 0; \end{cases} \qquad \delta(x) = \frac{d}{dx}\theta(x).$$

The delta function in n dimensions, denoted $\delta^{(n)}(\mathbf{x})$, is zero everywhere except at $\mathbf{x} = 0$ and satisfies

$$\int d^n x \, \delta^{(n)}(\mathbf{x}) = 1.$$

In Fourier transforms the factors of 2π will always appear with the momentum integral. For example, in four dimensions,

$$f(x) = \int \frac{d^4 k}{(2\pi)^4} \, e^{-ik \cdot x} \, \tilde{f}(k);$$

$$\tilde{f}(k) = \int d^4 x \, e^{ik \cdot x} \, f(x).$$

(In three-dimensional transforms the signs in the exponents will be $+$ and $-$, respectively.) The tilde on $\tilde{f}(k)$ will sometimes be omitted when there is no potential for confusion. The other important factors of 2π to remember appear in the identity

$$\int d^4 x \, e^{ik \cdot x} = (2\pi)^4 \delta^{(4)}(k).$$

Electrodynamics

We use the Heaviside-Lorentz conventions, in which the factors of 4π appear in Coulomb's law and the fine-structure constant rather than in Maxwell's equations. Thus the Coulomb potential of a point charge Q is

$$\Phi = \frac{Q}{4\pi r},$$

and the fine-structure constant is

$$\alpha = \frac{e^2}{4\pi} = \frac{e^2}{4\pi \hbar c} \approx \frac{1}{137}.$$

The symbol e stands for the charge of the electron, a negative quantity (although the sign rarely matters). We generally work with the relativistic form of Maxwell's equations:

$$\epsilon^{\mu\nu\rho\sigma} \partial_\nu F_{\rho\sigma} = 0, \qquad \partial_\mu F^{\mu\nu} = e j^\nu,$$

where

$$A^\mu = (\Phi, \mathbf{A}), \qquad F_{\mu\nu} = \partial_\mu A_\nu - \partial_\nu A_\mu,$$

and we have extracted the e from the 4-vector current density j^μ.

Dirac Equation

Some of our conventions differ from those of Bjorken and Drell (1964, 1965) and other texts: We use a chiral basis for Dirac matrices, and relativistic normalization for Dirac spinors. These conventions are introduced in Sections 3.2 and 3.3, and are summarized in the Appendix.

Editor's Foreword

The problem of communicating in a coherent fashion recent developments in the most exciting and active fields of physics continues to be with us. The enormous growth in the number of physicists has tended to make the familiar channels of communication considerably less effective. It has become increasingly difficult for experts in a given field to keep up with the current literature; the novice can only be confused. What is needed is both a consistent account of a field and the presentation of a definite "point of view" concerning it. Formal monographs cannot meet such a need in a rapidly developing field, while the review article seems to have fallen into disfavor. Indeed, it would seem that the people most actively engaged in developing a given field are the people least likely to write at length about it.

Frontiers in Physics was conceived in 1961 in an effort to improve the situation in several ways. Leading physicists frequently give a series of lectures, a graduate seminar, or a graduate course in their special fields of interest. Such lectures serve to summarize the present status of a rapidly developing field and may well constitute the only coherent account available at the time. Often, notes on lectures exist (prepared by the lecturer, by graduate students, or by postdoctoral fellows) and are distributed in photocopied form on a limited basis. One of the principal purposes of the *Frontiers in Physics* series is to make such notes available to a wider audience of physicists.

As *Frontiers in Physics* has evolved, a second category of book, the informal text/monograph, an intermediate step between lecture notes and formal texts or monographs, has played an increasingly important role in the series. In an informal text or monograph an author has reworked his/her lecture notes to the point at which the manuscript represents a coherent summation of a newly developed field, complete with references and problems, suitable for either classroom teaching or individual study.

During the past two decades significant advances have been made in both the conceptual framework of quantum field theory and its application to condensed matter physics and elementary particle physics. Given the fact that the study of quantum field theory has become an essential part of the education of graduate students in physics, a textbook which makes these recent developments accessible to the novice, while not neglecting the basic concepts, is highly desirable. Michael Peskin and Daniel Schroeder have written just such a book, describing in lucid fashion quantum electrodynamics, renormalization, and non-Abelian gauge theories while offering the reader a taste of what is to come. It is therefore quite appropriate to include this very polished text/monograph in the *Frontiers in Physics* series, and it gives me pleasure to welcome them to the ranks of its authors.

Aspen, Colorado   David Pines
August 1995

Part I

Feynman Diagrams
and
Quantum Electrodynamics

Chapter 1

Invitation: Pair Production in e^+e^- Annihilation

The main purpose of Part I of this book is to develop the basic calculational method of quantum field theory, the formalism of Feynman diagrams. We will then apply this formalism to computations in Quantum Electrodynamics, the quantum theory of electrons and photons.

Quantum Electrodynamics (QED) is perhaps the best fundamental physical theory we have. The theory is formulated as a set of simple equations (Maxwell's equations and the Dirac equation) whose form is essentially determined by relativistic invariance. The quantum-mechanical solutions of these equations give detailed predictions of electromagnetic phenomena from macroscopic distances down to regions several hundred times smaller than the proton.

Feynman diagrams provide for this elegant theory an equally elegant procedure for calculation: Imagine a process that can be carried out by electrons and photons, draw a diagram, and then use the diagram to write the mathematical form of the quantum-mechanical amplitude for that process to occur.

In this first part of the book we will develop both the theory of QED and the method of Feynman diagrams from the basic principles of quantum mechanics and relativity. Eventually, we will arrive at a point where we can calculate observable quantities that are of great interest in the study of elementary particles. But to reach our goal of deriving this simple calculational method, we must first, unfortunately, make a serious detour into formalism. The three chapters that follow this one are almost completely formal, and the reader might wonder, in the course of this development, where we are going. We would like to partially answer that question in advance by discussing the physics of an especially simple QED process—one sufficiently simple that many of its features follow directly from physical intuition. Of course, this intuitive, bottom-up approach will contain many gaps. In Chapter 5 we will return to this process with the full power of the Feynman diagram formalism. Working from the top down, we will then see all of these difficulties swept away.

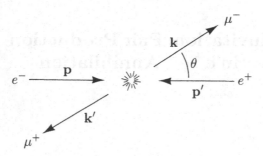

Figure 1.1. The annihilation reaction $e^+e^- \rightarrow \mu^+\mu^-$, shown in the center-of-mass frame.

The Simplest Situation

Since most particle physics experiments involve scattering, the most commonly calculated quantities in quantum field theory are scattering cross sections. We will now calculate the cross section for the simplest of all QED processes: the annihilation of an electron with its antiparticle, a positron, to form a pair of heavier leptons (such as muons). The existence of antiparticles is actually a prediction of quantum field theory, as we will discuss in Chapters 2 and 3. For the moment, though, we take their existence as given.

An experiment to measure this annihilation probability would proceed by firing a beam of electrons at a beam of positrons. The measurable quantity is the cross section for the reaction $e^+e^- \rightarrow \mu^+\mu^-$ as a function of the center-of-mass energy and the relative angle θ between the incoming electrons and the outgoing muons. The process is illustrated in Fig. 1.1. For simplicity, we work in the center-of-mass (CM) frame where the momenta satisfy $\mathbf{p}' = -\mathbf{p}$ and $\mathbf{k}' = -\mathbf{k}$. We also assume that the beam energy E is much greater than either the electron or the muon mass, so that $|\mathbf{p}| = |\mathbf{p}'| = |\mathbf{k}| = |\mathbf{k}'| = E \equiv E_{cm}/2$. (We use boldface type to denote 3-vectors and ordinary italic type to denote 4-vectors.)

Since both the electron and the muon have spin $1/2$, we must specify their spin orientations. It is useful to take the axis that defines the spin quantization of each particle to be in the direction of its motion; each particle can then have its spin polarized parallel or antiparallel to this axis. In practice, electron and positron beams are often unpolarized, and muon detectors are normally blind to the muon polarization. Hence we should average the cross section over electron and positron spin orientations, and sum the cross section over muon spin orientations.

For any given set of spin orientations, it is conventional to write the differential cross section for our process, with the μ^- produced into a solid angle $d\Omega$, as

$$\frac{d\sigma}{d\Omega} = \frac{1}{64\pi^2 E_{cm}^2} \cdot |\mathcal{M}|^2 \,. \tag{1.1}$$

The factor E_{cm}^{-2} provides the correct dimensions for a cross section, since in our units $(\text{energy})^{-2} \sim (\text{length})^2$. The quantity $\mathcal{M}$ is therefore dimensionless; it is the quantum-mechanical amplitude for the process to occur (analogous to the scattering amplitude f in nonrelativistic quantum mechanics), and we must now address the question of how to compute it from fundamental theory. The other factors in the expression are purely a matter of convention. Equation (1.1) is actually a special case, valid for CM scattering when the final state contains two massless particles, of a more general formula (whose form cannot be deduced from dimensional analysis) which we will derive in Section 4.5.

Now comes some bad news and some good news.

The bad news is that even for this simplest of QED processes, the exact expression for $\mathcal{M}$ is not known. Actually this fact should come as no surprise, since even in nonrelativistic quantum mechanics, scattering problems can rarely be solved exactly. The best we can do is obtain a formal expression for $\mathcal{M}$ as a perturbation series in the strength of the electromagnetic interaction, and evaluate the first few terms in this series.

The good news is that Feynman has invented a beautiful way to organize and visualize the perturbation series: the method of *Feynman diagrams*. Roughly speaking, the diagrams display the flow of electrons and photons during the scattering process. For our particular calculation, the lowest-order term in the perturbation series can be represented by a single diagram, shown in Fig. 1.2. The diagram is made up of three types of components: external lines (representing the four incoming and outgoing particles), internal lines (representing "virtual" particles, in this case one virtual photon), and vertices. It is conventional to use straight lines for fermions and wavy lines for photons. The arrows on the straight lines denote the direction of negative charge flow, not momentum. We assign a 4-momentum vector to each external line, as shown. In this diagram, the momentum q of the one internal line is determined by momentum conservation at either of the vertices: $q = p + p' = k + k'$. We must also associate a spin state (either "up" or "down") with each external fermion.

According to the *Feynman rules*, each diagram can be translated directly into a contribution to $\mathcal{M}$. The rules assign a short algebraic factor to each element of a diagram, and the product of these factors gives the value of the corresponding term in the perturbation series. Getting the resulting expression for $\mathcal{M}$ into a form that is usable, however, can still be nontrivial. We will develop much useful technology for doing such calculations in subsequent chapters. But we do not have that technology yet, so to get an answer to our particular problem we will use some heuristic arguments instead of the actual Feynman rules.

Recall that in quantum-mechanical perturbation theory, a transition amplitude can be computed, to first order, as an expression of the form

$$\langle \text{final state}| \, H_I \, |\text{initial state}\rangle , \qquad (1.2)$$

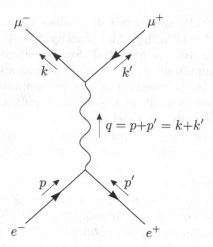

Figure 1.2. Feynman diagram for the lowest-order term in the $e^+e^- \to$ $\mu^+\mu^-$ cross section. At this order the only possible intermediate state is a photon (γ).

where H_I is the "interaction" part of the Hamiltonian. In our case the initial state is $|e^+e^-\rangle$ and the final state is $\langle\mu^+\mu^-|$. But our interaction Hamiltonian couples electrons to muons only through the electromagnetic field (that is, photons), not directly. So the first-order result (1.2) vanishes, and we must go to the second-order expression

$$\mathcal{M} \sim \langle\mu^+\mu^-|\,H_I\,|\gamma\rangle^\mu\,\langle\gamma|\,H_I\,|e^+e^-\rangle_\mu. \tag{1.3}$$

This is a heuristic way of writing the contribution to $\mathcal{M}$ from the diagram in Fig. 1.2. The external electron lines correspond to the factor $|e^+e^-\rangle$; the external muon lines correspond to $\langle\mu^+\mu^-|$. The vertices correspond to H_I, and the internal photon line corresponds to the operator $|\gamma\rangle\,\langle\gamma|$. We have added vector indices (μ) because the photon is a vector particle with four components. There are four possible intermediate states, one for each component, and according to the rules of perturbation theory we must sum over intermediate states. Note that since the sum in (1.3) takes the form of a 4-vector dot product, the amplitude $\mathcal{M}$ will be a Lorentz-invariant scalar as long as each half of (1.3) is a 4-vector.

Let us try to guess the form of the vector $\langle\gamma|\,H_I\,|e^+e^-\rangle_\mu$. Since H_I couples electrons to photons with a strength e (the electron charge), the matrix element should be proportional to e. Now consider one particular set of initial and final spin orientations, shown in Fig. 1.3. The electron and muon have spins parallel to their directions of motion; they are "right-handed". The antiparticles, similarly, are "left-handed". The electron and positron spins add up to one unit of angular momentum in the $+z$ direction. Since H_I should conserve angular momentum, the photon to which these particles couple must have the correct polarization vector to give it this same angular momentum:

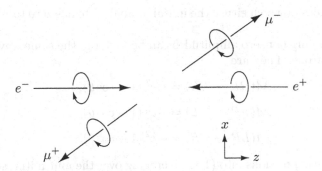

Figure 1.3. One possible set of spin orientations. The electron and the negative muon are right-handed, while the positron and the positive muon are left-handed.

$\epsilon^\mu = (0, 1, i, 0)$. Thus we have

$$\langle \gamma | H_I | e^+ e^- \rangle^\mu \propto e\, (0, 1, i, 0). \tag{1.4}$$

The muon matrix element should, similarly, have a polarization corresponding to one unit of angular momentum along the direction of the μ^- momentum **k**. To obtain the correct vector, rotate (1.4) through an angle θ in the xz-plane:

$$\langle \gamma | H_I | \mu^+ \mu^- \rangle^\mu \propto e\, (0, \cos\theta, i, -\sin\theta). \tag{1.5}$$

To compute the amplitude $\mathcal{M}$, we complex-conjugate this vector and dot it into (1.4). Thus we find, for this set of spin orientations,

$$\mathcal{M}(RL \rightarrow RL) = -e^2\, (1 + \cos\theta)\,. \tag{1.6}$$

Of course we cannot determine the overall factor by this method, but in (1.6) it happens to be correct, thanks to the conventions adopted in (1.1). Note that the amplitude vanishes for $\theta = 180°$, just as one would expect: A state whose angular momentum is in the $+z$ direction has no overlap with a state whose angular momentum is in the $-z$ direction.

Next consider the case in which the electron and positron are both right-handed. Now their total spin angular momentum is zero, and the argument is more subtle. We might expect to obtain a longitudinally polarized photon with a Clebsch-Gordan coefficient of $1/\sqrt{2}$, just as when we add angular momenta in three dimensions, $|\uparrow\downarrow\rangle = (1/\sqrt{2})(|j = 1, m = 0\rangle + |j = 0, m = 0\rangle)$. But we are really adding angular momenta in the four-dimensional Lorentz group, so we must take into account not only spin (the transformation properties of states under rotations), but also the transformation properties of states under boosts. It turns out, as we shall discuss in Chapter 3, that the Clebsch-Gordan coefficient that couples a 4-vector to the state $|e_R^- e_R^+\rangle$ of massless fermions is zero. (For the record, the state is a superposition of scalar and antisymmetric tensor pieces.) Thus the amplitude $\mathcal{M}(RR \rightarrow RL)$ is zero, as are the eleven other amplitudes in which either the initial or final state has zero total angular momentum.

The remaining nonzero amplitudes can be found in the same way that we found the first one. They are

$$\mathcal{M}(RL \to LR) = -e^2 \left(1 - \cos\theta\right),$$

$$\mathcal{M}(LR \to RL) = -e^2 \left(1 - \cos\theta\right), \tag{1.7}$$

$$\mathcal{M}(LR \to LR) = -e^2 \left(1 + \cos\theta\right).$$

Inserting these expressions into (1.1), averaging over the four initial-state spin orientations, and summing over the four final-state spin orientations, we find

$$\frac{d\sigma}{d\Omega} = \frac{\alpha^2}{4E_{\text{cm}}^2}\left(1 + \cos^2\theta\right), \tag{1.8}$$

where $\alpha = e^2/4\pi \simeq 1/137$. Integrating over the angular variables θ and ϕ gives the total cross section,

$$\sigma_{\text{total}} = \frac{4\pi\alpha^2}{3E_{\text{cm}}^2}. \tag{1.9}$$

Results (1.8) and (1.9) agree with experiments to about 10%; almost all of the discrepancy is accounted for by the next term in the perturbation series, corresponding to the diagrams shown in Fig. 1.4. The qualitative features of these expressions—the angular dependence and the sharp decrease with energy—are obvious in the actual data. (The properties of these results are discussed in detail in Section 5.1.)

Embellishments and Questions

We obtained the angular distribution predicted by Quantum Electrodynamics for the reaction $e^+e^- \to \mu^+\mu^-$ by applying angular momentum arguments, with little appeal to the underlying formalism. However, we used the simplifying features of the high-energy limit and the center-of-mass frame in a very strong way. The analysis we have presented will break down when we relax any of our simplifying assumptions. So how does one perform general QED calculations? To answer that question we must return to the Feynman rules.

As mentioned above, the Feynman rules tell us to draw the diagram(s) for the process we are considering, and to associate a short algebraic factor with each piece of each diagram. Figure 1.5 shows the diagram for our reaction, with the various assignments indicated.

For the internal photon line we write $-ig_{\mu\nu}/q^2$, where $g_{\mu\nu}$ is the usual Minkowski metric tensor and q is the 4-momentum of the virtual photon. This factor corresponds to the operator $|\gamma\rangle\langle\gamma|$ in our heuristic expression (1.3).

For each vertex we write $-ie\gamma^\mu$, corresponding to H_I in (1.3). The objects γ^μ are a set of four 4×4 constant matrices. They do the "addition of angular

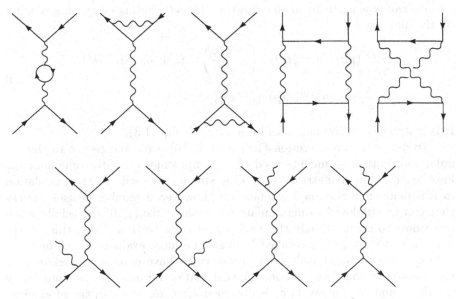

Figure 1.4. Feynman diagrams that contribute to the α^3 term in the $e^+e^- \to \mu^+\mu^-$ cross section.

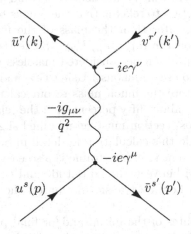

Figure 1.5. Diagram of Fig. 1.2, with expressions corresponding to each vertex, internal line, and external line.

momentum" for us, coupling a state of two spin-1/2 particles to a vector particle.

The external lines carry expressions for four-component column-spinors u, v, or row-spinors $\bar{u}$, $\bar{v}$. These are essentially the momentum-space wavefunctions of the initial and final particles, and correspond to $|e^+e^-\rangle$ and $\langle\mu^+\mu^-|$ in (1.3). The indices s, s', r, and r' denote the spin state, either up or down.

We can now write down an expression for $\mathcal{M}$, reading everything straight off the diagram:

$$
\begin{aligned}
\mathcal{M} &= \bar{v}^{s'}(p')\left(-ie\gamma^\mu\right)u^s(p)\left(\frac{-ig_{\mu\nu}}{q^2}\right)\bar{u}^r(k)\left(-ie\gamma^\nu\right)v^{r'}(k') \\
&= \frac{ie^2}{q^2}\left(\bar{v}^{s'}(p')\gamma^\mu u^s(p)\right)\left(\bar{u}^r(k)\gamma_\mu v^{r'}(k')\right).
\end{aligned}
\tag{1.10}
$$

It is instructive to compare this in detail with Eq. (1.3).

To derive the cross section (1.8) from (1.10), we could return to the angular momentum arguments used above, supplemented with some concrete knowledge about γ matrices and Dirac spinors. We will do the calculation in this manner in Section 5.2. There are, however, a number of useful tricks that can be employed to manipulate expressions like (1.10), especially when one wants to compute only the *unpolarized* cross section. Using this "Feynman trace technology" (so-called because one must evaluate traces of products of γ-matrices), it isn't even necessary to have explicit expressions for the γ-matrices and Dirac spinors. The calculation becomes almost completely mindless, and the answer (1.8) is obtained after less than a page of algebra. But since the Feynman rules and trace technology are so powerful, we can also relax some of our simplifying assumptions. To conclude this section, let us discuss several ways in which our calculation could have been more difficult.

The easiest restriction to relax is that the muons be massless. If the beam energy is not much greater than the mass of the muon, all of our predictions should depend on the ratio m_μ/E_{cm}. (Since the electron is 200 times lighter than the muon, it can be considered massless whenever the beam energy is large enough to create muons.) Using Feynman trace technology, it is extremely easy to restore the muon mass to our calculation. The amount of algebra is increased by about fifty percent, and the relation (1.1) between the amplitude and the cross section must be modified slightly, but the answer is worth the effort. We do this calculation in detail in Section 5.1.

Working in a different reference frame is also easy; the only modification is in the relation (1.1) between the amplitude and the cross section. Or one can simply perform a Lorentz transformation on the CM result, boosting it to a different frame.

When the spin states of the initial and/or final particles are known and we still wish to retain the muon mass, the calculation becomes somewhat cumbersome but no more difficult in principle. The trace technology can be generalized to this case, but it is often easier to evaluate expression (1.10) directly, using the explicit values of the spinors u and v.

Next one could compute cross sections for different processes. The process $e^+e^- \to e^+e^-$, known as *Bhabha scattering*, is more difficult because there is a second allowed diagram (see Fig. 1.6). The amplitudes for the two diagrams must first be added, then squared.

Other processes contain photons in the initial and/or final states. The

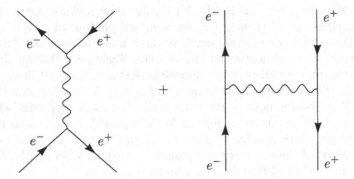

Figure 1.6. The two lowest-order diagrams for Bhabha scattering, $e^+e^- \to e^+e^-$.

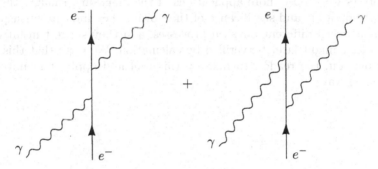

Figure 1.7. The two lowest-order diagrams for Compton scattering.

paradigm example is Compton scattering, for which the two lowest-order diagrams are shown in Fig. 1.7. The Feynman rules for external photon lines and for internal electron lines are no more complicated than those we have already seen. We discuss Compton scattering in detail in Section 5.5.

Finally we could compute higher-order terms in the perturbation series. Thanks to Feynman, the diagrams are at least easy to draw; we have seen those that contribute to the next term in the $e^+e^- \to \mu^+\mu^-$ cross section in Fig. 1.4. Remarkably, the algorithm that assigns algebraic factors to pieces of the diagrams holds for all higher-order contributions, and allows one to evaluate such diagrams in a straightforward, if tedious, way. The computation of the full set of nine diagrams is a serious chore, at the level of a research paper.

In this book, starting in Chapter 6, we will analyze much of the physics that arises from higher-order Feynman diagrams such as those in Fig. 1.4. We will see that the last four of these diagrams, which involve an additional photon in the final state, are necessary because no detector is sensitive enough to notice the presence of extremely low-energy photons. Thus a final state containing such a photon cannot be distinguished from our desired final state of just a muon pair.

The other five diagrams in Fig. 1.4 involve intermediate states of several virtual particles rather than just a single virtual photon. In each of these diagrams there will be one virtual particle whose momentum is not determined by conservation of momentum at the vertices. Since perturbation theory requires us to sum over all possible intermediate states, we must integrate over all possible values of this momentum. At this step, however, a new difficulty appears: The loop-momentum integrals in the first three diagrams, when performed naively, turn out to be infinite. We will provide a fix for this problem, so that we get finite results, by the end of Part I. But the question of the physical origin of these divergences cannot be dismissed so lightly; that will be the main subject of Part II of this book.

We have discussed Feynman diagrams as an algorithm for performing computations. The chapters that follow should amply illustrate the power of this tool. As we expose more applications of the diagrams, though, they begin to take on a life and significance of their own. They indicate unsuspected relations between different physical processes, and they suggest intuitive arguments that might later be verified by calculation. We hope that this book will enable you, the reader, to take up this tool and apply it in novel and enlightening ways.

Chapter 2

The Klein-Gordon Field

2.1 The Necessity of the Field Viewpoint

Quantum field theory is the application of quantum mechanics to dynamical systems of *fields*, in the same sense that the basic course in quantum mechanics is concerned mainly with the quantization of dynamical systems of *particles*. It is a subject that is absolutely essential for understanding the current state of elementary particle physics. With some modification, the methods we will discuss also play a crucial role in the most active areas of atomic, nuclear, and condensed-matter physics. In Part I of this book, however, our primary concern will be with elementary particles, and hence *relativistic* fields.

Given that we wish to understand processes that occur at very small (quantum-mechanical) scales and very large (relativistic) energies, one might still ask why we must study the quantization of *fields*. Why can't we just quantize relativistic particles the way we quantized nonrelativistic particles?

This question can be answered on a number of levels. Perhaps the best approach is to write down a single-particle relativistic wave equation (such as the Klein-Gordon equation or the Dirac equation) and see that it gives rise to negative-energy states and other inconsistencies. Since this discussion usually takes place near the end of a graduate-level quantum mechanics course, we will not repeat it here. It is easy, however, to understand why such an approach cannot work. We have no right to assume that any relativistic process can be explained in terms of a single particle, since the Einstein relation $E = mc^2$ allows for the creation of particle-antiparticle pairs. Even when there is not enough energy for pair creation, multiparticle states appear, for example, as intermediate states in second-order perturbation theory. We can think of such states as existing only for a very short time, according to the uncertainty principle $\Delta E \cdot \Delta t = \hbar$. As we go to higher orders in perturbation theory, arbitrarily many such "virtual" particles can be created.

The necessity of having a multiparticle theory also arises in a less obvious way, from considerations of causality. Consider the amplitude for a free particle to propagate from $\mathbf{x}_0$ to $\mathbf{x}$:

$$U(t) = \langle \mathbf{x} | e^{-iHt} | \mathbf{x}_0 \rangle .$$

In nonrelativistic quantum mechanics we have $E = \mathbf{p}^2/2m$, so

$$U(t) = \langle \mathbf{x} | e^{-i(\mathbf{p}^2/2m)t} | \mathbf{x}_0 \rangle$$

$$= \int \frac{d^3p}{(2\pi)^3} \langle \mathbf{x} | e^{-i(\mathbf{p}^2/2m)t} | \mathbf{p} \rangle \langle \mathbf{p} | \mathbf{x}_0 \rangle$$

$$= \frac{1}{(2\pi)^3} \int d^3p \, e^{-i(\mathbf{p}^2/2m)t} \cdot e^{i\mathbf{p} \cdot (\mathbf{x} - \mathbf{x}_0)}$$

$$= \left(\frac{m}{2\pi i t} \right)^{3/2} e^{im(\mathbf{x} - \mathbf{x}_0)^2/2t}.$$

This expression is nonzero for all x and t, indicating that a particle can propagate between any two points in an arbitrarily short time. In a relativistic theory, this conclusion would signal a violation of causality. One might hope that using the relativistic expression $E = \sqrt{p^2 + m^2}$ would help, but it does not. In analogy with the nonrelativistic case, we have

$$U(t) = \langle \mathbf{x} | e^{-it\sqrt{\mathbf{p}^2 + m^2}} | \mathbf{x}_0 \rangle$$

$$= \frac{1}{(2\pi)^3} \int d^3p \, e^{-it\sqrt{\mathbf{p}^2 + m^2}} \cdot e^{i\mathbf{p} \cdot (\mathbf{x} - \mathbf{x}_0)}$$

$$= \frac{1}{2\pi^2 |\mathbf{x} - \mathbf{x}_0|} \int_0^\infty dp \, p \, \sin(p|\mathbf{x} - \mathbf{x}_0|) e^{-it\sqrt{p^2 + m^2}}.$$

This integral can be evaluated explicitly in terms of Bessel functions.* We will content ourselves with looking at its asymptotic behavior for $x^2 \gg t^2$ (well outside the light-cone), using the method of stationary phase. The phase function $px - t\sqrt{p^2 + m^2}$ has a stationary point at $p = imx/\sqrt{x^2 - t^2}$. We may freely push the contour upward so that it goes through this point. Plugging in this value for p, we find that, up to a rational function of x and t,

$$U(t) \sim e^{-m\sqrt{x^2 - t^2}}.$$

Thus the propagation amplitude is small but nonzero outside the light-cone, and causality is still violated.

Quantum field theory solves the causality problem in a miraculous way, which we will discuss in Section 2.4. We will find that, in the multiparticle field theory, the propagation of a particle across a spacelike interval is indistinguishable from the propagation of an *antiparticle* in the opposite direction (see Fig. 2.1). When we ask whether an observation made at point x_0 can affect an observation made at point x, we will find that the amplitudes for particle and antiparticle propagation exactly cancel—so causality is preserved.

Quantum field theory provides a natural way to handle not only multiparticle states, but also transitions between states of different particle number. It solves the causality problem by introducing antiparticles, then goes on to

---

*See Gradshteyn and Ryzhik (1980), #3.914.

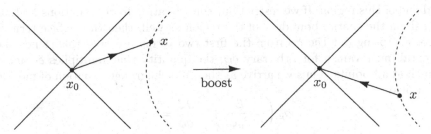

Figure 2.1. Propagation from x_0 to x in one frame looks like propagation from x to x_0 in another frame.

explain the relation between spin and statistics. But most important, it provides the tools necessary to calculate innumerable scattering cross sections, particle lifetimes, and other observable quantities. The experimental confirmation of these predictions, often to an unprecedented level of accuracy, is our real reason for studying quantum field theory.

2.2 Elements of Classical Field Theory

In this section we review some of the formalism of classical field theory that will be necessary in our subsequent discussion of quantum field theory.

Lagrangian Field Theory

The fundamental quantity of classical mechanics is the action, S, the time integral of the Lagrangian, L. In a local field theory the Lagrangian can be written as the spatial integral of a Lagrangian density, denoted by $\mathcal{L}$, which is a function of one or more fields $\phi(x)$ and their derivatives $\partial_\mu \phi$. Thus we have

$$S = \int L\,dt = \int \mathcal{L}(\phi, \partial_\mu \phi)\,d^4x. \tag{2.1}$$

Since this is a book on field theory, we will refer to $\mathcal{L}$ simply as the Lagrangian.

The principle of least action states that when a system evolves from one given configuration to another between times t_1 and t_2, it does so along the "path" in configuration space for which S is an extremum (normally a minimum). We can write this condition as

$$
\begin{aligned}
0 = \delta S \\
= \int d^4x \left\{ \frac{\partial \mathcal{L}}{\partial \phi} \delta\phi + \frac{\partial \mathcal{L}}{\partial(\partial_\mu \phi)} \delta(\partial_\mu \phi) \right\} \\
= \int d^4x \left\{ \frac{\partial \mathcal{L}}{\partial \phi} \delta\phi - \partial_\mu \left(\frac{\partial \mathcal{L}}{\partial(\partial_\mu \phi)} \right) \delta\phi + \partial_\mu \left(\frac{\partial \mathcal{L}}{\partial(\partial_\mu \phi)} \delta\phi \right) \right\}.
\end{aligned}
\tag{2.2}
$$

The last term can be turned into a surface integral over the boundary of the four-dimensional spacetime region of integration. Since the initial and final field configurations are assumed given, $\delta\phi$ is zero at the temporal beginning and end of this region. If we restrict our consideration to deformations $\delta\phi$ that vanish on the spatial boundary of the region as well, then the surface term is zero. Factoring out the $\delta\phi$ from the first two terms, we note that, since the integral must vanish for arbitrary $\delta\phi$, the quantity that multiplies $\delta\phi$ must vanish at all points. Thus we arrive at the Euler-Lagrange equation of motion for a field,

$$\partial_\mu \left(\frac{\partial \mathcal{L}}{\partial(\partial_\mu \phi)} \right) - \frac{\partial \mathcal{L}}{\partial \phi} = 0. \tag{2.3}$$

If the Lagrangian contains more than one field, there is one such equation for each.

Hamiltonian Field Theory

The Lagrangian formulation of field theory is particularly suited to relativistic dynamics because all expressions are explicitly Lorentz invariant. Nevertheless we will use the Hamiltonian formulation throughout the first part of this book, since it will make the transition to quantum mechanics easier. Recall that for a discrete system one can define a conjugate momentum $p \equiv \partial L / \partial \dot{q}$ (where $\dot{q} = \partial q / \partial t$) for each dynamical variable q. The Hamiltonian is then $H \equiv \sum p \dot{q} - L$. The generalization to a continuous system is best understood by pretending that the spatial points $\mathbf{x}$ are discretely spaced. We can define

$$\begin{aligned} p(\mathbf{x}) &\equiv \frac{\partial L}{\partial \dot{\phi}(\mathbf{x})} = \frac{\partial}{\partial \dot{\phi}(\mathbf{x})} \int \mathcal{L}(\phi(\mathbf{y}), \dot{\phi}(\mathbf{y})) \, d^3 y \\ &\sim \frac{\partial}{\partial \dot{\phi}(\mathbf{x})} \sum_{\mathbf{y}} \mathcal{L}(\phi(\mathbf{y}), \dot{\phi}(\mathbf{y})) d^3 y \\ &= \pi(\mathbf{x}) d^3 x, \end{aligned}$$

where

$$\pi(\mathbf{x}) \equiv \frac{\partial \mathcal{L}}{\partial \dot{\phi}(\mathbf{x})} \tag{2.4}$$

is called the *momentum density* conjugate to $\phi(\mathbf{x})$. Thus the Hamiltonian can be written

$$H = \sum_{\mathbf{x}} p(\mathbf{x}) \dot{\phi}(\mathbf{x}) - L.$$

Passing to the continuum, this becomes

$$H = \int d^3 x \left[\pi(\mathbf{x}) \dot{\phi}(\mathbf{x}) - \mathcal{L} \right] \equiv \int d^3 x \, \mathcal{H}. \tag{2.5}$$

We will rederive this expression for the Hamiltonian density $\mathcal{H}$ near the end of this section, using a different method.

As a simple example, consider the theory of a single field $\phi(x)$, governed by the Lagrangian

$$\begin{aligned} \mathcal{L} &= \tfrac{1}{2}\dot{\phi}^2 - \tfrac{1}{2}(\nabla\phi)^2 - \tfrac{1}{2}m^2\phi^2 \\ &= \tfrac{1}{2}(\partial_\mu\phi)^2 - \tfrac{1}{2}m^2\phi^2. \end{aligned} \tag{2.6}$$

For now we take ϕ to be a real-valued field. The quantity m will be interpreted as a mass in Section 2.3, but for now just think of it as a parameter. From this Lagrangian the usual procedure gives the equation of motion

$$\left(\frac{\partial^2}{\partial t^2} - \nabla^2 + m^2\right)\phi = 0 \qquad \text{or} \qquad (\partial^\mu \partial_\mu + m^2)\phi = 0, \tag{2.7}$$

which is the well-known Klein-Gordon equation. (In this context it is a *classical* field equation, like Maxwell's equations—not a quantum-mechanical wave equation.) Noting that the canonical momentum density conjugate to $\phi(x)$ is $\pi(x) = \dot{\phi}(x)$, we can also construct the Hamiltonian:

$$H = \int d^3x \, \mathcal{H} = \int d^3x \left[\tfrac{1}{2}\pi^2 + \tfrac{1}{2}(\nabla\phi)^2 + \tfrac{1}{2}m^2\phi^2\right]. \tag{2.8}$$

We can think of the three terms, respectively, as the energy cost of "moving" in time, the energy cost of "shearing" in space, and the energy cost of having the field around at all. We will investigate this Hamiltonian much further in Sections 2.3 and 2.4.

Noether's Theorem

Next let us discuss the relationship between symmetries and conservation laws in classical field theory, summarized in *Noether's theorem*. This theorem concerns continuous transformations on the fields ϕ, which in infinitesimal form can be written

$$\phi(x) \to \phi'(x) = \phi(x) + \alpha\Delta\phi(x), \tag{2.9}$$

where α is an infinitesimal parameter and $\Delta\phi$ is some deformation of the field configuration. We call this transformation a symmetry if it leaves the equations of motion invariant. This is insured if the action is invariant under (2.9). More generally, we can allow the action to change by a surface term, since the presence of such a term would not affect our derivation of the Euler-Lagrange equations of motion (2.3). The Lagrangian, therefore, must be invariant under (2.9) up to a 4-divergence:

$$\mathcal{L}(x) \to \mathcal{L}(x) + \alpha\partial_\mu \mathcal{J}^\mu(x), \tag{2.10}$$

for some $\mathcal{J}^\mu$. Let us compare this expectation for $\Delta\mathcal{L}$ to the result obtained by varying the fields:

$$\alpha\Delta\mathcal{L} = \frac{\partial\mathcal{L}}{\partial\phi}(\alpha\Delta\phi) + \left(\frac{\partial\mathcal{L}}{\partial(\partial_\mu\phi)}\right)\partial_\mu(\alpha\Delta\phi)$$

$$= \alpha\,\partial_\mu\left(\frac{\partial\mathcal{L}}{\partial(\partial_\mu\phi)}\Delta\phi\right) + \alpha\left[\frac{\partial\mathcal{L}}{\partial\phi} - \partial_\mu\left(\frac{\partial\mathcal{L}}{\partial(\partial_\mu\phi)}\right)\right]\Delta\phi. \tag{2.11}$$

The second term vanishes by the Euler-Lagrange equation (2.3). We set the remaining term equal to $\alpha \partial_\mu \mathcal{J}^\mu$ and find

$$\partial_\mu j^\mu(x) = 0, \qquad \text{for} \quad j^\mu(x) = \frac{\partial \mathcal{L}}{\partial(\partial_\mu \phi)} \Delta\phi - \mathcal{J}^\mu. \qquad (2.12)$$

(If the symmetry involves more than one field, the first term of this expression for $j^\mu(x)$ should be replaced by a sum of such terms, one for each field.) This result states that the current $j^\mu(x)$ is conserved. For each continuous symmetry of $\mathcal{L}$, we have such a conservation law.

The conservation law can also be expressed by saying that the charge

$$Q \equiv \int\limits_{\text{all space}} j^0 \, d^3x \qquad (2.13)$$

is a constant in time. Note, however, that the formulation of field theory in terms of a local Lagrangian density leads directly to the local form of the conservation law, Eq. (2.12).

The easiest example of such a conservation law arises from a Lagrangian with only a kinetic term: $\mathcal{L} = \frac{1}{2}(\partial_\mu \phi)^2$. The transformation $\phi \to \phi + \alpha$, where α is a constant, leaves $\mathcal{L}$ unchanged, so we conclude that the current $j^\mu = \partial^\mu \phi$ is conserved. As a less trivial example, consider the Lagrangian

$$\mathcal{L} = |\partial_\mu \phi|^2 - m^2 |\phi|^2, \qquad (2.14)$$

where ϕ is now a *complex*-valued field. You can easily show that the equation of motion for this Lagrangian is again the Klein-Gordon equation, (2.7). This Lagrangian is invariant under the transformation $\phi \to e^{i\alpha}\phi$; for an infinitesimal transformation we have

$$\alpha\Delta\phi = i\alpha\phi; \qquad \alpha\Delta\phi^* = -i\alpha\phi^*. \qquad (2.15)$$

(We treat ϕ and ϕ^* as independent fields. Alternatively, we could work with the real and imaginary parts of ϕ.) It is now a simple matter to show that the conserved Noether current is

$$j^\mu = i\big[(\partial^\mu \phi^*)\phi - \phi^*(\partial^\mu \phi)\big]. \qquad (2.16)$$

(The overall constant has been chosen arbitrarily.) You can check directly that the divergence of this current vanishes by using the Klein-Gordon equation. Later we will add terms to this Lagrangian that couple ϕ to an electromagnetic field. We will then interpret j^μ as the electromagnetic current density carried by the field, and the spatial integral of j^0 as its electric charge.

Noether's theorem can also be applied to spacetime transformations such as translations and rotations. We can describe the infinitesimal translation

$$x^\mu \to x^\mu - a^\mu$$

alternatively as a transformation of the field configuration

$$\phi(x) \to \phi(x+a) = \phi(x) + a^\mu \partial_\mu \phi(x).$$

The Lagrangian is also a scalar, so it must transform in the same way:

$$\mathcal{L} \to \mathcal{L} + a^\mu \partial_\mu \mathcal{L} = \mathcal{L} + a^\nu \partial_\mu (\delta^\mu_\nu \mathcal{L}).$$

Comparing this equation to (2.10), we see that we now have a nonzero $\mathcal{J}^\mu$. Taking this into account, we can apply the theorem to obtain four separately conserved currents:

$$T^\mu_{\ \nu} \equiv \frac{\partial \mathcal{L}}{\partial(\partial_\mu \phi)} \partial_\nu \phi - \mathcal{L}\delta^\mu_\nu. \tag{2.17}$$

This is precisely the *stress-energy tensor*, also called the *energy-momentum tensor*, of the field ϕ. The conserved charge associated with time translations is the Hamiltonian:

$$H = \int T^{00} \, d^3x = \int \mathcal{H} \, d^3x. \tag{2.18}$$

By computing this quantity for the Klein-Gordon field, one can recover the result (2.8). The conserved charges associated with spatial translations are

$$P^i = \int T^{0i} \, d^3x = -\int \pi \partial_i \phi \, d^3x, \tag{2.19}$$

and we naturally interpret this as the (physical) momentum carried by the field (not to be confused with the canonical momentum).

2.3 The Klein-Gordon Field as Harmonic Oscillators

We begin our discussion of *quantum* field theory with a rather formal treatment of the simplest type of field: the real Klein-Gordon field. The idea is to start with a classical field theory (the theory of a classical scalar field governed by the Lagrangian (2.6)) and then "quantize" it, that is, reinterpret the dynamical variables as operators that obey canonical commutation relations.[†] We will then "solve" the theory by finding the eigenvalues and eigenstates of the Hamiltonian, using the harmonic oscillator as an analogy.

The classical theory of the real Klein-Gordon field was discussed briefly (but sufficiently) in the previous section; the relevant expressions are given in Eqs. (2.6), (2.7), and (2.8). To quantize the theory, we follow the same procedure as for any other dynamical system: We promote ϕ and π to operators, and impose suitable commutation relations. Recall that for a discrete system of one or more particles the commutation relations are

$$[q_i, p_j] = i\delta_{ij};$$

$$[q_i, q_j] = [p_i, p_j] = 0.$$

[†]This procedure is sometimes called *second quantization*, to distinguish the resulting Klein-Gordon equation (in which ϕ is an operator) from the old one-particle Klein-Gordon equation (in which ϕ was a wavefunction). In this book we never adopt the latter point of view; we start with a classical equation (in which ϕ is a classical field) and quantize it exactly once.

For a continuous system the generalization is quite natural; since $\pi(\mathbf{x})$ is the momentum *density*, we get a Dirac delta function instead of a Kronecker delta:

$$[\phi(\mathbf{x}), \pi(\mathbf{y})] = i\delta^{(3)}(\mathbf{x} - \mathbf{y});$$

$$[\phi(\mathbf{x}), \phi(\mathbf{y})] = [\pi(\mathbf{x}), \pi(\mathbf{y})] = 0. \tag{2.20}$$

(For now we work in the Schrödinger picture where ϕ and π do not depend on time. When we switch to the Heisenberg picture in the next section, these "equal time" commutation relations will still hold provided that both operators are considered at the same time.)

The Hamiltonian, being a function of ϕ and π, also becomes an operator. Our next task is to find the spectrum from the Hamiltonian. Since there is no obvious way to do this, let us seek guidance by writing the Klein-Gordon equation in Fourier space. If we expand the classical Klein-Gordon field as

$$\phi(\mathbf{x}, t) = \int \frac{d^3p}{(2\pi)^3} \, e^{i\mathbf{p}\cdot\mathbf{x}} \, \phi(\mathbf{p}, t)$$

(with $\phi^*(\mathbf{p}) = \phi(-\mathbf{p})$ so that $\phi(\mathbf{x})$ is real), the Klein-Gordon equation (2.7) becomes

$$\left[\frac{\partial^2}{\partial t^2} + \left(|\mathbf{p}|^2 + m^2 \right) \right] \phi(\mathbf{p}, t) = 0. \tag{2.21}$$

This is the same as the equation of motion for a simple harmonic oscillator with frequency

$$\omega_{\mathbf{p}} = \sqrt{|\mathbf{p}|^2 + m^2}. \tag{2.22}$$

The simple harmonic oscillator is a system whose spectrum we already know how to find. Let us briefly recall how it is done. We write the Hamiltonian as

$$H_{\text{SHO}} = \tfrac{1}{2}p^2 + \tfrac{1}{2}\omega^2\phi^2.$$

To find the eigenvalues of H_{SHO}, we write ϕ and p in terms of ladder operators:

$$\phi = \frac{1}{\sqrt{2\omega}}(a + a^\dagger); \qquad p = -i\sqrt{\frac{\omega}{2}}(a - a^\dagger). \tag{2.23}$$

The canonical commutation relation $[\phi, p] = i$ is equivalent to

$$[a, a^\dagger] = 1. \tag{2.24}$$

The Hamiltonian can now be rewritten

$$H_{\text{SHO}} = \omega(a^\dagger a + \tfrac{1}{2}).$$

The state $|0\rangle$ such that $a|0\rangle = 0$ is an eigenstate of H with eigenvalue $\tfrac{1}{2}\omega$, the zero-point energy. Furthermore, the commutators

$$[H_{\text{SHO}}, a^\dagger] = \omega a^\dagger, \qquad [H_{\text{SHO}}, a] = -\omega a$$

make it easy to verify that the states

$$|n\rangle \equiv (a^\dagger)^n |0\rangle$$

are eigenstates of H_{SHO} with eigenvalues $(n + \frac{1}{2})\omega$. These states exhaust the spectrum.

We can find the spectrum of the Klein-Gordon Hamiltonian using the same trick, but now each Fourier mode of the field is treated as an independent oscillator with its own a and $a^\dagger$. In analogy with (2.23) we write

$$\phi(\mathbf{x}) = \int \frac{d^3p}{(2\pi)^3} \frac{1}{\sqrt{2\omega_\mathbf{p}}} \left(a_\mathbf{p} e^{i\mathbf{p}\cdot\mathbf{x}} + a_\mathbf{p}^\dagger e^{-i\mathbf{p}\cdot\mathbf{x}} \right); \tag{2.25}$$

$$\pi(\mathbf{x}) = \int \frac{d^3p}{(2\pi)^3} (-i)\sqrt{\frac{\omega_\mathbf{p}}{2}} \left(a_\mathbf{p} e^{i\mathbf{p}\cdot\mathbf{x}} - a_\mathbf{p}^\dagger e^{-i\mathbf{p}\cdot\mathbf{x}} \right). \tag{2.26}$$

The inverse expressions for $a_\mathbf{p}$ and $a_\mathbf{p}^\dagger$ in terms of ϕ and π are easy to derive but rarely needed. In the calculations below we will find it useful to rearrange (2.25) and (2.26) as follows:

$$\phi(\mathbf{x}) = \int \frac{d^3p}{(2\pi)^3} \frac{1}{\sqrt{2\omega_\mathbf{p}}} \left(a_\mathbf{p} + a_{-\mathbf{p}}^\dagger \right) e^{i\mathbf{p}\cdot\mathbf{x}}; \tag{2.27}$$

$$\pi(\mathbf{x}) = \int \frac{d^3p}{(2\pi)^3} (-i)\sqrt{\frac{\omega_\mathbf{p}}{2}} \left(a_\mathbf{p} - a_{-\mathbf{p}}^\dagger \right) e^{i\mathbf{p}\cdot\mathbf{x}}. \tag{2.28}$$

The commutation relation (2.24) becomes

$$[a_\mathbf{p}, a_{\mathbf{p}'}^\dagger] = (2\pi)^3 \delta^{(3)}(\mathbf{p} - \mathbf{p}'), \tag{2.29}$$

from which you can verify that the commutator of ϕ and π works out correctly:

$$[\phi(\mathbf{x}), \pi(\mathbf{x}')] = \int \frac{d^3p\, d^3p'}{(2\pi)^6} \frac{-i}{2} \sqrt{\frac{\omega_{\mathbf{p}'}}{\omega_\mathbf{p}}} \left([a_{-\mathbf{p}}^\dagger, a_{\mathbf{p}'}] - [a_\mathbf{p}, a_{-\mathbf{p}'}^\dagger] \right) e^{i(\mathbf{p}\cdot\mathbf{x}+\mathbf{p}'\cdot\mathbf{x}')}$$

$$= i\delta^{(3)}(\mathbf{x} - \mathbf{x}'). \tag{2.30}$$

(If computations such as this one and the next are unfamiliar to you, please work them out carefully; they are quite easy after a little practice, and are fundamental to the formalism of the next two chapters.)

We are now ready to express the Hamiltonian in terms of ladder operators. Starting from its expression (2.8) in terms of ϕ and π, we have

$$H = \int d^3x \int \frac{d^3p\, d^3p'}{(2\pi)^6} e^{i(\mathbf{p}+\mathbf{p}')\cdot\mathbf{x}} \left\{ -\frac{\sqrt{\omega_\mathbf{p}\omega_{\mathbf{p}'}}}{4} \left(a_\mathbf{p} - a_{-\mathbf{p}}^\dagger \right)\left(a_{\mathbf{p}'} - a_{-\mathbf{p}'}^\dagger \right) \right.$$

$$\left. + \frac{-\mathbf{p}\cdot\mathbf{p}' + m^2}{4\sqrt{\omega_\mathbf{p}\omega_{\mathbf{p}'}}} \left(a_\mathbf{p} + a_{-\mathbf{p}}^\dagger \right)\left(a_{\mathbf{p}'} + a_{-\mathbf{p}'}^\dagger \right) \right\}$$

$$= \int \frac{d^3p}{(2\pi)^3} \omega_\mathbf{p} \left(a_\mathbf{p}^\dagger a_\mathbf{p} + \frac{1}{2}[a_\mathbf{p}, a_\mathbf{p}^\dagger] \right). \tag{2.31}$$

The second term is proportional to $\delta(0)$, an infinite c-number. It is simply the sum over all modes of the zero-point energies $\omega_\mathbf{p}/2$, so its presence is completely expected, if somewhat disturbing. Fortunately, this infinite energy shift cannot be detected experimentally, since experiments measure only energy *differences* from the ground state of H. We will therefore ignore this infinite constant term in all of our calculations. It is possible that this energy shift of the ground state could create a problem at a deeper level in the theory; we will discuss this matter in the Epilogue.

Using this expression for the Hamiltonian in terms of $a_{\mathbf{p}}$ and $a_{\mathbf{p}}^{\dagger}$, it is easy to evaluate the commutators

$$[H, a_{\mathbf{p}}^{\dagger}] = \omega_{\mathbf{p}} a_{\mathbf{p}}^{\dagger}; \qquad [H, a_{\mathbf{p}}] = -\omega_{\mathbf{p}} a_{\mathbf{p}}. \qquad (2.32)$$

We can now write down the spectrum of the theory, just as for the harmonic oscillator. The state $|0\rangle$ such that $a_{\mathbf{p}}|0\rangle = 0$ for all $\mathbf{p}$ is the ground state or *vacuum*, and has $E = 0$ after we drop the infinite constant in (2.31). All other energy eigenstates can be built by acting on $|0\rangle$ with creation operators. In general, the state $a_{\mathbf{p}}^{\dagger} a_{\mathbf{q}}^{\dagger} \cdots |0\rangle$ is an eigenstate of H with energy $\omega_{\mathbf{p}} + \omega_{\mathbf{q}} + \cdots$. These states exhaust the spectrum.

Having found the spectrum of the Hamiltonian, let us try to interpret its eigenstates. From (2.19) and a calculation similar to (2.31) we can write down the total momentum operator,

$$\mathbf{P} = -\int d^3x \, \pi(\mathbf{x}) \nabla\phi(\mathbf{x}) = \int \frac{d^3p}{(2\pi)^3} \, \mathbf{p} \, a_{\mathbf{p}}^{\dagger} a_{\mathbf{p}}. \qquad (2.33)$$

So the operator $a_{\mathbf{p}}^{\dagger}$ creates momentum $\mathbf{p}$ and energy $\omega_{\mathbf{p}} = \sqrt{|\mathbf{p}|^2 + m^2}$. Similarly, the state $a_{\mathbf{p}}^{\dagger} a_{\mathbf{q}}^{\dagger} \cdots |0\rangle$ has momentum $\mathbf{p} + \mathbf{q} + \cdots$. It is quite natural to call these excitations *particles*, since they are discrete entities that have the proper relativistic energy-momentum relation. (By a *particle* we do not mean something that must be localized in space; $a_{\mathbf{p}}^{\dagger}$ creates particles in momentum eigenstates.) From now on we will refer to $\omega_{\mathbf{p}}$ as $E_{\mathbf{p}}$ (or simply E), since it really is the energy of a particle. Note, by the way, that the energy is always positive: $E_{\mathbf{p}} = +\sqrt{|\mathbf{p}|^2 + m^2}$.

This formalism also allows us to determine the statistics of our particles. Consider the two-particle state $a_{\mathbf{p}}^{\dagger} a_{\mathbf{q}}^{\dagger} |0\rangle$. Since $a_{\mathbf{p}}^{\dagger}$ and $a_{\mathbf{q}}^{\dagger}$ commute, this state is identical to the state $a_{\mathbf{q}}^{\dagger} a_{\mathbf{p}}^{\dagger} |0\rangle$ in which the two particles are interchanged. Moreover, a single mode $\mathbf{p}$ can contain arbitrarily many particles (just as a simple harmonic oscillator can be excited to arbitrarily high levels). Thus we conclude that Klein-Gordon particles obey *Bose-Einstein statistics*.

We naturally choose to normalize the vacuum state so that $\langle 0|0\rangle = 1$. The one-particle states $|\mathbf{p}\rangle \propto a_{\mathbf{p}}^{\dagger} |0\rangle$ will also appear quite often, and it is worthwhile to adopt a convention for their normalization. The simplest normalization $\langle \mathbf{p}|\mathbf{q}\rangle = (2\pi)^3 \delta^{(3)}(\mathbf{p} - \mathbf{q})$ (which many books use) is not Lorentz invariant, as we can demonstrate by considering the effect of a boost in the 3-direction. Under such a boost we have $p_3' = \gamma(p_3 + \beta E)$, $E' = \gamma(E + \beta p_3)$. Using the delta function identity

$$\delta\big(f(x) - f(x_0)\big) = \frac{1}{|f'(x_0)|} \delta(x - x_0), \qquad (2.34)$$

we can compute

$$\delta^{(3)}(\mathbf{p} - \mathbf{q}) = \delta^{(3)}(\mathbf{p}' - \mathbf{q}') \cdot \frac{dp_3'}{dp_3}$$

$$= \delta^{(3)}(\mathbf{p}' - \mathbf{q}')\gamma\left(1 + \beta\frac{dE}{dp_3}\right)$$

$$= \delta^{(3)}(\mathbf{p}' - \mathbf{q}')\frac{\gamma}{E}(E + \beta p_3)$$

$$= \delta^{(3)}(\mathbf{p}' - \mathbf{q}')\frac{E'}{E}.$$

The problem is that volumes are not invariant under boosts; a box whose volume is V in its rest frame has volume V/γ in a boosted frame, due to Lorentz contraction. But from the above calculation, we see that the quantity $E_{\mathbf{p}}\delta^{(3)}(\mathbf{p} - \mathbf{q})$ *is* Lorentz invariant. We therefore define

$$|\mathbf{p}\rangle = \sqrt{2E_{\mathbf{p}}}\, a_{\mathbf{p}}^{\dagger}\, |0\rangle, \tag{2.35}$$

so that

$$\langle \mathbf{p}|\mathbf{q}\rangle = 2E_{\mathbf{p}}(2\pi)^3\delta^{(3)}(\mathbf{p} - \mathbf{q}). \tag{2.36}$$

(The factor of 2 is unnecessary, but is convenient because of the factor of 2 in Eq. (2.25).)

On the Hilbert space of quantum states, a Lorentz transformation Λ will be implemented as some unitary operator $U(\Lambda)$. Our normalization condition (2.35) then implies that

$$U(\Lambda)|\mathbf{p}\rangle = |\Lambda\mathbf{p}\rangle. \tag{2.37}$$

If we prefer to think of this transformation as acting on the operator $a_{\mathbf{p}}^{\dagger}$, we can also write

$$U(\Lambda)\, a_{\mathbf{p}}^{\dagger}\, U^{-1}(\Lambda) = \sqrt{\frac{E_{\Lambda\mathbf{p}}}{E_{\mathbf{p}}}}\, a_{\Lambda\mathbf{p}}^{\dagger}. \tag{2.38}$$

With this normalization we must divide by $2E_{\mathbf{p}}$ in other places. For example, the completeness relation for the one-particle states is

$$(\mathbf{1})_{\text{1-particle}} = \int \frac{d^3p}{(2\pi)^3}\, |\mathbf{p}\rangle\, \frac{1}{2E_{\mathbf{p}}}\, \langle\mathbf{p}|, \tag{2.39}$$

where the operator on the left is simply the identity within the subspace of one-particle states, and zero in the rest of the Hilbert space. Integrals of this form will occur quite often; in fact, the integral

$$\int \frac{d^3p}{(2\pi)^3}\frac{1}{2E_{\mathbf{p}}} = \int \frac{d^4p}{(2\pi)^4}(2\pi)\delta(p^2 - m^2)\Big|_{p^0 > 0} \tag{2.40}$$

is a Lorentz-invariant 3-momentum integral, in the sense that if $f(p)$ is Lorentz-invariant, so is $\int d^3p\, f(p)/(2E_{\mathbf{p}})$. The integration can be thought of

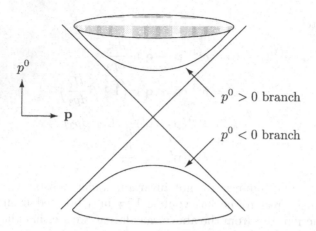

Figure 2.2. The Lorentz-invariant 3-momentum integral is over the upper branch of the hyperboloid $p^2 = m^2$.

as being over the $p^0 > 0$ branch of the hyperboloid $p^2 = m^2$ in 4-momentum space (see Fig. 2.2).

Finally let us consider the interpretation of the state $\phi(\mathbf{x})|0\rangle$. From the expansion (2.25) we see that

$$\phi(\mathbf{x})|0\rangle = \int \frac{d^3p}{(2\pi)^3}\frac{1}{2E_\mathbf{p}}e^{-i\mathbf{p}\cdot\mathbf{x}}|\mathbf{p}\rangle \tag{2.41}$$

is a linear superposition of single-particle states that have well-defined momentum. Except for the factor $1/2E_\mathbf{p}$, this is the same as the familiar nonrelativistic expression for the eigenstate of position $|\mathbf{x}\rangle$; in fact the extra factor is nearly constant for small (nonrelativistic) $\mathbf{p}$. We will therefore put forward the same interpretation, and claim that the operator $\phi(\mathbf{x})$, acting on the vacuum, *creates a particle at position* $\mathbf{x}$. This interpretation is further confirmed when we compute

$$\langle 0|\,\phi(\mathbf{x})\,|\mathbf{p}\rangle = \langle 0|\int \frac{d^3p'}{(2\pi)^3}\frac{1}{\sqrt{2E_{\mathbf{p}'}}}\left(a_{\mathbf{p}'}e^{i\mathbf{p}'\cdot\mathbf{x}} + a^\dagger_{\mathbf{p}'}e^{-i\mathbf{p}'\cdot\mathbf{x}}\right)\sqrt{2E_\mathbf{p}}\,a^\dagger_\mathbf{p}\,|0\rangle$$

$$= e^{i\mathbf{p}\cdot\mathbf{x}}. \tag{2.42}$$

We can interpret this as the position-space representation of the single-particle wavefunction of the state $|\mathbf{p}\rangle$, just as in nonrelativistic quantum mechanics $\langle\mathbf{x}|\mathbf{p}\rangle \propto e^{i\mathbf{p}\cdot\mathbf{x}}$ is the wavefunction of the state $|\mathbf{p}\rangle$.

2.4 The Klein-Gordon Field in Space-Time

In the previous section we quantized the Klein-Gordon field in the Schrödinger picture, and interpreted the resulting theory in terms of relativistic particles. In this section we will switch to the Heisenberg picture, where it will be easier to discuss time-dependent quantities and questions of causality. After a few preliminaries, we will return to the question of acausal propagation raised in Section 2.1. We will also derive an expression for the *Klein-Gordon propagator*, a crucial part of the Feynman rules to be developed in Chapter 4.

In the Heisenberg picture, we make the operators ϕ and π time-dependent in the usual way:

$$\phi(x) = \phi(\mathbf{x}, t) = e^{iHt} \phi(\mathbf{x}) e^{-iHt}, \tag{2.43}$$

and similarly for $\pi(x) = \pi(\mathbf{x}, t)$. The Heisenberg equation of motion,

$$i \frac{\partial}{\partial t} \mathcal{O} = [\mathcal{O}, H], \tag{2.44}$$

allows us to compute the time dependence of ϕ and π:

$$i \frac{\partial}{\partial t} \phi(\mathbf{x}, t) = \left[\phi(\mathbf{x}, t), \int d^3 x' \left\{ \tfrac{1}{2} \pi^2(\mathbf{x}', t) + \tfrac{1}{2} \left(\nabla \phi(\mathbf{x}', t) \right)^2 + \tfrac{1}{2} m^2 \phi^2(\mathbf{x}', t) \right\} \right]$$

$$= \int d^3 x' \left(i \delta^{(3)}(\mathbf{x} - \mathbf{x}') \pi(\mathbf{x}', t) \right)$$

$$= i \pi(\mathbf{x}, t);$$

$$i \frac{\partial}{\partial t} \pi(\mathbf{x}, t) = \left[\pi(\mathbf{x}, t), \int d^3 x' \left\{ \tfrac{1}{2} \pi^2(\mathbf{x}', t) + \tfrac{1}{2} \phi(\mathbf{x}', t) \left(-\nabla^2 + m^2 \right) \phi(\mathbf{x}', t) \right\} \right]$$

$$= \int d^3 x' \left(-i \delta^{(3)}(\mathbf{x} - \mathbf{x}') \left(-\nabla^2 + m^2 \right) \phi(\mathbf{x}', t) \right)$$

$$= -i \left(-\nabla^2 + m^2 \right) \phi(\mathbf{x}, t).$$

Combining the two results gives

$$\frac{\partial^2}{\partial t^2} \phi = \left(\nabla^2 - m^2 \right) \phi, \tag{2.45}$$

which is just the Klein-Gordon equation.

We can better understand the time dependence of $\phi(x)$ and $\pi(x)$ by writing them in terms of creation and annihilation operators. First note that

$$H a_{\mathbf{p}} = a_{\mathbf{p}} (H - E_{\mathbf{p}}),$$

and hence

$$H^n a_{\mathbf{p}} = a_{\mathbf{p}} (H - E_{\mathbf{p}})^n,$$

for any n. A similar relation (with $-$ replaced by $+$) holds for $a_{\mathbf{p}}^\dagger$. Thus we have derived the identities

$$e^{iHt} a_{\mathbf{p}} e^{-iHt} = a_{\mathbf{p}} e^{-iE_{\mathbf{p}}t}, \qquad e^{iHt} a_{\mathbf{p}}^\dagger e^{-iHt} = a_{\mathbf{p}}^\dagger e^{iE_{\mathbf{p}}t}, \tag{2.46}$$

which we can use on expression (2.25) for $\phi(\mathbf{x})$ to find the desired expression for the Heisenberg operator $\phi(x)$, according to (2.43). (We will always use the symbols $a_{\mathbf{p}}$ and $a_{\mathbf{p}}^{\dagger}$ to represent the time-independent, Schrödinger-picture ladder operators.) The result is

$$\phi(\mathbf{x},t) = \int \frac{d^3p}{(2\pi)^3} \frac{1}{\sqrt{2E_{\mathbf{p}}}} \left(a_{\mathbf{p}} e^{-ip\cdot x} + a_{\mathbf{p}}^{\dagger} e^{ip\cdot x}\right)\Big|_{p^0 = E_{\mathbf{p}}};$$

$$\pi(\mathbf{x},t) = \frac{\partial}{\partial t}\phi(\mathbf{x},t). \tag{2.47}$$

It is worth mentioning that we can perform the same manipulations with $\mathbf{P}$ instead of H to relate $\phi(\mathbf{x})$ to $\phi(0)$. In analogy with (2.46), one can show

$$e^{-i\mathbf{P}\cdot\mathbf{x}} a_{\mathbf{p}} e^{i\mathbf{P}\cdot\mathbf{x}} = a_{\mathbf{p}} e^{i\mathbf{p}\cdot\mathbf{x}}, \qquad e^{-i\mathbf{P}\cdot\mathbf{x}} a_{\mathbf{p}}^{\dagger} e^{i\mathbf{P}\cdot\mathbf{x}} = a_{\mathbf{p}}^{\dagger} e^{-i\mathbf{p}\cdot\mathbf{x}}, \tag{2.48}$$

and therefore

$$\phi(x) = e^{i(Ht - \mathbf{P}\cdot\mathbf{x})}\phi(0)e^{-i(Ht - \mathbf{P}\cdot\mathbf{x})}$$

$$= e^{iP\cdot x}\phi(0)e^{-iP\cdot x}, \tag{2.49}$$

where $P^{\mu} = (H, \mathbf{P})$. (The notation here is confusing but standard. Remember that $\mathbf{P}$ is the momentum operator, whose eigenvalue is the total momentum of the system. On the other hand, $\mathbf{p}$ is the momentum of a single Fourier mode of the field, which we interpret as the momentum of a particle in that mode. For a one-particle state of well-defined momentum, $\mathbf{p}$ is the eigenvalue of $\mathbf{P}$.)

Equation (2.47) makes explicit the dual particle and wave interpretations of the quantum field $\phi(x)$. On the one hand, $\phi(x)$ is written as a Hilbert space operator, which creates and destroys the particles that are the quanta of field excitation. On the other hand, $\phi(x)$ is written as a linear combination of solutions ($e^{ip\cdot x}$ and $e^{-ip\cdot x}$) of the Klein-Gordon equation. Both signs of the time dependence in the exponential appear: We find both $e^{-ip^0 t}$ and $e^{+ip^0 t}$, although p^0 is always positive. If these were single-particle wavefunctions, they would correspond to states of positive and negative energy; let us refer to them more generally as *positive-* and *negative-frequency* modes. The connection between the particle creation operators and the waveforms displayed here is always valid for free quantum fields: A positive-frequency solution of the field equation has as its coefficient the operator that *destroys* a particle in that single-particle wavefunction. A negative-frequency solution of the field equation, being the Hermitian conjugate of a positive-frequency solution, has as its coefficient the operator that *creates* a particle in that positive-energy single-particle wavefunction. In this way, the fact that relativistic wave equations have both positive- and negative-frequency solutions is reconciled with the requirement that a sensible quantum theory contain only positive excitation energies.

Causality

Now let us return to the question of causality raised at the beginning of this chapter. In our present formalism, still working in the Heisenberg picture, the amplitude for a particle to propagate from y to x is $\langle 0| \phi(x)\phi(y) |0\rangle$. We will call this quantity $D(x - y)$. Each operator ϕ is a sum of a and $a^\dagger$ operators, but only the term $\langle 0| a_\mathbf{p}a_\mathbf{q}^\dagger |0\rangle = (2\pi)^3\delta^{(3)}(\mathbf{p} - \mathbf{q})$ survives in this expression. It is easy to check that we are left with

$$D(x - y) = \langle 0| \phi(x)\phi(y) |0\rangle = \int \frac{d^3p}{(2\pi)^3} \frac{1}{2E_\mathbf{p}} e^{-ip\cdot(x-y)}. \qquad (2.50)$$

We have already argued in (2.40) that integrals of this form are Lorentz invariant. Let us now evaluate this integral for some particular values of $x - y$.

First consider the case where the difference $x - y$ is purely in the time-direction: $x^0 - y^0 = t$, $\mathbf{x} - \mathbf{y} = 0$. (If the interval from y to x is timelike, there is always a frame in which this is the case.) Then we have

$$\begin{aligned} D(x - y) &= \frac{4\pi}{(2\pi)^3} \int_0^\infty dp \, \frac{p^2}{2\sqrt{p^2 + m^2}} e^{-i\sqrt{p^2+m^2}\,t} \\ &= \frac{1}{4\pi^2} \int_m^\infty dE \, \sqrt{E^2 - m^2} \, e^{-iEt} \\ &\underset{t\to\infty}{\sim} e^{-imt}. \end{aligned} \qquad (2.51)$$

Next consider the case where $x - y$ is purely spatial: $x^0 - y^0 = 0$, $\mathbf{x} - \mathbf{y} = \mathbf{r}$. The amplitude is then

$$\begin{aligned} D(x - y) &= \int \frac{d^3p}{(2\pi)^3} \frac{1}{2E_\mathbf{p}} e^{i\mathbf{p}\cdot\mathbf{r}} \\ &= \frac{2\pi}{(2\pi)^3} \int_0^\infty dp \, \frac{p^2}{2E_\mathbf{p}} \frac{e^{ipr} - e^{-ipr}}{ipr} \\ &= \frac{-i}{2(2\pi)^2 r} \int_{-\infty}^\infty dp \, \frac{p\,e^{ipr}}{\sqrt{p^2 + m^2}}. \end{aligned}$$

The integrand, considered as a complex function of p, has branch cuts on the imaginary axis starting at $\pm im$ (see Fig. 2.3). To evaluate the integral we push the contour up to wrap around the upper branch cut. Defining $\rho = -ip$, we obtain

$$\frac{1}{4\pi^2 r} \int_m^\infty d\rho \, \frac{\rho\,e^{-\rho r}}{\sqrt{\rho^2 - m^2}} \underset{r\to\infty}{\sim} e^{-mr}. \qquad (2.52)$$

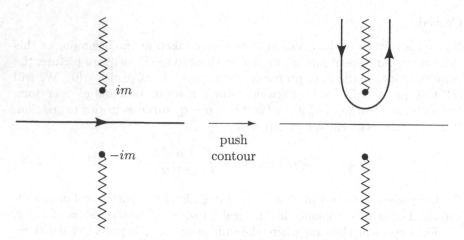

Figure 2.3. Contour for evaluating propagation amplitude $D(x-y)$ over a spacelike interval.

So again we find that outside the light-cone, the propagation amplitude is exponentially vanishing but nonzero.

To really discuss causality, however, we should ask not whether particles can propagate over spacelike intervals, but whether a *measurement* performed at one point can affect a measurement at another point whose separation from the first is spacelike. The simplest thing we could try to measure is the field $\phi(x)$, so we should compute the commutator $[\phi(x), \phi(y)]$; if this commutator vanishes, one measurement cannot affect the other. In fact, if the commutator vanishes for $(x-y)^2 < 0$, causality is preserved quite generally, since commutators involving any function of $\phi(x)$, including $\pi(x) = \partial\phi/\partial t$, would also have to vanish. Of course we know from Eq. (2.20) that the commutator vanishes for $x^0 = y^0$; now let's do the more general computation:

$$
\begin{aligned}
[\phi(x), \phi(y)] &= \int \frac{d^3p}{(2\pi)^3} \frac{1}{\sqrt{2E_\mathbf{p}}} \int \frac{d^3q}{(2\pi)^3} \frac{1}{\sqrt{2E_\mathbf{q}}} \\
&\quad \times \left[\left(a_\mathbf{p} e^{-ip\cdot x} + a_\mathbf{p}^\dagger e^{ip\cdot x}\right), \left(a_\mathbf{q} e^{-iq\cdot y} + a_\mathbf{q}^\dagger e^{iq\cdot y}\right) \right] \\
&= \int \frac{d^3p}{(2\pi)^3} \frac{1}{2E_\mathbf{p}} \left(e^{-ip\cdot(x-y)} - e^{ip\cdot(x-y)} \right) \\
&= D(x-y) - D(y-x). \quad\quad\quad (2.53)
\end{aligned}
$$

When $(x-y)^2 < 0$, we can perform a Lorentz transformation on the second term (since each term is separately Lorentz invariant), taking $(x-y) \to -(x-y)$, as shown in Fig. 2.4. The two terms are therefore equal and cancel to give zero; causality is preserved. Note that if $(x-y)^2 > 0$ there is no continuous Lorentz transformation that takes $(x-y) \longrightarrow -(x-y)$. In this case, by Eq. (2.51), the amplitude is (fortunately) nonzero, roughly $(e^{-imt} - e^{imt})$ for the special case $\mathbf{x} - \mathbf{y} = 0$. Thus we conclude that no measurement in the

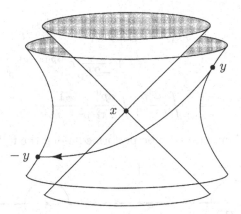

Figure 2.4. When $x - y$ is spacelike, a continuous Lorentz transformation can take $(x - y)$ to $-(x - y)$.

Klein-Gordon theory can affect another measurement outside the light-cone.

Causality is maintained in the Klein-Gordon theory just as suggested at the end of Section 2.1. To understand this mechanism properly, however, we should broaden the context of our discussion to include a *complex* Klein-Gordon field, which has distinct particle and antiparticle excitations. As was mentioned in the discussion of Eq. (2.15), we can add a conserved charge to the Klein-Gordon theory by considering the field $\phi(x)$ to be complex- rather than real-valued. When the complex scalar field theory is quantized (see Problem 2.2), $\phi(x)$ will create positively charged particles and destroy negatively charged ones, while $\phi^\dagger(x)$ will perform the opposite operations. Then the commutator $[\phi(x), \phi^\dagger(y)]$ will have nonzero contributions, which must delicately cancel outside the light-cone to preserve causality. The two contributions have the spacetime interpretation of the two terms in (2.53), but with charges attached. The first term will represent the propagation of a negatively charged particle from y to x. The second term will represent the propagation of a positively charged particle from x to y. In order for these two processes to be present and give canceling amplitudes, both of these particles must exist, and they must have the same mass. In quantum field theory, then, causality requires that every particle have a corresponding antiparticle with the same mass and opposite quantum numbers (in this case electric charge). For the real-valued Klein-Gordon field, the particle is its own antiparticle.

The Klein-Gordon Propagator

Let us study the commutator $[\phi(x), \phi(y)]$ a little further. Since it is a c-number, we can write $[\phi(x), \phi(y)] = \langle 0| [\phi(x), \phi(y)] |0\rangle$. This can be rewritten as a four-dimensional integral as follows, assuming for now that $x^0 > y^0$:

$$\langle 0| [\phi(x), \phi(y)] |0\rangle = \int \frac{d^3p}{(2\pi)^3} \frac{1}{2E_{\mathbf{p}}} \left(e^{-ip\cdot(x-y)} - e^{ip\cdot(x-y)}\right)$$

$$= \int \frac{d^3p}{(2\pi)^3} \left\{ \frac{1}{2E_{\mathbf{p}}} e^{-ip\cdot(x-y)} \bigg|_{p^0=E_{\mathbf{p}}} \right. $$

$$\left. + \frac{1}{-2E_{\mathbf{p}}} e^{-ip\cdot(x-y)} \bigg|_{p^0=-E_{\mathbf{p}}} \right\}$$

$$\underset{x^0>y^0}{=} \int \frac{d^3p}{(2\pi)^3} \int \frac{dp^0}{2\pi i} \frac{-1}{p^2-m^2} e^{-ip\cdot(x-y)}. \qquad (2.54)$$

In the last step the p^0 integral is to be performed along the following contour:

For $x^0 > y^0$ we can close the contour below, picking up both poles to obtain the previous line of (2.54). For $x^0 < y^0$ we may close the contour above, giving zero. Thus the last line of (2.54), together with the prescription for going around the poles, is an expression for what we will call

$$D_R(x-y) \equiv \theta(x^0-y^0) \langle 0| \left[\phi(x),\phi(y)\right] |0\rangle . \qquad (2.55)$$

To understand this quantity better, let's do another computation:

$$\begin{aligned}
(\partial^2 + m^2)D_R(x-y) &= \left(\partial^2\theta(x^0-y^0)\right) \langle 0| \left[\phi(x),\phi(y)\right] |0\rangle \\
&\quad + 2\left(\partial_\mu\theta(x^0-y^0)\right)\left(\partial^\mu \langle 0| \left[\phi(x),\phi(y)\right] |0\rangle\right) \\
&\quad + \theta(x^0-y^0)\left(\partial^2 + m^2\right) \langle 0| \left[\phi(x),\phi(y)\right] |0\rangle \\
&= -\delta(x^0-y^0) \langle 0| \left[\pi(x),\phi(y)\right] |0\rangle \\
&\quad + 2\delta(x^0-y^0) \langle 0| \left[\pi(x),\phi(y)\right] |0\rangle + 0 \\
&= -i\delta^{(4)}(x-y).
\end{aligned} \qquad (2.56)$$

This says that $D_R(x-y)$ is a Green's function of the Klein-Gordon operator. Since it vanishes for $x^0 < y^0$, it is the *retarded* Green's function.

If we had not already derived expression (2.54), we could find it by Fourier transformation. Writing

$$D_R(x-y) = \int \frac{d^4p}{(2\pi)^4} e^{-ip\cdot(x-y)} \tilde{D}_R(p), \qquad (2.57)$$

we obtain an algebraic expression for $\tilde{D}_R(p)$:

$$(-p^2 + m^2)\tilde{D}_R(p) = -i.$$

Thus we immediately arrive at the result

$$D_R(x-y) = \int \frac{d^4p}{(2\pi)^4} \frac{i}{p^2-m^2} e^{-ip\cdot(x-y)}. \qquad (2.58)$$

The p^0-integral of (2.58) can be evaluated according to four different contours, of which that used in (2.54) is only one. In Chapter 4 we will find that a different pole prescription,

is extremely useful; it is called the *Feynman prescription*. A convenient way to remember it is to write

$$D_F(x - y) \equiv \int \frac{d^4p}{(2\pi)^4} \frac{i}{p^2 - m^2 + i\epsilon} e^{-ip\cdot(x-y)}, \qquad (2.59)$$

since the poles are then at $p^0 = \pm(E_\mathbf{p} - i\epsilon)$, displaced properly above and below the real axis. When $x^0 > y^0$ we can perform the p^0 integral by closing the contour below, obtaining exactly the propagation amplitude $D(x - y)$ (2.50). When $x^0 < y^0$ we close the contour above, obtaining the same expression but with x and y interchanged. Thus we have

$$D_F(x - y) = \begin{cases} D(x - y) & \text{for } x^0 > y^0 \\ D(y - x) & \text{for } x^0 < y^0 \end{cases}$$

$$= \theta(x^0 - y^0) \langle 0| \phi(x)\phi(y) |0\rangle + \theta(y^0 - x^0) \langle 0| \phi(y)\phi(x) |0\rangle$$

$$\equiv \langle 0| T\phi(x)\phi(y) |0\rangle. \qquad (2.60)$$

The last line defines the "time-ordering" symbol T, which instructs us to place the operators that follow in order with the latest to the left. By applying $(\partial^2 + m^2)$ to the last line, you can verify directly that D_F is a Green's function of the Klein-Gordon operator.

Equations (2.59) and (2.60) are, from a practical point of view, the most important results of this chapter. The Green's function $D_F(x - y)$ is called the *Feynman propagator* for a Klein-Gordon particle, since it is, after all, a propagation amplitude. Indeed, the Feynman propagator will turn out to be part of the Feynman rules: $D_F(x-y)$ (or $\tilde{D}_F(p)$) is the expression that we will attach to internal lines of Feynman diagrams, representing the propagation of virtual particles.

Nevertheless we are still a long way from being able to do any real calculations, since so far we have talked only about the *free* Klein-Gordon theory, where the field equation is linear and there are no interactions. Individual particles live in their isolated modes, oblivious to each others' existence and to the existence of any other species of particles. In such a theory there is no hope of making any observations, by scattering or any other means. On the other hand, the formalism we have developed is extremely important, since the free theory forms the basis for doing perturbative calculations in the interacting theory.

Particle Creation by a Classical Source

There is one type of interaction, however, that we are already equipped to handle. Consider a Klein-Gordon field coupled to an external, classical source field $j(x)$. That is, consider the field equation

$$(\partial^2 + m^2)\phi(x) = j(x), \tag{2.61}$$

where $j(x)$ is some fixed, known function of space and time that is nonzero only for a finite time interval. If we start in the vacuum state, what will we find after $j(x)$ has been turned on and off again?

The field equation (2.61) follows from the Lagrangian

$$\mathcal{L} = \tfrac{1}{2}(\partial_\mu \phi)^2 - \tfrac{1}{2}m^2\phi^2 + j(x)\phi(x). \tag{2.62}$$

But if $j(x)$ is turned on for only a finite time, it is easiest to solve the problem using the field equation directly. Before $j(x)$ is turned on, $\phi(x)$ has the form

$$\phi_0(x) = \int \frac{d^3p}{(2\pi)^3} \frac{1}{\sqrt{2E_\mathbf{p}}} \left(a_\mathbf{p} e^{-ip\cdot x} + a_\mathbf{p}^\dagger e^{ip\cdot x}\right).$$

If there were no source, this would be the solution for all time. With a source, the solution of the equation of motion can be constructed using the retarded Green's function:

$$\phi(x) = \phi_0(x) + i \int d^4y\, D_R(x-y)j(y)$$

$$= \phi_0(x) + i \int d^4y \int \frac{d^3p}{(2\pi)^3} \frac{1}{2E_\mathbf{p}} \theta(x^0 - y^0)$$

$$\times \left(e^{-ip\cdot(x-y)} - e^{ip\cdot(x-y)}\right)j(y). \tag{2.63}$$

If we wait until all of j is in the past, the theta function equals 1 in the whole domain of integration. Then $\phi(x)$ involves only the Fourier transform of j,

$$\tilde{j}(p) = \int d^4y\, e^{ip\cdot y} j(y),$$

evaluated at 4-momenta p such that $p^2 = m^2$. It is natural to group the positive-frequency terms together with $a_\mathbf{p}$ and the negative-frequency terms with $a_\mathbf{p}^\dagger$; this yields the expression

$$\phi(x) = \int \frac{d^3p}{(2\pi)^3} \frac{1}{\sqrt{2E_\mathbf{p}}} \left\{\left(a_\mathbf{p} + \frac{i}{\sqrt{2E_\mathbf{p}}}\tilde{j}(p)\right)e^{-ip\cdot x} + \text{h.c.}\right\}. \tag{2.64}$$

You can now guess (or compute) the form of the Hamiltonian after $j(x)$ has acted: Just replace $a_\mathbf{p}$ with $(a_\mathbf{p} + i\tilde{j}(p)/\sqrt{2E_\mathbf{p}})$ to obtain

$$H = \int \frac{d^3p}{(2\pi)^3} E_\mathbf{p} \left(a_\mathbf{p}^\dagger - \frac{i}{\sqrt{2E_\mathbf{p}}}\tilde{j}^*(p)\right)\left(a_\mathbf{p} + \frac{i}{\sqrt{2E_\mathbf{p}}}\tilde{j}(p)\right).$$

The energy of the system after the source has been turned off is

$$\langle 0| \, H \, |0 \rangle = \int \frac{d^3 p}{(2\pi)^3} \frac{1}{2} |\tilde{\jmath}(p)|^2, \tag{2.65}$$

where $|0\rangle$ still denotes the ground state of the free theory. We can interpret these results in terms of particles by identifying $|\tilde{\jmath}(p)|^2/2E_{\mathbf{p}}$ as the probability density for creating a particle in the mode p. Then the total number of particles produced is

$$\int dN = \int \frac{d^3 p}{(2\pi)^3} \frac{1}{2E_{\mathbf{p}}} |\tilde{\jmath}(p)|^2. \tag{2.66}$$

Only those Fourier components of $j(x)$ that are in resonance with on-mass-shell (i.e., $p^2 = m^2$) Klein-Gordon waves are effective at creating particles.

We will return to this subject in Problem 4.1. In Chapter 6 we will study the analogous problem of photon creation by an accelerated electron (bremsstrahlung).

Problems

2.1 Classical electromagnetism (with no sources) follows from the action

$$S = \int d^4 x \left(-\tfrac{1}{4} F_{\mu\nu} F^{\mu\nu} \right), \qquad \text{where } F_{\mu\nu} = \partial_\mu A_\nu - \partial_\nu A_\mu.$$

(a) Derive Maxwell's equations as the Euler-Lagrange equations of this action, treating the components $A_\mu(x)$ as the dynamical variables. Write the equations in standard form by identifying $E^i = -F^{0i}$ and $\epsilon^{ijk} B^k = -F^{ij}$.

(b) Construct the energy-momentum tensor for this theory. Note that the usual procedure does not result in a symmetric tensor. To remedy that, we can add to $T^{\mu\nu}$ a term of the form $\partial_\lambda K^{\lambda\mu\nu}$, where $K^{\lambda\mu\nu}$ is antisymmetric in its first two indices. Such an object is automatically divergenceless, so

$$\widehat{T}^{\mu\nu} = T^{\mu\nu} + \partial_\lambda K^{\lambda\mu\nu}$$

is an equally good energy-momentum tensor with the same globally conserved energy and momentum. Show that this construction, with

$$K^{\lambda\mu\nu} = F^{\mu\lambda} A^\nu,$$

leads to an energy-momentum tensor $\widehat{T}$ that is symmetric and yields the standard formulae for the electromagnetic energy and momentum densities:

$$\mathcal{E} = \tfrac{1}{2}(E^2 + B^2); \qquad \mathbf{S} = \mathbf{E} \times \mathbf{B}.$$

2.2 **The complex scalar field.** Consider the field theory of a complex-valued scalar field obeying the Klein-Gordon equation. The action of this theory is

$$S = \int d^4 x \left(\partial_\mu \phi^* \partial^\mu \phi - m^2 \phi^* \phi \right).$$

It is easiest to analyze this theory by considering $\phi(x)$ and $\phi^*(x)$, rather than the real and imaginary parts of $\phi(x)$, as the basic dynamical variables.

(a) Find the conjugate momenta to $\phi(x)$ and $\phi^*(x)$ and the canonical commutation relations. Show that the Hamiltonian is

$$H = \int d^3x \left(\pi^*\pi + \nabla\phi^* \cdot \nabla\phi + m^2\phi^*\phi \right).$$

Compute the Heisenberg equation of motion for $\phi(x)$ and show that it is indeed the Klein-Gordon equation.

(b) Diagonalize H by introducing creation and annihilation operators. Show that the theory contains two sets of particles of mass m.

(c) Rewrite the conserved charge

$$Q = \int d^3x \frac{i}{2} \left(\phi^*\pi^* - \pi\phi \right)$$

in terms of creation and annihilation operators, and evaluate the charge of the particles of each type.

(d) Consider the case of two complex Klein-Gordon fields with the same mass. Label the fields as $\phi_a(x)$, where $a = 1, 2$. Show that there are now four conserved charges, one given by the generalization of part (c), and the other three given by

$$Q^i = \int d^3x \frac{i}{2} \left(\phi_a^*(\sigma^i)_{ab}\pi_b^* - \pi_a(\sigma^i)_{ab}\phi_b \right),$$

where σ^i are the Pauli sigma matrices. Show that these three charges have the commutation relations of angular momentum $(SU(2))$. Generalize these results to the case of n identical complex scalar fields.[‡]

2.3   Evaluate the function

$$\langle 0| \phi(x)\phi(y) |0\rangle = D(x - y) = \int \frac{d^3p}{(2\pi)^3} \frac{1}{2E_{\mathbf{p}}} e^{-ip\cdot(x-y)},$$

for $(x - y)$ spacelike so that $(x - y)^2 = -r^2$, explicitly in terms of Bessel functions.

[‡]With some additional work you can show that there are actually six conserved charges in the case of two complex fields, and $n(2n - 1)$ in the case of n fields, corresponding to the generators of the rotation group in four and $2n$ dimensions, respectively. The extra symmetries often do not survive when nonlinear interactions of the fields are included.

Chapter 3

The Dirac Field

Having exhaustively treated the simplest relativistic field equation, we now move on to the second simplest, the Dirac equation. You may already be familiar with the Dirac equation in its original incarnation, that is, as a single-particle quantum-mechanical wave equation.* In this chapter our viewpoint will be quite different. First we will rederive the Dirac equation as a *classical* relativistic field equation, with special emphasis on its relativistic invariance. Then, in Section 3.5, we will quantize the Dirac field in a manner similar to that used for the Klein-Gordon field.

3.1 Lorentz Invariance in Wave Equations

First we must address a question that we swept over in Chapter 2: What do we mean when we say that an equation is "relativistically invariant"? A reasonable definition is the following: If ϕ is a field or collection of fields and $\mathcal{D}$ is some differential operator, then the statement "$\mathcal{D}\phi = 0$ is relativistically invariant" means that if $\phi(x)$ satisfies this equation, and we perform a rotation or boost to a different frame of reference, then the transformed field, in the new frame of reference, satisfies the same equation. Equivalently, we can imagine physically rotating or boosting all particles or fields by a common angle or velocity; again, the equation $\mathcal{D}\phi = 0$ should be true after the transformation. We will adopt this "active" point of view toward transformations in the following analysis.

The Lagrangian formulation of field theory makes it especially easy to discuss Lorentz invariance. An equation of motion is automatically Lorentz invariant by the above definition if it follows from a Lagrangian that is a Lorentz *scalar*. This is an immediate consequence of the principle of least action: If boosts leave the Lagrangian unchanged, the boost of an extremum in the action will be another extremum.

---

*This subject is covered, for example, in Schiff (1968), Chapter 13; Baym (1969), Chapter 23; Sakurai (1967), Chapter 3. Although the present chapter is self-contained, we recommend that you also study the single-particle Dirac equation at some point.

As an example, consider the Klein-Gordon theory. We can write an arbitrary Lorentz transformation as

$$x^\mu \to x'^\mu = \Lambda^\mu{}_\nu x^\nu, \tag{3.1}$$

for some 4×4 matrix Λ. What happens to the Klein-Gordon field $\phi(x)$ under this transformation? Think of the field ϕ as measuring the local value of some quantity that is distributed through space. If there is an accumulation of this quantity at $x = x_0$, $\phi(x)$ will have a maximum at x_0. If we now transform the original distribution by a boost, the new distribution will have a maximum at $x = \Lambda x_0$. This is illustrated in Fig. 3.1(a). The corresponding transformation of the field is

$$\phi(x) \to \phi'(x) = \phi(\Lambda^{-1}x). \tag{3.2}$$

That is, the transformed field, evaluated at the boosted point, gives the same value as the original field evaluated at the point before boosting.

We should check that this transformation leaves the form of the Klein-Gordon Lagrangian unchanged. According to (3.2), the mass term $\frac{1}{2}m^2\phi^2(x)$ is simply shifted to the point $(\Lambda^{-1}x)$. The transformation of $\partial_\mu \phi(x)$ is

$$\partial_\mu \phi(x) \to \partial_\mu \big(\phi(\Lambda^{-1}x)\big) = (\Lambda^{-1})^\nu{}_\mu (\partial_\nu \phi)(\Lambda^{-1}x). \tag{3.3}$$

Since the metric tensor $g^{\mu\nu}$ is Lorentz invariant, the matrices Λ^{-1} obey the identity

$$(\Lambda^{-1})^\rho{}_\mu (\Lambda^{-1})^\sigma{}_\nu g^{\mu\nu} = g^{\rho\sigma}. \tag{3.4}$$

Using this relation, we can compute the transformation law of the kinetic term of the Klein-Gordon Lagrangian:

$$
\begin{aligned}
(\partial_\mu \phi(x))^2 &\to g^{\mu\nu}\big(\partial_\mu \phi'(x)\big)\big(\partial_\nu \phi'(x)\big) \\
&= g^{\mu\nu}\big[(\Lambda^{-1})^\rho{}_\mu \partial_\rho \phi\big]\big[(\Lambda^{-1})^\sigma{}_\nu \partial_\sigma \phi\big](\Lambda^{-1}x) \\
&= g^{\rho\sigma}\big(\partial_\rho \phi\big)\big(\partial_\sigma \phi\big)(\Lambda^{-1}x) \\
&= (\partial_\mu \phi)^2(\Lambda^{-1}x).
\end{aligned}
$$

Thus, the whole Lagrangian is simply transformed as a scalar:

$$\mathcal{L}(x) \to \mathcal{L}(\Lambda^{-1}x). \tag{3.5}$$

The action S, formed by integrating $\mathcal{L}$ over spacetime, is Lorentz invariant. A similar calculation shows that the equation of motion is invariant:

$$
\begin{aligned}
(\partial^2 + m^2)\phi'(x) &= \big[(\Lambda^{-1})^\nu{}_\mu \partial_\nu (\Lambda^{-1})^{\sigma\mu}\partial_\sigma + m^2\big]\phi(\Lambda^{-1}x) \\
&= (g^{\nu\sigma}\partial_\nu \partial_\sigma + m^2)\phi(\Lambda^{-1}x) \\
&= 0.
\end{aligned}
$$

The transformation law (3.2) used for ϕ is the simplest possible transformation law for a field. It is the only possibility for a field that has just one component. But we know examples of multiple-component fields that transform in more complicated ways. The most familiar case is that of a vector field,

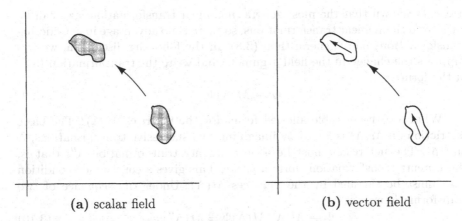

(a) scalar field   **(b)** vector field

Figure 3.1. When a rotation is performed on a vector field, it affects the *orientation* of the vector as well as the location of the region containing the configuration.

such as the 4-current density $j^\mu(x)$ or the vector potential $A^\mu(x)$. In this case, the quantity that is distributed in spacetime also carries an orientation, which must be rotated or boosted. As shown in Fig. 3.1(b), the orientation must be rotated forward as the point of evaluation of the field is changed:

under 3-dimensional rotations, $\quad V^i(x) \to R^{ij} V^j(R^{-1}x);$

under Lorentz transformations, $\quad V^\mu(x) \to \Lambda^\mu{}_\nu V^\nu(\Lambda^{-1}x).$

Tensors of arbitrary rank can be built out of vectors by adding more indices, with correspondingly more factors of Λ in the transformation law. Using such vector and tensor fields we can write a variety of Lorentz-invariant equations, for example, Maxwell's equations,

$$\partial^\mu F_{\mu\nu} = 0 \qquad \text{or} \qquad \partial^2 A_\nu - \partial_\nu \partial^\mu A_\mu = 0, \tag{3.6}$$

which follow from the Lagrangian

$$\mathcal{L}_{\text{Maxwell}} = -\tfrac{1}{4}(F_{\mu\nu})^2 = -\tfrac{1}{4}(\partial_\mu A_\nu - \partial_\nu A_\mu)^2. \tag{3.7}$$

In general, any equation in which each term has the same set of uncontracted Lorentz indices will naturally be invariant under Lorentz transformations.

This method of tensor notation yields a large class of Lorentz-invariant equations, but it turns out that there are still more. How do we find them? We could try to systematically find all possible transformation laws for a field. Then it would not be hard to write invariant Lagrangians. For simplicity, we will restrict our attention to linear transformations, so that, if Φ_a is an n component multiplet, the Lorentz transformation law is given by an $n \times n$ matrix $M(\Lambda)$:

$$\Phi_a(x) \to M_{ab}(\Lambda)\Phi_b(\Lambda^{-1}x). \tag{3.8}$$

It can be shown that the most general nonlinear transformation laws can be built from these linear transformations, so there is no advantage in considering transformations more general than (3.8). In the following discussion, we will suppress the change in the field argument and write the transformation (3.8) in the form

$$\Phi \to M(\Lambda)\Phi. \tag{3.9}$$

What are the possible allowed forms for the matrices $M(\Lambda)$? The basic restriction on $M(\Lambda)$ is found by imagining two successive transformations, Λ and Λ'. The net result must be a new Lorentz transformation Λ''; that is, the Lorentz transformations form a *group*. This gives a consistency condition that must be satisfied by the matrices $M(\Lambda)$: Under the sequence of two transformations,

$$\Phi \to M(\Lambda')M(\Lambda)\Phi = M(\Lambda'')\Phi, \tag{3.10}$$

for $\Lambda'' = \Lambda'\Lambda$. Thus the correspondence between the matrices M and the transformations Λ must be preserved under multiplication. In mathematical language, we say that the matrices M must form an n-dimensional *representation* of the Lorentz group. So our question now is rephrased in mathematical language: What are the (finite-dimensional) matrix representations of the Lorentz group?

Before answering this question for the Lorentz group, let us consider a simpler group, the rotation group in three dimensions. This group has representations of every dimensionality n, familiar in quantum mechanics as the matrices that rotate the n-component wavefunctions of particles of different spins. The dimensionality is related to the spin quantum number s by $n = 2s + 1$. The most important nontrivial representation is the two-dimensional representation, corresponding to spin 1/2. The matrices of this representation are the 2×2 unitary matrices with determinant 1, which can be expressed as

$$U = e^{-i\theta^i\sigma^i/2}, \tag{3.11}$$

where θ^i are three arbitrary parameters and σ^i are the Pauli sigma matrices.

For any continuous group, the transformations that lie infinitesimally close to the identity define a vector space, called the *Lie algebra* of the group. The basis vectors for this vector space are called the *generators* of the Lie algebra, or of the group. For the rotation group, the generators are the angular momentum operators J^i, which satisfy the commutation relations

$$[J^i, J^j] = i\epsilon^{ijk}J^k. \tag{3.12}$$

The finite rotation operations are formed by exponentiating these operators: In quantum mechanics, the operator

$$R = \exp[-i\theta^i J^i] \tag{3.13}$$

gives the rotation by an angle $|\theta|$ about the axis $\hat{\theta}$. The commutation relations of the operators J^i determine the multiplication laws of these rotation operators. Thus, a set of matrices satisfying the commutation relations (3.12) produces, through exponentiation as in (3.13), a representation of the rotation group. In the example given in the previous paragraph, the representation of the angular momentum operators

$$J^i \rightarrow \frac{\sigma^i}{2} \tag{3.14}$$

produces the representation of the rotation group given in Eq. (3.11). It is generally true that one can find matrix representations of a continuous group by finding matrix representations of the generators of the group (which must satisfy the proper commutation relations), then exponentiating these infinitesimal transformations.

For our present problem, we need to know the commutation relations of the generators of the group of Lorentz transformations. For the rotation group, one can work out the commutation relations by writing the generators as differential operators; from the expression

$$\mathbf{J} = \mathbf{x} \times \mathbf{p} = \mathbf{x} \times (-i\nabla), \tag{3.15}$$

the angular momentum commutation relations (3.12) follow straightforwardly. The use of the cross product in (3.15) is special to the case of three dimensions. However, we can also write the operators as an antisymmetric tensor,

$$J^{ij} = -i(x^i \nabla^j - x^j \nabla^i),$$

so that $J^3 = J^{12}$ and so on. The generalization to four-dimensional Lorentz transformations is now quite natural:

$$J^{\mu\nu} = i(x^\mu \partial^\nu - x^\nu \partial^\mu). \tag{3.16}$$

We will soon see that these six operators generate the three boosts and three rotations of the Lorentz group.

To determine the commutation rules of the Lorentz algebra, we can now simply compute the commutators of the differential operators (3.16). The result is

$$[J^{\mu\nu}, J^{\rho\sigma}] = i(g^{\nu\rho} J^{\mu\sigma} - g^{\mu\rho} J^{\nu\sigma} - g^{\nu\sigma} J^{\mu\rho} + g^{\mu\sigma} J^{\nu\rho}). \tag{3.17}$$

Any matrices that are to represent this algebra must obey these same commutation rules.

Just to see that we have this right, let us look at one particular representation (which we will simply pull out of a hat). Consider the 4×4 matrices

$$(\mathcal{J}^{\mu\nu})_{\alpha\beta} = i(\delta^\mu{}_\alpha \delta^\nu{}_\beta - \delta^\mu{}_\beta \delta^\nu{}_\alpha). \tag{3.18}$$

(Here μ and ν label which of the six matrices we want, while α and β label components of the matrices.) You can easily verify that these matrices satisfy the commutation relations (3.17). In fact, they are nothing but the matrices that act on ordinary Lorentz 4-vectors. To see this, parametrize an infinitesimal transformation as follows:

$$V^\alpha \to \left(\delta^\alpha_{\ \beta} - \frac{i}{2}\omega_{\mu\nu}(\mathcal{J}^{\mu\nu})^\alpha_{\ \beta}\right)V^\beta, \tag{3.19}$$

where V is a 4-vector and $\omega_{\mu\nu}$, an antisymmetric tensor, gives the infinitesimal angles. For example, consider the case $\omega_{12} = -\omega_{21} = \theta$, with all other components of ω equal to zero. Then Eq. (3.19) becomes

$$V \to \begin{pmatrix} 1 & 0 & 0 & 0 \\ 0 & 1 & -\theta & 0 \\ 0 & \theta & 1 & 0 \\ 0 & 0 & 0 & 1 \end{pmatrix} V, \tag{3.20}$$

which is just an infinitesimal rotation in the xy-plane. You can also verify that setting $\omega_{01} = -\omega_{10} = \beta$ gives

$$V \to \begin{pmatrix} 1 & \beta & 0 & 0 \\ \beta & 1 & 0 & 0 \\ 0 & 0 & 1 & 0 \\ 0 & 0 & 0 & 1 \end{pmatrix} V, \tag{3.21}$$

an infinitesimal boost in the x-direction. The other components of ω generate the remaining boosts and rotations in a similar manner.

3.2 The Dirac Equation

Now that we have seen one finite-dimensional representation of the Lorentz group, the logical next step would be to develop the formalism for finding all other representations. Although this is not very difficult to do (see Problem 3.1), it is hardly necessary for our purposes, since we are mainly interested in the representation(s) corresponding to spin 1/2.

We can find such a representation using a trick due to Dirac: Suppose that we had a set of four $n \times n$ matrices γ^μ satisfying the anticommutation relations

$$\{\gamma^\mu, \gamma^\nu\} \equiv \gamma^\mu\gamma^\nu + \gamma^\nu\gamma^\mu = 2g^{\mu\nu} \times \mathbf{1}_{n\times n} \qquad \text{(Dirac algebra)}. \tag{3.22}$$

Then we could immediately write down an n-dimensional representation of the Lorentz algebra. Here it is:

$$S^{\mu\nu} = \frac{i}{4}[\gamma^\mu, \gamma^\nu]. \tag{3.23}$$

By repeated use of (3.22), it is easy to verify that these matrices satisfy the commutation relations (3.17).

This computation goes through in any dimensionality, with Lorentz or Euclidean metric. In particular, it should work in three-dimensional Euclidean space, and in fact we can simply write

$$\gamma^j \equiv i\sigma^j \qquad \text{(Pauli sigma matrices)},$$

so that $\{\gamma^i, \gamma^j\} = -2\delta^{ij}$.

The factor of i in the first line and the minus sign in the second line are purely conventional. The matrices representing the Lorentz algebra are then

$$S^{ij} = \tfrac{1}{2}\epsilon^{ijk}\sigma^k, \tag{3.24}$$

which we recognize as the two-dimensional representation of the rotation group.

Now let us find Dirac matrices γ^μ for four-dimensional Minkowski space. It turns out that these matrices must be at least 4×4. (There is no fourth 2×2 matrix, for example, that anticommutes with the three Pauli sigma matrices.) Further, all 4×4 representations of the Dirac algebra are unitarily equivalent.[†] We thus need only write one explicit realization of the Dirac algebra. One representation, in 2×2 block form, is

$$\gamma^0 = \begin{pmatrix} 0 & 1 \\ 1 & 0 \end{pmatrix}; \qquad \gamma^i = \begin{pmatrix} 0 & \sigma^i \\ -\sigma^i & 0 \end{pmatrix}. \tag{3.25}$$

This representation is called the *Weyl* or *chiral* representation. We will find it an especially convenient choice, and we will use it exclusively throughout this book. (Be careful, however, since many field theory textbooks choose a different representation, in which γ^0 is diagonal. Furthermore, books that use chiral representations often make a different choice of sign conventions.)

In our representation, the boost and rotation generators are

$$S^{0i} = \frac{i}{4}[\gamma^0, \gamma^i] = -\frac{i}{2}\begin{pmatrix} \sigma^i & 0 \\ 0 & -\sigma^i \end{pmatrix}, \tag{3.26}$$

and

$$S^{ij} = \frac{i}{4}[\gamma^i, \gamma^j] = \frac{1}{2}\epsilon^{ijk}\begin{pmatrix} \sigma^k & 0 \\ 0 & \sigma^k \end{pmatrix} \equiv \frac{1}{2}\epsilon^{ijk}\Sigma^k. \tag{3.27}$$

A four-component field ψ that transforms under boosts and rotations according to (3.26) and (3.27) is called a *Dirac spinor*. Note that the rotation generator S^{ij} is just the three-dimensional spinor transformation matrix (3.24) replicated twice. The boost generators S^{0i} are not Hermitian, and thus our implementation of boosts is not unitary (this was also true of the vector representation (3.18)). In fact the Lorentz group, being "noncompact", has no faithful, finite-dimensional representations that are unitary. But that does not matter to us, since ψ is not a wavefunction; it is a classical field.

[†]This statement and the preceding one follow from the general theory of the representations of the Lorentz group derived in Problem 3.1.

Now that we have the transformation law for ψ, we should look for an appropriate field equation. One possibility is simply the Klein-Gordon equation:

$$(\partial^2 + m^2)\psi = 0. \qquad (3.28)$$

This works because the spinor transformation matrices (3.26) and (3.27) operate only in the "internal" space; they go right through the differential operator. But it is possible to write a stronger, first-order equation, which implies (3.28) but contains additional information. To do this we need to know one more property of the γ matrices. With a short computation you can verify that

$$[\gamma^\mu, S^{\rho\sigma}] = (\mathcal{J}^{\rho\sigma})^\mu{}_\nu \gamma^\nu,$$

or equivalently,

$$\left(1 + \tfrac{i}{2}\omega_{\rho\sigma}S^{\rho\sigma}\right)\gamma^\mu\left(1 - \tfrac{i}{2}\omega_{\rho\sigma}S^{\rho\sigma}\right) = \left(1 - \tfrac{i}{2}\omega_{\rho\sigma}\mathcal{J}^{\rho\sigma}\right)^\mu{}_\nu \gamma^\nu.$$

This equation is just the infinitesimal form of

$$\Lambda_{\frac{1}{2}}^{-1}\gamma^\mu\Lambda_{\frac{1}{2}} = \Lambda^\mu{}_\nu\gamma^\nu, \qquad (3.29)$$

where

$$\Lambda_{\frac{1}{2}} = \exp\left(-\frac{i}{2}\omega_{\mu\nu}S^{\mu\nu}\right) \qquad (3.30)$$

is the spinor representation of the Lorentz transformation Λ (compare (3.19)). Equation (3.29) says that the γ matrices are invariant under simultaneous rotations of their vector and spinor indices (just like the σ^i under spatial rotations). In other words, we can "take the vector index μ on γ^μ seriously," and dot γ^μ into ∂_μ to form a Lorentz-invariant differential operator.

We are now ready to write down the Dirac equation. Here it is:

$$(i\gamma^\mu\partial_\mu - m)\psi(x) = 0. \qquad (3.31)$$

To show that it is Lorentz invariant, write down the Lorentz-transformed version of the left-hand side and calculate:

$$\begin{aligned}
\left[i\gamma^\mu\partial_\mu - m\right]\psi(x) &\to \left[i\gamma^\mu(\Lambda^{-1})^\nu{}_\mu\partial_\nu - m\right]\Lambda_{\frac{1}{2}}\psi(\Lambda^{-1}x) \\
&= \Lambda_{\frac{1}{2}}\Lambda_{\frac{1}{2}}^{-1}\left[i\gamma^\mu(\Lambda^{-1})^\nu{}_\mu\partial_\nu - m\right]\Lambda_{\frac{1}{2}}\psi(\Lambda^{-1}x) \\
&= \Lambda_{\frac{1}{2}}\left[i\Lambda_{\frac{1}{2}}^{-1}\gamma^\mu\Lambda_{\frac{1}{2}}(\Lambda^{-1})^\nu{}_\mu\partial_\nu - m\right]\psi(\Lambda^{-1}x) \\
&= \Lambda_{\frac{1}{2}}\left[i\Lambda^\mu{}_\sigma\gamma^\sigma(\Lambda^{-1})^\nu{}_\mu\partial_\nu - m\right]\psi(\Lambda^{-1}x) \\
&= \Lambda_{\frac{1}{2}}\left[i\gamma^\nu\partial_\nu - m\right]\psi(\Lambda^{-1}x) \\
&= 0.
\end{aligned}$$

To see that the Dirac equation implies the Klein-Gordon equation, act on the left with $(-i\gamma^\mu \partial_\mu - m)$:

$$0 = (-i\gamma^\mu \partial_\mu - m)(i\gamma^\nu \partial_\nu - m)\psi$$
$$= (\gamma^\mu \gamma^\nu \partial_\mu \partial_\nu + m^2)\psi$$
$$= (\tfrac{1}{2}\{\gamma^\mu, \gamma^\nu\}\partial_\mu \partial_\nu + m^2)\psi$$
$$= (\partial^2 + m^2)\psi.$$

To write down a Lagrangian for the Dirac theory, we must figure out how to multiply two Dirac spinors to form a Lorentz scalar. The obvious guess, $\psi^\dagger \psi$, does not work. Under a Lorentz boost this becomes $\psi^\dagger \Lambda^\dagger_{\frac{1}{2}} \Lambda_{\frac{1}{2}} \psi$; if the boost matrix were unitary, we would have $\Lambda^\dagger_{\frac{1}{2}} = \Lambda^{-1}_{\frac{1}{2}}$ and everything would be fine. But $\Lambda_{\frac{1}{2}}$ is not unitary, because the generators (3.26) are not Hermitian. The solution is to define

$$\bar{\psi} \equiv \psi^\dagger \gamma^0. \tag{3.32}$$

Under an infinitesimal Lorentz transformation parametrized by $\omega_{\mu\nu}$, we have $\bar{\psi} \to \psi^\dagger (1 + \tfrac{i}{2}\omega_{\mu\nu}(S^{\mu\nu})^\dagger)\gamma^0$. The sum over μ and ν has six distinct nonzero terms. In the rotation terms, where μ and ν are both nonzero, $(S^{\mu\nu})^\dagger = S^{\mu\nu}$ and $S^{\mu\nu}$ commutes with γ^0. In the boost terms, where μ or ν is 0, $(S^{\mu\nu})^\dagger = -(S^{\mu\nu})$ but $S^{\mu\nu}$ anticommutes with γ^0. Passing the γ^0 to the left therefore removes the dagger from $S^{\mu\nu}$, yielding the transformation law

$$\bar{\psi} \to \bar{\psi}\Lambda^{-1}_{\frac{1}{2}}, \tag{3.33}$$

and therefore the quantity $\bar{\psi}\psi$ is a Lorentz scalar. Similarly you can show (with the aid of (3.29)) that $\bar{\psi}\gamma^\mu\psi$ is a Lorentz vector.

The correct, Lorentz-invariant Dirac Lagrangian is therefore

$$\mathcal{L}_{\text{Dirac}} = \bar{\psi}(i\gamma^\mu \partial_\mu - m)\psi. \tag{3.34}$$

The Euler-Lagrange equation for $\bar{\psi}$ (or $\psi^\dagger$) immediately yields the Dirac equation in the form (3.31); the Euler-Lagrange equation for ψ gives the same equation, in Hermitian-conjugate form:

$$-i\partial_\mu \bar{\psi}\gamma^\mu - m\bar{\psi} = 0. \tag{3.35}$$

Weyl Spinors

From the block-diagonal form of the generators (3.26) and (3.27), it is apparent that the Dirac representation of the Lorentz group is *reducible*.[‡] We can form two 2-dimensional representations by considering each block separately, and writing

$$\psi = \begin{pmatrix} \psi_L \\ \psi_R \end{pmatrix}. \tag{3.36}$$

[‡]If we had used a different representation of the gamma matrices, the reducibility would not be manifest; this is essentially the reason for using the chiral representation.

The two-component objects ψ_L and ψ_R are called left-handed and right-handed *Weyl spinors*. You can easily verify that their transformation laws, under infinitesimal rotations $\boldsymbol{\theta}$ and boosts $\boldsymbol{\beta}$, are

$$\psi_L \to (1 - i\boldsymbol{\theta} \cdot \tfrac{\boldsymbol{\sigma}}{2} - \boldsymbol{\beta} \cdot \tfrac{\boldsymbol{\sigma}}{2})\psi_L;$$
$$\psi_R \to (1 - i\boldsymbol{\theta} \cdot \tfrac{\boldsymbol{\sigma}}{2} + \boldsymbol{\beta} \cdot \tfrac{\boldsymbol{\sigma}}{2})\psi_R. \tag{3.37}$$

These transformation laws are connected by complex conjugation; using the identity

$$\sigma^2 \boldsymbol{\sigma}^* = -\boldsymbol{\sigma}\sigma^2, \tag{3.38}$$

it is not hard to show that the quantity $\sigma^2 \psi_L^*$ transforms like a right-handed spinor.

In terms of ψ_L and ψ_R, the Dirac equation is

$$(i\gamma^\mu \partial_\mu - m)\psi = \begin{pmatrix} -m & i(\partial_0 + \boldsymbol{\sigma} \cdot \boldsymbol{\nabla}) \\ i(\partial_0 - \boldsymbol{\sigma} \cdot \boldsymbol{\nabla}) & -m \end{pmatrix} \begin{pmatrix} \psi_L \\ \psi_R \end{pmatrix} = 0. \tag{3.39}$$

The two Lorentz group representations ψ_L and ψ_R are mixed by the mass term in the Dirac equation. But if we set $m = 0$, the equations for ψ_L and ψ_R decouple:

$$i(\partial_0 - \boldsymbol{\sigma} \cdot \boldsymbol{\nabla})\psi_L = 0;$$
$$i(\partial_0 + \boldsymbol{\sigma} \cdot \boldsymbol{\nabla})\psi_R = 0. \tag{3.40}$$

These are called the *Weyl equations*; they are especially important when treating neutrinos and the theory of weak interactions.

It is possible to clean up this notation slightly. Define

$$\sigma^\mu \equiv (1, \boldsymbol{\sigma}), \qquad \bar{\sigma}^\mu \equiv (1, -\boldsymbol{\sigma}), \tag{3.41}$$

so that

$$\gamma^\mu = \begin{pmatrix} 0 & \sigma^\mu \\ \bar{\sigma}^\mu & 0 \end{pmatrix}. \tag{3.42}$$

(The bar on $\bar{\sigma}$ has absolutely nothing to do with the bar on $\bar{\psi}$.) Then the Dirac equation can be written

$$\begin{pmatrix} -m & i\sigma \cdot \partial \\ i\bar{\sigma} \cdot \partial & -m \end{pmatrix} \begin{pmatrix} \psi_L \\ \psi_R \end{pmatrix} = 0, \tag{3.43}$$

and the Weyl equations become

$$i\bar{\sigma} \cdot \partial \psi_L = 0; \qquad i\sigma \cdot \partial \psi_R = 0. \tag{3.44}$$

3.3 Free-Particle Solutions of the Dirac Equation

To get some feel for the physics of the Dirac equation, let us now discuss its plane-wave solutions. Since a Dirac field ψ obeys the Klein-Gordon equation, we know immediately that it can be written as a linear combination of plane waves:

$$\psi(x) = u(p)e^{-ip\cdot x}, \qquad \text{where } p^2 = m^2. \tag{3.45}$$

For the moment we will concentrate on solutions with positive frequency, that is, $p^0 > 0$. The column vector $u(p)$ must obey an additional constraint, found by plugging (3.45) into the Dirac equation:

$$(\gamma^\mu p_\mu - m)u(p) = 0. \tag{3.46}$$

It is easiest to analyze this equation in the rest frame, where $p = p_0 = (m, \mathbf{0})$; the solution for general p can then be found by boosting with $\Lambda_{\frac{1}{2}}$. In the rest frame, Eq. (3.46) becomes

$$(m\gamma^0 - m)u(p_0) = m\begin{pmatrix} -1 & 1 \\ 1 & -1 \end{pmatrix}u(p_0) = 0,$$

and the solutions are

$$u(p_0) = \sqrt{m}\begin{pmatrix} \xi \\ \xi \end{pmatrix}, \tag{3.47}$$

for any numerical two-component spinor ξ. We conventionally normalize ξ so that $\xi^\dagger \xi = 1$; the factor $\sqrt{m}$ has been inserted for future convenience. We can interpret the spinor ξ by looking at the rotation generator (3.27): ξ transforms under rotations as an ordinary two-component spinor of the rotation group, and therefore determines the spin orientation of the Dirac solution in the usual way. For example, when $\xi = \binom{1}{0}$, the particle has spin up along the 3-direction.

Notice that after applying the Dirac equation, we are free to choose only two of the four components of $u(p)$. This is just what we want, since a spin-1/2 particle has only two physical states—spin up and spin down. (Of course we are being a bit premature in talking about *particles* and *spin*. We will *prove* that the spin angular momentum of a Dirac particle is $\hbar/2$ when we quantize the Dirac theory in Section 3.5; for now, just notice that there are two possible solutions $u(p)$ for any momentum p.)

Now that we have the general form of $u(p)$ in the rest frame, we can obtain $u(p)$ in any other frame by boosting. Consider a boost along the 3-direction. First we should remind ourselves of what the boost does to the 4-momentum vector. In infinitesimal form,

$$\begin{pmatrix} E \\ p^3 \end{pmatrix} = \left[1 + \eta \begin{pmatrix} 0 & 1 \\ 1 & 0 \end{pmatrix} \right] \begin{pmatrix} m \\ 0 \end{pmatrix},$$

where η is some infinitesimal parameter. For finite η we must write

$$\begin{pmatrix} E \\ p^3 \end{pmatrix} = \exp\left[\eta \begin{pmatrix} 0 & 1 \\ 1 & 0 \end{pmatrix}\right] \begin{pmatrix} m \\ 0 \end{pmatrix}$$

$$= \left[\cosh\eta \begin{pmatrix} 1 & 0 \\ 0 & 1 \end{pmatrix} + \sinh\eta \begin{pmatrix} 0 & 1 \\ 1 & 0 \end{pmatrix}\right] \begin{pmatrix} m \\ 0 \end{pmatrix} \tag{3.48}$$

$$= \begin{pmatrix} m\cosh\eta \\ m\sinh\eta \end{pmatrix}.$$

The parameter η is called the *rapidity*. It is the quantity that is additive under successive boosts.

Now apply the same boost to $u(p)$. According to Eqs. (3.26) and (3.30),

$$u(p) = \exp\left[-\tfrac{1}{2}\eta \begin{pmatrix} \sigma^3 & 0 \\ 0 & -\sigma^3 \end{pmatrix}\right] \sqrt{m} \begin{pmatrix} \xi \\ \xi \end{pmatrix}$$

$$= \left[\cosh(\tfrac{1}{2}\eta) \begin{pmatrix} 1 & 0 \\ 0 & 1 \end{pmatrix} - \sinh(\tfrac{1}{2}\eta) \begin{pmatrix} \sigma^3 & 0 \\ 0 & -\sigma^3 \end{pmatrix}\right] \sqrt{m} \begin{pmatrix} \xi \\ \xi \end{pmatrix}$$

$$= \begin{pmatrix} e^{\eta/2}\left(\frac{1-\sigma^3}{2}\right) + e^{-\eta/2}\left(\frac{1+\sigma^3}{2}\right) & 0 \\ 0 & e^{\eta/2}\left(\frac{1+\sigma^3}{2}\right) + e^{-\eta/2}\left(\frac{1-\sigma^3}{2}\right) \end{pmatrix} \sqrt{m} \begin{pmatrix} \xi \\ \xi \end{pmatrix}$$

$$= \begin{pmatrix} \left[\sqrt{E+p^3}\left(\frac{1-\sigma^3}{2}\right) + \sqrt{E-p^3}\left(\frac{1+\sigma^3}{2}\right)\right]\xi \\ \left[\sqrt{E+p^3}\left(\frac{1+\sigma^3}{2}\right) + \sqrt{E-p^3}\left(\frac{1-\sigma^3}{2}\right)\right]\xi \end{pmatrix}. \tag{3.49}$$

The last line can be simplified to give

$$u(p) = \begin{pmatrix} \sqrt{p\cdot\sigma}\,\xi \\ \sqrt{p\cdot\bar\sigma}\,\xi \end{pmatrix}, \tag{3.50}$$

where it is understood that in taking the square root of a matrix, we take the positive root of each eigenvalue. This expression for $u(p)$ is not only more compact, but is also valid for an arbitrary direction of $\mathbf{p}$. When working with expressions of this form, it is often useful to know the identity

$$(p\cdot\sigma)(p\cdot\bar\sigma) = p^2 = m^2. \tag{3.51}$$

You can then verify directly that (3.50) is a solution of the Dirac equation in the form of (3.43).

In practice it is often convenient to work with specific spinors ξ. A useful choice here would be eigenstates of σ^3. For example, if $\xi = \begin{pmatrix} 1 \\ 0 \end{pmatrix}$ (spin up along the 3-axis), we get

$$u(p) = \begin{pmatrix} \sqrt{E-p^3}\begin{pmatrix} 1 \\ 0 \end{pmatrix} \\ \sqrt{E+p^3}\begin{pmatrix} 1 \\ 0 \end{pmatrix} \end{pmatrix} \xrightarrow[\text{large boost}]{} \sqrt{2E}\begin{pmatrix} 0 \\ \begin{pmatrix} 1 \\ 0 \end{pmatrix} \end{pmatrix}, \tag{3.52}$$

while for $\xi = \binom{0}{1}$ (spin down along the 3-axis) we have

$$u(p) = \begin{pmatrix} \sqrt{E+p^3}\binom{0}{1} \\ \sqrt{E-p^3}\binom{0}{1} \end{pmatrix} \xrightarrow[\text{large boost}]{} \sqrt{2E}\begin{pmatrix} \binom{0}{1} \\ 0 \end{pmatrix}. \tag{3.53}$$

In the limit $\eta \to \infty$ the states degenerate into the two-component spinors of a massless particle. (We now see the reason for the factor of $\sqrt{m}$ in (3.47): It keeps the spinor expressions finite in the massless limit.)

The solutions (3.52) and (3.53) are eigenstates of the *helicity* operator,

$$h \equiv \hat{p} \cdot \mathbf{S} = \frac{1}{2}\hat{p}_i \begin{pmatrix} \sigma^i & 0 \\ 0 & \sigma^i \end{pmatrix}. \tag{3.54}$$

A particle with $h = +1/2$ is called *right-handed*, while one with $h = -1/2$ is called *left-handed*. The helicity of a massive particle depends on the frame of reference, since one can always boost to a frame in which its momentum is in the opposite direction (but its spin is unchanged). For a massless particle, which travels at the speed of light, one cannot perform such a boost.

The extremely simple form of $u(p)$ for a massless particle in a helicity eigenstate makes the behavior of such a particle easy to understand. In Chapter 1, it enabled us to guess the form of the $e^+e^- \to \mu^+\mu^-$ cross section in the massless limit. In subsequent chapters we will often do a mindless calculation first, then look at helicity eigenstates in the high-energy limit to understand what we have done.

Incidentally, we are now ready to understand the origin of the notation ψ_L and ψ_R for Weyl spinors. The solutions of the Weyl equations are states of definite helicity, corresponding to left- and right-handed particles, respectively. The Lorentz invariance of helicity (for a massless particle) is manifest in the notation of Weyl spinors, since ψ_L and ψ_R live in different representations of the Lorentz group.

It is convenient to write the normalization condition for $u(p)$ in a Lorentz-invariant way. We saw above that $\psi^\dagger\psi$ is not Lorentz invariant. Similarly,

$$u^\dagger u = \left(\xi^\dagger\sqrt{p\cdot\sigma},\, \xi^\dagger\sqrt{p\cdot\bar{\sigma}}\right) \cdot \begin{pmatrix} \sqrt{p\cdot\sigma}\,\xi \\ \sqrt{p\cdot\bar{\sigma}}\,\xi \end{pmatrix}$$

$$= 2E_{\mathbf{p}}\xi^\dagger\xi. \tag{3.55}$$

To make a Lorentz scalar we define

$$\bar{u}(p) = u^\dagger(p)\gamma^0. \tag{3.56}$$

Then by an almost identical calculation,

$$\bar{u}u = 2m\xi^\dagger\xi. \tag{3.57}$$

This will be our normalization condition, once we also require that the two-component spinor ξ be normalized as usual: $\xi^\dagger\xi = 1$. It is also conventional to choose basis spinors ξ^1 and ξ^2 (such as $\binom{1}{0}$ and $\binom{0}{1}$) that are orthogonal. For a massless particle Eq. (3.57) is trivial, so we must write the normalization condition in the form of (3.55).

Let us summarize our discussion so far. The general solution of the Dirac equation can be written as a linear combination of plane waves. The positive-frequency waves are of the form

$$\psi(x) = u(p)e^{-ip\cdot x}, \qquad p^2 = m^2, \qquad p^0 > 0. \tag{3.58}$$

There are two linearly independent solutions for $u(p)$,

$$u^s(p) = \begin{pmatrix} \sqrt{p\cdot\sigma}\,\xi^s \\ \sqrt{p\cdot\bar\sigma}\,\xi^s \end{pmatrix}, \qquad s = 1,2 \tag{3.59}$$

which we normalize according to

$$\bar u^r(p)u^s(p) = 2m\delta^{rs} \qquad \text{or} \qquad u^{r\dagger}(p)u^s(p) = 2E_{\mathbf p}\delta^{rs}. \tag{3.60}$$

In exactly the same way, we can find the negative-frequency solutions:

$$\psi(x) = v(p)e^{+ip\cdot x}, \qquad p^2 = m^2, \qquad p^0 > 0. \tag{3.61}$$

(Note that we have chosen to put the $+$ sign into the exponential, rather than having $p^0 < 0$.) There are two linearly independent solutions for $v(p)$,

$$v^s(p) = \begin{pmatrix} \sqrt{p\cdot\sigma}\,\eta^s \\ -\sqrt{p\cdot\bar\sigma}\,\eta^s \end{pmatrix}, \qquad s = 1,2 \tag{3.62}$$

where η^s is another basis of two-component spinors. These solutions are normalized according to

$$\bar v^r(p)v^s(p) = -2m\delta^{rs} \qquad \text{or} \qquad v^{r\dagger}(p)v^s(p) = +2E_{\mathbf p}\delta^{rs}. \tag{3.63}$$

The u's and v's are also orthogonal to each other:

$$\bar u^r(p)v^s(p) = \bar v^r(p)u^s(p) = 0. \tag{3.64}$$

Be careful, since $u^{r\dagger}(p)v^s(p) \neq 0$ and $v^{r\dagger}(p)u^s(p) \neq 0$. However, note that

$$u^{r\dagger}(\mathbf p)v^s(-\mathbf p) = v^{r\dagger}(-\mathbf p)u^s(\mathbf p) = 0, \tag{3.65}$$

where we have changed the sign of the 3-momentum in one factor of each spinor product.

Spin Sums

In evaluating Feynman diagrams, we will often wish to sum over the polarization states of a fermion. We can derive the relevant completeness relations with a simple calculation:

$$\sum_{s=1,2} u^s(p)\bar u^s(p) = \sum_s \begin{pmatrix} \sqrt{p\cdot\sigma}\,\xi^s \\ \sqrt{p\cdot\bar\sigma}\,\xi^s \end{pmatrix} \left(\xi^{s\dagger}\sqrt{p\cdot\bar\sigma}, \; \xi^{s\dagger}\sqrt{p\cdot\sigma} \right)$$

$$= \begin{pmatrix} \sqrt{p\cdot\sigma}\sqrt{p\cdot\bar\sigma} & \sqrt{p\cdot\sigma}\sqrt{p\cdot\sigma} \\ \sqrt{p\cdot\bar\sigma}\sqrt{p\cdot\bar\sigma} & \sqrt{p\cdot\bar\sigma}\sqrt{p\cdot\sigma} \end{pmatrix}$$

$$= \begin{pmatrix} m & p \cdot \sigma \\ p \cdot \bar{\sigma} & m \end{pmatrix}.$$

In the second line we have used

$$\sum_{s=1,2} \xi^s \xi^{s\dagger} = 1 = \begin{pmatrix} 1 & 0 \\ 0 & 1 \end{pmatrix}.$$

Thus we arrive at the desired formula,

$$\sum_s u^s(p) \bar{u}^s(p) = \gamma \cdot p + m. \tag{3.66}$$

Similarly,

$$\sum_s v^s(p) \bar{v}^s(p) = \gamma \cdot p - m. \tag{3.67}$$

The combination $\gamma \cdot p$ occurs so often that Feynman introduced the notation $\not{p} \equiv \gamma^\mu p_\mu$. We will use this notation frequently from now on.

3.4 Dirac Matrices and Dirac Field Bilinears

We saw in Section 3.2 that the quantity $\bar{\psi}\psi$ is a Lorentz scalar. It is also easy to show that $\bar{\psi}\gamma^\mu \psi$ is a 4-vector—we used this fact in writing down the Dirac Lagrangian (3.34). Now let us ask a more general question: Consider the expression $\bar{\psi}\Gamma\psi$, where Γ is any 4×4 constant matrix. Can we decompose this expression into terms that have definite transformation properties under the Lorentz group? The answer is yes, if we write Γ in terms of the following basis of sixteen 4×4 matrices, defined as antisymmetric combinations of γ-matrices:

1	1 of these
γ^μ	4 of these
$\gamma^{\mu\nu} = \frac{1}{2}[\gamma^\mu, \gamma^\nu] \equiv \gamma^{[\mu}\gamma^{\nu]} \equiv -i\sigma^{\mu\nu}$	6 of these
$\gamma^{\mu\nu\rho} = \gamma^{[\mu}\gamma^\nu \gamma^{\rho]}$	4 of these
$\gamma^{\mu\nu\rho\sigma} = \gamma^{[\mu}\gamma^\nu \gamma^\rho \gamma^{\sigma]}$	1 of these
	16 total

The Lorentz-transformation properties of these matrices are easy to determine. For example,

$$\bar{\psi}\gamma^{\mu\nu}\psi \rightarrow \left(\bar{\psi}\Lambda_{\frac{1}{2}}^{-1}\right)\left(\tfrac{1}{2}[\gamma^\mu, \gamma^\nu]\right)\left(\Lambda_{\frac{1}{2}}\psi\right)$$

$$= \tfrac{1}{2}\bar{\psi}\left(\Lambda_{\frac{1}{2}}^{-1}\gamma^\mu \Lambda_{\frac{1}{2}}\Lambda_{\frac{1}{2}}^{-1}\gamma^\nu \Lambda_{\frac{1}{2}} - \Lambda_{\frac{1}{2}}^{-1}\gamma^\nu \Lambda_{\frac{1}{2}}\Lambda_{\frac{1}{2}}^{-1}\gamma^\mu \Lambda_{\frac{1}{2}}\right)\psi$$

$$= \Lambda^\mu{}_\alpha \Lambda^\nu{}_\beta \bar{\psi}\gamma^{\alpha\beta}\psi.$$

Each set of matrices transforms as an antisymmetric tensor of successively higher rank.

The last two sets of matrices can be simplified by introducing an additional gamma matrix,

$$\gamma^5 \equiv i\gamma^0\gamma^1\gamma^2\gamma^3 = -\frac{i}{4!}\epsilon^{\mu\nu\rho\sigma}\gamma_\mu\gamma_\nu\gamma_\rho\gamma_\sigma. \qquad (3.68)$$

Then $\gamma^{\mu\nu\rho\sigma} = -i\epsilon^{\mu\nu\rho\sigma}\gamma^5$ and $\gamma^{\mu\nu\rho} = +i\epsilon^{\mu\nu\rho\sigma}\gamma_\sigma\gamma^5$. The matrix γ^5 has the following properties, all of which can be verified using (3.68) and the anticommutation relations (3.22):

$$(\gamma^5)^\dagger = \gamma^5; \qquad (3.69)$$

$$(\gamma^5)^2 = 1; \qquad (3.70)$$

$$\{\gamma^5, \gamma^\mu\} = 0. \qquad (3.71)$$

This last property implies that $[\gamma^5, S^{\mu\nu}] = 0$. Thus the Dirac representation must be reducible, since eigenvectors of γ^5 whose eigenvalues are different transform without mixing (this criterion for reducibility is known as Schur's lemma). In our basis,

$$\gamma^5 = \begin{pmatrix} -1 & 0 \\ 0 & 1 \end{pmatrix} \qquad (3.72)$$

in block-diagonal form. So a Dirac spinor with only left- (right-) handed components is an eigenstate of γ^5 with eigenvalue -1 $(+1)$, and indeed these spinors do transform without mixing, as we saw explicitly in Section 3.2.

Let us now rewrite our table of 4×4 matrices, and introduce some standard terminology:

1	scalar	1
γ^μ	vector	4
$\sigma^{\mu\nu} = \frac{i}{2}[\gamma^\mu, \gamma^\nu]$	tensor	6
$\gamma^\mu\gamma^5$	pseudo-vector	4
γ^5	pseudo-scalar	1
		16

The terms *pseudo-vector* and *pseudo-scalar* arise from the fact that these quantities transform as a vector and scalar, respectively, under continuous Lorentz transformations, but with an additional sign change under parity transformations (as we will discuss in Section 3.6).

From the vector and pseudo-vector matrices we can form two currents out of Dirac field bilinears:

$$j^\mu(x) = \bar{\psi}(x)\gamma^\mu\psi(x); \qquad j^{\mu 5}(x) = \bar{\psi}(x)\gamma^\mu\gamma^5\psi(x). \qquad (3.73)$$

Let us compute the divergences of these currents, assuming that ψ satisfies the Dirac equation:

$$\partial_\mu j^\mu = (\partial_\mu \bar\psi)\gamma^\mu \psi + \bar\psi \gamma^\mu \partial_\mu \psi$$
$$= (im\bar\psi)\psi + \bar\psi(-im\psi) \tag{3.74}$$
$$= 0.$$

Thus j^μ is always conserved if $\psi(x)$ satisfies the Dirac equation. When we couple the Dirac field to the electromagnetic field, j^μ will become the electric current density. Similarly, one can compute

$$\partial_\mu j^{\mu 5} = 2im\bar\psi \gamma^5 \psi. \tag{3.75}$$

If $m = 0$, this current (often called the *axial vector current*) is also conserved. It is then useful to form the linear combinations

$$j^\mu_L = \bar\psi \gamma^\mu \left(\frac{1-\gamma^5}{2}\right)\psi, \qquad j^\mu_R = \bar\psi \gamma^\mu \left(\frac{1+\gamma^5}{2}\right)\psi. \tag{3.76}$$

When $m = 0$, these are the electric current densities of left-handed and right-handed particles, respectively, and are separately conserved.

The two currents $j^\mu(x)$ and $j^{\mu 5}(x)$ are the Noether currents corresponding to the two transformations

$$\psi(x) \to e^{i\alpha}\psi(x) \qquad \text{and} \qquad \psi(x) \to e^{i\alpha\gamma^5}\psi(x).$$

The first of these is a symmetry of the Dirac Lagrangian (3.34). The second, called a *chiral transformation*, is a symmetry of the derivative term in $\mathcal{L}$ but not the mass term; thus, Noether's theorem confirms that the axial vector current is conserved only if $m = 0$.

Products of Dirac bilinears obey interchange relations, known as *Fierz identities*. We will discuss only the simplest of these, which will be needed several times later in the book. This simplest identity is most easily written in terms of the two-component Weyl spinors introduced in Eq. (3.36).

The core of the relation is the identity for the 2×2 matrices σ^μ defined in Eq. (3.41):

$$(\sigma^\mu)_{\alpha\beta}(\sigma_\mu)_{\gamma\delta} = 2\epsilon_{\alpha\gamma}\epsilon_{\beta\delta}. \tag{3.77}$$

(Here α, β, etc. are spinor indices, and ϵ is the antisymmetric symbol.) One can understand this relation by noting that the indices α, γ transform in the Lorentz representation of ψ_L, while β, δ transform in the separate representation of ψ_R, and the whole quantity must be a Lorentz invariant. Alternatively, one can just verify the 16 components of (3.77) explicitly.

By sandwiching identity (3.77) between the right-handed portions (i.e., lower half) of Dirac spinors u_1, u_2, u_3, u_4, we find the identity

$$(\bar u_{1R}\sigma^\mu u_{2R})(\bar u_{3R}\sigma_\mu u_{4R}) = 2\epsilon_{\alpha\gamma}\bar u_{1R\alpha}\bar u_{3R\gamma}\epsilon_{\beta\delta}u_{2R\beta}u_{4R\delta}$$
$$= -(\bar u_{1R}\sigma^\mu u_{4R})(\bar u_{3R}\sigma_\mu u_{2R}). \tag{3.78}$$

This nontrivial relation says that the product of bilinears in (3.78) is anti-symmetric under the interchange of the labels 2 and 4, and also under the interchange of 1 and 3. Identity (3.77) also holds for $\bar{\sigma}^\mu$, and so we also find

$$(\bar{u}_{1L}\bar{\sigma}^\mu u_{2L})(\bar{u}_{3L}\bar{\sigma}_\mu u_{4L}) = -(\bar{u}_{1L}\bar{\sigma}^\mu u_{4L})(\bar{u}_{3L}\bar{\sigma}_\mu u_{2L}). \qquad (3.79)$$

It is sometimes useful to combine the Fierz identity (3.78) with the identity linking σ^μ and $\bar{\sigma}^\mu$:

$$\epsilon_{\alpha\beta}(\sigma^\mu)_{\beta\gamma} = (\bar{\sigma}^{\mu T})_{\alpha\beta}\epsilon_{\beta\gamma}. \qquad (3.80)$$

This relation is also straightforward to verify explicitly. By the use of (3.80), (3.79), and the relation

$$\bar{\sigma}^\mu\sigma_\mu = 4, \qquad (3.81)$$

we can, for example, simplify horrible products of bilinears such as

$$\begin{aligned}(\bar{u}_{1L}\bar{\sigma}^\mu\sigma^\nu\bar{\sigma}^\lambda u_{2L})(\bar{u}_{3L}\bar{\sigma}_\mu\sigma_\nu\bar{\sigma}_\lambda u_{4L}) &= 2\epsilon_{\alpha\gamma}\bar{u}_{1L\alpha}\bar{u}_{3L\gamma}\epsilon_{\beta\delta}(\sigma^\nu\bar{\sigma}^\lambda u_{2L})_\beta(\sigma_\nu\bar{\sigma}_\lambda u_{4L})_\delta \\ &= 2\epsilon_{\alpha\gamma}\bar{u}_{1L\alpha}\bar{u}_{3L\gamma}\epsilon_{\beta\delta}u_{2L\beta}(\sigma^\lambda\bar{\sigma}^\nu\sigma_\nu\bar{\sigma}_\lambda u_{4L})_\delta \\ &= 2\cdot(4)^2\cdot\epsilon_{\alpha\gamma}\bar{u}_{1L\alpha}\bar{u}_{3L\gamma}\epsilon_{\beta\delta}u_{2L\beta}u_{4L\delta} \\ &= 16(\bar{u}_{1L}\bar{\sigma}^\mu u_{2L})(\bar{u}_{3L}\bar{\sigma}_\mu u_{4L}). \qquad (3.82)\end{aligned}$$

There are also Fierz rearrangement identities for 4-component Dirac spinors and 4×4 Dirac matrices. To derive these, however, it is useful to take a more systematic approach. Problem 3.6 presents a general method and gives some examples of its application.

3.5 Quantization of the Dirac Field

We are now ready to construct the quantum theory of the free Dirac field. From the Lagrangian

$$\mathcal{L} = \bar{\psi}(i\slashed{\partial} - m)\psi = \bar{\psi}(i\gamma^\mu\partial_\mu - m)\psi, \qquad (3.83)$$

we see that the canonical momentum conjugate to ψ is $i\psi^\dagger$, and thus the Hamiltonian is

$$H = \int d^3x\, \bar{\psi}(-i\gamma\cdot\nabla + m)\psi = \int d^3x\, \psi^\dagger[-i\gamma^0\gamma\cdot\nabla + m\gamma^0]\psi. \qquad (3.84)$$

If we define $\alpha = \gamma^0\gamma$, $\beta = \gamma^0$, you may recognize the quantity in brackets as the Dirac Hamiltonian of one-particle quantum mechanics:

$$h_D = -i\alpha\cdot\nabla + m\beta. \qquad (3.85)$$

How Not to Quantize the Dirac Field:
A Lesson in Spin and Statistics

To quantize the Dirac field in analogy with the Klein-Gordon field we would impose the canonical commutation relations

$$[\psi_a(\mathbf{x}), \psi_b^\dagger(\mathbf{y})] = \delta^{(3)}(\mathbf{x} - \mathbf{y})\delta_{ab}, \qquad \text{(equal times)} \qquad (3.86)$$

where a and b denote the spinor components of ψ. This already looks peculiar: If $\psi(x)$ were real-valued, the left-hand side would be antisymmetric under $\mathbf{x} \leftrightarrow \mathbf{y}$, while the right-hand side is symmetric. But ψ is complex, so we do not have a contradiction yet. In fact, we will soon find that much worse problems arise when we impose commutation relations on the Dirac field. But it is instructive to see how far we can get, in order to better understand the relation between spin and statistics. So let us press on; just remember that the next few pages will eventually turn out to be a blind alley.

Our first task is to find a representation of the commutation relations in terms of creation and annihilation operators that diagonalizes H. From the form of the Hamiltonian (3.84), it will clearly be helpful to expand $\psi(x)$ in a basis of eigenfunctions of h_D. We know these eigenfunctions already from our calculations in Section 3.3. There we found that

$$\left[i\gamma^0 \partial_0 + i\boldsymbol{\gamma} \cdot \boldsymbol{\nabla} - m\right] u^s(p) e^{-ip \cdot x} = 0,$$

so $u^s(\mathbf{p})e^{i\mathbf{p} \cdot \mathbf{x}}$ are eigenfunctions of h_D with eigenvalues $E_{\mathbf{p}}$. Similarly, the functions $v^s(\mathbf{p})e^{-i\mathbf{p} \cdot \mathbf{x}}$ (or equivalently, $v^s(-\mathbf{p})e^{+i\mathbf{p} \cdot \mathbf{x}}$) are eigenfunctions of h_D with eigenvalues $-E_{\mathbf{p}}$. These form a complete set of eigenfunctions, since for any $\mathbf{p}$ there are two u's and two v's, giving us four eigenvectors of the 4×4 matrix h_D.

Expanding ψ in this basis, we obtain

$$\psi(\mathbf{x}) = \int \frac{d^3p}{(2\pi)^3} \frac{1}{\sqrt{2E_{\mathbf{p}}}} e^{i\mathbf{p} \cdot \mathbf{x}} \sum_{s=1,2} \left(a_{\mathbf{p}}^s u^s(\mathbf{p}) + b_{-\mathbf{p}}^s v^s(-\mathbf{p}) \right), \qquad (3.87)$$

where $a_{\mathbf{p}}^s$ and $b_{\mathbf{p}}^s$ are operator coefficients. (For now we work in the Schrödinger picture, where ψ does not depend on time.) Postulate the commutation relations

$$\left[a_{\mathbf{p}}^r, a_{\mathbf{q}}^{s\dagger}\right] = \left[b_{\mathbf{p}}^r, b_{\mathbf{q}}^{s\dagger}\right] = (2\pi)^3 \delta^{(3)}(\mathbf{p} - \mathbf{q}) \delta^{rs}. \qquad (3.88)$$

It is then easy to verify the commutation relations (3.86) for ψ and $\psi^\dagger$:

$$[\psi(\mathbf{x}), \psi^\dagger(\mathbf{y})] = \int \frac{d^3p\, d^3q}{(2\pi)^6} \frac{1}{\sqrt{2E_{\mathbf{p}} 2E_{\mathbf{q}}}} e^{i(\mathbf{p} \cdot \mathbf{x} - \mathbf{q} \cdot \mathbf{y})}$$

$$\times \sum_{r,s} \left(\left[a_{\mathbf{p}}^r, a_{\mathbf{q}}^{s\dagger}\right] u^r(\mathbf{p}) \bar{u}^s(\mathbf{q}) + \left[b_{-\mathbf{p}}^r, b_{-\mathbf{q}}^{s\dagger}\right] v^r(-\mathbf{p}) \bar{v}^s(-\mathbf{q}) \right) \gamma^0$$

$$= \int \frac{d^3p}{(2\pi)^3} \frac{1}{2E_{\mathbf{p}}} e^{i\mathbf{p} \cdot (\mathbf{x} - \mathbf{y})}$$

$$\times \left[(\gamma^0 E_{\mathbf{p}} - \boldsymbol{\gamma} \cdot \mathbf{p} + m) + (\gamma^0 E_{\mathbf{p}} + \boldsymbol{\gamma} \cdot \mathbf{p} - m) \right] \gamma^0$$

$$= \delta^{(3)}(\mathbf{x} - \mathbf{y}) \times \mathbf{1}_{4 \times 4}. \qquad (3.89)$$

In the second step we have used the spin sum completeness relations (3.66) and (3.67).

We are now ready to write H in terms of the a's and b's. After another short calculation (making use of the orthogonality relations (3.60), (3.63), and (3.65)), we find

$$H = \int \frac{d^3p}{(2\pi)^3} \sum_s \left(E_{\mathbf{p}} a_{\mathbf{p}}^{s\dagger} a_{\mathbf{p}}^s - E_{\mathbf{p}} b_{\mathbf{p}}^{s\dagger} b_{\mathbf{p}}^s \right). \tag{3.90}$$

Something is terribly wrong with the second term: By creating more and more particles with $b^\dagger$, we can lower the energy indefinitely. (It would not have helped to rename $b \leftrightarrow b^\dagger$, since doing so would ruin the commutation relation (3.89).)

We seem to be in rather deep trouble, but again let's press on, and investigate the causality of this theory. To do this we should compute $[\psi(x), \psi^\dagger(y)]$ (or more conveniently, $[\psi(x), \bar{\psi}(y)]$) at non-equal times and hope to get zero outside the light-cone. First we must switch to the Heisenberg picture and restore the time-dependence of ψ and $\bar{\psi}$. Using the relations

$$e^{iHt} a_{\mathbf{p}}^s e^{-iHt} = a_{\mathbf{p}}^s e^{-iE_{\mathbf{p}}t}, \qquad e^{iHt} b_{\mathbf{p}}^s e^{-iHt} = b_{\mathbf{p}}^s e^{+iE_{\mathbf{p}}t}, \tag{3.91}$$

we immediately have

$$\begin{aligned}
\psi(x) &= \int \frac{d^3p}{(2\pi)^3} \frac{1}{\sqrt{2E_{\mathbf{p}}}} \sum_s \left(a_{\mathbf{p}}^s u^s(p) e^{-ip\cdot x} + b_{\mathbf{p}}^s v^s(p) e^{ip\cdot x} \right); \\
\bar{\psi}(x) &= \int \frac{d^3p}{(2\pi)^3} \frac{1}{\sqrt{2E_{\mathbf{p}}}} \sum_s \left(a_{\mathbf{p}}^{s\dagger} \bar{u}^s(p) e^{ip\cdot x} + b_{\mathbf{p}}^{s\dagger} \bar{v}^s(p) e^{-ip\cdot x} \right).
\end{aligned} \tag{3.92}$$

We can now calculate the general commutator:

$$\begin{aligned}
\left[\psi_a(x), \bar{\psi}_b(y) \right] &= \int \frac{d^3p}{(2\pi)^3} \frac{1}{2E_{\mathbf{p}}} \sum_s \left(u_a^s(p) \bar{u}_b^s(p) e^{-ip\cdot(x-y)} + v_a^s(p) \bar{v}_b^s(p) e^{ip\cdot(x-y)} \right) \\
&= \int \frac{d^3p}{(2\pi)^3} \frac{1}{2E_{\mathbf{p}}} \left((\not{p} + m)_{ab} e^{-ip\cdot(x-y)} + (\not{p} - m)_{ab} e^{ip\cdot(x-y)} \right) \\
&= (i\not{\partial}_x + m)_{ab} \int \frac{d^3p}{(2\pi)^3} \frac{1}{2E_{\mathbf{p}}} \left(e^{-ip\cdot(x-y)} - e^{ip\cdot(x-y)} \right) \\
&= (i\not{\partial}_x + m)_{ab} [\phi(x), \phi(y)].
\end{aligned}$$

Since $[\phi(x), \phi(y)]$ (the commutator of a real Klein-Gordon field) vanishes outside the light-cone, this quantity does also.

There is something odd, however, about this solution to the causality problem. Let $|0\rangle$ be the state that is annihilated by all the $a_{\mathbf{p}}^s$ and $b_{\mathbf{p}}^s$: $a_{\mathbf{p}}^s |0\rangle = b_{\mathbf{p}}^s |0\rangle = 0$. Then

$$\begin{aligned}
\left[\psi_a(x), \bar{\psi}_b(y) \right] &= \langle 0 | \left[\psi_a(x), \bar{\psi}_b(y) \right] | 0 \rangle \\
&= \langle 0 | \psi_a(x) \bar{\psi}_b(y) | 0 \rangle - \langle 0 | \bar{\psi}_b(y) \psi_a(x) | 0 \rangle,
\end{aligned}$$

just as for the Klein-Gordon field. But in the Klein-Gordon case, we got one term of the commutator from each of these two pieces: the propagation of a particle from y to x was canceled by the propagation of an antiparticle from x to y outside the light-cone. Here both terms come from the first piece, $\langle 0| \psi(x)\bar{\psi}(y) |0\rangle$, since the second piece is zero. The cancellation is between positive-energy particles and negative-energy particles, both propagating from y to x.

This observation can actually lead us to a resolution of the negative-energy problem. One of the assumptions we made in quantizing the Dirac theory must have been incorrect. Let us therefore forget about the postulated commutation relations (3.86) and (3.88), and see whether we can find a way for positive-energy particles to propagate in both directions. We will also have to drop our definition of the vacuum $|0\rangle$ as the state that is annihilated by all $a_{\mathbf{p}}^s$ and $b_{\mathbf{p}}^s$. We will, however, retain the expressions (3.92) for $\psi(x)$ and $\bar{\psi}(x)$ as Heisenberg operators, since if $\psi(x)$ and $\bar{\psi}(x)$ solve the Dirac equation, they must be decomposable into such plane-wave solutions.

First consider the propagation amplitude $\langle 0| \psi(x)\bar{\psi}(y) |0\rangle$, which is to represent a positive-energy particle propagating from y to x. In this case we want the (Heisenberg) state $\bar{\psi}(y) |0\rangle$ to be made up of only positive-energy, or negative-frequency components (since a Heisenberg state $\Psi_H = e^{+iHt}\Psi_S$). Thus only the $a_{\mathbf{p}}^{s\dagger}$ term of $\bar{\psi}(y)$ can contribute, which means that $b_{\mathbf{p}}^{s\dagger}$ must annihilate the vacuum. Similarly $\langle 0| \psi(x)$ can contain only positive-frequency components. Thus we have

$$
\langle 0| \psi(x)\bar{\psi}(y) |0\rangle = \langle 0| \int \frac{d^3p}{(2\pi)^3} \frac{1}{\sqrt{2E_{\mathbf{p}}}} \sum_r a_{\mathbf{p}}^r u^r(p)e^{-ipx}
$$
$$
\times \int \frac{d^3q}{(2\pi)^3} \frac{1}{\sqrt{2E_{\mathbf{q}}}} \sum_s a_{\mathbf{q}}^{s\dagger} \bar{u}^s(q)e^{iqy} |0\rangle .
$$
(3.93)

We can say something about the matrix element $\langle 0| a_{\mathbf{p}}^r a_{\mathbf{q}}^{s\dagger} |0\rangle$ even without knowing how to interchange $a_{\mathbf{p}}^r$ and $a_{\mathbf{q}}^{s\dagger}$, by using translational and rotational invariance. If the ground state $|0\rangle$ is to be invariant under translations, we must have $|0\rangle = e^{i\mathbf{P}\cdot\mathbf{x}} |0\rangle$. Furthermore, since $a_{\mathbf{q}}^{s\dagger}$ creates momentum $\mathbf{q}$, we can use Eq. (2.48) to compute

$$
\langle 0| a_{\mathbf{p}}^r a_{\mathbf{q}}^{s\dagger} |0\rangle = \langle 0| a_{\mathbf{p}}^r a_{\mathbf{q}}^{s\dagger} e^{i\mathbf{P}\cdot\mathbf{x}} |0\rangle
$$
$$
= e^{i(\mathbf{p}-\mathbf{q})\cdot\mathbf{x}} \langle 0| e^{i\mathbf{P}\cdot\mathbf{x}} a_{\mathbf{p}}^r a_{\mathbf{q}}^{s\dagger} |0\rangle
$$
$$
= e^{i(\mathbf{p}-\mathbf{q})\cdot\mathbf{x}} \langle 0| a_{\mathbf{p}}^r a_{\mathbf{q}}^{s\dagger} |0\rangle .
$$

This says that if $\langle 0| a_{\mathbf{p}}^r a_{\mathbf{q}}^{s\dagger} |0\rangle$ is to be nonzero, $\mathbf{p}$ must equal $\mathbf{q}$. Similarly, it can be shown that rotational invariance of $|0\rangle$ implies $r = s$. (This should be intuitively clear, and can be checked after we discuss the angular momentum operator later in this section.) From these considerations we conclude that the matrix element can be written

$$\langle 0 | a_{\mathbf{p}}^r a_{\mathbf{q}}^{s\dagger} | 0 \rangle = (2\pi)^3 \delta^{(3)}(\mathbf{p} - \mathbf{q}) \delta^{rs} \cdot A(\mathbf{p}),$$

where $A(\mathbf{p})$ is so far undetermined. Note, however, that if the norm of a state is always positive (as it should be in any self-respecting Hilbert space), $A(\mathbf{p})$ must be greater than zero. We can now go back to (3.93), and write

$$\langle 0 | \psi(x) \bar{\psi}(y) | 0 \rangle = \int \frac{d^3p}{(2\pi)^3} \frac{1}{2E_{\mathbf{p}}} \sum_s u^s(p) \bar{u}^s(p) A(\mathbf{p}) e^{-ip(x-y)}$$

$$= \int \frac{d^3p}{(2\pi)^3} \frac{1}{2E_{\mathbf{p}}} (\slashed{p} + m) A(\mathbf{p}) e^{-ip(x-y)}.$$

This expression is properly invariant under boosts only if $A(\mathbf{p})$ is a Lorentz scalar, i.e., $A(\mathbf{p}) = A(p^2)$. Since $p^2 = m^2$, A must be a constant. So finally we obtain

$$\langle 0 | \psi_a(x) \bar{\psi}_b(y) | 0 \rangle = (i\slashed{\partial}_x + m)_{ab} \int \frac{d^3p}{(2\pi)^3} \frac{1}{2E_{\mathbf{p}}} e^{-ip(x-y)} \cdot A. \qquad (3.94)$$

Similarly, in the amplitude $\langle 0 | \bar{\psi}(y) \psi(x) | 0 \rangle$, we want the only contributions to be from the positive-frequency terms of $\bar{\psi}(y)$ and the negative-frequency terms of $\psi(x)$. So $a_{\mathbf{p}}^s$ still annihilates the vacuum, but $b_{\mathbf{p}}^s$ does not. Then by arguments identical to those given above, we have

$$\langle 0 | \bar{\psi}_b(y) \psi_a(x) | 0 \rangle = -(i\slashed{\partial}_x + m)_{ab} \int \frac{d^3p}{(2\pi)^3} \frac{1}{2E_{\mathbf{p}}} e^{ip(x-y)} \cdot B, \qquad (3.95)$$

where B is another positive constant. The minus sign is important; it comes from the completeness relation (3.67) for $\sum v \bar{v}$ and the sign of x in the exponential factor. It implies that we cannot have $\langle 0 | [\psi(x), \bar{\psi}(y)] | 0 \rangle = 0$ outside the light-cone: The two terms (3.94) and (3.95) would indeed cancel if $A = -B$, but this is impossible since A and B must both be positive.

The solution, however, is now at hand. By setting $A = B = 1$, it is easy to obtain (outside the light-cone)

$$\langle 0 | \psi_a(x) \bar{\psi}_b(y) | 0 \rangle = - \langle 0 | \bar{\psi}_b(y) \psi_a(x) | 0 \rangle.$$

That is, the spinor fields *anticommute* at spacelike separation. This is enough to preserve causality, since all reasonable observables (such as energy, charge, and particle number) are built out of an *even* number of spinor fields; for any such observables $\mathcal{O}_1$ and $\mathcal{O}_2$, we still have $[\mathcal{O}_1(x), \mathcal{O}_2(y)] = 0$ for $(x - y)^2 < 0$.

And remarkably, postulating *anti*commutation relations for the Dirac field solves the negative energy problem. The equal-time anticommutation relations will be

$$\{\psi_a(\mathbf{x}), \psi_b^\dagger(\mathbf{y})\} = \delta^{(3)}(\mathbf{x} - \mathbf{y}) \delta_{ab};$$

$$\{\psi_a(\mathbf{x}), \psi_b(\mathbf{y})\} = \{\psi_a^\dagger(\mathbf{x}), \psi_b^\dagger(\mathbf{y})\} = 0. \qquad (3.96)$$

We can expand $\psi(\mathbf{x})$ in terms of $a_{\mathbf{p}}^s$ and $b_{\mathbf{p}}^s$ as before (Eq. (3.87)). The creation and annihilation operators must now obey

$$\{a_{\mathbf{p}}^r, a_{\mathbf{q}}^{s\dagger}\} = \{b_{\mathbf{p}}^r, b_{\mathbf{q}}^{s\dagger}\} = (2\pi)^3 \delta^{(3)}(\mathbf{p} - \mathbf{q})\delta^{rs} \qquad (3.97)$$

(with all other anticommutators equal to zero) in order that (3.96) be satisfied. Another computation gives the Hamiltonian,

$$H = \int \frac{d^3p}{(2\pi)^3} \sum_s \left(E_{\mathbf{p}} a_{\mathbf{p}}^{s\dagger} a_{\mathbf{p}}^s - E_{\mathbf{p}} b_{\mathbf{p}}^{s\dagger} b_{\mathbf{p}}^s \right),$$

which is the same as before; $b_{\mathbf{p}}^{s\dagger}$ still creates negative energy. However, the relation $\{b_{\mathbf{p}}^r, b_{\mathbf{q}}^{s\dagger}\} = (2\pi)^3\delta^{(3)}(\mathbf{p} - \mathbf{q})\delta^{rs}$ is symmetric between $b_{\mathbf{p}}^r$ and $b_{\mathbf{q}}^{s\dagger}$. So let us simply redefine

$$\tilde{b}_{\mathbf{p}}^s \equiv b_{\mathbf{p}}^{s\dagger}; \qquad \tilde{b}_{\mathbf{p}}^{s\dagger} \equiv b_{\mathbf{p}}^s. \qquad (3.98)$$

These of course obey exactly the same anticommutation relations, but now the second term in the Hamiltonian is

$$-E_{\mathbf{p}} b_{\mathbf{p}}^{s\dagger} b_{\mathbf{p}}^s = +E_{\mathbf{p}} \tilde{b}_{\mathbf{p}}^{s\dagger} \tilde{b}_{\mathbf{p}}^s - (\text{const}).$$

If we choose $|0\rangle$ to be the state that is annihilated by $a_{\mathbf{p}}^s$ and $\tilde{b}_{\mathbf{p}}^s$, then all excitations of $|0\rangle$ have positive energy.

What happened? To better understand this trick, let us abandon the field theory for a moment and consider a theory with a single pair of b and $b^\dagger$ operators obeying $\{b, b^\dagger\} = 1$ and $\{b, b\} = \{b^\dagger, b^\dagger\} = 0$. Choose a state $|0\rangle$ such that $b|0\rangle = 0$. Then $b^\dagger|0\rangle$ is a new state; call it $|1\rangle$. This state satisfies $b|1\rangle = |0\rangle$ and $b^\dagger|1\rangle = 0$. So b and $b^\dagger$ act on a Hilbert space of only two states, $|0\rangle$ and $|1\rangle$. We might say that $|0\rangle$ represents an "empty" state, and that $b^\dagger$ "fills" the state. But we could equally well call $|1\rangle$ the empty state and say that $b = \tilde{b}^\dagger$ fills it. The two descriptions are completely equivalent, until we specify some observable that allows us to distinguish the states physically. In our case the correct choice is to take the state of lower energy to be the empty one. And it is less confusing to put the dagger on the operator that creates positive energy. That is exactly what we have done.

Note, by the way, that since $(\tilde{b}^\dagger)^2 = 0$, the state cannot be filled twice. More generally, the anticommutation relations imply that any multiparticle state is antisymmetric under the interchange of two particles: $a_{\mathbf{p}}^\dagger a_{\mathbf{q}}^\dagger |0\rangle = -a_{\mathbf{q}}^\dagger a_{\mathbf{p}}^\dagger |0\rangle$. Thus we conclude that if the ladder operators obey *anticommutation* relations, the corresponding particles obey *Fermi-Dirac* statistics.

We have just shown that in order to insure that the vacuum has only positive-energy excitations, we must quantize the Dirac field with anticommutation relations; under these conditions the particles associated with the Dirac field obey Fermi-Dirac statistics. This conclusion is part of a more general result, first derived by Pauli*: Lorentz invariance, positive energies, positive norms, and causality together imply that particles of integer spin obey Bose-Einstein statistics, while particles of half-odd-integer spin obey Fermi-Dirac statistics.

The Quantized Dirac Field

Let us now summarize the results of the quantized Dirac theory in a systematic way. Since the dust has settled, we should clean up our notation: From now on we will write $\tilde{b}_{\mathbf{p}}$ (the operator that lowers the energy of a state) simply as $b_{\mathbf{p}}$, and $\tilde{b}_{\mathbf{p}}^{\dagger}$ as $b_{\mathbf{p}}^{\dagger}$. All the expressions we will need in our later work are listed below; corresponding expressions above, where they differ, should be forgotten.

First we write the field operators:

$$\psi(x) = \int \frac{d^3p}{(2\pi)^3} \frac{1}{\sqrt{2E_{\mathbf{p}}}} \sum_s \left(a_{\mathbf{p}}^s u^s(p) e^{-ip\cdot x} + b_{\mathbf{p}}^{s\dagger} v^s(p) e^{ip\cdot x} \right); \quad (3.99)$$

$$\bar{\psi}(x) = \int \frac{d^3p}{(2\pi)^3} \frac{1}{\sqrt{2E_{\mathbf{p}}}} \sum_s \left(b_{\mathbf{p}}^s \bar{v}^s(p) e^{-ip\cdot x} + a_{\mathbf{p}}^{s\dagger} \bar{u}^s(p) e^{ip\cdot x} \right). \quad (3.100)$$

The creation and annihilation operators obey the anticommutation rules

$$\{a_{\mathbf{p}}^r, a_{\mathbf{q}}^{s\dagger}\} = \{b_{\mathbf{p}}^r, b_{\mathbf{q}}^{s\dagger}\} = (2\pi)^3 \delta^{(3)}(\mathbf{p} - \mathbf{q}) \delta^{rs}, \quad (3.101)$$

with all other anticommutators equal to zero. The equal-time anticommutation relations for ψ and $\psi^{\dagger}$ are then

$$\{\psi_a(\mathbf{x}), \psi_b^{\dagger}(\mathbf{y})\} = \delta^{(3)}(\mathbf{x} - \mathbf{y})\delta_{ab};$$
$$\{\psi_a(\mathbf{x}), \psi_b(\mathbf{y})\} = \{\psi_a^{\dagger}(\mathbf{x}), \psi_b^{\dagger}(\mathbf{y})\} = 0. \quad (3.102)$$

The vacuum $|0\rangle$ is defined to be the state such that

$$a_{\mathbf{p}}^s |0\rangle = b_{\mathbf{p}}^s |0\rangle = 0. \quad (3.103)$$

The Hamiltonian can be written

$$H = \int \frac{d^3p}{(2\pi)^3} \sum_s E_{\mathbf{p}} \left(a_{\mathbf{p}}^{s\dagger} a_{\mathbf{p}}^s + b_{\mathbf{p}}^{s\dagger} b_{\mathbf{p}}^s \right), \quad (3.104)$$

where we have dropped the infinite constant term that comes from anticommuting $b_{\mathbf{p}}^s$ and $b_{\mathbf{p}}^{s\dagger}$. From this we see that the vacuum is the state of lowest energy, as desired. The momentum operator is

$$\mathbf{P} = \int d^3x\, \psi^{\dagger}(-i\boldsymbol{\nabla})\psi = \int \frac{d^3p}{(2\pi)^3} \sum_s \mathbf{p} \left(a_{\mathbf{p}}^{s\dagger} a_{\mathbf{p}}^s + b_{\mathbf{p}}^{s\dagger} b_{\mathbf{p}}^s \right). \quad (3.105)$$

---

*W. Pauli, *Phys. Rev.* **58**, 716 (1940), reprinted in Schwinger (1958). A rigorous treatment is given by R. F. Streater and A. S. Wightman, *PCT, Spin and Statistics, and All That* (Benjamin/Cummings, Reading, Mass., 1964).

Thus both $a_{\mathbf{p}}^{s\dagger}$ and $b_{\mathbf{p}}^{s\dagger}$ create particles with energy $+E_{\mathbf{p}}$ and momentum $\mathbf{p}$. We will refer to the particles created by $a_{\mathbf{p}}^{s\dagger}$ as *fermions* and to those created by $b_{\mathbf{p}}^{s\dagger}$ as *antifermions*.

The one-particle states

$$|\mathbf{p}, s\rangle \equiv \sqrt{2E_{\mathbf{p}}}\, a_{\mathbf{p}}^{s\dagger}\, |0\rangle \qquad (3.106)$$

are defined so that their inner product

$$\langle \mathbf{p}, r | \mathbf{q}, s \rangle = 2E_{\mathbf{p}}(2\pi)^3 \delta^{(3)}(\mathbf{p} - \mathbf{q})\delta^{rs} \qquad (3.107)$$

is Lorentz invariant. This implies that the operator $U(\Lambda)$ that implements Lorentz transformations on the states of the Hilbert space is unitary, even though for boosts, $\Lambda_{\frac{1}{2}}$ is not unitary.

It will be reassuring to do a consistency check, to see that $U(\Lambda)$ implements the right transformation on $\psi(x)$. So calculate

$$U\psi(x)U^{-1} = U \int \frac{d^3p}{(2\pi)^3} \frac{1}{\sqrt{2E_{\mathbf{p}}}} \sum_s \left(a_{\mathbf{p}}^s u^s(p)e^{-ipx} + b_{\mathbf{p}}^{s\dagger} v^s(p)e^{ipx} \right) U^{-1}. \quad (3.108)$$

We can concentrate on the first term; the second is completely analogous. Equation (3.106) implies that $a_{\mathbf{p}}^s$ transforms according to

$$U(\Lambda)a_{\mathbf{p}}^s U^{-1}(\Lambda) = \sqrt{\frac{E_{\Lambda\mathbf{p}}}{E_{\mathbf{p}}}}\, a_{\Lambda\mathbf{p}}^s, \qquad (3.109)$$

assuming that the axis of spin quantization is parallel to the boost or rotation axis. To use this relation to evaluate (3.108), rewrite the integral as

$$\int \frac{d^3p}{(2\pi)^3} \frac{1}{\sqrt{2E_{\mathbf{p}}}} a_{\mathbf{p}}^s = \int \frac{d^3p}{(2\pi)^3} \frac{1}{2E_{\mathbf{p}}} \cdot \sqrt{2E_{\mathbf{p}}} a_{\mathbf{p}}^s.$$

The second factor is transformed in a simple way by U, and the first is a Lorentz-invariant integral. Thus, if we apply (3.109) and make the substitution $\tilde{p} = \Lambda p$, Eq. (3.108) becomes

$$U(\Lambda)\psi(x)U^{-1}(\Lambda) = \int \frac{d^3\tilde{p}}{(2\pi)^3} \frac{1}{2E_{\tilde{\mathbf{p}}}} \sum_s u^s(\Lambda^{-1}\tilde{p})\sqrt{2E_{\tilde{\mathbf{p}}}} a_{\tilde{\mathbf{p}}}^s e^{-i\tilde{p}\cdot\Lambda x} + \cdots.$$

But $u^s(\Lambda^{-1}\tilde{p}) = \Lambda_{\frac{1}{2}}^{-1} u^s(\tilde{p})$, so indeed we have

$$U(\Lambda)\psi(x)U^{-1}(\Lambda) = \int \frac{d^3\tilde{p}}{(2\pi)^3} \frac{1}{\sqrt{2E_{\tilde{\mathbf{p}}}}} \sum_s \Lambda_{\frac{1}{2}}^{-1} u^s(\tilde{p}) a_{\tilde{\mathbf{p}}}^s e^{-i\tilde{p}\cdot\Lambda x} + \cdots$$

$$= \Lambda_{\frac{1}{2}}^{-1}\psi(\Lambda x). \qquad (3.110)$$

This result says that the transformed field creates and destroys particles at the point Λx, as it must. Note, however, that this transformation appears to be in the wrong direction compared to Eq. (3.2), where the transformed field ϕ was evaluated at $\Lambda^{-1}x$. The difference is that in Section 3.1 we imagined that we transformed a pre-existing field distribution that was measured by $\phi(x)$. Here, we are transforming the action of $\phi(x)$ in creating or destroying particles. These two ways of implementing the Lorentz transformation work in opposite directions. Notice, though, that the matrix acting on ψ and the transformation of the coordinate x have the correct relative orientation, consistent with Eq. (3.8).

Next we should discuss the spin of a Dirac particle. We expect Dirac fermions to have spin 1/2; now we can demonstrate this property from our formalism. We have already shown that the particles created by $a_{\mathbf{p}}^{s\dagger}$ and $b_{\mathbf{p}}^{s\dagger}$ each come in two "spin" states: $s = 1, 2$. But we haven't proved yet that this "spin" has anything to do with angular momentum. To do this, we must write down the angular momentum operator.

Recall that we found the linear momentum operator in Section 2.2 by looking for the conserved quantity associated with translational invariance. We can find the angular momentum operator in a similar way as a consequence of rotational invariance. Under a rotation (or any Lorentz transformation), the Dirac field ψ transforms (in our original convention) according to

$$\psi(x) \rightarrow \psi'(x) = \Lambda_{\frac{1}{2}}\psi(\Lambda^{-1}x).$$

To apply Noether's theorem we must compute the change in the field at a fixed point, that is,

$$\delta\psi = \psi'(x) - \psi(x) = \Lambda_{\frac{1}{2}}\psi(\Lambda^{-1}x) - \psi(x).$$

Consider for definiteness an infinitesimal rotation of coordinates by an angle θ about the z-axis. The parametrization of this transformation is given just below Eq. (3.19): $\omega_{12} = -\omega_{21} = \theta$. Using the same parameters in Eq. (3.30), we find

$$\Lambda_{\frac{1}{2}} \approx 1 - \tfrac{i}{2}\omega_{\mu\nu}S^{\mu\nu} = 1 - \tfrac{i}{2}\theta\Sigma^3.$$

We can now compute

$$\delta\psi(x) = \left(1 - \tfrac{i}{2}\theta\Sigma^3\right)\psi(t, x + \theta y, y - \theta x, z) - \psi(x)$$

$$= -\theta\left(x\partial_y - y\partial_x + \tfrac{i}{2}\Sigma^3\right)\psi(x) \equiv \theta\Delta\psi.$$

The time-component of the conserved Noether current is then

$$j^0 = \frac{\partial\mathcal{L}}{\partial(\partial_0\psi)}\Delta\psi = -i\bar{\psi}\gamma^0\left(x\partial_y - y\partial_x + \tfrac{i}{2}\Sigma^3\right)\psi.$$

Similar expressions hold for rotations about the x- and y-axes, so the angular momentum operator is

$$\mathbf{J} = \int d^3x\, \psi^\dagger\left(\mathbf{x} \times (-i\boldsymbol{\nabla}) + \tfrac{1}{2}\boldsymbol{\Sigma}\right)\psi. \tag{3.111}$$

For nonrelativistic fermions, the first term of (3.111) gives the orbital angular momentum. The second term therefore gives the spin angular momentum.

Unfortunately, the division of (3.111) into spin and orbital parts is not so straightforward for relativistic fermions, so it is not simple to write a general expression for this quantity in terms of ladder operators.

To prove that a Dirac particle has spin 1/2, however, it suffices to consider particles at rest. We would like to apply J_z to the state $a_0^{s\dagger}|0\rangle$ and show that this state is an eigenvector. This is most easily done using a trick: Since J_z must annihilate the vacuum, $J_z a_0^{s\dagger}|0\rangle = [J_z, a_0^{s\dagger}]|0\rangle$. The commutator is nonzero only for the terms in J_z that have annihiliation operators at $\mathbf{p} = 0$. For these terms, the orbital part of (3.111) does not contribute. To write the spin term of (3.111) in terms of ladder operators, use expansions (3.99) and (3.100), evaluated at $t = 0$:

$$J_z = \int d^3 x \int \frac{d^3 p\, d^3 p'}{(2\pi)^6} \frac{1}{\sqrt{2E_{\mathbf{p}}\, 2E_{\mathbf{p}'}}} e^{-i\mathbf{p}'\cdot\mathbf{x}} e^{i\mathbf{p}\cdot\mathbf{x}}$$

$$\times \sum_{r,r'} \left(a_{\mathbf{p}'}^{r'\dagger} u^{r'\dagger}(\mathbf{p}') + b_{-\mathbf{p}'}^{r'} v^{r'\dagger}(-\mathbf{p}') \right) \frac{\Sigma^3}{2} \left(a_{\mathbf{p}}^r u^r(\mathbf{p}) + b_{-\mathbf{p}}^{r\dagger} v^r(-\mathbf{p}) \right).$$

Taking the commutator with $a_0^{s\dagger}$, the only nonzero term has the structure $[a_{\mathbf{p}}^{r\dagger} a_{\mathbf{p}}^r, a_0^{s\dagger}] = (2\pi)^3 \delta^{(3)}(\mathbf{p}) a_0^{r\dagger} \delta^{rs}$; the other three terms in the commutator either vanish or annihilate the vacuum. Thus we find

$$J_z a_0^{s\dagger}|0\rangle = \frac{1}{2m} \sum_r \left(u^{s\dagger}(0) \frac{\Sigma^3}{2} u^r(0) \right) a_0^{r\dagger}|0\rangle = \sum_r \left(\xi^{s\dagger} \frac{\sigma^3}{2} \xi^r \right) a_0^{r\dagger}|0\rangle,$$

where we have used the explicit form (3.47) of $u(0)$ to obtain the last expression. The sum over r is accomplished most easily by choosing the spinors ξ^r to be eigenstates of σ^3. We then find that for $\xi^s = \binom{1}{0}$, the one-particle state is an eigenstate of J_z with eigenvalue $+1/2$, while for $\xi^s = \binom{0}{1}$, it is an eigenstate of J_z with eigenvalue $-1/2$. This result is exactly what we expect for electrons.

An analogous calculation determines the spin of a zero-momentum antifermion. But in this case, since the order of the b and $b^\dagger$ terms in J_z is reversed, we get an extra minus sign from evaluating $[b_{\mathbf{p}} b_{\mathbf{p}}^\dagger, b_0^\dagger] = -[b_{\mathbf{p}}^\dagger b_{\mathbf{p}}, b_0^\dagger]$. Thus for positrons, the association between the spinors η^s and the spin angular momentum is reversed: $\binom{1}{0}$ corresponds to spin $-1/2$, while $\binom{0}{1}$ corresponds to spin $+1/2$. This reversal of sign agrees with the prediction of Dirac hole theory. From that viewpoint, a positron is the absence of a negative-energy electron. If the missing electron had positive J_z, its absence has negative J_z.

In summary, the angular momentum of zero-momentum fermions is given by

$$J_z a_0^{s\dagger}|0\rangle = \pm\tfrac{1}{2} a_0^{s\dagger}|0\rangle, \qquad J_z b_0^{s\dagger}|0\rangle = \mp\tfrac{1}{2} b_0^{s\dagger}|0\rangle, \qquad (3.112)$$

where the upper sign is for $\xi^s = \binom{1}{0}$ and the lower sign is for $\xi^s = \binom{0}{1}$.

There is one more important conserved quantity in the Dirac theory. In Section 3.4 we saw that the current $j^\mu = \bar\psi\gamma^\mu\psi$ is conserved. The charge associated with this current is

$$Q = \int d^3x\, \psi^\dagger(x)\psi(x) = \int \frac{d^3p}{(2\pi)^3} \sum_s \left(a_{\mathbf{p}}^{s\dagger}a_{\mathbf{p}}^s + b_{-\mathbf{p}}^s b_{-\mathbf{p}}^{s\dagger} \right),$$

or, if we ignore another infinite constant,

$$Q = \int \frac{d^3p}{(2\pi)^3} \sum_s \left(a_{\mathbf{p}}^{s\dagger}a_{\mathbf{p}}^s - b_{\mathbf{p}}^{s\dagger}b_{\mathbf{p}}^s \right). \tag{3.113}$$

So $a_{\mathbf{p}}^{s\dagger}$ creates fermions with charge $+1$, while $b_{\mathbf{p}}^{s\dagger}$ creates antifermions with charge -1. When we couple the Dirac field to the electromagnetic field, we will see that Q is none other than the electric charge (up to a constant factor that depends on which type of particle we wish to describe; e.g., for electrons, the electric charge is Qe).

In Quantum Electrodynamics we will use the spinor field ψ to describe electrons and positrons. The particles created by $a_{\mathbf{p}}^{s\dagger}$ are electrons; they have energy $E_{\mathbf{p}}$, momentum $\mathbf{p}$, spin $1/2$ with polarization appropriate to ξ^s, and charge $+1$ (in units of e). The particles created by $b_{\mathbf{p}}^{s\dagger}$ are positrons; they have energy $E_{\mathbf{p}}$, momentum $\mathbf{p}$, spin $1/2$ with polarization opposite to that of ξ^s, and charge -1. The state $\psi_\alpha(x)\,|0\rangle$ contains a positron at position x, whose polarization corresponds to the spinor component chosen. Similarly, $\bar\psi_\alpha(x)\,|0\rangle$ is a state of one electron at position x.

The Dirac Propagator

Calculating propagation amplitudes for the Dirac field is by now a straight-forward exercise:

$$\langle 0|\, \psi_a(x)\bar\psi_b(y)\,|0\rangle = \int \frac{d^3p}{(2\pi)^3}\frac{1}{2E_{\mathbf{p}}} \sum_s u_a^s(p)\bar u_b^s(p)e^{-ip\cdot(x-y)}$$

$$= (i\not\partial_x + m)_{ab} \int \frac{d^3p}{(2\pi)^3}\frac{1}{2E_{\mathbf{p}}}e^{-ip\cdot(x-y)}, \tag{3.114}$$

$$\langle 0|\, \bar\psi_b(y)\psi_a(x)\,|0\rangle = \int \frac{d^3p}{(2\pi)^3}\frac{1}{2E_{\mathbf{p}}} \sum_s v_a^s(p)\bar v_b^s(p)e^{-ip\cdot(y-x)}$$

$$= -(i\not\partial_x + m)_{ab} \int \frac{d^3p}{(2\pi)^3}\frac{1}{2E_{\mathbf{p}}}e^{-ip\cdot(y-x)}. \tag{3.115}$$

Just as we did for the Klein-Gordon equation, we can construct Green's functions for the Dirac equation obeying various boundary conditions. For example, the retarded Green's function is

$$S_R^{ab}(x-y) \equiv \theta(x^0 - y^0)\,\langle 0|\, \{\psi_a(x), \bar\psi_b(y)\}\,|0\rangle. \tag{3.116}$$

It is easy to verify that

$$S_R(x - y) = (i\partial\!\!\!/_x + m)D_R(x - y), \tag{3.117}$$

since on the right-hand side the term involving $\partial_0\theta(x^0 - y^0)$ vanishes. Using (3.117) and the fact that $\partial\!\!\!/\partial\!\!\!/ = \partial^2$, we see that S_R is a Green's function of the Dirac operator:

$$(i\partial\!\!\!/_x - m)S_R(x - y) = i\delta^{(4)}(x - y) \cdot 1_{4\times4}. \tag{3.118}$$

The Green's function of the Dirac operator can also be found by Fourier transformation. Expanding $S_R(x - y)$ as a Fourier integral and acting on both sides with $(i\partial\!\!\!/_x - m)$, we find

$$i\delta^{(4)}(x - y) = \int \frac{d^4p}{(2\pi)^4}(p\!\!\!/ - m)e^{-ip\cdot(x-y)}\widetilde{S}_R(p), \tag{3.119}$$

and hence

$$\widetilde{S}_R(p) = \frac{i}{p\!\!\!/ - m} = \frac{i(p\!\!\!/ + m)}{p^2 - m^2}. \tag{3.120}$$

To obtain the retarded Green's function, we must evaluate the p^0 integral in (3.120) along the contour shown on page 30. For $x^0 > y^0$ we close the contour below, picking up both poles to obtain the sum of (3.114) and (3.115). For $x^0 < y^0$ we close the contour above and get zero.

The Green's function with Feynman boundary conditions is defined by the contour shown on page 31:

$$S_F(x - y) = \int \frac{d^4p}{(2\pi)^4} \frac{i(p\!\!\!/ + m)}{p^2 - m^2 + i\epsilon} e^{-ip\cdot(x-y)}$$

$$= \begin{cases} \langle 0| \psi(x)\bar{\psi}(y) |0\rangle & \text{for } x^0 > y^0 \text{ (close contour below)} \\ -\langle 0| \bar{\psi}(y)\psi(x) |0\rangle & \text{for } x^0 < y^0 \text{ (close contour above)} \end{cases}$$

$$\equiv \langle 0| T\psi(x)\bar{\psi}(y) |0\rangle, \tag{3.121}$$

where we have chosen to define the time-ordered product of spinor fields with an additional minus sign when the operators are interchanged. This minus sign is extremely important in the quantum field theory of fermions; we will meet it again in Section 4.7.

As with the Klein-Gordon theory, the expression (3.121) for the Feynman propagator is the most useful result of this chapter. When we do perturbative calculations with Feynman diagrams, we will associate the factor $\widetilde{S}_F(p)$ with each internal fermion line.

3.6 Discrete Symmetries of the Dirac Theory

In the last section we discussed the implementation of continuous Lorentz transformations on the Hilbert space of the Dirac theory. We found that for each transformation Λ there was a unitary operator $U(\Lambda)$, which induced the correct transformation on the fields:

$$U(\Lambda)\psi(x)U^{-1}(\Lambda) = \Lambda_{\frac{1}{2}}^{-1}\psi(\Lambda x). \tag{3.122}$$

In this section we will discuss the analogous operators that implement various discrete symmetries on the Dirac field.

In addition to continuous Lorentz transformations, there are two other spacetime operations that are potential symmetries of the Lagrangian: *parity* and *time reversal*. Parity, denoted by P, sends $(t, \mathbf{x}) \rightarrow (t, -\mathbf{x})$, reversing the handedness of space. Time reversal, denoted by T, sends $(t, \mathbf{x}) \rightarrow (-t, \mathbf{x})$, interchanging the forward and backward light-cones. Neither of these operations can be achieved by a continuous Lorentz transformation starting from the identity. Both, however, preserve the Minkowski interval $x^2 = t^2 - \mathbf{x}^2$. In standard terminology, the continuous Lorentz transformations are referred to as the proper, orthochronous Lorentz group, $\mathbf{L}_+^\uparrow$. Then the full Lorentz group breaks up into four disconnected subsets, as shown below.

$$\mathbf{L}_+^\uparrow \qquad \overset{P}{\longleftrightarrow} \qquad \mathbf{L}_-^\uparrow = P\mathbf{L}_+^\uparrow \qquad \text{"orthochronous"}$$

$$T\Big\updownarrow \qquad\qquad\qquad \Big\updownarrow T$$

$$\mathbf{L}_+^\downarrow = T\mathbf{L}_+^\uparrow \quad \underset{P}{\longleftrightarrow} \quad \mathbf{L}_-^\downarrow = PT\mathbf{L}_+^\uparrow \qquad \text{"nonorthochronous"}$$

$$\text{"proper"} \qquad\quad \text{"improper"}$$

At the same time that we discuss P and T, it will be convenient to discuss a third (non-spacetime) discrete operation: *charge conjugation*, denoted by C. Under this operation, particles and antiparticles are interchanged.

Although any relativistic field theory must be invariant under $\mathbf{L}_+^\uparrow$, it need not be invariant under P, T, or C. What is the status of these symmetry operations in the real world? From experiment, we know that three of the forces of Nature— the gravitational, electromagnetic, and strong interactions—are symmetric with respect to P, C, and T. The weak interactions violate C and P separately, but preserve CP and T. But certain rare processes (all so far observed involve neutral K mesons) also show CP and T violation. All observations indicate that the combination CPT is a perfect symmetry of Nature.

The currently accepted theoretical model of the weak interactions is the Glashow-Weinberg-Salam gauge theory, described in Chapter 20. This theory violates C and P in the strongest possible way. It is actually a surprise (though not quite an accident) that C and P happen to be quite good symmetries in the most readily observable processes. On the other hand, no one knows a really beautiful theory that violates CP. In the current theory, when there are three (or more) fermion generations, there is room for a parameter that, if nonzero, causes CP violation. But the value of this parameter is no better understood than the value of the electron mass; the physical origin of CP violation remains a mystery. We will discuss this question further in Section 20.3.

Parity

With this introduction, let us now discuss the action of P, T, and C on Dirac particles and fields. First consider parity. The operator P should reverse the momentum of a particle without flipping its spin:

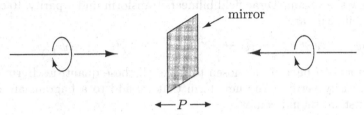

Mathematically, this means that P should be implemented by a unitary operator (properly called $U(P)$, but we'll just call it P) which, for example, transforms the state $a_{\mathbf{p}}^{s\dagger} |0\rangle$ into $a_{-\mathbf{p}}^{s\dagger} |0\rangle$. In other words, we want

$$Pa_{\mathbf{p}}^s P = \eta_a a_{-\mathbf{p}}^s \quad \text{and} \quad Pb_{\mathbf{p}}^s P = \eta_b b_{-\mathbf{p}}^s, \quad (3.123)$$

where η_a and η_b are possible phases. These phases are restricted by the condition that two applications of the parity operator should return observables to their original values. Since observables are built from an even number of fermion operators, this requires η_a^2, $\eta_b^2 = \pm 1$.

Just as a continuous Lorentz transformation is implemented on the Dirac field as the 4×4 constant matrix $\Lambda_{\frac{1}{2}}$, the parity transformation should also be represented by a 4×4 constant matrix. To find this matrix, and to determine η_a and η_b, we compute the action of P on $\psi(x)$. Using (3.123), we have

$$P\psi(x)P = \int \frac{d^3p}{(2\pi)^3} \frac{1}{\sqrt{2E_{\mathbf{p}}}} \sum_s \left(\eta_a a_{-\mathbf{p}}^s u^s(p) e^{-ipx} + \eta_b^* b_{-\mathbf{p}}^{s\dagger} v^s(p) e^{ipx} \right). \quad (3.124)$$

Now change variables to $\tilde{p} = (p^0, -\mathbf{p})$. Note that $p \cdot x = \tilde{p} \cdot (t, -\mathbf{x})$. Also $\tilde{p} \cdot \sigma = p \cdot \bar{\sigma}$ and $\tilde{p} \cdot \bar{\sigma} = p \cdot \sigma$. This allows us to write

$$u(p) = \begin{pmatrix} \sqrt{p \cdot \sigma}\, \xi \\ \sqrt{p \cdot \bar{\sigma}}\, \xi \end{pmatrix} = \begin{pmatrix} \sqrt{\tilde{p} \cdot \bar{\sigma}}\, \xi \\ \sqrt{\tilde{p} \cdot \sigma}\, \xi \end{pmatrix} = \gamma^0 u(\tilde{p});$$

$$v(p) = \begin{pmatrix} \sqrt{p \cdot \sigma}\, \xi \\ -\sqrt{p \cdot \bar{\sigma}}\, \xi \end{pmatrix} = \begin{pmatrix} \sqrt{\tilde{p} \cdot \bar{\sigma}}\, \xi \\ -\sqrt{\tilde{p} \cdot \sigma}\, \xi \end{pmatrix} = -\gamma^0 v(\tilde{p}).$$

So (3.124) becomes

$$P\psi(x)P = \int \frac{d^3\tilde{p}}{(2\pi)^3} \frac{1}{\sqrt{2E_{\tilde{p}}}} \sum_s \left(\eta_a a_{\tilde{p}}^s \gamma^0 u^s(\tilde{p}) e^{-i\tilde{p}(t,-\mathbf{x})} \right.$$
$$\left. - \eta_b^* b_{\tilde{p}}^{s\dagger} \gamma^0 v^s(\tilde{p}) e^{i\tilde{p}(t,-\mathbf{x})} \right).$$

This should equal some constant matrix times $\psi(t, -\mathbf{x})$, and indeed it works if we make $\eta_b^* = -\eta_a$. This implies

$$\eta_a \eta_b = -\eta_a \eta_a^* = -1. \tag{3.125}$$

Thus we have the parity transformation of $\psi(x)$ in its final form,

$$P\psi(t, \mathbf{x})P = \eta_a \gamma^0 \psi(t, -\mathbf{x}). \tag{3.126}$$

It will be very important (for example, in writing down Lagrangians) to know how the various Dirac field bilinears transform under parity. Recall that the five bilinears are

$$\bar{\psi}\psi, \qquad \bar{\psi}\gamma^\mu\psi, \qquad i\bar{\psi}[\gamma^\mu, \gamma^\nu]\psi, \qquad \bar{\psi}\gamma^\mu\gamma^5\psi, \qquad i\bar{\psi}\gamma^5\psi. \tag{3.127}$$

The factors of i have been chosen to make all these quantities Hermitian, as you can easily verify. (Any new term that we add to a Lagrangian must be real.) First we should compute

$$P\bar{\psi}(t, \mathbf{x})P = P\psi^\dagger(t, \mathbf{x})P\gamma^0 = \left(P\psi(t, \mathbf{x})P\right)^\dagger\gamma^0 = \eta_a^*\bar{\psi}(t, -\mathbf{x})\gamma^0. \tag{3.128}$$

Then the scalar bilinear transforms as

$$P\bar{\psi}\psi P = |\eta_a|^2 \bar{\psi}(t, -\mathbf{x})\gamma^0\gamma^0\psi(t, -\mathbf{x}) = +\bar{\psi}\psi(t, -\mathbf{x}), \tag{3.129}$$

while for the vector we obtain

$$P\bar{\psi}\gamma^\mu\psi P = \bar{\psi}\gamma^0\gamma^\mu\gamma^0\psi(t, -\mathbf{x}) = \begin{cases} +\bar{\psi}\gamma^\mu\psi(t, -\mathbf{x}) & \text{for } \mu = 0, \\ -\bar{\psi}\gamma^\mu\psi(t, -\mathbf{x}) & \text{for } \mu = 1, 2, 3. \end{cases} \tag{3.130}$$

Note that the vector acquires the same minus sign on the spatial components as does the vector x^μ. Similarly, the transformations of the pseudo-scalar and pseudo-vector are

$$Pi\bar{\psi}\gamma^5\psi P = i\bar{\psi}\gamma^0\gamma^5\gamma^0\psi(t, -\mathbf{x}) = -i\bar{\psi}\gamma^5\psi(t, -\mathbf{x}); \tag{3.131}$$

$$P\bar{\psi}\gamma^\mu\gamma^5\psi P = \bar{\psi}\gamma^0\gamma^\mu\gamma^5\gamma^0\psi(t, -\mathbf{x}) = \begin{cases} -\bar{\psi}\gamma^\mu\gamma^5\psi & \text{for } \mu = 0, \\ +\bar{\psi}\gamma^\mu\gamma^5\psi & \text{for } \mu = 1, 2, 3. \end{cases} \tag{3.132}$$

Just as we anticipated in Section 3.4, the "pseudo" signifies an extra minus sign in the parity transformation. (The transformation properties of $i\bar{\psi}[\gamma^\mu, \gamma^\nu]\psi = 2\bar{\psi}\sigma^{\mu\nu}\psi$ are reserved for Problem 3.7.) Note that the transformation properties of fermion bilinears were independent of η_a, so there would have been no loss of generality in setting $\eta_a = -\eta_b = 1$ from the beginning.

However, the relative minus sign (3.125) between the parity transformations of a fermion and an antifermion has important consequences. Consider a fermion-antifermion state, $a_\mathbf{p}^{s\dagger}b_\mathbf{q}^{s'\dagger}|0\rangle$. Applying P, we find $P\left(a_\mathbf{p}^{s\dagger}b_\mathbf{q}^{s'\dagger}|0\rangle\right) = -\left(a_{-\mathbf{p}}^{s\dagger}b_{-\mathbf{q}}^{s'\dagger}|0\rangle\right)$. Thus a state containing a fermion-antifermion pair gets an extra (-1) under parity. This information is most useful in the context of bound states, in which the fermion and antifermion momenta are integrated with the Schrödinger wavefunction to produce a system localized in space. We consider such states in detail in Section 5.3, but here we should remark that if the spatial wavefunction is symmetric under $\mathbf{x} \to -\mathbf{x}$, the state has *odd* parity, while if it is antisymmetric under $\mathbf{x} \to -\mathbf{x}$, the state has *even* parity. The $L = 0$ bound states, for example, have odd parity; the $J = 0$ state transforms as a pseudo-scalar, while the three $J = 1$ states transform as the spatial components of a vector. These properties show up in selection rules for decays of positronium and quark-antiquark systems (see Problem 3.8).

Time Reversal

Now let us turn to the implementation of time reversal. We would like T to take the form of a unitary operator that sends $a_{\mathbf{p}}$ to $a_{-\mathbf{p}}$ (and similarly for $b_{\mathbf{p}}$) and $\psi(t, \mathbf{x})$ to $\psi(-t, \mathbf{x})$ (times some constant matrix). These properties, however, are extremely difficult to achieve, since we saw above that sending $a_{\mathbf{p}}$ to $a_{-\mathbf{p}}$ instead sends $(t, \mathbf{x})$ to $(t, -\mathbf{x})$ in the expansion of ψ. The difficulty is even more apparent when we impose the constraint that time reversal should be a symmetry of the free Dirac theory, $[T, H] = 0$. Then

$$\psi(t, \mathbf{x}) = e^{iHt}\psi(\mathbf{x})e^{-iHt}$$

$$\Rightarrow T\psi(t, \mathbf{x})T = e^{iHt}[T\psi(\mathbf{x})T]e^{-iHt}$$

$$\Rightarrow T\psi(t, \mathbf{x})T|0\rangle = e^{iHt}[T\psi(\mathbf{x})T]|0\rangle ,$$

assuming that $H|0\rangle = 0$. The right-hand side is a sum of negative-frequency terms only. But if T is to reverse the time dependence of $\psi(t, \mathbf{x})$, then the left-hand side is (up to a constant matrix) $\psi(-t, \mathbf{x})|0\rangle = e^{-iHt}\psi(\mathbf{x})|0\rangle$, which is a sum of positive-frequency terms. Thus we have proved that T cannot be implemented as a linear unitary operator.

What can we do? The way out is to retain the unitarity condition $T^\dagger = T^{-1}$, but have T act on c-numbers as well as operators, as follows:

$$T(\text{c-number}) = (\text{c-number})^*T. \qquad (3.133)$$

Then even if $[T, H] = 0$, the time dependence of all exponential factors is reversed: $Te^{+iHt} = e^{-iHt}T$. Since all time evolution in quantum mechanics is performed with such exponential factors, this effectively changes the sign of t. Note that the operation of complex conjugation is nonlinear; T is referred to as an *antilinear* or *antiunitary* operator.

In addition to reversing the momentum of a particle, T should also flip the spin:

To quantify this, we must find a mathematical operation that flips a spinor ξ.

In the earlier parts of this chapter, we denoted the spin state of a fermion by a label $s = 1, 2$. In the remainder of this section, we will associate s with the physical spin component of the fermion along a specific axis. If this axis has polar coordinates θ, ϕ, the two-component spinors with spin up and spin down along this axis are

$$\xi(\uparrow) = \begin{pmatrix} \cos\frac{\theta}{2} \\ e^{i\phi}\sin\frac{\theta}{2} \end{pmatrix}, \qquad \xi(\downarrow) = \begin{pmatrix} -e^{-i\phi}\sin\frac{\theta}{2} \\ \cos\frac{\theta}{2} \end{pmatrix}.$$

Let $\xi^s = (\xi(\uparrow), \xi(\downarrow))$ for $s = 1, 2$. Also define

$$\xi^{-s} = -i\sigma^2(\xi^s)^*. \tag{3.134}$$

This quantity is the flipped spinor; from the explicit formulae,

$$\xi^{-s} = (\xi(\downarrow), -\xi(\uparrow)). \tag{3.135}$$

The form of the spin reversal relation follows more generally from the identity $\sigma\sigma^2 = \sigma^2(-\sigma^*)$. This equation implies that, if ξ satisfies $\mathbf{n}\cdot\sigma\xi = +\xi$ for some axis $\mathbf{n}$, then

$$(\mathbf{n}\cdot\sigma)(-i\sigma^2\xi^*) = -i\sigma^2(-\mathbf{n}\cdot\sigma)^*\xi^* = i\sigma^2(\xi^*) = -(-i\sigma^2\xi^*).$$

Notice that, with this convention for the spin flip, two successive spin flips return a spin to (-1) times the original state.

We now associate the various fermion spin states with these spinors. The electron annihilation operator $a_{\mathbf{p}}^s$ destroys an electron whose spinor $u^s(p)$ contains ξ^s. The positron annihilation operator $b_{\mathbf{p}}^s$ destroys a positron whose spinor $v^s(p)$ contains ξ^{-s}:

$$v^s(p) = \begin{pmatrix} \sqrt{p\cdot\sigma}\,\xi^{-s} \\ -\sqrt{p\cdot\bar{\sigma}}\,\xi^{-s} \end{pmatrix}. \tag{3.136}$$

As in Eq. (3.135), we define

$$a_{\mathbf{p}}^{-s} = (a_{\mathbf{p}}^2, -a_{\mathbf{p}}^1), \qquad b_{\mathbf{p}}^{-s} = (b_{\mathbf{p}}^2, -b_{\mathbf{p}}^1). \tag{3.137}$$

We can now work out the relation between the Dirac spinors u and v and their time reversals. Define $\tilde{p} = (p^0, -\mathbf{p})$. This vector satisfies the identity $\sqrt{\tilde{p}\cdot\sigma}\,\sigma^2 = \sigma^2\sqrt{p\cdot\sigma^*}$; to prove this, expand the square root as in (3.49). For some choice of spin and momentum, associated with the Dirac spinor $u^s(p)$, let $u^{-s}(\tilde{p})$ be the spinor with the reversed momentum and flipped spin. These quantities are related by

$$u^{-s}(\tilde{p}) = \begin{pmatrix} \sqrt{\tilde{p}\cdot\sigma}\,(-i\sigma^2\xi^{s*}) \\ \sqrt{\tilde{p}\cdot\bar{\sigma}}\,(-i\sigma^2\xi^{s*}) \end{pmatrix} = \begin{pmatrix} -i\sigma^2\sqrt{p\cdot\sigma^*}\,\xi^{s*} \\ -i\sigma^2\sqrt{p\cdot\bar{\sigma}^*}\,\xi^{s*} \end{pmatrix}$$

$$= -i\begin{pmatrix} \sigma^2 & 0 \\ 0 & \sigma^2 \end{pmatrix} [u^s(p)]^* = -\gamma^1\gamma^3 [u^s(p)]^*.$$

Similarly, for $v^s(p)$,

$$v^{-s}(\tilde{p}) = -\gamma^1\gamma^3 [v^s(p)]^*;$$

in this relation, v^{-s} contains $\xi^{-(-s)} = -\xi^s$.

Using the notation of Eq. (3.137), we define the time reversal transformation of fermion annihilation operators as follows:

$$Ta_{\mathbf{p}}^s T = a_{-\mathbf{p}}^{-s}, \qquad Tb_{\mathbf{p}}^s T = b_{-\mathbf{p}}^{-s}. \tag{3.138}$$

(An additional overall phase would have no effect on the rest of our discussion and is omitted for simplicity.) Relations (3.138) allow us to compute the action of T on the fermion field $\psi(x)$:

$$
\begin{aligned}
T\psi(t,\mathbf{x})T &= \int \frac{d^3p}{(2\pi)^3} \frac{1}{\sqrt{2E_{\mathbf{p}}}} \sum_s T\Big(a_{\mathbf{p}}^s u^s(p)e^{-ipx} + b_{\mathbf{p}}^{s\dagger} v^s(p)e^{ipx}\Big)T \\
&= \int \frac{d^3p}{(2\pi)^3} \frac{1}{\sqrt{2E_{\mathbf{p}}}} \sum_s \Big(a_{-\mathbf{p}}^{-s} [u^s(p)]^* e^{ipx} + b_{-\mathbf{p}}^{-s\dagger} [v^s(p)]^* e^{-ipx}\Big) \\
&= (\gamma^1\gamma^3) \int \frac{d^3\tilde{p}}{(2\pi)^3} \frac{1}{\sqrt{2E_{\tilde{p}}}} \sum_s \Big(a_{\tilde{p}}^{-s} u^{-s}(\tilde{p})e^{i\tilde{p}(t,-\mathbf{x})} \\
&\qquad\qquad\qquad\qquad\qquad\qquad + b_{\tilde{p}}^{-s\dagger} v^{-s}(\tilde{p})e^{-i\tilde{p}(t,-\mathbf{x})}\Big) \\
&= (\gamma^1\gamma^3)\psi(-t,\mathbf{x}). \tag{3.139}
\end{aligned}
$$

In the last step we used $\tilde{p} \cdot (t,-\mathbf{x}) = -\tilde{p} \cdot (-t,\mathbf{x})$. Just as for parity, we have derived a simple transformation law for the fermion field $\psi(x)$. The relative minus sign in the transformation laws for particle and antiparticle is present here as well, implicit in the twice-flipped spinor in v^{-s}.

Now we can check the action of T on the various bilinears. First we need

$$T\bar{\psi}T = (T\psi T)^\dagger (\gamma^0)^* = \psi^\dagger(-t,\mathbf{x})[\gamma^1\gamma^3]^\dagger \gamma^0 = \bar{\psi}(-t,\mathbf{x})[-\gamma^1\gamma^3]. \tag{3.140}$$

Then the transformation of the scalar bilinear is

$$T\bar{\psi}\psi(t,\mathbf{x})T = \bar{\psi}(-\gamma^1\gamma^3)(\gamma^1\gamma^3)\psi(-t,\mathbf{x}) = +\bar{\psi}\psi(-t,\mathbf{x}). \tag{3.141}$$

The pseudo-scalar acquires an extra minus sign when T goes through the i:

$$Ti\bar{\psi}\gamma^5\psi T = -i\bar{\psi}(-\gamma^1\gamma^3)\gamma^5(\gamma^1\gamma^3)\psi = -i\bar{\psi}\gamma^5\psi(-t,\mathbf{x}).$$

For the vector, we must separately compute each of the four cases $\mu = 0,1,2,3$. After a bit of work you should find

$$
\begin{aligned}
T\bar{\psi}\gamma^\mu\psi T &= \bar{\psi}(-\gamma^1\gamma^3)(\gamma^\mu)^*(\gamma^1\gamma^3)\psi \\
&= \begin{cases} +\bar{\psi}\gamma^\mu\psi(-t,\mathbf{x}) & \text{for } \mu = 0; \\ -\bar{\psi}\gamma^\mu\psi(-t,\mathbf{x}) & \text{for } \mu = 1,2,3. \end{cases} \tag{3.142}
\end{aligned}
$$

This is exactly the tranformation property we want for vectors such as the current density. You can verify that the pseudo-vector transforms in exactly the same way under time-reversal.

Charge Conjugation

The last of the three discrete symmetries is the particle-antiparticle symmetry C. There will be no problem in implementing C as a unitary linear operator. Charge conjugation is conventionally defined to take a fermion with a given spin orientation into an antifermion with the same spin orientation. Thus, a convenient choice for the transformation of fermion annihilation operators is

$$ Ca_{\mathbf{p}}^s C = b_{\mathbf{p}}^s; \qquad Cb_{\mathbf{p}}^s C = a_{\mathbf{p}}^s. \qquad (3.143) $$

Again, we ignore possible additional phases for simplicity.

Next we want to work out the action of C on $\psi(x)$. First we need a relation between $v^s(p)$ and $u^s(p)$. Using (3.136), and (3.134),

$$ \left(v^s(p) \right)^* = \left(\begin{array}{c} \sqrt{p \cdot \sigma}(-i\sigma^2 \xi^*) \\ -\sqrt{p \cdot \bar{\sigma}}(-i\sigma^2 \xi^*) \end{array} \right)^* = \left(\begin{array}{c} -i\sigma^2 \sqrt{p \cdot \bar{\sigma}^*} \xi^* \\ i\sigma^2 \sqrt{p \cdot \sigma^*} \xi^* \end{array} \right)^* = \left(\begin{array}{cc} 0 & -i\sigma^2 \\ i\sigma^2 & 0 \end{array} \right) \left(\begin{array}{c} \sqrt{p \cdot \sigma} \xi \\ \sqrt{p \cdot \bar{\sigma}} \xi \end{array} \right), $$

where ξ stands for ξ^s. That is,

$$ u^s(p) = -i\gamma^2 \left(v^s(p) \right)^*, \qquad v^s(p) = -i\gamma^2 \left(u^s(p) \right)^*. \qquad (3.144) $$

If we substitute (3.144) into the expression for the fermion field operator, and then transform this operator with C, we find

$$ C\psi(x)C = \int \frac{d^3p}{(2\pi)^3} \frac{1}{\sqrt{2E_{\mathbf{p}}}} \sum_s \left(-i\gamma^2 b_{\mathbf{p}}^s \left(v^s(p) \right)^* e^{-ipx} - i\gamma^2 a_{\mathbf{p}}^{s\dagger} \left(u^s(p) \right)^* e^{ipx} \right) $$

$$ = -i\gamma^2 \psi^*(x) = -i\gamma^2 (\psi^\dagger)^T = -i(\bar{\psi}\gamma^0\gamma^2)^T. \qquad (3.145) $$

Note that C is a linear unitary operator, even though it takes $\psi \to \psi^*$.

Once again, we would like to know how C acts on fermion bilinears. First we need

$$ C\bar{\psi}(x)C = C\psi^\dagger C \gamma^0 = (-i\gamma^2\psi)^T \gamma^0 = (-i\gamma^0\gamma^2\psi)^T. \qquad (3.146) $$

Working out the transformations of bilinears is a bit tricky, and it helps to write in spinor indices. For the scalar,

$$ C\bar{\psi}\psi C = (-i\gamma^0\gamma^2\psi)^T (-i\bar{\psi}\gamma^0\gamma^2)^T = -\gamma^0_{ab}\gamma^2_{bc}\psi_c \bar{\psi}_d \gamma^0_{de}\gamma^2_{ea} $$

$$ = +\bar{\psi}_d \gamma^0_{de}\gamma^2_{ea}\gamma^0_{ab}\gamma^2_{bc}\psi_c = -\bar{\psi}\gamma^2\gamma^0\gamma^0\gamma^2\psi \qquad (3.147) $$

$$ = +\bar{\psi}\psi. $$

(The minus sign in the third step is from fermion anticommutation.) The pseudo-scalar is no more difficult:

$$ Ci\bar{\psi}\gamma^5\psi C = i(-i\gamma^0\gamma^2\psi)^T \gamma^5 (-i\bar{\psi}\gamma^0\gamma^2)^T = i\bar{\psi}\gamma^5\psi. \qquad (3.148) $$

We must do each component of the vector and pseudo-vector separately. Noting that γ^0 and γ^2 are symmetric matrices while γ^1 and γ^3 are antisymmetric, we eventually find

$$C\bar{\psi}\gamma^\mu\psi C = -\bar{\psi}\gamma^\mu\psi; \tag{3.149}$$

$$C\bar{\psi}\gamma^\mu\gamma^5\psi C = +\bar{\psi}\gamma^\mu\gamma^5\psi. \tag{3.150}$$

Although the operator C interchanges ψ and $\bar{\psi}$, it does not actually change the order of the creation and annihilation operators. Thus, if $\bar{\psi}\gamma^0\psi$ is defined to subtract the infinite constant noted above Eq. (3.113), this constant does not reappear in the process of conjugation by C.

Summary of C, P, and T

The transformation properties of the various fermion bilinears under C, P, and T are summarized in the table below. Here we use the shorthand $(-1)^\mu \equiv 1$ for $\mu = 0$ and $(-1)^\mu \equiv -1$ for $\mu = 1, 2, 3$.

	$\bar{\psi}\psi$	$i\bar{\psi}\gamma^5\psi$	$\bar{\psi}\gamma^\mu\psi$	$\bar{\psi}\gamma^\mu\gamma^5\psi$	$\bar{\psi}\sigma^{\mu\nu}\psi$	∂_μ
P	$+1$	-1	$(-1)^\mu$	$-(-1)^\mu$	$(-1)^\mu(-1)^\nu$	$(-1)^\mu$
T	$+1$	-1	$(-1)^\mu$	$(-1)^\mu$	$-(-1)^\mu(-1)^\nu$	$-(-1)^\mu$
C	$+1$	$+1$	-1	$+1$	-1	$+1$
CPT	$+1$	$+1$	-1	-1	$+1$	-1

We have included the transformation properties of the tensor bilinear (see Problem 3.7), and also of the derivative operator.

Notice first that the free Dirac Lagrangian $\mathcal{L}_0 = \bar{\psi}(i\gamma^\mu\partial_\mu - m)\psi$ is invariant under C, P, and T separately. We can build more general quantum systems that violate any of these symmetries by adding to $\mathcal{L}_0$ some perturbation $\delta\mathcal{L}$. But $\delta\mathcal{L}$ must be a Lorentz scalar, and the last line of the table shows that all Lorentz scalar combinations of $\bar{\psi}$ and ψ are invariant under the combined symmetry CPT. Actually, it is quite generally true that one cannot build a Lorentz-invariant quantum field theory with a Hermitian Hamiltonian that violates CPT.[†]

Problems

3.1 Lorentz group. Recall from Eq. (3.17) the Lorentz commutation relations,

$$[J^{\mu\nu}, J^{\rho\sigma}] = i(g^{\nu\rho}J^{\mu\sigma} - g^{\mu\rho}J^{\nu\sigma} - g^{\nu\sigma}J^{\mu\rho} + g^{\mu\sigma}J^{\nu\rho}).$$

  (a) Define the generators of rotations and boosts as

$$L^i = \tfrac{1}{2}\epsilon^{ijk}J^{jk}, \qquad K^i = J^{0i},$$

[†]This theorem and the spin-statistics theorem are proved with great care in Streater and Wightman, *op. cit.*

where $i, j, k = 1, 2, 3$. An infinitesimal Lorentz transformation can then be written

$$\Phi \rightarrow (1 - i\boldsymbol{\theta} \cdot \mathbf{L} - i\boldsymbol{\beta} \cdot \mathbf{K})\Phi.$$

Write the commutation relations of these vector operators explicitly. (For example, $[L^i, L^j] = i\epsilon^{ijk}L^k$.) Show that the combinations

$$\mathbf{J_+} = \tfrac{1}{2}(\mathbf{L} + i\mathbf{K}) \quad \text{and} \quad \mathbf{J_-} = \tfrac{1}{2}(\mathbf{L} - i\mathbf{K})$$

commute with one another and separately satisfy the commutation relations of angular momentum.

(b) The finite-dimensional representations of the rotation group correspond precisely to the allowed values for angular momentum: integers or half-integers. The result of part (a) implies that all finite-dimensional representations of the Lorentz group correspond to pairs of integers or half integers, (j_+, j_-), corresponding to pairs of representations of the rotation group. Using the fact that $\mathbf{J} = \boldsymbol{\sigma}/2$ in the spin-1/2 representation of angular momentum, write explicitly the transformation laws of the 2-component objects transforming according to the $(\tfrac{1}{2}, 0)$ and $(0, \tfrac{1}{2})$ representations of the Lorentz group. Show that these correspond precisely to the transformations of ψ_L and ψ_R given in (3.37).

(c) The identity $\sigma^T = -\sigma^2 \sigma \sigma^2$ allows us to rewrite the ψ_L transformation in the unitarily equivalent form

$$\psi' \rightarrow \psi'(1 + i\boldsymbol{\theta} \cdot \frac{\boldsymbol{\sigma}}{2} + \boldsymbol{\beta} \cdot \frac{\boldsymbol{\sigma}}{2}),$$

where $\psi' = \psi_L^T \sigma^2$. Using this law, we can represent the object that transforms as $(\tfrac{1}{2}, \tfrac{1}{2})$ as a 2×2 matrix that has the ψ_R transformation law on the left and, simultaneously, the transposed ψ_L transformation on the right. Parametrize this matrix as

$$\begin{pmatrix} V^0 + V^3 & V^1 - iV^2 \\ V^1 + iV^2 & V^0 - V^3 \end{pmatrix}.$$

Show that the object V^μ transforms as a 4-vector.

3.2 Derive the *Gordon identity*,

$$\bar{u}(p')\gamma^\mu u(p) = \bar{u}(p') \left[\frac{p'^\mu + p^\mu}{2m} + \frac{i\sigma^{\mu\nu} q_\nu}{2m} \right] u(p),$$

where $q = (p' - p)$. We will put this formula to use in Chapter 6.

3.3 Spinor products. (This problem, together with Problems 5.3 and 5.6, introduces an efficient computational method for processes involving massless particles.) Let k_0^μ, k_1^μ be fixed 4-vectors satisfying $k_0^2 = 0$, $k_1^2 = -1$, $k_0 \cdot k_1 = 0$. Define basic spinors in the following way: Let u_{L0} be the left-handed spinor for a fermion with momentum k_0. Let $u_{R0} = \not{k}_1 u_{L0}$. Then, for any p such that p is lightlike ($p^2 = 0$), define

$$u_L(p) = \frac{1}{\sqrt{2p \cdot k_0}} \not{p} u_{R0} \quad \text{and} \quad u_R(p) = \frac{1}{\sqrt{2p \cdot k_0}} \not{p} u_{L0}.$$

This set of conventions defines the phases of spinors unambiguously (except when p is parallel to k_0).

(a) Show that $\not{k}_0 u_{R0} = 0$. Show that, for any lightlike p, $\not{p} u_L(p) = \not{p} u_R(p) = 0$.

(b) For the choices $k_0 = (E, 0, 0, -E)$, $k_1 = (0, 1, 0, 0)$, construct u_{L0}, u_{R0}, $u_L(p)$, and $u_R(p)$ explicitly.

(c) Define the *spinor products* $s(p_1, p_2)$ and $t(p_1, p_2)$, for p_1, p_2 lightlike, by

$$s(p_1, p_2) = \bar{u}_R(p_1) u_L(p_2), \qquad t(p_1, p_2) = \bar{u}_L(p_1) u_R(p_2).$$

Using the explicit forms for the u_λ given in part (b), compute the spinor products explicitly and show that $t(p_1, p_2) = (s(p_2, p_1))^*$ and $s(p_1, p_2) = -s(p_2, p_1)$. In addition, show that

$$|s(p_1, p_2)|^2 = 2p_1 \cdot p_2.$$

Thus the spinor products are the square roots of 4-vector dot products.

3.4   Majorana fermions. Recall from Eq. (3.40) that one can write a relativistic equation for a massless 2-component fermion field that transforms as the upper two components of a Dirac spinor (ψ_L). Call such a 2-component field $\chi_a(x)$, $a = 1, 2$.

(a) Show that it is possible to write an equation for $\chi(x)$ as a massive field in the following way:

$$i\bar{\sigma} \cdot \partial \chi - im\sigma^2 \chi^* = 0.$$

That is, show, first, that this equation is relativistically invariant and, second, that it implies the Klein-Gordon equation, $(\partial^2 + m^2)\chi = 0$. This form of the fermion mass is called a Majorana mass term.

(b) Does the Majorana equation follow from a Lagrangian? The mass term would seem to be the variation of $(\sigma^2)_{ab}\chi_a^*\chi_b^*$; however, since σ^2 is antisymmetric, this expression would vanish if $\chi(x)$ were an ordinary c-number field. When we go to quantum field theory, we know that $\chi(x)$ will become an anticommuting quantum field. Therefore, it makes sense to develop its classical theory by considering $\chi(x)$ as a classical anticommuting field, that is, as a field that takes as values *Grassmann numbers* which satisfy

$$\alpha\beta = -\beta\alpha \qquad \text{for any } \alpha, \beta.$$

Note that this relation implies that $\alpha^2 = 0$. A Grassmann field $\xi(x)$ can be expanded in a basis of functions as

$$\xi(x) = \sum_n \alpha_n \phi_n(x),$$

where the $\phi_n(x)$ are orthogonal c-number functions and the α_n are a set of independent Grassmann numbers. Define the complex conjugate of a product of Grassmann numbers to reverse the order:

$$(\alpha\beta)^* \equiv \beta^*\alpha^* = -\alpha^*\beta^*.$$

This rule imitates the Hermitian conjugation of quantum fields. Show that the classical action,

$$S = \int d^4x \left[\chi^\dagger i\bar{\sigma} \cdot \partial \chi + \frac{im}{2}(\chi^T\sigma^2\chi - \chi^\dagger\sigma^2\chi^*) \right],$$

(where $\chi^\dagger = (\chi^*)^T$) is real $(S^* = S)$, and that varying this S with respect to χ and χ^* yields the Majorana equation.

(c) Let us write a 4-component Dirac field as

$$\psi(x) = \begin{pmatrix} \psi_L \\ \psi_R \end{pmatrix},$$

and recall that the lower components of ψ transform in a way equivalent by a unitary transformation to the complex conjugate of the representation ψ_L. In this way, we can rewrite the 4-component Dirac field in terms of two 2-component spinors:

$$\psi_L(x) = \chi_1(x), \qquad \psi_R(x) = i\sigma^2 \chi_2^*(x).$$

Rewrite the Dirac Lagrangian in terms of χ_1 and χ_2 and note the form of the mass term.

(d) Show that the action of part (c) has a global symmetry. Compute the divergences of the currents

$$J^\mu = \chi^\dagger \bar\sigma^\mu \chi, \qquad J^\mu = \chi_1^\dagger \bar\sigma^\mu \chi_1 - \chi_2^\dagger \bar\sigma^\mu \chi_2,$$

for the theories of parts (b) and (c), respectively, and relate your results to the symmetries of these theories. Construct a theory of N free massive 2-component fermion fields with $O(N)$ symmetry (that is, the symmetry of rotations in an N-dimensional space).

(e) Quantize the Majorana theory of parts (a) and (b). That is, promote $\chi(x)$ to a quantum field satisfying the canonical anticommutation relation

$$\{\chi_a(\mathbf{x}), \chi_b^\dagger(\mathbf{y})\} = \delta_{ab}\delta^{(3)}(\mathbf{x} - \mathbf{y}),$$

construct a Hermitian Hamiltonian, and find a representation of the canonical commutation relations that diagonalizes the Hamiltonian in terms of a set of creation and annihilation operators. (Hint: Compare $\chi(x)$ to the top two components of the quantized Dirac field.)

3.5 Supersymmetry. It is possible to write field theories with continuous symmetries linking fermions and bosons; such transformations are called *supersymmetries*.

(a) The simplest example of a supersymmetric field theory is the theory of a free complex boson and a free Weyl fermion, written in the form

$$\mathcal{L} = \partial_\mu \phi^* \partial^\mu \phi + \chi^\dagger i\bar\sigma \cdot \partial\chi + F^* F.$$

Here F is an auxiliary complex scalar field whose field equation is $F = 0$. Show that this Lagrangian is invariant (up to a total divergence) under the infinitesimal tranformation

$$\delta\phi = -i\epsilon^T \sigma^2 \chi,$$
$$\delta\chi = \epsilon F + \sigma \cdot \partial\phi \sigma^2 \epsilon^*,$$
$$\delta F = -i\epsilon^\dagger \bar\sigma \cdot \partial\chi,$$

where the parameter ϵ_a is a 2-component spinor of Grassmann numbers.

(b) Show that the term

$$\Delta\mathcal{L} = [m\phi F + \tfrac{1}{2}im\chi^T \sigma^2 \chi] + (\text{complex conjugate})$$

is also left invariant by the transformation given in part (a). Eliminate F from the complete Lagrangian $\mathcal{L} + \Delta\mathcal{L}$ by solving its field equation, and show that the fermion and boson fields ϕ and χ are given the same mass.

(c) It is possible to write supersymmetric nonlinear field equations by adding cubic and higher-order terms to the Lagrangian. Show that the following rather general field theory, containing the field (ϕ_i, χ_i), $i = 1, \ldots, n$, is supersymmetric:

$$\mathcal{L} = \partial_\mu \phi_i^* \partial^\mu \phi_i + \chi_i^\dagger i \bar{\sigma} \cdot \partial \chi_i + F_i^* F_i$$

$$+ \left(F_i \frac{\partial W[\phi]}{\partial \phi_i} + \frac{i}{2} \frac{\partial^2 W[\phi]}{\partial \phi_i \partial \phi_j} \chi_i^T \sigma^2 \chi_j + \text{c.c.} \right),$$

where $W[\phi]$ is an arbitrary function of the ϕ_i, called the *superpotential*. For the simple case $n = 1$ and $W = g\phi^3/3$, write out the field equations for ϕ and χ (after elimination of F).

3.6   Fierz transformations. Let u_i, $i = 1, \ldots, 4$, be four 4-component Dirac spinors. In the text, we proved the Fierz rearrangement formulae (3.78) and (3.79). The first of these formulae can be written in 4-component notation as

$$\bar{u}_1 \gamma^\mu \left(\frac{1+\gamma^5}{2} \right) u_2 \bar{u}_3 \gamma_\mu \left(\frac{1+\gamma^5}{2} \right) u_4 = -\bar{u}_1 \gamma^\mu \left(\frac{1+\gamma^5}{2} \right) u_4 \bar{u}_3 \gamma_\mu \left(\frac{1+\gamma^5}{2} \right) u_2.$$

In fact, there are similar rearrangement formulae for any product

$$(\bar{u}_1 \Gamma^A u_2)(\bar{u}_3 \Gamma^B u_4),$$

where Γ^A, Γ^B are any of the 16 combinations of Dirac matrices listed in Section 3.4.

(a) To begin, normalize the 16 matrices Γ^A to the convention

$$\text{tr}[\Gamma^A \Gamma^B] = 4\delta^{AB}.$$

This gives $\Gamma^A = \{1, \gamma^0, i\gamma^j, \ldots\}$; write all 16 elements of this set.

(b) Write the general Fierz identity as an equation

$$(\bar{u}_1 \Gamma^A u_2)(\bar{u}_3 \Gamma^B u_4) = \sum_{C,D} C^{AB}{}_{CD} (\bar{u}_1 \Gamma^C u_4)(\bar{u}_3 \Gamma^D u_2),$$

with unknown coefficients $C^{AB}{}_{CD}$. Using the completeness of the 16 Γ^A matrices, show that

$$C^{AB}{}_{CD} = \frac{1}{16} \text{tr}[\Gamma^C \Gamma^A \Gamma^D \Gamma^B].$$

(c) Work out explicitly the Fierz transformation laws for the products $(\bar{u}_1 u_2)(\bar{u}_3 u_4)$ and $(\bar{u}_1 \gamma^\mu u_2)(\bar{u}_3 \gamma_\mu u_4)$.

3.7   This problem concerns the discrete symmetries P, C, and T.

(a) Compute the transformation properties under P, C, and T of the antisymmetric tensor fermion bilinears, $\bar{\psi}\sigma^{\mu\nu}\psi$, with $\sigma^{\mu\nu} = \frac{i}{2}[\gamma^\mu, \gamma^\nu]$. This completes the table of the transformation properties of bilinears at the end of the chapter.

(b) Let $\phi(x)$ be a complex-valued Klein-Gordon field, such as we considered in Problem 2.2. Find unitary operators P, C and an antiunitary operator T (all defined in terms of their action on the annihilation operators $a_\mathbf{p}$ and $b_\mathbf{p}$ for the Klein-Gordon particles and antiparticles) that give the following tranformations of the Klein-Gordon field:

$$P\,\phi(t,\mathbf{x})\,P \;=\; \phi(t,-\mathbf{x});$$

$$T\,\phi(t,\mathbf{x})\,T \;=\; \phi(-t,\mathbf{x});$$

$$C\,\phi(t,\mathbf{x})\,C \;=\; \phi^*(t,\mathbf{x}).$$

Find the transformation properties of the components of the current

$$J^\mu = i(\phi^* \partial^\mu \phi - \partial^\mu \phi^* \phi)$$

under P, C, and T.

(c) Show that any Hermitian Lorentz-scalar local operator built from $\psi(x)$, $\phi(x)$, and their conjugates has $CPT = +1$.

3.8   **Bound states.** Two spin-1/2 particles can combine to a state of total spin either 0 or 1. The wavefunctions for these states are odd and even, respectively, under the interchange of the two spins.

(a) Use this information to compute the quantum numbers under P and C of all electron-positron bound states with S, P, or D wavefunctions.

(b) Since the electron-photon coupling is given by the Hamiltonian

$$\Delta H = \int d^3x \; e A_\mu j^\mu,$$

where j^μ is the electric current, electrodynamics is invariant to P and C if the components of the vector potential have the same P and C parity as the corresponding components of j^μ. Show that this implies the following surprising fact: The spin-0 ground state of positronium can decay to 2 photons, but the spin-1 ground state must decay to 3 photons. Find the selection rules for the annihilation of higher positronium states, and for 1-photon transitions between positronium levels.

Chapter 4

Interacting Fields and Feynman Diagrams

4.1 Perturbation Theory—Philosophy and Examples

We have now discussed in some detail the quantization of two free field theories that give approximate descriptions of many of the particles found in Nature. Up to this point, however, free-particle states have been eigenstates of the Hamiltonian; we have seen no interactions and no scattering. In order to obtain a closer description of the real world, we must include new, nonlinear terms in the Hamiltonian (or Lagrangian) that will couple different Fourier modes (and the particles that occupy them) to one another. To preserve causality, we insist that the new terms may involve only products of fields at the same spacetime point: $[\phi(x)]^4$ is fine, but $\phi(x)\phi(y)$ is not allowed. Thus the terms describing the interactions will be of the form

$$H_{\text{int}} = \int d^3x\, \mathcal{H}_{\text{int}}\big[\phi(x)\big] = -\int d^3x\, \mathcal{L}_{\text{int}}\big[\phi(x)\big].$$

For now we restrict ourselves to theories in which $\mathcal{H}_{\text{int}}$ ($= -\mathcal{L}_{\text{int}}$) is a function only of the fields, not of their derivatives.

In this chapter we will discuss three important examples of interacting field theories. The first is "phi-fourth" theory,

$$\mathcal{L} = \frac{1}{2}(\partial_\mu \phi)^2 - \frac{1}{2}m^2\phi^2 - \frac{\lambda}{4!}\phi^4, \tag{4.1}$$

where λ is a dimensionless *coupling constant*. (A ϕ^3 interaction would be a bit simpler, but then the energy would not be positive-definite unless we added a higher even power of ϕ as well.) Although we are introducing this theory now for purely pedagogical reasons (since it is the simplest of all interacting quantum theories), models of the real world do contain ϕ^4 interactions; the most important example in particle physics is the self-interaction of the Higgs field in the standard electroweak theory. In Part II, we will see that ϕ^4 theory also arises in statistical mechanics. The equation of motion for ϕ^4 theory is

$$(\partial^2 + m^2)\phi = -\frac{\lambda}{3!}\phi^3, \tag{4.2}$$

which cannot be solved by Fourier analysis as the free Klein-Gordon equation could. In the quantum theory we impose the equal-time commutation relations

$$[\phi(\mathbf{x}), \pi(\mathbf{y})] = i\delta^{(3)}(\mathbf{x} - \mathbf{y}),$$

which are unaffected by $\mathcal{L}_{\text{int}}$. (Note, however, that if $\mathcal{L}_{\text{int}}$ contained $\partial_\mu \phi$, the definition of $\pi(\mathbf{x})$ would change.) It is an easy exercise to write down the Hamiltonian of this theory and find the Heisenberg equation of motion for the operator $\phi(x)$; the result is the same as the classical equation of motion (4.2), just as it was in the free theory.

Our second example of an interacting field theory will be Quantum Electrodynamics:

$$\mathcal{L}_{\text{QED}} = \mathcal{L}_{\text{Dirac}} + \mathcal{L}_{\text{Maxwell}} + \mathcal{L}_{\text{int}}$$
$$= \bar{\psi}(i\partial\!\!\!/ - m)\psi - \tfrac{1}{4}(F_{\mu\nu})^2 - e\bar{\psi}\gamma^\mu\psi A_\mu, \tag{4.3}$$

where A_μ is the electromagnetic vector potential, $F_{\mu\nu} = \partial_\mu A_\nu - \partial_\nu A_\mu$ is the electromagnetic field tensor, and $e = -|e|$ is the electron charge. (To describe a fermion of charge Q, replace e with Q. If we wish to consider several species of charged particles at once, we simply duplicate $\mathcal{L}_{\text{Dirac}}$ and $\mathcal{L}_{\text{int}}$ for each additional species.) That such a simple Lagrangian can account for nearly all observed phenomena from macroscopic scales down to 10^{-13} cm is rather astonishing. In fact, the QED Lagrangian can be written even more simply:

$$\mathcal{L}_{\text{QED}} = \bar{\psi}(i D\!\!\!\!/ - m)\psi - \tfrac{1}{4}(F_{\mu\nu})^2, \tag{4.4}$$

where D_μ is the *gauge covariant derivative*,

$$D_\mu \equiv \partial_\mu + ieA_\mu(x). \tag{4.5}$$

A crucial property of the QED Lagrangian is that it is invariant under the gauge transformation

$$\psi(x) \rightarrow e^{i\alpha(x)}\psi(x), \qquad A_\mu \rightarrow A_\mu - \frac{1}{e}\partial_\mu\alpha(x), \tag{4.6}$$

which is realized on the Dirac field as a *local* phase rotation. This invariance under local phase rotations has a fundamental geometrical significance, which motivates the term *covariant derivative*. For our present purposes, though, it is sufficient just to recognize (4.6) as a symmetry of the theory.

The equations of motion follow from (4.3) by the canonical procedure. The Euler-Lagrange equation for ψ is

$$(i D\!\!\!\!/ - m)\psi(x) = 0, \tag{4.7}$$

which is just the Dirac equation coupled to the electromagnetic field by the *minimal coupling* prescription, $\partial \rightarrow D$. The Euler-Lagrange equation for A_ν is

$$\partial_\mu F^{\mu\nu} = e\bar{\psi}\gamma^\nu\psi = ej^\nu. \tag{4.8}$$

These are the inhomogeneous Maxwell equations, with the current density $j^\nu = \bar{\psi}\gamma^\nu\psi$ given by the conserved Dirac vector current (3.73). As with ϕ^4 theory, the equations of motion can also be obtained as the Heisenberg equations of motion for the operators $\psi(x)$ and $A_\mu(x)$. This is easy to verify for $\psi(x)$; we have not yet discussed the quantization of the electromagnetic field.

In fact, we will not discuss canonical quantization of the electromagnetic field at all in this book. It is an awkward subject, essentially because of gauge invariance. Note that since $\dot{A}^0$ does not appear in the Lagrangian (4.3), the momentum conjugate to A^0 is identically zero. This contradicts the canonical commutation relation $[A^0(\mathbf{x}), \pi^0(\mathbf{y})] = i\delta(\mathbf{x} - \mathbf{y})$. One solution is to quantize in Coulomb gauge, where $\nabla \cdot \mathbf{A} = 0$ and A^0 is a constrained, rather than dynamical, variable; but then manifest Lorentz invariance is sacrificed. Alternatively, one can quantize the field in Lorentz gauge, $\partial_\mu A^\mu = 0$. It is then possible to modify the Lagrangian, adding an $\dot{A}^0$ term. One obtains the commutation relations $[A^\mu(\mathbf{x}), \dot{A}^\nu(\mathbf{y})] = -ig^{\mu\nu}\delta(\mathbf{x} - \mathbf{y})$, essentially the same as four Klein-Gordon fields. But the extra minus sign in $[A^0, \dot{A}^0]$ leads to another (surmountable) difficulty: states created by $a_\mathbf{p}^{0\dagger}$ have negative norm.*

The Feynman rules for calculating scattering amplitudes that involve photons are derived more easily in the functional integral formulation of field theory, to be discussed in Chapter 9. That method has the added advantage of generalizing readily to the case of non-Abelian gauge fields, as we will see in Part III. In the present chapter we will simply guess the Feynman rules for photons. This will actually be quite easy after we derive the rules for an analogous but simpler theory, *Yukawa theory*:

$$\mathcal{L}_{\text{Yukawa}} = \mathcal{L}_{\text{Dirac}} + \mathcal{L}_{\text{Klein-Gordon}} - g\bar{\psi}\psi\phi. \tag{4.9}$$

This will be our third example. It is similar to QED, but with the photon replaced by a scalar particle ϕ. The interaction term contains a dimensionless coupling constant g, analogous to the electron charge e. Yukawa originally invented this theory to describe nucleons (ψ) and pions (ϕ). In modern particle theory, the Standard Model contains Yukawa interaction terms coupling the scalar Higgs field to quarks and leptons; most of the free parameters in the Standard Model are Yukawa coupling constants.

Having written down our three paradigm interactions, let us pause a moment to discuss what other interactions could be found in Nature. At first it might seem that the list would be infinite; even for a scalar theory we could write down interactions of the form ϕ^n for any n. But remarkably, one simple and reasonable axiom eliminates all but a few of the possible interactions. That axiom is that the theory be *renormalizable*, and it arises as follows. Higher-order terms in perturbation theory, as mentioned in Chapter 1, will involve

---

*Excellent treatments of both quantization procedures are readily available. For Coulomb gauge quantization, see Bjorken and Drell (1965), Chapter 14; for Lorentz gauge quantization, see Mandl and Shaw (1984), Chapter 5.

integrals over the 4-momenta of intermediate ("virtual") particles. These integrals are often formally divergent, and it is generally necessary to impose some form of cut-off procedure; the simplest is just to cut off the integral at some large but finite momentum Λ. At the end of the calculation one takes the limit $\Lambda \to \infty$, and hopes that physical quantities turn out to be independent of Λ. If this is indeed the case, the theory is said to be *renormalizable*. Suppose, however, that the theory includes interactions whose coupling constants have the dimensions of mass to some *negative* power. Then to obtain a dimensionless scattering amplitude, this coupling constant must be multiplied by some quantity of positive mass dimension, and it turns out that this quantity is none other than Λ. Such a term diverges as $\Lambda \to \infty$, so the theory is not renormalizable.

We will discuss these matters in detail in Chapter 10. For now we merely note that any theory containing a coupling constant with negative mass dimension is not renormalizable. A bit of dimensional analysis then allows us to throw out nearly all candidate interactions. Since the action $S = \int \mathcal{L}\, d^4x$ is dimensionless, $\mathcal{L}$ must have dimension $(\text{mass})^4$ (or simply dimension 4). From the kinetic terms of the various free Lagrangians, we note that the scalar and vector fields ϕ and A^μ have dimension 1, while the spinor field ψ has dimension $3/2$. We can now tabulate all of the allowed renormalizable interactions.

For theories involving only scalars, the allowed interaction terms are

$$\mu\phi^3 \quad \text{and} \quad \lambda\phi^4.$$

The coupling constant μ has dimension 1, while λ is dimensionless. Terms of the form ϕ^n for $n > 4$ are not allowed, since their coupling constants would have dimension $4 - n$. Of course, more interesting theories can be obtained by including several scalar fields, real or complex (see Problem 4.3).

Next we can add spinor fields. Spinor self-interactions are not allowed, since ψ^3 (besides violating Lorentz invariance) already has dimension $9/2$. Thus the only allowable new interaction is the Yukawa term,

$$g\bar{\psi}\psi\phi,$$

although similar interactions can also be constructed out of Weyl and Majorana spinors.

When we add vector fields, many new interactions are possible. The most familiar is the vector-spinor interaction of QED,

$$e\bar{\psi}\gamma^\mu\psi A_\mu.$$

Again it is easy to construct similar terms out of Weyl and Majorana spinors. Less important is the *scalar QED* Lagrangian,

$$\mathcal{L} = |D_\mu\phi|^2 - m^2|\phi|^2, \quad \text{which contains} \quad eA^\mu\phi\partial_\mu\phi^*, \ e^2|\phi|^2A^2.$$

This is our first example of a derivative interaction; quantization of this theory will be much easier with the functional integral formalism, so we postpone its discussion until Chapter 9. Other possible Lorentz-invariant terms involving vectors are

$$A^2(\partial_\mu A^\mu) \quad \text{and} \quad A^4.$$

Although it is far from obvious, these terms lead to inconsistencies unless their coupling constants are precisely chosen on the basis of a special type of symmetry, which must involve several vector fields. This symmetry underlies the *non-Abelian gauge theories*, which will be the main subject of Part III. A mass term $\frac{1}{2}m^2A^2$ for vector fields is also inconsistent, except in the special case where it is added to QED; in any case, it breaks (Abelian or non-Abelian) gauge invariance.

This exhausts the list of possible Lagrangians involving scalar, spinor, and vector particles. It is interesting to note that the currently accepted models of the strong, weak, and electromagnetic interactions include *all* of the types of interactions listed above. The three paradigm interactions to be studied in this chapter cover nearly half of the possibilities; we will study the others in detail later in this book.

The assumption that realistic theories must be renormalizable is certainly convenient, since a nonrenormalizable theory would have little predictive power. However, one might still ask *why* Nature has been so kind as to use only renormalizable interactions. One might have expected that the true theory of Nature would be a quantum theory of a much more general type. But it can be shown that, however complicated a fundamental theory appears at very high energies, the low-energy approximation to this theory that we see in experiments should be a renormalizable quantum field theory. We will demonstrate this in Section 12.1.

At a more practical level, the preceding analysis highlights a great difference in methodology between nonrelativistic quantum mechanics and relativistic quantum field theory. Since the potential $V(\mathbf{r})$ that appears in the Schrödinger equation is completely arbitrary, nonrelativistic quantum mechanics puts no limits on what interactions can be found in the real world. But we have just seen that quantum field theory imposes very tight constraints on Nature (or vice versa). Taken literally, our discussion implies that the only tasks left for particle physicists are to enumerate the elementary particles that exist and to measure their masses and coupling constants. While this viewpoint is perhaps overly arrogant, the fact that it is even thinkable is surely a sign that particle physicists are on the right track toward a fundamental theory.

Given a set of particles and couplings, we must still work out the experimental consequences. How do we analyze the quantum mechanics of an interacting field theory? It would be nice if we could explicitly solve at least a few examples (that is, find the exact eigenvalues and eigenvectors as we did for the free theories) to get a feel for the properties of interacting theories. Unfortunately, this is easier said than done. No exactly solvable interacting field theories are known in more than two spacetime dimensions, and even there the solvable models involve special symmetries and considerable technical complication.[†] Studying these theories would be interesting, but hardly worth the effort at this stage. Instead we will fall back on a much simpler and more generally applicable approach: Treat the interaction term H_{int} as a perturbation, compute its effects as far in perturbation theory as is practicable, and hope that the coupling constant is small enough that this gives a reasonable approximation to the exact answer. In fact, the perturbation series we obtain will turn out to be very simple in structure; through the use of *Feynman diagrams* it will be possible at least to visualize the effects of interactions to arbitrarily high order.

This simplification of the perturbation series for relativistic field theories was the great advance of Tomonaga, Schwinger, and Feynman. To achieve this simplification, each, independently, found a way to reformulate quantum mechanics to remove the special role of time, and then applied his new viewpoint to recast each term of the perturbation expansion as a spacetime process. We will develop quantum field theory from a spacetime viewpoint, using Feynman's method of *functional integration*, in Chapter 9. In the present chapter we follow a more pedestrian line of analysis, first developed by Dyson, to derive the spacetime picture of perturbation theory from the conventional machinery of quantum mechanics.[‡]

4.2 Perturbation Expansion of Correlation Functions

Let us then begin the study of perturbation theory for interacting fields, aiming toward a formalism that will allow us to visualize the perturbation series as spacetime processes. Although we will not need to reformulate quantum mechanics, we will rederive time-dependent perturbation theory in a form that is convenient for our purposes. Ultimately, of course, we want to calculate scattering cross sections and decay rates. For now, however, let us be less ambitious and try to calculate a simpler (but more abstract) quantity, the *two-point correlation function*, or *two-point Green's function*,

$$\langle \Omega | T\phi(x)\phi(y) | \Omega \rangle , \qquad (4.10)$$

in ϕ^4 theory. We introduce the notation $|\Omega\rangle$ to denote the ground state of the interacting theory, which is generally different from $|0\rangle$, the ground state of the free theory. The time-ordering symbol T is inserted for later convenience. The correlation function can be interpreted physically as the amplitude for propagation of a particle or excitation between y and x. In the free theory, it

---

[†]A brief survey of exactly solvable quantum field theories is given in the Epilogue.

[‡]For a historical account of the contributions of Tomonaga, Schwinger, Feynman, and Dyson, see Schweber (1994).

is simply the Feynman propagator:

$$\langle 0| T\phi(x)\phi(y) |0\rangle_{\text{free}} = D_F(x-y) = \int \frac{d^4p}{(2\pi)^4} \frac{i\,e^{-ip\cdot(x-y)}}{p^2 - m^2 + i\epsilon}. \qquad (4.11)$$

We would like to know how this expression changes in the interacting theory. Once we have analyzed the two-point correlation function, it will be easy to generalize our results to higher correlation functions in which more than two field operators appear. In Sections 4.3 and 4.4 we will continue the analysis of correlation functions, eventually developing the formalism of Feynman diagrams for evaluating them perturbatively. Then in Sections 4.5 and 4.6 we will learn how to calculate cross sections and decay rates using the same techniques.

To attack this problem, we write the Hamiltonian of ϕ^4 theory as

$$H = H_0 + H_{\text{int}} = H_{\text{Klein–Gordon}} + \int d^3x \, \frac{\lambda}{4!} \phi^4(\mathbf{x}). \qquad (4.12)$$

We want an expression for the two-point correlation function (4.10) as a power series in λ. The interaction Hamiltonian H_{int} enters (4.10) in two places: first, in the definition of the Heisenberg field,

$$\phi(x) = e^{iHt}\phi(\mathbf{x})e^{-iHt}; \qquad (4.13)$$

and second, in the definition of $|\Omega\rangle$. We must express both $\phi(x)$ and $|\Omega\rangle$ in terms of quantities we know how to manipulate: free field operators and the free theory vacuum $|0\rangle$.

It is easiest to begin with $\phi(x)$. At any fixed time t_0, we can of course expand ϕ as before in terms of ladder operators:

$$\phi(t_0, \mathbf{x}) = \int \frac{d^3p}{(2\pi)^3} \frac{1}{\sqrt{2E_{\mathbf{p}}}} \left(a_{\mathbf{p}} e^{i\mathbf{p}\cdot\mathbf{x}} + a_{\mathbf{p}}^\dagger e^{-i\mathbf{p}\cdot\mathbf{x}} \right).$$

Then to obtain $\phi(t, \mathbf{x})$ for $t \neq t_0$, we just switch to the Heisenberg picture as usual:

$$\phi(t, \mathbf{x}) = e^{iH(t-t_0)}\phi(t_0, \mathbf{x})e^{-iH(t-t_0)}.$$

For $\lambda = 0$, H becomes H_0 and this reduces to

$$\phi(t, \mathbf{x})\big|_{\lambda=0} = e^{iH_0(t-t_0)}\phi(t_0, \mathbf{x})e^{-iH_0(t-t_0)} \equiv \phi_I(t, \mathbf{x}). \qquad (4.14)$$

When λ is small, this expression will still give the most important part of the time dependence of $\phi(x)$, and thus it is convenient to give this quantity a name: the *interaction picture* field, $\phi_I(t, \mathbf{x})$. Since we can diagonalize H_0, it is easy to construct ϕ_I explicitly:

$$\phi_I(t, \mathbf{x}) = \int \frac{d^3p}{(2\pi)^3} \frac{1}{\sqrt{2E_{\mathbf{p}}}} \left(a_{\mathbf{p}} e^{-ip\cdot x} + a_{\mathbf{p}}^\dagger e^{ip\cdot x} \right)\bigg|_{x^0=t-t_0}. \qquad (4.15)$$

This is just the familiar expression for the free field from Chapter 2.

The problem now is to express the full Heisenberg field ϕ in terms of ϕ_I. Formally, it is just

$$\phi(t, \mathbf{x}) = e^{iH(t-t_0)}e^{-iH_0(t-t_0)}\phi_I(t, \mathbf{x})e^{iH_0(t-t_0)}e^{-iH(t-t_0)}$$
$$\equiv U^\dagger(t, t_0)\phi_I(t, \mathbf{x})U(t, t_0), \qquad (4.16)$$

where we have defined the unitary operator

$$U(t, t_0) = e^{iH_0(t-t_0)}e^{-iH(t-t_0)}, \qquad (4.17)$$

known as the interaction picture propagator or time-evolution operator. We would like to express $U(t, t_0)$ entirely in terms of ϕ_I, for which we have an explicit expression in terms of ladder operators. To do this, we note that $U(t, t_0)$ is the unique solution, with initial condition $U(t_0, t_0) = 1$, of a simple differential equation (the Schrödinger equation):

$$i\frac{\partial}{\partial t}U(t, t_0) = e^{iH_0(t-t_0)}\big(H - H_0\big)e^{-iH(t-t_0)}$$
$$= e^{iH_0(t-t_0)}\big(H_{\text{int}}\big)e^{-iH(t-t_0)}$$
$$= e^{iH_0(t-t_0)}\big(H_{\text{int}}\big)e^{-iH_0(t-t_0)}e^{iH_0(t-t_0)}e^{-iH(t-t_0)}$$
$$= H_I(t)\, U(t, t_0), \qquad (4.18)$$

where

$$H_I(t) = e^{iH_0(t-t_0)}\big(H_{\text{int}}\big)e^{-iH_0(t-t_0)} = \int d^3x\, \frac{\lambda}{4!}\phi_I^4 \qquad (4.19)$$

is the interaction Hamiltonian written in the interaction picture. The solution of this differential equation for $U(t, t_0)$ should look something like $U \sim \exp(-iH_I t)$; this would be our desired formula for U in terms of ϕ_I. Doing it more carefully, we will show that the actual solution is the following power series in λ:

$$U(t, t_0) = 1 + (-i)\int_{t_0}^{t} dt_1\, H_I(t_1) + (-i)^2 \int_{t_0}^{t} dt_1 \int_{t_0}^{t_1} dt_2\, H_I(t_1)H_I(t_2)$$
$$+ (-i)^3 \int_{t_0}^{t} dt_1 \int_{t_0}^{t_1} dt_2 \int_{t_0}^{t_2} dt_3\, H_I(t_1)H_I(t_2)H_I(t_3) + \cdots. \qquad (4.20)$$

To verify this, just differentiate: Each term gives the previous one times $-iH_I(t)$. The initial condition $U(t, t_0) = 1$ for $t = t_0$ is obviously satisfied.

Note that the various factors of H_I in (4.20) stand in *time order*, later on the left. This allows us to simplify the expression considerably, using the time-ordering symbol T. The H_I^2 term, for example, can be written

$$\int_{t_0}^{t} dt_1 \int_{t_0}^{t_1} dt_2\, H_I(t_1)H_I(t_2) = \frac{1}{2}\int_{t_0}^{t} dt_1 \int_{t_0}^{t} dt_2\, T\{H_I(t_1)H_I(t_2)\}. \qquad (4.21)$$

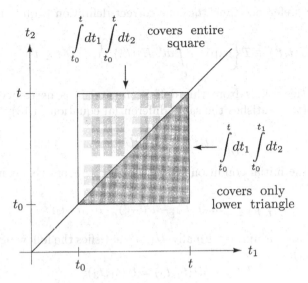

Figure 4.1. Geometric interpretation of Eq. (4.21).

The double integral on the right-hand side just counts everything twice, since in the $t_1 t_2$-plane, the integrand $T\{H_I(t_1)H_I(t_2)\}$ is symmetric about the line $t_1 = t_2$ (see Fig. 4.1).

A similar identity holds for the higher terms:

$$\int_{t_0}^{t} dt_1 \int_{t_0}^{t_1} dt_2 \cdots \int_{t_0}^{t_{n-1}} dt_n \, H_I(t_1) \cdots H_I(t_n) = \frac{1}{n!} \int_{t_0}^{t} dt_1 \cdots dt_n \, T\{H_I(t_1) \cdots H_I(t_n)\}.$$

This case is a little harder to visualize, but it is not hard to convince oneself that it is true. Using this identity, we can now write $U(t,t_0)$ in an extremely compact form:

$$U(t,t_0) = 1 + (-i)\int_{t_0}^{t} dt_1 \, H_I(t_1) + \frac{(-i)^2}{2!}\int_{t_0}^{t} dt_1 \, dt_2 \, T\{H_I(t_1)H_I(t_2)\} + \cdots$$

$$\equiv T\left\{ \exp\left[-i\int_{t_0}^{t} dt' \, H_I(t')\right] \right\}, \tag{4.22}$$

where the time-ordering of the exponential is just defined as the Taylor series with each term time-ordered. When we do real computations we will keep only the first few terms of the series; the time-ordered exponential is just a compact way of writing and remembering the correct expression.

We now have control over $\phi(t, \mathbf{x})$; we have written it entirely in terms of ϕ_I, as desired. Before moving on to consider $|\Omega\rangle$, however, it is convenient to generalize the definition of U, allowing its second argument to take on values other than our "reference time" t_0. The correct definition is quite natural:

$$U(t,t') \equiv T\left\{\exp\left[-i\int_{t'}^{t} dt'' \, H_I(t'')\right]\right\}. \qquad (t \geq t') \qquad (4.23)$$

Several properties follow from this definition, and it is necessary to verify them. First, $U(t,t')$ satisfies the same differential equation (4.18),

$$i\frac{\partial}{\partial t}U(t,t') = H_I(t)U(t,t'), \qquad (4.24)$$

but now with the initial condition $U = 1$ for $t = t'$. From this equation you can show that

$$U(t,t') = e^{iH_0(t-t_0)}e^{-iH(t-t')}e^{-iH_0(t'-t_0)}, \qquad (4.25)$$

which proves that U is unitary. Finally, $U(t,t')$ satisfies the following identities (for $t_1 \geq t_2 \geq t_3$):

$$U(t_1,t_2)U(t_2,t_3) = U(t_1,t_3);$$
$$U(t_1,t_3)\left[U(t_2,t_3)\right]^\dagger = U(t_1,t_2). \qquad (4.26)$$

Now we can go on to discuss $|\Omega\rangle$. Since $|\Omega\rangle$ is the ground state of H, we can isolate it by the following procedure. Imagine starting with $|0\rangle$, the ground state of H_0, and evolving through time with H:

$$e^{-iHT}|0\rangle = \sum_n e^{-iE_n T}|n\rangle\langle n|0\rangle,$$

where E_n are the eigenvalues of H. We must assume that $|\Omega\rangle$ has some overlap with $|0\rangle$, that is, $\langle\Omega|0\rangle \neq 0$ (if this were not the case, H_I would in no sense be a small perturbation). Then the above series contains $|\Omega\rangle$, and we can write

$$e^{-iHT}|0\rangle = e^{-iE_0 T}|\Omega\rangle\langle\Omega|0\rangle + \sum_{n\neq 0}e^{-iE_n T}|n\rangle\langle n|0\rangle,$$

where $E_0 \equiv \langle\Omega|H|\Omega\rangle$. (The zero of energy will be defined by $H_0|0\rangle = 0$.) Since $E_n > E_0$ for all $n \neq 0$, we can get rid of all the $n \neq 0$ terms in the series by sending T to ∞ in a slightly imaginary direction: $T \to \infty(1 - i\epsilon)$. Then the exponential factor $e^{-iE_n T}$ dies slowest for $n = 0$, and we have

$$|\Omega\rangle = \lim_{T\to\infty(1-i\epsilon)}\left(e^{-iE_0 T}\langle\Omega|0\rangle\right)^{-1}e^{-iHT}|0\rangle. \qquad (4.27)$$

Since T is now very large, we can shift it by a small constant:

$$|\Omega\rangle = \lim_{T\to\infty(1-i\epsilon)}\left(e^{-iE_0(T+t_0)}\langle\Omega|0\rangle\right)^{-1}e^{-iH(T+t_0)}|0\rangle$$

$$= \lim_{T\to\infty(1-i\epsilon)}\left(e^{-iE_0(t_0-(-T))}\langle\Omega|0\rangle\right)^{-1}e^{-iH(t_0-(-T))}e^{-iH_0(-T-t_0)}|0\rangle$$

$$= \lim_{T\to\infty(1-i\epsilon)}\left(e^{-iE_0(t_0-(-T))}\langle\Omega|0\rangle\right)^{-1}U(t_0,-T)|0\rangle. \qquad (4.28)$$

In the second line we have used $H_0 |0\rangle = 0$. Ignoring the c-number factor in front, this expression tells us that we can get $|\Omega\rangle$ by simply evolving $|0\rangle$ from time $-T$ to time t_0 with the operator U. Similarly, we can express $\langle\Omega|$ as

$$\langle\Omega| = \lim_{T\to\infty(1-i\epsilon)} \langle 0| U(T, t_0) \left(e^{-iE_0(T-t_0)} \langle 0| \Omega\rangle\right)^{-1}. \quad (4.29)$$

Let us put together the pieces of the two-point correlation function. For the moment, assume that $x^0 > y^0 > t_0$. Then

$$\langle\Omega| \phi(x)\phi(y) |\Omega\rangle = \lim_{T\to\infty(1-i\epsilon)} \left(e^{-iE_0(T-t_0)} \langle 0| \Omega\rangle\right)^{-1} \langle 0| U(T, t_0)$$
$$\times \left[U(x^0, t_0)\right]^\dagger \phi_I(x) U(x^0, t_0) \left[U(y^0, t_0)\right]^\dagger \phi_I(y) U(y^0, t_0)$$
$$\times U(t_0, -T) |0\rangle \left(e^{-iE_0(t_0-(-T))} \langle\Omega| 0\rangle\right)^{-1}$$
$$= \lim_{T\to\infty(1-i\epsilon)} \left(|\langle 0| \Omega\rangle|^2 e^{-iE_0(2T)}\right)^{-1}$$
$$\times \langle 0| U(T, x^0)\phi_I(x) U(x^0, y^0)\phi_I(y) U(y^0, -T) |0\rangle. \quad (4.30)$$

This is starting to look simple, except for the awkward factor in front. To get rid of it, divide by 1 in the form

$$1 = \langle\Omega| \Omega\rangle = \left(|\langle 0| \Omega\rangle|^2 e^{-iE_0(2T)}\right)^{-1} \langle 0| U(T, t_0) U(t_0, -T) |0\rangle.$$

Then our formula, still for $x^0 > y^0$, becomes

$$\langle\Omega| \phi(x)\phi(y) |\Omega\rangle = \lim_{T\to\infty(1-i\epsilon)} \frac{\langle 0| U(T, x^0)\phi_I(x) U(x^0, y^0)\phi_I(y) U(y^0, -T) |0\rangle}{\langle 0| U(T, -T) |0\rangle}.$$

Now note that all fields on both sides of this expression are in time order. If we had considered the case $y^0 > x^0$ this would still be true. Thus we arrive at our final expression, now valid for any x^0 and y^0:

$$\langle\Omega| T\{\phi(x)\phi(y)\} |\Omega\rangle = \lim_{T\to\infty(1-i\epsilon)} \frac{\langle 0| T\left\{\phi_I(x)\phi_I(y) \exp\left[-i \int_{-T}^{T} dt\, H_I(t)\right]\right\} |0\rangle}{\langle 0| T\left\{\exp\left[-i \int_{-T}^{T} dt\, H_I(t)\right]\right\} |0\rangle}.$$
$$(4.31)$$

The virtue of considering the time-ordered product is clear: It allows us to put everything inside one large T-operator. A similar formula holds for higher correlation functions of arbitrarily many fields; for each extra factor of ϕ on the left, put an extra factor of ϕ_I on the right. So far this expression is exact. But it is ideally suited to doing perturbative calculations; we need only retain as many terms as desired in the Taylor series expansions of the exponentials.

4.3 Wick's Theorem

We have now reduced the problem of calculating correlation functions to that of evaluating expressions of the form

$$\langle 0| T\{\phi_I(x_1)\phi_I(x_2)\cdots\phi_I(x_n)\} |0\rangle,$$

that is, vacuum expectation values of time-ordered products of finite (but arbitrary) numbers of free field operators. For $n = 2$ this expression is just the Feynman propagator. For higher n you could evaluate this object by brute force, plugging in the expansion of ϕ_I in terms of ladder operators. In this section and the next, however, we will see how to simplify such calculations immensely.

Consider again the case of two fields, $\langle 0| T\{\phi_I(x)\phi_I(y)\} |0\rangle$. We already know how to calculate this quantity, but now we would like to rewrite it in a form that is easy to evaluate and also generalizes to the case of more than two fields. To do this we first decompose $\phi_I(x)$ into positive- and negative-frequency parts:

$$\phi_I(x) = \phi_I^+(x) + \phi_I^-(x), \tag{4.32}$$

where

$$\phi_I^+(x) = \int \frac{d^3p}{(2\pi)^3} \frac{1}{\sqrt{2E_\mathbf{p}}} a_\mathbf{p} e^{-ip\cdot x}; \qquad \phi_I^-(x) = \int \frac{d^3p}{(2\pi)^3} \frac{1}{\sqrt{2E_\mathbf{p}}} a_\mathbf{p}^\dagger e^{+ip\cdot x}.$$

This decomposition can be done for any free field. It is useful because

$$\phi_I^+(x) |0\rangle = 0 \quad \text{and} \quad \langle 0| \phi_I^-(x) = 0.$$

For example, consider the case $x^0 > y^0$. The time-ordered product of two fields is then

$$T\phi_I(x)\phi_I(y) \underset{x^0>y^0}{=} \phi_I^+(x)\phi_I^+(y) + \phi_I^+(x)\phi_I^-(y) + \phi_I^-(x)\phi_I^+(y) + \phi_I^-(x)\phi_I^-(y)$$

$$= \phi_I^+(x)\phi_I^+(y) + \phi_I^-(y)\phi_I^+(x) + \phi_I^-(x)\phi_I^+(y) + \phi_I^-(x)\phi_I^-(y)$$

$$+ \left[\phi_I^+(x), \phi_I^-(y)\right]. \tag{4.33}$$

In every term except the commutator, all the $a_\mathbf{p}$'s are to the right of all the $a_\mathbf{p}^\dagger$'s. Such a term (e.g., $a_\mathbf{p}^\dagger a_\mathbf{q}^\dagger a_\mathbf{k} a_\mathbf{l}$) is said to be in *normal order*, and has vanishing vacuum expectation value. Let us also define the normal ordering symbol $N()$ to place whatever operators it contains in normal order, for example,

$$N\left(a_\mathbf{p} a_\mathbf{k}^\dagger a_\mathbf{q}\right) \equiv a_\mathbf{k}^\dagger a_\mathbf{p} a_\mathbf{q}. \tag{4.34}$$

The order of $a_\mathbf{p}$ and $a_\mathbf{q}$ on the right-hand side makes no difference since they commute.*

---

*In the literature one often sees the notation $:\phi_1\phi_2:$ instead of $N(\phi_1\phi_2)$.

If we had instead considered the case $y^0 > x^0$, we would get the same four normal-ordered terms as in (4.33), but this time the final commutator would be $[\phi_I^+(y), \phi_I^-(x)]$. Let us therefore define one more quantity, the *contraction* of two fields, as follows:

$$\overline{\phi(x)\phi(y)} \equiv \begin{cases} [\phi^+(x), \phi^-(y)] & \text{for } x^0 > y^0; \\ [\phi^+(y), \phi^-(x)] & \text{for } y^0 > x^0. \end{cases} \tag{4.35}$$

This quantity is exactly the Feynman propagator:

$$\overline{\phi(x)\phi(y)} = D_F(x-y). \tag{4.36}$$

(From here on we will often drop the subscript I for convenience; contractions will always involve interaction-picture fields.)

The relation between time-ordering and normal-ordering is now extremely simple to express, at least for two fields:

$$T\{\phi(x)\phi(y)\} = N\{\phi(x)\phi(y) + \overline{\phi(x)\phi(y)}\}. \tag{4.37}$$

But now that we have all this new notation, the generalization to arbitrarily many fields is also easy to write down:

$$\begin{aligned} T\{\phi(x_1)\phi(x_2)\cdots\phi(x_m)\} \\ = N\{\phi(x_1)\phi(x_2)\cdots\phi(x_m) + \text{all possible contractions}\}. \end{aligned} \tag{4.38}$$

This identity is known as *Wick's theorem*, and we will prove it in a moment. For $m = 2$ it is identical to (4.37). The phrase *all possible contractions* means there will be one term for each possible way of contracting the m fields in pairs. Thus for $m = 4$ we have (writing $\phi(x_a)$ as ϕ_a for brevity)

$$\begin{aligned} T\{\phi_1\phi_2\phi_3\phi_4\} = N\{ &\phi_1\phi_2\phi_3\phi_4 + \overline{\phi_1\phi_2}\phi_3\phi_4 + \overline{\phi_1\phi_2\phi_3}\phi_4 + \overline{\phi_1\phi_2\phi_3\phi_4} \\ &+ \phi_1\overline{\phi_2\phi_3}\phi_4 + \phi_1\overline{\phi_2\phi_3\phi_4} + \phi_1\phi_2\overline{\phi_3\phi_4} \\ &+ \overline{\phi_1\phi_2}\,\overline{\phi_3\phi_4} + \overline{\phi_1\overline{\phi_2\phi_3}\phi_4} + \overline{\phi_1\phi_2\phi_3\phi_4}\}. \end{aligned} \tag{4.39}$$

When the contraction symbol connects two operators that are not adjacent, we still define it to give a factor of D_F. For example,

$$N\{\overline{\phi_1\phi_2\phi_3}\phi_4\} \qquad \text{means} \qquad D_F(x_1 - x_3)\cdot N\{\phi_2\phi_4\}.$$

In the vacuum expectation value of (4.39), any term in which there remain uncontracted operators gives zero (since $\langle 0| N(\text{any operator}) |0\rangle = 0$). Only the three fully contracted terms in the last line survive, and they are all c-numbers, so we have

$$\begin{aligned} \langle 0| T\{\phi_1\phi_2\phi_3\phi_4\} |0\rangle = \ &D_F(x_1 - x_2)D_F(x_3 - x_4) \\ &+ D_F(x_1 - x_3)D_F(x_2 - x_4) \\ &+ D_F(x_1 - x_4)D_F(x_2 - x_3). \end{aligned} \tag{4.40}$$

Now let us prove Wick's theorem. Naturally the proof is by induction on m, the number of fields. We have already proved the case $m = 2$. So assume the theorem is true for $m - 1$ fields, and let's try to prove it for m fields. Without loss of generality, we can restrict ourselves to the case $x_1^0 \geq x_2^0 \geq \cdots x_m^0$; if this is not the case we can just relabel the points, without affecting either side of (4.38). Then applying Wick's theorem to $\phi_2 \cdots \phi_m$, we have

$$
\begin{aligned}
T\{\phi_1 \cdots \phi_m\} &= \phi_1 \cdots \phi_m \\
&= \phi_1 N\left\{\phi_2 \cdots \phi_m + \begin{pmatrix} \text{all contractions} \\ \text{not involving } \phi_1 \end{pmatrix}\right\} \\
&= (\phi_1^+ + \phi_1^-) N\left\{\phi_2 \cdots \phi_m + \begin{pmatrix} \text{all contractions} \\ \text{not involving } \phi_1 \end{pmatrix}\right\}. \quad (4.41)
\end{aligned}
$$

We want to move the ϕ_1^+ and ϕ_1^- inside the $N\{\}$. For the ϕ_1^- term this is easy: Just move it in, since (being on the left) it is already in normal order. The term with ϕ_1^+ must be put in normal order by commuting it to the right past all the other ϕ's. Consider, for example, the term with no contractions:

$$
\begin{aligned}
\phi_1^+ N(\phi_2 \cdots \phi_m) &= N(\phi_2 \cdots \phi_m)\phi_1^+ + [\phi_1^+, N(\phi_2 \cdots \phi_m)] \\
&= N(\phi_1^+ \phi_2 \cdots \phi_m) \\
&\quad + N([\phi_1^+, \phi_2^-]\phi_3 \cdots \phi_m + \phi_2[\phi_1^+, \phi_3^-]\phi_4 \cdots \phi_m + \cdots) \\
&= N\big(\phi_1^+ \phi_2 \cdots \phi_m + \overset{\frown}{\phi_1 \phi_2}\phi_3 \cdots \phi_m + \phi_1\phi_2\overset{\frown}{\phi_3} \cdots + \cdots\big).
\end{aligned}
$$

The first term in the last line combines with part of the ϕ_1^- term from (4.41) to give $N\{\phi_1\phi_2 \cdots \phi_m\}$, so we now have the first term on the right-hand side of Wick's theorem, as well as all possible terms involving a single contraction of ϕ_1 with another field. Similarly, a term in (4.41) involving one contraction will produce all possible terms involving both that contraction and a contraction of ϕ_1 with one of the other fields. Doing this with all the terms of (4.41), we eventually get all possible contractions of all the fields, including ϕ_1. Thus the induction step is complete, and Wick's theorem is proved.

4.4 Feynman Diagrams

Wick's theorem allows us to turn any expression of the form

$$
\langle 0| T\{\phi_I(x_1)\phi_I(x_2) \cdots \phi_I(x_n)\} |0\rangle
$$

into a sum of products of Feynman propagators. Now we are ready to develop a diagrammatic interpretation of such expressions. Consider first the case of four fields, all at different spacetime points, which we worked out in Eq. (4.40). Let us represent each of the points x_1 through x_4 by a dot, and each factor

$D_F(x-y)$ by a line joining x to y. Then Eq. (4.40) can be represented as the sum of three diagrams (called *Feynman diagrams*):

$$\langle 0| T\{\phi_1\phi_2\phi_3\phi_4\} |0\rangle = \qquad + \qquad + \qquad \tag{4.42}$$

Although this isn't exactly a measurable quantity, the diagrams do suggest an interpretation: Particles are created at two spacetime points, each propagates to one of the other points, and then they are annihilated. This can happen in three ways, corresponding to the three ways to connect the points in pairs, as shown in the three diagrams. The total amplitude for the process is the sum of the three diagrams.

Things get more interesting when the expression contains more than one field at the same spacetime point. So let us now return to the evaluation of the two-point function $\langle \Omega| T\{\phi(x)\phi(y)\} |\Omega\rangle$, and put formula (4.31) to use. We will ignore the denominator until the very end of this section. The numerator, with the exponential expanded as a power series, is

$$\langle 0| T\Big\{\phi(x)\phi(y) + \phi(x)\phi(y)\Big[-i\int dt\, H_I(t)\Big] + \cdots\Big\} |0\rangle. \tag{4.43}$$

The first term gives the free-field result, $\langle 0| T\{\phi(x)\phi(y)\} |0\rangle = D_F(x-y)$. The second term, in ϕ^4 theory, is

$$\langle 0| T\Big\{\phi(x)\phi(y)\,(-i)\int dt \int d^3z\, \frac{\lambda}{4!}\phi^4\Big\} |0\rangle$$

$$= \langle 0| T\Big\{\phi(x)\phi(y)\Big(\frac{-i\lambda}{4!}\Big)\int d^4z\, \phi(z)\phi(z)\phi(z)\phi(z)\Big\} |0\rangle.$$

Now apply Wick's theorem. We get one term for every way of contracting the six ϕ operators with each other in pairs. There are 15 ways to do this, but (fortunately) only two of them are really different. If we contract $\phi(x)$ with $\phi(y)$, then there are three ways to contract the four $\phi(z)$'s with each other, and all three give identical expressions. The other possibility is to contract $\phi(x)$ with one of the $\phi(z)$'s (four choices), $\phi(y)$ with one of the others (three choices), and the remaining two $\phi(z)$'s with each other (one choice). There are twelve ways to do this, and all give identical expressions. Thus we have

$$\langle 0| T\Big\{\phi(x)\phi(y)\,(-i)\int dt \int d^3z\, \frac{\lambda}{4!}\phi^4\Big\} |0\rangle$$

$$= 3\cdot\Big(\frac{-i\lambda}{4!}\Big) D_F(x-y)\int d^4z\, D_F(z-z)D_F(z-z) \tag{4.44}$$

$$+ 12\cdot\Big(\frac{-i\lambda}{4!}\Big) \int d^4z\, D_F(x-z)D_F(y-z)D_F(z-z).$$

We can understand this expression better if we represent each term as a Feynman diagram. Again we draw each contraction D_F as a line, and each point as a dot. But this time we must distinguish between the "external" points, x and y, and the "internal" point z; each internal point is associated with a factor of $(-i\lambda)\int d^4z$. We will worry about the constant factors in a moment. Using these rules, we see that the above expression (4.44) is equal to the sum of two diagrams:

$$\left(\begin{array}{c} \underset{x}{\bullet}\!\!-\!\!-\!\!-\!\!\underset{y}{\bullet}\,\, \underset{z}{}\!\!\bigotimes \end{array} \right) \;+\; \left(\begin{array}{c} \underset{x}{\bullet}\!\!-\!\!-\!\!\underset{z}{\bullet}\!\!-\!\!-\!\!\underset{y}{\bullet} \end{array} \right)$$

We refer to the lines in these diagrams as *propagators*, since they represent the propagation amplitude D_F. Internal points, where four lines meet, are called *vertices*. Since $D_F(x-y)$ is the amplitude for a free Klein-Gordon particle to propagate between x and y, the diagrams actually interpret the analytic formula as a process of particle creation, propagation, and annihilation which takes place in spacetime.

Now let's try a more complicated contraction, from the λ^3 term in the expansion of the correlation function:

$$\langle 0|\,\phi(x)\phi(y)\,\tfrac{1}{3!}\Big(\tfrac{-i\lambda}{4!}\Big)^3\int d^4z\,\phi\phi\phi\phi\,\int d^4w\,\phi\phi\phi\phi\,\int d^4u\,\phi\phi\phi\phi\,|0\rangle$$

$$= \frac{1}{3!}\left(\frac{-i\lambda}{4!}\right)^3\int d^4z\,d^4w\,d^4u\,\,D_F(x-z)D_F(z-z)D_F(z-w)$$

$$\times D_F(w-y)D_F^2(w-u)D_F(u-u). \qquad (4.45)$$

The number of "different" contractions that give this same expression is large:

$3!$	$\times$	$4\cdot 3$	$\times$	$4\cdot 3\cdot 2$	$\times$	$4\cdot 3$	$\times$	$1/2$
interchange of vertices		placement of contractions into z vertex		placement of contractions into w vertex		placement of contractions into u vertex		interchange of w–u contractions

The product of these combinatoric factors is 10,368, roughly 1/13 of the total of 135,135 possible full contractions of the 14 operators. The structure of this particular contraction can be represented by the following "cactus" diagram:

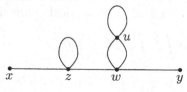

It is conventional, for obvious reasons, to let this one diagram represent the sum of all 10,368 identical terms.

In practice one always draws the diagram first, using it as a mnemonic device for writing down the analytic expression. But then the question arises, What is the overall constant? We could, of course, work it out as above: We could associate a factor $\int d^4z(-i\lambda/4!)$ with each vertex, put in the $1/n!$ from the Taylor series, and then do the combinatorics by writing out the product of fields as in (4.45) and counting. But the $1/n!$ from the Taylor series will almost always cancel the $n!$ from interchanging the vertices, so we can just forget about both of these factors. Furthermore, the generic vertex has four lines coming in from four different places, so the various placements of these contractions into $\phi\phi\phi\phi$ generates a factor of 4! (as in the w vertex above), which cancels the denominator in $(-i\lambda/4!)$. It is therefore conventional to associate the expression $\int d^4z(-i\lambda)$ with each vertex. (This was the reason for the factor of 4! in the ϕ^4 coupling.)

In the above diagram, this scheme gives a constant that is too large by a factor of $8 = 2 \cdot 2 \cdot 2$, the *symmetry factor* of the diagram. Two factors of 2 come from lines that start and end on the same vertex: The diagram is symmetric under the interchange of the two ends of such a line. The other factor of 2 comes from the two propagators connecting w to u: The diagram is symmetric under the interchange of these two lines with each other. A third possible type of symmetry is the equivalence of two vertices. To get the correct overall constant for a diagram, we divide by its symmetry factor, which is in general the number of ways of interchanging components without changing the diagram.

Most people never need to evaluate a diagram with a symmetry factor greater than 2, so there's no need to worry too much about these technicalities. But here are a few examples, to make some sense out of the above rules:

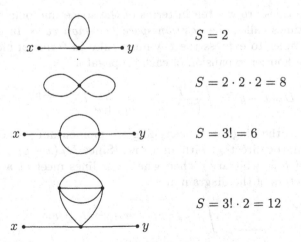

$$S = 2$$

$$S = 2 \cdot 2 \cdot 2 = 8$$

$$S = 3! = 6$$

$$S = 3! \cdot 2 = 12$$

When in doubt, you can always determine the symmetry factor by counting equivalent contractions, as we did above.

We are now ready to summarize our rules for calculating the numerator of our expression (4.31) for $\langle\Omega| T\phi(x)\phi(y) |\Omega\rangle$:

$$\langle 0| T\left\{\phi_I(x)\phi_I(y) \exp\left[-i\int dt\, H_I(t)\right]\right\} |0\rangle = \left(\begin{array}{c}\text{sum of all possible diagrams}\\ \text{with two external points}\end{array}\right),$$

where each diagram is built out of propagators, vertices, and external points. The rules for associating analytic expressions with pieces of diagrams are called the *Feynman rules*. In ϕ^4 theory the rules are:

1. For each propagator, $\quad x \bullet\!\!\!-\!\!\!-\!\!\!-\!\!\!-\!\!\!-\!\!\!\bullet y \quad = D_F(x-y)$;

2. For each vertex, $\qquad\qquad$ $\quad = (-i\lambda)\int d^4z$;

3. For each external point, $\quad x\bullet\!\!\!-\!\!\!-\!\!\!-\!\!\!-\!\!\!- \quad = 1$;

4. Divide by the symmetry factor.

One way to interpret these rules is to think of the vertex factor $(-i\lambda)$ as the amplitude for the emission and/or absorption of particles at a vertex. The integral $\int d^4z$ instructs us to sum over all points where this process can occur. This is just the *superposition* principle of quantum mechanics: When a process can happen in alternative ways, we *add* the amplitudes for each possible way. To compute each individual amplitude, the Feynman rules tell us to *multiply* the amplitudes (propagators and vertex factors) for each independent part of the process.

Since these rules are written in terms of the spacetime points x, y, etc., they are sometimes called the *position-space Feynman rules*. In most calculations, it is simpler to express the Feynman rules in terms of momenta, by introducing the Fourier expansion of each propagator:

$$D_F(x-y) = \int \frac{d^4p}{(2\pi)^4} \frac{i}{p^2 - m^2 + i\epsilon} e^{-ip\cdot(x-y)}. \qquad (4.46)$$

Represent this in the diagram by assigning a 4-momentum p to each propagator, indicating the direction with an arrow. (Since $D_F(x-y) = D_F(y-x)$, the direction of p is arbitrary.) Then when four lines meet at a vertex, the z-dependent factors of the diagram are

$\qquad\longrightarrow\qquad$ $\int d^4z\, e^{-ip_1 z} e^{-ip_2 z} e^{-ip_3 z} e^{+ip_4 z}$

$$= (2\pi)^4 \delta^{(4)}(p_1 + p_2 + p_3 - p_4). \qquad (4.47)$$

In other words, momentum is conserved at each vertex. The delta functions from the vertices can now be used to perform some of the momentum integrals from the propagators. We are left with the following *momentum-space Feynman rules*:

1. For each propagator, $\longrightarrow$ $= \dfrac{i}{p^2 - m^2 + i\epsilon};$

   p

2. For each vertex, $\times$ $= -i\lambda;$

3. For each external point, $x \bullet\!\!\longleftarrow\!\!\longrightarrow$ $= e^{-ip\cdot x};$

   p

4. Impose momentum conservation at each vertex;

5. Integrate over each undetermined momentum: $\displaystyle\int \dfrac{d^4p}{(2\pi)^4};$

6. Divide by the symmetry factor.

Again, we can interpret each factor as the amplitude for that part of the process, with the integrations coming from the superposition principle. The exponential factor for an external point is just the amplitude for a particle at that point to have the needed momentum, or, depending on the direction of the arrow, for a particle with a certain momentum to be found at that point.

This nearly completes our discussion of the computation of correlation functions, but there are still a few loose ends. First, what happened to the large time T that was taken to $\infty(1 - i\epsilon)$? We glossed over it completely in this section, starting with Eq. (4.43). The place to put it back is Eq. (4.47), where instead of just integrating over d^4z, we should have

$$\lim_{T \to \infty(1-i\epsilon)} \int_{-T}^{T} dz^0 \int d^3z \, e^{-i(p_1+p_2+p_3-p_4)\cdot z}.$$

The exponential blows up as $z^0 \to \infty$ or $z^0 \to -\infty$ unless its argument is purely imaginary. To achieve this, we can take each p^0 to have a small imaginary part: $p^0 \propto (1 + i\epsilon)$. But this is precisely what we do in following the Feynman boundary conditions for computing D_F: We integrate along a contour that is rotated slightly away from the real axis, so that $p^0 \propto (1+i\epsilon)$:

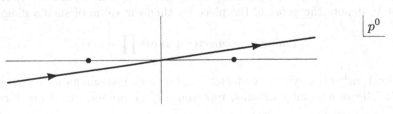

The explicit dependence on T seems to disappear when we take the limit $T \to \infty$ in the previous equation. But consider the diagram

$$p_1 \qquad (4.48)$$

The delta function for the left-hand vertex is $(2\pi)^4 \delta^{(4)}(p_1 + p_2)$, so momentum conservation at the right-hand vertex is automatically satisfied, and we get $(2\pi)^4 \delta^{(4)}(0)$ there. This awkward factor is easy to understand by going back to position space. It is simply the integral of a constant over $d^4 w$:

$$\int d^4 w \, (\text{const}) \propto (2T) \cdot (\text{volume of space}). \qquad (4.49)$$

This just tells us that the spacetime process (4.48) can happen at any place in space, and at any time between $-T$ and T. Every *disconnected* piece of a diagram, that is, every piece that is not connected to an external point, will have one such $(2\pi)^4 \delta^{(4)}(0) = 2T \cdot V$ factor.

The contributions to the correlation function coming from such diagrams can be better understood with the help of a very pretty identity, *the exponentiation of the disconnected diagrams*. It works as follows. A typical diagram has the form

$$(4.50)$$

with a piece connected to x and y, and several disconnected pieces. (Since each vertex has an even number of lines coming into it, x and y must be connected to each other.) Label the various possible disconnected pieces by V_i:

$$(4.51)$$

The elements V_i are connected internally, but disconnected from external points. Suppose that a diagram (such as (4.50)) has n_i pieces of the form V_i, for each i, in addition to its one piece that is connected to x and y. (In any given diagram, only finitely many of the n_i will be nonzero.) If we also let V_i denote the *value* of the piece V_i, then the value of such a diagram is

$$(\text{value of connected piece}) \cdot \prod_i \frac{1}{n_i!} \left(V_i\right)^{n_i}.$$

The $1/n_i!$ is the symmetry factor coming from interchanging the n_i copies of V_i. The sum of all diagrams, representing the numerator of our formula for the two-point correlation function, is then

$$\sum_{\substack{\text{all possible} \\ \text{connected} \\ \text{pieces}}} \sum_{\text{all } \{n_i\}} \left(\begin{array}{c} \text{value of} \\ \text{connected piece} \end{array} \right) \times \left(\prod_i \frac{1}{n_i!} (V_i)^{n_i} \right),$$

where "all $\{n_i\}$" means "all ordered sets $\{n_1, n_2, n_3, \ldots\}$ of nonnegative integers." The sum of the connected pieces factors out of this expression, giving

$$= \left(\sum \text{connected} \right) \times \sum_{\text{all } \{n_i\}} \left(\prod_i \frac{1}{n_i!} (V_i)^{n_i} \right),$$

where $\left(\sum \text{connected} \right)$ is an abbreviation for the sum of the values of all connected pieces. It is not too hard to see that the rest of the expression can also be factored (try working backwards):

$$= \left(\sum \text{connected} \right) \times \left(\sum_{n_1} \frac{1}{n_1!} V_1^{n_1} \right) \left(\sum_{n_2} \frac{1}{n_2!} V_2^{n_2} \right) \left(\sum_{n_3} \frac{1}{n_3!} V_3^{n_3} \right) \cdots$$

$$= \left(\sum \text{connected} \right) \times \prod_i \left(\sum_{n_i} \frac{1}{n_i!} V_i^{n_i} \right)$$

$$= \left(\sum \text{connected} \right) \times \prod_i \exp(V_i)$$

$$= \left(\sum \text{connected} \right) \times \exp \left(\sum_i V_i \right). \tag{4.52}$$

We have just shown that the sum of *all* diagrams is equal to the sum of all *connected* diagrams, times the exponential of the sum of all *disconnected* diagrams. (We should really say "pieces" rather than "diagrams" on the right-hand side of the equality, but from now on we will often just call a single piece a "diagram.") Pictorially, the identity is

$$\lim_{T \to \infty(1-i\epsilon)} \langle 0| T \left\{ \phi_I(x) \phi_I(y) \exp \left[-i \int_{-T}^{T} dt \, H_I(t) \right] \right\} |0\rangle$$

$$\times \exp \left[\; \right]. \tag{4.53}$$

Now consider the *denominator* of our formula (4.31) for the two-point function. By an argument identical to the above, it is just

$$\langle 0| T\left\{ \exp\left[-i \int_{-T}^{T} dt \, H_I(t) \right] \right\} |0\rangle = \exp\left[\; \text{} \; \right],$$

which cancels the exponential of the disconnected diagrams in the numerator. This is the final simplification of the formula, which now reads

$$\langle \Omega | T[\phi(x)\phi(y)] |\Omega \rangle$$

= sum of all connected diagrams with two external points

$$= \;\; \text{} \;\; . \tag{4.54}$$

We have come a long way from our original formula, Eq. (4.31).

Having gotten rid of the disconnected diagrams in our formula for the correlation function, we might pause a moment to go back and interpret them physically. The place to look is Eq. (4.30), which can be written

$$\lim_{T\to\infty(1-i\epsilon)} \langle 0| T\left\{ \phi_I(x)\phi_I(y) \exp\left[-i \int_{-T}^{T} dt \, H_I(t) \right] \right\} |0\rangle$$

$$= \langle \Omega | T\phi(x)\phi(y) |\Omega \rangle \cdot \lim_{T\to\infty(1-i\epsilon)} \left(|\langle 0|\Omega \rangle|^2 e^{-iE_0(2T)} \right).$$

Looking only at the T-dependent parts of both sides, this implies

$$\exp\left[\sum_i V_i \right] \propto \exp\left[-iE_0(2T) \right]. \tag{4.55}$$

Since each disconnected diagram V_i contains a factor of $(2\pi)^4 \delta^{(4)}(0) = 2T \cdot V$, this gives us a formula for the energy density of the vacuum (relative to the zero of energy set by $H_0 |0\rangle = 0$):

$$\frac{E_0}{\text{volume}} = i\left[\; \text{} \; \right] \Big/ \left[(2\pi)^4 \delta^{(4)}(0) \right]. \tag{4.56}$$

We should emphasize that the right-hand side is independent of T and (volume); in particular it is reassuring to see that E_0 is proportional to the volume of space. In Chapter 11 we will find that this formula is actually useful.

This completes our present analysis of the two-point correlation function. The generalization to higher correlation functions is easy:

$$\langle \Omega | T[\phi(x_1) \cdots \phi(x_n)] |\Omega \rangle = \left(\begin{array}{c} \text{sum of all connected diagrams} \\ \text{with } n \text{ external points} \end{array} \right). \tag{4.57}$$

The disconnected diagrams exponentiate, factor, and cancel as before, by the same argument. There is now a potential confusion in terminology, however. By "disconnected" we mean "disconnected from all external points"—exactly the same diagrams as in (4.51). (They are sometimes called "vacuum bubbles" or "vacuum to vacuum transitions".) In higher correlation functions, diagrams can also be disconnected in another sense. Consider, for example, the four-point function:

$$\langle \Omega | T \phi_1 \phi_2 \phi_3 \phi_4 | \Omega \rangle$$

$$
\begin{aligned}
= \quad & \overline{} \; + \; \big| \; \big| \; + \; \times \; + \; \frac{\circ}{} \; + \; \circlearrowleft \big| \; + \; \cdots \\
& + \; \times \; + \; \circlearrowleft \big| \; + \; \frac{\circ\circ}{} \; + \; \cdots \\
& + \; \times \; + \; \cdots \; + \; \bowtie \; + \; \cdots .
\end{aligned}
\tag{4.58}
$$

In many of these diagrams, external points are disconnected *from each other*. Such diagrams do not exponentiate or factor; they contribute to the amplitude just as do the fully connected diagrams (in which any point can be reached from any other by traveling along the lines).

Note that in ϕ^4 theory, all correlation functions of an odd number of fields vanish, since it is impossible to draw an allowed diagram with an odd number of external points. We can also see this by going back to Wick's theorem: The interaction Hamiltonian H_I contains an even number of fields, so all terms in the perturbation expansion of an odd correlation function will contain an odd number of fields. But it is impossible to fully contract an odd number of fields in pairs, and only fully contracted terms have nonvanishing vacuum expectation value.

4.5 Cross Sections and the S-Matrix

We now have an extremely beautiful formula, Eq. (4.57), for computing an extremely abstract quantity: the n-point correlation function. Our next task is to find equally beautiful ways of computing quantities that can actually be measured: cross sections and decay rates. In this section, after briefly reviewing the definitions of these objects, we will relate them (via a rather technical but fairly careful derivation) to a more primitive quantity, the S-matrix. In the next section we will learn how to compute the matrix elements of the S-matrix using Feynman diagrams.

The Cross Section

The experiments that probe the behavior of elementary particles, especially in the relativistic regime, are scattering experiments. One collides two beams of particles with well-defined momenta, and observes what comes out. The likelihood of any particular final state can be expressed in terms of the *cross section*, a quantity that is intrinsic to the colliding particles, and therefore allows comparison of two different experiments with different beam sizes and intensities.

The cross section is defined as follows. Consider a target, at rest, of particles of type $\mathcal{A}$, with density ρ_A (particles per unit volume). Aim at this target a bunch of particles of type $\mathcal{B}$, with number density ρ_B and velocity v:

Let ℓ_A and ℓ_B be the lengths of the bunches of particles. Then we expect the total number of scattering events (or scattering events of any particular desired type) to be proportional to ρ_A, ρ_B, ℓ_A, ℓ_B, and the cross-sectional area A common to the two bunches. The *cross section*, denoted by σ, is just the total number of events (of whatever type desired) divided by all of these quantities:

$$\sigma \equiv \frac{\text{Number of scattering events}}{\rho_A \, \ell_A \, \rho_B \, \ell_B \, A}. \tag{4.59}$$

The definition is symmetric between the $\mathcal{A}$'s and $\mathcal{B}$'s, so of course we could have taken the $\mathcal{B}$'s to be at rest, or worked in any other reference frame.

The cross section has units of area. In fact, it is the effective area of a chunk taken out of one beam, by each particle in the other beam, that subsequently becomes the final state we are interested in.

In real beams, ρ_A and ρ_B are not constant; the particle density is generally larger at the center of the beam than at the edges. We will assume, however, that both the range of the interaction between the particles and the width of the individual particle wavepackets are small compared to the beam diameter. We can then consider ρ_A and ρ_B to be constant in what follows, and remember that, to compute the event rate in an actual accelerator, one must integrate over the beam area:

$$\text{Number of events} = \sigma \, \ell_A \, \ell_B \int d^2x \, \rho_A(x) \, \rho_B(x). \tag{4.60}$$

If the densities are constant, or if we use this formula to compute an effective area A of the beams, then we have simply

$$\text{Number of events} = \frac{\sigma N_A N_B}{A}, \tag{4.61}$$

where N_A and N_B are the total numbers of $\mathcal{A}$ and $\mathcal{B}$ particles.

Cross sections for many different processes may be relevant to a single scattering experiment. In e^+e^- collisions, for example, one can measure the cross sections for production of $\mu^+\mu^-$, $\tau^+\tau^-$, $\mu^+\mu^-\gamma$, $\mu^+\mu^-\gamma\gamma$, etc., and countless processes involving hadron production, not to mention simple e^+e^- scattering. Usually, of course, we wish to measure not only what the final-state particles are, but also the momenta with which they come out. In this case our definition (4.59) of σ still works, but if we specify the exact momenta desired, σ will be infinitesimal. The solution is to define the *differential cross section, $d\sigma/(d^3p_1 \cdots d^3p_n)$*. It is simply the quantity that, when integrated over any small $d^3p_1 \cdots d^3p_n$, gives the cross section for scattering into that region of final-state momentum space. The various final-state momenta are not all independent: Four components will always be constrained by 4-momentum conservation. In the simplest case, where there are only two final-state particles, this leaves only two unconstrained momentum components, usually taken to be the angles θ and ϕ of the momentum of one of the particles. Integrating $d\sigma/(d^3p_1 d^3p_2)$ over the four constrained momentum components then leaves us with the usual differential cross section $d\sigma/d\Omega$.

A somewhat simpler measurable quantity is the *decay rate* Γ of an unstable particle $\mathcal{A}$ (assumed to be at rest) into a specified final state (of two or more particles). It is defined as

$$\Gamma \equiv \frac{\text{Number of decays per unit time}}{\text{Number of } \mathcal{A} \text{ particles present}}. \tag{4.62}$$

The lifetime τ of the particle is then the reciprocal of the sum of its decay rates into all possible final states. (The particle's half-life is $\tau \cdot \ln 2$.)

In nonrelativistic quantum mechanics, an unstable atomic state shows up in scattering experiments as a *resonance*. Near the resonance energy E_0, the scattering amplitude is given by the Breit-Wigner formula

$$f(E) \propto \frac{1}{E - E_0 + i\Gamma/2}. \tag{4.63}$$

The cross section therefore has a peak of the form

$$\sigma \propto \frac{1}{(E - E_0)^2 + \Gamma^2/4}.$$

The width of the resonance peak is equal to the decay rate of the unstable state.

The Breit-Wigner formula (4.63) also applies in relativistic quantum mechanics. In particular, it gives the scattering amplitude for processes in which initial particles combine to form an unstable particle, which then decays. The unstable particle, viewed as an excited state of the vacuum, is a direct analogue of the unstable nonrelativistic atomic state. If we call the 4-momentum of the unstable particle p and its mass m, we can make a relativistically invariant generalization of (4.63):

$$\frac{1}{p^2 - m^2 + im\Gamma} \approx \frac{1}{2E_{\mathbf{p}}(p^0 - E_{\mathbf{p}} + i(m/E_{\mathbf{p}})\Gamma/2)}. \tag{4.64}$$

The decay rate of the unstable particle in a general frame is $(m/E_{\mathbf{p}})\Gamma$, in accord with relativistic time dilation. Although the two expressions in (4.64) are equal in the vicinity of the resonance, the left-hand side, which is manifestly Lorentz invariant, is much more convenient.

The S-Matrix

How, then, do we calculate a cross section? We must set up wavepackets representing the initial-state particles, evolve this initial state for a very long time with the time-evolution operator $\exp(-iHt)$ of the interacting field theory, and overlap the resulting final state with wavepackets representing some desired set of final-state particles. This gives the probability amplitude for producing that final state, which is simply related to the cross section. We will find that, in the limit where the wavepackets are very narrow in momentum space, the amplitude depends only on the momenta of the wavepackets, not on the details of their shapes.[†]

A wavepacket representing some desired state $|\phi\rangle$ can be expressed as

$$|\phi\rangle = \int \frac{d^3k}{(2\pi)^3} \frac{1}{\sqrt{2E_\mathbf{k}}} \phi(\mathbf{k}) |\mathbf{k}\rangle, \qquad (4.65)$$

where $\phi(\mathbf{k})$ is the Fourier transform of the spatial wavefunction, and $|\mathbf{k}\rangle$ is a one-particle state of momentum $\mathbf{k}$ in the interacting theory. In the free theory, we would have $|\mathbf{k}\rangle = \sqrt{2E_\mathbf{k}} a_\mathbf{k}^\dagger |0\rangle$. The factor of $\sqrt{2E_\mathbf{k}}$ converts our relativistic normalization of $|\mathbf{k}\rangle$ to the conventional normalization in which the sum of all probabilities adds up to 1:

$$\langle\phi|\phi\rangle = 1 \quad \text{if} \quad \int \frac{d^3k}{(2\pi)^3} |\phi(\mathbf{k})|^2 = 1. \qquad (4.66)$$

The probability we wish to compute is then

$$\mathcal{P} = \Big| \langle \underbrace{\phi_1 \phi_2 \cdots}_{\text{future}} | \underbrace{\phi_\mathcal{A} \phi_\mathcal{B}}_{\text{past}} \rangle \Big|^2, \qquad (4.67)$$

where $|\phi_\mathcal{A}\phi_\mathcal{B}\rangle$ is a state of two wavepackets constructed in the far past and $\langle\phi_1\phi_2\cdots|$ is a state of several wavepackets (one for each final-state particle) constructed in the far future. The wavepackets are localized in space, so each can be constructed independently of the others. States constructed in this way are called *in* and *out* states. Note that we use the Heisenberg picture: States are time-independent, but the name we give a state depends on the eigenvalues or expectation values of time-dependent operators. Thus states with different names constructed at different times have a nontrivial overlap, which depends on the time dependence of the operators.

If we set up $|\phi_\mathcal{A}\phi_\mathcal{B}\rangle$ in the remote past, and then take the limit in which the wavepackets $\phi_i(\mathbf{k}_i)$ become concentrated about definite momenta $\mathbf{p}_i$, this defines an *in* state $|\mathbf{p}_\mathcal{A}\mathbf{p}_\mathcal{B}\rangle_{\text{in}}$ with definite initial momenta. It is useful to view $|\phi_\mathcal{A}\phi_\mathcal{B}\rangle$ as a linear superposition of such states. It is important, however, to

---

[†]Much of this section is based on the treatment of nonrelativistic scattering given in Taylor (1972), Chapters 2, 3, and 17. We concentrate on the additional complications of the relativistic theory, glossing over many subtleties, common to both cases, which Taylor explains carefully.

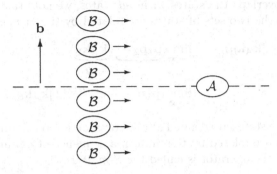

Figure 4.2. Incident wavepackets are uniformly distributed in impact parameter **b**.

take into account the transverse displacement of the wavepacket ϕ_B relative to ϕ_A in position space (see Fig. 4.2). Although we could leave this implicit in the form of $\phi_B(\mathbf{k}_B)$, we instead adopt the convention that our reference momentum-space wavefunctions are collinear (that is, have impact parameter $\mathbf{b} = 0$), and write $\phi_B(\mathbf{k}_B)$ with an explicit factor $\exp(-i\mathbf{b}\cdot\mathbf{k}_B)$ to account for the spatial translation. Then, since ϕ_A and ϕ_B are constructed independently at different locations, we can write the initial state as

$$|\phi_A\phi_B\rangle_{\text{in}} = \int \frac{d^3k_A}{(2\pi)^3} \int \frac{d^3k_B}{(2\pi)^3} \frac{\phi_A(\mathbf{k}_A)\phi_B(\mathbf{k}_B)e^{-i\mathbf{b}\cdot\mathbf{k}_B}}{\sqrt{(2E_A)(2E_B)}} |\mathbf{k}_A\mathbf{k}_B\rangle_{\text{in}}. \tag{4.68}$$

We could expand $_{\text{out}}\langle\phi_1\phi_2\cdots|$ in terms of similarly defined *out* states of definite momentum formed in the asymptotic future:[‡]

$$_{\text{out}}\langle\phi_1\phi_2\cdots| = \left(\prod_f \int \frac{d^3p_f}{(2\pi)^3} \frac{\phi_f(\mathbf{p}_f)}{\sqrt{2E_f}}\right) {}_{\text{out}}\langle\mathbf{p}_1\mathbf{p}_2\cdots|.$$

It is much easier, however, to use the *out* states of definite momentum as the final states in the probability amplitude (4.67), and to multiply by the various normalization factors after squaring the amplitude. This is physically reasonable as long as the detectors of final-state particles mainly measure momentum—that is, they do not resolve positions at the level of de Broglie wavelengths.

We can now relate the probability of scattering in a real experiment to an idealized set of transition amplitudes between the asymptotically defined *in* and *out* states of definite momentum,

$$_{\text{out}}\langle\mathbf{p}_1\mathbf{p}_2\cdots|\mathbf{k}_A\mathbf{k}_B\rangle_{\text{in}}. \tag{4.69}$$

[‡]Here and below, the product symbol applies (symbolically) to the integral as well as the other factors in parentheses; the integrals apply to what is outside the parentheses as well.

To compute the overlap of *in* states with *out* states, we note that the conventions for defining the two sets of states are related by time translation:

$$\text{out}\langle \mathbf{p}_1\mathbf{p}_2\cdots|\mathbf{k}_A\mathbf{k}_B\rangle_\text{in} = \lim_{T\to\infty}\underbrace{\langle \mathbf{p}_1\mathbf{p}_2\cdots}_{T}\,|\,\underbrace{\mathbf{k}_A\mathbf{k}_B\rangle}_{-T}$$

$$= \lim_{T\to\infty}\langle \mathbf{p}_1\mathbf{p}_2\cdots|\,e^{-iH(2T)}\,|\mathbf{k}_A\mathbf{k}_B\rangle. \tag{4.70}$$

In the last line, the states are defined at any common reference time. Thus, the *in* and *out* states are related by the limit of a sequence of unitary operators. This limiting unitary operator is called the *S-matrix*:

$$\text{out}\langle \mathbf{p}_1\mathbf{p}_2\cdots|\mathbf{k}_A\mathbf{k}_B\rangle_\text{in} \equiv \langle \mathbf{p}_1\mathbf{p}_2\cdots|\,S\,|\mathbf{k}_A\mathbf{k}_B\rangle. \tag{4.71}$$

The S-matrix has the following structure: If the particles in question do not interact at all, S is simply the identity operator. Even if the theory contains interactions, the particles have some probability of simply missing one another. To isolate the interesting part of the S-matrix—that is, the part due to interactions—we define the *T-matrix* by

$$S = \mathbf{1} + iT. \tag{4.72}$$

Next we note that the matrix elements of S should reflect 4-momentum conservation. Thus S or T should always contain a factor $\delta^{(4)}(k_A + k_B - \sum p_f)$. Extracting this factor, we define the *invariant matrix element* $\mathcal{M}$, by

$$\langle \mathbf{p}_1\mathbf{p}_2\cdots|\,iT\,|\mathbf{k}_A\mathbf{k}_B\rangle = (2\pi)^4\delta^{(4)}\big(k_A + k_B - \textstyle\sum p_f\big)\cdot i\mathcal{M}(k_A, k_B \to p_f). \tag{4.73}$$

We have written this expression in terms of 4-momenta p and k, but of course all 4-momenta are on mass-shell: $p^0 = E_\mathbf{p}$, $k^0 = E_\mathbf{k}$. (Note that our entire treatment is specific to the case where the initial state contains only two particles. For 3→many or many→many interactions, one can invent analogous constructions, but we will not consider such complicated experiments in this book.)

The matrix element $\mathcal{M}$ is analogous to the scattering amplitude f of one-particle quantum mechanics. It is useful because it allows us to separate all the physics that depends on the details of the interaction Hamiltonian ("dynamics") from all the physics that doesn't ("kinematics"). In the next section we will discuss how to compute $\mathcal{M}$ using Feynman diagrams. But first, we must figure out how to reconstruct the cross section σ from $\mathcal{M}$.

To do this, let us calculate, in terms of $\mathcal{M}$, the probability for the initial state $|\phi_A\phi_B\rangle$ to scatter and become a final state of n particles whose momenta lie in a small region $d^3p_1\cdots d^3p_n$. In our normalization, this probability is

$$\mathcal{P}(AB \to 1\,2\ldots n) = \left(\prod_f \frac{d^3p_f}{(2\pi)^3}\frac{1}{2E_f}\right)\big|\text{out}\langle \mathbf{p}_1\cdots \mathbf{p}_n\,|\,\phi_A\phi_B\rangle_\text{in}\big|^2. \tag{4.74}$$

For a single target ($\mathcal{A}$) particle and many incident ($\mathcal{B}$) particles with different impact parameters $\mathbf{b}$, the number of scattering events is

$$N = \sum_{\substack{\text{all incident} \\ \text{particles } i}} P_i = \int d^2b \, n_{\mathcal{B}} \, \mathcal{P}(\mathbf{b}),$$

where $n_{\mathcal{B}}$ is the number density (particles per unit area) of $\mathcal{B}$ particles. Since we are assuming that this number density is constant over the range of the interaction, $n_{\mathcal{B}}$ can be taken outside the integral. The cross section is then

$$\sigma = \frac{N}{n_{\mathcal{B}} N_{\mathcal{A}}} = \frac{N}{n_{\mathcal{B}} \cdot 1} = \int d^2b \, \mathcal{P}(\mathbf{b}). \qquad (4.75)$$

Deriving a simple expression for σ in terms of $\mathcal{M}$ is now a fairly straightforward calculation. Combining (4.75), (4.74), and (4.68), we have (writing $d\sigma$ rather than σ since this is an infinitesimal quantity)

$$d\sigma = \left(\prod_f \frac{d^3 p_f}{(2\pi)^3} \frac{1}{2E_f} \right) \int d^2b \left(\prod_{i=\mathcal{A},\mathcal{B}} \int \frac{d^3 k_i}{(2\pi)^3} \frac{\phi_i(\mathbf{k}_i)}{\sqrt{2E_i}} \int \frac{d^3 \bar{k}_i}{(2\pi)^3} \frac{\phi_i^*(\bar{\mathbf{k}}_i)}{\sqrt{2\bar{E}_i}} \right)$$

$$\times \, e^{i\mathbf{b}\cdot(\bar{\mathbf{k}}_{\mathcal{B}} - \mathbf{k}_{\mathcal{B}})} \, \left({}_{\text{out}}\langle\{\mathbf{p}_f\}|\{\mathbf{k}_i\}\rangle_{\text{in}} \right) \left({}_{\text{out}}\langle\{\mathbf{p}_f\}|\{\bar{\mathbf{k}}_i\}\rangle_{\text{in}} \right)^*, \quad (4.76)$$

where we have used $\bar{k}_{\mathcal{A}}$ and $\bar{k}_{\mathcal{B}}$ as dummy integration variables in the second half of the squared amplitude. The d^2b integral can be performed to give a factor of $(2\pi)^2 \delta^{(2)}(k_{\mathcal{B}}^\perp - \bar{k}_{\mathcal{B}}^\perp)$. We get more delta functions by writing the final two factors of (4.76) in terms of $\mathcal{M}$. Assuming that we are not interested in the trivial case of forward scattering where no interaction takes place, we can drop the **1** in Eq. (4.72) and write these factors as

$$\left({}_{\text{out}}\langle\{\mathbf{p}_f\}|\{\mathbf{k}_i\}\rangle_{\text{in}} \right) = i\mathcal{M}(\{k_i\} \to \{p_f\}) \, (2\pi)^4 \delta^{(4)}(\textstyle\sum k_i - \sum p_f);$$

$$\left({}_{\text{out}}\langle\{\mathbf{p}_f\}|\{\bar{\mathbf{k}}_i\}\rangle_{\text{in}} \right)^* = -i\mathcal{M}^*(\{\bar{k}_i\} \to \{p_f\}) \, (2\pi)^4 \delta^{(4)}(\textstyle\sum \bar{k}_i - \sum p_f).$$

We can use the second of these delta functions, together with the $\delta^{(2)}(k_{\mathcal{B}}^\perp - \bar{k}_{\mathcal{B}}^\perp)$, to perform all six of the $\bar{k}$ integrals in (4.76). Of the six integrals, only those over $\bar{k}_{\mathcal{A}}^z$ and $\bar{k}_{\mathcal{B}}^z$ require some work:

$$\int d\bar{k}_{\mathcal{A}}^z d\bar{k}_{\mathcal{B}}^z \, \delta(\bar{k}_{\mathcal{A}}^z + \bar{k}_{\mathcal{B}}^z - \textstyle\sum p_f^z) \, \delta(\bar{E}_{\mathcal{A}} + \bar{E}_{\mathcal{B}} - \sum E_f)$$

$$= \int d\bar{k}_{\mathcal{A}}^z \, \delta\left(\sqrt{\bar{k}_{\mathcal{A}}^2 + m_{\mathcal{A}}^2} + \sqrt{\bar{k}_{\mathcal{B}}^2 + m_{\mathcal{B}}^2} - \textstyle\sum E_f \right) \Big|_{\bar{k}_{\mathcal{B}}^z = \sum p_f^z - \bar{k}_{\mathcal{A}}^z}$$

$$= \frac{1}{\left| \dfrac{\bar{k}_{\mathcal{A}}^z}{\bar{E}_{\mathcal{A}}} - \dfrac{\bar{k}_{\mathcal{B}}^z}{\bar{E}_{\mathcal{B}}} \right|} \equiv \frac{1}{|v_{\mathcal{A}} - v_{\mathcal{B}}|}. \qquad (4.77)$$

In the last line and in the rest of Eq. (4.76) it is understood that the constraints $\bar{k}_{\mathcal{A}}^z + \bar{k}_{\mathcal{B}}^z = \sum p_f^z$ and $\bar{E}_{\mathcal{A}} + \bar{E}_{\mathcal{B}} = \sum E_f$ now apply (in addition to the constraints $\bar{k}_{\mathcal{A}}^\perp = k_{\mathcal{A}}^\perp$ and $\bar{k}_{\mathcal{B}}^\perp = k_{\mathcal{B}}^\perp$ coming from the other four integrals).

The difference $|v_A - v_B|$ is the relative velocity of the beams as viewed from the laboratory frame.

Now recall that the initial wavepackets are localized in momentum space, centered on $\mathbf{p}_A$ and $\mathbf{p}_B$. This means that we can evaluate all factors that are smooth functions of $\mathbf{k}_A$ and $\mathbf{k}_B$ at $\mathbf{p}_A$ and $\mathbf{p}_B$, pulling them outside the integrals. These factors include E_A, E_B, $|v_A - v_B|$, and $\mathcal{M}$—everything except the remaining delta function. After doing this, we arrive at the expression

$$d\sigma = \left(\prod_f \frac{d^3 p_f}{(2\pi)^3} \frac{1}{2E_f} \right) \frac{|\mathcal{M}(p_A, p_B \to \{p_f\})|^2}{2E_A 2E_B |v_A - v_B|} \int \frac{d^3 k_A}{(2\pi)^3} \int \frac{d^3 k_B}{(2\pi)^3} \tag{4.78}$$

$$\times \, |\phi_A(\mathbf{k}_A)|^2 |\phi_B(\mathbf{k}_B)|^2 (2\pi)^4 \delta^{(4)} (k_A + k_B - \textstyle\sum p_f).$$

To simplify this formula further, we should think a bit more about the properties of real particle detectors. We have already noted that real detectors project mainly onto eigenstates of momentum. But real detectors have finite resolution; that is, they sum incoherently over momentum bites of finite size. Normally, the measurement of the final-state momentum is not of such high quality that it can resolve the small variation of this momentum that results from the momentum spread of the initial wavepackets ϕ_A, ϕ_B. In that case, we may treat even the momentum vector $k_A + k_B$ inside the delta function as being well approximated by its central value $p_A + p_B$. With this further approximation, we can perform the integrals over k_A and k_B using the normalization condition (4.66). This produces the final form of the relation between S-matrix elements and cross sections,

$$d\sigma = \frac{1}{2E_A 2E_B |v_A - v_B|} \left(\prod_f \frac{d^3 p_f}{(2\pi)^3} \frac{1}{2E_f} \right) \tag{4.79}$$

$$\times \, |\mathcal{M}(p_A, p_B \to \{p_f\})|^2 \, (2\pi)^4 \delta^{(4)} (p_A + p_B - \textstyle\sum p_f).$$

All dependence on the shapes of the wavepackets has disappeared.

The integral over final-state momenta in (4.79) has the structure

$$\int d\Pi_n = \left(\prod_f \int \frac{d^3 p_f}{(2\pi)^3} \frac{1}{2E_f} \right) (2\pi)^4 \delta^{(4)} (P - \textstyle\sum p_f), \tag{4.80}$$

with P the total initial 4-momentum. This integral is manifestly Lorentz invariant, since it is built up from invariant 3-momentum integrals constrained by a 4-momentum delta function. This integral is known as *relativistically invariant n-body phase space*. Of the other ingredients in (4.79), the matrix element $\mathcal{M}$ is also Lorentz invariant. The Lorentz transformation property of (4.79) therefore comes entirely from the prefactor

$$\frac{1}{E_A E_B |v_A - v_B|} = \frac{1}{|E_B p_A^z - E_A p_B^z|} = \frac{1}{|\epsilon_{\mu x y \nu} p_A^\mu p_B^\nu|}.$$

This is not Lorentz invariant, but it is invariant to boosts along the z-direction. In fact, this expression has exactly the transformation properties of a cross-sectional area.

For the special case of two particles in the final state, we can simplify the general expression (4.79) by partially evaluating the phase-space integrals in the center-of-mass frame. Label the momenta of the two final particles p_1 and p_2. We first choose to integrate all three components of $\mathbf{p}_2$ over the delta functions enforcing 3-momentum conservation. This sets $\mathbf{p}_2 = -\mathbf{p}_1$ and converts the integral over two-body phase space to the form

$$\int d\Pi_2 = \int \frac{dp_1\, p_1^2\, d\Omega}{(2\pi)^3\, 2E_1\, 2E_2} (2\pi)\delta(E_{\mathrm{cm}} - E_1 - E_2), \qquad (4.81)$$

where $E_1 = \sqrt{p_1^2 + m_1^2}$, $E_2 = \sqrt{p_1^2 + m_2^2}$, and E_{cm} is the total initial energy. Integrating over the final delta function gives

$$\int d\Pi_2 = \int d\Omega\, \frac{p_1^2}{16\pi^2 E_1 E_2}\left(\frac{p_1}{E_1} + \frac{p_1}{E_2}\right)^{-1}$$
$$= \int d\Omega\, \frac{1}{16\pi^2} \frac{|\mathbf{p}_1|}{E_{\mathrm{cm}}}. \qquad (4.82)$$

For reactions symmetric about the collision axis, two-body phase space can be written simply as an integral over the polar angle in the center-of-mass frame:

$$\int d\Pi_2 = \int d\cos\theta\, \frac{1}{16\pi} \frac{2|\mathbf{p}_1|}{E_{\mathrm{cm}}}. \qquad (4.83)$$

The last factor tends to 1 at high energy.

Applying this simplification to (4.79), we find the following form of the cross section for two final-state particles:

$$\left(\frac{d\sigma}{d\Omega}\right)_{\mathrm{CM}} = \frac{1}{2E_A 2E_B |v_A - v_B|} \frac{|\mathbf{p}_1|}{(2\pi)^2\, 4E_{\mathrm{cm}}} \big|\mathcal{M}(p_A, p_B \to p_1, p_2)\big|^2. \qquad (4.84)$$

In the special case where all four particles have identical masses (including the commonly seen limit $m \to 0$), this reduces to the formula quoted in Chapter 1,

$$\left(\frac{d\sigma}{d\Omega}\right)_{\mathrm{CM}} = \frac{|\mathcal{M}|^2}{64\pi^2 E_{\mathrm{cm}}^2} \qquad \text{(all four masses identical).} \qquad (4.85)$$

To conclude this section, we should derive a formula for the differential decay rate, $d\Gamma$, in terms of $\mathcal{M}$. The correct expression is only a slight modification of (4.79), and is quite easy to guess: Just remove from (4.79) the factors that do not make sense when the initial state consists of a single particle. The definition of Γ assumes that the decaying particle is at rest, so the normalization factor $(2E_A)^{-1}$ becomes $(2m_A)^{-1}$. (In any other frame, this factor would give the usual time dilation.) Thus the decay rate formula is

$$d\Gamma = \frac{1}{2m_A}\left(\prod_f \frac{d^3 p_f}{(2\pi)^3} \frac{1}{2E_f}\right) \big|\mathcal{M}(m_A \to \{p_f\})\big|^2\, (2\pi)^4 \delta^{(4)}(p_A - \textstyle\sum p_f). \qquad (4.86)$$

Unfortunately, the meaning of this formula is far from clear. Since an unstable particle cannot be sent into the infinitely distant past, our definition (4.73) of $\mathcal{M}(m_{\mathcal{A}} \to \{p_f\})$ in terms of the S-matrix makes no sense in this context. Nevertheless formula (4.86) is correct, when $\mathcal{M}$ is computed according to the Feynman rules for S-matrix elements that we will present in the following section. We postpone the further discussion of these matters, and the proof of Eq. (4.86), until Section 7.3. Until then, an intuitive notion of $\mathcal{M}$ as a transition amplitude should suffice.

Equations (4.79) and (4.86) are completely general, whether or not the final state contains several identical particles. (The computation of $\mathcal{M}$, of course, will be quite different when identical particles are present, but that is another matter.) When integrating either of these formulae to obtain a *total* cross section or decay rate, however, we must be careful to avoid counting the same final state several times. If there are n identical particles in the final state, we must either restrict the integration to inequivalent configurations, or divide by $n!$ after integrating over all sets of momenta.

4.6 Computing S-Matrix Elements from Feynman Diagrams

Now that we have formulae for cross sections and decay rates in terms of the invariant matrix element $\mathcal{M}$, the only remaining task is to find a way of computing $\mathcal{M}$ for various processes in various interacting field theories. In this section we will write down (and try to motivate) a formula for $\mathcal{M}$ in terms of Feynman diagrams. We postpone the actual proof of this formula until Section 7.2, since the proof is somewhat technical and will be much easier to understand after we have seen how the formula is used.

Recall from its definition, Eq. (4.71), that the S-matrix is simply the time-evolution operator, $\exp(-iHt)$, in the limit of very large t:

$$\langle \mathbf{p}_1 \mathbf{p}_2 \cdots | S | \mathbf{k}_{\mathcal{A}} \mathbf{k}_{\mathcal{B}} \rangle = \lim_{T \to \infty} \langle \mathbf{p}_1 \mathbf{p}_2 \cdots | e^{-iH(2T)} | \mathbf{k}_{\mathcal{A}} \mathbf{k}_{\mathcal{B}} \rangle . \qquad (4.87)$$

To compute this quantity we would like to replace the external plane-wave states in (4.87), which are eigenstates of H, with their counterparts in the unperturbed theory, which are eigenstates of H_0. We successfully made such a replacement for the vacuum state $|\Omega\rangle$ in Eq. (4.27):

$$|\Omega\rangle = \lim_{T \to \infty(1-i\epsilon)} \left(e^{-iE_0 T} \langle \Omega | 0 \rangle \right)^{-1} e^{-iHT} | 0 \rangle .$$

This time we would like to find a relation of the form

$$|\mathbf{k}_{\mathcal{A}} \mathbf{k}_{\mathcal{B}}\rangle \propto \lim_{T \to \infty(1-i\epsilon)} e^{-iHT} |\mathbf{k}_{\mathcal{A}} \mathbf{k}_{\mathcal{B}}\rangle_0, \qquad (4.88)$$

where we have omitted some unknown phases and overlap factors like those in (4.27). To find such a relation would not be easy. In (4.27), we used the fact that the vacuum was the state of absolute lowest energy. Here we can use only the much weaker statement that the external states with well-separated initial and final particles have the lowest energy consistent with the predetermined nonzero values of momentum. The problem is a deep one, and it is associated with one of the most fundamental difficulties of field theory, that interactions affect not only the scattering of distinct particles but also the form of the single-particle states themselves.

If the formula (4.88) could somehow be justified, we could use it to rewrite the right-hand side of (4.87) as

$$
\lim_{T\to\infty(1-i\epsilon)} {}_0\langle \mathbf{p}_1 \cdots \mathbf{p}_n | e^{-iH(2T)} | \mathbf{p}_A \mathbf{p}_B \rangle_0
$$

$$
\propto \lim_{T\to\infty(1-i\epsilon)} {}_0\langle \mathbf{p}_1 \cdots \mathbf{p}_n | T\left(\exp\left[-i\int_{-T}^{T} dt\, H_I(t)\right]\right) | \mathbf{p}_A \mathbf{p}_B \rangle_0.
$$

$$(4.89)$$

In the evaluation of vacuum expectation values, the awkward proportionality factors between free and interacting vacuum states cancelled out of the final formula, Eq. (4.31). In the present case those factors are so horrible that we have not even attempted to write them down; we only hope that a similar dramatic cancellation will take place here. In fact such a cancellation does take place, although it is not easy to derive this conclusion from our present approach. Up to one small modification (which is unimportant for our present purposes), the formula for the nontrivial part of the S-matrix can be simplified to the following form:

$$
\langle \mathbf{p}_1 \cdots \mathbf{p}_n | iT | \mathbf{p}_A \mathbf{p}_B \rangle
$$

$$
= \lim_{T\to\infty(1-i\epsilon)} \left({}_0\langle \mathbf{p}_1 \cdots \mathbf{p}_n | T\left(\exp\left[-i\int_{-T}^{T} dt\, H_I(t)\right]\right) | \mathbf{p}_A \mathbf{p}_B \rangle_0 \right)_{\substack{\text{connected,}\\ \text{amputated}}}
$$

$$(4.90)$$

The attributes "connected" and "amputated" refer to restrictions on the class of possible Feynman diagrams; these terms will be defined in a moment. We will prove Eq. (4.90) in Section 7.2. In the remainder of this section, we will explain this formula and motivate the new restrictions that we have added.

First we must learn how to represent the matrix element in (4.90) as a sum of Feynman diagrams. Let us evaluate the first few terms explicitly, in ϕ^4 theory, for the case of two particles in the final state. The first term is

$$
{}_0\langle \mathbf{p}_1 \mathbf{p}_2 | \mathbf{p}_A \mathbf{p}_B \rangle_0 = \sqrt{2E_1 2E_2 2E_A 2E_B}\, \langle 0 | a_1 a_2 a_A^\dagger a_B^\dagger | 0 \rangle
$$

$$
= 2E_A 2E_B (2\pi)^6 \Big(\delta^{(3)}(\mathbf{p}_A - \mathbf{p}_1)\delta^{(3)}(\mathbf{p}_B - \mathbf{p}_2)
$$

$$(4.91)$$

$$
+ \delta^{(3)}(\mathbf{p}_A - \mathbf{p}_2)\delta^{(3)}(\mathbf{p}_B - \mathbf{p}_1) \Big).
$$

The delta functions force the final state to be identical to the initial state, so this term is part of the '**1**' in $S = \mathbf{1} + iT$, and does not contribute to the scattering matrix element $\mathcal{M}$. We can represent it diagrammatically as

The next term in $\langle \mathbf{p}_1 \mathbf{p}_2 | S | \mathbf{p}_A \mathbf{p}_B \rangle$ is

$$
\begin{aligned}
{}_0\langle \mathbf{p}_1 \mathbf{p}_2 | T \Big(-i\frac{\lambda}{4!} \int d^4x\, \phi_I^4(x) \Big) | \mathbf{p}_A \mathbf{p}_B \rangle_0 \\
= {}_0\langle \mathbf{p}_1 \mathbf{p}_2 | N \Big(-i\frac{\lambda}{4!} \int d^4x\, \phi_I^4(x) + \text{contractions} \Big) | \mathbf{p}_A \mathbf{p}_B \rangle_0,
\end{aligned}
\tag{4.92}
$$

using Wick's theorem. Since the external states are not $|0\rangle$, terms that are not fully contracted do not necessarily vanish; we can use an annihilation operator from $\phi_I(x)$ to annihilate an initial-state particle, or a creation operator from $\phi_I(x)$ to produce a final-state particle. For example,

$$
\begin{aligned}
\phi_I^+(x)|\mathbf{p}\rangle_0 &= \int \frac{d^3k}{(2\pi)^3} \frac{1}{\sqrt{2E_\mathbf{k}}}\, a_\mathbf{k} e^{-ik\cdot x}\, \sqrt{2E_\mathbf{p}}\, a_\mathbf{p}^\dagger |0\rangle \\
&= \int \frac{d^3k}{(2\pi)^3} \frac{1}{\sqrt{2E_\mathbf{k}}}\, e^{-ik\cdot x} \sqrt{2E_\mathbf{p}} (2\pi)^3 \delta^{(3)}(\mathbf{k}-\mathbf{p}) |0\rangle \\
&= e^{-ip\cdot x} |0\rangle.
\end{aligned}
\tag{4.93}
$$

An uncontracted ϕ_I operator inside the N-product of (4.92) has two terms: ϕ_I^+ on the far right and ϕ_I^- on the far left. We get one contribution to the S-matrix element for each way of commuting the a of ϕ_I^+ past an initial-state $a^\dagger$, and one contribution for each way of commuting the $a^\dagger$ of ϕ_I^- past a final-state a. It is natural, then, to define the contractions of field operators with external states as follows:

$$
\overbrace{\phi_I(x)|\mathbf{p}\rangle} = e^{-ip\cdot x}|0\rangle; \qquad \overbrace{\langle \mathbf{p}|\phi_I(x)} = \langle 0| e^{+ip\cdot x}.
\tag{4.94}
$$

To evaluate an S-matrix element such as (4.92), we simply write down all possible full contractions of the ϕ_I operators *and* the external-state momenta.

To see that this prescription is correct, let us evaluate (4.92) in detail. The N-product contains terms of the form

$$
\phi\phi\phi; \qquad \overset{\frown}{\phi\phi}\phi; \qquad \overset{\frown}{\phi\phi}\,\overset{\frown}{\phi\phi}.
\tag{4.95}
$$

The last term, in which the ϕ operators are fully contracted with each other, is equal to a vacuum bubble diagram times the value of (4.91) calculated above:

$$-i\frac{\lambda}{4!}\int d^4x \,\,_0\langle \mathbf{p}_1\mathbf{p}_2|\overset{\frown}{\phi}\overset{\frown}{\phi}\phi\phi|\mathbf{p}_A\mathbf{p}_B\rangle_0$$

(4.96)

This is just another contribution to the trivial part of the S-matrix, so we ignore it.

Next consider the second term of (4.95), in which two of the four ϕ operators are contracted. The normal-ordered product of the remaining two fields looks like $(a^\dagger a^\dagger + 2a^\dagger a + aa)$. As we commute these operators past the a's and $a^\dagger$'s of the initial and final states, we find that only a term with an equal number of a's and $a^\dagger$'s can survive. In the language of contractions, this says that one of the ϕ's must be contracted with an initial-state $|\mathbf{p}\rangle$, the other with a final-state $\langle\mathbf{p}|$. The uncontracted $|\mathbf{p}\rangle$ and $\langle\mathbf{p}|$ give a delta function as in (4.91). To represent these quantities diagrammatically, we introduce *external lines* to our Feynman rules:

$$\overset{\frown}{\phi_I(x)|\mathbf{p}\rangle} = \underset{x \quad p}{\longrightarrow} \qquad \overset{\frown}{\langle\mathbf{p}|\phi_I(x)} = \underset{p \quad x}{\longleftarrow} \qquad (4.97)$$

Feynman diagrams for S-matrix elements will always contain external lines, rather than the external points of diagrams for correlation functions. The second term of (4.95) thus yields four diagrams:

The integration $\int d^4x$ produces a momentum-conserving delta function at each vertex (including the external momenta), so these diagrams again describe trivial processes in which the initial and final states are identical. This illustrates a general principle: Only *fully connected* diagrams, in which all external lines are connected to each other, contribute to the T-matrix.

Finally, consider the term of (4.95) in which none of the ϕ operators are contracted with each other. Our prescription tells us to contract two of the ϕ's with $|\mathbf{p}_A\mathbf{p}_B\rangle$ and the other two with $\langle\mathbf{p}_1\mathbf{p}_2|$. There are 4! ways to do this.

Thus we obtain the diagram

$$= (4!) \cdot \left(-i\frac{\lambda}{4!}\right) \int d^4x \, e^{-i(p_A + p_B - p_1 - p_2)\cdot x} \tag{4.98}$$

$$= -i\lambda \, (2\pi)^4 \delta^{(4)}(p_A + p_B - p_1 - p_2).$$

This is exactly of the form $i\mathcal{M}(2\pi)^4\delta^{(4)}(p_A + p_B - p_1 - p_2)$, with $\mathcal{M} = -\lambda$.

Before continuing our discussion of Feynman diagrams for S-matrix elements, we should certainly pause to turn this result into a cross section. For scattering in the center-of-mass frame, we can simply plug $|\mathcal{M}|^2 = \lambda^2$ into Eq. (4.85) to obtain

$$\left(\frac{d\sigma}{d\Omega}\right)_{\text{CM}} = \frac{\lambda^2}{64\pi^2 E_{\text{cm}}^2}. \tag{4.99}$$

We have just computed our first quantum field theory cross section. It is a rather dull result, having no angular dependence at all. (This situation will be remedied when we consider fermions in the next section.) Integrating over $d\Omega$, and dividing by 2 since there are two identical particles in the final state, we find the total cross section,

$$\sigma_{\text{total}} = \frac{\lambda^2}{32\pi E_{\text{cm}}^2}. \tag{4.100}$$

In practice, one would probably use this result to measure the value of λ.

Returning to our general discussion, let us consider some higher-order contributions to the T-matrix for the process $A, B \to 1, 2$. If we ignore, for the moment, the "connected and amputated" prescription, we have the formula

$$\langle \mathbf{p_1 p_2}| \, iT \, |\mathbf{p_A p_B}\rangle \overset{?}{=} \quad + \quad + \quad + \quad + \cdots$$

$$+ \quad + \cdots + \quad + \cdots$$

$$+ \quad + \cdots \tag{4.101}$$

plus diagrams in which the four external lines are not all connected to each other. We have already seen that this last class of diagrams gives no contribution to the T-matrix. The first diagram shown in (4.101) gives the lowest-order contribution to T, which we calculated above. The next three diagrams give expected corrections to this amplitude, involving creation and annihilation of additional "virtual" particles.

The diagrams in the second line of (4.101) contain disconnected "vacuum bubbles". By the same argument as at the end of Section 4.4, the disconnected pieces exponentiate to an overall phase factor giving the shift of the energy of the interacting vacuum state upon which the scattering takes place. Thus they are irrelevant to S. We have now seen that only fully connected diagrams give sensible contributions to S-matrix elements.

The last diagram is more problematical; let us evaluate it. After integrating over the two vertex positions, we obtain

$$
\begin{aligned}
&= \frac{1}{2} \int \frac{d^4 p'}{(2\pi)^4} \frac{i}{p'^2 - m^2} \int \frac{d^4 k}{(2\pi)^4} \frac{i}{k^2 - m^2} \\
&\quad \times (-i\lambda)(2\pi)^4 \delta^{(4)}(p_A + p' - p_1 - p_2) \\
&\quad \times (-i\lambda)(2\pi)^4 \delta^{(4)}(p_B - p').
\end{aligned}
\tag{4.102}
$$

We can integrate over p' using the second delta function. It tells us to evaluate

$$
\frac{1}{p'^2 - m^2}\bigg|_{p'=p_B} = \frac{1}{p_B^2 - m^2} = \frac{1}{0}.
$$

We get infinity, since p_B, being the momentum of an external particle, is on-shell: $p_B^2 = m^2$. This is a disaster. Clearly, our formula for S makes sense only if we exclude diagrams of this form, that is, diagrams with loops connected to only one external leg. Fortunately, this is physically reasonable: In the same way that the vacuum bubble diagrams represent the evolution of $|0\rangle$ into $|\Omega\rangle$, these external leg corrections,

represent the evolution of $|\mathbf{p}\rangle_0$ into $|\mathbf{p}\rangle$, the single-particle state of the interacting theory. Since these corrections have nothing to do with the scattering process, we should exclude them from the computation of S.

For a general diagram with external legs, we define *amputation* in the following way. Starting from the tip of each external leg, find the last point at which the diagram can be cut by removing a single propagator, such that this operation separates the leg from the rest of the diagram. Cut there. For example:

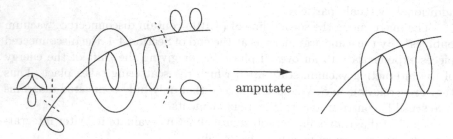

Let us summarize our prescription for calculating scattering amplitudes. Our formula for S-matrix elements, Eq. (4.90), can be rewritten

$$iM \cdot (2\pi)^4 \delta^{(4)}(p_A + p_B - \sum p_f)$$
$$= \begin{pmatrix} \text{sum of all connected, amputated Feynman} \\ \text{diagrams with } p_A,\ p_B \text{ incoming, } p_f \text{ outgoing} \end{pmatrix}. \qquad (4.103)$$

By 'connected', we now mean fully connected, that is, with no vacuum bubbles, and all external legs connected to each other. The Feynman rules for scattering amplitudes in ϕ^4 theory are, in position space,

1. For each propagator, $x \bullet\!\!-\!\!\!-\!\!\!-\!\!\!-\!\!\!-\!\!\bullet y$ $= D_F(x - y);$

2. For each vertex, $\times_{x}$ $= (-i\lambda) \int d^4x;$

3. For each external line, $\underset{x \qquad p}{\longrightarrow}$ $= e^{-ip\cdot x};$

4. Divide by the symmetry factor.

Notice that the factor for an ingoing line is just the amplitude for that particle to be found at the vertex it connects to, i.e., the particle's wavefunction. Similarly, the factor for an outgoing line is the amplitude for a particle produced at the vertex to have the desired final momentum.

Just as with the Feynman rules for correlation functions, it is usually simpler to introduce the momentum-space representation of the propagators, carry out the vertex integrals to obtain momentum-conserving delta functions, and use these delta functions to evaluate as many momentum integrals as possible. In a scattering amplitude, however, there will always be an overall delta function, which can be used to cancel the one on the left-hand side of Eq. (4.103). We are then left with

$$iM = \text{sum of all connected, amputated diagrams}, \qquad (4.104)$$

where the diagrams are evaluated according to the following rules:

1.   For each propagator, $\qquad \underset{p}{\longrightarrow} \qquad = \dfrac{i}{p^2 - m^2 + i\epsilon};$

2.   For each vertex, $\qquad \qquad \qquad \qquad = -i\lambda;$

3.   For each external line, $\qquad \qquad = 1;$

4.   Impose momentum conservation at each vertex;

5.   Integrate over each undetermined loop momentum: $\displaystyle\int \dfrac{d^4p}{(2\pi)^4};$

6.   Divide by the symmetry factor.

This is our final version of the Feynman rules for ϕ^4 theory; these rules are also listed in the Appendix, for reference.

Actually, Eq. (4.103) still isn't quite correct. One more modification is necessary, involving the proportionality factors that were omitted from Eq. (4.89). But the modification affects only diagrams containing loops, so we postpone its discussion until Chapters 6 and 7, where we first evaluate such diagrams. We will prove the corrected formula (4.103) in Section 7.2, by relating S-matrix elements to correlation functions, for which we have actually derived a formula in terms of Feynman diagrams.

4.7 Feynman Rules for Fermions

So far in this chapter we have discussed only ϕ^4 theory, in order to avoid unnecessary complication. We are now ready to generalize our results to theories containing fermions.

Our treatment of correlation functions in Section 4.2 generalizes without difficulty. Lorentz invariance requires that the interaction Hamiltionian H_I be a product of an even number of spinor fields, so no difficulties arise in defining the time-ordered exponential of H_I.

To apply Wick's theorem, however, we must generalize the definitions of the time-ordering and normal-ordering symbols to include fermions. We saw at the end of Section 3.5 that the time-ordering operator T acting on two spinor fields is most conveniently defined with an additional minus sign:

$$T\big(\psi(x)\bar\psi(y)\big) \equiv \begin{cases} \psi(x)\bar\psi(y) & \text{for } x^0 > y^0; \\ -\bar\psi(y)\psi(x) & \text{for } x^0 < y^0. \end{cases} \qquad (4.105)$$

With this definition, the Feynman propagator for the Dirac field is

$$S_F(x-y) = \int \frac{d^4p}{(2\pi)^4} \frac{i(\not p + m)}{p^2 - m^2 + i\epsilon} e^{-ip\cdot(x-y)} = \langle 0| \, T\psi(x)\bar\psi(y) \,|0\rangle. \qquad (4.106)$$

For products of more than two spinor fields, we generalize this definition in the natural way: The time-ordered product picks up one minus sign for each interchange of operators that is necessary to put the fields in time order. For example,

$$T(\psi_1\psi_2\psi_3\psi_4) = (-1)^3\psi_3\psi_1\psi_4\psi_2 \qquad \text{if } x_3^0 > x_1^0 > x_4^0 > x_2^0.$$

The definition of the normal-ordered product of spinor fields is analogous: Put in an extra minus sign for each fermion interchange. The anticommutation properties make it possible to write a normal-ordered product in several ways, but with our conventions these are completely equivalent:

$$N(a_{\mathbf{p}}a_{\mathbf{q}}a_{\mathbf{r}}^\dagger) = (-1)^2 a_{\mathbf{r}}^\dagger a_{\mathbf{p}}a_{\mathbf{q}} = (-1)^3 a_{\mathbf{r}}^\dagger a_{\mathbf{q}}a_{\mathbf{p}}.$$

Using these definitions, it is not hard to generalize Wick's theorem. Consider first the case of two Dirac fields, say $T[\psi(x)\bar\psi(y)]$. In analogy with (4.37), define the contraction of two fields by

$$T[\psi(x)\bar\psi(y)] = N[\psi(x)\bar\psi(y)] + \overbrace{\psi(x)\bar\psi(y)}. \qquad (4.107)$$

Explicitly, for the Dirac field,

$$\overbrace{\psi(x)\bar\psi(y)} \equiv \left\{ \begin{array}{ll} \{\psi^+(x), \bar\psi^-(y)\} & \text{for } x^0 > y^0 \\ -\{\bar\psi^+(y), \psi^-(x)\} & \text{for } x^0 < y^0 \end{array} \right\} = S_F(x-y); \qquad (4.108)$$

$$\overbrace{\psi(x)\psi(y)} = \overbrace{\bar\psi(x)\bar\psi(y)} = 0. \qquad (4.109)$$

Define contractions under the normal-ordering symbol to include minus signs for operator interchanges:

$$N(\overbrace{\psi_1\psi_2\bar\psi_3}\bar\psi_4) = -\overbrace{\psi_1\bar\psi_3}\, N(\psi_2\bar\psi_4) = -S_F(x_1-x_3)\, N(\psi_2\bar\psi_4). \qquad (4.110)$$

With these conventions, Wick's theorem takes the same form as before:

$$T[\psi_1\bar\psi_2\psi_3\cdots] = N[\psi_1\bar\psi_2\psi_3\cdots + \text{all possible contractions}]. \qquad (4.111)$$

The proof is essentially unchanged from the bosonic case, since all extra minus signs are accounted for by the above definitions.

Yukawa Theory

Writing down the Feynman rules for fermion correlation functions would now be easy, but instead let's press on and discuss scattering processes. For definiteness, we begin by analyzing the Yukawa theory:

$$H = H_{\text{Dirac}} + H_{\text{Klein–Gordon}} + \int d^3x\, g\bar\psi\psi\phi. \qquad (4.112)$$

This is a simplified model of Quantum Electrodynamics. In this section we will carefully work out the rules of calculation for Yukawa theory, so that in the next section we can guess the rules for QED without too much difficulty.

To be even more specific, consider the two-particle scattering reaction

$$\text{fermion}(p) + \text{fermion}(k) \longrightarrow \text{fermion}(p') + \text{fermion}(k').$$

The leading contribution comes from the H_I^2 term of the S-matrix:

$$_0\langle \mathbf{p}', \mathbf{k}' | T\Big(\frac{1}{2!} (-ig) \int d^4x\, \bar{\psi}_I \psi_I \phi_I \ (-ig) \int d^4y\, \bar{\psi}_I \psi_I \phi_I \Big) | \mathbf{p}, \mathbf{k} \rangle_0. \qquad (4.113)$$

To evaluate this expression, use Wick's theorem to reduce the T-product to an N-product of contractions, then act the uncontracted fields on the initial- and final-state particles. Represent this latter process as the contraction

$$\overline{\psi_I(x)|\mathbf{p}, s\rangle} = \int \frac{d^3p'}{(2\pi)^3} \frac{1}{\sqrt{2E_{\mathbf{p}'}}} \sum_{s'} a_{\mathbf{p}'}^{s'} u^{s'}(p') e^{-ip'\cdot x} \sqrt{2E_{\mathbf{p}}} a_{\mathbf{p}}^{s\dagger} |0\rangle \qquad (4.114)$$

$$= e^{-ip\cdot x} u^s(p) |0\rangle .$$

Similar expressions hold for the contraction of $\bar{\psi}_I$ with a final-state fermion, and for contractions of ψ_I and $\bar{\psi}_I$ with antifermion states. Note that ψ_I can be contracted with a fermion on the right or an antifermion on the left; the opposite is true for $\bar{\psi}_I$.

We can write a typical contribution to the matrix element (4.113) as the contraction

$$\langle \mathbf{p}', \mathbf{k}' | \tfrac{1}{2!} (-ig) \textstyle\int d^4x\, \bar{\psi}\psi\phi \ (-ig) \int d^4y\, \bar{\psi}\psi\phi\, | \mathbf{p}, \mathbf{k} \rangle. \qquad (4.115)$$

Up to a possible minus sign, the value of this quantity is

$$(-ig)^2 \int \frac{d^4q}{(2\pi)^4} \frac{i}{q^2 - m_\phi^2} (2\pi)^4 \delta^{(4)}(p'-p+q)$$

$$\times (2\pi)^4 \delta^{(4)}(k'-k-q)\bar{u}(p')u(p)\bar{u}(k')u(k).$$

(We have dropped the factor $1/2!$ because there is a second, identical term that comes from interchanging x and y in (4.115).) Using either delta function to perform the integral, we find that this expression takes the form $i\mathcal{M}(2\pi)^4\delta^{(4)}(\Sigma p)$, with

$$i\mathcal{M} = \frac{-ig^2}{q^2 - m_\phi^2} \bar{u}(p')u(p)\bar{u}(k')u(k). \qquad (4.116)$$

When writing it in this way, we must remember to impose the constraints $p - p' = q = k' - k$.

Instead of working from (4.115), we could draw a Feynman diagram:

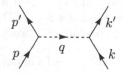

We denote scalar particles by dashed lines, and fermions by solid lines. The S-matrix element could then be obtained directly from the following momentum-space Feynman rules.

1. Propagators:

$$\contraction{}{\phi}{(x)}{\phi} \phi(x)\phi(y) = \quad \cdots\cdots\!\rightarrow\!\cdots\cdots \quad = \frac{i}{q^2 - m_\phi^2 + i\epsilon}$$

$$\contraction{}{\psi}{(x)}{\bar\psi} \psi(x)\bar\psi(y) = \quad \longrightarrow \quad = \frac{i(\not{p} + m)}{p^2 - m^2 + i\epsilon}$$

2. Vertices:

$$\quad = -ig$$

3. External leg contractions:

$$\phi\,|\mathbf{q}\rangle = \quad = 1 \qquad\qquad \langle\mathbf{q}|\,\phi = \quad = 1$$

$$\psi\,|\mathbf{p},s\rangle = \quad = u^s(p) \qquad \langle\mathbf{p},s|\,\bar\psi = \quad = \bar{u}^s(p)$$

$$\text{fermion} \qquad\qquad\qquad\qquad\qquad \text{fermion}$$

$$\bar\psi\,|\mathbf{k},s\rangle = \quad = \bar{v}^s(k) \qquad \langle\mathbf{k},s|\,\psi = \quad = v^s(k)$$

$$\text{antifermion} \qquad\qquad\qquad\qquad \text{antifermion}$$

4. Impose momentum conservation at each vertex.

5. Integrate over each undetermined loop momentum.

6. Figure out the overall sign of the diagram.

Several comments are in order regarding these rules.

First, note that the $1/n!$ from the Taylor series of the time-ordered exponential is always canceled by the $n!$ ways of interchanging vertices to obtain the same contraction. The diagrams of Yukawa theory never have symmetry factors, since the three fields ($\bar\psi\psi\phi$) in H_I cannot substitute for one another in contractions.

Second, the direction of the momentum on a fermion line is always significant. On external lines, as for bosons, the direction of the momentum is always ingoing for initial-state particles and outgoing for final-state particles. This follows immediately from the expansions of ψ and $\bar{\psi}$, where the annihilation operators $a_{\mathbf{p}}$ and $b_{\mathbf{p}}$ both multiply $e^{-ip\cdot x}$ and the creation operators $a_{\mathbf{p}}^{\dagger}$ and $b_{\mathbf{p}}^{\dagger}$ both multiply $e^{+ip\cdot x}$. On internal fermion lines (propagators), the momentum must be assigned in the direction of particle-number flow (for electrons, this is the direction of negative charge flow). This requirement is most easily seen by working out an example from first principles. Consider the annihilation of a fermion and an antifermion into two bosons:

$$\text{(diagram)} = \langle \mathbf{k,k'}| \int d^4x\,\phi\bar{\psi}\psi \int d^4y\,\phi\bar{\psi}\psi\,|\mathbf{p,p'}\rangle$$

$$\sim \int d^4x \int d^4y\, e^{ik'\cdot x}\,\bar{v}(p')e^{-ip'\cdot x} \int \frac{d^4q}{(2\pi)^4}\,\frac{i(\slashed{q}+m)}{q^2-m^2}\,e^{-iq\cdot(x-y)}\,u(p)e^{-ip\cdot y}\,e^{ik\cdot y}.$$

The integrals over x and y give delta functions that force q to flow from y to x, as shown. On internal boson lines the direction of the momentum is irrelevant and may be chosen for convenience, since $D_F(x-y) = D_F(y-x)$.

It is conventional to draw arrows on fermion lines, as shown, to represent the direction of particle-number flow. The momentum assigned to a fermion propagator then flows in the direction of this arrow. For external antiparticles, however, the momentum flows opposite to the arrow; it helps to show this explicitly by drawing a second arrow next to the line.

Third, note that in our examples the Dirac indices contract together along the fermion lines. This will also happen in more complicated diagrams:

$$\text{(diagram)} \sim \bar{u}(p_3) \cdot \frac{i(\slashed{p}_2 + m)}{p_2^2 - m^2} \cdot \frac{i(\slashed{p}_1 + m)}{p_1^2 - m^2} \cdot u(p_0). \qquad (4.117)$$

Finally, let's take a moment to worry about fermion minus signs. Return to the example of the fermion-fermion scattering process. We adopt a sign convention for the initial and final states:

$$|\mathbf{p,k}\rangle \sim a_{\mathbf{p}}^{\dagger}a_{\mathbf{k}}^{\dagger}\,|0\rangle\,, \qquad \langle\mathbf{p',k'}| \sim \langle 0|\,a_{\mathbf{k'}}a_{\mathbf{p'}}\,, \qquad (4.118)$$

so that $(|p,k\rangle)^{\dagger} = \langle p,k|$. Then the contraction

$$\langle\mathbf{p',k'}|(\bar{\psi}\psi)_x\,(\bar{\psi}\psi)_y|\mathbf{p,k}\rangle \sim \langle 0|\,a_{\mathbf{k'}}a_{\mathbf{p'}}\,\bar{\psi}_x\psi_x\,\bar{\psi}_y\psi_y\,a_{\mathbf{p}}^{\dagger}a_{\mathbf{k}}^{\dagger}\,|0\rangle$$

can be untangled by moving $\bar{\psi}_y$ two spaces to the left, and so picks up a factor of $(-1)^2 = +1$. But note that in the contraction

$$\langle\mathbf{p',k'}|(\bar{\psi}\psi)_x\,(\bar{\psi}\psi)_y|\mathbf{p,k}\rangle \sim \langle 0|\,a_{\mathbf{k'}}a_{\mathbf{p'}}\,\bar{\psi}_x\psi_x\,\bar{\psi}_y\psi_y\,a_{\mathbf{p}}^{\dagger}a_{\mathbf{k}}^{\dagger}\,|0\rangle\,,$$

it is sufficient to move the $\bar{\psi}_y$ one space to the left, giving a factor of -1. This contraction corresponds to the diagram

The full result, to lowest order, for the S-matrix element for this process is therefore

$$
i\mathcal{M} = \quad\rangle\!\cdots\cdots\!\langle \quad + \quad \rangle\!\!\!\!\times\!\!\!\!\langle
$$

$$
= (-ig^2)\left(\bar{u}(p')u(p)\frac{1}{(p'-p)^2 - m_\phi^2}\bar{u}(k')u(k) \right. \tag{4.119}
$$

$$
\left. - \bar{u}(p')u(k)\frac{1}{(p'-k)^2 - m_\phi^2.}\bar{u}(k')u(p) \right).
$$

The minus sign difference between these diagrams is a reflection of Fermi statistics. Turning this expression into an explicit cross section would require some additional work; we postpone such calculations until Chapter 5, when we can work with QED instead of the less interesting Yukawa theory.

In complicated diagrams, one can often simplify the determination of the minus signs by noting that the product $(\bar{\psi}\psi)$, or any other pair of fermions, *commutes* with any operator. Thus,

$$
\cdots(\bar{\psi}\psi)_x(\bar{\psi}\psi)_y(\bar{\psi}\psi)_z(\bar{\psi}\psi)_w\cdots = \cdots(+1)(\bar{\psi}\psi)_x(\bar{\psi}\psi)_z(\bar{\psi}\psi)_y(\bar{\psi}\psi)_w\cdots
$$

$$
= \cdots S_F(x-z)S_F(z-y)S_F(y-w)\cdots.
$$

But note that in a closed loop of n fermion propagators we have

$$
= \bar{\psi}\psi\ \bar{\psi}\psi\ \bar{\psi}\psi\ \bar{\psi}\psi
$$

$$
= (-1)\ \mathrm{tr}\left[\psi\ \bar{\psi}\psi\ \bar{\psi}\psi\ \bar{\psi}\psi\ \bar{\psi}\right]
$$

$$
= (-1)\ \mathrm{tr}\left[S_F\ S_F\ S_F\ S_F\right]. \tag{4.120}
$$

A closed fermion loop always gives a factor of -1 and the *trace* of a product of Dirac matrices.

The Yukawa Potential

We now have all the formal rules we need to compute scattering amplitudes in Yukawa theory. Before going on to discuss QED, let us briefly descend from abstraction to concrete physics, and consider one very simple application of these rules: the scattering of *distinguishable* fermions, in the nonrelativistic limit. By comparing the amplitude for this process to the Born approximation formula from nonrelativistic quantum mechanics, we can determine the potential $V(r)$ created by the Yukawa interaction.

If the two interacting particles are distinguishable, only the first diagram in (4.119) contributes. To evaluate the amplitude in the nonrelativistic limit, we keep terms only to lowest order in the 3-momenta. Thus, up to $\mathcal{O}(\mathbf{p}^2, \mathbf{p}'^2, \dots)$,

$$
\begin{aligned}
p &= (m, \mathbf{p}), & k &= (m, \mathbf{k}), \\
p' &= (m, \mathbf{p}'), & k' &= (m, \mathbf{k}').
\end{aligned}
\tag{4.121}
$$

Using these expressions, we have

$$
(p' - p)^2 = -|\mathbf{p}' - \mathbf{p}|^2 + \mathcal{O}(\mathbf{p}^4),
$$

$$
u^s(p) = \sqrt{m} \begin{pmatrix} \xi^s \\ \xi^s \end{pmatrix}, \quad \text{etc.,}
$$

where ξ^s is a two-component constant spinor normalized to $\xi^{s'\dagger}\xi^s = \delta^{ss'}$. The spinor products in (4.119) are then

$$
\begin{aligned}
\bar{u}^{s'}(p') u^s(p) &= 2m\xi^{s'\dagger}\xi^s = 2m\delta^{ss'}; \\
\bar{u}^{r'}(k') u^r(k) &= 2m\xi^{r'\dagger}\xi^r = 2m\delta^{rr'}.
\end{aligned}
\tag{4.122}
$$

So our first physical conclusion is that the spin of each particle is separately conserved in this nonrelativistic scattering interaction—a pleasing result.

Putting together the pieces of the scattering amplitude (4.119), we find

$$
i\mathcal{M} = \frac{ig^2}{|\mathbf{p}' - \mathbf{p}|^2 + m_\phi^2} 2m\delta^{ss'} 2m\delta^{rr'}.
\tag{4.123}
$$

This should be compared with the Born approximation to the scattering amplitude in nonrelativistic quantum mechanics, written in terms of the potential function $V(\mathbf{x})$:

$$
\langle p'|iT|p \rangle = -i\widetilde{V}(\mathbf{q})\,(2\pi)\delta(E_{\mathbf{p}'} - E_{\mathbf{p}}), \qquad (\mathbf{q} = \mathbf{p}' - \mathbf{p}).
\tag{4.124}
$$

So apparently, for the Yukawa interaction,

$$
\widetilde{V}(\mathbf{q}) = \frac{-g^2}{|\mathbf{q}|^2 + m_\phi^2}.
\tag{4.125}
$$

(The factors of $2m$ in (4.123) arise from our relativistic normalization conventions, and must be dropped when comparing to (4.124), which assumes conventional nonrelativistic normalization of states. The additional $\delta^{(3)}(\mathbf{p}' - \mathbf{p})$ goes away when we integrate over the momentum of the target.)

Inverting the Fourier transform to find $V(\mathbf{x})$ requires a short calculation:

$$
\begin{aligned}
V(\mathbf{x}) &= \int \frac{d^3q}{(2\pi)^3} \frac{-g^2}{|\mathbf{q}|^2 + m_\phi^2} e^{i\mathbf{q}\cdot\mathbf{x}} \\
&= \frac{-g^2}{4\pi^2} \int_0^\infty dq\, q^2 \frac{e^{iqr} - e^{-iqr}}{iqr} \frac{1}{q^2 + m_\phi^2} \\
&= \frac{-g^2}{4\pi^2 ir} \int_{-\infty}^\infty dq \frac{q\, e^{iqr}}{q^2 + m_\phi^2}.
\end{aligned}
\tag{4.126}
$$

The contour of this integral can be closed above in the complex plane, and we pick up the residue of the simple pole at $q = +im_\phi$. Thus we find

$$
V(r) = -\frac{g^2}{4\pi} \frac{1}{r} e^{-m_\phi r},
\tag{4.127}
$$

an *attractive* "Yukawa potential", with range $1/m_\phi = \hbar/m_\phi c$, the Compton wavelength of the exchanged boson. Yukawa made this potential the basis for his theory of the nuclear force, and worked backwards from the range of the force (about 1 fm) to predict the mass (about 200 MeV) of the required boson, the pion.

What happens if instead we scatter particles off of *anti*particles? For the process

$$
f_1(p)\bar{f}_2(k) \longrightarrow f_1(p')\bar{f}_2(k'),
$$

we need to evaluate (nonrelativistically)

$$
\bar{v}^s(k)v^{s'}(k') \approx m(\xi^{s\dagger}, -\xi^{s\dagger}) \begin{pmatrix} 0 & 1 \\ 1 & 0 \end{pmatrix} \begin{pmatrix} \xi^{s'} \\ -\xi^{s'} \end{pmatrix} = -2m\delta^{ss'}.
\tag{4.128}
$$

We must also work out the fermion minus sign. Using $|\mathbf{p}, \mathbf{k}\rangle = a_{\mathbf{p}}^\dagger b_{\mathbf{k}}^\dagger |0\rangle$ and $\langle \mathbf{p}', \mathbf{k}'| = \langle 0| b_{\mathbf{k}'} a_{\mathbf{p}'}$, we can write the contracted matrix element as

$$
\langle \mathbf{p}', \mathbf{k}'| \overline{\psi}\psi\; \overline{\psi}\psi\, |\mathbf{p}, \mathbf{k}\rangle = \langle 0| b_{\mathbf{k}'} a_{\mathbf{p}'}\; \overline{\psi}\psi\; \overline{\psi}\psi\; a_{\mathbf{p}}^\dagger b_{\mathbf{k}}^\dagger |0\rangle .
$$

To untangle the contractions requires three operator interchanges, so there is an overall factor of -1. This cancels the extra minus sign in (4.128), and therefore we see that the Yukawa potential between a fermion and an antifermion is also attractive, and identical in strength to that between two fermions.

The remaining case to consider is scattering of two antifermions. It should not be surprising that the potential is again attractive; there is an additional minus sign from changing the other $\bar{u}u$ into $\bar{v}v$, and the number of interchanges necessary to untangle the contractions is even. Thus we conclude that the Yukawa potential is *universally attractive*, whether it is between a pair of fermions, a pair of antifermions, or one of each.

4.8 Feynman Rules for Quantum Electrodynamics

Now we are ready to step from Yukawa theory to Quantum Electrodynamics. To do this, we replace the scalar particle ϕ with a vector particle A_μ, and replace the Yukawa interaction Hamiltonian with

$$H_{\text{int}} = \int d^3x \, e\bar{\psi}\gamma^\mu\psi A_\mu. \tag{4.129}$$

How do the Feynman rules change? The answer, though difficult to prove, is easy to guess. In addition to the fermion rules from the previous section, we have

New vertex: μ $= -ie\gamma^\mu$

Photon propagator: $\mu \longleftarrow q \; \nu$ $= \dfrac{-ig_{\mu\nu}}{q^2 + i\epsilon}$

External photon lines: $A_\mu \, |\mathbf{p}\rangle = \underset{\longleftarrow p}{} \mu = \epsilon_\mu(p)$

$\langle \mathbf{p}| \, A_\mu = \mu \underset{\longleftarrow p}{} = \epsilon_\mu^*(p)$

Photons are conventionally drawn as wavy lines. The symbol $\epsilon_\mu(p)$ stands for the *polarization vector* of the initial- or final-state photon.

To justify these rules, recall that in Lorentz gauge (which we employ to retain explicit relativistic invariance) the field equation for A_μ is

$$\partial^2 A_\mu = 0. \tag{4.130}$$

Thus each component of A separately obeys the Klein-Gordon equation (with $m = 0$). The momentum-space solutions of this equation are $\epsilon_\mu(p)e^{-ip\cdot x}$, where $p^2 = 0$ and $\epsilon_\mu(p)$ is any 4-vector. The interpretation of ϵ as the polarization vector of the field should be familiar from classical electromagnetism. If we expand the quantized electromagnetic field in terms of classical solutions of the wave equation, as we did for the Klein-Gordon field, we find

$$A_\mu(x) = \int \frac{d^3p}{(2\pi)^3} \frac{1}{\sqrt{2E_\mathbf{p}}} \sum_{r=0}^{3} \left(a_\mathbf{p}^r \epsilon_\mu^r(p) e^{-ip\cdot x} + a_\mathbf{p}^{r\dagger} \epsilon_\mu^{r*}(p) e^{ip\cdot x} \right), \tag{4.131}$$

where $r = 0, 1, 2, 3$ labels a basis of polarization vectors. The external line factors in the Feynman rules above follow immediately from this expansion, just as we obtained u's and v's as the external line factors for Dirac particles. The only subtlety is that we must restrict initial- and final-state photons to be transversely polarized: Their polarization vectors are always of the form $\epsilon^\mu = (0, \boldsymbol{\epsilon})$, where $\mathbf{p} \cdot \boldsymbol{\epsilon} = 0$. For $\mathbf{p}$ along the z-axis, the right- and left-handed polarization vectors are $\epsilon^\mu = (0, 1, \pm i, 0)/\sqrt{2}$.

The form of the QED vertex factor is also easy to justify, by simply looking at the interaction Hamiltonian (4.129). Note that the γ matrix in a QED amplitude will sit between spinors or other γ matrices, with the Dirac indices contracted along the fermion line. Note also that this interaction term is specific to the case of an electron (and its antiparticle, the positron). In general, for a Dirac particle with electric charge $Q|e|$,

$$= -iQ|e|\gamma^\mu.$$

For example, an electron has $Q = -1$, an up quark has $Q = +2/3$, and a down quark has $Q_d = -1/3$.

There is no easy way to derive the form of the photon propagator, so for now we will settle for a plausibility argument. Since the electromagnetic field in Lorentz gauge obeys the massless Klein-Gordon equation, it should come as no surprise that the photon propagator is nearly identical to the massless Klein-Gordon propagator. The factor of $-g_{\mu\nu}$, however, requires explanation. Lorentz invariance dictates that the photon propagator be an isotropic second-rank tensor that can dot together the γ^μ and γ^ν from the vertices at each end. The simplest candidate is $g^{\mu\nu}$. To understand the overall sign of the propagator, evaluate its Fourier transform:

$$\int \frac{d^4q}{(2\pi)^4} \frac{-ig_{\mu\nu}}{q^2 + i\epsilon} e^{-iq\cdot(x-y)} = \int \frac{d^3q}{(2\pi)^3} \frac{1}{2|\mathbf{q}|} e^{-iq\cdot(x-y)} \cdot (-g_{\mu\nu}). \qquad (4.132)$$

Presumably this is equal to $\langle 0| T[A_\mu(x)A_\nu(y)] |0\rangle$. Now set $\mu = \nu$, and take the limit $x^0 \to y^0$ from the positive direction. Then this quantity becomes the norm of the state $A_\mu(x)|0\rangle$, which should be positive. We see that our choice of signs in the propagator implies that the three states created by A_i, with $i = 1, 2, 3$, indeed have positive norm. These states include all real (non-virtual) photons, which always have spacelike polarizations. Unfortunately, because $g_{\mu\nu}$ is not positive definite, the states created by A_0 inevitably have negative norm. This is potentially a serious problem for any theory with vector particles. For Quantum Electrodynamics, we will show in Section 5.5 that the negative-norm states created by A_0 are never produced in physical processes. In Section 9.4 we will give a careful derivation of the photon propagator.

The Coulomb Potential

As a simple application of these Feynman rules, and to better understand the sign of the propagator, let us repeat the nonrelativistic scattering calculation of the previous section, this time for QED. The leading-order contribution is

$$i\mathcal{M} = \quad = (-ie)^2 \bar{u}(p')\gamma^\mu u(p)\frac{-ig_{\mu\nu}}{(p'-p)^2}\bar{u}(k')\gamma^\nu u(k). \quad (4.133)$$

In the nonrelativistic limit,

$$\bar{u}(p')\gamma^0 u(p) = u^\dagger(p')u(p) \approx +2m\xi'^\dagger\xi.$$

You can easily verify that the other terms, $\bar{u}(p')\gamma^i u(p)$, vanish if $p = p' = 0$; they can therefore be neglected compared to $\bar{u}(p')\gamma^0 u(p)$ in the nonrelativistic limit. Thus we have

$$i\mathcal{M} \approx \frac{+ie^2}{-|\mathbf{p}'-\mathbf{p}|^2}(2m\xi'^\dagger\xi)_p(2m\xi'^\dagger\xi)_k \cdot g_{00}$$

$$= \frac{-ie^2}{|\mathbf{p}'-\mathbf{p}|^2}(2m\xi'^\dagger\xi)_p(2m\xi'^\dagger\xi)_k. \quad (4.134)$$

Comparing this to the Yukawa case (4.123), we see that there is an extra factor of -1; the potential is a *repulsive* Yukawa potential with $m = 0$, that is, a repulsive Coulomb potential:

$$V(r) = \frac{e^2}{4\pi r} = \frac{\alpha}{r}, \quad (4.135)$$

where $\alpha = e^2/4\pi \approx 1/137$ is the fine-structure constant.

For particle-antiparticle scattering, note first that

$$\bar{v}(k)\gamma^0 v(k') = v^\dagger(k)v(k') \approx +2m\xi^\dagger\xi'.$$

The presence of the γ^0 eliminates the minus sign that we found in the Yukawa case. The nonrelativistic scattering amplitude is therefore

$$i\mathcal{M} = \quad = (-1) \cdot \frac{-ie^2}{|\mathbf{p}'-\mathbf{p}|^2}(+2m\xi'^\dagger\xi)_p(+2m\xi^\dagger\xi')_k, \quad (4.136)$$

where the (-1) is the same fermion minus sign we saw in the Yukawa case. This is an *attractive* potential. Similarly, for antifermion-antifermion scattering one finds a repulsive potential. We have just verified that in quantum field theory, when a vector particle is exchanged, like charges repel while unlike charges attract.

Note that the repulsion in fermion-fermion scattering came entirely from the extra factor $-g_{00} = -1$ in the vector boson propagator. A tensor boson, such as the graviton, would have a propagator

$$
\mu\, \nu \sim\!\!\!\sim\!\!\!\sim\!\!\!\sim\, \rho\, \sigma = \frac{1}{2}\Big((-g_{\mu\rho})(-g_{\nu\sigma}) + (-g_{\mu\sigma})(-g_{\nu\rho}) \Big) \left(\frac{i}{q^2 + i\epsilon} \right),
$$

which in nonrelativistic collisions gives a factor $(-g_{00})^2 = +1$; this will result in a universally attractive potential. It is reassuring to see that quantum field theory does indeed reproduce the obvious features of the electric and gravitational forces:

Exchanged particle	ff and $\bar{f}\bar{f}$	$f\bar{f}$
scalar (Yukawa)	attractive	attractive
vector (electricity)	repulsive	attractive
tensor (gravity)	attractive	attractive

Problems

4.1   Let us return to the problem of the creation of Klein-Gordon particles by a classical source. Recall from Chapter 2 that this process can be described by the Hamiltonian

$$
H = H_0 + \int d^3x \left(-j(t,\mathbf{x})\phi(x) \right),
$$

where H_0 is the free Klein-Gordon Hamiltonian, $\phi(x)$ is the Klein-Gordon field, and $j(x)$ is a c-number scalar function. We found that, if the system is in the vacuum state before the source is turned on, the source will create a mean number of particles

$$
\langle N \rangle = \int \frac{d^3p}{(2\pi)^3} \frac{1}{2E_{\mathbf{p}}} |\tilde{\jmath}(p)|^2.
$$

In this problem we will verify that statement, and extract more detailed information, by using a perturbation expansion in the strength of the source.

(a) Show that the probability that the source creates *no* particles is given by

$$
P(0) = \left| \langle 0|\, T\Big\{ \exp[\,i \int d^4x\, j(x)\phi_I(x)]\Big\}\, |0\rangle \right|^2.
$$

(b) Evaluate the term in P(0) of order j^2, and show that $P(0) = 1 - \lambda + \mathcal{O}(j^4)$, where λ equals the expression given above for $\langle N \rangle$.

(c) Represent the term computed in part (b) as a Feynman diagram. Now represent the whole pertubation series for $P(0)$ in terms of Feynman diagrams. Show that this series exponentiates, so that it can be summed exactly: $P(0) = \exp(-\lambda)$.

(d) Compute the probability that the source creates one particle of momentum k. Perform this computation first to $\mathcal{O}(j)$ and then to all orders, using the trick of part (c) to sum the series.

(e) Show that the probability of producing n particles is given by

$$P(n) = (1/n!)\lambda^n \exp(-\lambda).$$

This is a *Poisson distribution*.

(f) Prove the following facts about the Poisson distribution:

$$\sum_{n=0}^{\infty} P(n) = 1; \qquad \langle N \rangle = \sum_{n=0}^{\infty} n\, P(n) = \lambda.$$

The first identity says that the $P(n)$'s are properly normalized probabilities, while the second confirms our proposal for $\langle N \rangle$. Compute the mean square fluctuation $\langle (N - \langle N \rangle)^2 \rangle$.

4.2   Decay of a scalar particle. Consider the following Lagrangian, involving two real scalar fields Φ and ϕ:

$$\mathcal{L} = \tfrac{1}{2}(\partial_\mu \Phi)^2 - \tfrac{1}{2}M^2\Phi^2 + \tfrac{1}{2}(\partial_\mu \phi)^2 - \tfrac{1}{2}m^2\phi^2 - \mu\Phi\phi\phi.$$

The last term is an interaction that allows a Φ particle to decay into two ϕ's, provided that $M > 2m$. Assuming that this condition is met, calculate the lifetime of the Φ to lowest order in μ.

4.3   Linear sigma model. The interactions of pions at low energy can be described by a phenomenological model called the *linear sigma model*. Essentially, this model consists of N real scalar fields coupled by a ϕ^4 interaction that is symmetric under rotations of the N fields. More specifically, let $\Phi^i(x)$, $i = 1, \ldots, N$ be a set of N fields, governed by the Hamiltonian

$$H = \int d^3x \left(\tfrac{1}{2}(\Pi^i)^2 + \tfrac{1}{2}(\nabla\Phi^i)^2 + V(\Phi^2) \right),$$

where $(\Phi^i)^2 = \Phi \cdot \Phi$, and

$$V(\Phi^2) = \tfrac{1}{2}m^2(\Phi^i)^2 + \tfrac{\lambda}{4}\left((\Phi^i)^2\right)^2$$

is a function symmetric under rotations of Φ. For (classical) field configurations of $\Phi^i(x)$ that are constant in space and time, this term gives the only contribution to H; hence, V is the field potential energy.

(What does this Hamiltonian have to do with the strong interactions? There are two types of light quarks, u and d. These quarks have identical strong interactions, but different masses. If these quarks are massless, the Hamiltonian of the strong interactions is invariant to unitary transformations of the 2-component object (u, d):

$$\begin{pmatrix} u \\ d \end{pmatrix} \;\rightarrow\; \exp(i\boldsymbol{\alpha} \cdot \boldsymbol{\sigma}/2) \begin{pmatrix} u \\ d \end{pmatrix}.$$

This transformation is called an *isospin* rotation. If, in addition, the strong interactions are described by a vector "gluon" field (as is true in QCD), the strong interaction Hamiltonian is invariant to the isospin rotations done separately on the left-handed and right-handed components of the quark fields. Thus, the complete symmetry of QCD with two massless quarks is $SU(2) \times SU(2)$. It happens that $SO(4)$, the group of rotations in 4 dimensions, is isomorphic to $SU(2) \times SU(2)$, so for $N = 4$, the linear sigma model has the same symmetry group as the strong interactions.)

(a) Analyze the linear sigma model for $m^2 > 0$ by noticing that, for $\lambda = 0$, the Hamiltonian given above is exactly N copies of the Klein-Gordon Hamiltonian. We can then calculate scattering amplitudes as perturbation series in the parameter λ. Show that the propagator is

$$\overbracket{\Phi^i(x)\,\Phi^j(y)} = \delta^{ij}\,D_F(x-y),$$

where D_F is the standard Klein-Gordon propagator for mass m, and that there is one type of vertex given by

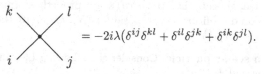

$$= -2i\lambda(\delta^{ij}\delta^{kl} + \delta^{il}\delta^{jk} + \delta^{ik}\delta^{jl}).$$

(That is, the vertex between two Φ^1s and two Φ^2s has the value $(-2i\lambda)$; that between four Φ^1s has the value $(-6i\lambda)$.) Compute, to leading order in λ, the differential cross sections $d\sigma/d\Omega$, in the center-of-mass frame, for the scattering processes

$$\Phi^1\Phi^2 \to \Phi^1\Phi^2, \qquad \Phi^1\Phi^1 \to \Phi^2\Phi^2, \qquad \text{and} \qquad \Phi^1\Phi^1 \to \Phi^1\Phi^1$$

as functions of the center-of-mass energy.

(b) Now consider the case $m^2 < 0$: $m^2 = -\mu^2$. In this case, V has a local maximum, rather than a minimum, at $\Phi^i = 0$. Since V is a potential energy, this implies that the ground state of the theory is not near $\Phi^i = 0$ but rather is obtained by shifting Φ^i toward the minimum of V. By rotational invariance, we can consider this shift to be in the Nth direction. Write, then,

$$\Phi^i(x) = \pi^i(x), \quad i = 1,\ldots,N-1,$$

$$\Phi^N(x) = v + \sigma(x),$$

where v is a constant chosen to minimize V. (The notation π^i suggests a pion field and should not be confused with a canonical momentum.) Show that, in these new coordinates (and substituting for v its expression in terms of λ and μ), we have a theory of a massive σ field and $N-1$ *massless* pion fields, interacting through cubic and quartic potential energy terms which all become small as $\lambda \to 0$. Construct the Feynman rules by assigning values to the propagators and vertices:

$$\overbracket{\sigma\ \sigma} = \ \Longrightarrow$$

$$\overbracket{\pi^i\ \pi^j} = \ i \longrightarrow j$$

(c) Compute the scattering amplitude for the process

$$\pi^i(p_1)\,\pi^j(p_2) \ \to \ \pi^k(p_3)\,\pi^l(p_4)$$

to leading order in λ. There are now four Feynman diagrams that contribute:

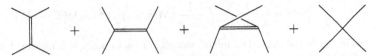

Show that, at threshold ($\mathbf{p}_i = 0$), these diagrams sum to *zero*. (Hint: It may be easiest to first consider the specific process $\pi^1 \pi^1 \to \pi^2 \pi^2$, for which only the first and fourth diagrams are nonzero, before tackling the general case.) Show that, in the special case $N = 2$ (1 species of pion), the term of $\mathcal{O}(p^2)$ also cancels.

(d) Add to V a symmetry-breaking term,

$$\Delta V = -a\Phi^N,$$

where a is a (small) constant. (In QCD, a term of this form is produced if the u and d quarks have the same nonvanishing mass.) Find the new value of v that minimizes V, and work out the content of the theory about that point. Show that the pion acquires a mass such that $m_\pi^2 \sim a$, and show that the pion scattering amplitude at threshold is now nonvanishing and also proportional to a.

4.4   Rutherford scattering. The cross section for scattering of an electron by the Coulomb field of a nucleus can be computed, to lowest order, without quantizing the electromagnetic field. Instead, treat the field as a given, classical potential $A_\mu(x)$. The interaction Hamiltonian is

$$H_I = \int d^3x \, e\bar{\psi}\gamma^\mu\psi \, A_\mu,$$

where $\psi(x)$ is the usual quantized Dirac field.

(a) Show that the T-matrix element for electron scattering off a localized classical potential is, to lowest order,

$$\langle p'|iT|p\rangle = -ie\,\bar{u}(p')\gamma^\mu u(p) \cdot \tilde{A}_\mu(p' - p),$$

where $\tilde{A}_\mu(q)$ is the four-dimensional Fourier transform of $A_\mu(x)$.

(b) If $A_\mu(x)$ is time independent, its Fourier transform contains a delta function of energy. It is then natural to define

$$\langle p'|iT|p\rangle \equiv i\mathcal{M} \cdot (2\pi)\delta(E_f - E_i),$$

where E_i and E_f are the initial and final energies of the particle, and to adopt a new Feynman rule for computing $\mathcal{M}$:

$$= -ie\gamma^\mu\tilde{A}_\mu(\mathbf{q}),$$

where $\tilde{A}_\mu(\mathbf{q})$ is the three-dimensional Fourier transform of $A_\mu(x)$. Given this definition of $\mathcal{M}$, show that the cross section for scattering off a time-independent, localized potential is

$$d\sigma = \frac{1}{v_i} \frac{1}{2E_i} \frac{d^3 p_f}{(2\pi)^3} \frac{1}{2E_f} |\mathcal{M}(p_i \to p_f)|^2 (2\pi)\delta(E_f - E_i),$$

where v_i is the particle's initial velocity. This formula is a natural modification of (4.79). Integrate over $|p_f|$ to find a simple expression for $d\sigma/d\Omega$.

(c) Specialize to the case of electron scattering from a Coulomb potential ($A^0 = Ze/4\pi r$). Working in the nonrelativistic limit, derive the Rutherford formula,

$$\frac{d\sigma}{d\Omega} = \frac{\alpha^2 Z^2}{4m^2 v^4 \sin^4(\theta/2)}.$$

(With a few calculational tricks from Section 5.1, you will have no difficulty evaluating the general cross section in the relativistic case; see Problem 5.1.)

Chapter 5

Elementary Processes of
Quantum Electrodynamics

Finally, after three long chapters of formalism, we are ready to perform some real relativistic calculations, to begin working out the predictions of Quantum Electrodynamics. First we will return to the process considered in Chapter 1, the annihilation of an electron-positron pair into a pair of heavier fermions. We will study this paradigm process in extreme detail in the next three sections, then do a few more simple QED calculations in Sections 5.4 and 5.5. The problems at the end of the chapter treat several additional QED processes. More complete surveys of QED can be found in the books of Jauch and Rohrlich (1976) and of Berestetskii, Lifshitz, and Pitaevskii (1982).

5.1 $e^+e^- \rightarrow \mu^+\mu^-$: Introduction

The reaction $e^+e^- \rightarrow \mu^+\mu^-$ is the simplest of all QED processes, but also one of the most important in high-energy physics. It is fundamental to the understanding of all reactions in e^+e^- colliders, and is in fact used to calibrate such machines. The related process $e^+e^- \rightarrow q\bar{q}$ (a quark-antiquark pair) is extraordinarily useful in determining the properties of elementary particles.

In this section we will compute the *unpolarized* cross section for $e^+e^- \rightarrow \mu^+\mu^-$, to lowest order. In Chapter 1 we used elementary arguments to guess the answer (Eq. (1.8)) in the limit where all the fermions are massless. We now relax that restriction and retain the muon mass in the calculation. Retaining the electron mass as well would be easy but pointless, since the ratio $m_e/m_\mu \approx 1/200$ is much smaller than the fractional error introduced by neglecting higher-order terms in the perturbation series.

Using the Feynman rules from Section 4.8, we can at once draw the diagram and write down the amplitude for our process:

$$= \bar{v}^{s'}(p')(-ie\gamma^\mu)u^s(p)\left(\frac{-ig_{\mu\nu}}{q^2}\right)\bar{u}^r(k)(-ie\gamma^\nu)v^{r'}(k').$$

Rearranging this slightly and leaving the spin superscripts implicit, we have

$$i\mathcal{M}\big(e^-(p)e^+(p') \to \mu^-(k)\mu^+(k')\big) = \frac{ie^2}{q^2}\Big(\bar{v}(p')\gamma^\mu u(p)\Big)\Big(\bar{u}(k)\gamma_\mu v(k')\Big). \quad (5.1)$$

This answer for the amplitude $\mathcal{M}$ is simple, but not yet very illuminating.

To compute the differential cross section, we need an expression for $|\mathcal{M}|^2$, so we must find the complex conjugate of $\mathcal{M}$. A bi-spinor product such as $\bar{v}\gamma^\mu u$ can be complex-conjugated as follows:

$$\big(\bar{v}\gamma^\mu u\big)^* = u^\dagger (\gamma^\mu)^\dagger (\gamma^0)^\dagger v = u^\dagger (\gamma^\mu)^\dagger \gamma^0 v = u^\dagger \gamma^0 \gamma^\mu v = \bar{u}\gamma^\mu v.$$

(This is another advantage of the 'bar' notation.) Thus the squared matrix element is

$$|\mathcal{M}|^2 = \frac{e^4}{q^4}\Big(\bar{v}(p')\gamma^\mu u(p)\bar{u}(p)\gamma^\nu v(p')\Big)\Big(\bar{u}(k)\gamma_\mu v(k')\bar{v}(k')\gamma_\nu u(k)\Big). \quad (5.2)$$

At this point we are still free to specify any particular spinors $u^s(p)$, $\bar{v}^{s'}(p')$, and so on, corresponding to any desired spin states of the fermions. In actual experiments, however, it is difficult (though not impossible) to retain control over spin states; one would have to prepare the initial state from polarized materials and/or analyze the final state using spin-dependent multiple scattering. In most experiments the electron and positron beams are unpolarized, so the measured cross section is an *average* over the electron and positron spins s and s'. Muon detectors are normally blind to polarization, so the measured cross section is a *sum* over the muon spins r and r'.

The expression for $|\mathcal{M}|^2$ simplifies considerably when we throw away the spin information. We want to compute

$$\frac{1}{2}\sum_s \frac{1}{2}\sum_{s'} \sum_r \sum_{r'} |\mathcal{M}(s,s' \to r,r')|^2.$$

The spin sums can be performed using the completeness relations from Section 3.3:

$$\sum_s u^s(p)\bar{u}^s(p) = \not{p} + m; \qquad \sum_s v^s(p)\bar{v}^s(p) = \not{p} - m. \quad (5.3)$$

Working with the first half of (5.2), and writing in spinor indices so we can freely move the v next to the $\bar{v}$, we have

$$\sum_{s,s'} \bar{v}^{s'}_a(p')\gamma^\mu_{ab} u^s_b(p)\bar{u}^s_c(p)\gamma^\nu_{cd} v^{s'}_d(p') = (\not{p}' - m)_{da}\gamma^\mu_{ab}(\not{p} + m)_{bc}\gamma^\nu_{cd}$$

$$= \mathrm{trace}\big[(\not{p}' - m)\gamma^\mu(\not{p} + m)\gamma^\nu\big].$$

Evaluating the second half of (5.2) in the same way, we arrive at the desired simplification:

$$\frac{1}{4}\sum_{\mathrm{spins}} |\mathcal{M}|^2 = \frac{e^4}{4q^4}\,\mathrm{tr}\Big[(\not{p}'-m_e)\gamma^\mu(\not{p}+m_e)\gamma^\nu\Big]\,\mathrm{tr}\Big[(\not{k}+m_\mu)\gamma_\mu(\not{k}'-m_\mu)\gamma_\nu\Big].$$

$$(5.4)$$

The spinors u and v have disappeared, leaving us with a much cleaner expression in terms of γ matrices. This trick is very general: Any QED amplitude involving external fermions, when squared and summed or averaged over spins, can be converted in this way to traces of products of Dirac matrices.

Trace Technology

This last step would hardly be an improvement if the traces had to be laboriously computed by brute force. But Feynman found that they could be worked out easily by appealing to the algebraic properties of the γ matrices. Since the evaluation of such traces occurs so often in QED calculations, it is worthwhile to pause and attack the problem systematically, once and for all.

We would like to evaluate traces of products of n gamma matrices, where $n = 0, 1, 2, \ldots$. (For the present problem we need $n = 2, 3, 4$.) The $n = 0$ case is fairly easy: $\operatorname{tr} \mathbf{1} = 4$. The trace of one γ matrix is also easy. From the explicit form of the matrices in the chiral representation, we have

$$\operatorname{tr} \gamma^\mu = \operatorname{tr} \begin{pmatrix} 0 & \sigma^\mu \\ \bar{\sigma}^\mu & 0 \end{pmatrix} = 0.$$

It is useful to prove this result in a more abstract way, which generalizes to an arbitrary odd number of γ matrices:

$$\begin{aligned}
\operatorname{tr} \gamma^\mu &= \operatorname{tr} \gamma^5 \gamma^5 \gamma^\mu && \text{since } (\gamma^5)^2 = 1 \\
&= -\operatorname{tr} \gamma^5 \gamma^\mu \gamma^5 && \text{since } \{\gamma^\mu, \gamma^5\} = 0 \\
&= -\operatorname{tr} \gamma^5 \gamma^5 \gamma^\mu && \text{using cyclic property of trace} \\
&= -\operatorname{tr} \gamma^\mu.
\end{aligned}$$

Since the trace of γ^μ is equal to minus itself, it must vanish. For n γ-matrices we would get n minus signs in the second step (as we move the second γ^5 all the way to the right), so the trace must vanish if n is odd.

To evaluate the trace of two γ matrices, we again use the anticommutation properties and the cyclic property of the trace:

$$\begin{aligned}
\operatorname{tr} \gamma^\mu \gamma^\nu &= \operatorname{tr}(2g^{\mu\nu} \cdot \mathbf{1} - \gamma^\nu \gamma^\mu) && \text{(anticommutation)} \\
&= 8g^{\mu\nu} - \operatorname{tr} \gamma^\mu \gamma^\nu && \text{(cyclicity)}
\end{aligned}$$

Thus $\operatorname{tr} \gamma^\mu \gamma^\nu = 4g^{\mu\nu}$. The trace of any even number of γ matrices can be evaluated in the same way: Anticommute the first γ matrix all the way to the right, then cycle it back to the left. Thus for the trace of four γ matrices, we have

$$\begin{aligned}
\operatorname{tr}(\gamma^\mu \gamma^\nu \gamma^\rho \gamma^\sigma) &= \operatorname{tr}(2g^{\mu\nu} \gamma^\rho \gamma^\sigma - \gamma^\nu \gamma^\mu \gamma^\rho \gamma^\sigma) \\
&= \operatorname{tr}(2g^{\mu\nu} \gamma^\rho \gamma^\sigma - \gamma^\nu 2g^{\mu\rho} \gamma^\sigma + \gamma^\nu \gamma^\rho 2g^{\mu\sigma} - \gamma^\nu \gamma^\rho \gamma^\sigma \gamma^\mu).
\end{aligned}$$

Using the cyclic property on the last term and bringing it to the left-hand side, we find

$$\text{tr}(\gamma^\mu\gamma^\nu\gamma^\rho\gamma^\sigma) = g^{\mu\nu}\,\text{tr}\,\gamma^\rho\gamma^\sigma - g^{\mu\rho}\,\text{tr}\,\gamma^\nu\gamma^\sigma + g^{\mu\sigma}\,\text{tr}\,\gamma^\nu\gamma^\rho$$
$$= 4(g^{\mu\nu}g^{\rho\sigma} - g^{\mu\rho}g^{\nu\sigma} + g^{\mu\sigma}g^{\nu\rho}).$$

In this manner one can always reduce a trace of n γ-matrices to a sum of traces of $(n-2)$ γ-matrices. The case $n = 6$ is easy to work out, but has fifteen terms (the number of ways of grouping the six indices in pairs to make terms of the form $g^{\mu\nu}g^{\rho\sigma}g^{\alpha\beta}$). Fortunately, we will not need it in this book. (If you ever do need to evaluate such complicated traces, it may be easier to learn to use one of the several computer programs that can perform symbolic manipulations on Dirac matrices.)

Starting in Section 5.2, we will often need to evaluate traces involving γ^5. Since $\gamma^5 = i\gamma^0\gamma^1\gamma^2\gamma^3$, the trace of γ^5 times any odd number of other γ matrices is zero. It is also easy to show that the trace of γ^5 itself is zero:

$$\text{tr}\,\gamma^5 = \text{tr}(\gamma^0\gamma^0\gamma^5) = -\text{tr}(\gamma^0\gamma^5\gamma^0) = -\text{tr}(\gamma^0\gamma^0\gamma^5) = -\text{tr}\,\gamma^5.$$

The same trick works for $\text{tr}(\gamma^\mu\gamma^\nu\gamma^5)$, if we insert two factors of γ^α for some α different from both μ and ν. The first nonvanishing trace involving γ^5 contains four other γ matrices. In this case the trick still works unless every γ matrix appears, so $\text{tr}(\gamma^\mu\gamma^\nu\gamma^\rho\gamma^\sigma\gamma^5) = 0$ unless $(\mu\nu\rho\sigma)$ is some permutation of (0123). From the anticommutation rules it also follows that interchanging any two of the indices simply changes the sign of the trace, so $\text{tr}(\gamma^\mu\gamma^\nu\gamma^\rho\gamma^\sigma\gamma^5)$ must be proportional to $\epsilon^{\mu\nu\rho\sigma}$. The overall constant turns out to be $-4i$, as you can easily check by plugging in $(\mu\nu\rho\sigma) = (0123)$.

Here is a summary of the trace theorems, for convenient reference:

$$\text{tr}(\mathbf{1}) = 4$$
$$\text{tr}(\text{any odd \# of }\gamma\text{'s}) = 0$$
$$\text{tr}(\gamma^\mu\gamma^\nu) = 4g^{\mu\nu}$$
$$\text{tr}(\gamma^\mu\gamma^\nu\gamma^\rho\gamma^\sigma) = 4(g^{\mu\nu}g^{\rho\sigma} - g^{\mu\rho}g^{\nu\sigma} + g^{\mu\sigma}g^{\nu\rho}) \qquad (5.5)$$
$$\text{tr}(\gamma^5) = 0$$
$$\text{tr}(\gamma^\mu\gamma^\nu\gamma^5) = 0$$
$$\text{tr}(\gamma^\mu\gamma^\nu\gamma^\rho\gamma^\sigma\gamma^5) = -4i\epsilon^{\mu\nu\rho\sigma}$$

Expressions resulting from use of the last formula can be simplified by means of the identities

$$\epsilon^{\alpha\beta\gamma\delta}\epsilon_{\alpha\beta\gamma\delta} = -24$$
$$\epsilon^{\alpha\beta\gamma\mu}\epsilon_{\alpha\beta\gamma\nu} = -6\delta^\mu_{\ \nu} \qquad (5.6)$$
$$\epsilon^{\alpha\beta\mu\nu}\epsilon_{\alpha\beta\rho\sigma} = -2(\delta^\mu_{\ \rho}\delta^\nu_{\ \sigma} - \delta^\mu_{\ \sigma}\delta^\nu_{\ \rho})$$

All of these can be derived by first appealing to symmetry arguments, then evaluating one special case to determine the overall constant.

Another useful identity allows one to reverse the order of all the γ matrices inside a trace:

$$\text{tr}(\gamma^\mu \gamma^\nu \gamma^\rho \gamma^\sigma \cdots) = \text{tr}(\cdots \gamma^\sigma \gamma^\rho \gamma^\nu \gamma^\mu). \tag{5.7}$$

To prove this relation, consider the matrix $C \equiv \gamma^0 \gamma^2$ (essentially the charge-conjugation operator). This matrix satisfies $C^2 = 1$ and $C\gamma^\mu C = -(\gamma^\mu)^T$. Thus if there are n γ-matrices inside the trace,

$$\begin{aligned}
\text{tr}(\gamma^\mu \gamma^\nu \cdots) &= \text{tr}(C\gamma^\mu C \, C\gamma^\nu C \cdots) \\
&= (-1)^n \, \text{tr}\big[(\gamma^\mu)^T (\gamma^\nu)^T \cdots\big] \\
&= \text{tr}(\cdots \gamma^\nu \gamma^\mu),
\end{aligned}$$

since the trace vanishes unless n is even. It is easy to show that the reversal identity (5.7) is also valid when the trace contains one or more factors of γ^5.

When two γ matrices inside a trace are dotted together, it is easiest to eliminate them before evaluating the trace. For example,

$$\gamma^\mu \gamma_\mu = g_{\mu\nu} \gamma^\mu \gamma^\nu = \tfrac{1}{2} g_{\mu\nu} \{\gamma^\mu, \gamma^\nu\} = g_{\mu\nu} g^{\mu\nu} = 4. \tag{5.8}$$

The following *contraction identities*, all easy to prove using the anticommutation relations, can be used when other γ matrices lie in between:

$$\begin{aligned}
\gamma^\mu \gamma^\nu \gamma_\mu &= -2\gamma^\nu \\
\gamma^\mu \gamma^\nu \gamma^\rho \gamma_\mu &= 4g^{\nu\rho} \\
\gamma^\mu \gamma^\nu \gamma^\rho \gamma^\sigma \gamma_\mu &= -2\gamma^\sigma \gamma^\rho \gamma^\nu
\end{aligned} \tag{5.9}$$

Note the reversal of order in the last identity.

All of the γ matrix identities proved in this section are collected for reference in the Appendix.

Unpolarized Cross Section

We now return to the evaluation of the squared matrix element, Eq. (5.4). The electron trace is

$$\text{tr}\big[(\not{p}' - m_e)\gamma^\mu(\not{p} + m_e)\gamma^\nu\big] = 4\big[p'^\mu p^\nu + p'^\nu p^\mu - g^{\mu\nu}(p \cdot p' + m_e^2)\big].$$

The terms with only one factor of m vanish, since they contain an odd number of γ matrices. Similarly, the muon trace is

$$\text{tr}\big[(\not{k} + m_\mu)\gamma_\mu(\not{k}' - m_\mu)\gamma_\nu\big] = 4\big[k_\mu k'_\nu + k_\nu k'_\mu - g_{\mu\nu}(k \cdot k' + m_\mu^2)\big].$$

From now on we will set $m_e = 0$, as discussed at the beginning of this section. Dotting these expressions together and collecting terms, we get the simple result

$$\frac{1}{4} \sum_{\text{spins}} |\mathcal{M}|^2 = \frac{8e^4}{q^4} \big[(p \cdot k)(p' \cdot k') + (p \cdot k')(p' \cdot k) + m_\mu^2(p \cdot p')\big]. \tag{5.10}$$

To obtain a more explicit formula we must specialize to a particular frame of reference and express the vectors p, p', k, k', and q in terms of the basic kinematic variables—energies and angles—in that frame. In practice, the choice of frame will be dictated by the experimental conditions. In this book, we will usually make the simplest choice of evaluating cross sections in the center-of-mass frame. For this choice, the initial and final 4-momenta for $e^+e^- \rightarrow \mu^+\mu^-$ can be written as follows:

$$k = (E, \mathbf{k})$$

$$p = (E, E\hat{z})$$

$$p' = (E, -E\hat{z})$$

$$k' = (E, -\mathbf{k})$$

$$|\mathbf{k}| = \sqrt{E^2 - m_\mu^2}$$

$$\mathbf{k} \cdot \hat{z} = |\mathbf{k}| \cos\theta$$

To compute the squared matrix element we need

$$q^2 = (p + p')^2 = 4E^2; \qquad\qquad p \cdot p' = 2E^2;$$

$$p \cdot k = p' \cdot k' = E^2 - E|\mathbf{k}|\cos\theta; \qquad p \cdot k' = p' \cdot k = E^2 + E|\mathbf{k}|\cos\theta.$$

We can now rewrite Eq. (5.10) in terms of E and θ:

$$\frac{1}{4}\sum_{\text{spins}}|\mathcal{M}|^2 = \frac{8e^4}{16E^4}\left[E^2(E - |\mathbf{k}|\cos\theta)^2 + E^2(E + |\mathbf{k}|\cos\theta)^2 + 2m_\mu^2 E^2\right]$$

$$= e^4\left[\left(1 + \frac{m_\mu^2}{E^2}\right) + \left(1 - \frac{m_\mu^2}{E^2}\right)\cos^2\theta\right]. \tag{5.11}$$

All that remains is to plug this expression into the cross-section formula derived in Section 4.5. Since there are only two particles in the final state and we are working in the center-of-mass frame, we can use the simplified formula (4.84). For our problem $|v_A - v_B| = 2$ and $E_A = E_B = E_{\text{cm}}/2$, so we have

$$\frac{d\sigma}{d\Omega} = \frac{1}{2E_{\text{cm}}^2}\frac{|\mathbf{k}|}{16\pi^2 E_{\text{cm}}} \cdot \frac{1}{4}\sum_{\text{spins}}|\mathcal{M}|^2$$

$$= \frac{\alpha^2}{4E_{\text{cm}}^2}\sqrt{1 - \frac{m_\mu^2}{E^2}}\left[\left(1 + \frac{m_\mu^2}{E^2}\right) + \left(1 - \frac{m_\mu^2}{E^2}\right)\cos^2\theta\right]. \tag{5.12}$$

Integrating over $d\Omega$, we find the total cross section:

$$\sigma_{\text{total}} = \frac{4\pi\alpha^2}{3E_{\text{cm}}^2}\sqrt{1 - \frac{m_\mu^2}{E^2}}\left(1 + \frac{1}{2}\frac{m_\mu^2}{E^2}\right). \tag{5.13}$$

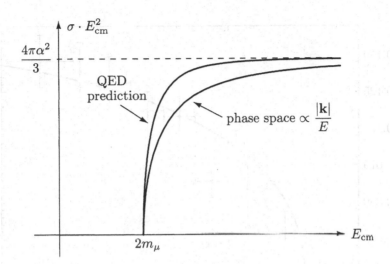

Figure 5.1. Energy dependence of the total cross section for $e^+e^- \rightarrow \mu^+\mu^-$, compared to "phase space" energy dependence.

In the high-energy limit where $E \gg m_\mu$, these formulae reduce to those given in Chapter 1:

$$\frac{d\sigma}{d\Omega} \xrightarrow[E \gg m_\mu]{} \frac{\alpha^2}{4E_{\text{cm}}^2}(1 + \cos^2\theta);$$

$$\sigma_{\text{total}} \xrightarrow[E \gg m_\mu]{} \frac{4\pi\alpha^2}{3E_{\text{cm}}^2}\left(1 - \frac{3}{8}\left(\frac{m_\mu}{E}\right)^4 - \cdots\right). \tag{5.14}$$

Note that these expressions have the correct dimensions of cross sections. In the high-energy limit, E_{cm} is the only dimensionful quantity in the problem, so dimensional analysis dictates that $\sigma_{\text{total}} \propto E_{\text{cm}}^{-2}$. Since we knew from the beginning that $\sigma_{\text{total}} \propto \alpha^2$, we only had to work to get the factor of $4\pi/3$.

The energy dependence of the total cross-section formula (5.13) near threshold is shown in Fig. 5.1. Of course the cross section is zero for $E_{\text{cm}} < 2m_\mu$. It is interesting to compare the shape of the actual curve to the shape one would obtain if $|\mathcal{M}|^2$ did not depend on energy, that is, if all the energy dependence came from the phase-space factor $|\mathbf{k}|/E$. To test Quantum Electrodynamics, an experiment must be able to resolve deviations from the naive phase-space prediction. Experimental results from pair production of both μ and τ leptons confirm that these particles behave as QED predicts. Figure 5.2 compares formula (5.13) to experimental measurements of the $\tau^+\tau^-$ threshold.

Before discussing our result further, let us pause to summarize how we obtained it. The method extends in a straightforward way to the calculation of unpolarized cross sections for other QED processes. The general procedure is as follows:

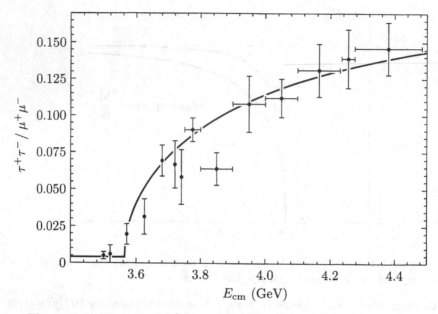

Figure 5.2. The ratio $\sigma(e^+e^- \to \tau^+\tau^-)/\sigma(e^+e^- \to \mu^+\mu^-)$ of measured cross sections near the threshold for $\tau^+\tau^-$ pair-production, as measured by the DELCO collaboration, W. Bacino, et. al., *Phys. Rev. Lett.* **41**, 13 (1978). Only a fraction of τ decays are included, hence the small overall scale. The curve shows a fit to the theoretical formula (5.13), with a small energy-independent background added. The fit yields $m_\tau = 1782^{+2}_{-7}$ MeV.

1. Draw the diagram(s) for the desired process.
2. Use the Feynman rules to write down the amplitude $\mathcal{M}$.
3. Square the amplitude and average or sum over spins, using the completeness relations (5.3). (For processes involving photons in the final state there is an analogous completeness relation, derived in Section 5.5.)
4. Evaluate traces using the trace theorems (5.5); collect terms and simplify the answer as much as possible.
5. Specialize to a particular frame of reference, and draw a picture of the kinematic variables in that frame. Express all 4-momentum vectors in terms of a suitably chosen set of variables such as E and θ.
6. Plug the resulting expression for $|\mathcal{M}|^2$ into the cross-section formula (4.79), and integrate over phase-space variables that are not measured to obtain a differential cross section in the desired form. (In our case these integrations were over the constrained momenta $\mathbf{k}'$ and $|\mathbf{k}|$, and were performed in the derivation of Eq. (4.84).)

While other calculations (especially those involving loop diagrams) often require additional tricks, nearly every QED calculation will involve the basic procedures outlined here.

Production of Quark-Antiquark Pairs

The asymptotic energy dependence of the $e^+e^- \to \mu^+\mu^-$ cross-section formula sets the scale for all e^+e^- annihilation cross sections. A particularly important example is the cross section for

$$e^+e^- \to \text{hadrons},$$

that is, the total cross section for production of any number of strongly interacting particles.

In our current understanding of the strong interactions, given by the theory called Quantum Chromodynamics (QCD), all hadrons are composed of Dirac fermions called *quarks*. Quarks appear in a variety of types, called *flavors*, each with its own mass and electric charge. A quark also carries an additional quantum number, *color*, which takes one of three values. Color serves as the "charge" of QCD, as we will discuss in Chapter 17.

According to QCD, the simplest e^+e^- process that ends in hadrons is

$$e^+e^- \to q\bar{q},$$

the annihilation of an electron and a positron, through a virtual photon, into a quark-antiquark pair. After they are created, the quarks interact with one another through their strong forces, producing more quark pairs. Eventually the quarks and antiquarks combine to form some number of mesons and baryons.

To adapt our results for muon production to handle the case of quarks, we must make three modifications:

1. Replace the muon charge e with the quark charge $Q|e|$.

2. Count each quark three times, one for each color.

3. Include the effects of the strong interactions of the produced quark and antiquark.

The first two changes are easy to make. For the first, it is simply necessary to know the masses and charges of each flavor of quark. For u, c, and t quarks we have $Q = 2/3$, while for d, s, and b quarks we have $Q = -1/3$. The cross-section formulae are proportional to the square of the charge of the final-state particle, so we can simply insert a factor of Q^2 into any of these formulae to obtain the cross section for production of any particular variety of quark. Counting colors is necessary because experiments measure only the total cross section for production of all three colors. (The hadrons that are actually detected are colorless.) In any case, this counting is easy: Just multiply the answer by 3.

If you know a little about the strong interaction, however, you might think this is all a big joke. Surely the third modification is extremely difficult to make, and will drastically alter the predictions of QED. The amazing truth is that in the high-energy limit, the effect of the strong interaction on the quark production process can be completely neglected. As we will discuss in Part III, the only effect of the strong interaction (in this limit) is to dress up the final-state quarks into bunches of hadrons. This simplification is due to a phenomenon called *asymptotic freedom*; it played a crucial role in the identification of Quantum Chromodynamics as the correct theory of the strong force.

Thus in the high-energy limit, we expect the cross section for the reaction $e^+e^- \to q\bar{q}$ to approach $3 \cdot Q^2 \cdot 4\pi\alpha^2/3E_{cm}^2$. It is conventional to define

$$1 \text{ unit of R} \equiv \frac{4\pi\alpha^2}{3E_{cm}^2} = \frac{86.8 \text{ nbarns}}{(E_{cm} \text{ in GeV})^2}. \tag{5.15}$$

The value of a cross section in units of R is therefore its ratio to the asymptotic value of the $e^+e^- \to \mu^+\mu^-$ cross section predicted by Eq. (5.14). Experimentally, the easiest quantity to measure is the total rate for production of all hadrons. Asymptotically, we expect

$$\sigma(e^+e^- \to \text{hadrons}) \xrightarrow[E_{cm} \to \infty]{} 3 \cdot \left(\sum_i Q_i^2\right) R, \tag{5.16}$$

where the sum runs over all quarks whose masses are smaller than $E_{cm}/2$. When $E_{cm}/2$ is in the vicinity of one of the quark masses, the strong interactions cause large deviations from this formula. The most dramatic such effect is the appearance of *bound states* just below $E_{cm} = 2m_q$, manifested as very sharp spikes in the cross section.

Experimental measurements of the cross section for e^+e^- annihilation to hadrons between 2.5 and 40 GeV are shown in Fig. 5.3. The data shows three distinct regions: a low-energy region in which u, d, and s quark pairs are produced; a region above the threshold for production of c quark pairs; and a region also above the threshold for b quark pairs. The prediction (5.16) is shown as a set of solid lines; it agrees quite well with the data in each region, as long as the energy is well away from the thresholds where the high-energy approximation breaks down. The dotted curves show an improved theoretical prediction, including higher-order corrections from QCD, which we will discuss in Section 17.2. This explanation of the e^+e^- annihilation cross section is a remarkable success of QCD. In particular, experimental verification of the factor of 3 in (5.16) is one piece of evidence for the existence of color.

The angular dependence of the differential cross section is also observed experimentally.* At high energy the hadrons appear in *jets*, clusters of several hadrons all moving in approximately the same direction. In most cases there are two jets, with back-to-back momenta, and these indeed have the angular dependence $(1 + \cos^2 \theta)$.

---

*The basic features of hadron production in high-energy e^+e^- annihilation are reviewed by P. Duinker, *Rev. Mod. Phys.* **54**, 325 (1982).

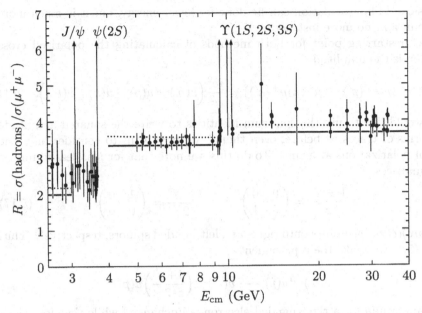

Figure 5.3. Experimental measurements of the total cross section for the reaction $e^+e^- \rightarrow$ hadrons, from the data compilation of M. Swartz, *Phys. Rev.* **D**53, 5268 (1996). Complete references to the various experiments are given there. The measurements are compared to theoretical predictions from Quantum Chromodynamics, as explained in the text. The solid line is the simple prediction (5.16).

5.2 $e^+e^- \rightarrow \mu^+\mu^-$: **Helicity Structure**

The unpolarized cross section for a reaction is generally easy to calculate (and to measure) but hard to understand. Where does the $(1 + \cos^2 \theta)$ angular dependence come from? We can answer this question by computing the $e^+e^- \rightarrow \mu^+\mu^-$ cross section for each set of spin orientations separately.

First we must choose a basis of polarization states. To get a simple answer in the high-energy limit, the best choice is to quantize each spin along the direction of the particle's motion, that is, to use states of definite helicity. Recall that in the massless limit, the left- and right-handed helicity states of a Dirac particle live in different representations of the Lorentz group. We might therefore expect them to behave independently, and in fact they do.

In this section we will compute the polarized $e^+e^- \rightarrow \mu^+\mu^-$ cross sections, using the helicity basis, in two different ways: first, by using trace technology but with the addition of helicity projection operators to project out the desired left- or right-handed spinors; and second, by plugging explicit expressions for these spinors directly into our formula for the amplitude $\mathcal{M}$. Throughout this section we work in the high-energy limit where all fermions are effectively massless. (The calculation can be done for lower energy, but it is much more difficult and no more instructive.)[†]

Our starting point for both methods of calculating the polarized cross section is the amplitude

$$i\mathcal{M}\big(e^-(p)e^+(p') \to \mu^-(k)\mu^+(k')\big) = \frac{ie^2}{q^2}\big(\bar{v}(p')\gamma^\mu u(p)\big)\big(\bar{u}(k)\gamma_\mu v(k')\big). \quad (5.1)$$

We would like to use the spin sum identities to write the squared amplitude in terms of traces as before, even though we now want to consider only one set of polarizations at a time. To do this, we note that for massless fermions, the matrices

$$\frac{1+\gamma^5}{2} = \begin{pmatrix} 0 & 0 \\ 0 & 1 \end{pmatrix}, \qquad \frac{1-\gamma^5}{2} = \begin{pmatrix} 1 & 0 \\ 0 & 0 \end{pmatrix} \qquad (5.17)$$

are *projection operators* onto right- and left-handed spinors, respectively. Thus if in (5.1) we make the replacement

$$\bar{v}(p')\gamma^\mu u(p) \longrightarrow \bar{v}(p')\gamma^\mu\Big(\frac{1+\gamma^5}{2}\Big)u(p),$$

the amplitude for a right-handed electron is unchanged while that for a left-handed electron becomes zero. Note that since

$$\bar{v}(p')\gamma^\mu\Big(\frac{1+\gamma^5}{2}\Big)u(p) = v^\dagger(p')\Big(\frac{1+\gamma^5}{2}\Big)\gamma^0\gamma^\mu u(p), \qquad (5.18)$$

this same replacement imposes the requirement that $v(p')$ also be a right-handed spinor. Recall from Section 3.5, however, that the right-handed spinor $v(p')$ corresponds to a *left*-handed positron. Thus we see that the annihilation amplitude vanishes when both the electron and the positron are right-handed. In general, the amplitude vanishes (in the massless limit) unless the electron and positron have opposite helicity, or equivalently, unless their spinors have the same helicity.

Having inserted this projection operator, we are now free to sum over the electron and positron spins in the squared amplitude; of the four terms in the sum, only one (the one we want) is nonzero. The electron half of $|\mathcal{M}|^2$, for a right-handed electron and a left-handed positron, is then

$$\sum_{\text{spins}}\Big|\bar{v}(p')\gamma^\mu\Big(\frac{1+\gamma^5}{2}\Big)u(p)\Big|^2 = \sum_{\text{spins}}\bar{v}(p')\gamma^\mu\Big(\frac{1+\gamma^5}{2}\Big)u(p)\,\bar{u}(p)\gamma^\nu\Big(\frac{1+\gamma^5}{2}\Big)v(p')$$

$$= \text{tr}\Big[\not{p}'\gamma^\mu\Big(\frac{1+\gamma^5}{2}\Big)\not{p}\gamma^\nu\Big(\frac{1+\gamma^5}{2}\Big)\Big]$$

$$= \text{tr}\Big[\not{p}'\gamma^\mu\not{p}\gamma^\nu\Big(\frac{1+\gamma^5}{2}\Big)\Big]$$

[†]The general formalism for S-matrix elements between states of definite helicity is presented in a beautiful paper of M. Jacob and G. C. Wick, *Ann. Phys.* **7**, 404 (1959).

$$= 2\left(p'^\mu p^\nu + p'^\nu p^\mu - g^{\mu\nu}p\cdot p' - i\epsilon^{\alpha\mu\beta\nu}p'_\alpha p_\beta\right). \quad (5.19)$$

The indices in this expression are to be dotted into those of the muon half of the squared amplitude. For a right-handed μ^- and a left-handed μ^+, an identical calculation yields

$$\sum_{\text{spins}}\left|\bar{u}(k)\gamma_\mu\left(\frac{1+\gamma^5}{2}\right)v(k')\right|^2 = 2\left(k_\mu k'_\nu + k_\nu k'_\mu - g_{\mu\nu}k\cdot k' - i\epsilon_{\rho\mu\sigma\nu}k^\rho k'^\sigma\right). \quad (5.20)$$

Dotting (5.19) into (5.20), we find that the squared matrix element for $e_R^- e_L^+ \to \mu_R^- \mu_L^+$ in the center-of-mass frame is

$$\begin{aligned}
|\mathcal{M}|^2 &= \frac{4e^4}{q^4}\left[2(p\cdot k)(p'\cdot k') + 2(p\cdot k')(p'\cdot k) - \epsilon^{\alpha\mu\beta\nu}\epsilon_{\rho\mu\sigma\nu}p'_\alpha p_\beta k^\rho k'^\sigma\right] \\
&= \frac{8e^4}{q^4}\left[(p\cdot k)(p'\cdot k') + (p\cdot k')(p'\cdot k) - (p\cdot k)(p'\cdot k') + (p\cdot k')(p'\cdot k)\right] \\
&= \frac{16e^4}{q^4}(p\cdot k')(p'\cdot k) \\
&= e^4(1+\cos\theta)^2. \quad (5.21)
\end{aligned}$$

Plugging this result into (4.85) gives the differential cross section,

$$\frac{d\sigma}{d\Omega}\left(e_R^- e_L^+ \to \mu_R^- \mu_L^+\right) = \frac{\alpha^2}{4E_{\text{cm}}^2}(1+\cos\theta)^2. \quad (5.22)$$

There is no need to repeat the entire calculation to obtain the other three nonvanishing helicity amplitudes. For example, the squared amplitude for $e_R^- e_L^+ \to \mu_L^- \mu_R^+$ is identical to (5.20) but with γ^5 replaced by $-\gamma^5$ on the left-hand side, and thus $\epsilon_{\rho\mu\sigma\nu}$ replaced by $-\epsilon_{\rho\mu\sigma\nu}$ on the right-hand side. Propagating this sign though (5.21), we easily see that

$$\frac{d\sigma}{d\Omega}\left(e_R^- e_L^+ \to \mu_L^- \mu_R^+\right) = \frac{\alpha^2}{4E_{\text{cm}}^2}(1-\cos\theta)^2. \quad (5.23)$$

Similarly,

$$\begin{aligned}
\frac{d\sigma}{d\Omega}\left(e_L^- e_R^+ \to \mu_R^- \mu_L^+\right) &= \frac{\alpha^2}{4E_{\text{cm}}^2}(1-\cos\theta)^2; \\
\frac{d\sigma}{d\Omega}\left(e_L^- e_R^+ \to \mu_L^- \mu_R^+\right) &= \frac{\alpha^2}{4E_{\text{cm}}^2}(1+\cos\theta)^2. \quad (5.24)
\end{aligned}$$

(These two results actually follow from the previous two by parity invariance.) The other twelve helicity cross sections (for instance, $e_L^- e_R^+ \to \mu_L^- \mu_L^+$) are zero, as we saw from Eq. (5.18). Adding up all sixteen contributions, and dividing by 4 to average over the electron and positron spins, we recover the unpolarized cross section in the massless limit, Eq. (5.14).

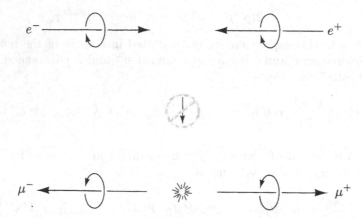

Figure 5.4. Conservation of angular momentum requires that if the z-component of angular momentum is measured, it must have the same value as initially.

Note that the cross section (5.22) for $e_R^- e_L^+ \to \mu_R^- \mu_L^+$ vanishes at $\theta = 180°$. This is just what we would expect, since for $\theta = 180°$, the total angular momentum of the final state is opposite to that of the initial state (see Figure 5.4).

This completes our first calculation of the polarized $e^+ e^- \to \mu^+ \mu^-$ cross sections. We will now redo the calculation in a manner that is more straightforward, more enlightening, and no more difficult. We will calculate the amplitude $\mathcal{M}$ (rather than the squared amplitude) directly, using explicit values for the spinors and γ matrices. This method does have its drawbacks: It forces us to specialize to a particular frame of reference much sooner, so manifest Lorentz invariance is lost. More pragmatically, it is very cumbersome except in the nonrelativistic and ultra-relativistic limits.

Consider again the amplitude

$$\mathcal{M} = \frac{e^2}{q^2} \Big(\bar{v}(p') \gamma^\mu u(p) \Big) \Big(\bar{u}(k) \gamma_\mu v(k') \Big). \tag{5.25}$$

In the high-energy limit, our general expressions for Dirac spinors become

$$u(p) = \begin{pmatrix} \sqrt{p \cdot \sigma}\, \xi \\ \sqrt{p \cdot \bar{\sigma}}\, \xi \end{pmatrix} \xrightarrow[E \to \infty]{} \sqrt{2E} \begin{pmatrix} \frac{1}{2}(1 - \hat{p} \cdot \boldsymbol{\sigma}) \xi \\ \frac{1}{2}(1 + \hat{p} \cdot \boldsymbol{\sigma}) \xi \end{pmatrix};$$

$$v(p) = \begin{pmatrix} \sqrt{p \cdot \sigma}\, \xi \\ -\sqrt{p \cdot \bar{\sigma}}\, \xi \end{pmatrix} \xrightarrow[E \to \infty]{} \sqrt{2E} \begin{pmatrix} \frac{1}{2}(1 - \hat{p} \cdot \boldsymbol{\sigma}) \xi \\ -\frac{1}{2}(1 + \hat{p} \cdot \boldsymbol{\sigma}) \xi \end{pmatrix}. \tag{5.26}$$

A right-handed spinor satisfies $(\hat{p} \cdot \boldsymbol{\sigma})\xi = +\xi$, while a left-handed spinor has $(\hat{p} \cdot \boldsymbol{\sigma})\xi = -\xi$. (Remember once again that for antiparticles, the handedness of the spinor is the opposite of the handedness of the particle.) We must evaluate expressions of the form $\bar{v}\gamma^\mu u$, so we need

$$\gamma^0 \gamma^\mu = \begin{pmatrix} 0 & 1 \\ 1 & 0 \end{pmatrix} \begin{pmatrix} 0 & \sigma^\mu \\ \bar{\sigma}^\mu & 0 \end{pmatrix} = \begin{pmatrix} \bar{\sigma}^\mu & 0 \\ 0 & \sigma^\mu \end{pmatrix}. \tag{5.27}$$

Thus we see explicitly that the amplitude is zero when one of the spinors is left-handed and the other is right-handed. In the language of Chapter 1, the Clebsch-Gordan coefficients that couple the vector photon to the product of such spinors are zero; those coefficients are just the off-block-diagonal elements of the matrix $\gamma^0\gamma^\mu$ (in the chiral representation).

Let us choose p and p' to be in the $\pm z$-directions, and first consider the case where the electron is right-handed and the positron is left-handed:

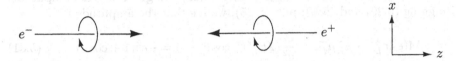

Thus for the electron we have $\xi = \binom{1}{0}$, corresponding to spin up in the z-direction, while for the positron we have $\xi = \binom{0}{1}$, also corresponding to (physical) spin up in the z-directon. Both particles have $(\hat{p}\cdot\boldsymbol{\sigma})\xi = +\xi$, so the spinors are

$$u(p) = \sqrt{2E} \begin{pmatrix} 0 \\ 0 \\ 1 \\ 0 \end{pmatrix} ; \qquad v(p') = \sqrt{2E} \begin{pmatrix} 0 \\ 0 \\ 0 \\ -1 \end{pmatrix}. \qquad (5.28)$$

The electron half of the matrix element is therefore

$$\bar{v}(p')\gamma^\mu u(p) = 2E\,(0,\ -1)\sigma^\mu \begin{pmatrix} 1 \\ 0 \end{pmatrix} = -2E\,(0,1,i,0). \qquad (5.29)$$

We can interpret this expression by saying that the virtual photon has circular polarization in the $+z$-direction; its polarization vector is $\epsilon_+ = (1/\sqrt{2})(\hat{x}+i\hat{y})$.

Next we must calculate the muon half of the matrix element. Let the μ^- be emitted at an angle θ to the z-axis, and consider first the case where it is right-handed (and the μ^+ is therefore left-handed):

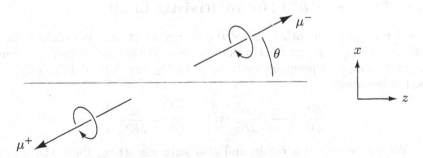

To calculate $\bar{u}(k)\gamma^\mu v(k')$ we could go back to expressions (5.26), but then it would be necessary to find the correct spinors ξ corresponding to polarization along the muon momentum. It is much easier to use a trick: Since any expression of the form $\bar{\psi}\gamma^\mu\psi$ transforms like a 4-vector, we can just rotate the result

(5.29). Rotating that vector by an angle θ in the xz-plane, we find

$$\bar{u}(k)\gamma^\mu v(k') = [\bar{v}(k')\gamma^\mu u(k)]^*$$
$$= [-2E\,(0, \cos\theta, i, -\sin\theta)]^* \qquad (5.30)$$
$$= -2E\,(0, \cos\theta, -i, -\sin\theta).$$

This vector can also be interpreted as the polarization of the virtual photon; when it has a nonzero overlap with (5.29), we get a nonzero amplitude. Plugging (5.29) and (5.30) into (5.25), we see that the amplitude is

$$\mathcal{M}(e_R^- e_L^+ \to \mu_R^- \mu_L^+) = \frac{e^2}{q^2}(2E)^2(-\cos\theta - 1) = -e^2(1 + \cos\theta), \qquad (5.31)$$

in agreement with (1.6), and also with (5.21). The differential cross section for this set of helicities can now be obtained in the same way as above, yielding (5.22).

We can calculate the other three nonvanishing helicity amplitudes in an analogous manner. For a left-handed electron and a right-handed positron, we easily find

$$\bar{v}(p')\gamma^\mu u(p) = -2E\,(0, 1, -i, 0) \equiv -2E \cdot \sqrt{2}\,\epsilon_-^\mu.$$

Perform a rotation to get the vector corresponding to a left-handed μ^- and a right-handed μ^+:

$$\bar{u}(k)\gamma^\mu v(k') = -2E\,(0, \cos\theta, i, -\sin\theta).$$

Putting the pieces together in various ways yields the remaining amplitudes,

$$\mathcal{M}(e_L^- e_R^+ \to \mu_L^- \mu_R^+) = -e^2(1 + \cos\theta);$$
$$\mathcal{M}(e_R^- e_L^+ \to \mu_L^- \mu_R^+) = \mathcal{M}(e_L^- e_R^+ \to \mu_R^- \mu_L^+) = -e^2(1 - \cos\theta). \qquad (5.32)$$

5.3 $e^+e^- \to \mu^+\mu^-$: Nonrelativistic Limit

Now let us go to the other end of the energy spectrum, and discuss the reaction $e^+e^- \to \mu^+\mu^-$ in the extreme nonrelativistic limit. When E is barely larger than m_μ, our previous result (5.12) for the unpolarized differential cross section becomes

$$\frac{d\sigma}{d\Omega} \xrightarrow[|\mathbf{k}|\to 0]{} \frac{\alpha^2}{2E_{\text{cm}}^2}\sqrt{1 - \frac{m_\mu^2}{E^2}} = \frac{\alpha^2}{2E_{\text{cm}}^2}\frac{|\mathbf{k}|}{E}. \qquad (5.33)$$

We can recover this result, and also learn something about the spin dependence of the reaction, by evaluating the amplitude with explicit spinors. Once again we begin with the matrix element

$$\mathcal{M} = \frac{e^2}{q^2}\Big(\bar{v}(p')\gamma^\mu u(p)\Big)\Big(\bar{u}(k)\gamma_\mu v(k')\Big).$$

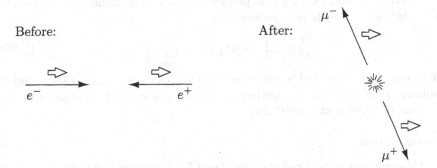

Figure 5.5. In the nonrelativistic limit the total spin of the system is conserved, and thus the muons are produced with both spins up along the z-axis.

The electron and positron are still very relativistic, so this expression will be simplest if we choose them to have definite helicity. Let the electron be right-handed, moving in the $+z$-direction, and the positron be left-handed, moving in the $-z$-direction. Then from Eq. (5.29) we have

$$\bar{v}(p')\gamma^\mu u(p) = -2E\,(0, 1, i, 0). \tag{5.34}$$

In the other half of the matrix element we should use the nonrelativistic expressions

$$u(k) = \sqrt{m}\begin{pmatrix} \xi \\ \xi \end{pmatrix}, \qquad v(k') = \sqrt{m}\begin{pmatrix} \xi' \\ -\xi' \end{pmatrix}. \tag{5.35}$$

Keep in mind, in the discussion of this section, that the spinor ξ' gives the flipped spin of the antiparticle. Leaving the muon spinors ξ and ξ' undetermined for now, we can easily compute

$$\bar{u}(k)\gamma^\mu v(k') = m(\xi^\dagger, \xi^\dagger)\begin{pmatrix} \sigma^\mu & 0 \\ 0 & \sigma^\mu \end{pmatrix}\begin{pmatrix} \xi' \\ -\xi' \end{pmatrix}$$

$$= \begin{cases} 0 & \text{for } \mu = 0, \\ -2m\xi^\dagger\sigma^i\xi' & \text{for } \mu = i. \end{cases} \tag{5.36}$$

To evaluate $\mathcal{M}$, we simply dot (5.34) into (5.36) and multiply by $e^2/q^2 = e^2/4m^2$. The result is

$$\mathcal{M}(e_R^- e_L^+ \to \mu^+\mu^-) = -2e^2\xi^\dagger\begin{pmatrix} 0 & 1 \\ 0 & 0 \end{pmatrix}\xi'. \tag{5.37}$$

Since there is no angular dependence in this expression, the muons are equally likely to come out in any direction. More precisely, they are emitted in an s-wave; their orbital angular momentum is zero. Angular momentum conservation therefore requires that the total spin of the final state equal 1, and indeed the matrix product gives zero unless both the muon and the antimuon have spin up along the z-axis (see Fig. 5.5).

To find the total rate for this process, we sum over muon spins to obtain $\mathcal{M}^2 = 4e^4$, which yields the cross section

$$\frac{d\sigma}{d\Omega}(e_R^- e_L^+ \to \mu^+\mu^-) = \frac{\alpha^2}{E_{cm}^2}\frac{|\mathbf{k}|}{E}. \tag{5.38}$$

The same expression holds for a left-handed electron and a right-handed positron. Thus the spin-averaged cross section is just $2 \cdot (1/4)$ times this expression, in agreement with (5.33).

Bound States

Until now we have considered the initial and final states of scattering processes to be states of isolated single particles. Very close to threshold, however, the Coulomb attraction of the muons should become an important effect. Just below threshold, we can still form $\mu^+\mu^-$ pairs in electromagnetic bound states.

The treatment of bound states in quantum field theory is a rich and complex subject, but one that lies mainly beyond the scope of this book.[‡] Fortunately, many of the familiar bound systems in Nature can be treated (at least to a good first approximation) as nonrelativistic systems, in which the internal motions are slow. The process of creating the constituent particles out of the vacuum is still a relativistic effect, requiring quantum field theory for its proper description. In this section we will develop a formalism for computing the amplitudes for creation and annihilation of two-particle, nonrelativistic bound states. We begin with a computation of the cross section for producing a $\mu^+\mu^-$ bound state in e^+e^- annihilation.

Consider first the case where the spins of the electron and positron both point up along the z-axis. From the preceding discussion we know that the resulting muons both have spin up, so the only type of bound state we can produce will have total spin 1, also pointing up. The amplitude for producing free muons in this configuration is

$$\mathcal{M}(\uparrow\uparrow \to \mathbf{k}_1\uparrow, \mathbf{k}_2\uparrow) = -2e^2, \tag{5.39}$$

independent of the momenta (which we now call $\mathbf{k}_1$ and $\mathbf{k}_2$) of the muons.

Next we need to know how to write a bound state in terms of free-particle states. For a general two-body system with equal constituent masses, the center-of-mass and relative coordinates are

$$\mathbf{R} = \tfrac{1}{2}(\mathbf{r}_1 + \mathbf{r}_2), \qquad \mathbf{r} = \mathbf{r}_1 - \mathbf{r}_2. \tag{5.40}$$

These have conjugate momenta

$$\mathbf{K} = \mathbf{k}_1 + \mathbf{k}_2, \qquad \mathbf{k} = \tfrac{1}{2}(\mathbf{k}_1 - \mathbf{k}_2). \tag{5.41}$$

The total momentum $\mathbf{K}$ is zero in the center-of-mass frame. If we know the force between the particles (for $\mu^+\mu^-$, it is just the Coulomb force), we can

---

[‡]Reviews of this subject can be found in Bodwin, Yennie, and Gregorio, *Rev. Mod. Phys.* **57**, 723 (1985), and in Sapirstein and Yennie, in Kinoshita (1990).

solve the nonrelativistic Schrödinger equation to find the Schrödinger wave-function, $\psi(\mathbf{r})$. The bound state is just a linear superposition of free states of definite $\mathbf{r}$ or $\mathbf{k}$, weighted by this wavefunction. For our purposes it is more convenient to build this superposition in momentum space, using the Fourier transform of $\psi(\mathbf{r})$:

$$\widetilde{\psi}(\mathbf{k}) = \int d^3x\, e^{i\mathbf{k}\cdot\mathbf{r}}\psi(\mathbf{r}); \qquad \int \frac{d^3k}{(2\pi)^3} \left|\widetilde{\psi}(\mathbf{k})\right|^2 = 1. \qquad (5.42)$$

If $\psi(\mathbf{r})$ is normalized conventionally, $\widetilde{\psi}(\mathbf{k})$ gives the amplitude for finding a particular value of $\mathbf{k}$. An explicit expression for a bound state with mass $M \approx 2m$, momentum $\mathbf{K} = 0$, and spin 1 oriented up is then

$$|B\rangle = \sqrt{2M} \int \frac{d^3k}{(2\pi)^3}\, \widetilde{\psi}(\mathbf{k})\, \frac{1}{\sqrt{2m}}\, \frac{1}{\sqrt{2m}}\, |\mathbf{k} \uparrow, -\mathbf{k} \uparrow\rangle. \qquad (5.43)$$

The factors of $(1/\sqrt{2m})$ convert our relativistically normalized free-particle states so that their integral with $\widetilde{\psi}(\mathbf{k})$ is a state of norm 1. (The factors should involve $\sqrt{2E_{\pm\mathbf{k}}}$, but for a nonrelativistic bound state, $|\mathbf{k}| \ll m$.) The outside factor of $\sqrt{2M}$ converts back to the relativistic normalization assumed by our formula for cross sections. These normalization factors could easily be modified to describe a bound state with nonzero total momentum $\mathbf{K}$.

Given this expression for the bound state, we can immediately write down the amplitude for its production:

$$\mathcal{M}(\uparrow\uparrow \rightarrow B) = \sqrt{2M} \int \frac{d^3k}{(2\pi)^3}\, \widetilde{\psi}^*(\mathbf{k})\, \frac{1}{\sqrt{2m}}\, \frac{1}{\sqrt{2m}}\, \mathcal{M}(\uparrow\uparrow \rightarrow \mathbf{k}\uparrow, -\mathbf{k}\uparrow). \ (5.44)$$

Since the free-state amplitude from (5.39) is independent of the momenta of the muons, the integral over $\mathbf{k}$ gives $\psi^*(0)$, the position-space wavefunction evaluated at the origin. It is quite natural that the amplitude for creation of a two-particle state from a pointlike virtual photon should be proportional to the value of the wavefunction at zero separation. Assembling the pieces, we find that the amplitude is simply

$$\mathcal{M}(\uparrow\uparrow \rightarrow B) = \sqrt{\frac{2}{M}}(-2e^2)\psi^*(0). \qquad (5.45)$$

In a moment we will compute the cross section from this amplitude. First, however, let us generalize this discussion to treat bound states with more general spin configurations. The analysis leading up to (5.37) will cast any S-matrix element for the production of nonrelativistic fermions with momenta $\mathbf{k}$ and $-\mathbf{k}$ into the form of a spin matrix element

$$i\mathcal{M}(\text{something} \rightarrow \mathbf{k}, \mathbf{k}') = \xi^\dagger [\Gamma(\mathbf{k})]\xi', \qquad (5.46)$$

where $\Gamma(\mathbf{k})$ is some 2×2 matrix. We now must replace the spinors with a nor-malized spin wavefunction for the bound state. In the example just completed, we replaced

$$\xi'\xi^\dagger \to \begin{pmatrix} 0 \\ 1 \end{pmatrix} \begin{pmatrix} 1 & 0 \end{pmatrix} = \begin{pmatrix} 0 & 0 \\ 1 & 0 \end{pmatrix}. \tag{5.47}$$

More generally, a spin-1 state is obtained by the replacement

$$\xi'\xi^\dagger \to \frac{1}{\sqrt{2}} \mathbf{n}^* \cdot \boldsymbol{\sigma}, \tag{5.48}$$

where $\mathbf{n}$ is a unit vector. Choosing $\mathbf{n} = (\hat{x} + i\hat{y})/\sqrt{2}$ gives back (5.47), while the choices $\mathbf{n} = (\hat{x} - i\hat{y})/\sqrt{2}$ and $\mathbf{n} = \hat{z}$ give the other two spin-1 states $\downarrow\downarrow$ and $(\uparrow\downarrow + \downarrow\uparrow)/\sqrt{2}$. (The relative minus sign in (5.48) for this last case comes from the rule (3.135) for the flipped spin.) Similarly, the spin-zero state $(\uparrow\downarrow - \downarrow\uparrow)/\sqrt{2}$ is given by the replacement

$$\xi'\xi^\dagger \to \frac{1}{\sqrt{2}} \mathbf{1}, \tag{5.49}$$

involving the 2×2 unit matrix. With these rules, we can convert an S-matrix element of the form (5.46) quite generally into an S-matrix element for production of a bound state at rest:

$$i\mathcal{M}(\text{something} \to B) = \sqrt{\frac{2}{M}} \int \frac{d^3k}{(2\pi)^3} \tilde{\psi}^*(\mathbf{k}) \, \text{tr}\left(\frac{\mathbf{n}^* \cdot \boldsymbol{\sigma}}{\sqrt{2}} \Gamma(\mathbf{k}) \right), \tag{5.50}$$

where the trace is taken over 2-component spinor indices. For a spin-0 bound state, replace $\mathbf{n} \cdot \boldsymbol{\sigma}$ by the unit matrix.

Vector Meson Production and Decay

Equation (5.45) can be straightforwardly converted into a cross section for production of $\mu^+\mu^-$ bound states in e^+e^- annihilation. To make it easier to extract all the physics in this equation, let us introduce polarization vectors for the initial and final spin configurations: $\boldsymbol{\epsilon}_+ = (\hat{x} + i\hat{y})/\sqrt{2}$, from Eq. (5.29), and $\mathbf{n}$, from Eq. (5.48). Then (5.45) can be rewritten in a more invariant form as

$$\mathcal{M}(e_R^- e_L^+ \to B) = \sqrt{\frac{2}{M}} (-2e^2) (\mathbf{n}^* \cdot \boldsymbol{\epsilon}_+) \psi^*(0). \tag{5.51}$$

The bound state spin polarization $\mathbf{n}$ is projected parallel to $\boldsymbol{\epsilon}_+$. Note that if the electrons are initially unpolarized, the cross section for production of B will involve the polarization average

$$\frac{1}{4}\left(|\mathbf{n}^* \cdot \boldsymbol{\epsilon}_+|^2 + |\mathbf{n}^* \cdot \boldsymbol{\epsilon}_-|^2 \right) = \frac{1}{4}\left(|n^x|^2 + |n^y|^2 \right). \tag{5.52}$$

Thus, the bound states produced will still be preferentially polarized along the e^+e^- collision axis.

Assuming an unpolarized electron beam, and summing (5.52) over the three possible directions of **n**, we find the following expression for the total cross section for production of the bound state:

$$\sigma(e^+e^- \rightarrow B) = \frac{1}{2}\frac{1}{2m}\frac{1}{2m}\int \frac{d^3K}{(2\pi)^3}\frac{1}{2E_{\mathbf{K}}}(2\pi)^4\delta^{(4)}(p+p'-K)\cdot\frac{2}{M}(4e^4)\frac{1}{2}|\psi(0)|^2.$$
(5.53)

Notice that the 1-body phase space integral can remove only three of the four delta functions. It is conventional to rewrite the last delta function using $\delta(P^0 - K^0) = 2K^0\delta(P^2 - K^2)$. Then

$$\sigma(e^+e^- \rightarrow B) = 64\pi^3\alpha^2\frac{|\psi(0)|^2}{M^3}\delta(E_{\text{cm}}^2 - M^2).$$
(5.54)

The last delta function enforces the constraint that the total center-of-mass energy must equal the bound-state mass; thus, the bound state is produced as a resonance in e^+e^- annihilation. If the bound state has a finite lifetime, this delta function will be broadened into a resonance peak. In practice, the intrinsic spread of the e^+e^- beam energy is often a more important broadening mechanism. In either case, (5.54) correctly predicts the area under the resonance peak.

If the bound state B can be produced from e^+e^-, it can also annihilate back to e^+e^-, or to any other sufficiently light lepton pair. According to (4.86), the total width for this decay mode is given by

$$\Gamma(B \rightarrow e^+e^-) = \frac{1}{2M}\int d\Pi_2\,|\mathcal{M}|^2,$$
(5.55)

where $\mathcal{M}$ is just the complex conjugate of the matrix element (5.51) we used to compute B production. Thus

$$\Gamma = \frac{1}{2M}\int\left(\frac{1}{8\pi}\frac{d\cos\theta}{2}\right)\frac{8e^4}{M}|\psi(0)|^2(|\mathbf{n}\cdot\boldsymbol{\epsilon}|^2 + |\mathbf{n}\cdot\boldsymbol{\epsilon}^*|^2).$$
(5.56)

This formula is summed over the possible final electron polarization states. It is easiest to evaluate by averaging over the three possible values of **n**. We thus obtain

$$\Gamma(B \rightarrow e^+e^-) = \frac{16\pi\alpha^2}{3}\frac{|\psi(0)|^2}{M^2}.$$
(5.57)

The formula for the decay width of B is very similar to that for the production cross section, and this is no surprise: Both calculations involve the square of the same matrix element, summed over initial and final polarizations. The two calculations differed only in how we formed the polarization averages, and in the phase-space factors. By this logic, the relation we have found between the two quantities,

$$\sigma(e^+e^- \rightarrow B) = 4\pi^2\cdot\frac{3\Gamma(B \rightarrow e^+e^-)}{M}\cdot\delta(E_{\text{cm}}^2 - M^2),$$
(5.58)

is very general and completely independent of the details of the matrix element computation. The factor 3 in (5.58) came from the orientation average for $\mathbf{n}$; for a spin-J bound state, this factor would be $(2J + 1)$.

The most famous application of this formalism is to bound states not of muons but of quarks: *quarkonium*. We saw the experimental evidence for $q\bar{q}$ bound states (the J/ψ and Υ, for example) in Fig. 5.3. (The resonance peaks are much too high and too narrow to show in the figure, but their sizes have been carefully measured.) Equations (5.54) and (5.57) must be multiplied by a color factor of 3 to give the production cross section and decay width for a spin-1 $q\bar{q}$ bound state. The value $\psi(0)$ of the $q\bar{q}$ wavefunction at the origin cannot be computed from first principles, but can be estimated from a nonrelativistic model of the $q\bar{q}$ spectrum with a phenomenologically chosen potential. Alternatively, we can use the formula

$$\Gamma(B(q\bar{q}) \to e^+e^-) = 16\pi\alpha^2 Q^2 \frac{|\psi(0)|^2}{M^2} \tag{5.59}$$

to measure $\psi(0)$ for a $q\bar{q}$ bound state. For example, the $1S$ spin-1 state of $s\bar{s}$, the ϕ meson, has an e^+e^- partial width of 1.4 keV and a mass of 1.02 GeV. From this we can infer $|\psi(0)|^2 = (1.2\,\text{fm})^{-3}$. This result is physically reasonable, since hadronic dimensions are typically ~ 1 fm.

Our viewpoint in this section has been quite different from that of earlier sections: Instead of computing everything from first principles, we have pieced together an approximate formula using a bit of quantum field theory and a bit of nonrelativistic quantum mechanics. In principle, however, we could treat bound states entirely in the relativistic formalism. Consider the annihilation of an e^+e^- pair to form a $\mu^+\mu^-$ bound state, which subsequently decays back into e^+e^-. In our present formalism we might represent this process by the diagram

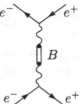

The net process is simply $e^+e^- \to e^+e^-$ (Bhabha scattering). What would happen if we tried to compute the Bhabha scattering cross section directly in QED perturbation theory? Obviously there is no $\mu^+\mu^-$ contribution in the tree-level diagrams:

As we go to higher orders in the perturbation series, however, we find (among others) the following set of diagrams:

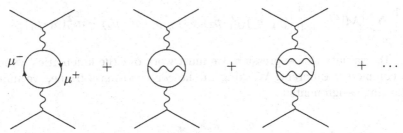

At most values of E_{cm}, these diagrams give only a small correction to the tree-level expression. But when E_{cm} is near the $\mu^+\mu^-$ threshold, the diagrams involving the exchange of photons within the muon loop contain the Coulomb interaction between the muons, and therefore become quite large. One must sum over all such diagrams, and it can be shown that this summation is equivalent to solving the nonrelativistic Schrödinger equation.* The final prediction is that the cross section contains a resonance peak, whose area is given by (5.54) and whose width is given by (5.57).

5.4 Crossing Symmetry

Electron-Muon Scattering

Now that we have completed our discussion of the process $e^+e^- \to \mu^+\mu^-$, let us consider a different but closely related QED process: electron-muon scattering, or $e^-\mu^- \to e^-\mu^-$. The lowest-order Feynman diagram is just the previous one turned on its side:

$$= \frac{ie^2}{q^2}\,\bar{u}(p_1')\gamma^\mu u(p_1)\,\bar{u}(p_2')\gamma_\mu u(p_2).$$

The relation between the processes $e^+e^- \to \mu^+\mu^-$ and $e^-\mu^- \to e^-\mu^-$ becomes clear when we compute the squared amplitude, averaged and summed over spins:

$$\frac{1}{4}\sum_{\text{spins}}|\mathcal{M}|^2 = \frac{e^4}{4q^4}\,\text{tr}\Big[(\slashed{p}'_1+m_e)\gamma^\mu(\slashed{p}_1+m_e)\gamma^\nu\Big]\,\text{tr}\Big[(\slashed{p}'_2+m_\mu)\gamma_\mu(\slashed{p}_2+m_\mu)\gamma_\nu\Big].$$

This is exactly the same as our result (5.4) for $e^+e^- \to \mu^+\mu^-$, with the replacements

$$p \to p_1, \qquad p' \to -p_1', \qquad k \to p_2', \qquad k' \to -p_2. \qquad (5.60)$$

---

*This analysis is carried out in Berestetskii, Lifshitz, and Pitaevskii (1982).

So instead of evaluating the traces from scratch, we can just make the same replacements in our previous result, Eq. (5.10). Setting $m_e = 0$, we find

$$\frac{1}{4} \sum_{\text{spins}} |\mathcal{M}|^2 = \frac{8e^4}{q^4} \left[(p_1 \cdot p_2')(p_1' \cdot p_2) + (p_1 \cdot p_2)(p_1' \cdot p_2') - m_\mu^2 (p_1 \cdot p_1') \right]. \quad (5.61)$$

To evaluate this expression, we must work out the kinematics, which will be completely different. Working in the center-of-mass frame, we make the following assignments:

The combinations we need are

$$p_1 \cdot p_2 = p_1' \cdot p_2' = k(E + k); \qquad p_1' \cdot p_2 = p_1 \cdot p_2' = k(E + k \cos \theta);$$

$$p_1 \cdot p_1' = k^2 (1 - \cos \theta); \qquad q^2 = -2p_1 \cdot p_1' = -2k^2 (1 - \cos \theta).$$

Our expression for the squared matrix element now becomes

$$\frac{1}{4} \sum_{\text{spins}} |\mathcal{M}|^2 = \frac{2e^4}{k^2 (1 - \cos \theta)^2} \left((E+k)^2 + (E+k \cos \theta)^2 - m_\mu^2 (1 - \cos \theta) \right). \quad (5.62)$$

To find the cross section from this expression, we use Eq. (4.84), which in the case where one particle is massless takes the simple form

$$\left(\frac{d\sigma}{d\Omega} \right)_{\text{CM}} = \frac{|\mathcal{M}|^2}{64\pi^2 (E + k)^2}. \quad (5.63)$$

Thus we have our result for unpolarized electron-muon scattering in the center-of-mass frame:

$$\frac{d\sigma}{d\Omega} = \frac{\alpha^2}{2k^2 (E+k)^2 (1 - \cos \theta)^2} \left((E+k)^2 + (E+k \cos \theta)^2 - m_\mu^2 (1 - \cos \theta) \right), \quad (5.64)$$

where $k = \sqrt{E^2 - m_\mu^2}$. In the high-energy limit where we can set $m_\mu = 0$, the differential cross section becomes

$$\frac{d\sigma}{d\Omega} = \frac{\alpha^2}{2E_{\text{cm}}^2 (1 - \cos \theta)^2} \left(4 + (1 + \cos \theta)^2 \right). \quad (5.65)$$

Note the singular behavior

$$\frac{d\sigma}{d\Omega} \propto \frac{1}{\theta^4} \qquad \text{as } \theta \to 0 \qquad (5.66)$$

of formulae (5.64) and (5.65). This singularity is the same as in the Rutherford formula (Problem 4.4). Such behavior is always present in Coulomb scattering; it arises from the nearly on-shell (that is, $q^2 \approx 0$) virtual photon.

Crossing Symmetry

The trick we made use of here, namely the relation between the two processes $e^+ e^- \to \mu^+ \mu^-$ and $e^- \mu^- \to e^- \mu^-$, is our first example of a type of relation known as *crossing symmetry*. In general, the S-matrix for any process involving a particle with momentum p in the initial state is equal to the S-matrix for an otherwise identical process but with an antiparticle of momentum $k = -p$ in the final state. That is,

$$\mathcal{M}\big(\phi(p) + \cdots \to \cdots\big) = \mathcal{M}\big(\cdots \to \cdots + \bar{\phi}(k)\big), \qquad (5.67)$$

where $\bar{\phi}$ is the antiparticle of ϕ and $k = -p$. (Note that there is no value of p for which p and k are both physically allowed, since the particle must have $p^0 > 0$ and the antiparticle must have $k^0 > 0$. So technically, we should say that either amplitude can be obtained from the other by analytic continuation.)

Relation (5.67) follows directly from the Feynman rules. The diagrams that contribute to the two amplitudes fall into a natural one-to-one correspondence, where corresponding diagrams differ only by changing the incoming ϕ into the outgoing $\bar{\phi}$. A typical pair of diagrams looks like this:

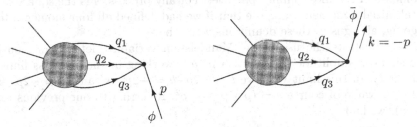

In the first diagram, the momenta q_i coming into the vertex from the rest of the diagram must add up to $-p$, while in the second diagram they must add up to k. Thus the two diagrams are equal, except for any possible difference in the external leg factors, if $p = -k$. If ϕ is a spin-zero boson, there is no external leg factor, so the identity is proved. If ϕ is a fermion, the analysis becomes more subtle, since the relation depends on the relative phase convention for the external spinors u and v. If we simply replace p by $-k$ in the fermion polarization sum, we find

$$\sum u(p)\bar{u}(p) = \not{p} + m = -(\not{k} - m) = -\sum v(k)\bar{v}(k). \qquad (5.68)$$

The minus sign can be compensated by changing our phase convention for $v(k)$. In practice, it is easiest to cancel by hand one minus sign for each crossed fermion. With appropriate conventions for the spinors $u(p)$ and $v(k)$, it is possible to prove the identity (5.67) without spin-averaging.

Mandelstam Variables

It is often useful to express scattering amplitudes in terms of variables that make it easy to apply crossing relations. For 2-body → 2-body processes, this can be done as follows. Label the four external momenta as

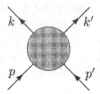

We now define three new quantities, the *Mandelstam variables*:

$$s = (p + p')^2 = (k + k')^2;$$
$$t = (k - p)^2 = (k' - p')^2; \qquad (5.69)$$
$$u = (k' - p)^2 = (k - p')^2.$$

The definitions of t and u appear to be interchangeable (by renaming $k \to k'$); it is conventional to define t as the squared difference of the initial and final momenta of the most similar particles. For any process, s is the square of the total initial 4-momentum. Note that if we had defined all four momenta to be ingoing, all signs in these definitions would be $+$.

To illustrate the use of the Mandelstam variables, let us first consider the squared amplitude for $e^+ e^- \to \mu^+ \mu^-$, working in the massless limit for simplicity. In this limit we have $t = -2p \cdot k = -2p' \cdot k'$ and $u = -2p \cdot k' = -2p' \cdot k$, while of course $s = (p + p')^2 = q^2$. Referring to our previous result (5.10), we find

$$\frac{1}{4} \sum_{\text{spins}} |\mathcal{M}|^2 = \frac{8e^4}{s^2} \left[\left(\frac{t}{2} \right)^2 + \left(\frac{u}{2} \right)^2 \right]. \qquad (5.70)$$

To convert to the process $e^- \mu^- \to e^- \mu^-$, we turn the diagram on its side and make use of the crossing relations, which become quite simple in terms of Mandelstam variables. For example, the crossing relations tell us to change the sign of p', the positron momentum, and reinterpret it as the momentum of the outgoing electron. Therefore $s = (p + p')^2$ becomes what we would now call t, the difference of the outgoing and incoming electron momenta. Similarly, t becomes s, while u remains unchanged. Thus for $e^- \mu^- \to e^- \mu^-$,

we can immediately write down

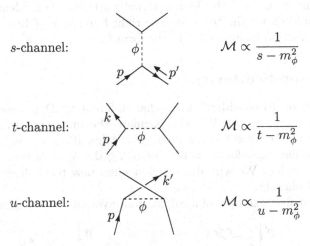

$$\frac{1}{4}\sum_{\text{spins}}|\mathcal{M}|^2 = \frac{8e^4}{t^2}\left[\left(\frac{s}{2}\right)^2 + \left(\frac{u}{2}\right)^2\right]. \qquad (5.71)$$

You can easily check that this agrees with (5.61) in the massless limit. Note that while (5.70) and (5.71) look quite similar, they are physically very different: The denominator of the first is just $s^2 = E_{\text{cm}}^4$, but that of the second involves t, which depends on angles and goes to zero as $\theta \to 0$.

When a 2-body $\to$ 2-body diagram contains only one virtual particle, it is conventional to describe that particle as being in a certain "channel". The channel can be read from the form of the Feynman diagram, and each channel leads to a characteristic angular dependence of the cross section:

s-channel: $\mathcal{M} \propto \dfrac{1}{s - m_\phi^2}$

t-channel: $\mathcal{M} \propto \dfrac{1}{t - m_\phi^2}$

u-channel: $\mathcal{M} \propto \dfrac{1}{u - m_\phi^2}$

In many cases, a single process will receive contributions from more than one channel; these must be added coherently. For example, the amplitude for *Bhabha scattering*, $e^+e^- \to e^+e^-$, is the sum of s- and t-channel diagrams; *Møller scattering*, $e^-e^- \to e^-e^-$, involves t- and u-channel diagrams.

To get a better feel for s, t, and u, let us evaluate them explicitly in the center-of-mass frame for particles all of mass m. The kinematics is as usual:

$k = (E, \mathbf{p})$

$p = (E, p\hat{z})$

θ

$p' = (E, -p\hat{z})$

$k' = (E, -\mathbf{p})$

Thus the Mandelstam variables are

$$s = (p + p')^2 = (2E)^2 = E_{cm}^2;$$

$$t = (k - p)^2 = -p^2 \sin^2 \theta - p^2(\cos \theta - 1)^2 = -2p^2(1 - \cos \theta); \qquad (5.72)$$

$$u = (k' - p)^2 = -p^2 \sin^2 \theta - p^2(\cos \theta + 1)^2 = -2p^2(1 + \cos \theta).$$

Thus we see that $t \to 0$ as $\theta \to 0$, while $u \to 0$ as $\theta \to \pi$. (When the masses are not all equal, the limiting values of t and u will shift slightly.)

Note from (5.72) that when all four particles have mass m, the sum of the Mandelstam variables is $s + t + u = 4E^2 - 4p^2 = 4m^2$. This is a special case of a more general relation, which is often quite useful:

$$s + t + u = \sum_{i=1}^{4} m_i^2, \qquad (5.73)$$

where the sum runs over the four external particles. This identity is easy to prove by adding up the terms on the right-hand side of Eqs. (5.69), and applying momentum conservation in the form $(p + p' - k - k')^2 = 0$.

5.5 Compton Scattering

We now move on to consider a somewhat different QED process: *Compton scattering*, or $e^- \gamma \to e^- \gamma$. We will calculate the unpolarized cross section for this reaction, to lowest order in α. The calculation will employ all the machinery we have developed so far, including the Mandelstam variables of the previous section. We will also develop some new technology for dealing with external photons.

This is our first example of a calculation involving two diagrams:

As usual, the Feynman rules tell us exactly how to write down an expression for $\mathcal{M}$. Note that since the fermion portions of the two diagrams are identical, there is no relative minus sign between the two terms. Using $\epsilon_\nu(k)$ and $\epsilon_\mu^*(k')$ to denote the polarization vectors of the initial and final photons, we have

$$i\mathcal{M} = \bar{u}(p')(-ie\gamma^\mu)\epsilon_\mu^*(k')\frac{i(\not{p} + \not{k} + m)}{(p + k)^2 - m^2}(-ie\gamma^\nu)\epsilon_\nu(k)u(p)$$

$$+ \bar{u}(p')(-ie\gamma^\nu)\epsilon_\nu(k)\frac{i(\not{p} - \not{k}' + m)}{(p - k')^2 - m^2}(-ie\gamma^\mu)\epsilon_\mu^*(k')u(p)$$

$$= -ie^2 \epsilon_\mu^*(k')\epsilon_\nu(k)\, \bar{u}(p')\left[\frac{\gamma^\mu(\not{p}+\not{k}+m)\gamma^\nu}{(p+k)^2-m^2} + \frac{\gamma^\nu(\not{p}-\not{k}'+m)\gamma^\mu}{(p-k')^2-m^2}\right]u(p).$$

We can make a few simplifications before squaring this expression. Since $p^2 = m^2$ and $k^2 = 0$, the denominators of the propagators are

$$(p+k)^2 - m^2 = 2p\cdot k \qquad \text{and} \qquad (p-k')^2 - m^2 = -2p\cdot k'.$$

To simplify the numerators, we use a bit of Dirac algebra:

$$(\not{p}+m)\gamma^\nu u(p) = (2p^\nu - \gamma^\nu\not{p} + \gamma^\nu m)u(p)$$
$$= 2p^\nu u(p) - \gamma^\nu(\not{p} - m)u(p)$$
$$= 2p^\nu u(p).$$

Using this trick on the numerator of each propagator, we obtain

$$i\mathcal{M} = -ie^2 \epsilon_\mu^*(k')\epsilon_\nu(k)\, \bar{u}(p')\left[\frac{\gamma^\mu\not{k}\gamma^\nu + 2\gamma^\mu p^\nu}{2p\cdot k} + \frac{-\gamma^\nu\not{k}'\gamma^\mu + 2\gamma^\nu p^\mu}{-2p\cdot k'}\right]u(p). \quad (5.74)$$

Photon Polarization Sums

The next step in the calculation will be to square this expression for $\mathcal{M}$ and sum (or average) over electron and photon polarization states. The sum over electron polarizations can be performed as before, using the identity $\Sigma u(p)\bar{u}(p) = \not{p} + m$. Fortunately, there is a similar trick for summing over photon polarization vectors. The correct prescription is to make the replacement

$$\sum_{\text{polarizations}} \epsilon_\mu^*\epsilon_\nu \longrightarrow -g_{\mu\nu}. \qquad (5.75)$$

The arrow indicates that this is not an actual equality. Nevertheless, the replacement is valid as long as both sides are dotted into the rest of the expression for a QED amplitude $\mathcal{M}$.

To derive this formula, let us consider an arbitrary QED process involving an external photon with momentum k:

$$= i\mathcal{M}(k) \equiv i\mathcal{M}^\mu(k)\epsilon_\mu^*(k). \qquad (5.76)$$

Since the amplitude always contains $\epsilon_\mu^*(k)$, we have extracted this factor and defined $\mathcal{M}^\mu(k)$ to be the rest of the amplitude $\mathcal{M}$. The cross section will be proportional to

$$\sum_\epsilon \left|\epsilon_\mu^*(k)\mathcal{M}^\mu(k)\right|^2 = \sum_\epsilon \epsilon_\mu^*\epsilon_\nu\, \mathcal{M}^\mu(k)\mathcal{M}^{\nu*}(k).$$

For simplicity, we orient k in the 3-direction: $k^\mu = (k, 0, 0, k)$. Then the two transverse polarization vectors, over which we are summing, can be chosen to be

$$\epsilon_1^\mu = (0, 1, 0, 0); \qquad \epsilon_2^\mu = (0, 0, 1, 0).$$

With these conventions, we have

$$\sum_\epsilon \left| \epsilon_\mu^*(k) \mathcal{M}^\mu(k) \right|^2 = \left| \mathcal{M}^1(k) \right|^2 + \left| \mathcal{M}^2(k) \right|^2. \tag{5.77}$$

Now recall from Chapter 4 that external photons are created by the interaction term $\int d^4x \, ej^\mu A_\mu$, where $j^\mu = \bar\psi \gamma^\mu \psi$ is the Dirac vector current. Therefore we expect $\mathcal{M}^\mu(k)$ to be given by a matrix element of the Heisenberg field j^μ:

$$\mathcal{M}^\mu(k) = \int d^4x \, e^{ik \cdot x} \, \langle f | j^\mu(x) | i \rangle, \tag{5.78}$$

where the initial and final states include all particles except the photon in question.

From the classical equations of motion, we know that the current j^μ is conserved: $\partial_\mu j^\mu(x) = 0$. Provided that this property still holds in the quantum theory, we can dot k_μ into (5.78) to obtain

$$k_\mu \mathcal{M}^\mu(k) = 0. \tag{5.79}$$

The amplitude $\mathcal{M}$ vanishes when the polarization vector $\epsilon_\mu(k)$ is replaced by k_μ. This famous relation is known as the *Ward identity*. It is essentially a statement of current conservation, which is a consequence of the gauge symmetry (4.6) of QED. We will give a formal proof of the Ward identity in Section 7.4, and discuss a number of subtle points skimmed over in this quick "derivation".

It is useful to check explicitly that the Compton amplitude given in (5.74) obeys the Ward identity. To do this, replace $\epsilon_\nu(k)$ by k_ν or $\epsilon_\mu^*(k')$ by k_μ', and manipulate the Dirac matrix products. In either case (after a bit of algebra) the terms from the two diagrams cancel each other to give zero.

Returning to our derivation of the polarization sum formula (5.75), we note that for $k^\mu = (k, 0, 0, k)$, the Ward identity takes the form

$$k \mathcal{M}^0(k) - k \mathcal{M}^3(k) = 0. \tag{5.80}$$

Thus $\mathcal{M}^0 = \mathcal{M}^3$, and we have

$$\sum_\epsilon \epsilon_\mu^* \epsilon_\nu \, \mathcal{M}^\mu(k) \mathcal{M}^{\nu*}(k) = |\mathcal{M}^1|^2 + |\mathcal{M}^2|^2$$

$$= |\mathcal{M}^1|^2 + |\mathcal{M}^2|^2 + |\mathcal{M}^3|^2 - |\mathcal{M}^0|^2$$

$$= -g_{\mu\nu} \, \mathcal{M}^\mu(k) \mathcal{M}^{\nu*}(k).$$

That is, we may sum over external photon polarizations by replacing $\sum \epsilon_\mu^* \epsilon_\nu$ with $-g_{\mu\nu}$.

Note that this proves (pending our general proof of the Ward identity) that the unphysical timelike and longitudinal photons can be consistently omitted from QED calculations, since in any event the squared amplitudes for producing these states cancel to give zero total probability. The negative norm of the timelike photon state, a property that troubled us in the discussion after Eq. (4.132), plays an essential role in this cancellation.

The Klein-Nishina Formula

The rest of the computation of the Compton scattering cross section is straightforward, although it helps to be somewhat organized. We want to average the squared amplitude over the initial electron and photon polarizations, and sum over the final electron and photon polarizations. Starting with expression (5.74) for $\mathcal{M}$, we find

$$\frac{1}{4}\sum_{\text{spins}} |\mathcal{M}|^2 = \frac{e^4}{4} g_{\mu\rho}\, g_{\nu\sigma} \cdot \text{tr}\left\{ (\not{p}'+m)\left[\frac{\gamma^\mu \not{k}\gamma^\nu + 2\gamma^\mu p^\nu}{2p\cdot k} + \frac{\gamma^\nu \not{k}'\gamma^\mu - 2\gamma^\nu p^\mu}{2p\cdot k'}\right]\right.$$

$$\left. \cdot (\not{p}+m)\left[\frac{\gamma^\sigma \not{k}\gamma^\rho + 2\gamma^\rho p^\sigma}{2p\cdot k} + \frac{\gamma^\rho \not{k}'\gamma^\sigma - 2\gamma^\sigma p^\rho}{2p\cdot k'}\right]\right\}$$

$$= \frac{e^4}{4}\left[\frac{\mathbf{I}}{(2p\cdot k)^2} + \frac{\mathbf{II}}{(2p\cdot k)(2p\cdot k')} + \frac{\mathbf{III}}{(2p\cdot k')(2p\cdot k)} + \frac{\mathbf{IV}}{(2p\cdot k')^2}\right], \quad (5.81)$$

where **I**, **II**, **III**, and **IV** are complicated traces. Note that **IV** is the same as **I** if we replace k with $-k'$. Also, since we can reverse the order of the γ matrices inside a trace (Eq. (5.7)), we see that **II = III**. Thus we must work only to compute **I** and **II**.

The first of the traces is

$$\mathbf{I} = \text{tr}\left[(\not{p}'+m)(\gamma^\mu \not{k}\gamma^\nu + 2\gamma^\mu p^\nu)(\not{p}+m)(\gamma_\nu \not{k}\gamma_\mu + 2\gamma_\mu p_\nu)\right].$$

There are 16 terms inside the trace, but half contain an odd number of γ matrices and therefore vanish. We must now evaluate the other eight terms, one at a time. For example,

$$\text{tr}\left[\not{p}'\gamma^\mu \not{k}\gamma^\nu \not{p}\gamma_\nu \not{k}\gamma_\mu\right] = \text{tr}\left[(-2\not{p}')\not{k}(-2\not{p})\not{k}\right]$$

$$= \text{tr}\left[4\not{p}'\not{k}(2p\cdot k - \not{k}\not{p})\right]$$

$$= 8p\cdot k \,\text{tr}[\not{p}'\not{k}]$$

$$= 32(p\cdot k)(p'\cdot k).$$

By similar use of the contraction identities (5.8) and (5.9), and other Dirac algebra such as $\not{p}\not{p} = p^2 = m^2$, each term in **I** can be reduced to a trace of no more than two γ matrices. When the smoke clears, we find

$$\mathbf{I} = 16\left(4m^4 - 2m^2 p\cdot p' + 4m^2 p\cdot k - 2m^2 p'\cdot k + 2(p\cdot k)(p'\cdot k)\right). \quad (5.82)$$

Although it is not obvious, this expression can be simplified further. To see how, introduce the Mandelstam variables:

$$s = (p + k)^2 = 2p \cdot k + m^2 = 2p' \cdot k' + m^2;$$

$$t = (p' - p)^2 = -2p \cdot p' + 2m^2 = -2k \cdot k'; \tag{5.83}$$

$$u = (k' - p)^2 = -2k' \cdot p + m^2 = -2k \cdot p' + m^2.$$

Recall from (5.73) that momentum conservation implies $s + t + u = 2m^2$. Writing everything in terms of s, t, and u, and using this identity, we eventually obtain

$$\mathbf{I} = 16\big(2m^4 + m^2(s - m^2) - \tfrac{1}{2}(s - m^2)(u - m^2)\big). \tag{5.84}$$

Sending $k \leftrightarrow -k'$, we can immediately write

$$\mathbf{IV} = 16\big(2m^4 + m^2(u - m^2) - \tfrac{1}{2}(s - m^2)(u - m^2)\big). \tag{5.85}$$

Evaluating the traces in the numerators $\mathbf{II}$ and $\mathbf{III}$ requires about the same amount of work as we have just done. The answer is

$$\mathbf{II} = \mathbf{III} = -8\big(4m^4 + m^2(s - m^2) + m^2(u - m^2)\big). \tag{5.86}$$

Putting together the pieces of the squared matrix element (5.81), and rewriting s and u in terms of $p \cdot k$ and $p \cdot k'$, we finally obtain

$$\frac{1}{4} \sum_{\text{spins}} |\mathcal{M}|^2 = 2e^4 \left[\frac{p \cdot k'}{p \cdot k} + \frac{p \cdot k}{p \cdot k'} + 2m^2 \Big(\frac{1}{p \cdot k} - \frac{1}{p \cdot k'} \Big) + m^4 \Big(\frac{1}{p \cdot k} - \frac{1}{p \cdot k'} \Big)^2 \right]. \tag{5.87}$$

To turn this expression into a cross section we must decide on a frame of reference and draw a picture of the kinematics. Compton scattering is most often analyzed in the "lab" frame, in which the electron is initially at rest:

$$k' = (\omega', \omega' \sin \theta, 0, \omega' \cos \theta)$$

Before:

After:

$$k = (\omega, \omega \hat{z}) \qquad p = (m, \mathbf{0})$$

$$\theta$$

$$p' = (E', \mathbf{p}')$$

We will express the cross section in terms of ω and θ. We can find ω', the energy of the final photon, using the following trick:

$$m^2 = (p')^2 = (p + k - k')^2 = p^2 + 2p \cdot (k - k') - 2k \cdot k'$$

$$= m^2 + 2m(\omega - \omega') - 2\omega\omega'(1 - \cos \theta),$$

hence, $$\frac{1}{\omega'} - \frac{1}{\omega} = \frac{1}{m}(1 - \cos \theta). \tag{5.88}$$

The last line is Compton's formula for the shift in the photon wavelength. For our purposes, however, it is more useful to solve for ω':

$$\omega' = \frac{\omega}{1 + \dfrac{\omega}{m}(1 - \cos\theta)}. \tag{5.89}$$

The phase space integral in this frame is

$$\int d\Pi_2 = \int \frac{d^3k'}{(2\pi)^3} \frac{1}{2\omega'} \frac{d^3p'}{(2\pi)^3} \frac{1}{2E'} (2\pi)^4 \delta^{(4)}(k' + p' - k - p)$$

$$= \int \frac{(\omega')^2 d\omega' \, d\Omega}{(2\pi)^3} \frac{1}{4\omega' E'}$$

$$\times 2\pi \, \delta(\omega' + \sqrt{m^2 + \omega^2 + (\omega')^2 - 2\omega\omega'\cos\theta} - \omega - m)$$

$$= \int \frac{d\cos\theta}{2\pi} \frac{\omega'}{4E'} \frac{1}{\left|1 + \dfrac{\omega' - \omega\cos\theta}{E'}\right|}$$

$$= \frac{1}{8\pi} \int d\cos\theta \frac{\omega'}{m + \omega(1 - \cos\theta)}$$

$$= \frac{1}{8\pi} \int d\cos\theta \frac{(\omega')^2}{\omega m}. \tag{5.90}$$

Plugging everything into our general cross-section formula (4.79) and setting $|v_A - v_B| = 1$, we find

$$\frac{d\sigma}{d\cos\theta} = \frac{1}{2\omega} \frac{1}{2m} \cdot \frac{1}{8\pi} \frac{(\omega')^2}{\omega m} \cdot \left(\frac{1}{4}\sum_{\text{spins}} |\mathcal{M}|^2\right).$$

To evaluate $|\mathcal{M}|^2$, we replace $p \cdot k = m\omega$ and $p \cdot k' = m\omega'$ in (5.87). The shortest way to write the final result is

$$\frac{d\sigma}{d\cos\theta} = \frac{\pi\alpha^2}{m^2} \left(\frac{\omega'}{\omega}\right)^2 \left[\frac{\omega'}{\omega} + \frac{\omega}{\omega'} - \sin^2\theta\right], \tag{5.91}$$

where ω'/ω is given by (5.89). This is the (spin-averaged) *Klein-Nishina formula*, first derived in 1929.[†]

In the limit $\omega \to 0$ we see from (5.89) that $\omega'/\omega \to 1$, so the cross section becomes

$$\frac{d\sigma}{d\cos\theta} = \frac{\pi\alpha^2}{m^2}(1 + \cos^2\theta); \qquad \sigma_{\text{total}} = \frac{8\pi\alpha^2}{3m^2}. \tag{5.92}$$

This is the familiar Thomson cross section for scattering of classical electromagnetic radiation by a free electron.

[†] O. Klein and Y. Nishina, *Z. Physik*, **52**, 853 (1929).

High-Energy Behavior

To analyze the high-energy behavior of the Compton scattering cross section, it is easiest to work in the center-of-mass frame. We can easily construct the differential cross section in this frame from the invariant expression (5.87). The kinematics of the reaction now looks like this:

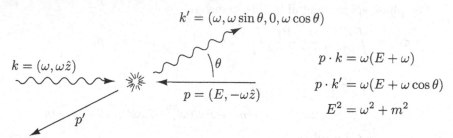

$$k' = (\omega, \omega \sin \theta, 0, \omega \cos \theta)$$

$$k = (\omega, \omega \hat{z})$$

$$p = (E, -\omega \hat{z})$$

$$p \cdot k = \omega(E + \omega)$$

$$p \cdot k' = \omega(E + \omega \cos \theta)$$

$$E^2 = \omega^2 + m^2$$

Plugging these values into (5.87), we see that for $\theta \approx \pi$, the term $p \cdot k / p \cdot k'$ becomes very large, while the other terms are all of $\mathcal{O}(1)$ or smaller. Thus for $E \gg m$ and $\theta \approx \pi$, we have

$$\frac{1}{4} \sum_{\text{spins}} |\mathcal{M}|^2 \approx 2e^4 \cdot \frac{p \cdot k}{p \cdot k'} = 2e^4 \cdot \frac{E + \omega}{E + \omega \cos \theta}. \tag{5.93}$$

The cross section in the CM frame is given by (4.84):

$$\begin{aligned}
\frac{d\sigma}{d \cos \theta} &= \frac{1}{2} \cdot \frac{1}{2E} \cdot \frac{1}{2\omega} \cdot \frac{\omega}{(2\pi)4(E + \omega)} \cdot \frac{2e^4(E + \omega)}{E + \omega \cos \theta} \\
&\approx \frac{2\pi \alpha^2}{2m^2 + s(1 + \cos \theta)}.
\end{aligned} \tag{5.94}$$

Notice that, since $s \gg m^2$, the denominator of (5.94) almost vanishes when the photon is emitted in the backward direction ($\theta \approx \pi$). In fact, the electron mass m could be neglected completely in this formula if it were not necessary to cut off this singularity. To integrate over $\cos \theta$, we can drop the electron mass term if we supply an equivalent cutoff near $\theta = \pi$. In this way, we can approximate the total Compton scattering cross section by

$$\int_{-1}^{1} d(\cos \theta) \frac{d\sigma}{d \cos \theta} \approx \frac{2\pi \alpha^2}{s} \int_{-1 + 2m^2/s}^{1} d(\cos \theta) \frac{1}{(1 + \cos \theta)}. \tag{5.95}$$

Thus, we find that the total cross section behaves at high energy as

$$\sigma_{\text{total}} = \frac{2\pi \alpha^2}{s} \log \left(\frac{s}{m^2} \right). \tag{5.96}$$

The main dependence α^2 / s follows from dimensional analysis. But the singularity associated with backward scattering of photons leads to an enhancement by an extra logarithm of the energy.

Let us try to understand the physics of this singularity. The singular term comes from the square of the u-channel diagram,

$$= -ie^2 \, \epsilon_\mu(k)\epsilon_\nu^*(k')\bar{u}(p')\gamma^\mu \frac{\not{p} - \not{k}' + m}{(p - k')^2 - m^2}\gamma^\nu u(p). \qquad (5.97)$$

The amplitude is large at $\theta \approx \pi$ because the denominator of the propagator is then small ($\sim m^2$) compared to s. To be more precise, define $\chi \equiv \pi - \theta$. We will be interested in values of χ that are somewhat larger than m/ω, but still small enough that we can approximate $1 - \cos\chi \approx \chi^2/2$. For χ in this range, the denominator is

$$(p - k')^2 - m^2 = -2p \cdot k' \approx -2\omega^2 \left(\frac{m^2}{2\omega^2} + 1 - \cos\chi\right) \approx -(\omega^2\chi^2 + m^2). \qquad (5.98)$$

This is small compared to s over a wide range of values for χ, hence the enhancement in the total cross section.

Looking back at (5.93), we see that for χ such that $m/\omega \ll \chi \ll 1$, the squared amplitude is proportional to $1/\chi^2$, and hence we expect $\mathcal{M} \propto 1/\chi$. But we have just seen that the denominator of $\mathcal{M}$ is proportional to χ^2, so there must be a compensating factor of χ in the numerator. We can understand the physical origin of that factor by looking at the amplitude for a particular set of electron and photon polarizations.

Suppose that the initial electron is right-handed. The dominant term of (5.97) comes from the term that involves $(\not{p} - \not{k}')$ in the numerator of the propagator. Since this term contains three γ-matrices in (5.97) between the $\bar{u}$ and the u, the final electron must also be right-handed. The amplitude is therefore

$$i\mathcal{M} = -ie^2\epsilon_\mu(k)\epsilon_\nu^*(k')u_R^\dagger(p')\sigma^\mu \frac{\bar{\sigma} \cdot (p - k')}{-(\omega^2\chi^2 + m^2)}\sigma^\nu u_R(p), \qquad (5.99)$$

where

$$u_R(p) = \sqrt{2E}\begin{pmatrix} 0 \\ 1 \end{pmatrix} \quad \text{and} \quad u_R(p') = \sqrt{2E}\begin{pmatrix} 1 \\ 0 \end{pmatrix}. \qquad (5.100)$$

If the initial photon is left-handed, with $\epsilon^\mu = (1/\sqrt{2})(0, 1, -i, 0)^\mu$, then

$$\sigma^\mu\epsilon_\mu(k) = \begin{pmatrix} 0 & 0 \\ -\sqrt{2} & 0 \end{pmatrix},$$

and the combination $u_R^\dagger(p')\sigma^\mu\epsilon_\mu(k)$ vanishes. The initial photon must therefore be right-handed. Similarly, the amplitude vanishes unless the final photon is right-handed. The kinematic situation for this set of polarizations is shown

Before:

After:

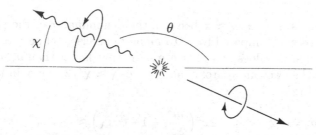

Figure 5.6. In the high-energy limit, the final photon is most likely to be emitted at backward angles. Since helicity is conserved, a unit of spin angular momentum is converted to orbital angular momentum.

in Fig. 5.6. Note that the total spin angular momentum of the final state is one unit less than that of the initial state.

Continuing with our calculation, let us consider the numerator of the propagator in (5.99). For χ in the range of interest, the dominant term is

$$-\sigma^1(p-k')^1 = \sigma^1 \cdot \omega\chi.$$

This is the factor of χ anticipated above. It indicates that the final state is a p-wave, as required by angular momentum conservation. Assembling all the pieces, we obtain

$$\mathcal{M}(e_R^-\gamma_R \to e_R^-\gamma_R) \approx e^2\sqrt{2E}\sqrt{2}\frac{\omega\chi}{(\omega^2\chi^2+m^2)}\sqrt{2E}\sqrt{2} \approx \frac{4e^2\chi}{\chi^2+m^2/\omega^2}.$$
(5.101)

We would find the same result in the case where all initial and final particles are left-handed.

Notice that for directly backward scattering, $\chi = 0$, the matrix element (5.101) vanishes due to the angular momentum zero in the numerator. Thus, at angles very close to backward, we should also take into account the mass term in the numerator of the propagator in (5.97). This term contains only two gamma matrices and so converts a right-handed electron into a left-handed electron. By an analysis similar to the one that led to Eq. (5.101), we can see that this amplitude is nonvanishing only when the initial photon is left-handed and the final photon is right-handed. Following this analysis in more detail, we find

$$\mathcal{M}(e_R^-\gamma_L \to e_L^-\gamma_R) \approx \frac{4e^2m/\omega}{\chi^2+m^2/\omega^2}.$$
(5.102)

The reaction with all four helicities reversed gives the same matrix element.

To compare this result to our previous calculations, we should add the contributions to the cross section from (5.101) and (5.102) and equal contributions for the reactions involving initial left-handed electrons, and divide by 4 to average over initial spins. The unpolarized differential cross section should then be

$$
\frac{d\sigma}{d\cos\theta} = \frac{1}{2}\frac{1}{2E}\frac{1}{2\omega}\frac{\omega}{(2\pi)\,4(E+\omega)}\left[\frac{8e^4\chi^2}{(\chi^2+m^2/\omega^2)^2} + \frac{8e^4 m^2/\omega^2}{(\chi^2+m^2/\omega^2)^2}\right]
$$

$$
= \frac{4\pi\alpha^2}{s(\chi^2+4m^2/s)}, \tag{5.103}
$$

which agrees precisely with Eq. (5.94).

The importance of the helicity-flip process (5.102) just at the kinematic endpoint has an interesting experimental consequence. Consider the process of *inverse* Compton scattering, a high-energy electron beam colliding with a low-energy photon beam (for example, a laser beam) to produce a high-energy photon beam. Let the electrons have energy E and the laser photons have energy ϖ, let the energy of the scattered photon be $E' = yE$, and assume for simplicity that $s = 4E\varpi \gg m^2$. Then the computation we have just done applies to this situation, with the highest energy photons resulting from scattering that is precisely backward in the center-of-mass frame. By computing $2k\cdot k'$ in the center-of-mass frame and in the lab frame, it is easy to show that the final photon energy is related to the center-of-mass scattering angle through

$$
y \approx \frac{1}{2}(1-\cos\theta) \approx 1 - \frac{\chi^2}{4}.
$$

Then Eq. (5.103) can be rewritten as a formula for the energy distribution of backscattered photons near the endpoint:

$$
\frac{d\sigma}{dy} = \frac{2\pi\alpha^2}{s((1-y)+m^2/s)^2}\left[(1-y) + \frac{m^2}{s}\right], \tag{5.104}
$$

where the first term in brackets corresponds to the helicity-conserving process and the second term to the helicity-flip process. Thus, for example, if a right-handed polarized laser beam is scattered from an unpolarized high-energy electron beam, most of the backscattered photons will be right-handed but the highest-energy photons will be left-handed. This effect can be used experimentally to measure the polarization of an electron beam or to create high-energy photon sources with adjustable energy distribution and polarization.

Pair Annihilation into Photons

We can still obtain one more result from the Compton-scattering amplitude. Consider the annihilation process

$$e^+ e^- \to 2\gamma,$$

given to lowest order by the diagrams

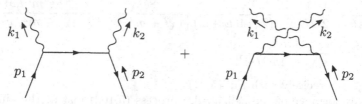

This process is related to Compton scattering by crossing symmetry; we can obtain the correct amplitude from the Compton amplitude by making the replacements

$$p \to p_1 \qquad p' \to -p_2 \qquad k \to -k_1 \qquad k' \to k_2.$$

Making these substitutions in (5.87), we find

$$\frac{1}{4} \sum_{\text{spins}} |\mathcal{M}|^2 = -2e^4 \left[\frac{p_1 \cdot k_2}{p_1 \cdot k_1} + \frac{p_1 \cdot k_1}{p_1 \cdot k_2} + 2m^2 \left(\frac{1}{p_1 \cdot k_1} + \frac{1}{p_1 \cdot k_2} \right) \right.$$
$$\left. - m^4 \left(\frac{1}{p_1 \cdot k_1} + \frac{1}{p_1 \cdot k_2} \right)^2 \right]. \tag{5.105}$$

The overall minus sign is the result of the crossing relation (5.68) and should be removed.

Now specialize to the center-of-mass frame. The kinematics is

$$k_1 = (E, E \sin\theta, 0, E \cos\theta)$$

$$e^- \qquad p_1 = (E, p\hat{z})$$

$$\theta$$

$$p_2 = (E, -p\hat{z}) \qquad e^+$$

$$k_2 = (E, -E \sin\theta, 0, -E \cos\theta)$$

A routine calculation yields the differential cross section,

$$\frac{d\sigma}{d\cos\theta} = \frac{2\pi\alpha^2}{s} \left(\frac{E}{p} \right) \left[\frac{E^2 + p^2 \cos^2\theta}{m^2 + p^2 \sin^2\theta} + \frac{2m^2}{m^2 + p^2 \sin^2\theta} - \frac{2m^4}{(m^2 + p^2 \sin^2\theta)^2} \right]. \tag{5.106}$$

In the high-energy limit, this becomes

$$\frac{d\sigma}{d\cos\theta} \xrightarrow[E \gg m]{} \frac{2\pi\alpha^2}{s} \left(\frac{1 + \cos^2\theta}{\sin^2\theta} \right), \tag{5.107}$$

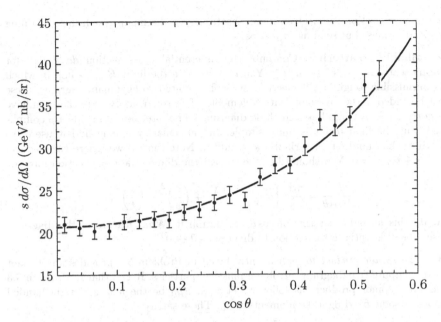

Figure 5.7. Angular dependence of the cross section for $e^+e^- \to 2\gamma$ at $E_{cm} = 29$ GeV, as measured by the HRS collaboration, M. Derrick, et. al., *Phys. Rev.* **D34**, 3286 (1986). The solid line is the lowest-order theoretical prediction, Eq. (5.107).

except when $\sin\theta$ is of order m/p or smaller. Note that since the two photons are identical, we count all possible final states by integrating only over $0 \le \theta \le \pi/2$. Thus the total cross section is computed as

$$\sigma_{\text{total}} = \int\limits_0^1 d(\cos\theta)\, \frac{d\sigma}{d\cos\theta}. \tag{5.108}$$

Figure 5.7 compares the asymptotic formula (5.107) for the differential cross section to measurements of e^+e^- annihilation into two photons at very high energy.

Problems

5.1   Coulomb scattering. Repeat the computation of Problem 4.4, part (c), this time using the full relativistic expression for the matrix element. You should find, for the spin-averaged cross section,

$$\frac{d\sigma}{d\Omega} = \frac{\alpha^2}{4|\mathbf{p}|^2\beta^2\sin^4(\theta/2)}\left(1 - \beta^2\sin^2\frac{\theta}{2}\right),$$

where $\mathbf{p}$ is the electron's 3-momentum and β is its velocity. This is the *Mott formula* for Coulomb scattering of relativistic electrons. Now derive it in a second way, by working out the cross section for electron-muon scattering, in the muon rest frame, retaining the electron mass but sending $m_\mu \to \infty$.

5.2   Bhabha scattering. Compute the differential cross section $d\sigma/d\cos\theta$ for Bhabha scattering, $e^+e^- \to e^+e^-$. You may work in the limit $E_{cm} \gg m_e$, in which it is permissible to ignore the electron mass. There are two Feynman diagrams; these must be added in the invariant matrix element before squaring. Be sure that you have the correct relative sign between these diagrams. The intermediate steps are complicated, but the final result is quite simple. In particular, you may find it useful to introduce the Mandelstam variables s, t, and u. Note that, if we ignore the electron mass, $s + t + u = 0$. You should be able to cast the differential cross section into the form

$$\frac{d\sigma}{d\cos\theta} = \frac{\pi\alpha^2}{s}\left[u^2\left(\frac{1}{s}+\frac{1}{t}\right)^2 + \left(\frac{t}{s}\right)^2 + \left(\frac{s}{t}\right)^2\right].$$

Rewrite this formula in terms of $\cos\theta$ and graph it. What feature of the diagrams causes the differential cross section to diverge as $\theta \to 0$?

5.3   The *spinor product* formalism introduced in Problem 3.3 provides an efficient way to compute tree diagrams involving massless particles. Recall that in Problem 3.3 we defined spinor products as follows: Let u_{L0}, u_{R0} be the left- and right-handed spinors at some fixed lightlike momentum k_0. These satisfy

$$u_{L0}\bar{u}_{L0} = \left(\frac{1-\gamma^5}{2}\right)\slashed{k}_0, \qquad u_{R0}\bar{u}_{R0} = \left(\frac{1+\gamma^5}{2}\right)\slashed{k}_0. \tag{1}$$

(These relations are just the projections onto definite helicity of the more standard formula $\sum u_0\bar{u}_0 = \slashed{k}_0$.) Then define spinors for any other lightlike momentum p by

$$u_L(p) = \frac{1}{\sqrt{2p\cdot k_0}}\slashed{p}\,u_{R0}, \qquad u_R(p) = \frac{1}{\sqrt{2p\cdot k_0}}\slashed{p}\,u_{L0}. \tag{2}$$

We showed that these spinors satisfy $\slashed{p}u(p) = 0$; because there is no m around, they can be used as spinors for either fermions or antifermions. We defined

$$s(p_1,p_2) = \bar{u}_R(p_1)u_L(p_2), \qquad t(p_1,p_2) = \bar{u}_L(p_1)u_R(p_2),$$

and, in a special frame, we proved the properties

$$t(p_1,p_2) = (s(p_2,p_1))^*, \quad s(p_1,p_2) = -s(p_2,p_1), \quad |s(p_1,p_2)|^2 = 2p_1\cdot p_2. \tag{3}$$

Now let us apply these results.

   (a) To warm up, give another proof of the last relation in Eq. (3) by using (1) to rewrite $|s(p_1,p_2)|^2$ as a trace of Dirac matrices, and then applying the trace calculus.

   (b) Show that, for any string of Dirac matrices,

$$\mathrm{tr}[\gamma^\mu\gamma^\nu\gamma^\rho\cdots] = \mathrm{tr}[\cdots\gamma^\rho\gamma^\nu\gamma^\mu]$$

   where $\mu,\nu,\rho,\ldots = 0,1,2,3$, or 5. Use this identity to show that

$$\bar{u}_L(p_1)\gamma^\mu u_L(p_2) = \bar{u}_R(p_2)\gamma^\mu u_R(p_1).$$

   (c) Prove the Fierz identity

$$\bar{u}_L(p_1)\gamma^\mu u_L(p_2)\,[\gamma_\mu]_{ab} = 2\left[u_L(p_2)\bar{u}_L(p_1) + u_R(p_1)\bar{u}_R(p_2)\right]_{ab},$$

where $a, b = 1, 2, 3, 4$ are Dirac indices. This can be done by justifying the following statements: The right-hand side of this equation is a Dirac matrix; thus, it can be written as a linear combination of the 16 Γ matrices discussed in Section 3.4. It satisfies

$$\gamma^5 [M] = -[M] \gamma^5,$$

thus, it must have the form

$$[M] = \left(\frac{1 - \gamma^5}{2} \right) \gamma_\mu V^\mu + \left(\frac{1 + \gamma^5}{2} \right) \gamma_\mu W^\mu$$

where V^μ and W^μ are 4-vectors. These 4-vectors can be computed by trace technology; for example,

$$V^\nu = \frac{1}{2} \text{tr}[\gamma^\nu \left(\frac{1 - \gamma^5}{2} \right) M].$$

(d) Consider the process $e^+ e^- \to \mu^+ \mu^-$, to the leading order in α, ignoring the masses of both the electron and the muon. Consider first the case in which the electron and the final muon are both right-handed and the positron and the final antimuon are both left-handed. (Use the spinor v_R for the antimuon and $\bar{u}_R$ for the positron.) Apply the Fierz identity to show that the amplitude can be evaluated directly in terms of spinor products. Square the amplitude and reproduce the result for

$$\frac{d\sigma}{d \cos \theta} (e^-_R e^+_L \to \mu^-_R \mu^+_L)$$

given in Eq. (5.22). Compute the other helicity cross sections for this process and show that they also reproduce the results found in Section 5.2.

(e) Compute the differential cross section for Bhabha scattering of massless electrons, helicity state by helicity state, using the spinor product formalism. The average over initial helicities, summed over final helicities, should reproduce the result of Problem 5.2. In the process, you should see how this result arises as the sum of definite-helicity contributions.

5.4 Positronium lifetimes.

(a) Compute the amplitude $\mathcal{M}$ for $e^+ e^-$ annihilation into 2 photons in the extreme nonrelativistic limit (i.e., keep only the term proportional to zero powers of the electron and positron 3-momentum). Use this result, together with our formalism for fermion-antifermion bound states, to compute the rate of annihilation of the $1S$ states of positronium into 2 photons. You should find that the spin-1 states of positronium do not annihilate into 2 photons, confirming the symmetry argument of Problem 3.8. For the spin-0 state of positronium, you should find a result proportional to the square of the $1S$ wavefunction at the origin. Inserting the value of this wavefunction from nonrelativistic quantum mechanics, you should find

$$\frac{1}{\tau} = \Gamma = \frac{\alpha^5 m_e}{2} \approx 8.03 \times 10^9 \text{ sec}^{-1}.$$

A recent measurement[‡] gives $\Gamma = 7.994 \pm .011 \, \text{nsec}^{-1}$; the 0.5% discrepancy is accounted for by radiative corrections.

(b) Computing the decay rates of higher-l positronium states is somewhat more difficult; in the rest of this problem, we will consider the case $l = 1$. First, work out the terms in the $e^+ e^- \to 2\gamma$ amplitude proportional to one power of the 3-momentum. (For simplicity, work in the center-of-mass frame.) Since

$$\int \frac{d^3 p}{(2\pi)^3} \, p^i \psi(\mathbf{p}) = i \frac{\partial}{\partial x^i} \psi(\mathbf{x}) \Big|_{\mathbf{x}=0},$$

this piece of the amplitude has overlap with P-wave bound states. Show that the $S = 1$, but not the $S = 0$ states, can decay to 2 photons. Again, this is a consequence of C.

(c) To compute the decay rates of these P-wave states, we need properly normalized state vectors. Denote the three P-state wavefunctions by

$$\psi_i = x^i f(|\mathbf{x}|), \qquad \text{normalized to} \quad \int d^3 x \, \psi_i^*(\mathbf{x}) \psi_j(\mathbf{x}) = \delta_{ij},$$

and their Fourier transforms by $\psi_i(\mathbf{p})$. Show that

$$|B(\mathbf{k})\rangle = \sqrt{2M} \int \frac{d^3 p}{(2\pi)^3} \, \psi_i(\mathbf{p}) \, a_{\mathbf{p}+\mathbf{k}/2}^\dagger \, \Sigma^i \, b_{-\mathbf{p}+\mathbf{k}/2}^\dagger \, |0\rangle$$

is a properly normalized bound-state vector if Σ^i denotes a set of three 2×2 matrices normalized to

$$\sum_i \text{tr}(\Sigma^{i\dagger} \Sigma^i) = 1.$$

To build $S = 1$ states, we should take each Σ^i to contain a Pauli sigma matrix. In general, spin-orbit coupling will split the multiplet of $S = 1$, $L = 1$ states according to the total angular momentum J. The states of definite J are given by

$$J = 0: \qquad \Sigma^i = \frac{1}{\sqrt{6}} \sigma^i,$$

$$J = 1: \qquad \Sigma^i = \frac{1}{2} \epsilon^{ijk} n^j \sigma^k,$$

$$J = 2: \qquad \Sigma^i = \frac{1}{\sqrt{3}} h^{ij} \sigma^j,$$

where $\mathbf{n}$ is a polarization vector satisfying $|\mathbf{n}|^2 = 1$ and h^{ij} is a traceless tensor, for which a typical value might be $h^{12} = 1$ and all other components zero.

(d) Using the expanded form for the $e^+ e^- \to 2\gamma$ amplitude derived in part (b) and the explicit form of the $S = 1$, $L = 1$, definite-J positronium states found in part (c), compute, for each J, the decay rate of the state into two photons.

[‡]D. W. Gidley et. al., *Phys. Rev. Lett.* **49**, 525 (1982).

5.5   Physics of a massive vector boson. Add to QED a massive photon field B_μ of mass M, which couples to electrons via

$$\Delta H = \int d^3x \, (g\bar{\psi}\gamma^\mu\psi B_\mu).$$

A massive photon in the initial or final state has three possible physical polarizations, corresponding to the three spacelike unit vectors in the boson's rest frame. These can be characterized invariantly, in terms of the boson's 4-momentum k^μ, as the three vectors $\epsilon_\mu^{(i)}$ satisfying

$$\epsilon^{(i)} \cdot \epsilon^{(j)} = -\delta^{ij}, \qquad k \cdot \epsilon^{(i)} = 0.$$

The four vectors $(k_\mu/M, \epsilon_\mu^{(i)})$ form a complete orthonormal basis. Because B_μ couples to the conserved current $\bar{\psi}\gamma^\mu\psi$, the Ward identity implies that k_μ dotted into the amplitude for B production gives zero; thus we can replace:

$$\sum_i \epsilon_\mu^{(i)} \epsilon_\nu^{(i)*} \quad \rightarrow \quad -g_{\mu\nu}.$$

This gives a generalization to massive bosons of the Feynman trick for photon polarization vectors and simplifies the calculation of B production cross sections. (Warning: This trick does not work (so simply) for "non-Abelian gauge fields".) Let's do a few of these computations, using always the approximation of ignoring the mass of the electron.

(a) Compute the cross section for the process $e^+e^- \to B$. Compute the lifetime of the B, assuming that it decays only to electrons. Verify the relation

$$\sigma(e^+e^- \to B) = \frac{12\pi^2}{M}\Gamma(B \to e^+e^-)\delta(M^2 - s)$$

discussed in Section 5.3.

(b) Compute the differential cross section, in the center-of-mass system, for the process $e^+e^- \to \gamma + B$. (This calculation goes over almost unchanged to the realistic process $e^+e^- \to \gamma + Z^0$; this allows one to measure the number of decays of the Z^0 into unobserved final states, which is in turn proportional to the number of neutrino species.)

(c) Notice that the cross section of part (b) diverges as $\theta \to 0$ or π. Let us analyze the region near $\theta = 0$. In this region, the dominant contribution comes from the t-channel diagram and corresponds intuitively to the emission of a photon from the electron line before e^+e^- annihilation into a B. Let us rearrange the formula in such a way as to support this interpretation. First, note that the divergence as $\theta \to 0$ is cut off by the electron mass: Let the electron momentum be $p^\mu = (E, 0, 0, k)$, with $k = (E^2 - m_e^2)^{1/2}$, and let the photon momentum be $k^\mu = (xE, xE\sin\theta, 0, xE\cos\theta)$. Show that the denominator of the propagator then never becomes smaller than $\mathcal{O}(m_e^2/s)$. Now integrate the cross section of part (b) over forward angles, cutting off the θ integral at $\theta^2 \sim (m_e^2/s)$ and keeping only the leading logarithmic term, proportional to $\log(s/m_e^2)$. Show that, in this approximation, the cross section for forward photon emission can be written

$$\sigma(e^+e^- \to \gamma + B) \approx \int dx \, f(x) \cdot \sigma(e^+e^- \to B \text{ at } E_{\text{cm}}^2 = (1-x)s),$$

where the annihilation cross section is evaluated for the collision of a positron of energy E and an electron of energy $(1 - x)E$, and the function $f(x)$, the *Weizsäcker-Williams distribution function*, is given by

$$f(x) = \frac{\alpha}{2\pi} \frac{1 + (1 - x)^2}{x} \cdot \log\left(\frac{s}{m_e^2}\right).$$

This function arises universally in processes in which a photon is emitted collinearly from an electron line, independent of the subsequent dynamics. We will meet it again, in another context, in Problem 6.2.

5.6   This problem extends the spinor product technology of Problem 5.3 to external photons.

(a) Let k be the momentum of a photon, and let p be another lightlike vector, chosen so that $p \cdot k \neq 0$. Let $u_R(p)$, $u_L(p)$ be spinors of definite helicity for fermions with the lightlike momentum p, defined according to the conventions of Problem 5.3. Define photon polarization vectors as follows:

$$\epsilon_+^\mu(k) = \frac{1}{\sqrt{4p \cdot k}} \bar{u}_R(k)\gamma^\mu u_R(p), \qquad \epsilon_-^\mu(k) = \frac{1}{\sqrt{4p \cdot k}} \bar{u}_L(k)\gamma^\mu u_L(p).$$

Use the identity

$$u_L(p)\bar{u}_L(p) + u_R(p)\bar{u}_R(p) = \not{p}$$

to compute the polarization sum

$$\epsilon_+^\mu \epsilon_+^{\nu*} + \epsilon_-^\mu \epsilon_-^{\nu*} = -g^{\mu\nu} + \frac{k^\mu p^\nu + k^\nu p^\mu}{p \cdot k}.$$

The second term on the right gives zero when dotted with any photon emission amplitude $\mathcal{M}^\mu$, so we have

$$|\epsilon_+ \cdot \mathcal{M}|^2 + |\epsilon_- \cdot \mathcal{M}|^2 = \mathcal{M}^\mu \mathcal{M}^{\nu*}(-g_{\mu\nu});$$

thus, we can use the vectors ϵ_+, ϵ_- to compute photon polarization sums.

(b) Using the polarization vectors just defined, and the spinor products and the Fierz identity from Problem 5.3, compute the differential cross section for a massless electron and positron to annihilate into 2 photons. Show that the result agrees with the massless limit derived in (5.107):

$$\frac{d\sigma}{d\cos\theta} = \frac{2\pi\alpha^2}{s}\left(\frac{1 + \cos^2\theta}{\sin^2\theta}\right)$$

in the center-of-mass frame. It follows from the result of part (a) that this answer is independent of the particular vector p used to define the polarization vectors; however, the calculation is greatly simplified by taking this vector to be the initial electron 4-vector.

Chapter 6

Radiative Corrections: Introduction

Now that we have acquired some experience at performing QED calculations, let us move on to some more complicated problems. Chapter 5 dealt only with *tree-level* processes, that is, with diagrams that contain no loops. But all such processes receive higher-order contributions, known as *radiative corrections*, from diagrams that do contain loops. Another source of radiative corrections in QED is *bremsstrahlung*, the emission of extra final-state photons during a reaction. In this chapter we will investigate both types of radiative corrections, and find that it is inconsistent to include one without also including the other.

Throughout this chapter, in order to illustrate these ideas in the simplest possible context, we will consider the process of electron scattering from another, very heavy, particle. We analyzed this process at tree level in Section 5.4 and Problem 5.1. At the next order in perturbation theory, we encounter the following four diagrams:

$$(6.1)$$

The order-α correction to the cross section comes from the interference term between these diagrams and the tree-level diagram. There are six additional one-loop diagrams involving the heavy particle in the loop, but they can be neglected in the limit where that particle is much heavier than the electron, since the mass appears in the denominator of the propagator. (Physically, the heavy particle accelerates less, and therefore radiates less, during the collision.)

Of the four diagrams in (6.1), the first (known as the *vertex correction*) is the most intricate and gives the largest variety of new effects. For example, it gives rise to an anomalous magnetic moment for the electron, which we will compute in Section 6.3.

The next two diagrams of (6.1) are *external leg corrections*. We will neglect them in this chapter because they are not amputated, as required by our formula (4.90) for S-matrix elements. We will discuss these diagrams in more detail when we prove that formula in Section 7.2.

The final diagram of (6.1) is called the *vacuum polarization*. Since it requires more computational machinery than the others, we will not evaluate this diagram until Section 7.5.

Our study of these corrections will be complicated by the fact that they are ill-defined. Each diagram of (6.1) involves an integration over the undetermined loop momentum, and in each case the integral is divergent in the $k \to \infty$ or *ultraviolet* region. Fortunately, the infinite parts of these integrals will always cancel out of expressions for observable quantities such as cross sections.

The first three diagrams of (6.1) also contain *infrared divergences*: infinities coming from the $k \to 0$ end of the loop-momentum integrals. We will see in Section 6.4 that these divergences are canceled when we also include the following *bremsstrahlung* diagrams:

$$\text{(6.2)}$$

These diagrams are divergent in the limit where the energy of the radiated photon tends to zero. In this limit, the photon cannot be observed by any physical detector, so it makes sense to add the cross section for producing these low-energy photons to the cross section for scattering without radiation. The bremsstrahlung diagrams are thus an essential part of the radiative correction, in this and any other QED process.

Our main goals in the present chapter are to understand bremsstrahlung of low-energy photons, the vertex correction diagram, and the cancellation of infrared divergences between these two types of radiative corrections.

6.1 Soft Bremsstrahlung

Let us begin our study of radiative corrections by analyzing the bremsstrahlung process. In this section we will first do a classical computation of the intensity of the low-frequency bremsstrahlung radiation when an electron undergoes a sudden acceleration. We will then compute a closely related quantity in quantum field theory: the cross section for emission of one very soft photon, given by diagrams (6.2). We would like to understand how the classical result arises as a limiting case of the quantum result.

Classical Computation

Suppose that a classical electron receives a sudden kick at time $t = 0$ and position $\mathbf{x} = 0$, causing its 4-momentum to change from p to p'. (An infinitely sudden change of momentum is of course an unrealistic idealization. The precise form of the trajectory during the acceleration does not affect the low-frequency radiation, however. Our calculation will be valid for radiation with a frequency less than the reciprocal of the scattering time.)

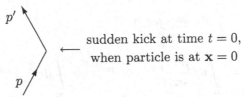

sudden kick at time $t = 0$,
when particle is at $\mathbf{x} = 0$

We can find the radiation field by writing down the current of this electron, and considering that current as a source for Maxwell's equations.

What is the current density of such a particle? For a charged particle at rest at $\mathbf{x} = 0$, the current would be

$$j^\mu(x) = (1,\mathbf{0})^\mu \cdot e\,\delta^{(3)}(\mathbf{x})$$

$$= \int dt\,(1,\mathbf{0})^\mu \cdot e\,\delta^{(4)}\big(x - y(t)\big), \qquad \text{with } y^\mu(t) = (t,\mathbf{0})^\mu.$$

From this we can guess the current for an arbitrary trajectory $y^\mu(\tau)$:

$$j^\mu(x) = e \int d\tau\, \frac{dy^\mu(\tau)}{d\tau}\, \delta^{(4)}\big(x - y(\tau)\big). \tag{6.3}$$

Note that this expression is independent of the precise way in which the curve $y^\mu(\tau)$ is parametrized: Changing variables from τ to $\sigma(\tau)$ gives a factor of $d\tau/d\sigma$ in the integration measure, which combines with $dy^\mu/d\tau$ via the chain rule to give $dy^\mu/d\sigma$. We can also prove from (6.3) that the current is automatically conserved: For any "test function" $f(x)$ that falls off at infinity, we have

$$\int d^4x\, f(x) \partial_\mu j^\mu(x) = \int d^4x\, f(x)\, e \int d\tau\, \frac{dy^\mu(\tau)}{d\tau}\, \partial_\mu \delta^{(4)}\big(x - y(\tau)\big)$$

$$= -e \int d\tau\, \frac{dy^\mu(\tau)}{d\tau}\, \frac{\partial}{\partial x^\mu} f(x)\Big|_{x=y(\tau)}$$

$$= -e \int d\tau\, \frac{d}{d\tau} f\big(y(\tau)\big)$$

$$= -e\, f\big(y(\tau)\big)\Big|_{-\infty}^{\infty} = 0.$$

For our process the trajectory is

$$y^\mu(\tau) = \begin{cases} (p^\mu/m)\tau & \text{for } \tau < 0; \\ (p'^\mu/m)\tau & \text{for } \tau > 0. \end{cases}$$

Thus the current can be written

$$j^\mu(x) = e \int_0^\infty d\tau \, \frac{p'^\mu}{m} \delta^{(4)}\left(x - \frac{p'}{m}\tau\right) + e \int_{-\infty}^0 d\tau \, \frac{p^\mu}{m} \delta^{(4)}\left(x - \frac{p}{m}\tau\right).$$

In a moment we will need to know the Fourier transform of this function. Inserting factors of $e^{-\epsilon\tau}$ and $e^{\epsilon\tau}$ to make the integrals converge, we have

$$\tilde{j}^\mu(k) = \int d^4x \, e^{ik\cdot x} j^\mu(x)$$

$$= e \int_0^\infty d\tau \, \frac{p'^\mu}{m} e^{i(k\cdot p'/m + i\epsilon)\tau} + e \int_{-\infty}^0 d\tau \, \frac{p^\mu}{m} e^{i(k\cdot p/m - i\epsilon)\tau}$$

$$= ie\left(\frac{p'^\mu}{k\cdot p' + i\epsilon} - \frac{p^\mu}{k\cdot p - i\epsilon}\right). \tag{6.4}$$

We are now ready to solve Maxwell's equations. In Lorentz gauge ($\partial^\mu A_\mu = 0$) we must solve $\partial^2 A^\mu = j^\mu$, or in Fourier space,

$$\tilde{A}^\mu(k) = -\frac{1}{k^2}\, \tilde{j}^\mu(k).$$

Plugging in (6.4), we obtain a formula for the vector potential:

$$A^\mu(x) = \int \frac{d^4k}{(2\pi)^4} \, e^{-ik\cdot x} \, \frac{-ie}{k^2}\left(\frac{p'^\mu}{k\cdot p' + i\epsilon} - \frac{p^\mu}{k\cdot p - i\epsilon}\right). \tag{6.5}$$

The k^0 integral can be performed as a contour integral in the complex plane. The locations of the poles are as follows:

We place the poles at $k^0 = \pm|\mathbf{k}|$ below the real axis so that (as we shall soon confirm) the radiation field will satisfy retarded boundary conditions.

For $t < 0$ we close the contour upward, picking up the pole at $k\cdot p = 0$, that is, $k^0 = \mathbf{k}\cdot\mathbf{p}/p^0$. The result is

$$A^\mu(x) = \int \frac{d^3k}{(2\pi)^3} \, e^{i\mathbf{k}\cdot\mathbf{x}} \, e^{-i(\mathbf{k}\cdot\mathbf{p}/p^0)t} \, \frac{(2\pi i)(+ie)}{(2\pi)k^2} \, \frac{p^\mu}{p^0}.$$

In the reference frame where the particle is initially at rest, its momentum vector is $p^\mu = (p^0, \mathbf{0})^\mu$ and the vector potential reduces to

$$A^\mu(x) = \int \frac{d^3k}{(2\pi)^3} e^{ik\cdot x} \frac{e}{|\mathbf{k}|^2} \cdot (1, \mathbf{0})^\mu.$$

This is just the Coulomb potential of an unaccelerated charge. As we would expect, there is no radiation field before the particle is scattered.

After scattering $(t > 0)$, we close the contour downward, picking up the three poles below the real axis. The pole at $k^0 = \mathbf{k} \cdot \mathbf{p}'/p'^0$ gives the Coulomb potential of the outgoing particle. Thus the other two poles are completely responsible for the radiation field. Their contribution gives

$$A_{\text{rad}}^\mu(x) = \int \frac{d^3k}{(2\pi)^3} \frac{-e}{2|\mathbf{k}|} \left\{ e^{-ik\cdot x} \left(\frac{p'^\mu}{k\cdot p'} - \frac{p^\mu}{k\cdot p} \right) + \text{c.c.} \right\} \bigg|_{k^0 = |\mathbf{k}|}$$

$$= \text{Re} \int \frac{d^3k}{(2\pi)^3} \mathcal{A}^\mu(\mathbf{k}) e^{-ik\cdot x}, \tag{6.6}$$

where the momentum-space amplitude $\mathcal{A}(\mathbf{k})$ is given by

$$\mathcal{A}^\mu(\mathbf{k}) = \frac{-e}{|\mathbf{k}|} \left(\frac{p'^\mu}{k\cdot p'} - \frac{p^\mu}{k\cdot p} \right). \tag{6.7}$$

(The condition $k^0 = |\mathbf{k}|$ is implicit here and in the rest of this calculation.)

To calculate the energy radiated, we must find the electric and magnetic fields. It is easiest to write $\mathbf{E}$ and $\mathbf{B}$ as the real parts of complex Fourier integrals, just as we did for A^μ:

$$\mathbf{E}(x) = \text{Re} \int \frac{d^3k}{(2\pi)^3} \boldsymbol{\mathcal{E}}(\mathbf{k}) e^{-ik\cdot x};$$

$$\mathbf{B}(x) = \text{Re} \int \frac{d^3k}{(2\pi)^3} \boldsymbol{\mathcal{B}}(\mathbf{k}) e^{-ik\cdot x}. \tag{6.8}$$

The momentum-space amplitudes $\boldsymbol{\mathcal{E}}(\mathbf{k})$ and $\boldsymbol{\mathcal{B}}(\mathbf{k})$ of the radiation fields are then simply

$$\boldsymbol{\mathcal{E}}(\mathbf{k}) = -i\mathbf{k}\mathcal{A}^0(\mathbf{k}) + ik^0 \boldsymbol{\mathcal{A}}(\mathbf{k});$$

$$\boldsymbol{\mathcal{B}}(\mathbf{k}) = i\mathbf{k} \times \boldsymbol{\mathcal{A}}(\mathbf{k}) = \hat{k} \times \boldsymbol{\mathcal{E}}(\mathbf{k}). \tag{6.9}$$

Using the explicit form (6.7) of $\mathcal{A}^\mu(\mathbf{k})$, you can easily check that the electric field is transverse: $\mathbf{k} \cdot \boldsymbol{\mathcal{E}}(\mathbf{k}) = 0$.

Having expressed the fields in this way, we can compute the energy radiated:

$$\text{Energy} = \tfrac{1}{2} \int d^3x \left(|\mathbf{E}(x)|^2 + |\mathbf{B}(x)|^2 \right). \tag{6.10}$$

The first term is

$$\tfrac{1}{8}\!\int\! d^3x \int\! \frac{d^3k}{(2\pi)^3} \int\! \frac{d^3k'}{(2\pi)^3} \left(\mathcal{E}(\mathbf{k})e^{-ikx} + \mathcal{E}^*(\mathbf{k})e^{ikx}\right)\cdot\left(\mathcal{E}(\mathbf{k}')e^{-ik'x} + \mathcal{E}^*(\mathbf{k}')e^{ik'x}\right)$$

$$= \tfrac{1}{8}\!\int\! \frac{d^3k}{(2\pi)^3} \left(\mathcal{E}(\mathbf{k})\cdot\mathcal{E}(-\mathbf{k})e^{-2ik^0t} + 2\mathcal{E}(\mathbf{k})\cdot\mathcal{E}^*(\mathbf{k}) + \mathcal{E}^*(\mathbf{k})\cdot\mathcal{E}^*(-\mathbf{k})e^{2ik^0t}\right).$$

A similar expression involving $\mathcal{B}(\mathbf{k})$ holds for the second term. Using (6.9) and the fact that $\mathcal{E}(\mathbf{k})$ is transverse, you can show that the time-dependent terms cancel between $\mathcal{E}$ and $\mathcal{B}$, while the remaining terms add to give

$$\text{Energy} = \tfrac{1}{2}\int \frac{d^3k}{(2\pi)^3}\, \mathcal{E}(\mathbf{k})\cdot\mathcal{E}^*(\mathbf{k}). \tag{6.11}$$

Since $\mathcal{E}(\mathbf{k})$ is transverse, let us introduce two transverse unit polarization vectors $\epsilon_\lambda(\mathbf{k})$, $\lambda = 1, 2$. We can then write the integrand as

$$\mathcal{E}(\mathbf{k})\cdot\mathcal{E}^*(\mathbf{k}) = \sum_{\lambda=1,2} \left|\epsilon_\lambda(\mathbf{k})\cdot\mathcal{E}(\mathbf{k})\right|^2 = |\mathbf{k}|^2 \sum_{\lambda=1,2} \left|\epsilon_\lambda(\mathbf{k})\cdot\mathcal{A}(\mathbf{k})\right|^2.$$

Using the explicit form of $\mathcal{A}(\mathbf{k})$ (6.7), we finally arrive at an expression for the energy radiated*:

$$\text{Energy} = \int \frac{d^3k}{(2\pi)^3} \sum_{\lambda=1,2} \frac{e^2}{2}\left|\epsilon_\lambda(\mathbf{k})\cdot\left(\frac{\mathbf{p}'}{k\cdot p'} - \frac{\mathbf{p}}{k\cdot p}\right)\right|^2. \tag{6.12}$$

We can freely change ϵ, $\mathbf{p}'$, and $\mathbf{p}$ into 4-vectors in this expression. Then, noting that substituting k^μ for ϵ^μ would give zero,

$$k_\mu\left(\frac{p'^\mu}{k\cdot p'} - \frac{p^\mu}{k\cdot p}\right) = 0,$$

we find that we can perform the sum over polarizations using the trick of Section 5.5, replacing $\sum \epsilon_\mu \epsilon_\nu^*$ by $-g_{\mu\nu}$. Our result then becomes

$$\text{Energy} = \int \frac{d^3k}{(2\pi)^3} \frac{e^2}{2} (-g_{\mu\nu})\left(\frac{p'^\mu}{k\cdot p'} - \frac{p^\mu}{k\cdot p}\right)\left(\frac{p'^\nu}{k\cdot p'} - \frac{p^\nu}{k\cdot p}\right)$$

$$= \int \frac{d^3k}{(2\pi)^3} \frac{e^2}{2}\left(\frac{2p\cdot p'}{(k\cdot p')(k\cdot p)} - \frac{m^2}{(k\cdot p')^2} - \frac{m^2}{(k\cdot p)^2}\right). \tag{6.13}$$

To make this formula more explicit, choose a frame in which $p^0 = p'^0 = E$. Then the momenta are

$$k^\mu = (k, \mathbf{k}), \qquad p^\mu = E(1, \mathbf{v}), \qquad p'^\mu = E(1, \mathbf{v}').$$

---

*This result is also derived in Jackson (1975), p. 703.

In such a frame our formula becomes

$$\text{Energy} = \frac{e^2}{(2\pi)^2} \int dk \; \mathcal{I}(\mathbf{v}, \mathbf{v}'), \tag{6.14}$$

where $\mathcal{I}(\mathbf{v}, \mathbf{v}')$ (which is essentially the differential intensity $d(\text{Energy})/dk$) is given by

$$\mathcal{I}(\mathbf{v}, \mathbf{v}') = \int \frac{d\Omega_{\hat{k}}}{4\pi} \left(\frac{2(1 - \mathbf{v} \cdot \mathbf{v}')}{(1 - \hat{k} \cdot \mathbf{v})(1 - \hat{k} \cdot \mathbf{v}')} - \frac{m^2/E^2}{(1 - \hat{k} \cdot \mathbf{v}')^2} - \frac{m^2/E^2}{(1 - \hat{k} \cdot \mathbf{v})^2} \right). \tag{6.15}$$

Since $\mathcal{I}(\mathbf{v}, \mathbf{v}')$ does not depend on k, we see that the integral over k in (6.14) is trivial but divergent. This divergence comes from our idealization of an infinitely sudden change in momentum. We expect our formula to be valid only for radiation whose frequency is less than the reciprocal of the scattering time. For a relativistic electron, another possible cutoff would take effect when individual photons carry away a sizable fraction of the electron's energy. In either case our formula is valid in the low-frequency limit, provided that we cut off the integral at some maximum frequency k_{max}. We then have

$$\text{Energy} = \frac{\alpha}{\pi} \cdot k_{\text{max}} \cdot \mathcal{I}(\mathbf{v}, \mathbf{v}'). \tag{6.16}$$

The integrand of $\mathcal{I}(\mathbf{v}, \mathbf{v}')$ peaks when $\hat{k}$ is parallel to either $\mathbf{v}$ or $\mathbf{v}'$:

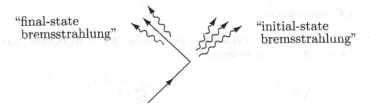

"final-state bremsstrahlung"   "initial-state bremsstrahlung"

In the extreme relativistic limit, most of the radiated energy comes from the two peaks in the first term of (6.15). Let us evaluate $\mathcal{I}(\mathbf{v}, \mathbf{v}')$ in this limit, by concentrating on the regions around these peaks. Break up the integral into a piece for each peak, and let $\theta = 0$ along the peak in each case. Integrate over a small region around $\theta = 0$, as follows:

$$\mathcal{I}(\mathbf{v}, \mathbf{v}') \approx \int\limits_{\hat{k}\cdot\mathbf{v}=\mathbf{v}'\cdot\mathbf{v}}^{\cos\theta=1} d\cos\theta \; \frac{1 - \mathbf{v} \cdot \mathbf{v}'}{(1 - v\cos\theta)(1 - \mathbf{v} \cdot \mathbf{v}')}$$

$$+ \int\limits_{\hat{k}\cdot\mathbf{v}'=\mathbf{v}'\cdot\mathbf{v}}^{\cos\theta=1} d\cos\theta \; \frac{1 - \mathbf{v} \cdot \mathbf{v}'}{(1 - \mathbf{v} \cdot \mathbf{v}')(1 - v'\cos\theta)}.$$

(The lower limits on the integrals are not critical; an equally good choice would be $\hat{k} \cdot \mathbf{v} = 1 - x(1 - \mathbf{v} \cdot \mathbf{v}')$, as long as x is neither too close to 0 nor too much bigger than 1. It is then easy to show that the leading term in the relativistic limit does not depend on x.) The integrals are easy to perform, and we obtain

$$\mathcal{I}(\mathbf{v}, \mathbf{v}') \approx \log\Big(\frac{1 - \mathbf{v}' \cdot \mathbf{v}}{1 - |\mathbf{v}|}\Big) + \log\Big(\frac{1 - \mathbf{v}' \cdot \mathbf{v}}{1 - |\mathbf{v}'|}\Big) = \log\Big(\frac{(E^2 - \mathbf{p} \cdot \mathbf{p}')^2}{E^2(E - |\mathbf{p}|)^2}\Big)$$

$$\approx 2\log\Big(\frac{p \cdot p'}{(E^2 - |\mathbf{p}|^2)/2}\Big) = 2\log\Big(\frac{-q^2}{m^2}\Big), \tag{6.17}$$

where $q^2 = (p' - p)^2$.

In conclusion, we have found that the radiated energy at low frequencies is given by

$$\text{Energy} = \frac{\alpha}{\pi} \int_0^{k_{\max}} dk\, \mathcal{I}(\mathbf{v}, \mathbf{v}') \xrightarrow[E \gg m]{} \frac{2\alpha}{\pi} \int_0^{k_{\max}} dk\, \log\Big(\frac{-q^2}{m^2}\Big). \tag{6.18}$$

If this energy is made up of photons, each photon contributes energy k. We would then expect

$$\text{Number of photons} = \frac{\alpha}{\pi} \int_0^{k_{\max}} dk\, \frac{1}{k} \mathcal{I}(\mathbf{v}, \mathbf{v}'). \tag{6.19}$$

We hope that a quantum-mechanical calculation will confirm this result.

Quantum Computation

Consider now the quantum-mechanical process in which one photon is radiated during the scattering of an electron:

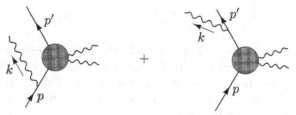

Let $\mathcal{M}_0$ denote the part of the amplitude that comes from the electron's interaction with the external field. Then the amplitude for the whole process is

$$iM = -ie\bar{u}(p')\Big(\mathcal{M}_0(p', p - k)\frac{i(\not{p} - \not{k} + m)}{(p - k)^2 - m^2}\gamma^\mu \epsilon_\mu^*(k)$$

$$+ \gamma^\mu \epsilon_\mu^*(k)\frac{i(\not{p}' + \not{k} + m)}{(p' + k)^2 - m^2}\mathcal{M}_0(p' + k, p)\Big)u(p). \tag{6.20}$$

Since we are interested in connecting with the classical limit, assume that the photon radiated is soft: $|\mathbf{k}| \ll |\mathbf{p}' - \mathbf{p}|$. Then we can approximate

$$\mathcal{M}_0(p', p - k) \approx \mathcal{M}_0(p' + k, p) \approx \mathcal{M}_0(p', p), \tag{6.21}$$

and we can ignore $\not{k}$ in the numerators of the propagators. The numerators can be further simplified with some Dirac algebra. In the first term we have

$$(\not{p} + m)\gamma^\mu \epsilon_\mu^* \, u(p) = [2p^\mu \epsilon_\mu^* + \gamma^\mu \epsilon_\mu^*(-\not{p} + m)] u(p)$$

$$= 2p^\mu \epsilon_\mu^* \, u(p).$$

Similarly, in the second term,

$$\bar{u}(p') \, \gamma^\mu \epsilon_\mu^* (\not{p}' + m) = \bar{u}(p') \, 2p'^\mu \epsilon_\mu^*.$$

The denominators of the propagators also simplify:

$$(p - k)^2 - m^2 = -2p \cdot k; \qquad (p' + k)^2 - m^2 = 2p' \cdot k.$$

So in the soft-photon approximation, the amplitude becomes

$$i\mathcal{M} = \bar{u}(p')\left[\mathcal{M}_0(p',p)\right]u(p) \cdot \left[e\left(\frac{p' \cdot \epsilon^*}{p' \cdot k} - \frac{p \cdot \epsilon^*}{p \cdot k}\right)\right]. \tag{6.22}$$

This is just the amplitude for elastic scattering (without bremsstrahlung), times a factor (in brackets) for the emission of the photon.

   The cross section for our process is also easy to express in terms of the elastic cross section; just insert an additional phase-space integration for the photon variable k. Summing over the two photon polarization states, we have

$$d\sigma\left(p \rightarrow p' + \gamma\right) = d\sigma(p \rightarrow p') \cdot \int \frac{d^3k}{(2\pi)^3} \frac{1}{2k} \sum_{\lambda=1,2} e^2 \left|\frac{p' \cdot \epsilon^{(\lambda)}}{p' \cdot k} - \frac{p \cdot \epsilon^{(\lambda)}}{p \cdot k}\right|^2. \tag{6.23}$$

Thus the differential probability of radiating a photon with momentum k, given that the electron scatters from p to p', is

$$d(\text{prob}) = \frac{d^3k}{(2\pi)^3} \sum_\lambda \frac{e^2}{2k} \left|\epsilon_\lambda \cdot \left(\frac{\mathbf{p}'}{p' \cdot k} - \frac{\mathbf{p}}{p \cdot k}\right)\right|^2. \tag{6.24}$$

This looks very familiar; if we multiply by the photon energy k to compute the expected energy radiated, we recover the classical expression (6.12).

   But there is a problem. Equation (6.24) is an expression not for the expected number of photons radiated, but for the probability of radiating a single photon. The problem becomes worse if we integrate over the photon momentum. As in (6.16), we can integrate only up to the energy at which our soft-photon approximations break down; a reasonable estimate for this energy is $|\mathbf{q}| = |\mathbf{p} - \mathbf{p}'|$. The integral is therefore

$$\text{Total probability} \approx \frac{\alpha}{\pi} \int_0^{|\mathbf{q}|} dk \, \frac{1}{k} \mathcal{I}(\mathbf{v}, \mathbf{v}'). \tag{6.25}$$

Since $\mathcal{I}(\mathbf{v}, \mathbf{v}')$ is independent of k, the integral diverges at its lower limit (where all our approximations are well justified). In other words, the total probability of radiating a very soft photon is infinite. This is the famous problem of infrared divergences in QED perturbation theory.

We can artificially make the integral in (6.25) well-defined by pretending that the photon has a very small mass μ. This mass would then provide a lower cutoff for the integral, allowing us to write the result of this section as

$$d\sigma\big(p \to p' + \gamma(k)\big) = d\sigma(p \to p') \cdot \frac{\alpha}{2\pi} \log\Big(\frac{-q^2}{\mu^2}\Big) \mathcal{I}(\mathbf{v}, \mathbf{v}')$$

$$\underset{-q^2 \to \infty}{\approx} d\sigma(p \to p') \cdot \frac{\alpha}{\pi} \log\Big(\frac{-q^2}{\mu^2}\Big) \log\Big(\frac{-q^2}{m^2}\Big).$$

(6.26)

The q^2 dependence of this result, known as the *Sudakov double logarithm*, is physical and will appear again in Section 6.4. The dependence on μ, however, presents a problem that we must solve. It is not hard to guess that the resolution of this problem will involve reinterpreting (6.24) as the expected number of radiated photons, rather than the probability of radiating a single photon. We will see in Sections 6.4 and 6.5 how this reinterpretation follows from the Feynman diagrams. To prepare for that discussion, however, we need to improve our understanding of the amplitude for scattering without radiation.

6.2 The Electron Vertex Function: Formal Structure

Having briefly discussed QED radiative corrections due to emission of photons (bremsstrahlung), let us now study the correction to electron scattering that comes from the presence of an additional *virtual* photon:

(6.27)

This will be our first experience with a Feynman diagram containing a loop. Such diagrams give rise to significant and profound complications in quantum field theory.

The result of computing this diagram will be rather complicated, so it will be useful to think ahead about what form we expect this correction to take and how to interpret its various possible terms. In this section, we will consider the general properties of vertex correction diagrams. We will see that the basic requirements of Lorentz invariance, the discrete symmetries of QED, and the Ward identity strongly constrain the form of the vertex.

Consider, then, the class of diagrams

where the gray circle indicates the sum of the lowest-order electron-photon vertex and all amputated loop corrections. We will call this sum of vertex diagrams $-ie\Gamma^\mu(p',p)$. Then, according to our master formula (4.103) for S-matrix elements, the amplitude for electron scattering from a heavy target is

$$i\mathcal{M} = ie^2\left(\bar{u}(p')\,\Gamma^\mu(p',p)\,u(p)\right)\frac{1}{q^2}\left(\bar{u}(k')\gamma_\mu u(k)\right). \tag{6.28}$$

More generally, the function $\Gamma^\mu(p',p)$ appears in the S-matrix element for the scattering of an electron from an external electromagnetic field. As in Problem 4.4, add to the Hamiltonian of QED the interaction

$$\Delta H_{\text{int}} = \int d^3x\, eA^{\text{cl}}_\mu j^\mu, \tag{6.29}$$

where $j^\mu(x) = \bar\psi(x)\gamma^\mu\psi(x)$ is the electromagnetic current and A^{cl}_μ is a fixed classical potential. In the leading order of perturbation theory, the S-matrix element for scattering from this field is

$$i\mathcal{M}\,(2\pi)\delta(p^{0\prime} - p^0) = -ie\bar{u}(p')\gamma^\mu u(p) \cdot \tilde{A}^{\text{cl}}_\mu(p' - p),$$

where $\tilde{A}^{\text{cl}}_\mu(q)$ is the Fourier transform of $A^{\text{cl}}_\mu(x)$. The vertex corrections modify this expression to

$$i\mathcal{M}\,(2\pi)\delta(p^{0\prime} - p^0) = -ie\bar{u}(p')\,\Gamma^\mu(p',p)\,u(p) \cdot \tilde{A}^{\text{cl}}_\mu(p' - p). \tag{6.30}$$

In writing (6.28) and (6.30), we have deliberately omitted the contribution of vacuum polarization diagrams, such as the fourth diagram of (6.1). The reason for this omission is that these diagrams should be considered corrections to the electromagnetic field itself, while the diagrams included in Γ^μ represent corrections to the electron's response to a given applied field.[†]

We can use general arguments to restrict the form of $\Gamma^\mu(p',p)$. To lowest order, $\Gamma^\mu = \gamma^\mu$. In general, Γ^μ is some expression that involves p, p', γ^μ, and constants such as m, e, and pure numbers. This list is exhaustive, since no other objects appear in the Feynman rules for evaluating the diagrams that contribute to Γ^μ. The only other object that could appear in any theory is $\epsilon^{\mu\nu\rho\sigma}$ (or equivalently, γ^5); but this is forbidden in any parity-conserving theory.

[†]To justify this statement, we must give a careful definition of an applied external field in a quantum field theory. We will do this in Chapter 11.

We can narrow down the form of Γ^μ considerably by appealing to Lorentz invariance. Since Γ^μ transforms as a vector (in the same sense that γ^μ does), it must be a linear combination of the vectors from the list above: γ^μ, p^μ, and p'^μ. Using the combinations $p' + p$ and $p' - p$ for convenience, we have

$$\Gamma^\mu = \gamma^\mu \cdot A + (p'^\mu + p^\mu) \cdot B + (p'^\mu - p^\mu) \cdot C. \qquad (6.31)$$

The coefficients A, B, and C could involve Dirac matrices dotted into vectors, that is, $\not{p}$ or $\not{p}'$. But since $\not{p} u(p) = m \cdot u(p)$ and $\bar{u}(p') \not{p}' = \bar{u}(p') \cdot m$, we can write the coefficients in terms of ordinary numbers without loss of generality. The only nontrivial scalar available is $q^2 = -2p' \cdot p + 2m^2$, so A, B, and C must be functions only of q^2 (and of constants such as m).

The list of allowed vectors can be further shortened by applying the Ward identity (5.79): $q_\mu \Gamma^\mu = 0$. (Note that our arguments for this identity in Section 5.5—and the proof in Section 7.4—do not require $q^2 = 0$.) Dotting q_μ into (6.31), we find that the second term vanishes, as does the first when sandwiched between $\bar{u}(p')$ and $u(p)$. The third term does not automatically vanish, so C must be zero.

We can make no further simplifications of (6.31) on general principles. It is conventional, however, to rewrite (6.31) by means of the Gordon identity (see Problem 3.2):

$$\bar{u}(p') \gamma^\mu u(p) = \bar{u}(p') \left[\frac{p'^\mu + p^\mu}{2m} + \frac{i\sigma^{\mu\nu} q_\nu}{2m} \right] u(p). \qquad (6.32)$$

This identity allows us to swap the $(p' + p)$ term for one involving $\sigma^{\mu\nu} q_\nu$. We write our final result as

$$\Gamma^\mu(p', p) = \gamma^\mu F_1(q^2) + \frac{i\sigma^{\mu\nu} q_\nu}{2m} F_2(q^2), \qquad (6.33)$$

where F_1 and F_2 are unknown functions of q^2 called *form factors*.

To lowest order, $F_1 = 1$ and $F_2 = 0$. In the next section we will compute the one-loop (order-α) corrections to the form factors, due to the vertex correction diagram (6.27). In principle, the form factors can be computed to any order in perturbation theory.

Since F_1 and F_2 contain complete information about the influence of an electromagnetic field on the electron, they should, in particular, contain the electron's gross electric and magnetic couplings. To identify the electric charge of the electron, we can use (6.30) to compute the amplitude for elastic Coulomb scattering of a nonrelativistic electron from a region of nonzero electrostatic potential. Set $A_\mu^{\text{cl}}(x) = (\phi(\mathbf{x}), \mathbf{0})$. Then $\tilde{A}_\mu^{\text{cl}}(q) = ((2\pi)\delta(q^0)\tilde{\phi}(\mathbf{q}), \mathbf{0})$. Inserting this into (6.30), we find

$$i\mathcal{M} = -ie\bar{u}(p') \, \Gamma^0(p', p) \, u(p) \cdot \tilde{\phi}(\mathbf{q}).$$

If the electrostatic field is very slowly varying over a large (perhaps macroscopic) region, $\tilde{\phi}(\mathbf{q})$ will be concentrated about $\mathbf{q} = 0$; then we can take the limit $\mathbf{q} \to 0$ in the spinor matrix element. Only the form factor F_1 contributes. Using the nonrelativistic limit of the spinors,

$$\bar{u}(p')\gamma^0 u(p) = u^\dagger(p')u(p) \approx 2m\xi'^\dagger\xi,$$

the amplitude for electron scattering from an electric field takes the form

$$i\mathcal{M} = -ieF_1(0)\tilde{\phi}(\mathbf{q}) \cdot 2m\xi'^\dagger\xi. \tag{6.34}$$

This is the Born approximation for scattering from a potential

$$V(\mathbf{x}) = eF_1(0)\phi(\mathbf{x}).$$

Thus $F_1(0)$ is the electric charge of the electron, in units of e. Since $F_1(0) = 1$ already in the leading order of perturbation theory, radiative corrections to $F_1(q^2)$ should vanish at $q^2 = 0$.

By repeating this analysis for an electron scattering from a static vector potential, we can derive a similar connection between the form factors and the electron's magnetic moment.[‡] Set $A_\mu^{cl}(x) = (0, \mathbf{A}^{cl}(\mathbf{x}))$. Then the amplitude for scattering from this field is

$$i\mathcal{M} = +ie\left[\bar{u}(p')\left(\gamma^i F_1 + \frac{i\sigma^{i\nu}q_\nu}{2m}F_2\right)u(p)\right]\tilde{A}_{cl}^i(\mathbf{q}). \tag{6.35}$$

The expression in brackets vanishes at $\mathbf{q} = 0$, so we must carefully extract from it a contribution linear in q^i. To do this, insert the nonrelativistic expansion of the spinors $u(p)$, keeping terms through first order in momenta:

$$u(p) = \begin{pmatrix} \sqrt{p\cdot\sigma}\,\xi \\ \sqrt{p\cdot\bar{\sigma}}\,\xi \end{pmatrix} \approx \sqrt{m}\begin{pmatrix} (1 - \mathbf{p}\cdot\boldsymbol{\sigma}/2m)\xi \\ (1 + \mathbf{p}\cdot\boldsymbol{\sigma}/2m)\xi \end{pmatrix}. \tag{6.36}$$

Then the F_1 term can be simplified as follows:

$$\bar{u}(p')\gamma^i u(p) = 2m\,\xi'^\dagger\left(\frac{\mathbf{p}'\cdot\boldsymbol{\sigma}}{2m}\sigma^i + \sigma^i\frac{\mathbf{p}\cdot\boldsymbol{\sigma}}{2m}\right)\xi.$$

Applying the identity $\sigma^i\sigma^j = \delta^{ij} + i\epsilon^{ijk}\sigma^k$, we find a spin-independent term, proportional to $(\mathbf{p}'+\mathbf{p})$, and a spin-dependent term, proportional to $(\mathbf{p}'-\mathbf{p})$. The first of these terms is the contribution of the operator $[\mathbf{p}\cdot\mathbf{A} + \mathbf{A}\cdot\mathbf{p}]$ in the standard kinetic energy term of nonrelativistic quantum mechanics. The second is the magnetic moment interaction we are seeking. Retaining only the latter term, we have

$$\bar{u}(p')\gamma^i u(p) = 2m\,\xi'^\dagger\left(\frac{-i}{2m}\epsilon^{ijk}q^j\sigma^k\right)\xi.$$

The F_2 term already contains an explicit factor of q, so we can evaluate it using the leading-order term of the expansion of the spinors. This gives

$$\bar{u}(p')\left(\frac{i}{2m}\sigma^{i\nu}q_\nu\right)u(p) = 2m\,\xi'^\dagger\left(\frac{-i}{2m}\epsilon^{ijk}q^j\sigma^k\right)\xi.$$

[‡]The following argument contains numerous factors of (-1) from raising and lowering spacelike indices. Be careful in verifying the algebra.

Thus, the complete term linear in q^j in the electron-photon vertex function is

$$\bar{u}(p')\left(\gamma^i F_1 + \frac{i\sigma^{i\nu}q_\nu}{2m}F_2\right)u(p) \underset{q\to 0}{\approx} 2m\,\xi'^\dagger\left(\frac{-i}{2m}\epsilon^{ijk}q^j\sigma^k\left[F_1(0)+F_2(0)\right]\right)\xi.$$

Inserting this expression into (6.35), we find

$$i\mathcal{M} = -i(2m)\cdot e\xi'^\dagger\left(\frac{-1}{2m}\sigma^k\left[F_1(0)+F_2(0)\right]\right)\xi\,\tilde{B}^k(\mathbf{q}),$$

where

$$\tilde{B}^k(\mathbf{q}) = -i\epsilon^{ijk}q^i\tilde{A}^j_{\rm cl}(\mathbf{q})$$

is the Fourier transform of the magnetic field produced by $\mathbf{A}^{\rm cl}(\mathbf{x})$.

Again we can interpret $\mathcal{M}$ as the Born approximation to the scattering of the electron from a potential well. The potential is just that of a magnetic moment interaction,

$$V(\mathbf{x}) = -\langle\boldsymbol{\mu}\rangle\cdot\mathbf{B}(\mathbf{x}),$$

where

$$\langle\boldsymbol{\mu}\rangle = \frac{e}{m}\left[F_1(0)+F_2(0)\right]\xi'^\dagger\frac{\boldsymbol{\sigma}}{2}\xi.$$

This expression for the magnetic moment of the electron can be rewritten in the standard form

$$\boldsymbol{\mu} = g\left(\frac{e}{2m}\right)\mathbf{S},$$

where $\mathbf{S}$ is the electron spin. The coefficient g, called the *Landé g-factor*, is

$$g = 2\left[F_1(0)+F_2(0)\right] = 2 + 2F_2(0). \tag{6.37}$$

Since the leading order of perturbation theory gives no F_2 term, QED predicts $g = 2 + \mathcal{O}(\alpha)$. The leading term is the standard prediction of the Dirac equation. In higher orders, however, we will find a nonzero F_2 and thus a small difference between the electron's magnetic moment and the Dirac value. We will compute the order-α contribution to this *anomalous magnetic moment* in the next section.

Since our derivation of the structure (6.33) for the vertex function used only general symmetry principles, we expect this formula to apply not only to the electron but to any fermion with electromagnetic interactions. For example, the electromagnetic scattering amplitude of the proton should also be described by two invariant functions of q^2. Since the proton is not an elementary particle, we should not expect the Dirac equation values $F_1 = 1$ and $F_2 = 0$ to be good approximations to the form factors of the proton. In fact, both proton form factors depend strongly on q^2. However, the description of the vertex function in term of form factors provides a useful summary of data on scattering at many energies and angles. The precise transcription between form factors and cross sections is worked out in Problem 6.1. In addition, the general constraints at $q^2 = 0$ that we have just derived apply to the proton: $F_1(0) = 1$, and $2F_2(0) = (g_p - 2)$, though the g-factor of the proton differs by 40% from the Dirac value.

6.3 The Electron Vertex Function: Evaluation

Now that we know what form the answer is to take (Eq. (6.33)), we are ready
to evaluate the one-loop contribution to the electron vertex function. Assign
momenta on the diagram as follows:

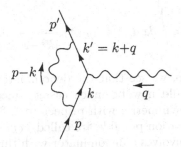

Applying the Feynman rules, we find, to order α, that $\Gamma^\mu = \gamma^\mu + \delta\Gamma^\mu$, where

$$\bar{u}(p')\delta\Gamma^\mu(p',p)u(p)$$

$$= \int \frac{d^4k}{(2\pi)^4}\, \frac{-ig_{\nu\rho}}{(k-p)^2+i\epsilon}\, \bar{u}(p')(-ie\gamma^\nu)\frac{i(\slashed{k}'+m)}{k'^2-m^2+i\epsilon}\,\gamma^\mu\, \frac{i(\slashed{k}+m)}{k^2-m^2+i\epsilon}(-ie\gamma^\rho)u(p)$$

$$= 2ie^2 \int \frac{d^4k}{(2\pi)^4}\, \frac{\bar{u}(p')\big[\slashed{k}\gamma^\mu\slashed{k}' + m^2\gamma^\mu - 2m(k+k')^\mu\big]u(p)}{((k-p)^2+i\epsilon)(k'^2-m^2+i\epsilon)(k^2-m^2+i\epsilon)}. \tag{6.38}$$

In the second line we have used the contraction identity $\gamma^\nu\gamma^\mu\gamma_\nu = -2\gamma^\mu$.
Note that the $+i\epsilon$ terms in the denominators cannot be dropped; they are
necessary for proper evaluation of the loop-momentum integral.

The integral looks impossible, and in fact it will not be easy. The evaluation of such integrals requires another piece of computational technology,
known as the method of *Feynman parameters* (although a very similar method
was introduced earlier by Schwinger).

Feynman Parameters

The goal of this method is to squeeze the three denominator factors of (6.38)
into a single quadratic polynomial in k, raised to the third power. We can then
shift k by a constant to complete the square in this polynomial and evaluate
the remaining spherically symmetric integral without difficulty. The price will
be the introduction of auxiliary parameters to be integrated over.

It is easiest to begin with the simpler case of two factors in the denominator. We would then use the identity

$$\frac{1}{AB} = \int_0^1 dx\, \frac{1}{[xA + (1-x)B]^2} = \int_0^1 dx\, dy\, \delta(x+y-1)\frac{1}{[xA + yB]^2}. \tag{6.39}$$

An example of its use might look like this:

$$\frac{1}{(k-p)^2(k^2-m^2)} = \int_0^1 dx\, dy\, \delta(x+y-1)\frac{1}{\left[x(k-p)^2 + y(k^2-m^2)\right]^2}$$

$$= \int_0^1 dx\, dy\, \delta(x+y-1)\frac{1}{\left[k^2-2xk\cdot p+xp^2-ym^2\right]^2}.$$

If we now let $\ell \equiv k - xp$, we see that the denominator depends only on ℓ^2. Integrating over d^4k would now be much easier, since $d^4k = d^4\ell$ and the integrand is spherically symmetric with respect to ℓ. The variables x and y that make this transformation possible are called *Feynman parameters*.

Our integral (6.38) involves a denominator with three factors, so we need a slightly better identity. By differentiating (6.39) with respect to B, it is easy to prove

$$\frac{1}{AB^n} = \int_0^1 dx\, dy\, \delta(x+y-1)\frac{ny^{n-1}}{[xA + yB]^{n+1}}. \tag{6.40}$$

But this still isn't quite good enough. The formula we need is

$$\frac{1}{A_1 A_2 \cdots A_n} = \int_0^1 dx_1 \cdots dx_n\, \delta(\textstyle\sum x_i-1)\frac{(n-1)!}{[x_1 A_1 + x_2 A_2 + \cdots x_n A_n]^n}. \tag{6.41}$$

The proof of this identity is by induction. The case $n = 2$ is just Eq. (6.39); the induction step is not difficult and involves the use of (6.40).

By repeated differentiation of (6.41), you can derive the even more general identity

$$\frac{1}{A_1^{m_1} A_2^{m_2} \cdots A_n^{m_n}} = \int_0^1 dx_1 \cdots dx_n\, \delta(\textstyle\sum x_i-1)\frac{\prod x_i^{m_i-1}}{[\sum x_i A_i]^{\Sigma m_i}}\frac{\Gamma(m_1 + \cdots + m_n)}{\Gamma(m_1)\cdots\Gamma(m_n)}. \tag{6.42}$$

This formula is true even when the m_i are not integers; in Section 10.5 we will apply it in such a case.

Evaluation of the Form Factors

Now let us apply formula (6.41) to the denominator of (6.38):

$$\frac{1}{((k-p)^2+i\epsilon)(k'^2-m^2+i\epsilon)(k^2-m^2+i\epsilon)} = \int_0^1 dx\, dy\, dz\, \delta(x+y+z-1)\frac{2}{D^3},$$

where the new denominator D is

$$D = x(k^2 - m^2) + y(k'^2 - m^2) + z(k - p)^2 + (x + y + z)i\epsilon$$

$$= k^2 + 2k \cdot (yq - zp) + yq^2 + zp^2 - (x + y)m^2 + i\epsilon. \tag{6.43}$$

In the second line we have used $x + y + z = 1$ and $k' = k + q$. Now shift k to complete the square:

$$\ell \equiv k + yq - zp.$$

After a bit of algebra we find that D simplifies to

$$D = \ell^2 - \Delta + i\epsilon,$$

where

$$\Delta \equiv -xyq^2 + (1 - z)^2 m^2. \tag{6.44}$$

Since $q^2 < 0$ for a scattering process, Δ is positive; we can think of it as an effective mass term.

Next we must express the numerator of (6.38) in terms of ℓ. This task is simplified by noting that since D depends only on the magnitude of ℓ,

$$\int \frac{d^4\ell}{(2\pi)^4} \frac{\ell^\mu}{D^3} = 0; \tag{6.45}$$

$$\int \frac{d^4\ell}{(2\pi)^4} \frac{\ell^\mu \ell^\nu}{D^3} = \int \frac{d^4\ell}{(2\pi)^4} \frac{\frac{1}{4} g^{\mu\nu} \ell^2}{D^3}. \tag{6.46}$$

The first identity follows from symmetry. To prove the second, note that the integral vanishes by symmetry unless $\mu = \nu$. Lorentz invariance therefore requires that we get something proportional to $g^{\mu\nu}$. To check the coefficient, contract each side with $g_{\mu\nu}$. Using these identities, we have

$$\text{Numerator} = \bar{u}(p')\Big[\slashed{k}\gamma^\mu \slashed{k}' + m^2\gamma^\mu - 2m(k + k')^\mu\Big]u(p)$$

$$\to \bar{u}(p')\Big[-\tfrac{1}{2}\gamma^\mu \ell^2 + (-y\slashed{q} + z\slashed{p})\gamma^\mu ((1 - y)\slashed{q} + z\slashed{p})$$

$$+ m^2\gamma^\mu - 2m((1 - 2y)q^\mu + 2zp^\mu)\Big]u(p).$$

(Remember that $k' = k + q$.)

Putting the numerator into a useful form is now just a matter of some tedious Dirac algebra (about a page or two). This is where our work in the last section pays off, since it tells us what kind of an answer to expect. We eventually want to group everything into two terms, proportional to γ^μ and $i\sigma^{\mu\nu}q_\nu$. The most straightforward way to accomplish this is to aim instead for an expression of the form

$$\gamma^\mu \cdot A + (p'^\mu + p^\mu) \cdot B + q^\mu \cdot C,$$

just as in (6.31). Attaining this form requires only the anticommutation relations (for example, $\slashed{p}\gamma^\mu = 2p^\mu - \gamma^\mu \slashed{p}$) and the Dirac equation ($\slashed{p}u(p) = m\,u(p)$)

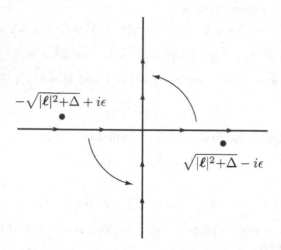

Figure 6.1. The contour of the ℓ^0 integration can be rotated as shown.

and $\bar{u}(p')\not{p}' = \bar{u}(p')\,m$; note that this implies $\bar{u}(p')\not{q}\,u(p) = 0$). It is also useful to remember that $x + y + z = 1$. When the smoke clears, we have

$$\text{Numerator} = \bar{u}(p')\Big[\gamma^\mu \cdot \big(-\tfrac{1}{2}\ell^2 + (1-x)(1-y)q^2 + (1-2z-z^2)m^2\big)$$

$$+ (p'^\mu + p^\mu)\cdot mz(z-1) + q^\mu \cdot m(z-2)(x-y)\Big]u(p).$$

The coefficient of q^μ must vanish according to the Ward identity, as discussed after Eq. (6.31). To see that it does, note from (6.44) that the denominator is symmetric under $x \leftrightarrow y$. The coefficient of q^μ is odd under $x \leftrightarrow y$ and therefore vanishes when integrated over x and y.

Still following our work in the previous section, we now use the Gordon identity (6.32) to eliminate $(p' + p)$ in favor of $i\sigma^{\mu\nu}q_\nu$. Our entire expression for the $\mathcal{O}(\alpha)$ contribution to the electron vertex then becomes

$$\bar{u}(p')\delta\Gamma^\mu(p',p)u(p) = 2ie^2\int\frac{d^4\ell}{(2\pi)^4}\int_0^1 dx\,dy\,dz\,\delta(x+y+z-1)\,\frac{2}{D^3}$$

$$\times\,\bar{u}(p')\Big[\gamma^\mu \cdot \big(-\tfrac{1}{2}\ell^2 + (1-x)(1-y)q^2 + (1-4z+z^2)m^2\big)$$

$$+ \frac{i\sigma^{\mu\nu}q_\nu}{2m}\big(2m^2z(1-z)\big)\Big]u(p), \qquad (6.47)$$

where as before,

$$D = \ell^2 - \Delta + i\epsilon, \qquad \Delta = -xyq^2 + (1-z)^2m^2 > 0.$$

The decomposition into form factors is now manifest.

With most of the work behind us, our main remaining task is to perform the momentum integral. It is not difficult to evaluate the ℓ^0 integral as a contour integral, then do the spatial integrals in spherical coordinates. We will use an even easier method, making use of a trick called *Wick rotation*. Note that if it were not for the minus signs in the Minkowski metric, we could perform the entire four-dimensional integral in four-dimensional "spherical" coordinates. To remove the minus signs, consider the contour of integration in the ℓ^0-plane (see Fig. 6.1). The locations of the poles, and the fact that the integrand falls off sufficiently rapidly at large $|\ell^0|$, allow us to rotate the contour counterclockwise by 90°. We then define a *Euclidean* 4-momentum variable ℓ_E:

$$\ell^0 \equiv i\ell_E^0; \qquad \ell = \ell_E. \tag{6.48}$$

Our rotated contour goes from $\ell_E^0 = -\infty$ to ∞. By simply changing variables to ℓ_E, we can now evaluate the integral in four-dimensional spherical coordinates.

Let us first evaluate

$$\int \frac{d^4\ell}{(2\pi)^4} \frac{1}{[\ell^2 - \Delta]^m} = \frac{i}{(-1)^m} \frac{1}{(2\pi)^4} \int d^4\ell_E \frac{1}{[\ell_E^2 + \Delta]^m}$$

$$= \frac{i(-1)^m}{(2\pi)^4} \int d\Omega_4 \int_0^\infty d\ell_E \frac{\ell_E^3}{[\ell_E^2 + \Delta]^m}.$$

(Here we need only the case $m = 3$, but the more general result will be useful for other loop calculations.) The factor $\int d\Omega_4$ is the surface "area" of a four-dimensional unit sphere, which happens to equal $2\pi^2$. (One way to compute this area is to use four-dimensional spherical coordinates,

$$x = (r\sin\omega\sin\theta\cos\phi,\ r\sin\omega\sin\theta\sin\phi,\ r\sin\omega\cos\theta,\ r\cos\omega).$$

The integration measure is then $d^4x = r^3 \sin^2\omega \sin\theta\, d\phi\, d\theta\, d\omega\, dr$.) The rest of the integral is straightforward, and we have

$$\int \frac{d^4\ell}{(2\pi)^4} \frac{1}{[\ell^2 - \Delta]^m} = \frac{i(-1)^m}{(4\pi)^2} \frac{1}{(m-1)(m-2)} \frac{1}{\Delta^{m-2}}. \tag{6.49}$$

Similarly,

$$\int \frac{d^4\ell}{(2\pi)^4} \frac{\ell^2}{[\ell^2 - \Delta]^m} = \frac{i(-1)^{m-1}}{(4\pi)^2} \frac{2}{(m-1)(m-2)(m-3)} \frac{1}{\Delta^{m-3}}. \tag{6.50}$$

Note that this second result is valid only when $m > 3$. When $m = 3$, the Wick rotation cannot be justified, and the integral is in any event divergent. But it is just this case that we need for (6.47).

We will eventually explore the physical meaning of this divergence, but for the moment we simply introduce an artificial prescription to make our integral finite. Go back to the original expression for the Feynman integral in

(6.38), and replace in the photon propagator

$$\frac{1}{(k-p)^2 + i\epsilon} \longrightarrow \frac{1}{(k-p)^2 + i\epsilon} - \frac{1}{(k-p)^2 - \Lambda^2 + i\epsilon}, \tag{6.51}$$

where Λ is a very large mass. The integrand is unaffected for small k (since Λ is large), but cuts off smoothly when $k \gtrsim \Lambda$. We can think of the second term as the propagator of a fictitious heavy photon, whose contribution is subtracted from that of the ordinary photon. In terms involving the heavy photon, the numerator algebra is unchanged and the denominator is altered by

$$\Delta \longrightarrow \Delta_\Lambda = -xyq^2 + (1-z)^2 m^2 + z\Lambda^2. \tag{6.52}$$

The integral (6.50) is then replaced with a convergent integral, which can be Wick-rotated and evaluated:

$$\int \frac{d^4\ell}{(2\pi)^4} \left(\frac{\ell^2}{[\ell^2 - \Delta]^3} - \frac{\ell^2}{[\ell^2 - \Delta_\Lambda]^3} \right) = \frac{i}{(4\pi)^2} \int\limits_0^\infty d\ell_E^2 \left(\frac{\ell_E^4}{[\ell_E^2 + \Delta]^3} - \frac{\ell_E^4}{[\ell_E^2 + \Delta_\Lambda]^3} \right)$$

$$= \frac{i}{(4\pi)^2} \log\left(\frac{\Delta_\Lambda}{\Delta} \right). \tag{6.53}$$

The convergent terms in (6.47) are modified by terms of order Λ^{-2}, which we ignore.

   This prescription for rendering Feynman integrals finite by introducing fictitious heavy particles is known as *Pauli-Villars regularization*. Please note that the fictitious photon has no physical significance, and that this method is only one of many for defining the divergent integrals. (We will discuss other methods in the next chapter; see especially Problem 7.2.) We must hope that the new parameter Λ will not appear in our final results for observable cross sections.

   Using formulae (6.49) and (6.53) to evaluate the integrals in (6.47), we obtain an explicit, though complicated, expression for the one-loop vertex correction:

$$\begin{aligned}
\left\{ \text{(vertex diagram)} \right\} &= \frac{\alpha}{2\pi} \int\limits_0^1 dx\, dy\, dz\, \delta(x+y+z-1) \\
&\times \bar{u}(p') \left(\gamma^\mu \left[\log\frac{z\Lambda^2}{\Delta} + \frac{1}{\Delta}\left((1-x)(1-y)q^2 + (1-4z+z^2)m^2 \right) \right] \right. \\
&\quad \left. + \frac{i\sigma^{\mu\nu}q_\nu}{2m}\left[\frac{1}{\Delta}2m^2 z(1-z) \right] \right) u(p). \tag{6.54}
\end{aligned}$$

The bracketed expressions are our desired corrections to the form factors.

Before we try to interpret this result, let us summarize the calculational methods we used. The techniques are common to all loop calculations:

1. Draw the diagram(s) and write down the amplitude.

2. Introduce Feynman parameters to combine the denominators of the propagators.

3. Complete the square in the new denominator by shifting to a new loop momentum variable, ℓ.

4. Write the numerator in terms of ℓ. Drop odd powers of ℓ, and rewrite even powers using identities like (6.46).

5. Perform the momentum integral by means of a Wick rotation and four-dimensional spherical coordinates.

The momentum integral in the last step will often be divergent. In that case we must define (or *regularize*) the integral using the Pauli-Villars prescription or some other device.

Now that we have parametrized the ultraviolet divergence in (6.54), let us try to interpret it. Notice that the divergence appears in the worst possible place: It corrects $F_1(q^2 = 0)$, which should (according to our discussion at the end of the previous section) be fixed at the value 1. But this is the only effect of the divergent term. We will therefore adopt a simple but completely *ad hoc* fix for this difficulty: Subtract from the above expression a term proportional to the zeroth-order vertex function $(\bar{u}(p')\gamma^\mu u(p))$, in such a way as to maintain the condition $F_1(0) = 1$. In other words, make the substitution

$$\delta F_1(q^2) \to \delta F_1(q^2) - \delta F_1(0) \tag{6.55}$$

(where δF_1 denotes the first-order correction to F_1). The justification of this procedure involves the minor correction to our S-matrix formula (4.103) mentioned in Section 4.6. In brief, the term we are subtracting corrects for our omission of the external leg correction diagrams of (6.1). We postpone the justification of this statement until Section 7.2.

There is also an infrared divergence in $F_1(q^2)$, coming from the $1/\Delta$ term. For example, at $q^2 = 0$ this term is

$$\int\limits_0^1 dx\, dy\, dz\, \delta(x+y+z-1)\, \frac{1-4z+z^2}{\Delta(q^2=0)} = \int\limits_0^1 dz \int\limits_0^{1-z} dy\, \frac{-2+(1-z)(3-z)}{m^2(1-z)^2}$$

$$= \int\limits_0^1 dz\, \frac{-2}{m^2(1-z)} + \text{finite terms.}$$

We can cure this disease by pretending that the photon has a small nonzero mass μ. Then in the denominator of the photon propagator, $(k-p)^2$ would become $(k-p)^2 - \mu^2$. This denominator was multiplied by z in (6.43), so the net effect is to add a term $z\mu^2$ to Δ. We will discuss the infrared divergence further in the next two sections.

With both of these provisional modifications, the form factors are

$$F_1(q^2) = 1 + \frac{\alpha}{2\pi} \int_0^1 dx\, dy\, dz\, \delta(x+y+z-1)$$

$$\times \left[\log\left(\frac{m^2(1-z)^2}{m^2(1-z)^2 - q^2xy} \right) + \frac{m^2(1-4z+z^2) + q^2(1-x)(1-y)}{m^2(1-z)^2 - q^2xy + \mu^2 z} \right.$$

$$\left. - \frac{m^2(1-4z+z^2)}{m^2(1-z)^2 + \mu^2 z} \right] + \mathcal{O}(\alpha^2); \tag{6.56}$$

$$F_2(q^2) = \frac{\alpha}{2\pi} \int_0^1 dx\, dy\, dz\, \delta(x+y+z-1) \left[\frac{2m^2 z(1-z)}{m^2(1-z)^2 - q^2xy} \right] + \mathcal{O}(\alpha^2). \tag{6.57}$$

Note that neither the ultraviolet nor the infrared divergence affects $F_2(q^2)$. We can therefore evaluate unambiguously

$$F_2(q^2 = 0) = \frac{\alpha}{2\pi} \int_0^1 dx\, dy\, dz\, \delta(x + y + z - 1) \frac{2m^2 z(1 - z)}{m^2(1 - z)^2}$$

$$= \frac{\alpha}{\pi} \int_0^1 dz \int_0^{1-z} dy\, \frac{z}{1 - z} = \frac{\alpha}{2\pi}. \tag{6.58}$$

Thus, we get a correction to the g-factor of the electron:

$$a_e \equiv \frac{g - 2}{2} = \frac{\alpha}{2\pi} \approx .0011614. \tag{6.59}$$

This result was first obtained by Schwinger in 1948.* Experiments give $a_e = .0011597$. Apparently, the unambiguous value that we obtained for $F_2(0)$ is also, up to higher orders in α, unambiguously correct.

Precision Tests of QED

Building on the success of the order-α QED prediction for a_e, successive generations of physicists have improved the accuracy of both the theoretical and the experimental determination of this quantity. The coefficients of the QED formula for a_e are now known through order α^4. The calculation of the order-α^2 and higher coefficients requires a systematic treatment of ultraviolet divergences.

These challenging theoretical calculations have been matched by increasingly imaginative experiments. The most recent measurement of a_e uses a technique, developed by Dehmelt and collaborators, in which individual electrons are trapped in a system of electrostatic and magnetostatic fields and

---

*J. Schwinger, *Phys. Rev.* **73**, 416L (1948).

excited to a spin resonance.[†] Today, the best theoretical and experimental values of a_e agree to eight significant figures.

High-order QED calculations have also been carried óut for several other quantities. These include transition energies in hydrogen and hydrogen-like atoms, the anomalous magnetic moment of the muon, and the decay rates of singlet and triplet positronium. Many of these quantities have also been measured to high precision. The full set of these comparisons gives a detailed test of the validity of QED in a variety of settings. The results of these precision tests are summarized in Table 6.1.

There is some subtlety in reporting the results of precision comparisons between QED theory and experiment, since theoretical predictions require an extremely precise value of α, which can only be obtained from another precision QED experiment. We therefore quote each comparison between theory and experiment as an independent determination of α. Each value of α is assigned an error that is the composite of the expected uncertainties from theory and experiment. QED is confirmed to the extent that the values of α from different sources agree.

The first nine entries in Table 6.1 refer to QED calculations in atomic physics settings. Of these, the hydrogen hyperfine splitting, measured using Ramsey's hydrogen maser, is the most precisely known quantity in physics. Unfortunately, the influence of the internal structure of the proton leads to uncertainties that limit the accuracy with which this quantity can be predicted theoretically. The same difficulty applies to the Lamb shift, the splitting between the $j = 1/2$ $2S$ and $2P$ levels of hydrogen. The most accurate QED tests now come from systems that involve no strongly interacting particles, the electron $g-2$ and the hyperfine splitting in the $e^- \mu^+$ atom, muonium. The last entry in this group gives a new method for determining α, by converting a very accurate measurement of the neutron Compton wavelength, using accurately known mass ratios, to a value of the electron mass. This can be combined with the known value of the Rydberg energy and accurate QED formulae to determine α. The only serious discrepancy among these numbers comes in the triplet positronium decay rate; however, there is some evidence that diagrams of relative order α^2 give a large correction to the value quoted in the table.

The next two entries are determinations of α from higher-order QED reactions at high-energy electron colliders. These high-energy experiments typically achieve only percent-level accuracy, but their results are consistent with the precise information available at lower energies.

Finally, the last two entries in the table give two independent measurements of α from exotic quantum interference phenomena in condensed-matter systems. These two effects provide a standard resistance and a standard frequency, respectively, which are believed to measure the charge of the electron

---

[†]R. Van Dyck, Jr., P. Schwinberg, and H. Dehmelt, *Phys. Rev. Lett.* **59**, 26 (1987).

Table 6.1. Values of α^{-1} Obtained from Precision QED Experiments

Low-Energy QED:

Electron $(g - 2)$	137.035 992 35 (73)
Muon $(g - 2)$	137.035 5 (1 1)
Muonium hyperfine splitting	137.035 994 (18)
Lamb shift	137.036 8 (7)
Hydrogen hyperfine splitting	137.036 0 (3)
$2\,^3S_1\text{-}1\,^3S_1$ splitting in positronium	137.034 (16)
$^1S_0$ positronium decay rate	137.00 (6)
$^3S_1$ positronium decay rate	136.971 (6)
Neutron compton wavelength	137.036 010 1 (5 4)

High-Energy QED:

$\sigma(e^+e^- \to e^+e^-e^+e^-)$	136.5 (2.7)
$\sigma(e^+e^- \to e^+e^-\mu^+\mu^-)$	139.9 (1.2)

Condensed Matter:

Quantum Hall effect	137.035 997 9 (3 2)
AC Josephson effect	137.035 977 0 (7 7)

Each value of α displayed in this table is obtained by fitting an experimental measurement to a theoretical expression that contains α as a parameter. The numbers in parentheses are the standard errors in the last displayed digits, including both theoretical and experimental uncertainties. This table is based on results presented in the survey of precision QED of Kinoshita (1990). That book contains a series of lucid reviews of the remarkable theoretical and experimental technology that has been developed for the detailed analysis of QED processes. The five most accurate values are updated as given by T. Kinoshita in *History of Original Ideas and Basic Discoveries in Particle Physics*, H. Newman and T. Ypsilantis, eds. (Plenum Press, New York, 1995). This latter paper also gives an interesting perspective on the future of precision QED experiments.

with corrections that are strictly zero for macroscropic systems.[‡]

The entire picture fits together well beyond any reasonable expectation. On the evidence presented in this table, QED is the most stringently tested— and the most dramatically successful—of all physical theories.

[‡]For a discussion of these effects, and their exact relation to α, see D. R. Yennie, *Rev. Mod. Phys.* **59**, 781 (1987).

6.4 The Electron Vertex Function: Infrared Divergence

Now let us confront the infrared divergence in our result (6.56) for $F_1(q^2)$. The dominant part, in the $\mu \to 0$ limit, is

$$F_1(q^2) \approx \frac{\alpha}{2\pi} \int\limits_0^1 dx\, dy\, dz\, \delta(x+y+z-1) \left[\frac{m^2(1-4z+z^2) + q^2(1-x)(1-y)}{m^2(1-z)^2 - q^2 xy + \mu^2 z} \right.$$
$$\left. - \frac{m^2(1-4z+z^2)}{m^2(1-z)^2 + \mu^2 z} \right]. \quad (6.60)$$

To understand this expression we must do some work to simplify it, extracting and evaluating the divergent part of the integral. Throughout this section we will retain only terms that diverge in the limit $\mu \to 0$.

First note that the divergence occurs in the corner of Feynman-parameter space where $z \approx 1$ (and therefore $x \approx y \approx 0$). In this region we can set $z = 1$ and $x = y = 0$ in the numerators of (6.60). We can also set $z = 1$ in the μ^2 terms in the denominators. Using the delta function to evaluate the x-integral, we then have

$$F_1(q^2) = \frac{\alpha}{2\pi} \int\limits_0^1 dz \int\limits_0^{1-z} dy \left[\frac{-2m^2 + q^2}{m^2(1-z)^2 - q^2 y(1-z-y) + \mu^2} - \frac{-2m^2}{m^2(1-z)^2 + \mu^2} \right].$$

(The lower limit on the z-integral is unimportant.) Making the variable changes

$$y = (1-z)\xi, \qquad w = (1-z),$$

this expression becomes

$$F_1(q^2) = \frac{\alpha}{2\pi} \int\limits_0^1 d\xi \, \tfrac{1}{2} \int\limits_0^1 d(w^2) \left[\frac{-2m^2 + q^2}{[m^2 - q^2\xi(1-\xi)]\,w^2 + \mu^2} - \frac{-2m^2}{m^2 w^2 + \mu^2} \right]$$

$$= \frac{\alpha}{4\pi} \int\limits_0^1 d\xi \left[\frac{-2m^2 + q^2}{m^2 - q^2\xi(1-\xi)} \log\!\left(\frac{m^2 - q^2\xi(1-\xi)}{\mu^2} \right) + 2\log\!\left(\frac{m^2}{\mu^2} \right) \right].$$

In the limit $\mu \to 0$ we can ignore the details of the numerators inside the logarithms; anything proportional to m^2 or q^2 is effectively the same. We therefore write

$$F_1(q^2) = 1 - \frac{\alpha}{2\pi} f_{\mathrm{IR}}(q^2) \log\!\left(\frac{-q^2 \text{ or } m^2}{\mu^2} \right) + \mathcal{O}(\alpha^2), \quad (6.61)$$

where the coefficient of the divergent logarithm is

$$f_{\mathrm{IR}}(q^2) = \int\limits_0^1 \left(\frac{m^2 - q^2/2}{m^2 - q^2\xi(1-\xi)} \right) d\xi - 1. \quad (6.62)$$

Since q^2 is negative and $\xi(1-\xi)$ has a maximum value of $1/4$, the first term is greater than 1 and hence $f_{IR}(q^2)$ is positive.

How does this infinite term affect the cross section for electron scattering off a potential? Since $F_1(q^2)$ is just the quantity that multiplies γ^μ in the matrix element, we can find the new cross section by making the replacement $e \to e \cdot F_1(q^2)$. The cross section for the process $\mathbf{p} \to \mathbf{p}'$ is therefore

$$\frac{d\sigma}{d\Omega} \simeq \left(\frac{d\sigma}{d\Omega}\right)_0 \cdot \left[1 - \frac{\alpha}{\pi}f_{IR}(q^2)\log\left(\frac{-q^2 \text{ or } m^2}{\mu^2}\right) + \mathcal{O}(\alpha^2)\right], \tag{6.63}$$

where the first factor is the tree-level result. Note that the $\mathcal{O}(\alpha)$ correction to the cross section is not only infinite, but negative. Something is terribly wrong.

To gain a better understanding of the divergence, let us evaluate the coefficient of the divergent logarithm, $f_{IR}(q^2)$, in the limit $-q^2 \to \infty$. In this limit, we find a second logarithm:

$$\int_0^1 d\xi \frac{-q^2/2}{-q^2\xi(1-\xi) + m^2} \simeq \frac{1}{2}\int_0^1 d\xi\frac{-q^2}{-q^2\xi + m^2} + \left(\begin{array}{c}\text{equal contribution}\\\text{from } \xi \approx 1\end{array}\right)$$

$$= \log\left(\frac{-q^2}{m^2}\right). \tag{6.64}$$

The form factor in this limit is therefore

$$F_1(-q^2 \to \infty) = 1 - \frac{\alpha}{2\pi}\log\left(\frac{-q^2}{m^2}\right)\log\left(\frac{-q^2}{\mu^2}\right) + \mathcal{O}(\alpha^2). \tag{6.65}$$

Note that the numerator in the second logarithm is $-q^2$, not m^2; this expression contains not only the correct coefficient of $\log(1/\mu^2)$, but also the correct coefficient of $\log^2(q^2)$.

The same double logarithm of $-q^2$ appeared in the cross section for soft bremsstrahlung, Eq. (6.26). This correspondence points to a resolution of the infrared divergence problem. Comparing (6.65) with (6.26), we find in the limit $-q^2 \to \infty$

$$\frac{d\sigma}{d\Omega}(p \to p') = \left(\frac{d\sigma}{d\Omega}\right)_0\left[1 - \frac{\alpha}{\pi}\log\left(\frac{-q^2}{m^2}\right)\log\left(\frac{-q^2}{\mu^2}\right) + \mathcal{O}(\alpha^2)\right];$$

$$\frac{d\sigma}{d\Omega}(p \to p' + \gamma) = \left(\frac{d\sigma}{d\Omega}\right)_0\left[\ + \frac{\alpha}{\pi}\log\left(\frac{-q^2}{m^2}\right)\log\left(\frac{-q^2}{\mu^2}\right) + \mathcal{O}(\alpha^2)\right].$$

$$\tag{6.66}$$

The separate cross sections are divergent, but their sum is independent of μ and therefore finite.

In fact, neither the elastic cross section nor the soft bremsstrahlung cross section can be measured individually; only their sum is physically observable. In any real experiment, a photon detector can detect photons only down to some minimum limiting energy E_ℓ. The probability that a scattering event occurs and this detector does not see a photon is the sum

$$\frac{d\sigma}{d\Omega}(p \to p') + \frac{d\sigma}{d\Omega}(p \to p' + \gamma(k < E_\ell)) \equiv \left(\frac{d\sigma}{d\Omega}\right)_{\text{measured}}. \qquad (6.67)$$

The divergent part of this "measured" cross section is

$$\left(\frac{d\sigma}{d\Omega}\right)_{\text{measured}} \approx \left(\frac{d\sigma}{d\Omega}\right)_0 \left[1 - \frac{\alpha}{\pi} f_{\text{IR}}(q^2) \log\left(\frac{-q^2 \text{ or } m^2}{\mu^2}\right)\right.$$

$$\left. + \frac{\alpha}{2\pi} \mathcal{I}(\mathbf{v}, \mathbf{v}') \log\left(\frac{E_\ell^2}{\mu^2}\right) + \mathcal{O}(\alpha^2)\right].$$

We have just seen that $\mathcal{I}(\mathbf{v}, \mathbf{v}') = 2 f_{\text{IR}}(q^2)$ when $-q^2 \gg m^2$. If the same relation holds for general q^2, the measured cross section becomes

$$\left(\frac{d\sigma}{d\Omega}\right)_{\text{measured}} \approx \left(\frac{d\sigma}{d\Omega}\right)_0 \left[1 - \frac{\alpha}{\pi} f_{\text{IR}}(q^2) \log\left(\frac{-q^2 \text{ or } m^2}{E_\ell^2}\right) + \mathcal{O}(\alpha^2)\right], \qquad (6.68)$$

which depends on the experimental conditions, but no longer on μ^2. The infrared divergences from soft bremsstrahlung and from $F_1(q^2)$ cancel each other, yielding a finite cross section for the quantity that can actually be measured.

We must still verify the identity $\mathcal{I}(\mathbf{v}, \mathbf{v}') = 2 f_{\text{IR}}(q^2)$ for arbitrary values of q^2. From (6.13) we have

$$\mathcal{I}(\mathbf{v}, \mathbf{v}') = \int \frac{d\Omega_{\mathbf{k}}}{4\pi} \left(\frac{2p \cdot p'}{(\hat{k} \cdot p')(\hat{k} \cdot p)} - \frac{m^2}{(\hat{k} \cdot p')^2} - \frac{m^2}{(\hat{k} \cdot p)^2}\right). \qquad (6.69)$$

The last two terms are easy to evaluate:

$$\int \frac{d\Omega_{\mathbf{k}}}{4\pi} \frac{1}{(\hat{k} \cdot p)^2} = \frac{1}{2} \int_{-1}^{1} d\cos\theta \frac{1}{(p^0 - p\cos\theta)^2} = \frac{1}{p^2} = \frac{1}{m^2}.$$

In the first term, we can combine the denominators with a Feynman parameter and perform the integral in the same way:

$$\int \frac{d\Omega_{\mathbf{k}}}{4\pi} \frac{1}{(\hat{k} \cdot p')(\hat{k} \cdot p)} = \int_0^1 d\xi \int \frac{d\Omega_{\mathbf{k}}}{4\pi} \frac{1}{\left[\xi \hat{k} \cdot p' + (1-\xi)\hat{k} \cdot p\right]^2}$$

$$= \int_0^1 d\xi \frac{1}{\left[\xi p' + (1-\xi)p\right]^2} = \int_0^1 d\xi \frac{1}{m^2 - \xi(1-\xi)q^2}.$$

(In the last step we have used $2p \cdot p' = 2m^2 - q^2$.) Putting all the terms of (6.69) together, we find

$$\mathcal{I}(\mathbf{v}, \mathbf{v}') = \int_0^1 \left(\frac{2m^2 - q^2}{m^2 - \xi(1-\xi)q^2}\right) d\xi - 2 = 2 f_{\text{IR}}(q^2), \qquad (6.70)$$

just what we need to cancel the infrared divergence.

Although Eq. (6.68) demonstrates the cancellation of the infrared divergence, this result has little practical use. An experimentalist would want to know the precise dependence on q^2, which we did not evaluate carefully. Recall from (6.65), however, that we were careful to obtain the correct coefficient of $\log^2(-q^2)$ in the limit $-q^2 \gg m^2$. In that limit, therefore, (6.68) becomes

$$\left(\frac{d\sigma}{d\Omega}\right)_{\text{measured}} \approx \left(\frac{d\sigma}{d\Omega}\right)_0 \left[1 - \frac{\alpha}{\pi} \log\left(\frac{-q^2}{m^2}\right) \log\left(\frac{-q^2}{E_\ell^2}\right) + \mathcal{O}(\alpha^2)\right]. \qquad (6.71)$$

This result is unambiguous and useful. Note that the $\mathcal{O}(\alpha)$ correction again involves the Sudakov double logarithm.

6.5 Summation and Interpretation of Infrared Divergences

The discussion of infrared divergences in the previous section suffices for removing the infinities from our bremsstrahlung and vertex-correction calculations. There are still, however, three points that we have not addressed:

1. We have not demonstrated the cancellation of infrared divergences beyond the leading order.

2. The correction to the measured cross section that we found after the infrared cancellation (Eqs. (6.68) and (6.71)) can be made arbitrarily negative by making photon detectors with a sufficiently low threshold E_ℓ.

3. We have not yet reproduced the classical result (6.19) for the number of photons radiated during a collision.

The solutions of the second and third problems will follow immediately from that of the first, to which we now turn.

A complete treatment of infrared divergences to all orders is beyond the scope of this book.* We will discuss here only the terms with the largest logarithmic enhancement at each order of perturbation theory. In general, these terms are of order

$$\left[\frac{\alpha}{\pi} \log\left(\frac{-q^2}{\mu^2}\right) \log\left(\frac{-q^2}{m^2}\right)\right]^n \qquad (6.72)$$

in the nth order of perturbation theory. Our final physical conclusions were first presented by Bloch and Nordsieck in a prescient paper written before the invention of relativistic perturbation theory.[†] We will follow a modern, and simplified, version of the analysis due to Weinberg.[‡]

---

*The definitive treatment is given in D. Yennie, S. Frautschi, and H. Suura, *Ann. Phys.* **13**, 379 (1961).

[†]F. Bloch and A. Nordsieck, *Phys. Rev.* **52**, 54 (1937).

[‡]S. Weinberg, *Phys. Rev.* **140**, B516 (1965).

Infrared divergences arise from photons with "soft" momenta: real photons with energy less than some cutoff E_ℓ, and virtual photons with (after Wick rotation) $k^2 < E_\ell^2$. A typical higher-order diagram will involve numerous real and virtual photons. But to find a divergence, we need more than a soft photon; we need a singular denominator in an electron propagator. Consider, for example, the following two diagrams:

The first diagram, in which the electron emits a soft photon followed by a hard photon, has no infrared divergence, since the momenta in both electron propagators are far from the mass shell. If the soft photon is emitted last, however, the denominator of the adjacent propagator is $(p' + k)^2 - m^2 = 2p' \cdot k$, which vanishes as $k \to 0$. Thus the second diagram does contain a divergence. We would like, then, to consider diagrams in which an arbitrary hard process, possibly involving emission of hard and soft photons, is modified by the addition of soft real and virtual photons on the electron legs:

Following Weinberg, we will add up the contributions of all such diagrams. The only new difficulty in this calculation will be in the combinatorics of counting all the ways in which the photons can appear.

First consider the outgoing electron line:

We attach n photons to the line, with momenta $k_1 \ldots k_n$. For the moment we do not care whether these are external photons, virtual photons connected to each other, or virtual photons connected to vertices on the incoming electron line. The Dirac structure of this diagram is

$$\bar{u}(p')(-ie\gamma^{\mu_1})\frac{i(p\!\!\!/' + k\!\!\!/_1 + m)}{2p'\cdot k_1}(-ie\gamma^{\mu_2})\frac{i(p\!\!\!/' + k\!\!\!/_1 + k\!\!\!/_2 + m)}{2p'\cdot(k_1 + k_2) + \mathcal{O}(k^2)}$$

$$\cdots (-ie\gamma^{\mu_n})\frac{i(p\!\!\!/' + k\!\!\!/_1 + \cdots + k\!\!\!/_n + m)}{2p'\cdot(k_1 + \cdots k_n) + \mathcal{O}(k^2)}(i\mathcal{M}_{\text{hard}})\cdots. \tag{6.73}$$

We will assume that all the k_i are small, dropping the $\mathcal{O}(k^2)$ terms in the denominators. We will also drop the $k\!\!\!/_i$ terms in the numerators, just as in our treatment of bremsstrahlung in Section 6.1. Also, as we did there, we can push the factors of $(p\!\!\!/' + m)$ to the left and use $\bar{u}(p')(-p\!\!\!/' + m) = 0$:

$$\bar{u}(p')\gamma^{\mu_1}(p\!\!\!/' + m)\gamma^{\mu_2}(p\!\!\!/' + m)\cdots = \bar{u}(p')2p'^{\mu_1}\gamma^{\mu_2}(p\!\!\!/' + m)\cdots$$

$$= \bar{u}(p')2p'^{\mu_1}2p'^{\mu_2}\cdots.$$

This turns expression (6.73) into

$$\bar{u}(p')\left(e\frac{p'^{\mu_1}}{p'\cdot k_1}\right)\left(e\frac{p'^{\mu_2}}{p'\cdot(k_1 + k_2)}\right)\cdots\left(e\frac{p'^{\mu_n}}{p'\cdot(k_1 + \cdots + k_n)}\right)\cdots. \tag{6.74}$$

Still working with only the outgoing electron line, we must now sum over all possible orderings of the momenta $k_1 \ldots k_n$. (This procedure will overcount when two of the photons are attached together to form a single virtual photon. We will deal with this overcounting later.) There are $n!$ different diagrams to sum, corresponding to the $n!$ permutations of the n photon momenta. Let π denote one such permutation, so that $\pi(i)$ is the number between 1 and n that i is taken to. (For example, if π denotes the permutation that takes $1 \rightarrow 3$, $2 \rightarrow 1$ and $3 \rightarrow 2$, then $\pi(1) = 3$, $\pi(2) = 1$, and $\pi(3) = 2$.)

Armed with this notation, we can perform the sum over permutations by means of the following identity:

$$\sum_{\substack{\text{all permu-}\\\text{tations } \pi}} \frac{1}{p\cdot k_{\pi(1)}}\frac{1}{p\cdot(k_{\pi(1)} + k_{\pi(2)})}\cdots\frac{1}{p\cdot(k_{\pi(1)} + k_{\pi(2)} + \cdots + k_{\pi(n)})}$$

$$= \frac{1}{p\cdot k_1}\frac{1}{p\cdot k_2}\cdots\frac{1}{p\cdot k_n}. \tag{6.75}$$

The proof of this formula proceeds by induction on n. For $n = 2$ we have

$$\sum_{\pi}\frac{1}{p\cdot k_{\pi(1)}}\frac{1}{p\cdot(k_{\pi(1)} + k_{\pi(2)})} = \frac{1}{p\cdot k_1}\frac{1}{p\cdot(k_1 + k_2)} + \frac{1}{p\cdot k_2}\frac{1}{p\cdot(k_2 + k_1)}$$

$$= \frac{1}{p\cdot k_1}\frac{1}{p\cdot k_2}.$$

For the induction step, notice that the last factor on the left-hand side of (6.75) is the same for every permutation π. Pulling this factor outside the sum, the left-hand side becomes

$$\text{LHS} = \frac{1}{p \cdot \sum k} \sum_\pi \frac{1}{p \cdot k_{\pi(1)}} \frac{1}{p \cdot (k_{\pi(1)} + k_{\pi(2)})} \cdots \frac{1}{p \cdot (k_{\pi(1)} + \cdots + k_{\pi(n-1)})}.$$

For any given π, the quantity being summed is independent of $k_{\pi(n)}$. Letting $i = \pi(n)$, we can now write

$$\sum_\pi = \sum_{i=1}^{n} \sum_{\pi'(i)},$$

where $\pi'(i)$ is the set of all permutations on the remaining $n - 1$ integers. Assuming by induction that (6.75) is true for $n - 1$, we have

$$\text{LHS} = \frac{1}{p \cdot \sum k} \sum_{i=1}^{n} \frac{1}{p \cdot k_1} \frac{1}{p \cdot k_2} \cdots \frac{1}{p \cdot k_{i-1}} \frac{1}{p \cdot k_{i+1}} \cdots \frac{1}{p \cdot k_n}.$$

If we now multiply and divide each term in this sum by $p \cdot k_i$, we easily obtain our desired result (6.75).

Applying (6.75) to (6.74), we find

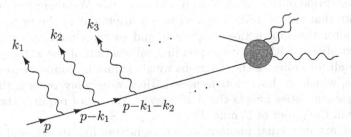

$$= \bar{u}(p') \left(e \frac{p'^{\mu_1}}{p' \cdot k_1} \right) \left(e \frac{p'^{\mu_2}}{p' \cdot k_2} \right) \cdots \left(e \frac{p'^{\mu_n}}{p' \cdot k_n} \right), \qquad (6.76)$$

where the blob denotes a sum over all possible orders of inserting the n photon lines.

A similar set of manipulations simplifies the sum over soft photon insertions on the initial electron line. There, however, the propagator momenta are $p - k_1$, $p - k_1 - k_2$, and so on:

We therefore get an extra minus sign in the factor for each photon, since $(p - \sum k)^2 - m^2 \approx -2p \cdot \sum k$.

Now consider diagrams containing a total of n soft photons, connected in any possible order to the initial or final electron lines. The sum over all such diagrams can be written

$$= \bar{u}(p') \, i\mathcal{M}_{\text{hard}} \, u(p)$$

$$\cdot e\Big(\frac{p'^{\mu_1}}{p' \cdot k_1} - \frac{p^{\mu_1}}{p \cdot k_1}\Big) \cdot e\Big(\frac{p'^{\mu_2}}{p' \cdot k_2} - \frac{p^{\mu_2}}{p \cdot k_2}\Big) \tag{6.77}$$

$$\cdots e\Big(\frac{p'^{\mu_n}}{p' \cdot k_n} - \frac{p^{\mu_n}}{p \cdot k_n}\Big).$$

By multiplying out all the factors, you can see that we get the correct term for each possible way of dividing the n photons between the two lines.

Next we must decide which photons are real and which are virtual.

We can make a virtual photon by picking two photon momenta k_i and k_j, setting $k_j = -k_i \equiv k$, multiplying by the photon propagator, and integrating over k. For each virtual photon we then obtain the expression

$$\frac{e^2}{2} \int \frac{d^4k}{(2\pi)^4} \frac{-i}{k^2 + i\epsilon} \Big(\frac{p'}{p' \cdot k} - \frac{p}{p \cdot k}\Big) \cdot \Big(\frac{p'}{-p' \cdot k} - \frac{p}{-p \cdot k}\Big) \equiv \mathbf{X}. \tag{6.78}$$

The factor of $1/2$ is required because our procedure has counted each Feynman diagram twice: interchanging k_i and k_j gives back the same diagram. It is possible to evaluate this expression by careful contour integration, but there is an easier way. Notice that this approximation scheme assigns to the diagram with one loop and no external photons the value

$$\bar{u}(p')\big(i\mathcal{M}_{\text{hard}}\big)u(p) \cdot \mathbf{X}.$$

Thus, $\mathbf{X}$ must be precisely the infrared limit of the one-loop correction to the form factor, as displayed in (6.61):

$$\mathbf{X} = -\frac{\alpha}{2\pi} f_{\text{IR}}(q^2) \log\Big(\frac{-q^2}{\mu^2}\Big). \tag{6.79}$$

A direct derivation of this result from (6.78) is given in Weinberg's paper cited above. Note that result (6.79) followed in our argument of the previous section only after the subtraction at $q^2 = 0$, and so we should worry whether (6.79) is consistent with the corresponding subtraction of the nth-order diagram. In addition, some of the diagrams we are summing contain external-leg corrections, which we have not discussed. Here we simply remark that neither of these subtleties affects the final answer; the proof requires the heavy machinery in the paper of Yennie, Frautschi, and Suura.

If there are m virtual photons we get m factors like (6.79), and also an additional symmetry factor of $1/m!$ since interchanging virtual photons with each other does not change the diagram. We can then sum over m to obtain the complete correction due to the presence of arbitrarily many soft virtual photons:

$$
\text{(diagram)} = \text{(diagram)} \times \sum_{m=0}^{\infty} \frac{\mathbf{X}^m}{m!} = \bar{u}(p')\big(i\mathcal{M}_{\text{hard}}\big)u(p)\exp(\mathbf{X}). \quad (6.80)
$$

If in addition to the m virtual photons we also emit a real photon, we must multiply by its polarization vector, sum over polarizations, and integrate the squared matrix element over the photon's phase space. This gives an additional factor

$$
\int \frac{d^3k}{(2\pi)^3} \frac{1}{2k} e^2(-g_{\mu\nu})\Big(\frac{p'^{\mu}}{p'\cdot k} - \frac{p^{\mu}}{p\cdot k}\Big)\Big(\frac{p'^{\nu}}{p'\cdot k} - \frac{p^{\nu}}{p\cdot k}\Big) \equiv \mathbf{Y} \quad (6.81)
$$

in the cross section. Assuming that the energy of the photon is greater than μ and less than E_ℓ (the detector threshold), this expression is simply

$$
\mathbf{Y} = \frac{\alpha}{\pi}\mathcal{I}(\mathbf{v},\mathbf{v}')\log\Big(\frac{E_\ell}{\mu}\Big) = \frac{\alpha}{\pi}f_{\text{IR}}(q^2)\log\Big(\frac{E_\ell^2}{\mu^2}\Big). \quad (6.82)
$$

If n real photons are emitted we get n such factors, and also a symmetry factor of $1/n!$ since there are n identical bosons in the final state. The cross section for emission of any number of soft photons is therefore

$$
\sum_{n=0}^{\infty} \frac{d\sigma}{d\Omega}(\mathbf{p}\to\mathbf{p}'+n\gamma) = \frac{d\sigma}{d\Omega}(\mathbf{p}\to\mathbf{p}')\cdot\sum_{n=0}^{\infty}\frac{1}{n!}\mathbf{Y}^n = \frac{d\sigma}{d\Omega}(\mathbf{p}\to\mathbf{p}')\cdot\exp(\mathbf{Y}).
$$
$$
(6.83)
$$

Combining our results for virtual and real photons gives our final result for the measured cross section, to all orders in α, for the process $\mathbf{p}\to\mathbf{p}'+$ (any number of photons with $k < E_\ell$):

$$
\Big(\frac{d\sigma}{d\Omega}\Big)_{\text{meas.}} = \Big(\frac{d\sigma}{d\Omega}\Big)_0 \times \exp(2\mathbf{X}) \times \exp(\mathbf{Y})
$$

$$
= \Big(\frac{d\sigma}{d\Omega}\Big)_0 \times \exp\Big[-\frac{\alpha}{\pi}f_{\text{IR}}(q^2)\log\Big(\frac{-q^2}{\mu^2}\Big)\Big] \times \exp\Big[\frac{\alpha}{\pi}f_{\text{IR}}(q^2)\log\Big(\frac{E_\ell^2}{\mu^2}\Big)\Big]
$$

$$
= \Big(\frac{d\sigma}{d\Omega}\Big)_0 \times \exp\Big[-\frac{\alpha}{\pi}f_{\text{IR}}(q^2)\log\Big(\frac{-q^2}{E_\ell^2}\Big)\Big]. \quad (6.84)
$$

The correction factor depends on the detector sensitivity E_ℓ, but is independent of the infrared cutoff μ. Note that if we expand this result to $\mathcal{O}(\alpha)$, we recover our earlier result (6.68). Now, however, the correction factor is controlled in magnitude—always between 0 and 1.

In the limit $-q^2 \gg m^2$, our result becomes

$$
\Big(\frac{d\sigma}{d\Omega}\Big)_{\text{meas.}} = \Big(\frac{d\sigma}{d\Omega}\Big)_0 \times \Big|\exp\Big[-\frac{\alpha}{2\pi}\log\Big(\frac{-q^2}{m^2}\Big)\log\Big(\frac{-q^2}{E_\ell^2}\Big)\Big]\Big|^2. \quad (6.85)
$$

In this limit, the probability of scattering without emitting a hard photon decreases faster than any power of q^2. The exponential correction factor, containing the Sudakov double logarithm, is known as the *Sudakov form factor*.

To conclude this section, let us calculate the probability, in the same approximation, that some hard scattering process is accompanied by the production of n soft photons, all with energies between E_- and E_+. The phase-space integral for these photons gives $\log(E_+/E_-)$ instead of $\log(E_\ell/\mu)$. If we assign photons with energy greater than E_+ to the "hard" part of the process, we find that the cross section is given by (6.84), times the additional factor

$$\text{Prob}(n\gamma \text{ with } E_- < E < E_+) = \frac{1}{n!}\left[\frac{\alpha}{\pi}f_{\text{IR}}(q^2)\log\left(\frac{E_+^2}{E_-^2}\right)\right]^n$$

$$\times \exp\left[-\frac{\alpha}{\pi}f_{\text{IR}}(q^2)\log\left(\frac{E_+^2}{E_-^2}\right)\right]. \qquad (6.86)$$

This expression has the form of a Poisson distribution,

$$P(n) = \frac{1}{n!}\lambda^n e^{-\lambda},$$

with

$$\lambda = \langle n \rangle = \frac{\alpha}{\pi}\log\left(\frac{E_+}{E_-}\right)\mathcal{I}(\mathbf{v}, \mathbf{v}').$$

This is precisely the semiclassical estimate of the number of radiated photons that we made in Eq. (6.19).

Problems

6.1   Rosenbluth formula. As discussed Section 6.2, the exact electromagnetic interaction vertex for a Dirac fermion can be written quite generally in terms of two form factors $F_1(q^2)$ and $F_2(q^2)$:

$$= \bar{u}(p')\left[\gamma^\mu F_1(q^2) + \frac{i\sigma^{\mu\nu}q_\nu}{2m}F_2(q^2)\right]u(p),$$

where $q = p' - p$ and $\sigma^{\mu\nu} = \frac{1}{2}i[\gamma^\mu, \gamma^\nu]$. If the fermion is a strongly interacting particle such as the proton, the form factors reflect the structure that results from the strong interactions and so are not easy to compute from first principles. However, these form factors can be determined experimentally. Consider the scattering of an electron with energy $E \gg m_e$ from a proton initially at rest. Show that the above expression for the vertex leads to the following expression (the Rosenbluth formula) for the elastic scattering cross section, computed to leading order in α but to all orders in the strong interactions:

$$\frac{d\sigma}{d\cos\theta} = \frac{\pi\alpha^2\left[(F_1^2 - \frac{q^2}{4m^2}F_2^2)\cos^2\frac{\theta}{2} - \frac{q^2}{2m^2}(F_1 + F_2)^2\sin^2\frac{\theta}{2}\right]}{2E^2[1 + \frac{2E}{m}\sin^2\frac{\theta}{2}]\sin^4\frac{\theta}{2}},$$

where θ is the lab-frame scattering angle and F_1 and F_2 are to be evaluated at the q^2 associated with elastic scattering at this angle. By measuring $(d\sigma/d\cos\theta)$ as a function of angle, it is thus possible to extract F_1 and F_2. Note that when $F_1 = 1$ and $F_2 = 0$, the Rosenbluth formula reduces to the Mott formula (in the massless limit) for scattering off a point particle (see Problem 5.1).

6.2   Equivalent photon approximation. Consider the process in which electrons of very high energy scatter from a target. In leading order in α, the electron is connected to the target by one photon propagator. If the initial and final energies of the electron are E and E', the photon will carry momentum q such that $q^2 \approx -2EE'(1 - \cos\theta)$. In the limit of forward scattering, whatever the energy loss, the photon momentum approaches $q^2 = 0$; thus the reaction is highly peaked in the forward direction. It is tempting to guess that, in this limit, the virtual photon becomes a real photon. Let us investigate in what sense that is true.

(a) The matrix element for the scattering process can be written as

$$\mathcal{M} = (-ie)\bar{u}(p')\gamma^\mu u(p)\frac{-ig_{\mu\nu}}{q^2}\widehat{\mathcal{M}}^\nu(q),$$

where $\widehat{\mathcal{M}}^\nu$ represents the (in general, complicated) coupling of the virtual photon to the target. Let us analyze the structure of the piece $\bar{u}(p')\gamma^\mu u(p)$. Let $q = (q^0, \mathbf{q})$, and define $\tilde{q} = (q^0, -\mathbf{q})$. We can expand the spinor product as:

$$\bar{u}(p')\gamma^\mu u(p) = A\cdot q^\mu + B\cdot\tilde{q}^\mu + C\cdot\epsilon_1^\mu + D\cdot\epsilon_2^\mu,$$

where A, B, C, D are functions of the scattering angle and energy loss and ϵ_i are two unit vectors transverse to $\mathbf{q}$. By dotting this expression with q_μ, show that the coefficient B is at most of order θ^2. This will mean that we can ignore it in the rest of the analysis. The coefficient A is large, but it is also irrelevant, since, by the Ward identity, $q^\mu\widehat{\mathcal{M}}_\mu = 0$.

(b) Working in the frame where $p = (E, 0, 0, E)$, compute explicitly

$$\bar{u}(p')\gamma\cdot\epsilon_i u(p)$$

using massless electrons, $u(p)$ and $u(p')$ spinors of definite helicity, and ϵ_1, ϵ_2 unit vectors parallel and perpendicular to the plane of scattering. We need this quantity only for scattering near the forward direction, and we need only the term of order θ. Note, however, that for ϵ in the plane of scattering, the small $\hat{3}$ component of ϵ also gives a term of order θ which must be taken into account.

(c) Now write the expression for the electron scattering cross section, in terms of $|\widehat{\mathcal{M}}^\mu|^2$ and the integral over phase space on the target side. This expression must be integrated over the final electron momentum p'. The integral over p'^3 is an integral over the energy loss of the electron. Show that the integral over $p'_\perp$ diverges logarithmically as $p'_\perp$ or $\theta \to 0$.

(d) The divergence as $\theta \to 0$ appears because we have ignored the electron mass in too many places. Show that reintroducing the electron mass in the expression for q^2,

$$q^2 = -2(EE' - pp'\cos\theta) + 2m^2,$$

cuts off the divergence and yields a factor of $\log(s/m^2)$ in its place.

(e) Assembling all the factors, and assuming that the target cross sections are independent of the photon polarization, show that the largest part of the electron-target scattering cross section is given by considering the electron to be the source of a beam of real photons with energy distribution ($x = E_\gamma/E$):

$$N_\gamma(x)dx = \frac{dx}{x} \frac{\alpha}{2\pi} [1 + (1-x)^2] \log\left(\frac{s}{m^2}\right).$$

This is the Weizsäcker-Williams *equivalent photon approximation*. This phenomenon allows us, for example, to study photon-photon scattering using e^+e^- collisions. Notice that the distribution we have found here is the same one that appeared in Problem 5.5 when we considered soft photon emission before electron scattering. It should be clear that a parallel general derivation can be constructed for that case.

6.3   Exotic contributions to $g-2$. Any particle that couples to the electron can produce a correction to the electron-photon form factors and, in particular, a correction to $g-2$. Because the electron $g-2$ agrees with QED to high accuracy, these corrections allow us to constrain the properties of hypothetical new particles.

(a) The unified theory of weak and electromagnetic interactions contains a scalar particle h called the *Higgs boson*, which couples to the electron according to

$$H_{\text{int}} = \int d^3x \, \frac{\lambda}{\sqrt{2}} \, h \, \bar\psi\psi.$$

Compute the contribution of a virtual Higgs boson to the electron $(g-2)$, in terms of λ and the mass m_h of the Higgs boson.

(b) QED accounts extremely well for the electron's anomalous magnetic moment. If $a = (g-2)/2$,

$$|a_{\text{expt.}} - a_{\text{QED}}| < 1 \times 10^{-10}.$$

What limits does this place on λ and m_h? In the simplest version of the electroweak theory, $\lambda = 3 \times 10^{-6}$ and $m_h > 60$ GeV. Show that these values are not excluded. The coupling of the Higgs boson to the muon is larger by a factor (m_μ/m_e): $\lambda = 6 \times 10^{-4}$. Thus, although our experimental knowledge of the muon anomalous magnetic moment is not as precise,

$$|a_{\text{expt.}} - a_{\text{QED}}| < 3 \times 10^{-8},$$

one can still obtain a stronger limit on m_h. Is it strong enough?

(c) Some more complex versions of this theory contain a pseudoscalar particle called the *axion*, which couples to the electron according to

$$H_{\text{int}} = \int d^3x \, \frac{i\lambda}{\sqrt{2}} \, a \, \bar\psi\gamma^5\psi.$$

The axion may be as light as the electron, or lighter, and may couple more strongly than the Higgs boson. Compute the contribution of a virtual axion to the $g-2$ of the electron, and work out the excluded values of λ and m_a.

Chapter 7

Radiative Corrections: Some Formal Developments

We cheated four times in the last three chapters,* stating (and sometimes motivating) a result but postponing its proof. Those results were:

1. The formula for decay rates in terms of S-matrix elements, Eq. (4.86).
2. The master formula for S-matrix elements in terms of Feynman diagrams, Eq. (4.103).
3. The Ward identity, Eq. (5.79).
4. The *ad hoc* subtraction to remove the ultraviolet divergence in the vertex-correction diagram, Eq. (6.55).

It is time now to return to these issues and give them a proper treatment. In Sections 7.2 through 7.4 we will derive all four of these results. The knowledge we gain along the way will help us interpret the three remaining loop corrections for electron scattering from a heavy target shown in (6.1): the external leg corrections and the vacuum polarization. We will evaluate the former in Section 7.1 and the latter in Section 7.5.

This chapter will be more abstract than the two preceding ones. Its main theme will be the singularities of Feynman diagrams viewed as analytic functions of their external momenta. We will find, however, that this apparently esoteric subject is rich in physical implications, and that it illuminates the relation between Feynman diagrams and the general principles of quantum theory.

7.1 Field-Strength Renormalization

In this section we will investigate the analytic structure of the two-point correlation function,

$$\langle \Omega | T\phi(x)\phi(y) | \Omega \rangle \qquad \text{or} \qquad \langle \Omega | T\psi(x)\bar{\psi}(y) | \Omega \rangle .$$

In a free field theory, the two-point function $\langle 0 | T\phi(x)\phi(y) | 0 \rangle$ has a simple interpretation: It is the amplitude for a particle to propagate from y to x. To what extent does this carry over into an interacting theory?

---

*A fifth cheat, postulating rather than deriving the photon propagator, will be remedied in Chapter 9.

Our analysis of the two-point function will rely only on general principles of relativity and quantum mechanics; it will not depend on the nature of the interactions or on an expansion in perturbation theory. We will, however, restrict our consideration to scalar fields. Similar results can be obtained for correlation functions of fields with spin; we will display the analogous result for Dirac fields at the end of the analysis.

To dissect the two-point function $\langle\Omega|\, T\phi(x)\phi(y)\,|\Omega\rangle$ we will insert the identity operator, in the form of a sum over a complete set of states, between $\phi(x)$ and $\phi(y)$. We choose these states to be eigenstates of the full interacting Hamiltonian, H. Since the momentum operator $\mathbf{P}$ commutes with H, we can also choose the states to be eigenstates of $\mathbf{P}$. But we can also make a stronger use of Lorentz invariance. Let $|\lambda_0\rangle$ be an eigenstate of H with momentum zero: $\mathbf{P}\,|\lambda_0\rangle = 0$. Then all the boosts of $|\lambda_0\rangle$ are also eigenstates of H, and these have all possible 3-momenta. Conversely, any eigenstate of H with definite momentum can be written as the boost of some zero-momentum eigenstate $|\lambda_0\rangle$. The eigenvalues of the 4-momentum operator $P^\mu = (H, \mathbf{P})$ organize themselves into hyperboloids, as shown in Fig. 7.1.

Recall from Chapter 2 that the completeness relation for the one-particle states is

$$(\mathbf{1})_{1-\text{particle}} = \int \frac{d^3p}{(2\pi)^3} \frac{1}{2E_{\mathbf{p}}}\, |\mathbf{p}\rangle\,\langle\mathbf{p}|\,. \tag{7.1}$$

We can write an analogous completeness relation for the entire Hilbert space with the aid of a bit of notation. Let $|\lambda_{\mathbf{p}}\rangle$ be the boost of $|\lambda_0\rangle$ with momentum $\mathbf{p}$, and assume that the states $|\lambda_{\mathbf{p}}\rangle$, like the one-particle states $|\mathbf{p}\rangle$, are relativistically normalized. Let $E_{\mathbf{p}}(\lambda) \equiv \sqrt{|\mathbf{p}|^2 + m_\lambda^2}$, where m_λ is the "mass" of the states $|\lambda_{\mathbf{p}}\rangle$, that is, the energy of the state $|\lambda_0\rangle$. Then the desired completeness relation is

$$\mathbf{1} = |\Omega\rangle\,\langle\Omega| + \sum_\lambda \int \frac{d^3p}{(2\pi)^3} \frac{1}{2E_{\mathbf{p}}(\lambda)}\, |\lambda_{\mathbf{p}}\rangle\,\langle\lambda_{\mathbf{p}}|\,, \tag{7.2}$$

where the sum runs over all zero-momentum states $|\lambda_0\rangle$.

We now insert this expansion between the operators in the two-point function. Assume for now that $x^0 > y^0$. Let us drop the uninteresting constant term $\langle\Omega|\,\phi(x)\,|\Omega\rangle\,\langle\Omega|\,\phi(y)\,|\Omega\rangle$. (This term is usually zero by symmetry; for higher-spin fields, it is zero by Lorentz invariance.) The two-point function is then

$$\langle\Omega|\,\phi(x)\phi(y)\,|\Omega\rangle = \sum_\lambda \int \frac{d^3p}{(2\pi)^3} \frac{1}{2E_{\mathbf{p}}(\lambda)}\, \langle\Omega|\,\phi(x)\,|\lambda_{\mathbf{p}}\rangle\,\langle\lambda_{\mathbf{p}}|\,\phi(y)\,|\Omega\rangle\,. \tag{7.3}$$

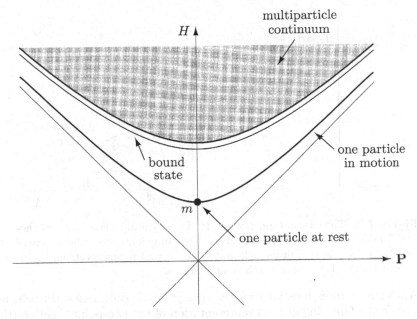

Figure 7.1. The eigenvalues of the 4-momentum operator $P^\mu = (H, \mathbf{P})$ occupy a set of hyperboloids in energy-momentum space. For a typical theory the states consist of one or more particles of mass m. Thus there is a hyperboloid of one-particle states and a continuum of hyperboloids of two-particle states, three-particle states, and so on. There may also be one or more bound-state hyperboloids below the threshold for creation of two free particles.

We can manipulate the matrix elements as follows:

$$\langle \Omega | \, \phi(x) \, | \lambda_{\mathbf{p}} \rangle = \langle \Omega | \, e^{iP \cdot x} \phi(0) e^{-iP \cdot x} \, | \lambda_{\mathbf{p}} \rangle$$

$$= \langle \Omega | \, \phi(0) \, | \lambda_{\mathbf{p}} \rangle \, e^{-ip \cdot x} \Big|_{p^0 = E_{\mathbf{p}}} \qquad (7.4)$$

$$= \langle \Omega | \, \phi(0) \, | \lambda_0 \rangle \, e^{-ip \cdot x} \Big|_{p^0 = E_{\mathbf{p}}}.$$

The last equality is a result of the Lorentz invariance of $\langle \Omega |$ and $\phi(0)$: Insert factors of $U^{-1}U$, where U is the unitary operator that implements a Lorentz boost from $\mathbf{p}$ to 0, and use $U\phi(0)U^{-1} = \phi(0)$. (For a field with spin, we would need to keep track of its nontrivial Lorentz transformation.) Introducing an integration over p^0, our expression for the two-point function (still for $x^0 > y^0$) becomes

$$\langle \Omega | \, \phi(x)\phi(y) \, | \Omega \rangle = \sum_\lambda \int \frac{d^4 p}{(2\pi)^4} \frac{i}{p^2 - m_\lambda^2 + i\epsilon} e^{-ip \cdot (x-y)} \big| \langle \Omega | \, \phi(0) \, | \lambda_0 \rangle \big|^2. \quad (7.5)$$

Note the appearance of the Feynman propagator, $D_F(x - y)$, but with m replaced by m_λ.

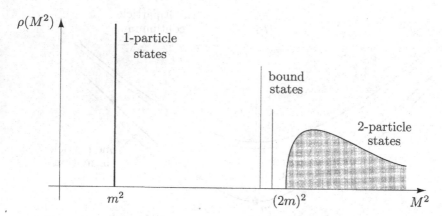

Figure 7.2. The spectral function $\rho(M^2)$ for a typical interacting field theory. The one-particle states contribute a delta function at m^2 (the square of the particle's mass). Multiparticle states have a continuous spectrum beginning at $(2m)^2$. There may also be bound states.

Analogous expressions hold for the case $y^0 > x^0$. Both cases can be summarized in the following general representation of the two-point function (the *Källén-Lehmann spectral representation*):

$$\langle \Omega | T\phi(x)\phi(y) |\Omega \rangle = \int_0^\infty \frac{dM^2}{2\pi}\, \rho(M^2)\, D_F(x - y; M^2), \qquad (7.6)$$

where $\rho(M^2)$ is a positive *spectral density* function,

$$\rho(M^2) = \sum_\lambda (2\pi)\delta(M^2 - m_\lambda^2)\big|\langle \Omega | \phi(0) |\lambda_0 \rangle\big|^2. \qquad (7.7)$$

The spectral density $\rho(M^2)$ for a typical theory is plotted in Fig. 7.2. Note that the one-particle states contribute an isolated delta function to the spectral density:

$$\rho(M^2) = 2\pi\, \delta(M^2 - m^2) \cdot Z + \text{(nothing else until } M^2 \gtrsim (2m)^2), \qquad (7.8)$$

where Z is some number given by the squared matrix element in (7.7). We refer to Z as the *field-strength renormalization*. The quantity m is the exact mass of a single particle—the exact energy eigenvalue at rest. This quantity will in general differ from the value of the mass parameter that appears in the Lagrangian. We will refer to the parameter in the Lagrangian as m_0, the *bare* mass, and refer to m as the *physical* mass of the ϕ boson. Only the physical mass m is directly observable.

The spectral decomposition (7.6) yields the following form for the Fourier

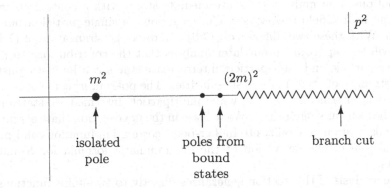

Figure 7.3. Analytic structure in the complex p^2-plane of the Fourier transform of the two-point function for a typical theory. The one-particle states contribute an isolated pole at the square of the particle mass. States of two or more free particles give a branch cut, while bound states give additional poles.

transform of the two-point function:

$$\int d^4x \, e^{ip\cdot x} \, \langle\Omega| \, T\phi(x)\phi(0) \, |\Omega\rangle = \int_0^\infty \frac{dM^2}{2\pi} \, \rho(M^2) \, \frac{i}{p^2-M^2+i\epsilon}$$

$$= \frac{iZ}{p^2-m^2+i\epsilon} + \int_{\sim 4m^2}^\infty \frac{dM^2}{2\pi} \, \rho(M^2) \, \frac{i}{p^2-M^2+i\epsilon}.$$

(7.9)

The analytic structure of this function in the complex p^2-plane is shown in Fig. 7.3. The first term gives an isolated simple pole at $p^2 = m^2$, while the second term contributes a branch cut beginning at $p^2 = (2m)^2$. If there are any two-particle bound states, these will appear as additional delta functions in $\rho(M^2)$ and thus as additional poles below the cut.

In Section 2.4, we found an explicit result for the two-point correlation function in the theory of a free scalar field:

$$\int d^4x \, e^{ip\cdot x} \, \langle 0| \, T\phi(x)\phi(0) \, |0\rangle = \frac{i}{p^2-m^2+i\epsilon}.$$

(7.10)

We interpreted this formula, for $x^0 > 0$, as the amplitude for a particle to propagate from 0 to x. Equation (7.9) shows that the two-point function takes a similar form in the most general theory of an interacting scalar field. The general expression is essentially a sum of scalar propagation amplitudes for states created from the vacuum by the field operator $\phi(0)$. There are two differences between (7.9) and (7.10). First, Eq. (7.9) contains the field-strength renormalization factor $Z = |\langle\lambda_0| \, \phi(0) \, |\Omega\rangle|^2$, the probability for $\phi(0)$ to create a given state from the vacuum. In (7.10), this factor is included implicitly, since $\langle p| \, \phi(0) \, |0\rangle = 1$ in free field theory. Second, Eq. (7.9) contains contributions from multiparticle intermediate states with a continuous mass spectrum. In free field theory, $\phi(0)$ can create only a single particle from the vacuum. With these two differences, (7.9) is a direct generalization of (7.10).

It will be important in our later analysis that the contributions to (7.9) from one-particle and multiparticle intermediate states can be distinguished by the strength of their analytic singularities. The poles in p^2 come only from one-particle intermediate states, while multiparticle intermediate states give weaker branch cut singularities. We will see in the next section that this rather formal observation generalizes to higher-point correlation functions and plays a crucial role in our derivation of the diagrammatic formula for S-matrix elements.

The analysis of this section generalizes directly to two-point functions of higher-spin fields. The main complication comes in the generalization of the manipulation (7.4), since now the field has a nontrivial transformation law under boosts. In general, several invariant spectral functions are required to represent the multiparticle states. But this complication does not affect the major result that a pole in p^2 can arise only from the contribution of a single-particle state created by the field operator. The two-point function of Dirac fields, for example, has the structure

$$
\int d^4x \, e^{ip \cdot x} \, \langle \Omega | \, T\psi(x)\bar{\psi}(0) \, | \Omega \rangle = \frac{iZ_2 \sum_s u^s(p)\bar{u}^s(p)}{p^2 - m^2 + i\epsilon} + \cdots
$$

$$
= \frac{iZ_2(\slashed{p} + m)}{p^2 - m^2 + i\epsilon} + \cdots,
$$

(7.11)

where the omitted terms give the multiparticle branch cut. As in the scalar case, the constant Z_2 is the probability for the quantum field to create or annihilate an exact one-particle eigenstate of H:

$$
\langle \Omega | \, \psi(0) \, | p, s \rangle = \sqrt{Z_2} \, u^s(p).
$$

(7.12)

(For an antiparticle, replace u with $\bar{v}$.) At the pole, the Dirac two-point function is exactly that of a free field with the physical mass, aside from the rescaling factor Z_2.

An Example: The Electron Self-Energy

This nonperturbative analysis of the two-point correlation function has been very different from our usual direct analysis of Feynman diagrams. But since this derivation was done in complete generality, the singularity structure of the two-point function that it implies ought also to be visible in a Feynman diagram computation. In the rest of this section we will explicitly check our results for the electron two-point function in QED.

The electron two-point function is equal to the sum of diagrams

$$
\langle \Omega | \, T\psi(x)\bar{\psi}(y) \, | \Omega \rangle = \underset{x \qquad\qquad y}{\xrightarrow{\hspace{2cm}}} \quad + \quad \underset{x \qquad\qquad y}{\overset{\frown}{\xrightarrow{\hspace{2cm}}}} \quad + \quad \cdots . \tag{7.13}
$$

Each of these diagrams, according to the Feynman rules for correlation functions, contains a factor of $e^{-ip\cdot(x-y)}$ for the two external points and an integration $\int (d^4p/(2\pi)^4)$ over the momentum p carried by the initial and final propagators. We will consistently omit these factors in this section; in other words, each diagram will denote the corresponding term in the Fourier transform of the two-point function.

The first diagram is just the free-field propagator:

$$\xrightarrow{\hspace{2cm}}_{\ p\ } \quad = \quad \frac{i(\not{p}+m_0)}{p^2 - m_0^2 + i\epsilon}. \tag{7.14}$$

Throughout this calculation, we will write m_0 instead of m as the mass in the electron propagator. This makes explicit the fact noted above that the mass appearing in the Lagrangian differs, in general, from the observable rest energy of a particle. However, if a perturbation expansion is applicable, the leading-order expression for the propagator should approximate the exact expression. Indeed, the function (7.14) has a pole, of just the form of (7.11), at $p^2 = m_0^2$. We therefore expect that the complete expression for the two-point function also has a pole of this form, at a slightly shifted location $m^2 = m_0^2 + \mathcal{O}(\alpha)$.

The second diagram in (7.13), called the *electron self-energy*, is somewhat more complicated:

$$= \frac{i(\not{p}+m_0)}{p^2 - m_0^2}\left[-i\Sigma_2(p)\right]\frac{i(\not{p}+m_0)}{p^2 - m_0^2}, \tag{7.15}$$

where

$$-i\Sigma_2(p) = (-ie)^2\int\frac{d^4k}{(2\pi)^4}\,\gamma^\mu\,\frac{i(\not{k}+m_0)}{k^2 - m_0^2 + i\epsilon}\,\gamma_\mu\,\frac{-i}{(p-k)^2 - \mu^2 + i\epsilon}. \tag{7.16}$$

(The notation Σ_2 indicates that this is the second-order (in e) contribution to a quantity Σ that we will define below.) The integral Σ_2 has an infrared divergence, which we have regularized by adding a small photon mass μ. Outside this integral, the diagram seems to have a double pole at $p^2 = m_0^2$. All in all, the form of this correction is quite unpleasant. But let us press on and try to evaluate $\Sigma_2(p)$ using the calculational techniques developed for the vertex correction in the Section 6.3.

First introduce a Feynman parameter to combine the two denominators:

$$\frac{1}{k^2 - m_0^2 + i\epsilon}\frac{1}{(p-k)^2 - \mu^2 + i\epsilon} = \int\limits_0^1 dx\,\frac{1}{\left[k^2 - 2xk\cdot p + xp^2 - x\mu^2 - (1-x)m_0^2 + i\epsilon\right]^2}.$$

Now complete the square and define a shifted momentum $\ell \equiv k - xp$. Dropping the term linear in ℓ from the numerator, we have

$$-i\Sigma_2(p) = -e^2 \int_0^1 dx \int \frac{d^4\ell}{(2\pi)^4} \frac{-2x\not{p} + 4m_0}{[\ell^2 - \Delta + i\epsilon]^2}, \qquad (7.17)$$

where $\Delta = -x(1-x)p^2 + x\mu^2 + (1-x)m_0^2$. The integral over ℓ is divergent. To evaluate it, we first regulate it using the Pauli-Villars procedure (6.51):

$$\frac{1}{(p-k)^2 - \mu^2 + i\epsilon} \rightarrow \frac{1}{(p-k)^2 - \mu^2 + i\epsilon} - \frac{1}{(p-k)^2 - \Lambda^2 + i\epsilon}.$$

The second term will have the same form as (7.17), but with μ replaced by Λ. As in Section 6.3, we now Wick-rotate and substitute the Euclidean variable $\ell_E^0 = -i\ell^0$. This gives

$$\int \frac{d^4\ell}{(2\pi)^4} \frac{1}{[\ell^2 - \Delta]^2} \rightarrow \frac{i}{(4\pi)^2} \int_0^\infty d\ell_E^2 \left(\frac{\ell_E^2}{[\ell_E^2 + \Delta]^2} - \frac{\ell_E^2}{[\ell_E^2 + \Delta_\Lambda]^2} \right)$$

$$= \frac{i}{(4\pi)^2} \log\left(\frac{\Delta_\Lambda}{\Delta} \right), \qquad (7.18)$$

where

$$\Delta_\Lambda = -x(1-x)p^2 + x\Lambda^2 + (1-x)m_0^2 \xrightarrow[\Lambda\to\infty]{} x\Lambda^2.$$

The final result is therefore

$$\Sigma_2(p) = \frac{\alpha}{2\pi} \int_0^1 dx\,(2m_0 - x\not{p}) \log\left(\frac{x\Lambda^2}{(1-x)m_0^2 + x\mu^2 - x(1-x)p^2} \right). \qquad (7.19)$$

Before discussing the divergences in this expression, let us work out its analytic behavior as a function of p^2. The logarithm in (7.19) has a branch cut when its argument becomes negative, and for any fixed x this will occur for sufficiently large p^2. More exactly, the cut begins at the point where

$$(1-x)m_0^2 + x\mu^2 - x(1-x)p^2 = 0.$$

Solving this equation for x, we find

$$x = \frac{1}{2} + \frac{m_0^2}{2p^2} - \frac{\mu^2}{2p^2} \pm \sqrt{\frac{(p^2 + m_0^2 - \mu^2)^2}{4p^4} - \frac{m_0^2}{p^2}}$$

$$= \frac{1}{2} + \frac{m_0^2}{2p^2} - \frac{\mu^2}{2p^2} \pm \frac{1}{2p^2} \sqrt{[p^2 - (m_0+\mu)^2][p^2 - (m_0-\mu)^2]}. \qquad (7.20)$$

The branch cut of $\Sigma_2(p^2)$ begins at the minimum value of p^2 such that this equation has a real solution for x between 0 and 1. This occurs when $p^2 = (m_0 + \mu)^2$, that is, at the threshold for creation of a two-particle (electron plus photon) state. In fact, it is a simple exercise in relativistic kinematics to show that the square root in (7.20), written in the form

$$k = \frac{1}{2\sqrt{p^2}} \sqrt{\left[p^2 - (m_0 + \mu)^2\right]\left[p^2 - (m_0 - \mu)^2\right]},$$

is precisely the momentum in the center-of-mass frame for two particles of mass m_0 and μ and total energy $\sqrt{p^2}$. It is natural that this momentum becomes real at the two-particle threshold. The location of the branch cut is exactly where we would expect from the Källén-Lehmann spectral representation.[†]

We have now located the two-particle branch cut predicted by the Källén-Lehmann representation, but we have not found the expected simple pole at $p^2 = m^2$. To find it we must actually include an infinite series of Feynman diagrams. Fortunately, this series will be easily summed.

Let us define a *one-particle irreducible* (1PI) diagram to be any diagram that cannot be split in two by removing a single line:

is 1PI, while is not.

Let $-i\Sigma(p)$ denote the sum of all 1PI diagrams with two external fermion lines:

$$= \text{...} + \text{...} + \text{...} + \cdots \qquad (7.21)$$

(Although each diagram has two external lines, the Feynman propagators for these lines are not to be included in the expression for $\Sigma(p)$.) To leading order in α we see that $\Sigma = \Sigma_2$.

The Fourier transform of the two-point function can now be written as

$$\int d^4x \, \langle\Omega|T\psi(x)\bar\psi(0)|\Omega\rangle \, e^{ip\cdot x} = \text{...}$$

$$= \text{...} + \text{...} + \text{...} + \cdots$$

$$= \frac{i(\not p + m_0)}{p^2 - m_0^2} + \frac{i(\not p + m_0)}{p^2 - m_0^2}(-i\Sigma)\frac{i(\not p + m_0)}{p^2 - m_0^2} + \cdots. \qquad (7.22)$$

[†]In real QED, $\mu = 0$ and the two-particle branch cut merges with the one-particle pole. This subtlety plays a role in the full treatment of the cancellation of infrared divergences, but it is beyond the scope of our present analysis.

The first diagram has a simple pole at $p^2 = m_0^2$. Each diagram in the second class has a double pole at $p^2 = m_0^2$. Each diagram in the third class has a triple pole. The behavior near $p^2 = m_0^2$ gets worse and worse as we include more and more diagrams. But fortunately, the sum of all the diagrams forms a geometric series. Note that $\Sigma(p)$ commutes with $\not{p}$, since $\Sigma(p)$ is a function only of pure numbers and $\not{p}$. In fact, we can consider $\Sigma(p)$ to be a function of $\not{p}$, writing $p^2 = (\not{p})^2$. Then we can rewrite each electron propagator as $i/(\not{p} - m_0)$ and express the above series as

$$\int d^4x \, \langle \Omega| T\psi(x)\bar{\psi}(0) |\Omega\rangle \, e^{ip\cdot x}$$

$$= \frac{i}{\not{p} - m_0} + \frac{i}{\not{p} - m_0}\left(\frac{\Sigma(\not{p})}{\not{p} - m_0}\right) + \frac{i}{\not{p} - m_0}\left(\frac{\Sigma(\not{p})}{\not{p} - m_0}\right)^2 + \cdots$$

$$= \frac{i}{\not{p} - m_0 - \Sigma(\not{p})}. \tag{7.23}$$

The full propagator has a simple pole, which is shifted away from m_0 by $\Sigma(\not{p})$.

The location of this pole, the physical mass m, is the solution of the equation

$$\left[\not{p} - m_0 - \Sigma(\not{p})\right]\big|_{\not{p}=m} = 0. \tag{7.24}$$

Notice that, if $\Sigma(\not{p})$ is defined by the convention (7.21), then a positive contribution to Σ yields a positive shift of the electron mass. Close to the pole, the denominator of (7.23) has the form

$$(\not{p} - m) \cdot \left(1 - \frac{d\Sigma}{d\not{p}}\bigg|_{\not{p}=m}\right) + \mathcal{O}((\not{p} - m)^2). \tag{7.25}$$

Thus the full electron propagator has a single-particle pole of just the form (7.11), with m given by (7.24) and

$$Z_2^{-1} = 1 - \frac{d\Sigma}{d\not{p}}\bigg|_{\not{p}=m}. \tag{7.26}$$

Our explicit calculation of Σ_2 allows us to compute the first corrections to m and Z_2. Let us begin with m. To order α, the mass shift is

$$\delta m = m - m_0 = \Sigma_2(\not{p} = m) \approx \Sigma_2(\not{p} = m_0). \tag{7.27}$$

Thus, using (7.19),

$$\delta m = \frac{\alpha}{2\pi}m_0 \int_0^1 dx \, (2-x) \log\left(\frac{x\Lambda^2}{(1-x)^2m_0^2 + x\mu^2}\right). \tag{7.28}$$

The mass shift is ultraviolet divergent; the divergent term has the form

$$\delta m \xrightarrow[\Lambda \to \infty]{} \frac{3\alpha}{4\pi}m_0 \log\left(\frac{\Lambda^2}{m_0^2}\right). \tag{7.29}$$

Is it a problem that m differs from m_0 by a divergent quantity? This question has two levels, those of concept and practice.

On the conceptual level, we should fully expect the mass of the electron to be modified by its coupling to the electromagnetic field. In classical electrodynamics, the rest energy of any charge is increased by the energy of its electrostatic field, and this energy shift diverges in the case of a point charge:

$$\int d^3r \, \tfrac{1}{2} |\mathbf{E}|^2 = \int d^3r \, \frac{1}{2} \left(\frac{e}{4\pi r^2} \right)^2 = \frac{\alpha}{2} \int \frac{dr}{r^2} \sim \alpha\Lambda. \tag{7.30}$$

In fact, it is puzzling why the divergence in (7.29) is so weak, logarithmic in Λ rather than linear as in (7.30). To understand this feature, suppose that m_0 were set to 0. Then the two helicity components of the electron field ψ_L and ψ_R would not be coupled by any term in the QED Hamiltonian. This would imply that perturbative corrections could never induce a coupling of ψ_L and ψ_R, nor, in particular, an electron mass term. In other words, δm must vanish when $m_0 = 0$. The mass shift must therefore be proportional to m_0, and so, by dimensional analysis, it can depend only logarithmically on Λ.

On a practical level, the infinite mass shift casts doubt on our perturbative calculations. For example, all of the theoretical results in Chapter 5 should technically involve m_0 rather than m. To compare theory to experiment we must eliminate m_0 in favor of m, using the relation $m_0 = m + \mathcal{O}(\alpha)$. Since the "small" $\mathcal{O}(\alpha)$ correction is infinite, the validity of this procedure is far from obvious. The validity of perturbation theory would be more plausible if we could compute Feynman diagrams using the propagator $i/(\not{p}-m)$, which has the correct pole location, instead of $i/(\not{p}-m_0)$. In Chapter 10 we will see how to rearrange the perturbation series in such a way that m_0 is systematically eliminated in favor of m and the zeroth-order propagator has its pole at the physical mass. In the remainder of this chapter, we will continue to simply replace m_0 by m in expressions for order-α corrections.

Finally, let us examine the perturbative correction to Z_2. From (7.26), we find that the order-α correction $\delta Z_2 = (Z_2 - 1)$ is

$$\delta Z_2 = \frac{d\Sigma_2}{d\not{p}} \bigg|_{\not{p}=m}$$

$$= \frac{\alpha}{2\pi} \int_0^1 dx \left[-x \log \frac{x\Lambda^2}{(1-x)^2 m^2 + x\mu^2} + 2(2-x) \frac{x(1-x)m^2}{(1-x)^2 m^2 + x\mu^2} \right]. \tag{7.31}$$

This expression is also logarithmically ultraviolet divergent. We will discuss the observability of this divergent term at the end of Section 7.2. However, it is interesting to note, even before that discussion, that (7.31) is very similar in form to the value of the *ad hoc* subtraction that we made in our calculation of the electron vertex correction in Section 6.3. From Eq. (6.56), the value of this subtraction was

$$\delta F_1(0) = \frac{\alpha}{2\pi} \int\limits_0^1 dx\, dy\, dz\, \delta(x+y+z-1)$$

$$\times \left[\log\left(\frac{z\Lambda^2}{(1-z)^2 m^2 + z\mu^2} \right) + \frac{(1-4z+z^2)m^2}{(1-z)^2 m^2 + z\mu^2} \right]$$

$$= \frac{\alpha}{2\pi} \int\limits_0^1 dz\,(1-z) \left[\log\left(\frac{z\Lambda^2}{(1-z)^2 m^2 + z\mu^2} \right) + \frac{(1-4z+z^2)m^2}{(1-z)^2 m^2 + z\mu^2} \right]. \quad (7.32)$$

Using the integration by parts

$$\int\limits_0^1 dz\,(1-2z) \log\left(\frac{\Lambda^2}{(1-z)^2 m^2 + z\mu^2} \right) = -\int\limits_0^1 dz\, z(1-z) \frac{2(1-z)m^2 - \mu^2}{(1-z)^2 m^2 + z\mu^2}$$

$$= \int\limits_0^1 dz \left[(1-z) - \frac{(1-z)(1-z^2)m^2}{(1-z)^2 m^2 + z\mu^2} \right],$$

it is not hard to show that $\delta F_1(0) + \delta Z_2 = 0$. This identity will play a crucial role in justifying the *ad hoc* subtraction of Section 6.3.

7.2 The LSZ Reduction Formula

In the last section we saw that the Fourier transform of the two-point correlation function, considered as an analytic function of p^2, has a simple pole at the mass of the one-particle state:

$$\int d^4x\, e^{ip \cdot x} \langle \Omega | T\phi(x)\phi(0) | \Omega \rangle \underset{p^2 \to m^2}{\sim} \frac{iZ}{p^2 - m^2 + i\epsilon}. \quad (7.33)$$

(Here and throughout this section we use the symbol $\sim$ to mean that the poles of both sides are identical; there are additional finite terms, given in this case by Eq. (7.9).) In this section we will generalize this result to higher correlation functions. We will derive a general relation between correlation functions and S-matrix elements first obtained by Lehmann, Symanzik, and Zimmermann and known as the *LSZ reduction formula*.[‡] This result, combined with our Feynman rules for computing correlation functions, will justify Eq. (4.103), our master formula for S-matrix elements in terms of Feynman diagrams. For simplicity, we will carry out the whole analysis for the case of scalar fields.

[‡]H. Lehmann, K. Symanzik, and W. Zimmermann, *Nuovo Cimento* **1**, 1425 (1955).

The strategy of the argument will be as follows. To calculate the S-matrix element for a 2-body $\to$ n-body scattering process, we begin with the correlation function of $n + 2$ Heisenberg fields. Fourier-transforming with respect to the coordinate of any one of these fields, we will find a pole of the form (7.33) in the Fourier-transform variable p^2. We will argue that the one-particle states associated with these poles are in fact asymptotic states, that is, states given by the limit of well-separated wavepackets as they become concentrated around definite momenta. Taking the limit in which all $n + 2$ external particles go on-shell, we can then interpret the coefficient of the multiple pole as an S-matrix element.

To begin, let us Fourier-transform the $(n + 2)$-point correlation function with respect to one argument x. We must then analyze the integral

$$\int d^4x \, e^{ip \cdot x} \, \langle \Omega | \, T\{\phi(x)\phi(z_1)\phi(z_2)\cdots\} \, |\Omega\rangle \, .$$

We would like to identify poles in the variable p^0. To do this, divide the integral over x^0 into three regions:

$$\int dx^0 = \int_{T_+}^{\infty} dx^0 + \int_{T_-}^{T_+} dx^0 + \int_{-\infty}^{T_-} dx^0 \, , \tag{7.34}$$

where T_+ is much greater than all the z_i^0 and T_- is much less than all the z_i^0. Call these three intervals regions I, II, and III. Since region II is bounded and the integrand depends on p^0 through the analytic function $\exp(ip^0 x^0)$, the contribution from this region will be analytic in p^0. However, regions I and III, which are unbounded, may develop singularities in p^0.

Consider first region I. Here x^0 is the latest time, so $\phi(x)$ stands first in the time ordering. Insert a complete set of intermediate states in the form of (7.2):

$$1 = \sum_{\lambda} \int \frac{d^3q}{(2\pi)^3} \frac{1}{2E_{\mathbf{q}}(\lambda)} \, |\lambda_{\mathbf{q}}\rangle \, \langle \lambda_{\mathbf{q}}| \, .$$

The integral over region I then becomes

$$\int_{T_+}^{\infty} dx^0 \int d^3x \, e^{ip^0 x^0} e^{-i\mathbf{p}\cdot\mathbf{x}} \sum_{\lambda} \int \frac{d^3q}{(2\pi)^3} \frac{1}{2E_{\mathbf{q}}(\lambda)} \, \langle \Omega | \, \phi(x) \, |\lambda_{\mathbf{q}}\rangle \tag{7.35}$$

$$\times \, \langle \lambda_{\mathbf{q}}| \, T\{\phi(z_1)\phi(z_2)\cdots\} \, |\Omega\rangle \, .$$

Using Eq. (7.4),

$$\langle \Omega | \, \phi(x) \, |\lambda_{\mathbf{q}}\rangle = \langle \Omega | \, \phi(0) \, |\lambda_0\rangle \, e^{-iq \cdot x} \Big|_{q^0 = E_{\mathbf{q}}(\lambda)},$$

and including a factor $e^{-\epsilon x^0}$ to insure that the integral is well defined, this integral becomes

$$\sum_\lambda \int_{T_+}^\infty dx^0 \int \frac{d^3q}{(2\pi)^3} \frac{1}{2E_{\mathbf{q}}(\lambda)} e^{ip^0 x^0} e^{-iq^0 x^0} e^{-\epsilon x^0} \langle\Omega|\,\phi(0)\,|\lambda_0\rangle\,(2\pi)^3\delta^{(3)}(\mathbf{p}-\mathbf{q})$$

$$\times \langle\lambda_{\mathbf{q}}|\,T\{\phi(z_1)\cdots\}\,|\Omega\rangle$$

$$= \sum_\lambda \frac{1}{2E_{\mathbf{p}}(\lambda)} \frac{ie^{i(p^0-E_{\mathbf{p}}+i\epsilon)T_+}}{p^0 - E_{\mathbf{p}}(\lambda) + i\epsilon} \langle\Omega|\,\phi(0)\,|\lambda_0\rangle\,\langle\lambda_{\mathbf{p}}|\,T\{\phi(z_1)\cdots\}\,|\Omega\rangle. \quad (7.36)$$

The denominator is just that of Eq. (7.5): $p^2 - m_\lambda^2$. There is an analytic singularity in p^0; as in Section 7.1, this singularity will be either a pole or a branch cut depending upon whether or not the rest energy m_λ is isolated. The one-particle state corresponds to an isolated energy value $p^0 = E_{\mathbf{p}} = \sqrt{|\mathbf{p}|^2 + m^2}$, and at this point Eq. (7.36) has a pole:

$$\int d^4x\, e^{ip\cdot x}\,\langle\Omega|\,T\{\phi(x)\phi(z_1)\cdots\}\,|\Omega\rangle$$

$$\underset{p^0\to+E_{\mathbf{p}}}{\sim} \frac{i}{p^2 - m^2 + i\epsilon}\sqrt{Z}\,\langle\mathbf{p}|\,T\{\phi(z_1)\cdots\}\,|\Omega\rangle. \quad (7.37)$$

The factor $\sqrt{Z}$ is the same field strength renormalization factor as in Eq. (7.8), since it replaces the same matrix element as in (7.7).

To evaluate the contribution from region III, we would put $\phi(x)$ last in the operator ordering, then insert a complete set of states between $T\{\phi(z_1)\cdots\}$ and $\phi(x)$. Repeating the above argument produces a pole as $p^0 \to -E_{\mathbf{p}}$:

$$\int d^4x\, e^{ip\cdot x}\,\langle\Omega|\,T\{\phi(x)\phi(z_1)\cdots\}\,|\Omega\rangle$$

$$\underset{p^0\to-E_{\mathbf{p}}}{\sim} \langle\Omega|\,T\{\phi(z_1)\cdots\}\,|-\mathbf{p}\rangle\sqrt{Z}\frac{i}{p^2 - m^2 + i\epsilon}. \quad (7.38)$$

Next we would like to Fourier-transform with respect to the remaining field coordinates. To keep the various external particles from interfering, however, we must isolate them from each other in space. Let us therefore repeat the preceding calculation using a wavepacket rather than a simple Fourier transform. In Eq. (7.35), replace

$$\int d^4x\, e^{ip^0 x^0} e^{-i\mathbf{p}\cdot\mathbf{x}} \rightarrow \int \frac{d^3k}{(2\pi)^3} \int d^4x\, e^{ip^0 x^0} e^{-i\mathbf{k}\cdot\mathbf{x}}\,\varphi(\mathbf{k}), \quad (7.39)$$

where $\varphi(\mathbf{k})$ is a narrow distribution centered on $\mathbf{k} = \mathbf{p}$. This distribution constrains x to lie within a band, whose spatial extent is that of the wavepacket, about the trajectory of a particle with momentum $\mathbf{p}$. With this modification, the right-hand side of (7.36) has a more complicated singularity structure:

$$\sum_\lambda \int \frac{d^3k}{(2\pi)^3}\,\varphi(\mathbf{k})\frac{1}{2E_{\mathbf{k}}(\lambda)}\frac{i}{p^0 - E_{\mathbf{k}}(\lambda) + i\epsilon} \langle\Omega|\,\phi(0)\,|\lambda_0\rangle\,\langle\lambda_{\mathbf{k}}|\,T\{\phi(z_1)\cdots\}\,|\Omega\rangle$$

$$\underset{p^0 \to +E_\mathbf{p}}{\sim} \int \frac{d^3k}{(2\pi)^3}\, \varphi(\mathbf{k}) \frac{i}{\tilde{p}^2 - m^2 + i\epsilon} \sqrt{Z}\, \langle \mathbf{k}|\, T\{\phi(z_1) \cdots\} \,|\Omega\rangle\,, \qquad (7.40)$$

where, in the second line, $\tilde{p} = (p_0, \mathbf{k})$. The one-particle singularity is now a branch cut, whose length is the width in momentum space of the wavepacket $\varphi(\mathbf{k})$. However, if $\varphi(\mathbf{k})$ defines the momentum narrowly, this branch cut is very short, and (7.40) has a well-defined limit in which $\varphi(\mathbf{k})$ tends to $(2\pi)^3 \delta^{(3)}(\mathbf{k} - \mathbf{p})$ and the singularity of (7.40) sharpens up to the pole of (7.36). The singularity due to single-particle states in the far past, Eq. (7.38), is modified in the same way.

Now consider integrating each of the coordinates in the $(n + 2)$-point correlation function against a wavepacket, to form*

$$\left(\prod_i \int \frac{d^3k_i}{(2\pi)^3} \int d^4x_i\, e^{i\tilde{p}_i \cdot x_i} \varphi_i(\mathbf{k}_i) \right) \langle \Omega|\, T\{\phi(x_1)\phi(x_2) \cdots\} \,|\Omega\rangle\,. \qquad (7.41)$$

The wavepackets should be chosen to overlap in a region around $x = 0$ and to separate in the far past and the far future. To analyze this integral, we choose a large positive time T_+ such that all of the wavepackets are well separated for $x_i^0 > T_+$, and we choose a large negative time T_- such that all of the wavepackets are well separated for $x_i^0 < T_-$. Then we can break up each of the integrals over x_i^0 into three regions as in (7.34). The integral of any x_i^0 over the bounded region II leads to an expression analytic in the corresponding energy p_0^i, so we can concentrate on the case in which all of the x_i^0 are placed at large past or future times.

For definiteness, consider the contribution in which only two of the time coordinates, x_1^0 and x_2^0, are in the future. In this case, the fields $\phi(x_1)$ and $\phi(x_2)$ stand to the left of the other fields in time order. Inserting a complete set of states $|\lambda_\mathbf{K}\rangle$, the integrations in (7.41) over the coordinates of these two fields take the form

$$\sum_\lambda \int \frac{d^3K}{(2\pi)^3} \frac{1}{2E_\mathbf{K}} \left(\prod_{i=1,2} \int \frac{d^3k_i}{(2\pi)^3} \int d^4x_i\, e^{i\tilde{p}_i \cdot x_i} \varphi_i(\mathbf{k}_i) \right)$$

$$\times \langle \Omega|\, T\{\phi(x_1)\phi(x_2)\} \,|\lambda_\mathbf{K}\rangle \langle \lambda_\mathbf{K}|\, T\{\phi(x_3) \cdots\} \,|\Omega\rangle\,.$$

The state $|\lambda_\mathbf{K}\rangle$ is annihilated by two field operators constrained to lie in distant wavepackets. It must therefore consist of two distinct excitations of the vacuum at two distinct locations. If these excitations are well separated,

---

*As in Section 4.5, the product symbol applies symbolically to the integrations as well as to the other factors within the parentheses; the x_i integrals apply to what lies outside the parentheses as well.

they should be independent of one another, so we can approximate

$$\sum_{\lambda} \int \frac{d^3K}{(2\pi)^3} \frac{1}{2E_\mathbf{K}} \langle\Omega|\, T\{\phi(x_1)\phi(x_2)\}\, |\lambda_\mathbf{K}\rangle \langle\lambda_\mathbf{K}|$$

$$= \sum_{\lambda_1,\lambda_2} \int \frac{d^3q_1}{(2\pi)^3} \frac{1}{2E_{\mathbf{q}1}} \int \frac{d^3q_2}{(2\pi)^3} \frac{1}{2E_{\mathbf{q}2}} \langle\Omega|\, \phi(x_1)\, |\lambda_{\mathbf{q}1}\rangle \langle\Omega|\, \phi(x_2)\, |\lambda_{\mathbf{q}2}\rangle \langle\lambda_{\mathbf{q}1}\lambda_{\mathbf{q}2}|.$$

The sums over λ_1 and λ_2 in this equation run over all zero-momentum states, but only single-particle states will contribute the poles we are looking for. In this case, the integrals over x_1^0 and $\mathbf{q}_1$ produce a sharp singularity in p_1^0 of the form of (7.40), and the integrals over x_2^0 and $\mathbf{q}_2$ produce the same singular behavior in p_2^0. The term in (7.41) with both singularities is

$$\left(\prod_{i=1,2} \int \frac{d^3k_i}{(2\pi)^3} \varphi_i(\mathbf{k}_i) \frac{i}{\tilde{p}_i^2 - m^2 + i\epsilon} \cdot \sqrt{Z}\right) \langle\mathbf{k}_1\mathbf{k}_2|\, T\{\phi(x_3)\cdots\}\, |\Omega\rangle.$$

In the limit in which the wavepackets tend to delta functions concentrated at definite momenta $\mathbf{p}_1$ and $\mathbf{p}_2$, this expression tends to

$$\left(\prod_{i=1,2} \frac{i}{p_i{}^2 - m^2 + i\epsilon} \cdot \sqrt{Z}\right)_{\text{out}}\langle\mathbf{p}_1\mathbf{p}_2|T\{\phi(x_3)\cdots\}\, |\Omega\rangle.$$

The state $\langle\mathbf{p}_1\mathbf{p}_2|$ is precisely an *out* state as defined in Section 4.5, since it is the definite-momentum limit of a state of particles constrained to well-separated wavepackets. Applying the same analysis to the times x_i^0 in the far past gives the result that the coefficient of the maximally singular term in the corresponding p_i^0 is a matrix element with an *in* state. This most singular term in (7.41) thus has the form

$$\left(\prod_{i=1,2} \frac{i}{p_i{}^2 - m^2 + i\epsilon} \cdot \sqrt{Z}\right)\left(\prod_{i=3,\ldots} \frac{i}{p_i{}^2 - m^2 + i\epsilon} \cdot \sqrt{Z}\right)_{\text{out}}\langle\mathbf{p}_1\mathbf{p}_2|-\mathbf{p}_3\cdots\rangle_{\text{in}}.$$

The last factor is just an S-matrix element.

We have now shown that we can extract the value of an S-matrix element by folding the corresponding vacuum expectation value of fields with wavepackets, extracting the leading singularities in the energies p_i^0, and then taking the limit as these wavepackets become delta functions of momenta. However, the computation would be made much simpler if we could do these operations in the reverse order—first letting the wavepackets become delta functions, returning us to the simple Fourier transform, and then extracting the singularities. In fact, the result for the leading singularity is not changed by this switch of the order of operations. It is, however, rather subtle to argue this point. Roughly, the explanation is the following: In the language of the analysis just completed, new singularities might arise because, in the Fourier transform, x_1 and x_2 can become close together in the far future. However, in this region, the exponential factor is close to $\exp[i(p_1+p_2) \cdot x_1]$, and thus the new singularities are single poles in the variable $(p_1^0 + p_2^0)$, rather than being products of poles in the two separate energy variables. A more careful argument (unfortunately, couched in a rather different language) can be found in the original paper of Lehmann, Symanzik, and Zimmermann cited at the beginning of this section.

Given the ability to make this reversal in the order of operations, we obtain a precise relation between Fourier transforms of correlation functions and S-matrix elements. This is the LSZ reduction formula:

$$\prod_1^n \int d^4x_i \, e^{ip_i \cdot x_i} \prod_1^m \int d^4y_j \, e^{-ik_j \cdot y_j} \, \langle \Omega | \, T\{\phi(x_1) \cdots \phi(x_n)\phi(y_1) \cdots \phi(y_m)\} \, | \Omega \rangle$$

$$\underset{\substack{\text{each } p_i^0 \to +E_{\mathbf{p}_i} \\ \text{each } k_j^0 \to +E_{\mathbf{k}_j}}}{\sim} \left(\prod_{i=1}^n \frac{\sqrt{Z}\,i}{p_i^2 - m^2 + i\epsilon} \right) \left(\prod_{j=1}^m \frac{\sqrt{Z}\,i}{k_j^2 - m^2 + i\epsilon} \right) \langle \mathbf{p}_1 \cdots \mathbf{p}_n | \, S \, | \mathbf{k}_1 \cdots \mathbf{k}_m \rangle .$$

$$(7.42)$$

The quantity Z that appears in this equation is exactly the field-strength renormalization constant, defined in Section 7.1 as the residue of the single-particle pole in the two-point function of fields. Each distinct particle will have a distinct renormalization factor Z, obtained from its own two-point function. For higher-spin fields, each factor of $\sqrt{Z}$ comes with a polarization factor such as $u^s(p)$, as in Eq. (7.12). The polarization s must be summed over in the second line of (7.42).

In words, the LSZ formula says that an S-matrix element can be computed as follows. Compute the appropriate Fourier-transformed correlation function, look at the region of momentum space where the external particles are near the mass shell, and identify the coefficient of the multiparticle pole. For fields with spin, one must also multiply by a polarization spinor (like $u^s(p)$) or vector (like $\epsilon^r(k)$) to project out the desired spin state.

Our next goal is to express this procedure in the language of Feynman diagrams. Let us analyze the relation between the diagrammatic expansion of the scalar field four-point function and the S-matrix element for 2-particle $\to$ 2-particle scattering. We will consider explicitly the fully connected Feynman diagrams contributing to the correlator. By a similar analysis, it is easy to confirm that disconnected diagrams should be disregarded because they do not have the singularity structure, with a product of four poles, indicated on the right-hand side of (7.42).

The exact four-point function

$$\left(\prod_1^2 \int d^4x_i \, e^{ip_i \cdot x_i} \right) \left(\prod_1^2 \int d^4y_i \, e^{-ik_i \cdot y_i} \right) \langle \Omega | \, T\{\phi(x_1)\phi(x_2)\phi(y_1)\phi(y_2)\} \, | \Omega \rangle$$

has the general form shown in Fig. 7.4. In this figure, we have indicated explicitly the diagrammatic corrections on each leg; the shaded circle in the center represents the sum of all amputated four-point diagrams.

We can sum up the corrections to each external leg just as we did for the electron propagator in the previous section. Let $-iM^2(p^2)$ denote the sum of

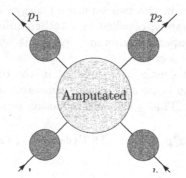

Figure 7.4. Structure of the exact four-point function in scalar field theory.

all one-particle-irreducible (1PI) insertions into the scalar propagator:

$$-iM^2(p^2) = \quad \text{⚬} \quad + \quad \text{⚬⚬} \quad + \quad \text{⚬} \quad + \cdots = \quad \text{—(1PI)—}$$

Then the exact propagator can be written as a geometric series and summed as in Eq. (7.23):

$$\text{⚫} = \text{——} + \text{—(1PI)—} + \text{—(1PI)—(1PI)—} + \cdots$$

$$= \frac{i}{p^2 - m_0^2} + \frac{i}{p^2 - m_0^2}(-iM^2)\frac{i}{p^2 - m_0^2} + \cdots$$

$$= \frac{i}{p^2 - m_0^2 - M^2(p^2)}. \tag{7.43}$$

Notice that, as in the case of the electron propagator, our sign convention for the 1PI self-energy $M^2(p^2)$ implies that a positive contribution to $M^2(p^2)$ corresponds to a positive shift of the scalar particle mass. If we expand each resummed propagator about the physical particle pole, we see that each external leg of the four-point amplitude contributes

$$\frac{i}{p^2 - m_0^2 - M^2} \underset{p^0 \to E_\mathbf{p}}{\sim} \frac{iZ}{p^2 - m^2} + \text{(regular)}. \tag{7.44}$$

Thus, the sum of diagrams contains a product of four poles:

$$\frac{iZ}{p_1^2 - m^2} \frac{iZ}{p_2^2 - m^2} \frac{iZ}{k_1^2 - m^2} \frac{iZ}{k_2^2 - m^2}.$$

This is exactly the singularity on the second line of (7.42). Comparing the coefficients of this product of poles, we find the relation

$$\langle \mathbf{p}_1 \mathbf{p}_2 | S | \mathbf{k}_1 \mathbf{k}_2 \rangle = \left(\sqrt{Z} \right)^4 \quad \text{(Amp.)} \quad ,$$

where the shaded circle is the sum of amputated four-point diagrams and Z is the field-strength renormalization factor.

An identical analysis can be applied to the Fourier transform of the $(n + 2)$-point correlator in a general field theory. The relation between S-matrix elements and Feynman diagrams then takes the form

$$\langle \mathbf{p}_1 \dots \mathbf{p}_n | S | \mathbf{k}_1 \mathbf{k}_2 \rangle = \left(\sqrt{Z} \right)^{n+2} \quad \text{(Amp.)} \quad . \tag{7.45}$$

(If the external particles are of different species, each has its own renormalization factor $\sqrt{Z}$; if the particles have nonzero spin, there will be additional polarization factors such as $u^s(k)$ on the right-hand side.) This is almost precisely the diagrammatic formula for the S-matrix that we wrote down in Section 4.6. The only new feature is the appearance of the renormalization factors $\sqrt{Z}$. The Z factors are irrelevant for calculations at the leading order of perturbation theory, but are important in the calculation of higher-order corrections.

Up to this point, we have performed only one full calculation of a higher-order correction, the computation of the order-α corrections to the electron form factors. We did not take into account the effects of the electron field-strength renormalization. Let us now add in this factor and examine its effects.

Since the expressions (6.28) and (6.30) for electron scattering from a heavy target were derived using our previous, incorrect formula for S-matrix elements, we should correct these formulae by inserting factors of $\sqrt{Z_2}$ for the initial and final electrons. Equation (6.33) for the structure of the exact vertex should then read

$$Z_2 \Gamma^\mu(p', p) = \gamma^\mu F_1(q^2) + \frac{i\sigma^{\mu\nu} q_\nu}{2m} F_2(q^2), \tag{7.46}$$

with $\Gamma^\mu(p', p)$ the sum of amputated electron-photon vertex diagrams.

We can use this equation to reevaluate the form factors to order α. Since $Z_2 = 1 + \mathcal{O}(\alpha)$ and F_2 begins in order α, our previous computation of F_2 is unaffected. To compute F_1, write the left-hand side of (7.46) as

$$Z_2 \Gamma^\mu = (1 + \delta Z_2)(\gamma^\mu + \delta \Gamma^\mu) = \gamma^\mu + \delta \Gamma^\mu + \gamma^\mu \cdot \delta Z_2,$$

where δZ_2 and $\delta \Gamma^\mu$ denote the order-α corrections to these quantities. Comparing to the right-hand side of (7.46), we see that $F_1(q^2)$ receives a new contribution equal to δZ_2. Now let $\delta F_1(q^2)$ denote the (unsubtracted) correction to the form factor that we computed in Section 6.3, and recall from the end of·Section 7.1 that $\delta Z_2 = -\delta F_1(0)$. Then

$$F_1(q^2) = 1 + \delta F_1(q^2) + \delta Z_2 = 1 + \left[\delta F_1(q^2) - \delta F_1(0)\right].$$

This is exactly the result we claimed, but did not prove, in Section 6.3. The inclusion of field-strength renormalization justifies the subtraction procedure that we applied on an *ad hoc* basis there.

At this level of analysis, it is difficult to see how the cancellation of divergences in F_1 will persist to higher orders. Worse, though we argued in Section 6.3 for the general result $F_1(0) = 1$, our verification of this result in order α seems to depend on a numerical coincidence.

We can state this problem carefully as follows: Define a second rescaling factor Z_1 by the relation

$$\Gamma^\mu(q = 0) = Z_1^{-1}\gamma^\mu, \tag{7.47}$$

where Γ^μ is the complete amputated vertex function. To find $F_1(0) = 1$, we must prove the identity $Z_1 = Z_2$, so that the vertex rescaling exactly compensates the electron field-strength renormalization. We will prove this identity to all orders in perturbation theory at the end Section 7.4.

We conclude our discussion of the LSZ reduction formula with one further formal observation. The LSZ formula distinguishes in and out particles only by the sign of the Fourier transform momentum p_i^0 or k_i^0. This means that, by analytically continuing the residue of the pole in p^2 from positive to negative p^0, we can convert the S-matrix element with $\phi(\mathbf{p})$ in the final state into the S-matrix element with the antiparticle $\phi^*(-\mathbf{p})$ in the initial state. This is exactly the statement of *crossing symmetry*, which we derived diagrammatically in Section 5.4:

$$\langle\cdots\phi(p)|\,S\,|\cdots\rangle\big|_{p=-k} = \langle\cdots|\,S\,|\phi^*(k)\cdots\rangle.$$

Since the proof of the LSZ formula does not depend on perturbation theory, we see that the crossing symmetry of the S-matrix is a general result of quantum theory, not merely a property of Feynman diagrams.

7.3 The Optical Theorem

In Section 7.1 we saw that the two-point correlation function of quantum fields, viewed as an analytic function of the momentum p^2, has branch cut singularities associated with multiparticle intermediate states. This conclusion should not be surprising to those familiar with nonrelativistic scattering theory, since it is already true there that the scattering amplitude, as a function of energy, has a branch cut on the positive real axis. The imaginary part of the scattering amplitude appears as a discontinuity across this branch cut. By the *optical theorem*, the imaginary part of the forward scattering amplitude is

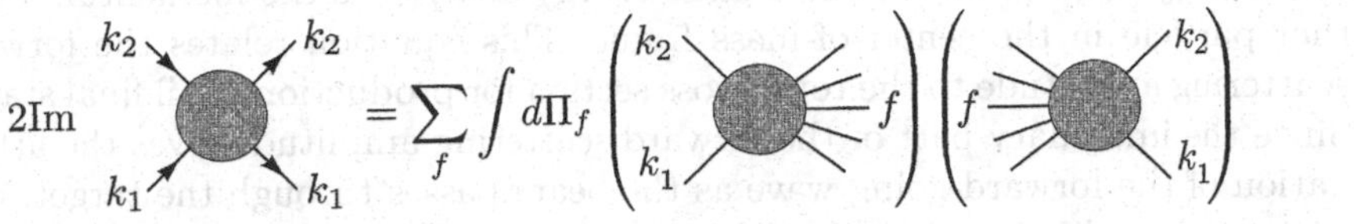

Figure 7.5. The optical theorem: The imaginary part of a forward scattering amplitude arises from a sum of contributions from all possible intermediate-state particles.

proportional to the total cross section. We will now prove the field-theoretic version of the optical theorem and illustrate how it arises in Feynman diagram calculations.

The optical theorem is a straightforward consequence of the unitarity of the S-matrix: $S^\dagger S = 1$. Inserting $S = 1 + iT$ as in (4.72), we have

$$-i(T - T^\dagger) = T^\dagger T. \tag{7.48}$$

Let us take the matrix element of this equation between two-particle states $|\mathbf{p}_1\mathbf{p}_2\rangle$ and $|\mathbf{k}_1\mathbf{k}_2\rangle$. To evaluate the right-hand side, insert a complete set of intermediate states:

$$\langle \mathbf{p}_1\mathbf{p}_2| T^\dagger T |\mathbf{k}_1\mathbf{k}_2\rangle = \sum_n \left(\prod_{i=1}^{n} \int \frac{d^3 q_i}{(2\pi)^3} \frac{1}{2E_i}\right) \langle \mathbf{p}_1\mathbf{p}_2| T^\dagger |\{\mathbf{q}_i\}\rangle \langle\{\mathbf{q}_i\}| T |\mathbf{k}_1\mathbf{k}_2\rangle.$$

Now express the T-matrix elements as invariant matrix elements $\mathcal{M}$ times 4-momentum-conserving delta functions. Identity (7.48) then becomes

$$- i\big[\mathcal{M}(k_1 k_2 \to p_1 p_2) - \mathcal{M}^*(p_1 p_2 \to k_1 k_2)\big]$$

$$= \sum_n \left(\prod_{i=1}^{n} \int \frac{d^3 q_i}{(2\pi)^3} \frac{1}{2E_i}\right) \mathcal{M}^*(p_1 p_2 \to \{q_i\})\mathcal{M}(k_1 k_2 \to \{q_i\})$$

$$\times (2\pi)^4 \delta^{(4)}(k_1 + k_2 - \sum_i q_i),$$

times an overall delta function $(2\pi)^4 \delta^{(4)}(k_1 + k_2 - p_1 - p_2)$. Let us abbreviate this identity as

$$-i\big[\mathcal{M}(a \to b) - \mathcal{M}^*(b \to a)\big] = \sum_f \int d\Pi_f\, \mathcal{M}^*(b \to f)\mathcal{M}(a \to f), \tag{7.49}$$

where the sum runs over all possible sets f of final-state particles. Although we have so far assumed that a and b are two-particle states, they could equally well be one-particle or multiparticle asymptotic states.

For the important special case of forward scattering, we can set $p_i = k_i$ to obtain a simpler identity, shown pictorially in Fig. 7.5. Supplying the kinematic factors required by (4.79) to build a cross section, we obtain the standard form of the optical theorem,

$$\operatorname{Im} \mathcal{M}(k_1, k_2 \to k_1, k_2) = 2E_{\mathrm{cm}} p_{\mathrm{cm}} \sigma_{\mathrm{tot}}(k_1, k_2 \to \text{anything}), \tag{7.50}$$

where E_{cm} is the total center-of-mass energy and p_{cm} is the momentum of either particle in the center-of-mass frame. This equation relates the forward scattering amplitude to the total cross section for production of all final states. Since the imaginary part of the forward scattering amplitude gives the attenuation of the forward-going wave as the beam passes through the target, it is natural that this quantity should be proportional to the probability of scattering. Identity (7.50) gives the precise connection.

The Optical Theorem for Feynman Diagrams

Let us now investigate how this identity for the imaginary part of an S-matrix element arises in the Feynman diagram expansion. It is easily checked (in QED, for example) that each diagram contributing to an S-matrix element $\mathcal{M}$ is purely real unless some denominators vanish, so that the $i\epsilon$ prescription for treating the poles becomes relevant. A Feynman diagram thus yields an imaginary part for $\mathcal{M}$ only when the virtual particles in the diagram go on-shell. We will now show how to isolate and compute this imaginary part.

For our present purposes, let us *define* $\mathcal{M}$ by the Feynman rules for perturbation theory. This allows us to consider $\mathcal{M}(s)$ as an analytic function of the complex variable $s = E_{cm}^2$, even though S-matrix elements are defined only for external particles with real momenta.

We first demonstrate that the appearance of an imaginary part of $\mathcal{M}(s)$ always requires a branch cut singularity. Let s_0 be the threshold energy for production of the lightest multiparticle state. For real s below s_0 the intermediate state cannot go on-shell, so $\mathcal{M}(s)$ is real. Thus, for real $s < s_0$, we have the identity

$$\mathcal{M}(s) = \big[\mathcal{M}(s^*)\big]^*. \qquad (7.51)$$

Each side of this equation is an analytic function of s, so it can be analytically continued to the entire complex s plane. In particular, near the real axis for $s > s_0$, Eq. (7.51) implies

$$\mathrm{Re}\,\mathcal{M}(s + i\epsilon) = \mathrm{Re}\,\mathcal{M}(s - i\epsilon);$$

$$\mathrm{Im}\,\mathcal{M}(s + i\epsilon) = -\,\mathrm{Im}\,\mathcal{M}(s - i\epsilon).$$

There is a branch cut across the real axis, starting at the threshold energy s_0; the discontinuity across the cut is

$$\mathrm{Disc}\,\mathcal{M}(s) = 2i\,\mathrm{Im}\,\mathcal{M}(s + i\epsilon).$$

Usually it is easier to compute the discontinuity of a diagram than to compute the imaginary part directly. The $i\epsilon$ prescription in the Feynman propagator indicates that physical scattering amplitudes should be evaluated above the cut, at $s + i\epsilon$.

We already saw in Section 7.1 that the electron self-energy diagram has a branch cut beginning at the physical electron-photon threshold. Let us now study more general one-loop diagrams, and show that their discontinuities give precisely the imaginary parts required by (7.49). The generalization of this result to multiloop diagrams has been proven by Cutkosky,[†] who showed in the process that the discontinuity of a Feynman diagram across its branch cut is always given by a simple set of *cutting rules.*[‡]

We begin by checking (7.49) in ϕ^4 theory. Since the right-hand side of (7.49) begins in order λ^2, we expect that $\operatorname{Im}\mathcal{M}$ should also receive its first contribution from higher-order diagrams. Consider, then, the order-λ^2 diagram

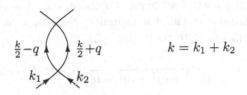

$$\frac{k}{2}-q \qquad \frac{k}{2}+q \qquad\qquad k = k_1 + k_2$$

$$k_1 \qquad k_2$$

with a loop in the s-channel. (It is easy to check that the corresponding t- and u-channel diagrams have no branch cut singularities for s above threshold.) The total momentum is $k = k_1 + k_2$, and for simplicity we have chosen the symmetrical routing of momenta shown above. The value of this Feynman diagram is

$$i\delta\mathcal{M} = \frac{\lambda^2}{2}\int\frac{d^4q}{(2\pi)^4}\frac{1}{(k/2-q)^2-m^2+i\epsilon}\frac{1}{(k/2+q)^2-m^2+i\epsilon}. \tag{7.52}$$

When this integral is evaluated using the methods of Section 6.3, the Wick rotation produces an extra factor of i, so that, below threshold, $\delta\mathcal{M}$ is purely real.

We would like to verify that the integral (7.52) has a discontinuity across the real axis in the physical region $k^0 > 2m$. It is easiest to identify this discontinuity by computing the integral for $k^0 < 2m$, then increasing k^0 by analytic continuation. It is not difficult to compute the integral directly using Feynman parameters (see Problem 7.1). However, it is illuminating to use a less direct approach, as follows.

Let us work in the center-of-mass frame, where $k = (k^0, \mathbf{0})$. Then the integrand of (7.52) has four poles in the integration variable q^0, at the locations

$$q^0 = \tfrac{1}{2}k^0 \pm (E_\mathbf{q} - i\epsilon), \qquad q^0 = -\tfrac{1}{2}k^0 \pm (E_\mathbf{q} - i\epsilon).$$

[†]R. E. Cutkosky, *J. Math. Phys.* **1**, 429 (1960).

[‡]These rules are simple only for singularities in the physical region. Away from the physical region, the singularities of three- and higher-point amplitudes can become quite intricate. This subject is reviewed in R. J. Eden, P. V. Landshoff, D. I. Olive, and J. C. Polkinghorne, *The Analytic S-Matrix* (Cambridge University Press, 1966).

Two of these poles lie above the real q^0 axis and two lie below, as shown:

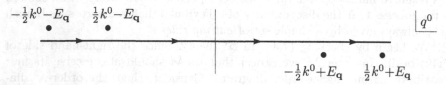

We will close the integration contour downward and pick up the residues of the poles in the lower half-plane. Of these, only the pole at $q^0 = -(1/2)k^0 + E_{\mathbf{q}}$ will contribute to the discontinuity. Note that picking up the residue of this pole is equivalent to replacing

$$\frac{1}{(k/2+q)^2 - m^2 + i\epsilon} \to -2\pi i \delta\big((k/2+q)^2 - m^2\big) \tag{7.53}$$

under the dq^0 integral.

The contribution of this pole yields the integral

$$i\delta\mathcal{M} = -2\pi i \frac{\lambda^2}{2} \int \frac{d^3q}{(2\pi)^4} \frac{1}{2E_{\mathbf{q}}} \frac{1}{(k^0 - E_{\mathbf{q}})^2 - E_{\mathbf{q}}^2}$$

$$= -2\pi i \frac{\lambda^2}{2} \frac{4\pi}{(2\pi)^4} \int\limits_m^\infty dE_{\mathbf{q}}\, E_{\mathbf{q}} |\mathbf{q}| \frac{1}{2E_{\mathbf{q}}} \frac{1}{k^0(k^0 - 2E_{\mathbf{q}})}. \tag{7.54}$$

The integrand in the second line has a pole at $E_{\mathbf{q}} = k^0/2$. When $k^0 < 2m$, this pole does not lie on the integration contour, so $\delta\mathcal{M}$ is manifestly real. When $k^0 > 2m$, however, the pole lies just above or below the contour of integration, depending upon whether k^0 is given a small positive or negative imaginary part:

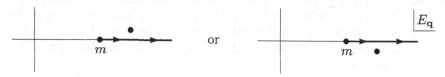

Thus the integral acquires a discontinuity between $k^2 + i\epsilon$ and $k^2 - i\epsilon$. To compute this discontinuity, apply

$$\frac{1}{k^0 - 2E_{\mathbf{q}} \pm i\epsilon} = P\frac{1}{k^0 - 2E_{\mathbf{q}}} \mp i\pi\delta(k^0 - 2E_{\mathbf{q}})$$

(where P denotes the principal value). The discontinuity is given by replacing the pole with a delta function. This in turn is equivalent to replacing the original propagator by a delta function:

$$\frac{1}{(k/2-q)^2 - m^2 + i\epsilon} \to -2\pi i \delta\big((k/2-q)^2 - m^2\big). \tag{7.55}$$

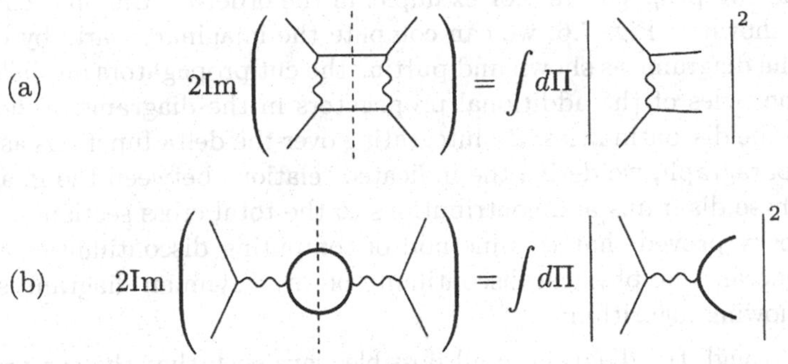

Figure 7.6. Two contributions to the optical theorem for Bhabha scattering.

Let us now retrace our steps and see what we have proved. Go back to the original integral (7.52), relabel the momenta on the two propagators as p_1, p_2 and substitute

$$\int \frac{d^4q}{(2\pi)^4} = \int \frac{d^4p_1}{(2\pi)^4} \int \frac{d^4p_2}{(2\pi)^4} (2\pi)^4 \delta^{(4)}(p_1 + p_2 - k).$$

We have shown that the discontinuity of the integral is computed by replacing each of the two propagators by a delta function:

$$\frac{1}{p_i^2 - m^2 + i\epsilon} \rightarrow -2\pi i \delta(p_i^2 - m^2). \tag{7.56}$$

The discontinuity of $\mathcal{M}$ comes only from the region of the d^4q integral in which the two delta functions are simultaneously satisfied. By integrating over the delta functions, we put the momenta p_i on shell and convert the integrals d^4p_i into integrals over relativistic phase space. What is left over in expression (7.52) is just the factor λ^2, the square of the leading-order scattering amplitude, and the symmetry factor $(1/2)$, which can be reinterpreted as the symmetry factor for identical bosons in the final state. Thus we have shown that, to order λ^2 on each side of the equation,

$$\text{Disc}\,\mathcal{M}(k) = 2i\,\text{Im}\,\mathcal{M}(k)$$

$$= \frac{i}{2} \int \frac{d^3p_1}{(2\pi)^3} \frac{1}{2E_1} \frac{d^3p_2}{(2\pi)^3} \frac{1}{2E_2} |\mathcal{M}(k)|^2 (2\pi)^4 \delta^{(4)}(p_1 + p_2 - k).$$

This explicitly verifies (7.49) to order λ^2 in ϕ^4 theory.

The preceding argument made no essential use of the fact that the two propagators in the diagram had equal masses, or of the fact that these propagators connected to a simple point vertex. Indeed, the analysis can be applied to an arbitrary one-loop diagram. Whenever, in the region of momentum integration of the diagram, two propagators can simultaneously go on-shell, we can follow the argument above to compute a nonzero discontinuity of $\mathcal{M}$. The value of this discontinuity is given by making the substitution (7.56) for each of the two propagators. For example, in the order-α^2 Bhabha scattering diagrams shown in Fig. 7.6, we can compute the imaginary parts by cutting through the diagrams as shown and putting the cut propagators on shell using (7.56). The poles of the additional propagators in the diagrams do not contribute to the discontinuities. By integrating over the delta functions as in the previous paragraph, we derive the indicated relations between the imaginary parts of these diagrams and contributions to the total cross section.

Cutkosky proved that this method of computing discontinuities is completely general. The physical discontinuity of any Feynman diagram is given by the following algorithm:

1. Cut through the diagram in all possible ways such that the cut propagators can simultaneously be put on shell.

2. For each cut, replace $1/(p^2-m^2+i\epsilon) \to -2\pi i\delta(p^2-m^2)$ in each cut propagator, then perform the loop integrals.

3. Sum the contributions of all possible cuts.

Using these *cutting rules*, it is possible to prove the optical theorem (7.49) to all orders in perturbation theory.

Unstable Particles

The cutting rules imply that the generalized optical theorem (7.49) is true not only for S-matrix elements, but for any amplitudes $\mathcal{M}$ that we can define in terms of Feynman diagrams. This fact is extremely useful for dealing with unstable particles, which never appear in asymptotic states.

Recall from Eq. (7.43) that the exact two-point function for a scalar particle has the form

$$\text{———} \bullet \text{———} = \frac{i}{p^2 - m_0^2 - M^2(p^2)}.$$

We defined the quantity $-iM^2(p^2)$ as the sum of all 1PI insertions into the boson propagator, but we can equally well think of it as the sum of all amputated diagrams for 1-particle $\to$ 1-particle "scattering". The LSZ formula then implies

$$\mathcal{M}(p \to p) = -ZM^2(p^2). \tag{7.57}$$

We can use this relation and the generalized optical theorem (7.49) to discuss the imaginary part of $M^2(p^2)$.

First consider the familiar case where the scalar boson is stable. In this case, there is no possible final state that can contribute to the right-hand side of (7.49). Thus $M^2(m^2)$ is real. The position of the pole in the propagator is determined by the equation $m^2 - m_0^2 - M^2(m^2) = 0$, which has a real-valued solution m. The pole therefore lies on the real p^2 axis, below the multiparticle branch cut.

Often, however, a particle can decay into two or more lighter particles. In this case $M^2(p^2)$ will acquire an imaginary part, so we must modify our definitions slightly. Let us define the particle's mass m by the condition

$$m^2 - m_0^2 - \text{Re}\, M^2(m^2) = 0. \tag{7.58}$$

Then the pole in the propagator is displaced from the real axis:

$$\sim \frac{iZ}{p^2 - m^2 - iZ\, \text{Im}\, M^2(p^2)}.$$

If this propagator appears in the s channel of a Feynman diagram, the cross section one computes, in the vicinity of the pole, will have the form

$$\sigma \propto \left| \frac{1}{s - m^2 - iZ\, \text{Im}\, M^2(s)} \right|^2. \tag{7.59}$$

This expression closely resembles the relativistic Breit-Wigner formula (4.64) for the cross section in the region of a resonance:

$$\sigma \propto \left| \frac{1}{p^2 - m^2 + im\Gamma} \right|^2. \tag{7.60}$$

The mass m defined by (7.58) is the position of the resonance. If $\text{Im}\, M^2(m^2)$ is small, so that the resonance in (7.59) is narrow, we can approximate $\text{Im}\, M^2(s)$ as $\text{Im}\, M^2(m^2)$ over the width of the resonance; then (7.59) will have precisely the Breit-Wigner form. In this case, we can identify

$$\Gamma = -\frac{Z}{m}\, \text{Im}\, M^2(m^2). \tag{7.61}$$

If the resonance is broad, it will show deviations from the Breit-Wigner shape, generally becoming narrower on the leading edge and broader on the trailing edge.

To compute $\text{Im}\, M^2$, and hence Γ, we could use the definition of M^2 as the sum of 1PI insertions into the propagator. The imaginary parts of the relevant loop diagrams give the decay rate. But the optical theorem (7.49), generalized to Feynman diagrams by the Cutkosky rules, simplifies this procedure. If we take (7.57) as the definition of the matrix element $\mathcal{M}(p \to p)$, and similarly define the decay matrix elements $\mathcal{M}(p \to f)$ through their Feynman diagram expansions, then (7.49) implies

$$Z\, \text{Im}\, M^2(p^2) = -\,\text{Im}\, \mathcal{M}(p \to p) = -\frac{1}{2} \sum_f \int d\Pi_f \, |\mathcal{M}(p \to f)|^2, \tag{7.62}$$

where the sum runs over all possible final states f. The decay rate is therefore

$$\Gamma = \frac{1}{2m} \sum_f \int d\Pi_f |\mathcal{M}(p \to f)|^2, \tag{7.63}$$

as quoted in Eq. (4.86).

We stress once again that our derivation of this equation applies only to the case of a long-lived unstable particle, so that $\Gamma \ll m$. For a broad resonance, the full energy dependence of $M^2(p^2)$ must be taken into account.

7.4 The Ward-Takahashi Identity

Of the loose ends listed at the beginning of this chapter, only one remains, the proof of the Ward identity. Recall from Section 5.5 that this identity states the following: If $\mathcal{M}(k) = \epsilon_\mu(k)\mathcal{M}^\mu(k)$ is the amplitude for some QED process involving an external photon with momentum k, then this amplitude vanishes if we replace ϵ_μ with k_μ:

$$k_\mu \mathcal{M}^\mu(k) = 0. \tag{7.64}$$

To prove this assertion, it is useful to prove a more general identity for QED correlation functions, called the *Ward-Takahashi identity*. To discuss this more general case we will let $\mathcal{M}$ denote a Fourier-transformed correlation function, in which the external momenta are not necessarily on-shell. The right-hand side of (7.64) will contain nonzero terms in this case; but when we apply the LSZ formula to extract an S-matrix element, those terms will not contribute.

We will prove the Ward-Takahashi identity order by order in α, by looking directly at the Feynman diagrams that contribute to $\mathcal{M}(k)$. The identity is generally not true for individual Feynman diagrams; we must sum over the diagrams for $\mathcal{M}(k)$ at any given order.

Consider a typical diagram for a typical amplitude $\mathcal{M}(k)$:

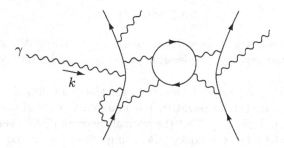

If we remove the photon $\gamma(k)$, we obtain a simpler diagram which is part of a simpler amplitude $\mathcal{M}_0$. If we reinsert the photon somewhere else inside the simpler diagram, we again obtain a contribution to $\mathcal{M}(k)$. The crucial observation is that by summing over all the diagrams that contribute to $\mathcal{M}_0$, then summing over all possible points of insertion in each of these diagrams, we obtain $\mathcal{M}(k)$. The Ward-Takahashi identity is true individually for each diagram contributing to $\mathcal{M}_0$, once we sum over insertion points; this is what we will prove.

When we insert our photon into one of the diagrams of $\mathcal{M}_0$, it must attach either to an electron line that runs out of the diagram to two external points, or to an internal electron loop. Let us consider each of these cases in turn.

First suppose that the electron line runs between external points. Before we insert our photon $\gamma(k)$, the line looks like this:

The electron propagators have momenta p, $p_1 = p + q_1$, $p_2 = p_1 + q_2$, and so on up to $p' = p_{n-1} + q_n$. If there are n vertices, we can insert our photon in $n + 1$ different places. Suppose we insert it after the ith vertex:

The electron propagators to the left of the new photon then have their momenta increased by k.

Let us now look at the values of these diagrams, with the polarization vector $\epsilon_\mu(k)$ replaced by k_μ. The product of k_μ with the new vertex is conveniently written:

$$-iek_\mu\gamma^\mu = -ie\big[(\not{p}_i + \not{k} - m) - (\not{p}_i - m)\big].$$

Multiplying by the adjacent electron propagators, we obtain

$$\frac{i}{\not{p}_i + \not{k} - m}(-ie\not{k})\frac{i}{\not{p}_i - m} = e\left(\frac{i}{\not{p}_i - m} - \frac{i}{\not{p}_i + \not{k} - m}\right). \tag{7.65}$$

The diagram with the photon inserted at position i therefore has the structure

$$= \cdots \left(\frac{i}{\not{p}_{i+1} + \not{k} - m}\right)\gamma^{\lambda_{i+1}}\left(\frac{i}{\not{p}_i - m} - \frac{i}{\not{p}_i + \not{k} - m}\right)\gamma^{\lambda_i}$$
$$\times \left(\frac{i}{\not{p}_{i-1} - m}\right)\gamma^{\lambda_{i-1}}\cdots.$$

Similarly, the diagram with the photon inserted at position $i - 1$ has the structure

$$= \cdots \left(\frac{i}{\not{p}_{i+1} + \not{k} - m}\right)\gamma^{\lambda_{i+1}}\left(\frac{i}{\not{p}_i + \not{k} - m}\right)\gamma^{\lambda_i}$$
$$\times \left(\frac{i}{\not{p}_{i-1} - m} - \frac{i}{\not{p}_{i-1} + \not{k} - m}\right)\gamma^{\lambda_{i-1}}\cdots.$$

Note that the first term of this expression cancels the second term of the previous expression. A similar cancellation occurs between any other pair of diagrams with adjacent insertions. When we sum over all possible insertion points along the line, everything cancels except the unpaired terms at the ends. The unpaired term coming from insertion after the last vertex (on the far left) is just e times the value of the original diagram; the other unpaired term, from insertion before the first vertex, is identical except for a minus sign and the replacement of p by $p + k$ everywhere. Diagrammatically, our result is

$$\sum_{\substack{\text{insertion} \\ \text{points}}} k_\mu \cdot \left(\mu \, \rightsquigarrow \underset{k}{\longrightarrow} \right) = e \left(\begin{array}{c} \end{array} - \begin{array}{c} \end{array} \right), \qquad (7.66)$$

where we have renamed $p' + k \rightarrow q$ for the sake of symmetry.

In each diagram on the left-hand side of (7.66), the momentum entering the electron line is p and the momentum exiting is q. According to the LSZ formula, we can extract from each diagram a contribution to an S-matrix element by taking the coefficient of the product of poles

$$\left(\frac{i}{\slashed{q} - m} \right) \left(\frac{i}{\slashed{p} - m} \right).$$

The terms on the right-hand side of (7.66) each contain one of these poles, but neither contains both poles. Thus the right-hand side of (7.66) contributes nothing to the S-matrix.*

To complete the proof of the Ward-Takahashi identity, we must consider the case in which our photon attaches to an internal electron loop. Before the insertion of the photon, a typical loop looks like this:

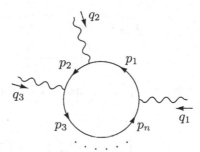

---

*This step of the argument is straightforward only if we have arranged the perturbation series so that the propagator contains m rather than m_0. We will do this in Chapter 10.

The electron propagators have momenta p_1, $p_1 + q_2 = p_2$, and so on up to p_n. Suppose now that we insert the photon $\gamma(k)$ between vertices i and $i+1$:

We now have an additional momentum k running around the loop from the new vertex; by convention, this momentum exits at vertex 1.

To evaluate the sum over all such insertions into the loop, apply identity (7.65) to each diagram. For the diagram in which the photon is inserted between vertices 1 and 2, we obtain

$$-e \int \frac{d^4 p_1}{(2\pi)^4} \, \text{tr}\left[\left(\frac{i}{\not{p}_n + \not{k} - m}\right)\gamma^{\lambda_n} \cdots \left(\frac{i}{\not{p}_2 + \not{k} - m}\right)\gamma^{\lambda_2} \right.$$
$$\left. \times \left(\frac{i}{\not{p}_1 - m} - \frac{i}{\not{p}_1 + \not{k} - m}\right)\gamma^{\lambda_1}\right].$$

The first term will be canceled by one of the terms from the diagram with the photon inserted between vertices 2 and 3. Similar cancellations take place between terms from other pairs of adjacent insertions. When we sum over all n insertion points we are left with

$$-e \int \frac{d^4 p_1}{(2\pi)^4} \, \text{tr}\left[\left(\frac{i}{\not{p}_n - m}\right)\gamma^{\lambda_n}\left(\frac{i}{\not{p}_{n-1} - m}\right)\gamma^{\lambda_{n-1}} \cdots \left(\frac{i}{\not{p}_1 - m}\right)\gamma^{\lambda_1}\right.$$
$$\left. -\left(\frac{i}{\not{p}_n + \not{k} - m}\right)\gamma^{\lambda_n}\left(\frac{i}{\not{p}_{n-1} + \not{k} - m}\right)\gamma^{\lambda_{n-1}} \cdots \left(\frac{i}{\not{p}_1 + \not{k} - m}\right)\gamma^{\lambda_1}\right].$$
$$\tag{7.67}$$

Shifting the integration variable from p_1 to $p_1 + k$ in the second term, we see that the two remaining terms cancel. Thus the diagrams in which the photon is inserted along a closed loop add up to zero.

We are now ready to assemble the pieces of the proof. Suppose that the amplitude $\mathcal{M}$ has $2n$ external electron lines, n incoming and n outgoing. Label the incoming momenta p_i and the outgoing momenta q_i:

$$\mathcal{M}(k; p_1 \cdots p_n; q_1 \cdots q_n) = \quad$$

(The amplitude can also involve an arbitrary number of additional external photons.) The amplitude $\mathcal{M}_0$ lacks the photon $\gamma(k)$ but is otherwise identical. To form $k_\mu \mathcal{M}^\mu$ from $\mathcal{M}_0$ we must sum over all diagrams that contribute to $\mathcal{M}_0$, and for each diagram, sum over all points at which the photon could be inserted. Summing over insertion points along an internal loop in any diagram gives zero. Summing over insertion points along a through-going line in any diagram gives a contribution of the form (7.66). Summing over *all* insertion points for any particular diagram, we obtain

where the shaded circle represents any particular diagram that contributes to $\mathcal{M}_0$. Summing over all such diagrams, we finally obtain

$$
\begin{aligned}
k_\mu \mathcal{M}^\mu(k; p_1 \cdots p_n; q_1 \cdots q_n) = e \sum_i \Big[& \mathcal{M}_0(p_1 \cdots p_n; q_1 \cdots (q_i - k) \cdots) \\
& - \mathcal{M}_0(p_1 \cdots (p_i + k) \cdots; q_1 \cdots q_n) \Big].
\end{aligned}
$$

(7.68)

This is the Ward-Takahashi identity for correlation functions in QED. We saw below (7.66) that the right-hand side does not contribute to the S-matrix; thus in the special case where $\mathcal{M}$ is an S-matrix element, Eq. (7.68) reduces to the Ward identity (7.64).

Before discussing this identity further, we should mention a potential flaw in the above proof. In order to find the necessary cancellation in Eq. (7.67), we had to shift the integration variable by a constant. If the integral diverges, however, this shift is not permissible. Similarly, there may be divergent loop-momentum integrals in the expressions leading to Eq. (7.66). Here there is no explicit shift in the proof, but in practice one does generally perform a shift while evaluating the integrals. In either case, ultraviolet divergences can potentially invalidate the Ward-Takahashi identity. We will see an example of this problem, as well as a general solution to it, in the next section.

The simplest example of the Ward-Takahashi identity involves, on the left-hand side, the three-point function with one entering and one exiting electron and one external photon:

$$k_\mu \cdot \left(\begin{matrix} p+k \\ \mu \sim\!\!\!\sim \bullet \\ k \quad p \end{matrix} \right) = e \left(\begin{matrix} p \\ \bullet \\ p \end{matrix} - \begin{matrix} p+k \\ \bullet \\ p+k \end{matrix} \right)$$

The quantities on the right-hand side are exact electron propagators, evaluated at p and $(p+k)$ respectively. Label these quantities $S(p)$ and $S(p+k)$; from Eq. (7.23),

$$S(p) = \frac{i}{\not{p} - m - \Sigma(p)}.$$

The full three-point amplitude on the left-hand side can be rewritten, just as in Eq. (7.44), as a product of full propagators for the entering and exiting electrons, times an amputated scattering diagram. In this case, the amputated function is just the vertex $\Gamma^\mu(p+k,p)$. Then the Ward-Takahashi identity reads:

$$S(p+k)\left[-iek_\mu\Gamma^\mu(p+k,p)\right]S(p) = e(S(p) - S(p+k)).$$

To simplify this equation, multiply, on the left and right respectively, by the Dirac matrices $S^{-1}(p+k)$ and $S^{-1}(p)$. This gives

$$-ik_\mu\Gamma^\mu(p+k,p) = S^{-1}(p+k) - S^{-1}(p). \qquad (7.69)$$

Often the term *Ward-Takahashi identity* is used to mean only this special case.

We can use identity (7.69) to obtain the general relation between the renormalization factors Z_1 and Z_2. We defined Z_1 in (7.47) by the relation

$$\Gamma^\mu(p+k,p) \to Z_1^{-1}\gamma^\mu \qquad \text{as} \quad k \to 0.$$

We defined Z_2 as the residue of the pole in $S(p)$:

$$S(p) \sim \frac{iZ_2}{\not{p} - m}.$$

Setting p near mass shell and expanding (7.69) about $k = 0$, we find for the first-order terms on the left and right

$$-iZ_1^{-1}\not{k} = -iZ_2^{-1}\not{k},$$

that is,

$$Z_1 = Z_2. \qquad (7.70)$$

Thus, the Ward-Takahashi identity guarantees the exact cancellation of infinite rescaling factors in the electron scattering amplitude that we found at the end of Section 7.2. When combined with the correct formal expression

(7.46) for the electron form factors, this identity guarantees that $F_1(0) = 1$ to all orders in perturbation theory.

Often, in the literature, the terms *Ward identity, current conservation,* and *gauge invariance* are used interchangeably. This is quite natural, since the Ward identity is the diagrammatic expression of the conservation of the electric current, which is in turn a consequence of gauge invariance. In this book, however, we will distinguish these three concepts. By *gauge invariance* we mean the fundamental symmetry of the Lagrangian; by *current conservation* we mean the equation of motion that follows from this symmetry; and by the *Ward identity* we mean the diagrammatic identity that imposes the symmetry on quantum mechanical amplitudes.

7.5 Renormalization of the Electric Charge

At the beginning of Chapter 6 we set out to study the order-α radiative corrections to electron scattering from a heavy target. We evaluated (at least in the classical limit) the bremsstrahlung diagrams,

and also the corrections due to virtual photons:

Our discussion of field-strength renormalization in this chapter has finally clarified the role of the last two diagrams. In particular, we have seen that the Ward identity, through the relation $Z_1 = Z_2$, insures that the sum of the virtual photon corrections vanishes as the momentum transfer q goes to zero.

There is one remaining type of radiative correction to this process:

This is the order-α *vacuum polarization* diagram, also known as the *photon self-energy*. It can be viewed as a modification to the photon structure by a virtual electron-positron pair. This diagram will alter the effective field $A^\mu(x)$ seen by the scattered electron. It can potentially shift the overall strength of this field, and can also change its dependence on x (or in Fourier space, on q). In this section we will compute this diagram, and see that it has both of these effects.

Overview of Charge Renormalization

Before beginning a detailed calculation, let's ask what kind of an answer we expect and what its interpretation will be. The interesting part of the diagram is the electron loop:

$$
\begin{aligned}
&= (-ie)^2(-1)\int \frac{d^4k}{(2\pi)^4}\,\mathrm{tr}\!\left[\gamma^\mu\frac{i}{\not k - m}\gamma^\nu\frac{i}{\not k + \not q - m}\right]\\
&\equiv i\Pi_2^{\mu\nu}(q). \qquad\qquad\qquad\qquad\qquad (7.71)
\end{aligned}
$$

(The fermion loop factor of (-1) was derived in Eq. (4.120).) More generally, let us define $i\Pi^{\mu\nu}(q)$ to be the sum of all 1-particle-irreducible insertions into the photon propagator,

$$
\mu \;\rightsquigarrow\; \boxed{1\text{PI}} \;\rightsquigarrow\; \nu \;\equiv\; i\Pi^{\mu\nu}(q), \qquad\qquad (7.72)
$$

so that $\Pi_2^{\mu\nu}(q)$ is the second-order (in e) contribution to $\Pi^{\mu\nu}(q)$.

The only tensors that can appear in $\Pi^{\mu\nu}(q)$ are $g^{\mu\nu}$ and $q^\mu q^\nu$. The Ward identity, however, tells us that $q_\mu\Pi^{\mu\nu}(q) = 0$. This implies that $\Pi^{\mu\nu}(q)$ is proportional to the projector $(g^{\mu\nu} - q^\mu q^\nu/q^2)$. Furthermore, we expect that $\Pi^{\mu\nu}(q)$ will not have a pole at $q^2 = 0$; the only obvious source of such a pole would be a single-massless-particle intermediate state, which cannot occur in any 1PI diagram.[†] It is therefore convenient to extract the tensor structure from $\Pi^{\mu\nu}$ in the following way:

$$
\Pi^{\mu\nu}(q) = (q^2 g^{\mu\nu} - q^\mu q^\nu)\Pi(q^2), \qquad\qquad (7.73)
$$

where $\Pi(q^2)$ is regular at $q^2 = 0$.

Using this notation, the exact photon two-point function is

$$
\begin{aligned}
&= \rightsquigarrow + \rightsquigarrow\!\boxed{1\text{PI}}\!\rightsquigarrow + \rightsquigarrow\!\boxed{1\text{PI}}\!\rightsquigarrow\!\boxed{1\text{PI}}\!\rightsquigarrow + \cdots\\
&= \frac{-ig_{\mu\nu}}{q^2} + \frac{-ig_{\mu\rho}}{q^2}\Big[i(q^2 g^{\rho\sigma} - q^\rho q^\sigma)\Pi(q^2)\Big]\frac{-ig_{\sigma\nu}}{q^2} + \cdots
\end{aligned}
$$

[†]One can prove that there is no such pole, but the proof is nontrivial. Schwinger has shown that, in two spacetime dimensions, the singularity in Π_2 due to a pair of massless fermions is a pole rather than a cut; this is a famous counterexample to our argument. There is no such problem in four dimensions.

$$= \frac{-ig_{\mu\nu}}{q^2} + \frac{-ig_{\mu\rho}}{q^2}\Delta_\nu^\rho\Pi(q^2) + \frac{-ig_{\mu\rho}}{q^2}\Delta_\sigma^\rho\Delta_\nu^\sigma\Pi^2(q^2) + \cdots,$$

where $\Delta_\nu^\rho \equiv \delta_\nu^\rho - q^\rho q_\nu/q^2$. Noting that $\Delta_\sigma^\rho\Delta_\nu^\sigma = \Delta_\nu^\rho$, we can simplify this expression to

$$\mu \enspace\underset{\sim}{}\!\!\!\bullet\!\!\!\underset{\sim}{}\enspace \nu = \frac{-ig_{\mu\nu}}{q^2} + \frac{-ig_{\mu\rho}}{q^2}\left(\delta_\nu^\rho - \frac{q^\rho q_\nu}{q^2}\right)\left(\Pi(q^2) + \Pi^2(q^2) + \cdots\right)$$

$$= \frac{-i}{q^2\left(1 - \Pi(q^2)\right)}\left(g_{\mu\nu} - \frac{q_\mu q_\nu}{q^2}\right) + \frac{-i}{q^2}\left(\frac{q_\mu q_\nu}{q^2}\right). \qquad (7.74)$$

In any S-matrix element calculation, at least one end of this exact propagator will connect to a fermion line. When we sum over all places along the line where it could connect, we must find, according to the Ward identity, that terms proportional to q_μ or q_ν vanish. For the purposes of computing S-matrix elements, therefore, we can abbreviate

$$\mu \enspace\underset{\sim}{}\!\!\!\bullet\!\!\!\underset{\sim}{}\enspace \nu = \frac{-ig_{\mu\nu}}{q^2\left(1 - \Pi(q^2)\right)}. \qquad (7.75)$$

Notice that as long as $\Pi(q^2)$ is regular at $q^2 = 0$, the exact propagator always has a pole at $q^2 = 0$. In other words, the photon remains absolutely massless at all orders in perturbation theory. This claim depends strongly on our use of the Ward identity in (7.73). If, for example, $\Pi^{\mu\nu}(q)$ contained a term $M^2 g^{\mu\nu}$ (with no compensating $q^\mu q^\nu$ term), the photon mass would be shifted to M.

The residue of the $q^2 = 0$ pole is

$$\frac{1}{1 - \Pi(0)} \equiv Z_3.$$

The amplitude for any low-q^2 scattering process will be shifted by this factor, relative to the tree-level approximation:

$$\text{or}\qquad \cdots\frac{e^2 g_{\mu\nu}}{q^2}\cdots \longrightarrow \cdots\frac{Z_3\, e^2 g_{\mu\nu}}{q^2}\cdots.$$

Since a factor of e lies at each end of the photon propagator, we can conveniently account for this shift by making the replacement $e \to \sqrt{Z_3}\, e$. This replacement is called *charge renormalization*; it is in many ways analogous to the mass renormalization introduced in Section 7.1. Note in particular that the "physical" electron charge measured in experiments is $\sqrt{Z_3}\, e$. We will therefore shift our notation and call this quantity simply e. From now on we will refer to the "bare" charge (the quantity that multiplies $A_\mu \bar{\psi} \gamma^\mu \psi$ in the Lagrangian) as e_0. We then have

$$\text{(physical charge)} = e = \sqrt{Z_3}\, e_0 = \sqrt{Z_3} \cdot \text{(bare charge)}. \qquad (7.76)$$

To lowest order, $Z_3 = 1$ and $e = e_0$.

In addition to this constant shift in the strength of the electric charge, $\Pi(q^2)$ has another effect. Consider a scattering process with nonzero q^2, and suppose that we have computed $\Pi(q^2)$ to leading order in α: $\Pi(q^2) \approx \Pi_2(q^2)$. The amplitude for the process will then involve the quantity

$$\frac{-ig_{\mu\nu}}{q^2} \left(\frac{e_0^2}{1 - \Pi(q^2)} \right) \underset{O(\alpha)}{=} \frac{-ig_{\mu\nu}}{q^2} \left(\frac{e^2}{1 - [\Pi_2(q^2) - \Pi_2(0)]} \right).$$

(Swapping e^2 for e_0^2 does not matter to lowest order.) The quantity in parentheses can be interpreted as a q^2-dependent electric charge. The full effect of replacing the tree-level photon propagator with the exact photon propagator is therefore to replace

$$\alpha_0 \to \alpha_{\text{eff}}(q^2) = \frac{e_0^2/4\pi}{1 - \Pi(q^2)} \underset{O(\alpha)}{=} \frac{\alpha}{1 - [\Pi_2(q^2) - \Pi_2(0)]}. \qquad (7.77)$$

(To leading order we could just as well bring the Π-terms into the numerator; but we will see in Chapter 12 that in this form, the expression is true to all orders when Π_2 is replaced by Π.)

Computation of Π_2

Having worked so hard to interpret $\Pi_2(q^2)$, we had better calculate it. Going back to (7.71), we have

$$i\Pi_2^{\mu\nu}(q) = -(-ie)^2 \int \frac{d^4k}{(2\pi)^4} \text{tr}\left[\gamma^\mu \frac{i(\slashed{k}+m)}{k^2 - m^2} \gamma^\nu \frac{i(\slashed{k}+\slashed{q}+m)}{(k+q)^2 - m^2} \right]$$

$$= -4e^2 \int \frac{d^4k}{(2\pi)^4} \frac{k^\mu(k+q)^\nu + k^\nu(k+q)^\mu - g^{\mu\nu}(k\cdot(k+q) - m^2)}{(k^2 - m^2)((k+q)^2 - m^2)}. \qquad (7.78)$$

We have written e and m instead of e_0 and m_0 for convenience, since the difference would give only an order-α^2 contribution to $\Pi^{\mu\nu}$.

Now introduce a Feynman parameter to combine the denominator factors:

$$\frac{1}{(k^2 - m^2)((k+q)^2 - m^2)} = \int_0^1 dx \, \frac{1}{(k^2 + 2xk\cdot q + xq^2 - m^2)^2}$$

$$= \int_0^1 dx \, \frac{1}{(\ell^2 + x(1-x)q^2 - m^2)^2},$$

where $\ell = k + xq$. In terms of ℓ, the numerator of (7.78) is

$$\text{Numerator} = 2\ell^\mu \ell^\nu - g^{\mu\nu}\ell^2 - 2x(1-x)q^\mu q^\nu + g^{\mu\nu}\left(m^2 + x(1-x)q^2\right)$$
$$+ \text{(terms linear in } \ell).$$

Performing a Wick rotation and substituting $\ell^0 = i\ell_E^0$, we obtain

$$i\Pi_2^{\mu\nu}(q) = -4ie^2 \int_0^1 dx \int \frac{d^4\ell_E}{(2\pi)^4}$$

$$\times \frac{-\frac{1}{2}g^{\mu\nu}\ell_E^2 + g^{\mu\nu}\ell_E^2 - 2x(1-x)q^\mu q^\nu + g^{\mu\nu}\left(m^2 + x(1-x)q^2\right)}{(\ell_E^2 + \Delta)^2},$$

(7.79)

where $\Delta = m^2 - x(1-x)q^2$. This integral is very badly ultraviolet divergent. If we were to cut it off at $\ell_E = \Lambda$, we would find for the leading term,

$$i\Pi_2^{\mu\nu}(q) \propto e^2 \Lambda^2 g^{\mu\nu},$$

with no compensating $q^\mu q^\nu$ term. This result violates the Ward identity; it would give the photon an infinite mass $M \propto e\Lambda$.

Our proof of the Ward identity has failed, in precisely the way anticipated at the end of the previous section: The shift of the integration variable in (7.67) is not permissible when the integral is divergent. In our present calculation, the failure of the Ward identity is catastrophic: It leads to an infinite photon mass,[‡] in conflict with experiment. Fortunately, there is a way to rescue the Ward identity.

In the above analysis we regulated the divergent integral in the most straightforward and most naive way: by cutting it off at a large momentum Λ. But other regulators are possible, and some will in fact preserve the Ward identity. In our computations of the vertex and electron self-energy diagrams, we used a Pauli-Villars regulator. This regulator preserved the relation $Z_1 = Z_2$, a consequence of the Ward identity; a naive cutoff does not (see Problem 7.2). We could fix our present computation by introducing Pauli-Villars fermions. Unfortunately, several sets of fermions are required, making the method rather complicated.[*] We will use a simpler method, *dimensional regularization*, due to 't Hooft and Veltman.[†] Dimensional regularization preserves the symmetries of QED and also of a wide class of more general theories.

The question of which regulator to use has no *a priori* answer in quantum field theory. Often the choice has no effect on the predictions of the theory.

[‡]We could still make the observed photon mass zero by adding a compensating infinite photon mass term to the Lagrangian. More generally, we could add terms to the Lagrangian to make the Ward identity valid for any n-point correlation function. This procedure would give the same results as the one we are about to follow, but would be much more complicated.

[*]This method is presented in Bjorken and Drell (1964), p. 154.

[†]G. 't Hooft and M. J. G. Veltman, *Nucl. Phys.* **B44**, 189 (1972).

When two regulators give different answers for observable quantities, it is generally because some symmetry (such as the Ward identity) is being violated by one (or both) of them. In these cases we take the symmetry to be fundamental and demand that it be preserved by the regulator. But the validity of this choice cannot be proven; we are adopting the symmetry as a new axiom.

Dimensional Regularization

The idea of dimensional regularization is very simple to state: Compute the Feynman diagram as an analytic function of the dimensionality of spacetime, d. For sufficiently small d, any loop-momentum integral will converge and therefore the Ward identity can be proved. The final expression for any observable quantity should have a well-defined limit as $d \to 4$.

Let us do a practice calculation to see how this technique works. We consider spacetime to have one time dimension and $(d-1)$ space dimensions. Then we can Wick-rotate Feynman integrals as before, to give integrals over a d-dimensional Euclidean space. A typical example is

$$\int \frac{d^d \ell_E}{(2\pi)^d} \frac{1}{(\ell_E^2 + \Delta)^2} = \int \frac{d\Omega_d}{(2\pi)^d} \cdot \int_0^\infty d\ell_E \frac{\ell_E^{d-1}}{(\ell_E^2 + \Delta)^2}. \tag{7.80}$$

The first factor in (7.80) contains the area of a unit sphere in d dimensions. To compute it, use the following trick:

$$(\sqrt{\pi})^d = \left(\int dx \, e^{-x^2} \right)^d = \int d^d x \, \exp\left(- \sum_{i=1}^d x_i^2 \right)$$

$$= \int d\Omega_d \int_0^\infty dx \, x^{d-1} e^{-x^2} = \left(\int d\Omega_d \right) \cdot \frac{1}{2} \int_0^\infty d(x^2) \, (x^2)^{\frac{d}{2}-1} e^{-(x^2)}$$

$$= \left(\int d\Omega_d \right) \cdot \frac{1}{2}\Gamma(d/2).$$

So the area of a d-dimensional unit sphere is

$$\int d\Omega_d = \frac{2\pi^{d/2}}{\Gamma(d/2)}. \tag{7.81}$$

This formula reproduces the familiar special cases:

d	$\Gamma(d/2)$	$\int d\Omega_d$
1	$\sqrt{\pi}$	2
2	1	2π
3	$\sqrt{\pi}/2$	4π
4	1	$2\pi^2$

The second factor in (7.80) is

$$\int_0^\infty d\ell \frac{\ell^{d-1}}{(\ell^2 + \Delta)^2} = \frac{1}{2} \int_0^\infty d(\ell^2) \frac{(\ell^2)^{\frac{d}{2}-1}}{(\ell^2 + \Delta)^2}$$

$$= \frac{1}{2} \left(\frac{1}{\Delta}\right)^{2-\frac{d}{2}} \int_0^1 dx \, x^{1-\frac{d}{2}} (1-x)^{\frac{d}{2}-1},$$

where we have substituted $x = \Delta/(\ell^2 + \Delta)$ in the second line. Using the definition of the beta function,

$$\int_0^1 dx \, x^{\alpha-1}(1-x)^{\beta-1} = B(\alpha, \beta) = \frac{\Gamma(\alpha)\Gamma(\beta)}{\Gamma(\alpha+\beta)}, \tag{7.82}$$

we can easily evaluate the integral over x. The final result for the d-dimensional integral is

$$\int \frac{d^d\ell_E}{(2\pi)^d} \frac{1}{(\ell_E^2 + \Delta)^2} = \frac{1}{(4\pi)^{d/2}} \frac{\Gamma(2-\frac{d}{2})}{\Gamma(2)} \left(\frac{1}{\Delta}\right)^{2-\frac{d}{2}}.$$

Since $\Gamma(z)$ has isolated poles at $z = 0, -1, -2, \ldots$, this integral has isolated poles at $d = 4, 6, 8, \ldots$. To find the behavior near $d = 4$, define $\epsilon = 4 - d$, and use the approximation[‡]

$$\Gamma(2-\tfrac{d}{2}) = \Gamma(\epsilon/2) = \frac{2}{\epsilon} - \gamma + \mathcal{O}(\epsilon), \tag{7.83}$$

where $\gamma \approx .5772$ is the Euler-Mascheroni constant. (This constant will always cancel in observable quantities.) The integral is then

$$\int \frac{d^d\ell_E}{(2\pi)^d} \frac{1}{(\ell_E^2 + \Delta)^2} \xrightarrow{d \to 4} \frac{1}{(4\pi)^2} \left(\frac{2}{\epsilon} - \log\Delta - \gamma + \log(4\pi) + \mathcal{O}(\epsilon)\right). \tag{7.84}$$

When we defined this integral with a Pauli-Villars regulator in Eq. (7.18), we found

$$\int \frac{d^4\ell_E}{(2\pi)^4} \frac{1}{(\ell_E^2 + \Delta)^2} \xrightarrow{\Lambda \to \infty} \frac{1}{(4\pi)^2} \left(\log\frac{x\Lambda^2}{\Delta} + \mathcal{O}(\Lambda^{-1})\right).$$

Thus the $1/\epsilon$ pole in dimensional regularization corresponds to a logarithmic divergence in the momentum integral. Note the curious fact that (7.84)

[‡]This expansion follows immediately from the infinite product representation

$$\frac{1}{\Gamma(z)} = ze^{\gamma z} \prod_{n=1}^\infty \left(1 + \frac{z}{n}\right) e^{-z/n}.$$

involves the logarithm of Δ, a dimensionful quantity. The scale of the logarithm is hidden in the $1/\epsilon$ term, and appears explicitly when the divergence is canceled.

You can easily verify the more general integration formulae,

$$\int \frac{d^d \ell_E}{(2\pi)^d} \frac{1}{(\ell_E^2 + \Delta)^n} = \frac{1}{(4\pi)^{d/2}} \frac{\Gamma(n-\frac{d}{2})}{\Gamma(n)} \left(\frac{1}{\Delta}\right)^{n-\frac{d}{2}}; \qquad (7.85)$$

$$\int \frac{d^d \ell_E}{(2\pi)^d} \frac{\ell_E^2}{(\ell_E^2 + \Delta)^n} = \frac{1}{(4\pi)^{d/2}} \frac{d}{2} \frac{\Gamma(n-\frac{d}{2}-1)}{\Gamma(n)} \left(\frac{1}{\Delta}\right)^{n-\frac{d}{2}-1}. \qquad (7.86)$$

In d dimensions, $g^{\mu\nu}$ obeys $g_{\mu\nu} g^{\mu\nu} = d$. Thus, if the numerator of a symmetric integrand contains $\ell^\mu \ell^\nu$, we should replace

$$\ell^\mu \ell^\nu \to \frac{1}{d} \ell^2 g^{\mu\nu}, \qquad (7.87)$$

in analogy with Eq. (6.46). In QED, the Dirac matrices can be manipulated as a set of d matrices satisfying

$$\{\gamma^\mu, \gamma^\nu\} = 2g^{\mu\nu}, \qquad \text{tr}[1] = 4. \qquad (7.88)$$

In manipulating Eq. (7.78), these rules give the same result as the purely four-dimensional rules. However, in the evaluation of other diagrams, there are additional contributions of order ϵ. In particular, the contraction identities (5.9) are modified in $d = 4 - \epsilon$ to

$$\begin{aligned} \gamma^\mu \gamma^\nu \gamma_\mu &= -(2-\epsilon)\gamma^\nu \\ \gamma^\mu \gamma^\nu \gamma^\rho \gamma_\mu &= 4g^{\nu\rho} - \epsilon\gamma^\nu \gamma^\rho \\ \gamma^\mu \gamma^\nu \gamma^\rho \gamma^\sigma \gamma_\mu &= -2\gamma^\sigma \gamma^\rho \gamma^\nu + \epsilon\gamma^\nu \gamma^\rho \gamma^\sigma. \end{aligned} \qquad (7.89)$$

These extra terms can contribute to the final value of the Feynman diagram if they multiply a factor ϵ^{-1} from a divergent integral. In QED at one-loop order, such extra terms appear in the vertex and self-energy diagrams but cancel when these diagrams are combined to compute an observable quantity.

Computation of Π_2, Continued

Now let us apply these dimensional regularization formulae to the momentum integral in (7.79). The unpleasant terms with ℓ^2 in the numerator give

$$\int \frac{d^d \ell_E}{(2\pi)^d} \frac{(-\frac{2}{d}+1)g^{\mu\nu}\ell_E^2}{(\ell_E^2 + \Delta)^2} = \frac{-1}{(4\pi)^{d/2}} (1-\tfrac{d}{2})\Gamma(1-\tfrac{d}{2}) \left(\frac{1}{\Delta}\right)^{1-\frac{d}{2}} g^{\mu\nu}$$

$$= \frac{1}{(4\pi)^{d/2}} \Gamma(2-\tfrac{d}{2}) \left(\frac{1}{\Delta}\right)^{2-\frac{d}{2}} \cdot (-\Delta g^{\mu\nu}).$$

We would have expected a pole at $d = 2$, since the quadratic divergence in 4 dimensions becomes a logarithmic divergence in 2 dimensions. But the pole cancels. The Ward identity is working.

Evaluating the remaining terms in (7.79) and using $\Delta = m^2 - x(1-x)q^2$, we obtain

$$i\Pi_2^{\mu\nu}(q) = -4ie^2 \int\limits_0^1 dx \, \frac{1}{(4\pi)^{d/2}} \frac{\Gamma(2-\frac{d}{2})}{\Delta^{2-d/2}}$$

$$\times \left[g^{\mu\nu}(-m^2 + x(1-x)q^2) + g^{\mu\nu}(m^2 + x(1-x)q^2) - 2x(1-x)q^\mu q^\nu \right]$$

$$= (q^2 g^{\mu\nu} - q^\mu q^\nu) \cdot i\Pi_2(q^2),$$

where

$$\Pi_2(q^2) = \frac{-8e^2}{(4\pi)^{d/2}} \int\limits_0^1 dx \, x(1-x) \frac{\Gamma(2-\frac{d}{2})}{\Delta^{2-d/2}} \tag{7.90}$$

$$\xrightarrow[d \to 4]{} -\frac{2\alpha}{\pi} \int\limits_0^1 dx \, x(1-x) \left(\frac{2}{\epsilon} - \log\Delta - \gamma + \log(4\pi) \right) \qquad (\epsilon = 4 - d).$$

With dimensional regularization, $\Pi_2^{\mu\nu}(q)$ indeed takes the form required by the Ward identity. But it is still logarithmically divergent.

We can now compute the order-α shift in the electric charge:

$$\frac{e^2 - e_0^2}{e_0^2} = \delta Z_3 \underset{O(\alpha)}{=} \Pi_2(0) \approx -\frac{2\alpha}{3\pi\epsilon}.$$

The bare charge is infinitely larger than the observed charge. But this difference is not observable. What can be observed is the q^2 dependence of the effective electric charge (7.77). This quantity depends on the difference

$$\hat{\Pi}_2(q^2) \equiv \Pi_2(q^2) - \Pi_2(0) = -\frac{2\alpha}{\pi} \int\limits_0^1 dx \, x(1-x) \log\left(\frac{m^2}{m^2 - x(1-x)q^2} \right), \tag{7.91}$$

which is independent of ϵ in the limit $\epsilon \to 0$. For the rest of this section we will investigate what physics this expression contains.

Interpretation of Π_2

First consider the analytic structure of $\hat{\Pi}_2(q^2)$. For $q^2 < 0$, as is the case when the photon propagator is in the t- or u-channel, $\hat{\Pi}_2(q^2)$ is manifestly real and analytic. But for an s-channel process, q^2 will be positive. The logarithm function has a branch cut when its argument becomes negative, that is, when

$$m^2 - x(1-x)q^2 < 0.$$

The product $x(1-x)$ is at most $1/4$, so $\hat{\Pi}_2(q^2)$ has a branch cut beginning at

$$q^2 = 4m^2,$$

at the threshold for creation of a real electron-positron pair.

Let us calculate the imaginary part of $\widehat{\Pi}_2$ for $q^2 > 4m^2$. For any fixed q^2, the x-values that contribute are between the points $x = \frac{1}{2} \pm \frac{1}{2}\beta$, where $\beta = \sqrt{1 - 4m^2/q^2}$. Since $\text{Im}[\log(-X \pm i\epsilon)] = \pm\pi$, we have

$$\text{Im}\left[\widehat{\Pi}_2(q^2 \pm i\epsilon)\right] = -\frac{2\alpha}{\pi}(\pm\pi)\int_{\frac{1}{2}-\frac{1}{2}\beta}^{\frac{1}{2}+\frac{1}{2}\beta} dx\, x(1-x)$$

$$= \mp 2\alpha \int_{-\beta/2}^{\beta/2} dy\,\left(\tfrac{1}{4} - y^2\right) \qquad (y \equiv x - \tfrac{1}{2})$$

$$= \mp\frac{\alpha}{3}\sqrt{1 - \frac{4m^2}{q^2}}\left(1 + \frac{2m^2}{q^2}\right). \tag{7.92}$$

This dependence on q^2 is exactly the same as in Eq. (5.13), the cross section for production of a fermion-antifermion pair. That is just what we would expect from the unitarity relation shown in Fig. 7.6(b); the cut through the diagram for forward Bhabha scattering gives the total cross section for $e^+e^- \to f\bar{f}$. The parameter β is precisely the velocity of the fermions in the center-of-mass frame.

Next let us examine how $\widehat{\Pi}_2(q^2)$ modifies the electromagnetic interaction, as determined by Eq. (7.77). In the nonrelativistic limit it makes sense to compute the potential $V(r)$. For the interaction between unlike charges, we have, in analogy with Eq. (4.126),

$$V(\mathbf{x}) = \int \frac{d^3q}{(2\pi)^3} e^{i\mathbf{q}\cdot\mathbf{x}} \frac{-e^2}{|\mathbf{q}|^2\left[1 - \widehat{\Pi}_2(-|\mathbf{q}|^2)\right]}. \tag{7.93}$$

Expanding $\widehat{\Pi}_2$ for $|q^2| \ll m^2$, we obtain

$$V(\mathbf{x}) = -\frac{\alpha}{r} - \frac{4\alpha^2}{15m^2}\delta^{(3)}(\mathbf{x}). \tag{7.94}$$

The correction term indicates that the electromagnetic force becomes much stronger at small distances. This effect can be measured in the hydrogen atom, where the energy levels are shifted by

$$\Delta E = \int d^3x\, |\psi(\mathbf{x})|^2 \cdot \left(-\frac{4\alpha^2}{15m^2}\delta^{(3)}(\mathbf{x})\right) = -\frac{4\alpha^2}{15m^2}|\psi(0)|^2.$$

The wavefunction $\psi(\mathbf{x})$ is nonzero at the origin only for s-wave states. For the $2S$ state, the shift is

$$\Delta E = -\frac{4\alpha^2}{15m^2} \cdot \frac{\alpha^3 m^3}{8\pi} = -\frac{\alpha^5 m}{30\pi} = -1.123 \times 10^{-7}\,\text{eV}.$$

This is a (small) part of the Lamb shift splitting listed in Table 6.1.

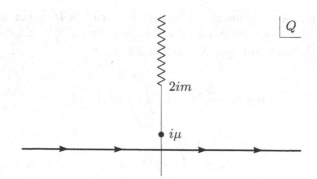

Figure 7.7. Contour for evaluating the effective strength of the electromagnetic interaction in the nonrelativistic limit. The pole at $Q = i\mu$ gives the Coulomb potential. The branch cut gives the order-α correction due to vacuum polarization.

The delta function in Eq. (7.94) is only an approximation; to find the true range of the correction term, we write Eq. (7.93) in the form

$$V(\mathbf{x}) = \frac{ie^2}{(2\pi)^2 r} \int\limits_{-\infty}^{\infty} dQ \, \frac{Q \, e^{iQr}}{Q^2 + \mu^2} \left[1 + \widehat{\Pi}_2(-Q^2) \right] \qquad (Q \equiv |\mathbf{q}|),$$

where we have inserted a photon mass μ to regulate the Coulomb potential. To perform this integral we push the contour upward (see Fig. 7.7). The leading contribution comes from the pole at $Q = i\mu$, giving the Coulomb potential, $-\alpha/r$. But there is an additional contribution from the branch cut, which begins at $Q = 2mi$. The real part of the integrand is the same on both sides of the cut, so the only contribution to the integral comes from the imaginary part of $\widehat{\Pi}_2$. Defining $q = -iQ$, we find that the contribution from the cut is

$$\delta V(r) = \frac{-e^2}{(2\pi)^2 r} \cdot 2 \int\limits_{2m}^{\infty} dq \, \frac{e^{-qr}}{q} \, \text{Im} \left[\widehat{\Pi}_2(q^2 - i\epsilon) \right]$$

$$= -\frac{\alpha}{r} \frac{2}{\pi} \int\limits_{2m}^{\infty} dq \, \frac{e^{-qr}}{q} \frac{\alpha}{3} \sqrt{1 - \frac{4m^2}{q^2}} \left(1 + \frac{2m^2}{q^2} \right).$$

When $r \gg 1/m$, this integral is dominated by the region where $q \approx 2m$. Approximating the integrand in this region and substituting $t = q - 2m$, we find

$$\delta V(r) = -\frac{\alpha}{r} \cdot \frac{2}{\pi} \int\limits_{0}^{\infty} dt \, \frac{e^{-(t+2m)r}}{2m} \frac{\alpha}{3} \sqrt{\frac{t}{m}} \left(\frac{3}{2} \right) + \mathcal{O}(t)$$

$$\approx -\frac{\alpha}{r} \cdot \frac{\alpha}{4\sqrt{\pi}} \frac{e^{-2mr}}{(mr)^{3/2}},$$

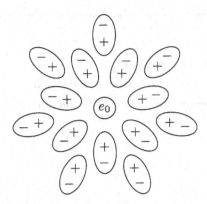

Figure 7.8. Virtual e^+e^- pairs are effectively dipoles of length $\sim 1/m$, which screen the bare charge of the electron.

so that

$$V(r) = -\frac{\alpha}{r}\left(1 + \frac{\alpha}{4\sqrt{\pi}}\frac{e^{-2mr}}{(mr)^{3/2}} + \cdots\right). \tag{7.95}$$

Thus the range of the correction term is roughly the electron Compton wavelength, $1/m$. Since hydrogen wavefunctions are nearly constant on this scale, the delta function in Eq. (7.94) was a good approximation. The radiative correction to $V(r)$ is called the *Uehling potential*.

We can interpret the correction as being due to screening. At $r \gtrsim 1/m$, virtual e^+e^- pairs make the vacuum a dielectric medium in which the apparent charge is less than the true charge (see Fig. 7.8). At smaller distances we begin to penetrate the polarization cloud and see the bare charge. This phenomenon is known as *vacuum polarization*.

Now consider the opposite limit: small distance or $-q^2 \gg m^2$. Equation (7.91) then becomes

$$\hat{\Pi}_2(q^2) \approx \frac{2\alpha}{\pi}\int_0^1 dx\, x(1-x)\left[\log\left(\frac{-q^2}{m^2}\right) + \log(x(1-x)) + \mathcal{O}\left(\frac{m^2}{q^2}\right)\right]$$

$$= \frac{\alpha}{3\pi}\left[\log\left(\frac{-q^2}{m^2}\right) - \frac{5}{3} + \mathcal{O}\left(\frac{m^2}{q^2}\right)\right].$$

The effective coupling constant in this limit is therefore

$$\alpha_{\text{eff}}(q^2) = \frac{\alpha}{1 - \frac{\alpha}{3\pi}\log\left(\frac{-q^2}{Am^2}\right)}, \tag{7.96}$$

where $A = \exp(5/3)$. The effective electric charge becomes much larger at small distances, as we penetrate the screening cloud of virtual electron-positron pairs.

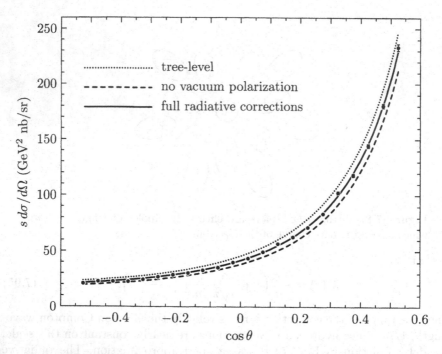

Figure 7.9. Differential cross section for Bhabha scattering, $e^+e^- \to e^+e^-$, at $E_{\mathrm{cm}} = 29$ GeV, as measured by the HRS collaboration, M. Derrick, et. al., *Phys. Rev.* **D34**, 3286 (1986). The upper curve is the order-α^2 prediction derived in Problem 5.2, plus a very small ($\sim$2%) correction due to the weak interaction. The lower curve includes all QED radiative corrections to order α^3 *except* the vacuum polarization contribution; note that these corrections depend on the experimental conditions, as explained in Chapter 6. The middle curve includes the vacuum polarization contribution as well, which increases the effective value of α^2 by about 10% at this energy.

The combined vacuum polarization effect of the electron and of heavier quarks and leptons causes the value of $\alpha_{\mathrm{eff}}(q^2)$ to increase by about 5% from $q = 0$ to $q = 30$ GeV, and this effect is observed in high-energy experiments. Figure 7.9 shows the cross section for Bhabha scattering at $E_{\mathrm{cm}} = 29$ GeV, and a comparison to QED with and without the vacuum polarization diagram.

We can write α_{eff} as a function of r by setting $q = 1/r$. The behavior of $\alpha_{\mathrm{eff}}(r)$ for all r is sketched in Fig. 7.10. The idea of a distance-dependent (or "scale-dependent" or "running") coupling constant will be a major theme of the rest of this book.

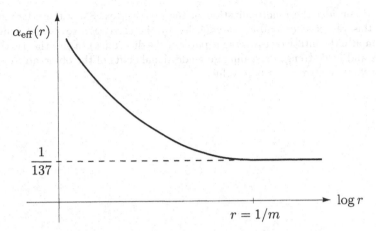

Figure 7.10. A qualitative sketch of the effective electromagnetic coupling constant generated by the one-loop vacuum polarization diagram, as a function of distance. The horizontal scale covers many orders of magnitude.

Problems

7.1   In Section 7.3 we used an indirect method to analyze the one-loop s-channel diagram for boson-boson scattering in ϕ^4 theory. To verify our indirect analysis, evaluate all three one-loop diagrams, using the standard method of Feynman parameters. Check the validity of the optical theorem.

7.2   Alternative regulators in QED. In Section 7.5, we saw that the Ward identity can be violated by an improperly chosen regulator. Let us check the validity of the identity $Z_1 = Z_2$, to order α, for several choices of the regulator. We have already verified that the relation holds for Pauli-Villars regularization.

(a) Recompute δZ_1 and δZ_2, defining the integrals (6.49) and (6.50) by simply placing an upper limit Λ on the integration over ℓ_E. Show that, with this definition, $\delta Z_1 \neq \delta Z_2$.

(b) Recompute δZ_1 and δZ_2, defining the integrals (6.49) and (6.50) by dimensional regularization. You may take the Dirac matrices to be 4×4 as usual, but note that, in d dimensions,

$$g_{\mu\nu}\gamma^\mu\gamma^\nu = d.$$

Show that, with this definition, $\delta Z_1 = \delta Z_2$.

7.3   Consider a theory of elementary fermions that couple both to QED and to a Yukawa field ϕ:

$$H_{\text{int}} = \int d^3x \, \frac{\lambda}{\sqrt{2}} \, \phi \bar\psi\psi + \int d^3x \, e \, A_\mu \bar\psi\gamma^\mu\psi.$$

(a) Verify that the contribution to Z_1 from the vertex diagram with a virtual ϕ equals the contribution to Z_2 from the diagram with a virtual ϕ. Use dimensional regularization. Is the Ward identity generally true in this theory?

(b) Now consider the renormalization of the $\phi\bar{\psi}\psi$ vertex. Show that the rescaling of this vertex at $q^2 = 0$ is *not* canceled by the correction to Z_2. (It suffices to compute the ultraviolet-divergent parts of the diagrams.) In this theory, the vertex and field-strength rescaling give additional shifts of the observable coupling constant relative to its bare value.

Final Project

Radiation of Gluon Jets

Although we have discussed QED radiative corrections at length in the last two chapters, so far we have made no attempt to compute a full radiatively corrected cross section. The reason is of course that such calculations are quite lengthy. Nevertheless it would be dishonest to pretend that one understands radiative corrections after computing only isolated effects as we have done. This "final project" is an attempt to remedy this situation. The project is the computation of one of the simplest, but most important, radiatively corrected cross sections. You should finish Chapter 6 before starting this project, but you need not have read Chapter 7.

Strongly interacting particles—pions, kaons, and protons—are produced in e^+e^- annihilation when the virtual photon creates a pair of quarks. If one ignores the effects of the strong interactions, it is easy to calculate the total cross section for quark pair production. In this final project, we will analyze the first corrections to this formula due to the strong interactions.

Let us represent the strong interactions by the following simple model: Introduce a new massless vector particle, the *gluon*, which couples universally to quarks:

$$\Delta H = \int d^3x \, g\bar{\psi}_{fi}\gamma^\mu\psi_{fi}B_\mu.$$

Here f labels the type ("flavor") of the quark (u, d, s, c, etc.) and $i = 1, 2, 3$ labels the color. The strong coupling constant g is independent of flavor and color. The electromagnetic coupling of quarks depends on the flavor, since the u and c quarks have charge $Q_f = +2/3$ while the d and s quarks have charge $Q_f = -1/3$. By analogy to α, let us define

$$\alpha_g = \frac{g^2}{4\pi}.$$

In this exercise, we will compute the radiative corrections to quark pair production proportional to α_g.

This model of the strong interactions of quarks does not quite agree with the currently accepted theory of the strong interactions, quantum chromodynamics (QCD). However, all of the results that we will derive here are also

259

correct in QCD with the replacement

$$\alpha_g \rightarrow \frac{4}{3}\alpha_s.$$

We will verify this claim in Chapter 17.

Throughout this exercise, you may ignore the masses of quarks. You may also ignore the mass of the electron, and average over electron and positron polarizations. To control infrared divergences, it will be necessary to assume that the gluons have a small nonzero mass μ, which can be taken to zero only at the end of the calculation. However (as we discussed in Problem 5.5), it is consistent to sum over polarization states of this massive boson by the replacement:

$$\sum \epsilon^\mu \epsilon^{\nu*} \rightarrow -g^{\mu\nu};$$

this also implies that we may use the propagator

$$B^\mu \overbrace{} B^\nu = \frac{-ig^{\mu\nu}}{k^2 - \mu^2 + i\epsilon}.$$

(a) Recall from Section 5.1 that, to lowest order in α and neglecting the effects of gluons, the total cross section for production of a pair of quarks of flavor f is

$$\sigma(e^+e^- \rightarrow \bar{q}q) = \frac{4\pi\alpha^2}{3s} \cdot 3Q_f^2.$$

Compute the diagram contributing to $e^+e^- \rightarrow \bar{q}q$ involving one virtual gluon. Reduce this expression to an integral over Feynman parameters, and renormalize it by subtraction at $q^2 = 0$, following the prescription used in Eq. (6.55). Notice that the resulting expression can be considered as a correction to $F_1(q^2)$ for the quark. Argue that, for massless quarks, to all orders in α_g, the total cross section for production of a quark pair unaccompanied by gluons is

$$\sigma(e^+e^- \rightarrow \bar{q}q) = \frac{4\pi\alpha^2}{3s} \cdot 3|F_1(q^2 = s)|^2,$$

with $F_1(q^2 = 0) = Q_f$.

(b) Before we attempt to evaluate the Feynman parameter integrals in part (a), let us put this contribution aside and study the process $e^+e^- \rightarrow \bar{q}qg$, quark pair production with an additional gluon emitted. Before we compute the cross section, it will be useful to work out some kinematics. Let q be the total 4-momentum of the reaction, let k_1 and k_2 be the 4-momenta of the final quark and antiquark, and let k_3 be the 4-momentum of the gluon. Define

$$x_i = \frac{2k_i \cdot q}{q^2}, \qquad i = 1, 2, 3;$$

this is the ratio of the center-of-mass energy of particle i to the maximum available energy. Then show (i) $\sum x_i = 2$, (ii) all other Lorentz scalars involving only the final-state momenta can be computed in terms of the x_i and the particle masses, and (iii) the complete integral over 3-body phase space can be written as

$$\int d\Pi_3 = \prod_i \int \frac{d^3 k_i}{(2\pi)^3} \frac{1}{2E_i} (2\pi)^4 \delta^{(4)}(q - \sum_i k_i) = \frac{q^2}{128\pi^3} \int dx_1\, dx_2.$$

Find the region of integration for x_1 and x_2 if the quark and antiquark are massless but the gluon has mass μ.

(c) Draw the Feynman diagrams for the process $e^+e^- \to \bar{q}qg$, to leading order in α and α_g, and compute the differential cross section. You may throw away the information concerning the correlation between the initial beam axis and the directions of the final particles. This is conveniently done as follows: The usual trace tricks for evaluating the square of the matrix element give for this process a result of the structure

$$\int d\Pi_3 \, \frac{1}{4} \sum |\mathcal{M}|^2 = L_{\mu\nu} \int d\Pi_3 \, H^{\mu\nu},$$

where $L_{\mu\nu}$ represents the electron trace and $H^{\mu\nu}$ represents the quark trace. If we integrate over all parameters of the final state except x_1 and x_2, which are scalars, the only preferred 4-vector characterizing the final state is q^μ. On the other hand, $H_{\mu\nu}$ satisfies

$$q^\mu H_{\mu\nu} = H_{\mu\nu} q^\nu = 0.$$

Why is this true? (There is an argument based on general principles; however, you might find it a useful check on your calculation to verify this property explicitly.) Since, after integrating over final-state vectors, $\int H^{\mu\nu}$ depends only on q^μ and scalars, it can only have the form

$$\int d\Pi_3 \, H^{\mu\nu} = \left(g^{\mu\nu} - \frac{q^\mu q^\nu}{q^2}\right) \cdot H,$$

where H is a scalar. With this information, show that

$$L_{\mu\nu} \int d\Pi_3 \, H^{\mu\nu} = \frac{1}{3} \left(g^{\mu\nu} L_{\mu\nu}\right) \cdot \int d\Pi_3 \left(g^{\rho\sigma} H_{\rho\sigma}\right).$$

Using this trick, derive the differential cross section

$$\frac{d\sigma}{dx_1 dx_2}(e^+e^- \to \bar{q}qg) = \frac{4\pi\alpha^2}{3s} \cdot 3Q_f^2 \cdot \frac{\alpha_g}{2\pi} \frac{x_1^2 + x_2^2}{(1-x_1)(1-x_2)}$$

in the limit $\mu \to 0$. If we assume that each original final-state particle is realized physically as a jet of strongly interacting particles, this formula gives the probability for observing three-jet events in e^+e^- annihilation and the kinematic distribution of these events. The form of the distribution in the x_i is an absolute prediction, and it agrees with experiment. The normalization of this distribution is a measure of the strong-interaction coupling constant.

(d) Now replace $\mu \neq 0$ in the formula of part (c) for the differential cross section, and carefully integrate over the region found in part (b). You may assume $\mu^2 \ll q^2$. In this limit, you will find infrared-divergent terms of order $\log(q^2/\mu^2)$ and also $\log^2(q^2/\mu^2)$, finite terms of order 1, and terms explicitly suppressed by powers of (μ^2/q^2). You may drop terms of the last type throughout this calculation. For the moment, collect and evaluate only the infrared-divergent terms.

(e) Now analyze the Feynman parameter integral obtained in part (a), again working in the limit $\mu^2 \ll q^2$. Note that this integral has singularities in the region of integration. These should be controlled by evaluating the integral for q spacelike and then analytically continuing into the physical region. That is, write $Q^2 = -q^2$, evaluate the integral for $Q^2 > 0$, and then carefully analytically continue the result to $Q^2 = -q^2 - i\epsilon$. Combine the result with the answer from part (d) to form the total cross section for $e^+e^- \to$ strongly interacting particles, to order α_g. Show that all infrared-divergent logarithms cancel out of this quantity, so that this total cross section is well-defined in the limit $\mu \to 0$.

(f) Finally, collect the terms of order 1 from the integrations of parts (d) and (e) and combine them. To evaluate certain of these terms, you may find the following formula useful:

$$\int\limits_0^1 dx \, \frac{\log(1-x)}{x} = -\frac{\pi^2}{6}.$$

(It is not hard to prove this.) Show that the total cross section is given, to this order in α_g, by

$$\sigma(e^+e^- \to \bar{q}q \text{ or } \bar{q}qg) = \frac{4\pi\alpha^2}{3s} \cdot 3Q_f^2 \cdot \left(1 + \frac{3\alpha_g}{4\pi}\right).$$

This formula gives a second way of measuring the strong-interaction coupling constant. The experimental results agree (within the current experimental errors) with the results obtained by the method of part (c). We will discuss the measurement of α_s more fully in Section 17.6.

Part II

Renormalization

Part II

Renormalization

Chapter 8

Invitation: Ultraviolet Cutoffs and Critical Fluctuations

The main purpose of Part II of this book is to develop a general theory of renormalization. This theory will explain the origin of ultraviolet divergences in field theory and will indicate when these divergences can be removed systematically. It will also give a way to convert the divergences of Feynman diagrams from a problem into a tool. We will apply this tool to study the asymptotic large- or small-momentum behavior of field theory amplitudes.

When we first encountered an ultraviolet divergence in the calculation of the one-loop vertex correction in Section 6.3, it seemed an aberration that ought to disappear before it caused us too much discomfort. In Chapter 7 we saw further examples of ultraviolet-divergent diagrams, enough to convince us that such divergences occur ubiquitously in Feynman diagram computations. Thus it is necessary for anyone studying field theory to develop a point of view toward these divergences. Most people begin with the belief that any theory that contains divergences must be nonsense. But this viewpoint is overly restrictive, since it excludes not only quantum field theory but even the classical electrodynamics of point particles.

With some experience, one might adopt a more permissive attitude of peaceful coexistence with the divergences: One can accept a theory with divergences, as long as they do not appear in physical predictions. In Chapter 7 we saw that all of the divergences that appear in the one-loop radiative corrections to electron scattering from a heavy target can be eliminated by consistently eliminating the bare values of the mass and charge of the electron in favor of their measured physical values. In Chapter 10, we will argue that all of the ultraviolet divergences of QED, in all orders of perturbation theory, can be eliminated in this way. Thus, as long as one is willing to consider the mass and charge of the electron as measured parameters, the predictions of QED perturbation theory will always be free of divergences. We will also show in Chapter 10 that QED belongs to a well-defined class of field theories in which all ultraviolet divergences are removed after a fixed small number of physical parameters are taken from experiment. These theories, called *renormalizable* quantum field theories, are the only ones in which perturbation theory gives well-defined predictions.

Ideally, though, one should take the further step of trying to understand

physically why the divergences appear and why their effects are more se-
vere in some theories than in others. This direct approach to the divergence
problem was pioneered in the 1960s by Kenneth Wilson. The crucial insights
needed to solve this problem emerged from a correspondence, discovered by
Wilson and others, between quantum field theory and the statistical physics
of magnets and fluids. Wilson's approach to renormalization is the subject
of Chapter 12. The present chapter gives a brief introduction to the issues
in condensed matter physics that have provided insight into the problem of
ultraviolet divergences.

Formal and Physical Cutoffs

Ultraviolet divergences signal that quantities calculated in a quantum field
theory depend on some very large momentum scale, the ultraviolet cutoff.
Equivalently, in position space, divergent quantities depend on some very
small distance scale.

The idea of a small-distance cutoff in the continuum description of a system occurs in classical field theories as well. Typically the cutoff is at the
scale of atomic distances, where the continuum description no longer applies.
However, the size of the cutoff manifests itself in certain parameters of the
continuum theory. In fluid dynamics, for instance, parameters such as the
viscosity and the speed of sound are of just the size one would expect by combining typical atomic radii and velocities. Similarly, in a magnet, the magnetic
susceptibility can be estimated by assuming that the energy cost of flipping
an electron spin is on the order of a tenth of an eV, as we would expect from
atomic physics. Each of these systems possesses a natural ultraviolet cutoff
at the scale of an atom; by understanding the physics at the atomic scale, we
can compute the parameters that determine the physics on larger scales.

In quantum field theory, however, we have no precise knowledge of the
fundamental physics at very short distance scales. Thus, we can only measure
parameters such as the physical charge and mass of the electron, not compute
them from first principles. The presence of ultraviolet divergences in the relations between these physical parameters and their bare values is a sign that
these parameters are controlled by the unknown short-distance physics.

Whether we know the fundamental physics at small distance scales or
not, we need two kinds of information in order to write an effective theory for
large-distance phenomena. First, we must know how many parameters from
the small distance scale are relevant to large-distance physics. Second, and
more importantly, we must know what degrees of freedom from the underlying
theory appear at large distances.

In fluid mechanics, it is something of a miracle, from the atomic point of
view, that any large-distance degrees of freedom even exist. Nevertheless, the
equations that express the transport of energy and mass over large distances
do have smooth, coherent solutions. The large-distance degrees of freedom are

the flows that transport these conserved quantities, and sound waves of long wavelength.

In quantum field theory, the large-distance physics involves only those particles that have masses that are very small compared to the fundamental cutoff scale. These particles and their dynamics are the quantum analogues of the large-scale flows in fluid mechanics. The simplest way to naturally arrange for such particles to appear is to make use of particles that naturally have zero mass. So far in this book, we have encountered two types of particles whose mass is precisely zero, the photon and the chiral fermion. (In Chapter 11 we will meet one further naturally massless particle, the *Goldstone boson*.) We might argue that QED exists as a theory on scales much larger than its cutoff because the photon is naturally massless and because the left- and right-handed electrons are very close to being chiral fermions.

There is another way that particles of zero or almost zero mass can arise in quantum field theory: We can simply tune the parameters of a scalar field theory so that the scalar particles have masses small compared to the cutoff. This method of introducing particles with small mass seems arbitrary and unnatural. Nevertheless, it has an analogue in statistical mechanics that is genuinely interesting in that discipline and can teach us some important lessons.

Normally, in a condensed matter system, the thermal fluctuations are correlated only over atomic distances. Under special circumstances, however, they can have much longer range. The clearest example of this phenomenon occurs in a ferromagnet. At high temperature, the electron spins in a magnet are disorganized and fluctuating; but at low temperature, these spins align to a fixed direction.* Let us think about how this alignment builds up as the temperature of the magnet is lowered. As the magnet cools from high temperature, clusters of correlated spins become larger and larger. At a certain point—the temperature of magnetization—the entire sample becomes a single large cluster with a well-defined macroscopic orientation. Just above this temperature, the magnet contains large clusters of spins with a common orientation, which in turn belong to still larger clusters, such that the orientations on the very largest scale are still randomized through the sample. This situation is illustrated in Fig. 8.1. Similar behavior occurs in the vicinity of any other second-order phase transition, for example, the order-disorder transition in binary alloys, the critical point in fluids, or the superfluid transition in Helium-4.

The natural description of these very long wavelength fluctuations is in terms of a fluctuating continuum field. At the lowest intuitive level, we might

---

*In a real ferromagnet, the long-range magnetic dipole-dipole interaction causes the state of uniform magnetization to break up into an array of magnetic domains. In this book, we will ignore this interaction and think of a magnetic spin as a pure orientation. It is this idealized system that is directly analogous to a quantum field theory.

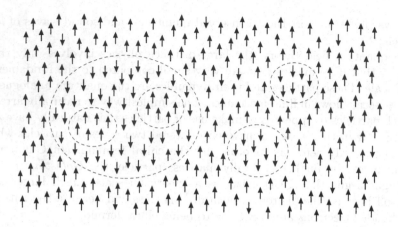

Figure 8.1. Clusters of oriented spins near the critical point of a ferromagnet.

substitute quantum for statistical fluctuations and try to describe this system as a quantum field theory. In Section 9.3 we will derive a somewhat more subtle relation that makes a precise connection between the statistical and the quantum systems. Through this connection, the behavior of any statistical system near a second-order phase transition can be translated into the behavior of a particular quantum field theory. This quantum field theory has a field with a mass that is very small compared to the basic atomic scale and that goes to zero precisely at the phase transition.

But this connection seems to compound the problem of ultraviolet divergences in quantum field theory: If the wealth of phase transitions observed in Nature generates a similar wealth of quantum field theories, how can we possibly define a quantum field theory without detailed reference to its origins in physics at the scale of its ultraviolet cutoff? Saying that a quantum field theory makes predictions independent of the cutoff would be equivalent to saying that the statistical fluctuations in the neighborhood of a critical point are independent of whether the system is a magnet, a fluid, or an alloy. But is this statement so obviously incorrect? By reversing the logic, we would find that quantum field theory makes a remarkably powerful prediction for condensed matter systems, a prediction of *universality* for the statistical fluctuations near a critical point. In fact, this prediction is verified experimentally.

A major theme of Part II of this book will be that these two ideas—cutoff independence in quantum field theory and universality in the theory of critical phenomena—are naturally the same idea, and that understanding either of these ideas gives insight into the other.

Landau Theory of Phase Transitions

To obtain a first notion of what could be universal in the phenomena of phase transitions, let us examine the simplest continuum theory of second-order phase transitions, due to Landau.

First we should review a little thermodynamics and clarify our nomenclature. In thermodynamics, a *first-order phase transition* is a point across which some thermodynamic variable (the density of a fluid, or the magnetization of a ferromagnet) changes discontinuously. At a phase transition point, two quite distinct thermodynamic states (liquid and gas, or magnetization parallel and antiparallel to a given axis) are in equilibrium. The thermodynamic quantity that changes discontinuously across the transition, and that characterizes the difference of the two competing phases, is called the *order parameter*. In most circumstances, it is possible to change a second thermodynamic parameter in such a way that the two competing states move closer together in the thermodynamic space, so that at some value of this parameter, these two states become identical and the discontinuity in the order parameter disappears. This endpoint of the line of first-order transitions is called a *second-order phase transition*, or, more properly, a *critical point*. Viewed from the other direction, a critical point is a point at which a single thermodynamic state bifurcates into two macroscopically distinct states. It is this bifurcation that leads to the long-ranged thermal fluctuations discussed in the previous section.

A concrete example of this behavior is exhibited by a ferromagnet. Let us assume for simplicity that the material we are discussing has a preferred axis of magnetization, so that at low temperature, the system will have its spins ordered either parallel or antiparallel to this axis. The total magnetization along this axis, M, is the order parameter. At low temperature, application of an external magnetic field H will favor one or the other of the two possible states. At $H = 0$, the two states will be in equilibrium; if H is changed from a small negative to a small positive value, the thermodynamic state and the value of M will change discontinuously. Thus, for any fixed (low) temperature, there is a first-order transition at $H = 0$. Now consider the effect of raising the temperature: The fluctuation of the spins increases and the value of $|M|$ decreases. At some temperature T_C the system ceases to be magnetized at $H = 0$. At this point, the first-order phase transition disappears and the two competing thermodynamic states coalesce. The system thus has a critical point at $T = T_C$. The location of these various transitions in the H-T plane is shown in Fig. 8.2.

Landau described this behavior by the use of the Gibbs free energy G; this is the thermodynamic potential that depends on M and T, such that

$$\left.\frac{\partial G}{\partial M}\right|_T = H. \tag{8.1}$$

He suggested that we concentrate our attention on the region of the critical

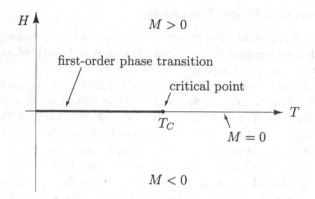

Figure 8.2. Phase diagram in the H-T plane for a uniaxial ferromagnet.

point: $T \approx T_C$, $M \approx 0$. Then it is reasonable to expand $G(M)$ as a Taylor series in M. For $H = 0$, we can write

$$G(M) = A(T) + B(T)M^2 + C(T)M^4 + \cdots . \qquad (8.2)$$

Because the system has a symmetry under $M \to -M$, $G(M)$ can contain only even powers of M. Since M is small, we will ignore the higher terms in the expansion. Given Eq. (8.2), we can find the possible values of M at $H = 0$ by solving

$$0 = \frac{\partial G}{\partial M} = 2B(T)M + 4C(T)M^3 . \qquad (8.3)$$

If B and C are positive, the only solution is $M = 0$. However, if $C > 0$ but B is negative below some temperature T_C, we have a nontrivial solution for $T < T_C$, as shown in Fig. 8.3. More concretely, approximate for $T \approx T_C$:

$$B(T) = b(T - T_C), \qquad C(T) = c . \qquad (8.4)$$

Then the solution to Eq. (8.3) is

$$M = \begin{cases} 0 & \text{for } T > T_C; \\ \pm\left[(b/2c)(T_C - T)\right]^{1/2} & \text{for } T < T_C. \end{cases} \qquad (8.5)$$

This is just the qualitative behavior that we expect at a critical point.

To find the value of M at nonzero external field, we could solve Eq. (8.1) with the left-hand side given by (8.2). An equivalent procedure is to minimize a new function, related to (8.2). Define

$$G(M, H) = A(T) + B(T)M^2 + C(T)M^4 - HM . \qquad (8.6)$$

Then the minimum of $G(M, H)$ with respect to M at fixed H gives the value of M that satisfies Eq. (8.1). The minimum is unique except when $H = 0$ and $T < T_C$, where we find the double minimum in the second line of (8.5). This is consistent with the phase diagram shown in Fig. 8.2.

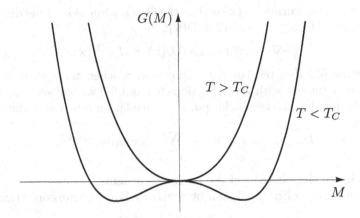

Figure 8.3. Behavior of the Gibbs free energy $G(M)$ in Landau theory, at temperatures above and below the critical temperature.

To study correlations in the vicinity of the phase transition, Landau generalized this description further by considering the magnetization M to be the integral of a local spin density:

$$M = \int d^3x \, s(\mathbf{x}). \tag{8.7}$$

Then the Gibbs free energy (8.6) becomes the integral of a local function of $s(\mathbf{x})$,

$$G = \int d^3x \left[\tfrac{1}{2} (\boldsymbol{\nabla} s)^2 + b(T - T_C)s^2 + cs^4 - Hs \right], \tag{8.8}$$

which must be minimized with respect to the field configuration $s(\mathbf{x})$. The first term is the simplest possible way to introduce the tendency of nearby spins to align with one another. We have rescaled $s(\mathbf{x})$ so that the coefficient of this term is set to $1/2$. In writing this free energy integral, we could even consider H to vary as a function of position. In fact, it is useful to do that; we can turn on $H(\mathbf{x})$ near $x = 0$ and see what response we find at another point.

The minimum of the free energy expression (8.8) with respect to $s(\mathbf{x})$ is given by the solution to the variational equation

$$0 = \delta G[s(\mathbf{x})] = -\nabla^2 s + 2b(T - T_C)s + 4cs^3 - H(\mathbf{x}). \tag{8.9}$$

For $T > T_C$, where the macroscopic magnetization vanishes and so $s(\mathbf{x})$ should be small, we can find the qualitative behavior by ignoring the s^3 term. Then $s(\mathbf{x})$ obeys a linear equation,

$$(-\nabla^2 + 2b(T - T_C))s(\mathbf{x}) = H(\mathbf{x}). \tag{8.10}$$

To study correlations of spins, we will set

$$H(\mathbf{x}) = H_0 \delta^{(3)}(\mathbf{x}). \tag{8.11}$$

The resulting configuration $s(\mathbf{x})$ is then the Green's function of the differential operator in Eq. (8.10), so we call it $D(\mathbf{x})$:

$$\left(-\nabla^2 + 2b(T - T_C)\right)D(\mathbf{x}) = H_0\delta^{(3)}(\mathbf{x}). \tag{8.12}$$

This Green's function tells us the response at $\mathbf{x}$ when the spin at $\mathbf{x} = 0$ is forced into alignment with H. In Sections 9.2 and 9.3 we will see that $D(\mathbf{x})$ is also proportional to the zero-field spin-spin correlation function in the thermal ensemble,

$$D(\mathbf{x}) \propto \langle s(\mathbf{x})s(0)\rangle \equiv \sum_{\text{all } s(\mathbf{x})} s(\mathbf{x})s(0)e^{-\mathbf{H}/kT}, \tag{8.13}$$

where $\mathbf{H}$ is the Hamiltonian of the magnetic system.

The solution to Eq. (8.12) can be found by Fourier transformation:

$$D(\mathbf{x}) = \int \frac{d^3k}{(2\pi)^3} \frac{H_0\, e^{i\mathbf{k}\cdot\mathbf{x}}}{|\mathbf{k}|^2 + 2b(T - T_C)}. \tag{8.14}$$

This is just the integral we encountered in our discussion of the Yukawa potential, Eq. (4.126). Evaluating it in the same way, we find

$$D(\mathbf{x}) = \frac{H_0}{4\pi}\frac{1}{r}e^{-r/\xi}, \tag{8.15}$$

where

$$\xi = \left[2b(T - T_C)\right]^{-1/2} \tag{8.16}$$

is the *correlation length*, the range of correlated spin fluctuations. Notice that this length diverges as $T \to T_C$.

The main results of this analysis, Eqs. (8.5) and (8.16), involve unknown constants b, c that depend on physics at the atomic scale. On the other hand, the power-law dependence in these formulae on $(T - T_C)$ follows simply from the structure of the Landau equations and is independent of any details of the microscopic physics. In fact, our derivation of this dependence did not even use the fact that G describes a ferromagnet; we assumed only that G can be expanded in powers of an order parameter and that G respects the reflection symmetry $M \to -M$. These assumptions apply equally well to many other types of systems: binary alloys, superfluids, and even (though the reflection symmetry is less obvious here) the liquid-gas transition. Landau theory predicts that, near the critical point, these systems show a universal behavior in the dependence of M, ξ, and other thermodynamic quantities on $(T - T_C)$.

Critical Exponents

The preceding treatment of the Landau theory of phase transitions emphasizes its similarity to classical field theory. We set up an appropriate free energy and found the thermodynamically preferred configuration by solving a classical variational equation. This gives only an approximation to the full statistical problem, analogous to the approximation of replacing quantum by classical dynamics in field theory. In Chapter 13, we will use methods of quantum field theory to account properly for the fluctuations about the preferred Landau thermodynamic state. These modifications turn out to be profound, and rather counterintuitive.

To describe the form of these modifications, let us write Eq. (8.15) more generally as

$$\langle s(\mathbf{x})s(0) \rangle = A \frac{1}{r^{1+\eta}} f(r/\xi), \tag{8.17}$$

where A is a constant and $f(y)$ is a function that satisfies $f(0) = 1$ and $f(y) \to 0$ as $y \to \infty$. Landau theory predicts that $\eta = 0$ and $f(y)$ is a simple exponential. This expression has a form strongly analogous to that of a Green's function in quantum field theory. The constant A can be absorbed into the field-strength renormalization of the field $s(\mathbf{x})$. The correlation length ξ is, in general, a complicated function of the atomic parameters, but in the continuum description we can simply trade these parameters for ξ. It is appropriate to consider ξ as a cutoff-independent, physical parameter, since it controls the large-distance behavior of a physical correlation. In fact, the analogy between Eq. (8.15) and the Yukawa potential suggests that we should identify ξ^{-1} with the physical mass in the associated quantum field theory. Then Eq. (8.17) gives a cutoff-independent, continuum representation of the statistical system.

If we were working in quantum field theory, we would derive corrections to Eq. (8.17) as a perturbation series in the parameter c multiplying the nonlinear term in (8.9). This would generalize the Landau result to

$$\langle s(\mathbf{x})s(0) \rangle = \frac{1}{r} F(r/\xi, c). \tag{8.18}$$

The perturbative corrections would depend on the properties of the continuum field theory. For example, $F(y, c)$ would depend on the number of components of the field $s(\mathbf{x})$, and its series expansion would differ depending on whether the magnetization formed along a preferred axis, in a preferred plane, or isotropically. For order parameters with many components, the expansion would also depend on higher discrete symmetries of the problem. However, we expect that systems described by the same Landau free energy (for example, a single-axis ferromagnet and a liquid-gas system) should have the same perturbation expansion when this expansion is written in terms of the physical mass and coupling. The complete universality of Landau theory then becomes a more limited concept, in which systems have the same large-distance correlations if their order parameters have the same symmetry. We might say that statistical systems divide into distinct *universality classes*, each with a characteristic large-scale behavior.

If this were the true behavior of systems near second-order phase transitions, it would already be a wonderful confirmation of the ideas required to formulate cutoff-independent quantum field theories. However, the true behavior of statistical systems is still another level more subtle. What one finds experimentally is a dependence of the form of Eq. (8.17), where the function

$f(y)$ is the same within each universality class. There is no need for an auxiliary parameter c. On the other hand, the exponent η takes a specific nonzero value in each universality class. Other power-law relations of Landau theory are also modified, in a specific manner for each universality class. For example, Eq. (8.5) is changed, for $T < T_C$, to

$$M \propto (T_C - T)^\beta, \qquad (8.19)$$

where the exponent β takes a fixed value for all systems in a given universality class. For three-dimensional single-axis magnets and for fluids, $\beta = 0.313$. The powers in these nontrivial scaling relations are called *critical exponents*.

The modification from Eq. (8.18) to Eq. (8.17) does not imperil the idea that a condensed matter system, in the vicinity of a second-order phase transition, has a well-defined, cutoff-independent, continuum behavior. However, we would like to understand why Eq. (8.17) should be expected as the correct representation. The answer to this question will come from a thorough analysis of the ultraviolet divergences of the corresponding quantum field theory. In Chapter 12, when we finally conclude our explication of the ultraviolet divergences, we will find that we have in hand the tools not only to justify Eq. (8.17), but also to calculate the values of the critical exponents using Feynman diagrams. In this way, we will uncover a beautiful application of quantum field theory to the domain of atomic physics. The success of this application will guide us, in Part III, to even more powerful tools, which we will need in the relativistic domain of elementary particles.

Chapter 9

Functional Methods

Feynman once said that* "every theoretical physicist who is any good knows six or seven different theoretical representations for exactly the same physics." Following his advice, we introduce in this chapter an alternative method of deriving the Feynman rules for an interacting quantum field theory: the method of *functional integration*.

Aside from Feynman's general principle, we have several specific reasons for introducing this formalism. It will provide us with a relatively easy derivation of our expression for the photon propagator, completing the proof of the Feynman rules for QED given in Section 4.8. The functional method generalizes more readily to other interacting theories, such as scalar QED (Problem 9.1), and especially the non-Abelian gauge theories (Part III). Since it uses the Lagrangian, rather than the Hamiltonian, as its fundamental quantity, the functional formalism explicitly preserves all symmetries of a theory. Finally, the functional approach reveals the close analogy between quantum field theory and statistical mechanics. Exploiting this analogy, we will turn Feynman's advice upside down and apply the same theoretical representation to two completely different areas of physics.

9.1 Path Integrals in Quantum Mechanics

We begin by applying the functional integral (or *path* integral) method to the simplest imaginable system: a nonrelativistic quantum-mechanical particle moving in one dimension. The Hamiltonian for this system is

$$H = \frac{p^2}{2m} + V(x).$$

Suppose that we wish to compute the amplitude for this particle to travel from one point (x_a) to another (x_b) in a given time (T). We will call this amplitude $U(x_a, x_b; T)$; it is the position representation of the Schrödinger time-evolution operator. In the canonical Hamiltonian formalism, U is given by

$$U(x_a, x_b; T) = \langle x_b | e^{-iHT/\hbar} | x_a \rangle. \tag{9.1}$$

---

*The Character of Physical Law (MIT Press, 1965), p. 168.

(For the next few pages we will display all factors of $\hbar$ explicitly.)

In the path-integral formalism, U is given by a very different-looking expression. We will first try to motivate that expression, then prove that it is equivalent to (9.1).

Recall that in quantum mechanics there is a *superposition principle*: When a process can take place in more than one way, its total amplitude is the coherent sum of the amplitudes for each way. A simple but nontrivial example is the famous double-slit experiment, shown in Fig. 9.1. The total amplitude for an electron to arrive at the detector is the sum of the amplitudes for the two paths shown. Since the paths differ in length, these two amplitudes generally differ, causing interference.

For a general system, we might therefore write the total amplitude for traveling from x_a to x_b as

$$U(x_a, x_b; T) = \sum_{\text{all paths}} e^{i \cdot (\text{phase})} = \int \mathcal{D}x(t)\, e^{i \cdot (\text{phase})}. \tag{9.2}$$

To be democratic, we have written the amplitude for each particular path as a pure phase, so that no path is inherently more important than any other. The symbol $\int \mathcal{D}x(t)$ is simply another way of writing "sum over all paths"; since there is one path for every function $x(t)$ that begins at x_a and ends at x_b, the sum is actually an integral over this continuous space of functions.

We can define this integral as part of a natural generalization of the calculus to spaces of functions. A function that maps functions to numbers is called a *functional*. The integrand in (9.2) is a functional, since it associates a complex amplitude with any function $x(t)$. The argument of a functional $F[x(t)]$ is conventionally written in square brackets rather than parentheses. Just as an ordinary function $y(x)$ can be integrated over a set of points x, a functional $F[x(t)]$ can be integrated over a set of functions $x(t)$; the measure of such a *functional integral* is conventionally written with a script capital $\mathcal{D}$, as in (9.2). A functional can also be differentiated with respect to its argument (a function), and this *functional derivative* is denoted by $\delta F / \delta x(t)$. We will develop more precise definitions of this new integral and derivative in the course of this section and the next.

What should we use for the "phase" in Eq. (9.2)? In the classical limit, we should find that only one path, the classical path, contributes to the total amplitude. We might therefore hope to evaluate the integral in (9.2) by the method of stationary phase, identifying the classical path $x_{\text{cl}}(t)$ by the stationary condition,

$$\frac{\delta}{\delta x(t)} \Big(\text{phase}[x(t)] \Big) \Big|_{x_{\text{cl}}} = 0.$$

But the classical path is the one that satisfies the principle of least action,

$$\frac{\delta}{\delta x(t)} \Big(S[x(t)] \Big) \Big|_{x_{\text{cl}}} = 0,$$

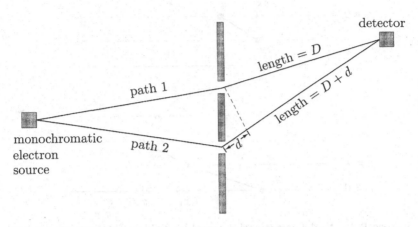

Figure 9.1. The double-slit experiment. Path 2 is longer than path 1 by an amount d, and therefore has a phase that is larger by $2\pi d/\lambda$, where $\lambda = 2\pi\hbar/p$ is the particle's de Broglie wavelength. Constructive interference occurs when $d = 0$, λ, ..., while destructive interference occurs when $d = \lambda/2$, $3\lambda/2$, ...

where $S = \int L\,dt$ is the classical action. It is tempting, therefore, to identify the phase with S, up to a constant. Since the stationary-phase approximation should be valid in the classical limit—that is, when $S \gg \hbar$—we will use $S/\hbar$ for the phase. Our final formula for the propagation amplitude is thus

$$\langle x_b | e^{-iHT/\hbar} | x_a \rangle = U(x_a, x_b; T) = \int \mathcal{D}x(t)\, e^{iS[x(t)]/\hbar}. \qquad (9.3)$$

We can easily verify that this formula gives the correct interference pattern in the double-slit experiment. The action for either path shown in Fig. 9.1 is just $(1/2)mv^2 t$, the kinetic energy times the time. For path 1 the velocity is $v_1 = D/t$, so the phase is $mD^2/2\hbar t$. For path 2 we have $v_2 = (D+d)/t$, so the phase is $m(D+d)^2/2\hbar t$. We must assume that $d \ll D$, so that $v_1 \approx v_2$ (i.e., the electrons have a well-defined velocity). The excess phase for path 2 is then $mDd/\hbar t \approx pd/\hbar$, where p is the momentum. This is exactly what we would expect from the de Broglie relation $p = h/\lambda$, so we must be doing something right.

To evaluate the functional integral more generally, we must define the symbol $\int \mathcal{D}x(t)$ in the case where the number of paths $x(t)$ is more than two (and, in fact, continuously infinite). We will use a brute-force definition, by discretization. Break up the time interval from 0 to T into many small pieces of duration ϵ, as shown in Fig. 9.2. Approximate a path $x(t)$ as a sequence of straight lines, one in each time slice. The action for this discretized path is

$$S = \int_0^T dt \left(\frac{m}{2}\dot{x}^2 - V(x) \right) \longrightarrow \sum_k \left[\frac{m}{2}\frac{(x_{k+1}-x_k)^2}{\epsilon} - \epsilon V\left(\frac{x_{k+1}+x_k}{2}\right) \right].$$

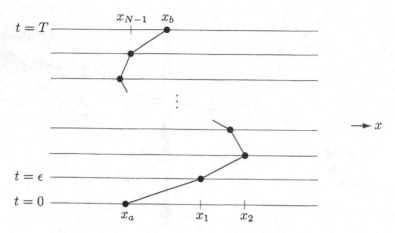

Figure 9.2. We define the path integral by dividing the time interval into small slices of duration ϵ, then integrating over the coordinate x_k of each slice.

We then define the path integral by

$$
\int \mathcal{D}x(t) \equiv \frac{1}{C(\epsilon)} \int \frac{dx_1}{C(\epsilon)} \int \frac{dx_2}{C(\epsilon)} \cdots \int \frac{dx_{N-1}}{C(\epsilon)} = \frac{1}{C(\epsilon)} \prod_k \int\limits_{-\infty}^{\infty} \frac{dx_k}{C(\epsilon)}, \tag{9.4}
$$

where $C(\epsilon)$ is a constant, to be determined later. (We have included one factor of $C(\epsilon)$ for each of the N time slices, for reasons that will be clear below.) At the end of the calculation we take the limit $\epsilon \to 0$. (As in Sections 4.5 and 6.2, the $\prod$ symbol is an instruction to write what follows once for each k.)

Using (9.4) as the definition of the right-hand side of (9.3), we will now demonstrate the validity of (9.3) for a general one-particle potential problem. To do this, we will show that the left- and right-hand sides of (9.3) are obtained by integrating the same differential equation, with the same initial condition. In the process, we will determine the constant $C(\epsilon)$.

To derive the differential equation satisfied by (9.4), consider the addition of the very last time slice in Fig. 9.2. According to (9.3) and the definition (9.4), we should have

$$
U(x_a, x_b; T) = \int\limits_{-\infty}^{\infty} \frac{dx'}{C(\epsilon)} \exp\left[\frac{i}{\hbar} \frac{m(x_b - x')^2}{2\epsilon} - \frac{i}{\hbar} \epsilon V\left(\frac{x_b + x'}{2}\right) \right] U(x_a, x'; T - \epsilon).
$$

The integral over x' is just the contribution to $\int \mathcal{D}x$ from the last time slice, while the exponential factor is the contribution to $e^{iS/\hbar}$ from that slice. All contributions from previous slices are contained in $U(x_a, x'; T - \epsilon)$.

As we send $\epsilon \to 0$, the rapid oscillation of the first term in the exponential constrains x' to be very close to x_b. We can therefore expand the above expression in powers of $(x'-x_b)$:

$$U(x_a, x_b; T) = \int_{-\infty}^{\infty} \frac{dx'}{C} \exp\left(\frac{i}{\hbar}\frac{m}{2\epsilon}(x_b-x')^2\right)\left[1 - \frac{i\epsilon}{\hbar}V(x_b) + \cdots\right]$$

$$\times \left[1 + (x'-x_b)\frac{\partial}{\partial x_b} + \frac{1}{2}(x'-x_b)^2 \frac{\partial^2}{\partial x_b^2} + \cdots\right]U(x_a, x_b; T-\epsilon).$$

$$(9.5)$$

We can now perform the x' integral by treating the exponential factor as a Gaussian. (Properly, we should introduce a small real term in the exponent for convergence; we will ignore this term until the next section, when we derive Feynman rules using functional methods.) Recall the Gaussian integration formulae

$$\int d\xi\, e^{-b\xi^2} = \sqrt{\frac{\pi}{b}}, \qquad \int d\xi\, \xi\, e^{-b\xi^2} = 0, \qquad \int d\xi\, \xi^2\, e^{-b\xi^2} = \frac{1}{2b}\sqrt{\frac{\pi}{b}}.$$

Applying these identities to (9.5), we find

$$U(x_a, x_b; T) = \left(\frac{1}{C}\sqrt{\frac{2\pi\hbar\epsilon}{-im}}\right)\left[1 - \frac{i\epsilon}{\hbar}V(x_b) + \frac{i\epsilon\hbar}{2m}\frac{\partial^2}{\partial x_b^2} + \mathcal{O}(\epsilon^2)\right]U(x_a, x_b, T-\epsilon).$$

This expression makes no sense in the limit $\epsilon \to 0$ unless the factor in parentheses is equal to 1. We can therefore identify the correct definition of C:

$$C(\epsilon) = \sqrt{\frac{2\pi\hbar\epsilon}{-im}}. \qquad (9.6)$$

Given this definition, we can compare terms of order ϵ and multiply by $i\hbar$ to obtain

$$i\hbar\frac{\partial}{\partial T}U(x_a, x_b; T) = \left[-\frac{\hbar^2}{2m}\frac{\partial^2}{\partial x_b^2} + V(x_b)\right]U(x_a, x_b; T)$$

$$= HU(x_a, x_b, T). \qquad (9.7)$$

This is the Schrödinger equation. But it is easy to show that the time-evolution operator U, as originally defined in (9.1), satisfies the same equation.

As $T \to 0$, the left-hand side of (9.3) tends to $\delta(x_a - x_b)$. Compare this to the value of (9.4) in the case of one time slice:

$$\frac{1}{C(\epsilon)}\exp\left[\frac{i}{\hbar}\frac{m(x_b - x_a)^2}{2\epsilon} + \mathcal{O}(\epsilon)\right].$$

This is just the peaked exponential of (9.5), and it also tends to $\delta(x_a - x_b)$ as $\epsilon \to 0$. Thus the left- and right-hand sides of (9.3) satisfy the same differential equation with the same initial condition. We conclude that the Hamiltonian definition of the time evolution operator (9.1) and the path-integral definition (9.3) are equivalent, at least for the case of this simple one-dimensional system.

To conclude this section, let us generalize our path-integral formula to more complicated quantum systems. Consider a very general quantum system, described by an arbitrary set of coordinates q^i, conjugate momenta p^i, and Hamiltonian $H(q,p)$. We will give a direct proof of the path-integral formula for transition amplitudes in this system.

The transition amplitude that we would like to compute is

$$U(q_a, q_b; T) = \langle q_b | e^{-iHT} | q_a \rangle. \tag{9.8}$$

(When q or p appears without a superscript, it will denote the set of all coordinates $\{q^i\}$ or momenta $\{p^i\}$. Also, for convenience, we now set $\hbar = 1$.) To write this amplitude as a functional integral, we first break the time interval into N short slices of duration ϵ. Thus we can write

$$e^{-iHT} = e^{-iH\epsilon} e^{-iH\epsilon} e^{-iH\epsilon} \cdots e^{-iH\epsilon} \qquad (N \text{ factors}).$$

The trick is to insert a complete set of intermediate states between each of these factors, in the form

$$\mathbf{1} = \left(\prod_i \int dq_k^i \right) |q_k\rangle \langle q_k|.$$

Inserting such factors for $k = 1 \ldots (N-1)$, we are left with a product of factors of the form

$$\langle q_{k+1} | e^{-iH\epsilon} | q_k \rangle \xrightarrow[\epsilon \to 0]{} \langle q_{k+1} | (1 - iH\epsilon + \cdots) | q_k \rangle. \tag{9.9}$$

To express the first and last factors in this form, we define $q_0 = q_a$ and $q_N = q_b$.

Now we must look inside H and consider what kinds of terms it might contain. The simplest kind of term to evaluate would be a function only of the coordinates, not of the momenta. The matrix element of such a term would be

$$\langle q_{k+1} | f(q) | q_k \rangle = f(q_k) \prod_i \delta(q_k^i - q_{k+1}^i).$$

It will be convenient to rewrite this as

$$\langle q_{k+1} | f(q) | q_k \rangle = f\left(\frac{q_{k+1} + q_k}{2} \right) \left(\prod_i \int \frac{dp_k^i}{2\pi} \right) \exp\left[i \sum_i p_k^i (q_{k+1}^i - q_k^i) \right],$$

for reasons that will soon be apparent.

Next consider a term in the Hamiltonian that is purely a function of the momenta. We introduce a complete set of momentum eigenstates to obtain

$$\langle q_{k+1} | f(p) | q_k \rangle = \left(\prod_i \int \frac{dp_k^i}{2\pi} \right) f(p_k) \exp\left[i \sum_i p_k^i (q_{k+1}^i - q_k^i) \right].$$

Thus if H contains only terms of the form $f(q)$ and $f(p)$, its matrix element can be written

$$\langle q_{k+1} | H(q,p) | q_k \rangle = \left(\prod_i \int \frac{dp_k^i}{2\pi} \right) H\left(\frac{q_{k+1} + q_k}{2}, p_k \right) \exp\left[i \sum_i p_k^i (q_{k+1}^i - q_k^i) \right]. \tag{9.10}$$

It would be nice if Eq. (9.10) were true even when H contains products of p's and q's. In general this formula must be false, since the order of a product pq matters on the left-hand side (where H is an operator) but not on the right-hand side (where H is just a function of the numbers p_k and q_k). But for one specific ordering, we can preserve (9.10). For example, the combination

$$\langle q_{k+1}| \tfrac{1}{4}\left(q^2 p^2 + 2qp^2 q + p^2 q^2\right)|q_k\rangle = \left(\frac{q_{k+1}+q_k}{2}\right)^2 \langle q_{k+1}|\,p^2\,|q_k\rangle$$

works out as desired, since the q's appear symmetrically on the left and right in just the right way. When this happens, the Hamiltonian is said to be *Weyl ordered*. Any Hamiltonian can be put into Weyl order by commuting p's and q's; in general this procedure will introduce some extra terms, and those extra terms must appear on the right-hand side of (9.10).

Assuming from now on that H is Weyl ordered, our typical matrix element from (9.9) can be expressed as

$$\langle q_{k+1}|\,e^{-i\epsilon H}\,|q_k\rangle = \left(\prod_i \int \frac{dp_k^i}{2\pi}\right)\exp\left[-i\epsilon H\left(\frac{q_{k+1}+q_k}{2},p_k\right)\right]$$

$$\times \exp\left[i\sum_i p_k^i(q_{k+1}^i - q_k^i)\right].$$

(We have again used the fact that ϵ is small, writing $1 - i\epsilon H$ as $e^{-i\epsilon H}$.) To obtain $U(q_a, q_b; T)$, we multiply N such factors, one for each k, and integrate over the intermediate coordinates q_k:

$$U(q_0, q_N; T) = \left(\prod_{i,k} \int dq_k^i \int \frac{dp_k^i}{2\pi}\right)$$

$$\times \exp\left[i\sum_k\left(\sum_i p_k^i(q_{k+1}^i-q_k^i) - \epsilon H\left(\frac{q_{k+1}+q_k}{2},p_k\right)\right)\right].$$

$$(9.11)$$

There is one momentum integral for each k from 0 to $N-1$, and one coordinate integral for each k from 1 to $N-1$. This expression is therefore the discretized form of

$$U(q_a, q_b; T) = \left(\prod_i \int \mathcal{D}q(t)\,\mathcal{D}p(t)\right)\exp\left[i\int_0^T dt\left(\sum_i p^i\dot{q}^i - H(q,p)\right)\right], \quad (9.12)$$

where the functions $q(t)$ are constrained at the endpoints, but the functions $p(t)$ are not. Note that the integration measure $\mathcal{D}q$ contains no peculiar constants, as it did in (9.4). The functional measure in (9.12) is just the product of the standard integral over phase space

$$\prod_i \int \frac{dq^i\,dp^i}{2\pi\hbar}$$

at each point in time. Equation (9.12) is the most general formula for computing transition amplitudes via functional integrals.

For a nonrelativistic particle, the Hamiltonian is simply $H = p^2/2m + V(q)$. In this case we can evaluate the p-integrals by completing the square in the exponent:

$$\int \frac{dp_k}{2\pi} \exp\left[i\big(p_k(q_{k+1}-q_k) - \epsilon p_k^2/2m\big)\right] = \frac{1}{C(\epsilon)} \exp\left[\frac{im}{2\epsilon}(q_{k+1} - q_k)^2\right],$$

where $C(\epsilon)$ is just the factor (9.6). Notice that we have one such factor for each time slice. Thus we recover expression (9.3), in discretized form, including the proper factors of C:

$$U(q_a, q_b; T) = \left(\frac{1}{C(\epsilon)} \prod_k \int \frac{dq_k}{C(\epsilon)}\right) \exp\left[i\sum_k \left(\frac{m}{2}\frac{(q_{k+1}-q_k)^2}{\epsilon} - \epsilon V\left(\frac{q_{k+1}+q_k}{2}\right)\right)\right]. \tag{9.13}$$

9.2 Functional Quantization of Scalar Fields

In this section we will apply the functional integral formalism to the quantum theory of a real scalar field $\phi(x)$. Our goal is to derive the Feynman rules for such a theory directly from functional integral expressions.

The general functional integral formula (9.12) derived in the last section holds for any quantum system, so it should hold for a quantum field theory. In the case of a real scalar field, the coordinates q^i are the field amplitudes $\phi(\mathbf{x})$, and the Hamiltonian is

$$H = \int d^3x \left[\tfrac{1}{2}\pi^2 + \tfrac{1}{2}(\nabla\phi)^2 + V(\phi)\right].$$

Thus our formula becomes

$$\langle\phi_b(\mathbf{x})| e^{-iHT} |\phi_a(\mathbf{x})\rangle = \int \mathcal{D}\phi\,\mathcal{D}\pi\, \exp\left[i\int_0^T d^4x\left(\pi\dot{\phi} - \tfrac{1}{2}\pi^2 - \tfrac{1}{2}(\nabla\phi)^2 - V(\phi)\right)\right],$$

where the functions $\phi(x)$ over which we integrate are constrained to the specific configurations $\phi_a(\mathbf{x})$ at $x^0 = 0$ and $\phi_b(\mathbf{x})$ at $x^0 = T$. Since the exponent is quadratic in π, we can complete the square and evaluate the $\mathcal{D}\pi$ integral to obtain

$$\langle\phi_b(\mathbf{x})| e^{-iHT} |\phi_a(\mathbf{x})\rangle = \int \mathcal{D}\phi\, \exp\left[i\int_0^T d^4x\,\mathcal{L}\right], \tag{9.14}$$

where

$$\mathcal{L} = \tfrac{1}{2}(\partial_\mu\phi)^2 - V(\phi)$$

is the Lagrangian density. The integration measure $\mathcal{D}\phi$ in (9.14) again involves an awkward constant, which we will not write explicitly.

The time integral in the exponent of (9.14) goes from 0 to T, as determined by our choice of what transition function to compute; in all other respects this formula is manifestly Lorentz invariant. Any other symmetries that the Lagrangian may have are also explicitly preserved by the functional integral. As we proceed in our study of quantum field theory, symmetries and their associated conservation laws will play an increasingly central role. We therefore propose to take a rash step: Abandon the Hamiltonian formalism, and take Eq. (9.14) to *define* the Hamiltonian dynamics. Any such formula corresponds to some Hamiltonian; to find it, one can always differentiate with respect to T and derive the Schrödinger equation as in the previous section. We thus consider the Lagrangian $\mathcal{L}$ to be the most fundamental specification of a quantum field theory. We will see next that one can use the functional integral to compute from $\mathcal{L}$ directly, without invoking the Hamiltonian at all.

Correlation Functions

To make direct use of the functional integral, we need a functional formula for computing correlation functions. To find such an expression, consider the object

$$\int \mathcal{D}\phi(x)\,\phi(x_1)\phi(x_2)\,\exp\left[i\int\limits_{-T}^{T}d^4x\,\mathcal{L}(\phi)\right], \qquad (9.15)$$

where the boundary conditions on the path integral are $\phi(-T, \mathbf{x}) = \phi_a(\mathbf{x})$ and $\phi(T, \mathbf{x}) = \phi_b(\mathbf{x})$ for some ϕ_a, ϕ_b. We would like to relate this quantity to the two-point correlation function, $\langle\Omega|\,T\phi_H(x_1)\phi_H(x_2)\,|\Omega\rangle$. (To distinguish operators from ordinary numbers, we write the Heisenberg picture operator with an explicit subscript: $\phi_H(x)$. Similarly, we will write $\phi_s(\mathbf{x})$ for the Schrödinger picture operator.)

First we break up the functional integral in (9.15) as follows:

$$\int \mathcal{D}\phi(x) = \int \mathcal{D}\phi_1(\mathbf{x}) \int \mathcal{D}\phi_2(\mathbf{x}) \int\limits_{\substack{\phi(x_1^0,\mathbf{x})=\phi_1(\mathbf{x}) \\ \phi(x_2^0,\mathbf{x})=\phi_2(\mathbf{x})}} \mathcal{D}\phi(x). \qquad (9.16)$$

The main functional integral $\int \mathcal{D}\phi(x)$ is now constrained at times x_1^0 and x_2^0 (in addition to the endpoints $-T$ and T), but we must integrate separately over the intermediate configurations $\phi_1(\mathbf{x})$ and $\phi_2(\mathbf{x})$. After this decomposition, the extra factors $\phi(x_1)$ and $\phi(x_2)$ in (9.15) become $\phi_1(\mathbf{x}_1)$ and $\phi_2(\mathbf{x}_2)$, and can be taken outside the main integral. The main integral then factors into three pieces, each being a simple transition amplitude according to (9.14). The times x_1^0 and x_2^0 automatically fall in order; for example, if $x_1^0 < x_2^0$, then (9.15) becomes

$$\int \mathcal{D}\phi_1(\mathbf{x}) \int \mathcal{D}\phi_2(\mathbf{x})\,\phi_1(\mathbf{x}_1)\phi_2(\mathbf{x}_2)\,\langle\phi_b|\,e^{-iH(T-x_2^0)}\,|\phi_2\rangle$$

$$\times \langle\phi_2|\,e^{-iH(x_2^0-x_1^0)}\,|\phi_1\rangle\,\langle\phi_1|\,e^{-iH(x_1^0+T)}\,|\phi_a\rangle\,.$$

We can turn the field $\phi_1(\mathbf{x}_1)$ into a Schrödinger operator using $\phi_S(\mathbf{x}_1)|\phi_1\rangle = \phi_1(\mathbf{x}_1)|\phi_1\rangle$. The completeness relation $\int \mathcal{D}\phi_1 |\phi_1\rangle \langle\phi_1| = 1$ then allows us to eliminate the intermediate state $|\phi_1\rangle$. Similar manipulations work for ϕ_2, yielding the expression

$$\langle\phi_b| e^{-iH(T-x_2^0)} \phi_S(\mathbf{x}_2) e^{-iH(x_2^0-x_1^0)} \phi_S(\mathbf{x}_1) e^{-iH(x_1^0+T)} |\phi_a\rangle .$$

Most of the exponential factors combine with the Schrödinger operators to make Heisenberg operators. In the case $x_1^0 > x_2^0$, the order of x_1 and x_2 would simply be interchanged. Thus expression (9.15) is equal to

$$\langle\phi_b| e^{-iHT} T\{\phi_H(x_1)\phi_H(x_2)\} e^{-iHT} |\phi_a\rangle . \tag{9.17}$$

This expression is almost equal to the two-point correlation function. To make it more nearly equal, we take the limit $T \to \infty(1 - i\epsilon)$. Just as in Section 4.2, this trick projects out the vacuum state $|\Omega\rangle$ from $|\phi_a\rangle$ and $|\phi_b\rangle$ (provided that these states have some overlap with $|\Omega\rangle$, which we assume). For example, decomposing $|\phi_a\rangle$ into eigenstates $|n\rangle$ of H, we have

$$e^{-iHT} |\phi_a\rangle = \sum_n e^{-iE_nT} |n\rangle \langle n|\phi_a\rangle \xrightarrow[T\to\infty(1-i\epsilon)]{} \langle\Omega|\phi_a\rangle e^{-iE_0\cdot\infty(1-i\epsilon)} |\Omega\rangle .$$

As in Section 4.2, we obtain some awkward phase and overlap factors. But these factors cancel if we divide by the same quantity as (9.15) but without the two extra fields $\phi(x_1)$ and $\phi(x_2)$. Thus we obtain the simple formula

$$\langle\Omega| T\phi_H(x_1)\phi_H(x_2) |\Omega\rangle = \lim_{T\to\infty(1-i\epsilon)} \frac{\int \mathcal{D}\phi\, \phi(x_1)\phi(x_2) \exp\left[i\int_{-T}^{T} d^4x\, \mathcal{L}\right]}{\int \mathcal{D}\phi \exp\left[i\int_{-T}^{T} d^4x\, \mathcal{L}\right]}. \tag{9.18}$$

This is our desired formula for the two-point correlation function in terms of functional integrals. For higher correlation functions, just insert additional factors of ϕ on both sides.

Feynman Rules

Our next task is to compute various correlation functions directly from the right-hand side of formula (9.18). In other words, we will now use (9.18) to derive the Feynman rules for a scalar field theory. We will begin by computing the two-point function in the free Klein-Gordon theory, then generalize to higher correlation functions in the free theory. Finally, we will consider ϕ^4 theory, in which we can perform a perturbation expansion to obtain the same Feynman rules as in Section 4.4.

Consider first a noninteracting real-valued scalar field:

$$S_0 = \int d^4x\, \mathcal{L}_0 = \int d^4x \left[\tfrac{1}{2}(\partial_\mu\phi)^2 - \tfrac{1}{2}m^2\phi^2\right]. \tag{9.19}$$

Since $\mathcal{L}_0$ is quadratic in ϕ, the functional integrals in (9.18) take the form of generalized, infinite-dimensional Gaussian integrals. We will therefore be able to evaluate the functional integrals exactly.

Since this is our first functional integral computation, we will do it in a very explicit, but ugly, way. We must first define the integral $\mathcal{D}\phi$ over field configurations. To do this, we use the method of Eq. (9.4) in considering the continuous integral as a limit of a large but finite number of integrals. We thus replace the variables $\phi(x)$ defined on a continuum of points by variables $\phi(x_i)$ defined at the points x_i of a square lattice. Let the lattice spacing be ϵ, let the four-dimensional spacetime volume be L^4, and define

$$\mathcal{D}\phi = \prod_i d\phi(x_i), \tag{9.20}$$

up to an irrelevant overall constant.

The field values $\phi(x_i)$ can be represented by a discrete Fourier series:

$$\phi(x_i) = \frac{1}{V} \sum_n e^{-ik_n \cdot x_i} \phi(k_n), \tag{9.21}$$

where $k_n^\mu = 2\pi n^\mu / L$, with n^μ an integer, $|k^\mu| < \pi/\epsilon$, and $V = L^4$. The Fourier coefficients $\phi(k)$ are complex. However, $\phi(x)$ is real, and so these coefficients must obey the constraint $\phi^*(k) = \phi(-k)$. We will consider the real and imaginary parts of the $\phi(k_n)$ with $k_n^0 > 0$ as independent variables. The change of variables from the $\phi(x_i)$ to these new variables $\phi(k_n)$ is a unitary transformation, so we can rewrite the integral as

$$\mathcal{D}\phi(x) = \prod_{k_n^0 > 0} d\,\mathrm{Re}\,\phi(k_n)\, d\,\mathrm{Im}\,\phi(k_n).$$

Later, we will take the limit $L \to \infty$, $\epsilon \to 0$. The effect of this limit is to convert discrete, finite sums over k_n to continuous integrals over k:

$$\frac{1}{V} \sum_n \to \int \frac{d^4 k}{(2\pi)^4}. \tag{9.22}$$

In the following discussion, this limit will produce Feynman perturbation theory in the form derived in Part I. We will not eliminate the infrared and ultraviolet divergences of Feynman diagrams that we encountered in Chapter 6, but at least the functional integral introduces no new types of singular behavior.

Having defined the measure of integration, we now compute the functional integral over ϕ. The action (9.19) can be rewritten in terms of the Fourier coefficients as

$$\int d^4x \left[\tfrac{1}{2}(\partial_\mu \phi)^2 - \tfrac{1}{2}m^2\phi^2 \right] = -\frac{1}{V} \sum_n \tfrac{1}{2}(m^2 - k_n^2)|\phi(k_n)|^2$$

$$= -\frac{1}{V} \sum_{k_n^0 > 0} (m^2 - k_n^2)\left[(\mathrm{Re}\,\phi_n)^2 + (\mathrm{Im}\,\phi_n)^2 \right],$$

where we have abbreviated $\phi(k_n)$ as ϕ_n in the second line. The quantity $(m^2 - k_n^2) = (m^2 + |\mathbf{k}_n|^2 - k_n^{02})$ is positive as long as k_n^0 is not too large. In the following discussion, we will treat this quantity as if it were positive. More precisely, we evaluate it by analytic continuation from the region where $|\mathbf{k}_n| > k_n^0$.

The denominator of formula (9.18) now takes the form of a product of Gaussian integrals:

$$
\begin{aligned}
\int \mathcal{D}\phi \, e^{iS_0} &= \left(\prod_{k_n^0 > 0} \int d\,\mathrm{Re}\,\phi_n \, d\,\mathrm{Im}\,\phi_n \right) \exp\left[-\frac{i}{V} \sum_{k_n^0 > 0} (m^2 - k_n^2)|\phi_n|^2 \right] \\
&= \prod_{k_n^0 > 0} \left(\int d\,\mathrm{Re}\,\phi_n \, \exp\left[-\frac{i}{V}(m^2 - k_n^2)(\mathrm{Re}\,\phi_n)^2 \right] \right) \\
&\qquad \times \left(\int d\,\mathrm{Im}\,\phi_n \, \exp\left[-\frac{i}{V}(m^2 - k_n^2)(\mathrm{Im}\,\phi_n)^2 \right] \right) \\
&= \prod_{k_n^0 > 0} \sqrt{\frac{-i\pi V}{m^2 - k_n^2}} \sqrt{\frac{-i\pi V}{m^2 - k_n^2}} \\
&= \prod_{\text{all } k_n} \sqrt{\frac{-i\pi V}{m^2 - k_n^2}}.
\end{aligned}
\tag{9.23}
$$

To justify using Gaussian integration formulae when the exponent appears to be purely imaginary, recall that the time integral in (9.18) is along a contour that is rotated clockwise in the complex plane: $t \to t(1 - i\epsilon)$. This means that we should change $k^0 \to k^0(1 + i\epsilon)$ in (9.21) and all subsequent equations; in particular, we should replace $(k^2 - m^2) \to (k^2 - m^2 + i\epsilon)$. The $i\epsilon$ term gives the necessary convergence factor for the Gaussian integrals. It also defines the direction of the analytic continuation that might be needed to define the square roots in (9.23).

To understand the result of (9.23), consider as an analogy the general Gaussian integral

$$
\left(\prod_k \int d\xi_k \right) \exp\left[-\xi_i B_{ij} \xi_j \right],
$$

where B is a symmetric matrix with eigenvalues b_i. To evaluate this integral we write $\xi_i = O_{ij} x_j$, where O is the orthogonal matrix of eigenvectors that diagonalizes B. Changing variables from ξ_i to the coefficients x_i, we have

$$
\left(\prod_k \int d\xi_k \right) \exp\left[-\xi_i B_{ij} \xi_j \right] = \left(\prod_k \int dx_k \right) \exp\left[-\sum_i b_i x_i^2 \right]
$$

$$
= \prod_i \left(\int dx_i \, \exp\left[-b_i x_i^2 \right] \right)
$$

$$= \prod_i \sqrt{\frac{\pi}{b_i}}$$

$$= \text{const} \times \left[\det B\right]^{-1/2}. \qquad (9.24)$$

The analogy is clearer if we perform an integration by parts to write the Klein-Gordon action as

$$S_0 = \tfrac{1}{2} \int d^4x \, \phi \, (-\partial^2 - m^2) \, \phi \; + \; (\text{surface term}).$$

Thus the matrix B corresponds to the operator $(m^2 + \partial^2)$, and we can formally write our result as

$$\int \mathcal{D}\phi \, e^{iS_0} = \text{const} \times \left[\det(m^2 + \partial^2)\right]^{-1/2}. \qquad (9.25)$$

This object is called a *functional determinant*. The actual result (9.23) looks quite ill-defined, and in fact all of these factors will cancel in Eq. (9.18). However, in many circumstances, the functional determinant itself has physical meaning. We will see examples of this in Sections 9.5 and 11.4.

Now consider the numerator of formula (9.18). We need to Fourier-expand the two extra factors of ϕ:

$$\phi(x_1)\phi(x_2) = \frac{1}{V} \sum_m e^{-ik_m \cdot x_1} \phi_m \, \frac{1}{V} \sum_l e^{-ik_l \cdot x_2} \phi_l.$$

Thus the numerator is

$$\frac{1}{V^2} \sum_{m,l} e^{-i(k_m \cdot x_1 + k_l \cdot x_2)} \left(\prod_{k_n^0 > 0} \int d\,\mathrm{Re}\,\phi_n \, d\,\mathrm{Im}\,\phi_n \right) \qquad (9.26)$$

$$\times \, (\mathrm{Re}\,\phi_m + i\,\mathrm{Im}\,\phi_m)(\mathrm{Re}\,\phi_l + i\,\mathrm{Im}\,\phi_l)$$

$$\times \, \exp\!\left[-\frac{i}{V} \sum_{k_n^0 > 0} (m^2 - k_n^2)\left[(\mathrm{Re}\,\phi_n)^2 + (\mathrm{Im}\,\phi_n)^2\right]\right].$$

For most values of k_m and k_l this expression is zero, since the extra factors of ϕ make the integrand odd. The situation is more complicated when $k_m = \pm k_l$. Suppose, for example, that $k_m^0 > 0$. Then if $k_l = +k_m$, the term involving $(\mathrm{Re}\,\phi_m)^2$ is nonzero, but is exactly canceled by the term involving $(\mathrm{Im}\,\phi_m)^2$. If $k_l = -k_m$, however, the relation $\phi(-k) = \phi^*(k)$ gives an extra minus sign on the $(\mathrm{Im}\,\phi_m)^2$ term, so the two terms add. When $k_m^0 < 0$ we obtain the same expression, so the numerator is

$$\text{Numerator} = \frac{1}{V^2} \sum_m e^{-ik_m \cdot (x_1 - x_2)} \left(\prod_{k_n^0 > 0} \frac{-i\pi V}{m^2 - k_n^2} \right) \frac{-iV}{m^2 - k_m^2 - i\epsilon}.$$

The factor in parentheses is identical to the denominator (9.23), while the rest of this expression is the discretized form of the Feynman propagator. Taking the continuum limit (9.22), we find

$$\langle 0| T\phi(x_1)\phi(x_2) |0\rangle = \int \frac{d^4k}{(2\pi)^4} \frac{i\,e^{-ik\cdot(x_1-x_2)}}{k^2 - m^2 + i\epsilon} = D_F(x_1-x_2). \qquad (9.27)$$

This is exactly right, including the $+i\epsilon$.

Next we would like to compute higher correlation functions in the free Klein-Gordon theory.

Inserting an extra factor of ϕ in (9.18), we see that the three-point function vanishes, since the integrand of the numerator is odd. All other odd correlation functions vanish for the same reason.

The four-point function has four factors of ϕ in the numerator. Fourier-expanding the fields, we obtain an expression similar to Eq. (9.26), but with a quadruple sum over indices that we will call m, l, p, and q. The integrand contains the product

$$(\mathrm{Re}\,\phi_m + i\,\mathrm{Im}\,\phi_m)(\mathrm{Re}\,\phi_l + i\,\mathrm{Im}\,\phi_l)(\mathrm{Re}\,\phi_p + i\,\mathrm{Im}\,\phi_p)(\mathrm{Re}\,\phi_q + i\,\mathrm{Im}\,\phi_q).$$

Again, most of the terms vanish because the integrand is odd. One of the nonvanishing terms occurs when $k_l = -k_m$ and $k_q = -k_p$. After the Gaussian integrations, this term of the numerator is

$$\frac{1}{V^4} \sum_{m,p} e^{-ik_m\cdot(x_1-x_2)} e^{-ik_p\cdot(x_3-x_4)} \left(\prod_{k_n^0>0} \frac{-i\pi V}{m^2-k_n^2} \right) \frac{-iV}{m^2-k_m^2-i\epsilon} \frac{-iV}{m^2-k_p^2-i\epsilon}$$

$$\xrightarrow[V\to\infty]{} \left(\prod_{k_n^0>0} \frac{-i\pi V}{m^2-k_n^2} \right) D_F(x_1 - x_2)D_F(x_3 - x_4).$$

The factor in parentheses is again canceled by the denominator. We obtain similar terms for each of the other two ways of grouping the four momenta in pairs. To keep track of the groupings, let us define the *contraction* of two fields as

$$\overline{\phi(x_1)\phi(x_2)} = \frac{\int \mathcal{D}\phi\, e^{iS_0}\phi(x_1)\phi(x_2)}{\int \mathcal{D}\phi\, e^{iS_0}} = D_F(x_1 - x_2). \qquad (9.28)$$

Then the four-point function is simply

$$\langle 0| T\phi_1\phi_2\phi_3\phi_4 |0\rangle = \text{sum of all full contractions}$$

$$= D_F(x_1 - x_2)D_F(x_3 - x_4)$$
$$+ D_F(x_1 - x_3)D_F(x_2 - x_4) \qquad (9.29)$$
$$+ D_F(x_1 - x_4)D_F(x_2 - x_3),$$

the same expression that we obtained using Wick's theorem in Eq. (4.40).

The same method allows us to compute still higher correlation functions. In each case the answer is just the sum of all possible full contractions of the fields. This result, identical to that obtained from Wick's theorem in Section 4.3, arises here from the simple rules of Gaussian integration.

We are now ready to move from the free Klein-Gordon theory to ϕ^4 theory. Add to $\mathcal{L}_0$ a ϕ^4 interaction:

$$\mathcal{L} = \mathcal{L}_0 - \frac{\lambda}{4!}\phi^4.$$

Assuming that λ is small, we can expand

$$\exp\left[i\int d^4x\,\mathcal{L}\right] = \exp\left[i\int d^4x\,\mathcal{L}_0\right]\left(1 - i\int d^4x\,\frac{\lambda}{4!}\phi^4 + \cdots\right).$$

Making this expansion in both the numerator and the denominator of (9.18), we see that each is (aside from the constant factor (9.23), which again cancels) expressed entirely in terms of free-field correlation functions. Moreover, since $i\int d^3x\,\mathcal{L}_{\rm int} = -iH_{\rm int}$, we obtain exactly the same expansion as in Eq. (4.31). We can express both the numerator and the denominator in terms of Feynman diagrams, with the fundamental interaction again given by the vertex

$$\times \quad = -i\lambda\,(2\pi)^4\delta^{(4)}(\textstyle\sum p). \tag{9.30}$$

All of the combinatorics work the same as in Section 4.4. In particular, the disconnected vacuum bubble diagrams exponentiate and factor from the numerator of (9.18), and are canceled by the denominator, just as in Eq. (4.31).

The vertex rule for ϕ^4 theory follows from the Lagrangian in an exceedingly simple way, and this simple procedure will turn out to be valid for other quantum field theories as well. Once the quadratic terms in the Lagrangian are properly understood and the propagators of the theory are computed, the vertices can be read directly from the Lagrangian as the coefficients of the cubic and higher-order terms.

Functional Derivatives and the Generating Functional

To conclude this section, we will now introduce a slicker, more formal, method for computing correlation functions. This method, based on an object called the *generating functional*, avoids the awkward Fourier expansions of the preceding derivation.

First we define the *functional derivative*, $\delta/\delta J(x)$, as follows. The functional derivative obeys the basic axiom (in four dimensions)

$$\frac{\delta}{\delta J(x)}J(y) = \delta^{(4)}(x - y) \quad \text{or} \quad \frac{\delta}{\delta J(x)}\int d^4y\,J(y)\phi(y) = \phi(x). \tag{9.31}$$

This definition is the natural generalization, to continuous functions, of the rule for discrete vectors,

$$\frac{\partial}{\partial x_i}x_j = \delta_{ij} \quad \text{or} \quad \frac{\partial}{\partial x_i}\sum_j x_j k_j = k_i.$$

To take functional derivatives of more complicated functionals we simply use the ordinary rules for derivatives of composite functions. For example,

$$\frac{\delta}{\delta J(x)} \exp\left[i \int d^4 y\, J(y)\phi(y)\right] = i\phi(x) \exp\left[i \int d^4 y\, J(y)\phi(y)\right]. \qquad (9.32)$$

When the functional depends on the derivative of J, we integrate by parts before applying the functional derivative:

$$\frac{\delta}{\delta J(x)} \int d^4 y\, \partial_\mu J(y) V^\mu(y) = -\partial_\mu V^\mu(x). \qquad (9.33)$$

The basic object of this formalism is the *generating functional* of correlation functions, $Z[J]$. (Some authors call it $W[J]$.) In a scalar field theory, $Z[J]$ is defined as

$$Z[J] \equiv \int \mathcal{D}\phi \, \exp\left[i \int d^4 x \big[\mathcal{L} + J(x)\phi(x)\big]\right]. \qquad (9.34)$$

This is a functional integral over ϕ in which we have added to $\mathcal{L}$ in the exponent a *source term*, $J(x)\phi(x)$.

Correlation functions of the Klein-Gordon field theory can be simply computed by taking functional derivatives of the generating functional. For example, the two-point function is

$$\langle 0| T\phi(x_1)\phi(x_2) |0\rangle = \frac{1}{Z_0} \left(-i\frac{\delta}{\delta J(x_1)}\right)\left(-i\frac{\delta}{\delta J(x_2)}\right) Z[J]\Big|_{J=0}, \qquad (9.35)$$

where $Z_0 = Z[J = 0]$. Each functional derivative brings down a factor of ϕ in the numerator of $Z[J]$; setting $J = 0$, we recover expression (9.18). To compute higher correlation functions we simply take more functional derivatives.

Formula (9.35) is useful because, in a free field theory, $Z[J]$ can be rewritten in a very explicit form. Consider the exponent of (9.34) in the free Klein-Gordon theory. Integrating by parts, we obtain

$$\int d^4 x \big[\mathcal{L}_0(\phi) + J\phi\big] = \int d^4 x \big[\tfrac{1}{2}\phi(-\partial^2 - m^2 + i\epsilon)\phi + J\phi\big]. \qquad (9.36)$$

(The $i\epsilon$ is a convergence factor for the functional integral, as we discussed below Eq. (9.23).) We can complete the square by introducing a shifted field,

$$\phi'(x) \equiv \phi(x) - i \int d^4 y\, D_F(x-y)J(y).$$

Making this substitution and using the fact that D_F is a Green's function of the Klein-Gordon operator, we find that (9.36) becomes

$$\int d^4 x \big[\mathcal{L}_0(\phi) + J\phi\big] = \int d^4 x \big[\tfrac{1}{2}\phi'(-\partial^2 - m^2 + i\epsilon)\phi'\big]$$

$$- \int d^4 x\, d^4 y\, \tfrac{1}{2} J(x)[-iD_F(x-y)]J(y).$$

More symbolically, we could write the change of variables as

$$\phi' \equiv \phi + (-\partial^2 - m^2 + i\epsilon)^{-1} J, \tag{9.37}$$

and the result

$$\int d^4x \left[\mathcal{L}_0(\phi) + J\phi \right] = \int d^4x \left[\tfrac{1}{2}\phi'(-\partial^2 - m^2 + i\epsilon)\phi' - \tfrac{1}{2}J(-\partial^2 - m^2 + i\epsilon)^{-1} J \right]. \tag{9.38}$$

Now change variables from ϕ to ϕ' in the functional integral of (9.34). This is just a shift, and so the Jacobian of the transformation is 1. The result is

$$\int \mathcal{D}\phi' \, \exp\left[i \int d^4x \, \mathcal{L}_0(\phi') \right] \, \exp\left[-i \int d^4x \, d^4y \, \tfrac{1}{2} J(x)[-iD_F(x-y)]J(y) \right].$$

The second exponential factor is independent of ϕ', while the remaining integral over ϕ' is precisely Z_0. Thus the generating functional of the free Klein-Gordon theory is simply

$$Z[J] = Z_0 \, \exp\left[-\tfrac{1}{2} \int d^4x \, d^4y \, J(x) D_F(x-y) J(y) \right]. \tag{9.39}$$

Let us use Eqs. (9.39) and (9.35) to compute some correlation functions. The two-point function is

$$\langle 0 | T\phi(x_1)\phi(x_2) | 0 \rangle$$

$$= -\frac{\delta}{\delta J(x_1)} \frac{\delta}{\delta J(x_2)} \exp\left[-\tfrac{1}{2} \int d^4x \, d^4y \, J(x)D_F(x-y)J(y) \right] \Big|_{J=0}$$

$$= -\frac{\delta}{\delta J(x_1)} \left[-\tfrac{1}{2} \int d^4y \, D_F(x_2-y)J(y) - \tfrac{1}{2} \int d^4x \, J(x)D_F(x-x_2) \right] \frac{Z[J]}{Z_0} \Big|_{J=0}$$

$$= D_F(x_1 - x_2). \tag{9.40}$$

Taking one derivative brings down two identical terms; the second derivative gives several terms, but only when it acts on the outside factor do we get a term that survives when we set $J = 0$.

It is instructive to work out the four-point function by this method as well. In order to fit the computation in a reasonable amount of space, let us abbreviate arguments of functions as subscripts: $\phi_1 \equiv \phi(x_1)$, $J_x \equiv J(x)$, $D_{x4} \equiv D_F(x-x_4)$, and so on. Repeated subscripts will be integrated over implicitly. The four-point function is then

$$\langle 0 | T\phi_1\phi_2\phi_3\phi_4 | 0 \rangle = \frac{\delta}{\delta J_1} \frac{\delta}{\delta J_2} \frac{\delta}{\delta J_3} \left[-J_x D_{x4} \right] e^{-\frac{1}{2} J_x D_{xy} J_y} \Big|_{J=0}$$

$$= \frac{\delta}{\delta J_1} \frac{\delta}{\delta J_2} \left[-D_{34} + J_x D_{x4} J_y D_{y3} \right] e^{-\frac{1}{2} J_x D_{xy} J_y} \Big|_{J=0}$$

$$= \frac{\delta}{\delta J_1} \left[D_{34} J_x D_{x2} + D_{24} J_y D_{y3} + J_x D_{x4} D_{23} \right] e^{-\frac{1}{2} J_x D_{xy} J_y} \Big|_{J=0}$$

$$= D_{34} D_{12} + D_{24} D_{13} + D_{14} D_{23}, \tag{9.41}$$

in agreement with (9.29). The rules for differentiating the exponential give rise to the same familiar pattern: We get one term for each possible way of contracting the four points in pairs, with a factor of D_F for each contraction.

The generating functional method used just above to construct the correlations of a free field theory can be used as well to represent the correlation functions of an interacting field theory. Formula (9.35) is independent of whether the theory is free or interacting. The factor $Z[J = 0]$ is nontrivial in the case of an interacting field theory, but it simply gives the denominator of Eq. (9.18), that is, the sum of vacuum diagrams. Again from this approach, the combinatoric issues in the evaluation of correlation functions are the same as in Section 4.4.

9.3 The Analogy Between Quantum Field Theory and Statistical Mechanics

Let us now pause from the technical aspects of this discussion to consider some implications of the formulae we have derived. To begin, let us summarize the formal conclusions of the previous section in the following way: For a field theory governed by the Lagrangian $\mathcal{L}$, the generating functional of correlation functions is

$$Z[J] = \int \mathcal{D}\phi \, \exp\left[i \int d^4x \, (\mathcal{L} + J\phi) \right].$$ (9.42)

The time variable of integration in the exponent runs from $-T$ to T, with $T \to \infty(1 - i\epsilon)$. A correlation function such as (9.18) is reproduced by writing

$$\langle 0| \, T\phi(x_1)\phi(x_2) \, |0\rangle = Z[J]^{-1} \left(-i\frac{\delta}{\delta J(x_1)} \right) \left(-i\frac{\delta}{\delta J(x_2)} \right) Z[J] \Big|_{J=0}.$$ (9.43)

The generating functional (9.42) is reminiscent of the partition function of statistical mechanics. It has the same general structure of an integral over all possible configurations of an exponential statistical weight. The source $J(x)$ plays the role of an external field. In fact, our method of computing correlation functions by differentiating with respect to $J(x)$ mimics the trick often used in statistical mechanics of computing correlation functions by differentiating with respect to such variables as the pressure or the magnetic field.

This analogy can be made more precise by manipulating the time variable of integration in (9.42). The derivation of the functional integral formula implied that the time integration was slightly tipped into the complex plane, in just the direction to permit the contour to be rotated clockwise onto the imaginary axis. We have already noted (below (9.23)) that the original infinitesimal rotation gives the correct $i\epsilon$ prescription to produce the Feynman propagator. The finite rotation is the analogue in configuration space of the Wick rotation of the time component of momentum illustrated in Fig. 6.1. Like the Wick rotation in a momentum integral, this Wick rotation of the time coordinate $t \to -ix^0$ produces a Euclidean 4-vector product:

$$x^2 = t^2 - |\mathbf{x}|^2 \to -(x^0)^2 - |\mathbf{x}|^2 = -|x_E|^2. \tag{9.44}$$

It is possible to show, by manipulating the expression for each Feynman diagram, that the analytic continuation of the time variables in any Green's function of a quantum field theory produces a correlation function invariant under the rotational symmetry of four-dimensional Euclidean space. This Wick rotation inside the functional integral demonstrates this same conclusion in a more general way.

To understand what we have achieved by this rotation, consider the example of ϕ^4 theory. The action of ϕ^4 theory coupled to sources is

$$\int d^4x \, (\mathcal{L} + J\phi) = \int d^4x \left[\frac{1}{2}(\partial_\mu \phi)^2 - \frac{1}{2}m^2\phi^2 - \frac{\lambda}{4!}\phi^4 + J\phi \right]. \tag{9.45}$$

After the Wick rotation (9.44), this expression takes the form

$$i \int d^4x_E(\mathcal{L}_E - J\phi) = i \int d^4x_E \left[\frac{1}{2}(\partial_{E\mu}\phi)^2 + \frac{1}{2}m^2\phi^2 + \frac{\lambda}{4!}\phi^4 - J\phi \right]. \tag{9.46}$$

This expression is identical in form to the expression (8.8) for the Gibbs free energy of a ferromagnet in the Landau theory. The field $\phi(x_E)$ plays the role of the fluctuating spin field $s(\mathbf{x})$, and the source $J(\mathbf{x})$ plays the role of an external magnetic field. Note that the new ferromagnet lives in four, rather than three, spatial dimensions.

The Wick-rotated generating functional $Z[J]$ becomes

$$Z[J] = \int \mathcal{D}\phi \, \exp\left[-\int d^4x_E \, (\mathcal{L}_E - J\phi) \right]. \tag{9.47}$$

The functional $\mathcal{L}_E[\phi]$ has the form of an energy: It is bounded from below and becomes large when the field ϕ has large amplitude or large gradients. The exponential, then, is a reasonable statistical weight for the fluctuations of ϕ. In this new form, $Z[J]$ is precisely the partition function describing the statistical mechanics of a macroscopic system, described approximately by treating the fluctuating variable as a continuum field.

The Green's functions of $\phi(x_E)$ after Wick rotation can be calculated from the functional integral (9.47) exactly as we computed Minkowski Green's functions in the previous section. For the free theory ($\lambda = 0$), a set of manipulations analogous to those that produced (9.27) or (9.40) gives the correlation function of ϕ as

$$\langle \phi(x_{E1})\phi(x_{E2}) \rangle = \int \frac{d^4k_E}{(2\pi)^4} \frac{e^{ik_E \cdot (x_{E1} - x_{E2})}}{k_E^2 + m^2}. \tag{9.48}$$

This is just the Feynman propagator evaluated in the spacelike region; according to Eq. (2.52), this function falls off as $\exp(-m|x_{E1} - x_{E2}|)$. That behavior is the four-dimensional analogue of the spin correlation function (8.15). We see that, in the Euclidean continuation of field theory Green's functions, the

Compton wavelength m^{-1} of the quanta becomes the correlation length of statistical fluctuations.

This correspondence between quantum field theory and statistical mechanics will play an important role in the developments of the next few chapters. In essence, it adds to our reserves of knowledge a completely new source of intuition about how field theory expectation values should behave. This intuition will be useful in imagining the general properties of loop diagrams and, as we have already discussed in Chapter 8, it will give important insights that will help us correctly understand the role of ultraviolet divergences in field theory calculations. In Chapter 13, we will see that field theory can also contribute to statistical mechanics by making profound predictions about the behavior of thermal systems from the properties of Feynman diagrams.

9.4 Quantization of the Electromagnetic Field

In Section 4.8 we stated without proof the Feynman rule for the photon propagator,

$$\underset{k \longrightarrow}{\wwwww} = \frac{-ig_{\mu\nu}}{k^2 + i\epsilon}. \tag{9.49}$$

Now that we have the functional integral quantization method at our command, let us apply it to the derivation of this expression.

Consider the functional integral

$$\int \mathcal{D}A\, e^{iS[A]}, \tag{9.50}$$

where $S[A]$ is the action for the free electromagnetic field. (The functional integral is over each of the four components: $\mathcal{D}A \equiv \mathcal{D}A^0 \mathcal{D}A^1 \mathcal{D}A^2 \mathcal{D}A^3$.) Integrating by parts and expanding the field as a Fourier integral, we can write the action as

$$S = \int d^4x \left[-\tfrac{1}{4}(F_{\mu\nu})^2 \right]$$

$$= \frac{1}{2} \int d^4x\, A_\mu(x)\big(\partial^2 g^{\mu\nu} - \partial^\mu \partial^\nu\big) A_\nu(x)$$

$$= \frac{1}{2} \int \frac{d^4k}{(2\pi)^4}\, \widetilde{A}_\mu(k)\big(-k^2 g^{\mu\nu} + k^\mu k^\nu\big)\widetilde{A}_\nu(-k). \tag{9.51}$$

This expression vanishes when $\widetilde{A}_\mu(k) = k_\mu\, \alpha(k)$, for any scalar function $\alpha(k)$. For this large set of field configurations the integrand of (9.50) is 1, and therefore the functional integral is badly divergent (there is no Gaussian damping). Equivalently, the equation

$$\big(\partial^2 g_{\mu\nu} - \partial_\mu \partial_\nu\big) D_F^{\nu\rho}(x - y) = i\delta_\mu^\rho\, \delta^{(4)}(x - y)$$

$$\text{or}\quad \big(-k^2 g_{\mu\nu} + k_\mu k_\nu\big)\widetilde{D}_F^{\nu\rho}(k) = i\delta_\mu^\rho, \tag{9.52}$$

which would define the Feynman propagator $D_F^{\nu\rho}$, has no solution, since the 4×4 matrix $(-k^2 g_{\mu\nu} + k_\mu k_\nu)$ is singular.

This difficulty is due to gauge invariance. Recall that $F_{\mu\nu}$, and hence $\mathcal{L}$, is invariant under a general gauge transformation of the form

$$A_\mu(x) \rightarrow A_\mu(x) + \frac{1}{e}\partial_\mu \alpha(x).$$

The troublesome modes are those for which $A_\mu(x) = \frac{1}{e}\partial_\mu \alpha(x)$, that is, those that are gauge-equivalent to $A_\mu(x) = 0$. The functional integral is badly defined because we are redundantly integrating over a continuous infinity of physically equivalent field configurations. To fix the problem, we would like to isolate the interesting part of the functional integral, which counts each physical configuration only once.

We can accomplish this by means of a trick, due to Faddeev and Popov.[†] Let $G(A)$ be some function that we wish to set equal to zero as a gauge-fixing condition; for example, $G(A) = \partial_\mu A^\mu$ corresponds to Lorentz gauge. We could constrain the functional integral to cover only the configurations with $G(A) = 0$ by inserting a functional delta function, $\delta\big(G(A)\big)$. (Think of this object as an infinite product of delta functions, one for each point x.) To do so legally, we insert 1 under the integral of (9.50), in the following form:

$$1 = \int \mathcal{D}\alpha(x)\, \delta\big(G(A^\alpha)\big)\, \det\left(\frac{\delta G(A^\alpha)}{\delta \alpha}\right), \tag{9.53}$$

where A^α denotes the gauge-transformed field,

$$A_\mu^\alpha(x) = A_\mu(x) + \frac{1}{e}\partial_\mu \alpha(x).$$

Equation (9.53) is the continuum generalization of the identity

$$1 = \left(\prod_i \int da_i\right) \delta^{(n)}\big(\mathbf{g}(\mathbf{a})\big)\, \det\left(\frac{\partial g_i}{\partial a_j}\right)$$

for discrete n-dimensional vectors. In Lorentz gauge we have $G(A^\alpha) = \partial^\mu A_\mu + (1/e)\partial^2 \alpha$, so the functional determinant $\det\big(\delta G(A^\alpha)/\delta\alpha\big)$ is equal to $\det(\partial^2/e)$. For the present discussion, the only relevant property of this determinant is that it is independent of A, so we can treat it as a constant in the functional integral.

After inserting (9.53), the functional integral (9.50) becomes

$$\det\left(\frac{\delta G(A^\alpha)}{\delta \alpha}\right) \int \mathcal{D}\alpha \int \mathcal{D}A\, e^{iS[A]}\, \delta\big(G(A^\alpha)\big).$$

Now change variables from A to A^α. This is a simple shift, so $\mathcal{D}A = \mathcal{D}A^\alpha$. Also, by gauge invariance, $S[A] = S[A^\alpha]$. Since A^α is now just a dummy

---

[†]L. D. Faddeev and V. N. Popov, *Phys. Lett.* **25B**, 29 (1967).

integration variable, we can rename it back to A, obtaining

$$\int \mathcal{D}A \, e^{iS[A]} = \det\left(\frac{\delta G(A^\alpha)}{\delta \alpha}\right) \int \mathcal{D}\alpha \int \mathcal{D}A \, e^{iS[A]} \, \delta(G(A)). \tag{9.54}$$

The functional integral over A is now restricted by the delta function to physically inequivalent field configurations, as desired. The divergent integral over $\alpha(x)$ simply gives an infinite multiplicative factor.

To go further we must specify a gauge-fixing function $G(A)$. We choose the general class of functions

$$G(A) = \partial^\mu A_\mu(x) - \omega(x), \tag{9.55}$$

where $\omega(x)$ can be any scalar function. Setting this $G(A)$ equal to zero gives a generalization of the Lorentz gauge condition. The functional determinant is the same as in Lorentz gauge, $\det(\delta G(A^\alpha)/\delta \alpha) = \det(\partial^2/e)$. Thus the functional integral becomes

$$\int \mathcal{D}A \, e^{iS[A]} = \det\left(\frac{1}{e}\partial^2\right)\left(\int \mathcal{D}\alpha\right)\int \mathcal{D}A \, e^{iS[A]} \, \delta(\partial^\mu A_\mu - \omega(x)).$$

This equality holds for any $\omega(x)$, so it will also hold if we replace the right-hand side with any properly normalized linear combination involving different functions $\omega(x)$. For our final trick, we will integrate over all $\omega(x)$, with a Gaussian weighting function centered on $\omega = 0$. The above expression is thus equal to

$$N(\xi) \int \mathcal{D}\omega \exp\left[-i\int d^4x \frac{\omega^2}{2\xi}\right]\det\left(\frac{1}{e}\partial^2\right)\left(\int \mathcal{D}\alpha\right)\int \mathcal{D}A \, e^{iS[A]} \, \delta(\partial^\mu A_\mu - \omega(x))$$

$$= N(\xi)\det\left(\frac{1}{e}\partial^2\right)\left(\int \mathcal{D}\alpha\right)\int \mathcal{D}A \, e^{iS[A]} \exp\left[-i\int d^4x \frac{1}{2\xi}(\partial^\mu A_\mu)^2\right],$$
$$\tag{9.56}$$

where $N(\xi)$ is an unimportant normalization constant and we have used the delta function to perform the integral over ω. We can choose ξ to be any finite constant. Effectively, we have added a new term $-(\partial^\mu A_\mu)^2/2\xi$ to the Lagrangian.

So far we have worked only with the denominator of our formula for correlation functions,

$$\langle\Omega| \, T \, \mathcal{O}(A) \, |\Omega\rangle = \lim_{T\to\infty(1-i\epsilon)} \frac{\int \mathcal{D}A \, \mathcal{O}(A) \exp\left[i\int_{-T}^{T} d^4x \, \mathcal{L}\right]}{\int \mathcal{D}A \exp\left[i\int_{-T}^{T} d^4x \, \mathcal{L}\right]}.$$

The same manipulations can also be performed on the numerator, provided that the operator $\mathcal{O}(A)$ is gauge invariant. (If it is not, the variable change from A to A^α preceding Eq. (9.54) does not work). Assuming that $\mathcal{O}(A)$ is gauge invariant, we find for its correlation function

$$\langle \Omega | \, T \, \mathcal{O}(A) \, | \Omega \rangle = \lim_{T \to \infty(1 - i\epsilon)} \frac{\int \mathcal{D}A \, \mathcal{O}(A) \exp\left[i \int_{-T}^{T} d^4x \left[\mathcal{L} - \frac{1}{2\xi}(\partial^\mu A_\mu)^2\right]\right]}{\int \mathcal{D}A \, \exp\left[i \int_{-T}^{T} d^4x \left[\mathcal{L} - \frac{1}{2\xi}(\partial^\mu A_\mu)^2\right]\right]}.$$

(9.57)

The awkward constant factors in (9.56) have canceled; the only trace left by this whole process is the extra ξ-term that is added to the action.

At the beginning of this section, in Eq. (9.52), we saw that we could not obtain a sensible photon propagator from the action $S[A]$. With the new ξ-term, however, that equation becomes

$$\left(-k^2 g_{\mu\nu} + (1 - \frac{1}{\xi})k_\mu k_\nu\right) \widetilde{D}_F^{\nu\rho}(k) = i\delta_\mu{}^\rho,$$

which has the solution

$$\widetilde{D}_F^{\mu\nu}(k) = \frac{-i}{k^2 + i\epsilon}\left(g^{\mu\nu} - (1-\xi)\frac{k^\mu k^\nu}{k^2}\right).$$

(9.58)

This is our desired expression for the photon propagator. The $i\epsilon$ term in the denominator arises exactly as in the Klein-Gordon case. Note the overall minus sign relative to the Klein-Gordon propagator, which was already evident in Eq. (9.52).

In practice one usually chooses a specific value of ξ when making computations. Two choices that are often convenient are

$$\xi = 0 \qquad \text{Landau gauge;}$$

$$\xi = 1 \qquad \text{Feynman gauge.}$$

So far in this book we have always used Feynman gauge.[‡]

The Faddeev-Popov procedure guarantees that the value of any correlation function of gauge-invariant operators computed from Feynman diagrams will be independent of the value of ξ used in the calculation (as long as the same value of ξ is used consistently). In the case of QED, it is not difficult to prove this ξ-independence directly. Notice in Eq. (9.58) that ξ multiplies a term in the photon propagator proportional to $k^\mu k^\nu$. According to the Ward-Takahashi identity (7.68), the replacement in a Green's function of any photon propagator by $k^\mu k^\nu$ yields zero, except for terms involving external off-shell fermions. These terms are equal and opposite for particle and antiparticle and vanish when the fermions are grouped into gauge-invariant combinations.

To complete our treatment of the quantization of the electromagnetic field, we need one additional ingredient. In Chapters 5 and 6, we computed S-matrix

---

[‡]Other choices of ξ may be useful in specific applications; for example, in certain problems of bound states in QED, the *Yennie gauge*, $\xi = 3$, produces a cancellation that is otherwise difficult to make explicit. See H. M. Fried and D. R. Yennie, *Phys. Rev.* **112**, 1391 (1958).

elements for QED from the correlation functions of non-gauge-invariant operators $\psi(x)$, $\bar{\psi}(x)$, and $A_\mu(x)$. We will now argue that the S-matrix elements are given correctly by this procedure. Since the S-matrix is defined between asymptotic states, we can compute S-matrix elements in a formalism in which the coupling constant is turned off adiabatically in the far past and far future. In the zero coupling limit, there is a clean separation between gauge-invariant and gauge-variant states. Single-particle states containing one electron, one positron, or one transversely polarized photon are gauge-invariant, while states with timelike and longitudinal photon polarizations transform under gauge motions. We can thus define a gauge-invariant S-matrix in the following way: Let S_{FP} be the S-matrix between general asymptotic states, computed from the Faddeev-Popov procedure. This matrix is unitary but not gauge-invariant. Let P_0 be a projection onto the subspace of the space of asymptotic states in which all particles are either electrons, positrons, or transverse photons. Then let

$$S = P_0 \, S_{\text{FP}} \, P_0 \,. \tag{9.59}$$

This S-matrix is gauge invariant by construction, because it is projected onto gauge-invariant states. It is not obvious that it is unitary. However, we addressed this issue in Section 5.5. We showed there that any matrix element $\mathcal{M}^\mu \epsilon_\mu^*$ for photon emission satisfies

$$\sum_{i=1,2} \epsilon_{i\mu}^* \epsilon_{i\nu} \mathcal{M}^\mu \mathcal{M}^{*\nu} = (-g_{\mu\nu}) \mathcal{M}^\mu \mathcal{M}^{*\nu}, \tag{9.60}$$

where the sum on the left-hand side runs only over transverse polarizations. The same argument applies if $\mathcal{M}^\mu$ and $\mathcal{M}^{*\nu}$ are distinct amplitudes, as long as they satisfy the Ward identity. This is exactly the information we need to see that

$$SS^\dagger = P_0 \, S_{\text{FP}} \, P_0 \, S_{\text{FP}}^\dagger \, P_0 = P_0 \, S_{\text{FP}} \, S_{\text{FP}}^\dagger \, P_0. \tag{9.61}$$

Now we can use the unitarity of S_{FP} to see that S is unitary, $SS^\dagger = 1$, on the subspace of gauge-invariant states. It is easy to check explicitly that the formula (9.59) for the S-matrix is independent of ξ: The Ward identity implies that any QED matrix element with all external fermions on-shell is unchanged if we add to the photon propagator $D^{\mu\nu}(q)$ any term proportional to q^μ.

9.5 Functional Quantization of Spinor Fields

The functional methods that we have used so far allow us to compute, using Eq. (9.18) or (9.35), correlation functions involving fields that obey canonical commutation relations. To generalize these methods to include spinor fields, which obey canonical anticommutation relations, we must do something different: We must represent even the classical fields by anticommuting numbers.

Anticommuting Numbers

We will define anticommuting numbers (also called *Grassmann numbers*) by giving algebraic rules for manipulating them. These rules are formal and might seem *ad hoc*. We will justify them by showing that they lead to the familiar quantum theory of the Dirac equation.

The basic feature of anticommuting numbers is that they *anticommute*. For any two such numbers θ and η,

$$\theta\eta = -\eta\theta. \tag{9.62}$$

In particular, the square of any Grassmann number is zero:

$$\theta^2 = 0.$$

(This fact makes algebra extremely easy.) A product $(\theta\eta)$ of two Grassmann numbers commutes with other Grassmann numbers. We will also wish to add Grassmann numbers, and to multiply them by ordinary numbers; these operations have all the properties of addition and scalar multiplication in any vector space.

The main thing we want to do with anticommuting numbers is integrate over them. To define functional integration, we do not need general definite integrals of these parameters, but only the analog of $\int_{-\infty}^{\infty} dx$. So let us define the integral of a general function f of a Grassmann variable θ, over the complete range of θ:

$$\int d\theta \, f(\theta) = \int d\theta \, (A + B\theta).$$

In general, $f(\theta)$ can be expanded in a Taylor series, which terminates after two terms since $\theta^2 = 0$. The integral should be linear in f; thus it must be a linear function of A and B. Its value is fixed by one additional property: In our analysis of bosonic functional integrals (for instance, in (9.38) and (9.54)), we made strong use of the invariance of the integral to shifts of the integration variable. We will see in Section 9.6 that this shift invariance of the functional integral plays a central role in the derivation of the quantum mechanical equations of motion and conservation laws, and thus must be considered a fundamental aspect of the formalism. We must, then, demand this same property for integrals over θ. Invariance under the shift $\theta \to \theta + \eta$ yields the condition

$$\int d\theta \, (A + B\theta) = \int d\theta \, ((A + B\eta) + B\theta).$$

The shift changes the constant term, but leaves the linear term unchanged. The only linear function of A and B that has this property is a constant (conventionally taken to be 1) times B, so we define*

$$\int d\theta \, (A + B\theta) = B. \tag{9.63}$$

When we perform a multiple integral over more than one Grassmann variable, an ambiguity in sign arises; we adopt the convention

$$\int d\theta \int d\eta \, \eta\theta = +1, \tag{9.64}$$

performing the innermost integral first.

Since the Dirac field is complex-valued, we will work primarily with complex Grassmann numbers, which can be built out of real and imaginary parts in the usual way. It is convenient to define complex conjugation to reverse the order of products, just like Hermitian conjugation of operators:

$$(\theta\eta)^* \equiv \eta^*\theta^* = -\theta^*\eta^*. \tag{9.65}$$

To integrate over complex Grassmann numbers, let us define

$$\theta = \frac{\theta_1 + i\theta_2}{\sqrt{2}}, \qquad \theta^* = \frac{\theta_1 - i\theta_2}{\sqrt{2}}.$$

We can now treat θ and θ^* as independent Grassmann numbers, and adopt the convention $\int d\theta^* d\theta \, (\theta\theta^*) = 1$.

Let us evaluate a Gaussian integral over a complex Grassmann variable:

$$\int d\theta^* \, d\theta \, e^{-\theta^* b\theta} = \int d\theta^* \, d\theta \, (1 - \theta^* b\theta) = \int d\theta^* \, d\theta \, (1 + \theta\theta^* b) = b. \tag{9.66}$$

If θ were an ordinary complex number, this integral would equal $2\pi/b$. The factor of 2π is unimportant; the main difference with anticommuting numbers is that the b comes out in the numerator rather than the denominator. However, if there is an additional factor of $\theta\theta^*$ in the integrand, we obtain

$$\int d\theta^* \, d\theta \, \theta\theta^* \, e^{-\theta^* b\theta} = 1 = \frac{1}{b} \cdot b. \tag{9.67}$$

The extra $\theta\theta^*$ introduces a factor of $(1/b)$, just as it does in an ordinary Gaussian integral.

To perform general Gaussian integrals in higher dimensions, we must first prove that an integral over complex Grassmann variables is invariant under unitary transformations. Consider a set of n complex Grassmann variables θ_i, and a unitary matrix U. If $\theta_i' = U_{ij}\theta_j$, then

$$\prod_i \theta_i' = \frac{1}{n!} \epsilon^{ij\dots l} \theta_i' \theta_j' \dots \theta_l'$$

---

*This definition is due to F. A. Berezin, *The Method of Second Quantization*, Academic Press, New York, 1966.

$$= \frac{1}{n!} \epsilon^{ij\cdots l} U_{ii'} \theta_i U_{jj'} \theta_{j'} \cdots U_{ll'} \theta_{l'}$$

$$= \frac{1}{n!} \epsilon^{ij\cdots l} U_{ii'} U_{jj'} \cdots U_{ll'} \epsilon^{i'j'\cdots l'} \left(\prod_i \theta_i \right)$$

$$= (\det U) \left(\prod_i \theta_i \right). \tag{9.68}$$

In a general integral

$$\left(\prod_i \int d\theta_i^* \, d\theta_i \right) f(\theta),$$

the only term of $f(\theta)$ that survives has exactly one factor of each θ_i and θ_i^*; it is proportional to $(\prod \theta_i)(\prod \theta_i^*)$. If we replace θ by $U\theta$, this term acquires a factor of $(\det U)(\det U)^* = 1$, so the integral is unchanged under the unitary transformation.

We can now evaluate a general Gaussian integral involving a Hermitian matrix B with eigenvalues b_i:

$$\left(\prod_i \int d\theta_i^* \, d\theta_i \right) e^{-\theta_i^* B_{ij} \theta_j} = \left(\prod_i \int d\theta_i^* \, d\theta_i \right) e^{-\Sigma_i \theta_i^* b_i \theta_i} = \prod_i b_i = \det B. \tag{9.69}$$

(If θ were an ordinary number, we would have obtained $(2\pi)^n/(\det B)$.) Similarly, you can show that

$$\left(\prod_i \int d\theta_i^* \, d\theta_i \right) \theta_k \theta_l^* \, e^{-\theta_i^* B_{ij} \theta_j} = (\det B)(B^{-1})_{kl}. \tag{9.70}$$

Inserting another pair $\theta_m \theta_n^*$ in the integrand would yield a second factor $(B^{-1})_{mn}$, and a second term in which the indices l and n are interchanged (the sum of all possible pairings). In general, except for the determinant being in the numerator rather than the denominator, Gaussian integrals over Grassmann variables behave exactly like Gaussian integrals over ordinary variables.

The Dirac Propagator

A Grassmann *field* is a function of spacetime whose values are anticommuting numbers. More precisely, we can define a Grassmann field $\psi(x)$ in terms of any set of orthonormal basis functions:

$$\psi(x) = \sum_i \psi_i \phi_i(x). \tag{9.71}$$

The basis functions $\phi_i(x)$ are ordinary c-number functions, while the coefficients ψ_i are Grassmann numbers. To describe the Dirac field, we take the ϕ_i to be a basis of four-component spinors.

We now have all the machinery needed to evaluate functional integrals, and hence correlation functions, involving fermions. For example, the Dirac two-point function is given by

$$\langle 0| T\psi(x_1)\bar\psi(x_2) |0\rangle = \frac{\int \mathcal{D}\bar\psi\, \mathcal{D}\psi\, \exp\left[i\int d^4x\, \bar\psi(i\slashed\partial - m)\psi\right]\psi(x_1)\bar\psi(x_2)}{\int \mathcal{D}\bar\psi\, \mathcal{D}\psi\, \exp\left[i\int d^4x\, \bar\psi(i\slashed\partial - m)\psi\right]}.$$

(We write $\mathcal{D}\bar\psi$ instead of $\mathcal{D}\psi^*$ for convenience; the two are unitarily equivalent. We also leave the limits on the time integrals implicit; they are the same as in Eq. (9.18), and will yield an $i\epsilon$ term in the propagator as usual.) The denominator of this expression, according to (9.69), is $\det(i\slashed\partial - m)$. The numerator, according to (9.70), is this same determinant times the inverse of the operator $-i(i\slashed\partial - m)$. Evaluating this inverse in Fourier space, we find the familiar result for the Feynman propagator,

$$\langle 0| T\psi(x_1)\bar\psi(x_2) |0\rangle = S_F(x_1 - x_2) = \int \frac{d^4k}{(2\pi)^4} \frac{ie^{-ik\cdot(x_1-x_2)}}{\slashed k - m + i\epsilon}. \tag{9.72}$$

Higher correlation functions of free Dirac fields can be evaluated in a similar manner. The answer is always just the sum of all possible full contractions of the operators, with a factor of S_F for each contraction, as we found from Wick's theorem in Chapter 4.

Generating Functional for the Dirac Field

As with the Klein-Gordon field, we can alternatively derive the Feynman rules for the free Dirac theory by means of a generating functional. In analogy with (9.34), we define the Dirac generating functional as

$$Z[\bar\eta, \eta] = \int \mathcal{D}\bar\psi\, \mathcal{D}\psi\, \exp\left[i\int d^4x\left[\bar\psi(i\slashed\partial - m)\psi + \bar\eta\psi + \bar\psi\eta\right]\right], \tag{9.73}$$

where $\eta(x)$ is a Grassmann-valued source field. You can easily shift $\psi(x)$ to complete the square, to derive the simpler expression

$$Z[\bar\eta, \eta] = Z_0 \cdot \exp\left[-\int d^4x\, d^4y\, \bar\eta(x)S_F(x - y)\eta(y)\right], \tag{9.74}$$

where, as before, Z_0 is the value of the generating functional with the external sources set to zero.

To obtain correlation functions, we will differentiate Z with respect to η and $\bar\eta$. First, however, we must adopt a sign convention for derivatives with respect to Grassmann numbers. If η and θ are anticommuting numbers, let us define

$$\frac{d}{d\eta}\theta\eta = -\frac{d}{d\eta}\eta\theta = -\theta. \tag{9.75}$$

Then referring to the definition (9.73) of Z, we see that the two-point function, for example, is given by

$$\langle 0| T\psi(x_1)\bar\psi(x_2) |0\rangle = Z_0^{-1}\left(-i\frac{\delta}{\delta\bar\eta(x_1)}\right)\left(+i\frac{\delta}{\delta\eta(x_2)}\right) Z[\bar\eta, \eta]\Big|_{\bar\eta,\eta=0}.$$

Plugging in formula (9.74) for $Z[\bar{\eta}, \eta]$ and carefully keeping track of the signs, we find that this expression is equal to the Feynman propagator, $S_F(x_1 - x_2)$. Higher correlation functions can be evaluated in a similar way.

QED

As we saw in Section 9.2 for the case of scalar fields, the functional integral method allows us to read the Feynman rules for vertices directly from the Lagrangian for an interacting field theory. For the theory of Quantum Electrodynamics, the full Lagrangian is

$$
\begin{aligned}
\mathcal{L}_{\text{QED}} &= \bar{\psi}(i\slashed{D} - m)\psi - \tfrac{1}{4}(F_{\mu\nu})^2 \\
&= \bar{\psi}(i\slashed{\partial} - m)\psi - \tfrac{1}{4}(F_{\mu\nu})^2 - e\bar{\psi}\gamma^\mu\psi A_\mu \\
&= \mathcal{L}_0 - e\bar{\psi}\gamma^\mu\psi A_\mu,
\end{aligned}
$$

where $D_\mu = \partial_\mu + ieA_\mu$ is the gauge-covariant derivative.

To evaluate correlation functions, we expand the exponential of the interaction term:

$$
\exp\left[i\int\mathcal{L}\right] = \exp\left[i\int\mathcal{L}_0\right]\left[1 - ie\int d^4x\,\bar{\psi}\gamma^\mu\psi A_\mu + \cdots\right].
$$

The two terms of the free Lagrangian yield the Dirac and electromagnetic propagators derived in this section and the last:

$$
\xrightarrow{\quad p \quad} = \int \frac{d^4p}{(2\pi)^4}\,\frac{i\,e^{-ip\cdot(x-y)}}{\slashed{p} - m + i\epsilon};
$$

$$
\rule{1cm}{0pt}\,\,{\sim\!\!\sim\!\!\sim\!\!\sim}\,\,{q \to} \rule{0.5cm}{0pt}= \int \frac{d^4q}{(2\pi)^4}\,\frac{-i\,g_{\mu\nu}\,e^{-iq\cdot(x-y)}}{q^2 + i\epsilon} \qquad \text{(Feynman gauge)}.
$$

The interaction term gives the QED vertex,

$$
\mu = -ie\gamma^\mu \int d^4x.
$$

As in Chapter 4, we can rearrange these rules, performing the integrations over vertex positions to obtain momentum-conserving delta functions, and using these delta functions to perform most of the propagator momentum integrals.

The only remaining aspect of the QED Feynman rules is the placement of various minus signs. These signs are also built into the functional integral; for example, interchanging θ_k and θ_l^* in Eq. (9.70) would introduce a factor of -1. We will see another example of a fermion minus sign in the computation that follows.

Functional Determinants

Throughout this chapter we have encountered expressions that we wrote formally as functional determinants. To end this section, let us investigate one of these objects more closely. We will find that, at least in this case, we can write the determinant explicitly as a sum of Feynman diagrams.

Consider the object

$$\int \mathcal{D}\bar{\psi}\,\mathcal{D}\psi \, \exp\left[i\int d^4x\, \bar{\psi}(i\slashed{D} - m)\psi\right], \tag{9.76}$$

where $D_\mu = \partial_\mu + ieA_\mu$ and $A_\mu(x)$ is a given external background field. Formally, this expression is a functional determinant:

$$= \det(i\slashed{D} - m) = \det(i\slashed{\partial} - m - e\slashed{A})$$

$$= \det(i\slashed{\partial} - m) \cdot \det\left(1 - \frac{i}{i\slashed{\partial} - m}(-ie\slashed{A})\right).$$

In the last form, the first term is an infinite constant. The second term contains the dependence of the determinant on the external field A. We will now show that this dependence is well defined and, in fact, is exactly equivalent to the sum of vacuum diagrams.

To demonstrate this, we need only apply standard identities from linear algebra. First notice that, if a matrix B has eigenvalues b_i, we can write its determinant as

$$\det B = \prod_i b_i = \exp\left[\sum_i \log b_i\right] = \exp\left[\mathrm{Tr}(\log B)\right], \tag{9.77}$$

where the logarithm of a matrix is defined by its power series. Applying this identity to our determinant, and writing out the power series of the logarithm, we obtain[†]

$$\det\left(1 - \frac{i}{i\slashed{\partial} - m}(-ie\slashed{A})\right) = \exp\left[\sum_{n=1}^{\infty} -\frac{1}{n}\,\mathrm{Tr}\left[\left(\frac{i}{i\slashed{\partial} - m}(-ie\slashed{A})\right)^n\right]\right]. \tag{9.78}$$

Alternatively, we can evaluate this determinant by returning to expression (9.76) and using Feynman diagrams. Expanding the interaction term, we obtain the vertex rule

$$= -ie\gamma^\mu \int d^4x\, A_\mu(x).$$

[†]We use $\mathrm{Tr}()$ to denote operator traces, and $\mathrm{tr}()$ to denote Dirac traces.

Our determinant is then equal to a sum of Feynman diagrams,

$$\det\left(1 - \frac{i(-ie\slashed{A})}{i\slashed{\partial} - m}\right) = 1 + \text{(diagram)} + \text{(diagram)} + \text{(diagram)} + \text{(diagram)} + \cdots$$

$$= \exp\left[\text{(diagram)} + \text{(diagram)} + \text{(diagram)} + \cdots\right]. \qquad (9.79)$$

The series exponentiates, since the disconnected diagrams are products of connected pieces (with appropriate symmetry factors when a piece is repeated). For example,

$$\text{(diagram)} = \frac{1}{2}\left(\text{(diagram)}\right)^2.$$

Now let us evaluate the nth diagram in the exponent of (9.79). There is a factor of -1 from the fermion loop, and a symmetry factor of $1/n$ since we could rotate the interactions around the diagram up to n times without changing it. (The factor is not $1/n!$, because the cyclic order of the interaction points is significant.) The diagram is therefore

$$\text{(diagram)} = -\frac{1}{n}\int dx_1 \cdots dx_n \, \text{tr}\left[(-ie\slashed{A}(x_1))S_F(x_2 - x_1)\cdots\right.$$
$$\left.(-ie\slashed{A}(x_n))S_F(x_1 - x_n)\right]$$

$$= -\frac{1}{n}\text{Tr}\left[\left(\frac{i}{i\slashed{\partial} - m}(-ie\slashed{A})\right)^n\right], \qquad (9.80)$$

in exact agreement with (9.78), including the minus sign and the symmetry factor.

The computation of functional determinants using Feynman diagrams is an important tool, as we will see in Chapter 11.

9.6 Symmetries in the Functional Formalism

We have now seen that the quantum field theoretic correlation functions of scalar, vector, and spinor fields can be computed from the functional integral, completely bypassing the construction of the Hamiltonian, the Hilbert space of states, and the equations of motion. The functional integral formalism makes the symmetries of the problem manifest; any invariance of the Lagrangian will be an invariance of the quantum dynamics.[‡] However, we would like to be able to appeal also to the conservation laws that follow from the quantum equations of motion, or to these equations of motion themselves. For example, the Ward identity, which played a major role in our discussion of photons in QED (Section 5.5), is essentially the conservation law of the electric charge current. Since, as we saw in Section 2.2, the conservation laws follow from symmetries of the Lagrangian, one might guess that it is not difficult to derive these conservation laws from the functional integral. In this section we will see how to do that. We will see that the functional integral gives, in a most direct way, a quantum generalization of Noether's theorem. This result will lead to the analogue of the Ward-Takahashi identity for any symmetry of a general quantum field theory.

Equations of Motion

To prepare for this discussion, we should determine how the quantum equations of motion follow from the functional integral formalism. As a first problem to study, let us examine the Green's functions of the free scalar field. To be specific, consider the three-point function:

$$\langle \Omega| T\phi(x_1)\phi(x_2)\phi(x_3) |\Omega\rangle = Z^{-1} \int \mathcal{D}\phi\, e^{i\int d^4x\, \mathcal{L}[\phi]}\phi(x_1)\phi(x_2)\phi(x_3), \quad (9.81)$$

where $\mathcal{L} = \frac{1}{2}(\partial_\mu\phi)^2 - \frac{1}{2}m^2\phi^2$ and Z is a shorthand for $Z[J=0]$, the functional integral over the exponential. In classical mechanics, we would derive the equations of motion by insisting that the action be stationary under an infinitesimal variation

$$\phi(x) \rightarrow \phi'(x) = \phi(x) + \epsilon(x). \quad (9.82)$$

The appropriate generalization is to consider (9.82) as an infinitesimal change of variables. A change of variables does not alter the value of the integral. Nor does a shift of the integration variable alter the measure: $\mathcal{D}\phi' = \mathcal{D}\phi$. Thus we can write

$$\int \mathcal{D}\phi\, e^{i\int d^4x\, \mathcal{L}[\phi]}\phi(x_1)\phi(x_2)\phi(x_3) = \int \mathcal{D}\phi\, e^{i\int d^4x\, \mathcal{L}[\phi']}\phi'(x_1)\phi'(x_2)\phi'(x_3),$$

[‡]There are some subtle exceptions to this rule, which we will treat in Chapter 19.

where $\phi' = \phi + \epsilon$. Expanding this equation to first order in ϵ, we find

$$0 = \int \mathcal{D}\phi \, e^{i \int d^4x \, \mathcal{L}} \left\{ \left(i \int d^4x \, \epsilon(x) \left[(-\partial^2 - m^2)\phi(x) \right] \phi(x_1)\phi(x_2)\phi(x_3) \right) \right.$$
$$\left. + \epsilon(x_1)\phi(x_2)\phi(x_3) + \phi(x_1)\epsilon(x_2)\phi(x_3) + \phi(x_1)\phi(x_2)\epsilon(x_3) \right\}.$$

(9.83)

The last three terms can be combined with the first by writing, for instance, $\epsilon(x_1) = \int d^4x \, \epsilon(x)\delta(x-x_1)$. Noting that the right-hand side must vanish for any possible variation $\epsilon(x)$, we then obtain

$$0 = \int \mathcal{D}\phi \, e^{i \int d^4x \, \mathcal{L}} \left[(\partial^2 + m^2)\phi(x) \, \phi(x_1)\phi(x_2)\phi(x_3) \right.$$
$$\left. + i\delta(x-x_1)\phi(x_2)\phi(x_3) + i\phi(x_1)\delta(x-x_2)\phi(x_3) + i\phi(x_1)\phi(x_2)\delta(x-x_3) \right].$$

(9.84)

A similar equation holds for any number of fields $\phi(x_i)$.

To see the implications of (9.84), let us specialize to the case of one field $\phi(x_1)$ in (9.81). Notice that the derivatives acting on $\phi(x)$ can be pulled outside the functional integral. Then, dividing (9.84) by Z yields the identity

$$(\partial^2 + m^2) \langle\Omega| T\phi(x)\phi(x_1) |\Omega\rangle = -i\delta(x - x_1). \tag{9.85}$$

The left-hand side of this relation is the Klein-Gordon operator acting on a correlation function of $\phi(x)$. The right-hand side is zero unless $x = x_1$; that is, the correlation function satisfies the Klein-Gordon equation except at the point where the arguments of the two ϕ fields coincide. The modification of the Klein-Gordon equation at this point is called a *contact term*. In this simple case, the modification is hardly unfamiliar to us; Eq. (9.85) merely says that the Feynman propagator is a Green's function of the Klein-Gordon operator, as we originally showed in Section 2.4. We saw there that the delta function arises when the time derivative in ∂^2 acts on the time-ordering symbol. We will see below that, quite generally in quantum field theory, the classical equations of motion for fields are satisfied by all quantum correlation functions of those fields, up to contact terms.

As an example, consider the identity that follows from (9.84) for an $(n+1)$-point correlation function of scalar fields:

$$(\partial^2 + m^2) \langle\Omega| T\phi(x)\phi(x_1) \cdots \phi(x_n) |\Omega\rangle$$
$$= \sum_{i=1}^{n} \langle\Omega| T\phi(x_1) \cdots \left(-i\delta(x - x_i) \right) \cdots \phi(x_n) |\Omega\rangle. \tag{9.86}$$

This identity says that the Klein-Gordon equation is obeyed by $\phi(x)$ inside any expectation value, up to contact terms associated with the time ordering. The result can also be derived from the Hamiltonian formalism using the methods of Section 2.4, or, using the special properties of free-field theory, by evaluating both sides of the equation using Wick's theorem.

As long as the functional measure is invariant under a shift of the integration variable, we can repeat this argument and obtain the quantum equations of motion for Green's functions for any theory of scalar, vector, and spinor fields. This is the reason why, in Eq. (9.63), we took the shift invariance to be the fundamental, defining property of the Grassmann integral.

For a general field theory of a field $\varphi(x)$, governed by the Lagrangian $\mathcal{L}[\varphi]$, the manipulations leading to (9.83) give the identity

$$0 = \int \mathcal{D}\varphi \, e^{i \int d^4 x \, \mathcal{L}} \left\{ i \int d^4 x \, \epsilon(x) \frac{\delta}{\delta\varphi(x)} \left(\int d^4 x' \mathcal{L} \right) \cdot \varphi(x_1)\varphi(x_2) \right.$$
$$\left. + \epsilon(x_1)\varphi(x_2) + \varphi(x_1)\epsilon(x_2) \right\}, \tag{9.87}$$

and similar identities for correlation functions of n fields. By the rule for functional differentiation (9.31), the derivative of the action is

$$\frac{\delta}{\delta\varphi(x)} \left(\int d^4 x' \mathcal{L} \right) = \frac{\partial \mathcal{L}}{\partial \varphi} - \partial_\mu \left(\frac{\partial \mathcal{L}}{\partial(\partial_\mu \varphi)} \right);$$

this is the quantity that equals zero by the Euler-Lagrange equation of motion (2.3) for φ. Formula (9.87) and its generalizations lead to the set of identities

$$\left\langle \left(\frac{\delta}{\delta\varphi(x)} \int d^4 x' \mathcal{L} \right) \varphi(x_1) \cdots \varphi(x_n) \right\rangle = \sum_{i=1}^{n} \langle \varphi(x_1) \cdots (i\delta(x - x_i)) \cdots \varphi(x_n) \rangle. \tag{9.88}$$

In this equation, the angle-brackets denote a time-ordered correlation function in which derivatives on $\varphi(x)$ are placed outside the time-ordering symbol, as in Eq. (9.86). Relation (9.88) states that the classical Euler-Lagrange equations of the field φ are obeyed for all Green's functions of φ, up to contact terms arising from the nontrivial commutation relations of field operators. These quantum equations of motion for Green's functions, including the proper contact terms, are called *Schwinger-Dyson equations*.

Conservation Laws

In classical field theory, Noether's theorem says that, to each symmetry of a local Lagrangian, there corresponds a conserved current. In Section 2.2 we proved Noether's theorem by subjecting the Lagrangian to an infinitesimal symmetry variation. In the spirit of the above discussion of equations of motion, we should find the quantum analogue of this theorem by subjecting the functional integral to an infinitesimal change of variables along the symmetry direction.

Again, it will be most instructive to begin with an example. Let us consider the theory of a free, complex-valued scalar field, with the Lagrangian

$$\mathcal{L} = |\partial_\mu \phi|^2 - m^2 |\phi|^2. \tag{9.89}$$

This Lagrangian is invariant under the transformation $\phi \to e^{i\alpha}\phi$. The classical consequences of this invariance were discussed in Section 2.2, below Eq. (2.14). To find the quantum formulae, consider the infinitesimal change of variables

$$\phi(x) \to \phi'(x) = \phi(x) + i\alpha(x)\phi(x). \tag{9.90}$$

Note that we have made the infinitesimal angle of rotation a function of x; the reason for this will be clear in a moment.

The measure of functional integration is invariant under the transformation (9.90), since this is a unitary transformation of the variables $\phi(x)$. Thus, for the case of two fields,

$$\int \mathcal{D}\phi\, e^{i\int d^4x\, \mathcal{L}[\phi]} \phi(x_1)\phi^*(x_2) = \int \mathcal{D}\phi\, e^{i\int d^4x\, \mathcal{L}[\phi']} \phi'(x_1)\phi'^*(x_2) \bigg|_{\phi'=(1+i\alpha)\phi}.$$

Expanding this equation to first order in α, we find

$$0 = \int \mathcal{D}\phi\, e^{i\int d^4x\, \mathcal{L}} \left\{ i\int d^4x \Big[(\partial_\mu\alpha)\cdot i(\phi\partial^\mu\phi^* - \phi^*\partial^\mu\phi)\Big]\phi(x_1)\phi^*(x_2) \right.$$
$$\left. + \Big[i\alpha(x_1)\phi(x_1)\Big]\phi^*(x_2) + \phi(x_1)\Big[-i\alpha(x_2)\phi^*(x_2)\Big] \right\}.$$

Notice that the variation of the Lagrangian contains only terms proportional to $\partial_\mu\alpha$, since the substitution (9.90) with a constant α leaves the Lagrangian invariant. To put this relation into a familiar form, integrate the term involving $\partial_\mu\alpha$ by parts. Then taking the coefficient of $\alpha(x)$ and dividing by Z gives

$$\langle \partial_\mu j^\mu(x)\phi(x_1)\phi^*(x_2)\rangle = (-i)\Big\langle \big(i\phi(x_1)\delta(x-x_1)\big)\phi^*(x_2)$$
$$+ \phi(x_1)\big(-i\phi^*(x_2)\delta(x-x_2)\big)\Big\rangle, \tag{9.91}$$

where

$$j^\mu = i(\phi\partial^\mu\phi^* - \phi^*\partial^\mu\phi) \tag{9.92}$$

is the Noether current identified in Eq. (2.16). As in Eq. (9.88), the correlation function denotes a time-ordered product with the derivative on $j^\mu(x)$ placed outside the time-ordering symbol. Relation (9.91) is the classical conservation law plus contact terms, that is, the Schwinger-Dyson equation associated with current conservation.

It is not much more difficult to discuss current conservation in more general situations. Consider a local field theory of a set of fields $\varphi_a(x)$, governed by a Lagrangian $\mathcal{L}[\varphi]$. An infinitesimal symmetry transformation on the fields φ_a will be of the general form

$$\varphi_a(x) \to \varphi_a(x) + \epsilon\Delta\varphi_a(x). \tag{9.93}$$

We assume that the action is invariant under this transformation. Then, as in Eq. (2.10), if the parameter ϵ is taken to be a constant, the Lagrangian must be invariant up to a total divergence:

$$\mathcal{L}[\varphi] \to \mathcal{L}[\varphi] + \epsilon\partial_\mu\mathcal{J}^\mu. \tag{9.94}$$

If the symmetry parameter ϵ depends on x, as in the analysis of the previous paragraph, the variation of the Lagrangian will be slightly more complicated:

$$\mathcal{L}[\varphi] \rightarrow \mathcal{L}[\varphi] + (\partial_\mu \epsilon)\Delta\varphi_a \frac{\partial\mathcal{L}}{\partial(\partial_\mu\varphi_a)} + \epsilon\partial_\mu\mathcal{J}^\mu.$$

Summation over the index a is understood. Then

$$\frac{\delta}{\delta\epsilon(x)} \int d^4x \, \mathcal{L}[\varphi + \epsilon\Delta\varphi] = -\partial_\mu j^\mu(x), \tag{9.95}$$

where j^μ is the Noether current of Eq. (2.12),

$$j^\mu = \frac{\partial\mathcal{L}}{\partial(\partial_\mu\varphi_a)}\Delta\varphi_a - \mathcal{J}^\mu. \tag{9.96}$$

Using result (9.95) and carrying through the steps leading up to (9.91), we find the Schwinger-Dyson equation:

$$\begin{aligned}
\langle\partial_\mu j^\mu(x)\varphi_a(x_1)\varphi_b(x_2)\rangle = (-i)\Big\langle &(\Delta\varphi_a(x_1)\delta(x-x_1))\varphi_b(x_2) \\
&+ \varphi_a(x_1)(\Delta\varphi_b(x_2)\delta(x-x_2))\Big\rangle.
\end{aligned} \tag{9.97}$$

A similar equation can be found for the correlator of $\partial_\mu j^\mu$ with n fields $\varphi(x)$. These give the full set of Schwinger-Dyson equations associated with the classical Noether theorem.

As an example of the use of this variational procedure to obtain the Noether current, consider the symmetry of the Lagrangian with respect to spacetime translations. Under the transformation

$$\varphi_a \rightarrow \varphi_a + a^\mu(x)\partial_\mu\varphi_a \tag{9.98}$$

the Lagrangian transforms as

$$\mathcal{L} \rightarrow \mathcal{L} + \partial_\nu a^\mu \partial_\mu\varphi_a \frac{\partial\mathcal{L}}{\partial(\partial_\nu\varphi_a)} + a^\mu\partial_\mu\mathcal{L}.$$

The variation of $\int d^4x\,\mathcal{L}$ with respect to a^μ then gives rise to the conservation equation for the energy-momentum tensor $\partial_\nu T^{\mu\nu} = 0$, with

$$T^{\mu\nu} = \frac{\partial\mathcal{L}}{\partial(\partial_\nu\varphi_a)}\partial^\mu\varphi_a - g^{\mu\nu}\mathcal{L}, \tag{9.99}$$

in agreement with Eq. (2.17).

The trick we have used in this section, that of considering a symmetry transformation whose parameter is a function of spacetime, is reminiscent of a technical feature of our earlier discussion introducing the Lagrangian of QED. In Eq. (4.6), we noted that the minimal coupling prescription for coupling the photon to charged fields produces a Lagrangian invariant not only under the global symmetry transformation with ϵ constant, but also under a transformation in which the symmetry parameter depends on x. In Chapter 15, we will draw these two ideas together in a general discussion of field theories with *local* symmetries.

The Ward-Takahashi Identity

As a final application of the methods of this section, let us derive the Schwinger-Dyson equations associated with the global symmetry of QED. Consider making, in the QED functional integral, the change of variables

$$\psi(x) \rightarrow (1 + ie\alpha(x))\psi(x), \tag{9.100}$$

without the corresponding term in the transformation law for A_μ (which would make the Lagrangian invariant under the transformation). The QED Lagrangian (4.3) then transforms according to

$$\mathcal{L} \rightarrow \mathcal{L} - e\partial_\mu \alpha \bar\psi \gamma^\mu \psi. \tag{9.101}$$

The transformation (9.100) thus leads to the following identity for the functional integral over two fermion fields:

$$0 = \int \mathcal{D}\bar\psi \mathcal{D}\psi \mathcal{D}A\, e^{i\int d^4 x\, \mathcal{L}} \left\{ -i \int d^4 x\, \partial_\mu \alpha(x) \left[j^\mu(x)\psi(x_1)\bar\psi(x_2) \right] \right.$$
$$\left. + (ie\alpha(x_1)\psi(x_1))\bar\psi(x_2) + \psi(x_1)(-ie\alpha(x_2)\bar\psi(x_2)) \right\}, \tag{9.102}$$

with $j^\mu = e\bar\psi\gamma^\mu\psi$. As in our other examples, an analogous equation holds for any number of fermion fields.

To understand the implications of this set of equations, consider first the specific case (9.102). Dividing this relation by Z, we find

$$i\partial_\mu \langle 0 | T j^\mu(x)\psi(x_1)\bar\psi(x_2) | 0 \rangle = -ie\delta(x - x_1) \langle 0 | T\psi(x_1)\bar\psi(x_2) | 0 \rangle$$
$$+ ie\delta(x - x_2) \langle 0 | T\psi(x_1)\bar\psi(x_2) | 0 \rangle. \tag{9.103}$$

To put this equation into a more familiar form, compute its Fourier transform by integrating:

$$\int d^4 x\, e^{-ik\cdot x} \int d^4 x_1\, e^{+iq\cdot x_1} \int d^4 x_2\, e^{-ip\cdot x_2}. \tag{9.104}$$

Then the amplitudes in (9.103) are converted to the amplitudes $\mathcal{M}(k; p; q)$ and $\mathcal{M}(p; q)$ defined below (7.67) in our discussion of the Ward-Takahashi identity. Indeed, (9.103) falls directly into the form

$$-ik_\mu \mathcal{M}^\mu(k; p; q) = -ie\mathcal{M}_0(p; q - k) + ie\mathcal{M}_0(p + k; q). \tag{9.105}$$

This is exactly the Ward-Takahashi identity for two external fermions, which we derived diagrammatically in Section 7.4. It is not difficult to check that the more general relations involving n fermion fields lead to the general Ward-Takahashi identity presented in (7.68). Because of this relation, the formula (9.97) associated with the arbitrary symmetry (9.93) is usually also referred to as a Ward-Takahashi identity, the one associated with the symmetry and its Noether current.

We have now arrived at a more general understanding of the terms on the right-hand side of the Ward-Takahashi identity. These are the contact terms that we now expect to find when we convert classical equations of motion to Schwinger-Dyson equations for quantum Green's functions. The functional integral formalism allows a simple and elegant derivation of these quantum-mechanical terms.

Problems

9.1   Scalar QED. This problem concerns the theory of a complex scalar field ϕ interacting with the electromagnetic field A^μ. The Lagrangian is

$$\mathcal{L} = -\tfrac{1}{4}F_{\mu\nu}^2 + (D_\mu\phi)^*(D^\mu\phi) - m^2\phi^*\phi,$$

where $D_\mu = \partial_\mu + ieA_\mu$ is the usual gauge-covariant derivative.

(a) Use the functional method of Section 9.2 to show that the propagator of the complex scalar field is the same as that of a real field:

$$----\!\blacktriangleleft----\cdot \atop p \qquad = \frac{i}{p^2 - m^2 + i\epsilon}.$$

Also derive the Feynman rules for the interactions between photons and scalar particles; you should find

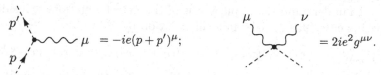

$$\mu \;\; = -ie(p+p')^\mu; \qquad\qquad = 2ie^2 g^{\mu\nu}.$$

(b) Compute, to lowest order, the differential cross section for $e^+e^- \to \phi\phi^*$. Ignore the electron mass (but not the scalar particle's mass), and average over the electron and positron polarizations. Find the asymptotic angular dependence and total cross section. Compare your results to the corresponding formulae for $e^+e^- \to \mu^+\mu^-$.

(c) Compute the contribution of the charged scalar to the photon vacuum polarization, using dimensional regularization. Note that there are two diagrams. To put the answer into the expected form,

$$\Pi^{\mu\nu}(q^2) = (g^{\mu\nu}q^2 - q^\mu q^\nu)\Pi(q^2),$$

it is useful to add the two diagrams at the beginning, putting both terms over a common denominator before introducing a Feynman parameter. Show that, for $-q^2 \gg m^2$, the charged boson contribution to $\Pi(q^2)$ is exactly 1/4 that of a virtual electron-positron pair.

9.2   Quantum statistical mechanics.

(a) Evaluate the quantum statistical partition function

$$Z \;=\; \mathrm{tr}[e^{-\beta H}]$$

(where $\beta = 1/kT$) using the strategy of Section 9.1 for evaluating the matrix elements of e^{-iHt} in terms of functional integrals. Show that one again finds a functional integral, over functions defined on a domain that is of length β and periodically connected in the time direction. Note that the Euclidean form of the Lagrangian appears in the weight.

(b) Evaluate this integral for a simple harmonic oscillator,

$$L_E = \tfrac{1}{2}\dot{x}^2 + \tfrac{1}{2}\omega^2 x^2,$$

by introducing a Fourier decomposition of x(t):

$$x(t) = \sum_n x_n \cdot \frac{1}{\sqrt{\beta}} e^{2\pi int/\beta}.$$

The dependence of the result on β is a bit subtle to obtain explicitly, since the measure for the integral over $x(t)$ depends on β in any discretization. However, the dependence on ω should be unambiguous. Show that, up to a (possibly divergent and β-dependent) constant, the integral reproduces exactly the familiar expression for the quantum partition function of an oscillator. [You may find the identity

$$\sinh z = z \cdot \prod_{n=1}^{\infty} \left(1 + \frac{z^2}{(n\pi)^2}\right)$$

useful.]

(c) Generalize this construction to field theory. Show that the quantum statistical partition function for a free scalar field can be written in terms of a functional integral. The value of this integral is given formally by

$$\left[\det(-\partial^2 + m^2)\right]^{-1/2},$$

where the operator acts on functions on Euclidean space that are periodic in the time direction with periodicity β. As before, the β dependence of this expression is difficult to compute directly. However, the dependence on m^2 is unambiguous. (More generally, one can usually evaluate the variation of a functional determinant with respect to any explicit parameter in the Lagrangian.) Show that the determinant indeed reproduces the partition function for relativistic scalar particles.

(d) Now let $\psi(t)$, $\bar{\psi}(t)$ be two Grassmann-valued coordinates, and define a fermionic oscillator by writing the Lagrangian

$$L_E = \bar{\psi}\dot{\psi} + \omega\bar{\psi}\psi.$$

This Lagrangian corresponds to the Hamiltonian

$$H = \omega\bar{\psi}\psi, \quad \text{with } \{\bar{\psi}, \psi\} = 1;$$

that is, to a simple two-level system. Evaluate the functional integral, assuming that the fermions obey antiperiodic boundary conditions: $\psi(t + \beta) = -\psi(t)$. (Why is this reasonable?) Show that the result reproduces the partition function of a quantum-mechanical two-level system, that is, of a quantum state with Fermi statistics.

(e) Define the partition function for the photon field as the gauge-invariant functional integral

$$Z = \int \mathcal{D}A \, \exp\left(-\int d^4x_E \left[\tfrac{1}{4}(F_{\mu\nu})^2\right]\right)$$

over vector fields A_μ that are periodic in the time direction with period β. Apply the gauge-fixing procedure discussed in Section 9.4 (working, for example, in Feynman gauge). Evaluate the functional determinants using the result of part (c) and show that the functional integral does give the correct quantum statistical result (including the correct counting of polarization states).

Chapter 10

Systematics of Renormalization

While computing radiative corrections in Chapters 6 and 7, we encountered three QED diagrams with ultraviolet divergences:

In each case we saw that the divergence could be regulated and canceled, yielding finite expressions for measurable quantities. In Chapter 8, we pointed out that such ultraviolet divergences occur commonly and, in fact, naturally in quantum field theory calculations. We sketched a physical interpretation of these divergences, with implications both in quantum field theory and in the statistical theory of phase transitions. In the next few chapters, we will convert this sketchy picture into a quantitative theory that allows precise calculations.

In this chapter, we begin this study by developing a classification of the ultraviolet divergences that can appear in a quantum field theory. Rather than stumbling across these divergences one by one and repairing them case by case, we now set out to determine once and for all which diagrams are divergent, and in which theories these divergences can be eliminated systematically. As examples we will consider both QED and scalar field theories.

10.1 Counting of Ultraviolet Divergences

In this section we will use elementary arguments to determine, tentatively, when a Feynman diagram contains an ultraviolet divergence. We begin by analyzing quantum electrodynamics.

First we introduce the following notation, to characterize a typical diagram in QED:

$$N_e = \text{number of external electron lines;}$$
$$N_\gamma = \text{number of external photon lines;}$$
$$P_e = \text{number of electron propagators;}$$
$$P_\gamma = \text{number of photon propagators;}$$
$$V = \text{number of vertices;}$$
$$L = \text{number of loops.}$$

(This analysis applies to correlation functions as well as scattering amplitudes. In the former case, propagators that are connected to external points should be counted as external lines, not as propagators.)

The expression corresponding to a typical diagram looks like this:

$$\sim \int \frac{d^4k_1\, d^4k_2 \cdots d^4k_L}{(\not{k}_i - m) \cdots (k_j^2) \cdots (k_n^2)}.$$

For each loop there is a potentially divergent 4-momentum integral, but each propagator aids the convergence of this integral by putting one or two powers of momentum into the denominator. Very roughly speaking, the diagram diverges unless there are more powers of momentum in the denominator than in the numerator. Let us therefore define the *superficial degree of divergence*, D, as the difference:

$$D \equiv (\text{power of } k \text{ in numerator}) - (\text{power of } k \text{ in denominator})$$
$$= 4L - P_e - 2P_\gamma. \tag{10.1}$$

Naively, we expect a diagram to have a divergence proportional to Λ^D, where Λ is a momentum cutoff, when $D > 0$. We expect a divergence of the form $\log \Lambda$ when $D = 0$, and no divergence when $D < 0$.

This naive expectation is often wrong, for one of three reasons (see Fig. 10.1). When a diagram contains a divergent subdiagram, its actual divergence may be worse than that indicated by D. When symmetries (such as the Ward identity) cause certain terms to cancel, the divergence of a diagram may be reduced or even eliminated. Finally, a trivial diagram with no propagators and no loops has $D = 0$ but no divergence.

Despite all of these complications, D is still a useful quantity. To see why, let us rewrite it in terms of the number of external lines (N_e, N_γ) and vertices (V). Note that the number of loop integrations in a diagram is

$$L = P_e + P_\gamma - V + 1, \tag{10.2}$$

since in our original Feynman rules each propagator has a momentum integral, each vertex has a delta function, and one delta function merely enforces overall momentum conservation. Furthermore, the number of vertices is

$$V = 2P_\gamma + N_\gamma = \tfrac{1}{2}(2P_e + N_e), \tag{10.3}$$

since each vertex involves exactly one photon line and two electron lines. (The propagators count twice since they have two ends on vertices.) Putting these relations together, we find that D can be expressed as

$$D = 4(P_e + P_\gamma - V + 1) - P_e - 2P_\gamma$$
$$= 4 - N_\gamma - \tfrac{3}{2}N_e, \tag{10.4}$$

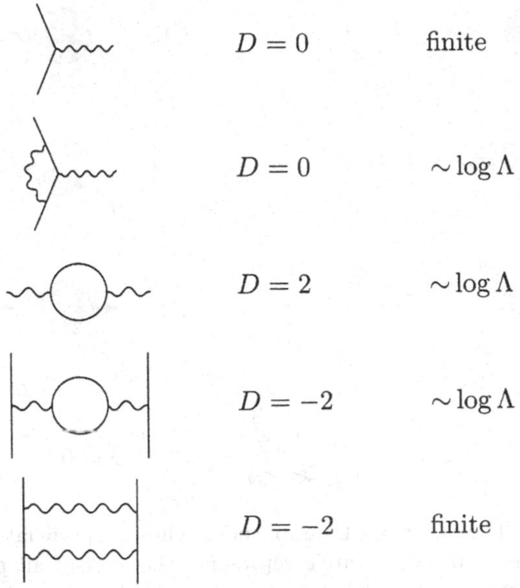

Figure 10.1. Some simple QED diagrams that illustrate the superficial degree of divergence. The first diagram is finite, even though $D = 0$. The third diagram has $D = 2$ but only a logarithmic divergence, due to the Ward identity (see Section 7.5). The fourth diagram diverges, even though $D < 0$, since it contains a divergent subdiagram. Only in the second and fifth diagrams does the superficial degree of divergence coincide with the actual degree of divergence.

independent of the number of vertices. The superficial degree of divergence of a QED diagram depends only on the number of external legs of each type.

According to result (10.4), only diagrams with a small number of external legs have $D \geq 0$; those seven types of diagrams are shown in Fig. 10.2. Since external legs do not enter the potentially divergent integral, we can restrict our attention to amputated diagrams. We can also restrict our attention to one-particle-irreducible diagrams, since reducible diagrams are simple products of the integrals corresponding to their irreducible parts. Thus the task of enumerating all of the divergent QED diagrams reduces to that of analyzing the seven types of amputated, one-particle-irreducible amplitudes shown in Fig. 10.2. Other diagrams may diverge, but only when they contain one of these seven as a subdiagram. Let us therefore consider each of these seven amplitudes in turn.

The zero-point function, Fig. 10.2a, is very badly divergent. But this object merely causes an unobservable shift of the vacuum energy; it never contributes to S-matrix elements.

To analyze the photon one-point function (Fig. 10.2b), note that the external photon must be attached to a QED vertex. Neglecting the external

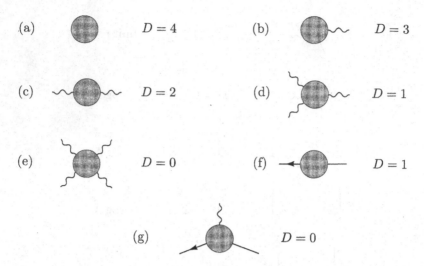

Figure 10.2. The seven QED amplitudes whose superficial degree of divergence (D) is ≥ 0. (Each circle represents the sum of all possible QED diagrams.) As explained in the text, amplitude (a) is irrelevant to scattering processes, while amplitudes (b) and (d) vanish because of symmetries. Amplitude (e) is nonzero, but its divergent parts cancel due to the Ward identity. The remaining amplitudes (c, f, and g) are all logarithmically divergent, even though $D > 0$ for (c) and (f).

photon propagator, this amplitude is therefore

$$= -ie \int d^4x \, e^{-iq \cdot x} \, \langle \Omega | \, T \, j_\mu(x) \, | \Omega \rangle , \qquad (10.5)$$

where $j^\mu = \bar\psi \gamma^\mu \psi$ is the electromagnetic current operator. But the vacuum expectation value of j^μ must vanish by Lorentz invariance, since otherwise it would be a preferred 4-vector.

The photon one-point function also vanishes for a second reason: charge-conjugation invariance. Recall that C is a symmetry of QED, so $C \, |\Omega\rangle = |\Omega\rangle$. But $j^\mu(x)$ changes sign under charge conjugation, $Cj^\mu(x)C^\dagger = -j^\mu(x)$, so its vacuum expectation value must vanish:

$$\langle \Omega | \, T j^\mu(x) \, | \Omega \rangle = \langle \Omega | \, C^\dagger C j^\mu(x) C^\dagger C \, | \Omega \rangle = - \langle \Omega | \, T j^\mu(x) \, | \Omega \rangle = 0.$$

The same argument applies to any vacuum expectation value of an odd number of electromagnetic currents. In particular, the photon three-point function, Fig. 10.2d, vanishes. (This result is known as Furry's theorem.) It is not hard to check explicitly that the photon one- and three-point functions vanish in the leading order of perturbation theory (see Problem 10.1).

The remaining amplitudes in Fig. 10.2 are all nonzero, so we must analyze their structures in more detail. Consider, for example, the electron self-energy

(Fig. 10.2f). This amplitude is a function of the electron momentum p, so let us expand it in a Taylor series about $p = 0$:

$$\text{(diagram)} = A_0 + A_1 p\!\!\!/ + A_2 p^2 + \cdots ,$$

where each coefficient is independent of p:

$$A_n = \frac{1}{n!} \frac{d^n}{dp\!\!\!/^n} \left(\text{(diagram)} \right) \Big|_{p\!\!\!/ = 0} .$$

(These coefficients are infrared divergent; to compute them explicitly we would need an infrared regulator, as in Chapter 6.) The diagrams contributing to the electron self-energy depend on p through the denominators of propagators. To compute the coefficients A_n, we differentiate these propagators, giving expressions like

$$\frac{d}{dp\!\!\!/} \left(\frac{1}{k\!\!\!/ + p\!\!\!/ - m} \right) = -\frac{1}{(k\!\!\!/ + p\!\!\!/ - m)^2} .$$

That is, each derivative with respect to the external momentum p lowers the superficial degree of divergence by 1. Since the constant term A_0 has (superficially) a linear divergence, A_1 can have only a logarithmic divergence; all the remaining A_n are finite. (This argument breaks down when the divergence is in a subdiagram, since then not all propagators involve the large momentum k. We will face this problem in Section 10.4.)

The electron self-energy amplitude has one additional subtlety. If the constant term A_0 were proportional to Λ (the ultraviolet cutoff), the electron mass shift would, according to the analysis in Section 7.1, also have a term proportional to Λ. But the electron mass shift must actually be proportional to m, since chiral symmetry would forbid a mass shift if m were zero. At worst, the constant term can be proportional to $m \log \Lambda$. We therefore expect the entire self-energy amplitude to have the form

$$\text{(diagram)} = a_0 m \log \Lambda + a_1 p\!\!\!/ \log \Lambda + \text{(finite terms)}, \qquad (10.6)$$

exactly what we found for the term of order α in Eq. (7.19).

Let us analyze the exact electron-photon vertex, Fig. 10.2g, in the same way. (Again we implicitly assume that infrared divergences have been regulated.) Expanding in powers of the three external momenta, we immediately see that only the constant term is divergent, since differentiating with respect to any external momentum would lower the degree of divergence to -1. This amplitude therefore contains only one divergent constant:

$$\text{(diagram)} \propto -ie\gamma^\mu \log \Lambda + \text{finite terms}. \qquad (10.7)$$

As discussed in Section 7.5, the photon self energy (Fig. 10.2c) is constrained by the Ward identity to have the form

$$\mu \sim\!\!\!\bigcirc\!\!\!\sim \nu = (g^{\mu\nu}q^2 - q^\mu q^\nu)\Pi(q^2). \tag{10.8}$$

Viewing this expression as a Taylor series in q, we see that the constant and linear terms both vanish, lowering the superficial degree of divergence from 2 to 0. The only divergence, therefore, is in the constant term of $\Pi(q^2)$, and this divergence is only logarithmic. This result is exactly what we found for the lowest-order contribution to $\Pi(q^2)$ in Eq. (7.90).

Finally, consider the photon-photon scattering amplitude, Fig. 10.2e. The Ward identity requires that if we replace any external photon by its momentum vector, the amplitude vanishes:

$$k^\mu \left(\begin{matrix} \mu & & \nu \\ & \bigcirc & \\ \rho & & \sigma \end{matrix} \right) = 0. \tag{10.9}$$

By exhaustion one can show that this condition is satisfied only if the amplitude is proportional to $(g^{\mu\nu}k^\sigma - g^{\mu\sigma}k^\nu)$, with a similar factor for each of the other three legs. Each of these factors involves one power of momentum, so all terms with less than four powers of momentum in the Taylor series of this amplitude must vanish. The first nonvanishing term has $D = 0 - 4 = -4$, and therefore this amplitude is finite.

In summary, we have found that there are only three "primitively" divergent amplitudes in QED: the three that we already found in Chapters 6 and 7. (Other amplitudes may also be divergent, but only because of diagrams that contain these primitive amplitudes as components.) Furthermore, the dependence of these divergent amplitudes on external momenta is extremely simple. If we expand each amplitude as a power series in its external momenta, there are altogether only four divergent coefficients in the expansions. In other words, QED contains only four divergent numbers. In the next section we will see how these numbers can be absorbed into unobservable Lagrangian parameters, so that observable scattering amplitudes are always finite.

For the remainder of this section, let us try to understand the superficial degree of divergence from a more general viewpoint. The theory of QED in four spacetime dimensions is rather special, so let us first generalize to QED in d dimensions. In this case, D is given by

$$D \equiv dL - P_e - 2P_\gamma, \tag{10.10}$$

since each loop contributes a d-dimensional momentum integral. Relations (10.2) and (10.3) still hold, so we can again rewrite D in terms of V, N_e, and N_γ. This time the result is

$$D = d + \left(\frac{d-4}{2}\right)V - \left(\frac{d-2}{2}\right)N_\gamma - \left(\frac{d-1}{2}\right)N_e. \qquad (10.11)$$

The cancellation of V in this expression is special to the case $d = 4$. For $d < 4$, diagrams with more vertices have a lower degree of divergence, so the total number of divergent *diagrams* is finite. For $d > 4$, diagrams with more vertices have a higher degree of divergence, so every amplitude becomes superficially divergent at a sufficiently high order in perturbation theory.

These three possible types of ultraviolet behavior will also occur in other quantum field theories. We will refer to them as follows:

Super-Renormalizable theory: Only a finite number of Feynman diagrams superficially diverge.

Renormalizable theory: Only a finite number of amplitudes superficially diverge; however, divergences occur at all orders in perturbation theory.

Non-Renormalizable theory: All amplitudes are divergent at a sufficiently high order in perturbation theory.

Using this nomenclature, we would say that QED is renormalizable in four dimensions, super-renormalizable in less than four dimensions, and non-renormalizable in more than four dimensions.

These superficial criteria give a correct picture of the true divergence structure of the theory for most cases that have been studied in detail. Examples are known in which the true behavior is better than this picture suggests, when powerful symmetries set to zero some or all of the superficially divergent amplitudes.* On the other hand, as we will explain in Section 10.4, it is always true that the divergences of superficially renormalizable theories can be absorbed into a finite number of Lagrangian parameters. For theories containing fields of spin 1 and higher, loop diagrams can produce additional problems, including violation of unitarity; we will discuss this difficulty in Chapter 16.

As another example of the counting of ultraviolet divergences, consider a pure scalar field theory, in d dimensions, with a ϕ^n interaction term:

$$\mathcal{L} = \frac{1}{2}(\partial_\mu\phi)^2 - \frac{1}{2}m^2\phi^2 - \frac{\lambda}{n!}\phi^n. \qquad (10.12)$$

Let N be the number of external lines in a diagram, P the number of propagators, and V the number of vertices. The number of loops in a diagram is $L = P - V + 1$. There are n lines meeting at each vertex, so $nV = N + 2P$.

---

*Some exotic four-dimensional field theories are actually free of divergences; see, for example, the article by P. West in *Shelter Island II*, R. Jackiw, N. N. Khuri, S. Weinberg, and E. Witten, eds. (MIT Press, Cambridge, 1985).

Combining these relations, we find that the superficial degree of divergence of a diagram is

$$D = dL - 2P$$

$$= d + \left[n\left(\frac{d-2}{2}\right) - d \right] V - \left(\frac{d-2}{2}\right) N. \tag{10.13}$$

In four dimensions a ϕ^4 coupling is renormalizable, while higher powers of ϕ are non-renormalizable. In three dimensions a ϕ^6 coupling becomes renormalizable, while ϕ^4 is super-renormalizable. In two spacetime dimensions any coupling of the form ϕ^n is super-renormalizable.

Expression (10.13) can also be derived in a somewhat different way, from dimensional analysis. In any quantum field theory, the action $S = \int d^d x \, \mathcal{L}$ must be dimensionless, since we work in units where $\hbar = 1$. In this system of units, the integral $d^d x$ has units $(\text{mass})^{-d}$, and so the Lagrangian has units $(\text{mass})^d$. Since all units can be expressed as powers of mass, it is unambiguous to say simply that the Lagrangian has "dimension d". Using this result, we can infer from the explicit form of (10.12) the dimensions of the field ϕ and the coupling constant λ. From the kinetic term in $\mathcal{L}$ we see that ϕ has dimension $(d-2)/2$. Note that the parameter m consistently has dimensions of mass. From the interaction term and the dimension of ϕ, we infer that the λ has dimension $d - n(d-2)/2$.

Now consider an arbitrary diagram with N external lines. One way that such a diagram could arise is from an interaction term $\eta \phi^N$ in the Lagrangian. The dimension of η would then be $d - N(d-2)/2$, and therefore we conclude that any (amputated) diagram with N external lines has dimension $d - N(d-2)/2$. In our theory with only the $\lambda \phi^n$ vertex, if the diagram has V vertices, its divergent part is proportional to $\lambda^V \Lambda^D$, where Λ is a high-momentum cutoff and D is the superficial degree of divergence. (This is the "generic" case; all the exceptions noted above also apply here.) Applying dimensional analysis, we find

$$d - N\left(\frac{d-2}{2}\right) = V\left[d - n\left(\frac{d-2}{2}\right) \right] + D,$$

in agreement with (10.13).

Note that the quantity that multiplies V in this expression is just the dimension of the coupling constant λ. This analysis can be carried out for QED and other field theories, with the same result. Thus we can characterize the three degrees of renormalizability in a second way:

Super-Renormalizable: Coupling constant has positive mass dimension.

Renormalizable: Coupling constant is dimensionless.

Non-Renormalizable: Coupling constant has negative mass dimension.

This is exactly the conclusion that we stated without proof in Section 4.1. In QED, the coupling constant e is dimensionless; thus QED is (at least superficially) renormalizable.

10.2 Renormalized Perturbation Theory

In the previous section we saw that a renormalizable quantum field theory contains only a small number of superficially divergent amplitudes. In QED, for example, there are three such amplitudes, containing four infinite constants. In Chapters 6 and 7 these infinities disappeared by the end of our computations: The infinity in the vertex correction diagram was canceled by the electron field-strength renormalization, while the infinity in the vacuum polarization diagram caused only an unobservable shift of the electron's charge. In fact, it is generally true that the divergences in a renormalizable quantum field theory never show up in observable quantities.

To obtain a finite result for an amplitude involving divergent diagrams, we have so far used the following procedure: Compute the diagrams using a regulator, to obtain an expression that depends on the bare mass (m_0), the bare coupling constant (e_0), and some ultraviolet cutoff (Λ). Then compute the physical mass (m) and the physical coupling constant (e), to whatever order is consistent with the rest of the calculation; these quantities will also depend on m_0, e_0, and Λ. To calculate an S-matrix element (rather than a correlation function), one must also compute the field-strength renormalization(s) Z (in accord with Eq. (7.45)). Combining all of these expressions, eliminate m_0 and e_0 in favor of m and e; this step is the "renormalization". The resulting expression for the amplitude should be finite in the limit $\Lambda \to \infty$.

The above procedure always works in a renormalizable quantum field theory. However, it can often be cumbersome, especially at higher orders in perturbation theory. In this section we will develop an alternative procedure which works more automatically. We will do this first for ϕ^4 theory, returning to QED in the next section.

The Lagrangian of ϕ^4 theory is

$$\mathcal{L} = \frac{1}{2}(\partial_\mu \phi)^2 - \frac{1}{2}m_0^2\phi^2 - \frac{\lambda_0}{4!}\phi^4.$$

We now write m_0 and λ_0, to emphasize that these are the bare values of the mass and coupling constant, not the values measured in experiments.

The superficial degree of divergence of a diagram with N external legs is, according to (10.13),

$$D = 4 - N.$$

Since the theory is invariant under $\phi \to -\phi$, all amplitudes with an odd number of external legs vanish. The only divergent amplitudes are therefore

(unobservable vacuum energy shift);

$\sim \Lambda^2 + p^2 \log \Lambda + (\text{finite terms})$;

$\sim \log \Lambda + (\text{finite terms})$.

Ignoring the vacuum diagram, these amplitudes contain three infinite constants. Our goal is to absorb these constants into the three unobservable parameters of the theory: the bare mass, the bare coupling constant, and the field strength. To accomplish this goal, it is convenient to reformulate the perturbation expansion so that these unobservable quantities do not appear explicitly in the Feynman rules.

First we will eliminate the shift in the field strength. Recall from Section 7.1 that the exact two-point function has the form

$$\int d^4x \, \langle \Omega | \, T\phi(x)\phi(0) \, |\Omega\rangle \, e^{ip\cdot x} = \frac{iZ}{p^2 - m^2} + (\text{terms regular at } p^2 = m^2),$$
(10.14)

where m is the physical mass. We can eliminate the awkward residue Z from this equation by rescaling the field:

$$\phi = Z^{1/2}\phi_r.$$
(10.15)

This transformation changes the values of correlation functions by a factor of $Z^{-1/2}$ for each field. Thus, in computing S-matrix elements, we no longer need the factors of Z in Eq. (7.45); a scattering amplitude is simply the sum of all connected, amputated diagrams, exactly as we originally guessed in Eq. (4.103).

The Lagrangian is much uglier after the rescaling:

$$\mathcal{L} = \frac{1}{2}Z(\partial_\mu\phi_r)^2 - \frac{1}{2}m_0^2 Z\phi_r^2 - \frac{\lambda_0}{4!}Z^2\phi_r^4.$$
(10.16)

The bare mass and coupling constant still appear in $\mathcal{L}$, but they can be eliminated as follows. Define

$$\delta_Z = Z - 1, \qquad \delta_m = m_0^2 Z - m^2, \qquad \delta_\lambda = \lambda_0 Z^2 - \lambda,$$
(10.17)

where m and λ are the physically measured mass and coupling constant. Then the Lagrangian becomes

$$\mathcal{L} = \frac{1}{2}(\partial_\mu\phi_r)^2 - \frac{1}{2}m^2\phi_r^2 - \frac{\lambda}{4!}\phi_r^4$$
$$+ \frac{1}{2}\delta_Z(\partial_\mu\phi_r)^2 - \frac{1}{2}\delta_m\phi_r^2 - \frac{\delta_\lambda}{4!}\phi_r^4.$$
(10.18)

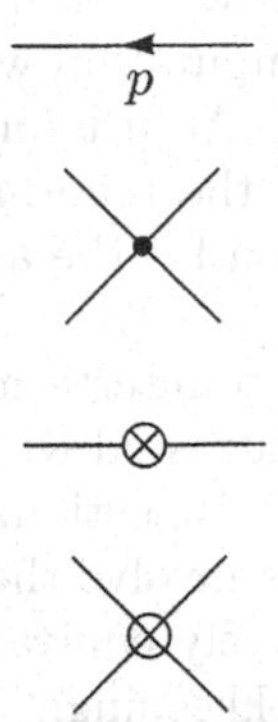

$$\xrightarrow{\quad p \quad} \qquad = \frac{i}{p^2 - m^2 + i\epsilon}$$

$$\times \qquad = -i\lambda$$

$$\longrightarrow\!\!\otimes\!\!\longrightarrow \qquad = i(p^2\delta_Z - \delta_m)$$

$$\bigotimes \qquad = -i\delta_\lambda$$

Figure 10.3. Feynman rules for ϕ^4 theory in renormalized perturbation theory.

The first line now looks like the familiar ϕ^4-theory Lagrangian, but is written in terms of the physical mass and coupling. The terms in the second line, known as *counterterms*, have absorbed the infinite but unobservable shifts between the bare parameters and the physical parameters. It is tempting to say that we have "added" these counterterms to the Lagrangian, but in fact we have merely split each term in (10.16) into two pieces.

The definitions in (10.17) are not useful unless we give precise definitions of the physical mass and coupling constant. Equation (10.14) defines m^2 as the location of the pole in the propagator. There is no obviously best definition of λ, but a perfectly good definition would be obtained by setting λ equal to the magnitude of the scattering amplitude at zero momentum. Thus we have the two defining relations,

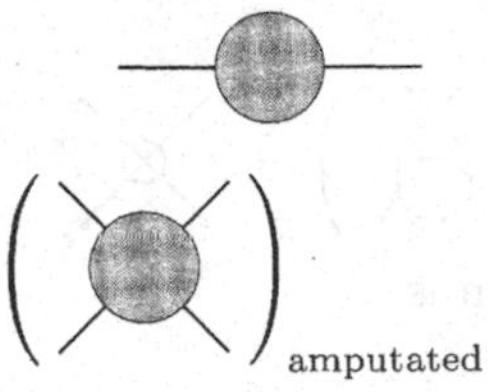

$$\xrightarrow{\quad\bullet\quad} \quad = \frac{i}{p^2 - m^2} + \text{(terms regular at } p^2 = m^2\text{)};$$

$$\left(\bigotimes\right)_{\text{amputated}} \quad = -i\lambda \qquad \text{at } s = 4m^2,\ t = u = 0. \qquad (10.19)$$

These equations are called *renormalization conditions*. (The first equation actually contains two conditions, specifying both the location of the pole and its residue.)

Our new Lagrangian, Eq. (10.18), gives a new set of Feynman rules, shown in Fig. 10.3. The propagator and the first vertex come from the first line of (10.18), and are identical to the old rules except for the appearance of the physical mass and coupling in place of the bare values. The counterterms in the second line of (10.18) give two new vertices (also called counterterms).

We can use these new Feynman rules to compute any amplitude in ϕ^4 theory. The procedure is as follows. Compute the desired amplitude as the sum of all possible diagrams created from the propagator and vertices shown in Fig. 10.3. The loop integrals in the diagrams will often diverge, so one must introduce a regulator. The result of this computation will be a function of the three unknown parameters δ_Z, δ_m, and δ_λ. Adjust (or "renormalize") these three parameters as necessary to maintain the renormalization conditions (10.19). After this adjustment, the expression for the amplitude should be finite and independent of the regulator.

This procedure, using Feynman rules with counterterms, is known as *renormalized perturbation theory*. It should be contrasted with the procedure we used in Part 1, outlined at the beginning of this section, which is called *bare perturbation theory* (since the Feynman rules involve the bare mass and coupling constant). The two methods are completely equivalent. The differences between them are purely a matter of bookkeeping. You will get the same answers using either procedure, so you may choose whichever you find more convenient. In general, renormalized perturbation theory is technically easier to use, especially for multiloop diagrams; however, bare perturbation theory is sometimes easier for complicated one-loop calculations. We will use renormalized perturbation theory in most of the rest of this book.

One-Loop Structure of ϕ^4 Theory

To make more sense of the renormalization procedure, let us carry it out explicitly at the one-loop level.

First consider the basic two-particle scattering amplitude,

If we define $p = p_1 + p_2$, then the second diagram is

$$= \frac{(-i\lambda)^2}{2} \int \frac{d^4k}{(2\pi)^4} \frac{i}{k^2 - m^2} \frac{i}{(k+p)^2 - m^2}$$

$$\equiv (-i\lambda)^2 \cdot iV(p^2). \tag{10.20}$$

Note that p^2 is equal to the Mandelstam variable s. The next two diagrams are identical, except that s will be replaced by t and u. The entire amplitude is therefore

$$i\mathcal{M} = -i\lambda + (-i\lambda)^2 \left[iV(s) + iV(t) + iV(u) \right] - i\delta_\lambda. \tag{10.21}$$

According to our renormalization condition (10.19), this amplitude should equal $-i\lambda$ at $s = 4m^2$ and $t = u = 0$. We must therefore set

$$\delta_\lambda = -\lambda^2 \big[V(4m^2) + 2V(0) \big]. \tag{10.22}$$

(At higher orders, δ_λ will receive additional contributions.)

We can compute $V(p^2)$ explicitly using dimensional regularization. The procedure is exactly the same as in Section 7.5: Introduce a Feynman parameter, shift the integration variable, rotate to Euclidean space, and perform the momentum integral. We obtain

$$V(p^2) = \frac{i}{2} \int_0^1 dx \int \frac{d^d k}{(2\pi)^d} \frac{1}{\big[k^2 + 2xk \cdot p + xp^2 - m^2 \big]^2}$$

$$= \frac{i}{2} \int_0^1 dx \int \frac{d^d \ell}{(2\pi)^d} \frac{1}{\big[\ell^2 + x(1-x)p^2 - m^2 \big]^2} \qquad (\ell = k + xp)$$

$$= -\frac{1}{2} \int_0^1 dx \int \frac{d^d \ell_E}{(2\pi)^d} \frac{1}{\big[\ell_E^2 - x(1-x)p^2 + m^2 \big]^2} \qquad (\ell_E^0 = -i\ell^0)$$

$$= -\frac{1}{2} \int_0^1 dx \, \frac{\Gamma(2 - \frac{d}{2})}{(4\pi)^{d/2}} \frac{1}{\big[m^2 - x(1-x)p^2 \big]^{2-d/2}}$$

$$\xrightarrow[d \to 4]{} -\frac{1}{32\pi^2} \int_0^1 dx \Big(\frac{2}{\epsilon} - \gamma + \log(4\pi) - \log \big[m^2 - x(1-x)p^2 \big] \Big), \tag{10.23}$$

where $\epsilon = 4 - d$. The shift in the coupling constant (10.22) is therefore

$$\delta_\lambda = \frac{\lambda^2}{2} \frac{\Gamma(2 - \frac{d}{2})}{(4\pi)^{d/2}} \int_0^1 dx \left(\frac{1}{[m^2 - x(1-x)4m^2]^{2-d/2}} + \frac{2}{[m^2]^{2-d/2}} \right)$$

$$\xrightarrow[d \to 4]{} \frac{\lambda^2}{32\pi^2} \int_0^1 dx \Big(\frac{6}{\epsilon} - 3\gamma + 3\log(4\pi) - \log \big[m^2 - x(1-x)4m^2 \big] - 2\log \big[m^2 \big] \Big). \tag{10.24}$$

These expressions are divergent as $d \to 4$. But if we combine them according to (10.21), we obtain the finite (if rather complicated) result,

$$i\mathcal{M} = -i\lambda - \frac{i\lambda^2}{32\pi^2} \int_0^1 dx \left[\log \Big(\frac{m^2 - x(1-x)s}{m^2 - x(1-x)4m^2} \Big) + \log \Big(\frac{m^2 - x(1-x)t}{m^2} \Big) \right.$$

$$\left. + \log \Big(\frac{m^2 - x(1-x)u}{m^2} \Big) \right]. \tag{10.25}$$

To determine δ_Z and δ_m we must compute the two-point function. As in Section 7.2, let us define $-iM^2(p^2)$ as the sum of all one-particle-irreducible insertions into the propagator:

$$\text{—(1PI)—} = -iM^2(p^2). \qquad (10.26)$$

Then the full two-point function is given by the geometric series,

$$\text{—⬤—} = \text{————} + \text{—(1PI)—} + \text{—(1PI)—(1PI)—} + \cdots$$

$$= \frac{i}{p^2 - m^2 - M^2(p^2)}. \qquad (10.27)$$

The renormalization conditions (10.19) require that the pole in this full propagator occur at $p^2 = m^2$ and have residue 1. These two conditions are equivalent, respectively, to

$$M^2(p^2)\big|_{p^2=m^2} = 0 \quad \text{and} \quad \frac{d}{dp^2}M^2(p^2)\big|_{p^2=m^2} = 0. \qquad (10.28)$$

(To check the latter condition, expand M^2 about $p^2 = m^2$ in Eq. (10.27).)
  Explicitly, to one-loop order,

$$-iM^2(p^2) = \text{⟲} + \text{—⊗—}$$

$$= -i\lambda \cdot \frac{1}{2} \cdot \int \frac{d^dk}{(2\pi)^d} \frac{i}{k^2 - m^2} + i(p^2\delta_Z - \delta_m)$$

$$= -\frac{i\lambda}{2} \frac{1}{(4\pi)^{d/2}} \frac{\Gamma(1-\frac{d}{2})}{(m^2)^{1-d/2}} + i(p^2\delta_Z - \delta_m). \qquad (10.29)$$

Since the first term is independent of p^2, the result is rather trivial: Setting

$$\delta_Z = 0 \quad \text{and} \quad \delta_m = -\frac{\lambda}{2(4\pi)^{d/2}} \frac{\Gamma(1-\frac{d}{2})}{(m^2)^{1-d/2}} \qquad (10.30)$$

yields $M^2(p^2) = 0$ for all p^2, satisfying both of the conditions in (10.28).
  The first nonzero contributions to $M^2(p^2)$ and δ_Z are proportional to λ^2, coming from the diagrams

$$\text{—◯—} + \text{⊗̲◯} + \text{—⊗—} \qquad (10.31)$$

The second diagram contains the δ_λ counterterm, which we have already computed. It cancels ultraviolet divergences in the first diagram that occur when one of the loop momenta is large and the other is small. The third diagram is again the $(p^2\delta_Z - \delta_m)$ counterterm, and is fixed to order λ^2 by requiring that the remaining divergences (when both loop momenta become large) cancel. In Section 10.4 we will see an explicit example of the interplay of various counterterms in a two-loop calculation.

The vanishing of δ_Z at one-loop order is a special feature of ϕ^4 theory, which does not occur in more general theories of scalar fields. The Yukawa theory described in Section 4.7 gives an explicit example of a one-loop correction for which this counterterm is required.

In the Yukawa theory, the scalar field propagator receives corrections at order g^2 from a fermion loop diagram and the two propagator counterterms. Using the Feynman rules on p. 118 to compute the loop diagram, we find

$$-iM^2(p^2) = \;\; \substack{k+p \\ \text{---}\bigcirc\text{---}} \;\; + \;\; \text{---}\otimes\text{---}$$

$$= -(-ig)^2 \int \frac{d^d k}{(2\pi)^d} \, \text{tr}\left[\frac{i(\slashed{k}+\slashed{p}+m_f)}{(k+p)^2-m_f^2} \frac{i(\slashed{k}+m_f)}{k^2-m_f^2}\right] + i(p^2\delta_Z - \delta_m)$$

$$= -4g^2 \int \frac{d^d k}{(2\pi)^d} \frac{k\cdot(p+k)+m_f^2}{((p+k)^2-m_f^2)(k^2-m_f^2)} + i(p^2\delta_Z - \delta_m), \quad (10.32)$$

where m_f is the mass of the fermion that couples to the Yukawa field. To evaluate the integral, combine denominators and shift as in Eq. (10.23). Then the first term in the last line becomes

$$-4g^2 \int_0^1 dx \int \frac{d^d \ell}{(2\pi)^d} \frac{\ell^2 - x(1-x)p^2 + m_f^2}{(\ell^2 + x(1-x)p^2 - m_f^2)^2}$$

$$= -4g^2 \int_0^1 dx \frac{-i}{(4\pi)^{d/2}} \left(\frac{\frac{d}{2}\Gamma(1-\frac{d}{2})}{\Delta^{1-d/2}} - \frac{\Delta\,\Gamma(2-\frac{d}{2})}{\Delta^{2-d/2}}\right)$$

$$= \frac{4ig^2(d-1)}{(4\pi)^{d/2}} \int_0^1 dx \frac{\Gamma(1-\frac{d}{2})}{\Delta^{1-d/2}}, \quad (10.33)$$

where $\Delta = m_f^2 - x(1-x)p^2$.

Now we can see that both of the counterterms δ_m and δ_Z must take nonzero values in order to satisfy the renormalization conditions (10.28). To determine δ_m, we subtract the value of the loop diagram at $p^2 = m^2$ as before, so that

$$\delta_m = \frac{4g^2(d-1)}{(4\pi)^{d/2}} \int_0^1 dx \frac{\Gamma(1-\frac{d}{2})}{[m_f^2 - x(1-x)m^2]^{1-d/2}} + m^2\delta_Z. \quad (10.34)$$

To determine δ_Z, we cancel also the first derivative with respect to p^2 of the loop integral (10.33). This gives

$$\delta_Z = -\frac{4g^2(d-1)}{(4\pi)^{d/2}} \int\limits_0^1 dx \, \frac{x(1-x)\Gamma(2-\frac{d}{2})}{[m_f^2 - x(1-x)m^2]^{2-d/2}}$$

$$\xrightarrow[d\to 4]{} -\frac{3g^2}{4\pi^2} \int\limits_0^1 dx \, x(1-x)\left(\frac{2}{\epsilon} - \gamma - \frac{2}{3} + \log(4\pi) - \log[m_f^2 - x(1-x)m^2]\right).$$

$$(10.35)$$

Thus, in Yukawa theory, the propagator corrections at one-loop order require a quadratically divergent mass renormalization and a logarithmically divergent field strength renormalization. This is the usual situation in scalar field theories.

10.3 Renormalization of Quantum Electrodynamics

The procedure we followed in the previous section, yielding a "renormalized" perturbation theory formulated in terms of physically measurable parameters, can be summarized as follows:

1. Absorb the field-strength renormalizations into the Lagrangian by rescaling the fields.

2. Split each term of the Lagrangian into two pieces, absorbing the infinite and unobservable shifts into counterterms.

3. Specify the renormalization conditions, which define the physical masses and coupling constants and keep the field-strength renormalizations equal to 1.

4. Compute amplitudes with the new Feynman rules, adjusting the counterterms as necessary to maintain the renormalization conditions.

Let us now use this procedure to construct a renormalized perturbation theory for Quantum Electrodynamics.

The original QED Lagrangian is

$$\mathcal{L} = -\tfrac{1}{4}(F_{\mu\nu})^2 + \bar{\psi}(i\slashed{\partial} - m_0)\psi - e_0\bar{\psi}\gamma^\mu\psi A_\mu.$$

Computing the electron and photon propagators with this Lagrangian, we would find expressions of the general form

$$= \frac{iZ_2}{\slashed{p} - m} + \cdots ; \qquad \qquad = \frac{-iZ_3 g_{\mu\nu}}{q^2} + \cdots .$$

(We found just such expressions in the explicit one-loop calculations of Chapter 7.) To absorb Z_2 and Z_3 into $\mathcal{L}$, and hence eliminate them from formula (7.45) for the S-matrix, we substitute $\psi = Z_2^{1/2}\psi_r$ and $A^\mu = Z_3^{1/2} A_r^\mu$. Then the Lagrangian becomes

$$\mathcal{L} = -\tfrac{1}{4}Z_3(F_r^{\mu\nu})^2 + Z_2\bar{\psi}_r(i\slashed{\partial} - m_0)\psi_r - e_0 Z_2 Z_3^{1/2}\bar{\psi}_r\gamma^\mu\psi_r A_{r\mu}. \qquad (10.36)$$

We can introduce the physical electric charge e, measured at large distances $(q = 0)$, by defining a scaling factor Z_1 as follows:[†]

$$e_0 Z_2 Z_3^{1/2} = e Z_1. \qquad (10.37)$$

If we let m be the physical mass (the location of the pole in the electron propagator), then we can split each term of the Lagrangian into two pieces as follows:

$$\mathcal{L} = -\tfrac{1}{4}(F_r^{\mu\nu})^2 + \bar{\psi}_r(i\slashed{\partial} - m)\psi_r - e\bar{\psi}_r \gamma^\mu \psi_r A_{r\mu}$$
$$- \tfrac{1}{4}\delta_3(F_r^{\mu\nu})^2 + \bar{\psi}_r(i\delta_2 \slashed{\partial} - \delta_m)\psi_r - e\delta_1 \bar{\psi}_r \gamma^\mu \psi_r A_{r\mu}, \qquad (10.38)$$

where

$$\delta_3 = Z_3 - 1, \qquad \delta_2 = Z_2 - 1,$$

$$\delta_m = Z_2 m_0 - m, \qquad \text{and} \qquad \delta_1 = Z_1 - 1 = (e_0/e)Z_2 Z_3^{1/2} - 1.$$

The Feynman rules for renormalized QED are shown in Fig. 10.4. In addition to the familiar propagators and vertex, there are three counterterm vertices. The ee and $ee\gamma$ counterterm vertices can be read directly from the Lagrangian (10.38). To derive the two-photon counterterm, integrate $-\tfrac{1}{4}(F_{\mu\nu})^2$ by parts to obtain $-\tfrac{1}{2}A_\mu(-\partial^2 g^{\mu\nu} + \partial^\mu \partial^\nu)A_\nu$; this gives the expression shown in the figure. In the remainder of the book, when we set up renormalized perturbation theory, we will drop the subscript r used here to distinguish the rescaled fields.

Each of the four counterterm coefficients must be fixed by a renormalization condition. The four conditions that we require have already been stated implicitly: Two of them fix the electron and photon field-strength renormalizations to 1, while the other two define the physical electron mass and charge. To write these conditions more explicitly, recall our notation from Chapters 6 and 7:

$$\mu \sim\!\!\!\!\!\!\text{1PI}\!\!\!\!\!\!\sim \nu \qquad = i\Pi^{\mu\nu}(q) = i(g^{\mu\nu}q^2 - q^\mu q^\nu)\Pi(q^2),$$

$$\blacktriangleleft\!\!\!\!\text{1PI}\!\!\!\!- \qquad = -i\Sigma(\slashed{p}), \qquad (10.39)$$

$$\left(\begin{array}{c} \text{amputated} \end{array}\right) \qquad = -ie\Gamma^\mu(p', p).$$

[†]Since we define e by the renormalization condition $\Gamma^\mu(q = 0) = \gamma^\mu$, the factor of Z_1 in the Lagrangian must cancel the multiplicative correction factor that arises from loop corrections. Therefore this definition of Z_1 is equivalent to that given in Eq. (7.47).

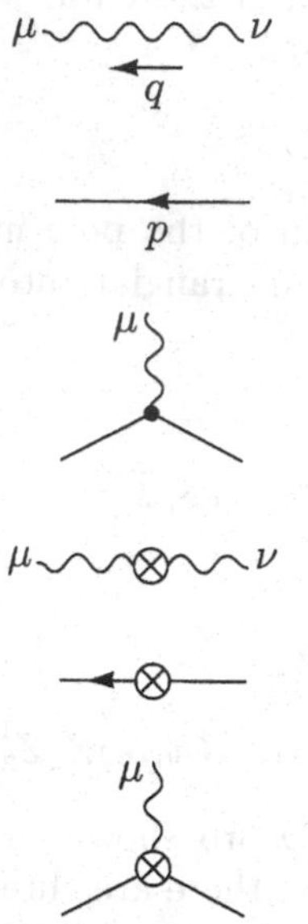

$$\mu \sim\!\!\sim\!\!\sim\!\!\sim\!\!\sim \nu \qquad = \frac{-ig_{\mu\nu}}{q^2 + i\epsilon} \qquad \text{(Feynman gauge)}$$

$$\xleftarrow[\;\;p\;\;]{} \qquad = \frac{i}{\not{p} - m + i\epsilon}$$

$$= -ie\gamma^\mu$$

$$\mu \sim\!\!\sim\!\!\otimes\!\!\sim\!\!\sim \nu \qquad = -i(g^{\mu\nu}q^2 - q^\mu q^\nu)\delta_3$$

$$\xleftarrow{}\!\!\otimes\!\!\xrightarrow{} \qquad = i(\not{p}\delta_2 - \delta_m)$$

$$= -ie\gamma^\mu \delta_1$$

Figure 10.4. Feynman rules for Quantum Electrodynamics in renormalized perturbation theory.

These amplitudes are now to be computed in renormalized perturbation theory; that is, we are now redefining $\Pi(q^2)$, $\Sigma(\not{p})$, and $\Gamma(p',p)$ to include counterterm vertices. Furthermore, the new definition of Γ involves the physical electron charge. With this notation, the four conditions are

$$\Sigma(\not{p} = m) = 0;$$

$$\left.\frac{d}{d\not{p}}\Sigma(\not{p})\right|_{\not{p}=m} = 0;$$

$$\Pi(q^2 = 0) = 0; \tag{10.40}$$

$$-ie\Gamma^\mu(p' - p = 0) = -ie\gamma^\mu.$$

The first condition fixes the electron mass at m, while the next two fix the residues of the electron and photon propagators at 1. Given these conditions, the final condition fixes the electron charge to be e.

One-Loop Structure of QED

The four conditions (10.40) allow us to determine the four counterterms in (10.38) in terms of the values of loop diagrams. In Chapters 6 and 7 we computed all of the diagrams required to carry out this determination to one-loop order. We will now collect these results and find explicit expressions for the renormalization constants of QED to order α. For overall consistency, we will use dimensional regularization to control ultraviolet divergences, and a photon mass μ to control infrared divergences. In Part I, we computed the vertex and self-energy diagrams using the Pauli-Villars regularization scheme, before introducing dimensional regularization. Now we have an opportunity to quote the values of these diagrams as computed with dimensional regularization.

The first two conditions involve the electron self-energy. We evaluated the one-loop diagram contributing to $\Sigma(p)$, using a Pauli-Villars regulator, in Section 7.1; the result is given in Eq. (7.19). If we re-evaluate the diagram in dimensional regularization, we find some additional terms in the Dirac algebra from the modified contraction identities (7.89). Taking these terms into account, we find for this diagram ($\epsilon = 4 - d$)

$$
-i\Sigma_2(p) = -i\frac{e^2}{(4\pi)^{d/2}} \int\limits_0^1 dx \, \frac{\Gamma(2-\frac{d}{2})}{((1-x)m^2 + x\mu^2 - x(1-x)p^2)^{2-d/2}}
$$

$$
\times \left((4-\epsilon)m - (2-\epsilon)x\not{p}\right). \qquad (10.41)
$$

Therefore, according to the first of conditions (10.40),

$$
m\delta_2 - \delta_m = \Sigma_2(m) = \frac{e^2 m}{(4\pi)^{d/2}} \int\limits_0^1 dx \, \frac{\Gamma(2-\frac{d}{2}) \cdot (4 - 2x - \epsilon(1-x))}{((1-x)^2 m^2 + x\mu^2)^{2-d/2}}. \qquad (10.42)
$$

Similarly, the second of conditions (10.40) determines δ_2:

$$
\delta_2 = \frac{d}{d\not{p}}\Sigma_2(m)
$$

$$
= -\frac{e^2}{(4\pi)^{d/2}} \int\limits_0^1 dx \, \frac{\Gamma(2-\frac{d}{2})}{((1-x)^2 m^2 + x\mu^2)^{2-d/2}}
$$

$$
\times \left[(2-\epsilon)x - \frac{\epsilon}{2}\frac{2x(1-x)m^2}{(1-x)^2 m^2 + x\mu^2}(4 - 2x - \epsilon(1-x))\right]. \qquad (10.43)
$$

Notice that the second term in the brackets gives a finite result as $\epsilon \to 0$, because it multiplies the divergent gamma function.

The third condition of (10.40) requires the value (7.90) of the photon self-energy diagram:

$$
\Pi_2(q^2) = -\frac{e^2}{(4\pi)^{d/2}} \int\limits_0^1 dx \, \frac{\Gamma(2-\frac{d}{2})}{(m^2 - x(1-x)q^2)^{2-d/2}}(8x(1-x)).
$$

Then

$$
\delta_3 = \Pi_2(0) = -\frac{e^2}{(4\pi)^{d/2}} \int\limits_0^1 dx \, \frac{\Gamma(2-\frac{d}{2})}{(m^2)^{2-d/2}}(8x(1-x)). \qquad (10.44)
$$

The last condition requires the value of the electron vertex function, computed in Section 6.3. Again, we will rework the diagram in dimensional regularization. Then the shift in the form factor $F_1(q^2)$ (6.56) becomes

$$\delta F_1(q^2) = \frac{e^2}{(4\pi)^{d/2}} \int dx\, dy\, dz\, \delta(x+y+z-1) \left[\frac{\Gamma(2-\frac{d}{2})}{\Delta^{2-d/2}} \frac{(2-\epsilon)^2}{2} \right.$$
$$\left. + \frac{\Gamma(3-\frac{d}{2})}{\Delta^{3-d/2}} \left(q^2[2(1-x)(1-y) - \epsilon xy] + m^2[2(1-4z+z^2) - \epsilon(1-z)^2] \right) \right],$$

$$(10.45)$$

where $\Delta = (1-z)^2 m^2 + z\mu^2 - xyq^2$ as before. The fourth renormalization condition then determines

$$\delta_1 = -\delta F_1(0) = -\frac{e^2}{(4\pi)^{d/2}} \int dz\, (1-z) \left[\frac{\Gamma(2-\frac{d}{2})}{((1-z)^2 m^2 + z\mu^2)^{2-d/2}} \frac{(2-\epsilon)^2}{2} \right.$$
$$\left. + \frac{\Gamma(3-\frac{d}{2})}{((1-z)^2 m^2 + z\mu^2)^{3-d/2}} [2(1-4z+z^2) - \epsilon(1-z)^2] m^2 \right].$$

$$(10.46)$$

Using an integration by parts similar to that following Eq. (7.32), one can show explicitly from (10.46) and (10.43) that $\delta_1 = \delta_2$, that is, that $Z_1 = Z_2$ to order α. As in our previous derivations, this formula follows from the Ward identity. The Lagrangian (10.38), with counterterms set to zero, is gauge invariant. If the regulator is also gauge invariant (and we do use dimensional regularization), this implies the Ward identity for diagrams without counterterm vertices. In particular, this implies that $\delta F_1(0) = -d\Sigma_2/d\slashed{p}|_m$. Then the counterterms δ_1 and δ_2, which are required to cancel these two factors, will be set equal.

By continuing this argument, it is straightforward to construct a full diagrammatic proof that $\delta_1 = \delta_2$, to all orders in renormalized perturbation theory, using the method we applied in Section 7.4 to prove the Ward-Takahashi identity in bare perturbation theory. With a generalization of the argument given there, one can show that the diagrammatic identity (7.68) holds for diagrams that include counterterm vertices in loops. Thus, if the counterterms δ_1 and δ_2 are determined up to order α^n, the unrenormalized vertex diagram at $q^2 = 0$ equals the derivative of the unrenormalized self-energy diagram on-shell in order α^{n+1}. To satisfy the renormalization conditions (10.40), we must then set the counterterms δ_1 and δ_2 equal to order α^{n+1}. This recursive argument gives yet another proof that $Z_1 = Z_2$ to all orders in QED perturbation theory.

The relation (10.37) between the bare and renormalized charge

$$e = \frac{Z_2}{Z_1} Z_3^{1/2} e_0 \tag{10.47}$$

gives a further physical interpretation of the identity $Z_1 = Z_2$. Using the identity, we can rewrite (10.47) as

$$e = \sqrt{Z_3}e_0,$$

which is just the relation (7.76) that we derived by a diagrammatic argument in Section 7.5. This says that the relation between the bare and renormalized electric charge depends only on the photon field strength renormalization, not on quantities particular to the electron. To see the importance of this observation, consider writing the renormalized quantum electrodynamics with two species of charged particles, say, electrons and muons. Then, in addition to (10.37), we will have a relation for the photon-muon vertex:

$$eZ_2'^{-1}Z_3^{-1/2} = e_0 Z_1'^{-1}, \tag{10.48}$$

where Z_1' and Z_2' are the vertex and field strength renormalizations for the muon. Each of these two constants depends on the mass of the muon, so (10.48) threatens to give a different relation between e_0 and e from the one written in (10.47). However, the Ward identity forces the factors Z_1' and Z_2' to cancel out of this relation, leaving over a universal electric charge which has the same value for all species.

10.4 Renormalization Beyond the Leading Order

In the last two sections we have developed an algorithm for computing scattering amplitudes to any order in a renormalizable field theory. We have seen explicitly that this algorithm yields finite results at the one-loop level in both ϕ^4 theory and QED. According to the naive analysis of Section 10.1, the algorithm should also work at higher orders. But that analysis ignored many of the intricacies of multiloop diagrams; specifically, it ignored the fact that diagrams can contain divergent subdiagrams.

When an otherwise finite diagram contains a divergent subdiagram, the treatment of the divergence is relatively straightforward. For example, the sum of diagrams

$$\tag{10.49}$$

is finite: The divergence in the photon propagator cancels just as when this propagator occurs in a tree diagram. The finite sum of the two propagator diagrams gives an integrand for the outer loop that falls off fast enough that this integral still converges.

A more difficult situation occurs when we have *nested* or *overlapping* divergences, that is, when two divergent loops share a propagator. Some examples of diagrams with overlapping divergences are

in ϕ^4 theory;

in QED.

To see the difficulty, consider the photon self-energy diagram:

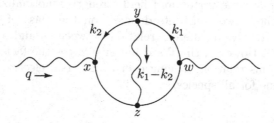

One contribution to this diagram comes from the region of momentum space where k_2 is very large. This means that, in position space, x, y, and z are very close together, while w can be farther away. In this region we can think of the virtual photon as giving a correction to the vertex at x. We saw in Section 6.3 that this vertex correction is logarithmically divergent, of the form

$$\mu \sim\!\!\!\!\!\!\!\!\!\underset{x}{\bigvee} \quad \sim -ie\gamma^\mu \cdot \alpha \log \Lambda^2$$

in the limit $\Lambda \to \infty$. Plugging this vertex into the rest of the diagram and integrating over k_1, we obtain an expression identical to the one-loop photon self-energy correction $\Pi_2(q^2)$, displayed in (7.90), multiplied by the additional logarithmic divergence:

$$\sim \alpha(g^{\mu\nu}q^2 - q^\mu q^\nu)\Pi_2(q^2) \cdot \alpha \log \Lambda^2$$

$$\tag{10.50}$$

$$\sim \alpha(g^{\mu\nu}q^2 - q^\mu q^\nu)(\log \Lambda^2 + \log q^2) \cdot \alpha \log \Lambda^2.$$

The $\log^2 \Lambda^2$ term comes from the region where both k_1 and k_2 are large, while the $\log q^2 \log \Lambda^2$ term comes from the region where k_2 is large but k_1 is small. Another such term would come from the region where k_1 is large but k_2 is small.

The appearance of terms proportional to $\Pi_2(q^2) \cdot \log \Lambda^2$ in the two-loop vacuum polarization diagram contradicts our naive argument, based on the criterion of the superficial degree of divergence, that the divergent terms of a Feynman integral are always simple polynomials in q^2. We will refer to divergences multiplying only polynomials in q^2 as *local divergences*, since their Fourier transforms back to position space are delta functions or derivatives of delta functions. We will call the new, nonpolynomial, term a *nonlocal* divergence. Fortunately, our derivation of the nonlocal divergent term gave this term a physical interpretation: It is a local divergence surrounded by an ordinary, nondivergent, quantum field theory process.

If this picture accurately describes all of the divergent terms of the two-loop diagram, we should expect that these divergences are canceled by two types of counterterm diagrams. First, we can build diagrams of order α^2 by inserting the order-α counterterm vertex into the one-loop vacuum polarization diagram:

These diagrams should cancel the nonlocal divergence in (10.50) and the corresponding contribution from the region where k_1 is large and k_2 is small. In fact, a detailed analysis shows that the sum of the original diagram and these two counterterm diagrams contains only local divergences. Once these diagrams are added, the only divergence that remains is a local one, which can be canceled by the diagram

that is, by adding an order-α^2 term to δ_3.

We can extend the lessons of this example to a general picture of the divergences of higher-loop Feynman diagrams and their cancellation. A given diagram may contain local divergences, as predicted by the analysis of Section 10.1. It may also contain nonlocal divergences due to divergent subgraphs embedded in loops carrying small momenta. These divergences are canceled by diagrams in which the divergent subgraphs are replaced by their counterterm vertices. One might still ask two questions: First, does this procedure remove all nonlocal divergences? Second, does this procedure preserve the finiteness of amplitudes, such as (10.49), that are not expected to be divergent by the superficial criteria of Section 10.1? To answer these questions requires an intricate study of nested Feynman integrals. The general analysis was begun by Bogoliubov and Parasiuk, completed by Hepp, and elegantly refined by

Zimmermann;[‡] they showed that the answer to both questions is yes. Their result, known as the BPHZ theorem, states that, for a general renormalizable quantum field theory, to any order in perturbation theory, all divergences are removed by the counterterm vertices corresponding to superficially divergent amplitudes. In other words, any superficially renormalizable quantum field theory is in fact rendered finite when one performs renormalized perturbation theory with the complete set of counterterms.

The proof of the BPHZ theorem is quite technical, and we will not include it in this book. Instead, we will investigate one detailed example of a two-loop calculation, which demonstrates explicitly the appearance and cancellation of nonlocal divergences.

10.5 A Two-Loop Example

To illustrate the issues discussed in the previous section, let us consider the two-loop contribution to the four-point function in ϕ^4 theory. There are 16 relevant diagrams, shown in Fig. 10.5. (There are also several diagrams involving the one-loop correction to the propagator. But each of these is exactly canceled by its counterterm, as we saw in Eq. (10.29), so we can just ignore them.) Fortunately, many of the diagrams are simply related to each other. Crossing symmetry reduces the number of distinct diagrams to only six,

$$\text{(10.51)}$$

where the last diagram denotes only the s-channel piece of the second-order vertex counterterm. If this sum of diagrams is finite, then simply replacing s with t or u gives a finite result for the remaining diagrams.

The value of the last diagram in (10.51) is just a constant, which we can freely adjust to absorb any divergent terms that are independent of the external momenta. Our goal, therefore, is to show that all momentum-dependent divergent terms cancel among the remaining five diagrams.

The fourth and fifth diagrams in (10.51) involve the one-loop vertex counterterm, which we computed in Eq. (10.24). Let us briefly recall that computation. We defined $iV(p^2)$ as the fundamental loop integral,

$$\text{(diagram)} = (-i\lambda)^2 \cdot iV(p^2) = (-i\lambda)^2 \left[-\frac{i}{2} \frac{\Gamma(2-\frac{d}{2})}{(4\pi)^{d/2}} \int_0^1 dx \frac{1}{\left[m^2 - x(1-x)p^2 \right]^{2-d/2}} \right].$$

$$\text{(10.52)}$$

[‡]N. N. Bogoliubov and O. S. Parasiuk, *Acta Math.* **97**, 227 (1957); K. Hepp, *Comm. Math. Phys.* **2**, 301 (1966); W. Zimmermann, in Deser, et. al. (1970).

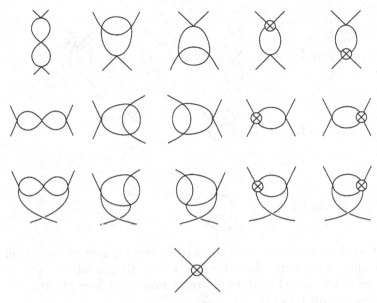

Figure 10.5. The two-loop contributions to the four-point function in ϕ^4 theory. Note that the diagrams in the first three lines are related to each other by crossing, being in the s-, t-, and u-channels, respectively. The last two diagrams in each of these lines involve the $\mathcal{O}(\lambda^2)$ vertex counterterm, while the final diagram is the $\mathcal{O}(\lambda^3)$ contribution to the vertex counterterm.

The counterterm, according to the renormalization condition (10.19), had to cancel the three one-loop diagrams (one for each channel) at threshold ($s = 4m^2$, $t = u = 0$); thus we found

$$\text{[diagram]} = -i\delta_\lambda = (-i\lambda)^2 \big[-iV(4m^2) - 2iV(0) \big].$$

For our present purposes it will be convenient to separate the two terms of this expression. Let us therefore define

$$\text{[diagram]}_s = (-i\lambda)^2 \cdot -iV(4m^2); \qquad \text{[diagram]}_{t+u} = (-i\lambda)^2 \cdot -2iV(0).$$

We can now divide the first five diagrams in (10.51) into three groups, as follows:

Group I:

Group II:

Group III:

We will find that all divergent terms that depend on momentum cancel separately within each group. Since Groups II and III are related by a simple interchange of initial and final momenta, it suffices to demonstrate this cancellation for Groups I and II.

Group I is actually quite easy, since each diagram factors into a product of objects we have already computed. Referring to Eq. (10.52), we have

$= (-i\lambda)^3 \cdot \left[iV(p^2)\right]^2;$

$= (-i\lambda)^3 \cdot iV(p^2) \cdot -iV(4m^2).$

The sum of all three diagrams is therefore

$$(-i\lambda)^3 \left(\left[iV(p^2)\right]^2 - 2iV(p^2)iV(4m^2) \right)$$

$$= (-i\lambda)^3 \left(-\left[V(p^2) - V(4m^2)\right]^2 + \left[V(4m^2)\right]^2 \right).$$

(10.53)

But the difference $V(p^2) - V(4m^2)$ is finite, as was required for the cancellation of divergences in the one-loop calculation:

$$V(p^2) - V(4m^2) = \frac{1}{32\pi^2} \int\limits_0^1 dx \, \log\left(\frac{m^2 - x(1-x)p^2}{m^2 - x(1-x)4m^2} \right).$$

The only remaining divergence is in the term $[V(4m^2)]^2$, which is independent of momentum and can therefore be absorbed into the second-order counterterm in (10.51).

Two general properties of result (10.53) are worth noting. First, the divergent piece (and hence the $\mathcal{O}(\lambda^3)$ vertex counterterm) is proportional to

$$\left[V(4m^2)\right]^2 \propto \left[\Gamma(2-\tfrac{d}{2})\right]^2 \xrightarrow[d\to 4]{} \left(\frac{2}{\epsilon}\right)^2 \qquad \text{for } d = 4 - \epsilon.$$

This is a double pole, in contrast to the simple pole we found for the one-loop counterterm. Higher-loop diagrams will similarly have higher-order poles, but in all cases the divergent terms are momentum-independent constants. Second, consider the large-momentum limit,

$$V(p^2) - V(4m^2) \underset{p^2\to\infty}{\sim} \log\frac{p^2}{m^2}.$$

The two-loop vertex is proportional to $\log^2(p^2/m^2)$. A diagram of this structure with n loops will have the form

$$\sim \lambda^{n+1}\left(\log\frac{p^2}{m^2}\right)^n.$$

This asymptotic behavior is actually a generic property of multiloop diagrams, which we will explore in more detail in Chapter 12.

Now consider the more difficult diagram, from Group II:

$$= (-i\lambda)^3 \int \frac{d^d k}{(2\pi)^d} \frac{i}{k^2 - m^2} \frac{i}{(k+p)^2 - m^2}\, iV\big((k+p_3)^2\big).$$

$$\tag{10.54}$$

In evaluating this diagram, we will combine denominators in the manner that makes it most straightforward to extract the divergent terms, at the price of complicating the evaluation of the finite parts. Another approach to the calculation of this diagram is discussed in Problem 10.4.

To begin the evaluation of (10.54), combine the pair of denominators shown explicitly, and substitute expression (10.52) for $V(p^2)$. This gives the expression

$$-\frac{\lambda^3}{2}\frac{\Gamma(2-\tfrac{d}{2})}{(4\pi)^{d/2}} \int_0^1 dx \int_0^1 dy \int \frac{d^d k}{(2\pi)^d} \frac{1}{\left[k^2 + 2yk\cdot p + yp^2 - m^2\right]^2}$$

$$\times \frac{1}{\left[m^2 - x(1-x)(k+p_3)^2\right]^{2-\frac{d}{2}}}. \tag{10.55}$$

It is possible to combine this pair of denominators by using the identity

$$\frac{1}{A^\alpha B^\beta} = \int_0^1 dw \, \frac{w^{\alpha-1}(1-w)^{\beta-1}}{\left[wA + (1-w)B\right]^{\alpha+\beta}} \frac{\Gamma(\alpha+\beta)}{\Gamma(\alpha)\Gamma(\beta)}. \tag{10.56}$$

This is a special case of the formula quoted in Section 6.3, Eq. (6.42). To prove it, change variables in the integral:

$$z \equiv \frac{wA}{wA + (1-w)B}, \quad (1-z) = \frac{(1-w)B}{wA + (1-w)B}, \quad dz = \frac{AB \, dw}{\left[wA + (1-w)B\right]^2},$$

so that

$$\int_0^1 dw \, \frac{w^{\alpha-1}(1-w)^{\beta-1}}{\left[wA + (1-w)B\right]^{\alpha+\beta}} = \frac{1}{A^\alpha B^\beta} \int_0^1 dz \, z^{\alpha-1}(1-z)^{\beta-1} = \frac{1}{A^\alpha B^\beta} B(\alpha, \beta),$$

where $B(\alpha, \beta)$ is the beta function, Eq. (7.82). The more general identity (6.42) can be proved by induction.

Applying identity (10.56) to (10.55), we obtain

$$= -\frac{\lambda^3}{2} \frac{\Gamma(4-\frac{d}{2})}{(4\pi)^{d/2}} \int_0^1 dx \int_0^1 dy \int_0^1 dw \int \frac{d^d k}{(2\pi)^d}$$

$$\times \frac{w^{1-\frac{d}{2}}(1-w)}{\left(w\left[m^2-x(1-x)(k+p_3)^2\right] + (1-w)\left[m^2-k^2-2yk\cdot p-yp^2\right]\right)^{4-\frac{d}{2}}}. \tag{10.57}$$

Completing the square in the denominator yields a polynomial of the form

$$-\left[(1-w) + wx(1-x)\right]\ell^2 - P^2 + m^2, \tag{10.58}$$

where ℓ is a shifted momentum variable and P^2 is a rather complicated function of p, p_3, and the various Feynman parameters. It will only be important for this analysis that, as $w \to 0$,

$$P^2(w) = y(1-y)p^2 + \mathcal{O}(w), \tag{10.59}$$

and this can be seen easily from (10.57). Changing variables to ℓ, Wick-rotating, and performing the integral, we eventually obtain

$$= -\frac{i\lambda^3}{2(4\pi)^d} \int_0^1 dx \int_0^1 dy \int_0^1 dw \, \frac{w^{1-\frac{d}{2}}(1-w)}{\left[1 - w + wx(1-x)\right]^{d/2}} \frac{\Gamma(4-d)}{(m^2 - P^2)^{4-d}}. \tag{10.60}$$

This expression has one obvious pole as $d \to 4$, coming from the gamma function. However, it also has a less obvious pole, coming from the zero end of the w integral. Let us write (10.60) as

$$\int\limits_0^1 dw\ w^{1-\frac{d}{2}}\ f(w),$$

where $f(w)$ incorporates all the factors not displayed explicitly. To isolate the pole at $w = 0$, we can add and subtract $f(0)$:

$$\int\limits_0^1 dw\ w^{1-\frac{d}{2}}\ f(w) = \int\limits_0^1 dw\ w^{1-\frac{d}{2}}\ f(0) + \int\limits_0^1 dw\ w^{1-\frac{d}{2}}\ [f(w) - f(0)]. \quad (10.61)$$

The second piece is

$$-\frac{i\lambda^3\Gamma(4{-}d)}{2(4\pi)^d}\int\limits_0^1 dx \int\limits_0^1 dy \int\limits_0^1 dw\ w^{1-\frac{d}{2}}$$

$$\times\left(\frac{(1{-}w)}{\left[1 - w + wx(1{-}x)\right]^{d/2}}\frac{1}{\left[m^2 - P^2(w)\right]^{4-d}} - \frac{1}{\left[m^2 - P^2(0)\right]^{4-d}}\right).$$

This term has only a simple pole as $d \to 4$; the residue of the pole is a momentum-independent constant, obtained by setting $d = 4$ everywhere except in $\Gamma(4{-}d)$. We can therefore absorb this divergence into the $\mathcal{O}(\lambda^3)$ vertex counterterm. (The finite part of this expression has a very complicated dependence on momentum, but we do not need to work this out to complete our argument.)

We are left with only the first term of (10.61). This expression contains only $P^2(0)$, which is given by (10.59). The w integral in this term is straightforward, and the x integral is trivial. With $\epsilon = 4 - d$, our remaining expression is

$$-\frac{i\lambda^3}{2(4\pi)^d}\left(\frac{2}{\epsilon}\right)\int\limits_0^1 dy\ \frac{\Gamma(\epsilon)}{\left[m^2 - y(1{-}y)p^2\right]^\epsilon}$$

$$\xrightarrow[d\to 4]{}\ -\frac{i\lambda^3}{2(4\pi)^4}\left(\frac{2}{\epsilon}\right)\int\limits_0^1 dy\left(\frac{1}{\epsilon} - \gamma + \log(4\pi) - \log\left[m^2 - y(1{-}y)p^2\right]\right),$$

$$(10.62)$$

where we have kept only the divergent terms in the second line. The logarithm, multiplied by the pole $2/\epsilon$, is the nonlocal divergence that we worried about in Section 10.4.

Fortunately, we must still add to this the "$t + u$" counterterm diagram of Group II. The computation of that diagram is by now a straightforward process:

$$
\begin{aligned}
\vcenter{\hbox{(diagram with label $t+u$)}} \quad &= (-i\lambda)^3 \cdot -2iV(0) \cdot iV(p^2) \\[1em]
&= \frac{i\lambda^3}{2(4\pi)^d} \int_0^1 dy\, \frac{\Gamma(2-\frac{d}{2})}{\left[m^2\right]^{2-d/2}} \frac{\Gamma(2-\frac{d}{2})}{\left[m^2 - y(1-y)p^2\right]^{2-d/2}} \\[1em]
&\xrightarrow[d\to 4]{} \frac{i\lambda^3}{2(4\pi)^4} \int_0^1 dy \left(\frac{2}{\epsilon} - \gamma + \log(4\pi) - \log m^2\right) \\[1em]
&\qquad\qquad \times \left(\frac{2}{\epsilon} - \gamma + \log(4\pi) - \log\left[m^2 - y(1-y)p^2\right]\right). \quad (10.63)
\end{aligned}
$$

(Again we have dropped finite terms from the last line.) This expression also contains a nonlocal divergence, given by the first pole times the second logarithm. It exactly cancels the nonlocal divergence in (10.62). The remaining terms are all either finite, or divergent but independent of momentum. This completes the proof that the two-loop contribution to the four-point function is finite.

The two features of the Group I diagrams appear here in Group II as well. The divergent pieces of (10.62) and (10.63) contain double poles that do not cancel, so we again find that the second-order vertex counterterm must contain a double pole. The finite pieces of (10.62) and (10.63) contain double logarithms, so we again find that the two-loop amplitude behaves as $\lambda^3 \log^2 p^2$ as $p \to \infty$.

Problems

10.1 One-loop structure of QED. In Section 10.1 we argued from general principles that the photon one-point and three-point functions vanish, while the four-point function is finite.

   (a) Verify directly that the one-loop diagram contributing to the one-point function vanishes. There are two Feynman diagrams contributing to the three-point function at one-loop order. Show that these cancel. Show that the diagrams contributing to any n-point photon amplitude, for n odd, cancel in pairs.

   (b) The photon four-point amplitude is a sum of six diagrams. Show explicitly that the potential logarithmic divergences of these diagrams cancel.

10.2 Renormalization of Yukawa theory. Consider the pseudoscalar Yukawa Lagrangian,
$$
\mathcal{L} = \tfrac{1}{2}(\partial_\mu \phi)^2 - \tfrac{1}{2}m^2\phi^2 + \bar\psi(i\slashed{\partial} - M)\psi - ig\bar\psi\gamma^5\psi\phi,
$$
where ϕ is a real scalar field and ψ is a Dirac fermion. Notice that this Lagrangian is invariant under the parity transformation $\psi(t,\mathbf{x}) \to \gamma^0\psi(t,-\mathbf{x})$, $\phi(t,\mathbf{x}) \to -\phi(t,-\mathbf{x})$,

in which the field ϕ carries odd parity.

(a) Determine the superficially divergent amplitudes and work out the Feynman rules for renormalized perturbation theory for this Lagrangian. Include all necessary counterterm vertices. Show that the theory contains a superficially divergent 4ϕ amplitude. This means that the theory cannot be renormalized unless one includes a scalar self-interaction,

$$\delta\mathcal{L} = \frac{\lambda}{4!}\phi^4,$$

and a counterterm of the same form. It is of course possible to set the renormalized value of this coupling to zero, but that is not a natural choice, since the counterterm will still be nonzero. Are any further interactions required?

(b) Compute the divergent part (the pole as $d \to 4$) of each counterterm, to the one-loop order of perturbation theory, implementing a sufficient set of renormalization conditions. You need not worry about finite parts of the counterterms. Since the divergent parts must have a fixed dependence on the external momenta, you can simplify this calculation by choosing the momenta in the simplest possible way.

10.3 Field-strength renormalization in ϕ^4 theory. The two-loop contribution to the propagator in ϕ^4 theory involves the three diagrams shown in (10.31). Compute the first of these diagrams in the limit of zero mass for the scalar field, using dimensional regularization. Show that, near $d = 4$, this diagram takes the form:

$$\underset{\displaystyle}{\bigcirc} \quad = -ip^2 \cdot \frac{\lambda^2}{12(4\pi)^4}\left[-\frac{1}{\epsilon} + \log p^2 + \cdots\right],$$

with $\epsilon = 4 - d$. The coefficient in this equation involves a Feynman parameter integral that can be evaluated by setting $d = 4$. Verify that the second diagram in (10.31) vanishes near $d = 4$. Thus the first diagram should contain a pole only at $\epsilon = 0$, which can be canceled by a field-strength renormalization counterterm.

10.4 Asymptotic behavior of diagrams in ϕ^4 theory. Compute the leading terms in the S-matrix element for boson-boson scattering in ϕ^4 theory in the limit $s \to \infty$, t fixed. Ignore all masses on internal lines, and keep external masses nonzero only as infrared regulators where these are needed. Show that

$$i\mathcal{M}(s,t) \sim -i\lambda - i\frac{\lambda^2}{(4\pi)^2}\log s - i\frac{5\lambda^3}{2(4\pi)^4}\log^2 s + \cdots.$$

Notice that ignoring the internal masses allows some pleasing simplifications of the Feynman parameter integrals.

Chapter 11

Renormalization and Symmetry

Now that we have determined the general structure of the ultraviolet divergences of quantum field theories, it would seem natural to continue investigating the implications of these divergences in Feynman diagram calculations. However, we will now put this issue aside until Chapter 12 and set off in what may seem an unrelated direction. In Chapter 8 and in Section 9.3, we noted the formal relation between quantum field theory and statistical mechanics. The closest formal analogue of a scalar field theory was seen to be the continuum description of a ferromagnet or some other system that allows a second-order phase transition. This analogy raises the possibility that in quantum field theory as well it may be possible for the field to take on a nonzero global value. As in a magnet, this global field might have a directional character, and thus violate a symmetry of the Lagrangian. In such a case, we say that the field theory has *hidden* or *spontaneously broken* symmetry. We devote this chapter to an analysis of this mechanism of symmetry violation.

Spontaneously broken symmetry is a central concept in the study of quantum field theory, for two reasons. First, it plays a major role in the applications of quantum field theory to Nature. In this book, we will see two very different examples of such applications: Chapter 13 will apply the theory of hidden symmetry to statistical mechanics, specifically to the behavior of thermodynamic variables near second-order phase transitions. Later, in Chapter 20, we will see that hidden symmetry is an essential ingredient in the theory of the weak interactions. Spontaneous symmetry breaking also finds applications in the theory of the strong interactions, and in the search for unified models of fundamental physics.

But spontaneous symmetry breaking is also interesting from a theoretical point of view. Quantum field theories with spontaneously broken symmetry contain ultraviolet divergences. Thus, it is natural to ask whether these divergences are constrained by the underlying symmetry of the theory. The answer to this question, first presented by Benjamin Lee,* will give us further insights into the nature of ultraviolet divergences and the meaning of renormalization.

---

*A beautiful summary of Lee's analysis is given in his lecture note volume: B. Lee, *Chiral Dynamics* (Gordon and Breach, New York, 1972).

11.1 Spontaneous Symmetry Breaking

We begin with an analysis of spontaneous symmetry breaking in classical field theory. Consider first the familiar ϕ^4 theory Lagrangian,

$$\mathcal{L} = \frac{1}{2}(\partial_\mu \phi)^2 - \frac{1}{2}m^2\phi^2 - \frac{\lambda}{4!}\phi^4,$$

but with m^2 replaced by a negative parameter, $-\mu^2$:

$$\mathcal{L} = \frac{1}{2}(\partial_\mu \phi)^2 + \frac{1}{2}\mu^2\phi^2 - \frac{\lambda}{4!}\phi^4. \tag{11.1}$$

This Lagrangian has a discrete symmetry: It is invariant under the operation $\phi \to -\phi$. The corresponding Hamiltonian is

$$H = \int d^3x \left[\frac{1}{2}\pi^2 + \frac{1}{2}(\boldsymbol{\nabla}\phi)^2 - \frac{1}{2}\mu^2\phi^2 + \frac{\lambda}{4!}\phi^4\right].$$

The minimum-energy classical configuration is a uniform field $\phi(x) = \phi_0$, with ϕ_0 chosen to minimize the potential

$$V(\phi) = -\frac{1}{2}\mu^2\phi^2 + \frac{\lambda}{4!}\phi^4$$

(see Fig. 11.1). This potential has two minima, given by

$$\phi_0 = \pm v = \pm\sqrt{\frac{6}{\lambda}}\,\mu. \tag{11.2}$$

The constant v is called the *vacuum expectation value* of ϕ.

   To interpret this theory, suppose that the system is near one of the minima (say the positive one). Then it is convenient to define

$$\phi(x) = v + \sigma(x), \tag{11.3}$$

and rewrite $\mathcal{L}$ in terms of $\sigma(x)$. Plugging (11.3) into (11.1), we find that the term linear in σ vanishes (as it must, since the minimum of the potential is at $\sigma = 0$). Dropping the constant term as well, we obtain the Lagrangian

$$\mathcal{L} = \frac{1}{2}(\partial_\mu \sigma)^2 - \frac{1}{2}(2\mu^2)\sigma^2 - \sqrt{\frac{\lambda}{6}}\,\mu\sigma^3 - \frac{\lambda}{4!}\sigma^4. \tag{11.4}$$

This Lagrangian describes a simple scalar field of mass $\sqrt{2}\mu$, with σ^3 and σ^4 interactions. The symmetry $\phi \to -\phi$ is no longer apparent; its only manifestation is in the relations among the three coefficients in (11.4), which depend in a special way on only two parameters. This is the simplest example of a spontaneously broken symmetry.

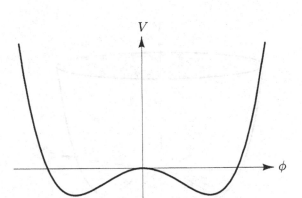

Figure 11.1. Potential for spontaneous symmetry breaking in the discrete case.

The Linear Sigma Model

A more interesting theory arises when the broken symmetry is continuous, rather than discrete. The most important example is a generalization of the preceding theory called the *linear sigma model*, which we considered briefly in Problem 4.3. We will study this model in detail throughout this chapter.

The Lagrangian of the linear sigma model involves a set of N real scalar field $\phi^i(x)$:

$$\mathcal{L} = \frac{1}{2}(\partial_\mu \phi^i)^2 + \frac{1}{2}\mu^2(\phi^i)^2 - \frac{\lambda}{4}\big[(\phi^i)^2\big]^2, \tag{11.5}$$

with an implicit sum over i in each factor $(\phi^i)^2$. Note that we have rescaled the coupling λ from the ϕ^4 theory Lagrangian to remove the awkward factors of 6 in the analysis above. The Lagrangian (11.5) is invariant under the symmetry

$$\phi^i \longrightarrow R^{ij}\phi^j \tag{11.6}$$

for any $N \times N$ orthogonal matrix R. The group of transformations (11.6) is just the rotation group in N dimensions, also called the N-dimensional *orthogonal group* or simply $O(N)$.

Again the lowest-energy classical configuration is a constant field ϕ_0^i, whose value is chosen to minimize the potential

$$V(\phi^i) = -\frac{1}{2}\mu^2(\phi^i)^2 + \frac{\lambda}{4}\big[(\phi^i)^2\big]^2$$

(see Fig. 11.2). This potential is minimized for any ϕ_0^i that satisfies

$$(\phi_0^i)^2 = \frac{\mu^2}{\lambda}.$$

This condition determines only the length of the vector ϕ_0^i; its direction is arbitrary. It is conventional to choose coordinates so that ϕ_0^i points in the

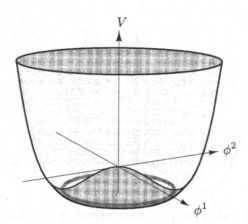

Figure 11.2. Potential for spontaneous breaking of a continuous $O(N)$ symmetry, drawn for the case $N = 2$. Oscillations along the trough in the potential correspond to the massless π fields.

Nth direction:

$$\phi_0^i = (0, 0, \ldots, 0, v), \qquad \text{where } v = \frac{\mu}{\sqrt{\lambda}}. \tag{11.7}$$

We can now define a set of shifted fields by writing

$$\phi^i(x) = \big(\pi^k(x), v + \sigma(x)\big), \qquad k = 1, \ldots, N-1. \tag{11.8}$$

(The notation, as in Problem 4.3, comes from the application of this formalism to pions in the case $N = 4$.)

It is now straightforward to rewrite the Lagrangian (11.5) in terms of the π and σ fields. The result is

$$\mathcal{L} = \frac{1}{2}(\partial_\mu \pi^k)^2 + \frac{1}{2}(\partial_\mu \sigma)^2 - \frac{1}{2}(2\mu^2)\sigma^2$$
$$- \sqrt{\lambda}\mu\sigma^3 - \sqrt{\lambda}\mu(\pi^k)^2\sigma - \frac{\lambda}{4}\sigma^4 - \frac{\lambda}{2}(\pi^k)^2\sigma^2 - \frac{\lambda}{4}\big[(\pi^k)^2\big]^2. \tag{11.9}$$

We obtain a massive σ field just as in (11.4), and also a set of $N-1$ massless π fields. The original $O(N)$ symmetry is hidden, leaving only the subgroup $O(N-1)$, which rotates the π fields among themselves. Referring to Fig. 11.2, we note that the massive σ field describes oscillations of ϕ^i in the radial direction, in which the potential has a nonvanishing second derivative. The massless π fields describe oscillations of ϕ^i in the tangential directions, along the trough of the potential. The trough is an $(N-1)$-dimensional surface, and all $N-1$ directions are equivalent, reflecting the unbroken $O(N-1)$ symmetry.

Goldstone's Theorem

The appearance of massless particles when a continuous symmetry is spontaneously broken is a general result, known as *Goldstone's theorem*. To state the theorem precisely, we must count the number of linearly independent continuous symmetry transformations. In the linear sigma model, there are no continuous symmetries for $N = 1$, while for $N = 2$ there is a single direction of rotation. A rotation in N dimensions can be in any one of $N(N-1)/2$ planes, so the $O(N)$-symmetric theory has $N(N-1)/2$ continuous symmetries. After spontaneous symmetry breaking there are $(N-1)(N-2)/2$ remaining symmetries, corresponding to rotations of the $(N-1)$ π fields. The number of *broken* symmetries is the difference, $N-1$.

Goldstone's theorem states that for every spontaneously broken continuous symmetry, the theory must contain a massless particle.[†] We have just seen that this theorem holds in the linear sigma model, at least at the classical level. The massless fields that arise through spontaneous symmetry breaking are called *Goldstone bosons*. Many light bosons seen in physics, such as the pions, may be interpreted (at least approximately) as Goldstone bosons. We conclude this section with a general proof of Goldstone's theorem for classical scalar field theories. The rest of this chapter is devoted to the quantum-mechanical analysis of theories with hidden symmetry. By the end of the chapter we will see that Goldstone bosons cannot acquire mass from any order of quantum corrections.

Consider, then, a theory involving several fields $\phi^a(x)$, with a Lagrangian of the form

$$\mathcal{L} = (\text{terms with derivatives}) - V(\phi). \tag{11.10}$$

Let ϕ_0^a be a constant field that minimizes V, so that

$$\left. \frac{\partial}{\partial \phi^a} V \right|_{\phi^a(x) = \phi_0^a} = 0.$$

Expanding V about this minimum, we find

$$V(\phi) = V(\phi_0) + \frac{1}{2}(\phi - \phi_0)^a (\phi - \phi_0)^b \left(\frac{\partial^2}{\partial \phi^a \partial \phi^b} V \right)_{\phi_0} + \cdots.$$

The coefficient of the quadratic term,

$$\left(\frac{\partial^2}{\partial \phi^a \partial \phi^b} V \right)_{\phi_0} = m_{ab}^2, \tag{11.11}$$

[†]J. Goldstone, *Nuovo Cim.* **19**, 154 (1961). An instructive four-page paper by J. Goldstone, A. Salam, and S. Weinberg, *Phys. Rev.* **127**, 965 (1962), gives three different proofs of the theorem.

is a symmetric matrix whose eigenvalues give the masses of the fields. These eigenvalues cannot be negative, since ϕ_0 is a minimum. To prove Goldstone's theorem, we must show that every continuous symmetry of the Lagrangian (11.10) that is not a symmetry of ϕ_0 gives rise to a zero eigenvalue of this mass matrix.

A general continuous symmetry transformation has the form

$$\phi^a \longrightarrow \phi^a + \alpha \Delta^a(\phi), \tag{11.12}$$

where α is an infinitesimal parameter and Δ^a is some function of all the ϕ's. Specialize to constant fields; then the derivative terms in $\mathcal{L}$ vanish and the potential alone must be invariant under (11.12). This condition can be written

$$V(\phi^a) = V\left(\phi^a + \alpha \Delta^a(\phi)\right) \qquad \text{or} \qquad \Delta^a(\phi) \frac{\partial}{\partial \phi^a} V(\phi) = 0.$$

Now differentiate with respect to ϕ^b, and set $\phi = \phi_0$:

$$0 = \left(\frac{\partial \Delta^a}{\partial \phi^b}\right)_{\phi_0} \left(\frac{\partial V}{\partial \phi^a}\right)_{\phi_0} + \Delta^a(\phi_0)\left(\frac{\partial^2}{\partial \phi^a \partial \phi^b} V\right)_{\phi_0}. \tag{11.13}$$

The first term vanishes since ϕ_0 is a minimum of V, so the second term must also vanish. If the transformation leaves ϕ_0 unchanged (i.e., if the symmetry is respected by the ground state), then $\Delta^a(\phi_0) = 0$ and this relation is trivial. A spontaneously broken symmetry is precisely one for which $\Delta^a(\phi_0) \neq 0$; in this case $\Delta^a(\phi_0)$ is our desired vector with eigenvalue zero, so Goldstone's theorem is proved.

11.2 Renormalization and Symmetry: An Explicit Example

Now let us investigate the quantum mechanics of a theory with spontaneously broken symmetry. Again we will use as our example the linear sigma model. The Lagrangian of this theory, written in terms of shifted fields, is given in Eq. (11.9). From this expression, we can read off the Feynman rules; these are shown in Fig. 11.3.

Using these Feynman rules, we can compute tree-level amplitudes without difficulty. Diagrams with loops, however, will often diverge. For the amplitude with N_e external legs, the superficial degree of divergence is

$$D = 4 - N_e,$$

just as in the discussion of ϕ^4 theory in Section 10.2. (Diagrams containing a three-point vertex will be less divergent than this expression indicates, because this vertex has a coefficient with dimensions of mass.) However, the symmetry constraints on the amplitudes are much weaker than in that earlier analysis. The linear sigma model has eight different superficially divergent amplitudes (see Fig. 11.4); several of these have $D > 0$ and therefore can contain

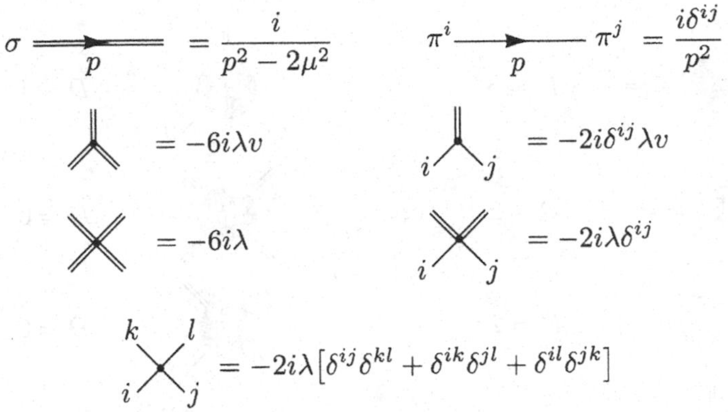

$$\sigma \overset{p}{=\!=\!=} \; = \frac{i}{p^2 - 2\mu^2} \qquad\qquad \pi^i \xrightarrow{p} \pi^j \; = \frac{i\delta^{ij}}{p^2}$$

$$\text{(vertex)} \; = -6i\lambda v \qquad\qquad \underset{i \quad j}{\text{(vertex)}} \; = -2i\delta^{ij}\lambda v$$

$$\text{(vertex)} \; = -6i\lambda \qquad\qquad \underset{i \quad j}{\text{(vertex)}} \; = -2i\lambda\delta^{ij}$$

$$\underset{i \quad j}{\overset{k \quad l}{\text{(vertex)}}} \; = -2i\lambda\big[\delta^{ij}\delta^{kl} + \delta^{ik}\delta^{jl} + \delta^{il}\delta^{jk}\big]$$

Figure 11.3. Feynman rules for the linear sigma model.

more than one infinite constant. Yet the number of bare parameters available to absorb these infinities is much smaller. If we follow the procedure of Section 10.2 to rewrite the original Lagrangian in terms of physical parameters and counterterms, we find only three counterterms:

$$\mathcal{L} = \frac{1}{2}(\partial_\mu \phi^i)^2 + \frac{1}{2}\mu^2(\phi^i)^2 - \frac{\lambda}{4}\big[(\phi^i)^2\big]^2$$
$$+ \frac{1}{2}\delta_Z(\partial_\mu \phi^i)^2 - \frac{1}{2}\delta_\mu(\phi^i)^2 - \frac{\delta_\lambda}{4}\big[(\phi^i)^2\big]^2. \tag{11.14}$$

Written in terms of π and σ fields, the second line takes the form

$$\frac{\delta_Z}{2}(\partial_\mu \pi^k)^2 - \frac{1}{2}(\delta_\mu + \delta_\lambda v^2)(\pi^k)^2 + \frac{\delta_Z}{2}(\partial_\mu \sigma)^2 - \frac{1}{2}(\delta_\mu + 3\delta_\lambda v^2)\sigma^2$$
$$- (\delta_\mu v + \delta_\lambda v^3)\sigma - \delta_\lambda v\sigma(\pi^k)^2 - \delta_\lambda v\sigma^3 \tag{11.15}$$
$$- \frac{\delta_\lambda}{4}\big[(\pi^k)^2\big]^2 - \frac{\delta_\lambda}{2}\sigma^2(\pi^k)^2 - \frac{\delta_\lambda}{4}\sigma^4.$$

The Feynman rules associated with these counterterms are shown in Fig. 11.5. There are now plenty of counterterms, but they still depend on only three renormalization parameters: δ_Z, δ_μ, and δ_λ. It would be a miracle if these three parameters were able to absorb all the infinities arising in the divergent amplitudes shown in Fig. 11.4.

If this miracle did not occur, that is, if the counterterms of (11.15) did not absorb all the infinities, we could still make this theory renormalizable by introducing new, symmetry-breaking terms in the Lagrangian. These would give rise to additional counterterms, which could be adjusted to render all amplitudes finite. If desired, we could set the physical values of the symmetry-breaking coupling constants to zero. The bare values of these constants, however, would still be nonzero, so the Lagrangian itself would no longer be invariant under the $O(N)$ symmetry. We would have to conclude that the symmetry is not consistent with quantum mechanics.

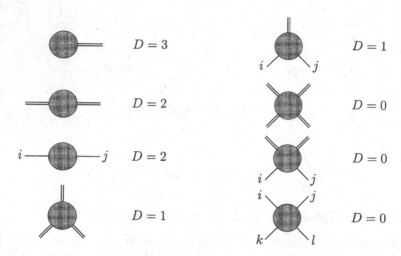

Figure 11.4. Divergent amplitudes in the linear sigma model.

$$\otimes\!\!=\ = -i(\delta_\mu v + \delta_\lambda v^3)$$

$$=\!\!\otimes\!\!=\ = i(\delta_Z p^2 - \delta_\mu - 3\delta_\lambda v^2)$$

$$i -\!\!\otimes\!\!- j\ = i\delta^{ij}(\delta_Z p^2 - \delta_\mu - \delta_\lambda v^2)$$

$$= -6i\delta_\lambda$$

$$i \qquad j \qquad = -2i\delta^{ij}\delta_\lambda$$

$$= -6i\delta_\lambda v$$

$$i \qquad j \qquad = -2i\delta^{ij}\delta_\lambda v$$

$$\begin{matrix} i & & j \\ & \otimes & \\ k & & l \end{matrix} \quad = -2i\delta_\lambda\left[\delta^{ij}\delta^{kl} + \delta^{ik}\delta^{jl} + \delta^{il}\delta^{jk}\right]$$

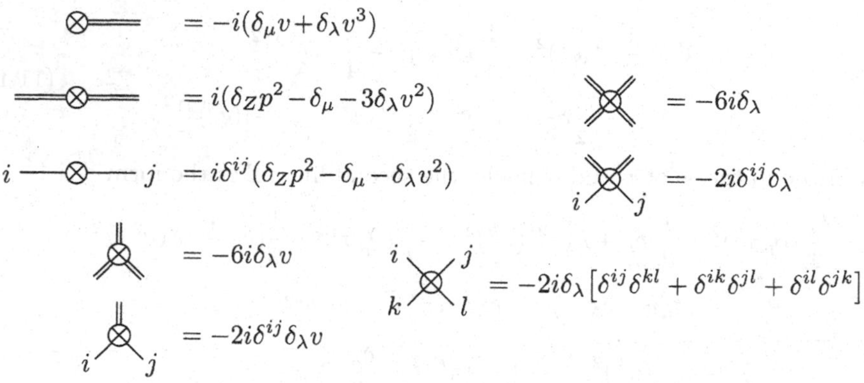

Figure 11.5. Feynman rules for counterterm vertices in the linear sigma model.

Fortunately, the miracle does occur. We will see below that the counterterms of (11.15), even though they contain only three adjustable parameters, are indeed sufficient to cancel all the infinities that occur in this theory. In this section we will demonstrate this cancellation explicitly at the one-loop level. The rest of this chapter is devoted to a more general discussion of these issues.

Renormalization Conditions

In the discussion to follow, we will keep track of only the divergent parts of Feynman diagrams. However, it will be useful to keep in mind a set of renormalization conditions that could, in principle, be used to determine also the finite parts of the counterterms. Since the counterterms contain three adjustable parameters, we need three conditions. We could take these to be the conditions (10.19) (implemented according to (10.28)), specifying the physical mass m of the σ field, its field strength, and the scattering amplitude at threshold. However, it is technically easier to replace one of these conditions with a constraint on the one-point amplitude for σ (the sum of *tadpole diagrams*):

$$\text{\Large\textbullet}\!\!-\!\!- = 0.$$

In QED the tadpole diagrams automatically vanish, as we saw in Eq. (10.5). In the linear sigma model, however, no symmetry forbids the appearance of a nonvanishing one-σ amplitude. This amplitude produces a vacuum expectation value of σ and so, since $\phi^N = v + \sigma$, shifts the vacuum expectation value of ϕ. Such a shift is quite acceptable, as long as it is finite after counterterms are properly added into the computation of the amplitude. However, it will simplify the bookkeeping to set up our conventions so that the relation

$$\langle \phi^N \rangle = \frac{\mu}{\sqrt{\lambda}} \tag{11.16}$$

is satisfied to all orders in perturbation theory. We will define λ, as in Eq. (10.19), as the scattering amplitude at threshold. Then Eq. (11.16) defines the parameter μ, so the mass m of the σ field will differ from the result of the classical equations $m^2 = 2\mu^2 = 2\lambda v^2$ by terms of order $(\lambda \mu^2)$. If indeed we can remove the divergences from the theory by adjusting three counterterms, these corrections will be finite and constitute a prediction of the quantum field theory.

To summarize, we will use the following renormalization conditions:

$$\text{(1PI)}\!\!-\!\!- = 0;$$

$$\frac{d}{dp^2}\left(-\!\!\text{(1PI)}\!\!- \right) = 0 \qquad \text{at } p^2 = m^2;$$

$$\text{Im}\left(\text{\Large$\times\!\!\!\bigcirc\!\!\!\times$} \right) = -6i\lambda \quad \text{at } s = 4m^2, \ t = u = 0. \tag{11.17}$$

In the last condition, the circle is the amputated four-point amplitude. Note that the last two conditions depend on the physical mass m of the σ particle. We must now show that these three conditions suffice to make all of the one-loop amplitudes of the linear sigma model finite.

The Vertex Counterterm

We begin by determining the counterterm δ_λ by computing the 4σ amplitude. The tree-level term comes from the 4σ vertex, and is just such as to satisfy (11.17). The one-loop contribution to this amplitude is the sum of diagrams:

$$\hspace{6cm}(11.18)$$

According to (11.17), we must adjust δ_λ so that this sum of diagrams vanishes at threshold. In this calculation, we will only keep track of the ultraviolet divergences. This greatly simplifies the analysis, because most of the diagrams in (11.18) are finite. All the diagrams with loops made of three or more propagators are finite, since they have at least six powers of the loop momentum in the denominator; for example,

$$\sim \int \frac{d^4 k}{(2\pi)^4} \frac{1}{k^2} \frac{1}{k^2} \frac{1}{k^2}.$$

Alternatively, we can see that this diagram is finite in the following way: Each three-point vertex carries a factor of μ, which has dimensions of mass. According to the dimensional analysis argument of Section 10.1, each such factor lowers the degree of divergence of a diagram by 1. Since the 4σ amplitude already has $D = 0$, any diagram containing a three-point vertex must be finite.

We are left with the first two diagrams of (11.18) and the four diagrams related to these by crossing. Let us evaluate the first diagram using dimensional regularization:

$$\text{} = \frac{1}{2} \cdot (-6i\lambda)^2 \cdot \int \frac{d^d k}{(2\pi)^d} \frac{i}{k^2 - 2\mu^2} \frac{i}{(k+p)^2 - 2\mu^2}$$

$$= 18\lambda^2 \int_0^1 dx \int \frac{d^d k}{(2\pi)^d} \frac{1}{[k^2 - \Delta]^2}$$

$$= 18\lambda^2 \int\limits_0^1 dx \, \frac{i}{(4\pi)^{d/2}} \Gamma(2-\tfrac{d}{2}) \left(\frac{1}{\Delta}\right)^{2-\frac{d}{2}}$$

$$= 18i\lambda^2 \frac{\Gamma(2-\tfrac{d}{2})}{(4\pi)^2} + \text{(finite terms)}. \tag{11.19}$$

Here Δ is a function of p and μ, whose exact form does not concern us. Since our objective is only to demonstrate the cancellation of the divergences, we will neglect finite terms here and throughout the rest of this section. The second diagram of (11.18) (with π's instead of σ's for the internal lines) is identical, except that each vertex factor is changed from $-6i\lambda$ to $-2i\lambda\delta^{ij}$. (Roman indices $i, j, \ldots$ run from 1 to $N-1$.) We therefore have

$= 2i\lambda^2(N-1) \dfrac{\Gamma(2-\tfrac{d}{2})}{(4\pi)^2} + \text{(finite terms)}. \tag{11.20}$

Since the infinite part of each of these diagrams is simply a momentum-independent constant, the infinite parts of the corresponding t- and u-channel diagrams must be identical. Therefore the infinite part of the 4σ vertex is just three times the sum of (11.19) and (11.20):

$+$ $+ \text{(crosses)} \sim 6i\lambda^2(N+8) \dfrac{\Gamma(2-\tfrac{d}{2})}{(4\pi)^2}, \tag{11.21}$

(In this section we use the $\sim$ symbol to indicate equality up to omitted finite corrections.) Applying the third condition of (11.17), we find that the counterterm δ_λ is given by

$$\delta_\lambda \sim \lambda^2(N+8) \frac{\Gamma(2-\tfrac{d}{2})}{(4\pi)^2}. \tag{11.22}$$

Once we have determined the value of δ_λ, we have fixed the counterterms for the two other four-point amplitudes. Are these amplitudes also made finite? Consider the amplitude with two σ's and two π's. This receives one-loop corrections from

$+$ $+$ $+$. $\tag{11.23}$

and from several diagrams with three-point vertices which, as argued earlier, are manifestly finite. Each of the diagrams in (11.23) contains a loop integral analogous to that in (11.19), whose infinite part is always $-i\Gamma(2-\tfrac{d}{2})/(4\pi)^2$.

The only differences are in the vertices and symmetry factors. For example, the infinite part of the first diagram of (11.23) is

$$\sim \frac{1}{2} \cdot (-6i\lambda)(-2i\lambda\delta^{ij}) \cdot \frac{-i}{(4\pi)^2}\Gamma(2-\tfrac{d}{2}) = 6i\lambda^2\delta^{ij}\frac{\Gamma(2-\tfrac{d}{2})}{(4\pi)^2}.$$

The second diagram is a bit more complicated:

$$\sim \frac{1}{2} \cdot (-2i\lambda\delta^{kl})\left(-2i\lambda(\delta^{ij}\delta^{kl} + \delta^{ik}\delta^{jl} + \delta^{il}\delta^{jk})\right) \cdot \frac{-i}{(4\pi)^2}\Gamma(2-\tfrac{d}{2})$$

$$= 2i\lambda^2(N+1)\delta^{ij}\frac{\Gamma(2-\tfrac{d}{2})}{(4\pi)^2}.$$

In the third diagram there is no symmetry factor:

$$\sim (-2i\lambda\delta^{il})(-2i\lambda\delta^{jl}) \cdot \frac{-i}{(4\pi)^2}\Gamma(2-\tfrac{d}{2}) = 4i\lambda^2\delta^{ij}\frac{\Gamma(2-\tfrac{d}{2})}{(4\pi)^2}.$$

The fourth diagram of (11.23) gives an identical expression, since it is the same as the third but with i and j interchanged. The sum of the four diagrams therefore gives, for the infinite part of the $\sigma\sigma\pi\pi$ vertex,

$$\sim 2i\lambda^2\,\delta^{ij}\,(N+8)\frac{\Gamma(2-\tfrac{d}{2})}{(4\pi)^2}. \qquad (11.24)$$

This divergent term is indeed canceled by the $\sigma\sigma\pi\pi$ counterterm, with the value of δ_λ given in (11.22).

The remaining four-point amplitude has four external π fields. The divergent one-loop diagrams are:

$$+ \qquad + \quad \text{(crosses)} \qquad\qquad (11.25)$$

These diagrams all have the same familiar form. The first is

$$\sim \frac{1}{2} \cdot (-2i\lambda\delta^{ij})(-2i\lambda\delta^{kl}) \cdot \frac{-i}{(4\pi)^2}\Gamma(2-\tfrac{d}{2}) = 2i\lambda^2\delta^{ij}\delta^{kl}\frac{\Gamma(2-\tfrac{d}{2})}{(4\pi)^2}.$$

The second diagram is more complicated:

$$\sim \frac{1}{2} \cdot \left(-2i\lambda(\delta^{ij}\delta^{mn}+\delta^{im}\delta^{jn}+\delta^{in}\delta^{jm})\right)$$

$$\cdot \left(-2i\lambda(\delta^{kl}\delta^{mn}+\delta^{km}\delta^{ln}+\delta^{kn}\delta^{lm})\right) \cdot \frac{-i}{(4\pi)^2}\Gamma(2-\tfrac{d}{2})$$

$$= 2i\lambda^2\left((N+3)\delta^{ij}\delta^{kl} + 2\delta^{ik}\delta^{jl} + 2\delta^{il}\delta^{jk}\right)\frac{\Gamma(2-\tfrac{d}{2})}{(4\pi)^2}.$$

For each of these diagrams there are two corresponding cross-channel diagrams, which differ only in the ways that the external indices $ijkl$ are paired together. For instance, the t-channel diagrams are identical to the s-channel diagrams, but with j and k interchanged. Adding all six diagrams, we find for the 4π vertex

$$\sim 2i\lambda^2 \left(\delta^{ij}\delta^{kl}+\delta^{ik}\delta^{jl}+\delta^{il}\delta^{jk}\right)(N+8)\frac{\Gamma(2-\tfrac{d}{2})}{(4\pi)^2}. \qquad (11.26)$$

Again, the value of δ_λ given in (11.22) gives a counterterm of the correct value and index structure to cancel this divergence.

The value of δ_λ that we have determined also fixes the counterterms for the three-point amplitudes. Thus we have no further freedom in canceling the divergences in the three-point amplitudes; we can only cross our fingers and hope these also come out finite. The 3σ amplitude is given by

$$\qquad (11.27)$$

The diagrams made of three three-point vertices are finite and play no role in the cancellation of divergences. Of the divergent diagrams in (11.27), the first has the form

$$= \frac{1}{2} \cdot (-6i\lambda)(-6i\lambda v) \cdot \int \frac{d^d k}{(2\pi)^d}\frac{i}{k^2-2\mu^2}\frac{i}{(k+p)^2-2\mu^2}$$

$$\sim 18i\lambda^2 v \frac{\Gamma(2-\tfrac{d}{2})}{(4\pi)^2}.$$

This is exactly the same as the corresponding diagram (11.19) for the 4σ vertex, except for the extra factor of v. The same is true of the other five divergent diagrams; thus,

$$\sim 6i\lambda^2 v(N+8)\frac{\Gamma(2-\frac{d}{2})}{(4\pi)^2}. \tag{11.28}$$

This is precisely canceled by the 3σ counterterm vertex in Fig. 11.5, with δ_λ given by (11.22).

There is a similar correspondence between the $\sigma\pi\pi$ amplitude and the $\sigma\sigma\pi\pi$ amplitude. The four divergent diagrams in the $\sigma\pi\pi$ amplitude are identical to those in (11.23), except that each has an external σ leg replaced by a factor of v. Referring to the $\sigma\pi\pi$ counterterm vertex in Fig. 11.5, we see that the cancellation of divergences will occur here as well.

What is happening? All the divergences we have seen so far are manifestations of the basic diagram

$$\tag{11.29}$$

with either four external particles or with one leg set to zero momentum and associated with the vacuum expectation value of ϕ. Since the $O(N)$ symmetry is broken, this diagram manifests itself in many different ways. But apparently, the divergent part of the diagram is unaffected by the symmetry breaking.

Two-Point and One-Point Amplitudes

To complete our investigation of the one-loop structure of this theory we must evaluate the two-point and one-point amplitudes. We first determine the counterterm δ_μ by applying the first renormalization condition in (11.17). At one-loop order, this condition reads

$$0 \; = \; \bigcirc\!\!\!- \;\; + \;\; \bigcirc\!\!\!- \;\; + \;\; \otimes\!\!\!- \;\; . \tag{11.30}$$

We will later need to make use of the finite part of the counterterm, so we will pay attention to the finite terms when we evaluate (11.30). The first diagram is

$$\bigcirc\!\!\!- \;\; = \frac{1}{2}(-6i\lambda v)\int\frac{d^d k}{(2\pi)^d}\frac{i}{k^2-2\mu^2} = -3i\lambda v\frac{\Gamma(1-\frac{d}{2})}{(4\pi)^{d/2}}\left(\frac{1}{2\mu^2}\right)^{1-\frac{d}{2}}.$$
$$\tag{11.31}$$

The second diagram involves a divergent integral over a massless propagator. To be sure that we understand how to treat this term, we will add a small mass ζ for the π field as an infrared regulator. Then the second diagram is

$$
\bigcirc\!\!-\!\!- = \frac{1}{2}(-2i\lambda v)\delta^{ij}\int\frac{d^dk}{(2\pi)^d}\frac{i\delta^{ij}}{k^2-\zeta^2}
$$

$$
= -i(N-1)\lambda v\frac{\Gamma(1-\frac{d}{2})}{(4\pi)^{d/2}}\left(\frac{1}{\zeta^2}\right)^{1-\frac{d}{2}}. \tag{11.32}
$$

Notice that, for $d > 2$, the diagram vanishes in the limit as $\zeta \to 0$; however, it has a pole at $d = 2$. Despite these strange features, we can add (11.32) to (11.31) and impose the condition that the tadpole diagrams be canceled by the counterterm from Fig. 11.5. This condition gives

$$
(\delta_\mu + v^2\delta_\lambda) = -\lambda\frac{\Gamma(1-\frac{d}{2})}{(4\pi)^{d/2}}\left(\frac{3}{(2\mu^2)^{1-d/2}} + \frac{N-1}{(\zeta^2)^{1-d/2}}\right). \tag{11.33}
$$

Now consider the 2σ amplitude. The one-particle-irreducible amplitude receives contributions from four one-loop diagrams and a counterterm:

$$
=\!\!\bigcirc\!\!= \; + \; =\!\!\bigcirc\!\!= \; + \; \underset{}{\bigcirc}\!\!\!\!\!\!\!\!\!\!\!\!-\!\!-\!\!- \; + \; \underset{}{\bigcirc}\!\!\!\!\!\!\!\!\!\!\!\!-\!\!-\!\!- \; + \; =\!\!\otimes\!\!= \; . \tag{11.34}
$$

It is convenient to write the counterterm vertex as

$$
-i(2v^2\delta_\lambda) - i(\delta_\mu + v^2\delta_\lambda) + ip^2\delta_Z. \tag{11.35}
$$

In a general renormalization scheme, the σ mass will also be shifted by the tadpole diagrams (and their counterterm):

$$
\underset{}{\bigcirc}\!\!\!\!\!\mid \; + \; \underset{}{\bigcirc}\!\!\!\!\!\mid \; + \; \underset{}{\otimes}\!\!\!\!\!\mid \; . \tag{11.36}
$$

However, the first renormalization condition in (11.17) forces these diagrams to cancel precisely. This is an example of the special simplicity of this renormalization condition.

The first two diagrams are again manifestations of the generic four-point diagram (11.29), now with two external legs replaced by the vacuum expectation value of ϕ. In analogy with the preceding calculations, we find for the first diagram

$$
=\!\!\bigcirc\!\!- \; \sim 18i\lambda^2 v^2\frac{\Gamma(2-\frac{d}{2})}{(4\pi)^2},
$$

and for the second diagram

$$
=\!\!\bigcirc\!\!- \; \sim 2i\lambda^2 v^2(N-1)\frac{\Gamma(2-\frac{d}{2})}{(4\pi)^2}.
$$

Using (11.22), we see that these two contributions are canceled by the first term of (11.35). The third and fourth diagrams of (11.34) contain precisely the same integrals as the tadpole diagrams of (11.30). Relation (11.33) implies that they are canceled by the second term in (11.35). Notice that there is no divergent term proportional to p^2 in any of the one-loop diagrams of (11.34). Thus the renormalization constant δ_Z is zero at the one-loop level, just as in ordinary ϕ^4 theory.

There remains only one potentially divergent amplitude—the $\pi\pi$ amplitude:

$$\text{(diagram)} + \text{(diagram)} + \text{(diagram)} + \text{(diagram)} . \tag{11.37}$$

In analogy with (11.31), the first diagram is

$$\text{(diagram)} = \frac{1}{2}(-2i\lambda\delta^{ij}) \int \frac{d^dk}{(2\pi)^d} \frac{i}{k^2 - 2\mu^2} = -i\lambda\delta^{ij} \frac{\Gamma(1-\frac{d}{2})}{(4\pi)^{d/2}} \left(\frac{1}{2\mu^2}\right)^{1-\frac{d}{2}}.$$

The second diagram is quite similar. As in (11.32), it is useful to introduce a small pion mass as an infrared regulator.

$$\text{(diagram)} = \frac{1}{2}(-2i\lambda(\delta^{ij}\delta^{kk}+\delta^{ik}\delta^{jk}+\delta^{ik}\delta^{jk})) \int \frac{d^dk}{(2\pi)^d} \frac{1}{k^2 - \zeta^2}$$

$$= -i\lambda(N+1)\delta^{ij} \frac{\Gamma(1-\frac{d}{2})}{(4\pi)^{d/2}} \left(\frac{1}{\zeta^2}\right)^{1-\frac{d}{2}}.$$

The third diagram is given by

$$\text{(diagram)} = (-2i\lambda v\delta^{ik})(-2i\lambda v\delta^{kj}) \int \frac{d^dk}{(2\pi)^d} \frac{i}{k^2 - \zeta^2} \frac{i}{(k+p)^2 - 2\mu^2}$$

$$= 4i\lambda^2 v^2 \delta^{ij} \frac{\Gamma(2-\frac{d}{2})}{(4\pi)^{d/2}} \int_0^1 dx \left(\frac{1}{2\mu^2 x + (1-x)\zeta^2 - p^2 x(1-x)}\right)^{2-\frac{d}{2}}.$$

The divergent part of this expression is independent of p, so to check the cancellation of the divergence, it suffices to set $p = 0$. It will be instructive to compute the complete amplitude at $p = 0$, including the finite terms. Adding the three loop diagrams and the counterterm, whose value is given by (11.33), we find

$$
-\!\!\bigoplus\!\!-\Big|_{p=0} = (-i\lambda\delta^{ij})\left\{\frac{\Gamma(1-\frac{d}{2})}{(4\pi)^{d/2}}\left(\frac{1}{(2\mu^2)^{1-d/2}}+\frac{N+1}{(\zeta^2)^{1-d/2}}\right)\right.
$$

$$
- 4\lambda v^2\frac{\Gamma(2-\frac{d}{2})}{(4\pi)^{d/2}}\int_0^1 dx\left(\frac{1}{2\mu^2 x+\zeta^2(1-x)}\right)^{2-\frac{d}{2}}
$$

$$
\left.-\frac{\Gamma(1-\frac{d}{2})}{(4\pi)^{d/2}}\left(\frac{3}{(2\mu^2)^{1-d/2}}+\frac{N-1}{(\zeta^2)^{1-d/2}}\right)\right\}.
$$

$$(11.38)$$

It is not hard to simplify this expression. The first and third lines can be combined to give

$$
2i\lambda\delta^{ij}\,\frac{\Gamma(1-\frac{d}{2})}{(4\pi)^{d/2}}\left[\frac{1}{(\zeta^2)^{1-d/2}}-\frac{1}{(2\mu^2)^{1-d/2}}\right].
$$

Near $d=2$ the quantity in brackets is proportional to $1-d/2$, and this factor cancels the pole in the gamma function. Thus the worst divergence cancels, leaving only a pole at $d=4$. Using the identity $\Gamma(x)=\Gamma(x+1)/x$, we can rewrite the above expression as

$$
2i\lambda\delta^{ij}\,\frac{\Gamma(2-\frac{d}{2})}{(4\pi)^{d/2}}\frac{1}{1-d/2}\left[\frac{\zeta^2}{(\zeta^2)^{2-d/2}}-\frac{2\mu^2}{(2\mu^2)^{2-d/2}}\right].
$$

$$(11.39)$$

The first term vanishes for $d>2$ and $\zeta\to 0$, and can be neglected. Meanwhile, the second line of expression (11.38) involves the elementary integral

$$
\int_0^1 dx\,(2\mu^2 x+(1-x)\zeta^2)^{\frac{d}{2}-2}=\frac{1}{d/2-1}\cdot\frac{(2\mu^2)^{d/2-1}-(\zeta^2)^{d/2-1}}{2\mu^2-\zeta^2}.
$$

This expression is also nonsingular at $d=2$ and reduces to

$$
\frac{1}{d/2-1}(2\mu^2)^{d/2-2}
$$

for $d>2$ and $\zeta\to 0$. Comparing this line with the remaining term from (11.39), and recalling that $\lambda v^2=\mu^2$, we find that the $\pi\pi$ amplitude is not only finite, but *vanishes* completely at $p=0$.

This result is very attractive. The $\pi\pi$ amplitude, at $p=0$, is precisely the mass shift δm_π^2 of the π field. We already knew that the π particles are massless at tree level—they are the $N-1$ massless bosons required by Goldstone's theorem. We have now verified that these bosons remain massless at the one-loop level in the linear sigma model; in other words, the first quantum corrections to the linear sigma model also respect Goldstone's theorem. At the end of this chapter, we will give a general argument that Goldstone's theorem is satisfied to all orders in perturbation theory.

11.3 The Effective Action

In the first section of this chapter, we analyzed spontaneous symmetry breaking in classical field theory. That analysis was geometrical: We found the vacuum state by finding the deepest well in a potential surface, and we proved Goldstone's theorem by showing that symmetry required the presence of a line of degenerate minima at the bottom of the well. But this geometrical picture was lost, or at least disguised, in the one-loop calculations of Section 11.2. It seems worthwhile to develop a formalism that will allow us to use geometrical arguments about spontaneous symmetry breaking even at the quantum level.

To define our goal somewhat better, consider the problem of determining the vacuum expectation value of the quantum field ϕ. This expectation value should be determined as a function of the parameters of the Lagrangian. At the classical level, it is easy to compute $\langle\phi\rangle$; one minimizes the potential energy. However, as we have seen in the previous section, this classical value can be altered by perturbative loop corrections. In fact, we saw that $\langle\phi\rangle$ could be shifted by a potentially divergent quantity, which we needed to control by renormalization.

It would be wonderful if, in the full quantum field theory, there were a function whose minimum gave the exact value of $\langle\phi\rangle$. This function would agree with the classical potential energy to lowest order in perturbation theory, but it would be modified in higher orders by quantum corrections. Presumably, these corrections would need renormalization to remove infinities. Nevertheless, after renormalization, this quantity should give the same relations between $\langle\phi\rangle$ and particle masses and couplings that we would find by direct Feynman diagram calculations. In this section, we will exhibit a function with these properties, called the *effective potential*. In Section 11.4 we will explain how to compute the effective potential in perturbation theory, in terms of renormalized masses and couplings. Then we will go on to use it as a tool in analyzing the renormalizability of theories with hidden symmetry.

To identify the effective potential, consider the analogy between quantum field theory and statistical mechanics set out in Section 9.3. In that section, we derived a correspondence between the correlation functions of a quantum field and those of a related statistical system, with quantum fluctuations being replaced by thermal fluctuations. At zero temperature the thermodynamic ground state is the state of lowest energy, but at nonzero temperature we still have a geometrical picture of the preferred thermodynamic state: It is the state that minimizes the Gibbs free energy. More explicitly, taking the example of a magnetic system, one defines the Helmholtz free energy $F(H)$ by

$$ Z(H) = e^{-\beta F(H)} = \int \mathcal{D}s \, \exp\left[-\beta \int dx \, (\mathcal{H}[s] - Hs(x))\right], \tag{11.40} $$

where H is the external magnetic field, $\mathcal{H}[s]$ is the spin energy density, and $\beta = 1/kT$. We can find the magnetization of the system by differentiating

$F(H)$:

$$-\frac{\partial F}{\partial H}\bigg|_{\beta \text{ fixed}} = \frac{1}{\beta} \frac{\partial}{\partial H} \log Z$$

$$= \frac{1}{Z} \int dx \int \mathcal{D}s \, s(x) \exp\left[-\beta \int dx \left(\mathcal{H}[s] - Hs\right)\right] \quad (11.41)$$

$$= \int dx \, \langle s(x) \rangle \equiv M.$$

The Gibbs free energy G is defined by the Legendre transformation

$$G = F + MH,$$

so that it satisfies

$$\frac{\partial G}{\partial M} = \frac{\partial F}{\partial M} + M \frac{\partial H}{\partial M} + H$$

$$= \frac{\partial H}{\partial M} \frac{\partial F}{\partial H} + M \frac{\partial H}{\partial M} + H \quad (11.42)$$

$$= H$$

(where all partial derivatives are taken with β fixed). If $H = 0$, the Gibbs free energy reaches an extremum at the corresponding value of M. The thermodynamically most stable state is the minimum of $G(M)$. Thus the function $G(M)$ gives a picture of the preferred thermodynamic state that is geometrical and at the same time includes all effects of thermal fluctuations.

By analogy, we can construct a similar quantity in a quantum field theory. For simplicity, we will work in this section only with a theory of one scalar field. All of the results generalize straightforwardly to systems with multiple scalar, spinor, and vector fields.

Consider a quantum field theory of a scalar field ϕ, in the presence of an external source J. As in Chapter 9, it is useful to take the external source to depend on x. Thus, we define an energy functional $E[J]$ by

$$Z[J] = e^{-iE[J]} = \int \mathcal{D}\phi \exp\left[i \int d^4x \left(\mathcal{L}[\phi] + J\phi\right)\right]. \quad (11.43)$$

The right-hand side of this equation is the functional integral representation of the amplitude $\langle \Omega| e^{-iHT} |\Omega\rangle$, where T is the time extent of the functional integration, in the presence of the source J. Thus, $E[J]$ is just the vacuum energy as a function of the external source. The functional $E[J]$ is the analogue of the Helmholtz free energy, and J is the analogue of the external magnetic field.

In principle, we could now Legendre-transform $E[J]$ with respect to a constant value of the source. However, since we have already developed a formalism for functional integration and differentiation, it will not be much more difficult to work with an external source $J(x)$ that depends on x in an arbitrary way. As we will see, this generalization yields additional relations which connect this formalism to our general study of renormalization theory.[‡]

Consider, then, the functional derivative of $E[J]$ with respect to $J(x)$:

$$\frac{\delta}{\delta J(x)} E[J] = i \frac{\delta}{\delta J(x)} \log Z = -\frac{\int \mathcal{D}\phi \, e^{i \int (\mathcal{L}+J\phi)} \phi(x)}{\int \mathcal{D}\phi \, e^{i \int (\mathcal{L}+J\phi)}}. \tag{11.44}$$

We abbreviate this relation as

$$\frac{\delta}{\delta J(x)} E[J] = - \langle \Omega | \, \phi(x) \, | \Omega \rangle_J \, ; \tag{11.45}$$

the right-hand side is the vacuum expectation value in the presence of a nonzero source $J(x)$. This relation is a functional analogue of Eq. (11.41): The functional derivative of $E[J]$ gives the expectation value of ϕ in the presence of the spatially varying source. We should treat this expectation value as the thermodynamic variable conjugate to $J(x)$. Thus we define the quantity $\phi_{\mathrm{cl}}(x)$, called the *classical field*, by

$$\phi_{\mathrm{cl}}(x) = \langle \Omega | \, \phi(x) \, | \Omega \rangle_J \, . \tag{11.46}$$

The classical field is related to $\phi(x)$ in the same way that the magnetization M is related to the local spin field $s(x)$: It is a weighted average over all possible fluctuations. Note that $\phi_{\mathrm{cl}}(x)$ depends on the external source $J(x)$, just as M depends on H.

Now, in analogy with the construction of the Gibbs free energy, define the Legendre transform of $E[J]$:

$$\Gamma[\phi_{\mathrm{cl}}] \equiv -E[J] - \int d^4y \, J(y)\phi_{\mathrm{cl}}(y). \tag{11.47}$$

This quantity is known as the *effective action*. In analogy with Eq. (11.42), we can now compute

$$\begin{aligned}
\frac{\delta}{\delta \phi_{\mathrm{cl}}(x)} \Gamma[\phi_{\mathrm{cl}}] &= -\frac{\delta}{\delta \phi_{\mathrm{cl}}(x)} E[J] - \int d^4y \, \frac{\delta J(y)}{\delta \phi_{\mathrm{cl}}(x)} \phi_{\mathrm{cl}}(y) - J(x) \\
&= -\int d^4y \, \frac{\delta J(y)}{\delta \phi_{\mathrm{cl}}(x)} \frac{\delta E[J]}{\delta J(y)} - \int d^4y \, \frac{\delta J(y)}{\delta \phi_{\mathrm{cl}}(x)} \phi_{\mathrm{cl}}(y) - J(x) \\
&= -J(x).
\end{aligned} \tag{11.48}$$

In the last step we have used Eq. (11.45).

For each of the thermodynamic quantities discussed at the beginning of this section, we have now defined an analogous quantity in quantum field theory. Table 11.1 summarizes these analogies.

[‡]This functional generalization of thermodynamics is due to C. DeDominicis and P. Martin, *J. Math. Phys.* **5**, 14 (1964), and was formulated for relativistic field theory by G. Jona-Lasinio, *Nuovo Cim.* **34A**, 1790 (1964).

Magnetic System	Quantum Field Theory
$\mathbf{x}$	$x = (t, \mathbf{x})$
$s(\mathbf{x})$	$\phi(x)$
H	$J(x)$
$\mathcal{H}(s)$	$\mathcal{L}(\phi)$
$Z(H)$	$Z[J]$
$F(H)$	$E[J]$
M	$\phi_{\mathrm{cl}}(x)$
$G(M)$	$-\Gamma[\phi_{\mathrm{cl}}]$

Table 11.1. Analogous quantities in a magnetic system and a scalar quantum field theory.

Relation (11.48) implies that, if the external source is set to zero, the effective action satisfies the equation

$$\frac{\delta}{\delta\phi_{\mathrm{cl}}(x)}\Gamma[\phi_{\mathrm{cl}}] = 0. \tag{11.49}$$

The solutions to this equation are the values of $\langle\phi(x)\rangle$ in the stable quantum states of the theory. For a translation-invariant vacuum state, we will find a solution in which ϕ_{cl} is independent of x. Sometimes, Eq. (11.49) will have additional solutions, corresponding to localized lumps of field held together by their self-interaction. In these states, called *solitons*, the solution $\phi_{\mathrm{cl}}(x)$ depends on x.

From here on we will assume, for the field theories we consider, that the possible vacuum states are invariant under translations and Lorentz transformations.* Then, for each possible vacuum state, the corresponding solution ϕ_{cl} will be a constant, independent of x, and the process of solving Eq. (11.49) reduces to that of solving an ordinary equation of one variable (ϕ_{cl}). Furthermore, we know that Γ is, in thermodynamic terms, an extensive quantity: It is proportional to the volume of the spacetime region over which the functional integral is taken. If T is the time extent of this region and V is its three-dimensional volume, we can write

$$\Gamma[\phi_{\mathrm{cl}}] = -(VT)\cdot V_{\mathrm{eff}}(\phi_{\mathrm{cl}}). \tag{11.50}$$

The coefficient V_{eff} is called the *effective potential*. The condition that $\Gamma[\phi_{\mathrm{cl}}]$ has an extremum then reduces to the simple equation

$$\frac{\partial}{\partial\phi_{\mathrm{cl}}}V_{\mathrm{eff}}(\phi_{\mathrm{cl}}) = 0. \tag{11.51}$$

---

*Certain condensed matter systems have ground states with preferred orientation; see, for example P. G. de Gennes, *The Physics of Liquid Crystals* (Oxford University Press, 1974).

Each solution of Eq. (11.51) is a translation-invariant state with $J = 0$. Equation (11.47) implies that $\Gamma = -E$ in this case, and therefore that $V_{\text{eff}}(\phi_{\text{cl}})$, evaluated at a solution to (11.51), is just the energy density of the corresponding state.

Figure 11.6 illustrates one possible shape for the function $V_{\text{eff}}(\phi)$. The local maxima (or, for systems of several fields ϕ^i, possible saddle points) are unstable configurations that cannot be realized as stationary states. The figure also contains a local minimum of V_{eff} that is not the absolute minimum; this is a metastable vacuum state, which can decay to the true vacuum by quantum-mechanical tunneling. The absolute minimum of V_{eff} is the state of lowest energy in the theory, and thus the true, stable, vacuum state. A system with spontaneously broken symmetry will have several minima of V_{eff}, all with the same energy by virtue of the symmetry. The choice of one among these vacua is the spontaneous symmetry breaking.

In drawing Fig. 11.6, we have assumed that we are computing the effective potential for a fixed constant background value of ϕ. Under some circumstances, this state does not give the true minimum energy configuration for states with a given expectation value of ϕ. This mismatch can occur in the following way: In a system for which the effective potential for constant background fields is given by Fig. 11.6, consider choosing a value of ϕ_{cl} that is intermediate between the locally stable vacuum states ϕ_1 and ϕ_3:

$$\phi_{\text{cl}} = x\phi_1 + (1 - x)\phi_3, \qquad 0 < x < 1. \tag{11.52}$$

The assumption of a constant background field gives a large value of the effective potential, as indicated in the figure. We can obtain a lower-energy configuration by considering states with macroscopic regions in which $\langle\phi\rangle = \phi_1$ and other regions in which $\langle\phi\rangle = \phi_3$, in such a way that the average value of $\langle\phi\rangle$ over the whole system is ϕ_{cl}. For such a configuration, the average vacuum energy is given by

$$V_{\text{eff}}(\phi_{\text{cl}}) = xV_{\text{eff}}(\phi_1) + (1 - x)V_{\text{eff}}(\phi_3), \tag{11.53}$$

as shown in Fig. 11.7. We have called the left-hand side of this equation $V_{\text{eff}}(\phi_{\text{cl}})$ because the result (11.53) would be the result of an exact evaluation of the functional integral definition of V_{eff} for values of ϕ_{cl} satisfying (11.52). The interpolation (11.53) is the field theoretic analogue of the Maxwell construction for the thermodynamic free energy. In general, for any ϕ_{cl}, ϕ_1, ϕ_3 satisfying (11.52), the estimate (11.53) will be an upper bound to the effective potential; we say that the effective potential is a *convex* function of ϕ_{cl}.[†]

Just as in thermodynamics, straightforward schemes for computing the effective potential do not take account of the possibility of phase separation and so lead to a structure of unstable and metastable configurations of the

---

[†]The convexity of the Gibbs free energy is a well-known exact result in statistical mechanics; see, for example, D. Ruelle, *Statistical Mechanics* (W. A. Benjamin, Reading, Mass., 1969).

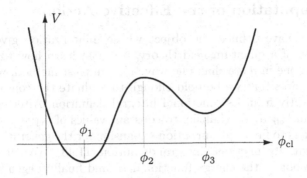

Figure 11.6. A possible form for the effective potential in a scalar field theory. The extrema of the effective potential occur at the points $\phi_{cl} = \phi_1, \phi_2, \phi_3$. The true vacuum state is the one corresponding to ϕ_1. The state ψ_2 is unstable. The state ϕ_3 is metastable, but it can decay to ϕ_1 by quantum-mechanical tunneling.

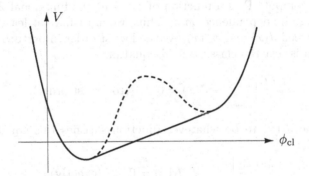

Figure 11.7. Exact convex form of the effective potential for the system of Fig. 11.6.

type shown in Fig. 11.6. The Maxwell construction must be performed by hand to yield the final form of $V_{eff}(\phi_{cl})$. Fortunately, the absolute minimum of V_{eff} is not affected by this nicety.

We have now solved the problem that we posed at the beginning of this section: The effective potential, defined by Eqs. (11.47) and (11.50), gives an easily visualized function whose minimization defines the exact vacuum state of the quantum field theory, including all effects of quantum corrections. It is not obvious from these definitions how to compute $V_{eff}(\phi_{cl})$. We will see how to do so in the next section, by direct evaluation of the functional integral.

11.4 Computation of the Effective Action

Now that we have defined the object whose minimization gives the exact vacuum state of a quantum field theory, we must learn how to compute it. This can be done in more than one way. The simplest method, which we will use here, requires that we be bold enough to evaluate the complete effective action Γ directly from its functional integral definition. After computing Γ, we can obtain V_{eff} by specializing to constant values of ϕ_{cl}.[‡]

Our plan is to find a perturbation expansion for the generating functional Z, starting with its functional integral definition (11.43). We will then take the logarithm to obtain the energy functional E, and finally Legendre-transform according to Eq. (11.47) to obtain Γ. We will use renormalized perturbation theory, so it is convenient to split the Lagrangian as we did in Eq. (10.18), into a piece depending on renormalized parameters and one containing the counterterms:

$$\mathcal{L} = \mathcal{L}_1 + \delta\mathcal{L}. \tag{11.54}$$

We wish to compute Γ as a function of ϕ_{cl}. But the functional $Z[J]$ depends on ϕ_{cl} through its dependence on J. Thus, we must find, at least implicitly, a relation between $J(x)$ and $\phi_{\text{cl}}(x)$. At the lowest order in perturbation theory, that relation is just the classical field equation:

$$\left.\frac{\delta\mathcal{L}}{\delta\phi}\right|_{\phi=\phi_{\text{cl}}} + J(x) = 0 \qquad \text{(to lowest order)}.$$

Let us define $J_1(x)$ to be whatever function satisfies this equation exactly, when $\mathcal{L} = \mathcal{L}_1$:

$$\left.\frac{\delta\mathcal{L}_1}{\delta\phi}\right|_{\phi=\phi_{\text{cl}}} + J_1(x) = 0 \qquad \text{(exactly)}. \tag{11.55}$$

We will think of the difference between J and J_1 as a counterterm, analogous to $\delta\mathcal{L}$, so we write

$$J(x) = J_1(x) + \delta J(x), \tag{11.56}$$

where δJ is determined, order by order in perturbation theory, by the original definition (11.46) of ϕ_{cl}, namely $\langle\phi(x)\rangle_J = \phi_{\text{cl}}(x)$.

Using this notation, we rewrite Eq. (11.43) as

$$e^{-iE[J]} = \int \mathcal{D}\phi \, e^{i\int d^4x(\mathcal{L}_1[\phi] + J_1\phi)} e^{i\int d^4x(\delta\mathcal{L}[\phi] + \delta J\phi)}. \tag{11.57}$$

The second exponential contains the counterterms; leave this aside for the moment. In the first exponential, expand the exponent about ϕ_{cl} by replacing

---

[‡]This method is due to R. Jackiw, *Phys. Rev.* **D9**, 1686 (1974).

$\phi(x) = \phi_{\text{cl}}(x) + \eta(x)$. This exponent takes the form

$$
\int d^4x (\mathcal{L}_1 + J_1\phi) = \int d^4x (\mathcal{L}_1[\phi_{\text{cl}}] + J_1\phi_{\text{cl}}) + \int d^4x\, \eta(x)\left(\frac{\delta\mathcal{L}_1}{\delta\phi} + J_1\right)
$$

$$
+ \frac{1}{2}\int d^4x\, d^4y\, \eta(x)\eta(y)\frac{\delta^2\mathcal{L}_1}{\delta\phi(x)\delta\phi(y)}
$$

$$
+ \frac{1}{3!}\int d^4x\, d^4y\, d^4z\, \eta(x)\eta(y)\eta(z)\frac{\delta^3\mathcal{L}_1}{\delta\phi(x)\delta\phi(y)\delta\phi(z)} + \cdots,
$$

(11.58)

where the various functional derivatives of $\mathcal{L}_1$ are evaluated at $\phi_{\text{cl}}(x)$. Notice that the term linear in η vanishes by the use of Eq. (11.55). The integral over η is thus a Gaussian integral, with the cubic and higher terms giving perturbative corrections.

We will describe a formal evaluation of this integral, following the prescriptions of Section 9.2. The ingredients in this evaluation will be the coefficients of Eq. (11.58), that is, the successive functional derivatives of $\mathcal{L}_1$. For the moment, please accept that these give well-defined operators. After presenting a general expression for $\Gamma[\phi_{\text{cl}}]$, we will carry out this calculation explicitly in a scalar field theory example. We will see in this example that the formal operators correspond to expressions familiar from Feynman diagram perturbation theory.

Let us, then, consider performing the integral over $\eta(x)$ using the expansion (11.58). Keeping only the terms up to quadratic order in η, and still neglecting the counterterms, we have a pure Gaussian integral, which can be evaluated in terms of a functional determinant:

$$
\int \mathcal{D}\eta\, \exp\left[i\left(\int (\mathcal{L}_1[\phi_{\text{cl}}] + J_1\phi_{\text{cl}}) + \frac{1}{2}\int \eta\frac{\delta^2\mathcal{L}_1}{\delta\phi\delta\phi}\eta\right)\right]
$$

$$
= \exp\left[i\int (\mathcal{L}_1[\phi_{\text{cl}}] + J_1\phi_{\text{cl}})\right] \cdot \left(\det\left[-\frac{\delta^2\mathcal{L}_1}{\delta\phi\delta\phi}\right]\right)^{-1/2}. \quad (11.59)
$$

This functional determinant will give us the lowest-order quantum correction to the effective action, and for many purposes it is unnecessary to go further in the expansion (11.58). Later we will see that if we do include the cubic and higher terms in η, these produce a Feynman diagram expansion of the functional integral (11.57) in which the propagator is the operator inverse

$$
-i\left(\frac{\delta^2\mathcal{L}_1}{\delta\phi\delta\phi}\right)^{-1} \quad (11.60)
$$

and the vertices are the third and higher functional derivatives of $\mathcal{L}_1$.

Finally, let us put back the effects of the second exponential in Eq. (11.57), that is, the counterterm Lagrangian. It is useful to expand this term about $\phi = \phi_{\text{cl}}$, writing it as

$$
\left(\delta\mathcal{L}[\phi_{\text{cl}}] + \delta J\phi_{\text{cl}}\right) + \left(\delta\mathcal{L}[\phi_{\text{cl}} + \eta] - \delta\mathcal{L}[\phi_{\text{cl}}] + \delta J\eta\right). \quad (11.61)
$$

The second term of (11.61) can be expanded as a Taylor series in η; the successive terms give counterterm vertices which can be included in the afore-mentioned Feynman diagrams. The first term is a constant with respect to the functional integral over η, and therefore gives additional terms in the exponent of Eq. (11.59).

Combining the integral (11.59) with the contributions from higher-order vertices and counterterms, one can obtain a complete expression for the functional integral (11.57). We will see in the example below that the Feynman diagrams representing the higher-order terms can be arranged to give the exponential of the sum of connected diagrams. Thus one obtains the following expression for $E[J]$:

$$
-iE[J] = i \int d^4x (\mathcal{L}_1[\phi_{\rm cl}] + J_1\phi_{\rm cl}) - \frac{1}{2} \log \det\Big[-\frac{\delta^2 \mathcal{L}_1}{\delta\phi\delta\phi}\Big]
$$
$$
+ \text{(connected diagrams)} + i \int d^4x (\delta\mathcal{L}[\phi_{\rm cl}] + \delta J\phi_{\rm cl}). \tag{11.62}
$$

From this equation, Γ follows directly: Using $J_1 + \delta J = J$ and the Legendre transform (11.47), we find

$$
\Gamma[\phi_{\rm cl}] = \int d^4x\, \mathcal{L}_1[\phi_{\rm cl}] + \frac{i}{2} \log \det\Big[-\frac{\delta^2 \mathcal{L}_1}{\delta\phi\delta\phi}\Big]
$$
$$
- i \cdot \text{(connected diagrams)} + \int d^4x\, \delta\mathcal{L}[\phi_{\rm cl}]. \tag{11.63}
$$

Notice that there are no terms remaining that depend explicitly on J; thus, Γ is expressed as a function of $\phi_{\rm cl}$, as it should be. The Feynman diagrams contributing to $\Gamma[\phi_{\rm cl}]$ have no external lines, and the simplest ones turn out to have two loops. The lowest-order quantum correction to Γ is given by the functional determinant, and this term is all that we will make use of in this book.

The last term of (11.63) provides a set of counterterms that can be used to satisfy the renormalization conditions on Γ and, in the process, to cancel divergences that appear in the evaluation of the functional determinant and the diagrams. We will show in the example below exactly how this cancellation works. The renormalization conditions will determine all of the counterterms in $\delta\mathcal{L}$. However, the formalism we have constructed contains a new counterterm δJ. That coefficient is determined by the following special criterion: In Eq. (11.55), we set up our analysis in such a way that, at the leading order, $\langle\phi\rangle = \phi_{\rm cl}$. Potentially, however, this relation could break down at higher orders: The quantity $\langle\phi\rangle$ could receive additional contributions from Feynman diagrams that might shift it from the value $\phi_{\rm cl}$. This will happen if there are nonzero tadpole diagrams that contribute to $\langle\eta\rangle$. But this amplitude also receives a contribution from the counterterm $(\delta J\eta)$ in (11.61). Thus we can maintain $\langle\eta\rangle = 0$, and in the process determine δJ to any order, by adjusting

δJ to satisfy the diagrammatic equation

$$\text{(diagram)} \quad + \quad \delta J \left(\otimes\!\!=\!\!=\!\!= \right) \quad = \quad 0 . \tag{11.64}$$

In practice, we will satisfy this condition by simply ignoring any one-particle-irreducible one-point diagram, since any such diagram will be canceled by adjustment of δJ. The removal of these tadpole diagrams, which we needed some effort to arrange in Section 11.2, is thus built in here as a natural part of the formalism.

The Effective Action in the Linear Sigma Model

In Eq. (11.63), we have given a complete, though not exactly transparent, evaluation of $\Gamma[\phi_{\rm cl}]$. Let us now clarify the meaning of this equation, and also put it to some good use, by computing $\Gamma[\phi_{\rm cl}]$ in the linear sigma model. We will see that the results that we obtained by brute-force perturbation theory in Section 11.2 emerge much more naturally from Eq. (11.63).

We begin again with the Lagrangian (11.5):

$$\mathcal{L} = \frac{1}{2}(\partial_\mu \phi^i)^2 + \frac{1}{2}\mu^2(\phi^i)^2 - \frac{\lambda}{4}\left[(\phi^i)^2\right]^2 . \tag{11.65}$$

Expand about the classical field: $\phi^i = \phi^i_{\rm cl} + \eta^i$. Because we expect to find a translation-invariant vacuum state, we will specialize to the case of a *constant* classical field. This will simplify some elements of the calculation below. In particular, according to Eq. (11.50), the final result will be proportional to the four-dimensional volume (VT) of the functional intergration. When this dependence is factored out, we will obtain a well-defined intensive expression for the effective potential. In any event, after this simplification, (11.65) takes the form

$$\mathcal{L} = \frac{1}{2}\mu^2(\phi^i_{\rm cl})^2 - \frac{\lambda}{4}\left[(\phi^i_{\rm cl})^2\right]^2 + (\mu^2 - \lambda(\phi^i_{\rm cl})^2)\phi^i_{\rm cl}\eta^i$$
$$+ \frac{1}{2}(\partial_\mu \eta^i)^2 + \frac{1}{2}\mu^2(\eta^i)^2 - \frac{\lambda}{2}\left[(\phi^i_{\rm cl})^2(\eta^i)^2 + 2(\phi^i_{\rm cl}\eta^i)^2\right] + \cdots . \tag{11.66}$$

According to Eq. (11.63), we should drop the term linear in η.

From the terms quadratic in η, we can read off

$$\frac{\delta^2 \mathcal{L}}{\delta\phi^i \delta\phi^j} = -\partial^2 \delta^{ij} + \mu^2 \delta^{ij} - \lambda\left[(\phi^k_{\rm cl})^2 \delta^{ij} + 2\phi^i_{\rm cl}\phi^j_{\rm cl}\right] . \tag{11.67}$$

Notice that this object has the general form of a Klein-Gordon operator. To clarify this relation, let us orient the coordinates so that $\phi^i_{\rm cl}$ points in the Nth direction,

$$\phi^i_{\rm cl} = (0, 0, \ldots, 0, \phi_{\rm cl}), \tag{11.68}$$

as we did in Eq. (11.7). Then the operator (11.67) is just equal to the Klein-Gordon operator $(-\partial^2 - m_i^2)$, where

$$m_i^2 = \begin{cases} \lambda\phi_{\rm cl}^2 - \mu^2 & \text{acting on } \eta^1, \ldots, \eta^{N-1}; \\ 3\lambda\phi_{\rm cl}^2 - \mu^2 & \text{acting on } \eta^N. \end{cases} \qquad (11.69)$$

The functional determinant in Eq. (11.63) is the product of the determinants of these Klein-Gordon operators:

$$\det \frac{\delta^2 \mathcal{L}}{\delta\phi\delta\phi} = \left[\det(\partial^2 + (\lambda\phi_{\rm cl}^2 - \mu^2))\right]^{N-1} [\det(\partial^2 + (3\lambda\phi_{\rm cl}^2 - \mu^2))]. \quad (11.70)$$

It is not difficult to obtain an explicit form for the determinant of a Klein-Gordon operator. To begin, use the trick of Eq. (9.77) to write

$$\log\det(\partial^2 + m^2) = \mathrm{Tr}\log(\partial^2 + m^2).$$

Now evaluate the trace of the operator as the sum of its eigenvalues:

$$\mathrm{Tr}\log(\partial^2 + m^2) = \sum_k \log(-k^2 + m^2)$$

$$= (VT)\int \frac{d^4 k}{(2\pi)^4} \log(-k^2 + m^2). \qquad (11.71)$$

In the second line, we have converted the sum over momenta to an integral. The factor (VT) is the four-dimensional volume of the functional integral; we have already noted that this is expected to appear as an overall factor in $\Gamma[\phi_{\rm cl}]$. This manipulation gives an integral that can be evaluated in dimensional regularization after a Wick rotation:

$$\int \frac{d^d k}{(2\pi)^d} \log(-k^2 + m^2) = i\int \frac{d^d k_E}{(2\pi)^d} \log(k_E^2 + m^2)$$

$$= -i\frac{\partial}{\partial\alpha} \int \frac{d^d k_E}{(2\pi)^d} \frac{1}{(k_E^2 + m^2)^\alpha}\bigg|_{\alpha=0}$$

$$= -i\frac{\partial}{\partial\alpha} \left(\frac{1}{(4\pi)^{d/2}} \frac{\Gamma(\alpha - \frac{d}{2})}{\Gamma(\alpha)} \frac{1}{(m^2)^{\alpha-d/2}}\right)\bigg|_{\alpha=0}$$

$$= -i\frac{\Gamma(-\frac{d}{2})}{(4\pi)^{d/2}} \frac{1}{(m^2)^{-d/2}}. \qquad (11.72)$$

In the last line, we have used $\Gamma(\alpha) \to 1/\alpha$ as $\alpha \to 0$. Thus,

$$\frac{1}{(VT)}\log\det(\partial^2 + m^2) = -i\frac{\Gamma(-\frac{d}{2})}{(4\pi)^{d/2}}(m^2)^{d/2}. \qquad (11.73)$$

Using this result to evaluate the determinant in Eq. (11.63), and choosing

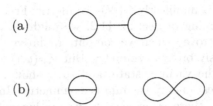

Figure 11.8. Feynman diagrams contributing to the evaluation of the effective potential of the $O(N)$ linear sigma model: (a) a diagram that is removed by (11.64); (b) the first nonzero diagrammatic corrections.

the counterterm Lagrangian as in Eq. (11.14), we find

$$
\begin{aligned}
V_{\text{eff}}(\phi_{\text{cl}}) &= -\frac{1}{(VT)}\Gamma[\phi_{\text{cl}}] \\
&= -\frac{1}{2}\mu^2\phi_{\text{cl}}^2 + \frac{\lambda}{4}\phi_{\text{cl}}^4 \\
&\quad -\frac{1}{2}\frac{\Gamma(-\frac{d}{2})}{(4\pi)^{d/2}}\left[(N-1)(\lambda\phi_{\text{cl}}^2 - \mu^2)^{d/2} + (3\lambda\phi_{\text{cl}}^2 - \mu^2)^{d/2}\right] \\
&\quad + \frac{1}{2}\delta_\mu\phi_{\text{cl}}^2 + \frac{1}{4}\delta_\lambda\phi_{\text{cl}}^4.
\end{aligned}
\tag{11.74}
$$

Here we have written ϕ_{cl}^2 as a shorthand for $(\phi_{\text{cl}}^i)^2$. Since the second line of this result is the leading radiative correction, we might expect that the result has the structure of a one-loop Feynman diagram. Indeed, we see that this expression contains Gamma functions and ultraviolet divergences similar to those that we found in the one-loop computations of Section 11.2. We will show below that this term in fact has exactly the same ultraviolet divergences that we found in Section 11.2. These divergences will be subtracted by the counterterms in the last line of Eq. (11.74).

Since the computation of the determinant in Eq. (11.63) gives the effect of one-loop corrections, we might expect the Feynman diagrams that contribute to Eq. (11.63) to begin in two-loop order. We can see this explicitly for the case of the $O(N)$ sigma model. The perturbation expansion described below Eq. (11.60) involves the propagator that is the inverse of Eq. (11.67):

$$
\langle \eta^i(k)\eta^j(-k)\rangle = \frac{i}{k^2 - m_i^2}\delta^{ij},
\tag{11.75}
$$

where m_i^2 is given by (11.69). The vertices are given by the terms of order η^3 and η^4 in the expansion of the Lagrangian. Combining these ingredients, we find that the leading Feynman diagrams contributing to the vacuum energy have the forms shown in Fig. 11.8. The diagram of Fig. 11.8(a) is actually canceled by the effects of the counterterm δJ, as shown in Eq. (11.64). Thus the leading diagrammatic contribution to the effective potential comes from the two-loop diagrams of Fig. 11.8(b).

The result (11.74) is manifestly $O(N)$-symmetric. From the question that we posed at the beginning of Section 11.2, we might have feared that this property would be destroyed when we compute radiative corrections about a state with spontaneously broken symmetry. But $V_{\text{eff}}(\phi_{\text{cl}})$ is the function that we minimize to find the vacuum state, and so it should properly be symmetric, even if the lowest-energy vacuum is asymmetric. In the formalism we have constructed here, there is no need to worry. Formula (11.63) is manifestly invariant, term by term, under the original $O(N)$ symmetry of the Lagrangian. Thus we must necessarily have arrived at an $O(N)$-symmetric result for $V_{\text{eff}}(\phi_{\text{cl}})$.

Before going on to determine δ_μ and δ_λ precisely, we might first check that the counterterms in Eq. (11.74) are sufficient to make the expression for $\Gamma[\phi_{\text{cl}}]$ finite. The factor $\Gamma(-d/2)$ has poles at $d = 0, 2, 4$. The pole at $d = 0$ is a constant, independent of ϕ_{cl}, and therefore without physical significance. The pole at $d = 2$ is an even quadratic polynomial in ϕ_{cl}. The pole at $d = 4$ is an even quartic polynomial in ϕ_{cl}. Thus Eq. (11.74) becomes a finite expression in the limit $d \to 2$ if we set

$$\delta_\mu = -\lambda(N+2)\frac{\Gamma(1-\frac{d}{2})}{(4\pi)} + \text{finite.}$$

The expression is finite as $d \to 4$ if we set

$$\delta_\mu = -\lambda\mu^2(N+2)\frac{\Gamma(2-\frac{d}{2})}{(4\pi)^2} + \text{finite;}$$

$$\delta_\lambda = \lambda^2(N+8)\frac{\Gamma(2-\frac{d}{2})}{(4\pi)^2} + \text{finite.} \tag{11.76}$$

These expressions agree with our earlier results from Section 11.2, Eqs. (11.33) and (11.22), in the limits $d \to 2$ and $d \to 4$ respectively.

The finite parts of δ_λ and δ_μ depend on the exact form of the renormalization conditions that are imposed. For example, in Section 11.2, we imposed the condition (11.16) that the vacuum expectation value of ϕ equals $\mu/\sqrt{\lambda}$ and the additional conditions in (11.17) on the scattering amplitude and field strength of the σ. Condition (11.16) is readily expressed in terms of the effective potential as

$$\frac{\partial V_{\text{eff}}}{\partial \phi_{\text{cl}}}(\phi_{\text{cl}} = \mu/\sqrt{\lambda}) = 0.$$

Using the connection between derivatives of Γ and one-particle-irreducible amplitudes, we could write the other two conditions as Fourier transforms to momentum space of functional derivatives of $\Gamma[\phi_{\text{cl}}]$. In this way, it is possible in principle to reconstruct the particular renormalization scheme used in Section 11.2.

However, if we want to visualize the modification of the lowest-order results that is induced by the quantum corrections, we can apply a renormalization scheme that can be implemented more easily. One such scheme, known as *minimal subtraction* (MS), is simply to remove the $(1/\epsilon)$ poles (for $\epsilon = 4 - d$) in potentially divergent quantities. Normally, though, these $(1/\epsilon)$ poles are accompanied by terms involving γ and $\log(4\pi)$. It is convenient, and no more arbitrary, to subtract these terms as well. In this prescription, known as *modified minimal subtraction* or $\overline{MS}$ ("em-ess-bar"), one replaces

$$\frac{\Gamma(2-\frac{d}{2})}{(4\pi)^{d/2}(m^2)^{2-d/2}} = \frac{1}{(4\pi)^2}\left(\frac{2}{\epsilon} - \gamma + \log(4\pi) - \log(m^2)\right)$$

$$\longrightarrow \frac{1}{(4\pi)^2}\left(-\log(m^2/M^2)\right), \qquad (11.77)$$

where M is an arbitrary mass parameter that we have introduced to make the final equation dimensionally correct. You should think of M as parametrizing a sequence of possible renormalization conditions. The $\overline{MS}$ renormalization scheme usually puts one-loop corrections in an especially simple form. The price of this simplicity is that it normally takes some effort to express physically measurable quantities in terms of the parameters of the $\overline{MS}$ expression.

To apply the $\overline{MS}$ renormalization prescription to (11.74), we need to expand the divergent terms in this equation in powers of ϵ. As an example, consider the $\overline{MS}$ regularization of expression (11.73):

$$\frac{\Gamma(-\frac{d}{2})}{(4\pi)^{d/2}}(m^2)^{d/2} = \frac{1}{\frac{d}{2}(\frac{d}{2}-1)}\frac{\Gamma(2-\frac{d}{2})}{(4\pi)^{d/2}}(m^2)^{d/2}$$

$$= \frac{m^4}{2(4\pi)^2}\left(\frac{2}{\epsilon} - \gamma + \log(4\pi) - \log(m^2) + \frac{3}{2}\right)$$

$$\longrightarrow \frac{m^4}{2(4\pi)^2}\left(-\log(m^2/M^2) + \frac{3}{2}\right). \qquad (11.78)$$

Modifying our result (11.74) in this way, we find

$$V_{\text{eff}} = -\frac{1}{2}\mu^2\phi_{\text{cl}}^2 + \frac{\lambda}{4}\phi_{\text{cl}}^4$$

$$+ \frac{1}{4}\frac{1}{(4\pi)^2}\bigg((N-1)(\lambda\phi_{\text{cl}}^2 - \mu^2)^2\big(\log[(\lambda\phi_{\text{cl}}^2 - \mu^2)/M^2] - \tfrac{3}{2}\big)$$

$$+ (3\lambda\phi_{\text{cl}}^2 - \mu^2)^2\big(\log[(3\lambda\phi_{\text{cl}}^2 - \mu^2)/M^2] - \tfrac{3}{2}\big)\bigg). \qquad (11.79)$$

The effective potential is thus modified to be slightly steeper at large values of ϕ_{cl} and more negative at smaller values, as shown in Fig. 11.9. For each set of values of μ, λ, and M, we can determine the preferred vacuum state by minimizing $V_{\text{eff}}(\phi)$ with respect to ϕ_{cl}. The correction to V_{eff} is undefined when the arguments of the logarithms become negative, but fortunately the minima of V_{eff} occur outside of this region, as is illustrated in the figure.

Before going on, we would like to raise two questions about this expression for the effective potential. The problems that we will raise occur generically in quantum field theory calculations, but expression (11.79) provides a concrete

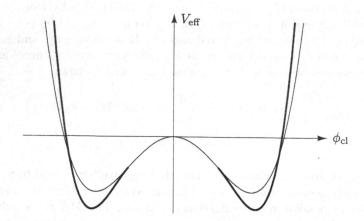

Figure 11.9. The effective potential for ϕ^4 theory ($N = 1$), with quantum corrections included as in Eq. (11.79). The lighter-weight curve shows the classical potential energy, for comparison.

illustration of these difficulties. Most of our discussion in the next two chapters will be devoted to building a formalism within which these questions can be answered.

First, it is troubling that, while our classical Lagrangian contained only two parameters, μ and λ, the result (11.79) depends on three parameters, of which one is the arbitrary mass scale M. A superficial reply to this complaint can be given as follows: Consider the change in $V_{\text{eff}}(\phi_{\text{cl}})$ that results from changing the value of M^2 to $M^2 + \delta M^2$. From the explicit form of (11.79), we can see that this change is compensated completely by shifting the values of μ and λ, according to

$$\lambda \to \lambda + \frac{\lambda^2}{(4\pi)^2}(N+8) \cdot \frac{\delta M^2}{M^2},$$

$$\mu^2 \to \mu^2 + \frac{\lambda\mu^2}{(4\pi)^2}(N+2) \cdot \frac{\delta M^2}{M^2}. \qquad (11.80)$$

Thus, a change in M^2 is completely equivalent to changes in the parameters μ and λ. It is not clear, however, why this should be true or how this fact helps us understand the dependence of our formulae on M^2.

The second problem arises from the fact that the one-loop correction in Eq. (11.79) includes a logarithm that can become large enough to compensate the small coupling constant λ. The problem is particularly clear in the limit $\mu^2 \to 0$; then Eq. (11.79) takes the form

$$V_{\text{eff}} = \frac{\lambda}{4}\phi_{\text{cl}}^4 + \frac{1}{4}\frac{\lambda^2}{(4\pi)^2}\phi_{\text{cl}}^4\left((N+8)\left(\log(\lambda\phi_{\text{cl}}^2/M^2) - \tfrac{3}{2}\right) + 9\log 3\right)$$

$$= \frac{1}{4}\phi_{\text{cl}}^4\left[\lambda + \frac{\lambda^2}{(4\pi)^2}\left((N+8)\left(\log(\lambda\phi_{\text{cl}}^2/M^2) - \tfrac{3}{2}\right) + 9\log 3\right)\right]. \qquad (11.81)$$

Where is the minimum of this potential? If we take this expression at face value, we find that $V_{\text{eff}}(\phi_{\text{cl}})$ passes through zero when ϕ_{cl} reaches a very small value of order

$$\phi_{\text{cl}}^2 \sim \frac{M^2}{\lambda} \cdot \exp\left[-\frac{(4\pi)^2}{(N+8)\lambda}\right],$$

and, near this point, attains a minimum with a nonzero value of ϕ_{cl}. But the zero occurs by the cancellation of the leading term against the quantum correction. In other words, perturbation theory breaks down completely before we can address the question of whether $V_{\text{eff}}(\phi_{\text{cl}})$, for $\mu^2 = 0$, has a symmetry-breaking minimum. It seems that our present tools are quite inadequate to resolve this case.

Although it is far from obvious, these two problems turn out to be related to each other. One of our major results in Chapter 12 will be an explanation of the interrelation of M^2, λ, and μ^2 displayed in Eq. (11.80). Then, in Chapter 13, we will use the insight we have gained from this analysis to solve completely the second problem of the appearance of large logarithms. Before beginning that study, however, there are a few issues we have yet to discuss in the more formal aspects of the renormalization of theories with spontaneously broken symmetry.

11.5 The Effective Action as a Generating Functional

Now that we have defined the effective action and computed it for one particular theory, let us return to our goal of understanding the renormalization of theories with hidden symmetry. In Section 11.6 we will use the effective action as a tool in achieving this goal. First, however, we must investigate in more detail the relation between the effective action and Feynman diagrams.

We saw in Section 9.2 that the functional derivatives of $Z[J]$ with respect to $J(x)$ produce the correlation functions of the scalar field (see, for example, Eq. (9.35)). In other words, $Z[J]$ is the *generating functional* of correlation functions. Our goal now is to show that $\Gamma[\phi_{\text{cl}}]$ is also such a generating functional; specifically, it is the generating functional of one-particle-irreducible (1PI) correlation functions. Since the 1PI correlation functions figure prominently in the theory of renormalization, this result will be central in the discussion of renormalization in the following section.

To begin, let us consider the functional derivatives not of $\Gamma[\phi_{\text{cl}}]$, but of $E[J] = i \log Z[J]$. The first derivative, given in Eq. (11.44), is precisely $- \langle \phi(x) \rangle$. The second derivative is

$$\frac{\delta^2 E[J]}{\delta J(x) \delta J(y)} = -\frac{i}{Z} \int \mathcal{D}\phi \, e^{i\int (\mathcal{L}+J\phi)} \phi(x)\phi(y)$$

$$+ \frac{i}{Z^2} \int \mathcal{D}\phi \, e^{i\int (\mathcal{L}+J\phi)} \phi(x) \cdot \int \mathcal{D}\phi \, e^{i\int (\mathcal{L}+J\phi)} \phi(y)$$

$$= -i\Big[\langle \phi(x)\phi(y) \rangle - \langle \phi(x) \rangle \langle \phi(y) \rangle \Big]. \tag{11.82}$$

If we were to compute the term $\langle \phi(x)\phi(y) \rangle$ from Feynman diagrams, there would be two types of contributions:

$$x \bullet\!\!-\!\!\bigcirc\!\!-\!\!\bullet y \;+\; x \bullet\!\!-\!\!\bigcirc \; \bigcirc\!\!-\!\!\bullet y, \tag{11.83}$$

where each circle corresponds to a sum of *connected* diagrams. The second term in the last line of Eq. (11.82) cancels the second, disconnected, term of (11.83). Thus the second derivative of $E[J]$ contains only those contributions to $\langle \phi(x)\phi(y) \rangle$ that come from connected Feynman diagrams. Let us call this object the *connected correlator*:

$$\frac{\delta^2 E[J]}{\delta J(x)\delta J(y)} = -i \,\langle \phi(x)\phi(y) \rangle_{\text{conn}} . \tag{11.84}$$

Similarly, the third functional derivative of $E[J]$ is

$$\frac{\delta^3 E[J]}{\delta J(x)\delta J(y)\delta J(z)} = \Big[\langle \phi(x)\phi(y)\phi(z) \rangle - \langle \phi(x)\phi(y) \rangle \,\langle \phi(z) \rangle - \langle \phi(x)\phi(z) \rangle \,\langle \phi(y) \rangle$$

$$- \langle \phi(y)\phi(z) \rangle \,\langle \phi(x) \rangle + 2 \,\langle \phi(x) \rangle \,\langle \phi(y) \rangle \,\langle \phi(z) \rangle \Big]$$

$$= \langle \phi(x)\phi(y)\phi(z) \rangle_{\text{conn}} . \tag{11.85}$$

In each successive derivative of $E[J]$ all contributions cancel except for those from fully connected diagrams. The general formula for n derivatives is

$$\frac{\delta^n E[J]}{\delta J(x_1)\cdots\delta J(x_n)} = (i)^{n+1} \,\langle \phi(x_1)\cdots\phi(x_n) \rangle_{\text{conn}} . \tag{11.86}$$

We therefore refer to $E[J]$ as the *generating functional of connected correlation functions*.

So much for $E[J]$. Now what about the functional derivatives of the effective action? Consider first the derivative of Eq. (11.48) with respect to $J(y)$:

$$\frac{\delta}{\delta J(y)} \frac{\delta\Gamma}{\delta\phi_{\text{cl}}(x)} = -\delta(x-y).$$

We can rewrite the left-hand side of this equation using the chain rule, to obtain

$$\delta(x-y) = -\int d^4 z \, \frac{\delta\phi_{\text{cl}}(z)}{\delta J(y)} \frac{\delta^2\Gamma}{\delta\phi_{\text{cl}}(z)\delta\phi_{\text{cl}}(x)}$$

$$= \int d^4 z \, \frac{\delta^2 E}{\delta J(y)\delta J(z)} \frac{\delta^2\Gamma}{\delta\phi_{\text{cl}}(z)\delta\phi_{\text{cl}}(x)}$$

$$= \left(\frac{\delta^2 E}{\delta J\delta J} \right)_{yz} \left(\frac{\delta^2\Gamma}{\delta\phi_{\text{cl}}\delta\phi_{\text{cl}}} \right)_{zx} . \tag{11.87}$$

In the second line we have used Eq. (11.45). The last line is an abstract representation of the second line, where we think of each of the second derivatives as an infinite-dimensional matrix, with the integral over z represented by matrix multiplication. What we have shown is that these two matrices are inverses of each other:

$$\left(\frac{\delta^2 E}{\delta J \delta J}\right) = \left(\frac{\delta^2 \Gamma}{\delta \phi_{\text{cl}} \delta \phi_{\text{cl}}}\right)^{-1}. \tag{11.88}$$

Now according to Eq. (11.84), the first of these matrices is $-i$ times the connected two-point function, that is, the exact propagator of the field ϕ. Let us call this propagator $D(x, y)$:

$$\left(\frac{\delta^2 E}{\delta J(x)\delta J(y)}\right) = -i\,\langle\phi(x)\phi(y)\rangle_{\text{conn}} \equiv -iD(x, y). \tag{11.89}$$

We will therefore refer to the other matrix (times $-i$) as the *inverse propagator*:

$$\left(\frac{\delta^2 \Gamma}{\delta \phi_{\text{cl}}(x)\delta \phi_{\text{cl}}(y)}\right) = iD^{-1}(x, y). \tag{11.90}$$

This provides an interpretation, of sorts, for the second functional derivative of the effective action. This interpretation becomes more concrete if we go to momentum space. On a translation-invariant vacuum state (one with ϕ_{cl} constant), the matrix $D(x, y)$ must be diagonal in momentum:

$$D(x, y) = \int \frac{d^4 p}{(2\pi)^4} e^{-ip\cdot(x-y)} \widetilde{D}(p). \tag{11.91}$$

We showed in Eq. (7.43) that the momentum-space propagator $\widetilde{D}(p)$ is a geometric series in one-particle-irreducible Feynman diagrams. The Fourier transform of $D^{-1}(x, y)$ then gives the inverse propagator:

$$\widetilde{D}^{-1}(p) = -i(p^2 - m^2 - M^2(p^2)), \tag{11.92}$$

where $M^2(p^2)$ is the sum of one-particle-irreducible two-point diagrams.

To evaluate higher derivatives of the effective action we again use the chain rule,

$$\frac{\delta}{\delta J(z)} = \int d^4 w \frac{\delta \phi_{\text{cl}}(w)}{\delta J(z)} \frac{\delta}{\delta \phi_{\text{cl}}(w)} = i\int d^4 w\, D(z, w)\frac{\delta}{\delta \phi_{\text{cl}}(w)}, \tag{11.93}$$

together with the standard rule for differentiating matrix inverses:

$$\frac{\partial}{\partial \alpha}M^{-1}(\alpha) = -M^{-1}\frac{\partial M}{\partial \alpha}M^{-1}. \tag{11.94}$$

Applying these identities to Eq. (11.88), we find (with some abbreviated notation)

$$\frac{\delta^3 E[J]}{\delta J_x \delta J_y \delta J_z} = i\int d^4 w\, D(z, w)\frac{\delta}{\delta \phi_w^{\text{cl}}}\left(\frac{\delta^2 \Gamma}{\delta \phi_x^{\text{cl}} \delta \phi_y^{\text{cl}}}\right)^{-1}$$

$$= i\int d^4 w\, D_{zw}\,(-1)\int d^4 u\int d^4 v(-iD_{xu})\frac{\delta^3 \Gamma}{\delta \phi_u^{\text{cl}} \delta \phi_v^{\text{cl}} \delta \phi_w^{\text{cl}}}(-iD_{vy})$$

$$= i \int d^4u \, d^4v \, d^4w \, D_{xu} D_{yv} D_{zw} \frac{\delta^3 \Gamma}{\delta \phi_u^{cl} \delta \phi_v^{cl} \delta \phi_w^{cl}}. \tag{11.95}$$

This relation is more clearly expressed diagrammatically. The left-hand side is the connected three-point function. If we extract exact propagators as indicated in (11.95), this decomposes as follows:

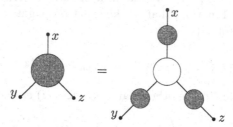

In this picture, each dark gray circle represents the sum of connected diagrams, while the light gray circle on the right-hand side represents the third derivative of $i\Gamma[\phi_{cl}]$. We see that the third derivative of $i\Gamma[\phi_{cl}]$ is just the connected correlation function with all three full propagators removed, that is, the *one-particle-irreducible* three-point function:

$$\frac{i\delta^3 \Gamma}{\delta \phi_{cl}(x) \phi_{cl}(y) \phi_{cl}(z)} = \langle \phi(x) \phi(y) \phi(z) \rangle_{1\text{PI}}.$$

By similar, if increasingly complicated, manipulations, one can derive the same relation for each successive derivative of Γ. For example, differentiating Eq. (11.95), we eventually find (using matrix notation with repeated indices implicitly integrated over)

$$\frac{-i\delta^4 E}{\delta J_w \delta J_x \delta J_y \delta J_z} = D_{sw} D_{xt} D_{yu} D_{zv} \left[\frac{i\delta^4 \Gamma}{\delta \phi_s^{cl} \delta \phi_t^{cl} \delta \phi_u^{cl} \delta \phi_v^{cl}} \right.$$
$$\left. + \frac{i\delta^3 \Gamma}{\delta \phi_s^{cl} \delta \phi_t^{cl} \delta \phi_r^{cl}} D_{qr} \frac{i\delta^3 \Gamma}{\delta \phi_q^{cl} \delta \phi_u^{cl} \delta \phi_v^{cl}} + (t \leftrightarrow u) + (t \leftrightarrow v) \right].$$

Since the left-hand side of this equation is the connected four-point function, we can rewrite it diagrammatically as

As above, the dark gray circles represent the sum of connected diagrams, while the light gray circles represent i times various derivatives of Γ. Subtracting the last three terms from each side removes all one-particle reducible pieces from the connected four-point function and so identifies the fourth derivative of $i\Gamma$ as the one-particle-irreducible four-point function. The general relation (for $n \geq 3$) is

$$\frac{\delta^n \Gamma[\phi_{\text{cl}}]}{\delta\phi_{\text{cl}}(x_1)\cdots\delta\phi_{\text{cl}}(x_n)} = -i\,\langle\phi(x_1)\cdots\phi(x_n)\rangle_{1\text{PI}}. \qquad (11.96)$$

In other words, the effective action is the generating functional of one-particle-irreducible correlation functions.

This conclusion implies that Γ contains the complete set of physical predictions of the quantum field theory. Let us review how this information is encoded. The vacuum state of the field theory is identified as the minimum of the effective potential. The location of the minimum determines whether the symmetries of the Lagrangian are preserved or spontaneously broken. The second derivative of Γ is the inverse propagator. The poles of the propagator, or the zeros of the inverse propagator, give the values of the particle masses. Thus the particle masses m^2 are determined as the values of p^2 that solve the equation

$$i\tilde{D}^{-1}(p^2) = \int d^4x\, e^{ip\cdot(x-y)}\frac{\delta^2\Gamma}{\delta\phi\delta\phi}(x,y) = 0. \qquad (11.97)$$

The higher derivatives of Γ are the one-particle-irreducible amplitudes. These can be connected by full propagators and joined together to construct four- and higher-point connected amplitudes, which give the S-matrix elements. Thus, from the knowledge of Γ, we can reconstruct the qualitative behavior of the quantum field theory, its pattern of symmetry-breaking, and then the quantitative details of its particles and their interactions.

11.6 Renormalization and Symmetry: General Analysis

In our analysis of the divergences of quantum field theories (especially in the paragraph below Eq. (10.4)), we noted that the basic divergences of Feynman integrals are associated with one-particle-irreducible diagrams. Thus we might expect that the effective action will be a useful object in discussing the renormalizability of quantum field theories, especially those with spontaneously broken symmetry. In this section we will make use of the effective action in precisely this way.

In Section 11.4, we saw in a particular example that the formalism for calculating the effective action provides the counterterms needed to remove the ultraviolet divergences, at least at the one-loop level. These counterterms were exactly those of the original Lagrangian. We will now argue that this set of counterterms is always sufficient—to all orders and for any renormalizable field theory—by applying the power-counting arguments of Section 10.1

directly to the computation of the effective action. We will use the language of scalar field theories, but the arguments can be generalized to theories of spinor and vector fields.

Consider first the computation of the effective potential for constant (x-independent) classical fields, in a field theory with an arbitrary number of fields ϕ^i. The effective potential has mass dimension 4, so we expect that $V_{\text{eff}}(\phi_{\text{cl}})$ will have divergent terms up to Λ^4. To understand these divergences, expand $V_{\text{eff}}(\phi_{\text{cl}})$ in a Taylor series:

$$V_{\text{eff}}(\phi_{\text{cl}}) = A_0 + A_2^{ij}\phi_{\text{cl}}^i\phi_{\text{cl}}^j + A_4^{ijk\ell}\phi_{\text{cl}}^i\phi_{\text{cl}}^j\phi_{\text{cl}}^k\phi_{\text{cl}}^\ell + \cdots.$$

In theories without a symmetry $\phi^i \to -\phi^i$, there might also be terms linear and cubic in ϕ^i; we omit these for simplicity. The coefficients A_0, A_2, A_4 have mass dimension, respectively, 4, 2, and 0; thus we expect them to contain Λ^4, Λ^2, and $\log \Lambda$ divergences, respectively. The power-counting analysis predicts that all higher terms in the Taylor series expansion should be finite. The constant term A_0 is independent of ϕ_{cl}; it has no physical significance. However, the divergences in A_2 and A_4 appear in physical quantities, since these coefficients enter the inverse propagator (11.90) and the irreducible four-point function (11.96) and therefore appear in the computation of S-matrix elements. There is one further coefficient in the effective action that has non-negative mass dimension by power counting; this is the coefficient of the term quadratic in $\partial_\mu \phi_{\text{cl}}^i$, which appears when the effective action is evaluated for a nonconstant background field:

$$\Delta\Gamma[\phi_{\text{cl}}] = \int d^4x \, B_2^{ij} \partial_\mu \phi_{\text{cl}}^i \partial^\mu \phi_{\text{cl}}^j. \tag{11.98}$$

All other coefficients in the Taylor expansion of the effective action in powers of ϕ_{cl}^i are finite by power counting.

We can now argue that the counterterms of the original Lagrangian suffice to remove the divergences that might appear in the computation of $\Gamma[\phi_{\text{cl}}]$. The argument proceeds in two steps. We first use the BPHZ theorem to argue that the divergences of Green's functions can be removed by adjusting a set of counterterms corresponding to the possible operators that can be added to the Lagrangian with coefficients of mass dimension greater than or equal to zero. The coefficients of these counterterms are in 1-to-1 correspondence with the coefficients A_2, A_4, and B_2 of the effective action. Next, we use the fact that the effective action is manifestly invariant to the original symmetry group of the model. This is true even if the vacuum state of the model has spontaneous symmetry breaking. This symmetry of the effective action follows from the analysis of Section 11.4, since the method we presented there for computing the effective action is manifestly invariant to the original symmetry of the Lagrangian. Combining these two results, we conclude that the effective action can always be made finite by adjusting the set of counterterms that are invariant to the original symmetry of the theory, even if this symmetry is spontaneously broken. By using the results of Section 11.5, which explain how to construct the Green's functions of the theory from the functional derivatives of the effective action, this conclusion of renormalizability extends to all the Green's functions of the theory.

To make this abstract argument more concrete, we will demonstrate in a simple example how the functional derivatives of the effective action yield a set of Feynman diagrams whose divergences correspond to symmetric counterterms. Let us, then, return once again to the $O(N)$-invariant linear sigma model and compute the second functional derivative of $\Gamma[\phi_{cl}]$. If the whole formalism we have constructed hangs together, we should be able to recognize the result as the Feynman diagram expansion of the inverse propagator, with divergences corresponding to the counterterms of $O(N)$-symmetric scalar field theory.

To begin, we write out expression (11.63) explicitly for this model:

$$\Gamma[\phi_{cl}] = \int d^4x \left(\frac{1}{2}(\partial_\mu^2 \phi_{cl}^i)^2 + \frac{1}{2}\mu^2(\phi_{cl}^i)^2 - \frac{\lambda}{4}((\phi_{cl}^i)^2)^2 + \frac{i}{2}\log\det[-i\mathcal{D}^{ij}] + \cdots \right),$$
(11.99)

where

$$-i\mathcal{D}^{ij} = -\frac{\delta^2 \mathcal{L}}{\delta\phi^i \delta\phi^j} = \partial^2 \delta^{ij} + \left(\lambda(\phi_{cl}^k(x))^2 - \mu^2\right)\delta^{ij} + 2\lambda\phi_{cl}^i(x)\phi_{cl}^j(x). \quad (11.100)$$

For constant ϕ_{cl}^i, $\mathcal{D}^{ij}$ is the operator that, acting on a given component of the scalar field, equals the Klein-Gordon operator with mass squared given by Eq. (11.69). This is the leading-order approximation to the inverse propagator of the linear sigma model.

To find the higher-order corrections to the inverse propagator, we must compute the second functional derivative of the quantum correction terms in $\Gamma[\phi_{cl}]$. From (11.99), we find

$$\frac{\delta^2\Gamma}{\delta\phi_{cl}^i(x)\delta\phi_{cl}^j(y)} = \frac{\delta^2\mathcal{L}}{\delta\phi_{cl}^i(x)\delta\phi_{cl}^j(y)} + \frac{i}{2}\frac{\delta^2}{\delta\phi_{cl}^i(x)\delta\phi_{cl}^j(y)}\log\det[-i\mathcal{D}] + \cdots.$$

The first term is just the Klein-Gordon operator $i\mathcal{D}^{ij}\delta(x - y)$. To compute the second term, use identity (9.77) for determinants of matrices:

$$\frac{\partial}{\partial\alpha}\log\det M(\alpha) = \frac{\partial}{\partial\alpha}\operatorname{tr}\log M(\alpha) = \operatorname{tr} M^{-1}\frac{\partial M}{\partial\alpha}. \quad (11.101)$$

Using this identity, we find

$$\frac{i}{2}\frac{\delta}{\delta\phi_{cl}^k(z)}\log\det[-i\mathcal{D}]$$

$$= i\left[\lambda\left(\phi_{cl}^k(z)\delta^{ij} + \phi_{cl}^i(z)\delta^{jk} + \phi_{cl}^j(z)\delta^{ik}\right)(i\mathcal{D}^{-1})^{ij}(z,z)\right]$$

$$= -\lambda\left(\phi_{cl}^k(z)\delta^{ij} + \phi_{cl}^i(z)\delta^{jk} + \phi_{cl}^j(z)\delta^{ik}\right)(\mathcal{D}^{-1})^{ij}(z,z). \quad (11.102)$$

The quantity $(\mathcal{D}^{-1})^{ij}(x, y)$ is the Klein-Gordon propagator. To differentiate a second time, we can use the identity (11.94); this yields

$$\frac{i}{2} \frac{\delta^2}{\delta\phi_{\text{cl}}^k(z)\delta\phi_{\text{cl}}^\ell(w)} \log \det[-i\mathcal{D}]$$

$$= -\lambda(\delta^{k\ell}\delta^{ij} + \delta^{ik}\delta^{j\ell} + \delta^{i\ell}\delta^{jk})(\mathcal{D}^{-1})^{ij}(z, z)\delta(z - w)$$

$$+ 2i\lambda^2(\phi_{\text{cl}}^k(z)\delta^{ij} + \phi_{\text{cl}}^i(z)\delta^{jk} + \phi_{\text{cl}}^j(z)\delta^{ik})(\mathcal{D}^{-1})^{im}(z, w)$$

$$\cdot (\phi_{\text{cl}}^\ell(w)\delta^{mn} + \phi_{\text{cl}}^m(w)\delta^{n\ell} + \phi_{\text{cl}}^n(w)\delta^{m\ell})(\mathcal{D}^{-1})^{nj}(w, z). \quad (11.103)$$

This is expected to be the formal correction to the inverse propagator at one-loop order, and indeed we can recognize in (11.103) the values of the one-loop diagrams

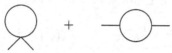

Notice how, in this derivation, every functional derivative on $\mathcal{D}^{-1}$ adds another propagator to the diagram and thus lowers the degree of divergence, in conformity with our general arguments in Section 10.1.

This example illustrates that the successive functional derivatives of $\Gamma[\phi_{\text{cl}}]$ are computed by a Feynman diagram expansion, with propagators and vertices that depend on the classical field. When the classical field is a constant, the propagators reduce to ordinary Klein-Gordon propagators and so the BPHZ theorem applies. All ultraviolet divergences can be removed from all of the amplitudes obtained by differentiating $\Gamma[\phi_{\text{cl}}]$ by the use of the most general set of mass, vertex, and field-strength renormalizations. At the same time, the perturbation theory is manifestly invariant to the symmetry of the original Lagrangian, and so the only divergences that appear—and thus the only counterterms required—are those that respect this symmetry. In general, then, all amplitudes of a renormalizable theory of scalar fields invariant under a symmetry group can be made finite using only the set of counterterms invariant to the symmetry. This gives a complete and quite satisfactory answer to the question posed at the beginning of Section 11.2.

The computation of the effective action in spatially varying background fields has not been analyzed at the level of rigor involved in the proof of the BPHZ theorem. However, it is expected that in this situation also, the standard set of counterterms for the symmetric theory should suffice. We can argue this intuitively by using the fact that the ultraviolet divergences of Feynman diagrams are local in spacetime. Thus, to understand the divergences of a computation in a background $\phi_{\text{cl}}(x)$ that is smoothly varying, we can divide spacetime into small boxes, in each of which $\phi_{\text{cl}}(x)$ is approximately constant, and expand in the derivatives $\partial_\mu\phi_{\text{cl}}(x)$. In this expansion in powers of $\partial_\mu\phi_{\text{cl}}(x)$, the Taylor series coefficients are functional derivatives of Γ in a constant background, which we know can be renormalized. The conclusion of this intuitive argument has been checked at the two-loop level for several nontrivial background field configurations.

Our general result on the renormalization of theories with spontaneously broken symmetry has an important implication for the physical predictions of these theories. In a renormalizable field theory, the most basic quantities of the theory cannot be predicted, because they are the quantities that must be specified as part of the definition of the theory. For example, in QED, the mass and charge of the electron must be adjusted from outside in order to define the theory. The predictions of QED are quantities that do not appear in the basic Lagrangian, for example, the anomalous magnetic moment of the electron. In renormalizable theories with spontaneously broken symmetry, however, the symmetry-breaking produces a large number of distinct masses and couplings, which depend on the relatively small number of parameters of the original symmetric theory. After the original parameters of the theory are fixed, any additional observable of the theory can be predicted unambiguously. For example, in the linear sigma model studied in this chapter, we took the values of the four-point coupling λ and the vacuum expectation value $\langle\phi\rangle$ as input parameters; we then calculated the mass of the σ particle in terms of these parameters in an unambiguous way.

There is a general argument that implies that, once we fix the parameters of the Lagrangian, we must find an unambiguous, finite formula for the σ mass in ϕ^4 theory, or, more generally, for any additional parameter of a renormalizable quantum field theory. In general, this parameter will be determined at the classical level in terms of the couplings in the Lagrangian. For the example of the σ mass in the linear sigma model, this classical relation is

$$m - \sqrt{2\lambda}\,\langle\phi\rangle = 0, \qquad (11.104)$$

where m is the mass of the σ and λ gives the four-ϕ scattering amplitude at threshold. In general, loop corrections will modify this relation, contributing some nonzero expression to the right-hand side of this equation. However, since Eq. (11.104) is valid at the classical level however the parameters of the Lagrangian are modified, it holds equally well when we add counterterms to the Lagrangian and then adjust these counterterms order by order. Thus, the counterterms must give zero contributions to the right-hand side of Eq. (11.104). Therefore, the perturbative corrections to Eq. (11.104) must be automatically ultraviolet-finite. A relation of this type, true at the classical level for all values of the couplings in the Lagrangian, but corrected by loop effects, is called a *zeroth-order natural relation*. The argument we have given implies that, for any such relation, the loop corrections are finite and constitute predictions of the quantum field theory. We will see another example of such a relation in Problem 11.2.

Goldstone's Theorem Revisited

As a final application of the effective action formalism, let us return to the question of whether Goldstone's theorem is valid in the presence of quantum corrections. Recall that we proved this theorem at the classical level at the end of Section 11.1: We showed in (11.13) that, if the Lagrangian has a continuous symmetry that is spontaneously broken, the matrix of second derivatives of the classical potential $V(\phi)$ has a corresponding zero eigenvalue. According to Eq. (11.11), this implies that the classical theory contains a massless scalar particle, associated with the spontaneously broken symmetry.

Using the effective action formalism, this argument can be repeated almost verbatim in the full quantum field theory. The effective potential $V_{\text{eff}}(\phi_{\text{cl}})$ encapsulates the full solution to the theory, including all orders of quantum corrections. At the same time, it satisfies the general properties of the classical potential: It is invariant to the symmetries of the theory, and its minimum gives the vacuum expectation value of ϕ. This means that the argument we gave in (11.13) works in exactly the same way for V_{eff} as it does for V: If a continuous symmetry of the original Lagrangian is spontaneously broken by $\langle\phi\rangle$, the matrix of second derivatives of $V_{\text{eff}}(\phi_{\text{cl}})$ has a zero eigenvalue along the symmetry direction.

We now argue that, just as at the classical level, the presence of such a zero eigenvalue implies the existence of a massless scalar particle. In our discussion of the general properties of the effective action, we showed that its second functional derivative is the inverse propagator, and that, through Eq. (11.97), this derivative yields the spectrum of masses in the quantum theory. Let us rewrite Eq. (11.97) for a theory that contains several scalar fields:

$$\int d^4x \, e^{ip \cdot (x-y)} \frac{\delta^2 \Gamma}{\delta \phi^i \delta \phi^j}(x, y) = 0. \qquad (11.105)$$

A particle of mass m corresponds to a zero eigenvalue of this matrix equation at $p^2 = m^2$. Now set $p = 0$. This implies that we differentiate $\Gamma[\phi_{\text{cl}}]$ with respect to constant fields. Thus, we can replace $\Gamma[\phi_{\text{cl}}]$ by its value with constant classical fields, which is just the effective potential. We find that the quantum field theory contains a scalar particle of zero mass when the matrix of second derivatives,

$$\frac{\partial^2 V_{\text{eff}}}{\partial \phi_{\text{cl}}^i \partial \phi_{\text{cl}}^j},$$

has a zero eigenvalue. This completes the proof of Goldstone's theorem.

This argument for Goldstone's theorem illustrates the power of the effective action formalism. The formalism gives a geometrical picture of spontaneous symmetry breaking that is valid to any order in quantum corrections. As a bonus, it is built up from objects that are renormalized in a simple way. This formalism will prove useful in understanding the applications of spontaneously broken symmetry that occur, in several different contexts, throughout the rest of this book.

Problems

11.1 Spin-wave theory.

(a) Prove the following wonderful formula: Let $\phi(x)$ be a free scalar field with propagator $\langle T\phi(x)\phi(0)\rangle = D(x)$. Then

$$\left\langle Te^{i\phi(x)}e^{-i\phi(0)}\right\rangle = e^{[D(x)-D(0)]}.$$

(The factor $D(0)$ gives a formally divergent adjustment of the overall normalization.)

(b) We can use this formula in Euclidean field theory to discuss correlation functions in a theory with spontaneously broken symmetry for $T < T_C$. Let us consider only the simplest case of a broken $O(2)$ or $U(1)$ symmetry. We can write the local spin density as a complex variable

$$s(x) = s^1(x) + is^2(x).$$

The global symmetry is the transformation

$$s(x) \rightarrow e^{-i\alpha}s(x).$$

If we assume that the physics freezes the modulus of $s(x)$, we can parametrize

$$s(x) = Ae^{i\phi(x)}$$

and write an effective Lagrangian for the field $\phi(x)$. The symmetry of the theory becomes the translation symmetry

$$\phi(x) \rightarrow \phi(x) - \alpha.$$

Show that (for $d > 0$) the most general renormalizable Lagrangian consistent with this symmetry is the free field theory

$$\mathcal{L} = \tfrac{1}{2}\rho(\vec{\nabla}\phi)^2.$$

In statistical mechanics, the constant ρ is called the *spin wave modulus*. A reasonable hypothesis for ρ is that it is finite for $T < T_C$ and tends to 0 as $T \rightarrow T_C$ from below.

(c) Compute the correlation function $\langle s(x)s^*(0)\rangle$. Adjust A to give a physically sensible normalization (assuming that the system has a physical cutoff at the scale of one atomic spacing) and display the dependence of this correlation function on x for $d = 1, 2, 3, 4$. Explain the significance of your results.

11.2 A zeroth-order natural relation. This problem studies an $N = 2$ linear sigma model coupled to fermions:

$$\mathcal{L} = \frac{1}{2}(\partial_\mu\phi^i)^2 + \frac{1}{2}\mu^2(\phi^i)^2 - \frac{\lambda}{4}((\phi^i)^2)^2 + \bar\psi(i\partial\!\!\!/)\psi - g\bar\psi(\phi^1 + i\gamma^5\phi^2)\psi \qquad (1)$$

where ϕ^i is a two-component field, $i = 1, 2$.

(a) Show that this theory has the following global symmetry:

$$\phi^1 \rightarrow \cos\alpha\,\phi^1 - \sin\alpha\,\phi^2,$$
$$\phi^2 \rightarrow \sin\alpha\,\phi^1 + \cos\alpha\,\phi^2, \qquad (2)$$
$$\psi \rightarrow e^{-i\alpha\gamma^5/2}\,\psi.$$

Show also that the solution to the classical equations of motion with the minimum energy breaks this symmetry spontaneously.

(b) Denote the vacuum expectation value of the field ϕ^i by v and make the change of variables

$$\phi^i(x) = (v + \sigma(x), \pi(x)). \qquad (3)$$

Write out the Lagrangian in these new variables, and show that the fermion acquires a mass given by

$$m_f = g \cdot v. \qquad (4)$$

(c) Compute the one-loop radiative correction to m_f, choosing renormalization conditions so that v and g (defined as the $\psi\psi\pi$ vertex at zero momentum transfer) receive no radiative corrections. Show that relation (4) receives nonzero corrections but that these corrections are *finite*. This is in accord with our general discussion in Section 11.6.

11.3 The Gross-Neveu model. The Gross-Neveu model is a model in two spacetime dimensions of fermions with a discrete chiral symmetry:

$$\mathcal{L} = \bar{\psi}_i i\partial\!\!\!/\psi_i + \tfrac{1}{2}g^2(\bar{\psi}_i\psi_i)^2$$

with $i = 1, \ldots, N$. The kinetic term of two-dimensional fermions is built from matrices γ^μ that satisfy the two-dimensional Dirac algebra. These matrices can be 2×2:

$$\gamma^0 = \sigma^2, \qquad \gamma^1 = i\sigma^1,$$

where σ^i are Pauli sigma matrices. Define

$$\gamma^5 = \gamma^0\gamma^1 = \sigma^3;$$

this matrix anticommutes with the γ^μ.

(a) Show that this theory is invariant with respect to

$$\psi_i \rightarrow \gamma^5\psi_i,$$

and that this symmetry forbids the appearance of a fermion mass.

(b) Show that this theory is renormalizable in 2 dimensions (at the level of dimensional analysis).

(c) Show that the functional integral for this theory can be represented in the following form:

$$\int \mathcal{D}\bar{\psi}\mathcal{D}\psi\; e^{i\int d^2x\,\mathcal{L}} = \int \mathcal{D}\bar{\psi}\mathcal{D}\psi\mathcal{D}\sigma\; \exp\left[i\int d^2x\left\{\bar{\psi}_i i\partial\!\!\!/\psi_i - \sigma\bar{\psi}_i\psi_i - \frac{1}{2g^2}\sigma^2\right\}\right],$$

where $\sigma(x)$ (not to be confused with a Pauli matrix) is a new scalar field with no kinetic energy terms.

(d) Compute the leading correction to the effective potential for σ by integrating over the fermion fields ψ_i. You will encounter the determinant of a Dirac operator; to evaluate this determinant, diagonalize the operator by first going to Fourier components and then diagonalizing the 2×2 Pauli matrix associated with each Fourier mode. (Alternatively, you might just take the determinant of this 2×2 matrix.) This 1-loop contribution requires a renormalization proportional to σ^2 (that is, a renormalization of g^2). Renormalize by minimal subtraction.

(e) Ignoring two-loop and higher-order contributions, minimize this potential. Show that the σ field acquires a vacuum expectation value which breaks the symmetry of part (a). Convince yourself that this result does not depend on the particular renormalization condition chosen.

(f) Note that the effective potential derived in part (e) depends on g and N according to the form

$$V_{\text{eff}}(\sigma_{\text{cl}}) = N \cdot f(g^2 N).$$

(The overall factor of N is expected in a theory with N fields.) Construct a few of the higher-order contributions to the effective potential and show that they contain additional factors of N^{-1} which suppress them if we take the limit $N \to \infty$, $(g^2 N)$ fixed. In this limit, the result of part (e) is unambiguous.

Chapter 12

The Renormalization Group

In the past two chapters, our main goal has been to determine when, and how, the cancellation of ultraviolet divergences in quantum field theory takes place. We have seen that, in a large class of field theories, the divergences appear only in the values of a few parameters: the bare masses and coupling constants, or, in renormalized perturbation theory, the counterterms. Aside from the shift in these parameters, virtual particles with very large momenta have no effect on computations in these theories.

The cancellation of ultraviolet divergences is essential if a theory is to yield quantitative physical predictions. But, at a deep level, the fact that high-momentum virtual quanta can have so little effect on a theory is quite surprising. One of the essential features of quantum field theory is locality, that is, the fact that fields at different spacetime points are independent degrees of freedom with independent quantum fluctuations. The quantum fluctuations at arbitrarily short distances appear in Feynman diagram computations as virtual quanta with arbitrarily high momenta. In a renormalizable theory, the loop integrals over virtual-particle momenta are always dominated by values comparable to the finite external particle momenta. But why? It is not easy to understand how the quantum fluctuations associated with extremely short distances can be so innocuous as to affect a theory only through the values of a few of its parameters.

This chapter begins with a physical picture, due to Kenneth Wilson, that explains this unusual and counterintuitive simplification. This picture generalizes the idea of the distance- or scale-dependent electric charge, introduced at the end of Chapter 7, and suggests that all of the parameters of a renormalizable field theory can usefully be thought of as scale-dependent entities. We will see that this scale dependence is described by simple differential equations, called *renormalization group* equations. The solutions of these equations will lead to physical predictions of a completely new type: predictions that, under certain circumstances, the correlation functions of a quantum field exhibit unusual but computable scaling laws as a function of their coordinates.

12.1 Wilson's Approach to Renormalization Theory

Wilson's method is based on the functional integral approach to field theory, in which the degrees of freedom of a quantum field are variables of integration. In this approach, one can study the origin of ultraviolet divergences by isolating the dependence of the functional integral on the short-distance degrees of freedom of the field.* In this section, we will illustrate this idea in the simplest example of ϕ^4 theory.

To make our analysis more concrete, we will drop the elegant but somewhat mysterious method of dimensional regularization in this section and instead use a sharp momentum cutoff. Since we will be working here only in ϕ^4 theory, we will not be concerned that this cutoff makes it difficult to satisfy Ward identities. Wilson's analysis can be adapted to QED and other situations where this subtlety is important, but the case of ϕ^4 theory is sufficient to give us the basic qualitative results of this approach.

In Section 9.2, we constructed the Green's functions of ϕ^4 theory in terms of a functional integral representation of the generating functional $Z[J]$. The basic integration variables are the Fourier components of the field $\phi(k)$, so $Z[J]$ is given concretely by the expression

$$Z[J] = \int \mathcal{D}\phi \, e^{i\int [\mathcal{L}+J\phi]} = \left(\prod_k \int d\phi(k) \right) e^{i\int [\mathcal{L}+J\phi]}. \tag{12.1}$$

To impose a sharp ultraviolet cutoff Λ, we restrict the number of the integration variables displayed in (12.1). That is, we integrate only over $\phi(k)$ with $|k| \leq \Lambda$, and set $\phi(k) = 0$ for $|k| > \Lambda$.

This modification of the functional integral suggests a method for assessing the influence of the quantum fluctuations at very short distances or very large momenta. In the functional integral representation, these fluctuations are represented by the integrals over the Fourier components of ϕ with momenta near the cutoff. Why not explicitly perform the integrals over these variables? Then we can compare the result to the original functional integral, and determine precisely the influence of these high-momentum modes on the physical predictions of the theory.

Before beginning this analysis, though, we must introduce one modification. At first sight, it seems most natural to define the ultraviolet cutoff in Minkowski space. However, a cutoff $k^2 \leq \Lambda^2$ is not completely effective in controlling large momenta, since in lightlike directions the components of k can be very large while k^2 remains small. We will therefore consider the cutoff to be imposed on the Euclidean momenta obtained after Wick rotation. Equivalently, we consider the Euclidean form of the functional integral, presented in Section 9.3, and restrict its variables $\phi(k)$, with k Euclidean, to $|k| \leq \Lambda$.

---

*Wilson's ideas are reviewed in K. G. Wilson and J. Kogut, *Phys. Repts.* **12C**, 75 (1974).

The transition to Euclidean space also brings us closer to the connection between renormalization theory and statistical mechanics advertised in Chapter 8. As we saw in Section 9.3, the Euclidean functional integral for ϕ^4 theory has precisely the same form as the continuum description of the statistical mechanics of a magnet. The field $\phi(x)$ is interpreted as the fluctuating spin field $s(x)$. A real magnet is built of atoms, and the atomic spacing provides a physical cutoff, a shortest distance over which fluctuations can take place. The cut-off functional integral models the effects of this atomic size in a crude way.

By pursuing this analogy, we can derive some physical intuition about the effects of the ultraviolet cutoff in a field theory. In a magnet, it is quite easy to visualize statistical fluctuations of the spins at the atomic scale. In fact, for values of the temperature away from any critical points, the statistical fluctuations are restricted to this scale; over distances of tens of atomic spacings, the magnet already shows its homogeneous macroscopic behavior. We have seen in Chapter 8 that we can approximate the correlation function of the spin field by the propagator of a Euclidean ϕ^4 theory. In this approximation,

$$\langle s(x)s(0)\rangle = \int \frac{d^4k}{(2\pi)^4} \frac{e^{ik\cdot x}}{k^2+m^2} \underset{|x|\to\infty}{\sim} e^{-m|x|}. \tag{12.2}$$

As long as the temperature is far from the critical temperature, the size of the "mass" m is determined by the one natural scale in the problem, the atomic spacing. Thus, we expect $m \approx \Lambda$. In our field theory calculations, we were specifically interested in the situation where $m \ll \Lambda$, and we adjusted the parameters of the theory to satisfy this condition. In describing a magnet, it appears that no such adjustment is called for.

However, we saw in Chapter 8 that there is one circumstance in which the correlations of the spin field are much longer than the atomic spacing, so that, indeed, $m \ll \Lambda$. When the spin system begins to magnetize, just in the vicinity of the critical point, the spins become correlated over arbitrarily long distances as the fluctuating spins attempt to choose their eventual direction of magnetization. To study these long-range correlations in a magnet, one must carefully adjust the temperature to bring the system into the vicinity of the phase transition. In the same way, we can imagine making a fine adjustment of the parameter m of ϕ^4 theory to bring the quantum field theory into a region of parameters where we do find correlations of the field $\phi(x)$ over distances much larger than $1/\Lambda$.

Integrating Over a Single Momentum Shell

With this introduction, we will now carry out the integration over the high-momentum degrees of freedom of ϕ. We begin by writing the functional integral (12.1) more explicitly for the case of ϕ^4 theory. We apply the cutoff prescription described earlier, and set $J = 0$ for simplicity. Then

$$Z = \int [\mathcal{D}\phi]_\Lambda \, \exp\left(-\int d^dx \left[\frac{1}{2}(\partial_\mu \phi)^2 + \frac{1}{2}m^2\phi^2 + \frac{\lambda}{4!}\phi^4\right]\right), \qquad (12.3)$$

where

$$[\mathcal{D}\phi]_\Lambda = \prod_{|k|<\Lambda} d\phi(k). \qquad (12.4)$$

In the Lagrangian of Eq. (12.3), m and λ are the bare parameters, and so there are no counterterms. As in our study of the superficial degree of divergence, it will be useful to carry out this analysis in an arbitrary spacetime dimension d.

We now divide the integration variables $\phi(k)$ into two groups. Choose a fraction $b < 1$. The variables $\phi(k)$ with $b\Lambda \leq |k| < \Lambda$ are the high-momentum degrees of freedom that we will integrate over. To label these degrees of freedom, let us define

$$\hat{\phi}(k) = \begin{cases} \phi(k) & \text{for } b\Lambda \leq |k| < \Lambda; \\ 0 & \text{otherwise.} \end{cases}$$

Next, let us define a new $\phi(k)$, which is identical to the old for $|k| < b\Lambda$ and zero for $|k| > b\Lambda$. Then we can replace the old ϕ in the Lagrangian with $\phi + \hat{\phi}$, and rewrite Eq. (12.3) as

$$Z = \int \mathcal{D}\phi \int \mathcal{D}\hat{\phi} \, \exp\left(-\int d^dx \left[\frac{1}{2}(\partial_\mu \phi + \partial_\mu \hat{\phi})^2 + \frac{1}{2}m^2(\phi + \hat{\phi})^2 + \frac{\lambda}{4!}(\phi + \hat{\phi})^4\right]\right)$$

$$= \int \mathcal{D}\phi \, e^{-\int \mathcal{L}(\phi)} \int \mathcal{D}\hat{\phi} \, \exp\left(-\int d^dx \left[\frac{1}{2}(\partial_\mu \hat{\phi})^2 + \frac{1}{2}m^2\hat{\phi}^2\right.\right.$$

$$\left.\left. + \lambda\left(\frac{1}{6}\phi^3\hat{\phi} + \frac{1}{4}\phi^2\hat{\phi}^2 + \frac{1}{6}\phi\hat{\phi}^3 + \frac{1}{4!}\hat{\phi}^4\right)\right]\right). \quad (12.5)$$

In the final expression we have gathered all terms independent of $\hat{\phi}$ into $\mathcal{L}(\phi)$. Note that quadratic terms of the form $\phi\hat{\phi}$ vanish, since Fourier components of different wavelengths are orthogonal.

The next few paragraphs will explain how to perform the integral over $\hat{\phi}$. This integration will transform (12.5) into an expression of the form

$$Z = \int [\mathcal{D}\phi]_{b\Lambda} \, \exp\left(-\int d^dx \, \mathcal{L}_{\text{eff}}\right), \qquad (12.6)$$

where $\mathcal{L}_{\text{eff}}(\phi)$ involves only the Fourier components $\phi(k)$ with $|k| < b\Lambda$. We will see that $\mathcal{L}_{\text{eff}}(\phi) = \mathcal{L}(\phi)$ plus corrections proportional to powers of λ. These correction terms compensate for the removal of the large-k Fourier components $\hat{\phi}$, by supplying the interactions among the remaining $\phi(k)$ that were previously mediated by fluctuations of the $\hat{\phi}$.

To carry out the integrals over the $\hat{\phi}(k)$, we use the same method that we applied in Section 9.2 to derive Feynman rules. In fact, we will see below that the new terms in $\mathcal{L}_{\text{eff}}$ can be written in a diagrammatic form. In this analysis, we treat the quartic terms in (12.5), all proportional to λ, as perturbations.

Since we are mainly interested in the situation $m^2 \ll \Lambda^2$, we will also treat the mass term $\frac{1}{2}m^2\hat{\phi}^2$ as a perturbation. Then the leading-order term in the portion of the Lagrangian involving $\hat{\phi}$ is

$$\int \mathcal{L}_0 = \frac{1}{2} \int_{b\Lambda \leq |k| < \Lambda} \frac{d^d k}{(2\pi)^d} \, \hat{\phi}^*(k) k^2 \hat{\phi}(k). \qquad (12.7)$$

This term leads to a propagator

$$\overbrace{\hat{\phi}(k)\hat{\phi}(p)} = \frac{\int \mathcal{D}\hat{\phi} \, e^{-\int \mathcal{L}_0} \hat{\phi}(k)\hat{\phi}(p)}{\int \mathcal{D}\hat{\phi} \, e^{-\int \mathcal{L}_0}} = \frac{1}{k^2} (2\pi)^d \delta^{(d)}(k+p) \Theta(k), \qquad (12.8)$$

where

$$\Theta(k) = \begin{cases} 1 & \text{if } b\Lambda \leq |k| < \Lambda; \\ 0 & \text{otherwise.} \end{cases} \qquad (12.9)$$

We will regard the remaining $\hat{\phi}$ terms in Eq. (12.5) as perturbations, and expand the exponential. The various contributions from these perturbations can be evaluated by using Wick's theorem with (12.8) as the propagator.

First consider the term that results from expanding to one power of the $\phi^2\hat{\phi}^2$ term in the exponent of (12.5). We find

$$-\int d^d x \, \frac{\lambda}{4} \phi^2 \overbrace{\hat{\phi}\hat{\phi}} = -\frac{1}{2} \int \frac{d^d k_1}{(2\pi)^d} \, \mu \, \phi(k_1)\phi(-k_1), \qquad (12.10)$$

where the coefficient μ is the result of contracting the two $\hat{\phi}$ fields:

$$\mu = \frac{\lambda}{2} \int_{b\Lambda \leq |k| < \Lambda} \frac{d^d k}{(2\pi)^d} \frac{1}{k^2} = \frac{\lambda}{(4\pi)^{d/2}\Gamma(\frac{d}{2})} \frac{1 - b^{d-2}}{d - 2} \Lambda^{d-2}. \qquad (12.11)$$

The term (12.10) could just as well have arisen from an expansion of the exponential

$$\exp\left(-\int d^d x \, \frac{1}{2}\mu\phi^2 + \cdots\right). \qquad (12.12)$$

We will soon see that the rest of the perturbation series also organizes itself into this form. The coefficient μ therefore gives a positive correction to the m^2 term in $\mathcal{L}$.

The higher orders of the perturbation theory in the correction terms can be worked out in a similar way. As in our derivation of the standard perturbation theory for ϕ^4 theory, it is useful to adopt a diagrammatic notation. Represent the propagator (12.8) by a double line. This propagator will connect pairs of fields $\hat{\phi}$ from the various quartic interactions. Represent the fields ϕ in these interactions, which are not integrated over, as single external lines.

Then, for example, the contribution of (12.10) corresponds to the following diagram:

At order λ^2, we will have, among other contributions, terms involving the contractions of two interaction terms $\lambda\phi^2\hat{\phi}^2$. Each term corresponds to a vertex connecting two single lines and two double lines. There are two possible contractions:

$$\left(\begin{array}{c} \text{\includegraphics{loop}} \end{array} \right)^2, \quad \text{\includegraphics{sunset}} \tag{12.13}$$

Of these, the first, which is a disconnected diagram, supplies the order-λ^2 term in the exponential (12.12). The second is a new contribution, which will become a correction to the ϕ^4 interaction in $\mathcal{L}(\phi)$.

Let us now evaluate this second contribution. For simplicity, we consider the limit in which the external momenta carried by the factors ϕ are very small compared to $b\Lambda$, so we can ignore them. Then this diagram has the value

$$-\frac{1}{4!}\int d^d x\, \zeta\, \phi^4, \tag{12.14}$$

where

$$\zeta = -4!\frac{2}{2!}\left(\frac{\lambda}{4}\right)^2 \int\limits_{b\Lambda\le |k|<\Lambda} \frac{d^d k}{(2\pi)^d}\left(\frac{1}{k^2}\right)^2 = \frac{-3\lambda^2}{(4\pi)^{d/2}\Gamma(\frac{d}{2})}\frac{(1-b^{d-4})}{d-4}\Lambda^{d-4}$$

$$\xrightarrow[d\to 4]{} -\frac{3\lambda^2}{16\pi^2}\log\frac{1}{b}. \tag{12.15}$$

The 2 in the numerator counts the two possible contractions; there are no additional combinatoric factors from counting external legs or vertices. In the analysis of ϕ^4 theory in Section 10.2, we encountered a similar diagram, integrated over a range of momenta from 0 to Λ, producing a logarithmic ultraviolet divergence. In Wilson's treatment this divergence is not a pathology but simply a sign that the diagram is receiving contributions from all momentum scales. Indeed, it receives an equal contribution from each logarithmic interval between the momentum scales m and Λ. We will see below that the (finite) contribution to this diagram from each momentum interval has a natural physical importance.

The diagrammatic perturbation theory we have described not only generates contributions proportional to ϕ^2 and ϕ^4 but also to higher powers of ϕ. For example, the following diagram generates a contribution to a ϕ^6 interaction:

$$\propto \frac{\lambda^2}{(p_1+p_2+p_3)^2}\Theta(p_1+p_2+p_3). \tag{12.16}$$

There are also derivative interactions, which arise when we no longer neglect the external momenta of the diagrams. A more exact treatment would Taylor-expand in these momenta; for instance, in addition to expression (12.14), we would obtain terms with two powers of external momenta, which we could rewrite as

$$-\frac{1}{4}\int d^dx\, \eta\, \phi^2(\partial_\mu\phi)^2. \qquad (12.17)$$

We would also find terms with four, six, and more powers of the momenta carried by the ϕ. In general, the procedure of integrating out the $\hat{\phi}$ generates all possible interactions of the fields ϕ and their derivatives.

The diagrammatic corrections can be simplified slightly by resumming them as an exponential. We have seen already in (12.13) that our diagrammatic expansion generates disconnected diagrams. By the same combinatoric argument that we used in Eq. (4.52), we can rewrite the sum of the series as the exponential of the sum of the connected diagrams. This leads precisely to expression (12.6), with

$$\mathcal{L}_{\text{eff}} = \frac{1}{2}(\partial_\mu\phi)^2 + \frac{1}{2}m^2\phi^2 + \frac{1}{4!}\lambda\phi^4 + (\text{sum of connected diagrams}). \quad (12.18)$$

The diagrammatic contributions include corrections to m^2 and λ, as well as all possible higher-dimension operators. We can now use the new Lagrangian $\mathcal{L}_{\text{eff}}(\phi)$ to compute correlation functions of the $\phi(k)$, or to compute S-matrix elements. Since the $\phi(k)$ include only momenta up to $b\Lambda$, the loop diagrams in such a calculation would be integrated only up to that lowered cutoff. The correction terms in (12.18) precisely compensate for this change.

One might well be puzzled by the appearance of higher-dimension operators in Eq. (12.18). We chose the original Lagrangian of ϕ^4 theory to contain only renormalizable interactions. At first sight, it is disturbing that all possible nonrenormalizable interactions appear when we integrate out the variables $\hat{\phi}$. However, we will see below that our procedure actually keeps the contributions of these nonrenormalizable interactions under control. In fact, our analysis will imply that the presence of nonrenormalizable interactions in the original Lagrangian, defined to be used with very large cutoff Λ, has negligible effect on physics at scales much less than Λ.

Renormalization Group Flows

Let us now make a more careful comparison of the new functional integral (12.6) and the one we started with (12.3). The most convenient way to do this is to rescale distances and momenta in (12.6) according to

$$k' = k/b, \qquad x' = xb, \qquad (12.19)$$

so that the variable k' is integrated over $|k'| < \Lambda$. Let us express the explicit form of (12.18) schematically as

$$\int d^d x \, \mathcal{L}_{\text{eff}} = \int d^d x \left[\frac{1}{2}(1 + \Delta Z)(\partial_\mu \phi)^2 + \frac{1}{2}(m^2 + \Delta m^2)\phi^2 \right.$$
$$\left. + \frac{1}{4!}(\lambda + \Delta\lambda)\phi^4 + \Delta C(\partial_\mu \phi)^4 + \Delta D\phi^6 + \cdots \right]. \tag{12.20}$$

In terms of the rescaled variable x', this becomes

$$\int d^d x \, \mathcal{L}_{\text{eff}} = \int d^d x' \, b^{-d} \left[\frac{1}{2}(1 + \Delta Z)b^2(\partial_\mu' \phi)^2 + \frac{1}{2}(m^2 + \Delta m^2)\phi^2 \right.$$
$$\left. + \frac{1}{4!}(\lambda + \Delta\lambda)\phi^4 + \Delta C b^4(\partial_\mu' \phi)^4 + \Delta D\phi^6 + \cdots \right]. \tag{12.21}$$

Throughout this analysis, we have treated all terms beyond the first as small perturbations. As long as the original couplings are small, this is still a valid approximation in treating (12.21).

The original functional integral led to the propagator (12.8). The new action (12.21) will give rise to exactly the same propagator, if we rescale the field ϕ according to

$$\phi' = \left[b^{2-d}(1 + \Delta Z) \right]^{1/2} \phi. \tag{12.22}$$

After this rescaling, the unperturbed action returns to its initial form, while the various perturbations undergo a transformation:

$$\int d^d x \, \mathcal{L}_{\text{eff}} = \int d^d x' \left[\frac{1}{2}(\partial_\mu' \phi')^2 + \frac{1}{2}m'^2 \phi'^2 \right.$$
$$\left. + \frac{1}{4!}\lambda'\phi'^4 + C'(\partial_\mu' \phi')^4 + D'\phi'^6 + \cdots \right]. \tag{12.23}$$

The new parameters of the Lagrangian are

$$
\begin{aligned}
m'^2 &= (m^2 + \Delta m^2)(1 + \Delta Z)^{-1} b^{-2}, \\
\lambda' &= (\lambda + \Delta\lambda)(1 + \Delta Z)^{-2} b^{d-4}, \\
C' &= (C + \Delta C)(1 + \Delta Z)^{-2} b^{d}, \\
D' &= (D + \Delta D)(1 + \Delta Z)^{-3} b^{2d-6},
\end{aligned}
\tag{12.24}
$$

and so on. (The original Lagrangian had $C = D = 0$, but the same equations would apply if the initial values of C and D were nonzero.) All of the corrections, Δm^2, $\Delta\lambda$, and so on, arise from diagrams and thus are small compared to the leading terms if perturbation theory is justified.

By combining the operation of integrating out high-momentum degrees of freedom with the rescaling (12.19), we have rewritten this operation as a transformation of the Lagrangian. Continuing this procedure, we could integrate over another shell of momentum space and transform the Lagrangian further. Successive integrations produce further iterations of the transformation (12.24). If we take the parameter b to be close to 1, so that the shells of momentum space are infinitesimally thin, the transformation becomes a continuous one. We can then describe the result of integrating over the high-momentum degrees of freedom of a field theory as a trajectory or a flow in the space of all possible Lagrangians.

For historical reasons, these continuously generated transformations of Lagrangians are referred to as the *renormalization group*. They do not form a group in the formal sense, because the operation of integrating out degrees of freedom is not invertible. On the other hand, they are most certainly connected to renormalization, as we will now see.

Imagine that we wish to compute a correlation function of fields whose momenta p_i are all much less than Λ. We could compute this correlation function perturbatively using either the original Lagrangian $\mathcal{L}$, or the effective Lagrangian $\mathcal{L}_{\text{eff}}$ obtained after integrating over all momentum shells down to the scale of the external momenta p_i. Both procedures must ultimately yield the same result. But in the first case, the effects of high-momentum fluctuations of the field do not show up until we compute loop diagrams. In the second case, these effects have already been absorbed into the new coupling constants $(m', \lambda', \text{etc.})$, so their influence can be seen directly from the Lagrangian. In the first procedure, the large shifts from the original (bare) parameters to the values appropriate to low-momentum processes appear suddenly in one-loop diagrams, and seem to invalidate the use of perturbation theory. In the second approach, these corrections are introduced slowly and systematically. A perturbative treatment is valid at every step as long as the effective coupling constants such as λ' remain small.

However, the parameters of the effective Lagrangian may be very different from those of the original Lagrangian, since we must iterate the transformation (12.24) many times to get from the large momentum Λ down to the momentum scale of typical experiments. Let us therefore look more closely at how the Lagrangian tends to vary under the renormalization group transformations.

The simplest case to consider is a Lagrangian in the vicinity of the point $m^2 = \lambda = C = D = \cdots = 0$, where all the perturbations vanish. We have defined our transformation so that this point is left unchanged; we say that the free-field Lagrangian

$$\mathcal{L}_0 = \tfrac{1}{2}(\partial_\mu \phi)^2 \tag{12.25}$$

is a *fixed point* of the renormalization group transformation.

In the vicinity of $\mathcal{L}_0$, we can ignore the terms Δm^2, $\Delta \lambda$, etc., in the iteration equations (12.24) and keep only those terms that are linear in the perturbations. This gives an especially simple transformation law:

$$m'^2 = m^2 b^{-2}, \quad \lambda' = \lambda b^{d-4}, \quad C' = C b^d, \quad D' = D b^{2d-6}, \quad \text{etc.} \tag{12.26}$$

Since $b < 1$, those parameters that are multiplied by negative powers of b

grow, while those that are multiplied by positive powers of b decay. If the Lagrangian contains growing coefficients, these will eventually carry it away from $\mathcal{L}_0$.

It is conventional to speak of the various terms in the effective Lagrangian as a set of local operators that can be added as perturbations to $\mathcal{L}_0$. We call the operators whose coefficients grow during the recursion procedure *relevant* operators. The coefficients that die away are associated with *irrelevant* operators. For example, the scalar field mass operator ϕ^2 is always relevant, while the ϕ^4 operator is relevant if $d < 4$. If the coefficient of some operator is multiplied by b^0 (for example, the operator ϕ^4 in $d = 4$), we call this operator *marginal*; to find out whether its coefficient grows or decays, we must include the effect of higher-order corrections.

In general, an operator with N powers of ϕ and M derivatives has a coefficient that transforms as

$$C'_{N,M} = b^{N(d/2-1)+M-d}C_{N,M}. \tag{12.27}$$

Notice that the coefficient is just $(d_{N,M}-d)$, where $d_{N,M}$ is the mass dimension of the operator as computed at the end of Section 10.1. In other words, relevant and marginal operators about the free theory $\mathcal{L}_0$ correspond precisely to super-renormalizable and renormalizable interaction terms in the power-counting analysis of Section 10.1.

We can also understand the evolution of coefficients near the free-field fixed point using straightforward dimensional analysis. An operator with mass dimension d_i has a coefficient with dimension $(\text{mass})^{d-d_i}$. The natural order of magnitude for this mass is the cutoff Λ. Thus, if $d_i < d$, the perturbation is increasingly important at low momenta. On the other hand, if $d_i > d$, the relative size of this term decreases as $(p/\Lambda)^{d_i-d}$ as the momentum $p \to 0$; thus the term is truly irrelevant.

We have now shown that, at least in the vicinity of the zero-coupling fixed point, an arbitrarily complicated Lagrangian at the scale of the cutoff degenerates to a Lagrangian containing only a finite number of renormalizable interactions. It is instructive to compare this result with the conclusions of Chapter 10. There we took the philosophy that the cutoff Λ should be disposed of by taking the limit $\Lambda \to \infty$ as quickly as possible. We found that this limit gives well-defined predictions only if the Lagrangian contains no parameters with negative mass dimension. From this viewpoint, it seemed exceedingly fortunate that QED, for example, contained no such parameters, since otherwise this theory would not yield well-defined predictions.

Wilson's analysis takes just the opposite point of view, that any quantum field theory is defined fundamentally with a cutoff Λ that has some physical significance. In statistical mechanical applications, this momentum scale is the inverse atomic spacing. In QED and other quantum field theories appropriate to elementary particle physics, the cutoff would have to be associated with some fundamental graininess of spacetime, perhaps a result of quantum fluctuations in gravity. We discuss some speculations on the nature of this

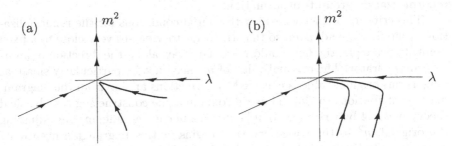

Figure 12.1. Renormalization group flows near the free-field fixed point in scalar field theory: (a) $d > 4$; (b) $d = 4$.

cutoff in the Epilogue. But whatever this scale is, it lies far beyond the reach of present-day experiments. The argument we have just given shows that this circumstance *explains* the renormalizability of QED and other quantum field theories of particle interactions. Whatever the Lagrangian of QED was at its fundamental scale, as long as its couplings are sufficiently weak, it must be described at the energies of our experiments by a renormalizable effective Lagrangian.

On the other hand, we should emphasize that these simple conclusions can be altered by sufficiently strong field theory interactions. Away from the free-field fixed point, the simple transformation laws (12.26) receive corrections proportional to higher powers of the coupling constants. If these corrections are large enough, they can halt or reverse the renormalization group flow. They could even create new fixed points, which would give new types of $\Lambda \to \infty$ limits.

To illustrate the possible influences of interactions in a relatively simple context, let us discuss the renormalization group flows near $\mathcal{L}_0$ for the specific case of ϕ^4 theory. It is instructive to consider the three cases $d > 4$, $d = 4$, and $d < 4$ in turn. When $d > 4$, the only relevant operator is the scalar field mass term. Then the renormalization group flows near $\mathcal{L}_0$ have the form shown in Fig. 12.1(a). The ϕ^4 interaction and possible higher-order interactions die away, while the mass term increases in importance.

In previous chapters, we have always discussed ϕ^4 theory in the limit in which the mass is small compared to the cutoff. Let us take a moment to rewrite this condition in the language of renormalization group flows. In the course of the flow, the effective mass term m'^2 becomes large and eventually comes to equal the current cutoff. For example, near the free-field fixed point, after n iterations, $m'^2 = m^2 b^{-2n}$, and eventually there is an n such that $m'^2 \sim \Lambda^2$. At this point, we have integrated out the entire momentum region between the original Λ and the effective mass of the scalar field. The mass term then suppresses the remaining quantum fluctuations. In general, the criterion that the scalar field mass is small compared to the cutoff is equivalent to the statement that $m'^2 \sim \Lambda^2$ only after a large number of iterations of the renormalization group transformation.

This criterion is met whenever the initial conditions for the renormalization group flow are adjusted so that the trajectory passes very close to a fixed point. In principle, the flow could begin far away, along the direction of an irrelevant operator. The original value of m^2 need not be particularly small, as long as this original value is canceled by corrections arising from the diagrammatic contributions to $\mathcal{L}_{\text{eff}}$. Thus we could imagine constructing a scalar field theory in $d > 4$ by writing a complicated nonlinear Lagrangian, but adjusting the original m^2 so the trajectory that begins at this Lagrangian eventually passes close to the free-field fixed point $\mathcal{L}_0$. In this case, the effective theory at momenta small compared to the cutoff should be extremely simple: It will be a free field theory with negligible nonlinear interaction. As will be discussed in the next chapter, this remarkable prediction has been verified in mathematical models of magnetic systems in more than four dimensions: Even though the original model is highly nonlinear, the correlation function of spins near the phase transition has the free-field form given by the higher-dimensional analogue of Eq. (12.2).

Next consider the case $d = 4$. For this case, Eq. (12.26) does not give enough information to tell us whether the ϕ^4 interaction is important or unimportant at large distances. So we must go back to the complete transformation law (12.24). The leading contribution to $\Delta\lambda$ is given by Eq. (12.15). The leading contribution to ΔZ is of order λ^2 and can be neglected. (This is just what happened with the first correction to δ_Z in Section 10.2.) Thus we find the transformation

$$\lambda' = \lambda - \frac{3\lambda^2}{16\pi^2}\log(1/b). \tag{12.28}$$

This says that λ slowly decreases as we integrate out high-momentum degrees of freedom.

The diagram contributing to the correction $\Delta\lambda$ has the same structure as the one-loop diagrams computed in Section 10.2. In fact, these are essentially the same diagrams, and differ only in whether the integrals are carried out iteratively or all at once. However, whereas the diagrams in Section 10.2 had ultraviolet divergences, the corresponding diagram in Wilson's approach is well defined and gives the coefficient of a simple evolution equation of the coupling constant. This transformation gives a first example of the reinterpretation of ultraviolet divergences that we will make in this chapter.

The transformation law (12.28) implies that the renormalization group flows near $\mathcal{L}_0$ have the form shown in Fig. 12.1(b), with one slowly decaying direction. If we follow the flows far enough, the behavior should again be that of a free field. This picture has the puzzling implication that four-dimensional interacting ϕ^4 theory does not exist in the limit in which the cutoff goes to infinity. We will discuss this result further—and explain why it nevertheless makes sense to use ϕ^4 theory as a model field theory—in Section 12.3.

Finally consider the case $d < 4$. Now λ becomes a relevant parameter.

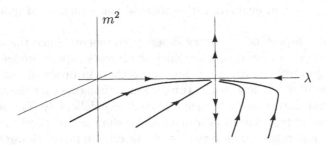

Figure 12.2. Renormalization group flows near the free-field fixed point in scalar field theory: $d < 4$.

The theory thus flows away from the free theory $\mathcal{L}_0$ as we integrate out degrees of freedom; at large distances, the ϕ^4 interaction becomes increasingly important. However, when λ becomes large, the nonlinear corrections such as that displayed in Eq. (12.28) must also be considered. If we include this specific effect in $d < 4$, we find the recursion formula

$$\lambda' = \left[\lambda - \frac{3\lambda^2}{(4\pi)^{d/2}\Gamma(\frac{d}{2})} \frac{b^{d-4}-1}{4-d}\Lambda^{d-4}\right]b^{d-4}. \tag{12.29}$$

This equation implies that there is a value of λ at which the increase due to rescaling is compensated by the decrease caused by the nonlinear effect. At this value, λ is unchanged when we integrate out degrees of freedom. The corresponding Lagrangian is a second fixed point of the renormalization group flow. In the limit $d \to 4$, the flow (12.29) tends to (12.28) and so the new fixed point merges with the free field fixed point. For d sufficiently close to 4, the new fixed point will share with $\mathcal{L}_0$ the property that the mass parameter m^2 is increased by the iteration. Then the mass operator will be a relevant operator near the new fixed point, so that the renormalization group flows will have the form shown in Fig. 12.2.

In this example, the new fixed point of the renormalization group had a Lagrangian with couplings weak enough that the transformation equations could be computed in perturbation theory. In principle, one could also find fixed points whose Lagrangians are strongly coupled, so that the renormalization group transformations cannot be understood by Feynman diagram analysis. Many examples of such fixed points are known in exactly solvable model field theories in two dimensions.[†] However, up to the present, all of the examples of quantum field theories that are important for physical applications have been found to be controlled either by the free field fixed point or by fixed points, like the one described in the previous paragraph, that approach the free-field fixed point in a specific limit. No one understands why this should be. This observation implies that Feynman diagram analysis has

---

[†]We mention some of these examples, and discuss other nonperturbative approaches to quantum field theory, in the Epilogue.

unexpected power in evaluating the physical consequences of quantum field theories.

One more aspect of ϕ^4 theory deserves comment. Since the mass term, $m^2\phi^2$, is a relevant operator, its coefficient diverges rapidly under the renormalization group flow. We have seen above that, in order to end up at the desired value of m^2 at low momentum, we must imagine that the value of m^2 in the original Lagrangian has been adjusted very delicately. This adjustment has a natural interpretation in a magnetic system as the need to sensitively adjust the temperature to be very close to the critical point. However, it seems quite artificial when applied to the quantum field theory of elementary particles, which purports to be a fundamental theory of Nature. This problem appears only for scalar fields, since for fermions the renormalization of the mass is proportional to the bare mass rather than being an arbitrary additive constant. Perhaps this is the reason why there seem to be no elementary scalar fields in Nature. We will return to this question in the Epilogue.

12.2 The Callan-Symanzik Equation

Wilson's picture of renormalization, as a flow in the space of possible Lagrangians, is beautifully intuitive, and gives us a deep understanding of why Nature should be describable in terms of renormalizable quantum field theories. In addition, however, this idea can be applied to extract further quantitative predictions from these theories. In the remainder of this chapter we will develop a formalism for extracting these predictions. Specifically, we will see that Wilson's picture leads to predictions for the form of the high- and low-momentum behavior of correlation functions. In the simplest cases, the correlation functions turn out to scale as powers of their external momenta, with power laws that do not appear at any fixed order of perturbation theory.

It is possible to derive these predictions directly from Wilson's procedure of integrating out slices in momentum space, as Wilson originally did. However, now that we understand the basic idea of renormalization group flows, it will be technically easier to work in the more familiar context of ordinary renormalized perturbation theory. The discussion of the previous section was physically motivated but technically complex. It involved awkward integrals over finite domains, and used the artificial parameter b, which must cancel out in any final results. Furthermore, we know from Section 7.5 that a cutoff regulator leads to even more trouble in QED, since it conflicts with the Ward identity. The discussion of the present section will be much more abstract and formal, but it will remove these technical problems. In this section and the next we will derive a flow equation for the coupling constant, similar to the one we derived in Section 12.1. To obtain the flows of the most general Lagrangians, we will need some additional tools, to be developed in Sections 12.4 and 12.5.

How can we hope to obtain information on renormalization group flows from the expressions for renormalized Green's functions, in which the cutoff has already been taken to infinity? We must first realize that renormalized quantum field theories correspond to a restricted class of the full set of possible Lagrangians that we considered in the previous section. In Wilson's language, a renormalized field theory with the cutoff taken arbitrarily large corresponds to a trajectory that takes an arbitrarily long time to evolve to a large value of the mass parameter. Such a trajectory must, then, pass arbitrarily close to a fixed point, which we will assume to be the weak-coupling fixed point. In the slow evolution past this fixed point, the irrelevant operators in the original Lagrangian die away, and we are left only with the relevant and marginal operators. The coefficients of these operators are in one-to-one correspondence with the parameters of the renormalizable field theory. Thus, in working with a renormalized field theory, we are throwing away information on the evolution of irrelevant perturbations, but keeping information on the flows of relevant and marginal perturbations.

The flows of these parameters cannot be determined from the cutoff dependence, because, in this framework, the cutoff has already been sent to infinity. However, we have an alternative, though more abstract, tool at our disposal. The parameters of a renormalized field theory are determined by a set of renormalization conditions, which are applied at a certain momentum scale (called the *renormalization scale*). By looking at how the parameters of the theory depend on the renormalization scale, we can recover the information contained in the renormalization group flows of the previous section.

We consider first the specific case of ϕ^4 theory in four dimensions, where the coupling constant λ is dimensionless and the corresponding operator is marginal. For simplicity, we will also assume that the mass term m^2 has been adjusted to zero, so that the theory sits just at its critical point. We will perform this analysis in Minkowski space, using spacelike reference momenta. However, the analysis would be essentially identical if carried out in Euclidean space. If we wish to consider renormalization group predictions at timelike momenta, we must consider the possibilities of new singularities which make the analysis more complicated. These include both physical thresholds and the Sudakov double logarithms discussed in Section 6.4. We postpone discussion of these complications until Chapters 17 and 18.

Renormalization Conditions

To define the theory properly, we must specify the renormalization conditions. In Chapter 10 we used a natural set of renormalization conditions (10.19) for ϕ^4 theory, defined in terms of the physical mass m. However, in a theory where $m = 0$, these conditions cannot be used because they lead to singularities in the counterterms. (Consider, for example, the limit $m^2 \to 0$ of Eq. (10.24).) To avoid such singularities, we choose an arbitrary momentum scale M and impose the renormalization conditions at a spacelike momentum p with $p^2 = -M^2$:

$$\overrightarrow{}\boxed{\text{1PI}}\overrightarrow{} = 0 \quad \text{at } p^2 = -M^2;$$

$$\frac{d}{dp^2}\left(\overrightarrow{}\boxed{\text{1PI}}\overrightarrow{} \right) = 0 \quad \text{at } p^2 = -M^2;$$

(12.30)

$$\begin{array}{c} p_3 \quad \quad p_4 \\ \\ p_1 \quad \quad p_2 \end{array} = -i\lambda$$

$$\text{at } (p_1 + p_2)^2 = (p_1 + p_3)^2 = (p_1 + p_4)^2 = -M^2.$$

The parameter M is called the *renormalization scale*. These conditions define the values of the two- and four-point Green's functions at a certain point and, in the process, remove all ultraviolet divergences. Speaking loosely, we say that we are "defining the theory at the scale M".

These new renormalization conditions take some getting used to. The second condition, in particular, implies that the two-point Green's function has a coefficient of 1 at the unphysical momentum $p^2 = -M^2$, rather than on shell (at $p^2 = 0$):

$$\langle \Omega | \, \phi(p)\phi(-p) \, | \Omega \rangle = \frac{i}{p^2} \quad \text{at } p^2 = -M^2.$$

Here ϕ is the renormalized field, related to the bare field ϕ_0 by a scale factor that we again call Z:

$$\phi = Z^{-1/2}\phi_0. \tag{12.31}$$

This Z, however, is not the residue of the physical pole in the two-point Green's function of bare fields, as it was in Chapters 7 and 10. Instead, we now have

$$\langle \Omega | \, \phi_0(p)\phi_0(-p) \, | \Omega \rangle = \frac{iZ}{p^2} \quad \text{at } p^2 = -M^2.$$

The Feynman rules for renormalized perturbation theory are the same as in Chapter 10, with the same relation between Z and the counterterm δ_Z,

$$\delta_Z = Z - 1.$$

Now, however, the counterterms δ_Z and δ_λ must be adjusted to maintain the new conditions (12.30).

The first renormalization condition in (12.30) holds the physical mass of the scalar field fixed at zero. We saw in Chapter 10 that, in ϕ^4 theory, the one-loop propagator correction is momentum-independent and is completely canceled by the mass renormalization counterterm. At two-loop order, however, the situation becomes more complicated, and the propagator corrections require both mass and field strength renormalizations. In more general scalar field theories, such as the Yukawa theory example considered at the end of Section 10.2, this complication arises already at one-loop order. Since the field strength renormalization counterterm will play an important role in the discussion below, it will be helpful to discuss briefly how we will treat this double subtraction.

The evaluation of propagator corrections has some special simplifications for the case of a massless scalar field, which we consider here, and specifically with the use of dimensional regularization. Consider, for example, the one-loop propagator correction in Yukawa theory. In Section 10.2 we found an expression of the form

$$\xrightarrow{\quad p \quad} \bigcirc \text{-----} \ \sim \frac{\Gamma(1-\frac{d}{2})}{\Delta^{1-d/2}}, \tag{12.32}$$

where Δ is a linear combination of the fermion mass m_f and p^2. If we compute the diagram using massless propagators only, Δ is proportional to p^2. Expression (12.32) has a pole at $d = 2$, corresponding to the quadratically divergent mass renormalization. However, the residue of this pole is independent of p^2, so we can completely cancel the pole with the mass counterterm δ_m. This allows us to analytically continue (12.32) to $d = 4$. Then this expression takes the form

$$-p^2 \Big(\frac{1}{2 - d/2} + \log \frac{1}{-p^2} + C \Big), \tag{12.33}$$

and gives no additional mass shift but only a field strength renormalization. The remaining divergence is canceled by the counterterm δ_Z. If we adopt the rule that we should simply continue expressions of the form (12.32) to $d = 4$, we can forget about the counterterm δ_m altogether.

In a regularization scheme with a momentum cutoff, the contributions to δ_m and δ_Z become tangled up with one another. Then it is more awkward to define the massless limit. In the following discussion, we will assume the use of dimensional regularization. However, to emphasize the physical role of the cutoff, we will write expressions of the form (12.33) as

$$-p^2 \Big(\log \frac{\Lambda^2}{-p^2} + C \Big). \tag{12.34}$$

The logarithmically divergent terms proportional to p^2 will agree with the divergences obtained with a momentum cutoff; the constant terms will not agree, but these will drop out of our final results.

In ϕ^4 theory, where the one-loop propagator correction is momentum-independent, the one-loop diagram is simply set to zero by this prescription. Then the preceding analysis applies to the two-loop and higher correction terms.

The generalization of the analysis of this section to massive scalar field theory requires some additional formalism, which we postpone to Section 12.5.

The Callan-Symanzik Equation

In the renormalization conditions (12.30), the renormalization scale M is arbitrary. We could just as well have defined the same theory at a different scale M'. By "the same theory", we mean a theory whose bare Green's functions,

$$\langle \Omega| T\phi_0(x_1)\phi_0(x_2)\cdots\phi_0(x_n) |\Omega\rangle,$$

are given by the same functions of the bare coupling constant λ_0 and the cutoff Λ. These functions make no reference to M. The dependence on M enters only when we remove the cutoff dependence by rescaling the fields and eliminating λ_0 in favor of the renormalized coupling λ. The renormalized Green's functions are numerically equal to the bare Green's functions, up to a rescaling by powers of the field strength renormalization Z:

$$\langle \Omega| T\phi(x_1)\phi(x_2)\cdots\phi(x_n) |\Omega\rangle = Z^{-n/2} \langle \Omega| T\phi_0(x_1)\phi_0(x_2)\cdots\phi_0(x_n) |\Omega\rangle. \tag{12.35}$$

The renormalized Green's functions could be defined equally well at another scale M', using a new renormalized coupling λ' and a new rescaling factor Z'.

Let us write more explicitly the effect of an infinitesimal shift of M. Let $G^{(n)}(x_1,\cdots,x_n)$ be the connected n-point function, computed in renormalized perturbation theory:

$$G^{(n)}(x_1,\cdots,x_n) = \langle \Omega| T\phi(x_1)\cdots\phi(x_n) |\Omega\rangle_{\text{connected}}. \tag{12.36}$$

Now suppose that we shift M by δM. There is a corresponding shift in the coupling constant and the field strength such that the bare Green's functions remain fixed:

$$M \to M + \delta M,$$
$$\lambda \to \lambda + \delta\lambda, \tag{12.37}$$
$$\phi \to (1 + \delta\eta)\phi.$$

Then the shift in any renormalized Green's function is simply that induced by the field rescaling,

$$G^{(n)} \to (1 + n\delta\eta)G^{(n)}.$$

If we think of $G^{(n)}$ as a function of M and λ, we can write this transformation as

$$dG^{(n)} = \frac{\partial G^{(n)}}{\partial M}\delta M + \frac{\partial G^{(n)}}{\partial \lambda}\delta\lambda = n\delta\eta G^{(n)}. \tag{12.38}$$

Rather than writing this relation in terms of $\delta\lambda$ and $\delta\eta$, it is conventional to define the dimensionless parameters

$$\beta \equiv \frac{M}{\delta M}\delta\lambda; \qquad \gamma \equiv -\frac{M}{\delta M}\delta\eta. \tag{12.39}$$

Making these substitutions in Eq. (12.38) and multiplying through by $M/\delta M$, we obtain

$$\left[M\frac{\partial}{\partial M} + \beta\frac{\partial}{\partial \lambda} + n\gamma\right]G^{(n)}(x_1,\cdots,x_n;M,\lambda) = 0. \tag{12.40}$$

The parameters β and γ are the same for every n, and must be independent of the x_i. Since the Green's function $G^{(n)}$ is renormalized, β and γ cannot depend on the cutoff, and hence, by dimensional analysis, these functions cannot depend on M. Therefore they are functions only of the dimensionless variable λ. We conclude that any Green's function of massless ϕ^4 theory must satisfy

$$\left[M\frac{\partial}{\partial M} + \beta(\lambda)\frac{\partial}{\partial \lambda} + n\gamma(\lambda)\right]G^{(n)}(\{x_i\}; M, \lambda) = 0. \qquad (12.41)$$

This relation is called the Callan-Symanzik equation.[‡] It asserts that there exist two universal functions $\beta(\lambda)$ and $\gamma(\lambda)$, related to the shifts in the coupling constant and field strength, that compensate for the shift in the renormalization scale M.

The preceding argument generalizes without difficulty to other massless theories with dimensionless couplings. In theories with multiple fields and couplings, there is a γ term for each field and a β term for each coupling. For example, we can define QED at zero electron mass by introducing a renormalization scale as in Eqs. (12.30). The renormalization conditions for the propagators are applied at $p^2 = -M^2$, and those for the vertex at a point where all three invariants are of order $-M^2$. Then the renormalized Green's functions of this theory satisfy the Callan-Symanzik equation

$$\left[M\frac{\partial}{\partial M} + \beta(e)\frac{\partial}{\partial e} + n\gamma_2(e) + m\gamma_3(e)\right]G^{(n,m)}(\{x_i\}; M, e) = 0, \qquad (12.42)$$

where n and m are, respectively, the number of electron and photon fields in the Green's function $G^{(n,m)}$ and γ_2 and γ_3 are the rescaling functions of the electron and photon fields.

Computation of β and γ

Before we work out the implications of the Callan-Symanzik equation, let us look more closely at the functions β and γ that appear in it. From their definitions (12.39), we see that they are proportional to the shift in the coupling constant and the shift in the field normalization, respectively, when the renormalization scale M is increased. The behavior of the coupling constant as a function of M is of particular interest, since it determines the strength of the interaction and the conditions under which perturbation theory is valid. We will see in the next section that the shift in the field strength is also reflected directly in the values of Green's functions.

The easiest way to compute the Callan-Symanzik functions is to begin with explicit perturbative expressions for some conveniently chosen Green's functions. If we insist that these expressions satisfy the Callan-Symanzik equation, we will obtain equations that can be solved for β and γ. Because the

---

[‡]C. G. Callan, *Phys. Rev.* **D2**, 1541 (1970), K. Symanzik, *Comm. Math. Phys.* **18**, 227 (1970).

M dependence of a renormalized Green's function originates in the counterterms that cancel its logarithmic divergences, we will find that the β and γ functions are simply related to these counterterms, or equivalently, to the coefficients of the divergent logarithms. The precise formulae that relate β and γ to the counterterms will depend on the specific renormalization prescription and other details of the calculational scheme. At one-loop order, however, the expressions for β and γ are simple and unambiguous.

As a first example, let us calculate the one-loop contributions to $\beta(\lambda)$ and $\gamma(\lambda)$ in massless ϕ^4 theory. We can simplify the analysis by working in momentum space rather than coordinate space. Our strategy will be to apply the Callan-Symanzik equation to the diagrammatic expressions for the two- and four-point Green's functions.

The two-point function is given by

In massless ϕ^4 theory, the one-loop propagator correction is completely canceled by the mass counterterm. Then the first nontrivial correction to the propagator comes from the two-loop diagram and its counterterm, and is of order λ^2. Meanwhile, the four-point function is given by

where we have omitted the canceled one-loop propagator corrections to the external legs. The diagrams of order λ^3 include nonvanishing two-loop propagator corrections to the external legs.

To calculate β, we apply the Callan-Symanzik equation to the four-point function:

$$\left[M\frac{\partial}{\partial M} + \beta(\lambda)\frac{\partial}{\partial \lambda} + 4\gamma(\lambda)\right]G^{(4)}(p_1,\ldots,p_4) = 0. \qquad (12.43)$$

Borrowing our result (10.21) from Section 10.2, we can write $G^{(4)}$ as

$$G^{(4)} = \left[-i\lambda + (-i\lambda)^2\left[iV(s) + iV(t) + iV(u)\right] - i\delta_\lambda\right] \cdot \prod_{i=1,\ldots,4} \frac{i}{p_i^2},$$

where $V(s)$ represents the loop integral in (10.20). Our renormalization condition (12.30) requires that the correction terms cancel at $s = t = u = -M^2$. The order-λ^2 vertex counterterm is therefore

$$\delta_\lambda = (-i\lambda)^2 \cdot 3V(-M^2) = \frac{3\lambda^2}{2(4\pi)^{d/2}} \int\limits_0^1 dx \frac{\Gamma(2-\frac{d}{2})}{(x(1-x)M^2)^{2-d/2}}. \qquad (12.44)$$

The last expression follows from setting $m = 0$ and $p^2 = -M^2$ in Eq. (10.23) for $V(p^2)$. In the limit as $d \to 4$, Eq. (12.44) becomes

$$\delta_\lambda = \frac{3\lambda^2}{2(4\pi)^2} \left[\frac{1}{2 - d/2} - \log M^2 + \text{finite} \right], \tag{12.45}$$

where the finite terms are independent of M. This counterterm gives $G^{(4)}$ its M dependence:

$$M \frac{\partial}{\partial M} G^{(4)} = \frac{3i\lambda^2}{(4\pi)^2} \prod_i \frac{i}{p_i^2}.$$

Let us assume for the moment that $\gamma(\lambda)$ has no term of order λ; we will justify this in the next paragraph. Then the Callan-Symanzik equation (12.43) can be satisfied to order λ^2 only if the β function of ϕ^4 theory is given by

$$\beta(\lambda) = \frac{3\lambda^2}{16\pi^2} + \mathcal{O}(\lambda^3). \tag{12.46}$$

Next, consider the Callan-Symanzik equation for the two-point function:

$$\left[M \frac{\partial}{\partial M} + \beta(\lambda) \frac{\partial}{\partial \lambda} + 2\gamma(\lambda) \right] G^{(2)}(p) = 0. \tag{12.47}$$

Since, to one-loop order, there are no propagator corrections to $G^{(2)}$, no dependence on M or λ is introduced to order λ. Thus the γ function is zero to this order:

$$\gamma = 0 + \mathcal{O}(\lambda^2). \tag{12.48}$$

This justifies the assumption made in the previous paragraph. The two-loop propagator correction is divergent, and its counterterm contains a term of order λ^2 which depends on M. This contributes to the first term in Eq. (12.47). Since β is of order λ^2 and the corrections to $G^{(2)}$ are of order λ^2, the leading contributions to the second term in (12.47) are of order λ^3. Thus γ acquires a nonzero contribution in order λ^2. This leading contribution to γ is computed in Problem 13.2.

The preceding example illustrates how β and γ can be calculated in more general theories with dimensionless couplings. In such theories, the M dependence of Green's functions enters through the field-strength and vertex counterterms, which are used to subtract the divergent logarithms. The lowest-order expressions for β and γ can be computed directly from these counterterms, or from the coefficients of the divergent logarithms.

In any renormalizable massless scalar field theory, the two-point Green's function has the generic form

$$G^{(2)}(p) = \underline{\qquad} \quad + \quad (\text{loop diagrams}) \quad + \quad \underline{\quad\otimes\quad} \quad + \quad \cdots$$

$$= \frac{i}{p^2} + \frac{i}{p^2} \left(A \log \frac{\Lambda^2}{-p^2} + \text{finite} \right) + \frac{i}{p^2} (ip^2 \delta_Z) \frac{i}{p^2} + \cdots. \tag{12.49}$$

The M dependence of this expression, to lowest order, comes entirely from the counterterm δ_Z. Applying the Callan-Symanzik equation to $G^{(2)}(p)$, and neglecting the β term (which is always smaller by at least one power of the coupling constant), we find

$$-\frac{i}{p^2}M\frac{\partial}{\partial M}\delta_Z + 2\gamma\frac{i}{p^2} = 0,$$

or

$$\gamma = \frac{1}{2}M\frac{\partial}{\partial M}\delta_Z \qquad \text{(to lowest order)}. \tag{12.50}$$

To make this result more explicit, note that the counterterm must be

$$\delta_Z = A\log\frac{\Lambda^2}{M^2} + \text{finite}$$

in order to cancel the divergent logarithm in $G^{(2)}$. Thus γ is simply the coefficient of the logarithm:

$$\gamma = -A \qquad \text{(to lowest order)}. \tag{12.51}$$

In most theories (e.g., Yukawa theory or QED), the first logarithmic divergence in δ_Z occurs at the one-loop level. However, even in ϕ^4 theory, formulae (12.50) and (12.51) are true for the first nonvanishing term in δ_Z, in this case the two-loop contribution.* By replacing the scalar field propagator (i/p^2) with a fermion propagator $(i/\not{p})$, we could repeat this argument line for line to compute the γ function for a fermion field in terms of its field strength counterterm δ_Z.

We can derive similar expressions for the β function of a generic dimensionless coupling constant g, associated with an n-point vertex. Taking propagator corrections into account, the full connected Green's function, to one-loop order, has the general form

$$G^{(n)} = \begin{pmatrix} \text{tree-level} \\ \text{diagram} \end{pmatrix} + \begin{pmatrix} \text{1PI loop} \\ \text{diagrams} \end{pmatrix} + \begin{pmatrix} \text{vertex} \\ \text{counterterm} \end{pmatrix} + \begin{pmatrix} \text{external leg} \\ \text{corrections} \end{pmatrix}$$

$$= \left(\prod_i \frac{i}{p_i^2}\right)\left[-ig - iB\log\frac{\Lambda^2}{-p^2} - i\delta_g + (-ig)\sum_i\left(A_i\log\frac{\Lambda^2}{-p_i^2} - \delta_{Zi}\right)\right]$$

$$+ \text{finite terms.} \tag{12.52}$$

In this expression, p_i are the momenta on the external legs, and p^2 represents a typical invariant built from these momenta. We assume that renormalization conditions are applied at a point where all such invariants are spacelike and of order $-M^2$. The M dependence of this expression comes from the counterterms δ_g and δ_{Zi}. Applying the Callan-Symanzik equation, we obtain

$$M\frac{\partial}{\partial M}\left(\delta_g - g\sum_i\delta_{Zi}\right) + \beta(g) + g\sum_i\frac{1}{2}M\frac{\partial}{\partial M}\delta_{Zi} = 0,$$

---

*At one loop, formula (12.33) implies that we can also identify A as the coefficient of $2/(4-d)$ in the 1PI self-energy, in the limit $d \to 4$. This relation changes in higher loops. However, Eq. (12.50) remains correct.

or

$$\beta(g) = M\frac{\partial}{\partial M}\left(-\delta_g + \frac{1}{2}g\sum_i \delta_{Zi}\right) \qquad \text{(to lowest order).} \tag{12.53}$$

To be more explicit, we note that

$$\delta_g = -B\log\frac{\Lambda^2}{M^2} + \text{finite.}$$

Thus the β function is just a combination of the coefficients of the divergent logarithms:

$$\beta(g) = -2B - g\sum_i A_i \qquad \text{(to lowest order).} \tag{12.54}$$

Notice that the finite parts of counterterms are independent of M and therefore never contribute to β or γ. This means that, to compute the leading terms in the Callan-Symanzik functions, we needn't be too precise in specifying renormalization conditions: Any momentum scale of order M^2 will yield the same results. The divergent parts of the counterterms can be estimated simply by setting all invariants inside of logarithms equal to M^2, as we did above in our expression for the n-point Green's function.

As in the computation of γ, this argument can be applied almost without change to coupling constants for fields with spin. In Yukawa theory, for example, we consider the three-point function with one incoming fermion, one outgoing fermion, and one scalar, with momenta p_1, p_2, and p_3, respectively. Then the tree-level expression for the three-point function is

$$\frac{i}{\not{p}_1}\frac{i}{\not{p}_2}\frac{1}{p_3^2}(-ig). \tag{12.55}$$

The one-loop corrections replace the quantity $(-ig)$ by the expression in brackets in Eq. (12.52). Then formulae (12.53) and (12.54) hold also for the β function of this theory.

Similar expressions also apply in QED, though there are a number of small complications. The first comes in computing the γ function for the photon propagator. In Eq. (7.74), we saw that the general form of the photon propagator in Feynman gauge is

$$D^{\mu\nu}(q) = D(q)\left(g^{\mu\nu} - \frac{q^\mu q^\nu}{q^2}\right) + \frac{-i}{q^2}\frac{q^\mu q^\nu}{q^2}. \tag{12.56}$$

The coefficient of the last term in (12.56) depends on the gauge. Fortunately, this term drops out of all gauge-invariant observables. Thus it makes sense to concentrate on the first term, projecting all external photons onto their transverse components. Projecting the photon propagator, we see that $D(q)$ satisfies the Callan-Symanzik equation. Since the corrections to this function have the form (12.49), the arguments following that formula are valid for photons as well as for electrons and scalars. Thus, to leading order,

$$\gamma_2 = \frac{1}{2}M\frac{\partial}{\partial M}\delta_2, \qquad \gamma_3 = \frac{1}{2}M\frac{\partial}{\partial M}\delta_3, \tag{12.57}$$

where δ_2 and δ_3 are the counterterms defined in Section 10.3.

Similarly, we may consider the three-point connected Green's function $\langle \bar{\psi}(p_1)\psi(p_2)A_\mu(q)\rangle$, projected onto transverse components of the photon. At leading order, this function equals

$$\frac{i}{\not{p}_1}(-ie\gamma^\mu)\frac{i}{\not{p}_2}\frac{-i}{q^2}\left(g^{\mu\nu} - \frac{q^\mu q^\nu}{q^2}\right).$$

The divergent one-loop corrections have the same form, with $(-ie)$ replaced by logarithmically divergent terms. Thus, Eq. (12.53) gives the lowest-order expression for the β function:

$$\beta(e) = M\frac{\partial}{\partial M}\left(-e\delta_1 + e\delta_2 + \frac{e}{2}\delta_3\right). \tag{12.58}$$

To find explicit expressions for the Callan-Symanzik functions of QED, we must write expressions for the counterterms δ_1, δ_2, δ_3. In Section 10.3, we evaluated these counterterms using on-shell renormalization conditions with massive fermions. We must now re-evaluate these terms for massless fermions and renormalization at $-M^2$. Fortunately, we need only evaluate the logarithmically divergent pieces of these counterterms, which are identical in the two cases. Reading from Eqs. (10.43) and (10.44), we find

$$\delta_1 = \delta_2 = -\frac{e^2}{(4\pi)^2}\frac{\Gamma(2-\frac{d}{2})}{(M^2)^{2-d/2}} + \text{finite},$$

$$\delta_3 = -\frac{e^2}{(4\pi)^2}\frac{4}{3}\frac{\Gamma(2-\frac{d}{2})}{(M^2)^{2-d/2}} + \text{finite}. \tag{12.59}$$

Using formulae (12.57) and (12.59), we obtain at leading order

$$\gamma_2(e) = \frac{e^2}{16\pi^2}, \qquad \gamma_3(e) = \frac{e^2}{12\pi^2}. \tag{12.60}$$

And from Eq. (12.58), we find

$$\beta(e) = \frac{e^3}{12\pi^2}. \tag{12.61}$$

It is important to remember that the expression we have used for δ_2 explicitly assumes the use of Feynman gauge. In fact, γ_2 depends on the gauge parameter, and this makes sense, because Green's functions of individual ψ and $\bar{\psi}$ fields are not gauge invariant. On the other hand, the QED vacuum polarization, and therefore γ_3 and β, are gauge invariant.

The Meaning of β and γ

We can obtain a deeper insight into the nature of β and γ by expressing them in terms of the parameters of bare perturbation theory: Z, λ_0, and Λ for the case of ϕ^4 theory.

First recall that the bare and renormalized field are related by

$$\phi(p) = Z(M)^{-1/2}\phi_0(p). \tag{12.62}$$

This equation expresses the dependence of the field rescaling on M. If M is increased by δM, the renormalized field is shifted by

$$\delta\eta = \frac{Z(M+\delta M)^{-1/2}}{Z(M)^{-1/2}} - 1.$$

Hence our original definition (12.39) of γ gives us immediately

$$\gamma(\lambda) = \frac{1}{2}\frac{M}{Z}\frac{\partial}{\partial M}Z. \tag{12.63}$$

Since $\delta_Z = Z - 1$ (Eq. (10.17)), this formula is in agreement with (12.50) to leading order. Formula (12.63), however, is an exact relation. This expression clarifies the relation of γ to the field strength rescaling. However, it obscures the fact that γ is independent of the cutoff Λ. To understand this aspect of γ, we have to go back to the original definition of this function in terms of renormalized Green's functions, whose cutoff independence follows from the renormalizability of the theory.

Similarly, we can find an instructive expression for β in terms of the parameters of bare perturbation theory. Our original definition of β in Eq. (12.39) made use of a quantity $\delta\lambda$, defined to be the shift of the renormalized coupling λ needed to preserve the values of the bare Green's functions when the renormalization point is shifted infinitesimally. Since the bare Green's functions depend on the bare coupling λ_0 and the cutoff, this definition can be rewritten as

$$\beta(\lambda) = M\frac{\partial}{\partial M}\lambda\bigg|_{\lambda_0,\Lambda}. \tag{12.64}$$

Thus the β function is the rate of change of the renormalized coupling at the scale M corresponding to a fixed bare coupling. Recalling our analysis in Section 12.1, it is tempting to associate $\lambda(M)$ with the coupling constant λ' obtained by integrating out degrees of freedom down to the scale M. With this correspondence, the β function is just the rate of the renormalization group flow of the coupling constant λ. A positive sign for the β function indicates a renormalized coupling that increases at large momenta and decreases at small momenta. We can see explicitly that this relation works for ϕ^4 theory, to leading order in λ, by comparing Eqs. (12.28) and (12.46). We will justify this correspondence further in the following section.

The equality of the exact formula (12.64) with the first-order formula (12.53) again follows from the counterterm definitions (10.17). As with (12.63), it is not obvious that this formula for $\beta(\lambda)$ is independent of Λ, but that fact again follows from renormalizability. Conversely, it is possible to prove the renormalizability of ϕ^4 theory by demonstrating, order by order in perturbation theory, that expressions (12.63) and (12.64) are independent of Λ.[†]

12.3 Evolution of Coupling Constants

Now that we have discussed all of the ingredients of the Callan-Symanzik equation, let us investigate its implications. We begin by finding the explicit solution to the Callan-Symanzik equation for the simplest situation, the two-point Green's function of a scalar field theory. This solution will clarify the physical implications of the equation. In particular, it will cement the relation suggested at the end of the previous section, which identifies the β function with the rate of the renormalization group flow of the coupling constant. We will then use this relation to discuss the qualitative features of the renormalization group flow in renormalizable field theories.

Solution of the Callan-Symanzik Equation

We would like to solve the Callan-Symanzik equation for the two-point Green's function, $G^{(2)}(p)$, in a theory with a single scalar field. Since $G^{(2)}(p)$ has dimensions of $(\text{mass})^{-2}$, we can express its dependence on p and M as

$$G^{(2)}(p) = \frac{i}{p^2} g(-p^2/M^2). \tag{12.65}$$

This equation allows us to trade the derivative with respect to M for a derivative with respect to p^2. For the remainder of this chapter, we will use the variable p to represent the magnitude of the spacelike momentum: $p = (-p^2)^{1/2}$. Then we can rewrite the Callan-Symanzik equation as

$$\left[p\frac{\partial}{\partial p} - \beta(\lambda)\frac{\partial}{\partial\lambda} + 2 - 2\gamma(\lambda) \right] G^{(2)}(p) = 0. \tag{12.66}$$

In free field theory, β and γ vanish and we recover the trivial result

$$G^{(2)}(p) = \frac{i}{p^2}. \tag{12.67}$$

In an interacting theory, β and γ are nonzero functions of λ. However, it is still possible to write the explicit solution to the Callan-Symanzik equation, using the method of characteristics. Equivalently (for those not well versed in the theory of partial differential equations), we will apply a lovely hydrodynamic-bacteriological analogy due to Sidney Coleman.[‡] Imagine a narrow pipe running in the x direction, containing a fluid whose velocity

---

[†]Callan has given a beautiful proof of the renormalizability of ϕ^4 theory, based on proving that the Callan-Symanzik equation holds order by order in λ, in his article in *Methods in Field Theory*, R. Balian and J. Zinn-Justin, eds. (North Holland, Amsterdam, 1976).

[‡]Coleman (1985), chap. 3.

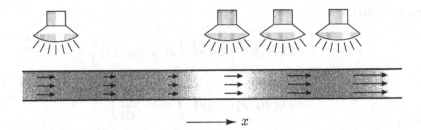

Figure 12.3. Coleman's bacteriological analogy to the Callan-Symanzik equation. The pipe is inhabited by bacteria with a given initial density $D_i(x)$. The growth rate (determined by the illumination) and flow velocity are given functions of x. The problem is to determine the density $D(t, x)$ at all subsequent times.

is $v(x)$, as shown in Fig. 12.3. The pipe is inhabited by bacteria, whose density is $D(t, x)$ and whose rate of growth is $\rho(x)$. Then the future behavior of the function $D(t, x)$ is governed by the differential equation

$$\left[\frac{\partial}{\partial t} + v(x)\frac{\partial}{\partial x} - \rho(x)\right]D(t, x) = 0. \tag{12.68}$$

The second term allows for the fact that the bacteria are swept along with the fluid, so their present density here determines their future density not here, but some distance ahead. This equation is identical to Eq. (12.66), with the replacements

$$\log(p/M) \leftrightarrow t,$$

$$\lambda \leftrightarrow x,$$

$$-\beta(\lambda) \leftrightarrow v(x), \tag{12.69}$$

$$2\gamma(\lambda)-2 \leftrightarrow \rho(x),$$

$$G^{(2)}(p, \lambda) \leftrightarrow D(t, x).$$

Now suppose we know the initial concentration of the bacteria: $D(t, x) = D_i(x)$ at time $t = 0$. Then we can determine the concentration of bacteria in a fluid element at the point x at any later time by computing the history of that fluid element and then integrating the rate of growth along that path. Consider the fluid element that is at x at the time t. We can find out where it was at time zero by integrating its motion backward in time. The position of this element at time $t = 0$ is given by $\bar{x}(t; x)$, which satisfies the differential equation

$$\frac{d}{dt'}\bar{x}(t'; x) = -v(\bar{x}), \qquad \text{with} \qquad \bar{x}(0; x) = x. \tag{12.70}$$

Then, immediately,

$$D(t,x) = D_i\big(\bar{x}(t;x)\big) \cdot \exp\left(\int_0^t dt'\, \rho(\bar{x}(t';x))\right)$$

$$= D_i\big(\bar{x}(t;x)\big) \cdot \exp\left(\int_{\bar{x}(t)}^x dx'\, \frac{\rho(x')}{v(x')}\right).$$
(12.71)

Now bring this solution back to our field theory problem by replacing each bacteriological parameter with its corresponding field theory parameter. The time $t = 0$ corresponds to $-p^2 = M^2$, and the initial concentration $D_i(x)$ becomes an unknown function $\hat{\mathcal{G}}(\lambda)$. Then

$$G^{(2)}(p,\lambda) = \hat{\mathcal{G}}\big(\bar{\lambda}(p;\lambda)\big) \cdot \exp\left(-\int_{p'=M}^{p'=p} d\log(p'/M) \cdot 2\big[1 - \gamma(\bar{\lambda}(p';\lambda))\big]\right), \quad (12.72)$$

where $\bar{\lambda}(p;\lambda)$ solves

$$\frac{d}{d\log(p/M)}\bar{\lambda}(p;\lambda) = \beta(\bar{\lambda}), \qquad \bar{\lambda}(M;\lambda) = \lambda. \tag{12.73}$$

This differential equation describes the flow of a modified coupling constant $\bar{\lambda}(p;\lambda)$ as a function of momentum. The rate of this flow is just the β function. Thus, this flow is strongly reminiscent of the dependence of the renormalized coupling on the renormalization scale given by Eq. (12.64). We will refer to $\bar{\lambda}(p)$ as the *running coupling constant*. Its equation (12.73) is often called the *renormalization group equation*.

One can check directly that (12.72) solves the Callan-Symanzik equation by using the identity

$$\int_\lambda^{\bar{\lambda}} \frac{d\lambda'}{\beta(\lambda')} = \int_{p'=M}^{p'=p} d\log(p'/M), \tag{12.74}$$

from which it follows that

$$\left(p\frac{\partial}{\partial p} - \beta(\lambda)\frac{\partial}{\partial \lambda}\right)\bar{\lambda} = 0. \tag{12.75}$$

A convenient way of writing the solution (12.72) is

$$G^{(2)}(p,\lambda) = \frac{i}{p^2}\mathcal{G}\big(\bar{\lambda}(p;\lambda)\big) \cdot \exp\left(2\int_M^p d\log(p'/M)\,\gamma(\bar{\lambda}(p';\lambda))\right), \tag{12.76}$$

in which $\mathcal{G}(\bar{\lambda})$ is a function that must be determined. This function cannot be determined from the general principles of renormalization theory. Instead, we must compute $G^{(2)}(p)$ as a perturbation series in λ and match terms to the expansion of (12.76) as a series in the same parameter. For the two-point function in ϕ^4 theory, this matching is rather trivial: $\mathcal{G}(\bar{\lambda}) = 1 + \mathcal{O}(\bar{\lambda}^2)$.

The preceding analysis can be applied to any family of Green's functions that are related by uniform rescaling of the momenta. Consider, for example, the connected four-point function of ϕ^4 theory evaluated at spacelike momenta p_i such that $p_i^2 = -P^2$, $p_i \cdot p_j = 0$, so that s, t, and u are of order $-P^2$. To leading order in perturbation theory, this function is given by

$$G^{(4)}(P) = \left(\frac{i}{P^2}\right)^4 (-i\lambda). \tag{12.77}$$

Using the fact that $G^{(4)}$ has dimensions of $(\text{mass})^{-8}$, we can exchange M for P in the Callan-Symanzik equation and write this equation as

$$\left[P\frac{\partial}{\partial P} - \beta(\lambda)\frac{\partial}{\partial \lambda} + 8 - 4\gamma(\lambda) \right] G^{(4)}(P; \lambda) = 0. \tag{12.78}$$

The solution to this equation is

$$G^{(4)}(P; \lambda) = \frac{1}{P^8}\mathcal{G}^{(4)}\left(\bar{\lambda}(p; \lambda)\right) \cdot \exp\left(4\int\limits_{M}^{p} d\log(p'/M)\,\gamma(\bar{\lambda}(p'; \lambda)) \right). \tag{12.79}$$

This formula must agree with (12.77) to leading order in λ; this matching requires that

$$\mathcal{G}^{(4)}\left(\bar{\lambda}(p; \lambda)\right) = -i\bar{\lambda} + \mathcal{O}(\bar{\lambda}^2). \tag{12.80}$$

We can now see the physical implication of the Callan-Symanzik equation. The ordinary Feynman perturbation series for a Green's function depends both on the coupling constant λ and on the dimensionless parameter $\log(-p^2/M^2)$. The perturbation theory can be badly behaved even when λ is small if the ratio p^2/M^2 is large. The solutions (12.76) and (12.79) reorganize this dependence into a function of the running coupling constant and an exponential scale factor. We consider these two pieces in turn.

The first factor in Eqs. (12.76) and (12.79) is a function of the running coupling constant, evaluated at the momentum scale p. If p were of order M, the renormalization scale, this function would essentially be the ordinary perturbative evaluation of the Green's function. The results (12.76) and (12.79) instruct us to make use of this same expression at the scale p, but to replace λ with a new coupling constant $\bar{\lambda}$ appropriate to that scale. Thus, the running coupling constant $\bar{\lambda}(p)$ is precisely the effective coupling constant of the renormalization group flow. This interpretation is particularly clear in the solution (12.79) for $G^{(4)}(P)$, since this function directly measures the strength of the ϕ^4 coupling constant.

The exponential factor in Eqs. (12.76) and (12.79) has an equally simple interpretation: It is the accumulated field strength rescaling of the correlation function from the reference point M to the actual momentum p at which the

Green's function is evaluated. This factor receives a multiplicative contribution from each intermediate scale between M and p. Each of these contributions is, appropriately, computed using the running coupling constant at that particular scale.

As a check on these formal arguments, we can use the explicit form of the β function of ϕ^4 theory found in Eq. (12.46) and the renormalization group equation (12.73) to evaluate the running coupling constant of ϕ^4 theory. This running coupling constant satisfies the differential equation

$$\frac{d}{d\log(p/M)}\bar{\lambda} = \frac{3\bar{\lambda}^2}{16\pi^2}, \qquad \text{with} \qquad \bar{\lambda}(M;\lambda) = \lambda. \qquad (12.81)$$

Integrating, we find

$$\left(\frac{3}{16\pi^2}\right)^{-1}\left[\frac{1}{\lambda} - \frac{1}{\bar{\lambda}}\right] = \log\frac{p}{M},$$

and thus,

$$\bar{\lambda}(p) = \frac{\lambda}{1 - (3\lambda/16\pi^2)\log(p/M)}. \qquad (12.82)$$

Many properties of the solution to the Callan-Symanzik equation are visible in this relation. First, the expansion of this formula for $\bar{\lambda}$ to order λ^2 agrees precisely with Eq. (12.28), the rate of the renormalization group flow from Wilson's method. Second, this expression for the running coupling constant goes to zero at a logarithmic rate as $p \to 0$. This coincides with our expectation that a positive value for the β function should imply an effective coupling that becomes stronger at large momenta and weaker at small momenta.

If we expand the running coupling constant $\bar{\lambda}(p)$ in powers of λ, we find that the successive powers of the coupling constant are multiplied by powers of logarithms,

$$\lambda^{n+1}(\log p/M)^n,$$

which become large and invalidate a simple perturbation expansion for p much greater or much less than M. We have seen this problem of large logarithms arising several times in our diagram calculations, and we have remarked on it specifically as a problem in the discussion following Eq. (11.81). We now see that the renormalization group gives a partial solution to this problem. In this example, and in many others that we will study, the Callan-Symanzik equation tells us how to sum these large logarithms into the running coupling constant and multiplicative rescalings. If the running coupling constant becomes large, as happens in ϕ^4 theory for $p \to \infty$, the perturbation expansion will break down anyway, and we will need more advanced methods. However, if the running coupling constant becomes small, as for ϕ^4 theory as $p \to 0$, we will have successfully organized the powers of logarithms into a meaningful and controlled expression. The specific problem posed at the end of Section 11.4 will be solved explicitly by this method in Section 13.2.

An Application to QED

For a more concrete application of the Callan-Symanzik equation, we can look again at the electromagnetic potential between static charges, $V(\mathbf{x})$, which we studied in Section 7.5. At very short distances or at large momenta, we can ignore the electron mass in the computation of QED corrections to this potential. In this approximation, the potential should obey the Callan-Symanzik equation of massless QED. We could write this equation either for $V(\mathbf{x})$ itself or for its Fourier transform; we choose to work in Fourier space in order to make contact more easily with the results of Section 7.5.

We define the massless limit of QED by specifying a renormalization scale M at which the renormalized coupling e_r is defined. If M is taken close to the electron mass m, at the point where the massless approximation is just becoming valid, then the value of e_r will be close to the physical electron charge e. The potential between static charges is a measurable energy, so its normalization is unambiguous and is not shifted from one renormalization point to another. Thus the Callan-Symanzik equation for the Fourier transform of the potential has no γ term, being simply

$$\left[M\frac{\partial}{\partial M} + \beta(e_r)\frac{\partial}{\partial e_r}\right]V(q; M, e_r) = 0. \tag{12.83}$$

The Fourier transform of the potential has dimensions of $(\text{mass})^{-2}$, so we can trade dependence on M for dependence on q as in the scalar field theory discussion above. This gives

$$\left[q\frac{\partial}{\partial q} - \beta(e_r)\frac{\partial}{\partial e_r} + 2\right]V(q; M, e_r) = 0. \tag{12.84}$$

Equation (12.84) is almost the same as Eq. (12.66), so we can immediately write down the solution as a special case of (12.76):

$$V(q, e_r) = \frac{1}{q^2}\mathcal{V}\big(\bar{e}(q; e_r)\big), \tag{12.85}$$

where $\bar{e}(q)$ is the solution of the renormalization group equation

$$\frac{d}{d\log(q/M)}\bar{e}(q; e_r) = \beta(\bar{e}), \qquad \bar{e}(M; e_r) = e_r. \tag{12.86}$$

By comparing this formula for $V(q)$ to the leading-order result

$$V(q) \approx \frac{e^2}{q^2},$$

we can identify $\mathcal{V}(\bar{e}) = \bar{e}^2 + \mathcal{O}(\bar{e}^4)$. Then

$$V(q, e_r) = \frac{\bar{e}^2(q; e_r)}{q^2}, \tag{12.87}$$

up to corrections that are suppressed by powers of e_r^2 and contain no compensatory large logarithms of q/M.

To turn Eq. (12.87) into a completely explicit formula, we need only solve the renormalization group equation (12.86). Using the QED β function (12.61), we can integrate (12.86) to find

$$\frac{12\pi^2}{2}\left(\frac{1}{e_r^2} - \frac{1}{\bar{e}^2}\right) = \log\frac{q}{M}.$$

This simplifies to

$$\bar{e}^2(q) = \frac{e_r^2}{1 - (e_r^2/6\pi^2)\log(q/M)}. \tag{12.88}$$

This result is almost identical to the formula for the effective electric charge that we found in Eq. (7.96). To cement the identification, set M to be of order the electron mass, $M^2 = Am^2$, and approximate e_r at this point by e, with $\alpha = e^2/4\pi$. Then Eq. (12.88) takes the form

$$\bar{\alpha}(q) = \frac{\alpha}{1 - (\alpha/3\pi)\log(-q^2/Am^2)}. \tag{12.89}$$

The particular choice $A = \exp(5/3)$ reproduces Eq. (7.96). Of course, we could not find this exact correspondence without the detailed one-loop calculation of Section 7.5. Nevertheless, our present analysis produces the correct asymptotic formula for the effective charge. Furthermore, our present formalism can be applied to any renormalizable quantum field theory; it does not rely on the special symmetries of QED that we exploited in Section 7.5.

Alternatives for the Running of Coupling Constants

Now that we have computed the behavior of the running coupling constant in two specific quantum field theories, let us consider more generally what behaviors of the running coupling constant are possible in principle. We continue to restrict our discussion to renormalizable theories in the massless limit, with a single dimensionless coupling constant λ.

By the arguments of the previous section, the Green's functions in any such theory obey a Callan-Symanzik equation. The solution of this equation depends on a running coupling constant, $\bar{\lambda}(p)$, which satisfies a differential equation

$$\frac{\partial}{\partial\log(p/M)}\bar{\lambda} = \beta(\bar{\lambda}), \tag{12.90}$$

in which the function $\beta(\lambda)$ is computable as a power series in the coupling constant. In the examples we have just discussed, the leading coefficient in this power series was positive. However, as a matter of principle, three behaviors are possible in the region of small λ:

      (1) $\beta(\lambda) > 0$;

      (2) $\beta(\lambda) = 0$;

      (3) $\beta(\lambda) < 0$.

Examples of quantum fields are known that exhibit each of these behaviors.

We have already seen how, in theories of the first class, the running coupling constant goes to zero in the infrared, leading to definite predictions about the small-momentum behavior of the theory. However, the running coupling constant becomes large in the region of high momenta. Thus the short-distance behavior of the theory cannot be computed using Feynman diagram perturbation theory. In fact, in the examples studied above, the coupling constant formally goes to infinity at a large but finite value of the momentum; thus it is not even clear that these theories possess a nontrivial limit $\Lambda \to \infty$. A Feynman diagram analysis is useful in such theories if one is mainly interested in large-distance or macroscopic behavior. In Chapter 13 we will use this observation to solve problems in the statistical mechanics of systems with critical points.

In theories of the second class, the coupling constant does not flow. In these theories, the running coupling constant is independent of the momentum scale, and thus equal to the bare coupling. This means that there can be no ultraviolet divergences in the relation of coupling constants. The only possible ultraviolet divergences in such theories are those associated with field rescaling, which automatically cancel in the computation of S-matrix elements. Such theories are called *finite* quantum field theories. Before the emergence of our modern understanding of renormalization, these theories would have been embraced as the solution to the problem of ultraviolet infinities. But in fact the known finite field theories in four dimensions are very special constructions—the so-called gauge theories with extended supersymmetry—with no known physical application.

In theories of the third class, the running coupling constant becomes large in the large-distance regime and becomes small at large momenta or short distances. Imagine, for instance, that the sign of the QED β function were reversed:

$$\beta(e) = -\tfrac{1}{2}Ce^3.$$ (12.91)

Then, following our earlier analysis, we would have

$$\bar{e}^2(p) = \frac{e^2}{1 + Ce^2 \log(p/M)}.$$ (12.92)

This coupling constant tends to zero at a logarithmic rate as the momentum scale increases. Such theories are called *asymptotically free*. In theories of this class, the short-distance behavior is completely solvable by Feynman diagram methods. Though ultraviolet divergences appear in every order of perturbation theory, the renormalization group tells us that the sum of these divergences is completely harmless. If we interpret these theories in terms of a bare coupling e_b and a finite cutoff Λ, the result (12.92) indicates that there is a smooth limit in which e_b tends to zero as Λ tends to infinity. Thus, asymptotically free theories give another, more sophisticated, resolution of the problem of ultraviolet divergences. In Chapter 17, we will see that asymptotic freedom plays an essential role in the formulation of a field theory that describes the strong interactions of elementary particle physics.

Now that we have enumerated the possibilities for the renormalization group flow in the region of weak coupling, let us turn our attention to the region of strong coupling. Here we will not be able to compute the β function quantitatively, but we can at least use the renormalization group equation to discuss qualitatively the possibilities for the coupling constant flow. All of our explicit solutions for running coupling constants—Eqs. (12.82), (12.88), and (12.92)—predict that the running coupling becomes infinite at a finite value of the momentum p. For example, according to Eq. (12.82), the running coupling constant of ϕ^4 theory should diverge at

$$p \sim M \exp\left(\frac{16\pi^2}{3\lambda}\right). \tag{12.93}$$

It is possible that this is the true behavior of the quantum field theory, but we have not proved this, because when the running coupling constant becomes large, the approximation we have made, ignoring the higher-order terms in the β function, is no longer valid. It is a logical possibility that the higher terms of the β function are negative, so that the β function has the form shown in Fig. 12.4(a). In this case the β function has a zero at a nonzero value λ_*. When $\bar{\lambda}$ approaches this value, the renormalization group flow slows to a halt; thus $\lambda = \lambda_*$ would be a nontrivial fixed point of the renormalization group. In this model, the running coupling constant $\bar{\lambda}$ tends to λ_* in the limit of large momentum.

For the specific case of ϕ^4 theory in four dimensions, we have strong evidence from numerical studies that there is no such nontrivial fixed point. However, we will soon demonstrate that there is a nontrivial fixed point in ϕ^4 theory in $d < 4$, and many more examples are known. It is thus worthwhile to explore the implications of a fixed point in the renormalization group flow.

For a β function of the form of Fig. 12.4(a), the β function behaves in the vicinity of the fixed point as

$$\beta \approx -B(\lambda - \lambda_*), \tag{12.94}$$

where B is a positive constant. For $\bar{\lambda}$ near λ_*,

$$\frac{d}{d\log p}\bar{\lambda} \approx -B(\bar{\lambda} - \lambda_*). \tag{12.95}$$

The solution of this equation is

$$\bar{\lambda}(p) = \lambda_* + C\left(\frac{M}{p}\right)^B. \tag{12.96}$$

Thus, $\bar{\lambda}$ indeed tends to λ_* as $p \to \infty$, and the rate of approach is governed by the slope of the β function at the fixed point.

This behavior has a dramatic consequence for the exact solution (12.72) of the Callan-Symanzik equation for $G(p)$. For p sufficiently large, the integral

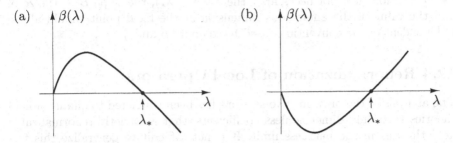

Figure 12.4. Possible forms of the β function with nontrivial zeros:
(a) ultraviolet-stable fixed point; (b) infrared-stable fixed point.

in the exponential factor in this equation will be dominated by values of p for which $\bar{\lambda}(p)$ is close to λ_*. Then

$$G(p) \approx \mathcal{G}(\lambda_*) \exp\left[-\left(\log \frac{p}{M}\right) \cdot 2\big(1 - \gamma(\lambda_*)\big)\right]$$

$$\approx C \cdot \left(\frac{1}{p^2}\right)^{1-\gamma(\lambda_*)}. \tag{12.97}$$

Thus the two-point correlation function returns to the form of a simple scaling law, but with a power law different from that expected by dimensional analysis. At the fixed point we have a scale-invariant quantum field theory in which the interactions of the theory affect the law of rescaling. The shift of the exponent $\gamma(\lambda_*)$ is called the *anomalous dimension* of the scalar field. By convention, the function $\gamma(\lambda)$ is often called the anomalous dimension even if there is no fixed point in the theory.

A similar behavior is possible in an asymptotically free theory. If the β function has the form shown in Fig. 12.4(b), the running coupling constant will tend to a fixed point λ_* as $p \to 0$. The two-point correlation function of fields $G(p)$ will tend to a power law as in (12.97) for asymptotically small momenta. The two cases shown in Figs. 12.4(a) and (b) are called, respectively, *ultraviolet-stable* and *infrared-stable* fixed points.

In the previous section, we saw that the leading-order expressions for the Callan-Symanzik functions β and γ are related in a simple way to the ultraviolet divergent parts of the one-loop counterterms. However, we noted that, in higher orders of perturbation theory, β and γ depend on the specific renormalization conventions used to define the Green's functions. Still, there are some properties of these functions that are independent of any convention. The coefficient of the logarithm in the denominator of such expressions as (12.82) or (12.89) can be determined unambiguously from experiments that measure this coupling constant. This confirms the convention independence of the first β-function coefficient. Experiments sensitive to the coupling constant can also determine the existence of a zero of the β function at strong coupling, and the rate of approach to this asymptote. Thus the existence of a zero of the β function (but not necessarily the value of λ_*), the slope B at the zero, and the value of the anomalous dimension at the fixed point should all be independent of the conventions used to compute β and γ.

12.4 Renormalization of Local Operators

The analysis of the previous two sections has been restricted to quantum field theories with only dimensionless coefficients, that is, strictly renormalizable field theories in the massless limit. It is not difficult to generalize this formalism to theories with mass terms and other operators whose coefficients have mass dimension. However, it is worthwhile to first devote some attention to an intermediate step, by analyzing the renormalization group properties of matrix elements of local operators. This is an interesting problem in its own right, and we will devote considerable space to the applications of this formalism in Chapter 18.

Matrix elements of local operators appear often in quantum field theory calculations. Typically one considers a set of interacting particles that couple weakly to an additional particle, which mediates new forces. Consider, for example, the theory of strongly interacting quarks perturbed by the effects of weak decay processes. The weak interaction is mediated by a massive vector boson, the W. Let us write the interaction of the quarks with the W very schematically as

$$\delta\mathcal{L} = \frac{g}{\sqrt{2}} W_\mu \bar{\psi}\gamma^\mu (1 - \gamma^5)\psi, \tag{12.98}$$

and assign the W boson the propagator

$$\frac{-ig^{\mu\nu}}{q^2 - m_W^2 + i\epsilon}. \tag{12.99}$$

(We will discuss this interaction more correctly in Section 18.2 and in Chapter 20.) Exchange of a W boson leads to the interaction shown in Fig. 12.5. For momentum transfers small compared to m_W, we can ignore the q^2 in the W propagator and write this interaction as the matrix element of the operator

$$\frac{g^2}{2m_W^2}\mathcal{O}(x), \quad \text{where} \quad \mathcal{O}(x) = \bar{\psi}\gamma^\mu(1-\gamma^5)\psi\,\bar{\psi}\gamma_\mu(1-\gamma^5)\psi. \tag{12.100}$$

In the spirit of Wilson's renormalization group procedure, we can say that, on distance scales larger than m_W^{-1}, the W boson can be integrated out, leaving over the interaction (12.100).

How would we analyze the effects of the operator (12.100) on strongly interacting particles composed of quarks and antiquarks? A useful way to begin is to compute the Green's function of the operator $\mathcal{O}$ together with fields that create and destroy quarks. If we approximate the theory of quarks by a theory of free fermions, it is easy to compute these Green's functions; for

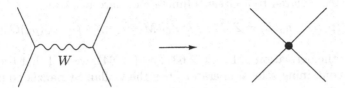

Figure 12.5. Interaction of quarks generated by the exchange of a W boson.

example:

$$\langle \psi(p_1)\bar{\psi}(-p_2)\psi(p_3)\bar{\psi}(-p_4)\mathcal{O}(0)\rangle$$
$$= S_F(p_1)\gamma^\mu(1-\gamma^5)S_F(p_2)\,S_F(p_3)\gamma_\mu(1-\gamma^5)S_F(p_4). \tag{12.101}$$

However, in an interacting field theory, the answer will be much more complicated. Some of these complications will involve the low-energy interactions of
quarks, and we will leave them outside of the present discussion. However, in
a renormalizable theory of quark interactions, one will also find that Green's
functions containing $\mathcal{O}$ have new ultraviolet divergences. The one-loop corrections to (12.101) will contain diagrams that evaluate to the right-hand side of
(12.101) times a divergent integral. These diagrams can be interpreted as field
strength renormalizations of the operator $\mathcal{O}$. As with correlation functions of
elementary fields, we can obtain finite and well-defined matrix elements of
local operators only if we establish conventions for the normalization of local operators and introduce operator rescalings in the form of counterterms,
order by order in perturbation theory, to preserve these conventions. More
specifically, in a massless, renormalizable field theory of the fermions ψ, we
should make the convention that Eq. (12.101) is exact at some spacelike normalization point for which $p_1^2 = p_2^2 = p_3^2 = p_4^2 = -M^2$. Then we should add a
counterterm of the form $\delta_{\mathcal{O}}\mathcal{O}(x)$, and adjust this counterterm at each order
of perturbation theory to insure that these relations are preserved. We refer
to the operator satisfying the normalization condition (12.101) at M^2 as $\mathcal{O}_M$.

The renormalized operator $\mathcal{O}_M$ is a rescaled version of the operator $\mathcal{O}_0$
built of bare fields,

$$\mathcal{O}_0(x) = \bar{\psi}_0\gamma^\mu(1-\gamma^5)\psi_0\bar{\psi}_0\gamma_\mu(1-\gamma^5)\psi_0. \tag{12.102}$$

As we did for the elementary fields, we can write this relation as

$$\mathcal{O}_0 = Z_{\mathcal{O}}(M)\mathcal{O}_M. \tag{12.103}$$

This allows us to write the generalization of the relation (12.35) between
Green's functions of bare and renormalized fields. Let us return to the language of scalar field theories and consider $\mathcal{O}(x)$ to be a local operator in a
scalar field theory. Define

$$G^{(n;1)}(p_1,\cdots,p_n;k) = \langle \phi(p_1)\cdots\phi(p_n)\mathcal{O}_M(k)\rangle. \tag{12.104}$$

Then $G^{(n;1)}$ is related to a Green's function of bare fields by

$$G^{(n;1)}(p_1, \cdots, p_n; k) = Z(M)^{-n/2} Z_{\mathcal{O}}(M)^{-1} \langle \phi_0(p_1) \cdots \phi_0(p_n) \mathcal{O}_0(k) \rangle. \tag{12.105}$$

Repeating the derivation of Eqs. (12.63) and (12.64), we find that the Green's functions containing a local operator obey the Callan-Symanzik equation

$$\left[M \frac{\partial}{\partial M} + \beta(\lambda) \frac{\partial}{\partial \lambda} + n\gamma(\lambda) + \gamma_{\mathcal{O}}(\lambda) \right] G^{(n;1)} = 0, \tag{12.106}$$

where

$$\gamma_{\mathcal{O}} = M \frac{\partial}{\partial M} \log Z_{\mathcal{O}}(M). \tag{12.107}$$

It often happens that a quantum field theory contains several operators with the same quantum numbers. For example, in quantum electrodynamics, the operators $\bar{\psi}[\gamma^\mu D^\nu + \gamma^\nu D^\mu]\psi$ and $F^{\mu\lambda} F^\nu{}_\lambda$ are both symmetric tensors with zero electric charge; in addition, both operators have mass dimension 4. Such operators, with the same quantum numbers and the same mass dimension, can be mixed by quantum corrections.* For such a set of operators $\{\mathcal{O}^i\}$, the relation of renormalized and bare operators must be generalized to

$$\mathcal{O}_0^i = Z_{\mathcal{O}}^{ij}(M) \mathcal{O}_M^j. \tag{12.108}$$

This relation in turn implies that the anomalous dimension function $\gamma_{\mathcal{O}}$ in the Callan-Symanzik equation must be generalized to a matrix,

$$\gamma_{\mathcal{O}}^{ij} = [Z_{\mathcal{O}}^{-1}(M)]^{ik} M \frac{\partial}{\partial M} [Z_{\mathcal{O}}(M)]^{kj}. \tag{12.109}$$

Most of our applications of (12.106) in Chapter 18 will require this generalization.

On the other hand, there are some operators for which the rescaling and anomalous dimensions are especially simple. If $\mathcal{O}$ is the quark number current $\bar{\psi}\gamma^\mu\psi$, its normalization is fixed once and for all because the associated charge

$$Q = \int d^3x\, \bar{\psi}\gamma^0\psi$$

is just the conserved integer number of quarks minus antiquarks in a given state. More generally, for any conserved current J^μ, $Z_J(M) = 1$ and $\gamma_J = 0$. The same argument applies to the energy-momentum tensor. Thus, in the QED example above, the specific linear combination

$$T^{\mu\nu} = \frac{1}{2}\bar{\psi}[\gamma^\mu D^\nu + \gamma^\nu D^\mu]\psi - F^{\mu\lambda} F^\nu{}_\lambda \tag{12.110}$$

receives no rescaling and no anomalous dimension. This linear combination of operators must be an eigenvector of the matrix γ^{ij} with eigenvalue zero.

---

*Our assumption that we are working in a massless field theory constrains the possibilities for operator mixing. In a massive field theory, operators of a given dimension can also mix with operators of lower dimension.

So far, our discussion of operator matrix elements has been rather abstract. To make it more concrete, we will construct a formula for computing $\gamma_\mathcal{O}$ to leading order from one-loop counterterms, and then apply this formula to a simple example in ϕ^4 theory.

To find a simple formula for $\gamma_\mathcal{O}$, we follow the same path that took us from Eq. (12.52) to the formula (12.53) for the β function. Consider an operator whose normalization condition is based on a Green's function with m scalar fields:

$$G^{(m;1)} = \langle \phi(p_1) \cdots \phi(p_m) \mathcal{O}_M(k) \rangle . \qquad (12.111)$$

To compute this Green's function to one-loop order, we find the set of diagrams:

The last diagram is the counterterm $\delta_\mathcal{O}$ needed to maintain the renormalization condition. Notice that the counterterm δ_Z also appears. If we insist that this sum of diagrams satisfies the Callan-Symanzik equation (12.106) to leading order in λ, we find, analogously to (12.53), the relation

$$\gamma_\mathcal{O}(\lambda) = M \frac{\partial}{\partial M} \left(-\delta_\mathcal{O} + \frac{m}{2} \delta_Z \right). \qquad (12.112)$$

As a specific example of the use of this formula, let us compute the anomalous dimension $\gamma_\mathcal{O}$ of the mass operator ϕ^2 in ϕ^4 theory. There is a small subtlety involved in this computation. The Feynman diagrams of ϕ^4 theory generate an additive mass renormalization, which must be removed by the mass counterterm at each order in perturbation theory. We would like to define the mass operator as a perturbation which we can add to the massless theory defined in this way. To clarify the distinction between the underlying mass, which is renormalized to zero, and the explicit mass perturbation, we will analyze a Green's function of ϕ^2 in which this operator carries a specific nonzero momentum. We thus choose to define the normalization of ϕ^2 by the convention

$$= \langle \phi(p)\phi(q)\phi^2(k) \rangle = \frac{i}{p^2} \frac{i}{q^2} \cdot 2 \qquad (12.113)$$

at $p^2 = q^2 = k^2 = -M^2$.

The one-particle-irreducible one-loop correction to (12.113) is

$$
\begin{aligned}
&\text{[diagram]} = \frac{i}{p^2}\frac{i}{q^2}\int \frac{d^d r}{(2\pi)^d}(-i\lambda)\frac{i}{r^2}\frac{i}{(k+r)^2} \\[2mm]
&\qquad\qquad = \frac{i}{p^2}\frac{i}{q^2}\left[-\frac{\lambda}{(4\pi)^2}\frac{\Gamma(2-\frac{d}{2})}{\Delta^{2-d/2}}\right],
\end{aligned}
\tag{12.114}
$$

where Δ is a function of the external momenta. At $-M^2$, this contribution must be canceled by a counterterm diagram,

$$
\text{[diagram]} = \frac{i}{p^2}\frac{i}{q^2}2\delta_{\phi^2}.
\tag{12.115}
$$

Thus, the counterterm must be

$$
\delta_{\phi^2} = \frac{\lambda}{2(4\pi)^2}\frac{\Gamma(2-\frac{d}{2})}{(M^2)^{2-d/2}}.
\tag{12.116}
$$

Since δ_Z is finite to order λ, this is the only contribution to (12.112), and we find

$$
\gamma_{\phi^2} = \frac{\lambda}{16\pi^2}.
\tag{12.117}
$$

This function can be used together with the γ and β functions of pure massless ϕ^4 theory to discuss the scaling of Green's functions that include the mass operator.

12.5 Evolution of Mass Parameters

Finally, we discuss the renormalization group for theories with masses. We note, though, that although we treat these masses as arbitrary parameters, we will continue to use renormalization conventions that are independent of mass, and we will often treat the masses as small parameters. This approach breaks down at momentum scales much less than the scale of masses, but it is sufficient, and simpler than alternative approaches, for most practical applications of the renormalization group.

In the previous section, we worked out the scaling of Green's functions containing one power of the mass operator. It is a small step to generalize this discussion to include an arbitrary number ℓ of mass operators; one simply finds the equation (12.106) with the coefficient ℓ in front of the term $\gamma_\mathcal{O}$. Now consider what would happen if we add the mass operator directly to the Lagrangian of the massless ϕ^4 theory, treating this operator as a perturbation. If $\mathcal{L}_M$ is the massless Lagrangian renormalized at the scale M, the new Lagrangian will be

$$
\mathcal{L}_M - \tfrac{1}{2}m^2\phi_M^2.
\tag{12.118}
$$

The Green's function of n scalar fields in the theory (12.118) could be expressed as a perturbation series in the mass parameter m^2. The coefficient of $(m^2)^\ell$ would be a joint correlation function of the n scalar fields with ℓ powers of ϕ_M^2, and would therefore satisfy the Callan-Symanzik equation (12.106) with the extra factor ℓ as noted above. In general, we can use the operator $m^2(\partial/\partial m^2)$ to count the number of insertions ℓ of ϕ^2. Then the Green's functions of the massive ϕ^4 theory, renormalized according to the mass-independent scheme, satisfy the equation

$$\left[M\frac{\partial}{\partial M} + \beta(\lambda)\frac{\partial}{\partial \lambda} + n\gamma(\lambda) + \gamma_{\phi^2}m^2\frac{\partial}{\partial m^2}\right]G^{(n)}(\{p_i\}; M, \lambda, m^2) = 0. \quad (12.119)$$

This argument extends to any perturbation of massless ϕ^4 theory. In the general case,

$$\mathcal{L}(C_i) = \mathcal{L}_M + C_i \mathcal{O}_M^i(x), \quad (12.120)$$

and the Green's functions of this perturbed theory satisfy

$$\left[M\frac{\partial}{\partial M} + \beta(\lambda)\frac{\partial}{\partial \lambda} + n\gamma(\lambda) + \sum_i \gamma_i(\lambda)C_i\frac{\partial}{\partial C_i}\right]G^{(n)}(\{p_i\}; M, \lambda, \{C_i\}) = 0.$$
$$(12.121)$$

To interpret this equation, it will help to make a slight change to bring the notation in line with our new viewpoint. Let d_i be the mass dimension of the operator $\mathcal{O}^i$. Then rewrite (12.120) by representing each coefficient C_i as a power of M and a dimensionless coefficient ρ_i:

$$\mathcal{L}(\rho_i) = \mathcal{L}_M + \rho_i M^{4-d_i}\mathcal{O}_M^i(x). \quad (12.122)$$

The size of each ρ_i indicates the importance of the corresponding operator at the scale M. This new convention introduces further explicit M dependence into the Green's functions, which is compensated by a rescaling of the ρ_i. Thus (12.121) must be modified to

$$\left[M\frac{\partial}{\partial M} + \beta\frac{\partial}{\partial \lambda} + n\gamma + \sum_i [\gamma_i(\lambda) + d_i - 4]\rho_i\frac{\partial}{\partial \rho_i}\right]G^{(n)}(\{p_i\}; M, \lambda, \{\rho_i\}) = 0.$$
$$(12.123)$$

The meaning of this equation becomes clearer if we define

$$\beta_i = (d_i - 4 + \gamma_i)\rho_i. \quad (12.124)$$

Then

$$\left[M\frac{\partial}{\partial M} + \beta\frac{\partial}{\partial \lambda} + \sum_i \beta_i\frac{\partial}{\partial \rho_i} + n\gamma\right]G^{(n)}(\{p_i\}; M, \lambda, \{\rho_i\}) = 0. \quad (12.125)$$

Now all of the coupling constants ρ_i appear on the same footing as λ. We can solve this generalized Callan-Symanzik equation using the same method as in Section 12.3, by introducing bacteria, which now live in a multidimensional velocity field (β, β_i). The solution will depend on a set of running coupling constants which obey the equations

$$\frac{d}{d\log(p/M)}\bar{\rho}_i = \beta_i(\bar{\rho}_i, \bar{\lambda}). \tag{12.126}$$

It is interesting to examine this flow of coupling constants for the case where all the dimensionless parameters λ, ρ_i are small, so that we are close to the free scalar field Lagrangian. In this situation, we can ignore the contribution of γ_i to β_i; then

$$\frac{d}{d\log(p/M)}\bar{\rho}_i = [d_i - 4 + \cdots]\bar{\rho}_i. \tag{12.127}$$

The solution to this equation is

$$\bar{\rho}_i = \rho_i\left(\frac{p}{M}\right)^{d_i-4}. \tag{12.128}$$

Operators with mass dimension greater than 4, corresponding to nonrenormalizable interactions, become less important as a power of p as $p \to 0$. This is exactly the behavior that we found in Eq. (12.27) using Wilson's method. Since we have now generalized the Callan-Symanzik equation to incorporate the most general perturbation of the free-field Lagrangian, it is pleasing that we recover the full structure of the Wilson flow of coupling constants. In addition, this more formal method gives us a way to compute the corrections to the Wilson flow due to $\lambda\phi^4$ interactions, order by order in λ, using Feynman diagrams.

We can move one step closer to the generality of Section 12.1 by moving from four dimensions to an arbitrary dimension d. We require only two small changes in the formalism. First, the operator ϕ^4 acquires a dimensionful coefficient when $d \neq 4$, and we must take account of this. We have seen in the discussion below Eq. (10.13) that a scalar field has mass dimension $(d-2)/2$. Thus, the operator ϕ^4 has mass dimension $(2d-4)$, and so its coefficient has dimension $4 - d$. To implement the renormalization group, we redefine λ so that this coefficient remains dimensionless in d dimensions. We treat the mass term similarly, replacing $m^2 \to \rho_m M^2$. Thus the expansion of the Lagrangian about the free scalar field theory $\mathcal{L}_0$ reads:

$$\mathcal{L} = \mathcal{L}_0 - \frac{1}{2}\rho_m M^2 \phi_M^2 - \frac{1}{4}\lambda M^{4-d}\phi_M^4 + \cdots. \tag{12.129}$$

The second required change in the formalism is that of recomputing the β and γ functions in the new dimension. To order λ, the result is surprisingly innocuous. Consider, for example, the computation of γ_{ϕ^2}, Eq. (12.114). This computation, which was performed in dimensional regularization, is essentially unchanged. For general values of d, the derivative of the counterterm δ_{ϕ^2} with respect to $\log M$ still involves the factor

$$M\frac{\partial}{\partial M}\left(\frac{\Gamma(2-\frac{d}{2})}{(M^2)^{2-d/2}}\right) = -2 + \mathcal{O}(4 - d). \tag{12.130}$$

This observation holds for all of the γ_i, and the β function is shifted only by the contribution of the mass dimension of λ. Thus, for d near 4,

$$\beta = (d - 4)\lambda + \beta^{(4)}(\lambda) + \cdots,$$

$$\beta_m = [-2 + \gamma_{\phi^2}^{(4)}]\rho_m + \cdots, \tag{12.131}$$

$$\beta_i = [d_i - d + \gamma_i^{(4)}]\rho_i + \cdots,$$

where the functions with a superscript (4) are the four-dimensional results obtained earlier in this section, and the omitted correction terms are of order $\lambda \cdot (d-4)$. The precise form of these corrections depends on the renormalization scheme.[†]

Using the explicit four-dimensional result (12.46) for β, we now find

$$\beta = -(4-d)\lambda + \frac{3\lambda^2}{16\pi^2}. \tag{12.132}$$

For $d \geq 4$, this function is positive and predicts that the coupling constant flows smoothly to zero at large distances. However, when $d < 4$, this $\beta(\lambda)$ has the form shown in Fig. 12.4(b). Thus it generates just the coupling constant flow that we discussed from Wilson's viewpoint below Eq. (12.29). At small values of λ, the coupling constant increases in importance with increasing distance, as dimensional analysis predicts. However, at larger λ, the coupling constant decreases as a result of its own nonlinear effects. These two tendencies come into balance at the zero of the beta function,

$$\lambda_* = \frac{16\pi^2}{3}(4 - d), \tag{12.133}$$

which gives a nontrivial fixed point of the renormalization group flows in scalar field theory for $d < 4$. If we formally consider values of d close to 4, this fixed point occurs in a region where the coupling constant is small and we can use Feynman diagrams to investigate its properties. This fixed point, which was discovered by Wilson and Fisher,[‡] has important consequences for statistical mechanics, which we will discuss in Chapter 13.

Critical Exponents: A First Look

As an application of the formalism of this section, let us calculate the renormalization group flow of the coefficient of the mass operator in ϕ^4 theory. This is found by integrating Eq. (12.126), using the value of β_m from (12.131):

$$\frac{d}{d\log p}\bar{\rho}_m = [-2 + \gamma_{\phi^2}(\bar{\lambda})]\bar{\rho}_m. \tag{12.134}$$

[†]This expansion is displayed to rather high order in E. Brezin, J. C. Le Gillou, and J. Zinn-Justin, *Phys. Rev.* **D9**, 1121 (1974).

[‡]K. G. Wilson and M. E. Fisher, *Phys. Rev. Lett.* **28**, 240 (1972).

For $\lambda = 0$, this equation gives the trivial relation

$$\bar{\rho}_m = \rho_m \left(\frac{M}{p}\right)^2. \tag{12.135}$$

If we recall that we originally defined $\rho_m = m^2/M^2$, this is just a complicated way of saying that, when p becomes of order m, the mass term becomes an important term in the Lagrangian. At this point, the correlations in the ϕ field begin to die away exponentially. The characteristic range of correlations, which in statistical mechanics would be called the correlation length ξ, is given by

$$\xi \sim p_0^{-1}, \qquad \text{where } \bar{\rho}_m(p_0) = 1. \tag{12.136}$$

If we evaluate this criterion, we find $\xi \sim (M^2 \rho_m)^{-1/2}$, that is, $\xi \sim m^{-1}$, as we would have expected.

However, the application of this criterion at the fixed point λ_* gives a much more interesting result. If we set $\bar{\lambda} = \lambda_*$, then Eq. (12.134) has the solution

$$\bar{\rho}_m = \rho_m \left(\frac{M}{p}\right)^{2-\gamma_{\phi^2}(\lambda_*)}. \tag{12.137}$$

This gives a nontrivial relation

$$\xi \sim \rho_m^{-\nu}, \tag{12.138}$$

where the exponent ν is given formally by the expression

$$\nu = \frac{1}{2 - \gamma_{\phi^2}(\lambda_*)}. \tag{12.139}$$

Using the results (12.133) and (12.117), we can evaluate this explicitly for d near 4:

$$\nu^{-1} = 2 - \frac{1}{3}(4 - d). \tag{12.140}$$

Wilson and Fisher showed that this expression can be extended to a systematic expansion of ν in powers of $\epsilon = (4 - d)$.

Because the exponent ν has an interpretation in statistical mechanics, it is directly measurable in the realistic case of three dimensions. In the statistical mechanical interpretation of scalar field theory, ρ_m is just the parameter that one must adjust finely to bring the system to the critical temperature. Thus ρ_m is proportional to the deviation from the critical temperature, $(T - T_C)$. Our field theoretic analysis thus implies that the correlation length in a magnet grows as $T \to T_C$ according to the scaling relation

$$\xi \sim (T - T_C)^{-\nu}. \tag{12.141}$$

It also gives a definite, and somewhat unusual, prediction for the value of ν. It predicts that ν is close to the value $1/2$ suggested by the Landau approximation studied in Chapter 8 (Eq. (8.16)), but that ν differs from this value by some systematic corrections.

A scaling behavior of the type (12.141) is observed in magnets, and it is known that several definite scaling laws occur, depending on the symmetry of the spin ordering. Magnets can be characterized by the number of fluctuating spin components: $N=1$ for magnets with a preferred axis, $N=2$ for magnets with a preferred plane, and $N=3$ for magnets that are isotropic in three-dimensional space. The experimental value of ν depends on this parameter. The ϕ^4 field theory discussed in this chapter contained only one fluctuating field; this is the analogue of a magnet with one spin component. In Chapter 11, we considered a generalization of ϕ^4 theory to a theory of N fields with $O(N)$ symmetry. We might guess that this system models magnets of general N.

If this correspondence is correct, Eq. (12.140) gives a prediction for the value of ν in magnets with a preferred axis. In Section 13.1, we will repeat the analysis leading to this equation in the $O(N)$-symmetric ϕ^4 theory and derive the formula

$$\nu^{-1} = 2 - \frac{N+2}{N+8}(4-d), \tag{12.142}$$

valid for general N to first order in $(4-d)$. For the cases $N = 1, 2, 3$ and $d = 3$, this formula predicts

$$\nu = 0.60, \ 0.63, \ 0.65. \tag{12.143}$$

For comparison, the best current determinations of ν in magnetic systems give*

$$\nu = 0.64, \ 0.67, \ 0.71 \tag{12.144}$$

for $N = 1, 2, 3$. The prediction (12.143) gives a reasonable first approximation to the experimental results.

The ability of quantum field theory to predict the critical exponents gives a concrete application both of the formal connection between quantum field theory and statistical mechanics and of the flows of coupling constants predicted by the renormalization group. However, there is another experimental aspect of critical behavior that is even more remarkable, and more persuasive. Critical behavior can be studied not only in magnets but also in fluids, binary alloys, superfluid helium, and a host of other systems. It has long been known that, for systems with this disparity of microscopic dynamics, the scaling exponents at the critical point depend only on the dimension N of the fluctuating variables and not on any other detail of the atomic structure. Fluids, binary alloys, and uniaxial magnets, for example, have the same critical exponents. To the untutored eye, this seems to be a miracle. But for a quantum field theorist, this conclusion is the natural outcome of the renormalization group idea, in which most details of the field theoretic interaction are described by operators that become irrelevant as the field theory finds its proper, simple, large-distance behavior.

---

*For further details, see Table 13.1 and the accompanying discussion.

Problems

12.1 Beta functions in Yukawa theory. In the pseudoscalar Yukawa theory studied in Problem 10.2, with masses set to zero,

$$\mathcal{L} = \frac{1}{2}(\partial_\mu \phi)^2 - \frac{\lambda}{4!}\phi^4 + \bar{\psi}(i\partial\!\!\!/)\psi - ig\bar{\psi}\gamma^5\psi\phi,$$

compute the Callan-Symanzik β functions for λ and g:

$$\beta_\lambda(\lambda, g), \qquad \beta_g(\lambda, g),$$

to leading order in coupling constants, assuming that λ and g^2 are of the same order. Sketch the coupling constant flows in the λ-g plane.

12.2 Beta function of the Gross-Neveu model. Compute $\beta(g)$ in the two-dimensional Gross-Neveu model studied in Problem 11.3,

$$\mathcal{L} = \bar{\psi}_i i\partial\!\!\!/\psi_i + \frac{1}{2}g^2(\bar{\psi}_i\psi_i)^2,$$

with $i = 1, \ldots, N$. You should find that this model is asymptotically free. How was that fact reflected in the solution to Problem 11.3?

12.3 Asymptotic symmetry. Consider the following Lagrangian, with two scalar fields ϕ_1 and ϕ_2:

$$\mathcal{L} = \frac{1}{2}((\partial_\mu \phi_1)^2 + (\partial_\mu \phi_2)^2) - \frac{\lambda}{4!}(\phi_1^4 + \phi_2^4) - \frac{2\rho}{4!}(\phi_1^2\phi_2^2).$$

Notice that, for the special value $\rho = \lambda$, this Lagrangian has an $O(2)$ invariance rotating the two fields into one another.

(a) Working in four dimensions, find the β functions for the two coupling constants λ and ρ, to leading order in the coupling constants.

(b) Write the renormalization group equation for the ratio of couplings ρ/λ. Show that, if $\rho/\lambda < 3$ at a renormalization point M, this ratio flows toward the condition $\rho = \lambda$ at large distances. Thus the $O(2)$ internal symmetry appears asymptotically.

(c) Write the β functions for λ and ρ in $4 - \epsilon$ dimensions. Show that there are nontrivial fixed points of the renormalization group flow at $\rho/\lambda = 0, 1, 3$. Which is the most stable? Sketch the pattern of coupling constant flows. This flow implies that the critical exponents are those of a symmetric two-component magnet.

Chapter 13

Critical Exponents and Scalar Field Theory

The idea of running coupling constants and renormalization-group flows gives us a new language with which to discuss the qualitative behavior of scalar field theory. In our first discussion of ϕ^4 theory, each value of the coupling constant—and, more generally, each form of the potential and each spacetime dimension—gave a separate problem to be explored. But in Chapter 12, we saw that ϕ^4 theories with different values of the coupling are connected by renormalization-group flows, and that the pattern of these flows changes continuously with the spacetime dimension. In this context, it makes sense to ask the very general question: How does ϕ^4 theory behave as a function of the dimension? This chapter will give a detailed answer to this question.

The central ingredient in our analysis will be the Wilson-Fisher fixed point discussed in Section 12.5. This fixed point exists in spacetime dimensions d with $d < 4$; in those dimensions it controls the renormalization group flows of massless ϕ^4 theory. The scalar field theory has manifest or spontaneously broken symmetry according to the sign of the mass parameter m^2. Near $m^2 = 0$, the theory exhibits scaling behavior with anomalous dimensions whose values are determined by the renormalization group equations. For $d > 4$, the Wilson-Fisher fixed point disappears, and only the free-field fixed point remains. Again, the theory exhibits two distinct phases, but now the behavior at the transition is determined by the renormalization group flows near the free-field fixed point, so the scaling laws are those that follow from simple dimensional analysis.

The continuation of these results to Euclidean space has important implications for the theory of phase transitions in magnets and fluids. As we discussed in the previous chapter, the ideas of the renormalization group imply that the power-law behaviors of thermodynamic quantities near a phase transition point are determined by the behavior of correlation functions in a Euclidean ϕ^4 theory. The results stated in the previous paragraph then imply the following conclusions for critical scaling laws: For statistical systems in a space of dimension $d > 4$, the scaling laws are just those following from simple dimensional analysis. These predictions are precisely those of Landau theory, which we discussed in Chapter 8. On the other hand, for $d < 4$, the critical scaling laws are modified, in a way that we can compute using the renormalization group.

In $d = 4$, we are on the boundary between the two types of scaling behavior. This corresponds to the situation in which ϕ^4 theory is precisely renormalizable. In this case, the dimensional analysis predictions are corrected, but only by logarithms. We will analyze this case specifically in Section 13.2.

Though it is not obvious, the case $d = 2$ provides another boundary. Here the transition to spontaneous symmetry breaking is described by a different quantum field theory, which becomes renormalizable in two dimensions. In Section 13.3, we will introduce that theory, called the *nonlinear sigma model*, and show how its renormalization group behavior merges smoothly with that of ϕ^4 theory. By combining all of the results of this chapter, we will obtain a quantitative understanding of the behavior of ϕ^4 theory, and of critical phenomena, over the whole range of spacetime dimensions.

13.1 Theory of Critical Exponents

At the end of Chapter 12, we used properties of the renormalization group for scalar field theory to make a prediction about the behavior of correlations near the critical point of a thermodynamic system. We argued that the range of correlations, the correlation length ξ, should increase to infinity as one approaches the critical point, according to the scaling law (12.141). The exponent in this equation, called ν, should depend only on the symmetry of the order parameter. We argued, further, that this exponent is related to the anomalous dimension of a local operator in ϕ^4 theory, and that it can be computed from Feynman diagrams. In this section, we will show that similar conclusions apply more generally to a large number of scaling laws associated with a critical point.

To begin, we will define systematically a set of *critical exponents*, exponents of scaling laws that describe the thermodynamic behavior in the vicinity of the critical point. We will then show, using the Callan-Symanzik equation, that all these exponents can be reduced to two basic anomalous dimensions. Finally, we will compare this remarkable prediction of quantum field theory to experiment.

In suggesting a set of critical scaling laws, we begin with the behavior of the correlation function of fluctuations of the ordering field. For definiteness, we will use the language appropriate to a magnet, as in Chapter 8. We will compute classical thermal expectation values as correlation functions in a Euclidean quantum field theory, as explained in Section 9.3. The fluctuating field will be called the spin field $s(x)$, its integral will be the magnetization M, and the external field that couples to $s(x)$ will be called the magnetic field H. (In deference to the magnetization, we will denote the renormalization scale in the Callan-Symanzik equation by μ in this section.)

Define the two-point correlation function by

$$G(x) = \langle s(x)s(0)\rangle, \tag{13.1}$$

or by the connected expectation value, if we are in the magnetized phase where

$\langle s(x) \rangle \neq 0$. Away from the critical point, $G(x)$ should decay exponentially, according to

$$G(x) \sim \exp[-|x|/\xi]. \tag{13.2}$$

The approach to the critical point is characterized by the parameter

$$t = \frac{T - T_C}{T_C}. \tag{13.3}$$

Then we expect that, as $t \to 0$, the correlation length should increase to infinity. Define the exponent ν, (12.141), by the formula

$$\xi \sim |t|^{-\nu}. \tag{13.4}$$

Just at $t = 0$, the correlation function should decay only as a power law. Define the exponent η by the formula

$$G(x) \sim \frac{1}{|x|^{d-2+\eta}}, \tag{13.5}$$

where d is the Euclidean space dimension.

The behaviors of thermodynamic quantities near the critical point define a number of additional exponents. Typically, the specific heat of the thermodynamic system diverges as $t \to 0$; define the exponent α by the formula for the specific heat at fixed external field $H = 0$:

$$C_H \sim |t|^{-\alpha}. \tag{13.6}$$

Since the ordering sets in at $t = 0$, the magnetization at zero field tends to zero as $t \to 0$ from below. Define the exponent β (not to be confused with the Callan-Symanzik function) by

$$M \sim |t|^{\beta}. \tag{13.7}$$

Even at $t = 0$ one has a nonzero magnetization at nonzero magnetic field. Write the law by which this magnetization tends to zero as $H \to 0$ as the relation

$$M \sim H^{1/\delta}. \tag{13.8}$$

Finally, the magnetic susceptibility diverges at the critical point; we write this divergence as the relation

$$\chi \sim |t|^{-\gamma}. \tag{13.9}$$

Equations (13.4)–(13.9) define a set of critical exponents α, β, γ, δ, ν, η, which can be measured experimentally for a variety of thermodynamic systems.*

In Chapter 12 we argued, following Wilson, that a thermodynamic system near its critical point can be described by a Euclidean quantum field theory. At the level of the atomic scale, the Lagrangian of this quantum field theory may be complicated; however, when we have integrated out the small-scale

---

*A variety of further critical exponents and relations are presented in M. E. Fisher, *Repts. Prog. Phys.* **30**, 615 (1967).

degrees of freedom, this Lagrangian simplifies. If we adjust a parameter of the theory to insure the presence of long-range correlations, the Lagrangian must closely approach a fixed point of the renormalization group. Generically, the Lagrangian will approach the fixed point with a single unstable or relevant direction, corresponding to the mass parameter of ϕ^4 theory. In $d < 4$, this is the Wilson-Fisher fixed point. In $d \geq 4$, it is the free-field fixed point. For definiteness, we will assume $d < 4$ in the following discussion.

Exponents of the Spin Correlation Function

In this setting, we can study the behavior of the spin-spin correlation function $G(x)$. By the argument just reviewed, $G(x)$ is proportional to the two-point correlation function of a Euclidean scalar field theory. The technology introduced in the previous chapter can be applied directly. The correlation function obeys the Callan-Symanzik equation (12.125),

$$\left[\mu\frac{\partial}{\partial\mu} + \sum_i \beta_i\frac{\partial}{\partial\rho_i} + 2\gamma\right]G(x;\mu,\{\rho_i\}) = 0. \tag{13.10}$$

Here we include the ϕ^4 coupling λ among the generalized couplings ρ_i.

By dimensional analysis, in d dimensions,

$$G(x) = \frac{1}{|x|^{d-2}}g(\mu|x|,\{\rho_i\}), \tag{13.11}$$

where g is an arbitrary function of the dimensionless parameters. (This is the Fourier transform of the statement that $G(p) \sim p^{-2}$ times a dimensionless function.) From this starting point, we can solve the Callan-Symanzik equation (13.10) by the method of Section 12.3, and find

$$G(x) = \frac{1}{|x|^{d-2}}h(\{\bar{\rho}_i(x)\}) \cdot \exp\left[-2\int\limits_{1/\mu}^{|x|} d\log(|x'|)\,\gamma(\{\bar{\rho}(x')\})\right], \tag{13.12}$$

where h is a dimensionless initial condition. The running coupling constants $\bar{\rho}_i$ obey the differential equation

$$\frac{d}{d\log(1/\mu|x|)}\bar{\rho}_i = \beta_i(\{\rho_j\}). \tag{13.13}$$

We studied the solution to this equation in Section 12.5. We saw there that, for flows that come to the vicinity of the Wilson-Fisher fixed point, the dimensionless coefficient of the mass operator grows as one moves toward large distances, while the other dimensionless parameters become small. Let λ_* be the location of the fixed point. Then we can write more explicitly

$$\bar{\rho}_m = \rho_m(\mu|x|)^{2-\gamma_{\phi^2}(\lambda_*)},$$
$$\bar{\rho}_i = \rho_i(\mu|x|)^{-A_i}, \tag{13.14}$$

where $A_i > 0$ for $i \neq m$. If the deviation of λ from the fixed point is treated as one of the ρ_i, by defining

$$\rho_\lambda = \lambda - \lambda_*, \tag{13.15}$$

this parameter also decreases in importance as a power of $|x|$, as we demonstrated in Eq. (12.96). In the language of Section 12.1, all of the parameters ρ_i multiply irrelevant operators, except for ρ_m, which multiplies a relevant operator.

To approach the critical point, we adjust the parameters of the underlying theory so that, at some scale $(1/\mu)$ near the atomic scale, $\rho_m \ll 1$. If ρ_m is adjusted by tuning the temperature of the thermodynamic system, then $\rho_m \sim t$. The critical scaling laws will be valid if there is a region of distance scales where $\bar{\rho}_m$ remains small while the other $\bar{\rho}_i$ can be neglected. The scaling laws can then be computed by evaluating the solution to the Callan-Symanzik equation with $\bar{\rho}_m$ given by (13.14) and the other $\bar{\rho}_i$ set equal to zero. The corrections to this approximation can be shown to be proportional to positive powers of t.

In this approximation, we should evaluate the function $\gamma(\lambda)$ in (13.12) at $\bar{\rho}_\lambda = 0$, that is, at the fixed point. Using this value and the solution for $\bar{\rho}_m$, Eq. (13.12) becomes

$$G(x) = \frac{1}{|x|^{d-2}} \cdot \frac{1}{(\mu|x|)^{2\gamma(\lambda_*)}} \cdot h\big(t(\mu|x|)^{2-\gamma_{\phi^2}(\lambda_*)}\big). \tag{13.16}$$

This equation implies the scaling laws (13.5) and (13.4): For the argument of h sufficiently small, $G(x)$ obeys Eq. (13.5), with

$$\eta = 2\gamma(\lambda_*). \tag{13.17}$$

At large distances, h must fall off exponentially, since this function is derived from a scalar field propagator. From the argument of h, we deduce that this exponential must be of the form

$$\exp[-|x|(\mu t^\nu)], \tag{13.18}$$

where, as in (12.139),

$$\nu = \frac{1}{2 - \gamma_{\phi^2}(\lambda_*)}. \tag{13.19}$$

This is precisely the scaling law (13.2), (13.4), with the identification of ν in terms of the anomalous dimension of the operator ϕ^2.

Exponents of Thermodynamic Functions

The thermodynamic critical scaling laws can be derived in a similar way, by studying the scaling behavior of macroscopic thermodynamic variables. These are derived from the Gibbs free energy, or, in the language of quantum fields, from the effective potential of the scalar field theory. Since the effective potential, and, more generally, the effective action, are constructed from correlation functions, these quantities should satisfy Callan-Symanzik equations. We will now construct those equations and then use them to identify the thermodynamic critical exponents.

In Eq. (11.96), we showed that the effective action Γ depends on the classical field ϕ_{cl} in such a way that the nth derivative of Γ with respect to ϕ_{cl} gives the one-particle-irreducible n-point function of the field theory. Thus we can reconstruct Γ from the 1PI functions by writing the Taylor series

$$\Gamma[\phi_{cl}] = i \sum_{2}^{\infty} \frac{1}{n!} \int dx_1 \cdots dx_n \, \phi_{cl}(x_1) \cdots \phi_{cl}(x_n) \, \Gamma^{(n)}(x_1, \ldots, x_n), \quad (13.20)$$

where the $\Gamma^{(n)}$ are the 1PI amplitudes.

To find the Callan-Symanzik equation satisfied by $\Gamma[\phi_{cl}]$, it is easiest to first work out the equation satisfied by $\Gamma^{(n)}$. We begin by considering the irreducible three-point function $\Gamma^{(3)}$. This function is defined as

$$\Gamma^{(3)}(p_1, p_2, p_3) = \frac{1}{G^{(2)}(p_1)G^{(2)}(p_2)G^{(2)}(p_3)} G^{(3)}(p_1, p_2, p_3). \quad (13.21)$$

Rescaling with factors $Z(\mu)$, we see that $\Gamma^{(3)}$ is related to the irreducible three-point function of bare fields by

$$\Gamma^{(3)}(p_1, p_2, p_3) = Z(\mu)^{+3/2}\Gamma_0^{(3)}(p_1, p_2, p_3).$$

Similarly, the irreducible n-point function is related to the corresponding function of bare fields by

$$\Gamma^{(n)} = Z(\mu)^{n/2}\Gamma_0^{(n)}. \quad (13.22)$$

This relation is identical in form to the corresponding relation for the full Green's functions, Eq. (12.35), except for the change of sign in the exponent. From this point, we can follow the logic used to derive the Callan-Symanzik equation for Green's functions, Eq. (12.41); the only difference is that the $n\gamma$ term enters with the opposite sign. Thus we find

$$\left[\mu\frac{\partial}{\partial\mu} + \beta(\lambda)\frac{\partial}{\partial\lambda} - n\gamma(\lambda)\right]\Gamma^{(n)}(\{p_i\}; \mu, \lambda) = 0. \quad (13.23)$$

To convert this to an equation for the effective action, note that, on the right-hand side of Eq. (13.20), the function $\Gamma^{(n)}$ is accompanied by n powers of the classical field. Then Eq. (13.23), integrated with n powers of ϕ_{cl} and summed over n, is equivalent to the equation

$$\left[\mu\frac{\partial}{\partial\mu} + \beta(\lambda)\frac{\partial}{\partial\lambda} - \gamma(\lambda)\int dx\,\phi_{cl}(x)\frac{\delta}{\delta\phi_{cl}(x)}\right]\Gamma([\phi_{cl}]; \mu, \lambda) = 0. \quad (13.24)$$

The operator multiplying $\gamma(\lambda)$ counts the number of powers of ϕ_{cl} in each term of the Taylor expansion. By specializing Eq. (13.24) to the case of constant ϕ_{cl}, we find the Callan-Symanzik equation for V_{eff}:

$$\left[\mu\frac{\partial}{\partial\mu} + \beta(\lambda)\frac{\partial}{\partial\lambda} - \gamma\phi_{cl}\frac{\partial}{\partial\phi_{cl}}\right]V_{eff}(\phi_{cl}, \mu, \lambda) = 0. \quad (13.25)$$

To apply Eq. (13.25) to the problem of critical exponents, we first convert this equation to the notation of statistical mechanics by replacing ϕ_{cl} with the magnetization M, the conjugate source J by H, and the effective potential V_{eff} by the Gibbs free energy $\mathbf{G}(M)$. At the same time, we will generalize λ to the full set of couplings ρ_i. Then (13.25) takes the form

$$\left[\mu\frac{\partial}{\partial\mu} + \sum_i \beta_i\frac{\partial}{\partial\rho_i} - \gamma M\frac{\partial}{\partial M}\right]\mathbf{G}(M, \mu, \{\rho_i\}) = 0. \qquad (13.26)$$

Now let us find the solution to this equation. As before, we begin from a statement of dimensional analysis. In d dimensions, the effective potential has mass dimension d, and a scalar field has mass dimension $(d-2)/2$. Thus

$$\mathbf{G}(M, \mu, \{\rho_i\}) = M^{2d/(d-2)}\hat{g}\big(M\mu^{-(d-2)/2}, \{\rho_i\}\big), \qquad (13.27)$$

where $\hat{g}$ is a new dimensionless function. Inserting (13.27) into (13.26), we see that $\hat{g}$ satisfies

$$\left[\sum_i \beta_i\frac{\partial}{\partial\rho_i} - \left(\frac{d-2}{2}+\gamma\right)M\frac{\partial}{\partial M} - d\frac{2}{d-2}\gamma\right]\hat{g}\big(M\mu^{-(d-2)/2}, \{\rho_i\}\big) = 0, \quad (13.28)$$

that is,

$$\left[M\frac{\partial}{\partial M} - \sum_i \frac{2\beta_i}{(d-2+2\gamma)}\frac{\partial}{\partial\rho_i} + \frac{4d\gamma}{(d-2)(d-2+2\gamma)}\right]\hat{g} = 0. \qquad (13.29)$$

Solving this equation, we find

$$\mathbf{G}(M) = M^{2d/(d-2)}\hat{h}(\{\bar{\rho}_i(M)\})$$

$$\times \exp\left[-\int_{\mu^{(d-2)/2}}^{M} d\log(M')\frac{4d\gamma}{(d-2)(d-2+2\gamma)}(\{\bar{\rho}(M')\})\right],$$

$$\qquad (13.30)$$

where the running coupling constants $\bar{\rho}_i$ obey

$$\frac{d}{d\log M}\bar{\rho}_i = \frac{2\beta_i(\{\bar{\rho}_i\})}{d-2+2\gamma(\{\bar{\rho}_i\})}. \qquad (13.31)$$

As in our discussion of the spin correlation function, we specialize to the critical region by assuming that we are on a renormalization group flow that passes close to the Wilson-Fisher fixed point. We again ignore the effects of irrelevant operators. Then we should set

$$\bar{\rho}_m = \rho_m\big(M\mu^{-(d-2)/2}\big)^{-2(2-\gamma_{\phi^2}(\lambda_*))/(d-2+2\gamma(\lambda_*))}, \qquad (13.32)$$

$$\bar{\rho}_i = 0 \qquad \text{for } i \neq m,$$

with $\rho_m \sim t$. In this approximation, the Gibbs free energy takes the form

$$\mathbf{G}(M, t) = M^{2d/(d-2)} \cdot \big(M\mu^{-(d-2)/2}\big)^{-4d\gamma(\lambda_*)/(d-2)(d-2+2\gamma(\lambda_*))}$$

$$\cdot \hat{h}\big(t(M\mu^{-(d-2)/2})^{-2(2-\gamma_{\phi^2}(\lambda_*))/(d-2+2\gamma(\lambda_*))}\big), \qquad (13.33)$$

where $\hat{h}$ is a smooth initial condition.

To simplify the form of the exponents in this expression, we anticipate some of the results below and replace

$$\beta = \frac{d - 2 + 2\gamma(\lambda_*)}{2(2 - \gamma_{\phi^2}(\lambda_*))},$$

$$\delta = \frac{2d}{d - 2 + 2\gamma(\lambda_*)} - 1 = \frac{d + 2 - 2\gamma(\lambda_*)}{d - 2 + 2\gamma(\lambda_*)}. \tag{13.34}$$

We must demonstrate that these new exponents indeed correspond to the ones we have defined in Eqs. (13.7) and (13.8). With these replacements (and ignoring the dependence on μ from here on), we find for **G** the scaling formula

$$\mathbf{G}(M, t) = M^{1+\delta}\hat{h}(tM^{-1/\beta}), \tag{13.35}$$

where $\hat{h}$ has a smooth limit as $t \to 0$. An equivalent way to represent this formula is

$$\mathbf{G}(M, t) = t^{\beta(1+\delta)}\hat{f}(Mt^{-\beta}). \tag{13.36}$$

The scaling laws for thermodynamic quantities follow immediately from these relations. Along the line $t = 0$, we find from (13.35) that

$$H = \frac{\partial \mathbf{G}}{\partial M} = \hat{h}(0)M^{\delta}, \tag{13.37}$$

which is precisely (13.8). Below the critical temperature, we find the nonzero value of the magnetization by minimizing **G** with respect to M. In the scaling region, this minimum occurs at the minimum m_0 of the function $\hat{f}(m)$ in (13.36). This leads to relation (13.7), in the form

$$Mt^{-\beta} = m_0. \tag{13.38}$$

If we work above T_C and in zero field, the minimum of $\hat{f}$ must occur at $M = 0$. Then

$$\mathbf{G}(t) \sim t^{\beta(1+\delta)}. \tag{13.39}$$

To compute the specific heat, we differentiate twice with respect to temperature; this gives the scaling law (13.6), with

$$2 - \alpha = \beta(1 + \delta) = \frac{d}{2 - \gamma_{\phi^2}(\lambda_*)}. \tag{13.40}$$

Finally, we must construct the scaling law for the magnetic susceptibility. From (13.36), the scaling law for H at nonzero t is

$$H = \frac{\partial \mathbf{G}}{\partial M} = t^{\beta\delta}\hat{f}'(Mt^{-\beta}). \tag{13.41}$$

The inverse of this relation is the scaling law

$$M = t^{\beta}\hat{c}(Ht^{-\beta\delta}). \tag{13.42}$$

The magnetic susceptibility at zero field is then

$$\chi = \left(\frac{\partial M}{\partial H}\right)_t = \hat{c}'(0)t^{-(\delta-1)\beta}. \tag{13.43}$$

Thus, we confirm Eq. (13.9), with the identification

$$\gamma = (\delta - 1)\beta = \frac{2(1 - \gamma(\lambda_*))}{2 - \gamma_{\phi^2}(\lambda_*)}. \tag{13.44}$$

We have now found explicit expressions for all of the various critical exponents in terms of the Callan-Symanzik functions. As the dimensionality d approaches 4 from below, the fixed point λ_* tends to zero. Then the six critical exponents approach the values that they would attain in simple dimensional analysis:

$$\eta = 0; \qquad \nu = \tfrac{1}{2}; \qquad \alpha = 0; \qquad \beta = \tfrac{1}{2};$$
$$\gamma = 1; \qquad \delta = 3. \tag{13.45}$$

It is no surprise that the values of η, ν and β given in (13.45) are those that we derived in Chapter 8 from the the Landau theory of critical phenomena. The other values can similarly be shown to follow from Landau theory. The renormalization group analysis tells us how to systematically correct the predictions of Landau theory to take proper account of the large-scale fluctuations of the spin field.

Notice that all of the exponents associated with thermodynamic quantities are constructed from the same ingredients as the exponents associated with the correlation function. From the field theory viewpoint, this is obvious, since all of the scaling laws in the field theory must ultimately follow from the anomalous dimensions of the operators $\phi(x)$ and $\phi^2(x)$, which are precisely $\gamma(\lambda_*)$ and $\gamma_{\phi^2}(\lambda_*)$. This result, however, has an interesting experimental consequence: It implies model-independent relations among critical exponents. For example, in any system with a critical point, this theory predicts

$$\alpha = 2 - d\nu, \qquad \beta = \tfrac{1}{2}(d - 2 + \eta)\nu. \tag{13.46}$$

These relations test the general framework of identifying a critical point with the fixed point of a renormalization group flow.

In addition, the field theoretic approach to critical phenomena predicts that critical exponents are *universal*, in the sense that they take the same values in condensed matter systems that approach the same scalar field fixed point in the limit $T \rightarrow T_C$.

Values of the Critical Exponents

Finally, scalar field theory actually predicts the values of $\gamma(\lambda_*)$ and $\gamma_{\phi^2}(\lambda_*)$, either from the expansion in powers of $\epsilon = 4 - d$ described in Section 12.5 or by direct expansion of the β and γ functions in powers of λ. We can use these expressions to generate quantitative predictions for the critical exponents. We gave an example of such a prediction at the end of Section 12.5, when we presented in Eq. (12.143) the first two terms of an expansion for ν. We now return to this question to give field-theoretic predictions for all of the critical exponents.

In our discussion at the end of Section 12.5, we remarked that magnets with different numbers of fluctuating spin components are observed to have different values for the critical exponents. An optimistic hypothesis would be that any thermodynamic system with N fluctuating spin components, or, more generally, N fluctuating thermodynamic variables at the critical point, would be described by the same fixed point field theory with N scalar fields. A natural candidate for this fixed point would be the Wilson-Fisher fixed point of the $O(N)$-symmetric ϕ^4 theory discussed in Chapter 11. We will now describe the computation of critical exponents for general values of N in this theory.

As a first step, we should compute the values of the functions $\beta(\lambda)$, $\gamma(\lambda)$, and $\gamma_{\phi^2}(\lambda)$ in four dimensions. This computation parallels the analysis done in Chapter 12 for ordinary ϕ^4 theory, so we will only indicate the changes that need to be made for this case. Just as in ordinary ϕ^4 theory, the propagator of the massless $O(N)$-symmetric theory receives no field strength corrections in one-loop order, and so the one-loop term in $\gamma(\lambda)$ again vanishes. In Problem 13.2, we compute the leading, two-loop, contribution to $\gamma(\lambda)$ in $O(N)$-symmetric ϕ^4 theory:

$$\gamma = (N+2)\frac{\lambda^2}{4(8\pi^2)^2} + \mathcal{O}(\lambda^3). \tag{13.47}$$

The one-loop contribution to the β function in ϕ^4 theory is derived from the one-loop vertex counterterm δ_λ, given in Eq. (12.44). For the $O(N)$-symmetric case, we computed the divergent part of the corresponding vertex counterterm in Section 11.2; from Eq. (11.22),

$$\delta_\lambda = \frac{\lambda^2}{(4\pi)^{d/2}}(N+8)\frac{\Gamma(2-\frac{d}{2})}{(M^2)^{2-d/2}} + \text{finite}. \tag{13.48}$$

Following the logic to Eq. (12.46), or using Eq. (12.54), we find

$$\beta = (N+8)\frac{\lambda^2}{8\pi^2} + \mathcal{O}(\lambda^3). \tag{13.49}$$

This reduces to the β function of ϕ^4 theory if we set $N = 1$ and replace $\lambda \to \lambda/6$, as indicated below Eq. (11.5). Finally, to compute γ_{ϕ^2}, we must repeat the computation done at the end of Section 12.4. If we consider, instead of (12.113), the Green's function $\langle\phi^i(p)\phi^j(q)\phi^2(k)\rangle$, and replace the vertex of ϕ^4 theory by the four-point vertex following from the Lagrangian (11.5), the factor $(-i\lambda)$ in the first line of (12.114) is replaced by

$$(-2i\lambda)[\delta^{ij}\delta^{k\ell} + \delta^{ik}\delta^{j\ell} + \delta^{i\ell}\delta^{jk}] \cdot \delta^{k\ell} = -2i\lambda(N+2)\delta^{ij}.$$

Then

$$\gamma_{\phi^2} = (N+2)\frac{\lambda}{8\pi^2} + \mathcal{O}(\lambda^2). \tag{13.50}$$

Next, we consider the same theory in $(4 - \epsilon)$ dimensions. The β function now becomes

$$\beta = -\epsilon\lambda + (N + 8)\frac{\lambda^2}{8\pi^2}, \qquad (13.51)$$

so there is a Wilson-Fisher fixed point at

$$\lambda_* = \frac{8\pi^2\epsilon}{N + 8}. \qquad (13.52)$$

At this fixed point, we find

$$\gamma(\lambda_*) = \frac{N + 2}{4(N + 8)^2}\epsilon^2 + \cdots, \qquad \gamma_{\phi^2}(\lambda_*) = \frac{N + 2}{N + 8}\epsilon + \cdots. \qquad (13.53)$$

From these two results, we can work out predictions for the whole set of critical exponents to order ϵ. As an example, inserting (13.53) into (13.19), we find

$$\nu^{-1} = 2 - \frac{N + 2}{N + 8}\epsilon + \mathcal{O}(\epsilon^2), \qquad (13.54)$$

as claimed at the end of Section 12.5.

In our discussion in Chapter 12, we claimed that the predictions of critical exponents are in rough agreement with experimental data. However, by computing to higher order, one can obtain a much more precise comparison of theory and experiment. The ϵ expansion of critical exponents has now been worked out through order ϵ^5. More impressively, the λ expansion for critical exponents in $d = 3$ has been worked out through order λ^9. By summing this perturbation series, it is possible to obtain very precise estimates of the anomalous dimensions $\gamma(\lambda_*)$ and $\gamma_{\phi^2}(\lambda_*)$ and, through them, precise predictions for the critical exponents.

A comparison of these values to direct determinations of the critical exponents is given in Table 13.1. The column labeled 'QFT' gives values of critical exponents obtained by anomalous dimension calculations using ϕ^4 perturbation theory in three dimensions. The column labeled 'Experiment' lists a selection of experimental determinations of the critical exponents in a variety of systems. These include the liquid-gas critical point in Xe, CO_2, and other fluids, the critical point in binary fluid mixtures with liquid-liquid phase separation, the order-disorder transition in the atomic arrangement of the Cu–Zn alloy β-brass, the superfluid transition in ^{4}He, and the order-disorder transitions in ferromagnets (EuO, EuS, Ni) and antiferromagnets (RbMnF$_3$). The agreement between experimental determinations of the exponents in different systems is a direct test of universality. For the case of systems with a single order parameter ($N = 1$), there is a remarkable diversity of physical systems that are characterized by the same critical exponents.

The column labeled 'Lattice' contains estimates of critical exponents in abstract lattice statistical mechanical models. For these simplified models, the statistical mechanical partition function can be calculated in an expansion for large temperature. With some effort, these expansions can be carried out to

Table 13.1. Values of Critical Exponents
for Three-Dimensional Statistical Systems

Exponent	Landau	QFT	Lattice	Experiment	
$N = 1$ Systems:					
γ	1.0	1.241 (2)	1.239 (3)	1.240 (7)	binary liquid
				1.22 (3)	liquid-gas
				1.24 (2)	β-brass
ν	0.5	0.630 (2)	0.631 (3)	0.625 (5)	binary liquid
				0.65 (2)	β-brass
α	0.0	0.110 (5)	0.103 (6)	0.113 (5)	binary liquid
				0.12 (2)	liquid-gas
β	0.5	0.325 (2)	0.329 (9)	0.325 (5)	binary liquid
				0.34 (1)	liquid-gas
η	0.0	0.032 (3)	0.027(5)	0.016 (7)	binary liquid
				0.04 (2)	β-brass
$N = 2$ Systems:					
γ	1.0	1.316 (3)	1.32 (1)		
ν	0.5	0.670 (3)	0.674 (6)	0.672 (1)	superfluid ^{4}He
α	0.0	−0.007 (6)	0.01 (3)	−0.013 (3)	superfluid ^{4}He
$N = 3$ Systems:					
γ	1.0	1.386 (4)	1.40 (3)	1.40 (3)	EuO, EuS
				1.33 (3)	Ni
				1.40 (3)	RbMnF$_3$
ν	0.5	0.705 (3)	0.711 (8)	0.70 (2)	EuO, EuS
				0.724 (8)	RbMnF$_3$
α	0.0	−0.115 (9)	−0.09 (6)	−0.011 (2)	Ni
β	0.5	0.365 (3)	0.37 (5)	0.37 (2)	EuO, EuS
				0.348 (5)	Ni
				0.316 (8)	RbMnF$_3$
η	0.0	0.033 (4)	0.041 (14)		

The values of critical exponents in the column 'QFT' are obtained by resumming
the perturbation series for anomalous dimensions at the Wilson-Fisher fixed point in
$O(N)$-symmetric ϕ^4 theory in three dimensions. The values in the column 'Lattice'
are based on analysis of high-temperature series expansions for lattice statistical mechanical models. The values in the column 'Experiment' are taken from experiments
on critical points in the systems described. In all cases, the numbers in parentheses are
the standard errors in the last displayed digits. This table is based on J. C. Le Guil-
lou and J. Zinn-Justin, *Phys. Rev.* **B21**, 3976 (1980), with some values updated from
J. Zinn-Justin (1993), Chapter 27. A full set of references for the last two columns can
be found in these sources.

15 terms or more. By resumming these series, one can obtain direct theoretical estimates of the critical exponents, with an accuracy comparable to that of the best experiments. The comparison between these values and experiment tests the identification of experimental systems with the simple Hamiltonians that were the starting point for our renormalization group analysis.

The agreement of all three types of determinations of the critical exponents presents an impressive picture. The picture is certainly not perfect, and a careful inspection of Table 13.1 reveals some significant discrepancies. But, in general, the evidence is compelling that quantum field theory provides the basic explanation for the thermodynamic critical behavior of a broad range of physical systems.

13.2 Critical Behavior in Four Dimensions

Now that we have discussed the general theory of critical exponents for $d < 4$, let us concentrate some attention on the case $d = 4$. This case obviously has special interest for the applications of quantum field theory to elementary particle physics. In addition, we now know that $d = 4$ lies on a boundary at which the Wilson-Fisher fixed point disappears. We would like to understand the special behavior of quantum field theory predictions at this boundary.

The most obvious difference between $d < 4$ and $d = 4$ is that, while in the former case the deviation of λ from the fixed point multiplies an irrelevant operator, in the case $d = 4$, λ multiplies a marginal operator. We have seen in Eq. (12.82) that, at small momenta or large distances, the running value of λ still approaches its fixed point, now located at $\lambda = 0$. However, this approach is described by a much slower function, not a power but only a logarithm of the distance scale. Thus it is normally not correct to ignore the deviation of λ from the fixed point. Including this effect, we find additional logarithmic terms, analogous to the dependence of correlation functions on $\log p$ that we already know characterizes a renormalizable field theory.

To give a nontrivial illustration of this logarithmic dependence, we return to a problem that we postponed at the end of Chapter 11. In Eq. (11.81), we obtained the expression for the effective potential of ϕ^4 theory to second order in λ, in the limit of vanishing mass parameter:

$$V_{\text{eff}} = \frac{1}{4}\phi_{\text{cl}}^4 \Big[\lambda + \frac{\lambda^2}{(4\pi)^2}\Big((N+8)\big(\log(\lambda\phi_{\text{cl}}^2/M^2) - \tfrac{3}{2}\big) + 9\log 3\Big)\Big]. \quad (13.55)$$

(Note that we now return to our standard notation, in which M is the renormalization scale and μ is a mass parameter.) This expression seemed to have a minimum for very small values of ϕ_{cl}, but only at values so small that

$$\big|\lambda \log(\lambda\phi_{\text{cl}}^2/M^2)\big| \sim 1. \quad (13.56)$$

Since, at the nth order of perturbation theory, one finds n powers of this logarithm, Eq. (13.56) implies that the higher-order terms in λ are not necessarily negligible. What we need is a technique that sums these terms.

This summation is provided by the Callan-Symanzik equation. From (13.24) or (13.25), the Callan-Symanzik equation for the effective potential in the massless limit of four-dimensional ϕ^4 theory is

$$\left[M \frac{\partial}{\partial M} + \beta(\lambda) \frac{\partial}{\partial \lambda} - \gamma \phi_{cl} \frac{\partial}{\partial \phi_{cl}} \right] V_{eff}(\phi_{cl}, M, \lambda) = 0. \tag{13.57}$$

As before, we can solve for V_{eff} by combining this equation with the predictions of dimensional analysis. In $d = 4$,

$$V_{eff}(\phi_{cl}, M, \lambda) = \phi_{cl}^4 v(\phi_{cl}/M, \lambda). \tag{13.58}$$

Then v satisfies

$$\left[\phi_{cl} \frac{\partial}{\partial \phi_{cl}} - \frac{\beta}{1+\gamma} \frac{\partial}{\partial \lambda} + \frac{4\gamma}{1+\gamma} \right] v = 0. \tag{13.59}$$

This equation for v can be solved by our standard methods, to give

$$v(\phi/M, \lambda) = v_0(\bar\lambda) \exp\left[-\int\limits_{M}^{\phi_{cl}} d\log \phi_{cl} \frac{4\gamma}{1+\gamma} (\bar\lambda(\phi_{cl})) \right], \tag{13.60}$$

where $\bar\lambda$ satisfies

$$\frac{d}{d\log(\phi_{cl}/M)} \bar\lambda = \frac{\beta(\bar\lambda)}{1+\gamma(\bar\lambda)}. \tag{13.61}$$

However, since we are working only to the order of the leading loop corrections, and since $\gamma(\lambda)$ is zero to this order, we can ignore the exponential in (13.60). In addition, we can ignore the denominator on the right-hand side of (13.61), so that this equation reduces to the more standard form of the equation (12.73) for the running coupling constant. Thus, using the leading-order Callan-Symanzik function, we find

$$V_{eff}(\phi_{cl}) = v_0(\bar\lambda(\phi_{cl}))\phi_{cl}^4. \tag{13.62}$$

The function v_0 in (13.62) is not determined by the Callan-Symanzik equation. To find this function, we compare (13.62) to the result (13.55) that we obtained from our explicit one-loop evaluation of the effective potential. The precise constraint is the following: After choosing the function $v_0(\bar\lambda)$, substitute for $\bar\lambda$ the solution (12.82) to the renormalization group equation,

$$\bar\lambda(\phi_{cl}) = \frac{\lambda}{1 - (\lambda/8\pi^2)(N+8)\log(\phi_{cl}/M)}. \tag{13.63}$$

Then expand the result in powers of λ and drop terms of order λ^3 and higher. If v_0 is chosen correctly, the result should agree with (13.55). Applying this criterion, we find the following result for the effective potential:

$$V_{eff}(\phi_{cl}) = \frac{1}{4}\phi_{cl}^4 \left[\bar\lambda + \frac{\bar\lambda^2}{(4\pi)^2} \left((N+8)(\log\bar\lambda - \tfrac{3}{2}) + 9\log 3 \right) \right], \tag{13.64}$$

where $\bar\lambda$ is given by (13.63).

The error in Eq. (13.64) comes in the determination of v_0 as a power series in $\bar{\lambda}$. Thus this error is of order $\bar{\lambda}^3$. As $\phi_{\mathrm{cl}} \to 0$, $\bar{\lambda} \to 0$, and so the representation (13.64) becomes more and more accurate. Thus this formula successfully sums the powers of the dangerous logarithm (13.56). Viewed as a function of ϕ_{cl}, (13.64) has its minimum at $\phi_{\mathrm{cl}} = 0$. Thus the apparent symmetry-breaking minimum of (13.55) is indeed an artifact of the incomplete perturbation expansion and disappears in a more complete treatment. This resolves the question that we raised in Section 11.4. We should note that, in more complicated examples, an apparent symmetry-breaking minimum of the effective potential found in the one-loop order of perturbation theory can survive a renormalization-group analysis. An example is given in the Final Project for Part II.

The procedure we have followed in this argument is called the *renormalization group improvement* of perturbation theory. The technique can be applied equally well to the computation of correlation functions and other predictions of Feynman diagram perturbation theory: One compares the solution of the Callan-Symanzik equation to the result of a straightforward perturbation theory computation to the same order in the coupling constant, choosing the undetermined function in the renormalization group solution in such a way as to reproduce the perturbation theory result. In this way, one finds a more compact formula in which large logarithms such as those in (13.56) are resummed into running coupling constants. This resummation produces the dependence of correlation functions on the logarithm of the mass scale that characterizes a field theory with a marginal or renormalizable perturbation.

In the case of ϕ^4 theory, the running coupling constant goes to zero at small momenta and becomes large at large momenta. Since the error term in improved perturbation theory is a power of $\bar{\lambda}$, the improved perturbation theory becomes accurate at small momenta but goes out of control at large momenta. This accords with our physical intuition: We would expect perturbation theory to be accurate only when the running coupling constant stays small.

In an asymptotically free theory, where the running coupling constant becomes small at large momenta, we can find accurate expressions for correlation functions at large momenta using renormalization-group-improved perturbation theory. In Chapters 17 and 18 we will use this idea as our major tool in analyzing the short-distance behavior of the strong interactions.

13.3 The Nonlinear Sigma Model

To complete our study of scalar field theory, we will discuss a nonlinear theory of scalar fields, whose structure is very different from that of ϕ^4 theory. This theory, called the *nonlinear sigma model*, was first proposed as an alternative description of spontaneous symmetry breaking. It will be interesting to us for three reasons. First, it provides a simple explicit example of an asymptotically free theory. Second, it will give us a second dimensional expansion with which we can study the Wilson-Fisher fixed point. Then we can see where the Wilson-Fisher fixed point goes in the space of Lagrangians for dimensions d well below 4. Finally, we will show that the nonlinear sigma model is exactly solvable in a limit that is different from the standard weak-coupling limit. This solution will give us further insight into the dependence of symmetry breaking on spacetime dimensionality.

The $d = 2$ Nonlinear Sigma Model

We begin our study in two dimensions. In $d = 2$, a scalar field is dimensionless; thus, any theory of scalar fields ϕ^i with a Lagrangian of the form

$$\mathcal{L} = f_{ij}(\{\phi^i\})\partial_\mu \phi^i \partial^\mu \phi^j \tag{13.65}$$

has dimensionless couplings and so is renormalizable. Since any function $f(\{\phi^i\})$ leads to a renormalizable theory, this class of scalar field theories contains an infinite number of marginal parameters. To restrict these possible parameters, we must impose some symmetries on the theory.

A simple choice is to take the scalar fields ϕ^i to form an N-component unit vector field $n^i(x)$, constrained to satisfy

$$\sum_{i=1}^{N} |n^i(x)|^2 = 1. \tag{13.66}$$

If we insist that the field theory has $O(N)$ symmetry, the function f in (13.65) can depend only on the invariant length of $\vec{n}(x)$, which is constrained by (13.66). Thus, the most general possible choice for f is a constant. Similarly, the only possible nonderivative interaction $g(\{n^i\})$ that one might add to (13.65) is a constant, and this would have no effect on the Green's functions of $\vec{n}$. With these restrictions, the most general Lagrangian one can build from $\vec{n}(x)$ with two derivatives and $O(N)$ symmetry is

$$\mathcal{L} = \frac{1}{2g^2} |\partial_\mu \vec{n}|^2. \tag{13.67}$$

This theory has a straightforward physical interpretation. It is a phenomenological description of a system with $O(N)$ symmetry spontaneously broken by the vacuum expectation value of a field that transforms as a vector of $O(N)$. Consider, for example, the situation in N-component ϕ^4 theory in its spontaneously broken phase. The field ϕ^i acquires a vacuum expectation value, which we can write in terms of a magnitude and a direction parameterized by a unit vector

$$\langle \phi^i \rangle = \phi_0 n^i(x). \tag{13.68}$$

The fluctuations of ϕ_0 correspond to a massive field, the field called σ in Chapter 11. The fluctuations of the direction of the unit vector $\vec{n}(x)$ correspond to the $N-1$ Goldstone bosons. Notice that $\vec{n}$ has N components subject to the one constraint (13.66), and so contains $N-1$ degrees of freedom. Formally, the nonlinear sigma model is the limit of ϕ^4 theory as the mass of the σ field is taken to infinity while ϕ_0 is held constant.

Despite this suggestive connection, we will first analyze the nonlinear sigma model on its own footing as an independent quantum field theory. It is convenient to solve the constraint and parametrize $\vec{n}$ by $N-1$ Goldstone boson fields π^k:

$$n^i = \left(\pi^1, \cdots, \pi^{N-1}, \sigma \right), \tag{13.69}$$

where, by definition,

$$\sigma = (1 - \pi^2)^{1/2}. \tag{13.70}$$

The configuration $\pi^k = 0$ corresponds to a uniform state of spontaneous symmetry breaking, oriented in the N direction. The representation (13.69) implies that

$$\left| \partial_\mu n^i \right|^2 = \left| \partial_\mu \vec{\pi} \right|^2 + \frac{(\vec{\pi} \cdot \partial_\mu \vec{\pi})^2}{1 - \pi^2}. \tag{13.71}$$

Then the Lagrangian (13.67) takes the form

$$\mathcal{L} = \frac{1}{2g^2} \left[\left| \partial_\mu \vec{\pi} \right|^2 + \frac{(\vec{\pi} \cdot \partial_\mu \vec{\pi})^2}{1 - \pi^2} \right]. \tag{13.72}$$

Notice that there is no mass term for the field $\vec{\pi}$, as required by Goldstone's theorem.

The perturbation theory for the π^k field can be read off straightforwardly by expanding the Lagrangian in powers of π^k:

$$\mathcal{L} = \frac{1}{2g^2} \left| \partial_\mu \vec{\pi} \right|^2 + \frac{1}{2g^2} (\vec{\pi} \cdot \partial_\mu \vec{\pi})^2 + \cdots. \tag{13.73}$$

This leads to the Feynman rules shown in Fig. 13.1, plus additional vertices with all even numbers of π^k fields. Since the Lagrangian (13.67) is the most general $O(N)$-symmetric Lagrangian with dimensionless coefficients that can be built out of these fields, the theory can be made finite by renormalization of the coupling constant g and $O(N)$-symmetric rescaling of the fields π^k and σ. In renormalized perturbation theory, there are divergences and counterterms

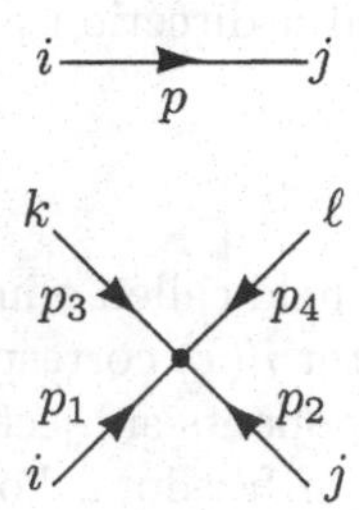

$$i \xrightarrow{} j \qquad = \frac{ig^2}{p^2}\delta^{ij}$$

$$
\begin{aligned}
&= -\frac{i}{g^2}\Big[(p_1 + p_2)\cdot(p_3 + p_4)\delta^{ij}\delta^{k\ell} \\
&\quad + (p_1 + p_3)\cdot(p_2 + p_4)\delta^{ik}\delta^{j\ell} \\
&\quad + (p_1 + p_4)\cdot(p_2 + p_3)\delta^{i\ell}\delta^{jk}\Big]
\end{aligned}
$$

Figure 13.1. Feynman rules for the nonlinear sigma model.

for each possible $2n$-π vertex; however, these counterterms are all related by the basic requirement that the bare Lagrangian preserve the $O(N)$ symmetry.

We now compute the Callan-Symanzik functions for this theory. Since the theory is renormalizable, its Green's functions obey the Callan-Symanzik equation for some functions β, γ. Explicitly,

$$\left[M\frac{\partial}{\partial M} + \beta(g)\frac{\partial}{\partial g} + n\gamma(g)\right]G^{(n)} = 0, \tag{13.74}$$

where $G^{(n)}$ is a Green's function of n fields π^k or σ. To identify the β and γ functions, to the leading order in perturbation theory, we compute two simple Green's functions to one-loop order and then see what forms are necessary if the Callan-Symanzik equation is to be satisfied.

The first Green's function we consider is

$$G^{(1)} = \langle\sigma(x)\rangle. \tag{13.75}$$

Expanding the definition (13.70), we find

$$\langle\sigma(0)\rangle = 1 - \tfrac{1}{2}\langle\pi^2(0)\rangle + \cdots \;\; = \;\; 1 \;\; - \;\; \tfrac{1}{2}\cdot \bigcirc \;. \tag{13.76}$$

To evaluate this formula, we use the propagator of Fig. 13.1 to compute

$$\langle\pi^k(0)\pi^\ell(0)\rangle = \bigcirc_{k\ell} = \int\frac{d^dk}{(2\pi)^d}\frac{ig^2}{k^2 - \mu^2}\delta^{k\ell}. \tag{13.77}$$

We have added a small mass μ as an infrared cutoff. Then

$$\langle\pi^k(0)\pi^\ell(0)\rangle = \frac{g^2}{(4\pi)^{d/2}}\frac{\Gamma(1-\frac{d}{2})}{(\mu^2)^{1-d/2}}\delta^{k\ell}. \tag{13.78}$$

Using this result in our expression for $\langle\sigma\rangle$ and then subtracting at the momentum scale M, we find

$$\langle\sigma\rangle = 1 - \frac{1}{2}(N-1)\frac{g^2}{(4\pi)^{d/2}}\Gamma(1-\tfrac{d}{2})\left(\frac{1}{(\mu^2)^{1-d/2}} - \frac{1}{(M^2)^{1-d/2}}\right) + \mathcal{O}(g^4)$$

$$\xrightarrow[d\to 2]{} 1 - \frac{g^2(N-1)}{8\pi}\log\frac{M^2}{\mu^2} + \mathcal{O}(g^4). \tag{13.79}$$

This expression satisfies the Callan-Symanzik equation to order g^2 only if

$$\gamma(g) = \frac{g^2(N-1)}{4\pi} + \mathcal{O}(g^4).$$ (13.80)

Next, consider the π^k two-point function,

$$\langle \pi^k(p)\pi^\ell(-p)\rangle = \underbrace{} + \underbrace{}_{\bigcirc} + \cdots$$

$$= \frac{ig^2}{p^2}\delta^{k\ell} + \frac{ig^2}{p^2}(-i\Pi^{k\ell})\frac{ig^2}{p^2} + \cdots.$$ (13.81)

In evaluating $\Pi^{k\ell}$ from the Feynman rules in Fig. 13.1, we again encounter the integral (13.77), and also the integral

$$\langle \partial_\mu \pi^k(0)\partial^\mu \pi^\ell(0)\rangle = \int \frac{d^d k}{(2\pi)^d} \frac{ig^2 k^2}{k^2 - \mu^2}\delta^{k\ell}$$

$$= -\frac{g^2}{(4\pi)^{d/2}}\frac{\frac{d}{2}\Gamma(-\frac{d}{2})}{(\mu^2)^{-d/2}}\delta^{k\ell}.$$ (13.82)

This formula has no pole at $d = 0$, and for $d > 0$ it is proportional to a positive power of μ^2; hence, we can set this contraction to zero. Then

$$\Pi^{k\ell}(p) = -\delta^{k\ell}p^2 \frac{1}{(4\pi)^{d/2}}\frac{\Gamma(1-\frac{d}{2})}{(\mu^2)^{1-d/2}}.$$ (13.83)

Subtracting at M as above and taking the limit $d \to 2$, we find

$$\langle \pi^k(p)\pi^\ell(-p)\rangle = \frac{ig^2}{p^2}\delta^{k\ell} + \frac{ig^2}{p^2}\left(+ip^2\frac{1}{4\pi}\log\frac{M^2}{\mu^2}\right)\frac{ig^2}{p^2}\delta^{k\ell} + \cdots.$$

$$= \frac{i}{p^2}\delta^{k\ell}\left(g^2 - \frac{g^4}{4\pi}\log\frac{M^2}{\mu^2} + \mathcal{O}(g^6)\right).$$ (13.84)

Applying the Callan-Symanzik equation to this result gives

$$\left[M\frac{\partial}{\partial M} + \beta(g)\frac{\partial}{\partial g} + 2\gamma(g)\right]\langle \pi^k(p)\pi^\ell(-p)\rangle = 0,$$

$$= \frac{i\delta^{k\ell}}{p^2}\left[-\frac{g^4}{2\pi} + \beta(g)\cdot 2g + 2g^2\gamma(g)\right].$$ (13.85)

Inserting the result (13.80) for $\gamma(g)$, we find

$$\beta(g) = -(N-2)\frac{g^3}{4\pi} + \mathcal{O}(g^5).$$ (13.86)

At $N = 2$ precisely, the beta function vanishes. This is not an accident but rather is a nontrivial check of our calculation. For $N = 2$, we can make the change of variables $\pi^1 = \sin\theta$; then $\sigma = \cos\theta$, and the Lagrangian takes the form

$$\mathcal{L} = \frac{1}{2g^2}(\partial_\mu \theta)^2.$$ (13.87)

This is a free field theory for the field $\theta(x)$, so it can have no renormalization group flow.

For $N > 2$, the β function is *negative*: This theory is asymptotically free. The running coupling constant $\bar{g}$ becomes small at small distances and grows large at large distances.

In quantum electrodynamics, we found an appealing physical picture for the sign of the coupling constant evolution. As we discussed in Section 7.5, the process of virtual pair creation makes the vacuum a dielectric medium, which screens electric charge. One would therefore expect the effective Coulomb interaction of charge to decrease at large distances and increase at small distances. It is easy to imagine that a similar screening phenomenon might occur in any quantum field theory. Thus, it is surprising that, in this theory, we have found by explicit calculation that the coupling constant evolution has the opposite sign. What is the physical explanation for this?

In fact, the original derivation of the asymptotic freedom of the nonlinear sigma model, due to Polyakov,[†] gave a clear physical argument for the sign of the evolution. Now that we have derived the β function by the automatic method of the Callan-Symanzik equation, let us review Polyakov's more physical derivation.

Polyakov analyzed the nonlinear sigma model using Wilson's momentum-slicing technique, which we discussed in Section 12.1. Consider, then, the nonlinear sigma model defined with a momentum cutoff in place of the dimensional regulator. As in Section 12.1, we work in Euclidean space with initial cutoff Λ.

The original integration variables are the Fourier components of the unit vector field $n^i(x)$. We wish to integrate out of the functional integral those Fourier components corresponding to momenta k in the range $b\Lambda \leq |k| < \Lambda$. If the remaining components are Fourier-transformed back to coordinate space, they describe a coarse-grained average of the original unit vector field. This averaged field can be rescaled so that it is again a unit vector at each point. Call this averaged and rescaled field $\tilde{n}^i$. Then we can write the relation of n^i and $\tilde{n}^i$ as follows:

$$n^i(x) = \tilde{n}^i(x)\left(1 - \phi^2\right)^{1/2} + \sum_{a=1}^{N-1} \phi_a(x)e_a^i(x). \tag{13.88}$$

In this equation, the vectors $\vec{e}_a(x)$ form a basis of unit vectors orthogonal to $\tilde{n}(x)$. In Polyakov's picture, $\tilde{n}(x)$ and the $\vec{e}_a(x)$ are slowly varying. On the other hand, the coefficients $\phi_a(x)$ contain only Fourier components in the range $b\Lambda \leq |k| < \Lambda$. These are the variables we integrate over to achieve the renormalization group transformation.

[†]A. M. Polyakov, *Phys. Lett.* **59B**, 79 (1975).

To set up the integral over ϕ_a, we first work out

$$\partial_\mu n^i = \partial_\mu \tilde{n}^i (1 - \phi^2)^{1/2} - \tilde{n}^i \left(\frac{\phi \cdot \partial_\mu \phi}{(1 - \phi^2)^{1/2}} \right) + \partial_\mu \phi_a e_a^i + \phi_a \partial_\mu e_a^i. \quad (13.89)$$

By the definition of $\tilde{n}$, $\vec{e}_a$, these vectors satisfy

$$|\tilde{n}|^2 = 1; \qquad \tilde{n} \cdot \vec{e}_a = 0. \quad (13.90)$$

Taking the derivative of these identities, we find

$$\tilde{n} \cdot \partial_\mu \tilde{n} = 0; \qquad \tilde{n} \cdot \partial_\mu \vec{e}_a + \partial_\mu \tilde{n} \cdot \vec{e}_a = 0. \quad (13.91)$$

Using the identities in (13.90) and (13.91), we can compute the Lagrangian of the nonlinear sigma model through terms quadratic in the ϕ_a:

$$\mathcal{L} = \frac{1}{2g^2} |\partial_\mu n^i|^2 = \frac{1}{2g^2} \Big[|\partial_\mu \tilde{n}^i|^2 (1 - \phi^2) + (\partial_\mu \phi_a)^2 + 2(\phi_a \partial^\mu \phi_b)(\vec{e}_u \cdot \partial_\mu \vec{e}_b)$$

$$+ \partial^\mu \phi_a \partial_\mu \tilde{n} \cdot \vec{e}_a + \phi_a \phi_b \partial^\mu \vec{e}_a \cdot \partial_\mu \vec{e}_b + \cdots \Big]. \quad (13.92)$$

We will consider the second term of (13.92) to be the zeroth-order Lagrangian for ϕ_a. Thus,

$$\mathcal{L}_0 = \frac{1}{2g^2} (\partial_\mu \phi_a)^2, \quad (13.93)$$

which gives the propagator

$$\langle \phi_a(p) \phi_b(-p) \rangle = \frac{g^2}{p^2} \delta_{ab}, \quad (13.94)$$

restricted to the momentum region $b\Lambda \leq |p| < \Lambda$. This propagator can be used to integrate the remaining terms of the Lagrangian over the ϕ_a. Borrowing the integrals from the derivation of (13.84), we can set

$$\langle \phi_a(0) \partial_\mu \phi_b(0) \rangle = \langle \partial^\mu \phi_a(0) \partial_\mu \phi_b(0) \rangle = 0 \quad (13.95)$$

and

$$\langle \phi_a(0) \phi_b(0) \rangle = \delta_{ab} \frac{g^2}{4\pi} \log \frac{\Lambda^2}{(b\Lambda)^2}. \quad (13.96)$$

Then, after the integral over ϕ, the new Lagrangian is given approximately by

$$\mathcal{L}_{\text{eff}} = \frac{1}{2g^2} \Big[|\partial_\mu \tilde{n}|^2 (1 - \langle \phi^2 \rangle) + \langle \phi_a \phi_b \rangle \partial_\mu \vec{e}_a \cdot \partial^\mu \vec{e}_b + \mathcal{O}(g^4) \Big], \quad (13.97)$$

where the expectation values of ϕ_a are given by (13.96).

To simplify this further, we must simplify the structure $(\partial_\mu \vec{e}_a)^2$ that appears in the second term of (13.97). Introduce a complete basis of vectors:

$$(\partial_\mu \vec{e}_a)^2 = (\tilde{n} \cdot \partial_\mu \vec{e}_a)^2 + (\vec{e}_c \cdot \partial_\mu \vec{e}_a)^2. \quad (13.98)$$

The second term on the right is a new structure, associated with the torsion of the coordinate system for e_a; it turns out to correspond to an irrelevant operator induced by the renormalization procedure. The first term, however, can be put into a familiar form by using the two identities (13.91):

$$(\tilde{n} \cdot \partial_\mu \vec{e}_a)^2 = (\vec{e}_a \cdot \partial_\mu \tilde{n})^2 = (\partial_\mu \tilde{n})^2. \tag{13.99}$$

Then

$$\begin{aligned}\mathcal{L}_{\text{eff}} &= \frac{1}{2g^2}\left[|\partial_\mu \tilde{n}|^2\left(1 - (N-1)\frac{g^2}{4\pi}\log\frac{1}{b^2} + \frac{g^2}{4\pi}\log\frac{1}{b^2}\right) + \cdots\right] \\ &= \frac{1}{2}\left(g^2 + \frac{g^4}{2\pi}(N-2)\log\frac{1}{b} + \cdots\right)^{-1}|\partial_\mu \tilde{n}|^2.\end{aligned} \tag{13.100}$$

The quantity in parentheses is the square of a running coupling constant. To the order of our calculation, this quantity satisfies

$$\frac{d}{d\log b}\bar{g} = -(N-2)\frac{\bar{g}^3}{4\pi}, \tag{13.101}$$

in agreement with (13.86).

In this calculation, the sign of the coupling constant renormalization comes from the fact that the effective length of the unit vector $\vec{n}$ is reduced by averaging over short-wavelength fluctuations. This lowers the effective action associated with a configuration in which the direction of $\vec{n}$ changes over a displacement Δx (see Fig. 13.2). Looking back at (13.67), we see that a decrease of the magnitude of $\mathcal{L}$ for the same configuration of $\vec{n}$ can be interpreted as an *increase* of the effective coupling. Thus the nonlinear sigma model is more strongly coupled, or, in terms of the physical configuration of the $\vec{n}$ field, more disordered, at large distances.

Our calculation implies that, if any two-dimensional statistical system apparently has spontaneously broken symmetry and Goldstone bosons, then, at large distances, the ordering disappears. This is an unexpected conclusion. However, this conclusion is in accord with a theorem proved by Mermin and Wagner[‡] that a two-dimensional system with a continuous symmetry cannot support an ordered state in which a symmetry-breaking field has a nonzero vacuum expectation value. This theorem applies to the case $N = 2$ as well as to $N > 2$. We have motivated this theorem in Problem 11.1.

The Nonlinear Sigma Model for $2 < d < 4$

We now extend the results of this analysis to dimensions $d > 2$. In general d, we will continue to define the action of the nonlinear sigma model by

$$\int d^d x\, \mathcal{L} = \int d^d x\, \frac{1}{2g^2}(\partial_\mu \vec{n})^2, \tag{13.102}$$

where $\vec{n}$ is still dimensionless, since it obeys the constraint $|\vec{n}|^2 = 1$. Thus g has the dimensions $(\text{mass})^{(2-d)/2}$. We define a dimensionless coupling by

---

[‡]N. D. Mermin and H. Wagner, *Phys. Rev. Lett.* **17**, 1133 (1966).

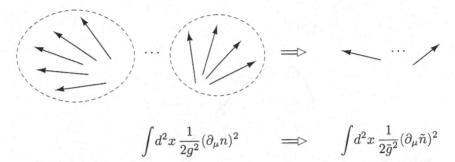

$$\int d^2x \, \frac{1}{2g^2}(\partial_\mu n)^2 \qquad \Longrightarrow \qquad \int d^2x \, \frac{1}{2\bar{g}^2}(\partial_\mu \tilde{n})^2$$

Figure 13.2. Averaging of the direction of $\vec{n}$, and its interpretation as an increase of the running coupling constant.

writing

$$T = g^2 M^{d-2}, \tag{13.103}$$

just as we did in (12.122). If (13.102) is viewed as the Boltzmann weight of a partition function, then T is a dimensionless variable proportional to the temperature.

From (13.103), we can find the β function for T in d dimensions, in analogy to Eq. (12.131):

$$\beta(T) = (d-2)T + 2g\beta^{(2)}(g), \tag{13.104}$$

where the factor of $2g$ in the second term comes from the definition $T \sim g^2$. Since $\vec{n}$ is dimensionless, the γ function is unchanged from the two-dimensional result when expressed in terms of dimensionless couplings. Thus, in $d = 2 + \epsilon$,

$$\beta(T) = +\epsilon T - (N-2)\frac{T^2}{2\pi};$$
$$\gamma(T) = (N-1)\frac{T}{4\pi}. \tag{13.105}$$

Notice that the β function for T has a nontrivial zero, which approaches $T = 0$ as $\epsilon \to 0$. This zero is located at

$$T_* = \frac{2\pi\epsilon}{N-2}. \tag{13.106}$$

The form of the β function is sketched in Fig. 13.3. In contrast to the Wilson-Fisher zero in $d = 4 - \epsilon$, discussed in Section 12.5, this is an ultraviolet-stable fixed point. The flows to the infrared go out from this fixed point. Since T is proportional to the temperature of the corresponding statistical system, $t \to 0$ is the state of complete order, while $t \gg 1$ is the state of complete disorder. This agrees with the intuition that accompanied Polyakov's derivation of the β function. The fixed point T_* corresponds to the critical temperature. Thus, the critical temperature tends to zero as $d \to 2$, in accord with the Mermin-Wagner theorem.

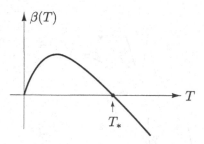

Figure 13.3. The form of the β function in the nonlinear sigma model for $d > 2$.

We can now compute the critical exponents of the nonlinear sigma model in an expansion in $\epsilon = d - 2$. The exponent η is given straightforwardly by

$$\eta = 2\gamma(T_*) = \frac{\epsilon}{N-2}. \tag{13.107}$$

To find the second exponent ν, we need to identify the relevant perturbation that corresponds to the renormalization group flow away from the fixed point for $T \neq T_C$. This is just the deviation of T from T_*:

$$\rho_T = T - T_*. \tag{13.108}$$

From the renormalization group equation for the running coupling constant, we find that the running ρ_T obeys

$$\frac{d}{d\log p}\bar{\rho}_T = \left[\frac{d}{dT}\beta(T)\Big|_{T=T_*}\right] \cdot \bar{\rho}_T. \tag{13.109}$$

The quantity in brackets is negative. As in Eqs. (12.134) and (12.137), we can identify this quantity with $(-1/\nu)$: At a momentum $p \ll M$,

$$\bar{\rho}_T(p) = \rho_T\left(\frac{p}{M}\right)^{-1/\nu}; \tag{13.110}$$

thus $\bar{\rho}(p)$ becomes of order 1 at a momentum that is the inverse of $\xi \sim (T-T_*)^{-\nu}$, as required. Using the explicit form of the β function from (13.105), we find

$$\nu = \frac{1}{\epsilon}, \tag{13.111}$$

independent of N to this order in ϵ. (Of course, these results apply only for $N \geq 3$.) The thermodynamic critical exponents can be found from (13.107) and (13.111) using the model-independent relations derived in Section 13.1. When the values found here for ν and η are extrapolated to $d = 3$ (that is, $\epsilon = 1$), the agreement with experiment is not spectacular, but the results at least suggest that the fixed point we have found here may be the continuation of the Wilson-Fisher fixed point to the vicinity of two dimensions.

Exact Solution for Large N

It is possible to obtain further insight into the nature of this fixed point by attacking the nonlinear sigma model using another approach. Since the nonlinear sigma model depends on a parameter N, the number of components of the unit vector, it is reasonable to ask how this model behaves as $N \to \infty$. We now show that if we take this limit holding $g^2 N$ fixed, we can obtain an exact solution to the model with nontrivial behavior.

The manipulations that lead to this solution are most clear if we work in Euclidean space, regarding the Lagrangian as the Boltzmann weight of a spin system. Then we must compute the functional integral

$$Z = \int \mathcal{D}n \, \exp\left[-\int d^d x \, \frac{1}{2g_0^2}(\partial_\mu n)^2\right] \cdot \prod_x \delta(n^2(x) - 1). \qquad (13.112)$$

Here g_0 is the bare value of the coupling constant, while the product of delta functions, one at each point, enforces the constraint. Introduce an integral representation of the delta functions; this requires a second functional integral over a Lagrange multiplier variable $\alpha(x)$:

$$Z = \int \mathcal{D}\alpha \mathcal{D}n \, \exp\left[-\int d^d x \, \frac{1}{2g_0^2}(\partial_\mu n)^2 - \frac{i}{2g_0^2}\int d^d x \, \alpha(n^2 - 1)\right]. \qquad (13.113)$$

Now the variable n is unconstrained and appears in the exponent only quadratically. Thus, we can integrate over n, to obtain

$$\begin{aligned} Z &= \int \mathcal{D}\alpha \, (\det[-\partial^2 + i\alpha(x)])^{-N/2} \exp\left[\frac{i}{2g_0^2}\int d^d x \, \alpha\right]. \\ &= \int \mathcal{D}\alpha \, \exp\left[-\frac{N}{2}\operatorname{tr}\log(-\partial^2 + i\alpha) + \frac{i}{2g_0^2}\int d^d x \, \alpha\right]. \end{aligned} \qquad (13.114)$$

Since we are taking the limit $N \to \infty$ with $g_0^2 N$ held fixed, both terms in the exponent are of order N. Thus it makes sense to evaluate the integral by steepest descents. This entails dominating the integral by the value of the function $\alpha(x)$ that minimizes the exponent. To determine this configuration, we compute the functional derivative of the exponent with respect to $\alpha(x)$. This gives the variational equation

$$\frac{N}{2}\langle x| \frac{1}{-\partial^2 + i\alpha} |x\rangle = \frac{1}{2g_0^2}. \qquad (13.115)$$

The left-hand side of this equation must be constant and real; thus, we should look for a solution in which $\alpha(x)$ is constant and pure imaginary. Write

$$\alpha(x) = -im^2; \qquad (13.116)$$

then m^2 obeys

$$N \int \frac{d^d k}{(2\pi)^d} \frac{1}{k^2 + m^2} = \frac{1}{g_0^2}. \qquad (13.117)$$

We will study this equation first in $d = 2$. If we define the integral in (13.117) with a momentum cutoff, we can evaluate this integral and find the equation for m:

$$\frac{N}{2\pi} \log \frac{\Lambda}{m} = \frac{1}{g_0^2}. \tag{13.118}$$

We can make this equation finite by the renormalization

$$\frac{1}{g_0^2} = \frac{1}{g^2} + \frac{N}{2\pi} \log \frac{\Lambda}{M}, \tag{13.119}$$

which introduces an arbitrary renormalization scale M. Then we can solve for m, to find

$$m = M \exp\left[-\frac{2\pi}{g^2 N}\right]. \tag{13.120}$$

This is a nonzero, $O(N)$-invariant mass term for the N unconstrained components of $\vec{n}$. In this solution, $\langle \vec{n} \rangle = 0$ and the symmetry is unbroken, for any value of g^2 or T.

The solution of the theory does depend on the arbitrary renormalization scale M; this dependence simply reflects the arbitrariness of the definition of the renormalized coupling constant. The statement that m follows unambiguously from an underlying theory with fixed bare coupling and cutoff is precisely the statement that m obeys the Callan-Symanzik equation with no overall rescaling:

$$\left[M\frac{\partial}{\partial M} + \beta(g^2)\frac{\partial}{\partial g}\right]m(g^2, M) = 0. \tag{13.121}$$

Using the large-N limit of (13.86),

$$\beta(g) = -\frac{g^3 N}{4\pi}, \tag{13.122}$$

it is easy to check that (13.121) is satisfied. Conversely, the validity of (13.121) with (13.122) tells us that Eq. (13.122) is an *exact* representation of the β function to all orders in $g^2 N$ in the limit of large N. The corrections to (13.122) are of order $(1/N)$ or, equivalently, of order g^2 with no compensatory factor of N. Equation (13.122) agrees with our earlier calculation (13.86) to this order.

Now let us redo this exercise in $d > 2$. In this case, the integral in (13.117) diverges as a power of the cutoff. Even when the dependence on Λ is removed by renormalization, this change in behavior leads to a change in the dependence of the integral on m, which has important physical implications.

It is not difficult to work out the integral in (13.117) as an expansion in (Λ/m). One finds:

$$\int \frac{d^d k}{(2\pi)^d} \frac{1}{k^2 + m^2} = \begin{cases} C_1 \Lambda^{d-2} - C_2 m^{d-2} + \cdots & \text{for } d < 4, \\ C_1 \Lambda^{d-2} - \tilde{C}_2 m^2 \Lambda^{d-4} + \cdots & \text{for } d > 4, \end{cases} \tag{13.123}$$

where C_1, C_2, $\tilde{C}_2$ are some functions of d. In particular,

$$C_1 = \left[2^{d-1}\pi^{(d+1)/2}\Gamma\left(\frac{d-1}{2}\right)(d-2)\right]^{-1}. \tag{13.124}$$

In $d > 4$, the first derivative of the integral with respect to m^2 is smooth as $m^2 \to 0$; this is the reason for the change in behavior.

In the case $d = 2$, the left-hand side of (13.117) covered the whole range from 0 to ∞ as m was varied; thus, we could always find a solution for any value of g_0^2. In $d > 2$, this is no longer true. Equation (13.117) can be solved for m only if Ng_0^2 is greater than the critical value

$$Ng_C^2 = \left(C_1\Lambda^{d-2}\right)^{-1}. \tag{13.125}$$

Just at the boundary, $m = 0$. For bare couplings weaker than (13.125), it is possible to lower the value of the effective action by giving one component of $\vec{n}$ a vacuum expectation value while keeping the other components massless. Thus (13.125) is the criterion for the second-order phase transition in this model. Equation (13.124) implies that the critical value of g_0^2, which is proportional to the critical temperature, goes to zero as $d \to 2$, in accord with our renormalization-group analysis.

In the symmetric phase of the nonlinear sigma model, the mass m determines the exponential fall-off of correlations, so $\xi = m^{-1}$. Thus we can determine the exponent ν by solving for the dependence of m on the deviation of g_0^2 from the critical temperature. Write

$$t = \frac{g_0^2 - g_C^2}{g_C^2}. \tag{13.126}$$

Then, in $2 < d < 4$, we can use (13.123) to solve (13.117) for m for small values of t. This gives

$$\frac{1}{Ng_C^4} \cdot t = C_2 m^{d-2}, \tag{13.127}$$

which implies $m \sim t^\nu$ with

$$\nu = \frac{1}{d-2}, \qquad 2 < d < 4. \tag{13.128}$$

Similarly,

$$\nu = \frac{1}{2}, \qquad d \geq 4. \tag{13.129}$$

The discontinuity in the dependence of ν on d is exactly what we predicted from renormalization group analysis. For $d > 4$, the value of ν goes over to the prediction of naive dimensional analysis. The value of ν given by (13.128) is in precise agreement with (13.111), in the expansion $\epsilon = d - 2$, and with the $N \to \infty$ limit of (12.142), in the expansion $\epsilon = 4 - d$. Apparently, all of our results for critical exponents mesh in a very satisfying way.

By combining all of our results, we arrive at a pleasing picture of the behavior of scalar field theory as a function of spacetime dimensionality. Above four dimensions, any scalar field interaction is irrelevant and the expected behavior is trivial. Just at four dimensions, the coupling constant tends to zero only logarithmically at large scale, giving rise to a renormalizable theory with predictions such as those in Section 13.2. Below four dimensions, the theory is intrinsically a theory of interacting scalar fields, dominated by the Wilson-Fisher fixed point. The coupling at this fixed point is small near four dimensions but grows large as the dimensionality decreases. Finally (for $N > 2$), as $d \to 2$, the fixed-point theory approaches the weak-coupling limit of a completely different Lagrangian with the same symmetries, the nonlinear sigma model.

This evolution of the behavior of the model as a function of d illustrates the main point of the previous two chapters: The qualitative behavior of a quantum field theory is determined not by the fundamental Lagrangian, but rather by the nature of the renormalization group flow and its fixed points. These, in turn, depend only on the basic symmetries that are imposed on the family of Lagrangians that flow into one another. This conclusion signals, at the deepest level, the importance of symmetry principles in determining the fundamental laws of physics.

Problems

13.1 Correction-to-scaling exponent. For critical phenomena in $4 - \epsilon$ dimensions, the irrelevant contributions that disappear most slowly are those associated with the deviation of the coupling constant λ from its fixed-point value. This gives the most important nonuniversal correction to the scaling laws derived in Section 13.1. By studying the solution of the Callan-Symanzik equation, show that if the bare value of λ differs slightly from λ_*, the Gibbs free energy receives a correction

$$G(M,t) \to G(M,t) \cdot \left(1 + (\lambda - \lambda_*)t^{\omega\nu}\hat{k}(tM^{-1/\beta})\right).$$

This formula defines a new critical exponent ω, called the *correction-to-scaling exponent*. Show that

$$\omega = \frac{d}{d\lambda}\beta\Big|_{\lambda_*} = \epsilon + \mathcal{O}(\epsilon^2).$$

13.2 The exponent η. By combining the result of Problem 10.3 with an appropriate renormalization prescription, show that the leading term in $\gamma(\lambda)$ in ϕ^4 theory is

$$\gamma = \frac{\lambda^2}{12(4\pi)^4}.$$

Generalize this result to the $O(N)$-symmetric ϕ^4 theory to derive Eq. (13.47). Compute the leading-order (ϵ^2) contribution to η.

13.3 The CP^N model. The nonlinear sigma model discussed in the text can be thought of as a quantum theory of fields that are coordinates on the unit sphere. A slightly more complicated space of high symmetry is complex projective space,

CP^N. This space can be defined as the space of $(N+1)$-dimensional complex vectors $(z_1, \ldots, z_{N+1})$ subject to the condition

$$\sum_j |z_j|^2 = 1,$$

with points related by an overall phase rotation identified, that is,

$$(e^{i\alpha} z_1, \ldots, e^{i\alpha} z_{N+1}) \quad \text{identified with} \quad (z_1, \ldots, z_{N+1}).$$

In this problem, we study the two-dimensional quantum field theory whose fields are coordinates on this space.

(a) One way to represent a theory of coordinates on CP^N is to write a Lagrangian depending on fields $z_j(x)$, subject to the constraint, which also has the local symmetry

$$z_j(x) \to e^{i\alpha(x)} z_j(x),$$

independently at each point x. Show that the following Lagrangian has this symmetry:

$$\mathcal{L} = \frac{1}{g^2} [|\partial_\mu z_j|^2 - |z_j^* \partial_\mu z_j|^2].$$

To prove the invariance, you will need to use the constraint on the z_j, and its consequence

$$z_j^* \partial_\mu z_j = -(\partial_\mu z_j^*) z_j.$$

Show that the nonlinear sigma model for the case $N = 3$ can be converted to the CP^N model for the case $N = 1$ by the substitution

$$n^i = z^* \sigma^i z,$$

where σ^i are the Pauli sigma matrices.

(b) To write the Lagrangian in a simpler form, introduce a scalar Lagrange multiplier λ which implements the constraint and also a vector Lagrange multiplier A_μ to express the local symmetry. More specifically, show that the Lagrangian of the CP^N model is obtained from the Lagrangian

$$\mathcal{L} = \frac{1}{g^2} [|D_\mu z_j|^2 - \lambda(|z_j|^2 - 1)],$$

where $D_\mu = (\partial_\mu + iA_\mu)$, by functionally integrating over the fields λ and A_μ.

(c) We can solve the CP^N model in the limit $N \to \infty$ by integrating over the fields z_j. Show that this integral leads to the expression

$$Z = \int \mathcal{D}A \mathcal{D}\lambda \exp\left[-N \operatorname{tr} \log(-D^2 - \lambda) + \frac{i}{g^2} \int d^2x\, \lambda \right],$$

where we have kept only the leading terms for $N \to \infty$, $g^2 N$ fixed. Using methods similar to those we used for the nonlinear sigma model, examine the conditions for minimizing the exponent with respect to λ and A_μ. Show that these conditions have a solution at $A_\mu = 0$ and $\lambda = m^2 > 0$. Show that, if g^2 is renormalized at the scale M, m can be written as

$$m = M \exp\left[-\frac{2\pi}{g^2 N} \right].$$

(d) Now expand the exponent about $A_\mu = 0$. Show that the first nontrivial term in this expansion is proportional to the vacuum polarization of massive scalar fields. Evaluate this expression using dimensional regularization, and show that it yields a standard kinetic energy term for A_μ. Thus the strange nonlinear field theory that we started with is finally transformed into a theory of $(N + 1)$ massive scalar fields interacting with a massless photon.

Final Project

The Coleman-Weinberg Potential

In Chapter 11 and Section 13.2 we discussed the effective potential for an $O(N)$-symmetric ϕ^4 theory in four dimensions. We computed the perturbative corrections to this effective potential, and used the renormalization group to clarify the behavior of the potential for small values of the scalar field mass. After all this work, however, we found that the qualitative dependence of the theory on the mass parameter was unchanged by perturbative corrections. The theory still possessed a second-order phase transition as a function of the mass. The loop corrections affected this picture only in providing some logarithmic corrections to the scaling behavior near the phase transition.

However, loop corrections are not always so innocuous. For some systems, they can change the structure of the phase transition qualitatively. This Final Project treats the simplest example of such a system, the *Coleman-Weinberg model*. The analysis of this model draws on a broad variety of topics discussed in Part II; it provides a quite nontrivial application of the effective potential formalism and the use of the renormalization group equation. The phenomenon displayed in this exercise reappears in many contexts, from displacive phase transitions in solids to the thermodynamics of the early universe.

This problem makes use of material in starred sections of the book, in particular, Sections 11.3, 11.4, and 13.2. Parts (a) and (e), however, depend only on the unstarred material of Part II. We recommend part (c) as excellent practice in the computation of renormalization group functions.

The Coleman-Weinberg model is the quantum electrodynamics of a scalar field in four dimensions, considered for small values of the scalar field mass. The Lagrangian is

$$\mathcal{L} = -\tfrac{1}{4}(F_{\mu\nu})^2 + (D_\mu\phi)^\dagger D^\mu\phi - m^2\phi^\dagger\phi - \tfrac{\lambda}{6}(\phi^\dagger\phi)^2,$$

where $\phi(x)$ is a complex-valued scalar field and $D_\mu\phi = (\partial_\mu + ieA_\mu)\phi$.

(a) Assume that $m^2 = -\mu^2 < 0$, so that the symmetry $\phi(x) \to e^{-i\alpha}\phi(x)$ is spontaneously broken. Write out the expression for $\mathcal{L}$, expanded around the broken-symmetry state by introducing

$$\phi(x) = \phi_0 + \frac{1}{\sqrt{2}}[\sigma(x) + i\,\pi(x)],$$

where ϕ_0, $\sigma(x)$, and π are real-valued. Show that the A_μ field acquires a mass. This mechanism of mass generation for vector fields is called the *Higgs mechanism*. We will study it in great detail in Chapter 20.

(b) Working in Landau gauge ($\partial^\mu A_\mu = 0$), compute the one-loop correction to the effective potential $V(\phi_{\rm cl})$. Show that it is renormalized by counterterms for m^2 and λ. Renormalize by minimal subtraction, introducing a renormalization scale M.

(c) In the result of part (b), take the limit $\mu^2 \to 0$. The result should be an effective potential that is scale-invariant up to logarithms containing M. Analyze this expression for λ very small, of order $(e^2)^2$. Show that, with this choice of coupling constants, $V(\phi_{\rm cl})$ has a symmetry-breaking minimum at a value of $\phi_{\rm cl}$ for which no logarithm is large, so that a straightforward perturbation theory analysis should be valid. Thus the $\mu^2 = 0$ theory, for this choice of coupling constants, still has spontaneously broken symmetry, due to the influence of quantum corrections.

(d) Sketch the behavior of $V(\phi_{\rm cl})$ as a function of m^2, on both sides of $m^2 = 0$, for the choice of coupling constants made in part (c).

(e) Compute the Callan-Symanzik β functions for e and λ. You should find

$$\beta_e = \frac{e^3}{48\pi^2}, \qquad \beta_\lambda = \frac{1}{24\pi^2}\left(5\lambda^2 - 18e^2\lambda + 54e^4\right).$$

Sketch the renormalization group flows in the (λ, e^2) plane. Show that every renormalization group trajectory passes through the region of coupling constants considered in part (c).

(f) Construct the renormalization-group-improved effective potential at $\mu^2 = 0$ by applying the results of part (e) to the calculation of part (c). Compute $\langle\phi\rangle$ and the mass of the σ particle as a function of λ, e^2, M. Compute the ratio m_σ/m_A to leading order in e^2, for $\lambda \ll e^2$.

(g) Include the effects of a nonzero m^2 in the analysis of part (f). Show that m_σ/m_A takes a minimum nonzero value as m^2 increases from zero, before the broken-symmetry state disappears entirely. Compute this value as a function of e^2, for $\lambda \ll e^2$.

(h) The Lagrangian of this problem (in its Euclidean form) is equivalent to the Landau free energy for a superconductor in d dimensions, coupled to an electromagnetic field. This expression is known as the Landau-Ginzburg free energy. Compute the β functions for this system and sketch the renormalization group flows for $d = 4 - \epsilon$. Describe the qualitative behavior you would expect for the superconducting phase transition in three dimensions. (For realistic superconductors, the value of e^2—after it is made dimensionless in the appropriate way—is very small. The effect you will find is expected to be important only for $|T - T_C|/T_C < 10^{-5}$.)

Part III

Non-Abelian Gauge Theories

Chapter 14

Invitation: The Parton Model
of Hadron Structure

In Part II of this book, we explored the structure of quantum field theories in a formal way. We developed sophisticated calculational algorithms (Chapter 10), derived a formalism for the extraction of scaling laws and asymptotic behavior (Chapter 12), and worked out some of the consequences of spontaneously broken symmetry (Chapter 11). Much of this formalism turned out to have unexpected applications in statistical mechanics. However, we have not yet investigated its implications for elementary particle physics. To do so, we must first ask which particular quantum field theories describe the interactions of elementary particles.

Since the mid-1970s, most high-energy physicists have agreed that the elementary particles that make up matter are a set of fermions, interacting primarily through the exchange of vector bosons. The elementary fermions include the *leptons* (the electron, its heavy counterparts μ and τ, and a neutral, almost massless neutrino corresponding to each of these species), and the *quarks*, whose bound states form the particles with nuclear interactions, mesons and baryons (collectively called hadrons). These fermions interact through three forces: the strong, weak, and electromagnetic interactions. Of these, the *strong interaction* is responsible for nuclear binding and the interactions of the constituents of nuclei, while the *weak interaction* is responsible for radioactive beta decay processes. The *electromagnetic interaction* is the familiar Quantum Electrodynamics, coupled minimally to all charged quarks and leptons. It is not clear that these three forces suffice to explain the most subtle properties of the elementary fermions—we will discuss this question in Chapter 20—but these three forces are certainly the most prominent. All three are now understood to be mediated by the exchange of vector bosons. The equations describing the electromagnetic interaction were discovered by Maxwell, and their quantum mechanical implications have been treated in detail in Part I. The correct theories of the weak and strong interactions were discovered much later.

By the late 1950s, studies of the helicity dependence of weak interaction cross sections and decay rates had shown that the weak interaction involves a coupling of vector currents built of quark and lepton fields.* It was thus natural to assume that the weak interaction is due to the exchange of very heavy vector bosons, and indeed, such bosons, the W and Z particles, were discovered in experiments at CERN in 1982. But a complete theory of the weak interaction must include not only the correct couplings of the bosons to fermions, but also the equations of motion of the boson fields themselves, the analogue, for the W and Z, of Maxwell's equations. Finding the correct form of these equations was not straightforward, because Maxwell's equations prohibit the generation of a mass for the vector particle. The proper reconciliation of the generalized Maxwell equations with the nonzero W and Z masses turned out to require incorporating into the theory a spontaneously broken symmetry. Chapters 20 and 21 treat this subject in some detail, describing the interplay of vector field theories with spontaneously broken symmetry. This interplay leads to new twists and new phenomena, beyond those discussed in our treatment of spontaneous symmetry breaking in Chapter 11. A complete theory of the weak interaction also requires the simultaneous incorporation of the electromagnetic interaction, forming a unified structure as first hypothesized by Glashow, Weinberg, and Salam.

On the other hand, it was for a long time completely obscure that a theory of exchanged vector bosons could correctly describe the strong interaction. Part of the mystery was that quarks do not exist as isolated species. Their existence, and eventually their quantum numbers, had to be deduced from the spectrum of observable strongly interacting particles. But, in addition, there were complications due to the fact that the strong interactions are strong. The Feynman diagram expansion assumes that the coupling constant is small; when the coupling becomes strong, a large number of diagrams are important (if the series converges at all) and it becomes impossible to pick out the contributions of the elementary interaction vertices. The crucial clue that the strong interactions have a vector character arose from what at first seemed to be just another mystery, the observation that the strong interactions turn themselves off when the momentum transfer is large, in a sense that we will now describe.

Almost Free Partons

In Section 5.1 we computed the cross section for the QED process $e^+e^- \rightarrow \mu^+\mu^-$. We then remarked that the corresponding cross section for e^+e^- annihilation into hadrons could be computed in the same way, using a simplistic model in which the quarks are treated as noninteracting fermions. This method gives a surprisingly accurate formula for the cross section, capturing its most important qualitative features. But we deferred the explanation of this puzzle: How can a model of noninteracting quarks represent the behavior of a force that, under other circumstances, is extremely *strong*?

---

*For an overview of weak interaction phenomenology, see Perkins (1987), Chapter 7, or any other modern particle physics text.

In fact, there are many circumstances in the study of the strong interaction at high energy in which this force has an unexpectedly weak effect. Historically, the first of these appeared in proton-proton collisions. At high energy, above 10 GeV or so in the center of mass, collisions of protons (or any other hadrons) produce large numbers of pions. One might have imagined that these pions would fill all of the allowed phase space, but, in fact, they are mainly produced with momenta almost collinear with the collision axis. The probability of producing a pion with a large component of momentum transverse to the collision axis falls off exponentially in the value of this transverse momentum, suppressing the production substantially for transverse momenta greater than a few hundred MeV.

This phenomenon of limited transverse momentum led to a picture of a hadron as a loosely bound assemblage of many components. In this picture, a proton struck by another proton would be torn into a cloud of pieces. These pieces would have momenta roughly collinear with the original momentum of the proton and would eventually reform into hadrons moving along the collision axis. By hypothesis, these pieces could not absorb a large momentum transfer. We can characterize this hypothesis mathematically as follows: In a high-energy collision, the momenta of the two initial hadrons are almost lightlike. The shattered pieces of the hadrons, arrayed along the collision axis, also have lightlike momenta parallel to the original momentum vectors. This final state can be produced by exchanging momenta q among the pieces in such a way that, though the components of q might be large, the invariant q^2 is always small. The ejection of a hadron at large transverse momentum would require large (spacelike) q^2, but such a process was very rare. Thus it was hypothesized that hadrons were loose clouds of constituents, like jelly, which could not absorb a large q^2.

This picture of hadronic structure was put to a crucial test in the late 1960s, in the SLAC-MIT deep inelastic scattering experiments.[†] In these experiments, a 20 GeV electron beam was scattered from a hydrogen target, and the scattering rate was measured for large deflection angles, corresponding to large invariant momentum transfers from the electron to a proton in the target. The large momentum transfer was delivered through the electromagnetic rather than the strong interaction, so that the amount of momentum delivered could be computed from the momentum of the scattered electron. In models in which hadrons were complex and softly bound, very low scattering rates were expected.

Instead, the SLAC-MIT experiments saw a substantial rate for hard scattering of electrons from protons. The total reaction rate was comparable to what would have been expected if the proton were an elementary particle scattering according to the simplest expectations from QED. However, only in rare cases did a single proton emerge from the scattering process. The largest part

---

[†]For a description of these experiments and their ramifications, see J. I. Friedman, H. W. Kendall, and R. E. Taylor, *Rev. Mod. Phys.* **63**, 573 (1991).

of the rate came from the *deep inelastic* region of phase space, in which the electromagnetic impulse shattered the proton and produced a system with a large number of hadrons.

How could one reconcile the presence of electromagnetic hard scattering processes with the virtual absence of hard scattering in strong interaction processes? To answer this question, Bjorken and Feynman advanced the following simple model, called the *parton model*: Assume that the proton is a loosely bound assemblage of a small number of constituents, called *partons*. These include quarks (and antiquarks), which are fermions carrying electric charge, and possibly other neutral species responsible for their binding. By assumption, these constituents are incapable of exchanging large momenta q^2 through the strong interactions. However, the quarks have the electromagnetic interactions of elementary fermions, so that an electron scattering from a quark can knock it out of the proton. The struck quark then exchanges momentum softly with the remainder of the proton, so that the pieces of the proton materialize as a jet of hadrons. The produced hadrons should be collinear with the direction of the original struck parton.

The parton model, incomplete though it is, imposes a strong constraint on the cross section for deep inelastic electron scattering. To derive this constraint, consider first the cross section for electron scattering from a single constituent quark. We discussed the related process of electron-muon scattering in Section 5.4, and we can borrow that result. Since we imagine the reaction to occur at very high energy, we will ignore all masses. The square of the invariant matrix element in the massless limit is written in a simple form in Eq. (5.71):

$$\frac{1}{4}\sum_{\text{spins}} |\mathcal{M}|^2 = \frac{8e^4 Q_i^2}{\hat{t}^2}\left(\frac{\hat{s}^2 + \hat{u}^2}{4}\right), \tag{14.1}$$

where $\hat{s}$, $\hat{t}$, $\hat{u}$ are the Mandelstam variables for the electron-quark collision and Q_i is the electric charge of the quark in units of $|e|$. Recall from Eq. (5.73) that, for a collision involving massless particles, $\hat{s} + \hat{t} + \hat{u} = 0$. Then the differential cross section in the center of mass system is

$$\begin{aligned}
\frac{d\sigma}{d\cos\theta_{\text{CM}}} &= \frac{1}{2\hat{s}}\frac{1}{16\pi}\frac{8e^4 Q_i^2}{\hat{t}^2}\left(\frac{\hat{s}^2 + \hat{u}^2}{4}\right) \\
&= \frac{\pi\alpha^2 Q_i^2}{\hat{s}}\left(\frac{\hat{s}^2 + (\hat{s}+\hat{t})^2}{\hat{t}^2}\right).
\end{aligned} \tag{14.2}$$

Or, since $\hat{t} = -\hat{s}(1 - \cos\theta_{\text{CM}})/2$,

$$\frac{d\sigma}{d\hat{t}} = \frac{2\pi\alpha^2 Q_i^2}{\hat{s}^2}\left(\frac{\hat{s}^2 + (\hat{s}+\hat{t})^2}{\hat{t}^2}\right). \tag{14.3}$$

To make use of this result, we must relate the invariants $\hat{s}$ and $\hat{t}$ to experimental observables of electron-proton inelastic scattering. The kinematic variables are shown in Fig. 14.1. The momentum transfer q from the electron

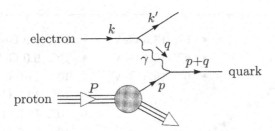

Figure 14.1. Kinematics of deep inelastic electron scattering in the parton model.

can be measured by measuring the final momentum and energy of the electron, without using any information from the hadronic products. Since q^μ is a spacelike vector, one conventionally expresses its invariant square in terms of a positive quantity Q, with

$$Q^2 \equiv -q^2. \tag{14.4}$$

Then the invariant $\hat{t}$ is simply $-Q^2$.

Expressing $\hat{s}$ in terms of measurable quantities is more difficult. If the collision is viewed from the electron-proton center of mass frame, and we visualize the proton as a loosely bound collection of partons (and continue to ignore masses), we can characterize a given parton by the fraction of the proton's total momentum that it carries. We denote this *longitudinal fraction* by the parameter ξ, with $0 < \xi < 1$. For each species i of parton, for example, up-type quarks with electric charge $Q_i = +2/3$, there will be a function $f_i(\xi)$ that expresses the probability that the proton contains a parton of type i and longitudinal fraction ξ. The expression for the total cross section for electron-proton inelastic scattering will contain an integral over the value of ξ for the struck parton. The momentum vector of the parton is then $p = \xi P$, where P is the total momentum of the proton. Thus, if k is the initial electron momentum,

$$\hat{s} = (p + k)^2 = 2p \cdot k = 2\xi P \cdot k = \xi s, \tag{14.5}$$

where s is the square of the electron-proton center of mass energy.

Remarkably, ξ can also be determined from measurements of only the electron momentum, if one makes the assumption that the electron-parton scattering is elastic. Since the scattered parton has a mass small compared to s and Q^2,

$$0 \approx (p + q)^2 = 2p \cdot q + q^2 = 2\xi P \cdot q - Q^2. \tag{14.6}$$

Thus

$$\xi = x, \qquad \text{where} \quad x \equiv \frac{Q^2}{2P \cdot q}. \tag{14.7}$$

From each scattered electron, one can determine the values of Q^2 and x for the scattering process. The parton model then predicts the event distribution

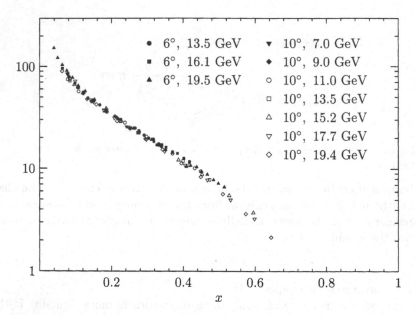

Figure 14.2. Test of Bjorken scaling using the e^-p deep inelastic scattering cross sections measured by the SLAC-MIT experiment, J. S. Poucher, et. al., *Phys. Rev. Lett.* **32**, 118 (1974). We plot $d^2\sigma/dxdQ^2$ divided by the factor (14.9) against x, for the various initial electron energies and scattering angles indicated. The data span the range 1 GeV^2 < Q^2 < 8 GeV^2.

in the x-Q^2 plane. Using the parton distribution functions $f_i(\xi)$, evaluated at $\xi = x$, and the cross-section formula (14.3), we find the distribution

$$\frac{d^2\sigma}{dxdQ^2} = \sum_i f_i(x)Q_i^2 \cdot \frac{2\pi\alpha^2}{Q^4}\left[1 + \left(1 - \frac{Q^2}{xs}\right)^2\right]. \qquad (14.8)$$

The distribution functions $f_i(x)$ depend on the details of the structure of the proton and it is not known how to compute them from first principles. But formula (14.8) still makes a striking prediction, that the deep inelastic scattering cross section, when divided by the factor

$$\frac{1 + (1 - Q^2/xs)^2}{Q^4} \qquad (14.9)$$

to remove the kinematic dependence of the QED cross section, gives a quantity that depends only on x and is independent of Q^2. This behavior is known as *Bjorken scaling*. Indeed, the data from the SLAC-MIT experiment exhibited Bjorken scaling to about 10% accuracy for values of Q above 1 GeV, as shown in Fig. 14.2.

Bjorken scaling is, essentially, the statement that the structure of the proton looks the same to an electromagnetic probe no matter how hard the proton is struck. In the frame of the proton, the energy of the exchanged virtual photon is

$$q^0 = \frac{P \cdot q}{m} = \frac{Q^2}{2xm},$$

(14.10)

where m is the proton mass. The reciprocal of this energy transfer is, roughly, the duration of the scattering process as seen by the components of the proton. This time should be compared to the reciprocal of the proton mass, which is the characteristic time over which the partons interact. The deep inelastic regime occurs when $q^0 \gg m$, that is, when the scattering is very rapid compared to the normal time scales of the proton. Bjorken scaling implies that, during such a rapid scattering process, interactions among the constituents of the proton can be ignored. We might imagine that the partons are approximately free particles over the very short times scales corresponding to energy transfers of a GeV or more, though they have strong interactions on longer time scales.

Asymptotically Free Partons

The picture of the proton structure implied by Bjorken scaling was beautifully simple, but it raised new, fundamental questions. In quantum field theory, fermions interact by exchanging virtual particles. These virtual particles can have arbitrarily high momenta, hence the fluctuations associated with them can occur on arbitrarily short time scales. Quantum field theory processes do not turn themselves off at short times to reveal free-particle equations. Thus the discovery of Bjorken scaling suggested a conflict between the observation of almost free partons and the basic principles of quantum field theory.

The resolution of this paradox came from the renormalization group. In Chapter 12 we saw that coupling constants vary with distance scale. In QED and ϕ^4 theory, we found that the couplings become strong at large momenta and weak at small momenta. However, we noted the possibility that, in some theories, the coupling constant could have the opposite behavior, becoming strong at small momenta or large times but weak at large momenta or short times. We referred to such behavior as *asymptotic freedom*. Section 13.3 discussed an example of an asymptotically free quantum field theory, the nonlinear sigma model in two dimensions. The problem posed in the previous paragraph would be resolved if there existed a suitable asymptotically free quantum field theory in four dimensions that could describe the interaction and binding of quarks. Then, at least to some level of approximation, the strong interaction described by this theory would turn off in large-momentum-transfer or short-time processes.

At the time of the discovery of Bjorken scaling, no asymptotically free field theories in four dimensions were known. Then, in the early 1970s, 't Hooft, Politzer, Gross, and Wilczek discovered a class of such theories. These are the *non-Abelian gauge theories*: theories of interacting vector bosons that can be constructed as generalizations of quantum electrodynamics. It was subsequently shown that these are the only asymptotically free field theories in four dimensions. This discovery gave the crucial clue for the construction of the fundamental theory of the strong interactions. Apparently, the quarks are bound together by interacting vector bosons (called *gluons*) of precisely this type.

However, these gauge theories cannot precisely reproduce the expectations of strict Bjorken scaling. The differences between the free parton model and the quantum field theory model with asymptotic freedom appear when one moves to a higher level of accuracy in measurements of deep inelastic scattering and other strong interaction processes involving large momentum transfer. In an asymptotically free quantum field theory, the coupling constant is still nonzero at any finite momentum transfer. In fact, the final evolution of the coupling to zero is very slow, logarithmic in momentum. Thus, at some level, one must find small corrections to Bjorken scaling, associated with the exchange or emission of high-momentum gluons. Similarly, the other qualitative simplifications of hadron physics at high momentum transfer—for example, the phenomenon of limited transverse momentum in hadron-hadron collisions—should be only approximate, receiving corrections due to gluon exchange and emission. Thus the predictions of an asymptotically free theory of the strong interaction are twofold. On one hand, such a theory predicts qualitative simplifications of behavior at high momentum. But, on the other hand, such a theory predicts a specific pattern of corrections to this behavior.

In fact, particle physics experiments of the 1970s revealed precisely this picture. Bjorken scaling was found to be only an approximate relation, showing violations that correspond to a slow evolution of the parton distributions $f_i(x)$ over a logarithmic scale in Q^2. The rate of particle production in hadron-hadron collisions was found to decrease only as a power rather than exponentially at very large values of the transverse momentum, and the particles produced at large transverse momentum were shown to be associated with jets of hadrons created by the soft evolution of a hard-scattered quark or gluon. Most remarkably, the forms of the cross sections found for these and other deviations from scaling did, finally, give direct evidence for the vector character of the elementary field that mediates the strong interaction.

We will review all of these phenomena in Chapter 17, as we study the particular gauge theory that describes the strong interactions. First, however, we must learn how to construct non-Abelian gauge theories and how to work out their predictions using Feynman diagrams. Throughout our analysis of these theories, the renormalization group will play an essential role. One of the very beautiful aspects of the study of non-Abelian gauge theories is the way in which the most powerful general ideas of quantum field theory acquire even more strength as they intertwine with the specific features of these particular, intricately built models. This interplay between general principles and the specific features of gauge theories will be the major theme of Part III of this book.

Chapter 15

Non-Abelian Gauge Invariance

So far in this book we have worked with a rather limited class of quantum fields and interactions, restricting our attention to scalar field theories, Yukawa theory, and Quantum Electrodynamics. It is hardly surprising that these theories are not sufficient to describe all of the known interactions of elementary particles. But what other theories are possible, given that the Lagrangian of a renormalizable theory can contain no terms of mass dimension higher than 4?

The most natural theories to try next would be ones with interactions among vector fields, of the form $A^\mu A^\nu \partial_\mu A_\nu$ or A^4. Sensible theories of this type are difficult to construct, however, because of the negative-norm states produced by the time component A^0 of the vector field operator. In Section 5.5 we saw that these negative-norm states cause no difficulty in QED: They are effectively canceled out by the longitudinal polarization states, by virtue of the Ward identity. The Ward identity, in turn, follows from the invariance of the QED Lagrangian under local gauge transformations. Perhaps, then, if we can generalize the principle of local gauge invariance, it will lead us to the construction of other sensible theories of vector particles.

The goal of this chapter is to do just that. First we will return briefly to the study of QED, this time taking the gauge symmetry to be fundamental and deriving the rest of the theory from this principle. Then, in Section 15.2, we will see that the gauge invariance of electrodynamics is only the most trivial example of an infinite-parameter symmetry, and that the more general examples lead to other interesting Lagrangians. These field theories, the first of which was constructed by Yang and Mills,* generalize electrodynamics in a profound way. They are theories of multiple vector particles, whose interactions are strongly constrained by the symmetry principle. In subsequent chapters we will study the quantization of these theories and their application to the real world of elementary particle physics.

---

*C. N. Yang and R. Mills, *Phys. Rev.* **96**, 191 (1954).

15.1 The Geometry of Gauge Invariance

In Section 4.1 we wrote down the Lagrangian of Quantum Electrodynamics and noted the curious fact that it is invariant under a very large group of transformations (4.6), allowing an independent symmetry transformation at every point in spacetime. This invariance is the famous *gauge symmetry* of QED. From the modern viewpoint, however, gauge symmetry is not an incidental curiosity, but rather the fundamental principle that determines the form of the Lagrangian. Let us now review the elements of the theory, taking the modern viewpoint.

We begin with the complex-valued Dirac field $\psi(x)$, and stipulate that our theory should be invariant under the transformation

$$\psi(x) \rightarrow e^{i\alpha(x)}\psi(x). \tag{15.1}$$

This is a phase rotation through an angle $\alpha(x)$ that varies arbitrarily from point to point. How can we write a Lagrangian that is invariant under this transformation? As long as we consider terms in the Lagrangian that have no derivatives, this is easy: We simply write the same terms that are invariant to global phase rotations. For example, the fermion mass term

$$m\bar{\psi}\psi(x)$$

is permitted by global phase invariance, and the local invariance gives no further restriction.

The difficulty arises when we try to write terms including derivatives. The derivative of $\psi(x)$ in the direction of the vector n^μ is defined by the limiting procedure

$$n^\mu \partial_\mu \psi = \lim_{\epsilon \to 0} \frac{1}{\epsilon} \left[\psi(x + \epsilon n) - \psi(x) \right]. \tag{15.2}$$

However, in a theory with local phase invariance, this definition is not very sensible, since the two fields that are subtracted, $\psi(x + \epsilon n)$ and $\psi(x)$, have completely different transformations under the symmetry (15.1). The quantity $\partial_\mu \psi$, in other words, has no simple tranformation law and no useful geometric interpretation.

In order to subtract the values of $\psi(x)$ at neighboring points in a meaningful way, we must introduce a factor that compensates for the difference in phase transformations from one point to the next. The simplest way to do this is to define a scalar quantity $U(y, x)$ that depends on the two points and has the transformation law

$$U(y, x) \rightarrow e^{i\alpha(y)}U(y, x)e^{-i\alpha(x)} \tag{15.3}$$

simultaneously with (15.1). At zero separation, we set $U(y, y) = 1$; in general, we can require $U(y, x)$ to be a pure phase: $U(y, x) = \exp[i\phi(y, x)]$. With this definition, the objects $\psi(y)$ and $U(y, x)\psi(x)$ have the same transformation law, and we can subtract them in a manner that is meaningful despite the local symmetry. Thus we can define a sensible derivative, called the *covariant derivative*, as follows:

$$n^\mu D_\mu \psi = \lim_{\epsilon \to 0} \frac{1}{\epsilon} \left[\psi(x + \epsilon n) - U(x + \epsilon n, x) \psi(x) \right]. \tag{15.4}$$

To make this definition explicit, we need an expression for the comparator $U(y, x)$ at infinitesimally separated points. If the phase of $U(y, x)$ is a continuous function of the positions y and x, then $U(y, x)$ can be expanded in the separation of the two points:

$$U(x + \epsilon n, x) = 1 - ie \, \epsilon n^\mu A_\mu(x) + \mathcal{O}(\epsilon^2). \tag{15.5}$$

Here we have arbitrarily extracted a constant e. The coefficient of the displacement ϵn^μ is a new vector field $A_\mu(x)$. Such a field, which appears as the infinitesimal limit of a comparator of local symmetry transformations, is called a *connection*. The covariant derivative then takes the form

$$D_\mu \psi(x) = \partial_\mu \psi(x) + ie A_\mu \psi(x). \tag{15.6}$$

By inserting (15.5) into (15.3), one finds that A_μ transforms under this local gauge transformation as

$$A_\mu(x) \to A_\mu(x) - \frac{1}{e} \partial_\mu \alpha(x). \tag{15.7}$$

To check that all of these expressions are consistent, we can transform $D_\mu \psi(x)$ according to Eqs. (15.1) and (15.7):

$$\begin{aligned} D_\mu \psi(x) &\to \left[\partial_\mu + ie \left(A_\mu - \frac{1}{e} \partial_\mu \alpha \right) \right] e^{i\alpha(x)} \psi(x) \\ &= e^{i\alpha(x)} \left(\partial_\mu + ie A_\mu \right) \psi(x) = e^{i\alpha(x)} D_\mu \psi(x). \end{aligned} \tag{15.8}$$

Thus the covariant derivative transforms in the same way as the field ψ, exactly as we constructed it to in the original definition (15.4).

We have now recovered most of the familiar ingredients of the QED Lagrangian. From our current viewpoint, however, the definition of the covariant derivative and the transformation law for the connection A_μ follow from the postulate of local phase rotation symmetry. Even the very existence of the vector field A_μ is a consequence of local symmetry: Without it we could not write an invariant Lagrangian involving derivatives of ψ.

More generally, our present analysis gives us a way to construct all possible Lagrangians that are invariant under the local symmetry. In any term with derivatives of ψ, replace these with covariant derivatives. According to Eq. (15.8), these transform in exactly the same manner as ψ itself. Therefore any combination of ψ and its covariant derivatives that is invariant under a global phase rotation (and only these combinations) will also be locally invariant.

To complete the construction of a locally invariant Lagrangian, we must find a kinetic energy term for the field A_μ: a locally invariant term that depends on A_μ and its derivatives, but not on ψ. This term can be constructed

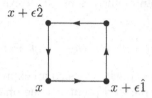

Figure 15.1. Construction of the field strength by comparisons around a small square in the $(1, 2)$ plane.

either integrally, from the comparator $U(y, x)$, or infinitesimally, from the covariant derivative.

Working from $U(y, x)$, we will need to extend our explicit formula (15.5) to the next term in the expansion in ϵ. Using the assumption that $U(y, x)$ is a pure phase and the restriction $(U(x, y))^\dagger = U(y, x)$, it follows that

$$U(x + \epsilon n, x) = \exp\left[-ie\epsilon n^\mu A_\mu(x + \tfrac{\epsilon}{2}n) + \mathcal{O}(\epsilon^3)\right]. \tag{15.9}$$

(Relaxing these restrictions introduces additional vector fields into the theory; this is an unnecessary complication.) Using this expansion for $U(y, x)$, we link together comparisons of the phase direction around a small square in spacetime. For definiteness, we take this square to lie in the $(1, 2)$-plane, as defined by the unit vectors $\hat{1}$, $\hat{2}$ (see Fig. 15.1). Define $\mathbf{U}(x)$ to be the product of the four comparisons around the corners of the loop:

$$\begin{aligned}\mathbf{U}(x) &\equiv U(x, x + \epsilon\hat{2})U(x + \epsilon\hat{2}, x + \epsilon\hat{1} + \epsilon\hat{2}) \\ &\quad \times U(x + \epsilon\hat{1} + \epsilon\hat{2}, x + \epsilon\hat{1})U(x + \epsilon\hat{1}, x).\end{aligned} \tag{15.10}$$

The transformation law (15.3) for U implies that $\mathbf{U}(x)$ is locally invariant. In the limit $\epsilon \to 0$, it will therefore give us a locally invariant function of A_μ. To find the form of this function, insert the expansion (15.9) to obtain

$$\begin{aligned}\mathbf{U}(x) = \exp\Big\{ &- ie\epsilon\left[-A_2(x + \tfrac{\epsilon}{2}\hat{2}) - A_1(x + \tfrac{\epsilon}{2}\hat{1} + \epsilon\hat{2})\right. \\ &\left. + A_2(x + \epsilon\hat{1} + \tfrac{\epsilon}{2}\hat{2}) + A_1(x + \tfrac{\epsilon}{2}\hat{1})\right] + \mathcal{O}(\epsilon^3)\Big\}.\end{aligned} \tag{15.11}$$

When we expand the exponent in powers of ϵ, this reduces to

$$\mathbf{U}(x) = 1 - ie^2 e\left[\partial_1 A_2(x) - \partial_2 A_1(x)\right] + \mathcal{O}(\epsilon^3). \tag{15.12}$$

Therefore the structure

$$F_{\mu\nu} = \partial_\mu A_\nu - \partial_\nu A_\mu \tag{15.13}$$

is locally invariant. Of course, $F_{\mu\nu}$ is the familiar electromagnetic field tensor, and its invariance under (15.7) can be checked directly. The preceding construction, however, shows us the geometrical origin of the structure of $F_{\mu\nu}$. Any function that depends on A_μ only through $F_{\mu\nu}$ and its derivatives is locally invariant. More general functions, such as the vector field mass term

$A_\mu A^\mu$, transform under (15.7) in ways that cannot be compensated and thus cannot appear in an invariant Lagrangian.

A related argument for the invariance of $F_{\mu\nu}$ can be made using the covariant derivative. We have seen above that, if a field has the local transfomation law (15.1), then its covariant derivative has the same transformation law. Thus the second covariant derivative of ψ also transforms according to (15.1). The same conclusion holds for the commutator of covariant derivatives:

$$[D_\mu, D_\nu]\psi(x) \rightarrow e^{i\alpha(x)}[D_\mu, D_\nu]\psi(x). \tag{15.14}$$

However, the commutator is not itself a derivative at all:

$$\begin{aligned}[D_\mu, D_\nu]\psi &= [\partial_\mu, \partial_\nu]\psi + ie([\partial_\mu, A_\nu] - [\partial_\nu, A_\mu])\psi - e^2[A_\mu, A_\nu]\psi \\ &= ie(\partial_\mu A_\nu - \partial_\nu A_\mu)\cdot\psi. \end{aligned} \tag{15.15}$$

That is,

$$[D_\mu, D_\nu] = ieF_{\mu\nu}. \tag{15.16}$$

On the right-hand side of (15.14), the factor $\psi(x)$ accounts for the entire transformation law, so the multiplicative factor $F_{\mu\nu}$ must be invariant. One can visualize the commutator of covariant derivatives as the comparison of comparisons across a small square; fundamentally, therefore, this argument is equivalent to that of the previous paragraph.

Whatever the method of proving the invariance of $F_{\mu\nu}$, we have now assembled all of the ingredients we need to write the most general locally invariant Lagrangian for the electron field ψ and its associated connection A_μ. This Lagrangian must be a function of ψ and its covariant derivatives, and of $F_{\mu\nu}$ and its derivatives, and must be invariant to global phase transformations. Up to operators of dimension 4, there are only four possible terms:

$$\mathcal{L}_4 = \bar{\psi}(i\slashed{D})\psi - \frac{1}{4}(F_{\mu\nu})^2 - c\epsilon^{\alpha\beta\mu\nu}F_{\alpha\beta}F_{\mu\nu} - m\bar{\psi}\psi. \tag{15.17}$$

By adjusting the normalization of the fields ψ and A_μ, we have set the coefficients of the first two terms to their standard values. This normalization of A_μ requires the arbitrary scale factor e in our original definition (15.5) of A_μ. The third term violates the discrete symmetries P and T, so we may exclude it if we postulate these symmetries.[†] Then $\mathcal{L}_4$ contains only two free parameters, the scale factor e and the coefficient m.

By using operators of dimension 5 and 6, we can form many additional gauge-invariant combinations:

$$\mathcal{L}_6 = ic_1\bar{\psi}\sigma_{\mu\nu}F^{\mu\nu}\psi + c_2(\bar{\psi}\psi)^2 + c_3(\bar{\psi}\gamma^5\psi)^2 + \cdots. \tag{15.18}$$

More allowed terms appear at each higher order in mass dimension. But all of these terms are nonrenormalizable interactions. In the language of Section 12.1, they are irrelevant to physics in four dimensions in the limit where the cutoff is taken to infinity.

[†]The general systematics of P, C, and T violation are discussed in Section 20.3.

We have now reached a remarkable conclusion. We began by postulating that the electron field obeys the local symmetry (15.1). From this postulate, we showed that there must be an electromagnetic vector potential. Further, the symmetry principle implies that the most general Lagrangian in four dimensions that is renormalizable (or relevant, in Wilson's sense) is the general form $\mathcal{L}_4$. If we insist that this Lagrangian also be invariant under time reversal or parity, we are led uniquely to the Maxwell-Dirac Lagrangian that is the basis of quantum electrodynamics.

15.2 The Yang-Mills Lagrangian

If the simple geometrical constructions of the previous section yield Maxwell's theory of electrodynamics, then surely it must be possible to construct other interesting theories by starting with more general geometrical principles. Yang and Mills proposed that the argument of the previous section could be generalized from local phase rotation invariance to invariance under any continuous symmetry group. In this section, we will introduce this generalization of local symmetry. For most of the discussion, we will consider our local symmetry to be the three-dimensional rotation group, $O(3)$ or $SU(2)$, since in this case the necessary group theory should be familiar. At the end of the section, we will generalize further to the case of an arbitrary local symmetry.

Consider, then, the following generalization of the phase rotation (15.1): Instead of a single fermion field, we start with a doublet of Dirac fields,

$$\psi = \begin{pmatrix} \psi_1(x) \\ \psi_2(x) \end{pmatrix}, \tag{15.19}$$

which transform into one another under abstract three-dimensional rotations as a two-component spinor:

$$\psi \rightarrow \exp\left(i\alpha^i \frac{\sigma^i}{2}\right)\psi. \tag{15.20}$$

Here σ^i are the Pauli sigma matrices, and, as usual, a sum over repeated indices is implied. It is important to distinguish this abstract transformation from a rotation in physical three-dimensional space; in their original paper, Yang and Mills considered (ψ_1, ψ_2) to be the proton-neutron doublet as it is transformed under isotopic spin. As in the case of a phase rotation, it is not hard to construct Lagrangians for ψ that are invariant to (15.20) as a global symmetry.

We now promote (15.20) to a local symmetry, by insisting that the Lagrangian be invariant to this transformation for α^i an arbitrary function of x. Write this transformation as

$$\psi(x) \rightarrow V(x)\psi(x), \qquad \text{where } V(x) = \exp\left(i\alpha^i(x)\frac{\sigma^i}{2}\right). \tag{15.21}$$

We can construct a suitable Lagrangian by applying the methods of the previous section. However, we will encounter a number of additional complications, due to the fact that there are now three orthogonal symmetry motions, which do not commute with one another. This feature is sufficiently important to earn a special name for theories that have it: We refer to the *Abelian* symmetry group of electrodynamics, and the *non-Abelian* symmetry group of the more general theories. The field theory associated with a noncommuting local symmetry is termed a *non-Abelian gauge theory*.

To construct a Lagrangian that is invariant under this new group of transformations, we must again define a covariant derivative that transforms in a simple way. Again we use the definition (15.4), but since ψ is now a two-component object, the comparator $U(y, x)$ must be a 2×2 matrix. The transformation law for $U(y, x)$ is now

$$U(y, x) \rightarrow V(y) \, U(y, x) \, V^\dagger(x), \qquad (15.22)$$

where $V(x)$ is as in (15.21), and again we set $U(y, y) = 1$. At points $x \neq y$ we can consistently restrict $U(y, x)$ to be a unitary matrix. Near $U = 1$, any such matrix can be expanded in terms of the Hermitian generators of $SU(2)$; thus for infinitesimal separation we can write

$$U(x + \epsilon n, x) = 1 + ig\epsilon n^\mu A_\mu^i \frac{\sigma^i}{2} + \mathcal{O}(\epsilon^2). \qquad (15.23)$$

Here g is a constant, extracted for later convenience. Inserting this expansion into the definition (15.4) of the covariant derivative, we find the following expression for the covariant derivative associated with local $SU(2)$ symmetry:

$$D_\mu = \partial_\mu - ig A_\mu^i \frac{\sigma^i}{2}. \qquad (15.24)$$

This covariant derivative requires three vector fields, one for each generator of the transformation group.

We can find the gauge transformation law of the connection A_μ^i by inserting the expansion (15.23) into the transformation law (15.22):

$$1 + ig\epsilon n^\mu A_\mu^i \frac{\sigma^i}{2} \rightarrow V(x + \epsilon n)\left(1 + ig\epsilon n^\mu A_\mu^i \frac{\sigma^i}{2}\right)V^\dagger(x). \qquad (15.25)$$

We must expand the right-hand side to order ϵ, taking care that the various Pauli matrices do not commute with one another. The expansion of $V(x + \epsilon n)$ is conveniently done using the identity

$$V(x + \epsilon n)V^\dagger(x) = \left[\left(1 + \epsilon n^\mu \frac{\partial}{\partial x^\mu} + \mathcal{O}(\epsilon^2)\right)V(x)\right]V^\dagger(x)$$

$$= 1 + \epsilon n^\mu \left(\frac{\partial}{\partial x^\mu}V(x)\right)V^\dagger(x) + \mathcal{O}(\epsilon^2) \qquad (15.26)$$

$$= 1 + \epsilon n^\mu V(x)\left(-\frac{\partial}{\partial x^\mu}V^\dagger(x)\right) + \mathcal{O}(\epsilon^2).$$

Then the terms proportional to ϵn^μ in (15.25) give the transformation

$$A^i_\mu(x)\frac{\sigma^i}{2} \to V(x)\left(A^i_\mu(x)\frac{\sigma^i}{2} + \frac{i}{g}\partial_\mu\right)V^\dagger(x). \qquad (15.27)$$

The derivative acts on $V^\dagger(x) = \exp(-i\alpha^i\sigma^i/2)$; it is not so easy to compute this derivative explicitly because the exponent does not necessarily commute with its derivative. For infinitesimal transformations we can expand $V(x)$ to first order in α. In this case we obtain

$$A^i_\mu\frac{\sigma^i}{2} \to A^i_\mu\frac{\sigma^i}{2} + \frac{1}{g}(\partial_\mu\alpha^i)\frac{\sigma^i}{2} + i\left[\alpha^i\frac{\sigma^i}{2}, A^j_\mu\frac{\sigma^j}{2}\right] + \cdots. \qquad (15.28)$$

The last term in this transformation law is new, and arises from the noncommutativity of the local transformations. By combining this relation with the infinitesimal form of the fermion transformation,

$$\psi \to \left(1 + i\alpha^i\frac{\sigma^i}{2}\right)\psi + \cdots, \qquad (15.29)$$

we can check the infinitesimal transformation of the covariant derivative:

$$D_\mu\psi \to \left(\partial_\mu - igA^i_\mu\frac{\sigma^i}{2} - i(\partial_\mu\alpha^i)\frac{\sigma^i}{2} + g\left[\alpha^i\frac{\sigma^i}{2}, A^j_\mu\frac{\sigma^j}{2}\right]\right)\left(1 + i\alpha^k\frac{\sigma^k}{2}\right)\psi$$

$$= \left(1 + i\alpha^i\frac{\sigma^i}{2}\right)D_\mu\psi, \qquad (15.30)$$

up to terms of order α^2. It is not difficult to check using (15.27) and (15.21) that, even for finite transformations, the covariant derivative has the same transformation law as the field on which it acts.

Using the covariant derivative, we can build the most general gauge-invariant Lagrangians involving ψ. But to write a complete Lagrangian, we must also find gauge-invariant terms that depend only on A^i_μ. To do this, we construct the analogue of the electromagnetic field tensor. We will use the second method of the previous section, working from the commutator of covariant derivatives. The transformation law of the covariant derivative implies that

$$[D_\mu, D_\nu]\psi(x) \to V(x)[D_\mu, D_\nu]\psi(x). \qquad (15.31)$$

At the same time, by writing out the commutator using formula (15.24), we can show, as in the Abelian case, that $[D_\mu, D_\nu]$ is not a differential operator but merely a multiplicative factor (now a matrix) acting on ψ. This time, however, there is a new feature: The last term in the expansion of the commutator no longer vanishes. Instead, we find

$$[D_\mu, D_\nu] = -igF^i_{\mu\nu}\frac{\sigma^i}{2}, \qquad (15.32)$$

with

$$F^i_{\mu\nu}\frac{\sigma^i}{2} = \partial_\mu A^i_\nu\frac{\sigma^i}{2} - \partial_\nu A^i_\mu\frac{\sigma^i}{2} - ig\left[A^i_\mu\frac{\sigma^i}{2}, A^j_\nu\frac{\sigma^j}{2}\right]. \qquad (15.33)$$

We can simplify this relation by applying the standard commutation relations of Pauli matrices:

$$\left[\frac{\sigma^i}{2}, \frac{\sigma^j}{2}\right] = i\epsilon^{ijk}\frac{\sigma^k}{2}. \tag{15.34}$$

Then

$$F^i_{\mu\nu} = \partial_\mu A^i_\nu - \partial_\nu A^i_\mu + g\epsilon^{ijk} A^j_\mu A^k_\nu. \tag{15.35}$$

The transformation law for the field strength follows from Eqs. (15.21) and (15.31):

$$F^i_{\mu\nu}\frac{\sigma^i}{2} \to V(x) F^j_{\mu\nu}\frac{\sigma^j}{2} V^\dagger(x). \tag{15.36}$$

The infinitesimal form is

$$F^i_{\mu\nu}\frac{\sigma^i}{2} \to F^i_{\mu\nu}\frac{\sigma^i}{2} + \left[i\alpha^i\frac{\sigma^i}{2}, F^j_{\mu\nu}\frac{\sigma^j}{2}\right]. \tag{15.37}$$

Notice that the field strength is no longer a gauge-invariant quantity. It cannot be, since there are now three field strengths, each associated with a given direction of rotation in the abstract space. However, it is easy to form gauge-invariant combinations of the field strengths. For example,

$$\mathcal{L} = -\frac{1}{2}\text{tr}\left[\left(F^i_{\mu\nu}\frac{\sigma^i}{2}\right)^2\right] = -\frac{1}{4}\left(F^i_{\mu\nu}\right)^2 \tag{15.38}$$

is a gauge-invariant kinetic energy term for A^i_μ. Notice that, in contrast to the case of electrodynamics, this Lagrangian contains cubic and quartic terms in A^i_μ. Thus, this Lagrangian describes a nontrivial, interacting field theory, called *Yang-Mills theory*. This is the simplest example of a non-Abelian gauge theory.

To construct a theory of Yang-Mills vector fields interacting with fermions, we simply add the gauge-field Lagrangian (15.38) to the familiar Dirac Lagrangian, with the ordinary derivative of ψ replaced by the covariant derivative. The result looks almost identical to the QED Lagrangian:

$$\mathcal{L} = \bar{\psi}(i\slashed{D})\psi - \frac{1}{4}(F^i_{\mu\nu})^2 - m\bar{\psi}\psi. \tag{15.39}$$

This is the famous Yang-Mills Lagrangian. Like that of QED, it depends on two parameters: the scale factor g (which is analogous to the electron charge) and the fermion mass m. By varying this Lagrangian, we find the classical equations of motion of the gauge theory. These are the Dirac equation for the fermion field and the equation

$$\partial^\mu F^i_{\mu\nu} + g\epsilon^{ijk}A^{j\mu}F^k_{\mu\nu} = -g\bar{\psi}\gamma_\nu\frac{\sigma^i}{2}\psi \tag{15.40}$$

for the vector field.

Everything that we have done for the $SU(2)$ symmetry transformation (15.20) generalizes easily to any other continuous group of symmetries. The full range of possible symmetry groups is enumerated and classified in Section 15.4. For any such group, however, the general expressions for elements of the Lagrangian are quite similar. Consider any continuous group of transformations, represented by a set of $n \times n$ unitary matrices V. Then the basic fields $\psi(x)$ will form an n-plet, and transform according to

$$\psi(x) \rightarrow V(x)\psi(x), \qquad (15.41)$$

where the x dependence of V makes the transformation local. In infinitesimal form, $V(x)$ can be expanded in terms of a set of basic generators of the symmetry group, which can be represented as Hermitian matrices t^a:

$$V(x) = 1 + i\alpha^a(x)t^a + \mathcal{O}(\alpha^2). \qquad (15.42)$$

Now one can carry through the whole analysis from Eq. (15.22) to Eq. (15.33) for a general local symmetry group simply by replacing

$$\frac{\sigma^i}{2} \rightarrow t^a \qquad (15.43)$$

at each step of the analysis.

To generalize the explicit expression (15.35) for the field tensor, we need to know the commutation relations of the matrices t^a. It is conventional to write these in the standard form

$$[t^a, t^b] = if^{abc}t^c, \qquad (15.44)$$

where f^{abc} is a set of numbers called *structure constants*. This object replaces ϵ^{ijk} in Eq. (15.34). It is conventional to choose a basis for the matrices t^a such that f^{abc} is completely antisymmetric; we will prove that this is always possible in Section 15.4.

We can now recapitulate all of our results as follows. The covariant derivative associated with the general transformation (15.41) is

$$D_\mu = \partial_\mu - igA_\mu^a t^a; \qquad (15.45)$$

it contains one vector field for each independent generator of the local symmetry. The infinitesimal tranformation laws for ψ and A_μ^a are

$$\psi \rightarrow (1 + i\alpha^a t^a)\psi;$$
$$A_\mu^a \rightarrow A_\mu^a + \frac{1}{g}\partial_\mu\alpha^a + f^{abc}A_\mu^b\alpha^c. \qquad (15.46)$$

The finite transformation of A_μ^a has exactly the form of (15.27):

$$A_\mu^a(x)t^a \rightarrow V(x)\left(A_\mu^a(x)t^a + \frac{i}{g}\partial_\mu\right)V^\dagger(x). \qquad (15.47)$$

These transformation laws imply that the covariant derivative of ψ has the same transformation law as ψ itself. The field tensor is defined by

$$[D_\mu, D_\nu] = -igF_{\mu\nu}^a t^a, \qquad (15.48)$$

or more explicitly,

$$F_{\mu\nu}^a = \partial_\mu A_\nu^a - \partial_\nu A_\mu^a + g f^{abc} A_\mu^b A_\nu^c. \tag{15.49}$$

This quantity has the infinitesimal transformation

$$F_{\mu\nu}^a \to F_{\mu\nu}^a - f^{abc}\alpha^b F_{\mu\nu}^c. \tag{15.50}$$

From Eqs. (15.46) and (15.50), one can show that any globally symmetric function of ψ, $F_{\mu\nu}^a$, and their covariant derivatives is also locally symmetric, and is therefore a candidate for a term in a gauge-invariant Lagrangian. However, there are very few permissible terms up to dimension 4. The most general gauge-invariant Lagrangian that is renormalizable and conserves P and T is again given by Eq. (15.39). The corresponding classical equation of motion is

$$\partial^\mu F_{\mu\nu}^a + g f^{abc} A^{b\mu} F_{\mu\nu}^c = -g j_\nu^a, \tag{15.51}$$

where

$$j_\nu^a = \bar{\psi}\gamma_\nu t^a \psi \tag{15.52}$$

is the global symmetry current of the fermion field.

Notice that the nonlinear terms in the Yang-Mills Lagrangian (15.39) appear in the covariant derivative, where they are proportional to t^a, and in the field tensor, where they are proportional to f^{abc}. Thus the form of the interactions in a non-Abelian gauge theory is dictated by the local symmetry. The nonlinear interactions of the vector field with itself are proportional to the commutators of symmetry generators and thus explicitly require the non-Abelian nature of the symmetry group.

15.3 The Gauge-Invariant Wilson Loop

In both of the previous sections we made use of the *comparator*, $U(y,x)$, which converts the fermion gauge transformation law at point x to that at point y. So far, in writing expressions for this object, it has sufficed to assume that x and y are infinitesimally separated. However, it is worthwhile to think further about the comparator in the case where x and y are far apart. This discussion will give us further insights into the geometry of gauge invariance, and will reveal some additional useful functions of the gauge field which we will put to work in Chapter 19.

We first return to the Abelian theory and expand upon our discussion of $U(y,x)$ in that context. In Eq. (15.10) we constructed a product of comparators on a path that wound around a small square. We showed that this product $\mathbf{U}(x)$ is not trivial, even though we eventually return to the starting point; rather, we found that $\mathbf{U}(x)$ differs from 1 by a term proportional to the electromagnetic field strength and to the area of the square. This is a particular case of a general conclusion: The comparator between two points x and y at finite separation depends on the path taken from x to y.

To explain this statement, it is useful to reverse some of the logic of Section 15.1. We begin from the connection A_μ, which we assume to have the transformation law (15.7), and construct $U(z, y)$ as a function of A_μ that transforms according to (15.3). It is not difficult to verify that the expression

$$U_P(z, y) = \exp\left[-ie \int_P dx^\mu A_\mu(x)\right] \qquad (15.53)$$

meets this criterion if the integral is taken along any path P that runs from y to z. This object $U_P(z, y)$ is called the *Wilson line*.[‡] Expression (15.53) gives an explicit realization of the abstract comparator $U(z, y)$ for points at finite separation.

A crucial property of the Wilson line is that it depends on the path P. If P is a closed path that returns to y, we obtain the *Wilson loop*,

$$U_P(y, y) = \exp\left[-ie \oint_P dx^\mu A_\mu(x)\right]. \qquad (15.54)$$

This quantity is a nontrivial function of A_μ that is, by construction, locally gauge invariant. In fact, all gauge-invariant functions of A_μ can be thought of as combinations of Wilson loops for various choices of the path P. To motivate this claim, we use Stokes's theorem to rewrite the Wilson loop as

$$U_P(y, y) = \exp\left[-i\frac{e}{2} \int_\Sigma d\sigma^{\mu\nu} F_{\mu\nu}\right], \qquad (15.55)$$

where Σ is a surface that spans the closed loop P, $d\sigma^{\mu\nu}$ is an area element on this surface, and $F_{\mu\nu}$ is the field tensor (15.13). This relation between the Wilson loop and the field strength is illustrated in Fig. 15.2. Since the Wilson loop is gauge invariant, this argument gives one more way to visualize the gauge invariance of the field strength. Conversely, since (almost) all gauge-invariant functions of A_μ can be built up from $F_{\mu\nu}$, this expression gives weight to the statement that $U_P(y, y)$ is the most general gauge invariant.

Both the Wilson line and the Wilson loop can be generalized to the non-Abelian case. Here, however, additional subtleties arise when we consider exponentials of noncommuting matrices. Let us first construct the Wilson line, which now transforms according to Eq. (15.22). It is not correct to make a straightforward rewriting of (15.53) with the integral of $A_\mu^a t^a$ in the exponent, since these matrices do not necessarily commute at different points. Instead, we must order these matrices in a particular way. We will now give the correct ordering prescription and then prove its transformation law.

Let s be a parameter of the path P, running from 0 at $x = y$ to 1 at $x = z$. Then define the Wilson line as the power-series expansion of the exponential, with the matrices in each term ordered so that higher values of s stand to the

---

[‡]This path-dependent phase was used long before Wilson's work, in Schwinger's early papers on QED, and in Y. Aharonov and D. Böhm, *Phys. Rev.* **115**, 485 (1959).

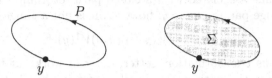

Figure 15.2. The Wilson loop integral is taken around an arbitrary loop. It can also be expressed as a flux integral of the field strength over a surface spanning the loop.

left. This prescription is called *path-ordering* and is denoted by the symbol $P\{\}$. Thus the Wilson line is written

$$U_P(z, y) = P\left\{ \exp\left[ig \int_0^1 ds \frac{dx^\mu}{ds} A_\mu^a(x(s))t^a \right] \right\}. \tag{15.56}$$

This expression is similar to the time-ordered exponential that we wrote for the interaction-picture propagator in Eq. (4.23). Pursuing this analogy, one can show that this expression for U_P is the solution of a differential equation similar to (4.24):

$$\frac{d}{ds} U_P(x(s), y) = \left(ig \frac{dx^\mu}{ds} A_\mu^a(x(s))t^a \right) U_P(x(s), y). \tag{15.57}$$

(Here we consider U_P to be a continuous function of the parameter s, rather than fixing $s = 1$ at the endpoint.)

To show that expression (15.56) is the correct generalization of the Wilson line, we must show that it satisfies the correct gauge transformation law (15.22). This follows from the differential equation (15.57), which can be rewritten as

$$\frac{dx^\mu}{ds} D_\mu U_P(x, y) = 0. \tag{15.58}$$

Now let A^V represent the gauge transform of a field configuration A, and use these arguments to denote explicitly the dependence of gauge functions on the gauge field. We would like to show that

$$U_P(z, y, A^V) = V(z)U_P(z, y, A)V^\dagger(y), \tag{15.59}$$

which is equivalent to (15.22). In (15.30) we proved, in its infinitesimal version, the relation

$$D_\mu(A^V) V(x) = V(x) D_\mu(A). \tag{15.60}$$

This relation implies that the right-hand side of (15.59) satisfies (15.58) for the gauge field A^V if $U_P(z, y, A)$ satisfies this equation for the gauge field A. But the solution of a first-order differential equation with a fixed boundary condition is unique. Thus, if $U_P(z, y)$ is defined to be the solution of (15.57) or (15.58), it indeed has the transformation law (15.59).

The Wilson line associated with a closed path returning to y transforms only with the gauge parameter at y; however, it is not a gauge invariant:

$$U_P(y,y) \to V(y)U_P(y,y)V^\dagger(y). \qquad (15.61)$$

To understand this transformation better, one can work out the expression for $U_P(x,x)$, where the path is the small square in the $(1,2)$ plane shown in Fig. 15.1. In addition to the terms in Eq. (15.11), there are additional corrections of order ϵ^2 coming from products of $(A_\mu^a t^a)$ factors from pairs of sides, which sum up to a commutator of these factors. One finds

$$U_P(x,x) = 1 + ig\epsilon^2 F_{12}^a(x)t^a + \mathcal{O}(\epsilon^3), \qquad (15.62)$$

where $F_{\mu\nu}^a$ is given by the full expression in (15.49). If we then expand the transformation law (15.61) in powers of ϵ, the term of order ϵ^2 is the transformation law of $F_{\mu\nu}^a$ given in Eq. (15.36).

To convert the Wilson line for a closed path into a true gauge invariant, take the trace. By cyclic invariance, (15.61) implies

$$\operatorname{tr} U_P(x,x) \to \operatorname{tr} U_P(x,x). \qquad (15.63)$$

Thus for a non-Abelian gauge theory, we define the Wilson loop to be the trace of the Wilson line around a closed path.

Let us evaluate $\operatorname{tr} U_P(x,x)$ more explicitly for the case of an $SU(2)$ gauge group. If $U(\epsilon)$ is any 2×2 unitary matrix that tends to 1 as $\epsilon \to 0$, we can expand it in ϵ as follows:

$$
\begin{aligned}
U(\epsilon) &= \exp\left[i(\epsilon\beta^i + \epsilon^2\gamma^i + \cdots)\frac{\sigma^i}{2}\right] \\
&= 1 + i(\epsilon\beta^i + \epsilon^2\gamma^i + \cdots)\frac{\sigma^i}{2} - \frac{1}{2}(\epsilon\beta^i \cdot \epsilon\beta^j + \cdots)\frac{\sigma^i}{2}\frac{\sigma^j}{2} + \cdots.
\end{aligned}
\qquad (15.64)
$$

Then, since the Pauli matrices are traceless and satisfy $\operatorname{tr}[\sigma^i\sigma^j] = 2\delta^{ij}$,

$$\operatorname{tr} U(\epsilon) = 2 - \frac{1}{4}\epsilon^2(\beta^i)^2 + \mathcal{O}(\epsilon^3). \qquad (15.65)$$

Applying this formula to Eq. (15.62), we find

$$\operatorname{tr} U_P(x,x) = 2 - \frac{1}{4}g^2\epsilon^4(F_{12}^i)^2 + \mathcal{O}(\epsilon^5). \qquad (15.66)$$

Thus the gauge invariance of $(F_{\mu\nu}^i)^2$ can be derived from a geometrical argument, just as in the Abelian case. Using the notation that will be introduced in the next section, one can show that the same argument goes through for any gauge group.

15.4 Basic Facts about Lie Algebras

At the end of Section 15.2 we saw that the class of non-Abelian gauge theories is very large. To work with these theories most efficiently, it is worthwhile to pause and consider the general properties of the continuous groups on which they are based. In this section we will enumerate all the possible groups that can be used to construct non-Abelian gauge theories. We will then compute some numerical factors, built out of group transformation matrices, that are needed in performing explicit calculations in quantized gauge theories.*

To a mathematician, a group is made up of abstract entities that obey certain algebraic rules. In quantum mechanics, however, we are interested specifically in groups of unitary operators that act on the vector space of quantum states. We focus our attention on continuously generated groups, that is, groups that contain elements arbitrarily close to the identity, such that the general element can be reached by the repeated action of these infinitesimal elements. Then any infinitesimal group element g can be written

$$g(\alpha) = 1 + i\alpha^a T^a + \mathcal{O}(\alpha^2). \tag{15.67}$$

The coefficients of the infinitesimal group parameters α^a are Hermitian operators T^a, called the *generators* of the symmetry group. A continuous group with this structure is called a *Lie group*.

The set of generators T^a must span the space of infinitesimal group transformations, so the commutator of generators T^a must be a linear combination of generators. Thus the commutation relations of the operators T^a can be written

$$[T^a, T^b] = if^{abc}T^c; \tag{15.68}$$

the numbers f^{abc} are called *structure constants*. The vector space spanned by the generators, with the additional operation of commutation, is called a *Lie Algebra*.

The commutation relations (15.68) and the identity

$$[T^a, [T^b, T^c]] + [T^b, [T^c, T^a]] + [T^c, [T^a, T^b]] = 0 \tag{15.69}$$

imply that the structure constants obey

$$f^{ade} f^{bcd} + f^{bde} f^{cad} + f^{cde} f^{abd} = 0, \tag{15.70}$$

called the *Jacobi identity*. From the mathematician's viewpoint (considering the generators to be abstract entities rather than Hermitian operators), the

---

*In this section we will state, without proof, some general results from the theory of continuous groups. There are many excellent books that review these mathematical results systematically. Among these, we recommend especially Cahn (1984), for a brief but incisive discussion, and S. Helgason, *Differential Geometry, Lie Groups, and Symmetric Spaces* (Academic Press, 1978), which gives an elegant and rigorous account. R. Slansky, *Phys. Repts.* **79**, 1 (1981), has compiled an especially useful set of tables of group-theoretic identities relevant to the construction of non-Abelian gauge theories.

Jacobi identity is an axiom that must be satisfied in order for a given set of commutation rules to define a Lie algebra.

The commutation relations of the Lie algebra completely determine the group multiplication law of an associated Lie group sufficiently close to the identity. For large enough transformations, additional global questions come into play; to give a familiar example, $SU(2)$ and $O(3)$ have the same commutation relations but different global structure. However, the Lagrangian of a non-Abelian gauge theory depends only on the Lie algebra of the local symmetry group, so we will ignore these global questions from here on.

Classification of Lie Algebras

For the application to gauge theories, the local symmetry is normally a unitary transformation of a set of fields. Thus we are primarily interested in Lie algebras that have finite-dimensional Hermitian representations, leading to finite-dimensional unitary representations of the corresponding Lie group. We will also assume that the number of generators is finite. Such Lie algebras are called *compact*, because these conditions imply that the Lie group is a finite-dimensional compact manifold.

If one of the generators T^a commutes with all of the others, it generates an independent continuous Abelian group. Such a group, which has the structure of the group of phase rotations

$$\psi \to e^{i\alpha}\psi, \tag{15.71}$$

we call $U(1)$. If the algebra contains no such commuting elements, so that the group contains no $U(1)$ factors, then we call the algebra *semi-simple*. If, in addition, the Lie algebra cannot be divided into two mutually commuting sets of generators, the algebra is *simple*. A general Lie algebra is the direct sum of non-Abelian simple components and additional Abelian generators.

Surprisingly, the basic conditions that a Lie algebra be compact and simple turn out to be extremely restrictive. In one of the triumphs of nineteenth-century mathematics, Killing and Cartan classified all possible compact simple Lie algebras. Almost all of these algebras belong to one of three infinite families, with only five exceptions. The three infinite families are the algebras corresponding to the so-called *classical groups*, whose structures are conveniently defined in terms of particular matrix representations. The definitions of the three families of classical groups are as follows:

1. *Unitary transformations of N-dimensional vectors.* Let ξ and η be complex N-vectors. A general linear transformation then has the form

$$\eta_a \to U_{ab}\eta_b, \qquad \xi_a \to U_{ab}\xi_b. \tag{15.72}$$

We say that this transformation is *unitary* if it preserves the inner product $\eta_a^*\xi_a$. The pure phase transformations

$$\xi_a \to e^{i\alpha}\xi_a \tag{15.73}$$

form a $U(1)$ subgroup which commutes with all other unitary transformations; we remove this subgroup to form a simple Lie group, called $SU(N)$; it consists of all $N \times N$ unitary transformations satisfying $\det(U) = 1$. The generators of $SU(N)$ are represented by $N \times N$ Hermitian matrices t^a, subject to the condition that they be orthogonal to the generator of (15.73):

$$\text{tr}[t^a] = 0. \tag{15.74}$$

There are $N^2 - 1$ independent matrices satisfying these conditions.

2. *Orthogonal transformations of N-dimensional vectors.* This is the subgroup of unitary $N \times N$ transformations that preserves the symmetric inner product

$$\eta_a E_{ab} \xi_b, \qquad \text{with } E_{ab} = \delta_{ab}. \tag{15.75}$$

This is the usual vector product, and so this group is the rotation group in N dimensions, $SO(N)$. (Adding the reflection gives the group $O(N)$.) There is an independent rotation corresponding to each plane in N dimensions, so $SO(N)$ has $N(N-1)/2$ generators.

3. *Symplectic transformations of N-dimensional vectors.* This is the subgroup of unitary $N \times N$ transformations, for N even, that preserves the antisymmetric inner product

$$\eta_a E_{ab} \xi_b, \qquad \text{with } E_{ab} = \begin{pmatrix} 0 & 1 \\ -1 & 0 \end{pmatrix}, \tag{15.76}$$

where the elements of the matrix are $N/2 \times N/2$ blocks. This group is called $Sp(N)$; it has $N(N+1)/2$ generators.

Beyond these three families, there are five more *exceptional* Lie algebras, denoted in Cartan's classification system as G_2, F_4, E_6, E_7, and E_8. Of these, E_6 and E_8 have been applied as local symmetry groups in interesting unified models of the fundamental interactions. However, we will not consider these exceptional groups further in this book. In fact, most of our examples will involve only $SU(N)$ groups.

Representations

Once we have specified the local symmetry group, the fields that appear in the Lagrangian most naturally transform according to a finite-dimensional unitary representation of this group. Thus we might next ask how to systematically find all such representations of any given Lie group. Recall that for the group $SU(2)$, the representations can be constructed directly from the commutation relations, using the raising and lowering operators J_+ and J_-. This construction can be generalized to find the finite-dimensional representations of any compact Lie algebra. In this book, however, we will work with relatively simple representations whose structure we can work out by less formal methods.

Before discussing representations of Lie algebras, we should review some general aspects of group representations. Given a symmetry group G, a finite-dimensional unitary representation of the group's Lie algebra is a set of $d \times d$ Hermitian matrices t^a that satisfy the commutation relations (15.68). The size d is the *dimension* of the representation. An arbitrary representation can generally be decomposed by finding a basis in which all representation matrices are simultaneously block-diagonal. Through this change of basis, we can write the representation as the direct sum of *irreducible* representations. We denote the representation matrices in the irreducible representation r by t_r^a.

It is standard practice to adopt a normalization convention for the matrices t_r^a, based on traces of their products. If the Lie algebra is semi-simple, the matrices t_r^a themselves are traceless. Consider, however, the trace of the product of two generator matrices:

$$\mathrm{tr}[t_r^a t_r^b] \equiv D^{ab}. \tag{15.77}$$

As long as the generator matrices are Hermitian, the matrix D^{ab} is positive definite. Let us choose a basis for the generators T^a so that this matrix is proportional to the identity. It can be shown that, once this is done for one irreducible representation, it is true for all irreducible representations. Thus, in this basis,

$$\mathrm{tr}[t_r^a t_r^b] = C(r)\delta^{ab}, \tag{15.78}$$

where $C(r)$ is a constant for each representation r. Equation (15.78) and the commutation relations (15.68) yield the following representation of the structure constants:

$$f^{abc} = -\frac{i}{C(r)}\, \mathrm{tr}\left\{ [t_r^a, t_r^b] t_r^c \right\}. \tag{15.79}$$

This equation implies that f^{abc} is totally antisymmetric.

For each irreducible representation r of G, there is an associated *conjugate* representation $\bar{r}$. The representation r yields the infinitesimal transformation

$$\phi \rightarrow (1 + i\alpha^a t_r^a)\phi. \tag{15.80}$$

The complex conjugate of this transformation,

$$\phi^* \rightarrow (1 - i\alpha^a (t_r^a)^*)\phi^*, \tag{15.81}$$

must also be the infinitesimal element of a representation of G. Thus the conjugate representation to r has representation matrices

$$t_{\bar{r}}^a = -(t_r^a)^* = -(t_r^a)^T. \tag{15.82}$$

Since $\phi^*\phi$ is invariant to unitary transformations, it is possible to combine fields transforming in the representations r and $\bar{r}$ to form a group invariant.

It is possible that the representation $\bar{r}$ may be equivalent to r, if there is a unitary transformation U such that $t_{\bar{r}}^a = U t_r^a U^\dagger$. If so, the representation r is *real*. In this case, there is a matrix G_{ab} such that, if η and ξ belong to the representation r, the combination $G_{ab}\eta_a\xi_b$ is an invariant. It is sometimes useful to distinguish the case in which G_{ab} is symmetric from that in which G_{ab} is antisymmetric. In the former case the representation is *strictly real*; in the latter case it is *pseudoreal*. Both cases occur already in $SU(2)$: The invariant combination of two vectors is $v_a w_a$, so the vector is a real representation; the invariant combination of two spinors is $\epsilon^{\alpha\beta}\eta_\alpha \xi_\beta$, so the spinor is a pseudoreal representation.

With this language we can discuss the simplest representations of the classical groups. In $SU(N)$, the basic irreducible representation (often called the *fundamental* representation) is the N-dimensional complex vector. For $N > 2$ this representation is complex, so that there is a second, inequivalent, representation $\overline{N}$. (In $SU(2)$ this representation is the pseudoreal spinor representation.) In $SO(N)$, the basic N-dimensional vector is a (strictly) real representation. In $Sp(N)$, the N-dimensional vector is a pseudoreal representation.

Another irreducible representation, present for any simple Lie algebra, is the one to which the generators of the algebra belong. This representation is called the *adjoint representation* and denoted by $r = G$. The representation matrices are given by the structure constants:

$$(t_G^b)_{ac} = i f^{abc}. \tag{15.83}$$

With this definition, the statement that t_G^a satisfies the Lie algebra

$$\left([t_G^b, t_G^c]\right)_{ae} = i f^{bcd}(t_G^d)_{ae} \tag{15.84}$$

is just a rewriting of the Jacobi identity (15.70). Since the structure constants are real and antisymmetric, $t_G^a = -(t_G^a)^*$; thus the adjoint representation is always a real representation. From the descriptions of the Lie groups given above, the dimension of the adjoint representation $d(G)$ is given, for the classical groups, by

$$d(G) = \begin{cases} N^2 - 1 & \text{for } SU(N), \\ N(N-1)/2 & \text{for } SO(N), \\ N(N+1)/2 & \text{for } Sp(N). \end{cases} \tag{15.85}$$

The identification of f^{abc} as a representation matrix allows us to gain further insight into some of the quantities introduced in Section 15.2. The covariant derivative acting on a field in the adjoint representation is

$$\begin{aligned} (D_\mu \phi)_a &= \partial_\mu \phi_a - i g A_\mu^b (t_G^b)_{ac} \phi_c \\ &= \partial_\mu \phi_a + g f^{abc} A_\mu^b \phi_c. \end{aligned} \tag{15.86}$$

Thus we can recognize the infinitesimal form of the gauge transformation of the vector field in (15.46) as the motion

$$A_\mu^a \rightarrow A_\mu^a + \frac{1}{g}(D_\mu \alpha)^a. \tag{15.87}$$

The gauge field equation of motion (15.51) can be rewritten as

$$(D^\mu F_{\mu\nu})^a = -gj^a_\nu. \tag{15.88}$$

In both of these expressions, the arbitrary-looking terms involving f^{abc} arise naturally as part of a covariant derivative. An additional identity follows from considering the antisymmetric double commutator of covariant derivatives,

$$\epsilon^{\mu\nu\lambda\sigma}[D_\nu, [D_\lambda, D_\sigma]].$$

This quantity vanishes by its total antisymmetry, in the same way as (15.69). This result can be reduced to the identity

$$\epsilon^{\mu\nu\lambda\sigma}(D_\nu F_{\lambda\sigma})^a = 0. \tag{15.89}$$

This equation, called the *Bianchi identity* of a non-Abelian gauge theory, is the analogue of the homogeneous Maxwell equations in electrodynamics.

The Casimir Operator

In $SU(2)$, we characterize representations by the eigenvalue of the total spin J^2. In fact, for any simple Lie algebra, the operator

$$T^2 = T^a T^a \tag{15.90}$$

(with the repeated index summed, as always) commutes with all group generators:

$$\begin{aligned} [T^b, T^a T^a] &= (if^{bac}T^c)T^a + T^a(if^{bac}T^c) \\ &= if^{bac}\{T^c, T^a\}, \end{aligned} \tag{15.91}$$

which vanishes by the antisymmetry of f^{abc}. In other words, T^2 is an invariant of the algebra; this implies that T^2 takes a constant value on each irreducible representation. Thus, the matrix representation of T^2 is proportional to the unit matrix:

$$t^a_r t^a_r = C_2(r) \cdot \mathbf{1}, \tag{15.92}$$

where $\mathbf{1}$ is the $d(r) \times d(r)$ unit matrix and $C_2(r)$ is a constant, called the *quadratic Casimir operator*, for each representation. For the adjoint representation, Eq. (15.92) is more conveniently written as

$$f^{acd} f^{bcd} = C_2(G)\delta^{ab}. \tag{15.93}$$

Casimir operators appear very often in computations in non-Abelian gauge theories. Furthermore, the related invariant $C(r)$ given by (15.78) is simply related to the Casimir operator: If we contract (15.78) with δ^{ab} and evaluate the left-hand side using (15.92), we find

$$d(r)C_2(r) = d(G)C(r). \tag{15.94}$$

Thus it will be useful for us to compute $C_2(r)$ for the simplest $SU(N)$ representations.

For $SU(2)$, the fundamental two-dimensional representation is the spinor representation, which is given in terms of Pauli matrices by

$$t_2^a = \frac{\sigma^a}{2}. \tag{15.95}$$

These satisfy $\text{tr}[t_2^a t_2^b] = \frac{1}{2}\delta^{ab}$. We will choose the generators of $SU(N)$ so that three of these are the generators (15.95), acting on the first two components of the N-vector ξ. Then, for any matrices of the fundamental representation,

$$\text{tr}[t_N^a t_N^b] = \frac{1}{2}\delta^{ab}. \tag{15.96}$$

This convention fixes the values of $C(r)$ and $C_2(r)$ for all of the irreducible representations of $SU(N)$. For the fundamental representations N and $\overline{N}$, $C(N)$ is given directly by (15.96), and $C_2(N)$ follows from (15.94). We find

$$C(N) = \frac{1}{2}, \qquad C_2(N) = \frac{N^2 - 1}{2N}. \tag{15.97}$$

To compute the Casimir operator for the adjoint representation, we build up this representation from the product of the N and $\overline{N}$. Let us first discuss the product of irreducible representations more generally. The direct product of two representations r_1, r_2 is a representation of dimension $d(r_1) \cdot d(r_2)$. An object that transforms according to this representation can be written as a tensor Ξ_{pq}, in which the first index transforms according to r_1, the second according to r_2. In general, such a product can be decomposed into a direct sum of irreducible representations; symbolically, we write

$$r_1 \times r_2 = \sum_i r_i. \tag{15.98}$$

The representation matrices in the representation $r_1 \times r_2$ are

$$t_{r_1 \times r_2}^a = t_{r_1}^a \otimes 1 + 1 \otimes t_{r_2}^a, \tag{15.99}$$

where the first matrix of each product acts on the first index of Ξ_{pq} and the second matrix acts on the second index.

The Casimir operator in the product representation is

$$(t_{r_1 \times r_2}^a)^2 = (t_{r_1}^a)^2 \otimes 1 + 2t_{r_1}^a \otimes t_{r_2}^a + 1 \otimes (t_{r_2}^a)^2.$$

Take the trace; since the matrices t_r^a are traceless, the trace of the second term on the right is zero. Then

$$\text{tr}(t_{r_1 \times r_2}^a)^2 = (C_2(r_1) + C_2(r_2))d(r_1)d(r_2). \tag{15.100}$$

On the other hand, the decomposition (15.98) implies

$$\text{tr}(t_{r_1 \times r_2}^a)^2 = \sum C_2(r_i)d(r_i). \tag{15.101}$$

Equating (15.100) and (15.101), we find a useful identity for $C_2(r)$.

Now apply this identity to the product of the N and $\overline{N}$ representations of $SU(N)$. In this case, the tensor Ξ_{pq} can contain a term proportional to the invariant δ_{pq}. The remaining $(N^2 - 1)$ independent components of Ξ_{pq}

transform as a general traceless $N \times N$ tensor; the matrices that effect these transformations make up the adjoint representation of $SU(N)$. In this case Eq. (15.98) becomes explicitly

$$N \times \overline{N} = 1 + (N^2 - 1). \tag{15.102}$$

For this decomposition, Eqs. (15.100) and (15.101) imply the identity

$$\left(2 \cdot \frac{N^2 - 1}{2N}\right) N^2 = 0 + C_2(G) \cdot (N^2 - 1). \tag{15.103}$$

Thus, for $SU(N)$,

$$C_2(G) = C(G) = N. \tag{15.104}$$

Some additional examples of the computation of quadratic Casimir operators are given in Problem 15.5. However, the examples we have discussed in this section, combined with the basic group-theoretic concepts that we have reviewed, already provide enough material to carry out the most important computations of physical interest in non-Abelian gauge theories.

Problems

15.1 Brute-force computations in $SU(3)$. The standard basis for the fundamental representation of $SU(3)$ is

$$t^1 = \frac{1}{2}\begin{pmatrix} 0 & 1 & 0 \\ 1 & 0 & 0 \\ 0 & 0 & 0 \end{pmatrix}, \quad t^2 = \frac{1}{2}\begin{pmatrix} 0 & -i & 0 \\ i & 0 & 0 \\ 0 & 0 & 0 \end{pmatrix}, \quad t^3 = \frac{1}{2}\begin{pmatrix} 1 & 0 & 0 \\ 0 & -1 & 0 \\ 0 & 0 & 0 \end{pmatrix},$$

$$t^4 = \frac{1}{2}\begin{pmatrix} 0 & 0 & 1 \\ 0 & 0 & 0 \\ 1 & 0 & 0 \end{pmatrix}, \quad t^5 = \frac{1}{2}\begin{pmatrix} 0 & 0 & -i \\ 0 & 0 & 0 \\ i & 0 & 0 \end{pmatrix},$$

$$t^6 = \frac{1}{2}\begin{pmatrix} 0 & 0 & 0 \\ 0 & 0 & 1 \\ 0 & 1 & 0 \end{pmatrix}, \quad t^7 = \frac{1}{2}\begin{pmatrix} 0 & 0 & 0 \\ 0 & 0 & -i \\ 0 & i & 0 \end{pmatrix}, \quad t^8 = \frac{1}{2\sqrt{3}}\begin{pmatrix} 1 & 0 & 0 \\ 0 & 1 & 0 \\ 0 & 0 & -2 \end{pmatrix}.$$

(a) Explain why there are exactly eight matrices in the basis.

(b) Evaluate all the commutators of these matrices, to determine the structure constants of $SU(3)$. Show that, with the normalizations used here, f^{abc} is totally antisymmetric. (This exercise is tedious; you may wish to check only a representative sample of the commutators.)

(c) Check the orthogonality condition (15.78), and evaluate the constant $C(r)$ for this representation.

(d) Compute the quadratic Casimir operator $C_2(r)$ directly from its definition (15.92), and verify the relation (15.94) between $C_2(r)$ and $C(r)$.

15.2 Write down the basis matrices of the adjoint representation of $SU(2)$. Compute $C(G)$ and $C_2(G)$ directly from their definitions (15.78) and (15.92).

15.3 Coulomb potential.

(a) Using functional integration, compute the expectation value of the Wilson loop in pure quantum electrodynamics without fermions. Show that

$$\langle U_P(z,z) \rangle = \exp\left[-e^2 \oint_P dx^\mu \oint_P dy^\nu g_{\mu\nu} \frac{1}{8\pi^2(x-y)^2}\right],$$

with x and y integrated around the closed curve P.

(b) Consider the Wilson loop of a rectangular path of (spacelike) width R and (timelike) length T, $T \gg R$. Compute the expectation value of the Wilson loop in this limit and compare to the general expression for time evolution,

$$\langle U_P \rangle = \exp[-iE(R)T],$$

where $E(R)$ is the energy of the electromagnetic sources corresponding to the Wilson loop. Show that the potential energy of these sources is just the Coulomb potential, $V(R) = -e^2/4\pi R$.

(c) Assuming that the propagator of the non-Abelian gauge field is given by the Feynman gauge expression

$$\langle A_\mu^a(x) A_\mu^b(y) \rangle = \int \frac{d^4p}{(2\pi)^4} \frac{-ig_{\mu\nu}\delta^{ab}}{p^2} e^{-ip\cdot(x-y)},$$

compute the expectation value of a non-Abelian Wilson loop to order g^2. The result will depend on the representation r of the gauge group in which one chooses the matrices that appear in the exponential. Show that, to this order, the Coulomb potential of the non-Abelian gauge theory is $V(R) = -g^2 C_2(r)/4\pi R$.

15.4 Scalar propagator in a gauge theory. Consider the equation for the Green's function of the Klein-Gordon equation:

$$(\partial^2 + m^2)D_F(x,y) = -i\delta^{(4)}(x-y).$$

We can find an interesting representation for this Green's function by writing

$$D_F(x,y) = \int_0^\infty dT\, D(x,y,T),$$

where $D(x,y,T)$ satisfies the Schrödinger equation

$$\left[i\frac{\partial}{\partial T} - (\partial^2 + m^2)\right] D(x,y,T) = i\delta(T)\delta^{(4)}(x-y).$$

Now, represent $D(x,y,T)$ using the functional integral solution of the Schrödinger equation presented in Section 9.1.

(a) Using the explicit formula of the propagator of the Schrödinger equation, show that this integral formula gives the standard expression for the Feynman propagator.

(b) Using the method just described, show that the expression

$$D_F(x,y) = \int_0^\infty dT \int \mathcal{D}x \, \exp\left[i \int dt \, \frac{1}{2}\left(\left(\frac{dx^\mu}{dt}\right)^2 - m^2 \right) - ie \int dt \, \frac{dx^\mu}{dt} A_\mu(x) \right]$$

is a functional integral representation for the scalar field propagator in an arbitrary background electromagnetic field. Show, in particular, that the functional integral satisfies the relevant Schrödinger equation. Notice that this integral depends on A_μ through the Wilson line.

(c) Generalize this expression to a non-Abelian gauge theory. Show that the functional integral solves the relevant Schrödinger equation only if the group matrices in the exponential for the Wilson line are path-ordered.

15.5 Casimir operator computations. An alternative strategy for computing the quadratic Casimir operator is to compute $C(r)$ in the formula

$$\text{tr}[t_r^a t_r^b] = C(r)\delta^{ab}$$

by choosing t^a and t^b to lie in an $SU(2)$ subgroup of the gauge group.

(a) Under an $SU(2)$ subgroup of a general group G, an irreducible representation r of G will decompose into a sum of representations of $SU(2)$:

$$r \to \sum j_i,$$

where the j_i are the spins of $SU(2)$ representations. Show that

$$3C(r) = \sum_i j_i(j_i + 1)(2j_i + 1).$$

(b) Under an $SU(2)$ subgroup of $SU(N)$, the fundamental representation N transforms as a 2-component spinor $(j = \frac{1}{2})$ and $(N-2)$ singlets. Use this relation to check the formula $C(N) = \frac{1}{2}$. Show that the adjoint representation of $SU(N)$ decomposes into one spin 1, $2(N-2)$ spin-$\frac{1}{2}$'s, plus singlets, and use this decomposition to check that $C(G) = N$.

(c) Symmetric and antisymmetric 2-index tensors form irreducible representations of $SU(N)$. Compute $C_2(r)$ for each of these representations. The direct sum of these representations is the product representation $N \times N$. Verify that your results for $C_2(r)$ satisfy the identity for product representations that follows from Eqs. (15.100) and (15.101).

Chapter 16

Quantization of Non-Abelian Gauge Theories

The previous chapter showed how to construct Lagrangians with non-Abelian gauge symmetry. However, this is only the first step in the process of relating the idea of non-Abelian gauge invariance to the real interactions of particle physics. We must next work out the rules for computing Feynman diagrams containing the non-Abelian gauge vector particles, then use these rules to compute scattering amplitudes and cross sections. This chapter will develop the technology needed for such calculations.

Alongside this technical discussion, we will study how the gauge symmetry affects the Feynman amplitudes. In any theory with a local symmetry, some degrees of freedom of the fields that appear in the Lagrangian are *unphysical*, in the sense that they can be adjusted arbitrarily by gauge transformations. In electrodynamics, the components of the field $A_\mu(k)$ proportional to k^μ lie along the symmetry directions. We saw in Section 9.4 that this fact has two important consequences. First, the propagator of the field A_μ is ambiguous; there are multiple expressions for the propagator, which follow equally well from the QED Lagrangian. Second, the vertices of electrodynamics are such that this ambiguity makes no difference in the calculation of cross sections. For example, Eq. (9.58) displays a continuous family of photon propagators, one for each value of the continuous parameter ξ; but we saw immediately that all dependence of S-matrix elements on ξ is eliminated by the Ward identity. Non-Abelian gauge theories contain similar ambiguities and cancellations, but, as we will see in this chapter, the structure of the cancellations is more intricate.

An additional goal of this chapter is to compute the Callan-Symanzik β function, and hence determine the behavior of the running coupling constant, for non-Abelian gauge theories. As discussed in Chapter 14, these theories are in fact *asymptotically free*: The coupling constant becomes weak at large momenta. This result indicates the applicability of non-Abelian gauge theory to model the strong interactions. We will be able to derive this result once we have determined the correct Feynman rules for non-Abelian gauge theories.

16.1 Interactions of Non-Abelian Gauge Bosons

Most of the Feynman rules for non-Abelian gauge theory can be read directly from the Yang-Mills Lagrangian, following the method of Section 9.2. However, when we quantized the electromagnetic field in Section 9.4, we saw that the functional integral over a gauge field must be defined carefully, and that the subtle aspects of this construction can introduce new ingredients into the quantum theory. In this section we will see how far we can go in the non-Abelian theory by ignoring these subtleties. In Section 16.2 we will carry out a more proper derivation of the Feynman rules, through a careful analysis of the functional integral.

Feynman Rules for Fermions and Gauge Bosons

The Yang-Mills Lagrangian, as derived in the previous chapter, is

$$\mathcal{L} = -\frac{1}{4}(F_{\mu\nu}^a)^2 + \bar{\psi}(i\slashed{D} - m)\psi, \tag{16.1}$$

where the index a is summed over the generators of the gauge group G, and the fermion multiplet ψ belongs to an irreducible representation r of G. The field strength is

$$F_{\mu\nu}^a = \partial_\mu A_\nu^a - \partial_\nu A_\mu^a + g f^{abc} A_\mu^b A_\nu^c, \tag{16.2}$$

where f^{abc} are the structure constants of G. The covariant derivative is defined in terms of the representation matrices t_r^a by

$$D_\mu = \partial_\mu - ig A_\mu^a t_r^a. \tag{16.3}$$

From now on we will drop the subscript r except where it is needed for clarity.

The Feynman rules for this Lagrangian can be derived from a functional integral over the fields ψ, $\bar{\psi}$, and A_μ^a. Imagine expanding the functional integral in perturbation theory, starting with the free Lagrangian, at $g = 0$. The free theory contains a number of free fermions equal to the dimension $d(r)$ of the representation r, and a number of free vector bosons equal to the number $d(G)$ of generators of G. Using the methods of Section 9.5, it is straightforward to derive the fermion propagator

$$\langle \psi_{i\alpha}(x)\bar{\psi}_{j\beta}(y) \rangle = \int \frac{d^4k}{(2\pi)^4}\left(\frac{i}{\slashed{k} - m}\right)_{\alpha\beta} \delta_{ij}\, e^{-ik\cdot(x-y)}, \tag{16.4}$$

where α, β are Dirac indices and i, j are indices of the symmetry group: $i, j = 1, \ldots, d(r)$. In analogy with electrodynamics, we would guess that the propagator of the vector fields is

$$\langle A_\mu^a(x)A_\nu^b(y) \rangle = \int \frac{d^4k}{(2\pi)^4}\left(\frac{-ig_{\mu\nu}}{k^2}\right)\delta^{ab}\, e^{-ik\cdot(x-y)}, \tag{16.5}$$

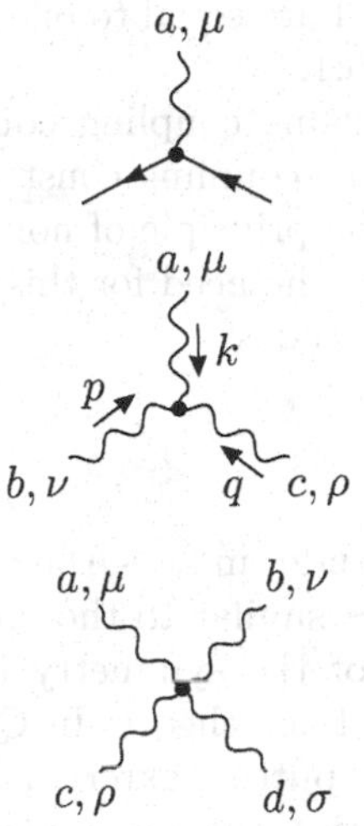

$$= \quad ig\gamma^\mu t^a$$

$$=\quad \begin{aligned} gf^{abc}\big[&g^{\mu\nu}(k-p)^\rho\\ &+g^{\nu\rho}(p-q)^\mu\\ &+g^{\rho\mu}(q-k)^\nu\big]\end{aligned}$$

$$=\quad \begin{aligned} -ig^2\big[&f^{abe}f^{cde}(g^{\mu\rho}g^{\nu\sigma}-g^{\mu\sigma}g^{\nu\rho})\\ &+f^{ace}f^{bde}(g^{\mu\nu}g^{\rho\sigma}-g^{\mu\sigma}g^{\nu\rho})\\ &+f^{ade}f^{bce}(g^{\mu\nu}g^{\rho\sigma}-g^{\mu\rho}g^{\nu\sigma})\big]\end{aligned}$$

Figure 16.1. Feynman rules for fermion and gauge boson vertices of a non-Abelian gauge theory.

with $a, b = 1, \ldots, d(G)$. We will derive this formula in the next section.

To find the vertices, we write out the nonlinear terms in (16.1). If $\mathcal{L}_0$ is the free field Lagrangian, then

$$\mathcal{L} = \mathcal{L}_0 + gA_\lambda^a\bar{\psi}\gamma^\lambda t^a\psi - gf^{abc}(\partial_\kappa A_\lambda^a)A^{\kappa b}A^{\lambda c}$$
$$- \tfrac{1}{4}g^2(f^{eab}A_\kappa^a A_\lambda^b)(f^{ecd}A^{\kappa c}A^{\lambda d}). \tag{16.6}$$

The first of the three nonlinear terms gives the fermion-gauge boson vertex

$$ig\gamma^\mu t^a; \tag{16.7}$$

this is a matrix that acts on the Dirac and gauge indices of the fermions. The second nonlinear term leads to a three gauge boson vertex. To work out this vertex, we first choose a definite convention for the external momenta and Lorentz and gauge indices. A suitable convention is shown in Fig. 16.1, with all momenta pointing inward. Consider first contracting the external gauge particle with momentum k to the first factor of A_μ^a, the gauge particle with momentum p to the second, and the gauge particle of momentum q to the third. The derivative contributes a factor $(-ik_\kappa)$ if the momentum points into the diagram. Then this contribution is

$$-igf^{abc}(-ik^\nu)g^{\mu\rho}. \tag{16.8}$$

In all, there are 3! possible contractions, which alternate in sign according to the total antisymmetry of f^{abc}. The sum of these is exhibited in Fig. 16.1. Finally, the last term of (16.6) leads to a four gauge boson vertex. Following the conventions of Fig. 16.1, one possible contraction gives the contribution

$$-ig^2f^{eab}f^{ecd}g^{\mu\rho}g^{\nu\sigma}. \tag{16.9}$$

There are 4! possible contractions, of which sets of 4 are equal to one another. The sum of these contributions is shown in Fig. 16.1.

Notice that all of these vertices involve the same coupling constant g. We derived the vertices, and thus the equality of the coupling constants, as a part of our construction of the Lagrangian from the principle of non-Abelian gauge invariance. However, it is also possible to see the need for this equality *a posteriori*, from the properties of Feynman amplitudes.

Equality of Coupling Constants

One property that we expect from Feynman amplitudes in non-Abelian gauge theories is that they should satisfy Ward identities similar to those of QED. These Ward identities express the conservation of the symmetry currents, which follows already from the global symmetry of the theory. In QED, the simplest form of the Ward identity was obtained by putting external electrons and positrons on shell. In non-Abelian gauge theories, the gauge bosons also carry charge and so these must also be put on shell to remove contact terms. With all external particles on shell, the amplitude for production of a gauge boson should obey

$$k^{\mu}\left(\text{—} \right) = 0. \tag{16.10}$$

This identity is not only an indication of the local gauge symmetry, but is physically important in its own right. Like the photon, the non-Abelian gauge boson has only two physical polarization states. In QED, the on-shell Ward identity expressed the fact that the orthogonal, unphysical polarization states are not produced in scattering processes. The on-shell Ward identity will play a similar role in the non-Abelian case.

Let us check the Ward identity in a simple case, the lowest-order diagrams contributing to fermion-antifermion annihilation into a pair of gauge bosons. In order g^2, there are three diagrams, shown in Fig. 16.2. The first two diagrams are similar to the QED diagrams that we studied in Section 5.5; they sum to

$$i\mathcal{M}^{\mu\nu}_{1,2}\epsilon^{*}_{\mu}(k_1)\epsilon^{*}_{\nu}(k_2) = (ig)^2 \bar{v}(p_+)\left\{ \gamma^{\mu}t^a \frac{i}{\not{p} - \not{k}_2 - m}\gamma^{\nu}t^b \right.$$

$$\left. + \gamma^{\nu}t^b \frac{i}{\not{k}_2 - \not{p}_+ - m}\gamma^{\mu}t^a \right\} u(p)\, \epsilon^{*}_{\mu}(k_1)\epsilon^{*}_{\nu}(k_2).$$

$$\tag{16.11}$$

The vectors $\epsilon(k_i)$ are the gauge boson polarization vectors; for physical polarizations, these satisfy $k^{\mu}_i \epsilon_{\mu}(k_i) = 0$. To check the Ward identity (16.10), we

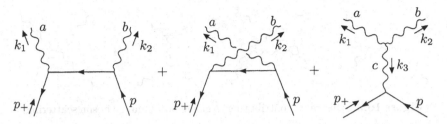

Figure 16.2. Diagrams contributing to fermion-antifermion annihilation to two gauge bosons.

replace $\epsilon_\nu^*(k_2)$ in (16.11) by $k_{2\nu}$. This gives

$$i\mathcal{M}_{1,2}^{\mu\nu}\epsilon_{1\mu}^* k_{2\nu} = (ig)^2 \bar{v}(p_+)\left\{\gamma^\mu t^a \frac{i}{\not{p} - \not{k}_2 - m} \not{k}_2 t^b \right.$$
$$\left. + \not{k}_2 t^b \frac{i}{\not{k}_2 - \not{p}_+ - m}\gamma^\mu t^a \right\} u(p)\, \epsilon_{1\mu}^*. \tag{16.12}$$

Since

$$(\not{p} - m)u(p) = 0 \qquad \text{and} \qquad \bar{v}(p_+)(-\not{p}_+ - m) = 0, \tag{16.13}$$

we can add these quantities to $\not{k}_2$ in the first and second terms of (16.12), to cancel the denominators. This gives

$$i\mathcal{M}_{1,2}^{\mu\nu}\epsilon_{1\mu}^* k_{2\nu} = (ig)^2 \bar{v}(p_+)\left\{-i\gamma^\mu [t^a, t^b]\right\} u(p)\, \epsilon_{1\mu}^*. \tag{16.14}$$

In the Abelian case, this expression would vanish. In the non-Abelian case, however, the residual term is nonzero and depends on the commutator of gauge group generators:

$$i\mathcal{M}_{1,2}^{\mu\nu}\epsilon_{1\mu}^* k_{2\nu} = -g^2 \bar{v}(p_+)\gamma^\mu u(p)\, \epsilon_{1\mu}^* \cdot f^{abc}t^c. \tag{16.15}$$

We need to find another contribution to cancel this term. Notice, however, that this term has the group index structure of a fermion-gauge boson vertex $(g\gamma^\mu t^c)$ multiplied by a three gauge boson vertex (gf^{abc}). This is just the structure of the third diagram in Fig. 16.2.

To check that the cancellation works, let us evaluate the third diagram:

$$i\mathcal{M}_3^{\mu\nu}\epsilon_{1\mu}^* \epsilon_{2\nu}^* = ig\bar{v}(p_+)\gamma_\rho t^c u(p)\frac{-i}{k_3^2}\epsilon_\mu^*(k_1)\epsilon_\nu^*(k_2)$$
$$\times gf^{abc}\left[g^{\mu\nu}(k_2 - k_1)^\rho + g^{\nu\rho}(k_3 - k_2)^\mu + g^{\rho\mu}(k_1 - k_3)^\nu\right],$$

with $k_3 = -k_1 - k_2$. If we replace $\epsilon_\nu^*(k_2)$ with $k_{2\nu}$, then eliminate k_2 using momentum conservation, the expression in brackets simplifies as follows:

$$\epsilon_\nu^*(k_2)\left[g^{\mu\nu}(k_2 - k_1)^\rho + g^{\nu\rho}(k_3 - k_2)^\mu + g^{\rho\mu}(k_1 - k_3)^\nu\right]$$
$$\to k_2^\mu(k_2 - k_1)^\rho + k_2^\rho(k_3 - k_2)^\mu + g^{\rho\mu}(k_1 - k_3)\cdot k_2 \tag{16.16}$$
$$= g^{\rho\mu}k_3^2 - k_3^\rho k_3^\mu - g^{\rho\mu}k_1^2 + k_1^\rho k_1^\mu.$$

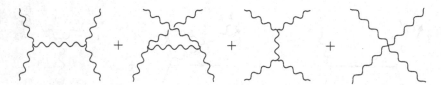

Figure 16.3. Diagrams contributing to gauge boson-gauge boson scattering.

Let us assume that the other gauge boson, with momentum k_1, is on shell ($k_1^2 = 0$), and that it has transverse polarization ($k_1^\mu \epsilon_\mu(k_1) = 0$). Then the third and fourth terms in the last line vanish. Furthermore, the term $k_3^\rho k_3^\mu$ vanishes when it is contracted with the fermion current. In the remaining term, the factor k_3^2 cancels the gauge boson propagator, and we are left with

$$i\mathcal{M}_3^{\mu\nu} \epsilon_{1\mu}^* k_{2\nu} = +g^2 \bar{v}(p_+) \gamma^\mu u(p) \, \epsilon_{1\mu}^* \cdot f^{abc} t^c, \tag{16.17}$$

which precisely cancels (16.15).

Notice that this cancellation takes place only if the value of the coupling constant in the three-boson vertex is identical to that in the fermion-boson vertex. In a similar way, the Ward identity cannot be satisfied among the diagrams for boson-boson scattering, shown in Fig. 16.3, unless the coupling constant g in the four-boson vertex is identical to that in the three-boson vertex. Thus, the coupling constants of all three nonlinear terms in the Yang-Mills Lagrangian *must* be equal in order to preserve the Ward identity and avoid the production of bosons with unphysical polarization states. Conversely, the non-Abelian gauge symmetry guarantees that these couplings *are* equal. The symmetry thus accomplishes exactly what we hoped it would in our discussion at the beginning of Chapter 15, giving us a consistent theory of physical vector particle interactions.

A Flaw in the Argument

The preceding argument has one serious deficiency. At the final stage, we needed to assume that the second gauge boson was transverse. However, one might have expected that this information would come out of the argument rather than having to be put in. In QED, the Feynman diagrams predict that, when an electron and a positron annihilate to form two photons, only the physical transverse polarization states of the photons are produced. Amplitudes to produce other photon polarizations cancel each other to yield zero, as we saw in Eq. (5.80). This statement is not true for the non-Abelian gauge theory Feynman rules that we have worked with so far.

To state the discrepancy more concretely, we introduce some notation. Let $k^\mu = (k^0, \mathbf{k})$ be a lightlike vector: $k^2 = 0$. Then there are two purely spatial vectors orthogonal to $\mathbf{k}$. If k is the momentum of a vector boson, these are the two transverse polarizations. To construct an orthogonal basis, we must include also the longitudinal polarization state, with polarization vector parallel to $\mathbf{k}$, and the timelike polarization state. It is most convenient to work with the two lightlike linear combinations of these states, with polarization vectors parallel to the vectors k^μ and $\tilde{k}^\mu = (k^0, -\mathbf{k})$. These two unphysical polarization states of a massless vector particle can be written as follows:

$$\epsilon_\mu^+(k) = \left(\frac{k^0}{\sqrt{2}|\mathbf{k}|}, \frac{\mathbf{k}}{\sqrt{2}|\mathbf{k}|} \right); \qquad \epsilon_\mu^-(k) = \left(\frac{k^0}{\sqrt{2}|\mathbf{k}|}, -\frac{\mathbf{k}}{\sqrt{2}|\mathbf{k}|} \right). \tag{16.18}$$

We will refer to $\epsilon^+(k)$ and $\epsilon^-(k)$ as the *forward* and *backward* lightlike polarization vectors. Denote the two transverse polarization states $\epsilon_{i\mu}^T(k)$, for $i = 1, 2$. These four polarization vectors obey the orthogonality relations

$$\epsilon_i^T \cdot \epsilon_j^{*T} = -\delta_{ij}, \qquad \epsilon^+ \cdot \epsilon_i^T = \epsilon^- \cdot \epsilon_i^T = 0,$$
$$(\epsilon^+)^2 = (\epsilon^-)^2 = 0, \qquad \epsilon^+ \cdot \epsilon^- = 1. \tag{16.19}$$

They also satisfy the completeness relation

$$g_{\mu\nu} = \epsilon_\mu^- \epsilon_\nu^{+*} + \epsilon_\mu^+ \epsilon_\nu^{-*} - \sum_i \epsilon_{i\mu}^T \epsilon_{i\nu}^{T*}. \tag{16.20}$$

Using this notation, we can express concretely the gap in the argument for the Ward identity. The Feynman diagrams of Fig. 16.2 apparently predict that there is a nonzero amplitude to produce a forward-polarized gauge boson together with a backward-polarized gauge boson. For this case, we substitute $\epsilon_\mu^{-*}(k_1)$ and $\epsilon_\nu^{+*}(k_2)$ for the two polarization vectors. Then the term proportional to $k_1^\rho k_1^\mu$ in Eq. (16.16) no longer vanishes; it now yields

$$i\mathcal{M} = ig\bar{v}(p_+)\gamma_\rho t^c u(p) \frac{-i}{k_3^2} \epsilon_\mu^{-*}(k_1) \cdot \frac{1}{\sqrt{2}|\mathbf{k}_2|} \cdot gf^{abc}k_1^\rho k_1^\mu$$
$$= ig\bar{v}(p_+)\gamma_\rho t^c u(p) \frac{-i}{k_3^2} \cdot gf^{abc}k_1^\rho \cdot \frac{|\mathbf{k}_1|}{|\mathbf{k}_2|}. \tag{16.21}$$

Can we simply ignore this totally unphysical process? We are free to do so in calculations of leading-order amplitudes, but the process will come back to haunt us in loop diagrams. Recall from Section 7.3 how the optical theorem (7.49) links the imaginary part of a loop diagram to the square of a corresponding scattering amplitude, obtained by cutting the diagram across the loop. If we apply the optical theorem to the diagram shown in Fig. 16.4, we obtain a paradox. In the gauge boson loop on the left-hand side we can replace the $g_{\mu\nu}$ factors in the propagators with sums over all four polarization vectors (16.20). The theorem thus implies that all four polarizations, even the unphysical ones, should be included for the final-state gauge bosons on the right-hand side. We are faced with a choice of allowing the production of unphysical states or violating the optical theorem. A third alternative, equally unattractive, would be to discard our expression (16.5) for the gauge boson propagator. Clearly, we are missing some crucial element of the quantum-mechanical structure of non-Abelian gauge theories.

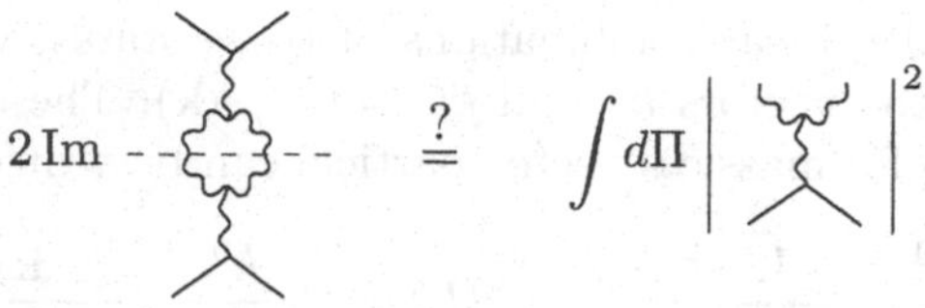

$$2\,\mathrm{Im}\ \ \text{---}\!\!\begin{array}{c}\text{(diagram)}\end{array}\!\!\text{---}\ \ \overset{?}{=}\ \ \int d\Pi\left|\begin{array}{c}\text{(diagram)}\end{array}\right|^2$$

Figure 16.4. A paradox for the optical theorem in gauge theories.

16.2 The Faddeev-Popov Lagrangian

It is not surprising that we have found a problem with our Feynman rules for non-Abelian gauge theories, since we were not very careful in deriving them. In particular, we did not actually derive expression (16.5) for the gauge field propagator. In this section we will remedy this by going through a formal derivation of this expression. We will find that, although expression (16.5) is indeed correct, it is incomplete: It must be supplemented by additional rules of a completely new type.

To define the functional integral for a theory with non-Abelian gauge invariance, we will use the Faddeev-Popov method, as introduced in Section 9.4 to quantize the electromagnetic field. Our present discussion will follow Section 9.4 closely. However, as we have by now come to expect, the case of non-Abelian local symmetry brings with it new tricks and surprises.

First consider the quantization of the pure gauge theory, without fermions. To derive the Feynman rules, we must define the functional integral

$$\int \mathcal{D}A\, \exp\left[i\int d^4x\left(-\tfrac{1}{4}(F^a_{\mu\nu})^2\right)\right]. \tag{16.22}$$

As in the Abelian case, the Lagrangian is unchanged along the infinite number of directions in the space of field configurations corresponding to local gauge transformations. To compute the functional integral we must factor out the integrations along these directions, constraining the remaining integral to a much smaller space.

As in electrodynamics, we will constrain the gauge directions by applying a gauge-fixing condition $G(A) = 0$ at each point x. Following Faddeev and Popov, we can introduce this constraint by inserting into the functional integral the identity (9.53):

$$1 = \int \mathcal{D}\alpha(x)\, \delta\big(G(A^\alpha)\big)\, \det\!\left(\frac{\delta G(A^\alpha)}{\delta\alpha}\right). \tag{16.23}$$

Here A^α is the gauge field A transformed through a finite gauge transformation as in (15.47):

$$(A^\alpha)^a_\mu t^a = e^{i\alpha^a t^a}\big[A^b_\mu t^b + \tfrac{i}{g}\partial_\mu\big]e^{-i\alpha^c t^c}. \tag{16.24}$$

In evaluating the determinant, the infinitesimal form of this transformation will be more useful:

$$(A^\alpha)^a_\mu = A^a_\mu + \frac{1}{g}\partial_\mu\alpha^a + f^{abc}A^b_\mu\alpha^c = A^a_\mu + \frac{1}{g}D_\mu\alpha^a, \qquad (16.25)$$

where D_μ is the covariant derivative (15.86) acting on a field in the adjoint representation. Note that, as long as the gauge-fixing function $G(A)$ is linear, the functional derivative $\delta G(A^\alpha)/\delta\alpha$ is independent of α.

Since the Lagrangian is gauge invariant, we can replace A by A^α in the exponential of (16.22). Then, as in the Abelian case, we can interchange the order of the functional integrals over A and α, and then change variables in the inner integral from A to $A' = A^\alpha$. The transformation (16.24) looks more complicated than in the Abelian case, but it is nothing more than a linear shift of the A^a_μ, followed by a unitary rotation of the various components of the symmetry multiplet $A^a_\mu(x)$ at each point. Both of these operations preserve the measure

$$\mathcal{D}A = \prod_x \prod_{a,\mu} dA^a_\mu. \qquad (16.26)$$

Thus $\mathcal{D}A = \mathcal{D}A'$, under the integral over α. Just as in the Abelian case, the integral over gauge motions α can be factored out of the functional integral into an overall normalization, leaving us with

$$\int \mathcal{D}A\, e^{iS[A]} = \left(\int \mathcal{D}\alpha\right)\int \mathcal{D}A\, e^{iS[A]}\,\delta\big(G(A)\big)\,\det\!\left(\frac{\delta G(A^\alpha)}{\delta\alpha}\right). \qquad (16.27)$$

This normalization factor cancels in the computation of correlation functions of gauge-invariant operators.

From this point, the derivation of the gauge boson propagator proceeds as for the photon propagator. We choose the generalized Lorentz gauge condition

$$G(A) = \partial^\mu A^a_\mu(x) - \omega^a(x), \qquad (16.28)$$

with a Gaussian weight for ω^a as in Eq. (9.56). The manipulations of Section 9.4 then lead to the class of gauge field propagators

$$\langle A^a_\mu(x)A^b_\nu(y)\rangle = \int \frac{d^4k}{(2\pi)^4}\frac{-i}{k^2+i\epsilon}\left(g_{\mu\nu} - (1-\xi)\frac{k_\mu k_\nu}{k^2}\right)\delta^{ab}e^{-ik\cdot(x-y)}, \qquad (16.29)$$

with a freely adjustable gauge parameter ξ. Our guess (16.5) corresponds to the choice $\xi = 1$, called the *Feynman-'t Hooft gauge*.

So far, this whole derivation parallels the case of electrodynamics. Here, however, there is one more nontrivial ingredient. In QED, the determinant in Eq. (16.23) was independent of A, so this quantity could be treated as just another contribution to the normalization factor. In the non-Abelian case this is no longer true. Using the infinitesimal form (16.25) of the gauge transformation, we can evaluate

$$\frac{\delta G(A^\alpha)}{\delta\alpha} = \frac{1}{g}\partial^\mu D_\mu, \qquad (16.30)$$

acting on a field in the adjoint representation; this operator depends on A. The functional determinant of (16.30) thus contributes new terms to the Lagrangian.

Faddeev and Popov chose to represent this determinant as a functional integral over a new set of anticommuting fields belonging to the adjoint representation:

$$\det\left(\frac{1}{g}\partial^\mu D_\mu\right) = \int \mathcal{D}c\,\mathcal{D}\bar{c}\,\exp\left[i\int d^4x\,\bar{c}(-\partial^\mu D_\mu)c\right]. \tag{16.31}$$

We derived this formal identity in Eq. (9.69), using our rules for fermionic functional integrals. (The factor of $1/g$ is absorbed into the normalization of the fields c and $\bar{c}$.) But to give the correct identity, c and $\bar{c}$ must be anticommuting fields that are scalars under Lorentz transformations. The quantum excitations of these fields have the wrong relation between spin and statistics to be physical particles. However, we can nevertheless treat these excitations as additional particles in the computation of Feynman diagrams. These new fields and their particle excitations are called *Faddeev-Popov ghosts*.

If we temporarily suppress our curiosity about the physical interpretation of the ghosts, we can work out their Feynman rules. We write the ghost Lagrangian more explicitly as

$$\mathcal{L}_{\text{ghost}} = \bar{c}^a\left(-\partial^2\delta^{ac} - g\partial^\mu f^{abc}A_\mu^b\right)c^c. \tag{16.32}$$

The first term gives a ghost propagator,

$$\langle c^a(x)\bar{c}^b(y)\rangle = \int \frac{d^4k}{(2\pi)^4}\frac{i}{k^2}\delta^{ab}e^{-ik\cdot(x-y)}. \tag{16.33}$$

In a diagram, this propagator carries an arrow that shows the flow of ghost number, as in Fig. 16.5. In the interaction term of (16.32), the derivative stands to the left of the gauge field; this implies that this derivative is evaluated with the momentum coming out of the vertex along the ghost line. The explicit Feynman rule is shown in Fig. 16.5. As with the other vertices we have encountered, the coupling constant g that appears in this vertex must be equal to the coupling constant g in the three-boson vertex in order to avoid upsetting the Ward identities.

There are no further subtleties in the construction of the perturbation theory for non-Abelian gauge theories. In particular, it is straightforward to include fermions. The final Lagrangian, including all of the effects of Faddeev-Popov gauge fixing, is

$$\mathcal{L} = -\frac{1}{4}(F_{\mu\nu}^a)^2 - \frac{1}{2\xi}(\partial^\mu A_\mu^a)^2 + \bar{\psi}(i\slashed{D} - m)\psi + \bar{c}^a(-\partial^\mu D_\mu^{ac})c^c. \tag{16.34}$$

This Lagrangian leads to the propagator (16.29), and to the set of Feynman rules for vertices shown in Figs. 16.1 and 16.5.

The argument we have just completed suffices to derive the Feynman diagram expansion of any correlation function of gauge-invariant operators in a non-Abelian gauge theory. At the end of Section 9.4, we explained that the

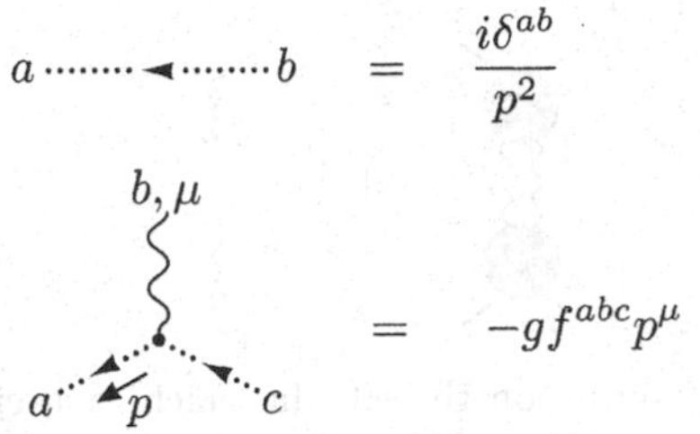

Figure 16.5. Feynman rules for Faddeev-Popov ghosts.

Faddeev-Popov gauge-fixing technique also gives the correct gauge-invariant expressions for S-matrix elements. This remains true in the non-Abelian case. However, the argument given in Section 9.4 relied upon the cancellation in QED of the emission probabilities for timelike and longitudinal photons, and we have already found that this cancellation does not go through in the non-Abelian case. In Section 16.4 we will construct a more sophisticated argument, in which the Faddeev-Popov ghosts play an essential role, that will correctly generalize our previous argument to non-Abelian gauge theories.

16.3 Ghosts and Unitarity

We might now ask whether the new ingredients that we found in the previous section, the Faddeev-Popov ghosts, can resolve the paradox that we encountered at the end of Section 16.1. There we saw that the first diagram in Fig. 16.6 contains a nonzero contribution to its imaginary part that does not correspond to a possible final state with physical gauge boson polarizations. We will now compute this contribution more carefully. We must then add a new potential contribution from the ghosts, shown as the second diagram in Fig. 16.6.

Let us call the amplitude for fermion-fermion annihilation into gauge bosons, which we studied in Section 16.1,

$$i\mathcal{M}^{\mu\nu}\epsilon_\mu^*(k_1)\epsilon_\nu^*(k_2); \tag{16.35}$$

the amplitude for two gauge bosons to convert to a fermion-antifermion pair will be, correspondingly, $\mathcal{M}'$. Then, following the Cutkosky rules of Section 7.3, we find the imaginary part of the first diagram in Fig. 16.6 by replacing the cut gauge boson propagator with momentum k_i by

$$-ig_{\mu\nu}\cdot(-2\pi i)\delta(k_i^2). \tag{16.36}$$

Replacing both propagators gives two delta functions, turning the four-dimensional integrals over the gauge boson momenta into three-dimensional phase space integrals, as in the example in Section 7.3. We are thus left with the expression

$$\tfrac{1}{2}(i\mathcal{M}^{\mu\nu})g_{\mu\rho}g_{\nu\sigma}(i\mathcal{M}'^{\rho\sigma}), \tag{16.37}$$

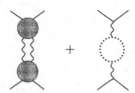

Figure 16.6. The diagram on the left, in which each circle represents the sum of the three contributions of Fig. 16.2, gives a possible problem for the optical theorem. The ghost diagram on the right cancels the anomalous terms.

integrated over the phase space of two massless particles. The factor $1/2$ is a symmetry factor for the Feynman diagram or, equivalently, a correction to the phase space integral for identical particles.

Now introduce the representation (16.20) for $g_{\mu\rho}$ and $g_{\nu\sigma}$. The pieces that involve only transverse polarizations correspond to the expected imaginary parts necessary to satisfy the optical theorem. We need not consider these terms further. The cross terms between physical and unphysical polarizations vanish: We showed in Section 16.1 that

$$i\mathcal{M}^{\mu\nu}\epsilon_\mu^{T*}(k_1)\epsilon_\nu^{+*}(k_2) = 0. \tag{16.38}$$

The same identity holds if $\mathcal{M}$ is replaced by $\mathcal{M}'$, and if ϵ^+ is replaced by ϵ^-. Furthermore, the amplitude vanishes if both polarization vectors are forward or both are backward. The only surviving terms are the cross terms between forward and backward polarization, which yield the expression

$$\tfrac{1}{2}\big[(i\mathcal{M}^{\mu\nu}\epsilon_\mu^{-*}\epsilon_\nu^{+*})(i\mathcal{M}'^{\rho\sigma}\epsilon_\rho^+\epsilon_\sigma^-) + (i\mathcal{M}^{\mu\nu}\epsilon_\mu^{+*}\epsilon_\nu^{-*})(i\mathcal{M}'^{\rho\sigma}\epsilon_\rho^-\epsilon_\sigma^+)\big], \tag{16.39}$$

integrated over phase space. We worked out the value of the first factor in Eq. (16.21), and the contraction with $\mathcal{M}'$ is very similar. Substituting these results, expression (16.39) becomes

$$\frac{1}{2}\left(ig\bar{v}(p_+)\gamma_\mu t^c u(p) \cdot \frac{-i}{(k_1+k_2)^2} \cdot gf^{abc}k_1^\mu\right)$$
$$\times \left(ig\bar{u}(p')\gamma_\rho t^d v(p'_+) \cdot \frac{-i}{(k_1+k_2)^2} \cdot gf^{abd}(-k_2)^\rho\right) + (k_1 \leftrightarrow k_2). \tag{16.40}$$

Using the identity

$$\bar{v}(p_+)\gamma_\mu(k_1+k_2)^\mu u(p) = \bar{v}(p_+)\gamma_\mu(p+p_+)^\mu u(p) = 0, \tag{16.41}$$

we see that the two terms added in (16.40) are equal.

Now add the contribution from the Faddeev-Popov ghosts. Using the Feynman rules in Fig. 16.5, we can assemble the amplitude for fermion-antifermion annihilation into a pair of ghosts:

$$i\mathcal{M}_{\text{ghost}} = ig\bar{v}(p_+)\gamma_\mu t^c u(p) \cdot \frac{-i}{(k_1+k_2)^2} \cdot gf^{abc}k_1^\mu. \tag{16.42}$$

This is precisely the first half of expression (16.40). Similarly, the amplitude for the ghost-antighost pair to annihilate into fermions is equal to the second half of (16.40). Finally, since Faddeev-Popov ghost fields anticommute, we must supply a factor of -1 for each ghost loop. Thus the ghost contribution exactly cancels the contribution of unphysical gauge boson polarizations to the Cutkosky cut of the diagrams in Fig. 16.6.

This example illustrates a general physical interpretation of Faddeev-Popov ghosts. These "particles" serve as negative degrees of freedom to cancel the effects of the unphysical timelike and longitudinal polarization states of the gauge bosons. The simplest effect of the ghosts can already be seen from the determinants that appear when one integrates over the gauge and ghost fields in the Faddeev-Popov Lagrangian (16.34). In a general dimension d, working in Feynman gauge and at zero coupling for simplicity, the functional integral over the gauge and ghost fields in (16.34) yields

$$\left(\det[-\partial^2]\right)^{-d/2} \cdot \left(\det[-\partial^2]\right)^{+1}. \tag{16.43}$$

The second determinant, which appears with a positive exponent because the ghost fields anticommute, cancels the contribution to the first determinant of two components of the field A_μ. This physical effect was illustrated, using the language of Section 9.4, in Problem 9.2.

16.4 BRST Symmetry

To show how this cancellation extends to the complete interacting theory, Becchi, Rouet, Stora, and Tyutin introduced as a beautiful formal tool a new symmetry of the gauge-fixed Lagrangian (16.34), which involves the ghost in an essential way.* This *BRST symmetry* has a continuous parameter that is an anticommuting number. To write the symmetry in its simplest form, let us rewrite the Faddeev-Popov Lagrangian by introducing a new (commuting) scalar field B^a:

$$\mathcal{L} = -\frac{1}{4}(F_{\mu\nu}^a)^2 + \bar{\psi}(i\slashed{D} - m)\psi + \frac{\xi}{2}(B^a)^2 + B^a\partial^\mu A_\mu^a + \bar{c}^a(-\partial^\mu D_\mu^{ac})c^c. \tag{16.44}$$

The new field B^a has a quadratic term without derivatives, so it is not a normal propagating field. The functional integral over B^a can be done by completing the square in (16.44); this procedure brings us back precisely to Eq. (16.34). A field of this type, which appears in the functional integral but has no independent dynamics, is called an *auxiliary field*.

---

*C. Becchi, A. Rouet, and R. Stora, *Ann. Phys.* **98**, 287 (1976); I. V. Tyutin, Lebedev Institute preprint (1975, unpublished); M. Z. Iofa and I. V. Tyutin, *Theor. Math. Phys.* **27**, 316 (1976).

Now let ϵ be an infinitesimal anticommuting parameter, and consider the following infinitesimal transformation of the fields in (16.44):

$$\delta A_\mu^a = \epsilon D_\mu^{ac} c^c$$

$$\delta \psi = ig\epsilon c^a t^a \psi$$

$$\delta c^a = -\tfrac{1}{2} g\epsilon f^{abc} c^b c^c \qquad (16.45)$$

$$\delta \bar{c}^a = \epsilon B^a$$

$$\delta B^a = 0.$$

The transformation of the fields A_μ^a and ψ is a local gauge transformation whose parameter is proportional to the ghost field: $\alpha^a(x) = g\epsilon c^a(x)$. Thus, the first two terms of (16.44) are invariant to (16.45). The third term is trivially invariant. The transformation of A_μ^a in the fourth term cancels the transformation of $\bar{c}^a$ in the last term. Finally, we must examine the transformation of the last ingredient in (16.44):

$$\begin{aligned}
\delta(D_\mu^{ac} c^c) &= D_\mu^{ac} \delta c^c + g f^{abc} \delta A_\mu^b c^c \\
&= -\tfrac{1}{2} g\epsilon \partial_\mu (f^{abc} c^b c^c) - \tfrac{1}{2} g^2 \epsilon f^{abc} f^{cde} A_\mu^b c^d c^e \qquad (16.46) \\
&\quad + g\epsilon f^{abc} (\partial_\mu c^b) c^c + g^2 \epsilon f^{abc} f^{bde} A_\mu^d c^e c^c.
\end{aligned}$$

The two terms of order g manifestly cancel. By using the anticommuting nature of the ghost fields and exchanging the names of indices, we can write the remaining two terms as

$$-\tfrac{1}{2} \epsilon g^2 f^{abc} f^{cde} \left(A_\mu^b c^d c^e + A_\mu^d c^e c^b + A_\mu^e c^b c^d \right), \qquad (16.47)$$

which vanishes by the Jacobi identity (15.70). Apparently, the BRST transformation (16.45) is a global symmetry of the gauge-fixed Lagrangian (16.44), for any value of the gauge parameter ξ.

The BRST transformation has one more remarkable feature, which is a natural consequence of its anticommuting nature. Let $Q\phi$ be the BRST transformation of the field ϕ: $\delta\phi = \epsilon Q\phi$. For example, $QA_\mu^a = D_\mu^{ac} c^c$. Then, for any field, the BRST variation of $Q\phi$ vanishes:

$$Q^2 \phi = 0. \qquad (16.48)$$

The vanishing of (16.46) proves this identity for the second BRST variation of the gauge field. For the ghost field,

$$Q^2 c^a = \tfrac{1}{2} g^2 f^{abc} f^{bde} c^c c^d c^e, \qquad (16.49)$$

which vanishes by the Jacobi identity. It is straightforward to check that the second BRST variations of the other fields in (16.44) also vanish.

To describe the implications of identity (16.48), we now consider studying the effective theory (16.44) in the Hamiltonian picture after canonical quantization. Because the Lagrangian has the continuous symmetry (16.45), the theory will have a conserved current, and the integral of the time component of this current will be a conserved charge Q that commutes with H. The action of Q on field configurations will be just that described in the previous paragraph. The relation (16.48) is equivalent to the operator identity

$$Q^2 = 0. \tag{16.50}$$

We say that the BRST operator Q is *nilpotent*.

A nilpotent operator that commutes with H divides the eigenstates of H into three subspaces. Many eigenstates of H must be annihilated by Q so that (16.50) can be satisfied. Let $\mathcal{H}_1$ be the subspace of states that are *not* annihilated by Q. Let $\mathcal{H}_2$ be the subspace of states of the form

$$|\psi_2\rangle = Q\,|\psi_1\rangle\,, \tag{16.51}$$

where $|\psi_1\rangle$ is in $\mathcal{H}_1$. According to (16.50), acting Q again on these states gives zero. Finally, let $\mathcal{H}_0$ be the subspace of states $|\psi_0\rangle$ that satisfy $Q\,|\psi_0\rangle = 0$ but that cannot be written in the form (16.51). The subspace $\mathcal{H}_2$ is quite peculiar, because any two states in this subspace have zero inner product:

$$\langle\psi_{2a}|\psi_{2b}\rangle = \langle\psi_{1a}|\,Q\,|\psi_{2b}\rangle = 0 \tag{16.52}$$

by (16.50). By the same argument, the states of $\mathcal{H}_2$ have zero inner product with the states of $\mathcal{H}_0$.

These considerations seem extremely abstract, but they have a direct physical correspondence.[†] To see this, consider single-particle states of the non-Abelian gauge theory in the limit of zero coupling. According to the transformation (16.45), Q converts the forward component of A^a_μ to a ghost field; equivalently, Q converts a single forward-polarized gauge boson to a ghost. At $g = 0$, Q annihilates the one-ghost state. At the same time, Q converts the antighost state to a quantum of B^a. To identify this state, note that the Lagrangian (16.44) implies the classical field equation

$$\xi B^a = -\partial^\mu A^a_\mu. \tag{16.53}$$

Thus the quanta of the field B^a are those quanta of A^a_μ with polarization vectors such that $k^\mu\epsilon_\mu(k) \neq 0$; these are the backward-polarized gauge bosons.

We have now seen that, among the single-particle states of the gauge theory, forward gauge bosons and antighosts belong to $\mathcal{H}_1$, ghosts and backward gauge bosons belong to $\mathcal{H}_2$, and transverse gauge bosons belong to $\mathcal{H}_0$. More generally, it can be shown that asymptotic states containing ghosts,

---

[†]The following argument is presented only at an intuitive level. For a rigorous discussion, see T. Kugo and I. Ojima, *Suppl. Prog. Theor. Phys.* **66**, 1 (1979).

antighosts, or gauge bosons of unphysical polarization always belong to $\mathcal{H}_1$ or $\mathcal{H}_2$, while the asymptotic states in $\mathcal{H}_0$ are those with only transversely polarized gauge bosons. The BRST operator thus gives a precise relation between the unphysical gauge boson polarization states and the ghosts and antighosts as positive and negative degrees of freedom.

In Section 9.4, we argued that the Faddeev-Popov prescription gave the correct, gauge-invariant result for a certain subclass of S-matrix elements, from which we could compute the physical scattering cross sections of transversely polarized gauge bosons. These S-matrix elements were constructed by putting operators in the far past to create transversely polarized gauge bosons, adiabatically turning on the gauge coupling, adiabatically turning off the gauge coupling, and then placing operators in the far future to annihilate gauge bosons with transverse polarization. However, this argument had a possible problem: If the states created as collections of transversely polarized bosons in the far past could evolve into states that contained gauge bosons of other polarizations in the far future, the S-matrix projected between transverse gauge boson states would not be unitary. This problem would also lead to the technical problem discussed in the previous section: The Cutkosky cuts of diagrams contributing to S-matrix elements would have nonzero contributions from unphysical polarizations. In Section 9.4, we used an argument special to the Abelian case to show that these problems do not arise in QED. In the non-Abelian case, the removal of unphysical gauge boson polarizations is more subtle, and we have seen that it involves the ghosts in an essential way. To resolve this subtle problem, we apply the principle of BRST symmetry.

Let $|A; \text{tr}\rangle$ be an external state that contains no ghosts or antighosts and only gauge bosons with transverse polarization. We wish to show that the S-matrix projected onto such states is unitary:

$$\sum_C \langle A; \text{tr}| S^\dagger |C; \text{tr}\rangle \langle C; \text{tr}| S |B; \text{tr}\rangle = \langle A; \text{tr}| \mathbf{1} |B; \text{tr}\rangle. \tag{16.54}$$

As we explained above, the physical states $|A; \text{tr}\rangle$ belong to—and, in fact, span—the subspace $\mathcal{H}_0$ defined by the BRST operator. In particular, all of these states are annihilated by Q. Since Q commutes with the Hamiltonian, the time evolution of any such state must also produce a state annihilated by Q. Thus,

$$Q \cdot S |A; \text{tr}\rangle = 0. \tag{16.55}$$

This implies that the states $S |A; \text{tr}\rangle$ must be linear combinations of states in $\mathcal{H}_0$ and $\mathcal{H}_2$. However, states in $\mathcal{H}_2$ have zero inner product with one another and with states in $\mathcal{H}_0$. Thus the inner product of any two states of the form $S |A; \text{tr}\rangle$ comes only from the overlap of the components in $\mathcal{H}_0$, so we can write

$$\langle A; \text{tr}| S^\dagger \cdot S |B; \text{tr}\rangle = \sum_C \langle A; \text{tr}| S^\dagger |C; \text{tr}\rangle \langle C; \text{tr}| S |B; \text{tr}\rangle. \tag{16.56}$$

Since the full S-matrix is unitary, this relation implies that the restricted S-matrix is also unitary, Eq. (16.54). In addition, (16.56) implies that the sum of the Cutkosky cuts of diagrams contributing to the S-matrix in a given order is equal to the sum of the cuts involving transverse gauge bosons only. Thus, the cancellation between diagrams that produce pairs of gauge bosons with unphysical polarizations and those that produce ghosts is a general property that persists to all orders in perturbation theory.

Since the BRST transformation generates a continuous symmetry, it generates a set of Ward identities. These identities are similar in structure to the Ward identities of the non-Abelian gauge symmetry, since the BRST symmetry contains a gauge transformation whose parameter is the ghost field. However, the identities that follow from BRST symmetry are simpler. We will not study the Ward identities of non-Abelian gauge theory further in this book. However, when one discusses the renormalization of gauge theories at a higher level, the central identities among renormalization constants that follow from the Ward identities are most easily derived using the BRST symmetry.[‡]

16.5 One-Loop Divergences of Non-Abelian Gauge Theory

Now that we have discussed the general properties of tree-level diagrams in non-Abelian gauge theories, we turn our attention to diagrams with loops. As always in quantum field theory, some of these loop diagrams will diverge, and we must take care to treat the divergent integrals correctly.

The Lagrangian of a non-Abelian gauge theory (15.39) contains no interactions of dimension higher than 4. Therefore, by the general arguments of Chapter 10, this Lagrangian is renormalizable, in the sense that the divergences can be removed by a finite number of counterterms. However, in non-Abelian gauge theories, as in QED, the gauge symmetries of the theory imply stronger restrictions on the structure of the divergences. In QED, provided that we use a gauge-invariant regulator, there are only four possible divergent coefficients, which are subtracted by the counterterms for the electromagnetic vertex (δ_1), for the electron and photon field strength $(\delta_2$ and $\delta_3)$, and for the electron mass (δ_m). In particular, the possibility of a photon mass renormalization is excluded by gauge invariance. Furthermore, the two counterterms δ_1 and δ_2 are equal to one another, and cancel in the evaluation of the electron-photon vertex function, as a consequence of the Ward identity. Non-Abelian gauge symmetries imply similar restrictions on the divergences of Feynman diagrams. In this section, we will illustrate some of these restrictions through examples of one-loop diagrams.

[‡]An introduction to the Ward identities of the BRST symmetry is given by Taylor (1976).

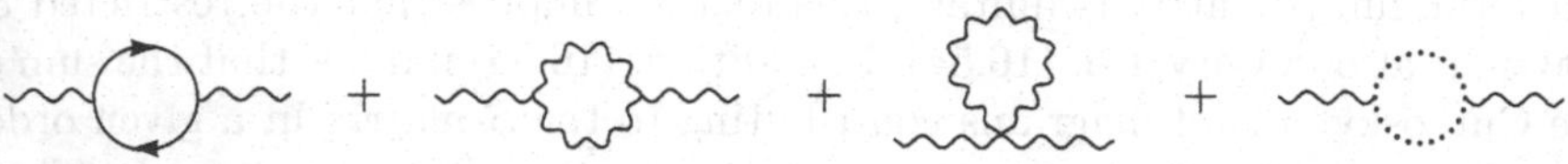

Figure 16.7. Contributions to the gauge boson self-energy in order g^2.

The Gauge Boson Self-Energy

In QED, the strongest constraints of gauge invariance come in the evaluation of the photon self-energy. The Ward identity implies the relation

$$q^\mu \left(\text{~~~} \right) = 0, \tag{16.57}$$

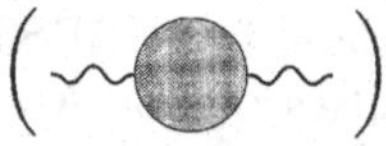

which in turn implies that the photon self-energy diagrams have the structure

$$\text{~~~} = i(q^2 g^{\mu\nu} - q^\mu q^\nu)\Pi(q^2). \tag{16.58}$$

The only divergence possible is a logarithmically divergent contribution to $\Pi(q^2)$. In non-Abelian gauge theories, (16.57) still holds, so the self-energy again has the Lorentz structure (16.58). However, the cancellations that lead to this structure are more complex. Here we will exhibit these cancellations by computing the gauge boson self-energy in detail at the one-loop level. In order to preserve gauge invariance, we will use dimensional regularization.

The contributions of order g^2 to the gauge boson self-energy are shown in Fig. 16.7. (In addition to these 1PI diagrams, there are three "tadpole" diagrams; but these automatically vanish, as in QED, by the argument given below Eq. (10.5).) The fermion loop diagram can be considered separately from the other diagrams, since in principle we could include any number of fermions in the theory. We will see below that the contributions of the three remaining diagrams interlock in an essential way.

Let us first calculate the fermion loop diagram. The Feynman rule for the vertices in this diagram is identical to the QED Feynman rule, except for the addition of a group matrix t^a that acts on the fermion gauge group indices. The value of this diagram is therefore the same as in QED, Eq. (7.90), multiplied by a trace over group matrices:

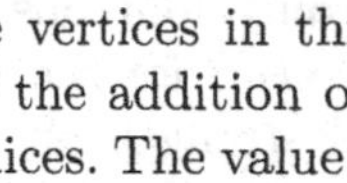

$$\text{~~~} = \text{tr}[t^a t^b]\, i(q^2 g^{\mu\nu} - q^\mu q^\nu)$$

$$\times \frac{-g^2}{(4\pi)^{d/2}} \int\limits_0^1 dx\, 8x(1-x) \frac{\Gamma(2-\frac{d}{2})}{(m^2 - x(1-x)q^2)^{2-d/2}}.$$

The value of the trace is given by Eq. (15.78): $\text{tr}[t^a t^b] = C(r)\delta^{ab}$. In a theory with several species of fermions, there would be a diagram of this type for each species. We will be mainly interested in the divergent part of this diagram, which is independent of the fermion mass. If there are n_f species of fermions, all in the same representation r, then the total contribution of fermion loop diagrams takes the form

$$\sum_{\text{fermions}} \left(\rule{0pt}{1.2em}\text{—}\!\bigcirc\!\text{—} \right) \tag{16.59}$$

$$= i(q^2 g^{\mu\nu} - q^\mu q^\nu)\delta^{ab} \left(\frac{-g^2}{(4\pi)^2} \cdot \frac{4}{3} n_f C(r)\Gamma(2-\tfrac{d}{2}) + \cdots \right).$$

Now consider the three diagrams from the pure gauge sector. The contribution of these diagrams depends on the gauge; we will use Feynman-'t Hooft gauge, $\xi = 1$.

Using the three-gauge-boson vertex from Fig. 16.1, we can write the first of the three diagrams as

$$= \frac{1}{2} \int \frac{d^4 p}{(2\pi)^4} \frac{-i}{p^2} \frac{-i}{(p+q)^2} g^2 f^{acd} f^{bcd} N^{\mu\nu}, \tag{16.60}$$

where the numerator structure is

$$N^{\mu\nu} = \left[g^{\mu\rho}(q-p)^\sigma + g^{\rho\sigma}(2p+q)^\mu + g^{\sigma\mu}(-p-2q)^\rho \right]$$
$$\times \left[\delta^\nu_{\ \rho}(p-q)_\sigma + g_{\rho\sigma}(-2p-q)^\nu + \delta^\nu_{\ \sigma}(p+2q)_\rho \right].$$

The overall factor of $1/2$ is a symmetry factor. The contraction of structure constants can be evaluated using Eq. (15.93): $f^{acd} f^{bcd} = C_2(G)\delta^{ab}$.

To simplify the expression further, combine denominators in the standard way:

$$\frac{1}{p^2} \frac{1}{(p+q)^2} = \int_0^1 dx \frac{1}{((1-x)p^2 + x(p+q)^2)^2} = \int_0^1 dx \frac{1}{(P^2 - \Delta)^2}, \tag{16.61}$$

where $P = p + xq$ and $\Delta = -x(1-x)q^2$. Then (16.60) can be rewritten

$$= -\frac{g^2}{2} C_2(G)\delta^{ab} \int_0^1 dx \int \frac{d^4 P}{(2\pi)^4} \frac{1}{(P^2 - \Delta)^2} N^{\mu\nu}.$$

The numerator structure can be simplified by eliminating p in favor of P, discarding terms linear in P^μ (which integrate symmetrically to zero), and replacing $P^\mu P^\nu$ with $g^{\mu\nu} P^2/d$ (also by symmetry):

$$\begin{aligned}
N^{\mu\nu} &= -g^{\mu\nu}\left[(2q+p)^2 + (q-p)^2\right] - d(q+2p)^\mu(q+2p)^\nu \\
&\quad + \left[(2q+p)^\mu(q+2p)^\nu + (q-p)^\mu(2q+p)^\nu - (q+2p)^\mu(q-p)^\nu \right.\\
&\qquad \left. + (\mu \leftrightarrow \nu)\right] \\
&\to -g^{\mu\nu}P^2 \cdot 6(1-\tfrac{1}{d}) - g^{\mu\nu}q^2\left[(2-x)^2 + (1+x)^2\right] \\
&\quad + q^\mu q^\nu\left[(2-d)(1-2x)^2 + 2(1+x)(2-x)\right].
\end{aligned}$$

The final step in the evaluation is to Wick-rotate and apply the integration formulae (7.85) and (7.86). This brings the diagram into the following form:

$$\begin{aligned}
&= \frac{ig^2}{(4\pi)^{d/2}} C_2(G)\delta^{ab} \int_0^1 dx\, \frac{1}{\Delta^{2-d/2}} \\
&\quad \times \Big(\Gamma(1-\tfrac{d}{2})\, g^{\mu\nu}q^2\left[\tfrac{3}{2}(d-1)x(1-x)\right] \\
&\qquad + \Gamma(2-\tfrac{d}{2})\, g^{\mu\nu}q^2\left[\tfrac{1}{2}(2-x)^2 + \tfrac{1}{2}(1+x)^2\right] \\
&\qquad - \Gamma(2-\tfrac{d}{2})\, q^\mu q^\nu\left[(1-\tfrac{d}{2})(1-2x)^2 + (1+x)(2-x)\right]\Big).
\end{aligned}$$

(16.62)

Next consider the diagram with a four-gauge-boson vertex. Using the vertex Feynman rule in Fig. 16.1, we find

$$\begin{aligned}
&= \frac{1}{2}\int \frac{d^4p}{(2\pi)^4}\, \frac{-ig_{\rho\sigma}}{p^2}\delta^{cd}\,(-ig^2) \\
&\quad \times \Big[f^{abe}f^{cde}(g^{\mu\rho}g^{\nu\sigma} - g^{\mu\sigma}g^{\nu\rho}) \\
&\qquad + f^{ace}f^{bde}(g^{\mu\nu}g^{\rho\sigma} - g^{\mu\sigma}g^{\nu\rho}) \\
&\qquad + f^{ade}f^{bce}(g^{\mu\nu}g^{\rho\sigma} - g^{\mu\rho}g^{\nu\sigma})\Big].
\end{aligned}$$

(16.63)

The factor $1/2$ in the first line is a symmetry factor. The first combination of structure constants in the vertex factor vanishes by antisymmetry; the second and third can be reduced by the use of Eq. (15.93). We then find simply

$$= -g^2 C_2(G)\delta^{ab} \int \frac{d^4p}{(2\pi)^4}\, \frac{1}{p^2} \cdot g^{\mu\nu}(d-1).$$

(16.64)

In dimensional regularization, the integral over p gives a pole at $d=2$ but yields zero as $d \to 4$. We could simply discard this diagram and trust that the pole at $d=2$ is canceled by the other two diagrams. It is instructive, however, and no more difficult, to demonstrate the cancellation explicitly. To do so, we can force the integral to look like that of the previous diagram, multiplying the integrand by 1 in the form $(q+p)^2/(q+p)^2$. We then combine denominators as before, and eliminate p in favor of the shifted variable $P = p + xq$. After dropping the term linear in P, we obtain

$$= -g^2 C_2(G)\delta^{ab} \int_0^1 dx \int \frac{d^4 P}{(2\pi)^4} \frac{1}{(P^2-\Delta)^2} g^{\mu\nu}(d-1)[P^2+(1-x)^2 q^2].$$

We can now Wick-rotate and integrate over P to obtain

$$= \frac{ig^2}{(4\pi)^{d/2}} C_2(G)\delta^{ab} \int_0^1 dx \frac{1}{\Delta^{2-d/2}}$$

$$\times \left(-\Gamma(1-\tfrac{d}{2}) g^{\mu\nu} q^2 \left[\tfrac{1}{2} d(d-1)x(1-x) \right] \right.$$

$$\left. - \Gamma(2-\tfrac{d}{2}) g^{\mu\nu} q^2 \left[(d-1)(1-x)^2 \right] \right).$$

(16.65)

Expressions (16.62) and (16.65), by themselves, do not add to any reasonable value: The pole at $d = 2$ does not cancel, and the sum does not have a transverse Lorentz structure. To bring the gauge boson self-energy into its desired form, we must include the diagram with a ghost loop. According to the rules shown in Fig. 16.5, this diagram is

$$= (-1) \int \frac{d^4 p}{(2\pi)^4} \frac{i}{p^2} \frac{i}{(p+q)^2} g^2 f^{dac}(p+q)^\mu f^{cbd} p^\nu. \quad (16.66)$$

There is no symmetry factor in this case, but there is a factor of -1 because the ghost fields anticommute. The ghost diagram can be simplified using the same set of tricks that we applied to the previous two: combine denominators, shift the integral to P, Wick-rotate, and integrate over P using dimensional regularization. The result is

$$= \frac{ig^2}{(4\pi)^{d/2}} C_2(G)\delta^{ab} \int_0^1 dx \frac{1}{\Delta^{2-d/2}}$$

$$\times \left(-\Gamma(1-\tfrac{d}{2}) g^{\mu\nu} q^2 \left[\tfrac{1}{2} x(1-x) \right] \right.$$

$$\left. + \Gamma(2-\tfrac{d}{2}) q^\mu q^\nu \left[x(1-x) \right] \right).$$

(16.67)

Now we are ready to put these results together. In the sum of the three diagrams, the coefficient of $\Gamma(1-\tfrac{d}{2}) g^{\mu\nu} q^2 x(1-x)$ is

$$\tfrac{1}{2}(3d - 3 - d^2 + d - 1) = (1 - \tfrac{d}{2})(d - 2). \quad (16.68)$$

The first factor cancels the pole of the gamma function at $d = 2$. Thus, the sum of the three diagrams has no quadratic divergence and no gauge boson mass renormalization. Notice that the ghost diagram plays an essential role in this cancellation.

After the pole at $d = 2$ is canceled, $\Gamma(1-\frac{d}{2})$ becomes $\Gamma(2-\frac{d}{2})$. This term therefore combines with the others that are proportional to $\Gamma(2-\frac{d}{2})g^{\mu\nu}q^2$, to give a total coefficient of

$$(d-2)x(1-x) + \tfrac{1}{2}(2-x)^2 + \tfrac{1}{2}(1+x)^2 - (d-1)(1-x)^2. \tag{16.69}$$

Since the best way to simplify this expression is not obvious, let us put it aside and work first with the coefficient of $\Gamma(2-\frac{d}{2})q^\mu q^\nu$:

$$-(1-\tfrac{d}{2})(1-2x)^2 - (1+x)(2-x) + x(1-x) = -(1-\tfrac{d}{2})(1-2x)^2 - 2.$$

If the total self-energy is to be proportional to $(g^{\mu\nu}q^2 - q^\mu q^\nu)$, it must be possible to reduce expression (16.69) to this same form (times -1). To do so, note that Δ is symmetric with respect to $x \leftrightarrow (1-x)$, and therefore we can substitute $(1-x)$ for x in any term of the numerator. In particular, terms that are linear in x can be transformed as follows:

$$x \to \tfrac{1}{2}x + \tfrac{1}{2}(1-x) = \tfrac{1}{2}.$$

In the end, the sum of the three pure-gauge diagrams simplifies to

$$\frac{ig^2}{(4\pi)^{d/2}} C_2(G)\delta^{ab} \int_0^1 dx \, \frac{\Gamma(2-\frac{d}{2})}{\Delta^{2-d/2}} (g^{\mu\nu}q^2 - q^\mu q^\nu)\left[(1-\tfrac{d}{2})(1-2x)^2 + 2\right]. \tag{16.70}$$

This expression is manifestly transverse, as required by the Ward identity of the non-Abelian gauge theory. For future reference, we record the ultraviolet divergent part of (16.70):

$$\tag{16.71}$$

$$= i(q^2 g^{\mu\nu} - q^\mu q^\nu)\delta^{ab}\left(\frac{-g^2}{(4\pi)^2} \cdot \left(-\frac{5}{3}\right) C_2(G)\Gamma(2-\tfrac{d}{2}) + \cdots\right).$$

As we noted above, the result (16.70) depends on the gauge used in the calculation. In any gauge, the boson self-energy is transverse and free of quadratic divergences. However, the coefficient of the transverse Lorentz structure may depend on ξ. It turns out that, for a general value of ξ, the coefficient of the ultraviolet divergence in (16.71) is modified according to

$$-\frac{5}{3} \to -\left(\frac{13}{6} - \frac{\xi}{2}\right). \tag{16.72}$$

The fact that the boson self-energy depends on the gauge does not contradict the general theorem that S-matrix elements are independent of ξ. The full set of one-loop corrections to a gauge theory S-matrix element always involves a number of different radiative corrections to vertices and propagators; the gauge dependence cancels in an intricate fashion among these various terms.

The β Function

The simplest calculation that involves a gauge-invariant combination of radiative corrections is the computation of the leading term of the Callan-Symanzik β function of a non-Abelian gauge theory. The invariance of the leading term of β could be argued intuitively, by saying that the coupling constant of the gauge theory should not evolve to large values in one scheme of calculation while it stays small in another scheme. In Section 17.2 we will demonstrate this result more cleanly by showing that the leading coefficient of the β function can be extracted from a physical cross section and so must be gauge independent. (Surprisingly, this conclusion actually applies to the first *two* coefficients of the β function, written as a power series in g.)

Recall from Section 12.2 that the β function gives the rate at which the renormalized coupling constant changes as the renormalization scale M is increased. Since Green's functions depend on M through the counterterms that subtract ultraviolet divergences, β can be computed from the counterterms that enter an appropriately chosen Green's function. For example, in Eq. (12.58), we saw that the β function of QED can be computed from the counterterms for the electron-photon vertex, the electron self-energy, and the photon self-energy. The same derivation goes through in the case of a non-Abelian gauge theory. Thus, to lowest order,

$$\beta(g) = gM \frac{\partial}{\partial M} \left(-\delta_1 + \delta_2 + \tfrac{1}{2}\delta_3 \right), \qquad (16.73)$$

with the conventions for the counterterm vertices shown in Fig. 16.8. In QED, the first two terms cancel by the Ward identity, so β depends only on δ_3. In the non-Abelian case, all three terms contribute. The most difficult to compute is δ_3, but we have nearly done so already by computing the gauge-boson self-energy diagrams. Let us now complete this calculation of the β function of non-Abelian gauge theory.

In order for the counterterm δ_3 to cancel the divergence of Eqs. (16.59) and (16.71), it must be of the form

$$\delta_3 = \frac{g^2}{(4\pi)^2} \frac{\Gamma(2-\frac{d}{2})}{(M^2)^{2-d/2}} \left[\frac{5}{3}C_2(G) - \frac{4}{3}n_f C(r) \right], \qquad (16.74)$$

where M is the renormalization scale. Depending on the precise renormalization conditions used, there may be additional finite contributions to δ_3, but these do not contribute to the β function (to one-loop order). Similarly, the finite parts of δ_2 and δ_1 will depend on the details of the renormalization scheme. However, as we saw in Section 12.2, the one-loop contribution to the β function is the same in any scheme in which amplitudes are renormalized at a point where all momentum invariants are of the same order M^2. In dimensional regularization, a logarithmic divergence always takes the form $\Gamma(2-\frac{d}{2})/\Delta^{2-d/2}$, where Δ is some combination of momentum invariants. Thus, to compute the β function, we can simply set $\Delta = M^2$ in such expressions.

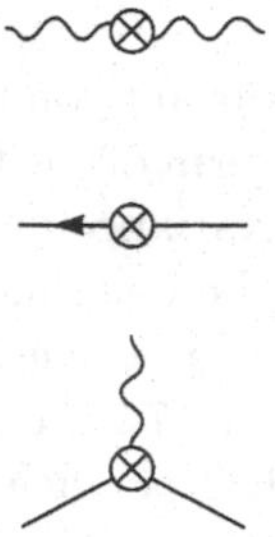

$$\text{(wavy line with crossed circle)} = -i\left(k^2 g^{\mu\nu} - k^\mu k^\nu\right)\delta^{ab}\delta_3$$

$$\text{(fermion line with crossed circle)} = i\not{p}\,\delta_2$$

$$\text{(vertex with crossed circle)} = ig t^a \gamma^\mu \delta_1$$

Figure 16.8. Counterterms needed for computing fermion interactions in a non-Abelian gauge theory.

Figure 16.9. Diagrams whose divergences are subtracted by the counterterms δ_2 and δ_1.

To complete the computation of the β function, we must compute δ_2 and δ_1 to the same level of approximation. The fermion self-energy counterterm δ_2 cancels the divergence proportional to $\not{k}$ in the first diagram of Fig. 16.9. In Feynman-'t Hooft gauge, the value of this diagram is

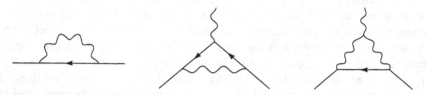

$$= \int \frac{d^4 p}{(2\pi)^4} (ig)^2 \gamma^\mu t^a \frac{i(\not{p}+\not{k})}{(p+k)^2} \gamma_\mu t^a \frac{-i}{p^2}. \qquad (16.75)$$

Since the divergence in the field strength renormalization is independent of the fermion mass, we have simplified (16.75) by setting the mass to zero. The product of group matrices equals the quadratic Casimir operator, by definition (15.92). The Dirac matrix structure can be reduced using a contraction identity (7.89). The rest of the calculation follows the same steps as for the boson self-energy diagrams:

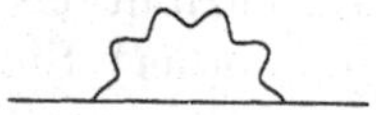

$$= g^2 C_2(r)(d-2) \int \frac{d^4 p}{(2\pi)^4} \frac{(\not{p}+\not{k})}{(p+k)^2 p^2}$$

$$= g^2 C_2(r)(d-2) \int_0^1 dx \int \frac{d^4 P}{(2\pi)^4} \frac{(1-x)\not{k}}{(P^2 - \Delta)^2}$$

$$= \frac{ig^2}{(4\pi)^{d/2}} C_2(r)\not{k} \int_0^1 dx\,(1-x)(d-2)\frac{\Gamma(2-\frac{d}{2})}{\Delta^{2-d/2}}$$

$$= \frac{ig^2}{(4\pi)^2} \not{k} C_2(r)\Gamma(2-\tfrac{d}{2}) + \cdots . \tag{16.76}$$

(Here $P = p + xk$ and $\Delta = -x(1-x)k^2$.)

The divergent part of this expression must be canceled by the second counterterm diagram of Fig. 16.8. Thus, if the renormalization scale is M, the counterterm must be

$$\delta_2 = -\frac{g^2}{(4\pi)^2} \frac{\Gamma(2-\tfrac{d}{2})}{(M^2)^{2-d/2}} \cdot C_2(r), \tag{16.77}$$

plus finite terms. We note that, like δ_3, δ_2 depends on the gauge; for example, δ_2 has no one-loop divergence in Landau gauge ($\xi = 0$).

To determine δ_1, we must compute the second and third diagrams of Fig. 16.9. The second diagram, computed in Feynman-'t Hooft gauge and for massless fermions, is

$$= \int \frac{d^4p}{(2\pi)^4} g^3 \, t^b t^a t^b \frac{\gamma^\nu (\not{p} + \not{k}')\gamma^\mu (\not{p} + \not{k})\gamma_\nu}{(p + k')^2 (p + k)^2 p^2} . \tag{16.78}$$

The gauge group matrices can be simplified according to

$$
\begin{aligned}
t^b t^a t^b &= t^b t^b t^a + t^b [t^a, t^b] \\
&= C_2(r) t^a + i t^b f^{abc} t^c \\
&= C_2(r) t^a + \tfrac{1}{2} i f^{abc} \cdot i f^{bcd} t^d \\
&= [C_2(r) - \tfrac{1}{2} C_2(G)] t^a .
\end{aligned}
\tag{16.79}
$$

In the third line we have used the antisymmetry of f^{abc} to rewrite the matrix product as a commutator; in the last line we have used Eq. (15.93).

The diagrams computed earlier in this section had positive superficial degrees of divergence, so we needed to extract their logarithmic divergences carefully. The integral in (16.78), however, is superficially logarithmically divergent, and so the coefficient of this divergence can be extracted easily by considering the limit in which the integration variable p is much greater than any external momentum. In this limit, the diagram is estimated as follows:

$$\sim g^3 [C_2(r) - \tfrac{1}{2} C_2(G)] t^a \int \frac{d^4p}{(2\pi)^4} \frac{\gamma^\nu \not{p} \gamma^\mu \not{p} \gamma_\nu}{p^2 \cdot p^2 \cdot p^2} . \tag{16.80}$$

If we replace $p^\rho p^\sigma$ by $g^{\rho\sigma} p^2/d$ in the numerator of (16.80), this expression simplifies easily:

$$\sim g^3[C_2(r) - \tfrac{1}{2}C_2(G)]t^a(2-d)^2\frac{1}{d}\gamma^\mu \int \frac{d^4p}{(2\pi)^4}\frac{1}{(p^2)^2} \tag{16.81}$$

$$\sim \frac{ig^3}{(4\pi)^2}[C_2(r) - \tfrac{1}{2}C_2(G)]t^a\gamma^\mu\left(\Gamma(2-\tfrac{d}{2}) + \cdots\right).$$

This estimate gives the correct coefficient of the divergent term. It drops completely the finite terms in the vertex function, but we do not need these to compute the β function.

The third diagram of Fig. 16.9 can be analyzed in the same way. Its value, in Feynman-'t Hooft gauge and for massless fermions, is

$$= \int \frac{d^4p}{(2\pi)^4}(ig\gamma_\nu t^b)\frac{i\not p}{p^2}(ig\gamma_\rho t^c)\frac{-i}{(k'-p)^2}\frac{-i}{(k-p)^2}$$

$$\times gf^{abc}\left[g^{\mu\nu}(2k'-k-p)^\rho + g^{\nu\rho}(-k'-k+2p)^\mu \right.$$
$$\left. + g^{\rho\mu}(2k-k'-p)^\nu\right]. \tag{16.82}$$

The gauge matrix product can be reduced as follows:

$$f^{abc}t^bt^c = \frac{1}{2}f^{abc}\cdot if^{bcd}t^d = \frac{i}{2}C_2(G)t^a.$$

Again we can determine the logarithmic divergence of this diagram by neglecting all external momenta in comparison with p. A straightforward calculation then yields

$$\sim \frac{g^3}{2}C_2(G)t^a \int \frac{d^4p}{(2\pi)^4}\gamma_\nu\not p\gamma_\rho\frac{g^{\mu\nu}p^\rho - 2g^{\nu\rho}p^\mu + g^{\rho\mu}p^\nu}{(p^2)^3}$$

$$\sim \frac{g^3}{2}C_2(G)t^a\frac{1}{d}\int \frac{d^4p}{(2\pi)^4}\frac{1}{(p^2)^2}\left[\gamma^\mu\gamma^\rho\gamma_\rho - 2\gamma^\rho\gamma^\mu\gamma_\rho + \gamma^\sigma\gamma_\sigma\gamma^\mu\right]$$

$$\sim \frac{ig^3}{(4\pi)^2}\frac{3}{2}C_2(G)\,t^a\gamma^\mu\left(\Gamma(2-\tfrac{d}{2}) + \cdots\right). \tag{16.83}$$

In the second line we have again replaced $p^\rho p^\sigma$ with $g^{\rho\sigma}p^2/d$.

The sum of the divergences in results (16.81) and (16.83) must be canceled by the third counterterm diagram in Fig 16.8. With a renormalization scale of M, we find

$$\delta_1 = -\frac{g^2}{(4\pi)^2}\frac{\Gamma(2-\tfrac{d}{2})}{(M^2)^{2-d/2}}[C_2(r) + C_2(G)]. \tag{16.84}$$

Notice that δ_1 is not equal to δ_2, as would have been true in the Abelian case; here δ_1 has an extra term, proportional to $C_2(G)$.

We are now ready to compute the β function. Plugging the three counterterms (16.74), (16.77), and (16.84) into our formula (16.73), we find

$$\beta(g) = (-2)\frac{g^3}{(4\pi)^2}\left[(C_2(r) + C_2(G)) - C_2(r) + \frac{1}{2}\left(\frac{5}{3}C_2(G) - \frac{4}{3}n_fC(r)\right)\right];$$

that is,

$$\beta(g) = -\frac{g^3}{(4\pi)^2}\left[\frac{11}{3}C_2(G) - \frac{4}{3}n_fC(r)\right]. \tag{16.85}$$

Notice that, at least for small values of n_f, the β function is *negative* and so non-Abelian gauge theories are asymptotically free. This is a result of exceptional physical importance, first discovered by 't Hooft, Politzer, and Gross and Wilczek.* We will discuss the physical interpretation of this result further in Section 16.7, and in the next several chapters. However, for the rest of this section, we will resist the temptation to pursue the physics and instead complete our technical analysis of the divergences of non-Abelian gauge theories.

Relations among Counterterms

In the analysis just completed, we computed the β function of a non-Abelian gauge theory from the divergences of the fermion vertex and field strength renormalizations. One might visualize that we were computing the running of the coupling constant at the fermion-gauge boson vertex. Alternatively, we could have studied the divergences of the three-gauge-boson vertex or the four-gauge-boson vertex, and thus computed the running of these coupling constants. However, we saw already in Section 16.1 that non-Abelian gauge invariance knits together these separate coupling constants and requires their equality. Thus we might expect that these different calculations should produce the same value of the β function.

To clarify this issue, let us carefully enumerate all the counterterms that appear in a non-Abelian gauge theory. We start from the Lagrangian (16.34), regarded as a combination of bare fields and a bare coupling constant. In the following discussion, we denote bare quantities by the subscript 0. Then,

$$\begin{aligned}\mathcal{L} = &-\frac{1}{4}(\partial_\mu A^a_{0\nu} - \partial_\nu A^a_{0\mu})^2 + \bar{\psi}_0(i\slashed{\partial} - m_0)\psi_0 - \bar{c}^a_0\partial^2 c^a_0 \\ &+ g_0 A^a_{0\mu}\bar{\psi}_0\gamma^\mu t^a\psi_0 - g_0 f^{abc}(\partial_\mu A^a_{0\nu})A^{b\mu}_0 A^{c\nu}_0 \\ &- \frac{1}{4}g_0^2(f^{eab}A^a_{0\mu}A^b_{0\nu})(f^{ecd}A^{c\mu}_0 A^{d\nu}_0) - g_0\bar{c}^a_0 f^{abc}\partial^\mu A^b_{0\mu}c^c_0.\end{aligned} \tag{16.86}$$

We choose $\xi = \infty$ for simplicity. We now rescale the fields to the renormalized field strengths by extracting the factors Z_2, Z_3, Z^c_2 for the fermions, gauge

---

*G. 't Hooft, unpublished; H. D. Politzer, *Phys. Rev. Lett.* **30**, 1346 (1973); D. J. Gross and F. Wilczek, *Phys. Rev. Lett.* **30**, 1343 (1973).

bosons, and ghosts, and shift the coupling to the renormalized coupling g. The Lagrangian then takes the form

$$\mathcal{L} = \mathcal{L}_{\text{ren}} + \mathcal{L}_{\text{c.t.}},$$

where $\mathcal{L}_{\text{ren}}$ is the Lagrangian (16.34) and $\mathcal{L}_{\text{c.t.}}$ takes the form

$$
\begin{aligned}
\mathcal{L}_{\text{c.t.}} = &-\frac{1}{4}\delta_3(\partial_\mu A_\nu^a - \partial_\nu A_\mu^a)^2 + \bar{\psi}(i\delta_2 \slashed{\partial} - \delta_m)\psi - \delta_2^c \bar{c}^a \partial^2 c^a \\
&+ g\delta_1 A_\mu^a \bar{\psi}\gamma^\mu \psi - g\delta_1^{3g} f^{abc}(\partial_\mu A_\nu^a)A_\mu^b A_\nu^c \\
&-\frac{1}{4}g^2\delta_1^{4g}(f^{eab}A_\mu^a A_\nu^b)(f^{ecd}A_\mu^c A_\nu^d) - g\delta_1^c \bar{c}^a f^{abc}\partial^\mu A_\mu^b c^c,
\end{aligned}
\tag{16.87}
$$

with the counterterms defined by

$$
\delta_2 = Z_2 - 1, \qquad \delta_3 = Z_3 - 1, \qquad \delta_2^c = Z_2^c - 1, \qquad \delta_m = Z_2 m_0 - m,
$$

$$
\delta_1 = \frac{g_0}{g}Z_2(Z_3)^{1/2} - 1, \qquad \delta_1^{3g} = \frac{g_0}{g}(Z_3)^{3/2} - 1,
$$

$$
\delta_1^{4g} = \frac{g_0^2}{g^2}(Z_3)^2 - 1, \qquad \delta_1^c = \frac{g_0}{g}Z_2^c(Z_3)^{1/2} - 1.
\tag{16.88}
$$

Notice that these eight counterterms depend on five underlying parameters; thus, there are three relations among them. The situation is very similar to that for the scalar theories with spontaneously broken symmetry that we studied in Chapter 11. The underlying symmetry of the theory—here, local gauge invariance—implies relations among the divergent amplitudes of the theory and among the counterterms required to cancel them. In the present case, a set of five renormalization conditions uniquely specifies all of the counterterms in a way that removes all divergences from the theory.

This program is especially simple at one-loop order. In this case we can expand g_0/g and the various Z factors about 1, keeping only the leading-order contribution to each counterterm. Then the three relations among the counterterms can be written

$$
\delta_1 - \delta_2 = \delta_1^{3g} - \delta_3 = \tfrac{1}{2}(\delta_1^{4g} - \delta_3) = \delta_1^c - \delta_2^c.
\tag{16.89}
$$

It is instructive to check explicitly that the values of δ_1^{3g}, δ_1^{4g}, and δ_1^c determined from (16.89) indeed remove the divergences of the corresponding vertex diagrams; this is the subject of Problem 16.3. Using relations (16.89), it is easy to show that the one-loop calculation of the β function will yield the same value, whichever gauge boson vertex is used in the computation. More generally, consider a non-Abelian gauge theory with many different species of particles, bosons and fermions, which couple to the gauge field. Then, to one-loop order, the quantity

$$
\delta_1^i - \delta_2^i,
$$

where δ_1^i is the vertex counterterm for species i and δ_2^i is the corresponding field strength counterterm, takes a universal value. This value is gauge dependent, so that the gauge dependence of its divergent part cancels the gauge dependence of δ_3 in the computation of the β function.

In our discussion of the counterterms of QED at the end of Section 10.3, we remarked that the relation between δ_1 and δ_2 insured that all electrically charged species see a common universal value of the coupling constant e. In non-Abelian gauge theories, the relations (16.89) and their higher-loop generalizations preserve the universality of the non-Abelian couplings. In QED, we were able to obtain an even stronger relation, $\delta_1 = \delta_2$ or $Z_1 = Z_2$, from the absolute normalization of the matrix elements of the vector current. However, in non-Abelian gauge theories, the corresponding vector current $j^{\mu a} = \bar{\psi}\gamma^\mu t^a \psi$ transforms under local gauge transformations in the adjoint representation. Thus the Faddeev-Popov prescription cannot be used to compute matrix elements of this current unambiguously, and thus the normalization of these matrix elements is not preserved by the perturbation theory.

16.6 Asymptotic Freedom: The Background Field Method

In the previous section, we saw that the β function of a non-Abelian gauge theory with a sufficiently small number of fermions is negative. This result is important enough that it is worthwhile to derive it twice. The preceding derivation was straightforward but not very illuminating. In this section we give a second derivation of the same result, which is more abstract but much cleaner and more transparent.

The method of this section reflects the spirit of Wilson's idea of integrating out the high-momentum degrees of freedom, while taking proper care to preserve gauge invariance. We will compute the effective action of a non-Abelian gauge theory for a fixed, slowly varying, classical background gauge field $A_\mu^a(x)$. By adopting a canonical normalization of this field, we can interpret the coefficient of the effective action as a running coupling constant. This method is analogous to Polyakov's method for computing the β function of the nonlinear sigma model, presented in Section 13.3.

Background Field Perturbation Theory

To set up the computation, rescale the gauge field $gA_\mu^a \to A_\mu^a$. In this normalization, the gauge coupling is removed from the covariant derivative and moved to the coefficient of the gauge field kinetic energy term. We thus start from the Lagrangian

$$\mathcal{L} = -\frac{1}{4g^2}(F_{\mu\nu}^a)^2 + \bar{\psi}(i\not{D})\psi, \tag{16.90}$$

with

$$D_\mu = \partial_\mu - iA_\mu^a t^a,$$
$$F_{\mu\nu}^a = \partial_\mu A_\nu^a - \partial_\nu A_\mu^a + f^{abc} A_\mu^b A_\nu^c, \tag{16.91}$$

and the fermion mass set to zero for simplicity. The transformation laws of A_μ^a and ψ are also independent of the coupling constant:

$$\delta A_\mu^a = \partial_\mu \alpha^a + f^{abc} A_\mu^b \alpha^c, \qquad \delta\psi = i\alpha^a t^a \psi. \tag{16.92}$$

On the other hand, the coupling constant g will appear in the gauge field propagator.

Next, split the gauge field into a classical background field and a fluctuating quantum field:

$$A_\mu^a \to A_\mu^a + \mathcal{A}_\mu^a. \tag{16.93}$$

We will treat the classical part A_μ^a as a fixed field configuration and the fluctuating part $\mathcal{A}_\mu^a$ as the integration variable of the functional integral. From here on, we will use the symbol D_μ to denote the covariant derivative with respect to the background field: $D_\mu = \partial_\mu - iA_\mu^a t^a$. Then

$$\bar\psi(i\slashed{D})\psi \to \bar\psi(i\slashed{D})\psi + \mathcal{A}_\mu^a \bar\psi\gamma^\mu t^a\psi. \tag{16.94}$$

The Yang-Mills field strength decomposes as follows:

$$\begin{aligned}
F_{\mu\nu}^a \to &\; \partial_\mu A_\nu^a - \partial_\nu A_\mu^a + f^{abc} A_\mu^b A_\nu^c \\
&+ \partial_\mu \mathcal{A}_\nu^a - \partial_\nu \mathcal{A}_\mu^a + f^{abc}(A_\mu^b \mathcal{A}_\nu^c - A_\nu^b \mathcal{A}_\mu^c) + f^{abc} \mathcal{A}_\mu^b \mathcal{A}_\nu^c \\
= &\; F_{\mu\nu}^a + D_\mu \mathcal{A}_\nu^a - D_\nu \mathcal{A}_\mu^a + f^{abc} \mathcal{A}_\mu^b \mathcal{A}_\nu^c,
\end{aligned} \tag{16.95}$$

where, in the last line, $F_{\mu\nu}^a$ is the field strength of the classical field, and D_μ is the covariant derivative in the adjoint representation, Eq. (15.86). Notice that, both in (16.94) and in (16.95), the derivative ∂_μ appears only as a part of the covariant derivative with respect to the background field.

If the background field A_μ^a is regarded as fixed, the Lagrangian has a local gauge symmetry implemented by transformations on $\mathcal{A}_\mu^a$:

$$\mathcal{A}_\mu^a \to \mathcal{A}_\mu^a + D_\mu \alpha^a + f^{abc} \mathcal{A}_\mu^b \alpha^c. \tag{16.96}$$

To define the functional integral, we must gauge-fix using the Faddeev-Popov procedure. We choose a gauge-fixing condition that is covariant with respect to the background gauge field:

$$G^a(A) = D^\mu \mathcal{A}_\mu^a - \omega^a, \tag{16.97}$$

instead of (16.28). The Faddeev-Popov determinant involves the variation of this operator with respect to the gauge transformation (16.96). As in Section 16.2, we can promote the gauge-fixing term to the exponent, to quantize the theory in the background field analogue of Feynman-'t Hooft gauge. Then the gauge-fixed Lagrangian is

$$\mathcal{L}_{\text{FP}} = -\frac{1}{4g^2}\left(F_{\mu\nu}^a + D_\mu A_\nu^a - D_\nu A_\mu^a * f^{abc} A_\mu^b A_\nu^c\right)^2 - \frac{1}{2g^2}(D^\mu A_\mu^a)^2$$
$$+ \bar{\psi}(i\slashed{D} + A_\mu^a \gamma^\mu t^a)\psi - \bar{c}^a(D^2)^{ac}c^c - \bar{c}^a D^\mu(f^{abc} A_\mu^b c^c). \tag{16.98}$$

The Lagrangian (16.98) is gauge-fixed, but it is invariant under a local symmetry that transforms both $\mathcal{A}_\mu^a$ and the background field A_μ^a:

$$\begin{aligned} A_\mu^a &\to A_\mu^a + D_\mu \beta^a \\ \mathcal{A}_\mu^a &\to \mathcal{A}_\mu^a - f^{abc}\beta^b \mathcal{A}_\mu^c \\ \psi &\to \psi + i\beta^a t^a \psi \\ c^a &\to c^a - f^{abc}\beta^b c^c. \end{aligned} \tag{16.99}$$

Under this transformation, $\mathcal{A}_\mu^a$ transforms as a matter field in the adjoint representation, while A_μ^a carries the part of the local gauge transformation proportional to $\partial_\mu \beta^a$. To prove that (16.99) is a symmetry of (16.98), we need only note that (16.98) is globally invariant, and that A_μ^a appears in (16.98) only as a part of the covariant derivative and the field strength. The transformation (16.99) is also a symmetry of the functional measure. Thus, if we functionally integrate over $\mathcal{A}_\mu^a$, ψ, and c^a to compute the effective action, the result must be invariant to local gauge transformations of A_μ^a. This observation greatly simplifies the analysis of the effective action.

One-Loop Correction to the Effective Action

Let us now compute the effective action, using the method of Section 11.4. To compute $\Gamma[A_\mu^a]$ to one-loop order, we drop terms linear in the fluctuating field $\mathcal{A}_\mu^a$ and then integrate over the terms quadratic in $\mathcal{A}_\mu^a$ and the fermion and ghost fields. This produces functional determinants, which we can evaluate into an appropriate form for an effective action.

To carry out this program, we must work out the terms in (16.98) quadratic in each of the various fields. The terms quadratic in $\mathcal{A}_\mu^a$ are:

$$\mathcal{L}_\mathcal{A} = -\frac{1}{2g^2}\{\tfrac{1}{2}(D_\mu \mathcal{A}_\nu^a - D_\nu \mathcal{A}_\mu^a)^2 + F^{a\mu\nu}f^{abc}\mathcal{A}_\mu^b \mathcal{A}_\nu^c + (D^\mu \mathcal{A}_\mu^a)^2\}. \tag{16.100}$$

After integrating by parts, we can rewrite this as

$$\mathcal{L}_\mathcal{A} = -\frac{1}{2g^2}\{\mathcal{A}_\mu^a[-(D^2)^{ab}g^{\mu\nu} + (D^\nu D^\mu)^{ab} - (D^\mu D^\nu)^{ab}]\mathcal{A}_\nu^b - \mathcal{A}_\mu^a f^{abc} F^{b\mu\nu}\mathcal{A}_\nu^c\}. \tag{16.101}$$

The term in brackets contains the commutator of covariant derivatives. This can be simplified using (15.48); the result combines with the last term to give

$$\mathcal{L}_\mathcal{A} = -\frac{1}{2g^2}\{\mathcal{A}_\mu^a[-(D^2)^{ac}g^{\mu\nu} - 2f^{abc}F^{b\mu\nu}]\mathcal{A}_\nu^c\}. \tag{16.102}$$

The first term is part of a covariant d'Alembertian operator. The second term seems quite special, but we can put it into a form that will be convenient later as follows: First, we recognize that $F_{\mu\nu}^b$ is contracted with a group generator in the adjoint representation. Next, we introduce the matrix (3.18) that is the generator of Lorentz transformations on 4-vectors:

$$(\mathcal{J}^{\rho\sigma})_{\alpha\beta} = i(\delta_\alpha^\rho \delta_\beta^\sigma - \delta_\alpha^\sigma \delta_\beta^\rho). \tag{16.103}$$

With these replacements, we can write (16.102) in the form

$$\mathcal{L}_\mathcal{A} = -\frac{1}{2g^2}\{\mathcal{A}_\mu^a[-(D^2)^{ac}g^{\mu\nu} + 2(\tfrac{1}{2}F_{\rho\sigma}^b\mathcal{J}^{\rho\sigma})^{\mu\nu}(t_G^b)^{ac}]\mathcal{A}_\nu^c\}. \tag{16.104}$$

The object in brackets can be considered as a generalized d'Alembertian for fluctuations on the background field.

Next, we reduce the quadratic terms in fermion fields in a similar way. The quadratic Lagrangian for the fermion field is

$$\mathcal{L}_\psi = \bar{\psi}(i\not{D})\psi. \tag{16.105}$$

Integrating over the fermion fields, we find the determinant of the operator $(i\not{D})$. This is conveniently expressed as the square root of the determinant of the operator

$$\begin{aligned}
(i\not{D})^2 &= -\gamma^\mu\gamma^\nu D_\mu D_\nu \\
&= (-\tfrac{1}{2}\{\gamma^\mu, \gamma^\nu\} - \tfrac{1}{2}[\gamma^\mu, \gamma^\nu])D_\mu D_\nu \\
&= -D^2 + 2i(\tfrac{i}{4}[\gamma^\mu, \gamma^\nu])D_\mu D_\nu.
\end{aligned} \tag{16.106}$$

In the last line, the commutator of Dirac matrices forms the generator of Lorentz transformations in the spinor representation, $S^{\mu\nu}$ (3.23). Since this object is antisymmetric in its indices, the product $D_\mu D_\nu$ that is contracted with it can be replaced by half of their commutator. Then (16.106) takes the form

$$(i\not{D})^2 = -D^2 + 2(\tfrac{1}{2}F_{\rho\sigma}^b S^{\rho\sigma})t^b, \tag{16.107}$$

where t^a is now given in the representation of the fermions. This is just the d'Alembertian in (16.104), rewritten for the new set of spin and gauge quantum numbers. If the theory contains n_f species of fermions, the fermionic functional integral gives the determinant of (16.107) raised to the power $n_f/2$.

The quadratic term in ghosts is simply

$$\mathcal{L}_c = \bar{c}^a[-(D^2)^{ab}]c^b; \tag{16.108}$$

This contains the same d'Alembertian operator written for the case of spin zero.

To summarize all of these results, we define the general covariant background-field d'Alembertian as

$$\Delta_{r,j} = -D^2 + 2(\tfrac{1}{2}F_{\rho\sigma}^b\mathcal{J}^{\rho\sigma})t^b, \tag{16.109}$$

acting on a field of representation r and spin j. The square of the covariant derivative gives the normal, convective, minimal coupling of the particle described by $\Delta_{r,j}$ to the gauge field. The additional term is a magnetic moment interaction with the gauge field, whose strength corresponds to a g-factor $g = 2$. Using this general expression, we can write the effective action for the classical fields A^a_μ, to one-loop order, as

$$
\begin{aligned}
e^{i\Gamma[A]} &= \int \mathcal{D}A\mathcal{D}\psi\mathcal{D}c \, \exp\left[i\int d^4x\,(\mathcal{L}_{\text{FP}} + \mathcal{L}_{\text{c.t.}})\right] \\
&= \exp\left[i\int d^4x\,(-\frac{1}{4g^2}(F^a_{\mu\nu})^2 + \mathcal{L}_{\text{c.t.}})\right] \\
&\quad \cdot \left(\det\Delta_{G,1}\right)^{-1/2}\left(\det\Delta_{r,1/2}\right)^{+n_f/2}\left(\det\Delta_{G,0}\right)^{+1},
\end{aligned}
\tag{16.110}
$$

where $\mathcal{L}_{\text{c.t.}}$ is the counterterm Lagrangian and the three determinants are the results of evaluating the gauge field, fermion, and ghost functional integrals. Additional loop corrections to the effective action are suppressed by another factor of g^2.

Since each integral contributing to (16.110) is invariant to (16.99), each determinant will be a gauge-invariant functional of A^a_μ. If we expand the determinants in powers of the background field, we should then find a series of terms that begins

$$
\log\det\Delta_{r,j} = i\int d^4x\,\left(\frac{1}{4}\mathbf{C}_{r,j}(F^a_{\mu\nu})^2 + \cdots\right),
\tag{16.111}
$$

where the succeeding terms contain higher-dimension gauge-invariant operators. The coefficient $\mathbf{C}_{r,j}$ can depend on the representation r and the spin j. This first term of the expansion modifies the zeroth-order effective action according to

$$
\frac{1}{4g^2}(F^a_{\mu\nu})^2 \rightarrow \frac{1}{4}\left(\frac{1}{g^2} + \frac{1}{2}\mathbf{C}_{G,1} - \mathbf{C}_{G,0} - \frac{n_f}{2}\mathbf{C}_{r,1/2}\right)(F^a_{\mu\nu})^2.
\tag{16.112}
$$

The factors $\mathbf{C}_{r,j}$ are dimensionless but, since they arise from a one-loop computation, we should expect that they are logarithmically divergent:

$$
\mathbf{C}_{r,j} = c_{r,j}\log\frac{\Lambda^2}{k^2} + \cdots,
\tag{16.113}
$$

where k is a momentum characterizing the variation of the background field. The counterterm δ_3 removes the divergence; if we impose a renormalization condition at the scale M, then the addition of (16.113) and its counterterm gives the result (16.112) with the replacement

$$
\mathbf{C}_{r,j} = c_{r,j}\log\frac{M^2}{k^2} + \cdots.
\tag{16.114}
$$

Then the original fixed coupling constant in the effective action is replaced by a running coupling constant

$$\frac{1}{g^2(k^2)} = \frac{1}{g^2} + \left(\frac{1}{2}c_{G,1} - c_{G,0} - \frac{n_f}{2}c_{r,1/2}\right)\log\frac{M^2}{k^2}, \tag{16.115}$$

or

$$g^2(k^2) = \frac{g^2}{1 - \left(\frac{1}{2}c_{G,1} - c_{G,0} - \frac{n_f}{2}c_{r,1/2}\right)g^2\log k^2/M^2}. \tag{16.116}$$

By comparing this form to Eq. (12.88), we see that this running coupling constant is the solution to the renormalization group equation for the β function

$$\beta(g) = \left(\frac{1}{2}c_{G,1} - c_{G,0} - \frac{n_f}{2}c_{r,1/2}\right)g^3. \tag{16.117}$$

Thus, by calculating the $c_{r,j}$, we can directly obtain the leading coefficient of the β function.

Computation of the Functional Determinants

To compute $c_{r,j}$, we must work out the first term in the expansion of the determinant in powers of the external field. To expand the determinant, we proceed as in the example in Section 9.5. Write

$$\Delta_{r,j} = -\partial^2 + \Delta^{(1)} + \Delta^{(2)} + \Delta^{(\mathcal{J})}, \tag{16.118}$$

where

$$\Delta^{(1)} = i[\partial^\mu A_\mu^a t^a + A_\mu^a t^a \partial^\mu]$$
$$\Delta^{(2)} = A^{a\mu} t^a A_\mu^b t^b \tag{16.119}$$
$$\Delta^{(\mathcal{J})} = 2(\tfrac{1}{2}F_{\rho\sigma}^b \mathcal{J}^{\rho\sigma})t^b.$$

The pieces $\Delta^{(1)}$ and $\Delta^{(\mathcal{J})}$ contain one power of the external field; $\Delta^{(2)}$ contains two powers of A_μ^a. Treating these terms as perturbations, we write

$$\log\det\Delta_{r,j} = \log\det[-\partial^2 + (\Delta^{(1)} + \Delta^{(2)} + \Delta^{(\mathcal{J})})]$$
$$= \log\det[-\partial^2] + \log\det[1 + (-\partial^2)^{-1}(\Delta^{(1)} + \Delta^{(2)} + \Delta^{(\mathcal{J})})]$$
$$= \log\det[-\partial^2] + \mathrm{tr}\log[1 + (-\partial^2)^{-1}(\Delta^{(1)} + \Delta^{(2)} + \Delta^{(\mathcal{J})})]$$
$$= \log\det[-\partial^2] + \mathrm{tr}[(-\partial^2)^{-1}(\Delta^{(1)} + \Delta^{(2)} + \Delta^{(\mathcal{J})}) + \cdots].$$
$$\tag{16.120}$$

The first term of the right in (16.120) is an irrelevant constant. The terms in this expansion that are linear in A_μ^a vanish by gauge invariance (or, more explicitly, because $\mathrm{tr}[t^a] = 0$). The quadratic terms in A_μ^a must organize themselves into the structure of (16.111), plus terms with higher derivatives.

Figure 16.10. Terms quadratic in the external field in the expansion of $\log \det \Delta_{r,j}$. The special vertex arises from the $F^{\rho\sigma} \mathcal{J}_{\rho\sigma}$ coupling.

The terms in (16.111) quadratic in A_μ^a can be written in Fourier space as

$$\log \det \Delta_{r,j} = \frac{i}{2} \int \frac{d^4k}{(2\pi)^4} A_\mu^a(-k) A_\nu^b(k) (k^2 g^{\mu\nu} - k^\mu k^\nu) \cdot [\mathbf{C}_{r,j} + \mathcal{O}(k^2)].$$

$$(16.121)$$

We will now compute these terms explicitly from (16.120) and bring them into the form of (16.121). The terms with two powers of A_μ^a in the expansion (16.120) are those with one power of $\Delta^{(2)}$ or two powers of $\Delta^{(1)}$ or $\Delta^{(\mathcal{J})}$. Further, terms linear in $\Delta^{(\mathcal{J})}$ are proportional to $\text{tr}[\mathcal{J}^{\rho\sigma}] = 0$, so the cross term between these two structures vanishes. The three remaining contributions correspond to the Feynman diagrams shown in Fig. 16.10.

The term involving two powers of $\Delta^{(1)}$ is

$$-\tfrac{1}{2} \text{tr}\big[(-\partial^2)^{-1} \Delta^{(1)} (-\partial^2)^{-1} \Delta^{(1)}\big] = \quad \text{}$$

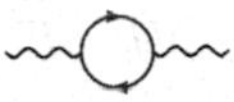

$$= -\tfrac{1}{2} \int \frac{d^4k}{(2\pi)^4} A_\mu^a A_\nu^b \int \frac{d^4p}{(2\pi)^4} \, \text{tr} \, \frac{1}{p^2} (2p+k)^\mu t^a \frac{1}{(p+k)^2} (2p+k)^\nu t^b,$$

$$(16.122)$$

where the trace is now simply a trace over gauge and spin indices. The factor $1/2$ comes from the expansion of the logarithm. The term involving one power of $\Delta^{(2)}$ is

$$\text{tr}\big[(-\partial^2)^{-1} \Delta^{(2)}\big] = \quad \text{}$$

$$(16.123)$$

$$= \int \frac{d^4k}{(2\pi)^4} A_\mu^a A_\nu^b \int \frac{d^4p}{(2\pi)^4} \, \text{tr} \, \frac{1}{p^2} g^{\mu\nu} t^a t^b.$$

As Fig. 16.10 suggests, these two contributions are precisely proportional to the contribution of a scalar particle to the QED vacuum polarization, times the factor

$$\text{tr}[t^a t^b] = C(r) d(j) \delta^{ab}, \qquad (16.124)$$

where $d(j)$ is the number of spin components. The values of the diagrams can be worked out using the methods of the previous section (or simply recalled from Problem 9.1). One finds that the two diagrams together sum up to the gauge-invariant form (16.121), to give

$$\tfrac{1}{2} \int \frac{d^4k}{(2\pi)^4} A_\mu^a(-k) A_\nu^b(k) (k^2 g^{\mu\nu} - k^\mu k^\nu) \cdot \Big[i \frac{C(r) d(j)}{3(4\pi)^2} \Gamma(2-\tfrac{d}{2}) + \cdots \Big]. \quad (16.125)$$

The term involving two powers of $\Delta^{(\mathcal{J})}$ is

$$-\tfrac{1}{2}\operatorname{tr}\left[(-\partial^2)^{-1}\Delta^{(\mathcal{J})}(-\partial^2)^{-1}\Delta^{(\mathcal{J})}\right] = \quad\text{(diagram)}$$

$$= -\tfrac{1}{2}\int\frac{d^4k}{(2\pi)^4}A_\mu^a A_\nu^b\int\frac{d^4p}{(2\pi)^4}\operatorname{tr}\frac{1}{p^2}(2ik_\rho g_{\mu\sigma}\mathcal{J}^{\rho\sigma})t^a\frac{1}{(p+k)^2}(-2ik_\alpha g_{\nu\beta}\mathcal{J}^{\alpha\beta})t^b.$$
$$(16.126)$$

To evaluate this, define $C(j)$ as the trace over spin indices

$$\operatorname{tr}[\mathcal{J}^{\rho\sigma}\,\mathcal{J}^{\alpha\beta}] = (g^{\rho\alpha}g^{\sigma\beta} - g^{\rho\beta}g^{\sigma\alpha})C(j). \qquad (16.127)$$

It is straightforward to work out from the explicit expressions that

$$C(j) = \begin{cases} 0 & \text{scalars;} \\ 1 & \text{Dirac spinors;} \\ 2 & \text{4-vectors.} \end{cases} \qquad (16.128)$$

Then (16.126) can be evaluated as

$$-\frac{1}{2}\int\frac{d^4k}{(2\pi)^4}A_\mu^a A_\nu^b\int\frac{d^4p}{(2\pi)^4}\frac{1}{p^2}\frac{1}{(p+k)^2}(k^2 g^{\mu\nu} - k^\mu k^\nu)4C(r)C(j)$$

$$= \frac{1}{2}\int\frac{d^4k}{(2\pi)^4}A_\mu^a(-k)A_\nu^b(k)(k^2 g^{\mu\nu} - k^\mu k^\nu)\left(-i\frac{4C(r)C(j)}{(4\pi)^2}\Gamma(2-\tfrac{d}{2}) + \cdots\right).$$
$$(16.129)$$

Adding (16.125) and (16.129), we find that the coefficient $\mathbf{C}_{r,j}$ in (16.111) is given by

$$\mathbf{C}_{r,j} = \frac{1}{(4\pi)^2}\left[\tfrac{1}{3}d(j) - 4C(j)\right]C(r)\Gamma(2-\tfrac{d}{2}). \qquad (16.130)$$

Thus,

$$c_{r,j} = \frac{1}{(4\pi)^2}\left[\tfrac{1}{3}d(j) - 4C(j)\right]C(r), \qquad (16.131)$$

or explicitly,

$$c_{r,j} = \frac{C(r)}{(4\pi)^2}\cdot\begin{cases} +1/3 & \text{scalars;} \\ -8/3 & \text{Dirac spinors;} \\ -20/3 & \text{4-vectors.} \end{cases} \qquad (16.132)$$

Notice that, whenever the magnetic moment term is nonzero, it dominates, and that its coefficient is opposite in sign from the convective term.

Inserting the values from (16.132) into (16.117), we find

$$\beta(g) = -\frac{g^3}{(4\pi)^2}\left(\frac{11}{3}C_2(G) - \frac{4}{3}n_f C(r)\right). \qquad (16.133)$$

We thus confirm the conclusion of the previous section, that non-Abelian gauge theories with sufficiently few fermions are asymptotically free.

16.7 Asymptotic Freedom: A Qualitative Explanation

In the previous two sections[†] we twice calculated the β function in non-Abelian gauge theory:

$$\beta(g) = -\frac{g^3}{(4\pi)^2}\left(\frac{11}{3}C_2(G) - \frac{4}{3}n_f C(r)\right). \tag{16.134}$$

Here n_f is the number of fermion species in representation r, $C(r)$ is the constant appearing in the orthogonality relation (15.78) for the representation matrices, and $C_2(G)$ is the quadratic Casimir operator of the adjoint representation of the group, defined in Eq. (15.92). In an $SU(N)$ gauge theory with fermions in the fundamental representation, this result becomes

$$\beta(g) = -\frac{g^3}{(4\pi)^2}\left(\frac{11}{3}N - \frac{2}{3}n_f\right). \tag{16.135}$$

The overall minus sign implies that, for sufficiently small n_f, non-Abelian gauge theories are asymptotically free. In this case the running coupling constant tends to zero at large momenta, according to Eq. (12.92):

$$g^2(k) = \frac{g^2}{1 + \frac{g^2}{(4\pi)^2}\left(\frac{11}{3}N - \frac{2}{3}n_f\right)\log(k^2/M^2)}. \tag{16.136}$$

The asymptotic freedom of non-Abelian gauge theories is a surprising conclusion. When we first encountered the running of the electromagnetic coupling in Section 7.5, we found it easy to understand the direction of the coupling constant flow: The vacuum acquires a dielectric property due to virtual electron-positron pair creation, causing the effective electric charge to decrease at large distances. In non-Abelian gauge theories, according to Eq. (16.134), the fermions still produce such an effect. Furthermore, since the non-Abelian gauge bosons are charged, they should produce an additional screening effect. But according to Eq. (16.134), the net effect of the non-Abelian gauge bosons is *opposite* in sign. Apparently there must be other, competing, effects, which overcome the effect of screening and change the sign of the β function.

The precise form of these effects depends on the gauge. They are simplest to describe in the Coulomb gauge, for which the gauge fixing condition is

$$\partial_i A^{ai} = 0. \tag{16.137}$$

We will not work out the full quantization in this gauge; rather, we will just describe its qualitative features. As in electrodynamics, the field quanta in Coulomb gauge are described in a non-Lorentz-covariant manner as transversely polarized photons. There are no timelike or longitudinal photons and

---

[†]Section 16.7 draws on the main result of 16.5 and 16.6, but does not depend on these earlier sections. However, even if you have not read Section 16.5, you may wish to skim pages 522 through 531 to get an overview of how the β function can be calculated.

no propagating ghosts. However, there is a Coulomb potential, described by the field A^{a0}, which obeys a constraint equation analogous to Gauss's law. Not surprisingly, in the non-Abelian case, Gauss's law takes the gauge-covariant form

$$D_i E^{ai} = g\rho^a, \qquad (16.138)$$

where $E^{ai} = F^{a0i}$ and ρ^a is the charge density of the global symmetry current of the fermions. Recall from Eq. (15.86) that the covariant derivative acting on a field in the adjoint representation is

$$(D_\mu \phi)^a = \partial_\mu \phi^a + g f^{abc} A_\mu^b \phi^c.$$

To analyze the consequences of Eq. (16.138), we will choose an example as simple and explicit as possible. Let the gauge group be $SU(2)$, so that $a = 1, 2, 3$ and $f^{abc} = \epsilon^{abc}$. Let us compute the Coulomb potential of a point charge of magnitude $+1$ with the orientation $a = 1$. We will solve for E^{ai} using an iteration process, putting the gauge-field term of the covariant derivative on the right-hand side of the equation:

$$\partial_i E^{ai} = g\delta^{(3)}(\mathbf{x})\delta^{a1} + g\epsilon^{abc} A^{bi} E^{ci}. \qquad (16.139)$$

The second term on the right shows that, in a non-Abelian gauge theory, a region containing vector potentials and electric fields that are parallel in physical space and perpendicular in the group space is a source of electric field.

The implication of Eq. (16.139) is worked out pictorially in Fig. 16.11. The leading term on the right-hand side of (16.139) implies a $1/r^2$ electric field of type $a = 1$ radiating from $\mathbf{x} = 0$. Somewhere in space, this electric field will cross with a bit of vector potential A^{ai} arising as a fluctuation of the vacuum. For definiteness, let us assume that this fluctuation has $a = 2$ and points in some diagonal direction, as shown in Fig. 16.11(a). Then the second term on the right-hand side of Eq. (16.139) is negative for $a = 3$: There is a sink of the field E^{3i} at this location, as shown in Fig. 16.11(b). These new fields are, in two locations, parallel or antiparallel to the original A^{ai} field fluctuation. Looking again at the second term of Eq. (16.139), we see that there is a source of electric field with $a = 1$ closer to the origin, and a sink of electric field with $a = 1$ farther away. This is an induced electric dipole in the vacuum, shown in Fig. 16.11(c). But look at the signs: This dipole is oriented toward the original charge, and thus serves to amplify rather than screen it! The effect of the original charge thus becomes stronger at larger distances.

The competition between this antiscreening effect and the screening due to virtual pairs of gauge bosons must be worked out quantitatively. When this is done,[‡] one finds that the antiscreening effect is 12 times larger.

In this argument, it is a set of dynamical features peculiar to the non-Abelian gauge theory that enables the coupling constant to be amplified rather

---

[‡]T. Appelquist, M. Dine, and I. Muzinich, *Phys. Lett.* **69B**, 231 (1977).

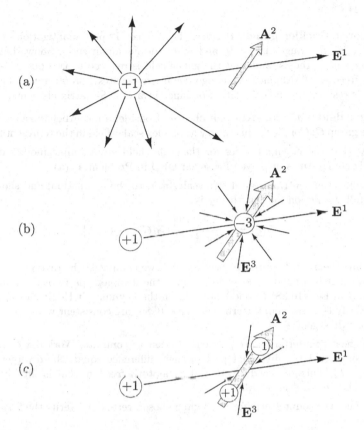

Figure 16.11. The effect of vacuum fluctuations on the Coulomb field of an $SU(2)$ gauge theory. In (a), a fluctuation $\mathbf{A}^2$ occurs on top of the $1/r^2$ field $\mathbf{E}^1$. These combined fields generate a sink of the field $\mathbf{E}^3$, as shown in (b). The $\mathbf{E}^3$ field, in turn, combines with $\mathbf{A}^2$ to create an effective $\mathbf{E}^1$ dipole, shown in (c). The dipole points toward the original charge, enhancing its field at large distances.

than screened at large distances. This suggests that asymptotic freedom might be a special property of non-Abelian gauge theories. Although the statement can be proved only by exhausting other cases, it does actually turn out to be true: Among renormalizable quantum field theories in four spacetime dimensions, only the non-Abelian gauge theories are asymptotically free.* We have already seen in Chapter 14 that asymptotic freedom was suggested experimentally as a property of the strong interactions. In the following chapter we will build a model of the strong interactions out of a non-Abelian gauge theory and explore its properties in detail.

---

*S. Coleman and D. J. Gross, *Phys. Rev. Lett.* **31**, 851 (1973).

Problems

16.1 Arnowitt-Fickler gauge. Perform the Faddeev-Popov quantization of Yang-Mills theory in the gauge $A^{3a} = 0$, and write the Feynman rules. Show that there are no propagating ghosts, and that the gauge field is reduced to two positive-metric degrees of freedom. (Although the gauge condition violates Lorentz invariance, this symmetry is restored in the calculation of gauge-invariant S-matrix elements.)

16.2 Scalar field with non-Abelian charge. Consider a non-Abelian gauge theory with gauge group G. Couple to this theory a complex scalar field in the representation r.

   (a) Show that the Feynman rules for the scalar field are a simple modification of the Feynman rules displayed for scalar QED in Problem 9.1(a).

   (b) Compute the contribution of this scalar field to the β function, and show that the full β function for this theory is

$$\beta(g) = -\frac{g^3}{(4\pi)^2}\left(\frac{11}{3}C_2(G) - \frac{1}{3}C(r)\right).$$

16.3 Counterterm relations. In Section 16.5, we computed the divergent parts of δ_1, δ_2, and δ_3. It is a good exercise to compute the divergent parts of the remaining counterterms in Eq. (16.88) to one-loop order in the Feynman-'t Hooft gauge, and to explicitly verify that the counterterm relations (16.89) are consistent with the removal of ultraviolet divergences.

   (a) The ghost counterterms are particularly easy to compute. Work out δ_1^c and δ_2^c, and show that the divergent part of their difference equals the divergent part of $\delta_1 - \delta_2$. This gives a derivation of asymptotic freedom that is slightly easier than the one in Section 16.5.

   (b) Compute the counterterm for the 3-gauge-boson vertex and verify the first equality in (16.89).

   (c) Compute the counterterm for the 4-gauge-boson vertex and find, when the smoke clears, the second relation in (16.89).

Chapter 17

Quantum Chromodynamics

The key to constructing a realistic model of the strong interactions is the phenomenon of asymptotic freedom. Chapter 14 described the experimental discovery of this phenomenon, while Chapter 16 presented the theoretical proof that non-Abelian gauge theories are asymptotically free. We are now ready to explore the consequences of these discoveries.

We will begin by arguing that the most natural candidate for a model of the strong interactions is the non-Abelian gauge theory with gauge group $SU(3)$, coupled to fermions (quarks) in the fundamental representation. This theory is known as *Quantum Chromodynamics*, or QCD. After some general discussion of QCD in Section 17.1, we will investigate a number of specific QCD scattering processes in Sections 17.2 through 17.4. The most interesting application of QCD, however, is of a somewhat more sophisticated nature; it comes in the prediction of a pattern of slow violations of the Bjorken scaling relation discussed in Chapter 14. Section 17.5 develops the additional theoretical tools that are needed to understand these violations.

Although this chapter includes many references to experiments, we remind the reader that, for QCD as for QED or critical phenomena, this book is primarily a textbook of theoretical methods rather than a review and interpretation of experimental data. The details of experimental techniques and results on strong interaction physics are reviewed in a number of excellent texts (see the bibliography). We hope that this chapter will give the theoretical foundation necessary to illuminate and interpret these results.

17.1 From Quarks to QCD

Our current theoretical picture of the strong interactions began with the identification of the elementary fermions that make up the proton and other hadrons. As the properties of these fermions became better understood, the nature of their interactions became tightly constrained, in a way that led eventually to a unique candidate theory. In order to appreciate the uniqueness of this theory, we begin this chapter with a simplified history of how it arose.

In 1963, Gell-Mann and Zweig proposed a model that explained the spectrum of strongly interacting particles in terms of elementary constituents called *quarks*. Mesons were expected to be quark-antiquark bound states. Indeed, the lightest mesons have just the correct quantum numbers to justify this interpretation; they are spin-0 and spin-1 states of odd parity, just as we found for fermion-antifermion bound states of zero orbital angular momentum in Chapter 3. Baryons were interpreted as bound states of three quarks. To explain the electric charges and other quantum numbers of hadrons, Gell-Mann and Zweig needed to assume three species of quarks, up (u), down (d), and strange (s). Additional hadrons discovered since that time require the existence of three more species: charm (c), bottom (b), and top (t). To make baryons with integer charges, the quarks needed to be assigned fractional electric charge: $+2/3$ for u, c, t, and $-1/3$ for d, s, b. Then, for example, the proton would be a bound state of uud, while the neutron would be a bound state of udd. The six types of quarks are conventionally referred to as *flavors*.

The quark model had great success in predicting new hadronic states, and in explaining the strengths of electromagnetic and weak-interaction transitions among different hadrons. In particular, the quark model naturally incorporates the most important symmetry relations among strongly interacting particles. If one assumes that the u and d quarks have identical masses and interactions, the $SU(2)$ group that acts as a unitary rotation of u and d states,

$$\begin{pmatrix} u \\ d \end{pmatrix} \to U \begin{pmatrix} u \\ d \end{pmatrix}, \tag{17.1}$$

should be a symmetry of the strong interactions. Indeed, both in nuclear and in elementary particle physics, the quantum states form multiplets of this $SU(2)$ symmetry, called *isotopic spin* or *isospin*. Similarly, since the strange quark is only a little heavier than the u and d quarks, it makes sense to consider the symmetry of unitary transformations of the triplet (u, d, s). Gell-Mann and Ne'eman showed that the elementary particles naturally fill out irreducible representations of this $SU(3)$ symmetry.

Despite the phenomenological success of the original quark model, it had two serious problems. First, despite considerable effort, free particles with fractional charge could not be found. Second, the spectrum of baryons required the assumption that the wavefunction of the three quarks be totally symmetric under the interchange of the quark spin and flavor quantum numbers, contradicting the expectation that quarks, which must have spin $1/2$, should obey Fermi-Dirac statistics. The need for this symmetry is most clearly illustrated in the fact that one of the lightest excited states of the nucleon is a spin-$3/2$ particle with charge $+2$, the Δ^{++}. This particle is readily interpreted as a uuu bound state with zero orbital angular momentum and all three quark spins parallel.

To reconcile the baryon spectrum with the spin-statistics theorem, Han and Nambu, Greenberg, and Gell-Mann proposed that quarks carry an additional, unobserved quantum number, called *color*. They introduced the *ad hoc* assumption that baryon wavefunctions must be totally antisymmetric in color quantum numbers. Then, if the quark wavefunctions are totally symmetric in spin and flavor, they are totally antisymmetric overall, in agreement with Fermi-Dirac statistics. The simplest model of color would be to assign quarks to the fundamental representation of a new, internal $SU(3)$ global symmetry. Suppressing for a moment the spin and flavor quantum numbers, we can represent quarks by q_i, where $i = 1, 2, 3$ is the color index. Thus quarks transform under the fundamental, or "3", representation of the color $SU(3)$ symmetry. Antiquarks, $\bar{q}^i$, transform in the $\bar{3}$ representation. The inner product of a 3 and a $\bar{3}$ is an invariant of $SU(3)$. One can also form an invariant by using the totally antisymmetric combination of three 3's, ϵ_{ijk}: This object transforms under a unitary transformation according to

$$\epsilon_{ijk} \rightarrow U_{ii'}U_{jj'}U_{kk'}\epsilon_{i'j'k'} = (\det U)\epsilon_{ijk}, \qquad (17.2)$$

and so is invariant under $SU(3)$ transformations, which have $\det U = 1$. Under the postulate that all hadron wavefunctions must be invariant under $SU(3)$ symmetry transformations, these two types of combinations are the only simple ones allowed:

$$\bar{q}^i q_i, \qquad \epsilon^{ijk} q_i q_j q_k, \qquad \epsilon_{ijk} \bar{q}^i \bar{q}^j \bar{q}^k. \qquad (17.3)$$

That is, the assumption that physical hadrons are singlets under color implies that the only possible light hadrons are the mesons, baryons, and antibaryons.

Like the original quark model, the color hypothesis was phenomenologically successful but raised additional questions: Why should quarks have this seemingly superfluous property, and what mechanism insures that all hadron wavefunctions are color singlets? The answers to these questions came not from hadron spectroscopy, but from the deep-inelastic scattering experiments described in Chapter 14 and the ensuing search for a theory of parton binding with the property of asymptotic freedom. When it was discovered that non-Abelian gauge theories have this property, all that remained was to identify the correct gauge group and fermion representation. Since the color symmetry had no other obvious physical role, it was natural to identify this symmetry with the gauge group, with the colors being the gauge quantum numbers of the quarks. This reasoning resulted in a model of the strong interactions as a system of quarks, of the various flavors, each assigned to the fundamental representation of the local gauge group $SU(3)$. The quanta of the $SU(3)$ gauge field are called *gluons*, and the theory is known as Quantum Chromodynamics, or QCD.

In this book, we will mainly investigate the properties of QCD in the high-energy regime, where the coupling constant has become small. However, we should point out that one can also study QCD in the regime of strong coupling, using an approximation scheme introduced by Wilson in which the continuum gauge theory is replaced by a discrete statistical mechanical system on a four-dimensional Euclidean lattice. Using this approximation, Wilson showed that, for sufficiently strong coupling, QCD exhibits *confinement of color*: The only finite-energy asymptotic states of the theory are those that are singlets of color $SU(3)$. Thus the *ad hoc* assumption that explains the

Figure 17.1. Gauge electric field configuration associated with the separation of color sources in a strong-coupling gauge theory.

spectrum of hadrons turns out to be a consequence of the non-Abelian gauge theory coupling to color. If one attempts to separate a color-singlet state into colored components—for example, to dissociate a meson into a quark and an antiquark—a tube of gauge field forms between the two sources, as shown in Fig. 17.1. In a non-Abelian gauge theory with sufficiently strong coupling, this tube has fixed radius and energy density, so the energy cost of separating color sources grows proportionally to the separation. A force law of this type can consistently be weak at short distances and strong at long distances, accounting for the fact that isolated quarks are not observed. We will discuss the large-distance, strong-coupling limit of gauge theories further in the Epilogue.

The short-distance limit of Quantum Chromodynamics can be readily studied using the Feynman diagram technology that we have developed in previous chapters. Here asymptotic freedom makes the coupling weak, and there is a sensible diagrammatic perturbation theory that begins from the model of free quarks and gluons. The following sections treat the elementary interactions among quarks and gluons that can be observed in high-energy experiments.

17.2 e^+e^- Annihilation into Hadrons

The simplest reaction involving quarks is the production of quark pairs in e^+e^- annihilation, a process that we treated already in Section 5.1. There we analyzed this process only at the most elementary level, viewing it as a pure QED reaction in which free quarks are created by a virtual photon. The diagram for this lowest-order process is shown in Fig. 17.2(a). The computation of the total cross section includes a sum over the various color states of the quark fields, and so provides a confirmation that the number of allowed colors is 3. Combining the color factor with the square of the quark electric charges, we found (Eq. (5.16))

$$\sigma(e^+e^- \rightarrow \text{hadrons}) = \sigma_0 \cdot 3 \cdot \sum_f Q_f^2, \qquad (17.4)$$

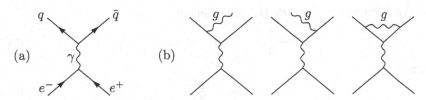

Figure 17.2. Diagrams contributing to the process $e^+e^- \to$ hadrons in QCD: (a) the leading-order diagram; (b) corrections of order α_s.

where σ_0 is the QED cross section for $e^+e^- \to \mu^+\mu^-$,

$$\sigma_0 = \frac{4\pi\alpha^2}{3s}, \tag{17.5}$$

and the sum in (17.4) is taken over quark flavors. This formula assumes that the center of mass energy is high enough that we can ignore the quark masses.

When we couple the quarks to an $SU(3)$ gauge theory, we add many important processes that affect both the value of this cross section and the final states that it includes. Some of the most important effects cannot be discussed within the context of perturbation theory. In particular, though the leading diagram contains free quarks, the particles that emerge from the reaction are color-singlet mesons and baryons. However, we will find that QCD perturbation theory with quarks and gluons does make a number of important predictions for e^+e^- annihilation to hadrons. The ideas that we develop in working out these predictions will also apply to many other strong-interaction processes.

Total Cross Section

The leading corrections to the rate of e^+e^- annihilation due to gluon exchange and emission are shown in Fig. 17.2(b). These are precisely the diagrams computed in the Final Project of Part I. The first two diagrams give a cross section of order g^2, where g is the $SU(3)$ gauge coupling, to produce a gluon in addition to the quark and antiquark. The third diagram must be summed in the amplitude with the leading diagram to produce a correction to the rate of $q\bar{q}$ production without gluon emission. In Part I, we computed these two contributions as if the strong interactions were an Abelian gauge theory. To obtain the corresponding expressions in QCD, we need only multiply the Abelian formulae by the group theory factor

$$\mathrm{tr}[t^a t^a] = C_2(r) \cdot \mathrm{tr}[1] = \frac{4}{3} \cdot 3, \tag{17.6}$$

where we have used Eq. (15.97) to evaluate $C_2(r)$ for the fundamental representation of $SU(3)$. The factor of 3 is the same color sum that appears in Eq. (17.4). Thus we can obtain the correct formulae for QCD from those of the Final Project by making the replacement

$$g^2 \to \frac{4}{3}g^2, \qquad \text{or} \qquad \alpha_g \to \frac{4}{3}\alpha_s, \tag{17.7}$$

where

$$\alpha_s = \frac{g^2}{4\pi} \tag{17.8}$$

is the strong-interaction analogue of the fine-structure constant.

The end result of the Final Project of Part I was a formula for the total cross section to produce hadrons in e^+e^- annihilation. If we replace α_g with $(4/3)\alpha_s$, that result becomes

$$\sigma(e^+e^- \to \text{hadrons}) = \sigma_0 \cdot \left(3\sum_f Q_f^2\right) \cdot \left[1 + \frac{\alpha_s}{\pi} + \mathcal{O}(\alpha_s^2)\right]. \tag{17.9}$$

This is actually the sum of the rates for two elementary processes, $e^+e^- \to q\bar{q}$ (including the correction from the third diagram of Fig. 17.2(b)) and $e^+e^- \to q\bar{q}g$. Although the rate for each of these processes is divergent as the gluon mass is taken to zero, that divergence cancels when they are combined. This is another example of the phenomenon of infrared divergence cancellations that we studied for the example of electron scattering in Sections 6.4 and 6.5. There we showed that the dressing of the final state by the emission of soft and collinear photons does not affect the overall scattering rate. Here, we see again that infrared divergences cancel in the total rate, although the sum over real and virtual gluon corrections leaves over a simple numerical correction.

It is not difficult to understand the cancellation of infrared logarithms intuitively. The original process $e^+e^- \to q\bar{q}$ is extremely rapid: Since the virtual photon is off-shell by an amount $q^2 = s$, the quarks are created in a time $1/\sqrt{s}$. However, the emission of collinear gluons, and the virtual corrections associated with the exchange of soft gluons, occur over a much longer time scale. In the diagrams with gluon emission, the virtual quark or antiquark is off-mass-shell by an amount $p_{\perp g}^2$, where $p_{\perp g}$ is the transverse momentum of the gluon relative to the $q\bar{q}$ system. Thus this virtual state survives for a time $1/p_{\perp g}$ before it decays. Such a slow process cannot affect the probability that a $q\bar{q}$ pair was produced; it can only affect the properties of the final state into which the $q\bar{q}$ system will evolve. By this logic, the only perturbative corrections that can affect the total cross section are those for which $p_{\perp g} \sim \sqrt{s}$. Another way to express this conclusion would be to argue that, after contributions from the infrared-sensitive regions have canceled, the contributions that remain come from the region of large real or virtual gluon momenta. By either argument, formula (17.9) should be a meaningful prediction of QCD perturbation theory, even though it involves an integral over the region of soft gluon emission.

The Running of α_s

Formula (17.9) depends on the QCD coupling constant α_s, which must be defined at some renormalization point M. This is in contrast to the QED coupling constant, which we defined in a natural way by on-shell renormalization. In QCD, we would like to avoid discussing on-shell quarks, since these are strongly interacting particles that are significantly affected by nonperturbative forces. The use of an arbitrary renormalization point M allows us to avoid this problem. We will define α_s by renormalization conditions imposed at a large momentum scale M where the coupling is small; this value of α_s can then be used to predict the results of scattering processes with any large momentum transfer.

However, the use of renormalization at a scale M in a computation involving momentum invariants of order P^2 involves some subtlety when P^2 and M^2 are very different. In our discussion of Section 12.3, we saw that, in this circumstance, Feynman diagrams with n loops typically contain correction terms proportional to $(\alpha_s \log(P^2/M^2))^n$. Fortunately, we can absorb these corrections into the lowest-order terms by using the renormalization group to replace the fixed renormalized coupling with a running coupling constant.

To illustrate how this analysis applies to QCD, let us examine the implications of the Callan-Symanzik equation for the e^+e^- annihilation total cross section σ, viewed as a function of s, a renormalization scale M, and the value of α_s at the renormalization scale. Like the QED potential (12.87), the e^+e^- total cross section is an observable quantity and so its normalization is independent of any conventions. It therefore obeys a Callan-Symanzik equation with $\gamma = 0$:

$$\left[M\frac{\partial}{\partial M} + \beta(g)\frac{\partial}{\partial g}\right]\sigma(s, M, \alpha_s) = 0. \tag{17.10}$$

By dimensional analysis, we can write

$$\sigma = \frac{c}{s}f\left(\frac{s}{M^2}, \alpha_s\right), \tag{17.11}$$

were c is a dimensionless constant. Then the Callan-Symanzik equation implies that f depends on its arguments only through the running coupling constant $\alpha_s(Q) = \bar{g}^2/4\pi$, evaluated at $Q^2 = s$. The coupling constant $\bar{g}$ is defined to satisfy the renormalization group equation

$$\frac{d}{d\log(Q/M)}\bar{g} = \beta(\bar{g}), \tag{17.12}$$

with initial condition $\alpha_s(M) = \alpha_s$. For QCD with three colors and n_f approximately massless quarks, the β function is given by Eq. (16.135):

$$\beta(g) = -\frac{b_0 g^3}{(4\pi)^2}, \quad \text{with } b_0 = 11 - \frac{2}{3}n_f. \tag{17.13}$$

Then the solution of the renormalization group equation is

$$\alpha_s(Q) = \frac{\alpha_s}{1 + (b_0 \alpha_s / 2\pi) \log(Q/M)}. \tag{17.14}$$

The explicit dependence of σ on α_s can be found by matching the successive terms in the expansion of $f(\alpha_s(\sqrt{s}))$ to the terms in the perturbative expansion. To the order of the first corrections, we find simply

$$\sigma = \sigma_0 \cdot \left(3 \sum_f Q_f^2\right) \cdot \left[1 + \frac{\alpha_s(\sqrt{s})}{\pi} + + \mathcal{O}(\alpha_s^2(\sqrt{s}))\right]. \tag{17.15}$$

Thus the Callan-Symanzik equation instructs us to replace the fixed renormalized coupling α_s with the running coupling constant $\alpha_s(Q)$, evaluated at $Q^2 = s$.

Because the fixed coupling α_s depends on the arbitrary renormalization point M, it is sometimes useful to remove it from our formulae completely. To do this, we define a mass scale conventionally called Λ (not to be confused with an ultraviolet cutoff!) satisfying

$$1 = g^2 (b_0/8\pi^2) \log(M/\Lambda). \tag{17.16}$$

Then Eq. (17.14) can be rearranged into the form

$$\alpha_s(Q) = \frac{2\pi}{b_0 \log(Q/\Lambda)}. \tag{17.17}$$

This formula is the clearest expression of the statement that $\alpha_s(Q)$ becomes small as $(\log(Q))^{-1}$ for large Q. The momentum scale Λ is the scale at which α_s becomes strong as Q^2 is decreased.

Experimental measurements of the rate of this reaction and others yield a value of $\Lambda \approx 200$ MeV. QCD perturbation theory is valid only when Q is somewhat larger than this, say above $Q = 1$ GeV, where $\alpha_s(Q) \approx 0.4$. The strong interactions become strong at distances larger than $\sim 1/\Lambda$, which is roughly the size of the light hadrons.

Although the example of the $e^+ e^-$ annihilation cross section is especially simple, since it depends on only one momentum invariant, similar conclusions carry over to other QCD predictions. In analyzing strong-interaction processes that are sensitive to the quark and gluon substructure, we will find leading-order formulae for the reaction cross sections that depend on the renormalized coupling α_s. To make these expressions satisfy the Callan-Symanzik equation, we must replace this fixed coupling with the running coupling constant $\alpha_s(Q)$, evaluated with Q of the order of the momentum invariants of the reaction. Since the running coupling constant depends only logarithmically on Q, we need not worry about choosing Q precisely. If we guess the proper scale incorrectly by a factor of 2, this induces an error in $\alpha_s(Q)$ that is of order $(\log(Q))^{-2} \sim \alpha_s^2(Q)$. Conversely, this ambiguity would be resolved by computing to the next order in α_s.

Before concluding this formal treatment of the e^+e^- annihilation cross section, we should add one qualification. At the beginning of Section 12.2, we remarked that renormalization group predictions can be complicated by the appearance of physical thresholds and their associated singularities, and so we stated these predictions only for when the relevant momentum invariant P^2 was large and spacelike. In the present chapter, we will be concerned with cross sections for quark and gluon reactions, evaluated on-shell. This introduces additional complications of principle. For example, in order to apply the Callan-Symanzik equation to $\sigma(s)$, we needed to know that this quantity contains no infrared divergences whose regulator might provide another dimensionful parameter. Throughout this chapter we will assume that similiar cancellations of divergences associated with soft and collinear gluons occur in the processes of interest to us. The complete proof of these cancellations in QCD can be carried through, but it is rather technical.* In some cases, an alternative point of view is possible; one can justify the use of the renormalization group to analyze an on-shell amplitude by relating it to Green's functions evaluated in the spacelike region. This method of analysis, which brings its own insights, will be the main subject of Chapter 18.

Gluon Emission and Jet Production

A second result of the Final Project of Part I was a formula for the differential cross section for $q\bar{q}$ production with gluon emission. Transcribing this formula to QCD using (17.7) gives the following result: Let x_1, x_2, x_3 be the ratios of the quark, antiquark, and gluon energies to the electron beam energy. These satisfy $0 < x_i < 1$ and $x_1 + x_2 + x_3 = 2$. Then the cross section for $e^+e^- \to q\bar{q}g$ is given by

$$\frac{d\sigma}{dx_1 dx_2}(e^+e^- \to q\bar{q}g) = \sigma_0 \cdot \left(3\sum_f Q_f^2\right) \cdot \frac{2\alpha_s}{3\pi} \frac{x_1^2 + x_2^2}{(1-x_1)(1-x_2)}. \quad (17.18)$$

This cross section is singular as x_1 or x_2 approaches 1. The limit $x_1 \to 1$ corresponds to configurations in which the quark has the maximum possible energy, while the antiquark and the gluon go off in the opposite direction, sharing the remaining energy. Then the antiquark and gluon have almost collinear lightlike momentum vectors and so form a system of very small invariant mass. Similarly, the limit $x_2 \to 1$ corresponds to configurations in which the quark and gluon are collinear. These singularities are responsible for the divergence of the integrated cross section in the limit of vanishing gluon mass.

How should we interpret these singularities? In our general treatment of bremsstrahlung in Section 6.1, we saw that the emission of a photon by

---

*For a review of the theorems justifying the formulae of perturbative QCD, see J. C. Collins and D. E. Soper, *Ann. Rev. Nucl. Sci.* **37**, 383 (1987).

a scattered electron is enhanced, for collinear radiation, by a factor of order $\log(q^2/m^2)$, where m is the mass of the electron. Thus the total rate for emitting a collinear photon formally diverges in the limit of zero mass. The same conclusion holds for the emission of gluons by quarks. A divergence that appears for collinear emissions in the limit of zero mass is called a *mass singularity*. In QED, we saw that the mass singularity signals a real physical effect of strong collinear radiation when q^2 is large. In QCD, we might expect strong gluon radiation in this limit, but we must think carefully about how this radiation appears experimentally. Whether a collinear gluon is radiated or not, the quark and antiquark emerging from the reaction will undergo further soft interactions with the other products. These processes must continue, producing quark-antiquark pairs and emitting and absorbing gluons, until all colored particles are collected into color-singlet hadrons. Thus the presence of one or more collinear gluons will have no noticeable effect on the final state, which consists of two back-to-back jets of hadrons. For this reason, formula (17.18) is of no use when the gluon transverse momentum is less than the typical scale of soft gluon interactions, roughly 1 GeV.

When the gluon is emitted with substantial transverse momentum with respect to the $q\bar{q}$ axis, however, it is not possible for subsequent soft exchanges to recall or reverse this transverse momentum. In this case, the $q\bar{q}g$ system evolves into a system of three distinct jets of hadrons. Thus, sufficiently far from the collinear regions, we can interpret Eq. (17.18) as the cross section for producing events with three hadronic jets having energies x_1, x_2, x_3 times the electron beam energy.

By an analysis similar to that given above for the total cross section, we can improve Eq. (17.18) by replacing the fixed coupling constant α_s with a running $\alpha_s(Q)$. A reasonable choice for Q is the transverse momentum of the gluon, $p_{\perp g}$, described below Eq. (17.9). If this transverse momentum is too small, however, $\alpha_s(Q)$ will be large, and our leading-order formula will break down. This is a second reason why we cannot use formula (17.18) when the transverse momentum of the gluon is less than about 1 GeV.

On the other hand, when the gluon transverse momentum is much larger than 1 GeV, there is no reason to distrust QCD perturbation theory. Soft processes cannot disturb the three-jet nature of the hadronic state, and asymptotic freedom insures that the coupling constant is small, so that the leading order of perturbation theory will be a good approximation.

The three-jet cross section (17.18) is a good example of the type of prediction that one obtains from the use of perturbation theory in QCD. We describe a strong-interaction process by the invariant momentum transfer Q it gives to hadronic constituents. QCD perturbation theory makes predictions about the flow of energy and momentum in such a reaction into the final system of jets of hadrons. If Q is small, perturbation theory is invalid, and we obtain no useful prediction. However, if Q is large, the asymptotic freedom of QCD implies that Feynman diagrams for quarks and gluons will correctly predict the behavior of the final system of hadronic jets.

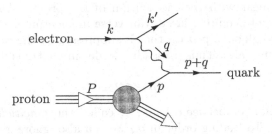

Figure 17.3. Deep inelastic scattering in QCD. The diagram shows the flow of momentum when a high-energy electron scatters from a quark taken from tł e wavefunction of the proton.

17.3 Deep Inelastic Scattering

After e^+e^- annihilation into hadrons, the next simplest reaction involving strongly interacting particles is electron scattering from a proton, or from some other hadron. At the most elementary level, this reaction can be viewed as the electromagnetic scattering of an electron from a quark inside the proton.[†] A way to visualize the process is shown in Fig. 17.3. Call the proton momentum P, and the initial quark momentum p. Call the initial and final momenta of the electron k and k'. If the final electron momentum is measured, one can deduce the momentum $q = k - k'$ transferred by the virtual photon to the hadronic system. The vector q is spacelike, and one conventionally writes $q^2 = -Q^2$.

If the invariant momentum transfer Q^2 is large, the quark is ejected from the proton in a manner that cannot be balanced by subsequent soft processes. These soft processes will, however, create gluons and quark-antiquark pairs that eventually neutralize the color and cause the struck quark to materialize as a jet of hadrons in the direction of the momentum transfer from the electron. Typically the total invariant mass of the final hadronic system is large, since the struck quark carries a large momentum with respect to the other "spectator" quarks. In this case, the process is referred to as *deep inelastic scattering*.

To derive a first approximation to the cross section for electron-proton scattering, we consider this reaction from a frame in which the electron and proton are moving rapidly toward each other, for example, the electron-proton center-of-mass frame. We assume that the center-of-mass energy is large enough that we can ignore the proton mass in working out the kinematics. Then the proton has an almost lightlike momentum along the collision axis. The constituents of the proton also have lightlike momenta, which

---

[†]The electron could just as well be a muon; all the same formulae apply in this case. Leptons can also scatter from quarks via the neutral-current weak interaction, as we will see in Chapter 20.

are almost collinear with the momentum of the proton. This is because a constituent cannot acquire a large transverse momentum except through exchange of a hard gluon, a process that is suppressed by the smallness of α_s at large momentum scales. Thus, to leading order in QCD perturbation theory, we can write

$$p = \xi P, \tag{17.19}$$

where ξ is a number between 0 and 1, called the *longitudinal fraction* of the constituent. To leading order in α_s we can also ignore gluon emission or exchanges during the collision process. The cross section for electron-proton scattering is then given by the cross section for electron-quark scattering at given longitudinal fraction ξ, multiplied by the probability that the proton contains a quark at that value of ξ, integrated over ξ.

But the probability that the proton contains a certain constituent with a certain momentum fraction cannot be computed using QCD perturbation theory, since it depends on the soft processes that determine the structure of the proton as a bound state of quarks and gluons. We will therefore consider this probability to be an unknown function, to be determined from experiment. Eventually, we will need to make use of such a probability function for each species of quark, antiquark, and gluon that can be found in the wavefunction of the proton. Collectively, these constituents are called *partons*. For each parton species f, we write the probability of finding a constituent of the proton of type f at longitudinal fraction ξ as

$$\left(\begin{array}{c} \text{probability of finding constituent } f \\ \text{with longitudinal fraction } \xi \end{array} \right) = f_f(\xi)d\xi. \tag{17.20}$$

The functions $f_f(\xi)$ are called the *parton distribution functions*. Using this notation, the cross section for electron-proton inelastic scattering is given to leading order in α_s by the expression

$$\sigma\big(e^-(k)p(P) \to e^-(k') + X\big)$$

$$= \int\limits_0^1 d\xi \sum_f f_f(\xi)\, \sigma\big(e^-(k)q_f(\xi P) \to e^-(k') + q_f(p')\big), \tag{17.21}$$

where X stands for any hadronic final state. The sum in (17.21) contains contributions from constituent antiquarks as well as constituent quarks.

Equation (17.21) is equivalent to the formula (14.8) that we constructed for this reaction in Chapter 14. Now we see that this formula is justified by the smallness of the QCD coupling constant at large momentum scales. It is important to remember, however, that (17.21) is not the complete prediction of QCD, but only the first term of an expansion in α_s; this level of approximation is called the *parton model*. The higher-order QCD corrections to Eq. (17.21) will involve modifications both to the electron-quark scattering cross section and to the parton distribution functions. The most important of these corrections are discussed in Section 17.5.

In the same way, all other reactions of the proton that involve large momentum transfer also have parton model descriptions. In QCD, all of these reaction cross sections are computed from scattering amplitudes for quarks and gluons. The initial motion of the partons for any process is described by the same parton distribution functions $f_f(\xi)$ that appear in deep inelastic scattering.

Let us now work out the explicit leading-order formula for the deep inelastic scattering cross section, reviewing the analysis in Chapter 14. In Eq. (14.3), we wrote the leading-order differential cross section for the parton-level process,

$$\frac{d\sigma}{d\hat{t}}(e^-q \to e^-q) = \frac{2\pi\alpha^2 Q_f^2}{\hat{s}^2}\left[\frac{\hat{s}^2 + \hat{u}^2}{\hat{t}^2}\right]. \tag{17.22}$$

In general, we will use the symbols $\hat{s}$, $\hat{t}$, $\hat{u}$ to denote the Mandelstam variables for two-body scattering processes at the parton level. These variables must be related to observable properties of the hadronic system or the scattered electron. For massless initial and final particles,

$$\hat{s} + \hat{t} + \hat{u} = 0.$$

In the case of deep inelastic scattering,

$$\hat{t} = -Q^2$$

and

$$\hat{s} = 2p \cdot k = 2\xi P \cdot k = \xi s.$$

Thus the cross section for deep inelastic scattering at fixed Q^2 is given by

$$\frac{d\sigma}{dQ^2} = \int_0^1 d\xi \sum_f f_f(\xi) Q_f^2 \frac{2\pi\alpha^2}{Q^4}\left[1 + \left(1 - \frac{Q^2}{\xi s}\right)^2\right]\theta(\xi s - Q^2). \tag{17.23}$$

The final factor expresses the kinematic constraint $\hat{s} \geq |\hat{t}|$. Expression (17.23) should be an accurate first approximation to the deep inelastic scattering cross section when Q^2 is large. In that case, the corrections to this formula from hard gluon emissions and exchanges will be of order $\alpha_s(Q^2)$.

We also showed in Chapter 14 that the measurement of the scattered electron momentum k' and thus the momentum transfer q uniquely determines an allowed value of ξ for elastic electron-quark scattering. This value is given by Eq. (14.7):

$$\xi = x, \qquad \text{where} \quad x \equiv \frac{Q^2}{2P \cdot q}. \tag{17.24}$$

When (17.23) is expressed as a doubly differential cross section in x and Q^2, it becomes the simple product of a parton-level cross section and a sum of parton distribution functions evaluated at $\xi = x$. In the literature, the symbol x is often used interchangably with ξ, and we will follow this practice from here on.

It is especially convenient to represent the cross section in terms of dimensionless combinations of kinematic variables. One of these should be x; a good choice for the other is

$$y \equiv \frac{2P \cdot q}{2P \cdot k} = \frac{2P \cdot q}{s}. \tag{17.25}$$

In the frame in which the proton is at rest,

$$y = \frac{q^0}{k^0}, \tag{17.26}$$

that is, y is the fraction of the incident electron's energy that is transferred to the hadronic system. On the other hand, since $p = \xi P$, we can evaluate y in terms of parton variables:

$$y = \frac{2p \cdot (k - k')}{2p \cdot k} = \frac{\hat{s} + \hat{u}}{\hat{s}}, \tag{17.27}$$

so that

$$\frac{\hat{u}}{\hat{s}} = -(1 - y). \tag{17.28}$$

From Eq. (17.26) or (17.28), we see that $y \le 1$. The kinematically allowed region in the (x, y) plane is thus

$$0 \le x \le 1, \qquad 0 \le y \le 1.$$

To express Eq. (17.23) in terms of x and y, we need the formula

$$Q^2 = xys, \tag{17.29}$$

which follows from Eqs. (17.24) and (17.25), and the change of variables

$$d\xi \, dQ^2 = dx \, dQ^2 = \frac{dQ^2}{dy} dx \, dy = xs \, dx \, dy.$$

Then the differential cross section becomes

$$\frac{d^2\sigma}{dx\,dy}(e^- p \to e^- X) = \left(\sum_f x f_f(x) Q_f^2\right) \frac{2\pi\alpha^2 s}{Q^4} \left[1 + (1-y)^2\right]. \tag{17.30}$$

The factor $1/Q^4$ comes from the square of the virtual photon propagator. Once this factor is removed, the dependence on x and y completely factorizes. Each half of this relation contains physical information. The fact that the parton distributions $f_f(x)$ depend only on x and are independent of Q^2 is the statement of Bjorken scaling. This tells us that the initial distribution of partons is independent of the details of the hard scattering. The y dependence of the cross section comes from the underlying parton cross section. In Chapter 5, we saw that the elementary QED cross sections, viewed in the high-energy limit, reflect the helicities of the interacting particles. The behavior $[1 + (1-y)^2]$ in (17.30) is known as the *Callan-Gross relation*; it is specific to the scattering of electrons from massless fermions. This relation gave evidence that the partons involved in deep inelastic scattering were fermions, at a time when the relation between partons and quarks was still unclear.

Deep Inelastic Neutrino Scattering

Because the sum over quark flavors factorizes in Eq. (17.30), one cannot determine the individual parton distribution functions $f_f(x)$ from electron scattering experiments alone. One can, however, obtain more detailed information on the structure of the proton through deep inelastic *neutrino* scattering.

Neutrinos have zero electric charge and so do not interact by photon exchange, but they do interact with quarks through weak interactions. We will discuss the weak interactions in detail in Chapter 20; for the moment, let us adopt a simplified description that concentrates on the elementary process shown in Fig. 17.4. In this process, a neutrino converts to the associated charged lepton,[‡] exchanging a virtual massive vector boson, the W^+. This boson couples to a quark current that converts a d quark to a u quark. The effect of this exchange process is to provide a different, but completely characterized, method for injecting a hard momentum transfer q. The amplitude for this process is described by the effective Lagrangian

$$\Delta\mathcal{L} = \frac{g^2}{2}\frac{1}{m_W^2}\left[\bar{\ell}\gamma^\mu\left(\frac{1-\gamma^5}{2}\right)\nu\right]\left[\bar{u}\gamma_\mu\left(\frac{1-\gamma^5}{2}\right)d\right] + \text{h.c.}, \qquad (17.31)$$

where ℓ, ν, d, u are the fermion fields associated with the charged lepton, the neutrino, and the d and u quarks, and g is the weak interaction coupling constant. The factor $1/m_W^2$ comes from the W boson propagator, considered in the limit $q^2 \ll m_W^2$. The first two factors are often written in terms of the *Fermi constant* G_F, defined as

$$\frac{G_F}{\sqrt{2}} = \frac{g^2}{8m_W^2}. \qquad (17.32)$$

This constant gives the strength of the weak interactions at energies much less than m_W. The crucial property of the weak interactions, shown explicitly in (17.31), is that the W boson couples only to the left-handed helicity states of relativistic fermions. The deeper significance of this property will be discussed in Chapter 20.

For technical reasons, it is easiest to do neutrino deep inelastic scattering using muon neutrinos, which convert to muons after emitting a W boson. It is equally feasible to scatter muon antineutrinos from nuclear targets, and, as we will see, it is interesting to compare the effects of neutrinos and antineutrinos. Since the proton contains a small admixture of the heavier quarks s, c, these also give small contributions to neutrino deep inelastic scattering. However, we will ignore these contributions in our discussion.

[‡]There is also a *neutral-current* weak interaction in which the neutrino remains a neutrino; see Problem 20.4.

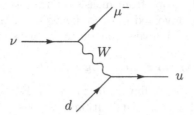

Figure 17.4. The elementary neutrino scattering process mediated by the weak interaction.

The cross section for neutrino deep inelastic scattering is given by a formula analogous to (17.21), with the photon-exchange cross section replaced by one resulting from W exchange. It is straightforward to work out this cross section directly. However, we can also obtain the result from Eq. (17.22), if we look back to Chapter 5 and recall how the structure of this equation arises from the various helicity contributions. In (17.22), the factor $\hat{t}^2$ in the denominator came from the photon propagator; this factor is replaced by m_W^4 in the weak interaction case. The factor $[\hat{s}^2 + \hat{u}^2]$ came from the Dirac matrix algebra. We saw in Section 5.2 that the first term is the contribution of left-handed electrons scattering from left-handed fermions or right-handed electrons scattering from right-handed fermions, and that the second term arises from the other helicity combinations. For the case of neutrino-quark scattering, the interaction (17.31) allows only the scattering of left-handed neutrinos from left-handed quarks, so only the $\hat{s}^2$ term appears. To determine the overall normalization of the cross section, we note that, since the neutrinos are produced in weak interactions, they always have left-handed polarization, so no polarization average should be done. On the other hand, we must still average over the polarization of the initial quark. In all, we find

$$\frac{d\sigma}{d\hat{t}}(\nu d \rightarrow \mu^- u) = \frac{\pi g^4}{2(4\pi)^2 \hat{s}^2}\left[\frac{\hat{s}^2}{m_W^4}\right] = \frac{G_F^2}{\pi}. \tag{17.33}$$

It is easy to check this formula by explicit computation starting from (17.31).

The cross section for the scattering of antineutrinos from quarks can be worked out in the same way. Note that this reaction involves the exchange of a W^-, and so converts u quarks to d quarks. However, the u quarks must still be left-handed. The only modification from the previous paragraph comes in the fact that the antineutrinos that couple to the interaction (17.31) are right-handed, so the cross section comes from the term in (17.22) proportional to $\hat{u}^2$:

$$\frac{d\sigma}{d\hat{t}}(\bar{\nu} u \rightarrow \mu^+ d) = \frac{\pi g^4}{2(4\pi)^2 \hat{s}^2}\left[\frac{\hat{u}^2}{m_W^4}\right] = \frac{G_F^2}{\pi}(1-y)^2. \tag{17.34}$$

Again, it is easy to verify this formula directly. The cross section for neutrino-antiquark scattering, converting a $\bar{u}$ into a $\bar{d}$, is also given by Eq. (17.34),

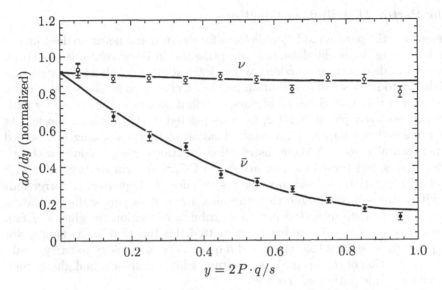

Figure 17.5. The distribution in y of neutrino and anti-neutrino deep inelastic scattering from an iron target, as measured by the CDHS experiment, J. G. H. de Groot, et. al., *Z. Phys.* **C1**, 143 (1979). The solid curves are fits to the form $A + B(1-y)^2$.

while the cross section for antineutrino-antiquark scattering, converting a $\bar{d}$ into a $\bar{u}$, is given by Eq. (17.33).

To convert these parton-level cross sections to physical cross sections, we combine them with the parton distribution functions. The kinematics is exactly the same as in the case of electron scattering. Thus, following the arguments that led to Eq. (17.30), we obtain the expressions

$$
\begin{aligned}
\frac{d^2\sigma}{dxdy}(\nu p \to \mu^- X) &= \frac{G_F^2 s}{\pi}\left[xf_d(x) + xf_{\bar{u}}(x)\cdot(1-y)^2\right], \\
\frac{d^2\sigma}{dxdy}(\bar{\nu} p \to \mu^+ X) &= \frac{G_F^2 s}{\pi}\left[xf_u(x)\cdot(1-y)^2 + xf_{\bar{d}}(x)\right].
\end{aligned}
\tag{17.35}
$$

According to these relations, deep inelastic neutrino scattering allows one to map separately the parton distribution functions for u and d quarks and antiquarks in the proton.

In addition, Eq. (17.35) makes a dramatic qualitative prediction: To the extent that a proton (or neutron) is made of quarks with very few additional quark-antiquark pairs, the deep inelastic neutrino scattering cross section should be constant in y, while the antineutrino scattering cross section should fall off as $(1-y)^2$. The measured y dependence of these deep inelastic cross sections is shown in Fig. 17.5. The qualitative behavior predicted by the parton description is clearly evident; the discrepancy from the strict prediction can be accounted for by a small antiquark component in the nucleon wavefunction.

The Parton Distribution Functions

Given that the parton model predictions for electron and neutrino deep inelastic scattering do fit the data, one can make use of these relations to extract the parton distribution functions and so learn something about the structure of the proton.* A set of distribution functions, chosen to fit all available data, is shown in Fig. 17.6. Since all of these distributions, especially those for antiquarks, peak sharply at small x, we have plotted $x f_f(x)$ for each species. As we remarked in Chapter 14, a small violation of Bjorken scaling is observed experimentally, so that these distribution functions change slowly with Q^2. The figure shows these functions at $Q^2 = 4$ GeV^2. We will see in Section 17.5 that this violation of Bjorken scaling is an effect of higher-order corrections in QCD; we will also argue that the measurement of this scaling violation allows one to determine the parton distribution function for gluons, $f_g(x)$. Anticipating this result, we have also plotted this function in the figure. Not surprisingly, one finds that the u and d quarks are most likely to carry a substantial fraction of the proton's momentum, while antiquarks and gluons tend to have small longitudinal fractions.

Since the parton distributions are the probabilities of finding various proton constituents, they must be normalized in a way that reflects the quantum numbers of the proton. The proton is a bound state of uud, plus some admixture of quark-antiquark pairs. Thus it should contain an excess of two u quarks and one d quark over the corresponding antiquarks. These considerations imply the constraints

$$\int_0^1 dx \left[f_u(x) - f_{\bar{u}}(x) \right] = 2, \qquad \int_0^1 dx \left[f_d(x) - f_{\bar{d}}(x) \right] = 1. \qquad (17.36)$$

So far we have discussed the parton distributions only for the proton. Similar considerations, however, apply to any other hadron. Each hadron has its own set of parton distribution functions; these obey sum rules analogous to (17.36) but reflecting the particular quantum numbers of the hadron. The parton distribution functions should also reflect the symmetries that link different hadrons. For example, since the neutron can be generated, to a few percent accuracy, by interchanging the role of u and d quarks in the proton, its distribution functions obey

$$f_u^n(x) = f_d(x), \qquad f_d^n(x) = f_u(x), \qquad f_{\bar{u}}^n(x) = f_{\bar{d}}, \qquad \text{etc.} \qquad (17.37)$$

In these equations, and henceforth, a distribution function without a special label refers to the proton. The parton distribution functions of the antiproton are given by the exact relations

$$f_u^{\bar{p}}(x) = f_{\bar{u}}(x), \qquad f_{\bar{u}}^{\bar{p}}(x) = f_u(x), \qquad \text{etc.} \qquad (17.38)$$

---

*A detailed discussion of the extraction of parton distribution functions from data can be found in G. Sterman, et. al., *Rev. Mod. Phys.* **67**, 157 (1995).

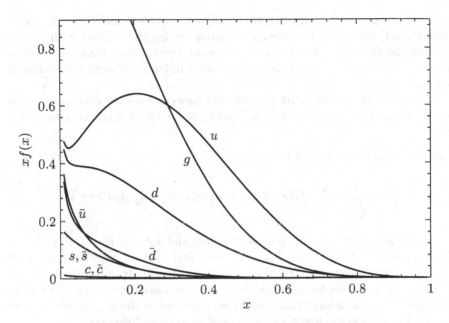

Figure 17.6. Parton distribution functions $xf_f(x)$ for quarks, antiquarks, and gluons in the proton, at $Q^2 = 4 \text{ GeV}^2$. These distributions are obtained from a fit to deep inelastic scattering data performed by the CTEQ collaboration (CTEQ2L), described in J. Botts, et. al., *Phys. Lett.* **B304**, 159 (1993).

In any case, the total amount of momentum carried by the partons must sum to the total momentum of the hadron. This implies

$$\int_0^1 dx\, x\big[f_u(x) + f_d(x) + f_{\bar{u}}(x) + f_{\bar{d}}(x) + f_g(x)\big] = 1. \qquad (17.39)$$

The distribution functions of quarks and antiquarks in the proton, as extracted from deep inelastic scattering data, contribute only about *half* of the total value required for this integral. The remaining energy-momentum must be carried by the gluons.

17.4 Hard-Scattering Processes in Hadron Collisions

If one collides hadrons with other hadrons at very high energy, most of the collisions will involve only soft interactions of the consituent quarks and gluons. Such interactions cannot be treated using perturbative QCD, because α_s is large when the momentum transfer is small. In some collisions, however, two quarks or gluons will exchange a large momentum $p_\perp$ perpendicular to the collision axis. Then, as in deep inelastic scattering, the elementary interaction takes place very rapidly compared to the internal time scale of the hadron wavefunctions, so the lowest-order QCD prediction should accurately describe the process. Again, we should find a parton-model formula that is built from a leading-order subprocess cross section, integrated with parton distribution functions. For the case of proton-proton scattering, these functions will be the same ones that are measured in lepton-proton deep inelastic scattering.

For example, if the hard parton-level process involves quark-antiquark scattering into a final state Y, the leading-order QCD prediction takes the form

$$\sigma(p(P_1) + p(P_2) \rightarrow Y + X)$$

$$= \int_0^1 dx_1 \int_0^1 dx_2 \sum_f f_f(x_1) f_{\bar{f}}(x_2) \cdot \sigma(q_f(x_1 P) + \bar{q}_f(x_2 P) \rightarrow Y), \qquad (17.40)$$

where the sum runs over all species of quarks and antiquarks—u, d, $\bar{u}$, $\bar{d}$, ... (Here again, X denotes any hadronic final state.) The same formula, with appropriately modified distribution functions, applies to any other hadron-hadron collision. This formula will be a good first approximation if, by some invariant measure, a large momentum is transfered in the $q\bar{q}$ reaction. In this section we will discuss several examples of processes of this type.

Lepton Pair Production

The simplest example to analyze is the reaction in which a high-mass lepton pair $\ell^+\ell^-$ emerges from $q\bar{q}$ annihilation in a proton-proton collision. This reaction, called the *Drell-Yan process*, is illustrated in Fig. 17.7. In this case, the underlying $q\bar{q}$ reaction is described by an elementary QED cross section. To the leading order in QCD, the cross section that we require, $\sigma(q\bar{q} \rightarrow \ell^+\ell^-)$, is simply related to the cross section $\sigma(e^+e^- \rightarrow q\bar{q})$ given in (17.4). The only difference between the two calculations is that we must average rather than sum over the color orientations of the quark and antiquark. This gives two extra factors of $1/3$. Thus,

$$\sigma(q_f \bar{q}_f \rightarrow \ell^+\ell^-) = \frac{1}{3} Q_f^2 \cdot \frac{4\pi\alpha^2}{3\hat{s}}. \qquad (17.41)$$

If both final-state lepton momenta are observed, it is possible to reconstruct the total 4-momentum q of the virtual photon. It is also possible to determine the longitudinal fractions of the initial quark and antiquark, as we will now show. Let

$$M^2 = q^2 \qquad (17.42)$$

be the square of the invariant mass of the Drell-Yan pair. (Do not confuse this quantity M with the renormalization scale.) If the initial partons have small transverse momentum, the transverse momentum of the virtual photon will also be small. Its longitudinal momentum, however, will in general be

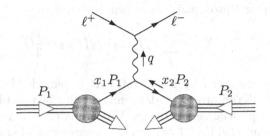

Figure 17.7. The Drell-Yan process: $pp \to \ell^+\ell^- +$ anything.

substantial. We parametrize this using the *rapidity*, Y, of the virtual photon, as defined in Eq. (3.48):

$$q^0 = M \cosh Y, \qquad (17.43)$$

where q^0 is measured in the pp center of mass frame. We will express the longitudinal fractions of the quarks, and hence the Drell-Yan cross section, in terms of the observables M^2 and Y.

In the pp center of mass frame, the proton momenta take the explicit forms

$$P_1 = (E, 0, 0, E), \qquad P_2 = (E, 0, 0, -E),$$

where E satisfies $s = 4E^2$. Ignoring their small transverse momenta, we can write the constituent quark and antiquark momenta as x_1 and x_2 times these vectors, so that

$$q = x_1 P_1 + x_2 P_2 = \big((x_1+x_2)E, 0, 0, (x_1-x_2)E\big). \qquad (17.44)$$

By computing the invariant square of this vector we find

$$M^2 = x_1 x_2 s. \qquad (17.45)$$

Similarly, comparing (17.43) with (17.44), we find

$$\cosh Y = \frac{x_1 + x_2}{2\sqrt{x_1 x_2}} = \frac{1}{2}\left(\sqrt{\frac{x_1}{x_2}} + \sqrt{\frac{x_2}{x_1}}\right),$$

which implies

$$\exp Y = \sqrt{\frac{x_1}{x_2}}. \qquad (17.46)$$

These equations can be inverted to determine x_1 and x_2:

$$x_1 = \frac{M}{\sqrt{s}} e^Y, \qquad x_2 = \frac{M}{\sqrt{s}} e^{-Y}. \qquad (17.47)$$

Relations (17.45) and (17.46) let us convert the integral in Eq. (17.40) into an integral over the parameters M^2, Y of the produced leptons. The Jacobian of the change of variables is

$$\frac{\partial(M^2, Y)}{\partial(x_1, x_2)} = \begin{vmatrix} x_2 s & x_1 s \\ 1/2x_1 & -1/2x_2 \end{vmatrix} = s = \frac{M^2}{x_1 x_2}.$$

The cross section for lepton pair production is therefore

$$\frac{d^2\sigma}{dM^2dY}(pp \to e^+e^- + X) = \sum_f x_1 f_f(x_1)\, x_2 f_{\bar{f}}(x_2) \cdot \frac{1}{3}Q_f^2 \cdot \frac{4\pi\alpha^2}{3M^4}, \quad (17.48)$$

where x_1 and x_2 are given by Eq. (17.47). It is remarkable that the cross section for the Drell-Yan process is determined point by point by information derivable from deep inelastic scattering. Unfortunately, the relation between the two processes implied by (17.48) receives a correction of order $\alpha_s(M)$ that turns out to be numerically large, and which must be included to check this prediction against experiment.

General Kinematics of Pair Production

In deriving Eq. (17.48), we used the total cross section (17.41) for the parton-level process, integrated over the angular distribution of the outgoing leptons. In principle, we could have retained the angular information and derived a triply differential distribution. This would be the most complete prediction possible for a two-body parton-level reaction. It will be useful to work out the kinematics of such reactions, taking a more general viewpoint. In the generic situation, a parton of type 1 from proton 1 scatters from a parton of type 2 from proton 2, yielding partons of types 3 and 4, with a squared momentum transfer $\hat{t}$. This generic process is shown in Fig. 17.8. In the Drell-Yan process, partons 3 and 4 are leptons. But these partons could also be quarks or gluons, which materialize as hadronic jets. We assume that all partons can be treated as massless. In parton variables, the cross section for this process is

$$\frac{d^3\sigma}{dx_1 dx_2 d\hat{t}}(pp \to 3 + 4 + X) = f_1(x_1)f_2(x_2)\frac{d\sigma}{d\hat{t}}(1 + 2 \to 3 + 4). \quad (17.49)$$

Let us now translate this formula to observable parameters of the final state.

In the leading order of QCD, the transverse momenta of partons 3 and 4 must be equal and opposite, but their longitudinal momenta are not constrained. We will take the three parameters of the final state to be the common magnitude of the parton transverse momenta $p_\perp$ and the *longitudinal rapidities* y_3, y_4 of the final-state partons, defined by the formulae

$$E_i = p_\perp \cosh y_i; \qquad p_{i\|} = p_\perp \sinh y_i. \quad (17.50)$$

The longitudinal rapidity y_i gives the boost of the particle i from the frame where it has zero longitudinal momentum.[†] Recall from Section 3.3 that rapidities simply add under collinear boosts. The transverse momentum is invariant under longitudinal boosts. Thus, $(y_3, y_4, p_\perp)$ is a set of variables with convenient Lorentz transformation properties with respect to boosts along the

---

[†]In the literature on hadron collisions, y_i is usually called simply the rapidity, with the restriction to longitudinal boosts being understood.

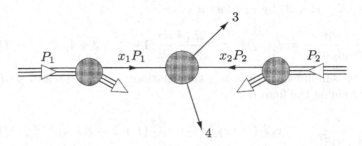

Figure 17.8. A generic two-body parton scattering process.

collision axis. We will now see that these three parameters are related in a straightforward way to the underlying variables x_1, x_2, $\hat{t}$.

Consider the center of mass frame of the colliding partons. The total energy in this frame is $\sqrt{\hat{s}}$. Let us use a subscript $*$ to denote other quantities measured in this frame, for instance, θ_* for the parton scattering angle. Then

$$p_{3\|*} = \tfrac{1}{2}\sqrt{\hat{s}}\cos\theta_*, \qquad p_{3\perp*} = \tfrac{1}{2}\sqrt{\hat{s}}\sin\theta_*, \tag{17.51}$$

and p_{4*} is oriented just oppositely. This frame is also the center of mass frame of partons 3 and 4, so

$$y_{3*} = -y_{4*} \equiv y_*. \tag{17.52}$$

Since rapidities transform by shifts, we can solve for y_* and for the rapidity Y by which one must boost to reach this frame:

$$y_* = \tfrac{1}{2}(y_3 - y_4), \qquad Y = \tfrac{1}{2}(y_3 + y_4). \tag{17.53}$$

The scattering angle θ_* is determined from y_* by combining (17.51) with the relation $E_* = p_\perp \cosh y_*$:

$$\frac{1}{\sin\theta_*} = \cosh y_*. \tag{17.54}$$

Then the Mandelstam variables

$$\hat{s} = \frac{4p_\perp^2}{\sin^2\theta_*}, \qquad \hat{t} = -\tfrac{1}{2}\hat{s}(1 - \cos\theta_*) \tag{17.55}$$

can be expressed as

$$\hat{s} = 4p_\perp^2\cosh^2 y_*, \qquad \hat{t} = -2p_\perp^2\cosh y_* \, e^{-y_*}. \tag{17.56}$$

We can combine the first of these expressions with (17.47) to determine x_1 and x_2:

$$x_1 = \frac{2p_\perp}{\sqrt{s}}\cosh y_* \, e^Y, \qquad x_2 = \frac{2p_\perp}{\sqrt{s}}\cosh y_* \, e^{-Y}. \tag{17.57}$$

To translate the cross section (17.49) to the final parton observables, we need the Jacobian

$$\frac{\partial(x_1, x_2, \hat{t})}{\partial(y_3, y_4, p_\perp)} = \frac{8p_\perp^3}{s}\cosh^2 y_* = \frac{2p_\perp\hat{s}}{s}. \tag{17.58}$$

Multiplying Eq. (17.49) by this factor gives

$$\frac{d^3\sigma}{dy_3 dy_4 dp_\perp} = f_1(x_1) f_2(x_2) \frac{2p_\perp \hat{s}}{s} \frac{d\sigma}{d\hat{t}} (1 + 2 \rightarrow 3 + 4). \tag{17.59}$$

This can be simplified a bit using the relations $\hat{s} = x_1 x_2 s$ and $p_\perp dp_\perp = d^2 p_\perp / 2\pi$, yielding the final result:

$$\frac{d^4\sigma}{dy_3 dy_4 d^2 p_\perp} = x_1 f_1(x_1)\, x_2 f_2(x_2)\, \frac{1}{\pi} \frac{d\sigma}{d\hat{t}} (1 + 2 \rightarrow 3 + 4). \tag{17.60}$$

In this formula, x_1, x_2, and the Mandelstam variables of the parton subprocess are given by Eqs. (17.57) and (17.56).

This result gives us the complete distribution of final-state leptons or jets in any two-body reaction of partons. For example, to find the distribution of final-state leptons in the Drell-Yan process, we would insert into this formula the differential cross section for quark annihilation into leptons,

$$\frac{d\sigma}{d\hat{t}}(q_f \bar{q}_f \rightarrow \ell^+ \ell^-) = \frac{1}{3} Q_f^2 \cdot \frac{2\pi\alpha^2}{\hat{s}^2} \frac{\hat{t}^2 + \hat{u}^2}{\hat{s}^2}. \tag{17.61}$$

The formula can be applied equally well to other two-body parton reactions, if we know the relevant parton-level differential cross sections.

Jet Pair Production

The most common two-body parton reactions are those of QCD, involving quarks, gluons, or both. Unfortunately, it is very difficult to distinguish hadronic jets initiated by gluons from those initiated by quarks. It is even more difficult to determine experimentally whether the initial partons in a hard-scattering process were quarks or gluons. Thus, the predictions of QCD for hard-scattering processes are most often quoted as cross sections for jet production in hadronic collisions, summing over all possible reactions of quarks, antiquarks, and gluons. In any event, to derive these predictions, we must work out the basic parton-parton cross sections.

The simple two-body scattering processes of quarks, antiquarks, and gluons are the elementary processes of QCD perturbation theory, in the same sense that the reactions studied in Chapter 5 are the elementary processes of QED perturbation theory. They are the basic hadronic hard-scattering reactions that appear in QCD at the leading order in α_s. In the remainder of this section, we will write down the cross section formulae for the various possible quark and gluon subprocesses. All of these cross sections will be of order α_s^2. In practice, this α_s should be evaluated at a typical momentum transfer of the reaction, for example, at $Q^2 = \hat{t}$.

The simplest subprocess is the scattering of different species of quarks, for example, $u + d \rightarrow u + d$. At order α_s^2, this process occurs via the Feynman diagram shown in Fig. 17.9. This process is analogous to electron-muon

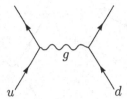

Figure 17.9. Feynman diagram contributing to $ud \to ud$.

scattering in QED, for which we wrote the cross section in Eq. (17.22):

$$\frac{d\sigma}{d\hat{t}}(e^-\mu \to e^-\mu) = \frac{2\pi\alpha^2}{\hat{s}^2}\left[\frac{\hat{s}^2 + \hat{u}^2}{\hat{t}^2}\right].$$

(17.62)

To convert this to the cross section for quark scattering in QCD, we need only replace the QED coupling e^2 by g^2 times an $SU(3)$ group theory factor. The QCD diagram contains the factor

$$(t^a)_{i'i}(t^a)_{j'j},$$

where i, i' are the initial and final colors of the u quark and j, j' are the initial and final colors of the d quark. To compute the cross section, we must square this factor, sum over final colors, and average over initial colors. This gives the factor

$$\frac{1}{3} \cdot \frac{1}{3} \cdot \mathrm{tr}[t^b t^a] \cdot \mathrm{tr}[t^b t^a] = \frac{1}{9}[C(r)]^2 \delta^{ab}\delta^{ab} = \frac{1}{9} \cdot \frac{1}{4} \cdot 8 = \frac{2}{9},$$

(17.63)

where we have used Eq. (15.78) and $C(r) = 1/2$ for the fundamental representation of $SU(3)$. Thus for ud scattering,

$$\frac{d\sigma}{d\hat{t}}(ud \to ud) = \frac{4\pi\alpha_s^2}{9\hat{s}^2}\left[\frac{\hat{s}^2 + \hat{u}^2}{\hat{t}^2}\right].$$

(17.64)

The same formula applies for the scattering of any two different quarks, or, by crossing, to the scattering of a quark and an antiquark of a different species. Crossing from the t to the s channel gives the cross section for $q\bar{q}$ annihilation into a different species:

$$\frac{d\sigma}{d\hat{t}}(u\bar{u} \to d\bar{d}) = \frac{4\pi\alpha_s^2}{9\hat{s}^2}\left[\frac{\hat{t}^2 + \hat{u}^2}{\hat{s}^2}\right].$$

(17.65)

The scattering of a quark with an antiquark of the same species is more complicated, since now there are two Feynman diagrams, shown in Fig. 17.10, which interfere with one another. The analogous QED process is Bhabha scattering, $e^+e^- \to e^+e^-$, for which we worked out the cross section in Problem 5.2:

$$\frac{d\sigma}{d\hat{t}}(e^+e^- \to e^+e^-) = \frac{2\pi\alpha^2}{\hat{s}^2}\left[\left(\frac{\hat{s}}{\hat{t}}\right)^2 + \left(\frac{\hat{t}}{\hat{s}}\right)^2 + \hat{u}^2\left(\frac{1}{\hat{s}} + \frac{1}{\hat{t}}\right)^2\right].$$

(17.66)

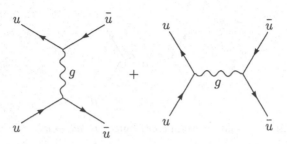

Figure 17.10. Feynman diagrams contributing to $u\bar{u} \to u\bar{u}$.

However, it is not quite straightforward to transcribe this to QCD, because different terms receive different color factors.

This process is most easily analyzed using initial and final states of definite helicity. For massless fermions, helicity is conserved, so the reaction $e_R^+ e_L^- \to e_L^+ e_R^-$ can receive a contribution only from the s-channel diagram, while $e_R^+ e_R^- \to e_R^+ e_R^-$ can receive a contribution only from the t-channel diagram. The corresponding cross sections are

$$\frac{d\sigma}{d\hat{t}}(e_R^+ e_L^- \to e_L^+ e_R^-) = \frac{4\pi\alpha^2}{\hat{s}^2}\left(\frac{\hat{t}}{\hat{s}}\right)^2,$$

$$\frac{d\sigma}{d\hat{t}}(e_R^+ e_R^- \to e_R^+ e_R^-) = \frac{4\pi\alpha^2}{\hat{s}^2}\left(\frac{\hat{s}}{\hat{t}}\right)^2. \tag{17.67}$$

The cross section for $e_R^+ e_L^- \to e_L^+ e_L^-$ must vanish. The fourth possible process involving e_R^+ receives contributions from both s- and t-channel diagrams. Computing this contribution explicitly, one finds

$$\frac{d\sigma}{d\hat{t}}(e_R^+ e_L^- \to e_R^+ e_L^-) = \frac{4\pi\alpha^2}{\hat{s}^2}\hat{u}^2\left(\frac{1}{\hat{t}} + \frac{1}{\hat{s}}\right)^2; \tag{17.68}$$

the cross term in the square is the interference term between the two diagrams. The invariance of QED under parity implies that the values of all of these cross sections remain identical when all helicities are reversed. It is easy to check that the spin-averaged cross section is indeed given by (17.66).

To convert Eq. (17.66) to a QCD cross section averaged over colors, we can assign the color factor (17.63) to the square of any individual diagram. However, the cross term between the two diagrams in Fig. 17.10 receives a different color factor:

$$\left(\frac{1}{3}\right)^2 \cdot (t^a)_{i'i}(t^a)_{jj'} \cdot (t^b)_{j'i'}(t^b)_{ij} = \frac{1}{9}\operatorname{tr}[t^a t^b t^a t^b]. \tag{17.69}$$

To evaluate this factor, we can make use of Eq. (16.79):

$$t^a t^b t^a t^b = \left(C_2(r) - \frac{1}{2}C_2(G)\right)t^a t^a = \left(\frac{4}{3} - \frac{3}{2}\right)\frac{4}{3} = -\frac{2}{9}.$$

So the color factor (17.69) equals $-2/27$.

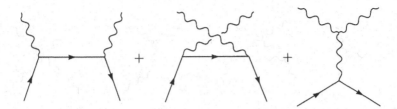

Figure 17.11. Feynman diagrams contributing to $q\bar{q} \rightarrow gg$.

Assembling the color factors and the helicity cross sections, we find the following result for the $u\bar{u}$ scattering cross section:

$$\frac{d\sigma}{d\hat{t}}(u\bar{u} \rightarrow u\bar{u}) = \frac{4\pi\alpha_s^2}{9\hat{s}^2}\left[\frac{\hat{s}^2 + \hat{u}^2}{\hat{t}^2} + \frac{\hat{t}^2 + \hat{u}^2}{\hat{s}^2} - \frac{2}{3}\frac{\hat{u}^2}{\hat{s}\hat{t}}\right]. \tag{17.70}$$

By crossing between the s and u channels, we find the corresponding cross section for $uu \rightarrow uu$:

$$\frac{d\sigma}{d\hat{t}}(uu \rightarrow uu) = \frac{4\pi\alpha_s^2}{9\hat{s}^2}\left[\frac{\hat{u}^2 + \hat{s}^2}{\hat{t}^2} + \frac{\hat{t}^2 + \hat{s}^2}{\hat{u}^2} - \frac{2}{3}\frac{\hat{s}^2}{\hat{u}\hat{t}}\right]. \tag{17.71}$$

The process $\bar{u}\bar{u} \rightarrow \bar{u}\bar{u}$ has the same cross section. This completes our catalogue of cross sections for the scattering of quarks and antiquarks.

We turn next to processes that involve both quarks and gluons. We will begin with the reaction $q\bar{q} \rightarrow gg$. This is the analogue of the QED annihilation of e^+e^- to $\gamma\gamma$, discussed in Section 5.5. The QED cross section is

$$\frac{d\sigma}{d\hat{t}}(e^+e^- \rightarrow \gamma\gamma) = \frac{2\pi\alpha^2}{\hat{s}^2}\left[\frac{\hat{u}}{\hat{t}} + \frac{\hat{t}}{\hat{u}}\right]. \tag{17.72}$$

Since the photons are identical particles, this expression should be integrated over only half of the 4π solid angle.

The QCD reaction is considerably more complicated. As we saw in Section 16.1, this process receives contributions from three Feynman diagrams, shown in Fig. 17.11. These contributions must be summed over the transverse polarization states of the gluons. If one chooses instead to evaluate the sum over gluon polarizations by the replacement

$$\sum_\epsilon \epsilon^\mu \epsilon^{*\nu} \rightarrow -g^{\mu\nu}, \tag{17.73}$$

we saw in Section 16.3 that one must also include the (negative) cross section for $q\bar{q}$ annihilation to a ghost-antighost pair.

The leading behavior of the $q\bar{q} \rightarrow gg$ cross section as $\hat{t}$ or $\hat{u} \rightarrow 0$ is not so hard to evaluate. In either case, only the single diagram with the corresponding kinematic singularity contributes. The color factor associated with the square of either of these diagrams is the square of

$$(t^a)_{ij}(t^b)_{jk},$$

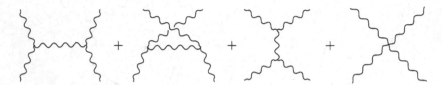

Figure 17.12. Feynman diagrams contributing to $gg \to gg$.

summed over the gluon colors a, b and averaged over the q and $\bar{q}$ colors i, k. This gives

$$\left(\frac{1}{3}\right)^2 \cdot \text{tr}[t^a t^b t^b t^a] = \frac{1}{9} \cdot 3\big(C_2(r)\big)^2 = \frac{16}{27}. \tag{17.74}$$

Thus the most singular terms are given by the QED result, with α replaced by α_s, multiplied by 16/27. The complete evaluation of the cross section is left for Problem 17.3; the result is

$$\frac{d\sigma}{d\hat{t}}(q\bar{q} \to gg) = \frac{32\pi\alpha_s^2}{27\hat{s}^2}\left[\frac{\hat{u}}{\hat{t}} + \frac{\hat{t}}{\hat{u}} - \frac{9}{4}\Big(\frac{\hat{t}^2 + \hat{u}^2}{\hat{s}^2}\Big)\right]. \tag{17.75}$$

The cross sections for the remaining quark-gluon processes can be obtained from this result by crossing. The result for the inverse reaction $gg \to q\bar{q}$ involves the same squared matrix element as (17.75); the only difference is that we average over gluon rather than quark colors, giving a relative factor of $(3/8)^2$. Thus,

$$\frac{d\sigma}{d\hat{t}}(gg \to q\bar{q}) = \frac{\pi\alpha_s^2}{6\hat{s}^2}\left[\frac{\hat{u}}{\hat{t}} + \frac{\hat{t}}{\hat{u}} - \frac{9}{4}\Big(\frac{\hat{t}^2 + \hat{u}^2}{\hat{s}^2}\Big)\right]. \tag{17.76}$$

For the reaction $qg \to qg$, cross the s and t channels in Eq. (17.75) and multiply by 3/8 since there is one gluon color average. This gives

$$\frac{d\sigma}{d\hat{t}}(qg \to qg) = \frac{4\pi\alpha_s^2}{9\hat{s}^2}\left[-\frac{\hat{u}}{\hat{s}} - \frac{\hat{s}}{\hat{u}} + \frac{9}{4}\Big(\frac{\hat{s}^2 + \hat{u}^2}{\hat{t}^2}\Big)\right]. \tag{17.77}$$

The cross section for $\bar{q}g \to \bar{q}g$ is identical.

The final elementary process of QCD is gluon-gluon scattering. This has no QED analogue, and is rather tedious to evaluate. There are four leading-order diagrams, shown in Fig. 17.12. We discuss this process also in Problem 17.3. The final result for the spin- and color-averaged cross section is

$$\frac{d\sigma}{d\hat{t}}(gg \to gg) = \frac{9\pi\alpha_s^2}{2\hat{s}^2}\left[3 - \frac{\hat{t}\hat{u}}{\hat{s}^2} - \frac{\hat{s}\hat{u}}{\hat{t}^2} - \frac{\hat{s}\hat{t}}{\hat{u}^2}\right]. \tag{17.78}$$

The various parton cross sections listed in this section can be combined with the parton distribution functions to predict the cross section for jet production in hadron-hadron collisions. As an example, we show in Fig. 17.13 a comparison of the invariant mass ($\hat{s}$) distribution predicted for parton-parton

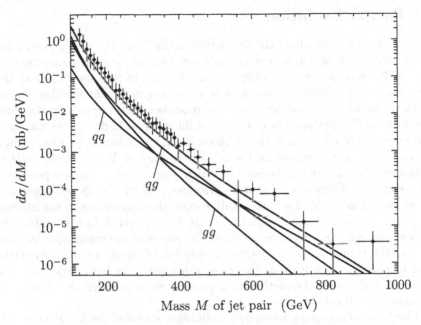

Figure 17.13. Two-jet invariant mass distribution in $p\bar{p}$ collisions at $E_{\rm cm} =$ 1.8 TeV, as measured by the CDF collaboration, F. Abe, et. al., *Phys. Rev.* **D48**, 998 (1993). The measurement is compared to a leading-order QCD calculation using the CTEQ structure functions described in Fig. 17.6. The three lower curves show the invariant mass distributions for the three components of the theoretical prediction: quark-quark (and antiquark) scattering, quark-gluon scattering, and gluon-gluon scattering.

scattering with the invariant mass distribution of two-jet events observed in high-energy $p\bar{p}$ collisions. The overall normalization of the theoretical prediction is uncertain by about a factor of 2 due to the ambiguity of the choice of Q^2 used to evaluate $\alpha_s(Q^2)$ in the parton cross sections, and due to similar ambiguities in deriving parton distributions from deep inelastic scattering cross sections. This uncertainty is reduced to about 30% when corrections of order α_s are included. Still, it is remarkable that the lowest-order QCD prediction tracks the observed distribution as a function of the two-jet invariant mass as it falls by six orders of magnitude. Thus, for the jet production cross section, as for hard processes involving leptons, QCD indeed gives a reasonable description of the behavior of the strong interactions at large momentum transfer.

17.5 Parton Evolution

Now that we have examined the predictions of QCD at the leading order for several strong interaction processes, we should investigate the corrections to these predictions at the next order in α_s. We saw in Section 17.2 that the corrections from individual diagrams may contain mass singularities, singularities associated with collinear emission processes which appear in the limit of zero mass. For the process of e^+e^- annihilation to hadrons, we saw that these mass singularities, and the infrared divergences from soft gluon emission, cancel in the expression for the total cross section. It can be shown that this is a general feature of processes in which quarks and gluons are produced in the collision of leptons or photons. However, when quarks or gluons appear in the initial state of a parton subprocess, the corrections to the process will, in general, have mass singularities that do not cancel. In this section we will demonstrate this effect and work out its physical interpretation. We will find that these singular terms predict a violation of Bjorken scaling by terms depending on the logarithm of the momentum scale. In fact, they lead to a precise set of differential equations that govern the momentum dependence of the parton distributions.

The basic phenomena associated with mass singularities in QCD are already present in the physics of collinear photon emission in QED at high energies, and so it is most straightforward to begin by studying that case. In this section, we will show that collinear photon emission leads to an analogue of a parton distribution function for the electron. We will derive a differential equation describing this distribution function, first constructed by Gribov and Lipatov. Finally, we will generalize this equation to QCD, following the construction of Altarelli and Parisi.[‡]

In Chapters 5 and 6, we studied several examples of QED processes that involved diagrams with t- or u-channel singularities. In these cases, we found that the total cross section was generally enhanced by an extra factor $\log(s/m^2)$ in the high-energy limit. For example, in Eq. (5.95) we saw that the u-channel exchange diagram in Compton scattering, Fig. 17.14(a), leads to an integral that, in the high-energy limit, takes the form

$$\int \frac{d\cos\theta}{(1+\cos\theta)}.$$

The singularity as $\cos\theta \to -1$ is cut off by the electron mass, leading to the logarithmic enhancement factor. Thus the collinear photon emission costs a factor that is not α but rather $\alpha\log(s/m^2)$. Emission of multiple collinear photons, as in Fig. 17.14(b), gives contributions of order $(\alpha\log(s/m^2))^n$. To improve the accuracy of perturbation theory, it would be useful to find a procedure for summing these terms to all orders in α. In QCD, the corresponding

---

[‡]V. N. Gribov and L. Lipatov, *Sov. J. Nucl. Phys.* **15**, 438 (1972); G. Altarelli and G. Parisi, *Nucl. Phys.* **B126**, 298 (1977). We also strongly recommend reading the papers of J. Kogut and L. Susskind, *Phys. Rev.* **D9**, 697, 3391 (1974).

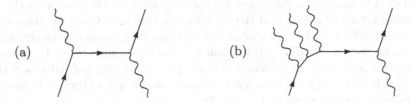

Figure 17.14. Diagrams with mass singularities associated with collinear photon emission: (a) leading order; (b) higher order.

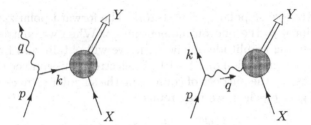

Figure 17.15. General form of diagrams with mass singularities in QED.

factor for collinear gluon emission would be

$$\alpha_s(Q^2) \log \frac{Q^2}{\mu^2},$$

where μ is the momentum scale where nonperturbative QCD effects become important. Comparing with Eq. (17.17), we see that this product is of order 1. Thus, in this case, the resummation of large logarithms is essential if we are to make any quantitative predictions.

In QED, diagrams with mass singularities associated with one collinear emission are of one of the forms shown in Fig. 17.15. In each case, the circle represents a scattering process with large momentum transfer. The mass singularity appears when the denominator of the intermediate propagator vanishes, that is, when the intermediate state is almost on-shell. Thus, it is natural to consider the first diagram in Fig. 17.15 to be a transition to a real photon and an almost-real electron, followed by the interaction of the electron with the remaining particles in the amplitude. The second diagram should have a similar interpretation with an almost-real photon in the intermediate state.

The only subtlety comes in defining the polarization of the intermediate-state particle. For the case of an intermediate-state electron, the numerator of the propagator is

$$\not{k} = \sum_s u^s(k)\bar{u}^s(k). \tag{17.79}$$

Thus, when $k^2 \to 0$, the photon emission vertex and the remaining part of the amplitude are contracted with on-shell polarization spinors for a massless electron. The analogous statement for the diagram with the photon in the intermediate state would be that the electron emission vertex and the remaining photon amplitude should be contracted with physical transverse polarization vectors for the intermediate-state photon. Since the numerator of the photon propagator is $g^{\mu\nu}$, it is not obvious that the photon propagator reduces in this way. But it is true. To see this, use the expansion for $g^{\mu\nu}$ in terms of massless polarization vectors given in Eq. (16.20):

$$g^{\mu\nu} = \epsilon_+^\mu \epsilon_-^{\nu *} + \epsilon_-^\mu \epsilon_+^{\nu *} - \sum_i \epsilon_{Ti}^\mu \epsilon_{Ti}^{\nu *}. \tag{17.80}$$

Here ϵ_{Ti}^μ are transverse polarization vectors. The forward polarization vector ϵ_+^μ is proportional to the photon momentum q^μ. When we contract ϵ_+^μ with the QED scattering amplitude on the right, we will obtain zero by the Ward identity, and the contraction of $\epsilon_+^{*\nu}$ with the electron emission vertex similarly gives zero. Thus, for the purpose of computing the singular term as the photon momentum q goes on-shell, we may replace

$$\frac{-ig^{\mu\nu}}{q^2} \rightarrow \frac{+i}{q^2} \sum_i \epsilon_{Ti}^\mu \epsilon_{Ti}^{\nu *} \tag{17.81}$$

and evaluate the photon emission and absorption amplitudes with transverse polarization vectors.

Matrix Element for Electron Splitting

By replacing the numerator of the intermediate propagator with a sum over polarization vectors, we decouple the photon or electron emission vertex from the rest of the diagram. We will now evaluate this vertex explicitly between physical polarization states of massless particles. The kinematics is shown in Fig. 17.16. The two final particles should be almost collinear, with a small relative transverse momentum. We can choose the incident electron momentum to lie along the $\hat{3}$ axis and the outgoing momenta to lie in the $\hat{1}$-$\hat{3}$ plane. Let z be the fraction of the energy of the initial electron that is carried off by the photon. Then the three 4-momenta can be written as

$$p = (p, 0, 0, p),$$

$$q \approx (zp, p_\perp, 0, zp), \tag{17.82}$$

$$k \approx ((1-z)p, -p_\perp, 0, (1-z)p).$$

These three vectors satisfy $p^2 = q^2 = k^2 = 0$, up to terms of order $p_\perp^2$.

   In the process where a real photon is emitted, we should have p^2 and q^2 exactly zero, and k^2 slightly off-shell by an amount of order $p_\perp^2$. We will need to know the value of k^2, which appears in the virtual electron propagator. So let us modify Eqs. (17.82) to satisfy the condition $q^2 = 0$ up to terms of order

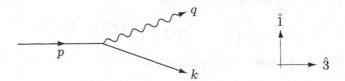

Figure 17.16. Kinematics of the vertex for emission of a collinear electron or photon.

$p_\perp^4$, rewriting q and k as

$$q = (zp, p_\perp, 0, zp - \frac{p_\perp^2}{2zp}),$$

$$(17.83)$$

$$k = ((1-z)p, -p_\perp, 0, (1-z)p + \frac{p_\perp^2}{2zp}).$$

With this modification,

$$k^2 = -p_\perp^2 - 2(1-z)\frac{p_\perp^2}{2z} + \mathcal{O}(p_\perp^4).$$

Thus, if the photon is real and the electron is virtual, we have

$$q^2 = 0, \qquad k^2 = -\frac{p_\perp^2}{z}. \qquad (17.84)$$

Reciprocally, in the process with a real electron and a virtual photon,

$$k^2 = 0, \qquad q^2 = -\frac{p_\perp^2}{(1-z)}. \qquad (17.85)$$

These more accurate expressions will be needed only in the propagator of the virtual particle. The matrix element of the electron-photon vertex begins in order $p_\perp$, so it is not significantly affected by the modification of (17.82) to (17.83), and is the same (to lowest order) no matter which particle is virtual.

We now calculate the matrix elements of the QED vertex between massless states of definite helicity. If the initial electron is left-handed, the final electron must also be left-handed, by helicity conservation. Then the photon emission vertex is given by

$$i\mathcal{M} = \bar{u}_L(k)(-ie\gamma_\mu)u_L(p)\epsilon_T^{*\mu}(q), \qquad (17.86)$$

where the photon polarization vector may be either left- or right-handed. Recalling the helicity-basis expressions

$$\gamma_\mu = \begin{pmatrix} 0 & \sigma_\mu \\ \bar{\sigma}_\mu & 0 \end{pmatrix}, \qquad u_L(p) = \sqrt{2p^0} \begin{pmatrix} \xi(p) \\ 0 \end{pmatrix} \quad \text{(for } m = 0),$$

we can write more explicitly

$$i\mathcal{M} = ie\sqrt{2(1-z)p}\sqrt{2p}\,\xi^\dagger(k)\sigma^i\xi(p)\,\epsilon_T^{*i}(q). \qquad (17.87)$$

To order $p_\perp$, the left-handed spinors are

$$\xi(p) = \begin{pmatrix} 0 \\ 1 \end{pmatrix}, \qquad \xi(k) = \begin{pmatrix} p_\perp/2(1-z)p \\ 1 \end{pmatrix}. \tag{17.88}$$

The polarization vectors for the photon are

$$\epsilon_L^{*i}(q) = \frac{1}{\sqrt{2}}\left(1, i, -\frac{p_\perp}{zp}\right), \qquad \epsilon_R^{*i}(q) = \frac{1}{\sqrt{2}}\left(1, -i, -\frac{p_\perp}{zp}\right). \tag{17.89}$$

Notice that, when these vectors are contracted with the Pauli matrix in Eq. (17.87), the first two components of the right-handed polarization vector give $(\sigma^1 - i\sigma^2) = 2\sigma^-$, which annihilates $\xi(p)$. The only remaining term comes from the $i = 3$ component, and we find

$$i\mathcal{M}(e_L^- \to e_L^- \gamma_R) = ie\frac{\sqrt{2(1-z)}}{z}p_\perp. \tag{17.90}$$

For the left-handed photon polarization, there is an additional contribution from the first two components of ϵ_L^*. These add to

$$i\mathcal{M}(e_L^- \to e_L^- \gamma_L) = ie\frac{\sqrt{2(1-z)}}{z(1-z)}p_\perp. \tag{17.91}$$

Parity invariance implies that the values of the matrix elements are unchanged if all helicities are flipped; this immediately gives the required matrix elements for the case of an initial e_R^-. The squared matrix element, averaged over initial helicities, is therefore

$$\frac{1}{2}\sum_{\text{pols.}} |\mathcal{M}|^2 = \frac{2e^2p_\perp^2}{z(1-z)}\left[\frac{1 + (1-z)^2}{z}\right]. \tag{17.92}$$

The first term in the brackets comes from a photon with spin parallel to the electron spin; the second term comes from a photon with spin opposite to the electron spin.

The Equivalent Photon Approximation

Now we have all the pieces needed to compute the cross sections for the processes shown in Fig. 17.15. We first consider the process with a virtual photon. Call the initial state on the right-hand side of the diagram X and the final state Y, and let $\mathcal{M}_{\gamma X}$ represent the matrix element for the scattering of the photon from X. We will assume for simplicity that X is unpolarized, so that the scattering cross section does not depend on the virtual photon polarization. Then the complete diagram gives a cross section

$$\sigma = \frac{1}{(1+v_X)2p2E_X}\int\frac{d^3k}{(2\pi)^3}\frac{1}{2k^0}\int d\Pi_Y \left[\frac{1}{2}\sum|\mathcal{M}|^2\right]\left(\frac{1}{q^2}\right)^2|\mathcal{M}_{\gamma X}|^2, \tag{17.93}$$

where v_X is the velocity of X and $\int d\Pi_Y$ is the phase space integral over Y.

The integral has a singularity when k is collinear with the incident electron momentum p. To isolate the singularity, substitute for k^0 and q^2 from Eqs.

(17.82) and (17.85) and rewrite the integral over k as

$$d^3k = dk^3 d^2k_\perp = pdz \cdot \pi dp_\perp^2. \tag{17.94}$$

Then the cross section can be expressed as

$$\sigma = \int \frac{pdzdp_\perp^2}{16\pi^2(1-z)p} \Big[\frac{1}{2}\sum|\mathcal{M}|^2\Big]\frac{(1-z)^2}{p_\perp^4}\frac{z}{(1+v_X)2zp2E_X}\int d\Pi_Y |\mathcal{M}_{\gamma X}|^2$$

$$= \int \frac{dzdp_\perp^2}{16\pi^2(1-z)} \Big[\frac{1}{2}\sum|\mathcal{M}|^2\Big]\frac{z(1-z)^2}{p_\perp^4} \cdot \sigma(\gamma X \to Y). \tag{17.95}$$

Finally, insert the spin-averaged electron emission vertex (17.92), to obtain

$$\sigma = \int \frac{dzdp_\perp^2}{16\pi^2}\frac{z(1-z)}{p_\perp^4}\frac{2e^2p_\perp^2}{z(1-z)}\Big[\frac{1+(1-z)^2}{z}\Big] \cdot \sigma(\gamma X \to Y)$$

$$= \int_0^1 dz \int \frac{dp_\perp^2}{p_\perp^2}\frac{\alpha}{2\pi}\Big[\frac{1+(1-z)^2}{z}\Big] \cdot \sigma(\gamma X \to Y). \tag{17.96}$$

The integral over $p_\perp^2$ runs from momentum transfers of order s down to the electron mass m^2, which cuts off the singularity. Thus, our final result is

$$\sigma(e^-X \to e^-Y) = \int_0^1 dz \frac{\alpha}{2\pi}\log\frac{s}{m^2}\Big[\frac{1+(1-z)^2}{z}\Big]\cdot\sigma(\gamma X \to Y). \tag{17.97}$$

The cross section on the right-hand side is computed for a real, transversely polarized photon of momentum zp. The factor $\log(s/m^2)$ represents the mass singularity. This formula is the Weizsäcker-Williams *equivalent photon approximation*, which we encountered earlier in Problems 5.5 and 6.2.

Formula (17.97) takes on a new significance when it is juxtaposed with the QCD predictions of the previous two sections. This QED formula has just the same form as a parton model expression, with the Weizsäcker-Williams distribution function

$$f_\gamma(z) = \frac{\alpha}{2\pi}\log\frac{s}{m^2}\Big[\frac{1+(1-z)^2}{z}\Big] \tag{17.98}$$

playing the role of the probability to find a photon of longitudinal fraction z in the incident electron.

The Electron Distribution

The first diagram of Fig. 17.15, with an emitted photon and a virtual electron, can be treated in the same way. The analogue of (17.93) is

$$\sigma = \frac{1}{(1+v_X)2p2E_X}\int\frac{d^3q}{(2\pi)^3}\frac{1}{2q^0}\int d\Pi_Y\Big[\frac{1}{2}\sum|\mathcal{M}|^2\Big]\Big(\frac{1}{k^2}\Big)^2|\mathcal{M}_{e^-X}|^2.$$

Following the steps that led to (17.97), we find

$$\sigma(e^- X \to \gamma Y) = \int \frac{dz\,dp_\perp^2}{16\pi^2 z} \left[\frac{1}{2} \sum |\mathcal{M}|^2 \right] \frac{z^2}{p_\perp^4} \cdot (1-z)\sigma(e^- X \to Y)$$

$$= \int \frac{dz\,dp_\perp^2}{16\pi^2} \frac{z(1-z)}{p_\perp^4} \frac{2e^2 p_\perp^2}{z(1-z)} \left[\frac{1+(1-z)^2}{z} \right] \cdot \sigma(e^- X \to Y)$$

$$= \int_0^1 dz \int \frac{dp_\perp^2}{p_\perp^2} \frac{\alpha}{2\pi} \left[\frac{1+(1-z)^2}{z} \right] \cdot \sigma(e^- X \to Y), \qquad (17.99)$$

where the intermediate electron carries a longitudinal fraction $(1-z)$.

It is tempting to substitute $x = (1-z)$ and interpret the factor multiplying the cross section under the integral in (17.99) as the parton distribution for finding an electron parton in the electron. This would give

$$f_e^{(1)}(x) = \frac{\alpha}{2\pi} \log \frac{s}{m^2} \left[\frac{1+x^2}{1-x} \right]. \qquad (17.100)$$

However, this expression is not adequate. Most obviously, it does not take into account the processes without radiation, in which the electron remains an electron at the full energy. This is easily remedied by considering (17.100) as the order-α correction to the most naive expectation,

$$f_e^{(0)}(x) = \delta(1-x), \qquad (17.101)$$

in which we consider the electron to contain only an electron at the full energy. Unfortunately, the sum of (17.101) and (17.100) still does not give an adequate description of the electron distribution, for two reasons. First, Eq. (17.100) diverges near $x = 1$, and we need a prescription for treating this singularity. Second, while Eq. (17.100) takes into account the virtual electrons moved to longitudinal fraction x from $x = 1$ by radiation, it does not take into account the concomitant loss of electrons from the delta function peak at $x = 1$.

The divergence of (17.100) at $x = 1$ corresponds to the emission of soft photons. We saw in Section 6.5 that the emission of soft photons does not affect the rate of a QED reaction. Order by order in α, one finds that infrared-divergent positive contributions to the total rate from the emission of soft photons are balanced by negative contributions from diagrams with soft virtual photons. In the present example, the negative contribution must decrease the weight of the process in which no photon is emitted. Thus, to order α, the parton distribution for electrons in the electron should have the form

$$f_e(x) = \delta(1-x) + \frac{\alpha}{2\pi} \log \frac{s}{m^2} \left(\frac{1+x^2}{(1-x)} - A\delta(1-x) \right). \qquad (17.102)$$

The coefficient A results from diagrams with virtual photons that we have not computed. However, the effect of these diagrams is easy to understand; they subtract from the delta function the probability that has been moved to lower x by radiation, so that the integral over the full term of order α is zero. Another way of expressing this criterion is that A is determined by the condition that the electron contain exactly one electron parton,

$$\int_0^1 dx\, f_e(x) = 1. \tag{17.103}$$

(This equation will be modified below, when we include pair-creation processes.)

It is not so clear how to integrate over the singular denominator in (17.100) to determine A explicitly. It is conventional to define a distribution that can be integrated by subtracting a delta function from the singular term. Define the distribution

$$\frac{1}{(1-x)_+} \tag{17.104}$$

to agree with the function $1/(1-x)$ for all values of x less than 1, and to have a singularity at $x = 1$ such that the integral of this distribution with any smooth function $f(x)$ gives

$$\int_0^1 dx\, \frac{f(x)}{(1-x)_+} = \int_0^1 dx\, \frac{f(x) - f(1)}{(1-x)}. \tag{17.105}$$

Less formally,

$$\frac{1}{(1-x)_+} = \lim_{\epsilon \to 0}\left[\frac{1}{(1-x)}\theta(1-x-\epsilon) - \delta(1-x)\int_0^{1-\epsilon} dx'\, \frac{1}{(1-x')}\right]. \tag{17.106}$$

The more formal definition (17.105) is often easier to use in practice.

Using this definition, we can bring a piece of the delta function into the singular term of (17.102) by changing the denominator $(1-x)$ to $(1-x)_+$. Then, to normalize (17.102), we need the integral

$$\int_0^1 dx\, \frac{1+x^2}{(1-x)_+} = \int_0^1 dx\, \frac{x^2-1}{(1-x)} = -\frac{3}{2}.$$

Our final form of the electron distribution, to order α, is

$$f_e(x) = \delta(1-x) + \frac{\alpha}{2\pi}\log\frac{s}{m^2}\left[\frac{1+x^2}{(1-x)_+} + \frac{3}{2}\delta(1-x)\right]. \tag{17.107}$$

This distribution is now properly normalized, but it is still highly singular near $x = 1$. Thus, we should expect higher-order corrections to the electron distribution function to be important in this region. We must, then, think about how to treat the emission of many collinear photons.

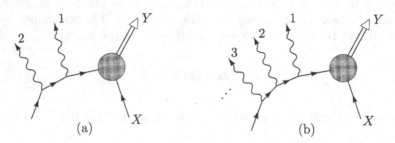

Figure 17.17. Higher-order diagrams with collinear photon emission: (a) two collinear photons; (b) many collinear photons.

Multiple Splittings

In fact, it is not difficult to extend the analysis we have just completed to account for emission of many collinear photons. Consider the process shown in Fig. 17.17(a), in which photon 1 is radiated with a transverse momentum $p_{1\perp}$ and photon 2 with a transverse momentum $p_{2\perp}$. The emission of photon 2 can be computed just as we did above. If $p_{2\perp} \ll p_{1\perp}$, the first virtual electron is very close to mass shell compared to $p_{1\perp}^2$ and so we can ignore its virtuality in computing the emission of the photon 1. The double photon emission gives a contribution of order

$$\left(\frac{\alpha}{2\pi}\right)^2 \int_{m^2}^{s} \frac{dp_{1\perp}^2}{p_{1\perp}^2} \int_{m^2}^{p_{1\perp}^2} \frac{dp_{2\perp}^2}{p_{2\perp}^2} = \frac{1}{2}\left(\frac{\alpha}{2\pi}\right)^2 \log^2 \frac{s}{m^2}.$$

In the opposite limit, $p_{2\perp} \gg p_{1\perp}$, there is no denominator of order $p_{1\perp}^2$, and so we do not find a double logarithm. Only in the case $p_{2\perp} \ll p_{1\perp}$ can the contribution of order α^2 compete with the contribution of order α.

This argument extends to the emission of arbitrarily many collinear photons, Fig. 17.17(b). The region of integration over the photon phase space corresponding to the ordering

$$p_{1\perp} \gg p_{2\perp} \gg p_{3\perp} \gg \cdots \tag{17.108}$$

gives a contribution that contains the factor

$$\frac{1}{n!}\left(\frac{\alpha}{2\pi}\right)^n \log^n \frac{s}{m^2}. \tag{17.109}$$

If the photon transverse momenta are ordered in any other way, the contribution from that region contains at least one less power of the large logarithm at the same order in α. If condition (17.108) holds, the virtual electron momenta are increasingly off-mass-shell as one proceeds from the outside of the diagram toward the hard collision. In this case, the electron momenta are said to be *strongly ordered*.

This set of conclusions has an interesting physical interpretation. Since the intermediate electrons are increasingly virtual as we go into the diagram, it is natural to interpret them as components of the physical electron when this particle is analyzed at successively smaller distance scales. The intermediate electron with $k^2 \sim p_\perp^2$ may be thought of as a constituent of the electron made visible when the wavefunction of the physical electron is probed with a resolution $\Delta r \sim (p_\perp)^{-1}$. In this picture, the electron seen at one resolution can be resolved at a finer scale into a more virtual electron and a number of photons.

From either the perspective of computing Feynman diagrams or the grander perspective of the electron structure, it is useful to imagine the splitting of the electron into a virtual electron plus photons to be a continuous evolution process as a function of the transverse momentum of the electron constituent. To describe this process mathematically, we introduce an explicit $p_\perp$ dependence of the electron and photon distribution functions. We define the functions $f_\gamma(x, Q)$ and $f_e(x, Q)$ to give the probabilities of finding a photon or an electron of longitudinal fraction x in the physical electron, taking into account collinear photon emissions with transverse momenta $p_\perp < Q$. If Q is slightly increased to $Q + \Delta Q$, we must take into account the possibility that an electron constituent in $f_e(x, Q)$ will radiate a photon with $Q < p_\perp < Q + \Delta Q$. The differential probability for an electron to split off a photon that carries away a fraction z of its energy is

$$\frac{\alpha}{2\pi} \frac{dp_\perp^2}{p_\perp^2} \frac{1 + (1-z)^2}{z}. \tag{17.110}$$

The new photon distribution can therefore be computed as follows:

$$f_\gamma(x, Q + \Delta Q)$$
$$= f_\gamma(x, Q) + \int_0^1 dx' \int_0^1 dz \left[\frac{\alpha}{2\pi} \frac{\Delta Q^2}{Q^2} \frac{1 + (1-z)^2}{z} \right] f_e(x', p_\perp) \delta(x - zx')$$

$$= f_\gamma(x, Q) + \frac{\Delta Q}{Q} \int_x^1 \frac{dz}{z} \left[\frac{\alpha}{\pi} \frac{1 + (1-z)^2}{z} \right] f_e\left(\frac{x}{z}, p_\perp\right). \tag{17.111}$$

Passing to a continuous evolution, we find that the function $f_\gamma(x, Q)$ is determined by the integral-differential equation

$$\frac{d}{d \log Q} f_\gamma(x, Q) = \int_x^1 \frac{dz}{z} \left[\frac{\alpha}{\pi} \frac{1 + (1-z)^2}{z} \right] f_e\left(\frac{x}{z}, Q\right). \tag{17.112}$$

Similarly, the distribution of component electrons in the physical electron will evolve with Q, reflecting the appearance of electrons at lower values of x due to photon radiation, and the disappearance of these electrons at higher x. The term in brackets in Eq. (17.107) gives a correct accounting of both effects for the radiation of a single photon. Thus, the electron distribution evolves according to

$$\frac{d}{d\log Q} f_e(x, Q) = \int\limits_x^1 \frac{dz}{z} \left[\frac{\alpha}{\pi}\left(\frac{1+z^2}{(1-z)_+} + \frac{3}{2}\delta(1-z)\right)\right] f_e\left(\frac{x}{z}, Q\right). \quad (17.113)$$

By integrating these integral-differential equations using appropriate initial conditions, we sum all of the logarithmically enhanced terms of the form (17.109). The initial conditions should be fixed at a point that will reproduce the correct denominator of the logarithms in Eqs. (17.98) and (17.107). Thus, we should set

$$f_e(x, Q) = \delta(1 - x), \qquad f_\gamma(x, Q) = 0, \quad (17.114)$$

at $Q^2 \sim m^2$.

The resulting distribution functions can be used to compute the cross sections for electron hard scattering from an arbitrary target. Then Eqs. (17.97) and (17.99) should be replaced by

$$\sigma(e^- X \to e^- + n\gamma + Y) = \int\limits_0^1 dx\, f_\gamma(x, Q)\sigma(\gamma X \to Y),$$

$$\quad (17.115)$$

$$\sigma(e^- X \to n\gamma + Y) = \int\limits_0^1 dx\, f_e(x, Q)\sigma(e^- X \to Y),$$

where the cross sections under the integrals are computed for a photon or electron carrying a fraction x of the original electron momentum, the functions $f_\gamma(x, Q)$, $f_e(x, Q)$ are the solutions to Eqs. (17.112) and (17.113), and the momentum Q is chosen as a characteristic momentum transfer of the γX or $e^- X$ subprocess.

Photon Splitting to Pairs

The evolution equations for $f_\gamma(x)$ and $f_e(x)$ need one more modification before they can be considered complete. As written, these equations account for the radiation of photons by electrons to all orders. However, they omit another process that is of the same order in α: the splitting of a photon into an electron-positron pair. We must include this process in our evolution equations, because the process shown in Fig. 17.18, for example, has the same logarithmic enhancement as that shown in Fig. 17.17(a).

We can compute the effects of photon splitting in the same way that we computed with effects of photon radiation. The basic kinematics of the process is very similar, as shown in Fig. 17.19; the only difference is that the photon is now in the initial state, while the final state consists of an almost collinear electron-positron pair. We need to work out the analogue of Eq. (17.92) for this process.

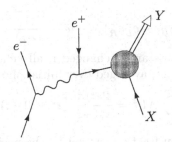

Figure 17.18. A process that involves e^+e^- pair creation enhanced by a collinear mass singularity.

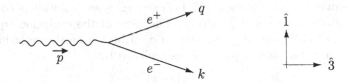

Figure 17.19. Kinematics of the vertex for photon conversion to a collinear electron-positron pair.

Consider the case in which the outgoing electron is left-handed. Then the outgoing positron must be right-handed, by helicity conservation; its spin wavefunction will contain a left-handed spinor. Let us take the electron momentum to be k, given by Eq. (17.82), and the positron momentum to be q. Then the vertex gives a matrix element

$$i\mathcal{M} = -ie\bar{u}_L(k)\gamma_\mu v_L(q)\epsilon_T^\mu(p), \qquad (17.116)$$

where the photon polarization vector can be either left- or right-handed. When we insert the explicit forms for the massless spinors, we obtain

$$i\mathcal{M} = ie\sqrt{2(1-z)p}\sqrt{2zp}\,\xi^\dagger(k)\sigma^i\xi(q)\cdot\epsilon_T^i(p),$$

where the electron and positron spinors are given, to order $p_\perp$, by

$$\xi(q) = \begin{pmatrix} -p_\perp/2zp \\ 1 \end{pmatrix}, \qquad \xi(k) = \begin{pmatrix} p_\perp/2(1-z)p \\ 1 \end{pmatrix}.$$

The polarization vectors for the photon are

$$\epsilon_L^i(p) = \frac{1}{\sqrt{2}}(1, -i, 0), \qquad \epsilon_R^i(p) = \frac{1}{\sqrt{2}}(1, i, 0).$$

Dotting these vectors into σ^i, we find for the polarized matrix elements

$$i\mathcal{M}(\gamma_L \to e_L^- e_R^+) = -ie\frac{\sqrt{2z(1-z)}}{z}p_\perp,$$

and

$$iM(\gamma_R \to e_L^- e_R^+) = +ie\frac{\sqrt{2z(1-z)}}{(1-z)}p_\perp.$$

Again, the matrix elements are unchanged if all helicities are flipped. Thus the squared matrix element, averaged over initial photon polarizations, is

$$\frac{1}{2}\sum_{\text{pols.}}|\mathcal{M}|^2 = \frac{2e^2 p_\perp^2}{z(1-z)}\left[z^2 + (1-z)^2\right], \tag{17.117}$$

where z is the momentum fraction carried by the positron. The first term in the brackets comes from processes in which the spin of the positron is parallel to the spin of the photon; the second term comes from processes where the electron spin is parallel to the photon spin.

The squared matrix element (17.117) generates an evolution of constituent photons into electrons and positrons. The form of the evolution equation is similar to (17.113), but with the photon distribution on the right-hand side, and with the expression in parentheses replaced by

$$\left(z^2 + (1-z)^2\right). \tag{17.118}$$

When we create an electron-positron pair, we must remove a photon; this requires a negative term in the evolution equation for the photon distribution (17.112) that contains a delta function multiplying the normalization of (17.118):

$$\int_0^1 dz\left(z^2 + (1-z)^2\right) = \frac{2}{3}. \tag{17.119}$$

Evolution Equations for QED

Including the effects of pair creation, we find the complete evolution equations for electron, positron, and photon distributions in QED. These equations, originally derived by Gribov and Lipatov, sum the leading logarithms from collinear singularities to all orders in α. The evolution equations take the form

$$\frac{d}{d\log Q}f_\gamma(x,Q) = \frac{\alpha}{\pi}\int_x^1 \frac{dz}{z}\left\{P_{\gamma\leftarrow e}(z)\left[f_e(\frac{x}{z},Q) + f_{\bar{e}}(\frac{x}{z},Q)\right]\right.$$

$$\left. + P_{\gamma\leftarrow\gamma}(z)f_\gamma(\frac{x}{z},Q)\right\},$$

$$\tag{17.120}$$

$$\frac{d}{d\log Q}f_e(x,Q) = \frac{\alpha}{\pi}\int_x^1 \frac{dz}{z}\left\{P_{e\leftarrow e}(z)f_e(\frac{x}{z},Q) + P_{e\leftarrow\gamma}(z)f_\gamma(\frac{x}{z},Q)\right\},$$

$$\frac{d}{d\log Q}f_{\bar{e}}(x,Q) = \frac{\alpha}{\pi}\int_x^1 \frac{dz}{z}\left\{P_{e\leftarrow e}(z)f_{\bar{e}}(\frac{x}{z},Q) + P_{e\leftarrow\gamma}(z)f_\gamma(\frac{x}{z},Q)\right\}.$$

The *splitting functions* $P_{i\leftarrow j}(z)$ are given by

$$P_{e\leftarrow e}(z) = \frac{1+z^2}{(1-z)_+} + \frac{3}{2}\delta(1-z),$$

$$P_{\gamma\leftarrow e}(z) = \frac{1+(1-z)^2}{z},$$

$$P_{e\leftarrow\gamma}(z) = z^2 + (1-z)^2, \tag{17.121}$$

$$P_{\gamma\leftarrow\gamma}(z) = -\frac{2}{3}\delta(1-z).$$

To obtain the distribution functions for an electron relevant to a given momentum transfer Q, we should integrate these equations with the initial conditions

$$f_e(x,Q) = \delta(1-x), \qquad f_{\bar{e}}(x,Q) = 0, \qquad f_\gamma(x,Q) = 0, \tag{17.122}$$

at $Q = m$. With different initial conditions, the same equations give the distribution functions for a physical positron or photon. The solutions to these equations are used as in Eq. (17.115) to compute cross sections involving processes induced by electrons, positrons, or photons that involve large momentum transfer.

The evolution equations (17.120) are constructed in such a way as to conserve electron number and longitudinal momentum. Thus, the basic sum rules (17.36) and (17.39) satisfied by the parton distributions of hadrons also apply to the QED distribution functions. Specifically, the distribution functions of the electron contain one net electron constituent,

$$\int_0^1 dx\left[f_e(x,Q) - f_{\bar{e}}(x,Q)\right] = 1, \tag{17.123}$$

and account for the total momentum of the physical electron,

$$\int_0^1 dx\, x\left[f_e(x,Q) + f_{\bar{e}}(x,Q) + f_\gamma(x,Q)\right] = 1. \tag{17.124}$$

It is an instructive exercise to verify explicitly, using Eq. (17.120), that the values of these integrals do not depend on Q.

The Altarelli-Parisi Equations

If we encounter mass singularities in QED associated with collinear photon emission, we must also encounter mass singularities in QCD associated with collinear gluon and quark emission. If we compute the corrections of order α_s to the leading-order parton cross sections discussed in Sections 17.3 and 17.4, using massless quarks and gluons, we will find that these correction terms diverge when we integrate over the collinear configurations. Thus the parton-model expressions, at least in their simplest form, break down already at the next-to-leading order in α_s.

However, assuming that the singularities of QCD are no worse than those of QED, the considerations of the previous section tell us how to treat these singular terms. In QED, we found it natural to include the large corrections associated with the mass singularities in the parton distributions rather than in the hard-scattering cross sections. Viewed in this way, the singular terms supply the kernel of an evolution equation for the parton distributions as a function of the logarithm of the momentum scale. Hard scattering with a momentum transfer Q probes the electron at a distance of order Q^{-1}. When the electron wavefunction is resolved to very small scales, it appears as a constituent electron, carrying only a fraction of the total longitudinal momentum, plus a number of constituent photons and electron-positron pairs. Any one of these constituents that carries a substantial fraction of the total electron momentum can initiate a hard-scattering process.

Precisely the same logic applies to the calculation of QCD cross sections. The contributions from the region of collinear gluon or quark emission should be associated with the parton distribution functions rather than with the hard-scattering cross sections. If we make this association, we find that the parton distributions are no longer independent of the momentum Q that characterizes the hard-scattering process; rather, they now evolve logarithmically with Q. For example, the basic equation (17.30) for deep-inelastic scattering will become

$$\frac{d^2\sigma}{dxdy}(e^- p \rightarrow e^- X) = \left(\sum_f x f_f(x, Q) Q_f^2\right) \cdot \frac{2\pi\alpha^2 s}{Q^4}[1 + (1-y)^2], \quad (17.125)$$

and so Bjorken scaling will be violated. Since this violation takes place only on a logarithmic scale in Q^2, it will be a subtle effect, and *approximate* Bjorken scaling will still be a prediction of QCD. But the violation of Bjorken scaling is inevitable, since QCD is a quantum field theory with degrees of freedom at all momentum scales. As we probe the proton wavefunction at increasingly short distances, we excite the high-momentum degrees of freedom and resolve the wavefunction into an increasing number of quarks, antiquarks, and gluons.

The evolution of the QED parton distributions, governed by Eq. (17.120), is characterized by the parameter α/π, so the parton distributions change by $\sim 1\%$ as Q is changed by a factor of 10. In QCD, the corresponding factor governing the rate of evolution should be $\alpha_s(Q)/\pi$. Thus, when Q is very small, the evolution is rapid and contributions of higher order in perturbation theory are important. Ultimately, the initial conditions for the evolution are determined by the form of the proton wavefunction at large distance scales, which cannot be calculated using Feynman diagrams. On the other hand, when Q is large, well above 1 GeV in practice, the evolution becomes slow and is dominated by the leading order in perturbation theory. In that case,

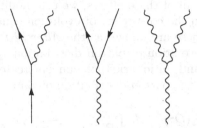

Figure 17.20. The three vertices that contribute to parton evolution in QCD.

QCD perturbation theory makes precise predictions for the form of the evolution of the parton distributions, and these predictions can be tested against experiment.

To derive the evolution equations of parton distributions in QCD we can use the same techniques and logic that we used above for QED. There is a subtlety, that the reduction of the gluon propagator to transverse polarization states in the limit $q^2 \to 0$, Eq. (17.81), cannot be proved so simply as in QED. However, the result is correct also in the non-Abelian case.* Once this technical point is resolved, the kinematics of collinear emission is exactly the same as in QED. Thus we find evolution equations of the same form as in QED, modified only by the replacement of α by α_s, the insertion of appropriate color factors, and the accounting of the effects of the three-gluon vertex.

Collinear emission processes in QCD involve the three vertices shown in Fig. 17.20. Of these, the first two have the same Lorentz structure as those shown in Figs. 17.16 and 17.19. The only difference, aside from the strength of the coupling constant, comes in the color indices. We will treat color just as we treated spin in the preceding analysis: We average over initial colors, and sum over final colors. Then the first vertex of Fig. 17.20, representing the splitting of a quark into a quark and a gluon, receives the color factor

$$\tfrac{1}{3} \operatorname{tr}[t^a t^a] = C_2(r) = \tfrac{4}{3}. \tag{17.126}$$

The second vertex, representing the splitting of a gluon into a quark-antiquark pair, receives the factor

$$\tfrac{1}{8} \operatorname{tr}[t^a t^a] = \tfrac{1}{2}. \tag{17.127}$$

The third vertex in Fig. 17.20 represents the splitting of a gluon to two gluons, an effect that is new to the non-Abelian case. It is straightforward to compute the contribution of this vertex to the evolution equations by taking the matrix elements of the vertex between transverse gluon states of definite helicity. This calculation is the subject of Problem 17.4.

---

*See, for example, J. Collins and D. Soper, in A. Mueller, *Quantum Chromodynamics* (World Scientific, Singapore, 1991).

By accounting for all of these effects, we can modify the QED evolution equations (17.120) into the correct set of evolution equations for parton distributions in QCD. These are known as the *Altarelli-Parisi equations*. They describe the coupled evolution of parton distributions $f_f(x, Q)$, $f_{\bar{f}}(x, Q)$ for each flavor of quark and antiquark that can be treated as massless at the scale Q, together with the parton distribution of gluons, $f_g(x, Q)$. Explicitly,

$$\frac{d}{d\log Q} f_g(x, Q) = \frac{\alpha_s(Q^2)}{\pi} \int_x^1 \frac{dz}{z} \left\{ P_{g\leftarrow q}(z) \sum_f [f_f(\frac{x}{z}, Q) + f_{\bar{f}}(\frac{x}{z}, Q)] \right.$$

$$\left. + P_{g\leftarrow g}(z) f_g(\frac{x}{z}, Q) \right\},$$

$$\frac{d}{d\log Q} f_f(x, Q) = \frac{\alpha_s(Q^2)}{\pi} \int_x^1 \frac{dz}{z} \left\{ P_{q\leftarrow q}(z) f_f(\frac{x}{z}, Q) + P_{q\leftarrow g}(z) f_g(\frac{x}{z}, Q) \right\},$$

$$\frac{d}{d\log Q} f_{\bar{f}}(x, Q) = \frac{\alpha_s(Q^2)}{\pi} \int_x^1 \frac{dz}{z} \left\{ P_{q\leftarrow q}(z) f_{\bar{f}}(\frac{x}{z}, Q) + P_{q\leftarrow g}(z) f_g(\frac{x}{z}, Q) \right\}.$$

$$(17.128)$$

The first three splitting functions can be taken from Eqs. (17.121), multiplied by the color factors computed in Eqs. (17.126) and (17.127):

$$P_{q\leftarrow q}(z) = \frac{4}{3} \left[\frac{1 + z^2}{(1-z)_+} + \frac{3}{2} \delta(1 - z) \right],$$

$$P_{g\leftarrow q}(z) = \frac{4}{3} \left[\frac{1 + (1-z)^2}{z} \right], \qquad (17.129)$$

$$P_{q\leftarrow g}(z) = \frac{1}{2} \left[z^2 + (1-z)^2 \right].$$

The fourth splitting function requires also the computation of Problem 17.4; the result is

$$P_{g\leftarrow g}(z) = 6 \left[\frac{(1-z)}{z} + \frac{z}{(1-z)_+} + z(1-z) + \left(\frac{11}{12} - \frac{n_f}{18} \right) \delta(1 - z) \right]. \quad (17.130)$$

The final term in this expression, which is proportional to n_f, the number of light quark flavors, is the subtraction term associated with gluon splitting into $q\bar{q}$ pairs. The Altarelli-Parisi equations describe the evolution of parton distributions for any hadron, or any hadronic constituent, up to corrections of order α_s that are not enhanced by large logarithms.

Our derivation of the Altarelli-Parisi equations respects the conservation laws of QCD for quark numbers and longitudinal momentum. Thus, the equations must respect the parton-model sum rules (17.36) and (17.39). As in the QED case, it is instructive to verify explicitly that these integrals are independent of Q.

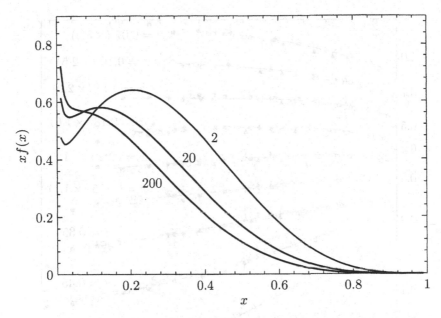

Figure 17.21. The u quark parton distribution function $xf_u(x, Q)$ at $Q =$ 2, 20, and 200 GeV, showing the effects of parton evolution according the Altarelli-Parisi equations. These curves are taken from the CTEQ fit to deep inelastic scattering data described in Fig. 17.6.

In QED, we could use the evolution equations to explicitly compute the structure function of the electron. In QCD, this is no longer possible, because the initial conditions required to integrate the equations are determined by the strong-coupling region of QCD and so are not known *a priori*. However, one can determine the initial conditions of the proton structure experimentally, by measuring the cross section for deep inelastic scattering at a given value of Q^2. One can then predict the structure functions, and thus the deep inelastic cross sections, at higher values of Q^2. There is one subtlety in this analysis: The gluon distribution is not directly measured in deep inelastic scattering, but it does enter the evolution equation for the quark distributions. Thus, some of the information on the Q^2 dependence of deep inelastic scattering simply goes into determining the gluon distribution. However, the gluon distribution is absolutely normalized by the momentum sum rule (17.39), so the evolution equations have predictive power even if this distribution must be fit from the data.

The Altarelli-Parisi equations predict a characteristic form for the evolution of parton distribution functions, shown in Fig. 17.21. Partons at high x tend to radiate and drop down to lower values of x. Meanwhile, new partons are formed at low x as products of this radiation. Thus, the parton distributions decrease at large x and increase much more rapidly at small x as Q^2 increases. We can picture the proton as having more and more constituents,

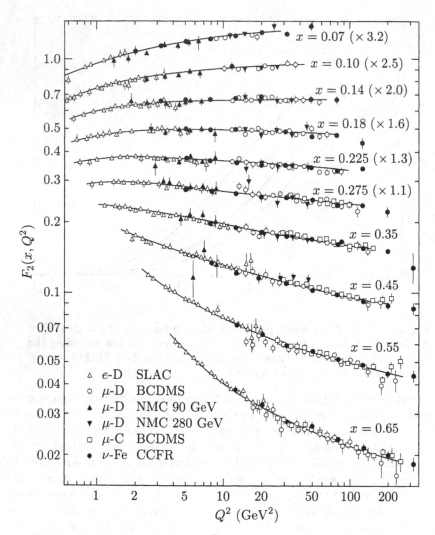

Figure 17.22. Dependence on Q^2 of the combination of quark distribution functions $F_2 = \sum_f x Q_f^2 f_f(x, Q^2)$ measured in deep inelastic electron-proton scattering. The various curves show the variation of F_2 for fixed values of x, and the comparison of this variation to a model evolved with the Altarelli-Parisi equations. The upper six data sets have been multiplied by the indicated factors to separate them on the plot. The data were compiled by M. Virchaux and R. Voss for the Particle Data Group, *Phys. Rev.* **D50**, 1173 (1994), Fig. 32.2. The complete references to the original experiments are given there.

which share its total momentum, as its wavefunction is probed on finer and finer distance scales.

Figure 17.22 shows the evolution of the combination of distribution functions that is measured in deep inelastic scattering, as a function of Q^2. We see the characteristic decrease of the distribution functions at large x and the increase at small x. The data are compared to a model evolved according to the Altarelli-Parisi equations; this model apparently describes the data quite well.

17.6 Measurements of α_s

Before concluding our introductory survey of QCD, we should summarize the quantitative verification of the theory. We discussed precision tests of QED in Section 6.3, bringing together various measurements of the coupling α; the best determinations agree to eight significant figures. Since QCD perturbation theory works only for hard-scattering processes, with uncertainties due to soft processes that are difficult to estimate, this theory has not been tested to such extreme accuracy. Nevertheless, it is interesting to bring together the best available determinations of α_s, to see how well they agree.

In order to compare values of α_s, it is necessary to express these using a common set of conventions. First, one must set the renormalization scale; a useful choice is the mass of the neutral weak boson Z^0: $m_Z = 91.19$ GeV. Second, one must fix the renormalization scheme that defines the QCD coupling constant at this scale. It has become conventional to use as a standard the bare coupling after regularization by modified minimal subtraction, Eq. (11.77). The resulting standard coupling constant is called $\alpha_{s\overline{MS}}(m_Z^2)$.

Measurements of α_s from a number of types of experiments are summarized in Table 17.1. In Section 17.2 we saw that one can obtain a value of α_s from the measurement of the total cross section for e^+e^- annihilation to hadrons or, equivalently, the ratio R of the number of observed hadronic and leptonic events. An independent measurement of α_s can be obtained from the fraction of e^+e^- annihilation events with three-jet final states or, equivalently, from the transverse momentum distribution of produced hadrons relative to the jet axis. A number of measurements of this type are collected and averaged under the heading 'Event shapes'. A similar measurement of α_s is obtained from the measurement of the transverse momentum spectrum of W bosons produced from quark-antiquark annihilation at high-energy $p\bar{p}$ colliders. The gluon radiative correction to the vertex in deep inelastic neutrino scattering can also be used to extract α_s. The rate of Bjorken scaling violation in deep inelastic scattering is controlled by α_s, and so this effect provides another α_s measurement. The decays of the lightest $b\bar{b}$ bound state Υ and the $c\bar{c}$ bound state ψ are governed by QCD and yield a measurement of α_s. Finally, the spectrum of $c\bar{c}$ and $b\bar{b}$ bound states can be computed numerically in terms of the QCD coupling constant, and the comparison with experiment gives a determination of α_s.

Table 17.1. Values of $\alpha_s(m_Z)$ Obtained from QCD Experiments

Process:	$\alpha_s(m_Z)$	Q (GeV)
Deep inelastic scattering	0.118 (6)	1.7
R in τ lepton decay	0.123 (4)	1.8
ψ, Υ spectroscopy	0.110 (6)	2.3
Transverse momentum of W production	0.121 (24)	4.
Deep inelastic scattering (evolution)	0.112 (4)	5.
Event shapes in e^+e^- annihilation	0.121 (6)	5.8,9.1
Rate for ψ, Υ decay	0.108 (10)	9.5
R in e^+e^- annihilation (20–65 GeV)	0.124 (21)	35.
R in Z^0 decay	0.124 (7)	91.2

The values of $\alpha_s(m_Z)$ displayed in this table are obtained by fitting experimental results to the theoretical expressions given by perturbative QCD using minimal subtraction. The values of α_s have been evolved to $Q = m_Z$ using the renormalization group equation. R refers to the ratio of cross sections or partial widths to hadrons versus leptons. The numbers in parentheses are the standard errors in the last displayed digits. The column labelled 'Q' gives an idea of the value of Q at which the measurement was made. (Typically, these measurements average over a range of Q, and that averaging is taken into account in the quoted values of α_s.) This table is based on the results compiled by I. Hinchliffe in his article for the Particle Data Group, *Phys. Rev.* **D50**, 1297 (1994). This article contains a full set of references and a discussion of the sources of uncertainty in these determinations.

The table shows the values of α_s extracted from each of these measurements, expressed in terms of the value in the reference conventions, $\alpha_{s\overline{\mathrm{MS}}}(m_Z^2)$. We see that several of the experiments determine α_s to an accuracy of 5%, and that the various determinations are consistent with one another at this level. In Fig. 17.23, we have plotted the original values of α_s represented in Table 17.1, before conversion to a common scale, versus the momentum scale Q at which each was obtained. This comparison gives a striking direct verification of the running of α_s.

At the beginning of this chapter, we wrote down a candidate for the fundamental theory of strong interactions using only a few simple principles: the existence of quarks and the identification of their quantum numbers, and the idea that the theory of the quark interactions should be an asymptotically free gauge theory. It is remarkable that these simple considerations have led us to a description of strong interactions that is quantitatively correct for a broad range of phenomena in the hard-scattering regime where asymptotic freedom can be used as a tool for calculation.

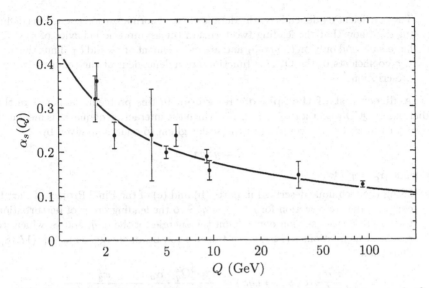

Figure 17.23. Measurements of α_s, plotted against the momentum scale Q at which the measurement was made. This figure was constructed by evolving the values of $\alpha_s(m_Z)$ listed in Table 17.1 back to the values of Q indicated in the table. The value for e^+e^- event shapes has been split into two points corresponding to experiments at the TRISTAN and LEP accelerators. These values are compared to the theoretical expectation from the renormalization group evolution with the initial condition $\alpha_s(m_Z) = 0.117$.

Problems

17.1 Two-loop renormalization group relations.

(a) In higher orders of perturbation theory, the expression for the QCD β function will be a series

$$\beta(g) = -\frac{b_0}{(4\pi)^2}g^3 - \frac{b_1}{(4\pi)^4}g^5 - \frac{b_2}{(4\pi)^6}g^7 + \cdots.$$

Integrate the renormalization group equation and show that the running coupling constant is now given by

$$\alpha_s(Q^2) = \frac{4\pi}{b_0}\left[\frac{1}{\log(Q^2/\Lambda^2)} - \frac{b_1}{b_0^2}\frac{\log\log(Q^2/\Lambda^2)}{(\log(Q^2/\Lambda^2))^2} + \cdots\right],$$

where the omitted terms decrease as $(\log(Q^2/\Lambda^2))^{-2}$.

(b) Combine this formula with the perturbation series for the e^+e^- annihilation cross section:

$$\sigma(e^+e^- \to \text{hadrons}) = \sigma_0 \cdot \left(3\sum_f Q_f^2\right) \cdot \left[1 + \frac{\alpha_s}{\pi} + a_2\left(\frac{\alpha_s}{\pi}\right)^2 + \mathcal{O}(\alpha_s^3)\right].$$

The coefficient a_2 depends on the details of the renormalization conditions defining α_s. Show that the leading two terms in the asymptotic behavior of $\sigma(s)$ for large s depend only on b_0 and b_1 and are independent of a_2 and b_2. Thus the first *two* coefficients of the QCD β function are independent of the renormalization prescription.

17.2 A direct test of the spin of the gluon. In this problem, we compare the predictions of QCD with those of a model in which the interaction of quarks is mediated by a scalar boson. Let the coupling of the scalar gluon to quarks be given by

$$\delta\mathcal{L} = gS\bar{q}q,$$

and define $\alpha_g = g^2/4\pi$.

(a) Using the technique described in parts (b) and (c) of the Final Project of Part I, compute the cross section for $e^+e^- \to q\bar{q}S$ to the leading order of perturbation theory. This cross section depends on the energies of the q, $\bar{q}$, and S, which we represent as fractions x_1, x_2, x_3 of the electron beam energy, as in Eq. (17.18). Show that

$$\frac{d^2\sigma}{dx_1 dx_2}(e^+e^- \to q\bar{q}S) = \frac{4\pi\alpha^2 Q_q^2}{3s} \cdot \frac{\alpha_g}{4\pi} \frac{x_3^2}{(1-x_q)(1-x_{\bar{q}})}.$$

(b) In practice, it is very difficult to tell quarks from gluons experimentally, since both particles appear as jets of hadrons. Therefore, let x_a be the largest of x_1, x_2, x_3, let x_b be the second largest, and let x_c be the smallest. Sum over the various possibilities to derive an expression for $d^2\sigma/dx_a dx_b$, both in QCD, using Eq. (17.18), and in the scalar gluon model. Show that these models can be distinguished by their distributions in the x_a, x_b plane.

17.3 Quark-gluon and gluon-gluon scattering.

(a) Compute the differential cross section

$$\frac{d\sigma}{d\hat{t}}(q\bar{q} \to gg)$$

for quark-antiquark annihilation in QCD to the leading order in α_s. This is most easily done by computing the amplitudes between states of definite quark and gluon helicity. Ignore all masses. Use explicit polarization vectors and spinors, for example,

$$\epsilon^\mu = \frac{1}{\sqrt{2}}(0, 1, i, 0)$$

for a right-handed gluon moving in the $+\hat{3}$ direction. You need only consider transversely polarized gluons. By helicity conservation, only the initial states $q_L\bar{q}_R$ and $q_R\bar{q}_L$ can contribute; by parity, these two states give identical cross sections. Thus it is necessary only to compute the amplitudes for the three processes

$$q_L\bar{q}_R \to g_R g_R,$$
$$q_L\bar{q}_R \to g_R g_L,$$
$$q_L\bar{q}_R \to g_L g_L.$$

In fact, by CP invariance, the first and third processes have equal cross sections. After computing the amplitudes, square them and combine them properly with color factors to construct the various helicity cross sections. Finally, combine these to form the total cross section averaged over initial spins and colors.

(b) Compute the differential cross section

$$\frac{d\sigma}{d\hat{t}}(gg \to gg)$$

for gluon-gluon scattering. There are 16 possible combinations of helicities, but many of them are related to each other by parity and crossing symmetry. All 16 can be built up from the three amplitudes for

$$g_R g_R \to g_R g_R,$$

$$g_R g_R \to g_R g_L,$$

$$g_R g_R \to g_L g_L.$$

Show that the last two of these amplitudes vanish. The first can be dramatically simplified using the Jacobi identity. When the smoke clears, only three of the 16 polarized gluon scattering cross sections are nonzero. Combine these to compute the spin- and color-averaged differential cross section.

17.4 The gluon splitting function. Compute the gluon splitting function (17.130) for the Altarelli-Parisi equations. To carry out this computation, first compute the matrix elements of the three-gluon vertex shown in Fig. 17.20 between gluon states of definite helicity. Combine these to derive the splitting function in the region $x < 1$. Then fix the singularity of the splitting function at $x = 1$ to give this function the correct overall normalization.

17.5 Photoproduction of heavy quarks. Consider the process of heavy quark pair photoproduction, $\gamma + p \to Q\bar{Q} + X$, for a heavy quark of mass M and electric charge Q. If M is large enough, any diagram contributing to this process must involve a large momentum transfer; thus a perturbative QCD analysis should apply. This idea applies in practice already for the production of c quark pairs. Work out the cross section to the leading order in QCD. Choose the parton subprocess that gives the leading contribution to this reaction, and write the parton-model expression for the cross section. You will need to compute the relevant subprocess cross section, but this can be taken directly from one of the QED calculations in Chapter 5. Then use this result to write an expression for the cross section for γ-proton scattering.

17.6 Behavior of parton distribution functions at small x. It is possible to solve the Altarelli-Parisi equations analytically for very small x, using some physically motivated approximations. This discussion is based on a paper of Ralston.[†]

(a) Show that the Q^2 dependence of the right-hand side of the A-P equations can be expressed by rewriting the equations as differential equations in

$$\xi = \log \log \left(\frac{Q^2}{\Lambda^2} \right),$$

where Λ is the value of Q^2 at which $\alpha_s(Q^2)$, evolved with the leading-order β function, formally goes to infinity.

[†]J. P. Ralston, *Phys. Lett.* **172B**, 430 (1986).

(b) Since the branching functions to gluons are singular as z^{-1} as $z \to 0$, it is reasonable to guess that the gluon distribution function will blow up approximately as x^{-1} as $x \to 0$. The resulting distribution

$$dx \, f_g(x) \sim \frac{dx}{x}$$

is approximately scale invariant, and so its form should be roughly preserved by the A-P equations. Let us, then, make the following two approximations: (1) the terms involving the gluon distribution completely dominate the right-hand sides of the A-P equations; and (2) the function

$$\tilde{g}(x, Q^2) = x f_g(x, Q^2)$$

is a slowly varying function of x. Using these approximations, and the limit $x \to 0$, show that the A-P equation for $f_g(x)$ can be converted to the following differential equation:

$$\frac{\partial^2}{\partial w \partial \xi} \tilde{g}(x, \xi) = \frac{12}{b_0} \tilde{g}(x, \xi),$$

where $w = \log(1/x)$ and $b = (11 - \frac{2}{3} n_f)$. Show that if $w\xi \gg 1$, this equation has the approximate solution

$$\tilde{g} = K(Q^2) \cdot \exp\left(\left[\frac{48}{b_0} w(\xi - \xi_0)\right]^{1/2}\right),$$

where $K(Q^2)$ is an initial condition.

(c) The quark distribution at very small x is mainly created by branching of gluons. Using the approximations of part (b), show that, for any flavor of quark, the right-hand side of the A-P equation for $f_q(x)$ can be approximately integrated to yield an equation for $\tilde{q}(x) = x f_q(x)$:

$$\frac{\partial}{\partial \xi} \tilde{q}(x, \xi) = \frac{2}{3b_0} \tilde{g}(x, \xi).$$

Show, again using $w\xi \gg 1$, that this equation has as its integral

$$\tilde{q} = \left(\frac{\xi - \xi_0}{27 b_0 w}\right)^{1/2} K(Q^2) \cdot \exp\left(\left[\frac{48}{b_0} w(\xi - \xi_0)\right]^{1/2}\right).$$

(d) Ralston suggested that the initial condition

$$K(Q^2) = 50.36(\exp(\xi - \xi_0) - 0.957) \cdot \exp\left[-7.597(\xi - \xi_0)^{1/2}\right],$$

with $Q_0^2 = 5$ GeV^2, $\Lambda = 0.2$ GeV, and $n_f = 5$, gave a reasonable fit to the known properties of parton distributions, extrapolated into the small x region. Use this function and the results above to sketch the behavior of the quark and gluon distributions at small x and large Q^2.

Chapter 18

Operator Products and Effective Vertices

Our analysis of QCD in Chapter 17 was founded on the principle of asymptotic freedom, which told us that strong interaction processes with large momentum transfer might reliably be treated in weak-coupling perturbation theory. So far, however, we have made little use in QCD of the more powerful tools of the renormalization group. In this chapter, we will work out some implications of the Callan-Symanzik equation in QCD. We will see that asymptotically free theories have their own characteristic scaling behavior, with corrections in the form of anomalous powers of logarithms of the momentum scale. Though these corrections are generally weaker than those in the scalar field theories studied in Chapter 13, they nevertheless have important qualitative effects on the strong interactions.

We begin by considering the scaling law for mass terms in QCD, taking over directly the formalism that we used to describe the mass term of ϕ^4 theory in Sections 12.4 and 12.5. Other applications, however, require a more powerful theoretical tool, the *operator product expansion*. Section 18.3 introduces a general description of products of operators in quantum field theory and explains how such operator products are constrained by the Callan-Symanzik equation. The last two sections use this tool to develop a new viewpoint toward deep inelastic scattering and other hard processes in QCD.

18.1 Renormalization of the Quark Mass Parameter

Up to this point, we have always assumed that quark masses are small enough that they can be ignored in high-energy processes. This is not always an adequate assumption even for the light quarks u, d, s; for the heavier quarks c, b, t, the masses can have very important effects. However, since isolated quarks do not exist, it is not possible to define the mass of a quark unambiguously. In the discussion to follow, we will consider the quark mass to be a parameter of QCD perturbation theory, defined by a renormalization prescription at some renormalization scale M.

Because we define the quark mass as we would a coupling constant, by a renormalization convention, we should expect that this parameter will run according to a renormalization group evolution, so that different values of the mass parameter apply to different processes. We say that our original prescription leads to an *effective* quark mass, which depends on the momentum scale at which it is evaluated. In this section, we will work out the leading dependence of this effective mass on the momentum scale.

The basic formalism for effective mass terms was set out in Section 12.5. To add a mass term to the QCD Lagrangian, we must first define the mass operator $(\bar{q}q)$ by a renormalization prescription at a scale M. Then we can define the quark mass by adding to the Lagrangian the term

$$\Delta \mathcal{L}_m = -m(\bar{q}q)_M. \tag{18.1}$$

In this discussion, we will assume that the quark mass m is small enough that we need only keep terms of leading order in m. We will also assume, for simplicity, that we have such a mass term for only one quark flavor.

In the zero-mass limit, Green's functions of the operator $(\bar{q}q)$ with quark fields,

$$
\begin{aligned}
G^{(n,k)}&(x_1,\ldots,x_n,y_1,\ldots,y_n,z_1,\ldots,z_k) \\
&= \langle q(x_1)\cdots q(x_n)\bar{q}(y_1)\cdots \bar{q}(y_n)\bar{q}q(z_1)\cdots \bar{q}q(z_k)\rangle,
\end{aligned} \tag{18.2}
$$

obey the Callan-Symanzik equation

$$\left[M\frac{\partial}{\partial M} + \beta\frac{\partial}{\partial g} + 2n\gamma + k\gamma_{\bar{q}q}\right]G^{(n,k)}(\{x_i\},\{y_i\},\{z_j\},g,M) = 0, \tag{18.3}$$

where γ is the anomalous dimension of the quark field and $\gamma_{\bar{q}q}$ is the anomalous dimension of the operator $\bar{q}q$. If we include the mass terms in the Lagrangian according to (18.1), the Green's function of n quark fields and n antiquark fields satisfies

$$\left[M\frac{\partial}{\partial M} + \beta\frac{\partial}{\partial g} + 2n\gamma + \gamma_{\bar{q}q}m\frac{\partial}{\partial m}\right]G^{(n)}(\{x_i\},\{y_i\},g,m,M) = 0. \tag{18.4}$$

The derivative with respect to m counts the number of times the mass operator is used. In Section 12.5, we traded the variable m, with the dimensions of mass, for a dimensionless variable. However, in QCD, it is just as convenient to consider the dimensionful parameter m as a coupling constant. The solution of the Callan-Symanzik equation will then contain a running mass parameter $\overline{m}(Q)$, which depends on a typical momentum Q of the Green's function. This parameter is defined as the solution to a renormalization group equation analogous to Eq. (12.126). For this case, the equation is

$$\frac{d}{d\log(Q/M)}\overline{m} = \gamma_{\bar{q}q}(\bar{g})\cdot\overline{m}, \tag{18.5}$$

with the initial condition

$$\overline{m}(M) = m. \tag{18.6}$$

The quantity $\overline{m}(Q)$ is the *effective mass*, which should be used to compute the mass effects on quark production or scattering processes with the momentum transfer Q.

To compute $\overline{m}(Q)$ explicitly, we need to work out the anomalous dimension of the mass operator $\gamma_{\bar{q}q}$. This can be done as explained in Section 12.4. We define the normalization of the operator explicitly by the prescription that the vertex function of $(\bar{q}q)$ between renormalized quark fields should satisfy

$$= 1 \qquad (18.7)$$

for $p^2 = q^2 = (p+q)^2 = -M^2$. To preserve (18.7), we will need a counter-term vertex $\delta_{\bar{q}q}$ with the structure of the operator insertion. Then, as in Eq. (12.112), the anomalous dimension is given to one-loop order by

$$\gamma_{\bar{q}q} = M \frac{\partial}{\partial M} (-\delta_{\bar{q}q} + \delta_2), \qquad (18.8)$$

where δ_2 is the counterterm for the quark field strength renormalization, defined in Fig. 16.8. Correlation functions of the gauge-invariant operator $(\bar{q}q)$ are gauge invariant, and so the various terms in the Callan-Symanzik equation for this function must sum to a gauge-invariant result. Since the leading coefficient of $\beta(g)$ is independent of the gauge and of other conventions, it follows from (18.3) that the leading coefficient of $\gamma_{\bar{q}q}$ is also convention independent. The counterterms δ_2 and $\delta_{\bar{q}q}$ both depend on the gauge. This argument shows that the gauge dependence must cancel in (18.8). In the calculation to follow, and in the other anomalous dimension calculations in this chapter, we will work consistently in Feynman-'t Hooft gauge.

We have already computed the divergent part of the counterterm δ_2 in Feynman-'t Hooft gauge in Section 16.4. Evaluating the group-theory factor in the result (16.77) for QCD, we find

$$\delta_2 = -\frac{4}{3} \frac{g^2}{(4\pi)^2} \frac{\Gamma(2-\frac{d}{2})}{(M^2)^{2-d/2}}. \qquad (18.9)$$

To compute $\delta_{\bar{q}q}$, we must work out the one-loop correction to the vertex (18.7). This is given by the diagram

$$= \int \frac{d^4k}{(2\pi)^4} (ig)^2 t^a \gamma^\mu \frac{i(\not{k}+\not{q})}{(k+q)^2} \cdot 1 \cdot \frac{i\not{k}}{k^2} t^a \gamma_\mu \frac{-i}{(k-p)^2}. \qquad (18.10)$$

In the expression for this diagram, the factor 1 represents the $\bar{q}q$ operator insertion. In the corresponding diagram for the renormalization of the quark number current $j^\nu = \bar{q}\gamma^\nu q$, this factor would be replaced by γ^ν. Since we need only the divergent part of the vertex renormalization (18.10), we can approximate the integrand by its value for large k. Then this diagram becomes

$$\sim \int \frac{d^4 k}{(2\pi)^4} (ig)^2 t^a \gamma^\mu \frac{i\slashed{k}}{k^2} \cdot 1 \cdot \frac{i\slashed{k}}{k^2} t^a \gamma_\mu \frac{-i}{k^2}$$

$$\sim -i\frac{4}{3}g^2 \int \frac{d^4 k}{(2\pi)^4} \frac{d \cdot k^2}{(k^2)^3} \qquad (18.11)$$

$$\sim \frac{4}{3}g^2 \cdot 4 \cdot \frac{1}{(4\pi)^2}\Gamma(2-\tfrac{d}{2}).$$

To preserve the normalization condition (18.7), we must add the counterterm

$$\delta_{\bar{q}q} = -4 \cdot \frac{4}{3}\frac{g^2}{(4\pi)^2}\frac{\Gamma(2-\frac{d}{2})}{(M^2)^{2-d/2}}. \qquad (18.12)$$

Assembling (18.8), (18.9), and (18.12), we find

$$\gamma_{\bar{q}q} = -8\frac{g^2}{(4\pi)^2}. \qquad (18.13)$$

As we have noted in the previous paragraph, the anomalous dimension γ_j of the quark number current can be found by a very similar calculation. This will give a good check on our formalism, since, as we have argued above Eq. (12.110), a conserved current is unambiguously normalized by its integral, the conserved charge, and so must have zero anomalous dimension. If we substitute γ^ν for 1 in (18.10) and use the same set of approximations to reduce the integral, we find in the numerator the Dirac matrix structure

$$\gamma^\mu \slashed{k}\gamma^\nu \slashed{k}\gamma_\mu = \frac{1}{d}k^2 \gamma^\mu \gamma^\lambda \gamma^\nu \gamma_\lambda \gamma_\mu$$

$$= \frac{1}{4}(-2)^2 k^2 \gamma^\nu. \qquad (18.14)$$

Then, instead of (18.12), we need the counterterm

$$\delta_j = -\frac{4}{3}\frac{g^2}{(4\pi)^2}\frac{\Gamma(2-\frac{d}{2})}{(M^2)^{2-d/2}}. \qquad (18.15)$$

Combining this result with (18.9), we find

$$\gamma_j = 0, \qquad (18.16)$$

in accord with our general arguments.

If we replace the gamma function in (18.11) by an explicit factor of $\log(\Lambda^2/Q^2)$, and then subtract the divergence using the counterterm (18.12), we find that the vertex diagram behaves as

$$= \frac{4}{3} \cdot 4\frac{g^2}{(4\pi)^2}\log\frac{M^2}{Q^2}. \qquad (18.17)$$

This diagram gives an enhancement at small external momenta. Some of this enhancement is associated with the (gauge-dependent) rescaling of the external quark fields; relation (18.8) tells us how to extract the piece of this

Figure 18.1. Diagrams giving the leading logarithmic contributions to the momentum dependence of the quark effective mass.

logarithm associated with the gauge-invariant enhancement of the effective mass. Thus, to order α_s,

$$\overline{m}(Q) = m \cdot \left(1 + 4\frac{g^2}{(4\pi)^2} \log \frac{M^2}{Q^2}\right). \tag{18.18}$$

To compute the momentum dependence of the effective mass more accurately, we must take two more features of the calculation into account. First, the quantity $(\alpha_s \log(M^2/Q^2))$ may become of order 1, and, in this case, we must take into account all leading logarithmic terms of the form $(\alpha_s \log(M^2/Q^2))^n$. Contributions of this type come from all the diagrams shown in Fig. 18.1. Second, the coupling constant α_s is itself a function of the momentum scale, giving a further enhancement to contributions from small Q. Both of these effects are properly accounted by solving the renormalization group equation (18.5). To the leading order in g^2, this equation takes the explicit form

$$\frac{d}{d\log(Q/M)}\overline{m} = -8\frac{g^2}{(4\pi)^2}m = -2\frac{\alpha_s(Q^2)}{\pi}\overline{m}. \tag{18.19}$$

Inserting the solution of the renormalization group equation for $\bar{g}$ in the form (17.17), we find

$$\frac{d}{d\log(Q/M)}\overline{m} = -\frac{8}{b_0 \log(Q^2/\Lambda^2)}\overline{m}, \tag{18.20}$$

where b_0 is the first coefficient of the QCD β function and Λ is now the QCD scale parameter defined in (17.16). The integral of this equation, satisfying the initial condition (18.6), is

$$\overline{m}(Q^2) = \left(\frac{\log(M^2/\Lambda^2)}{\log(Q^2/\Lambda^2)}\right)^{4/b_0} m. \tag{18.21}$$

Recall that $b_0 = 11 - \frac{2}{3}n_f$ in QCD. Another way to express (18.21) is by writing

$$\overline{m}(Q^2) = \left(\frac{\alpha_s(Q^2)}{\alpha_s(M^2)}\right)^{4/b_0} m. \tag{18.22}$$

Just as an illustration, take $n_f = 4$ and $\Lambda = 150$ MeV; then the effective masses of the light quarks increase by about a factor 2 from $Q = 100$ GeV to $Q = 1$ GeV.

The method we have just used for computing the QCD enhancement of the quark mass operator applies equally well to the matrix elements of any other gauge-invariant operator. We conclude this section by recapitulating the conclusions of the argument in their more general form.

Let $\mathcal{O}(x)$ be any gauge-invariant operator in QCD. As we saw for the mass term, the one-loop corrections to the matrix elements of this operator may contain enhancement or suppression terms proportional to $\alpha_s \log(M^2/Q^2)$, where Q is the momentum scale of a QCD process mediated by $\mathcal{O}(x)$ and M is the renormalization scale used to define the operator normalization. The part of these one-loop corrections specifically associated with the operator normalization is given by the anomalous dimension $\gamma_\mathcal{O}$. For an operator containing n quark or antiquark fields and k gluon fields,

$$\gamma_\mathcal{O} = M\frac{\partial}{\partial M}\left(-\delta_\mathcal{O} + \frac{n}{2}\delta_2 + \frac{k}{2}\delta_3\right), \tag{18.23}$$

where $\delta_\mathcal{O}$ is the counterterm needed to preserve the operator normalization condition and δ_2 and δ_3 are the counterterms for the quark and gluon field strength renormalization defined in Fig. 16.8. From (18.23), we can derive the explicit one-loop expression for $\gamma_\mathcal{O}$ in the form

$$\gamma_\mathcal{O} = -a_\mathcal{O}\frac{g^2}{(4\pi)^2}. \tag{18.24}$$

Using this result, we can solve the renormalization group equation for the coefficient of $\mathcal{O}(x)$ and find the QCD renormalization factor

$$\left(\frac{\log(M^2/\Lambda^2)}{\log(Q^2/\Lambda^2)}\right)^{a_\mathcal{O}/2b_0}, \tag{18.25}$$

where b_0 is the first coefficient of the QCD β function,

$$b_0 = 11 - \frac{2}{3}n_f. \tag{18.26}$$

The QCD renormalization (18.25) is an enhancement at small momenta if $a_\mathcal{O} > 0$.

In the remainder of this chapter, we will present further examples of this enhancement or suppression by QCD logarithms. In many cases, we will see that these factors lead to striking and nontrivial physical effects.

18.2 QCD Renormalization of the Weak Interaction

Our next example of the appearance of QCD enhancement factors occurs in the theory of the weak interactions of hadrons. In Section 17.3, we introduced the weak interaction coupling of quarks and leptons, which we described by an effective Lagrangian. For our analysis here, we will need to know a few more details of the structure of the weak interactions, so we begin this section by presenting these facts. The complete structure of the weak interactions of quarks and leptons will be discussed systematically in Chapter 20.

As we discussed in Section 17.3, the weak interactions among quarks and leptons are described by an effective Lagrangian resulting from the exchange of a virtual W vector boson. In (17.31), we wrote the effective vertex that couples quarks to leptons:

$$\Delta\mathcal{L} = \frac{g^2}{2m_W^2}\bar{\ell}\gamma^\mu\frac{(1-\gamma^5)}{2}\nu\,\bar{u}\gamma_\mu\frac{(1-\gamma^5)}{2}d + \text{h.c.} \qquad (18.27)$$

In this chapter, we will mainly be concerned with the effects of this interaction for momentum scales much larger than 1 GeV. Thus, we will ignore quark masses. All fermion fields that appear in the weak-interaction vertices are multiplied by the left-handed projector $\frac{1}{2}(1-\gamma^5)$. In the rest of this section, we will not write this projector explicitly; rather, we will denote the projection by a subscript L. We will also introduce the Fermi constant, given by (17.32). Then (18.27) can be rewritten as

$$\Delta\mathcal{L} = \frac{4G_F}{\sqrt{2}}\left(\bar{\ell}_L\gamma^\mu\nu_L\right)\left(\bar{u}_L\gamma_\mu d_L\right) + \text{h.c.} \qquad (18.28)$$

There is an analogous vertex that represents W exchange between pairs of quarks; this has the form

$$\Delta\mathcal{L} = \frac{4G_F}{\sqrt{2}}\left(\bar{d}_L\gamma^\mu u_L\right)\left(\bar{u}_L\gamma_\mu d_L\right) + \text{h.c.} \qquad (18.29)$$

However, for the discussion of this chapter, we will need to write a modified, and less approximate, expression. When we discuss the theory of weak interactions in detail in Chapter 20, we will learn that the charge $+2/3$ quarks (u,c,t) couple to the charge $-1/3$ quarks (d,s,b) through the weak interactions via a unitary rotation. Thus, for example, u couples to the combination

$$\cos\theta_c\,d + \sin\theta_c\,s, \qquad (18.30)$$

plus a small admixture of b, which we will ignore in this section. The mixing angle θ_c is called the *Cabibbo angle*. Because of this rotation, the weak interaction effective Lagrangian coupling quarks to quarks actually contains a number of terms, of which a particularly important one is

$$\Delta\mathcal{L} = \frac{4G_F}{\sqrt{2}}\cos\theta_c\sin\theta_c(\bar{d}_L\gamma^\mu u_L)(\bar{u}_L\gamma_\mu s_L). \qquad (18.31)$$

This term allows the s quarks to decay through the process $s \to u\bar{u}d$. Similarly, the rotation of (18.28) produces the effective interaction

$$\Delta\mathcal{L} = \frac{4G_F}{\sqrt{2}} \sin\theta_c (\bar{\ell}_L \gamma^\mu \nu_L)(\bar{u}_L \gamma_\mu s_L), \qquad (18.32)$$

which leads to the decay $s \to u\bar{\nu}\ell$. These weak interaction processes are referred to as *nonleptonic* and *semileptonic* decay processes, respectively. Similar expressions apply to the other heavy quarks.

Given that (18.31) and (18.32) describe the weak interaction coupling of the s quark at a fundamental level, we now discuss the modification of these couplings by QCD logarithms. We have seen in the previous section that QCD corrections have a profound effect in enhancing the strength of the quark mass term of the underlying Lagrangian. We will now investigate whether the strength of the weak interactions can receive a similar enhancement.

We first consider the semileptonic weak interaction operator (18.32). The leptonic fermion bilinear is not affected by QCD, so the QCD enhancement of this operator is just the same as that of its quark component

$$\bar{u}_L \gamma_\mu s_L. \qquad (18.33)$$

However, this operator is a current and so has $\gamma = 0$. In terms of diagrams, the logarithmic enhancement resulting from the diagram shown in Fig. 18.2 is canceled by the quark field-strength renormalization, as we saw already in our discussion of the current vertex is Section 18.1. The left-handed projector $\frac{1}{2}(1 - \gamma^5)$ commutes through the diagram and has no effect on the final result. The same remark applies to the semileptonic weak interaction that links u and d quarks. It implies, for that case, that the normalization of the cross sections for deep inelastic neutrino scattering given in (17.35) is not affected by QCD logarithms.

In the case of nonleptonic weak interactions, however, the effect of QCD is not so simple. Let us first compute the Feynman diagrams that give the leading corrections to the renormalization of the weak interaction vertex (18.31) and then, at a later stage, build up the renormalization group interpretation of these results.

At order α_s, the nonleptonic weak interaction vertex receives corrections from the diagrams shown in Fig. 18.3. Notice that the first diagram is precisely the current renormalization found in the semileptonic case. The second diagram gives the analogous renormalization of the second quark current. In the computation of γ, these two contributions cancel the contributions from the field-strength renormalization of the four quark fields. The remaining four diagrams of Fig. 18.3 are new contributions which contribute potentially large rescaling factors.

We now compute these diagrams, beginning with the third diagram in Fig. 18.3. As in the computation of Section 18.1, we are interested in the logarithmically divergent contribution associated with values of the loop momentum k much larger than the external momenta. The simplest way to extract this

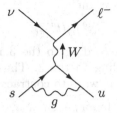

Figure 18.2. QCD correction to the strength of the semileptonic weak interaction vertex.

$$\text{\Large ⅀} \,\,+\,\, \text{\Large ⅀} \,\,+\,\, \text{\Large ⅀}g \,\,+\,\, g\text{\Large ⅀} \,\,+\,\, g\text{\Large ⅀} \,\,+\,\, \text{\Large ⅀}g$$

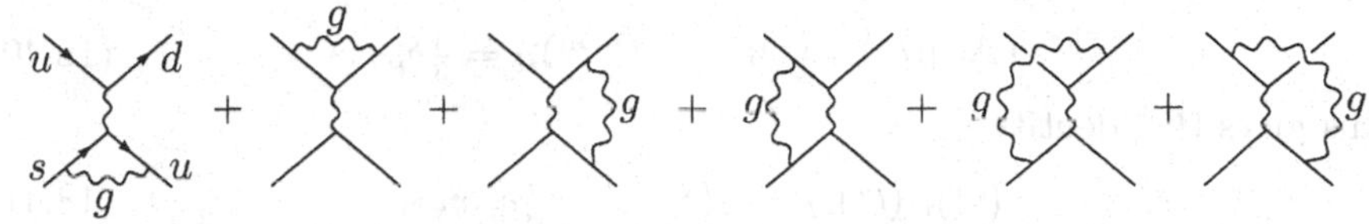

Figure 18.3. QCD corrections to the strength of the nonleptonic weak interaction vertex.

contribution is to compute each diagram in the approximation of zero external momentum. In writing the expression for these diagrams, we will omit the prefactor

$$\frac{4G_F}{\sqrt{2}} \cos\theta_c \sin\theta_c. \tag{18.34}$$

We will retain the quark fields to represent the external states, so that our final expressions will have the form of rescaled operators.

Using this notation, the third diagram in Fig. 18.3 has the value

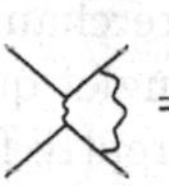

$$\text{⅀} = \int \frac{d^4 k}{(2\pi)^4} (ig)^2 \frac{-i}{k^2} \left(\bar{d}_L \gamma^\nu t^a \frac{i\slashed{k}}{k^2} \gamma^\mu u_L \right) \left(\bar{u}_L \gamma_\nu t^a \frac{-i\slashed{k}}{k^2} \gamma_\mu s_L \right). \tag{18.35}$$

Using the symmetry of the k integral, we extract the divergent piece:

$$\text{⅀} = ig^2 \int \frac{d^4 k}{(2\pi)^4} \frac{k^2/d}{(k^2)^3} \left(\bar{d}_L \gamma^\nu t^a \gamma^\lambda \gamma^\mu u_L \right) \left(\bar{u}_L \gamma_\nu t^a \gamma_\lambda \gamma_\mu s_L \right)$$

$$= -\frac{g^2}{4} \frac{\Gamma(2-\frac{d}{2})}{(4\pi)^2} \left(\bar{d}_L \gamma^\nu t^a \gamma^\lambda \gamma^\mu u_L \right) \left(\bar{u}_L \gamma_\nu t^a \gamma_\lambda \gamma_\mu s_L \right). \tag{18.36}$$

To put the product of quark fields into a more familiar form, we apply the Fierz transformation discussed at the end of Section 3.4. If the color matrices t^a were not present, the product of fermion fields would be exactly the one appearing in (3.82), and we would find

$$\left(\bar{d}_L \gamma^\nu \gamma^\lambda \gamma^\mu u_L \right) \left(\bar{u}_L \gamma_\nu \gamma_\lambda \gamma_\mu s_L \right) = 16 \bar{d}_L \gamma^\nu u_L \bar{u}_L \gamma_\nu s_L. \tag{18.37}$$

The matrices t^a redirect the color quantum numbers of the quark fields. To clarify this, we need the analogue of identity (3.77) for color. To find this identity, consider the color invariant

$$(t^a)_{ij}(t^a)_{k\ell}. \tag{18.38}$$

The indices i, k transform according to the 3 representation of color; the indices j, ℓ transform according to the $\bar{3}$. Thus, (18.38) must be a linear combination of the two possible ways to contract these indices,

$$A\delta_{i\ell}\delta_{kj} + B\delta_{ij}\delta_{k\ell}. \tag{18.39}$$

The constants A and B can be determined by contracting (18.38) and (18.39) with δ_{ij} and with δ_{jk} and adjusting A and B so that the contractions of (18.39) obey the identities

$$\mathrm{tr}[t^a](t^a)_{k\ell} = 0; \qquad (t^a t^a)_{i\ell} = \tfrac{4}{3}\delta_{i\ell}. \tag{18.40}$$

This gives the identity

$$(t^a)_{ij}(t^a)_{k\ell} = \tfrac{1}{2}\big(\delta_{i\ell}\delta_{kj} - \tfrac{1}{3}\delta_{ij}\delta_{k\ell}\big). \tag{18.41}$$

A similar relation holds for the generators of $SU(N)$ in the fundamental representation, with $(1/3)$ replaced by $(1/N)$ in that case.

Inserting (18.41) into (18.36), we find that the first term of the identity generates a new four-fermion operator,

$$\big(\bar{d}_{Li}\gamma^\nu\gamma^\lambda\gamma^\mu u_{Lj}\big)\big(\bar{u}_{Lj}\gamma_\nu\gamma_\lambda\gamma_\mu s_{Li}\big), \tag{18.42}$$

where i, j are color indices. Applying the Fierz rearrangement in (18.37), and then applying the additional rearrangement (3.79), we can convert this operator to the form

$$16(\bar{d}_{Li}\gamma^\nu u_{Lj})(\bar{u}_{Lj}\gamma_\nu s_{Li}) = 16(\bar{u}_{Lj}\gamma^\nu u_{Lj})(\bar{d}_{Li}\gamma_\nu s_{Li}). \tag{18.43}$$

The minus sign in (3.79) is compensated by a minus sign from interchanging the order of fermion fields. The final result is a product of color-singlet quark currents; however, the fields in these currents are associated differently from the original operator.

The final result of our evaluation of this diagram is

$$\includegraphics[] = -4g^2\frac{\Gamma(2-\frac{d}{2})}{(4\pi)^2}\big[\tfrac{1}{2}\bar{u}_L\gamma^\mu u_L\bar{d}_L\gamma_\mu s_L - \tfrac{1}{6}\bar{d}_L\gamma^\mu u_L\bar{u}_L\gamma_\mu s_L\big]. \tag{18.44}$$

The fourth diagram of Fig. 18.3 gives precisely the same contribution.

The evaluation of the last two diagrams in Fig. 18.3 is quite similar. The fifth diagram gives

$$\includegraphics[] = \int\frac{d^4k}{(2\pi)^4}(ig)^2\frac{-i}{k^2}\Big(\bar{d}_L\gamma^\nu t^a\frac{i\slashed{k}}{k^2}\gamma^\mu u_L\Big)\Big(\bar{u}_L\gamma_\mu t^a\frac{i\slashed{k}}{k^2}\gamma_\nu s_L\Big)$$

$$= -ig^2\int\frac{d^4k}{(2\pi)^4}\frac{k^2/d}{(k^2)^3}\big(\bar{d}_L\gamma^\nu t^a\gamma^\lambda\gamma^\mu u_L\big)\big(\bar{u}_L\gamma_\mu t^a\gamma_\lambda\gamma_\nu s_L\big)$$

$$= +\frac{g^2}{4}\frac{\Gamma(2-\frac{d}{2})}{(4\pi)^2}\big(\bar{d}_L\gamma^\nu t^a\gamma^\lambda\gamma^\mu u_L\big)\big(\bar{u}_L\gamma_\mu t^a\gamma_\lambda\gamma_\nu s_L\big). \tag{18.45}$$

The four-fermion operator can be simplified as follows, by the use of the Fierz identity (3.79):

$$\begin{aligned}
\left(\bar{d}_L \gamma^\nu \gamma^\lambda \gamma^\mu u_L\right)\left(\bar{u}_L \gamma_\mu \gamma_\lambda \gamma_\nu s_L\right) &= +\left(\bar{d}_L \gamma^\nu \gamma^\lambda \gamma^\mu \gamma_\lambda \gamma_\nu s_L\right)\left(\bar{u}_L \gamma_\mu u_L\right) \\
&= +(-2)^2\left(\bar{d}_L \gamma^\mu s_L\right)\left(\bar{u}_L \gamma_\mu u_L\right) \qquad (18.46) \\
&= 4(\bar{d}_L \gamma^\mu u_L)(\bar{u}_L \gamma_\mu s_L).
\end{aligned}$$

Again, we must reduce the product of color matrices using identity (18.41), and, again, the first term of this identity will require an additional Fierz transformation. The final result is

$$\text{(diagram)} = +g^2 \frac{\Gamma(2-\frac{d}{2})}{(4\pi)^2}\left[\tfrac{1}{2}\bar{u}_L \gamma^\mu u_L \bar{d}_L \gamma_\mu s_L - \tfrac{1}{6}\bar{d}_L \gamma^\mu u_L \bar{u}_L \gamma_\mu s_L\right]. \qquad (18.47)$$

The last diagram in Fig. 18.3 gives an identical contribution. The sum of the contributions from these four diagrams is

$$-3g^2 \frac{\Gamma(2-\frac{d}{2})}{(4\pi)^2}\left[\bar{u}_L \gamma^\mu u_L \bar{d}_L \gamma_\mu s_L - \tfrac{1}{3}\bar{d}_L \gamma^\mu u_L \bar{u}_L \gamma_\mu s_L\right]. \qquad (18.48)$$

The extraction of the ultraviolet-divergent pieces of the diagrams of Fig. 18.3 is part of our formal prescription for computing the Callan-Symanzik γ function of the weak interaction vertex. However, it is useful to pause at this point and ask about the physical significance of this divergence. The diagrams of Fig. 18.3 would not be divergent if we computed them in the underlying theory with W bosons. In writing the weak interaction as an effective local vertex, we approximated the W boson propagator by a constant, assuming that the momentum k that it carried was much less than m_W:

$$\frac{1}{k^2 - m_W^2} \rightarrow \frac{-1}{m_W^2}. \qquad (18.49)$$

The approximation we used to compute the QCD corrections to the effective vertex is valid only in the region of integration where $k^2 \ll m_W^2$. Outside this region we must use the full W propagator; this introduces an extra factor of k^2 in the denominator and makes the integral converge. Thus, in a direct calculation of the QCD correction, the ultraviolet-divergent terms in the evaluation of Fig. 18.3 would be replaced by logarithms cut off at m_W. The lower limit of the logarithm is set by the external momenta. In the decay of a K meson—the lightest hadron containing the s quark—these are of order m_K. Thus the correction given in (18.48) should be evaluated by replacing

$$g^2 \frac{\Gamma(2-\frac{d}{2})}{(4\pi)^2} \rightarrow \frac{\alpha_s}{4\pi} \log \frac{m_W^2}{m_K^2}. \qquad (18.50)$$

With this interpretation, we can rewrite (18.48) as the order-α_s correction to the leading-order weak interaction vertex. The effect of this correction is the rescaling and modification of the weak interaction operator:

$$\bar{d}_L\gamma^\mu u_L \bar{u}_L \gamma_\mu s_L \rightarrow$$

$$\left(1 + \frac{\alpha_s}{4\pi}\log\frac{m_W^2}{m_K^2}\right)\bar{d}_L\gamma^\mu u_L \bar{u}_L\gamma_\mu s_L - 3\left(\frac{\alpha_s}{4\pi}\log\frac{m_W^2}{m_K^2}\right)\bar{u}_L\gamma^\mu u_L \bar{d}_L\gamma_\mu s_L.$$

(18.51)

Notice that the QCD corrections not only rescale the normalization of the original operator but also introduce a new operator with a different structure. This calculation makes concrete the idea introduced in Section 12.4 that the diagrams that change the normalization of local operators may also mix together different operators with the same dimension and quantum numbers.

Since the value of the logarithm in (18.50) is about 10, the size of the leading QCD correction is of order 1 and so higher-order corrections are important. To sum the leading logarithmic corrections, we return to the renormalization group analysis. For clarity, define

$$\mathcal{O}^1 = \bar{d}_L\gamma^\mu u_L \bar{u}_L\gamma_\mu s_L; \qquad \mathcal{O}^2 = \bar{u}_L\gamma^\mu u_L \bar{d}_L\gamma_\mu s_L. \qquad (18.52)$$

We will use the subscript 0 to denote bare operators and the subscript M to denote operators obeying renormalization conditions at the scale M. From the diagrams of Fig. 18.3, we have found that the operator whose matrix elements have the quark structure of $\mathcal{O}^1$, properly normalized at the scale M, is given by

$$\mathcal{O}_M^1 = \mathcal{O}_0^1 + \delta^{11}\mathcal{O}_0^1 + \delta^{12}\mathcal{O}_0^2, \qquad (18.53)$$

where the δ^{ij} are counterterms,

$$\delta^{11} = -\frac{g^2}{(4\pi)^2}\frac{\Gamma(2-\frac{d}{2})}{(M^2)^{2-d/2}}; \qquad \delta^{12} = +3\frac{g^2}{(4\pi)^2}\frac{\Gamma(2-\frac{d}{2})}{(M^2)^{2-d/2}}. \qquad (18.54)$$

A reciprocal calculation gives $\mathcal{O}_M^2$ in terms of bare operators:

$$\mathcal{O}_M^2 = \mathcal{O}_0^2 + \delta^{21}\mathcal{O}_0^1 + \delta^{22}\mathcal{O}_0^2, \qquad (18.55)$$

with

$$\delta^{21} = \delta^{12}, \qquad \delta^{22} = \delta^{11}.$$

Then, in the manner than we discussed in Eq. (12.109), the operator rescaling of $\mathcal{O}^1$ and $\mathcal{O}^2$ is described in the Callan-Symanzik equation by a matrix γ^{ij} linking the two operators. Expanding this equation to first order in g^2, we see that this matrix is given to one-loop order by

$$\gamma^{ij} = M\frac{\partial}{\partial M}[-\delta^{ij}]. \qquad (18.56)$$

Thus we find

$$\gamma = \frac{g^2}{(4\pi)^2}\begin{pmatrix} -2 & 6 \\ 6 & -2 \end{pmatrix}, \qquad (18.57)$$

acting on the space of operators $\mathcal{O}^1$, $\mathcal{O}^2$.

The simplest way to deduce the physical effects of the rescaling described by (18.57) is to diagonalize this matrix and thus find a new basis of operators that are rescaled without mixing. For the matrix (18.57), the eigenoperators are easily seen to be

$$\mathcal{O}^{1/2} = \tfrac{1}{2}\left[\bar{d}_L\gamma^\mu u_L \bar{u}_L\gamma_\mu s_L - \bar{u}_L\gamma^\mu u_L \bar{d}_L\gamma_\mu s_L\right],$$
$$\mathcal{O}^{3/2} = \tfrac{1}{2}\left[\bar{d}_L\gamma^\mu u_L \bar{u}_L\gamma_\mu s_L + \bar{u}_L\gamma^\mu u_L \bar{d}_L\gamma_\mu s_L\right]. \tag{18.58}$$

The superscripts on these operators are their isospin quantum numbers. The operator $\mathcal{O}^{1/2}$ is antisymmetric under the interchange of the labels $\bar{d}$ and $\bar{u}$; thus, these two isospin-1/2 fields are combined to total isospin zero, and so the whole operator is isospin-1/2. This operator can mediate decays of the K meson that change the isospin by 1/2 unit, such as $K^0 \to \pi^+\pi^-$, but not processes that change the isospin by 3/2, such as $K^+ \to \pi^+\pi^0$. Experimentally, processes of the former type occur almost a thousand times faster (an observation called the $\Delta I = 1/2$ rule). Thus, it is interesting that the hard QCD corrections already make a distinction between these operators.

From the eigenvalues of (18.57), we obtain the Callan-Symanzik γ functions of the eigenoperators (18.58):

$$\gamma_{1/2} = -8\frac{g^2}{(4\pi)^2}; \qquad \gamma_{3/2} = +4\frac{g^2}{(4\pi)^2}. \tag{18.59}$$

According to Eqs. (18.24) and (18.25), this implies that the operator $\mathcal{O}^{1/2}$ receives an enhancement from hard QCD logarithms, while the operator $\mathcal{O}^{3/2}$ receives a suppression. More explicitly, we can write the operator that appears in the original nonleptonic weak interaction vertex (18.31) as

$$\left[\bar{d}_L\gamma^\mu u_L \bar{u}_L\gamma_\mu s_L\right]\big|_{m_W} = \left[\mathcal{O}^{1/2}\right]\big|_{m_W} + \left[\mathcal{O}^{3/2}\right]\big|_{m_W}. \tag{18.60}$$

As above, the subscript refers to the mass scale at which the operator is normalized. We now account for the QCD logarithms associated with evaluating the matrix element of this operator at a lower momentum scale, m_K, by replacing the operators on the right-hand side of (18.60) with operators renormalized at m_K, with the rescaling factor (18.25). This gives

$$\left[\bar{d}_L\gamma^\mu u_L \bar{u}_L\gamma_\mu s_L\right]\big|_{m_W} = \left(\frac{\log(m_W^2/\Lambda^2)}{\log(m_K^2/\Lambda^2)}\right)^{4/b_0}\left[\mathcal{O}^{1/2}\right]\big|_{m_K}$$
$$+ \left(\frac{\log(m_W^2/\Lambda^2)}{\log(m_K^2/\Lambda^2)}\right)^{-2/b_0}\left[\mathcal{O}^{3/2}\right]\big|_{m_K}, \tag{18.61}$$

where, again, $b_0 = 11 - \tfrac{2}{3}n_f$. This equation shows that, unlike the case of semileptonic weak interactions, the overall normalization of the effective Lagrangian for nonleptonic weak interactions is changed by QCD logarithms. In addition, the quark structure of the effective Lagrangian is altered.

Quantitatively, taking $n_f = 4$ and $\Lambda = 150$ MeV as an illustration, we find

$$\left[\bar{d}_L\gamma^\mu u_L \bar{u}_L \gamma_\mu s_L\right]\Big|_{m_W} = 2.1\left[\mathcal{O}^{1/2}\right]\Big|_{m_K} + 0.7\left[\mathcal{O}^{3/2}\right]\Big|_{m_K}. \tag{18.62}$$

Thus, the QCD logarithmic corrections from m_W to m_K give the $\Delta I = 1/2$ part of the effective vertex an enhancement of about a factor of 3.* The observed $\Delta I = 1/2$ rule in K decays requires a factor of 20 enhancement. However, part of this is expected to arise from the ratio of the matrix elements of the operators $\mathcal{O}_{m_K}^{1/2}$ and $\mathcal{O}_{m_K}^{3/2}$ between physical hadron states, which are determined by the soft, nonperturbative part of QCD dynamics.

18.3 The Operator Product Expansion

One way to describe the development of the previous section is to say that we studied an interaction that was fundamentally a product of currents by replacing this product of operators with a single local operator. We then derived the physical consequences of the original, composite, interaction by working out the QCD rescaling of this operator. The procedure of replacing a product of operators with a single effective vertex is useful in many contexts in quantum field theory. Thus, in this section, we will pause from our study of QCD to write out the general formalism governing this procedure.

Let us abstract the situation described in the previous section as follows: Consider a quantum field theory process that includes two operators $\mathcal{O}_1$, $\mathcal{O}_2$ separated by a small distance x, together with other fields $\phi(y_i)$ located much farther away, or together with external physical states. In the example above, the two operators are the quark currents that appear in the weak interaction vertex, and their separation x is a distance of order m_W^{-1}, the range of the W propagator. The external states, which contain K and π mesons, can be described by operators that create and destroy these particles. The amplitude for K decay by the weak interactions, or any more general process of this class, can then be extracted from the Green's function

$$G_{12}(x; y_1, \cdots, y_m) = \langle \mathcal{O}_1(x)\mathcal{O}_2(0)\phi(y_1)\cdots\phi(y_m)\rangle, \tag{18.63}$$

considered in the limit $x \to 0$, with the y_i fixed away from the origin. Here and in the following discussion, products of operators will be considered to be time-ordered, just as we would find by writing the product of fields under the functional integral.

The product of operators $\mathcal{O}_1(x)\mathcal{O}_2(0)$ can potentially create the most general local disturbance in the vicinity of the point 0. However, any such disturbance can be described as the effect of a local operator placed at 0. This

---

*M. K. Gaillard and B. W. Lee, *Phys. Rev. Lett.* **33**, 108 (1974); G. Altarelli and L. Maiani, *Phys. Lett.* **52B**, 351 (1974).

local operator must have the global symmetry quantum numbers of the product of $\mathcal{O}_1\mathcal{O}_2$, but it is otherwise unrestricted. It is useful to write this operator as a linear combination of operators from a standard basis. The coefficients in this linear combination can depend on the separation x. Typically, products of operators in quantum field theory are singular, so it is likely that some of the coefficients will have singularities as $x \to 0$. Combining these observations, Wilson proposed that the effects of the operator product could be computed by replacing the product of operators in (18.63) with a linear combination of local operators,

$$\mathcal{O}_1(x)\mathcal{O}_2(0) \to \sum_n C_{12}{}^n(x)\mathcal{O}_n(0), \qquad (18.64)$$

where the coefficients $C_{12}{}^n(x)$ are c-number functions. This *operator product expansion* (OPE) will depend only on the operators $\mathcal{O}_1$, $\mathcal{O}_2$, and their separation and will be independent of the identity and location of the other fields appearing in the Green's function.

The expansion (18.64) implies that the Green's function (18.63) can be expanded for small x as follows:

$$G_{12}(x; y_1, \cdots, y_m) = \sum_n C_{12}{}^n(x)G_n(y_1, \cdots y_m), \qquad (18.65)$$

where

$$G_n(y_1, \cdots, y_m) = \langle \mathcal{O}_n(0)\phi(y_1)\cdots\phi(y_m)\rangle, \qquad (18.66)$$

and all of the dependence on x is now carried by the OPE coefficient functions. In the example of the previous section, the final amplitudes depended in a rather involved way on the small separation of the two operators, through the dependence of the coefficients in (18.61) on m_W. From the viewpoint of the operator product expansion, this dependence is carried by the coefficient functions and is determined for all matrix elements when these are computed.

In Sections 18.1 and 18.2, we used the renormalization group to compute the enhancement or suppression factors for operator matrix elements. Thus it is natural to expect that the form of the operator product coefficients is also determined by the renormalization group. We will now work out this relation. To begin, we rewrite the expansion (18.64) more precisely. The operators that appear in this relation must be defined at some renormalization scale M. Then the operator product expansion reads:

$$[\mathcal{O}_1(x)]_M [\mathcal{O}_2(0)]_M = \sum_n C_{12}{}^n(x; M)[\mathcal{O}_n(0)]_M. \qquad (18.67)$$

Note that the coefficient functions can depend on M, since they must absorb the M-dependent operator rescalings. If we use the left-hand side of (18.67) to compute (18.63), this function obeys the Callan-Symanzik equation

$$\left[M\frac{\partial}{\partial M} + \beta\frac{\partial}{\partial g} + m\gamma + \gamma_1 + \gamma_2\right]G_{12}(x; y_1, \cdots, y_m; M) = 0. \qquad (18.68)$$

Similarly, with the operator $\mathcal{O}_n$ normalized at M, the Green's function (18.66) obeys

$$\left[M\frac{\partial}{\partial M} + \beta\frac{\partial}{\partial g} + m\gamma + \gamma_n\right]G_n(y_1,\cdots,y_m;M) = 0. \tag{18.69}$$

By applying (18.68) to the right-hand side of (18.65), we see that these relations are consistent only if the OPE coefficient functions obey the Callan-Symanzik equation

$$\left[M\frac{\partial}{\partial M} + \beta\frac{\partial}{\partial g} + \gamma_1 + \gamma_2 - \gamma_n\right]C_{12}{}^n(x;M) = 0. \tag{18.70}$$

We now solve this equation by our standard methods. First, let us apply dimensional analysis. If the operators $\mathcal{O}_1$, $\mathcal{O}_2$, $\mathcal{O}_n$ have dimensions d_1, d_2, d_n, the coefficient function $C_{12}{}^n(x)$ must have the dimensions of $(\text{mass})^{d_1+d_2-d_n}$. Thus,

$$C_{12}{}^n(x) = \left(\frac{1}{|x|}\right)^{d_1+d_2-d_n}\widetilde{C}(xM), \tag{18.71}$$

where $\widetilde{C}(xM)$ is a dimensionless function. This function is determined from (18.70) according to the method of Section 12.3. Thus,

$$C_{12}{}^n(x) = \left(\frac{1}{|x|}\right)^{d_1+d_2-d_n}c(\bar{g}(1/x))\exp\left[\int_{1/x}^{M} d\log M'(\gamma_n - \gamma_1 - \gamma_2)\right], \tag{18.72}$$

with $c(\bar{g})$ a dimensionless function of the running coupling constant at the separation scale $1/x$.

At a fixed point of the renormalization group, the γ functions would take definite values $\gamma_{j*} = \gamma_j(g_*)$. Then, the solution (18.72) can be evaluated as

$$C_{12}{}^n(x) = \left(\frac{1}{|x|}\right)^{d_1+d_2-d_n}c(g_*)\exp\left[\log(xM)\left(\gamma_{n*} - \gamma_{1*} - \gamma_{2*}\right)\right]. \tag{18.73}$$

Thus, in this case,

$$C_{12}{}^n(x) \sim \left(\frac{1}{|x|}\right)^{d_1^*+d_2^*-d_n^*}, \tag{18.74}$$

where

$$d_j^* = d_j + \gamma_j(g_*) \tag{18.75}$$

is the true scaling dimension of the operator $\mathcal{O}_j$ at the fixed point.

For the case of an asymptotically free theory, the scaling relation is complicated in the way that we worked out in Section 18.1. In the leading order of perturbation theory, the three γ functions take the form (18.24). Then the solution of (18.72) takes the form

$$C_{12}{}^n(x) \sim \left(\frac{1}{|x|}\right)^{d_1+d_2-d_n}\left(\frac{\log(1/|x|^2\Lambda^2)}{\log(M^2/\Lambda^2)}\right)^{(a_n-a_1-a_2)/2b_0} \tag{18.76}$$

In the example of Section 18.2, the original operators were currents with dimension 3 and $\gamma = 0$, at separation m_W^{-1}, and the final local operators had dimension 6. Thus, (18.76) does properly reproduce the dependence of (18.61) on m_W. Notice that the renormalization group dependence is less complicated for a product of currents, which have a fixed normalization independent of scale. This special case occurs often in applications of the operator product expansion.

We have written Eq. (18.70) without taking account of operator mixing. However, as we have already seen, operator mixing is often an essential part of the applications of the OPE. It is straightforward to include this effect by rewriting the analysis that leads to (18.70) using matrix-valued γ functions. For example, with operator mixing, the Callan-Symanzik equation for G_n will be modified to

$$\left[\delta_{np}\left(M\frac{\partial}{\partial M} + \beta\frac{\partial}{\partial g} + m\gamma\right) + \gamma_{np}\right]G_p(y_1,\cdots,y_m;M) = 0. \qquad (18.77)$$

With these changes, (18.70) becomes

$$\left[M\frac{\partial}{\partial M} + \beta\frac{\partial}{\partial g}\right]C_{12}{}^n(x;M) + \gamma_{1k}C_{k2}{}^n(x;M)$$
$$+ \gamma_{2k}C_{1k}{}^n(x;M) - \gamma_{kn}C_{12}{}^k(x;M) = 0. \qquad (18.78)$$

Notice that the first two γ matrices act on the OPE coefficient from the left, while the third acts from the right. In the case of a product of currents, the first two γ matrices vanish and (18.78) simplifies to

$$\left[M\frac{\partial}{\partial M} + \beta\frac{\partial}{\partial g}\right]C_{12}{}^n(x;M) - C_{12}{}^k(x;M)\gamma_{kn} = 0. \qquad (18.79)$$

This equation will play an important role in the analysis of Section 18.5.

18.4 Operator Analysis of e^+e^- Annihilation

It is not difficult to imagine that there is a connection between matrix elements in which currents are placed at short distances from one another and matrix elements in which currents deliver a hard momentum transfer. Thus we might expect that the idea of the operator product expansion will give us a new viewpoint from which to understand the theory of hard-scattering processes in QCD. In this section and the next, we will work out the relation of the operator product expansion to the perturbative QCD analysis of Chapter 17.

We begin by discussing the total cross section for e^+e^- annihilation to hadrons. Below Eq. (17.9), we argued that this total cross section could be computed in QCD perturbation theory, using a value of α_s corresponding to the scale of the total center of mass energy. However, this argument was a purely intuitive one, with many logical jumps. In this section, we will give a more rigorous argument to the same conclusion.

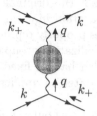

Figure 18.4. Diagrams whose imaginary part yields the total cross section for $e^+e^- \to$ hadrons.

In order to invoke the operator product expansion, we must write the total cross section for e^+e^- annihilation to hadrons as the matrix element of a product of currents. To do this, we use the optical theorem to relate the total e^+e^- scattering cross section to the forward scattering amplitude for $e^+e^- \to e^+e^-$. Ignoring the mass of the electron, we see from Eq. (7.49) that

$$\sigma(e^+e^-) = \frac{1}{2s} \operatorname{Im} \mathcal{M}(e^+e^- \to e^+e^-). \tag{18.80}$$

To compute the cross section for $e^+e^- \to$ hadrons, we consider in the computation of the imaginary part only the contributions from hadronic intermediate states. To leading order in α, but to all orders in the strong interactions, these contributions come from considering only diagrams of the form of Fig. 18.4, and taking the imaginary part of the hadronic contributions to the vacuum polarization.

The value of the diagrams shown in Fig. 18.4 is

$$i\mathcal{M} = (-ie)^2 \bar{u}(k)\gamma_\mu v(k_+) \frac{-i}{s} \left(i\Pi_h^{\mu\nu}(q) \right) \frac{-i}{s} \bar{v}(k_+)\gamma_\nu u(k), \tag{18.81}$$

where $s = q^2$ and $\Pi_h^{\mu\nu}(q)$ is the hadronic part of the vacuum polarization. By the Ward identity, this can be written

$$\Pi_h^{\mu\nu}(q) = (q^2 g^{\mu\nu} - q^\mu q^\nu)\Pi_h(q^2). \tag{18.82}$$

The $q^\mu q^\nu$ terms give zero when contracted with the external electron currents, so only the $g^{\mu\nu}$ term survives. To evaluate the electron spinor part of (18.81), we use the fact that, in this forward scattering amplitude, the initial and final momenta and spins are set equal. Then, averaging over the initial spin gives

$$\frac{1}{2} \cdot \frac{1}{2} \sum_{\text{spins}} \bar{u}(k)\gamma^\mu v(k_+)\bar{v}(k_+)\gamma_\mu u(k) = \frac{1}{4} \operatorname{tr}\left[\not{k}\gamma^\mu \not{k}_+ \gamma_\mu\right]$$

$$= \frac{1}{4} \cdot (-2) \cdot 4(k \cdot k_+) \tag{18.83}$$

$$= -s.$$

Thus, we find

$$\sigma(e^+e^- \to \text{hadrons}) = -\frac{4\pi\alpha}{s} \operatorname{Im} \Pi_h(s). \tag{18.84}$$

To check this result, we can look back to the one-loop value of Π in QED (7.91), or to the imaginary part of this expression given in Eq. (7.92):

$$\operatorname{Im}\Pi(s+i\epsilon) = -\frac{\alpha}{3}\sqrt{1-\frac{4m^2}{s}}\left(1+\frac{2m^2}{s}\right). \tag{18.85}$$

Combining (18.85) with (18.84), we obtain the correct leading-order cross section for production of a new heavy lepton in e^+e^- annihilation,

$$\sigma(e^+e^- \to L^+L^-) = \frac{4\pi\alpha^2}{3s}\sqrt{1-\frac{4m^2}{s}}\left(1+\frac{2m^2}{s}\right). \tag{18.86}$$

If we multiply (18.86) by a factor of 3 for color and sum over quark flavors with the squares of the quark charges, we obtain the leading-order prediction of QCD.

Now that we have relation (18.84), we complete the connection we wished to prove by noting that the hadronic vacuum polarization is simply a matrix element of a product of currents. Let J^μ be the electromagnetic current of quarks,

$$J^\mu = \sum_f Q_f \bar{q}_f \gamma^\mu q_f. \tag{18.87}$$

Then

$$i\Pi_h^{\mu\nu}(q) = -e^2 \int d^4x\, e^{iq\cdot x} \langle 0|\, T\{J^\mu(x)J^\nu(0)\}\,|0\rangle. \tag{18.88}$$

In the limit in which the point x approaches 0, we can reduce the product of currents by applying the operator product expansion. Since we will be taking the vacuum expectation value of the product, we need only list the contribution from operators that are gauge-invariant Lorentz scalars. Thus,

$$J_\mu(x)J_\nu(0) \sim C_{\mu\nu}{}^1(x)\cdot 1 + C_{\mu\nu}{}^{\bar{q}q}(x)\bar{q}q(0) + C_{\mu\nu}{}^{F^2}(x)(F_{\alpha\beta}^a)^2(0) + \cdots. \tag{18.89}$$

Note that we have included the operator 1 on the right-hand side, and the next possible operators in QCD have dimension 3 and 4, respectively. Since the operator $\bar{q}q$ violates chiral symmetry, its coefficient function must have an explicit factor of the quark mass. Thus, by dimensional analysis,

$$C_{\mu\nu}{}^1 \sim x^{-6}, \qquad C_{\mu\nu}{}^{\bar{q}q} \sim mx^{-2}, \qquad C_{\mu\nu}{}^{F^2} \sim x^{-2}, \tag{18.90}$$

and the higher terms in the series are less singular as $x \to 0$.

To compute $\Pi_h^{\mu\nu}(q)$, we need the Fourier transform of the product of currents. Assuming that this Fourier transform is indeed dominated by the limit of short distances, we can compute it by Fourier-transforming the individual OPE coefficients. Since the currents are conserved, the individual terms in the OPE must give zero when dotted with q^μ. Thus the transformed OPE takes the form

$$-e^2 \int d^4x \, e^{iq \cdot x} J^\mu(x) J^\nu(0)$$

$$= -ie^2(q^2 g^{\mu\nu} - q^\mu q^\nu)\left[c^1(q^2) \cdot 1 + c^{\bar{q}q}(q^2) \cdot m\bar{q}q + c^{F^2}(q^2) \cdot (F^a_{\alpha\beta})^2 + \cdots\right],$$
$$(18.91)$$

where the c^i are Lorentz-invariant c-number functions of q^2, and the factor of i at the beginning of the second line is inserted as a convenient convention. By dimensional analysis, we find

$$c^1 \sim (q^2)^0, \qquad c^{\bar{q}q} \sim (q^2)^{-2}, \qquad c^{F^2} \sim (q^2)^{-2}, \qquad (18.92)$$

and the higher terms are more irrelevant for large q.

The OPE coefficients $c^i(q^2)$ can be computed from Feynman diagrams. As shown in Fig. 18.5, the coefficient of the operator 1 is the sum of diagrams with no external legs other than the current insertions. The leading QCD diagram is just the simple vacuum polarization diagram, multiplied by the color factor 3 and the sum of the squares of the quark charges. Combining these factors with Eq. (7.91), we have

$$c^1(q^2) = -\left(3 \sum_f Q_f^2\right) \cdot \frac{\alpha}{3\pi} \log(-q^2). \qquad (18.93)$$

The corrections to this result are of order $\alpha_s(q^2)$. The higher coefficient functions are extracted from diagrams with more external legs. For example, the coefficient function of $(F^a_{\alpha\beta})^2$ is determined by diagrams with two external gluon legs.

Still assuming that the Fourier transform of the product of currents can be computed from the OPE for the region of large timelike q^2, we can complete our evaluation of the cross section for $e^+e^- \to$ hadrons by taking the vacuum expectation value of (18.91), extracting the imaginary parts of the coefficient functions, and substituting the result into (18.84). We find

$$\sigma(e^+e^- \to \text{hadrons}) = \frac{4\pi\alpha^2}{s}\left[\text{Im}\, c^1(q^2) + \text{Im}\, c^{\bar{q}q}(q^2) \langle 0| m\bar{q}q |0\rangle \right.$$
$$(18.94)$$
$$\left. + \text{Im}\, c^{F^2}(q^2) \langle 0| (F^a_{\alpha\beta})^2 |0\rangle + \cdots\right].$$

The first term of this series is just the result of summing perturbative QCD diagrams for the e^+e^- total cross section. The additional terms give corrections to this result which depend on soft hadronic matrix elements, but these corrections are explicitly suppressed at high energy by factors $(q^2)^{-2}$. (Incidentally, this expansion, which applies equally well in the absence of QCD interactions, explains why (18.86) contains no term of order s^{-2} when expanded for large s.) If we insert the leading-order expression (18.93) into (18.94), we obtain the familiar result

$$\sigma(e^+e^- \to \text{hadrons}) = \frac{4\pi\alpha^2}{s} \sum_f Q_f^2. \qquad (18.95)$$

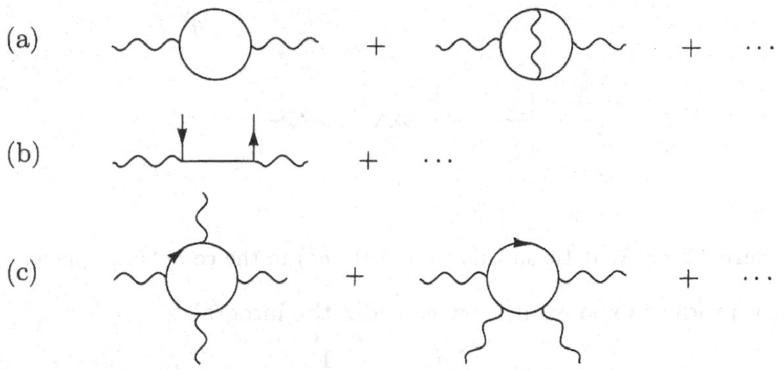

Figure 18.5. Feynman diagrams contributing the operator product coefficient, in the expansion of the product of currents, for the operator (a) 1; (b) $\bar{q}q$; (c) $(F^a_{\alpha\beta})^2$.

Our result (18.94) is pleasing, but the logic that led us to it was not correct. To compute the e^+e^- total cross section, we must compute $\Pi_h(q^2)$ in the region of large *timelike* momentum q, where the expectation value of the product of currents is dominated by intermediate states of high energy, involving large numbers of physical hadrons. Thus we need $\Pi_h(q^2)$ in precisely the region where it is not dominated by short-distance perturbations of the quark and gluon fields. To compute the product of currents from the short-distance expansion, we choose kinematic conditions such that the intermediate states that enter the computation of the product of currents are far off-shell, so that they cannot propagate far from the converging points x and 0. This condition is satisfied at large *spacelike* momentum, or, equivalently, at small spacelike separation. However, it seems at first sight that a computation in this region is useless for determination of the e^+e^- cross section.

Fortunately, there is a wonderful trick for relating the values of a quantum field theory amplitude in two well-separated kinematic regions. This trick, called the method of *dispersion relations*, makes use of the general analytic properties of the amplitude. Since (18.88) is the Fourier transform of a two-point correlation function, we know from the analysis of Section 7.1 that $\Pi_h(q^2)$ possesses a Källén-Lehmann spectral representation. Thus, $\Pi_h(q^2)$ is an analytic function of q^2 with a branch cut on the positive q^2 axis and no other singularities in the complex q^2 plane. This analytic structure is shown in Fig. 18.6. The discontinuity of $\Pi_h(q^2)$ across the branch cut is $(2i)$ times the imaginary part of Π_h and so is directly related to the total e^+e^- annihilation cross section.

With this additional knowledge about $\Pi_h(q^2)$, we can argue as follows. Let $q^2 = -Q_0^2$ be a value sufficiently far into the spacelike region of q that the Fourier transform of the product of currents can be computed from the

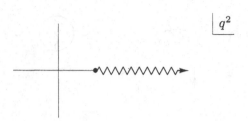

Figure 18.6. Analytic singularities of $\Pi_h(q^2)$ in the complex q^2 plane.

operator product expansion. Now consider the integral

$$I_n = -4\pi\alpha \oint \frac{dq^2}{2\pi i} \frac{1}{(q^2 + Q_0^2)^{n+1}} \Pi_h(q^2), \tag{18.96}$$

for $n \geq 1$, evaluated on a contour encircling $q^2 = -Q_0^2$. If we contract the contour onto the pole, we find

$$I_n = -4\pi\alpha \frac{1}{n!} \frac{d^n}{d(q^2)^n} \Pi_h \bigg|_{q^2=-Q_0^2}, \tag{18.97}$$

which can be computed by evaluating Π_h from the the operator product relation (18.91),

$$\Pi_h(q^2) = -e^2 \left[c^1(q^2) + c^{\bar{q}q}(q^2) \langle 0| m\bar{q}q |0\rangle + c^{F^2}(q^2) \langle 0| (F^a_{\alpha\beta})^2 |0\rangle + \cdots \right]. \tag{18.98}$$

On the other hand, we can evaluate the integral by distorting the contour to the form of Fig. 18.7. Since none of the coefficient functions grow faster than $(q^2)^0$ times logarithms as $q^2 \to \infty$, the contour at infinity can be neglected for $n \geq 1$. The piece of the contour that wraps around the branch cut gives

$$\begin{aligned}
I_n &= -4\pi\alpha \int \frac{dq^2}{2\pi i} \frac{1}{(q^2 + Q_0^2)^{n+1}} \, \text{Disc} \, \Pi_h(q^2) \\
&= -4\pi\alpha \int \frac{dq^2}{2\pi} \frac{1}{(q^2 + Q_0^2)^{n+1}} \frac{1}{i} 2i \, \text{Im} \, \Pi_h(q^2) \\
&= \frac{1}{\pi} \int_0^\infty ds \frac{s}{(s + Q_0^2)^{n+1}} \sigma(s).
\end{aligned} \tag{18.99}$$

This is an integral over the total cross section for $e^+e^- \to$ hadrons. By equating (18.97) and (18.99), we obtain a series of integral relations between the OPE coefficients, evaluated in QCD perturbation theory, and the observable cross section. These relations, which were first constructed by Novikov, Shifman, Voloshin, Vainshtein, and Zakharov, are known as the *ITEP sum rules*.[†]

[†]The theory of these sum rules is reviewed in V. A. Novikov, L. B. Okun, M. A. Shifman, A. I. Vainshtein, M. B. Voloshin, and V. I. Zakharov, *Phys. Repts.* **41**, 1 (1978).

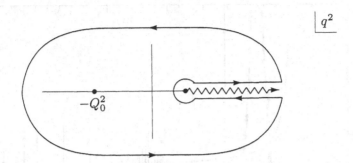

Figure 18.7. Contour of integration involved in the derivation of the ITEP sum rules for $\sigma(e^+e^- \to \text{hadrons})$.

Evaluating the sum rules with only the leading QCD expression for $c^1(q^2)$, we find

$$\int_0^\infty ds\, \frac{s}{(s+Q_0^2)^{n+1}}\sigma(s) = \frac{4\pi\alpha^2}{n(Q_0^2)^n}\sum_f Q_f^2 + \mathcal{O}(\alpha_s(Q_0^2)) + \mathcal{O}((Q_0^2)^{-2}). \quad (18.100)$$

The leading-order relation is consistent with the lowest-order cross section given in Eq. (18.95). The corrections come from higher orders of QCD perturbation theory, with α_s taken at the scale Q_0^2, and from the higher operator terms in the OPE.

If the correction terms in (18.100) converged to zero uniformly in n, we could invert the sum rules and derive from them our result (18.94). However, the true situation is more subtle. Because the derivatives in (18.97) emphasize terms with stronger q^2 variation, the correction terms in the ITEP sum rules are more and more important as n increases. Thus the most important deviations of the cross section from the prediction of QCD perturbation theory are oscillations about this prediction, which average out in the sum rules for low n. The comparison of theory and experiment is shown in Fig. 18.8. At large s, (18.94) is quite accurate. As s becomes smaller, however, the oscillations grow in size. Eventually, they come to dominate the total cross section as the resonances associated with quark-antiquark bound states.

18.5 Operator Analysis of Deep Inelastic Scattering

We now apply the operator product expansion to another example of a QCD hard-scattering process, deep inelastic electron scattering. In Chapter 17 we found that the predictions of QCD for deep inelastic scattering are precise but also intricate in structure. At a first level, QCD implies that deep inelastic scattering is described by the parton model, in which the incident electron scatters from quarks and antiquarks that carry fractions of the total momentum of the proton. These fractions are determined by parton distribution

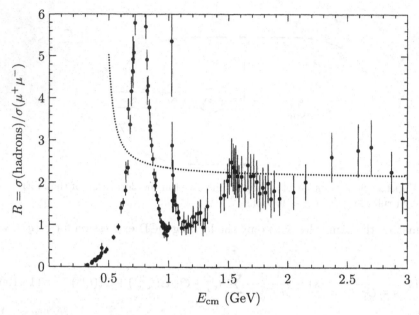

Figure 18.8. Experimental measurements of the total cross section for the reaction $e^+e^- \to$ hadrons at energies below 3 GeV, compared to the prediction of perturbative QCD for 3 quark flavors. The data are taken from the compilation of M. Swartz, *Phys. Rev.* **D**53, 5268 (1996). Complete references to the various results are given there.

functions, which reflect the form of the proton wavefunction and are determined by soft QCD dynamics. However, we saw in Section 17.5 that effects of QCD perturbation theory cause the parton distributions to change their form as a function of the momentum transfer Q^2. We will now show that much of this picture can be reconstructed from our new viewpoint, using the operator product expansion.

In the previous section, we derived the OPE relations for the e^+e^- annihilation cross section in three steps. First, we used the optical theorem to relate this cross section to a matrix element of a product of currents. Second, we applied the operator product expansion to the product of currents. Unfortunately, this expansion could be used only in an unphysical kinematic region. However, in the third step, we used the method of dispersion relations to connect this unphysical result to an integral over the cross section we wished to predict. In our discussion of deep inelastic scattering, we will go through these same three steps. To obtain our final result, we will need to add a fourth step, involving QCD operator rescaling.

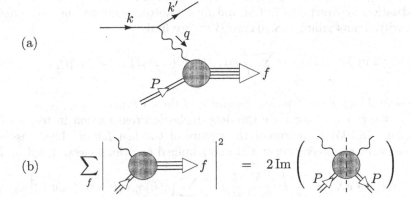

Figure 18.9. Computation of the cross section for deep inelastic electron scattering: (a) general structure of the amplitudes; (b) application of the optical theorem.

Kinematics of Deep Inelastic Scattering

We begin by writing a general expression for the deep inelastic scattering cross section. The matrix element for deep inelastic electron scattering to a final state f is computed as shown in Fig. 18.9(a):

$$i\mathcal{M}(ep \to ef) = (-ie)\bar{u}(k')\gamma_\mu u(k)\frac{-i}{q^2}(ie)\int d^4x\, e^{iq\cdot x} \langle f| J^\mu(x) |P\rangle, \quad (18.101)$$

where $J^\mu(x)$ is the quark electromagnetic current (18.87). The core of this expression is the hadronic matrix element of the current between the proton and some high-energy hadronic state. This matrix element must be squared and summed over possible final states. That sum can be computed, using the optical theorem, by relating it to the forward matrix element of two currents in the proton state, as shown in Fig. 18.9(b). Define

$$W^{\mu\nu} = i\int d^4x\, e^{iq\cdot x} \langle P| T\{J^\mu(x)J^\nu(0)\} |P\rangle, \quad (18.102)$$

averaged over the spin of the proton. This object is known as the *forward Compton amplitude*, since if it is evaluated at $q^2 = 0$ and contracted with physical polarization vectors, it gives the forward amplitude for photon-proton scattering:

$$i\mathcal{M}(\gamma p \to \gamma p) = (ie)^2\epsilon_\mu^*(q)\epsilon_\nu(q)(-iW^{\mu\nu}(P,q)). \quad (18.103)$$

However, in the following discussion we will need to analyze (18.102) for general spacelike q and for general polarization states.

The optical theorem for Compton scattering from a proton is

$$2\,\mathrm{Im}\,\mathcal{M}(\gamma p \to \gamma p) = \sum_f \int d\Pi_f |\mathcal{M}(\gamma p \to f)|^2. \quad (18.104)$$

In the generalization given in (7.49), this result extends to the more general situation in which the initial and final photon polarizations can differ arbitrarily. Transcribing (18.104) to $W^{\mu\nu}$, we find

$$2 \operatorname{Im} W^{\mu\nu}(P,q) = \sum_f \int d\Pi_f \, \langle P| \, J^\mu(-q) \, |f\rangle \, \langle f| \, J^\nu(q) \, |P\rangle, \qquad (18.105)$$

where $J^\mu(q)$ is the Fourier transform of the current.

We can now compute the deep inelastic cross section in terms of $W^{\mu\nu}$, using (18.105) to represent the square of the last factor. The cross section should be averaged over initial and summed over final electron spins. Thus,

$$\sigma(ep \to eX) = \frac{1}{2s} \int \frac{d^3 k'}{(2\pi)^3} \frac{1}{2k'} e^4 \frac{1}{2} \sum_{\text{spins}} \left[\bar{u}(k)\gamma_\mu u(k')\bar{u}(k')\gamma_\nu u(k) \right]$$
$$\cdot \left(\frac{1}{Q^2}\right)^2 \cdot 2 \operatorname{Im} W^{\mu\nu}(P,q). \qquad (18.106)$$

The electron spinor product can be evaluated as

$$\frac{1}{2} \sum_{\text{spins}} \left[\bar{u}(k)\gamma_\mu u(k')\bar{u}(k')\gamma_\nu u(k) \right] = \frac{1}{2} \operatorname{tr} \left[\slashed{k}\gamma_\mu \slashed{k}'\gamma_\nu \right]$$
$$= 2\left(k_\mu k'_\nu + k_\nu k'_\mu - g_{\mu\nu} k \cdot k' \right). \qquad (18.107)$$

It is useful to convert the integral over the final electron momentum k' and scattering angle θ to an integral over the dimensionless variables x and y that we introduced in Section 17.3. These variables are given in terms of the initial and final electron energies k and k' by

$$x = \frac{Q^2}{2P \cdot q} = \frac{2kk'(1 - \cos\theta)}{2m(k - k')}, \qquad y = \frac{2P \cdot q}{2P \cdot k} = \frac{k - k'}{k}. \qquad (18.108)$$

Then

$$\left| \frac{\partial(x,y)}{\partial(k', \cos\theta)} \right| = \frac{2k'}{2m(k - k')} = \frac{2k'}{ys}, \qquad (18.109)$$

and so

$$\int \frac{d^3 k'}{(2\pi)^3} \frac{1}{2k'} = \int \frac{2\pi dk' k' d\cos\theta}{(2\pi)^3 \cdot 2} = \int dx dy \frac{ys}{(4\pi)^2}. \qquad (18.110)$$

Using (18.107) and (18.110) to simplify (18.106), we find

$$\frac{d^2\sigma}{dxdy}(ep \to eX) = \frac{2\alpha^2 y}{(Q^2)^2} \left(k_\mu k'_\nu + k_\nu k'_\mu - g_{\mu\nu} k \cdot k' \right) \operatorname{Im} W^{\mu\nu}(P,q). \qquad (18.111)$$

To go further, we need to know something about the structure of $W^{\mu\nu}$. In the previous section, we used current conservation to write the matrix element of currents in terms of a single scalar function $\Pi_h(q^2)$, as in Eq. (18.82). In the case of the forward Compton amplitude, the Ward identity again requires

$$q_\mu W^{\mu\nu} = q_\nu W^{\mu\nu} = 0, \tag{18.112}$$

but now there are two possible tensors built from P and q that satisfy these constraints. Thus the forward Compton amplitude is written as an expression involving two scalar form factors:

$$W^{\mu\nu} = \left(-g^{\mu\nu} + \frac{q^\mu q^\nu}{q^2}\right)W_1 + \left(P^\mu - q^\mu \frac{P \cdot q}{q^2}\right)\left(P^\nu - q^\nu \frac{P \cdot q}{q^2}\right)W_2. \tag{18.113}$$

The scalar functions W_1, W_2 depend on the two invariants of the problem, $(P \cdot q)$ and q^2, or, alternatively, x and Q^2. If we insert (18.113) into (18.111) and use the fact that dotting q^μ with the lepton tensor gives zero, we find

$$\frac{d^2\sigma}{dx dy}(ep \rightarrow eX) = \frac{2\alpha^2 y}{(Q^2)^2}\left[2k \cdot Pk' \cdot P \operatorname{Im} W_2 + 2k \cdot k' \operatorname{Im} W_1\right]$$

$$= \frac{\alpha^2 y}{Q^4}\left[s^2(1-y) \operatorname{Im} W_2 + 2xys \operatorname{Im} W_1\right]. \tag{18.114}$$

Expression (18.114) is completely general and makes no assumptions about the nature of the strong interactions. It is also rather formal. However, we can easily get an idea of the relation of this formula to our earlier analysis by evaluating $W^{\mu\nu}$ in the parton model and working out the parton expressions for W_1 and W_2. In the parton model, we replace the proton matrix element in (18.102) by a sum of quark matrix elements, weighted with the parton distribution functions. Thus,

$$W^{\mu\nu} \approx i \int d^4 x \, e^{iq \cdot x} \int_0^1 d\xi \sum_f f_f(\xi) \cdot \frac{1}{\xi} \langle q_f(p)| T\{J^\mu(x) J^\nu(0)\} |q_f(p)\rangle \Big|_{p=\xi P}. \tag{18.115}$$

The factor $(1/\xi)$ in front of the matrix element gives the proper normalization of the proton state in terms of the quark states. The simplest way to understand this factor is to note that the kinematic prefactor $(1/2s)$ in (18.106) and in other expressions involving an initial-state proton becomes $(1/2\xi s)$, under the ξ integral, in the parton model.

We now evaluate the matrix element in (18.115) using noninteracting fermions. There are two Feynman diagrams, shown in Fig. 18.10. The first diagram on the right in Fig. 18.10 has the value

$$i \int_0^1 d\xi \sum_f f_f(\xi) \frac{1}{\xi} Q_f^2 \bar{u}(p) \gamma^\mu \frac{i(\not p + \not q)}{(p+q)^2 + i\epsilon} \gamma^\nu u(p); \tag{18.116}$$

the second diagram gives a contribution identical to this one after the interchange of q, μ with $(-q)$, ν. To evaluate (18.116), we average over the quark

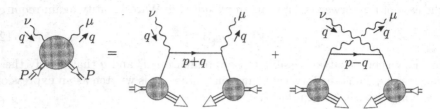

Figure 18.10. Evaluation of $W^{\mu\nu}$ in the parton model.

spin to find

$$
\int_0^1 d\xi \sum_f f_f(\xi) \frac{1}{\xi} \cdot \frac{1}{2} \operatorname{tr}\left[\not{p}\gamma^\mu(\not{p}+\not{q})\gamma^\nu\right] \frac{-1}{2p\cdot q + q^2 + i\epsilon}
$$

$$
= \int_0^1 d\xi \sum_f f_f(\xi) \frac{1}{\xi} \cdot 2\big(p^\mu(p+q)^\nu + p^\nu(p+q)^\mu - g^{\mu\nu} p\cdot(p+q)\big)
$$

$$
\cdot \frac{-1}{2\xi P\cdot q - Q^2 + i\epsilon}. \tag{18.117}
$$

The imaginary part of this expression, which we need to evaluate (18.114), comes from the last factor in (18.117):

$$
\operatorname{Im}\left(\frac{-1}{2\xi P\cdot q - Q^2 + i\epsilon}\right) = \pi\delta(2\xi P\cdot q - Q^2) = \frac{\pi}{ys}\delta(\xi - x). \tag{18.118}
$$

In the second diagram of Fig. 18.10, the two factors in the denominator have a relative $+$ sign, so this diagram has no imaginary part in the physical region for deep inelastic scattering. Thus, we find that in the parton model,

$$
\operatorname{Im} W^{\mu\nu} = \sum Q_f^2 f_f(x) \frac{1}{x}\frac{\pi}{ys}\big(4x^2 P^\mu P^\nu + 2x(P^\mu q^\nu + P^\nu q^\mu) - g^{\mu\nu} xys\big). \tag{18.119}
$$

By adding and subtracting terms proportional to $q^\mu q^\nu$, we can see that this expression is of the form (18.113), with

$$
\operatorname{Im} W_1 = \pi \sum_f Q_f^2 f_f(x), \qquad \operatorname{Im} W_2 = \frac{4\pi}{ys} \sum_f Q_f^2 x f_f(x). \tag{18.120}
$$

The parton model expressions for W_1 and W_2 obey the relation

$$
\operatorname{Im} W_1 = \frac{ys}{4x} \operatorname{Im} W_2. \tag{18.121}
$$

This is another form of the Callan-Gross relation, since the substitution of (18.121) into (18.114) gives

$$
\frac{d^2\sigma}{dxdy}(ep \to eX) = \frac{\alpha^2 ys^2}{2Q^4}\big[1 + (1-y)^2\big] \operatorname{Im} W_2, \tag{18.122}
$$

with the y dependence characteristic of free fermions, as in Eq. (17.125). Finally, substituting from (18.120) for the imaginary part of W_2, we recover this parton model expression precisely:

$$\frac{d^2\sigma}{dx\,dy}(ep \to eX) = \frac{2\pi\alpha^2 s}{Q^4}\Big(\sum_f Q_f^2 x f_f(x)\Big)\big[1 + (1-y)^2\big]. \qquad (18.123)$$

This equation will give us a reference point for comparison with more general expressions that we will derive as we continue our analysis.

Expansion of the Operator Product

Since the forward Compton amplitude is a matrix element of a product of currents, an alternative strategy for calculating $W^{\mu\nu}$ is to expand this product as a series of local operators. Like the parton model evaluation, this method makes use of asymptotic freedom. However, in this case, the assumption is applied more directly. The computation of the operator product coefficients will take place explicitly at a small distance of order $1/Q$, and so we can calculate these coefficients in a perturbation theory whose coupling constant is $\alpha_s(Q^2)$.

In the previous section, we computed the coefficients of operators that contribute to the vacuum expectation value of the product of currents by considering the various ways of contracting the quark fields in the product. Here, we should note that the operator 1 does not contribute to the Compton scattering amplitude. The leading contributions come from operators that can create and annihilate quarks in the proton wavefunction.

The most important terms in the operator product of two currents J^μ come from products of two quark currents $\bar{q}_f \gamma^\mu q_f$ with quarks of the same flavor. Therefore we will begin by studying the OPE of the individual quark currents. To zeroth order in α_s, the leading terms of the operator product of quark currents are given by

$$\bar{q}\gamma^\mu q(x)\,\bar{q}\gamma^\nu q(0)$$

$$(18.124)$$

$$= \bar{q}(x)\gamma^\mu \overline{q(x)\bar{q}(0)}\gamma^\nu q(0) + \bar{q}(x)\gamma^\mu \overline{q(x)\bar{q}(0)}\gamma^\nu q(0) + \cdots,$$

where the contractions should be evaluated as Feynman propagators for the quark fields. The terms with explicit contractions are singular as $x \to 0$; the remaining terms are nonsingular and thus less important in the short-distance limit. In the OPE of currents with quarks of different flavor, there are no corresponding singular terms; we will argue below that this conclusion is valid even beyond the leading order in α_s.

To evaluate $W^{\mu\nu}$, we must take the Fourier transform of the terms in (18.124), as indicated in (18.102). When we do this, we should remember that the propagators carry not only the Fourier transform momentum q but also whatever momentum is carried in through the quark fields. To take account of this, it is convenient to represent

$$\int d^4x\, e^{iq\cdot x}\, \bar{q}(x)\gamma^\mu \overbracket{q(x)\bar{q}(0)}\gamma^\nu q(0) = \bar{q}\gamma^\mu \frac{i(i\slashed{\partial}+\slashed{q})}{(i\partial+q)^2}\gamma^\nu q(0), \tag{18.125}$$

where the derivatives ∂ act to the right on the quark field. Notice that this contribution has the structure of the first diagram on the right in Fig. 18.10. Similarly, the second contraction indicated in (18.124) has the form of the second diagram in Fig. 18.10.

In the short-distance limit, the momentum q will be larger than any external momentum entering the quark fields. Thus we should expand

$$\frac{1}{(i\partial+q)^2} = \frac{-1}{Q^2 - 2iq\cdot\partial + \partial^2} = -\frac{1}{Q^2}\sum_{n=0}^{\infty}\Big(\frac{2iq\cdot\partial - \partial^2}{Q^2}\Big)^n. \tag{18.126}$$

We will argue below that the terms with ∂^2 in the numerator are unimportant and may be dropped. However, we should retain all powers of the ratio $(2iq\cdot\partial/Q^2)$. This ratio has Q^2 in the denominator and so is formally suppressed in the short-distance limit. However, in the parton model

$$\frac{2iq\cdot\partial}{Q^2} \to \frac{2q\cdot\xi P}{Q^2} = 1, \tag{18.127}$$

so, eventually, all of these terms must be equally important. We will see how this works in a moment.

The last step required to reduce the operator product (18.124) to a useful form is to reduce the product of Dirac matrices. We know from (18.113) that, after we average over the proton spin, $W^{\mu\nu}$ will be symmetric under the interchange of μ and ν. Thus, it does no harm to symmetrize the OPE. We can then reduce the product of three Dirac matrices to one by using the identity

$$\tfrac{1}{2}\big(\gamma^\mu\gamma^\alpha\gamma^\nu + \gamma^\nu\gamma^\alpha\gamma^\mu\big) = g^{\mu\alpha}\gamma^\nu + \gamma^\mu g^{\alpha\nu} - g^{\mu\nu}\gamma^\alpha, \tag{18.128}$$

which is easily proved from the anticommutation relations. By the use of (18.126) and (18.128), we can rewrite (18.125) as

$$-i\bar{q}\Big(\gamma^\mu(i\partial^\nu) + \gamma^\nu(i\partial^\mu) - ig^{\mu\nu}\slashed{\partial} + \gamma^\mu q^\nu + \gamma^\nu q^\mu - g^{\mu\nu}\slashed{q}\Big)\frac{1}{Q^2}\sum_{n=0}^{\infty}\Big(\frac{2iq\cdot\partial}{Q^2}\Big)^n q. \tag{18.129}$$

We can remove the term $(i\slashed{\partial})q$, which vanishes to leading order in α_s, since the quark field obeys the Dirac equation. To compute W_1 and W_2, we can also drop the terms with explicit factors of q^μ, since these will eventually be organized into the general form (18.113). Then, finally, (18.125) takes the form

$$\int d^4x\, e^{iq\cdot x}\bar{q}(x)\gamma^\mu \overbracket{q(x)\bar{q}(0)}\gamma^\nu q(0)$$
$$= -i\bar{q}\big(2\gamma^\mu(i\partial^\nu) - g^{\mu\nu}\slashed{q}\big)\frac{1}{Q^2}\sum_{n=0}^{\infty}\Big(\frac{2iq\cdot\partial}{Q^2}\Big)^n q, \tag{18.130}$$

symmetrized under $\mu \leftrightarrow \nu$.

The second term in (18.124) differs from the first by the interchange of the points x and 0 and the interchange of indices μ and ν. Its Fourier transform is thus given by (18.130) with the replacement $q \to -q$. The complete operator product therefore contains only terms even in q. All remaining contributions from the singular terms of the operator product contain the operator

$$\bar{q}\gamma^{\mu_1}(i\partial^{\mu_2})\cdots(i\partial^{\mu_k})q, \tag{18.131}$$

with an even number of indices, with these indices either identified with μ or ν or contracted with powers of q. To write the relevant terms of the operator product expansion, we will modify this operator in two ways. First, since the operator in (18.131) has n vector indices, it contains components that transform under many different irreducible representations of the Lorentz group. Each component has a different rescaling law under renormalization. However, we will see below that only the component of (18.131) with the highest spin is relevant to our analysis. This component is obtained by totally symmetrizing the indices $\mu_1, \ldots, \mu_n$ and then subtracting terms proportional to $g^{\mu_i\mu_j}$ so that the operator is traceless on all pairs of indices. We will retain only this component when we write out the operator product of currents. Second, the operator (18.131) does not transform simply under gauge transformations. Since the original currents J^μ were invariant to color gauge transformations, the operator product of two currents must be a sum of gauge-invariant operators. We can make (18.131) gauge-invariant by replacing each factor of $(i\partial^\mu)$ with a covariant derivative (iD^μ). This modification adds only terms proportional to the strong coupling constant g, so it has no effect on our derivation of the operator product coefficients.

Incorporating these changes, let us define a spin-n operator with quarks of flavor f as follows:

$$\mathcal{O}_f^{(n)\,\mu_1\cdots\mu_n} = \bar{q}_f\gamma^{\{\mu_1}(iD^{\mu_2})\cdots(iD^{\mu_n\}})q_f - \text{traces}, \tag{18.132}$$

with indices symmetrized and with appropriate subtractions. We can use these operators to write a final expression for the most singular part of the OPE of two currents J^μ. The leading terms in this operator product come from (18.130) and the corresponding contraction with $q \leftrightarrow -q$. Extracting the pieces of these expressions that contain the highest spin operators (18.132), we find

$$i\int d^4x\, e^{iq\cdot x} J^\mu(x) J^\nu(0)$$

$$= \sum_f Q_f^2 \left[4\sum_{n=2}^{\infty} \frac{(2q^{\mu_1})\cdots(2q^{\mu_{n-2}})}{(Q^2)^{n-1}} \mathcal{O}_f^{(n)\,\mu\nu\mu_1\cdots\mu_{n-2}} \right. \tag{18.133}$$

$$\left. - g^{\mu\nu}\sum_{n=2}^{\infty} \frac{(2q^{\mu_1})\cdots(2q^{\mu_n})}{(Q^2)^n} \mathcal{O}_f^{(n)\,\mu_1\cdots\mu_n} \right] + \cdots,$$

where the sums over n run over even integers only.

Expression (18.133) has been derived in the leading order in α_s. Higher-order Feynman diagrams will contribute corrections to the coefficient functions of order $\alpha_s(Q^2)$. These corrections will be important only if they are multiplied by large logarithms. If we consider the operators $\mathcal{O}_f^{(n)}$ appearing on the right-hand side to be normalized at the renormalization scale Q, there is no large ratio of momenta available to enhance the QCD corrections to the coefficient functions. Large logarithmic corrections may still arise at a later stage of the calculation, when we compute the matrix elements of the operators $\mathcal{O}_f^{(n)}$.

From the expansion (18.133), it is straightforward to compute an expansion for $W^{\mu\nu}$ by taking its expectation value in the proton state. To carry out this computation, we need to know the proton matrix elements of the operators $\mathcal{O}_f^{(n)}$. Notice that these matrix elements cannot depend on the direction of the momentum q^μ, since that dependence has been isolated in the coefficient functions. This means that only the proton momentum P^μ is available to carry the vector indices of the matrix element. We can therefore write the spin-averaged matrix element of $\mathcal{O}_f^{(n)}$ as

$$\langle P| \mathcal{O}_f^{(n)\,\mu_1\cdots\mu_n} |P\rangle = A_f^n \cdot 2P^{\mu_1}\cdots P^{\mu_n} - \text{traces}. \qquad (18.134)$$

The coefficients A_f^n are dimensionless. They are not quite pure numbers, because they depend on the renormalization scale of the operators, but we will treat them as constants in the next few paragraphs.

For the case $n = 1$, the operators $\mathcal{O}_f^{(1)}$ reduce simply to the quark flavor currents $\bar{q}\gamma^\mu q$; in this case the operators are normalized independently of any scale and the coefficients A_f^1 are truly constants. From our general discussion of form factors in Section 6.2, we know that the proton matrix element of a conserved flavor current at zero momentum transfer is given by

$$\langle P| \bar{q}_f\gamma^\mu q_f |P\rangle = \bar{u}(P)\gamma^\mu u(P)F_{f1}(0), \qquad (18.135)$$

where $F_{f1}(0)$ is equal to the value of the corresponding conserved charge in the proton state. For the quark currents, this charge is just the number of quarks (minus antiquarks) of flavor f in the state $|P\rangle$, which we will call N_f. Averaging (18.135) over the proton spin, we find

$$\langle P| \bar{q}_f\gamma^\mu q_f |P\rangle = 2P^\mu \cdot N_f. \qquad (18.136)$$

Thus, for $n = 1$,

$$A_f^1 = N_f = \begin{cases} 2 & f = u, \\ 1 & f = d. \end{cases} \qquad (18.137)$$

Similarly, $\mathcal{O}_f^{(2)}$ is the contribution of the quark flavor f to the energy-momentum tensor of QCD:

$$(T^{\mu\nu})_f = \bar{q}_f\gamma^{\{\mu}(iD^{\nu\}})q_f. \qquad (18.138)$$

Thus, A_f^2 is the fraction of the total energy-momentum of the proton that is carried by the quark flavor f.

When we evaluate the series for $W^{\mu\nu}$ using (18.133) and the expression (18.134) for the operator matrix elements, we find

$$W^{\mu\nu} = \sum_f Q_f^2 \left[8 \sum_n P^\mu P^\nu \frac{(2q \cdot P)^{n-2}}{(Q^2)^{n-1}} A_f^n - 2g^{\mu\nu} \sum_n \frac{(2q \cdot P)^n}{(Q^2)^n} A_f^n \right] + \cdots,$$

(18.139)

where the sums over n run over even integers from 2 to infinity. In addition to the corrections to the OPE omitted in (18.133), we have also dropped contributions from the trace terms in (18.134). This is quite appropriate: In each of these terms, two factors of the proton momentum $P^\alpha P^\beta$ are replaced by $g^{\alpha\beta} m_p^2$, were $m_p^2 = P^2$ is the proton mass. When the indices are contracted with powers of q, we obtain a term of order

$$m_p^2 Q^2 \ll (2q \cdot P)^2.$$

(18.140)

Since $(Q^2/2P \cdot q) = x$, which is held fixed in deep inelastic scattering as Q^2 becomes large, the contribution from the trace terms is suppressed by a factor m_p^2/Q^2, times powers of x.

In general, an operator of dimension d has a coefficient function in the operator product expansion of currents that has dimension $(\text{mass})^{6-d}$; in the Fourier transform of the OPE, this coefficient function will carry a suppression factor

$$\left(\frac{1}{Q} \right)^{d-2}.$$

(18.141)

However, if the operator has spin s, the operator matrix element will contribute s factors of the vector P^μ, so that, in the kinematic region of deep inelastic scattering, the contribution will be of order

$$\left(\frac{2P \cdot q}{Q^2} \right)^s \left(\frac{1}{Q} \right)^{d-s-2}.$$

(18.142)

Thus, the relative size of contributions from the OPE to deep inelastic scattering is controlled, not exactly by the dimension of the operator, but rather by the *twist*, defined as

$$t = d - s.$$

(18.143)

In our selection of the leading terms in the operator product expansion of currents, we have consistently kept the contribution of leading spin for each dimension or for each power of Q^{-1} in the coefficient. The operators $\mathcal{O}_f^{(n)}$ all have twist $t = 2$, which is the smallest possible value for QCD operators other than the operator 1.

In the operator product of two different flavor currents—for example, $\bar{u}\gamma^\mu u$ and $\bar{d}\gamma^\nu d$—the leading terms in the OPE have the quark structure $(\bar{u}\Gamma u \bar{d}\Gamma d)$ and thus have twist $t \geq 4$. Thus, to all orders in α_s, the cross terms in the operator product of currents J^μ are suppressed by at least a factor

$(1/Q^2)$ relative to the leading-twist terms presented in (18.133). If we neglect these suppressed terms, the expression for $W^{\mu\nu}$ separates, to all orders, into a sum of contributions

$$W^{\mu\nu} = \sum_f Q_f^2 W_f^{\mu\nu}, \tag{18.144}$$

where $W_f^{\mu\nu}$ is the matrix element of two quark flavor currents $\bar{q}_f \gamma^\mu q_f$.

We can read from (18.139) the following expressions for W_1 and W_2:

$$
\begin{aligned}
W_1 &= \sum_f Q_f^2 \sum_n 2 \frac{(2q \cdot P)^n}{(Q^2)^n} A_f^n, \\
W_2 &= \sum_f Q_f^2 \sum_n \frac{8}{Q^2} \frac{(2q \cdot P)^{n-2}}{(Q^2)^{n-2}} A_f^n,
\end{aligned}
\tag{18.145}
$$

where the sum over n in each line runs over even integers from 2 to infinity. Like (18.139), these expressions explicitly separate according to (18.144). It is noteworthy that the series (18.145) satisfy the Callan-Gross relation in the form (18.121), without further parton model input. However, this relation is corrected in order α_s due to the next-order contributions to the operator product coefficients.

Because the leading contributions to the deep inelastic form factors can be written as sums over quark flavors, it is tempting to reverse the logic of Eq. (18.120) and use these equations to define the parton distribution functions. In particular, let us define

$$x f_f^+(x, Q^2) = \frac{ys}{4\pi} \operatorname{Im} W_{2f}(x, Q^2), \tag{18.146}$$

where W_{2f} is the second form factor of $W_f^{\mu\nu}$, defined in (18.144), neglecting terms suppressed by powers of Q^2. In the parton model evaluation,

$$f_f^+(x) = f_f(x) + f_{\bar{f}}(x). \tag{18.147}$$

From (18.123) and the definition (18.146), we know that $f_f^+(x)$ enters in the correct way into the formula for the deep inelastic scattering cross section. However, parton distribution functions have other important properties, including the normalization conditions (17.36) and (17.39) and the evolution with Q^2 discussed in Section 17.6. We must now see whether we can derive these properties from (18.146) using the operator product expansion.

The Dispersion Integral

The operator product analysis has given us explicit expressions for W_1 and W_2 as a series in inverse powers of Q^2. In the following discussion, we will concentrate on the analysis of W_2. We must work out the relation of its series expansion to the observable deep inelastic scattering cross section. As in the discussion of Section 18.4, the OPE analysis naturally takes place in an unphysical kinematic region. To make the operator product expansion, we

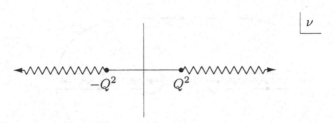

Figure 18.11. Analytic singularities of $W_2(\nu, Q^2)$ in the complex ν plane, for fixed Q^2.

needed to consider Q^2 to be larger than any other kinematic invariant. However, in the physical region for deep inelastic scattering, $2P \cdot q \geq Q^2$. We need a formula that connects these two distinct regions.

To state this problem more precisely, define

$$\nu = 2P \cdot q = ys; \tag{18.148}$$

in the frame in which the proton is at rest, $\nu = 2m_p q^0$. The form factor W_2 can be viewed as a function of ν and Q^2. Then, for fixed Q^2, the OPE gives a series expansion about the point $\nu = 0$, while the physical region for deep inelastic scattering is $\nu \geq Q^2$. Because this region is associated with a physical scattering process, $W_2(\nu, Q^2)$, viewed as an analytic function of ν for fixed Q^2, will have a branch cut along the real ν axis in this region. The discontinuity across this branch cut will be $(2i)$ times the imaginary part of W_2, which appears in the expression (18.123) for the deep inelastic cross section. Because expression (18.102) is symmetric under the interchange of (q, μ) and $(-q, \nu)$, W_2 must obey

$$W_2(-\nu, Q^2) = W_2(\nu, Q^2). \tag{18.149}$$

Thus, W_2 must also have a branch cut along the negative real axis, from $\nu = -Q^2$ to $-\infty$. The discontinuity across this cut gives the cross section for the u-channel process in which positive energy comes in through the second current and out through the first. Since $q^2 = -Q^2 < 0$, there is no possible physical t-channel process; thus W_2 has no further singularities in the complex ν plane. The analytic structure of $W_2(\nu, Q^2)$ is shown in Fig. 18.11.

Now consider the contour integral

$$I_n = \int \frac{d\nu}{2\pi i} \frac{1}{\nu^{n-1}} W_2(\nu, Q^2), \tag{18.150}$$

for n even, taken on a small circle surrounding the origin. This integral picks out the coefficient of ν^{n-2} in the series expansion for W_2. The OPE formula (18.145) gives us the leading contribution to this coefficient for large Q^2:

$$I_n = \sum_f Q_f^2 \frac{8}{(Q^2)^{n-1}} A_f^n. \tag{18.151}$$

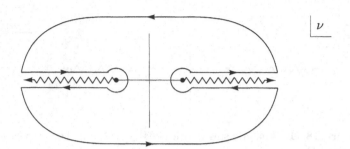

Figure 18.12. Contour of integration involved in the derivation of the moment sum rules for W_2.

The corrections to this formula are of order $\alpha_s(Q^2)$, from the evaluation of the OPE coefficient functions.

On the other hand, we can also distort the contour as shown in Fig. 18.12 and evaluate it as an integral over the discontinuities of W_2. By the symmetry (18.149), the two branch cuts give equal contributions. Thus,

$$I_n = 2 \int_{Q^2}^{\infty} \frac{d\nu}{2\pi i} \frac{1}{\nu^{n-1}} (2i) \operatorname{Im} W_2(\nu, Q^2). \tag{18.152}$$

Now change variables to $x = Q^2/\nu$. The integral becomes

$$I_n = \frac{8}{(Q^2)^{n-1}} \int_0^1 dx \, x^{n-2} \frac{\nu}{4\pi} \operatorname{Im} W_2. \tag{18.153}$$

When we equate (18.151) and (18.153) and relate $\operatorname{Im} W_2$ to the parton distributions $f_f^+(x)$ using (18.146), the relation we have derived splits into a series of sum rules,

$$\int_0^1 dx \, x^{n-1} f_f^+(x, Q^2) = A_f^n, \tag{18.154}$$

for n even. These relations are known as the *moment sum rules* for the deep inelastic form factors. They relate the x moments of the parton distribution functions, as defined by Eq. (18.146), to the proton matrix elements of twist-2 operators.

Because W_2 is a symmetric function of ν, the moment sum rules apply only for even n. However, in deep inelastic neutrino scattering, there is a third form factor in $W^{\mu\nu}$, associated with the interference term between the vector and axial vector parts of the weak interaction current. In Problem 18.2, we show that this form factor can be used to derive a set of sum rules for odd n:

$$\int_0^1 dx \, x^{n-1} f_f^-(x, Q^2) = A_f^n, \tag{18.155}$$

where A_f^n is the coefficient of the proton matrix element (18.134) for odd n, and $f_{\bar{f}}^-(x)$ is a form factor which, in the parton model, evaluates to

$$f_{\bar{f}}^-(x) = f_f(x) - f_{\bar{f}}(x). \tag{18.156}$$

Combining this information with the argument given below (18.136), we can see that the definition of the parton distribution functions from the deep inelastic form factors has the correct normalization. Using (18.137), we find

$$\int_0^1 dx\, f^-(x) = N_f, \tag{18.157}$$

the (net) number of quarks of flavor f in the proton. Similarly, (18.154) and (18.138) imply

$$\int_0^1 dx\, x f^+(x) = \langle x \rangle_f, \tag{18.158}$$

where $\langle x \rangle_f$ is the fraction of the total energy-momentum of the proton carried by quarks and antiquarks of flavor f.

Operator Rescaling

If the coefficients A_f^n were truly constants, relations (18.154) and (18.155) would be consistent with parton distribution functions that satisfy Bjorken scaling. However, as we remarked below (18.134), these factors actually depend on Q^2, since this is the normalization point of the operators in the operator product expansion (18.133). Since this dependence comes only through operator rescaling, it involves only logarithms of Q^2, and so contributes only a slow violation of Bjorken scaling. We can work out the Q^2 dependence of the parton distribution functions quantitatively by summing the leading logarithmic corrections to the matrix elements of the twist-2 operators.

To account for these corrections, let us first assume (incorrectly, as we will see below) that the twist-2 operators (18.132) are renormalized without operator mixing. Then the leading logarithmic corrections to the matrix element of the operator $\mathcal{O}_f^{(n)}$ would be summed by rescaling the operator normalized at Q to operators normalized at a standard reference point μ, of order 1 GeV. The relation between these conventions would be

$$\left[\mathcal{O}_f^{(n)}\right]_Q = \left(\frac{\log(Q^2/\Lambda^2)}{\log(\mu^2/\Lambda^2)}\right)^{a_f^n/2b_0} \left[\mathcal{O}_f^{(n)}\right]_\mu, \tag{18.159}$$

where a_f^n is the first coefficient of the γ function of $\mathcal{O}_f^{(n)}$. Then the factors A_f^n would depend on Q^2 according to

$$A_f^n(Q^2) = \left(\frac{\log(Q^2/\Lambda^2)}{\log(\mu^2/\Lambda^2)}\right)^{a_f^n/2b_0} A_f^n(\mu^2). \tag{18.160}$$

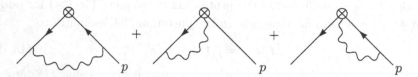

Figure 18.13. Diagrams contributing the anomalous dimension of the quark twist-2 operators.

This equation agrees with the scale dependence of operator product coefficients written in (18.76), for the special case of an operator product of currents, $a_1 = a_2 = 0$. To find the explicit form of the rescaling factor, we must compute a_f^n.

To compute the γ functions of the quark twist-2 operators, we must compute their counterterms for operator rescaling. These are determined by the diagrams shown in Fig. 18.13. It suffices to compute these diagrams with external momentum p entering through the quark line and zero external momentum injected into the operator. Under these conditions, the matrix element of the operator $\mathcal{O}_f^{(n)}$, in leading order, equals

$$\mathord{\text{\raisebox{0pt}{\(\otimes\)}}}\ p \quad = \gamma^{\mu_1} p^{\mu_2} \cdots p^{\mu_n}. \tag{18.161}$$

Here and at all later points in the discussion, we will treat the matrix elements of $\mathcal{O}_f^{(n)}$ as though they are symmetrized in the n indices and have all possible traces subtracted. We must now evaluate the diagrams of Fig. 18.13 and collect all terms that rescale this structure.

The first diagram of Fig. 18.13 is quite straightforward to evaluate:

$$\mathord{\text{\raisebox{0pt}{\(\otimes\)}}}\ k \;=\; \int \frac{d^4k}{(2\pi)^4} (ig)^2 \gamma^\lambda t^a \frac{i\slashed{k}}{k^2} \gamma^{\mu_1} k^{\mu_2} \cdots k^{\mu_n} \frac{i\slashed{k}}{k^2} \gamma_\lambda t^a \frac{-i}{(k-p)^2}$$

$$= -ig^2 C_2(r) \cdot (-2) \int \frac{d^4k}{(2\pi)^4} \frac{1}{(k^2)^2 (k-p)^2} \slashed{k}\gamma^{\mu_1} k^{\mu_2} \cdots k^{\mu_n} \slashed{k}. \tag{18.162}$$

We combine denominators using identity (6.40):

$$\frac{1}{(k^2)^2 (k-p)^2} = \int_0^1 dx\, \frac{2(1-x)}{(\mathbf{k}^2 - \Delta)^3}; \tag{18.163}$$

the quantities in the denominator on the right are $\mathbf{k} = k - xp$ and $\Delta = -x(1-x)p^2$. We must now shift the integral, substitute $k = \mathbf{k} + xp$ in the numerator, and pick out a term proportional to $(n-1)$ powers of p. If this term contains the factor $g^{\mu_i \mu_j}$, we may drop it, since it contributes to the coefficient of an operator of higher twist and since, in any event, it will be removed when we subtract traces. Thus, we must choose carefully which two factors of k we replace with $\mathbf{k}$ when we replace the others with (xp). The following choices, simplified using the rotational symmetry of the **k** integral, do not give useful contributions:

$$\mathbf{k}^{\mu_i}\mathbf{k}^{\mu_j} = \frac{1}{4}\mathbf{k}^2 g^{\mu_i\mu_j},$$

$$\not{k}\gamma^{\mu_1}\mathbf{k}^{\mu_j} \rightarrow \frac{1}{4}\mathbf{k}^2\gamma^{\mu_j}\gamma^{\mu_1} = \frac{1}{4}\mathbf{k}^2 g^{\mu_1\mu_j}. \tag{18.164}$$

In the second line, we have used the symmetry under $\mu_1 \leftrightarrow \mu_j$. The one remaining placement of the factors of **k** is

$$\not{k}\gamma^{\mu_1}\not{k} \rightarrow \frac{1}{4}k^2\gamma^\nu\gamma^{\mu_1}\gamma_\nu = -\frac{1}{2}k^2\gamma^{\mu_1}. \tag{18.165}$$

Thus (18.162) has the value

$$
\begin{aligned}
&= -ig^2\frac{4}{3}\int_0^1 dx \cdot 2(1-x)\int\frac{d^4k}{(2\pi)^4}\frac{k^2}{(k^2-\Delta)^3}\gamma^{\mu_1}(xp^{\mu_2})\cdots(xp^{\mu_n})\\
&= -i\frac{8}{3}g^2\int_0^1 dx(1-x)x^{n-1}\frac{i}{(4\pi)^2}\Gamma(2-\tfrac{d}{2})\gamma^{\mu_1}p^{\mu_2}\cdots p^{\mu_n}\\
&= \frac{4}{3}\frac{2}{n(n+1)}\frac{g^2}{(4\pi)^2}\Gamma(2-\tfrac{d}{2})\gamma^{\mu_1}p^{\mu_2}\cdots p^{\mu_n}.
\end{aligned}\tag{18.166}
$$

It is not so obvious that there are additional contributions to the rescaling of the operators $\mathcal{O}_f^{(n)}$. Note, however, that the covariant derivatives in (18.132) contain explicit factors of the gauge field,

$$iD^{\mu_j} = i\partial^{\mu_j} - gA^{a\mu_j}t^a, \tag{18.167}$$

and these may be contracted with gauge field vertices on the external legs. These contributions give rise to the second and third diagrams in Figure 18.13. The term in which two factors of $A^{a\mu}$ from (18.167) are contracted with one another is proportional to $G^{\mu_i\mu_j}$ and thus does not contribute to the rescaling of the leading-twist operators.

The contributions we have just described have the form of sums over j, where μ_j is the index of the derivative that includes the contraction. Then the second diagram of Fig. 18.3 is the sum over j of the following integral:

$$
\begin{aligned}
&= \int\frac{d^4k}{(2\pi)^4}(ig)\gamma_\lambda t^a\frac{i\not{k}}{k^2}\gamma^{\mu_1}k^{\mu_2}\cdots k^{\mu_j-1}\\
&\qquad\qquad \cdot(-gt^a g^{\lambda\mu_j})p^{\mu_j+1}\cdots p^{\mu_n}\frac{-i}{(k-p)^2}\\
&= ig^2 C_2(r)\int\frac{d^4k}{(2\pi)^4}\frac{1}{k^2(k-p)^2}\gamma^{\mu_j}\not{k}\gamma^{\mu_1}k^{\mu_2}\cdots k^{\mu_j-1}p^{\mu_j+1}\cdots p^{\mu_n}.
\end{aligned}\tag{18.168}
$$

Since μ_j and μ_1 are symmetrized, we can use (18.128) to rewrite

$$\gamma^{\mu_j} \not{k} \gamma^{\mu_1} \rightarrow k^{\mu_j} \gamma^{\mu_1} + \gamma^{\mu_j} k^{\mu_1} - g^{\mu_j \mu_1} \not{k}$$
$$\rightarrow 2\gamma^{\mu_1} k^{\mu_j}, \tag{18.169}$$

where, in the second line, the symmetrization of indices and subtraction of traces is understood. Now combine denominators. To obtain a term with $(n-1)$ factors of p, we must replace every factor k in the numerator of (18.168) with (xp). This gives

$$
\begin{aligned}
\text{(diagram)} &= ig^2 \frac{4}{3} \int_0^1 dx \int \frac{d^4k}{(2\pi)^4} \frac{1}{(k^2 - \Delta)^2} 2\gamma^{\mu_1} (xp^{\mu_2}) \cdots (xp^{\mu_j}) p^{\mu_{j+1}} \cdots p^{\mu_n} \\
&= ig^2 \frac{8}{3} \int_0^1 dx\, x^{j-1} \frac{i}{(4\pi)^2} \Gamma(2-\tfrac{d}{2}) \gamma^{\mu_1} p^{\mu_2} \cdots p^{\mu_n} \\
&= -\frac{4}{3} \frac{2}{j} \frac{g^2}{(4\pi)^2} \Gamma(2-\tfrac{d}{2}) \gamma^{\mu_1} p^{\mu_2} \cdots p^{\mu_n}.
\end{aligned}
\tag{18.170}
$$

This contribution must be summed over j from 2 to n. The third diagram of Fig. 18.13 makes an equal contribution.

Summing the rescaling factors from the three diagrams of Fig. 18.13, we find for the operator rescaling counterterm of $\mathcal{O}_f^{(n)}$

$$\delta_f = \frac{g^2}{(4\pi)^2} \frac{4}{3} \left[4 \sum_{j=2}^{n} \frac{1}{j} - \frac{2}{n(n+1)} \right] \frac{\Gamma(2-\tfrac{d}{2})}{(M^2)^{2-d/2}}. \tag{18.171}$$

From this result, we can derive the Callan-Symanzik γ function by the use of (18.23) and the field strength renormalization counterterm (18.9). We find

$$\gamma_f^n = \frac{8}{3} \frac{g^2}{(4\pi)^2} \left[1 + 4 \sum_{2}^{n} \frac{1}{j} - \frac{2}{n(n+1)} \right]. \tag{18.172}$$

Notice that this expression vanishes for $n = 1$, so that there is no rescaling of A_f^1, as required by (18.157). For $n > 1$, γ_f^n is positive and so its coefficient a_f^n is negative. This implies that the higher moments of the quark distribution functions are suppressed as Q^2 becomes large.

Operator Mixing

The QCD rescaling of the operators $\mathcal{O}_f^{(n)}$ is still more complicated because QCD contains additional twist-2 operators which can be built from gluon fields. These new operators are mixed with the quark twist-2 operators by the diagrams of Fig. 18.14.

For n even, the diagrams of Fig. 18.14 give the operators $\mathcal{O}_f^{(n)}$ matrix elements in the state of a gluon with momentum p. The tensor structure of

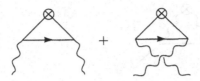

Figure 18.14. Diagrams that produce operator mixing between twist-2 quark and gluon operators.

this matrix element contains the term

$$g^{\alpha\beta}p^{\mu_1}\cdots p^{\mu_n}, \tag{18.173}$$

where α, β are the polarization indices of the external gluons. This structure arises from the operator

$$\mathcal{O}_g^{(n)\mu_1\cdots\mu_n} = -F^{\{\mu_1\nu}(iD^{\mu_2})\cdots(iD^{\mu_{n-1}})F^{\mu_n\}}{}_\nu - \text{traces}, \tag{18.174}$$

symmetrized on $\mu_1,\cdots\mu_n$, with traces subtracted. These operators have dimension $(n+2)$ and spin n, and thus have twist 2.

The gluon operators (18.174) are relevant only for n even. Using the manipulation

$$F^{\mu_1\nu}(iD^{\mu_2})\cdots F^{\mu_n}{}_\nu = i\partial^{\mu_2}\left(F^{\mu_1\nu}\cdots F^{\mu_n}{}_\nu\right) - (iD^{\mu_2})F^{\mu_1\nu}\cdots F^{\mu_n}{}_\nu, \tag{18.175}$$

we can transfer the covariant derivatives from one factor of $F^{\mu\nu}$ to the other, giving

$$\mathcal{O}_g^{(n)} = (-1)^n\mathcal{O}_g^{(n)} + \partial^{\mu_1}(\mathcal{O}'). \tag{18.176}$$

Thus, for n odd, the operator $\mathcal{O}_g^{(n)}$ is equal to a total derivative. The matrix elements of a total derivative are proportional to the momentum injected into this operator. Since zero momentum is injected in the calculation of the proton matrix elements of the OPE of currents, the operators $\mathcal{O}_g^{(n)}$ have no effect on the deep inelastic scattering cross section for n odd.

For n even, however, we must take account of the mixing of $\mathcal{O}_g^{(n)}$ with $\mathcal{O}_f^{(n)}$. The computation of the diagrams of Fig. 18.14 is quite similar to the other operator rescaling calculations we have done in this chapter, and so we reserve working out the details for Problem 18.3. We find that the diagrams of Fig. 18.14 contain a structure proportional to (18.173) with the coefficient

$$\frac{2(n^2+n+2)}{n(n+1)(n+2)}\frac{g^2}{(4\pi)^2}\Gamma(2-\tfrac{d}{2}). \tag{18.177}$$

From this computation, we find that the renormalized twist-2 quark operator, properly normalized at the scale M, is given in terms of bare operators by

$$\left[\mathcal{O}_f^{(n)}\right]_M = (1+\delta_f)\left[\mathcal{O}_f^{(n)}\right]_0 + (\delta_g)\left[\mathcal{O}_g^{(n)}\right]_0, \tag{18.178}$$

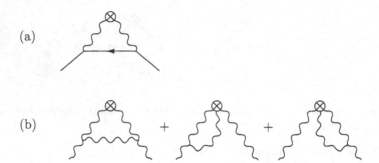

Figure 18.15. Diagrams contributing to the operator rescaling of twist-2 gluon operators: (a) contributions to gluon-quark mixing; (b) contributions to diagonal gluon operator renormalization.

where δ_f is given by (18.171) and

$$\delta_g = -\frac{g^2}{(4\pi)^2}\frac{2(n^2+n+2)}{n(n+1)(n+2)}\frac{\Gamma(2-\frac{d}{2})}{(M^2)^{2-d/2}}. \tag{18.179}$$

This equation gives us two elements of the anomalous dimension matrix of twist-2 operators.

The remaining elements of the γ matrix for twist-2 operators are generated by the diagrams shown in Fig. 18.15. The diagram of Fig. 18.15(a) gives the mixing of $\mathcal{O}_g^{(n)}$ back into $\mathcal{O}_f^{(n)}$. The diagrams of Fig. 18.15(b), combined with the counterterm δ_3 for gluon field strength rescaling, gives the diagonal anomalous dimension. The counterterm δ_3 is given explicitly, in Feynman-'t Hooft gauge, in (16.74). The remainder of this anomalous dimension computation is discussed in Problem 18.3.

To describe the complete anomalous dimension matrix, we begin by considering a strong interaction model with one quark flavor. In this case, there is one twist-two operator $\mathcal{O}_f^{(n)}$ which mixes with $\mathcal{O}_g^{(n)}$. These two operators mix through a 2×2 matrix

$$\gamma^n = -\frac{g^2}{(4\pi)^2}\begin{pmatrix} a_{ff}^n & a_{fg}^n \\ a_{gf}^n & a_{gg}^n \end{pmatrix}, \tag{18.180}$$

where

$$a_{ff}^n = -\frac{8}{3}\Big[1 + 4\sum_2^n \frac{1}{j} - \frac{2}{n(n+1)}\Big],$$

$$a_{fg}^n = 4\frac{n^2+n+2}{n(n+1)(n+2)},$$

$$a_{gf}^n = \frac{16}{3}\frac{n^2+n+2}{n(n^2-1)}, \tag{18.181}$$

$$a_{gg}^n = -6\Big[\frac{1}{3} + \frac{2}{9}n_f + 4\sum_2^n \frac{1}{j} - \frac{4}{n(n-1)} - \frac{4}{(n+1)(n+2)}\Big].$$

Notice that this matrix is not symmetric. In the last line, n_f is the number of quark flavors, equal to 1 in this case; this term comes from (16.74).

In the realistic case, QCD contains several quark flavors—u, d, s, and also c and b when we work at momenta sufficiently large that we can ignore the masses of these particles. Then the anomalous dimension matrix γ^n has size $(n_f + 1) \times (n_f + 1)$. The submatrix acting on quark operators is diagonal, with all of the diagonal entries being given by a_{ff}^n in (18.181). The quark-gluon and gluon-quark entries are all given by a_{fg}^n and a_{gf}^n, respectively, and are independent of the flavor. The gluon diagonal entry is given by a_{gg}^n in (18.181) with the realistic value of n_f. This means that the gluon operator mixes with only one linear combination of quark operators:

$$\sum_f \mathcal{O}_f^{(n)}; \tag{18.182}$$

the orthogonal linear combinations are simply rescaled, with the exponent given by a_{ff}^n or (18.172).

Let us now apply this analysis of operator mixing to the evaluation of the moment sum rules. For odd n, there is no operator mixing, and so the Q^2 dependence of the right-hand side of (18.155) is correctly given by the simple rescaling (18.160).

For even n, we must take operator mixing into account. The right-hand side of the sum rule (18.154) is the proton matrix element of a twist-2 operator normalized at the scale Q. Let us write an arbitrary linear combination of these operators as

$$c_i^n \left[\mathcal{O}_i^{(n)}\right]_Q, \tag{18.183}$$

where the index i runs over g and the various flavors f. To rescale this operator to a fixed reference momentum μ, we rewrite the coefficients in a basis of *left* eigenvectors of γ^n and rescale each eigenvector acccording to (18.159). In terms of the matrix a_{ij}^n of rescaling coefficients, we can write the rescaling abstractly as

$$c_i^n \left[\mathcal{O}_i^{(n)}\right]_Q = c_i^n \left\{ \left(\frac{\log(Q^2/\Lambda^2)}{\log(\mu^2/\Lambda^2)} \right)^{a^n/2b_0} \right\}_{ij} \left[\mathcal{O}_j^{(n)}\right]_\mu. \tag{18.184}$$

This rescaling, acting with c_i^n to the left of the matrix (a^n), is precisely the prescription required by Eq. (18.79).

Let us work this out explicitly for the case $n = 2$. The right-hand side of the moment sum rule (18.154) is given by the matrix element of $\mathcal{O}_f^{(2)}$. We rewrite this as

$$\mathcal{O}_f^{(2)} = \left(\mathcal{O}_f^{(2)} - \frac{1}{n_f} \sum_{f'} \mathcal{O}_{f'}^{(2)}\right) + \frac{1}{n_f} \sum_{f'} \mathcal{O}_{f'}^{(2)}. \tag{18.185}$$

The first term is simply rescaled; the second term mixes with the gluon operator $\mathcal{O}_g^{(2)}$. The anomalous dimension matrix acting on $(\sum_f \mathcal{O}_f, \mathcal{O}_g)$ for $n = 2$

has coefficients

$$\begin{pmatrix} a_{ff}^2 & a_{fg}^2 n_f \\ a_{gf}^2 & a_{gg}^2 \end{pmatrix} = \begin{pmatrix} -\frac{64}{9} & \frac{4}{3} n_f \\ \frac{64}{9} & -\frac{4}{3} n_f \end{pmatrix}. \tag{18.186}$$

The left eigenvectors of this matrix, and their corresponding eigenvalues, are

$$\begin{aligned} (1,1) & \rightarrow & a^2 = 0 \\ \left(\frac{16}{3}, -n_f\right) & \rightarrow & a^2 = -\frac{4}{3}\left(\frac{16}{3} + n_f\right). \end{aligned} \tag{18.187}$$

Notice that the first eigenvector gives a linear combination of operators $c_i^2 \mathcal{O}_i^{(2)}$ with zero anomalous dimension. This operator is in fact the total energy momentum tensor of QCD,

$$T^{\mu\nu} = \sum_f \mathcal{O}_f^{(2)\,\mu\nu} + \mathcal{O}_g^{(2)\,\mu\nu}, \tag{18.188}$$

which must have $\gamma = 0$. If we expand the second term in (18.185) in terms of the components (18.187), we can compute the full form of the operator rescaling. We find

$$\begin{aligned} [\mathcal{O}_f^{(2)}]_Q = {} & \frac{1}{16/3 + n_f} T \\ & + \frac{1}{n_f(\frac{16}{3} + n_f)} \left(\frac{\log(Q^2/\Lambda^2)}{\log(\mu^2/\Lambda^2)}\right)^{-\frac{4}{3}(\frac{16}{3} + n_f)/2b_0} \left[\frac{16}{3}\sum_f \mathcal{O}_f^{(2)} - n_f \mathcal{O}_g^{(2)}\right]_\mu \\ & + \left(\frac{\log(Q^2/\Lambda^2)}{\log(\mu^2/\Lambda^2)}\right)^{-32/9b_0} \left[\mathcal{O}_f^{(2)} - \frac{1}{n_f}\sum_{f'} \mathcal{O}_{f'}^{(2)}\right]_\mu, \end{aligned} \tag{18.189}$$

where T is the energy-momentum tensor (18.188). The right-hand side of the $n = 2$ moment sum rule is given by the coefficient of the proton matrix element of this operator. To evaluate this coefficient, we need to define gluon analogues of the A_f^n, by writing, analogously to (18.134),

$$\langle P | \mathcal{O}_g^{(n)\mu_1\cdots\mu_n} | P \rangle = A_g^n \cdot 2P^{\mu_1}\cdots P^{\mu_n} - \text{traces}. \tag{18.190}$$

For the case $n = 2$, we note in particular that

$$\langle P | T^{\mu\nu} | P \rangle = 2P^\mu P^\nu; \tag{18.191}$$

thus, (18.188) implies

$$\sum_f A_f^2 + A_g^2 = 1. \tag{18.192}$$

If we replace each operator in (18.189) by the corresponding coefficient $A_i^2(\mu)$, we will have an expression for the right-hand side of the $n = 2$ moment sum rule which makes its Q^2 dependence explicit.

Although expression (18.189) is rather complicated, it has a simple form in the extreme limit $Q^2 \to \infty$. At asymptotic Q^2, the last two terms of (18.189)

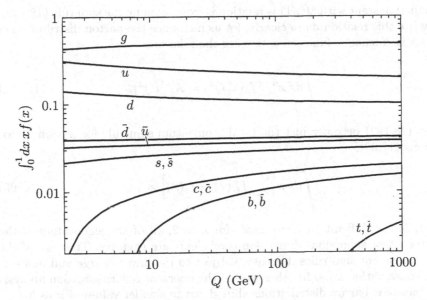

Figure 18.16. Fractions of the total energy-momentum of the proton carried by various parton species, as a function of Q, according to the CTEQ fit to deep inelastic scattering data described in Fig. 17.6. The Q dependence of the curves is calculated from the QCD evolution equations.

tend to zero, and the right-hand side of (18.189) becomes a fixed number times the energy-momentum tensor. Then, using (18.191), we can evaluate the $n = 2$ moment sum rule completely:

$$\int_0^1 dx\, x f_f^+(x) \to \frac{1}{16/3 + n_f}. \qquad (18.193)$$

In this extreme limit, we find that each quark flavor carries the same fixed fraction of the energy-momentum of the proton. By (18.192), the remainder is carried by the gluons. To illustrate, in a theory with $n_f = 4$, each quark flavor carries 3/28 of the total momentum of the proton, and the gluons carry the remaining 4/7. Figure 18.16 shows how slowly these asymptotic results are approached starting from realistic parton distributions.

Relation to the Altarelli-Parisi Equations

The operator mixing analysis just described gives predictions for the moments of parton distributions which imply that these integrals are Q^2 dependent. Of the various moment integrals that do not involve operator mixing, only the $n = 1$ integrals which give the flavor quantum numbers of the proton are constant as a function of Q^2. The rest decrease as powers of $\log Q^2$. Similarly, one linear combination of the matrix elements of $n = 2$ twist-2 operators remains constant with Q^2. This relation is expression by the sum rule (18.192). To write this relation more clearly, let us introduce the parton distribution of gluons as a smooth function satisfying the relations

$$\int_0^1 dx\, x^{n-1} f_g(x, Q^2) = A_g^{(n)}(Q^2).$$
(18.194)

Then (18.192) becomes just the total momentum sum rule for parton distributions (17.39):

$$\int_0^1 dx\, x \left[\sum_f f_f^+(x) + f_g(x) \right] = 1.$$
(18.195)

It is not difficult to verify that, for $n > 2$, all of the eigenvalues of the matrix a_{ij}^n of anomalous dimension coefficients are negative. Thus, all of the higher moment sum rules decrease, subject to the flavor charge and momentum conservation laws. In other words, the operator renormalization analysis predicts that parton distributions shift down to smaller values of x as $\log Q^2$ increases. It is pleasing that this is the same conclusion that we reached in Section 17.5, where we derived the Altarelli-Parisi equations to describe this evolution of the parton distributions.

Given that the operator analysis and the Altarelli-Parisi equations imply the same qualitative behavior for the parton distributions, how do these analyses compare quantitatively? To compare them directly, we should work out what predictions the Altarelli-Parisi equations make for the moments of the parton distribution functions. Let us begin with the simpler case of $f_f^-(x) = f_f(x) - f_{\bar f}(x)$.

To find the Altarelli-Parisi equation for this quantity, subtract the last two equations of (17.128). The term involving the gluon distribution cancels, and we find

$$\frac{d}{d\log Q^2} f_f^-(x) = \frac{\alpha_s(Q^2)}{2\pi} \int_x^1 \frac{dz}{z} P_{q \leftarrow q}(z) f_f^- \left(\frac{x}{z} \right).$$
(18.196)

Now define

$$M_{fn}^- = \int_0^1 dx\, x^{n-1} f_f^-(x).$$
(18.197)

This quantity obeys the differential equation

$$\frac{d}{d\log Q^2} M_{fn}^- = \frac{\alpha_s(Q^2)}{2\pi} \int_0^1 dx\, x^{n-1} \int_x^1 \frac{dz}{z} P_{q \leftarrow q}(z) f_f^- \left(\frac{x}{z} \right).$$
(18.198)

Interchange the order of integration on the right-hand side, and change variables to $y = x/z$:

$$\int_0^1 dx\, x^{n-1} \int_x^1 \frac{dz}{z} = \int_0^1 \frac{dz}{z} \int_0^z dx\, x^{n-1}$$

$$= \int_0^1 \frac{dz}{z} \int_0^1 dy\, y^{n-1} z^n$$

$$= \int_0^1 dz\, z^{n-1} \int_0^1 dy\, y^{n-1}. \qquad (18.199)$$

Then the right-hand side of the differential equation neatly factorizes:

$$\frac{d}{d\log Q^2} M_{nf}^- = \frac{\alpha_s(Q^2)}{2\pi} \left[\int_0^1 dz\, z^{n-1} P_{q \leftarrow q}(z) \right] \cdot \int_0^1 dy\, y^{n-1} f_f^-(y); \qquad (18.200)$$

the last factor is again M_{fn}^-. The coefficient in this relation is the nth moment of the splitting function $P_{q \leftarrow q}(z)$. We can compute this from the explicit form of this function given in (17.129):

$$\int_0^1 dz\, z^{n-1} P_{q \leftarrow q}(z) = \int_0^1 dz\, z^{n-1} \frac{4}{3} \left[\frac{1+z^2}{(1-z)_+} + \frac{3}{2}\delta(1-z) \right]. \qquad (18.201)$$

The integral over the distribution is done by using the definition (17.105):

$$\int_0^1 dz\, z^{n-1} \frac{1}{(1-z)_+} = \int_0^1 dz\, \frac{z^{n-1}-1}{(1-z)}$$

$$= \int_0^1 dz \left(-1 - z - \cdots - z^{n-2} \right)$$

$$= -\sum_1^{n-1} \frac{1}{j}. \qquad (18.202)$$

Then

$$\int_0^1 dz\, z^{n-1} P_{q \leftarrow q}(z) = -\frac{4}{3} \left[\sum_1^{n-1} \frac{1}{j} + \sum_1^{n+1} \frac{1}{j} - \frac{3}{2} \right] \qquad (18.203)$$

$$= -\frac{2}{3} \left[1 + 4\sum_2^n \frac{1}{j} - \frac{2}{n(n+1)} \right].$$

Remarkably, this is just $a_f^n/4$, as the anomalous dimension coefficient is given in (18.172) or (18.181). Thus, according to the Altarelli-Parisi equations, the nth moment of $f_f^-(x)$ obeys

$$\frac{d}{d\log Q^2} M_{fn}^- = \frac{\alpha_s(Q^2)}{8\pi} a_f^n \cdot M_{fn}^-. \tag{18.204}$$

To integrate this equation, we need the explicit form of $\alpha_s(Q^2)$. Inserting expression (17.17), we find

$$\frac{d}{d\log Q^2} M_{fn}^- = \frac{a_f^n}{2b_0} \frac{1}{\log(Q^2/\Lambda^2)} M_{fn}^-. \tag{18.205}$$

The solution of this equation, derived from the Altarelli-Parisi equations, is precisely the function (18.160) that we derived from the operator analysis of the moment sum rules for f_f^-.

It is not difficult to check that this conclusion is more general. By taking the nth moment of the full Altarelli-Parisi equations (17.128), we convert these equations to a set of ordinary differential equations for the moments. The linear combination of quark distribution functions

$$\sum_f (f_f(x) + f_{\bar{f}}(x)) \tag{18.206}$$

couples to the gluon distribution and leads to a 2×2 set of equations. All orthogonal linear combinations separate from the gluon distribution and thus have moments that obey equations identical to (18.205). To analyze the coupled equations, define

$$M_n^+ = \int_0^1 dx\, x^{n-1} \sum_f (f_f(x) + f_{\bar{f}}(x)); \qquad M_{gn} = \int_0^1 dx\, x^{n-1} f_g(x). \tag{18.207}$$

Then one can show, by the manipulations that led to (18.205), that the Altarelli-Parisi equations predict for these moments the set of coupled equations

$$\frac{d}{d\log Q^2} M_n^+ = \frac{1}{2b_0} \frac{1}{\log(Q^2/\Lambda^2)} \left[a_{ff}^n M_n^+ + a_{fg}^n M_{gn} \right],$$
$$\frac{d}{d\log Q^2} M_{gn} = \frac{1}{2b_0} \frac{1}{\log(Q^2/\Lambda^2)} \left[n_f a_{gf}^n M_n^+ + a_{gg}^n M_{gn} \right], \tag{18.208}$$

where the coefficients a_{ij}^n are proportional to the nth moments of the splitting functions given in (17.129) and (17.130). In all cases, one can see that these coefficients agree precisely with the corresponding coefficients in (18.181). Thus, the solution of these equations gives the same Q^2 dependence for the moments of parton distribution functions that we found from the operator analysis.

Remarkably, the analysis of parton splitting functions given in Chapter 17 and the analysis of operator renormalization factors given above have turned out to be two views of the same basic phenomenon. Both sets of equations express the manner in which the constituents of hadrons in QCD are resolved, layer by layer, by hard-scattering processes at successively higher values of the momentum transfer. Our understanding that a quark, when studied on a fine scale, is resolved into a set of quarks, antiquarks, and gluons indicates that we have gone far beyond the simple notions of one-particle relativistic mechanics. Our two complementary derivations of this idea reinforce its fundamental character as a prediction of quantum field theory. It is especially pleasing that, as we saw at the end of Section 17.5, Nature apparently accepts this prediction and makes this consequence of quantum field theory an essential part of the structure of hadrons.

Problems

18.1 Matrix element for proton decay. Some advanced theories of particle interactions include heavy particles X whose couplings violate the conservation of baryon number. Integrating out these particles produces an effective interaction that allows the proton to decay to a positron and a photon or a pion. This effective interaction is most easily written using the definite-helicity components of the quark and electron fields: If u_L, d_L, u_R, e_R are two-component spinors, then this effective interaction is

$$\Delta \mathcal{L} = \frac{2}{m_X^2} \epsilon_{abc} \epsilon^{\alpha\beta} \epsilon^{\gamma\delta} e_{R\alpha} u_{R\alpha\beta} u_{Lb\gamma} d_{Lc\delta}.$$

A typical value for the mass of the X boson is $m_X = 10^{16}$ GeV.

(a) Estimate, in order of magnitude, the value of the proton lifetime if the proton is allowed to decay through this interaction.

(b) Show that the three-quark operator in $\Delta \mathcal{L}$ has an anomalous dimension

$$\gamma = -4 \frac{g^2}{(4\pi)^2}.$$

Estimate the enhancement of the proton decay rate due to the leading QCD corrections.

18.2 Parity-violating deep inelastic form factor. In this problem, we first motivate the presence of additional deep inelastic form factors that are proportional to *differences* of quark and antiquark distribution functions. Then we define these functions formally and work out their properties.

(a) Analyze neutrino-proton scattering following the method used at the beginning of Section 18.5. Define

$$J_+^\mu = \bar{u}\gamma^\mu \left(\frac{1-\gamma^5}{2}\right) d, \qquad J_-^\mu = \bar{d}\gamma^\mu \left(\frac{1-\gamma^5}{2}\right) u.$$

Let

$$W^{\mu\nu(\nu)} = 2i \int d^4x \, e^{iq\cdot x} \, \langle P| \, T\{J_-^\mu(x) J_+^\nu(0)\} \, |P\rangle,$$

averaged over the proton spin. Show that the cross section for deep inelastic neutrino scattering can be computed from $W^{\mu\nu(\nu)}$ according to

$$\frac{d^2\sigma}{dxdy}(\nu p \to \mu^- X) = \frac{G_F^2 y}{2\pi^2}$$

$$\cdot \operatorname{Im}[(k_\mu k_\nu' + k_\nu k_\mu' - g_{\mu\nu} k \cdot k' - i\epsilon_{\mu\nu}{}^{\alpha\beta} k_\alpha' k_\beta) W^{\mu\nu(\nu)}(P,q)].$$

(b) Show that any term in $W^{\mu\nu(\nu)}$ proportional to q^μ or q^ν gives zero when contracted with the lepton momentum tensor in the formula above. Thus we can expand $W^{\mu\nu(\nu)}$ with three scalar form factors,

$$W^{\mu\nu(\nu)} = -g^{\mu\nu} W_1^{(\nu)} + P^\mu P^\nu W_2^{(\nu)} + i\epsilon^{\mu\nu\lambda\sigma} P_\lambda q_\sigma W_3^{(\nu)} + \cdots,$$

where the additional terms do not contribute to the deep inelastic cross section. Find the formula for the deep inelastic cross section in terms of the imaginary parts of $W_1^{(\nu)}$, $W_2^{(\nu)}$, and $W_3^{(\nu)}$.

(c) Evaluate the form factors $W_i^{(\nu)}$ in the parton model, and show that

$$\operatorname{Im} W_1^{(\nu)} = \pi(f_d(x) + f_{\bar{u}}(x)),$$

$$\operatorname{Im} W_2^{(\nu)} = \frac{4\pi}{ys} x(f_d(x) + f_{\bar{u}}(x)),$$

$$\operatorname{Im} W_3^{(\nu)} = \frac{2\pi}{ys}(f_d(x) - f_{\bar{u}}(x)).$$

Insert these expressions into the formula derived in part (b) and show that the result reproduces the first line of Eq. (17.35).

(d) This analysis motivates the following definition: For a single quark flavor f, let

$$J_{fL}^\mu = \bar{f}\gamma^\mu \left(\frac{1-\gamma^5}{2}\right) f.$$

Define

$$W_{fL}^{\mu\nu} = 2i \int d^4x \, e^{iq\cdot x} \langle P| T\{J_{fL}^\mu(x) J_{fL}^\nu(0)\} |P\rangle .$$

Decompose this tensor according to

$$W_{fL}^{\mu\nu} = -g^{\mu\nu} W_{1fL} + P^\mu P^\nu W_{2fL} + i\epsilon^{\mu\nu\lambda\sigma} P_\lambda q_\sigma W_{3fL} + \cdots,$$

where the remaining terms are proportional to q^μ or q^ν. Evaluate the W_{iL} in the parton model. Show that the quantities W_{1fL} and W_{2fL} reproduce the expressions for W_{1f} and W_{2f} given by Eqs. (18.120) and (18.144), and that W_{3fL} is given by

$$\operatorname{Im} W_{3fL} = \frac{2\pi}{ys}(f_f(x) - f_{\bar{f}}(x)).$$

(e) Compute the operator product of the currents in the expression for $W_{fL}^{\mu\nu}$, and write the terms in this product that involve twist-2 operators. Show that the expressions for W_{1fL} and W_{2fL} that follow from this analysis reproduce the expressions for W_{1f} and W_{2f} given by Eqs. (18.144) and (18.145). Find the corresponding expression for W_{3fL}.

(f) Define the parton distribution $f_{\bar{f}}$ by the relation

$$f_{\bar{f}}(x, Q^2) = \frac{ys}{2\pi} \operatorname{Im} W_{3fL}(x, Q^2).$$

Show that, by virtue of this definition, the distribution function $f_{\bar{f}}$ satisfies the sum rule (18.155) for odd n.

18.3 Anomalous dimensions of gluon twist-2 operators.

(a) Compute the divergent parts of the diagrams in Fig. 18.14, and use these to derive the second line of Eq. (18.181). Notice that this result holds only for n even. Show that the two diagrams cancel for n odd.

(b) Compute the divergent parts of the diagrams in Fig. 18.5, and use these to derive the third and fourth lines of Eq. (18.181).

18.4 Deep inelastic scattering from a photon. Consider the problem of deep-inelastic scattering of an electron from a photon. This process can actually be measured by analyzing the reaction $e^+e^- \to e^+e^- + X$ in the regime where the positron goes forward, with emission of a collinear photon, which then has a hard reaction with the electron. Let us analyze this process to leading order in QED and to leading-log order in QCD. To predict the photon structure functions, it is reasonable to integrate the renormalization group equations with the initial condition that the parton distribution for photons in the photon is $\delta(x - 1)$ at $Q^2 = (\frac{1}{2} \text{ GeV})^2$. Take $\Lambda - 150$ MeV. Assume for simplicity that there are four flavors of quarks, u, d, c, and s, with charges $2/3$, $-1/3$, $2/3$, $-1/3$, respectively, and that it is always possible to ignore the masses of these quarks.

(a) Use the Altarelli-Parisi equations to compute the parton distributions for quarks and antiquarks in the photon, to leading order in QED and to zeroth order in QCD. Compute also the probability that the photon remains a photon as a function of Q^2.

(b) Formulate the problem of computing the moments of W_2 for the photon as a problem in operator mixing. Compute the relevant anomalous dimension matrix γ. You should be able to assemble this matrix from familiar ingredients without doing further Feynman diagram computations.

(c) Compute the $n = 2$ moments of the photon structure functions as a function of Q^2.

(d) Describe qualitatively the evolution of the photon structure function as a function of x and Q^2.

Chapter 19

Perturbation Theory Anomalies

In many examples, we have seen that loop corrections can have an important effect on the predictions of quantum field theory. We have studied examples in which the relative importance of operators is shifted by radiative corrections, and in which the form of the interactions they mediate is altered. However, in specific circumstances, radiative corrections can have an even more significant effect: They can destroy symmetries of the classical equations of motion.

The most important effect of this type involves the chiral symmetries of theories with massless fermions. In Section 3.4, we saw that the massless Dirac Lagrangian has an enhanced symmetry associated with the separate number conservation of left- and right-handed fermions. This symmetry is generated by the axial vector current $j^{\mu 5} = \bar{\psi}\gamma^\mu \gamma^5 \psi$. Classically,

$$\partial_\mu j^{\mu 5} = 0 \tag{19.1}$$

for zero-mass fermions. This equation of motion is true not only in free fermion theory but also, as a classical field equation, in massless QED and QCD.

However, in this chapter, we will see that the true picture is not so simple. We will show that, in gauge theories, the conservation of the axial vector current is actually incompatible with gauge invariance, and that radiative corrections in gauge theories supply a nonzero operator that appears on the right-hand side of Eq. (19.1). This new conservation equation for the axial current has a number of remarkable consequences, which we will discuss in Sections 19.3 and 19.4.

19.1 The Axial Current in Two Dimensions

Eventually, we will want to analyze the current conservation equation for the axial current in massless QCD. However, this discussion will involve some technical complication, so we will first study the physics that violates axial current conservation in a context in which the calculations are relatively simple. A particularly simple model problem is that of two-dimensional massless QED.

The Lagrangian of two-dimensional QED is

$$\mathcal{L} = \bar{\psi}(i\slashed{D})\psi - \tfrac{1}{4}(F_{\mu\nu})^2, \tag{19.2}$$

with $\mu, \nu = 0, 1$ and $D_\mu = \partial_\mu + ieA_\mu$. The Dirac matrices must be chosen to satisfy the Dirac algebra

$$\{\gamma^\mu, \gamma^\nu\} = 2g^{\mu\nu}. \tag{19.3}$$

In two dimensions, this set of relations can be represented by 2×2 matrices; we choose

$$\gamma^0 = \begin{pmatrix} 0 & -i \\ i & 0 \end{pmatrix}, \qquad \gamma^1 = \begin{pmatrix} 0 & i \\ i & 0 \end{pmatrix}. \tag{19.4}$$

The Dirac spinors will be two-component fields.

The product of the Dirac matrices, which anticommutes with each of the γ^μ, is

$$\gamma^5 = \gamma^0\gamma^1 = \begin{pmatrix} 1 & 0 \\ 0 & -1 \end{pmatrix}. \tag{19.5}$$

Then, just as in four dimensions, there are two possible currents,

$$j^\mu = \bar{\psi}\gamma^\mu\psi, \qquad j^{\mu 5} = \bar{\psi}\gamma^\mu\gamma^5\psi, \tag{19.6}$$

and both are conserved if there is no mass term in the Lagrangian.

To make the conservation laws quite explicit, we label the components of the fermion field ψ in this spinor basis as

$$\psi = \begin{pmatrix} \psi_+ \\ \psi_- \end{pmatrix}. \tag{19.7}$$

The subscript indicates the γ^5 eigenvalue. Then, using the explicit representations (19.4) and (19.7), we can rewrite the fermionic part of (19.2) as

$$\mathcal{L} = \psi_+^\dagger i(D_0 + D_1)\psi_+ + \psi_-^\dagger i(D_0 - D_1)\psi_-. \tag{19.8}$$

In the free theory, the field equation of ψ_+ would be

$$i(\partial_0 + \partial_1)\psi_+ = 0; \tag{19.9}$$

the solutions to this equation are waves that move to the right in the one-dimensional space at the speed of light. We will thus refer to the particles associated with ψ_+ as *right-moving* fermions. The quanta associated with ψ_- are, similarly, *left-moving*. This distinction is analogous to the distinction between left- and right-handed particles which gives the physical interpretation of γ^5 in four dimensions. Since the Lagrangian (19.8) contains no terms that mix left- and right-moving fields, it seems obvious that the number currents for these fields are separately conserved. Thus,

$$\partial_\mu\left(\bar{\psi}\gamma^\mu\left(\frac{1-\gamma^5}{2}\right)\psi\right) = 0, \qquad \partial_\mu\left(\bar{\psi}\gamma^\mu\left(\frac{1+\gamma^5}{2}\right)\psi\right) = 0. \tag{19.10}$$

It is a curious property of two-dimensional spacetime that the vector and axial vector fermionic currents are not independent of each other. Let $\epsilon^{\mu\nu}$ be the totally antisymmetric symbol in two dimensions, with $\epsilon^{01} = +1$. Then the two-dimensional Dirac matrices obey the identity

$$\gamma^\mu \gamma^5 = -\epsilon^{\mu\nu} \gamma_\nu. \tag{19.11}$$

The currents $j^{\mu 5}$ and j^μ have the same relation. Thus we can study the properties of the axial vector current by using results that we have already derived for the vector current.

Vacuum Polarization Diagrams

In Section 7.5, we computed the lowest-order vacuum polarization of QED in dimensional regularization. In the limit of zero mass, we found, in Eq. (7.90),

$$i\Pi^{\mu\nu}(q) = -i(q^2 g^{\mu\nu} - q^\mu q^\nu) \frac{2e^2}{(4\pi)^{d/2}} \operatorname{tr}[1] \int\limits_0^1 dx\, x(1-x) \frac{\Gamma(2-\frac{d}{2})}{(-x(1-x)q^2)^{2-d/2}}, \tag{19.12}$$

where $\operatorname{tr}[1] = 4$ gives the convention for tracing over Dirac matrices given in Eq. (7.88). If we set $\operatorname{tr}[1] = 2$ to be consistent with (19.4) and then set $d = 2$ in (19.12), we find the finite and well-defined result

$$
\begin{aligned}
i\Pi^{\mu\nu}(q) &= i(q^2 g^{\mu\nu} - q^\mu q^\nu) \frac{2e^2}{4\pi} \cdot 2 \cdot \frac{1}{q^2} \\
&= i\left(g^{\mu\nu} - \frac{q^\mu q^\nu}{q^2}\right) \frac{e^2}{\pi}.
\end{aligned} \tag{19.13}
$$

Notice that this expression has the structure of a photon mass term; the photon receives the mass

$$m_\gamma^2 = \frac{e^2}{\pi}. \tag{19.14}$$

Schwinger showed that this result is exact, and that the photon of two-dimensional QED is a free massive boson.* In the discussion below Eq. (7.72), we pointed out that it is not possible for a vacuum polarization amplitude consistent with the Ward identity to generate a mass for the photon unless it also contains a pole at $q^2 = 0$. In two dimensions, such a pole can arise from the infrared behavior of the fermion-antifermion intermediate state, and we see this behavior explicitly in (19.13).

Once we have an explicit expression for the vacuum polarization, we can find the expectation value of the current induced by a background electromagnetic field. This quantity is generated by the diagram of Fig. 19.1, which gives

$$\int d^2x\, e^{iq\cdot x} \langle j^\mu(x) \rangle = \frac{i}{e}\left(i\Pi^{\mu\nu}(q)\right) A_\nu(q) = -\left(g^{\mu\nu} - \frac{q^\mu q^\nu}{q^2}\right) \cdot \frac{e}{\pi} A_\nu(q), \tag{19.15}$$

---

*J. Schwinger, *Phys. Rev.* **128**, 2425 (1962).

Figure 19.1. Computation of $\langle j^\mu \rangle$ in a background electromagnetic field.

where $A_\nu(q)$ is the Fourier transform of the background field. This quantity manifestly satisfies the current conservation relation $q_\mu \langle j^\mu(q) \rangle = 0$.

The identity (19.11) between the vector and axial vector currents allows us to derive from (19.15) the corresponding expectation value of $j^{\mu 5}$. We find

$$\langle j^{\mu 5}(q) \rangle = -\epsilon^{\mu\nu} \langle j_\nu(q) \rangle$$
$$= \epsilon^{\mu\nu} \frac{e}{\pi} \left(A_\nu(q) - \frac{q_\nu q^\lambda}{q^2} A_\lambda(q) \right). \tag{19.16}$$

If the axial vector current were conserved, this object would satisfy the Ward identity. Instead, we find

$$q_\mu \langle j^{\mu 5}(q) \rangle = \frac{e}{\pi} \epsilon^{\mu\nu} q_\mu A_\nu(q). \tag{19.17}$$

This is the Fourier transform of the field equation

$$\partial_\mu j^{\mu 5} = \frac{e}{2\pi} \epsilon^{\mu\nu} F_{\mu\nu}. \tag{19.18}$$

Apparently, the axial vector current is not conserved in the presence of electromagnetic fields, as the result of an anomalous behavior of its vacuum polarization diagram.

How could this happen? The Feynman diagrams formally satisfy the Ward identity both for the vector and for the axial vector current. The problem must come in the regularization of the vacuum polarization diagram. By dimensional analysis, we know that this diagram has the form

$$\text{ } = ie^2 \left(Ag^{\mu\nu} - B\frac{q^\mu q^\nu}{q^2} \right). \tag{19.19}$$

The coefficient B is a finite integral, and is, in any event, unambiguously determined by the low-energy structure of the theory since it is the residue of the pole in q^2. However, the integral A is logarithmically divergent, so its value depends on the regularization. Dimensional regularization automatically subtracts this integral to set $A = B$; then the vector current Ward identity is satisfied. But then we are led directly to (19.17). We could, alternatively, regularize the integral A so that $A = 0$. Working through the steps of the previous paragraph with this modification, we now find $q_\mu \langle j^{\mu 5}(q) \rangle = 0$, but

$$q_\mu \langle j^\mu(q) \rangle = \frac{e}{\pi} q^\nu A_\nu(q). \tag{19.20}$$

Though the result (19.17) is unpleasant, the result (19.20) would be a complete disaster, since it depends on the unphysical gauge degrees of freedom of the vector potential. We conclude that it is not possible to regularize two-dimensional QED so that, simultaneously, the theory is gauge invariant and the axial vector current is conserved. The price of requiring gauge invariance is the anomalous nonconservation of the axial current shown in (19.18).

The Axial Vector Current Operator Equation

To understand what happened to the axial current from another viewpoint, we now study the operator equation for the divergence of $j^{\mu 5}$. Varying the Lagrangian (19.2), we find the following equations of motion for the fermion fields:

$$\slashed{\partial}\psi = -ie\slashed{A}\psi, \qquad \partial_\mu \bar{\psi}\gamma^\mu = ie\bar{\psi}\slashed{A}. \tag{19.21}$$

By using these equations of motion in the most straightforward way, it is easy to conclude that $\partial_\mu j^{\mu 5} = 0$. However, a closer look at these manipulations reveals some subtleties, which alter the final conclusion.

The axial vector current is a composite operator built out of fermion fields. In the previous chapter we saw that products of local operators are often singular, so we will define the current by placing the two fermion fields at distinct points separated by a distance ϵ and then carefully taking the limit as the two fields approach each other. Explicitly, we define

$$j^{\mu 5} = \operatorname*{symm\,lim}_{\epsilon \to 0} \left\{ \bar{\psi}(x + \tfrac{\epsilon}{2})\gamma^\mu\gamma^5 \exp\left[-ie \int_{x-\epsilon/2}^{x+\epsilon/2} dz \cdot A(z)\right] \psi(x - \tfrac{\epsilon}{2}) \right\}. \tag{19.22}$$

Notice that, because we have placed ψ and $\bar{\psi}$ at different points, we must introduce a Wilson line (15.53) in order that the operator be locally gauge invariant. To give $j^{\mu 5}$ the correct transformation properties under Lorentz transformations, the limit $\epsilon \to 0$ should be taken symmetrically,

$$\operatorname*{symm\,lim}_{\epsilon \to 0} \left\{ \frac{\epsilon^\mu}{\epsilon^2} \right\} = 0, \qquad \operatorname*{symm\,lim}_{\epsilon \to 0} \left\{ \frac{\epsilon^\mu \epsilon^\nu}{\epsilon^2} \right\} = \frac{1}{d}g^{\mu\nu}, \tag{19.23}$$

with $d = 2$ in this case.

We now compute the divergence of the axial current defined as in (19.22):

$$\partial_\mu j^{\mu 5} = \operatorname*{symm\,lim}_{\epsilon \to 0} \left\{ \left(\partial_\mu \bar{\psi}(x + \tfrac{\epsilon}{2})\right)\gamma^\mu\gamma^5 \exp\left[-ie \int_{x-\epsilon/2}^{x+\epsilon/2} dz \cdot A(z)\right] \psi(x - \tfrac{\epsilon}{2}) \right.$$

$$+ \bar{\psi}(x + \tfrac{\epsilon}{2})\gamma^\mu\gamma^5 \exp\left[-ie \int_{x-\epsilon/2}^{x+\epsilon/2} dz \cdot A(z)\right] \left(\partial_\mu \psi(x - \tfrac{\epsilon}{2})\right)$$

$$\left. + \bar{\psi}(x + \tfrac{\epsilon}{2})\gamma^\mu\gamma^5 \left[-ie\epsilon^\nu \partial_\mu A_\nu(x)\right] \psi(x - \tfrac{\epsilon}{2}) \right\}. \tag{19.24}$$

Using the equations of motion (19.21), and keeping terms up to order ϵ, we can reduce this to

$$\partial_\mu j^{\mu 5} = \text{symm} \lim_{\epsilon \to 0} \left\{ \bar\psi(x + \tfrac{\epsilon}{2}) \left[ie\slashed{A}(x + \tfrac{\epsilon}{2}) - ie\slashed{A}(x - \tfrac{\epsilon}{2}) \right. \right.$$

$$\left. \left. - ie\epsilon^\nu \gamma^\mu \partial_\mu A_\nu(x) \right] \gamma^5 \psi(x - \tfrac{\epsilon}{2}) \right\}$$

$$= \text{symm} \lim_{\epsilon \to 0} \left\{ \bar\psi(x + \tfrac{\epsilon}{2}) \left[-ie\gamma^\mu \epsilon^\nu (\partial_\mu A_\nu - \partial_\nu A_\mu) \right] \gamma^5 \psi(x - \tfrac{\epsilon}{2}) \right\}.$$

$$(19.25)$$

Expression (19.25) seems to vanish in the limit $\epsilon \to 0$. However, we must take account of the fact that the product of the fermion operators is singular. In two dimensions, the contraction of fermion fields is

$$\overline{\psi(y)\bar\psi}(z) = \int \frac{d^2 k}{(2\pi)^2} e^{-ik\cdot(y-z)} \frac{i\slashed{k}}{k^2}$$

$$= -\slashed{\partial} \left(\frac{i}{4\pi} \log(y - z)^2 \right) \qquad (19.26)$$

$$= \frac{-i}{2\pi} \frac{\gamma^\alpha (y - z)_\alpha}{(y - z)^2}.$$

Thus,

$$\overline{\psi(x + \tfrac{\epsilon}{2}) \Gamma \psi}(x - \tfrac{\epsilon}{2}) = \frac{-i}{2\pi} \text{tr}\left[\frac{\gamma^\alpha \epsilon_\alpha}{\epsilon^2} \Gamma \right]. \qquad (19.27)$$

Notice that the result (19.27) contains an extra minus sign from the interchange of fermion operators.

Because the contraction of fermion fields is singular as $\epsilon \to 0$, the terms of order ϵ in the last line of (19.25) can give a finite contribution. Taking the contraction according to (19.27), we find

$$\partial_\mu j^{\mu 5} = \text{symm} \lim_{\epsilon \to 0} \left\{ \frac{-i}{2\pi} \text{tr}\left[\frac{\gamma^\alpha \epsilon_\alpha}{\epsilon^2} \gamma^\mu \gamma^5 \right] \cdot \left(-ie\epsilon^\nu F_{\mu\nu} \right) \right\}. \qquad (19.28)$$

In two dimensions, $\text{tr}[\gamma^\alpha \gamma^\mu \gamma^5] = 2\epsilon^{\alpha\mu}$. Thus,

$$\partial_\mu j^{\mu 5} = \frac{e}{2\pi} \text{symm} \lim_{\epsilon \to 0} \left\{ 2\frac{\epsilon_\mu \epsilon^\nu}{\epsilon^2} \right\} \epsilon^{\mu\alpha} F_{\nu\alpha}. \qquad (19.29)$$

Now take the symmetric limit according to the prescription (19.23). We find precisely the anomalous nonconservation equation (19.18). In this derivation, (19.18) appears as an operator relation, rather than in a simple matrix element. Notice that, as in our first derivation of this equation, the assumption of local gauge invariance played a crucial role. If we had defined the axial vector current by reversing the sign of the Wilson line in (19.22), a prescription that would have done violence to local gauge invariance, we would have found the various contributions canceling on the right-hand side of (19.29).

An Example with Fermion Number Nonconservation

To complete our discussion of the two-dimensional axial vector current, we will show that the nonconservation equation (19.18) also has a global aspect. In free fermion theory, the integral of the axial current conservation law gives

$$\int d^2x \, \partial_\mu j^{\mu 5} = N_R - N_L = 0. \tag{19.30}$$

This relation implies that the difference in the number of right-moving and left-moving fermions cannot be changed in any possible process. Combining this with the conservation law for the vector current, we conclude that the number of each type of fermion is separately conserved. From (19.8), we might conclude that these separate conservation laws hold also in two-dimensional QED. However, we have already found that we must be careful in making statements about the axial current.

In two-dimensional QED, the conservation equation for the axial current is replaced by the anomalous nonconservation equation (19.18). If the right-hand side of this equation were the total derivative of a quantity falling off sufficiently rapidly at infinity, its integral would vanish and we would still retain the global conservation law. In fact, $\epsilon^{\mu\nu} F_{\mu\nu}$ is a total derivative:

$$\epsilon^{\mu\nu} F_{\mu\nu} = 2\partial_\mu \left(\epsilon^{\mu\nu} A_\nu\right). \tag{19.31}$$

However, it is easy to imagine examples where the integral of this quantity does not vanish, for example, a world with a constant background electric field. In such a world, the conservation law (19.30) must be violated. But how can this happen?

Let us analyze this problem by thinking about fermions in one space dimension in a background A^1 field that is constant in space and has a very slow time dependence. We will assume that the system has a finite length L, with periodic boundary conditions. Notice that the constant A^1 field cannot be removed by a gauge transformation that satisfies the periodic boundary conditions. One way to see this is to note that the system gives a nonzero value to the Wilson line

$$\exp\left[-ie \int_0^L dx \, A_1(x)\right], \tag{19.32}$$

which forms a gauge-invariant closed loop due to the periodic boundary conditions.

Following the derivation of the three-dimensional Hamiltonian, Eq. (3.84), we find that the Hamiltonian of this one-dimensional system is

$$H = \int dx \, \psi^\dagger \left(-i\alpha^1 D_1\right)\psi, \tag{19.33}$$

where $\alpha = \gamma^0 \gamma^1 = \gamma^5$. In the components (19.7),

$$H = \int dx \left\{ -i\psi_+^\dagger (\partial_1 - ieA^1)\psi_+ + i\psi_-^\dagger (\partial_1 - ieA^1)\psi_- \right\}. \qquad (19.34)$$

For a constant A^1 field, it is easy to diagonalize this Hamiltonian. The eigenstates of the covariant derivatives are wavefunctions

$$e^{ik_n x}, \qquad \text{with} \quad k_n = \frac{2\pi n}{L}, \; n = -\infty, \ldots, \infty, \qquad (19.35)$$

to satisfy the periodic boundary conditions. Then the single-particle eigenstates of H have energies

$$
\begin{aligned}
\psi_+ : \qquad E_n &= +(k_n - eA^1), \\
\psi_- : \qquad E_n &= -(k_n - eA^1).
\end{aligned}
\qquad (19.36)
$$

Each type of fermion has an infinite tower of equally spaced levels. To find the ground state of H, we fill the negative energy levels and interpret holes created among these filled states as antiparticles.

Now adiabatically change the value of A^1. The fermion energy levels slowly shift in accord with the relations (19.36). If A^1 changes by the finite amount

$$\Delta A^1 = \frac{2\pi}{eL}, \qquad (19.37)$$

which brings the Wilson loop (19.32) back to its original value, the spectrum of H returns to its original form. In this process, each level of ψ_+ moves down to the next position, and each level of ψ_- moves up to the next position, as shown in Fig. 19.2. The occupation numbers of levels should be maintained in this adiabatic process. Thus, remarkably, one right-moving fermion disappears from the vacuum and one extra left-moving fermion appears. At the same time,

$$
\begin{aligned}
\int d^2x \left(\frac{e}{\pi} \epsilon^{\mu\nu} F_{\mu\nu} \right) &= \int dt\, dx\, \frac{e}{\pi} \partial_0 A_1 \\
&= \frac{e}{\pi} L(-\Delta A^1) \\
&= -2,
\end{aligned}
\qquad (19.38)
$$

where we have inserted (19.37) in the last line. Thus the integrated form of the anomalous nonconservation equation (19.18) is indeed satisfied:

$$N_R - N_L = \int d^2x \left(\frac{e}{2\pi} \epsilon^{\mu\nu} F_{\mu\nu} \right). \qquad (19.39)$$

Even in this simple example, we see that it is not possible to escape the question of ultraviolet regularization in analyzing the chiral conservation law. Right-moving fermions are lost and left-moving fermions appear from the depths of the fermionic spectrum, $E \to -\infty$. In computing the changes in the separate fermion numbers, we have assumed that the vacuum cannot change the charge it contains at large negative energies. This prescription is gauge invariant, but it leads to the nonconservation of the axial vector current.

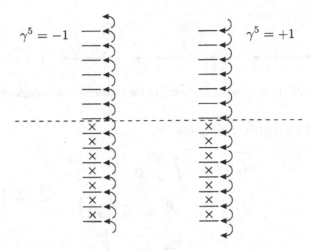

Figure 19.2. Effect on the vacuum state of the Hamiltonian H of one-dimensional QED due to an adiabatic change in the background A^1 field.

19.2 The Axial Current in Four Dimensions

All of the derivations we have just given for the two-dimensional axial current have analogues in four dimensions. In Eq. (3.40), we showed that, in the case of massless fermions, the four-dimensional Dirac equation splits neatly into separate equations for left- and right-handed fermions. If we couple the Dirac equation to a gauge field, we replace derivatives by covariant derivatives. This does not seem to affect the manifest separation between the two helicity components. Thus it seems clear that both the vector and axial vector currents should remain conserved. However, after the analysis we have just completed for the two-dimensional case, we know that we should not take these conservation laws for granted. We will now make a more careful analysis of the axial vector conservation law in four dimensions.

The Axial Vector Current Operator Equation

We begin with the case of massless four-dimensional QED. Of the three arguments that we gave in previous section for the two-dimensional axial current conservation law, the operator derivation generalizes most easily. The fermion field equations (19.21) are identical in the four-dimensional case. We can again adopt the gauge-invariant definition of the axial vector current (19.22). When we take the divergence of this current, all of the manipulations leading to Eq. (19.25) are still correct.

From this point, we must compute the singular terms in the operator product of the two fermion fields in the limit $\epsilon \to 0$. As in two dimensions, the leading term is given by contracting the two operators using a free-field

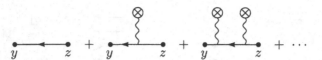

Figure 19.3. Expansion of $\psi(y)\bar{\psi}(z)$ in the presence of a background gauge field.

propagator. This contribution gives

$$\overline{\psi(y)\bar{\psi}}(z) = \int \frac{d^4k}{(2\pi)^4} e^{-ik\cdot(y-z)} \frac{i\not{k}}{k^2}$$

$$= -\not\partial\left(\frac{i}{4\pi^2}\frac{1}{(y-z)^2}\right)$$

$$= \frac{-i}{2\pi^2}\frac{\gamma^\alpha(y-z)_\alpha}{(y-z)^4}. \tag{19.40}$$

This is highly singular as $(y-z) \to 0$, but it gives zero when traced with $\gamma^\mu\gamma^5$. To find a nonzero result, we must consider terms of higher order in the expansion of the product of operators.

In a nonzero background gauge field, the contraction of fermion fields is given by the series of diagrams shown in Fig. 19.3. We have computed the leading term in this series in (19.40). The higher terms give less singular terms as $(y-z) \to 0$. The second term in the series is given by

$$\text{[diagram]} = \int \frac{d^4k}{(2\pi)^4}\frac{d^4p}{(2\pi)^4} e^{-i(k+p)\cdot y}e^{ik\cdot z}\frac{i(\not{k}+\not{p})}{(k+p)^2}\left(-ie\not{A}(p)\right)\frac{i\not{k}}{k^2}. \tag{19.41}$$

This contribution leads to

$$\langle\bar{\psi}(x+\tfrac{\epsilon}{2})\gamma^\mu\gamma^5\psi(x-\tfrac{\epsilon}{2})\rangle$$

$$= \int \frac{d^4k}{(2\pi)^4}\frac{d^4p}{(2\pi)^4} e^{ik\cdot\epsilon}e^{-ip\cdot x}\,\text{tr}\left[(-\gamma^\mu\gamma^5)\frac{i(\not{k}+\not{p})}{(k+p)^2}\left(-ie\not{A}(p)\right)\frac{i\not{k}}{k^2}\right]$$

$$= \int \frac{d^4k}{(2\pi)^4}\frac{d^4p}{(2\pi)^4} e^{ik\cdot\epsilon}e^{-ip\cdot x}\frac{4e\epsilon^{\mu\alpha\beta\gamma}(k+p)_\alpha A_\beta(p)k_\gamma}{k^2(k+p)^2}. \tag{19.42}$$

To evaluate the limit $\epsilon \to 0$, we can expand the integrand for large k. Then

$$\langle\bar{\psi}(x+\tfrac{\epsilon}{2})\gamma^\mu\gamma^5\psi(x-\tfrac{\epsilon}{2})\rangle \sim 4e\epsilon^{\mu\alpha\beta\gamma}\int\frac{d^4p}{(2\pi)^4}e^{-ip\cdot x}p_\alpha A_\beta(p)\int\frac{d^4k}{(2\pi)^4}e^{ik\cdot\epsilon}\frac{k_\gamma}{k^4}$$

$$= 4e\epsilon^{\alpha\beta\mu\gamma}\left(\partial_\alpha A_\beta(x)\right)\frac{\partial}{\partial\epsilon^\gamma}\left(\frac{i}{16\pi^2}\log\frac{1}{\epsilon^2}\right)$$

$$= 2e\epsilon^{\alpha\beta\mu\gamma}F_{\alpha\beta}(x)\left(\frac{-i}{8\pi^2}\frac{\epsilon_\gamma}{\epsilon^2}\right). \tag{19.43}$$

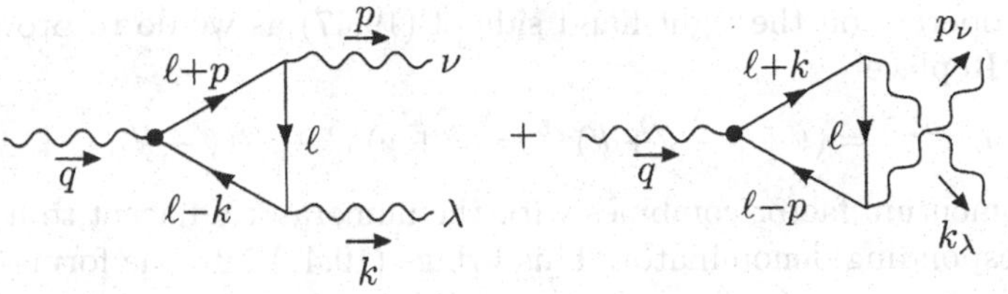

Figure 19.4. Diagrams contributing to the two-photon matrix element of the divergence of the axial vector current.

Substituting this expression into (19.25), we find

$$\partial_\mu j^{\mu 5} = \underset{\epsilon \to 0}{\text{symm lim}} \left\{ \frac{e}{4\pi^2} \epsilon^{\alpha\beta\mu\gamma} F_{\alpha\beta} \left(\frac{-i\epsilon_\gamma}{\epsilon^2} \right) (-ie\epsilon^\nu F_{\mu\nu}) \right\}. \tag{19.44}$$

Now take the symmetric limit $\epsilon \to 0$ in four dimensions. We find

$$\partial_\mu j^{\mu 5} = -\frac{e^2}{16\pi^2} \epsilon^{\alpha\beta\mu\nu} F_{\alpha\beta} F_{\mu\nu}. \tag{19.45}$$

This equation, which expresses the anomalous nonconservation of the four-dimensional axial current, is known as the Adler-Bell-Jackiw anomaly. Adler and Bardeen proved that this operator relation is actually correct to all orders in QED perturbation theory and receives no further radiative corrections.[†]

Triangle Diagrams

We can confirm the Adler-Bell-Jackiw relation by checking, in standard perturbation theory, that the divergence of the axial vector current has a nonzero matrix element to create two photons. To do this, we must analyze the matrix element

$$\int d^4x e^{-iq\cdot x} \langle p, k | j^{\mu 5}(x) | 0 \rangle = (2\pi)^4 \delta^{(4)}(p + k - q) \epsilon_\nu^*(p) \epsilon_\lambda^*(k) \mathcal{M}^{\mu\nu\lambda}(p, k). \tag{19.46}$$

The leading-order diagrams contributing to $\mathcal{M}^{\mu\nu\lambda}$ are shown in Fig. 19.4. The first diagram gives the contribution

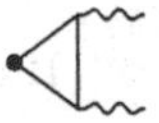

$$\text{\raisebox{-1ex}{}} = (-1)(-ie)^2 \int \frac{d^4\ell}{(2\pi)^4} \text{tr}\left[\gamma^\mu \gamma^5 \frac{i(\slashed{\ell} - \slashed{k})}{(\ell - k)^2} \gamma^\lambda \frac{i\slashed{\ell}}{\ell^2} \gamma^\nu \frac{i(\slashed{\ell} + \slashed{p})}{(\ell + p)^2} \right], \tag{19.47}$$

and the second diagram gives an identical contribution with (p, ν) and (k, λ) interchanged.

It is easy to give a formal argument that the matrix element of the divergence of the axial current vanishes at this order. Taking the divergence of the axial current in (19.46) is equivalent to dotting this quantity with iq_μ.

[†]S. Adler and W. A. Bardeen, *Phys. Rev.* **182**, 1517 (1969); S. Adler, in Deser, et. al. (1970).

Now we operate on the right-hand side of (19.47) as we do to prove a Ward identity. Replace

$$q_\mu \gamma^\mu \gamma^5 = (\ell\!\!\!/ + p\!\!\!/ - \ell\!\!\!/ + k\!\!\!/)\gamma^5 = (\ell\!\!\!/ + p\!\!\!/)\gamma^5 + \gamma^5(\ell\!\!\!/ - k\!\!\!/). \qquad (19.48)$$

Each momentum factor combines with the numerator adjacent to it to cancel the corresponding denominator. This brings (19.47) into the form

$$iq_\mu \cdot \;\;\begin{array}{c}\text{(diagram)}\end{array}\;\; = e^2 \int \frac{d^4\ell}{(2\pi)^4} \, \mathrm{tr}\left[\gamma^5 \frac{(\ell\!\!\!/ - k\!\!\!/)}{(\ell-k)^2}\gamma^\lambda \frac{\ell\!\!\!/}{\ell^2}\gamma^\nu + \gamma^5\gamma^\lambda \frac{\ell\!\!\!/}{\ell^2}\gamma^\nu \frac{(\ell\!\!\!/ + p\!\!\!/)}{(\ell+p)^2}\right]. \tag{19.49}$$

Now pass γ^ν through γ^5 in the second term and shift the integral over the first term according to $\ell \to (\ell+k)$:

$$iq_\mu \cdot \;\;\begin{array}{c}\text{(diagram)}\end{array}\;\; = e^2 \int \frac{d^4\ell}{(2\pi)^4} \, \mathrm{tr}\left[\gamma^5 \frac{\ell\!\!\!/}{\ell^2}\gamma^\lambda \frac{(\ell\!\!\!/ + k\!\!\!/)}{(\ell+k)^2}\gamma^\nu - \gamma^5 \frac{\ell\!\!\!/}{\ell^2}\gamma^\nu \frac{(\ell\!\!\!/ + p\!\!\!/)}{(\ell+p)^2}\gamma^\lambda\right]. \tag{19.50}$$

This expression is manifestly antisymmetric under the interchange of (p,ν) and (k,λ), so the contribution of the second diagram in Fig. 19.4 precisely cancels (19.47).

However, because this derivation involves a shift of the integration variable, we should look closely at whether this shift is allowed by the regularization. From (19.47), we see that the integral that must be shifted is divergent. If the diagram is regulated with a simple momentum cutoff, or even with Pauli-Villars regularization, it turns out that the shift leaves over a finite, nonzero term. In Chapter 7, we encountered a similar problem in our discussion of the QED vacuum polarization diagram. We evaded the problem there by using dimensional regularization. Dimensional regularization of the diagrams of Fig. 19.4 will automatically insure the validity of the QED Ward identities for the photon emission vertices,

$$p_\nu \mathcal{M}^{\mu\nu\lambda} = k_\lambda \mathcal{M}^{\mu\nu\lambda} = 0. \tag{19.51}$$

But in the analysis of the axial vector current, even dimensional regularization has an extra subtlety, because γ^5 is an intrinsically four-dimensional object. In their original paper on dimensional regularization,[‡] 't Hooft and Veltman suggested using the definition

$$\gamma^5 = i\gamma^0\gamma^1\gamma^2\gamma^3 \tag{19.52}$$

in d dimensions. This definition has the consequence that γ^5 anticommutes with γ^μ for $\mu = 0, 1, 2, 3$ but commutes with γ^μ for other values of μ.

In the evaluation of (19.47), the external indices and the momenta p, k, q all live in the physical four dimensions, but the loop momentum ℓ has components in all dimensions. Write

$$\ell = \ell_\parallel + \ell_\perp, \tag{19.53}$$

[‡]G. 't Hooft and M. J. G. Veltman, *Nucl. Phys.* **B44**, 189 (1972).

where the first term has nonzero components in dimensions $0, 1, 2, 3$ and the second term has nonzero components in the other $d-4$ dimensions. Because γ^5 commutes with the γ^μ in these extra dimensions, identity (19.48) is modified to

$$q_\mu \gamma^\mu \gamma^5 = (\slashed{\ell} + \slashed{k})\gamma^5 + \gamma^5(\slashed{\ell} - \slashed{p}) - 2\gamma^5 \slashed{\ell}_\perp. \qquad (19.54)$$

The first two terms cancel according to the argument given above; the shift in (19.50) is justified by the dimensional regularization. However, the third term of (19.54) gives an additional contribution:

$$iq_\mu \cdot \quad = e^2 \int \frac{d^4\ell}{(2\pi)^4} \, \text{tr}\left[-2\gamma^5 \slashed{\ell}_\perp \frac{(\slashed{\ell} - \slashed{k})}{(\ell - k)^2} \gamma^\lambda \frac{\slashed{\ell}}{\ell^2} \gamma^\nu \frac{(\slashed{\ell} + \slashed{p})}{(\ell + p)^2}\right]. \qquad (19.55)$$

To evaluate this contribution, combine denominators in the standard way, and shift the integration variable $\ell \to \ell + P$, where $P = xk - yp$. In expanding the numerator, we must retain one factor each of γ^ν, γ^λ, $\slashed{p}$, and $\slashed{k}$ to give a nonzero trace with γ^5. This leaves over one factor of $\slashed{\ell}_\perp$ and one factor of $\slashed{\ell}$ which must also be evaluated with components in extra dimensions in order to give a nonzero integral. The factors $\slashed{\ell}_\perp$ anticommute with the other Dirac matrices in the problem and thus can be moved to adjacent positions. Then we must evaluate the integral

$$\int \frac{d^4\ell}{(2\pi)^4} \frac{\slashed{\ell}_\perp \slashed{\ell}_\perp}{(\ell^2 - \Delta)^3}, \qquad (19.56)$$

where Δ is a function of k, p, and the Feynman parameters. Using

$$(\slashed{\ell}_\perp)^2 = \ell_\perp^2 \to \frac{(d-4)}{d}\ell^2 \qquad (19.57)$$

under the symmetrical integration, we can evaluate (19.56) as

$$\frac{i}{(4\pi)^{d/2}} \frac{(d-4)}{2} \frac{\Gamma(2-\frac{d}{2})}{\Gamma(3)\Delta^{2-d/2}} \xrightarrow{d \to 4} \frac{-i}{2(4\pi)^2}. \qquad (19.58)$$

Notice the behavior in which a logarithmically divergent integral contributes a factor $(d-4)$ in the denominator and allows an anomalous term, formally proportional to $(d-4)$, to give a finite contribution. The remainder of the algebra in the evaluation of (19.55) is straightforward. The terms involving the momentum shift P cancel, and we find

$$iq_\mu \cdot \quad = e^2 \left(\frac{-i}{2(4\pi)^2}\right) \text{tr}\left[2\gamma^5(-\slashed{k})\gamma^\lambda \slashed{p}\gamma^\nu\right]$$

$$= \frac{e^2}{4\pi^2} \epsilon^{\alpha\lambda\beta\nu} k_\alpha p_\beta. \qquad (19.59)$$

This term is symmetric under the interchange of (p, ν) with (k, λ), so the second diagram of Fig. 19.4 gives an equal contribution. Thus,

$$
\begin{aligned}
\langle p, k | \, \partial_\mu j^{\mu 5}(0) \, | 0 \rangle &= -\frac{e^2}{2\pi^2} \epsilon^{\alpha\nu\beta\lambda} (-ip_\alpha) \epsilon_\nu^*(p) (-ik_\beta) \epsilon_\lambda^*(k) \\
&= -\frac{e^2}{16\pi^2} \langle p, k | \, \epsilon^{\alpha\nu\beta\lambda} F_{\alpha\nu} F_{\beta\lambda}(0) \, | 0 \rangle,
\end{aligned}
\tag{19.60}
$$

as we would expect from the Adler-Bell-Jackiw anomaly equation.

Chiral Transformation of the Functional Integral

A third way of understanding the Adler-Bell-Jackiw anomaly comes from analyzing the conservation law for the axial vector current from the functional integral for the fermion field. In Section 9.6, we used the functional integral to derive the current conservation equations and the Ward identities associated with any symmetry of the Lagrangian. It is instructive to see how this argument breaks down when we apply it to the chiral symmetry of massless fermions.

We first review the standard derivation of the axial vector Ward identities following the method of Section 9.6. Starting from the fermionic functional integral

$$
Z = \int \mathcal{D}\psi \mathcal{D}\bar{\psi} \exp\left[i \int d^4x \, \bar{\psi}(i\slashed{D})\psi\right],
\tag{19.61}
$$

make the change of variables

$$
\begin{aligned}
\psi(x) &\to \psi'(x) = (1 + i\alpha(x)\gamma^5)\psi(x), \\
\bar{\psi}(x) &\to \bar{\psi}'(x) = \bar{\psi}(1 + i\alpha(x)\gamma^5).
\end{aligned}
\tag{19.62}
$$

Since the global chiral rotation, with constant α, is a symmetry of the Lagrangian, the only new terms in the Lagrangian that result from (19.62) contain derivatives of α. Thus,

$$
\begin{aligned}
\int d^4x \, \bar{\psi}'(i\slashed{D})\psi' &= \int d^4x \left[\bar{\psi}(i\slashed{D})\psi - \partial_\mu \alpha(x) \bar{\psi}\gamma^\mu\gamma^5\psi\right] \\
&= \int d^4x \left[\bar{\psi}(i\slashed{D})\psi + \alpha(x)\partial_\mu(\bar{\psi}\gamma^\mu\gamma^5\psi)\right].
\end{aligned}
\tag{19.63}
$$

Then, by varying the Lagrangian with respect to $\alpha(x)$, we derive the classical conservation equation for the axial current. By carrying out a similar manipulation on the functional expression for a correlation function, as in Eq. (9.102), we would derive the associated Ward identities.

In the argument just given, we assumed that the functional measure does not change when we change variables from $\psi'(x)$ to ψ. This seems reasonable, because the relation of ψ' and ψ in (19.62) looks like a unitary transformation. However, we should examine this point more closely.* First, we must carefully

---

*K. Fujikawa, *Phys. Rev. Lett.* **42**, 1195 (1979); *Phys. Rev.* **D21**, 2848 (1980).

define the functional measure. To do this, expand the fermion field in a basis of eigenstates of $\not{D}$. Define right and left eigenvectors of $\not{D}$ by

$$(i\not{D})\phi_m = \lambda_m \phi_m, \qquad \hat{\phi}_m(i\not{D}) = -iD_\mu \hat{\phi}_m \gamma^\mu = \lambda_m \hat{\phi}_m. \tag{19.64}$$

For zero background A_μ field, these eigenstates are Dirac wavefunctions of definite momentum; the eigenvalues satisfy

$$\lambda_m^2 = k^2 = (k^0)^2 - (\mathbf{k})^2. \tag{19.65}$$

For a fixed A_μ field, this is also the asymptotic form of the eigenvalues for large k. These eigenfunctions give us a basis that we can use to expand ψ and $\bar{\psi}$:

$$\psi(x) = \sum_m a_m \phi_m(x), \qquad \bar{\psi}(x) = \sum_m \hat{a}_m \hat{\phi}_m(x), \tag{19.66}$$

where a_m, $\hat{a}_m$ are anticommuting coefficients multiplying the c-number eigen-functions (19.64). The functional measure over ψ, $\bar{\psi}$ can then be defined as

$$\mathcal{D}\psi \mathcal{D}\bar{\psi} = \prod_m da_m d\hat{a}_m, \tag{19.67}$$

and the functional measure over ψ', $\bar{\psi}'$ can be defined in the same way.

If $\psi'(x) = (1 + i\alpha(x)\gamma^5)\psi(x)$, the expansion coefficients of ψ and ψ' are related by a infinitesimal linear transformation $(1 + C)$, computed as follows:

$$a_m' = \sum_n \int d^4x \, \phi_m^\dagger(x)(1 + i\alpha(x)\gamma^5)\phi_n(x)a_n = \sum_n (\delta_{mn} + C_{mn})a_n. \tag{19.68}$$

Then

$$\mathcal{D}\psi' \mathcal{D}\bar{\psi}' = \mathcal{J}^{-2} \cdot \mathcal{D}\psi \mathcal{D}\bar{\psi}, \tag{19.69}$$

where $\mathcal{J}$ is the Jacobian determinant of the transformation $(1 + C)$. The inverse of $\mathcal{J}$ appears in (19.69) as a result of the rule (9.63) or (9.69) for fermionic integration. To evaluate $\mathcal{J}$, we write

$$\mathcal{J} = \det(1 + C) = \exp\left[\operatorname{tr} \log(1 + C)\right] = \exp\left[\sum_n C_{nn} + \cdots\right], \tag{19.70}$$

and we can ignore higher order terms in the last line because C is infinitesimal. Thus,

$$\log \mathcal{J} = i \int d^4x \, \alpha(x) \sum_n \phi_n^\dagger(x)\gamma^5 \phi_n(x). \tag{19.71}$$

The coefficient of $\alpha(x)$ looks like $\operatorname{tr}[\gamma^5] = 0$. However, we must regularize the sum over eigenstates n in a gauge-invariant way. The natural choice is

$$\sum_n \phi_n^\dagger(x)\gamma^5 \phi_n(x) = \lim_{M \to \infty} \sum_n \phi_n^\dagger(x)\gamma^5 \phi_n(x)e^{\lambda_n^2/M^2}. \tag{19.72}$$

As (19.65) indicates, the sign of λ_n^2 will be negative at large momentum after a Wick rotation; thus, the sign in the exponent of the convergence factor is given correctly. We can write (19.72) in an operator form

$$\sum_n \phi_n^\dagger(x)\gamma^5\phi_n(x) = \lim_{M\to\infty}\sum_n \phi_n^\dagger(x)\gamma^5 e^{(i\slashed{D})^2/M^2}\phi_n(x)$$

$$= \lim_{M\to\infty}\langle x|\operatorname{tr}\big[\gamma^5 e^{(i\slashed{D})^2/M^2}\big]|x\rangle,$$

(19.73)

where, in the second line, we trace over Dirac indices.

To evaluate (19.73), we rewrite $(i\slashed{D})^2$ according to (16.107). In our present conventions, this equation reads

$$(i\slashed{D})^2 = -D^2 + \frac{e}{2}\sigma^{\mu\nu}F_{\mu\nu},$$

(19.74)

with $\sigma^{\mu\nu} = \frac{i}{2}[\gamma^\mu,\gamma^\nu]$. Since we are taking the limit $M\to\infty$, we can concentrate our attention on the asymptotic part of the spectrum, where momentum k is large and we can expand in powers of the gauge field. To obtain a nonzero trace with γ^5, we must bring down four Dirac matrices from the exponent. The leading term is given by expanding the exponent to order $(\sigma\cdot F)^2$, and then ignoring the background A_μ field in all other terms. This gives

$$\lim_{M\to\infty}\langle x|\operatorname{tr}\big[\gamma^5 e^{(-D^2+(e/2)\sigma\cdot F)/M^2}\big]|x\rangle$$

$$= \lim_{M\to\infty}\operatorname{tr}\Big[\gamma^5\frac{1}{2!}\Big(\frac{e}{2M^2}\sigma^{\mu\nu}F_{\mu\nu}(x)\Big)^2\Big]\langle x|e^{-\partial^2/M^2}|x\rangle.$$

(19.75)

The matrix element in (19.75) can be evaluated by a Wick rotation:

$$\langle x|e^{-\partial^2/M^2}|x\rangle = \lim_{x\to y}\int\frac{d^4k}{(2\pi)^4}e^{-ik\cdot(x-y)}e^{k^2/M^2}$$

$$= i\int\frac{d^4k_E}{(2\pi)^4}e^{-k_E^2/M^2}$$

$$= i\frac{M^4}{16\pi^2}.$$

(19.76)

Then (19.75) reduces to

$$\lim_{M\to\infty}\frac{-ie^2}{8\cdot16\pi^2}M^4\operatorname{tr}\Big[\gamma^5\gamma^\mu\gamma^\nu\gamma^\lambda\gamma^\sigma\frac{1}{(M^2)^2}F_{\mu\nu}F_{\lambda\sigma}(x)\Big]$$

$$= -\frac{e^2}{32\pi^2}\epsilon^{\alpha\beta\mu\nu}F_{\alpha\beta}F_{\mu\nu}(x).$$

(19.77)

Thus,

$$J = \exp\Big[-i\int d^4x\,\alpha(x)\Big(\frac{e^2}{32\pi^2}\epsilon^{\mu\nu\lambda\sigma}F_{\mu\nu}F_{\lambda\sigma}(x)\Big)\Big].$$

(19.78)

In all, we find that, after the change of variables (19.62), the functional integral (19.61) takes the form

$$Z = \int \mathcal{D}\psi \mathcal{D}\bar{\psi} \exp\left[i \int d^4x \left(\bar{\psi}(i\slashed{D})\psi + \alpha(x)\left\{ \partial_\mu j^{\mu 5} + \frac{e^2}{16\pi^2} \epsilon^{\mu\nu\lambda\sigma} F_{\mu\nu} F_{\lambda\sigma} \right\} \right)\right].$$
$$(19.79)$$

Varying the exponent with respect to $\alpha(x)$, we find precisely the Adler-Bell-Jackiw anomaly equation.

This derivation of the axial vector anomaly is especially interesting because it generalizes readily to any even dimensionality. The functional derivation always picks out for the right-hand side of the anomaly equation the pseudoscalar operator built from gauge fields that has the same dimension, d, as the divergence of the current. In two dimensions, this derivation leads immediately to (19.18). As long as d is even, we can always construct a matrix γ^5 that anticommutes with all of the Dirac matrices by taking their product. Then, the functional derivation leads straightforwardly to the result

$$\partial_\mu j^{\mu 5} = (-1)^{n+1} \frac{2e^n}{n!(4\pi)^n} \epsilon^{\mu_1\mu_2\cdots\mu_{2n}} F_{\mu_1\mu_2} \cdots F_{\mu_{2n-1}\mu_{2n}}, \qquad (19.80)$$

where $n = d/2$.

At the end of the previous section, we argued that the axial vector anomaly leads to global nonconservation of fermionic charges in a two-dimensional system with a macroscopic electric field. In the same way, the four-dimensional anomaly equation leads to global nonconservation of the number of left- and right-handed fermions in background fields in which the right-hand side of (19.45) is nonzero. These are field configurations with parallel electric and magnetic fields. In Problem 19.1, we work out an example of four-dimensional massless fermions in a simple situation of this type and show that the fermion numbers are indeed violated, in a manner similar to what we saw at the end of Section 19.1, in accord with the Adler-Bell-Jackiw anomaly.

19.3 Goldstone Bosons and Chiral Symmetries in QCD

The Adler-Bell-Jackiw anomaly has a number of important implications for QCD. To describe these, we must first discuss the chiral symmetries of QCD systematically. In this discussion, we will ignore all but the lightest quarks u and d. In many analyses of the low-energy structure of the strong interactions, one also treats the s quark as light; this gives results that naturally generalize the ones we will find below.

The fermionic part of the QCD Lagrangian is

$$\mathcal{L} = \bar{u}i\slashed{D}u + \bar{d}i\slashed{D}d - m_u\bar{u}u - m_d\bar{d}d. \qquad (19.81)$$

If the u and d quarks are very light, the last two terms are small and can be neglected. Let us study the implications of making this approximation. If we ignore the u and d masses, the Lagrangian (19.81) of course has isospin symmetry, the symmetry of an $SU(2)$ unitary transformation mixing the u and d fields. However, because the classical Lagrangian for massless fermions contains no coupling between left- and right-handed quarks, this Lagrangian actually is symmetric under the separate unitary transformations

$$\begin{pmatrix} u \\ d \end{pmatrix}_L \to U_L \begin{pmatrix} u \\ d \end{pmatrix}_L, \qquad \begin{pmatrix} u \\ d \end{pmatrix}_R \to U_R \begin{pmatrix} u \\ d \end{pmatrix}_R. \tag{19.82}$$

It is useful to separate the $U(1)$ and $SU(2)$ parts of these transformations; then the symmetry group of the classical, massless QCD Lagrangian is $SU(2) \times SU(2) \times U(1) \times U(1)$. Let Q denote the quark doublet, with chiral components

$$Q_L = \left(\frac{1-\gamma^5}{2}\right)\begin{pmatrix} u \\ d \end{pmatrix}, \qquad Q_R = \left(\frac{1+\gamma^5}{2}\right)\begin{pmatrix} u \\ d \end{pmatrix}. \tag{19.83}$$

Then we can write the currents associated with these symmetries as

$$\begin{aligned} j_L^\mu &= \bar{Q}_L \gamma^\mu Q_L, & j_R^\mu &= \bar{Q}_R \gamma^\mu Q_R, \\ j_L^{\mu a} &= \bar{Q}_L \gamma^\mu \tau^a Q_L, & j_R^{\mu a} &= \bar{Q}_R \gamma^\mu \tau^a Q_R, \end{aligned} \tag{19.84}$$

where $\tau^a = \sigma^a/2$ represent the generators of $SU(2)$. The sums of left- and right-handed currents give the baryon number and isospin currents

$$j^\mu = \bar{Q} \gamma^\mu Q, \qquad j^{\mu a} = \bar{Q} \gamma^\mu \tau^a Q. \tag{19.85}$$

The corresponding symmetries are the transformations (19.82) with $U_L = U_R$. The differences of the currents (19.84) give the corresponding axial vector currents $j^{\mu 5}$, $j^{\mu 5a}$:

$$j^{\mu 5} = \bar{Q} \gamma^\mu \gamma^5 Q, \qquad j^{\mu 5a} = \bar{Q} \gamma^\mu \gamma^5 \tau^a Q. \tag{19.86}$$

In the discussion to follow, we will derive conclusions about the strong interactions by assuming that the classical conservation laws for these currents are not spoiled by anomalies. We will show below that this assumption is correct for the isotriplet currents $j^{\mu 5a}$ but not for $j^{\mu 5}$.

The vector $SU(2) \times U(1)$ transformations are manifest symmetries of the strong interactions, and the associated currents lead to familiar conservation laws. What about the orthogonal, axial vector, transformations? These do not correspond to any obvious symmetry of the strong interactions. In 1960, Nambu and Jona-Lasinio hypothesized that these are accurate symmetries of the strong interactions that are spontaneously broken.[†] This idea has led to a correct and surprisingly detailed description of the properties of the strong interactions at low energy.

[†]Y. Nambu and G. Jona-Lasinio, *Phys. Rev.* **122**, 345 (1961).

Figure 19.5. A quark-antiquark pair with zero total momentum and angular momentum.

Spontaneous Breaking of Chiral Symmetry

Before we describe the consequences of spontaneously broken chiral symmetry, let us ask why we might expect the chiral symmetries to be spontaneously broken in the first place. In the theory of superconductivity, a small electron-electron attraction leads to the appearance of a condensate of electron pairs in the ground state of a metal. In QCD, quarks and antiquarks have strong attractive interactions, and, if these quarks are massless, the energy cost of creating an extra quark-antiquark pair is small. Thus we expect that the vacuum of QCD will contain a condensate of quark-antiquark pairs. These fermion pairs must have zero total momentum and angular momentum. Thus, as Fig. 19.5 shows, they must contain net chiral charge, pairing left-handed quarks with the antiparticles of right-handed quarks. The vacuum state with a quark pair condensate is characterized by a nonzero vacuum expectation value for the scalar operator

$$\langle 0| \, \overline{Q}Q \, |0\rangle = \langle 0| \, \overline{Q}_L Q_R + \overline{Q}_R Q_L \, |0\rangle \neq 0, \qquad (19.87)$$

which transforms under (19.82) with $U_L \neq U_R$. The expectation value signals that the vacuum mixes the two quark helicities. This allows the u and d quarks to acquire effective masses as they move through the vacuum. Inside quark-antiquark bound states, the u and d quarks would appear to move as if they had a sizable effective mass, even if they had zero mass in the original QCD Lagrangian.

The vacuum expectation value (19.87) signals the spontaneous breaking of the full symmetry group (19.82) down to the subgroup of vector symmetries with $U_L = U_R$. Thus there are four spontaneously broken continuous symmetries, associated with the four axial vector currents. At the end of Section 11.1, we proved Goldstone's theorem, which states that every spontaneously broken continuous symmetry of a quantum field theory leads to a massless particle with the quantum numbers of a local symmetry rotation. This means that, in QCD with massless u and d quarks, we should find four spin-zero particles with the correct quantum numbers to be created by the four axial vector currents.

The real strong interactions do not contain any massless particles, but they do contain an isospin triplet of relatively light mesons, the pions. These particles are known to have odd parity (as we expect if they are quark-antiquark bound states). Thus, they can be created by the axial isospin currents. We can parametrize the matrix element of $j^{\mu 5a}$ between the vacuum and an on-shell pion by writing

$$\langle 0| j^{\mu 5a}(x) |\pi^b(p)\rangle = -ip^\mu f_\pi \delta^{ab} e^{-ip\cdot x}, \tag{19.88}$$

where a, b are isospin indices and f_π is a constant with the dimensions of $(\text{mass})^1$. We show in Problem 19.2 that the value of f_π can be determined from the rate of π^+ decay through the weak interaction; one finds $f_\pi = 93$ MeV. For this reason, f_π is often called the *pion decay constant*. If we contract (19.88) with p_μ and use the conservation of the axial currents, we find that an on-shell pion must satisfy $p^2 = 0$, that is, it must be massless, as required by Goldstone's theorem.

If we now restore the quark mass terms in (19.81), the axial currents are no longer exactly conserved. The equation of motion of the quark field is now

$$i\slashed{D}Q = \mathbf{m}Q, \qquad -iD_\mu \bar{Q}\gamma^\mu = \bar{Q}\mathbf{m}, \tag{19.89}$$

where

$$\mathbf{m} = \begin{pmatrix} m_u & 0 \\ 0 & m_d \end{pmatrix} \tag{19.90}$$

is the quark mass matrix. Then one can readily compute

$$\partial_\mu j^{\mu 5a} = i\bar{Q}\{\mathbf{m}, \tau^a\}\gamma^5 Q. \tag{19.91}$$

Using this equation together with (19.88), we find

$$\langle 0| \partial_\mu j^{\mu 5a}(0) |\pi^b(p)\rangle = -p^2 f_\pi \delta^{ab} = \langle 0| i\bar{Q}\{\mathbf{m}, \tau^a\}\gamma^5 Q |\pi^b(p)\rangle. \tag{19.92}$$

The last expression is an invariant quantity times

$$\text{tr}\left[\{\mathbf{m}, \tau^a\}\tau^b\right] = \tfrac{1}{2}\delta^{ab}(m_u + m_d). \tag{19.93}$$

Thus, the quark mass terms give the pions masses of the form

$$m_\pi^2 = (m_u + m_d)\frac{M^2}{f_\pi}. \tag{19.94}$$

The mass parameter M has been estimated to be of order 400 MeV. Thus, to give the observed pion mass of 140 MeV, one needs only $(m_u + m_d) \sim 10$ MeV. This is a small perturbation on the strong interactions.

This argument has an interesting implication for the nature of the isospin symmetry of the strong interactions. In the limit in which the u and d quarks have zero mass in the Lagrangian, these quarks acquire large, equal effective masses from the vacuum with spontaneously broken chiral symmetry. As long as the masses m_u and m_d in the Lagrangian are small compared to the effective mass, the u and d quarks will behave inside hadrons as though they are approximately degenerate. Thus the isospin symmetry of the strong interactions need have nothing to do with a fundamental symmetry linking u and d; it follows for any arbitrary relation between m_u and m_d, provided that both of these parameters are much less than 300 MeV. Similarly, the approximate $SU(3)$ symmetry of the strong interactions follows if the fundamental mass of the s quark is also small compared to the strong interaction scale. The best

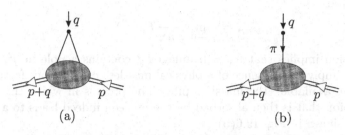

Figure 19.6. Matrix element of the axial isospin current in the nucleon: (a) kinematics of the amplitude; (b) contribution that leads to a pole in q^2.

current estimates of the mass ratios $m_u : m_d : m_s$ are in fact 1 : 2 : 40, so that the fundamental Lagrangian of the strong interactions shows no sign of flavor symmetry among the quark masses.[‡]

The identification of the pions as Goldstone bosons of broken chiral symmetry has a number of implications for hadronic matrix elements. Here we will give only one example. In the following argument, we will work in the limit of exact chiral symmetry, ignoring the small corrections from the u and d masses.

The matrix element of the axial isospin current in the nucleon, a quantity that enters the theory of neutron and nuclear β decay, can be written in terms of form factors as follows:

$$\langle N| j^{\mu 5 a}(q) |N\rangle = \bar{u}\Big[\gamma^\mu \gamma^5 F_1^5(q^2) + \frac{i\sigma^{\mu\nu} q_\nu}{2m}\gamma^5 F_2^5(q^2) + q^\mu \gamma^5 F_3^5(q^2)\Big]\tau^a u.$$

$$(19.95)$$

The kinematics of the vertex is shown in Fig. 19.6. Notice that there is one more possible form factor than in the vector case, Eq. (6.33). The value of F_1^5 at $q^2 = 0$ is not restricted by the value of any manifestly conserved charge. Conventionally, one writes simply

$$F_1^5(0) = g_A. \qquad (19.96)$$

However, we will now show that the value of this quantity can be computed.

If we ignore quark masses, the axial vector current in (19.95) is conserved, so the form factors satisfy

$$0 = \bar{u}(p')\Big[\slashed{q}\gamma^5 F_1^5(q^2) + q^2 \gamma^5 F_3^5(q^2)\Big]u(p)$$

$$= \bar{u}(p')\Big[(\slashed{p}' - \slashed{p})\gamma^5 F_1^5(q^2) + q^2 \gamma^5 F_3^5(q^2)\Big]u(p) \qquad (19.97)$$

$$= \bar{u}(p')\Big[2m_N \gamma^5 F_1^5(q^2) + q^2 \gamma^5 F_3^5(q^2)\Big]u(p).$$

[‡]The determination of the fundamental quark masses is reviewed by J. Gasser and H. Leutwyler, *Phys. Repts.* **87**, 77 (1982).

Thus, we find

$$g_A = \lim_{q^2 \to 0} \frac{q^2}{2m_N} F_3^5(q^2).$$ (19.98)

This equation implies that $g_A = 0$ unless F_3^5 contains a pole in q^2. Such a pole would imply the presence of a physical massless particle, but fortunately, there is one available—the massless pion. The process in which the current creates a pion that is then absorbed by the nucleon indeed leads to a pole in $F_3^5(q^2)$, as shown in Fig. 19.6(b).

Let us now compute this pole term and use it to determine g_A. The low-energy pion-nucleon interaction is conventionally parametrized by the Lagrangian

$$\Delta\mathcal{L} = ig_{\pi NN}\pi^a \overline{N}\gamma^5\sigma^a N.$$ (19.99)

The amplitude for the current $j^{\mu 5a}$ to create the pion is given by (19.88). Then the contribution of Fig. 19.6(b) to the current vertex is

$$-g_{\pi NN}\bar{u}(2\tau^a\gamma^5)u \cdot \frac{i}{q^2} \cdot (iq^\mu f_\pi).$$ (19.100)

Thus,

$$F_3^5(q^2) = \frac{1}{q^2} \cdot 2f_\pi g_{\pi NN}.$$ (19.101)

We find that g_A is given by a combination of f_π, the nucleon mass, and the pion-nucleon coupling constant:

$$g_A = \frac{f_\pi}{m_N}g_{\pi NN}.$$ (19.102)

This strange identity, called the *Goldberger-Treiman relation*, is satisfied experimentally to 5% accuracy.

The identification of the pion as the Goldstone boson of spontaneously broken chiral symmetry leads to numerous other predictions for current matrix elements and pion scattering amplitudes. In particular, the leading terms of the pion-pion and pion-nucleon scattering amplitudes at low energy can be computed directly in terms of f_π by arguments similar to one just given.*

Anomalies of Chiral Currents

Up to this point, we have discussed the chiral symmetries of QCD according to the classical current conservation equations. We must now ask whether these equations are affected by the Adler-Bell-Jackiw anomaly, and what the consequences of that modification are.

To begin, we study the modification of the chiral conservation laws due to the coupling of the quark currents to the gluon fields of QCD. The arguments given in the previous section go through equally well in the case of

---

*The detailed consequences of spontaneously broken chiral symmetry are worked out in a very clear manner in Georgi (1984).

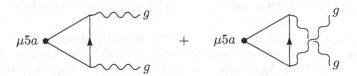

Figure 19.7. Diagrams that lead to an axial vector anomaly for a chiral current in QCD.

massless fermions coupling to a non-Abelian gauge field, so we expect that an axial vector current will receive an anomalous contribution from the diagrams shown in Fig. 19.7. The anomaly equation should be the Abelian result, supplemented by an appropriate group theory factor. In addition, since the axial current is gauge invariant, the anomaly must also be gauge invariant. That is, it must contain the full non-Abelian field strength, including its nonlinear terms. These terms are actually included in the functional derivation of the anomaly given at the end of Section 19.2.

For the axial currents of QCD, written in (19.86), we can read the group theory factors for the Adler-Bell-Jackiw anomaly from the diagrams of Fig. 19.7. For the axial isospin currents,

$$\partial_\mu j^{\mu 5a} = -\frac{g^2}{16\pi^2}\epsilon^{\alpha\beta\mu\nu}F^c_{\alpha\beta}F^d_{\mu\nu}\cdot\mathrm{tr}\left[\tau^a t^c t^d\right],\qquad(19.103)$$

where $F^c_{\mu\nu}$ is a gluon field strength, τ^a is an isospin matrix, t^c is a color matrix, and the trace is taken over colors and flavors. In this case, we find

$$\mathrm{tr}\left[\tau^a t^c t^d\right] = \mathrm{tr}[\tau^a]\,\mathrm{tr}[t^c t^d] = 0,\qquad(19.104)$$

since the trace of a single τ^a vanishes. Thus the conservation of the axial isospin currents is unaffected by the Adler-Bell-Jackiw anomaly of QCD. However, in the case of the isospin singlet axial current, the matrix τ^a is replaced by the matrix 1 on flavors, and we find

$$\partial_\mu j^{\mu 5} = -\frac{g^2 n_f}{32\pi^2}\epsilon^{\alpha\beta\mu\nu}F^c_{\alpha\beta}F^c_{\mu\nu},\qquad(19.105)$$

where n_f is the number of flavors; $n_f = 2$ in our current model.

Thus, the isospin singlet axial current is not in fact conserved in QCD. The divergence of this current is equal to a gluon operator with nontrivial matrix elements between hadron states. Some subtle questions remain concerning the effects of this operator. In particular, it can be shown, as we saw for the two-dimensional axial anomaly in Eq. (19.31), that the right-hand side of (19.105) is a total divergence. Nevertheless, again in accord with our experience in two dimensions, there are physically reasonable field configurations in which the four-dimensional integral of this term takes a nonzero value. This topic is discussed further at the end of Section 22.3. In any event, Eq. (19.105) indeed implies that QCD has no isosinglet axial symmetry and no associated Goldstone boson. This equation explains why the strong interactions contain no light isosinglet pseudoscalar meson with mass comparable to that of the pions.

Though the axial isospin currents have no axial anomaly from QCD interactions, they do have an anomaly associated with the coupling of quarks to electromagnetism. Again referring to the diagrams of Fig. 19.7, we see that the electromagnetic anomaly of the axial isospin currents is given by

$$\partial_\mu j^{\mu 5a} = -\frac{e^2}{16\pi^2}\epsilon^{\alpha\beta\mu\nu}F_{\alpha\beta}F_{\mu\nu} \cdot \text{tr}\left[\tau^a Q^2\right], \tag{19.106}$$

where $F_{\mu\nu}$ is the electromagnetic field strength, Q is the matrix of quark electric charges,

$$Q = \begin{pmatrix} \frac{2}{3} & 0 \\ 0 & -\frac{1}{3} \end{pmatrix}, \tag{19.107}$$

and the trace again runs over flavors and colors. Since the matrices in the trace do not depend on color, the color sum simply gives a factor of 3. The flavor trace is nonzero only for $a = 3$; in that case, the electromagnetic anomaly is

$$\partial_\mu j^{\mu 53} = -\frac{e^2}{32\pi^2}\epsilon^{\alpha\beta\mu\nu}F_{\alpha\beta}F_{\mu\nu}. \tag{19.108}$$

Because the current $j^{\mu 53}$ annihilates a π^0 meson, Eq. (19.108) indicates that the axial vector anomaly contributes to the matrix element for the decay $\pi^0 \to 2\gamma$. We will now show that, in fact, it gives the leading contribution to this amplitude. Again, we work in the limit of massless u and d quarks, so that the chiral symmetries are exact up to the effects of the anomaly.

Consider the matrix element of the axial current between the vacuum and a two-photon state:

$$\langle p, k| j^{\mu 53}(q) |0\rangle = \epsilon_\nu^* \epsilon_\lambda^* \mathcal{M}^{\mu\nu\lambda}(p, k). \tag{19.109}$$

This is the same matrix element (19.46) that we studied in QED perturbation theory in Section 19.2. Now, however, we will study the general properties of this matrix element by expanding it in form factors. In general, the amplitude can be decomposed by writing all possible tensor structures and applying the restrictions that follow from symmetry under the interchange of (p, ν) and (k, λ) and the QED Ward identities (19.51). This leaves three possible structures:

$$\mathcal{M}^{\mu\nu\lambda} = q^\mu \epsilon^{\nu\lambda\alpha\beta}p_\alpha k_\beta \mathcal{M}_1 + \left(\epsilon^{\mu\nu\alpha\beta}k^\lambda - \epsilon^{\mu\lambda\alpha\beta}p^\nu\right)k_\alpha p_\beta \mathcal{M}_2 \\ + \left[\left(\epsilon^{\mu\nu\alpha\beta}p^\lambda - \epsilon^{\mu\lambda\alpha\beta}k^\nu\right)k_\alpha p_\beta - \epsilon^{\mu\nu\lambda\sigma}(p-k)_\sigma p \cdot k\right]\mathcal{M}_3. \tag{19.110}$$

The second term satisfies (19.51) by virtue of the on-shell conditions $p^2 = k^2 = 0$.

Now contract (19.110) with (iq_μ) to take the divergence of the axial vector current. We find

$$iq_\mu \mathcal{M}^{\mu\nu\lambda} = iq^2 \epsilon^{\nu\lambda\alpha\beta}p_\alpha k_\beta \mathcal{M}_1 - i\epsilon^{\mu\nu\lambda\sigma}q_\mu(p-k)_\sigma p \cdot k\mathcal{M}_3; \tag{19.111}$$

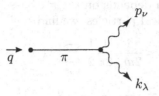

Figure 19.8. Contribution that leads to a pole in the axial vector current form factor $\mathcal{M}_1$.

the other terms automatically give zero. Using $q = p + k$, $q^2 = 2p \cdot k$, we can simplify this to

$$iq_\mu \mathcal{M}^{\mu\nu\lambda} = iq^2 \epsilon^{\nu\lambda\alpha\beta} p_\alpha k_\beta (\mathcal{M}_1 + \mathcal{M}_3). \tag{19.112}$$

The whole quantity is proportional to q^2 and apparently vanishes in the limit $q^2 \to 0$. This contrasts with the prediction of the axial vector anomaly. Taking the matrix element of the right-hand side of (19.108), we find

$$iq_\mu \mathcal{M}^{\mu\nu\lambda} = -\frac{e^2}{4\pi^2} \epsilon^{\nu\lambda\alpha\beta} p_\alpha k_\beta. \tag{19.113}$$

The conflict can be resolved if one of the form factors appearing in (19.112) contains a pole in q^2. Such a pole can arise through the process shown in Fig. 19.8, in which the current creates a π^0 meson which subsequently decays to two photons. The amplitude for the current to create the meson is given by (19.88). Let us parametrize the pion decay amplitude as

$$i\mathcal{M}(\pi^0 \to 2\gamma) = iA\epsilon_\nu^* \epsilon_\lambda^* \epsilon^{\nu\lambda\alpha\beta} p_\alpha k_\beta, \tag{19.114}$$

where A is a constant to be determined. Then the contribution of the process of Fig. 19.8 to the amplitude $\mathcal{M}^{\mu\nu\lambda}$ defined in (19.109) is

$$\left(iq^\mu f_\pi\right) \frac{i}{q^2} \left(iA\epsilon^{\nu\lambda\alpha\beta} p_\alpha k_\beta\right). \tag{19.115}$$

This is a contribution to the form factor $\mathcal{M}_1$,

$$\mathcal{M}_1 = \frac{-i}{q^2} f_\pi \cdot A, \tag{19.116}$$

plus terms regular at $q^2 = 0$. Now, by equating (19.112) to (19.113), we determine A in terms of the coefficient of the anomaly:

$$A = \frac{e^2}{4\pi^2} \frac{1}{f_\pi}. \tag{19.117}$$

From the decay matrix element (19.114), it is straightforward to work out the decay rate of the π^0. Note that, though we have worked out the decay matrix element in the limit of a massless π^0, we must supply the physically correct kinematics which depends on the π^0 mass. Including a factor $1/2$ for the phase space of identical particles, we find

$$\Gamma(\pi^0 \rightarrow 2\gamma) = \frac{1}{2m_\pi} \frac{1}{8\pi} \frac{1}{2} \sum_{\text{pols.}} |\mathcal{M}(\pi^0 \rightarrow 2\gamma)|^2$$

$$= \frac{1}{32\pi m_\pi} \cdot A^2 \cdot 2(p \cdot k)^2 \tag{19.118}$$

$$= A^2 \cdot \frac{m_\pi^3}{64\pi}.$$

Thus, finally,

$$\Gamma(\pi^0 \rightarrow 2\gamma) = \frac{\alpha^2}{64\pi^3} \frac{m_\pi^3}{f_\pi^2}. \tag{19.119}$$

This relation, which provides a direct measure of the coefficient of the Adler-Bell-Jackiw anomaly, is satisfied experimentally to an accuracy of a few percent.

19.4 Chiral Anomalies and Chiral Gauge Theories

Up to this point, we have coupled gauge fields to fermions in a parity-symmetric manner, replacing the derivative in the Dirac equation by a covariant derivative. This procedure couples the gauge field to the vector current of fermions. However, this procedure gives only a subset of the possible couplings of fermions to gauge bosons. In this section we will construct more general, parity-asymmetric, couplings and discuss their interplay with the axial vector anomaly.

We will focus primarily on theories of massless fermions. If the Lagrangian contains no fermion mass terms, it has no terms that mix the two helicity states of a Dirac fermion. Thus, in a theory that contains massless Dirac fermions ψ_i, we can write the kinetic energy term in the helicity basis (3.36) as

$$\mathcal{L} = \psi_{Li}^\dagger i\bar{\sigma} \cdot \partial \psi_{Li} + \psi_{Ri}^\dagger i\sigma \cdot \partial \psi_{Ri}. \tag{19.120}$$

There is no difficulty in coupling this system to a gauge field by assigning the left-handed fields ψ_{Li} to one representation of the gauge group G and assigning the right-handed fields to a different representation. For example, we might assign the left-handed fields to a representation r of G and take the right-handed fields to be invariant under G. This gives

$$\mathcal{L} = \psi_{Li}^\dagger i\bar{\sigma} \cdot D\psi_{Li} + \psi_{Ri}^\dagger i\sigma \cdot \partial \psi_{Ri}, \tag{19.121}$$

with $D_\mu = \partial_\mu - igA_\mu^a t_r^a$. In more conventional notation, (19.121) becomes

$$\mathcal{L} = \bar{\psi} i\gamma^\mu \Big(\partial_\mu - igA_\mu^a t_r^a \Big(\frac{1-\gamma^5}{2} \Big) \Big) \psi. \tag{19.122}$$

It is straightforward to verify that the classical Lagrangian (19.122) is invariant to the local gauge transformation

$$\psi \rightarrow \left(1 + i\alpha^a t^a \left(\frac{1-\gamma^5}{2}\right)\right)\psi,$$

$$A_\mu^a \rightarrow A_\mu^a + \frac{1}{g}\partial_\mu \alpha^a + f^{abc}A_\mu^b \alpha^c, \qquad (19.123)$$

which generalizes (15.46). Since the right-handed fields are free fields, we can even eliminate these fields and write a gauge-invariant Lagrangian for purely left-handed fermions.

The idea of gauge fields that couple only to left-handed fermions plays a central role in the construction of a theory of weak interactions. The coupling of the W boson to quarks and leptons described in (17.31) can be derived by assigning the left-handed components of quarks and leptons to doublets of an $SU(2)$ gauge symmetry

$$Q_L = \begin{pmatrix} u \\ d \end{pmatrix}_L, \qquad L_L = \begin{pmatrix} \nu \\ \ell \end{pmatrix}_L, \qquad (19.124)$$

and then identifying the W bosons as gauge fields that couple to this $SU(2)$ group. In this picture, it is the restriction of the symmetry to left-handed fields that leads to the helicity structure of the weak interaction effective Lagrangian. We will discuss a complete, explicit model of weak interactions, incorporating this idea, in the next chapter.

To work out the general properties of chirally coupled fermions, it is useful to rewrite their Lagrangian with one further transformation. Below Eq. (3.38), we noted that the quantity $\sigma^2 \psi_R^*$ transforms under Lorentz transformations as a left-handed field. Thus it is useful to rewrite the right-handed components in (19.120) as new left-handed fermions, by defining

$$\psi_{Li}' = \sigma^2 \psi_{Ri}^*, \qquad \psi_{Li}'^\dagger = \psi_{Ri}^T \sigma^2. \qquad (19.125)$$

This transformation relabels the right-handed fermions as antifermions and calls their left-handed antiparticles a new species of left-handed fermions. By using (3.38), we can rewrite the Lagrangian for the right-handed fermions as

$$\int d^4x\, \psi_{Ri}^\dagger i\sigma \cdot \partial\psi_{Ri} = \int d^4x\, \psi_{Li}'^\dagger i\bar{\sigma} \cdot \partial\psi_{Li}'. \qquad (19.126)$$

The minus sign from fermion interchange cancels the minus sign from integration by parts. Notice that, if the fermions are coupled to gauge fields in the representation r, this manipulation changes the covariant derivative as follows:

$$\psi_R^\dagger i\sigma \cdot \left(\partial - igA^a t_r^a\right)\psi_R = \psi_L'^\dagger i\bar{\sigma} \cdot \left(\partial + igA^a (t_r^a)^T\right)\psi_L'$$

$$= \psi_L'^\dagger i\bar{\sigma} \cdot \left(\partial - igA^a t_{\bar{r}}^a\right)\psi_L'. \qquad (19.127)$$

Thus the new fields ψ_L' belong to the conjugate representation to r, for which the representation matrices are given by (15.82). In this notation, QCD with

n_f flavors of massless fermions is rewritten as an $SU(3)$ gauge theory coupled to n_f massless fermions in the 3 and n_f massless fermions in the $\bar{3}$ representation of $SU(3)$. The most general gauge theory of massless fermions would simply assign left-handed fermions to an arbitrary, reducible representation R of the gauge group G. We have just seen that rewriting a system of Dirac fermions leads to $R = r \oplus \bar{r}$, a *real* representation in the sense described below (15.82). Conversely, if R is not a real representation, then the theory cannot be rewritten in terms of Dirac fermions and is intrinsically chiral.

The rewriting (19.125) transforms the mass term of the QCD Lagrangian as follows:

$$ m\bar{\psi}_i\psi_i = m\big(\psi_R^\dagger\psi_L + \text{h.c.}\big) = -m\big(\psi_{Li}'^T\sigma^2\psi_{Li} + \text{h.c.}\big). \tag{19.128} $$

This has the form of the *Majorana* mass term that we encountered in Problem 3.4. The most general mass term that can be built purely from left-handed fermion fields is

$$ \Delta\mathcal{L}_M = M_{ij}\psi_{Li}^T\sigma^2\psi_{Lj} + \text{h.c.} \tag{19.129} $$

The matrix M_{ij} is symmetric under $i \leftrightarrow j$, since the minus sign from the antisymmetry of σ^2 is compensated by a minus sign from fermion interchange. This mass term is gauge invariant if M_{ij} is invariant under G. For example, the mass term in (19.128) couples 3 and $\bar{3}$ indices together in an $SU(3)$ singlet combination. In general, a gauge-invariant mass term exists if the representation containing the fermions is *strictly real*, in the sense described below (15.82). In an intrinsically chiral theory, there is no possible gauge-invariant mass term. We will see in the next chapter that, in the gauge theory of the weak interactions, mass terms for the quarks and leptons are forbidden by gauge invariance. We will present a solution to this problem in Section 20.2.

At the classical level, there is no restriction on the representation R of the left-handed fermions. However, at the level of one-loop corrections, many possible choices become inconsistent due to the axial vector anomaly. In a gauge theory of left-handed massless fermions, consider computing the diagrams of Fig. 19.9, in which the external fields are non-Abelian gauge bosons and the marked vertex represents the gauge symmetry current

$$ j^{\mu a} = \bar{\psi}\gamma^\mu\Big(\frac{1-\gamma^5}{2}\Big)t^a\psi. \tag{19.130} $$

The gauge boson vertices also contain factors of $(1 - \gamma^5)/2$. The three projectors can be moved together into a single factor. Then, if we regularize this diagram as in Section 19.2, the term containing a γ^5 has an axial vector anomaly that leads to the relation

$$ \langle p, \nu, b; k, \lambda, c| \, \partial_\mu j^{\mu a} \, |0\rangle = \frac{g^2}{8\pi^2}\epsilon^{\alpha\nu\beta\lambda}p_\alpha k_\beta \cdot \mathcal{A}^{abc}, \tag{19.131} $$

where $\mathcal{A}^{abc}$ is a trace over group matrices in the representation R:

$$ \mathcal{A}^{abc} = \text{tr}\big[t^a\{t^b, t^c\}\big]. \tag{19.132} $$

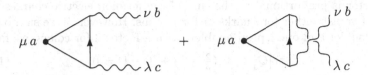

Figure 19.9. Diagrams contributing to the anomaly of a gauge symmetry current in a chiral gauge theory.

This equation implies that, unless $\mathcal{A}^{abc}$ vanishes, the current $j^{\mu a}$ is not conserved. The factor (19.132) is totally symmetric in (a, b, c), so this condition is independent of which current is treated as an external operator. As we described in Sections 19.1 and 19.2, we can change the regularization of the diagram so that the external current is conserved, but only at the price of violating the conservation of one of the other two currents in the diagram.

Since the whole construction of a theory with local gauge invariance is based on the existence of an exact global symmetry, the violation of the conservation of $j^{\mu a}$ does violence to the structure of the theory. For example, triangle diagrams of the form of Fig. 19.9 will now generate divergent gauge boson mass terms and will upset the delicate relations between three- and four-point vertices discussed in Chapter 16. These relations, following from the Ward identity, were necessary to insure the cancellation of unphysical states and the unitarity of the S-matrix. The only way to avoid this problem is to insist that $\mathcal{A}^{abc} = 0$ as a fundamental consistency condition for chiral gauge theories.[†] Gauge theories satisfying this condition are said to be *anomaly free*.

As an example of the application of this condition, consider the prototype weak interaction gauge theory that we presented in (19.124). If the two gauge bosons in Fig. 19.9 are $SU(2)$ gauge bosons and the current $j^{\mu a}$ is an $SU(2)$ gauge current, we would evaluate (19.132) by substituting $t^a = \tau^a = \sigma^a/2$ and using the relation $\{\sigma^b, \sigma^c\} = 2\delta^{bc}$. This gives

$$\mathcal{A}^{abc} = \frac{1}{8}\,\mathrm{tr}\big[\sigma^a \cdot 2\delta^{bc}\big] = 0, \tag{19.133}$$

so the consistency condition is satisfied. If the fermions in (19.124) also couple to electromagnetism, there is an additional consistency condition that we would find by taking the current in Fig. 19.9 to be the electromagnetic current. The factor $\mathcal{A}^{abc}$ for this case is

$$\mathrm{tr}\big[Q\{\tau^b, \tau^c\}\big], \tag{19.134}$$

where Q is the matrix of electric charges. If we simplify as in (19.133), the trace (19.134) becomes

$$\tfrac{1}{2}\,\mathrm{tr}\big[Q\big]\delta^{bc}. \tag{19.135}$$

[†]D. J. Gross and R. Jackiw, *Phys. Rev.* **D6**, 477 (1972).

This factor is proportional to the sum of the fermion electric charges, which does not vanish either for quarks or for leptons. However, if we sum over one quark doublet and one lepton doublet, with a factor 3 for colors, we find

$$\text{tr}[Q] = 3 \cdot (\tfrac{2}{3} - \tfrac{1}{3}) + (0 - 1) = 0. \tag{19.136}$$

Remarkably, the weak interaction gauge theory described by (19.124) can be consistently combined with QED only if the theory contains equal numbers of quark and lepton doublets.

We complete this section by working out more generally the condition that a chiral gauge theory be anomaly free. We will first derive some basic properties of the anomaly factor $\mathcal{A}^{abc}$ and then apply these to chiral gauge theories with simple gauge groups.

If the fermion representation R is real, R is equivalent to its conjugate reprsentation $\bar{R}$. Thus, as we described below (15.82), t_R^a is related by a unitary transformation to $t_{\bar{R}}^a = -(t_R^a)^T$. Since (19.132) is invariant to unitary transformations of the t^a, we can replace t_R^a by $t_{\bar{R}}^a$. Then

$$\begin{aligned}
\mathcal{A}^{abc} &= \text{tr}\big[(-t^a)^T\{(-t^b)^T, (-t^c)^T\}\big] \\
&= -\text{tr}\big[\{t^c, t^b\}t^a\big] \\
&= -\mathcal{A}^{abc}.
\end{aligned} \tag{19.137}$$

Thus, if R is real, the gauge theory is automatically anomaly free. As a special case, any gauge theory of Dirac fermions is anomaly free.

In more general circumstances, we can simplify the calculation of $\mathcal{A}^{abc}$ by noting that it is an invariant of the gauge group G that is totally symmetric with three indices in the adjoint representation. For some possible groups, a suitable invariant may not exist, and in those cases $\mathcal{A}^{abc}$ must vanish. For example, in $SU(2)$ the adjoint representation has spin 1. The symmetric product of two spin-1 multiplets gives spin 0 plus spin 2, with no spin-1 component. Thus, there is no symmetric tensor coupling two spin-1 indices to give a spin 1. The factor (19.132) must then vanish in any $SU(2)$ gauge theory. We saw this happen in an explicit example in Eq. (19.133).

In $SU(n)$ groups, $n \geq 3$, there is a unique symmetric invariant d^{abc} of the required type. It appears in the anticommutator of representation matrices of the fundamental representation:

$$\{t_n^a, t_n^b\} = \tfrac{1}{n}\delta^{ab} + d^{abc}t_n^c. \tag{19.138}$$

The uniqueness of this invariant implies that, in an $SU(n)$ gauge theory, any trace of the form of (19.132) is proportional to d^{abc}. For each representation r, we can define an *anomaly coefficient* $A(r)$ by

$$\text{tr}\big[t_r^a\{t_r^b, t_r^c\}\big] = \tfrac{1}{2}A(r)d^{abc}. \tag{19.139}$$

For the fundamental representation, we can see from (19.138) that $A(n) = 1$. It follows from the argument of (19.137) that

$$A(\bar{r}) = -A(r). \tag{19.140}$$

For higher representations, the anomaly coefficients can be worked out using methods similar to those we used in Section 15.4 to compute $C_2(r)$. For example, we show in Problem 19.3 that, if a and s are the $SU(n)$ representations corresponding to antisymmetric and symmetric two-index tensors, then

$$A(a) = n - 4, \qquad A(s) = n + 4. \tag{19.141}$$

An $SU(n)$ gauge theory is anomaly free if the anomaly coefficients of the various irreducible components of the fermion multiplet R sum to zero. For example, the $SU(n)$ gauge theory of left-handed fermions with representation content

$$R = a + (n - 4)\bar{n} \tag{19.142}$$

is anomaly free.

Of the various simple Lie groups listed below (15.72), only $SU(n)$, $SO(4n+2)$, and E_6 have complex representations. Of these, only $SU(n)$ and $SO(6)$, which has the same Lie algebra as $SU(4)$, have a symmetric invariant of the type required to build the anomaly. Gauge theories based on $SO(4n+2)$, $n \geq 2$, and on E_6 are automatically anomaly free. The groups $SO(10)$ and E_6 have been suggested as candidates for the *grand unified* gauge symmetry of particle physics, which we will discuss in Section 22.2.

There is one further constraint on the representation content of a chiral gauge theory, which comes from considering its coupling to gravity. It is possible to show that the diagrams of Fig. 19.9 give an anomaly contribution when computed with a gauge current $j^{\mu a}$ and external gravitational fields. The group-theory factor that multiplies this diagram is

$$\mathrm{tr}\left[t_R^a\right]. \tag{19.143}$$

This factor automatically vanishes if the gauge group is non-Abelian. However, if the gauge group of the theory contains $U(1)$ factors, the theory cannot be consistently coupled to gravity unless each of the $U(1)$ generators is traceless.[‡]

Once we have constructed a consistent chiral gauge theory, we have an additional problem of finding a prescription for calculating in this theory consistently. In a vector-like gauge theory, we can define ultraviolet-divergent diagrams with dimensional regularization. This guarantees that the divergent diagrams will be regulated in a way that respects the Ward identities of local gauge invariance. To generalize dimensional regularization to chiral gauge theories, we need to introduce a dimensional continuation of γ^5. The 't Hooft-Veltman definition used to define the chiral current in Section 19.2 is not satisfactory, because this definition does not manifestly respect the conservation of the gauge currents. A useful alternative procedure is to define γ^5 formally as an object that anticommutes with all of the γ^μ. This prescription gives unambiguous, gauge-invariant results for amplitudes that are not proportional to $\epsilon^{\mu\nu\lambda\sigma}$, at least through two-loop order. In Section 21.3, we will

---

[‡]L. Alvarez-Gaumé and E. Witten, *Nucl. Phys.* **B234**, 269 (1984).

use this prescription to compute loop diagrams in weak interaction theory. As a last resort, one can always compute with a non-gauge-invariant regulator and add non-gauge-invariant counterterms to the theory so that the gauge theory Ward identities remain valid.

19.5 Anomalous Breaking of Scale Invariance

There is one more important example of a symmetry that is an invariance at the classical level and is broken by quantum corrections. This is the classical scale invariance of a massless field theory with a dimensionless coupling constant. In Chapter 12, we saw that a quantum field theory with no classical dimensionful parameters still depends on a mass scale through the regularization of ultraviolet divergences, or, equivalently, through the running of coupling constants. We have already seen how to analyze this induced dependence on the renormalization scale using the Callan-Symanzik equation. In this section, we will show how the violation of classical scale invariance by quantum corrections can be described as a current conservation anomaly.

In this book we have avoided giving a careful treatment of the energy-momentum tensor of a quantum field theory. In Section 2.2, we used Noether's theorem to demonstrate that the invariance of a quantum field theory under spacetime translations implies the presence of a conserved tensor $T^{\mu\nu}$. In Section 9.6, we gave an alternative derivation of this result using the functional integral formalism. However, to discuss the theory of scale invariance, we will need some more detailed properties of the energy-momentum tensor. We will now simply state these properties and refer elsewhere for their derivations.[*]

The tensor $T^{\mu\nu}$ defined by expressions (2.17) and (9.99) is called the *canonical energy-momentum tensor*. The expressions that defined this tensor do not imply that $T^{\mu\nu}$ is symmetric. In fact, this tensor need not be symmetric, and, in a gauge theory, it need not be gauge-invariant. However, it is always possible to convert $T^{\mu\nu}$ into a symmetric and gauge-invariant tensor $\Theta^{\mu\nu}$ by the addition

$$\Theta^{\mu\nu} = T^{\mu\nu} + \partial_\lambda \Sigma^{\mu\nu\lambda}, \tag{19.144}$$

where $\Sigma^{\mu\nu\lambda}$ is antisymmetric under interchange of μ and λ. The form of the added term implies that $\Theta^{\mu\nu}$ is conserved if $T^{\mu\nu}$ is, and that the global energy-momentum four-vector is unchanged,

$$P^\nu = \int d^3x \, T^{0\nu} = \int d^3x \, \Theta^{0\nu}. \tag{19.145}$$

A scale transformation of a scalar field theory can be defined as a transformation of variables

$$\phi(x) \rightarrow e^{-D\sigma}\phi(xe^{-\sigma}), \tag{19.146}$$

---

[*]The conclusions presented in the next three paragraphs are derived with care in C. G. Callan, S. Coleman, and R. Jackiw, *Ann. Phys.* **59**, 42 (1970).

with $D = 1$, the canonical mass dimension of the field. The scale transformation is defined similarly in theories of fermion and gauge fields. If this transformation is an invariance of the classical Lagrangian, as it will be if there are no dimensionful couplings, this theory will possess a conserved current D^μ, called the *dilatation current*. The dilatation current has a simple relation to the symmetric energy-momentum tensor $\Theta^{\mu\nu}$: $D^\mu = \Theta^{\mu\nu} x_\nu$, so that

$$\partial_\mu D^\mu = \Theta^\mu{}_\mu. \tag{19.147}$$

The derivation of these results from Noether's theorem is not straightforward. There is a simpler derivation, which, however, uses formalism beyond the scope of this book. If the quantum field theory under consideration is coupled to gravity, then the energy-momentum tensor can be identified as the source of the gravitational field. This energy-momentum tensor can be found by varying the Lagrangian $\mathcal{L}_m$ of matter fields with respect to the spacetime metric $g_{\mu\nu}(x)$. This construction gives a manifestly symmetric and gauge-invariant tensor, which turns out to be $\Theta^{\mu\nu}$:

$$\Theta^{\mu\nu} = 2 \frac{\delta}{\delta g_{\mu\nu}(x)} \int d^4x \, \mathcal{L}_m. \tag{19.148}$$

A scale transformation can be represented as a change in the spacetime metric

$$g_{\mu\nu}(x) \rightarrow e^{2\sigma} g_{\mu\nu}(x). \tag{19.149}$$

Combining (19.148) and (19.149), we see that the change in the Lagrangian induced by this transformation is just the trace of $\Theta^{\mu\nu}$. This will be equal by Noether's theorem to the divergence of the corresponding current, giving us back Eq. (19.147).

In QED, it is not hard to guess the form of the symmetric energy-momentum tensor:

$$\Theta^{\mu\nu} = -F^{\mu\lambda} F^\nu{}_\lambda + \tfrac{1}{4} g^{\mu\nu} (F_{\lambda\sigma})^2 + \tfrac{1}{2}\bar\psi i(\gamma^\mu D^\nu + \gamma^\nu D^\mu)\psi - g^{\mu\nu}\bar\psi(i\slashed{D} - m)\psi. \tag{19.150}$$

This is a gauge-invariant symmetric tensor that leads to the familiar expression for the total energy,

$$H = \int d^3x \left\{ \tfrac{1}{2}(E^2 + B^2) + \psi^\dagger(-i\gamma^0\boldsymbol{\gamma} \cdot \boldsymbol{\nabla} + m)\psi \right\}. \tag{19.151}$$

For future reference, we note that these results are true at the classical level in any spacetime dimension d. In four dimensions, the trace of the gauge field terms cancels automatically. After using the Dirac equation, which is valid as an operator equation of motion, we find that the trace of $\Theta^{\mu\nu}$ is given by

$$\Theta^\mu{}_\mu = m\bar\psi\psi \tag{19.152}$$

and indeed vanishes, classically, if $m = 0$.

When quantum corrections are included, we know that a scale transformation is not a symmetry of the theory, since the same theory referred to a larger scale contains a different value of the renormalized coupling constant. The shift of the renormalized coupling is

$$g \rightarrow g + \sigma\beta(g), \tag{19.153}$$

and the corresponding change in the Lagrangian is

$$\sigma\beta(g)\frac{\partial}{\partial g}\mathcal{L}. \tag{19.154}$$

Thus, when quantum corrections are included, the equation for the dilatation current in a classically scale-invariant theory should read

$$\partial_\mu D^\mu = \Theta^\mu{}_\mu = \beta(g)\frac{\partial}{\partial g}\mathcal{L}. \tag{19.155}$$

In massless QED, we can write this formula most usefully by rescaling the gauge fields so that the coupling constant e is removed from the covariant derivative: $eA^\mu \rightarrow A^\mu$. Then e appears only in the term

$$\mathcal{L} = -\frac{1}{4e^2}(F_{\lambda\sigma})^2 + \cdots, \tag{19.156}$$

and Eq. (19.155) reads

$$\Theta^\mu{}_\mu = \frac{\beta(e)}{2e^3}(F_{\lambda\sigma})^2. \tag{19.157}$$

This relation, which says that the trace of the symmetric energy-momentum tensor takes a nonzero value as a result of quantum corrections, is known as the *trace anomaly*.

We should be able to check the trace anomaly equation (19.157) directly in perturbation theory. We now evaluate the trace of $\Theta^{\mu\nu}$ explicitly to one-loop order. The formalism we have set up is very similar to that of the background field calculation of the β function done in Section 16.6. As in that section, we will integrate over the fluctuating parts of quantum fields in the presence of a background field with a nonzero $F_{\mu\nu}$. Equation (19.157) predicts that this integration will lead to the expression

$$\langle \Theta^\mu{}_\mu \rangle = C \int \frac{d^4k}{(2\pi)^4} A_\mu(-k)(k^2 g^{\mu\nu} - k^\mu k^\nu)A_\nu(k), \tag{19.158}$$

where A_μ is the background field and the constant C is equal to $\beta(e)/e^3$.

Since we will be using dimensional regularization, we should begin by writing the trace of $\Theta^{\mu\nu}$ in d dimensions:

$$\Theta^\mu{}_\mu = -\frac{4-d}{4}(F_{\lambda\sigma})^2 + (1-d)\bar{\psi}i\slashed{D}\psi. \tag{19.159}$$

The one-loop matrix element of this quantity proportional to two powers of the background field arises from the three diagrams shown in Fig. 19.10. Since the second term on the right-hand side of (19.159) vanishes by the equation of motion of $\psi(x)$, one might expect that this term gives zero contribution to the trace. Indeed, it is easy to check that the first two diagrams in Fig.

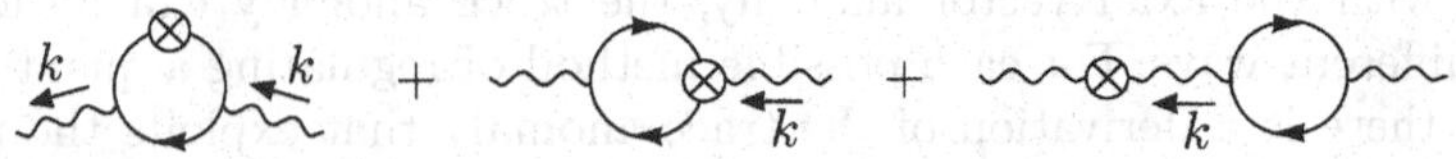

Figure 19.10. One-loop diagrams contributing to the anomalous trace of $\Theta^{\mu\nu}$.

19.10 cancel: These diagrams have the same structure, since the first has an extra propagator and an extra factor $\not{p}$ from the operator matrix element, and opposite overall signs.

The first term on the left-hand side of (19.159) is unexpected, since it apparently vanishes in four dimensions. However, the fermion loop diagram is divergent, and in dimensional regularization, this introduces a factor $1/(2 - d/2)$. As a result, this diagram gives a nonzero contribution to the operator matrix element. In massless QED, the fermion loop diagram has the value

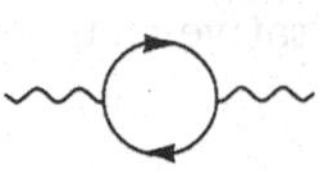

$$
\text{} = -i(k^2 g^{\mu\nu} - k^\mu k^\nu)\frac{4}{3(4\pi)^2}\left(\Gamma(2-\tfrac{d}{2}) + (\text{finite})\right). \quad (19.160)
$$

Then the complete expression for the third diagram in Fig. 19.10 is

$$
\int \frac{d^4 k}{(2\pi)^4}\, A_\mu(-k)\left(-2\frac{4-d}{4}\right)(k^2 g^{\mu\nu} - k^\mu k^\nu)\frac{-i}{k^2}\left(-ik^2 \frac{4}{3(4\pi)^2}\frac{1}{2-d/2}\right)A_\nu(k). \quad (19.161)
$$

This is of the form of (19.158), with

$$
C = \frac{1}{12\pi^2}, \quad (19.162)
$$

which is indeed the first β function coefficient in massless QED.

This discussion generalizes to QCD and other gauge theories. In a non-Abelian gauge theory, $\Theta^{\mu\nu}$ is given by the obvious generalization of (19.150) with the Abelian field-strength tensor $F_{\mu\nu}$ replaced by the non-Abelian expression $F^a_{\mu\nu}$. The trace of $\Theta^{\mu\nu}$ is again given by

$$
\Theta^\mu{}_\mu = -\frac{4-d}{4}(F^a_{\lambda\sigma})^2, \quad (19.163)
$$

plus terms that vanish by the equations of motion. In the background field gauge, the one-loop diagrams with the operator $\Theta^\mu{}_\mu$ inserted into the loop cancel as above. We saw in Section 16.6 that the two-point functions in this gauge sum to

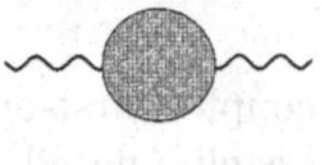

$$
\text{} = -i(k^2 g^{\mu\nu} - k^\mu k^\nu)\left[\frac{-b_0}{(4\pi)^2}\right]\left(\Gamma(2-\tfrac{d}{2}) + (\text{finite})\right), \quad (19.164)
$$

where $\beta(g) = -b_0 g^3/(4\pi)^2$. Following through the logic of the previous paragraph, we again find the result (19.158) with the identification of C as the first β function coefficient.

As with the axial vector anomaly, the trace anomaly can be found in many different ways. For each possible method of regulating a quantum field theory, there is a derivation of the trace anomaly that exploits the possible pathology of that particular regulator. For example, if one uses a Pauli-Villars regulator with heavy fermions to cancel the divergence of the QED fermion loop diagram, the heavy fermions Ψ contribute a term $M\bar{\Psi}\Psi$ to the trace of $\Theta^{\mu\nu}$. The loop diagram with this term inserted turns out to have a finite limit as $M \to \infty$, which precisely reproduces the trace anomaly. This computation is worked out in Problem 19.4.

As with the axial vector anomaly, each derivation of the anomaly with a different regulator, taken individually, seems artificial, as if there were a problem with the field theory that we are not quite clever enough to fix. Eventually, though, we are forced to conclude that the quantum field theory is trying to tell us something. The anomalous symmetries of the classical theory cannot be promoted to symmetries of the quantum theory. Instead, the anomalous conservation laws require profound and qualitative changes in the theory from the classical to the quantum level.

Problems

19.1 Fermion number nonconservation in parallel E and B fields.

(a) Show that the Adler-Bell-Jackiw anomaly equation leads to the following law for global fermion number conservation: If N_R and N_L are, respectively, the numbers of right- and left-handed massless fermions, then

$$\Delta N_R - \Delta N_L = -\frac{e^2}{2\pi^2} \int d^4x \, \mathbf{E} \cdot \mathbf{B}.$$

To set up a solvable problem, take the background field to be $A^\mu = (0,0,Bx^1,A)$, with B constant and A constant in space and varying only adiabatically in time.

(b) Show that the Hamiltonian for massless fermions represented in the components (3.36) is

$$H = \int d^3x \left[\psi_R^\dagger(-i\boldsymbol{\sigma} \cdot \mathbf{D})\psi_R - \psi_L^\dagger(-i\boldsymbol{\sigma} \cdot \mathbf{D})\psi_L \right],$$

with $D^i = \nabla^i - ieA^i$. Concentrate on the term in the Hamiltonian that involves right-handed fermions. To diagonalize this term, one must solve the eigenvalue equation $-i\boldsymbol{\sigma} \cdot \mathbf{D}\psi_R = E\psi_R$.

(c) The ψ_R eigenvectors can be written in the form

$$\psi_R = \begin{pmatrix} \phi_1(x^1) \\ \phi_2(x^1) \end{pmatrix} e^{i(k_2x^2 + k_3x^3)}.$$

The functions ϕ_1 and ϕ_2, which depend only on x^1, obey coupled first-order differential equations. Show that, when one of these functions is eliminated, the other obeys the equation of a simple harmonic oscillator. Use this observation to find the single-particle spectrum of the Hamiltonian. Notice that the eigenvalues do not depend on k_2.

(d) If the system of fermions is set up in a box with sides of length L and periodic boundary conditions, the momenta k_2 and k_3 will be quantized:

$$k_i = \frac{2\pi n_i}{L}.$$

By looking back to the harmonic oscillator equation in part (c), show that the condition that the center of the oscillation is inside the box leads to the condition

$$k_2 < eBL.$$

Combining these two conditions, we see that each level found in part (c) has a degeneracy of

$$\frac{eL^2 B}{2\pi}.$$

(e) Now consider the effect of changing the background A adiabatically by an amount (19.37). Show that the vacuum loses right-handed fermions. Repeating this analysis for the left-handed spectrum, one sees that the vacuum gains the same number of left-handed fermions. Show that these numbers are in accord with the global nonconservation law given in part (a).

19.2 Weak decay of the pion.

(a) In the effective Lagrangian for semileptonic weak interactions (18.28), the hadronic part of the operator is a left-handed current involving the u and d quarks. Show that this current is related to the quark currents of Section 19.3 as follows:

$$\bar{u}_L \gamma^\mu d_L = \frac{1}{2}(j^{\mu 1} + i j^{\mu 2} - j^{\mu 51} - i j^{\mu 52}),$$

where 1, 2 are isospin indices. Using this identification and (19.88), show that the amplitude for the decay $\pi^+ \to \ell^+ \nu$ is given by

$$i\mathcal{M} = G_F f_\pi \bar{u}(q)\not{p}(1 - \gamma^5)v(k),$$

where p, k, q are the momenta of the π^+, ℓ^+, ν.

(b) Compute the decay rate of the pion. Show that this rate vanishes in the limit of zero lepton mass, and that the relative rate of pion decay to muons and electrons is given by

$$\frac{\Gamma(\pi^+ \to e^+ \nu)}{\Gamma(\pi^+ \to \mu^+ \nu)} = \left(\frac{m_e}{m_\mu}\right)^2 \frac{(1 - m_e^2/m_\pi^2)^2}{(1 - m_\mu^2/m_\pi^2)^2} = 10^{-4}.$$

From the measured pion lifetime, $\tau_\pi = 2.6 \times 10^{-8}$ sec, and the pion and muon masses, $m_\pi = 140$ MeV, $m_\mu = 106$ MeV, determine the value of f_π.

19.3 Computation of anomaly coefficients.

(a) Consider a product $r_1 \times r_2$ of $SU(n)$ representations, which is decomposed into irreducible representations as in (15.98). Using the explicit form of the generators given in (15.99), show that the anomaly coefficients satisfy

$$d_1 A(r_2) + d_2 A(r_1) = \sum_i A(r_i).$$

(b) As we saw in Problem 15.5, the two-index symmetric and antisymmetric tensors form irreducible representations of $SU(n)$, which we will call s and a, respectively. In $SU(3)$, the representation a is three-dimensional. Show that it is equivalent to the $\bar{3}$. Compute the anomaly coefficients for a and s, making use of the identity in part (a).

(c) Since $SU(n)$ has a unique three-index symmetric tensor d^{abc} which is already nonvanishing in an $SU(3)$ subgroup, we can compute the anomaly coefficient in $SU(n)$ by restricting our attention to three generators in this subgroup. By decomposing $SU(n)$ representations into $SU(3)$ representations, compute the anomaly coefficients for a and s in $SU(n)$ and derive Eq. (19.141). Find the anomaly coefficient of the j-index totally antisymmetric tensor representation of $SU(n)$. Why does the result always vanish when $2j = n$?

19.4 Large fermion mass limits. In the text, we derived the Adler-Bell-Jackiw and trace anomalies by the use of dimensional regularization. As an alternative, one could imagine deriving these results using Pauli-Villars regularization. In that technique, one regularizes the value of a fermion loop integral by subtracting the value of the same loop diagram computed with fermions Ψ of large mass M. The parameter M plays the role of the cutoff and should be taken to infinity at the end of the calculation. The anomalies arise because some pieces of the diagrams computed for very heavy fermions do not disappear in the limit $M \to \infty$. These nontrivial $M \to \infty$ limits are interesting in their own right and can have physical applications (for example, in part (c) of the Final Project for Part III).

(a) Show that the Adler-Bell-Jackiw anomaly equation is equivalent to the following large-mass limit of a fermion matrix element between the vacuum and a two-photon state:

$$\lim_{M \to \infty} \left\{ \langle p, k| \, 2iM\bar{\Psi}\gamma^5\Psi \, |0\rangle \right\} = -\frac{e^2}{2\pi^2} \epsilon^{\alpha\nu\beta\lambda} p_\alpha \epsilon_\nu^*(p) k_\beta \epsilon_\lambda^*(k).$$

(b) Show that the trace anomaly, at one-loop order, is equivalent to the following large-mass limit:

$$\lim_{M \to \infty} \left\{ \langle p, k| \, M\bar{\Psi}\Psi \, |0\rangle \right\} = +\frac{e^2}{6\pi^2} [p \cdot k \, \epsilon^*(p) \cdot \epsilon^*(k) - p \cdot \epsilon^*(k) \, k \cdot \epsilon^*(p)].$$

(c) Show that the matrix element in part (a) is ultraviolet-finite before the $M \to \infty$ limit is taken. Evaluate the matrix element explicitly at one-loop order and verify the limit claimed in part (a).

(d) The matrix element in part (b) has a potential ultraviolet divergence. However, show that the coefficient of $(p \cdot \epsilon^*(k) k \cdot \epsilon^*(p))$ is ultraviolet-finite, and that the rest of the expression is determined by gauge invariance. Compute the full matrix element using dimensional regularization as a gauge-invariant regulator and verify the result claimed in part (b).

Chapter 20

Gauge Theories with Spontaneous Symmetry Breaking

In the course of this book, we have discussed three distinct fashions in which symmetries can be realized in a quantum field theory. The simplest case is a global symmetry that is manifest, leading to particle multiplets with restricted interactions. A second possibility is a global symmetry that is spontaneously broken. Then, as discussed in Chapter 11,* the symmetry currents are still conserved and interactions are similarly restricted, but the vacuum state does not respect the symmetry and the particles do not form obvious symmetry multiplets. Instead, such a theory contains massless particles, Goldstone bosons, one for each generator of the spontaneously broken symmetry. The third case is that of a local, or gauge, symmetry. As we saw in Chapter 15, such a symmetry requires the existence of a massless vector field for each symmetry generator, and the interactions among these fields are highly restricted.

It is now only natural to consider a fourth possibility: What happens if we include both local gauge invariance and spontaneous symmetry breaking in the same theory? In this chapter and the next, we will find that this combination of ingredients leads to new possibilities for the construction of quantum field theory models. We will see that spontaneous symmetry breaking requires gauge vector bosons to acquire mass. However, the interactions of these massive bosons are still constrained by the underlying gauge symmetry, and these constraints can have observable consequences.

In elementary particle physics, the principal application of spontaneously broken local symmetry is in the currently accepted model of weak interactions. This model, due to Glashow, Weinberg, and Salam, is introduced in Section 20.2. There we will see that it makes a number of precise and successful predictions for weak interaction phenomena. Remarkably, this model also unifies the weak interactions with electromagnetism in a single larger gauge theory.

---

*Section 11.1 is necessary background for the present chapter, but the rest of Chapter 11 is not.

20.1 The Higgs Mechanism

In this section we analyze some simple examples of gauge theories with spontaneous symmetry breaking. We begin with an Abelian gauge theory, and then study several examples of non-Abelian models.

An Abelian Example

As our first example, consider a complex scalar field coupled both to itself and to an electromagnetic field:

$$\mathcal{L} = -\tfrac{1}{4}(F_{\mu\nu})^2 + |D_\mu\phi|^2 - V(\phi), \tag{20.1}$$

with $D_\mu = \partial_\mu + ieA_\mu$. This Lagrangian is invariant under the local $U(1)$ transformation

$$\phi(x) \to e^{i\alpha(x)}\phi(x), \qquad A_\mu(x) \to A_\mu(x) - \frac{1}{e}\partial_\mu\alpha(x). \tag{20.2}$$

If we choose the potential in $\mathcal{L}$ to be of the form

$$V(\phi) = -\mu^2\phi^*\phi + \frac{\lambda}{2}(\phi^*\phi)^2, \tag{20.3}$$

with $\mu^2 > 0$, the field ϕ will acquire a vacuum expectation value and the $U(1)$ global symmetry will be spontaneously broken. The minimum of this potential occurs at

$$\langle\phi\rangle = \phi_0 = \left(\frac{\mu^2}{\lambda}\right)^{1/2}, \tag{20.4}$$

or at any other value related by the $U(1)$ symmetry (20.2).

Let us expand the Lagrangian (20.1) about the vacuum state (20.4). Decompose the complex field $\phi(x)$ as

$$\phi(x) = \phi_0 + \frac{1}{\sqrt{2}}\big(\phi_1(x) + i\phi_2(x)\big). \tag{20.5}$$

The potential (20.3) is rewritten

$$V(\phi) = -\frac{1}{2\lambda}\mu^4 + \frac{1}{2}\cdot 2\mu^2\phi_1^2 + \mathcal{O}(\phi_i^3), \tag{20.6}$$

so that the field ϕ_1 acquires the mass $m = \sqrt{2}\mu$ and ϕ_2 is the massless Goldstone boson. So far, this whole analysis follows that in Section 11.1.

But now consider how the kinetic energy term of ϕ is transformed. Inserting the expansion (20.5), we rewrite

$$|D_\mu\phi|^2 = \tfrac{1}{2}(\partial_\mu\phi_1)^2 + \tfrac{1}{2}(\partial_\mu\phi_2)^2 + \sqrt{2}e\phi_0\cdot A_\mu\partial^\mu\phi_2 + e^2\phi_0^2 A_\mu A^\mu + \cdots, \tag{20.7}$$

where we have omitted terms cubic and quartic in the fields A_μ, ϕ_1, and ϕ_2. The last term written explicitly in (20.7) is a photon mass term

$$\Delta\mathcal{L} = \tfrac{1}{2}m_A^2 A_\mu A^\mu, \tag{20.8}$$

where the mass

$$m_A^2 = 2e^2\phi_0^2 \tag{20.9}$$

arises from the nonvanishing vacuum expectation value of ϕ. Notice that the sign of this mass term is correct; the physical spacelike components of A^μ appear in (20.8) as

$$\Delta\mathcal{L} = -\tfrac{1}{2}m_A^2(A^i)^2,$$

with the correct sign for a potential energy term.

In Chapter 7, and again in Chapter 16 for the non-Abelian case, we argued that a gauge boson cannot obtain a mass, unless this mass term is associated with a pole in the vacuum polarization amplitude. There is a counterexample to this result in two-dimensional spacetime; there, as we saw in Section 19.1, a pole of the required form can arise from the infrared singularity generated by a massless fermion pair. However, in four dimensions, a pole in the vacuum polarization amplitude can be created only by a massless scalar particle. Typically, in situations with unbroken symmetry, no such particle is available.

However, a model with a spontaneously broken continuous symmetry must have massless Goldstone bosons. These scalar particles have the quantum numbers of the symmetry currents, and therefore have just the right quantum numbers to appear as intermediate states in the vacuum polarization. In the model we are now discussing, we can see this pole arise explicitly in the following way: The third term in Eq. (20.7) couples the gauge boson directly to the Goldstone boson ϕ_2; this gives a vertex of the form

$$\mu\,\, \rule{0pt}{0pt} \underset{k}{\longleftarrow} \quad = i\sqrt{2}e\phi_0(-ik^\mu) = m_A k^\mu. \tag{20.10}$$

If we also treat the mass term (20.8) as a vertex in perturbation theory, then the leading-order contributions to the vacuum polarization amplitude give the expression

$$\rule{0pt}{0pt} = \rule{0pt}{0pt} + \rule{0pt}{0pt}$$

$$= i m_A^2 g^{\mu\nu} + (m_A k^\mu)\frac{i}{k^2}(-m_A k^\nu) \tag{20.11}$$

$$= i m_A^2 \left(g^{\mu\nu} - \frac{k^\mu k^\nu}{k^2} \right).$$

The Goldstone boson supplies exactly the right pole to make the vacuum polarization amplitude properly transverse.

Although the Goldstone boson plays an important formal role in this theory, it does not appear as an independent physical particle. The easiest way to see this is to make a particular choice of gauge, called the *unitarity gauge*. Using the local $U(1)$ gauge symmetry (20.2), we can choose $\alpha(x)$ in such a way that $\phi(x)$ becomes real-valued at every point x. With this choice, the field ϕ_2 is removed from the theory. The Lagrangian (20.1) becomes

$$\mathcal{L} = -\tfrac{1}{4}(F_{\mu\nu})^2 + (\partial_\mu\phi)^2 + e^2\phi^2 A_\mu A^\mu - V(\phi). \tag{20.12}$$

If the potential $V(\phi)$ favors a nonzero vacuum expectation value of ϕ, the gauge field acquires a mass; it also retains a coupling to the remaining, physical field ϕ_1.

This mechanism, by which spontaneous symmetry breaking generates a mass for a gauge boson, was explored and generalized to the non-Abelian case by Higgs, Kibble, Guralnik, Hagen, Brout, and Englert, and is now known as the *Higgs mechanism*. However, this mechanism had an earlier application to the theory of superconductivity. In Chapter 8, we constructed the Landau description of a second-order phase transition. To describe a superconductor, Landau and Ginzburg coupled this theory to an external electromagnetic field; they obtained precisely the Lagrangian (20.1). Since the gauge field acquires a nonzero mass, external electromagnetic fields penetrate a superconductor only to the depth m_A^{-1}. This explains the *Meissner effect*, the observed exclusion of macroscopic magnetic fields from a superconductor.

The role of the Goldstone boson in the Higgs mechanism is intricate, and seems mysterious at this level of the discussion. We first saw that the involvement of the Goldstone boson is necessary, as a matter of principle, in order for the gauge boson to acquire a mass. We then saw that the Goldstone boson can be formally eliminated from the theory. However, we might argue that the Goldstone boson has not completely disappeared. A massless vector boson has only two physical polarization states; we saw in Chapter 16 that the longitudinal polarization state cannot be produced, and appears in the formalism only to cancel other unphysical contributions. However, a massive vector boson must have three physical polarization states: In its rest frame, it is a spin-1 object, which can make no distinction between transverse and longitudinal polarizations. It is tempting to say that the gauge boson acquired its extra degree of freedom by *eating* the Goldstone boson. In Sections 21.1 and 21.2 we will clarify this picture, by studying the quantization and gauge-fixing of spontaneously broken gauge theories.

Systematics of the Higgs Mechanism

The Higgs mechanism extends straightforwardly to systems with non-Abelian gauge symmetry. It is not difficult to derive the general relation by which a set of scalar field vacuum expectation values leads to the appearance of gauge boson masses. Let us work out this relation and then apply it in a number of examples.

Consider a system of scalar fields ϕ_i that appear in a Lagrangian invariant under a symmetry group G, represented by the transformation

$$\phi_i \rightarrow (1 + i\alpha^a t^a)_{ij}\phi_j. \tag{20.13}$$

It is convenient to write the ϕ_i as real-valued fields, for example, writing n complex fields as $2n$ real fields. Then the group representation matrices t^a must be pure imaginary and, since they are Hermitian, antisymmetric. Let us write

$$t^a_{ij} = iT^a_{ij}, \tag{20.14}$$

so that the T^a are real and antisymmetric.

If we promote the symmetry group G to a local gauge symmetry, the covariant derivative on the ϕ_i is

$$D_\mu \phi = (\partial_\mu - igA^a_\mu t^a)\phi = (\partial_\mu + gA^a_\mu T^a)\phi.$$

Then the kinetic energy term for the ϕ_i is

$$\tfrac{1}{2}(D_\mu \phi_i)^2 = \tfrac{1}{2}(\partial_\mu \phi_i)^2 + gA^a_\mu(\partial_\mu \phi_i T^a_{ij}\phi_j) + \tfrac{1}{2}g^2 A^a_\mu A^{b\mu}(T^a\phi)_i(T^b\phi)_i. \tag{20.15}$$

Now let the ϕ_i acquire vacuum expectation values

$$\langle \phi_i \rangle = (\phi_0)_i, \tag{20.16}$$

and expand the ϕ_i about these values. The last term in Eq. (20.15) contains a term with the structure of a gauge boson mass,

$$\Delta \mathcal{L} = \tfrac{1}{2}m^2_{ab}A^a_\mu A^{b\mu}, \tag{20.17}$$

with the mass matrix

$$m^2_{ab} = g^2(T^a\phi_0)_i(T^b\phi_0)_i. \tag{20.18}$$

This matrix is positive semidefinite, since any diagonal element, in any basis, has the form

$$m^2_{aa} = g^2(T^a\phi_0)^2 \geq 0 \qquad \text{(no sum)}.$$

Thus, generically, all of the gauge bosons will receive positive masses. However, it may be that some particular generator T^a of G leaves the vacuum invariant:

$$T^a\phi_0 = 0.$$

In that case, the generator T^a will give no contribution to (20.18), and the corresponding gauge boson will remain massless.

As in the Abelian case, the gauge boson propagator receives a contribution from the Goldstone bosons, which is necessary to make the vacuum polarization amplitude transverse. To compute this contribution, we need the vertex that mixes gauge bosons and Goldstone bosons. This comes from the second term of the Lagrangian (20.15). When we insert the vacuum expectation value of the scalar field (20.16), this term becomes

$$\Delta \mathcal{L} = gA^a_\mu \partial_\mu \phi_i(T^a\phi_0)_i. \tag{20.19}$$

This interaction term does not involve all of the components of ϕ—only those that are parallel to a vector $T^a\phi_0$ for some choice of T^a. These vectors represent the infinitesimal rotations of the vacuum; thus the components ϕ_i that appear in (20.19) are precisely the Goldstone bosons. Using the fact that these bosons are massless, we can compute the counterpart, for the non-Abelian case, of the Goldstone boson diagram in Eq. (20.11):

$$\mu \atop a \;\; \text{\wavy}\!-\!\bullet\!-\!\bullet\!-\!\text{\wavy} \;\; {\nu \atop b} = \sum_j \left(g k^\mu (T^a \phi_0)_j \right) \frac{i}{k^2} \left(-g k^\nu (T^b \phi_0)_j \right). \quad (20.20)$$

The sum runs over those components j with a nonzero projection onto the space spanned by the $T^a \phi_0$, or equally well, over all j. This diagram is therefore proportional to the mass matrix (20.18). Combining this expression with the contribution to the vacuum polarization from (20.17), we find a properly transverse result,

$$\text{\wavy}\!-\!\bigcirc\!-\!\text{\wavy} = i m_{ab}^2 \left(g^{\mu\nu} - \frac{k^\mu k^\nu}{k^2} \right), \quad (20.21)$$

where m_{ab}^2 is given by Eq. (20.18).

Non-Abelian Examples

Let us now apply this general formalism to some specific examples of non-Abelian gauge theories. Consider first a model with an $SU(2)$ gauge field coupled to a scalar field ϕ that transforms as a spinor of $SU(2)$. The covariant derivative acting on ϕ is

$$D_\mu \phi = (\partial_\mu - i g A_\mu^a \tau^a) \phi, \quad (20.22)$$

where $\tau^a = \sigma^a / 2$. The square of this expression is the scalar field kinetic energy term.

If ϕ acquires a vacuum expectation value, we can use the freedom of $SU(2)$ rotations to write this expectation value in the form

$$\langle \phi \rangle = \frac{1}{\sqrt{2}} \begin{pmatrix} 0 \\ v \end{pmatrix}. \quad (20.23)$$

Then the gauge boson masses arise from

$$|D_\mu \phi|^2 = \frac{1}{2} g^2 \begin{pmatrix} 0 & v \end{pmatrix} \tau^a \tau^b \begin{pmatrix} 0 \\ v \end{pmatrix} A_\mu^a A^{b\mu} + \cdots. \quad (20.24)$$

We can symmetrize the matrix product under the interchange of a and b; using $\{\tau^a, \tau^b\} = \frac{1}{2} \delta^{ab}$, we find the mass term

$$\Delta \mathcal{L} = \frac{g^2 v^2}{8} A_\mu^a A^{a\mu}. \quad (20.25)$$

All three gauge bosons receive the mass

$$m_A = \frac{gv}{2}, \quad (20.26)$$

signaling that all three generators of $SU(2)$ are broken equally well by (20.23).

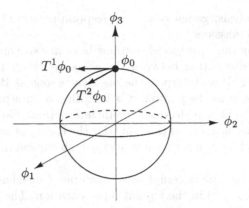

Figure 20.1. The space of configurations for a scalar field in the vector representation of $SU(2)$. When the $SU(2)$ symmetry is spontaneously broken, the allowed vacuum states lie on a spherical surface. If the vacuum expectation value ϕ_0 lies in the 3 direction, then the generator T^3 leaves ϕ_0 unchanged, while T^1 and T^2 rotate ϕ_0 in the directions shown.

What if we had taken ϕ to transform according to the vector representation of $SU(2)$? If we take ϕ to be a real-valued vector under $SU(2)$, we must assign it the covariant derivative

$$(D_\mu \phi)_a = \partial_\mu \phi_a + g\epsilon_{abc} A_\mu^b \phi_c. \tag{20.27}$$

Again, the ϕ kinetic energy term is the square of this object, and so, if ϕ acquires a vacuum expectation value, we find the gauge boson mass term

$$\Delta \mathcal{L} = \frac{1}{2}(D_\mu \phi)^2 = \frac{g^2}{2}\left(\epsilon_{abc} A_\mu^b (\phi_0)_c\right)^2 + \cdots. \tag{20.28}$$

If a vector of $SU(2)$ acquires an expectation value ϕ_0, we can choose our coordinates so that this vector points in any particular direction in the internal space. We will choose it to point in the 3 direction, as indicated in Fig. 20.1:

$$\langle \phi_c \rangle = (\phi_0)_c = V\delta_{c3}. \tag{20.29}$$

Inserting (20.29) into (20.28), we find

$$\Delta \mathcal{L} = \frac{g^2}{2} V^2 \left(\epsilon_{ab3} A_\mu^b\right)^2 = \frac{g^2}{2} V^2 \left((A_\mu^1)^2 + (A_\mu^2)^2\right). \tag{20.30}$$

The gauge bosons corresponding to the generators 1 and 2 acquire masses

$$m_1 = m_2 = gV, \tag{20.31}$$

while the boson corresponding to the generator 3 remains massless. It is easy to see the reason for this distinction by glancing at Fig. 20.1. The vacuum expectation value of ϕ_c destroys the symmetry of rotation about the axes 1 and 2, but it preserves the symmetry of rotation about the 3 axis. As we saw in our general analysis, gauge bosons corresponding to unbroken symmetry generators remain massless.

It is interesting that this model contains both massive and massless gauge bosons, with the distinction between these bosons created by spontaneous symmetry breaking. If we interpret the massive bosons as W bosons and the massless gauge boson as the photon, it is tempting to interpret this theory as a unified model of weak and electromagnetic interactions. Georgi and Glashow proposed this model as a serious candidate for the theory of weak interactions.[†] However, Nature chooses a different model, which we will discuss in the next section.

We turn next to a more complicated example. Consider an $SU(3)$ gauge theory with a scalar field in the adjoint representation. The covariant derivative of ϕ takes the form

$$D_\mu \phi_a = \partial_\mu \phi_a + g f_{abc} A_\mu^b \phi_c, \tag{20.32}$$

and so the gauge field masses arise from the term

$$\Delta \mathcal{L} = \frac{g^2}{2} \left(f_{abc} A_\mu^b \phi_c \right)^2. \tag{20.33}$$

We can write this more clearly by defining the quantity

$$\Phi = \phi_c t^c, \tag{20.34}$$

where t^c are the 3×3 traceless Hermitian matrices that represent the generators of $SU(3)$. Using this notation and the definition (15.68) of the structure constants, we can rewrite the mass term (20.33) as

$$\Delta \mathcal{L} = -g^2 \operatorname{tr} \left[[t^a, \Phi][t^b, \Phi] \right] A_\mu^a A^{b\mu}. \tag{20.35}$$

Now let Φ acquire a vacuum expectation value

$$\langle \Phi \rangle = \Phi_0. \tag{20.36}$$

Since Φ_0 is a traceless Hermitian matrix, we should analyze its effects by diagonalizing it. In principle, Φ_0 could have three arbitrary eigenvalues that sum to zero. However, when one minimizes explicit potential energy functions, one often finds expectation values that preserve some of the original symmetry. We will consider two examples.

First, Φ_0 might have the orientation

$$\Phi_0 = |\phi| \cdot \begin{pmatrix} 1 & & \\ & 1 & \\ & & -2 \end{pmatrix}. \tag{20.37}$$

[†]H. Georgi and S. L. Glashow, *Phys. Rev. Lett.* **28**, 1494 (1972).

This matrix commutes with the four $SU(3)$ generators

$$t^a = \begin{pmatrix} \tau^a & 0 \\ 0 & 0 \end{pmatrix}, \qquad t^8 = \frac{1}{2\sqrt{3}} \begin{pmatrix} 1 & & \\ & 1 & \\ & & -2 \end{pmatrix}. \tag{20.38}$$

Thus, the expectation value (20.37) breaks $SU(3)$ spontaneously to $SU(2) \times U(1)$ and leaves the gauge bosons corresponding to these four generators massless. The remaining four generators of $SU(3)$,

$$t^4 = \frac{1}{2} \begin{pmatrix} 0 & 0 & 1 \\ 0 & 0 & 0 \\ 1 & 0 & 0 \end{pmatrix}, \qquad t^5 = \frac{1}{2} \begin{pmatrix} 0 & 0 & -i \\ 0 & 0 & 0 \\ i & 0 & 0 \end{pmatrix},$$

$$t^6 = \frac{1}{2} \begin{pmatrix} 0 & 0 & 0 \\ 0 & 0 & 1 \\ 0 & 1 & 0 \end{pmatrix}, \qquad t^7 = \frac{1}{2} \begin{pmatrix} 0 & 0 & 0 \\ 0 & 0 & -i \\ 0 & i & 0 \end{pmatrix}, \tag{20.39}$$

acquire the masses

$$m^2 = \big(3g|\phi|\big)^2, \tag{20.40}$$

as one can check by substituting these matrices into Eq. (20.35).

Another possible orientation for Φ_0 is

$$\Phi_0 = |\phi| \cdot \begin{pmatrix} 1 & & \\ & -1 & \\ & & 0 \end{pmatrix}. \tag{20.41}$$

In this case, only t^3 and t^8 commute with Φ_0, so the original $SU(3)$ symmetry is broken down to $U(1) \times U(1)$. By substituting into (20.35), one can determine that the gauge bosons corresponding to the remaining generators of $SU(3)$ acquire the masses

$$t^1, t^2 : \quad m^2 = \big(2g|\phi|\big)^2, \qquad t^4, t^5, t^6, t^7 : \quad m^2 = \big(g|\phi|\big)^2. \tag{20.42}$$

Still larger symmetry groups offer a wider variety of symmetry-breaking patterns, and more complex mass matrices. We consider one further example in Problem 20.1.

Formal Description of the Higgs Mechanism

Up to this point, our study of the Higgs mechanism has been based on the explicit analysis of scalar field Lagrangians coupled to gauge fields. Scalar field theories provide the simplest examples of systems with spontaneous symmetry breaking, and the explicit calculations they allow are useful for visualization. But symmetries can be broken in other ways. In the theory of superconductivity, for example, the Abelian gauge invariance of electromagnetism is broken by pairs of electrons that condense in the ground state of a metal. In Section 19.3, we argued that, in the approximation that quark masses are very small, QCD possesses global symmetries that are spontaneously broken by a condensation of quark-antiquark pairs. In these examples, spontaneous symmetry breaking is the result of strong interactions beyond perturbation theory. We would like to understand whether these more general mechanisms of spontaneous symmetry breaking can also give mass to vector bosons, and, if so, how the masses can be calculated.

To carry out this analysis, we will need to abstract several ideas from the preceding discussion. First, we will discuss in general terms the relations between gauge bosons, Goldstone bosons, and global symmetry currents. Then we will use this information to construct the gauge boson mass matrix without making direct use of the Lagrangian.

Consider, first, an arbitrary quantum field theory $\mathcal{L}_0$ with a global symmetry G. In Section 9.6, we derived the Noether current corresponding to the G symmetry by varying the Lagrangian by a local gauge transformation with infinitesimal parameter $\alpha^a(x)$. Transforming with a constant α^a should leave $\mathcal{L}_0$ unchanged. Then the more general variation of $\mathcal{L}_0$ must take the form

$$\delta\mathcal{L}_0 = (\partial_\mu \alpha^a)J^{\mu a}, \tag{20.43}$$

for some set of vector operators $J^{\mu a}$ built from the fields of $\mathcal{L}_0$. The variational principle then tells us that

$$\partial_\mu J^{\mu a} = 0. \tag{20.44}$$

We can identify the $J^{\mu a}$ as the Noether currents of the global gauge symmetry.

We can now couple this globally symmetric theory to non-Abelian gauge fields, promoting the global symmetry to a local symmetry. To first order in g, the new Lagrangian should take the form

$$\mathcal{L} = \mathcal{L}_0 - gA_\mu^a J^{\mu a} + \mathcal{O}(A^2). \tag{20.45}$$

To check this, note that the transformation (20.43) compensates the variation due to a gauge transformation of A_μ^a, Eq. (15.46), to leading order in g. The terms of order A^2 and higher can in general be arranged to compensate the higher-order terms in the gauge transformation. Thus, matrix elements involving only one insertion of the gauge field can be evaluated using properties of the Noether currents of the original globally symmetric theory. Note in particular that the conservation law for these currents, Eq. (20.44), guarantees that the Ward identities for these matrix elements are satisfied.

If the global symmetry of the theory $\mathcal{L}_0$ is spontaneously broken, this theory will contain Goldstone bosons, which will stand in a special relation to the Noether currents. At long wavelength, the Goldstone bosons become infinitesimal symmetry rotations of the vacuum, $Q^a |0\rangle$, where Q^a is the global charge associated with $J^{\mu a}$. Thus, the operators $J^{\mu a}$ have the correct quantum numbers to create Goldstone bosons from the vacuum. Let $|\pi_k\rangle$ denote a Goldstone boson state. In general, there will be a current $J^{\mu a}$ that can create or destroy this boson; we can parametrize the corresponding matrix element as

$$\langle 0| J^{\mu a}(x) |\pi_k(p)\rangle = -ip^\mu F^a{}_k e^{-ip\cdot x}, \tag{20.46}$$

where p^μ is the on-shell momentum of the boson and $F^a{}_k$ is a matrix of constants. The elements $F^a{}_k$ vanish when a denotes a generator of an unbroken symmetry. Then the nonvanishing matrix elements of $F^a{}_k$ connect the currents of the spontaneously broken symmetries to their corresponding Goldstone bosons. Since the currents $J^{\mu a}$ are conserved, we find

$$0 = \partial_\mu \langle 0| \, J^{\mu a}(x) \, |\pi_k(p)\rangle = -p^2 F^a{}_k e^{-ip\cdot x}, \qquad (20.47)$$

which implies that the bosons with nonzero matrix elements (20.46) satisfy $p^2 = 0$ on shell and so are massless. This is another proof of Goldstone's theorem.[‡]

Since the scalar field theory that we examined earlier in this section should be a special case of this analysis, we should find there an example of the relation given in Eq. (20.46). Comparing Eqs. (20.15) and (20.45), we see that, for the scalar field theory,

$$J^{\mu a} = \partial_\mu \phi_i T^a_{ij} \phi_j, \qquad (20.48)$$

which is indeed the Noether current corresponding to the global gauge symmetry. Inserting the vacuum expectation value (20.16), we find

$$J^{\mu a} = \partial_\mu \phi_i (T^a \phi_0)_i, \qquad (20.49)$$

which leads to the set of matrix elements

$$\langle 0| \, J^{\mu a}(x) \, |\phi_i(p)\rangle = -ip^\mu (T^a \phi_0)_i e^{-ip\cdot x}. \qquad (20.50)$$

Using this relation, we can identify

$$F^a{}_i = T^a_{ij} \phi_{0j} \qquad (20.51)$$

for the Higgs mechanism in a weakly coupled scalar field theory. To be more precise, the index i runs over all components of the scalar field ϕ. However, we saw in the discussion below Eq. (20.19) that (20.51) is nonzero only for components ϕ_i that are Goldstone bosons, and only for symmetry generators a that are spontaneously broken. Thus the nonzero components of (20.51) form precisely the structure (20.46).

As a concrete illustration of the way that the objects $T^a \phi_0$ link spontaneously broken generators and Goldstone bosons, consider the situation of $SU(2)$ symmetry broken by a scalar field in the vector representation, as in Eq. (20.29) and Fig. 20.1. According to the figure, rotations about the $\hat{1}$ axis tip the vacuum expectation value of ϕ into the $\hat{2}$ direction, rotations about the $\hat{2}$ axis tip this expectation value into the $\hat{1}$ direction, and rotations about $\hat{3}$ leave $\langle\phi\rangle$ invariant. Thus the gauge generators T^1 and T^2 are spontaneously broken, and the scalar field components ϕ^2 and ϕ^1 are the corresponding Goldstone bosons. This accords with the result of computing the elements of $(T^a \phi_0)_i$ explicitly: Using $(T^a)_{bc} = \epsilon^{bac}$, we find

$$(T^a \phi_0)_b = \epsilon_{bac} \langle \phi^c \rangle = V \epsilon^{ba3}. \qquad (20.52)$$

[‡]A special case of this argument appeared in the discussion of Eq. (19.88).

Inserting this result into formula (20.50), we see that the current of each spontaneously broken symmetry creates and destroys its own Goldstone boson.

Now we can use this formalism to study the working of the Higgs mechanism in this general context. Consider the original theory $\mathcal{L}_0$ coupled to gauge bosons of G. To see how the Higgs mechanism operates, we must compute the vacuum polarization amplitude. This amplitude is required by the Ward identity to be transverse, so it is necessarily of the form

$$a \sim\!\!\!\bigcirc\!\!\!\sim b \; = i\left(g^{\mu\nu} - \frac{k^\mu k^\nu}{k^2}\right) \cdot (m_{ab}^2 + \mathcal{O}(k^2)). \qquad (20.53)$$

It is not easy to compute the nonsingular terms in (20.53) in this general situation, but it is straightforward to compute the singular term, which comes from contributions with an intermediate Goldstone boson. Combining Eqs. (20.45) and (20.46), we see that the amplitude for a gauge boson to convert to a Goldstone boson is

$$\underset{k}{\xleftarrow{\hspace{1.5cm}}}\!\!\!\bullet\!\!\sim\!\!\sim\!\!\sim_\mu \; = -gk^\mu F^a{}_j. \qquad (20.54)$$

Then the pole contribution to the vacuum polarization is

$$\sim\!\!\sim\!\!\bullet\!\!-\!\!\bullet\!\!\sim\!\!\sim \; = (gk^\mu F^a{}_j)\frac{i}{k^2}(-gk^\nu F^b{}_j). \qquad (20.55)$$

Comparing (20.55) with (20.21), we identify

$$m_{ab}^2 = g^2 F^a{}_j F^b{}_j. \qquad (20.56)$$

Notice that, in the case in which the symmetry is broken by a scalar field, this result reverts to (20.18). However, Eq. (20.56) applies to any theory of spontaneously broken symmetry, whether the symmetry breaking is apparent from the Lagrangian or whether it requires strong interactions or other nonperturbative effects. It is a general result, then, that any gauge boson coupled to the current of a spontaneously broken symmetry acquires a mass.

20.2 The Glashow-Weinberg-Salam Theory of Weak Interactions

We are now ready to write down the spontaneously broken gauge theory that gives the experimentally correct description of the weak interactions, a model introduced by Glashow, Weinberg, and Salam (GWS). Like the second $SU(2)$ model considered in the previous section, this model gives a unified description of weak and electromagnetic interactions, in which the massless photon corresponds to a particular combination of symmetry generators that remains unbroken.

Again we begin with a theory with $SU(2)$ gauge symmetry. To break the symmetry spontaneously, we introduce a scalar field in the spinor representation of $SU(2)$, as in Eq. (20.22). However, we know that this theory leads to a system with no massless gauge bosons. We therefore introduce an additional $U(1)$ gauge symmetry. We assign the scalar field a charge $+1/2$ under this $U(1)$ symmetry, so that its complete gauge transformation is

$$\phi \to e^{i\alpha^a \tau^a} e^{i\beta/2} \phi. \tag{20.57}$$

(Here $\tau^a = \sigma^a/2$.) If the field ϕ acquires a vacuum expectation value of the form

$$\langle \phi \rangle = \frac{1}{\sqrt{2}} \begin{pmatrix} 0 \\ v \end{pmatrix}, \tag{20.58}$$

then a gauge transformation with

$$\alpha^1 = \alpha^2 = 0, \qquad \alpha^3 = \beta \tag{20.59}$$

leaves $\langle \phi \rangle$ invariant. Thus, the theory will contain one massless gauge boson, corresponding to this particular combination of generators. The remaining three gauge bosons will acquire masses from the Higgs mechanism.

Gauge Boson Masses

It is straightforward to work out the details of the mass spectrum by using the methods of the previous section. The covariant derivative of ϕ is

$$D_\mu \phi = \left(\partial_\mu - ig A_\mu^a \tau^a - i\tfrac{1}{2} g' B_\mu \right) \phi, \tag{20.60}$$

where A_μ^a and B_μ are, respectively, the $SU(2)$ and $U(1)$ gauge bosons. Since the $SU(2)$ and $U(1)$ factors of the gauge group commute with one another, they can have different coupling constants, which we have called g and g'.

The gauge boson mass terms come from the square of Eq. (20.60), evaluated at the scalar field vacuum expectation value (20.58). The relevant terms are

$$\Delta \mathcal{L} = \frac{1}{2} (0 \quad v) \left(g A_\mu^a \tau^a + \frac{1}{2} g' B_\mu \right) \left(g A^{b\mu} \tau^b + \frac{1}{2} g' B^\mu \right) \begin{pmatrix} 0 \\ v \end{pmatrix}. \tag{20.61}$$

If we evaluate the matrix product explicitly, using $\tau^a = \sigma^a/2$, we find

$$\Delta \mathcal{L} = \frac{1}{2} \frac{v^2}{4} \left[g^2 (A_\mu^1)^2 + g^2 (A_\mu^2)^2 + (-g A_\mu^3 + g' B_\mu)^2 \right]. \tag{20.62}$$

There are three massive vector bosons, which we will notate as follows:

$$W_\mu^\pm = \frac{1}{\sqrt{2}} \left(A_\mu^1 \mp i A_\mu^2 \right) \qquad \text{with mass} \quad m_W = g \frac{v}{2};$$

$$Z_\mu^0 = \frac{1}{\sqrt{g^2 + g'^2}} \left(g A_\mu^3 - g' B_\mu \right) \qquad \text{with mass} \quad m_Z = \sqrt{g^2 + g'^2} \, \frac{v}{2}. \tag{20.63}$$

The fourth vector field, orthogonal to Z_μ^0, remains massless:

$$A_\mu = \frac{1}{\sqrt{g^2 + g'^2}} \left(g' A_\mu^3 + g B_\mu \right) \qquad \text{with mass} \quad m_A = 0. \tag{20.64}$$

We will identify this field with the electromagnetic vector potential.

From now on it will be more convenient to write all expressions in terms of these mass eigenstate fields. Consider, for instance, the coupling of the vector fields to fermions. For a fermion field belonging to a general $SU(2)$ representation, with $U(1)$ charge Y, the covariant derivative takes the form

$$D_\mu = \partial_\mu - igA_\mu^a T^a - ig'YB_\mu. \tag{20.65}$$

In terms of the mass eigenstate fields, this becomes

$$D_\mu = \partial_\mu - i\frac{g}{\sqrt{2}}\left(W_\mu^+ T^+ + W_\mu^- T^-\right) - i\frac{1}{\sqrt{g^2+g'^2}}Z_\mu\left(g^2 T^3 - g'^2 Y\right)$$
$$- i\frac{gg'}{\sqrt{g^2+g'^2}}A_\mu\left(T^3 + Y\right), \tag{20.66}$$

where $T^\pm = (T^1 \pm iT^2)$. The normalization is chosen so that, in the spinor representation of $SU(2)$,

$$T^\pm = \tfrac{1}{2}(\sigma^1 \pm i\sigma^2) = \sigma^\pm. \tag{20.67}$$

The last term of Eq. (20.66) makes explicit the fact that the massless gauge boson A_μ couples to the gauge generator $(T^3 + Y)$, which generates precisely the symmetry operation (20.59).

To put expression (20.66) into a more useful form, we should identify the coefficient of the electromagnetic interaction as the electron charge e,

$$e = \frac{gg'}{\sqrt{g^2+g'^2}}, \tag{20.68}$$

and identify the electric charge quantum number as

$$Q = T^3 + Y. \tag{20.69}$$

These substitutions, with $Q = -1$ for the electron, give the conventional form of the coupling of the electromagnetic field.

To simplify expression (20.66) further, we define the *weak mixing angle*, θ_w, to be the angle that appears in the change of basis from (A^3, B) to (Z^0, A):

$$\begin{pmatrix} Z^0 \\ A \end{pmatrix} = \begin{pmatrix} \cos\theta_w & -\sin\theta_w \\ \sin\theta_w & \cos\theta_w \end{pmatrix} \begin{pmatrix} A^3 \\ B \end{pmatrix},$$

that is,

$$\cos\theta_w = \frac{g}{\sqrt{g^2+g'^2}}, \qquad \sin\theta_w = \frac{g'}{\sqrt{g^2+g'^2}}. \tag{20.70}$$

Then, with the manipulation in the Z^0 coupling

$$g^2 T^3 - g'^2 Y = (g^2 + g'^2)T^3 - g'^2 Q,$$

we can rewrite the covariant derivative (20.66) in the form

$$D_\mu = \partial_\mu - i\frac{g}{\sqrt{2}}\left(W_\mu^+ T^+ + W_\mu^- T^-\right) - i\frac{g}{\cos\theta_w}Z_\mu\left(T^3 - \sin^2\theta_w Q\right) - ieA_\mu Q,$$

(20.71)

where

$$g = \frac{e}{\sin\theta_w}.$$

(20.72)

We see here that the couplings of all of the weak bosons are described by two parameters: the well-measured electron charge e, and a new parameter θ_w. The couplings induced by W and Z exchange will also involve the masses of these particles. However, these masses are not independent, since it follows from Eqs. (20.63) that

$$m_W = m_Z \cos\theta_w.$$

(20.73)

All effects of W and Z exchange processes, at least at tree level, can be written in terms of the three basic parameters e, θ_w, and m_W.

Coupling to Fermions

The covariant derivative (20.71) uniquely determines the coupling of the W and Z^0 fields to fermions, once the quantum numbers of the fermion fields are specified. To determine these quantum numbers, we must take account of the fact, mentioned in Section 17.3, that the W boson couples only to left-handed helicity states of quarks and leptons.

At the level of the classical Lagrangian, there is no difficulty in constructing theories in which the left- and right-handed components of a fermion field couple differently to gauge bosons.* Already in Section 3.2 we saw that the kinetic energy term for Dirac fermions splits into separate pieces for the left- and right-handed fields:

$$\bar{\psi}i\partial\!\!\!/\psi = \bar{\psi}_L i\partial\!\!\!/\psi_L + \bar{\psi}_R i\partial\!\!\!/\psi_R.$$

(20.74)

When we couple ψ to a gauge field, we can assign ψ_L and ψ_R to different representations of the gauge group. Then the two terms on the right-hand side of (20.74) will contain two different covariant derivatives, and these will imply two different sets of couplings.

In the GWS model, we can use this technique to insure that only the left-handed components of the quark and lepton fields couple to the W bosons. We assign the left-handed fermion fields to doublets of $SU(2)$, while making the right-handed fermion fields singlets under this group. Once we have specified the T^3 value for each fermion field, the value of Y that we must assign follows from Eq. (20.69). This means that the Y assignments will also be different for

---

*In Section 19.4, we argued that there is a possible problem with this strategy at the level of quantum corrections. We will check below whether the specific model we construct avoids this problem.

the left- and right-handed components of quarks and leptons. For the right-handed fields, $T^3 = 0$, and so we reproduce the standard electric charges by assigning Y to equal the electric charge. For example, for the right-handed u quark field, $Y = +2/3$; for e_R^-, $Y = -1$. For the left-handed fields,

$$E_L = \begin{pmatrix} \nu_e \\ e^- \end{pmatrix}_L, \qquad Q_L = \begin{pmatrix} u \\ d \end{pmatrix}_L, \tag{20.75}$$

the assignments $Y = -1/2$ and $Y = +1/6$, respectively, combine with $T^3 = \pm 1/2$ to give the correct electric charge assignments. Since the left- and right-handed fermions live in different representations of the fundamental gauge group, it is often useful to think of these components as distinct particles, which are mixed by the fermion mass terms.

In fact, the construction of fermion mass terms is a serious problem, because all possible such terms are forbidden by global gauge invariances. For example, we cannot write an electron mass term

$$\Delta\mathcal{L} = -m_e(\bar{e}_L e_R + \bar{e}_R e_L), \tag{20.76}$$

because the fields e_L and e_R belong to different $SU(2)$ representations and have different $U(1)$ charges. For the next few pages, we will ignore this problem by treating all fermion fields as massless. This description will suffice to analyze the structure of the weak interactions at high energies, where the quark and lepton masses can be ignored. At the end of this section we will return to the problem of writing quark and lepton mass terms in the GWS theory. The solution to this problem will reinforce the idea that the left- and right-handed fermion fields are fundamentally independent entities, mixed to form massive fermions by some subsidiary process.

If we ignore fermion masses, the Lagrangian for the weak interactions of quarks and leptons follows directly from the charge assignments given above. The fermion kinetic energy terms for e, ν, u, and d are

$$\mathcal{L} = \bar{E}_L(i\slashed{D})E_L + \bar{e}_R(i\slashed{D})e_R + \bar{Q}_L(i\slashed{D})Q_L + \bar{u}_R(i\slashed{D})u_R + \bar{d}_R(i\slashed{D})d_R. \tag{20.77}$$

In each term, the covariant derivative is given by Eq. (20.65), with T^a and Y evaluated in the particular representation to which that fermion field belongs. For example,

$$\bar{Q}_L(i\slashed{D})Q_L = \bar{Q}_L i\gamma^\mu \left(\partial_\mu - igA_\mu^a \tau^a - i\tfrac{1}{6}g'B_\mu\right)Q_L. \tag{20.78}$$

A right-handed neutrino would have zero coupling both to $SU(2)$ and to $U(1)$, so we have simply omitted this field from Eq. (20.77).

To work out the physical consequences of the fermion-vector boson couplings, we should write Eq. (20.77) in terms of the vector boson mass eigenstates, using the form of the covariant derivative given in Eq. (20.71). Equation (20.77) then takes the form

$$\begin{aligned} \mathcal{L} = & \bar{E}_L(i\slashed{\partial})E_L + \bar{e}_R(i\slashed{\partial})e_R + \bar{Q}_L(i\slashed{\partial})Q_L + \bar{u}_R(i\slashed{\partial})u_R + \bar{d}_R(i\slashed{\partial})d_R \\ & + g\left(W_\mu^+ J_W^{\mu+} + W_\mu^- J_W^{\mu-} + Z_\mu^0 J_Z^\mu\right) + eA_\mu J_{EM}^\mu, \end{aligned} \tag{20.79}$$

where

$$J_W^{\mu+} = \frac{1}{\sqrt{2}}\left(\bar{\nu}_L\gamma^\mu e_L + \bar{u}_L\gamma^\mu d_L\right);$$

$$J_W^{\mu-} = \frac{1}{\sqrt{2}}\left(\bar{e}_L\gamma^\mu \nu_L + \bar{d}_L\gamma^\mu u_L\right);$$

$$J_Z^\mu = \frac{1}{\cos\theta_w}\Big[\bar{\nu}_L\gamma^\mu\left(\tfrac{1}{2}\right)\nu_L + \bar{e}_L\gamma^\mu\left(-\tfrac{1}{2} + \sin^2\theta_w\right)e_L + \bar{e}_R\gamma^\mu\left(\sin^2\theta_w\right)e_R$$

$$+ \bar{u}_L\gamma^\mu\left(\tfrac{1}{2} - \tfrac{2}{3}\sin^2\theta_w\right)u_L + \bar{u}_R\gamma^\mu\left(-\tfrac{2}{3}\sin^2\theta_w\right)u_R$$

$$+ \bar{d}_L\gamma^\mu\left(-\tfrac{1}{2} + \tfrac{1}{3}\sin^2\theta_w\right)d_L + \bar{d}_R\gamma^\mu\left(\tfrac{1}{3}\sin^2\theta_w\right)d_R\Big];$$

$$J_{EM}^\mu = \bar{e}\gamma^\mu(-1)e + \bar{u}\gamma^\mu\left(+\tfrac{2}{3}\right)u + \bar{d}\gamma^\mu\left(-\tfrac{1}{3}\right)d. \tag{20.80}$$

Here we have used Eq. (20.67) to simplify the W boson currents. Notice that the current J_{EM}^μ associated with the photon field is indeed the standard electromagnetic current.

Anomaly Cancellation

As we have just seen, there is no difficulty in writing a Lagrangian that couples the GWS gauge bosons to fermions in a chiral fashion. However, these chiral couplings do present a potential problem that appears at the level of one-loop corrections. In Section 19.2, we saw that an axial current that is conserved at the level of the classical equations of motion can acquire a nonzero divergence through one-loop diagrams that couple this current to a pair of gauge bosons. The Feynman diagram that contains this anomalous contribution is a triangle diagram with the axial current and the two gauge currents at its vertices. In a gauge theory in which gauge bosons couple to a chiral current, the dangerous triangle diagrams appear in the one-loop corrections to the three-gauge-boson vertex function. The anomalous terms violate the Ward identity for this amplitude. Thus, as we argued in Section 19.4, theories in which gauge bosons couple to chiral currents can be gauge invariant only if the anomalous contribution somehow disappears. Fortunately, as we saw there, the anomalous terms can be arranged to cancel when one sums over all possible fermion species that can circulate in these diagrams.[†]

Within the GWS theory, the requirement from experiment that the weak interaction currents are left-handed forced us to choose a chiral gauge coupling. Now we must check that the anomalous terms from the triangle diagrams cancel as required. We will find that they do, but only through a subtle and rather magical interplay of the quantum numbers of quarks and leptons.

The anomalous term of a triangle diagram of three gauge bosons A_μ^a, A_ν^b,

---

[†]If you have not read Chapter 19, but you are willing to assume that the fermion triangle diagram contains a contribution that violates gauge invariance, you should still be able to follow the argument that follows.

and A_λ^c is proportional to the group theoretic invariant

$$\text{tr}\left[\gamma^5 t^a \{t^b, t^c\}\right], \tag{20.81}$$

where the trace is taken over all fermion species. The anticommutator comes from taking the sum of two triangle diagrams in which the fermions circle in opposite directions. The factor γ^5 registers the fact that the anomaly is associated with chiral currents; this factor equals -1 for left-handed fermions and $+1$ for right-handed fermions. In theories such as QED or QCD in which the gauge bosons couple equally to right- and left-handed species, the anomalies automatically cancel. This bookkeeping method is a special case of the more general method presented in Section 19.4.

To evaluate the anomalies of the GWS theory, it is easiest to work in the basis of $SU(2) \times U(1)$ gauge bosons, before the mixing to the photon and Z^0 mass eigenstates. It suffices to evaluate the triangle diagrams for massless fermions, so that right- and left-handed fermions have distinct quantum numbers. However, we must consider not only the anomalies of diagrams with three $SU(2) \times U(1)$ gauge bosons, but also diagrams with both weak-interaction gauge bosons and color $SU(3)$ gauge bosons of QCD. If we consider effects of gravity on the weak-interaction gauge theory, there is also a possibly anomalous diagram with one weak-interaction gauge boson and two gravitons. We can omit diagrams, such as the anomaly of three $SU(3)$ bosons or of one $SU(3)$ boson and two gravitons, in which all of the couplings are left-right symmetric. Then the full set of diagrams with possible anomalous terms is shown in Fig. 20.2. All of the possible anomalies must cancel if the Ward identities of the $SU(2) \times U(1)$ gauge theory are to be satisfied.

It is a special property of $SU(2)$ gauge theory that the anomaly of three $SU(2)$ gauge bosons always vanishes; this result follows from the property of Pauli sigma matrices $\{\sigma^a, \sigma^b\} = 2\delta^{ab}$, which implies that the trace (20.81) vanishes. The anomalies containing one $SU(3)$ boson or one $SU(2)$ boson are proportional to

$$\text{tr}[t^a] = 0 \qquad \text{or} \qquad \text{tr}[\tau^a] = 0. \tag{20.82}$$

The remaining nontrivial anomalies are those of one $U(1)$ boson with two $SU(2)$ bosons or two $SU(3)$ bosons, the anomaly of three $U(1)$ bosons, and the gravitational anomaly with one $U(1)$ gauge boson.

The anomaly of one $U(1)$ boson with two $SU(3)$ bosons is proportional to the group theory factor

$$\text{tr}[t^a t^b Y] = \tfrac{1}{2}\delta^{ab} \cdot \sum_q Y_q, \tag{20.83}$$

where the sum runs over left-handed quarks and right-handed quarks, with an extra (-1) for the left-handed contributions. Inserting the charge assignments given above for u_L, d_L, u_R, and d_R, we find

$$\sum_q Y_q = -2 \cdot \tfrac{1}{6} + (\tfrac{2}{3}) + (-\tfrac{1}{3}) = 0. \tag{20.84}$$

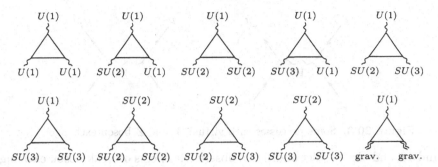

Figure 20.2. Possible gauge anomalies of weak interaction theory. All of these anomalies must vanish for the Glashow-Weinberg-Salam theory to be consistent.

Similarly, the anomaly of a $U(1)$ boson with two $SU(2)$ bosons is proportional to

$$\text{tr}[\tau^a \tau^b Y] = \tfrac{1}{2}\delta^{ab} \sum_{fL} Y_{fL}, \tag{20.85}$$

where the sum runs over the left-handed fermions E_L and Q_L. Thus,

$$\sum_{fL} Y_{fL} = -(-\tfrac{1}{2}) - 3 \cdot \tfrac{1}{6} = 0; \tag{20.86}$$

the factor 3 counts the color states of the quarks. The anomaly of three $U(1)$ gauge bosons is proportional to a sum involving left- and right-handed leptons and quarks:

$$\text{tr}[Y^3] = -2(-\tfrac{1}{2})^3 + (-1)^3 - 3\big[2(\tfrac{1}{6})^3 - (\tfrac{2}{3})^3 - (-\tfrac{1}{3})^3\big] = 0. \tag{20.87}$$

Finally, the gravitational anomaly with one $U(1)$ gauge boson is proportional to

$$\text{tr}[Y] = -2(-\tfrac{1}{2}) + (-1) - 3\big[2(\tfrac{1}{6}) - (\tfrac{2}{3}) - (-\tfrac{1}{3})\big] = 0. \tag{20.88}$$

   The Glashow-Weinberg-Salam theory is thus a chiral gauge theory that is completely free of axial vector anomalies among the gauge currents. However, the cancellation of anomalies requires that leptons and quarks appear in complete multiplets with the structure of $(E_L, e_R, Q_L, u_R, d_R)$. This set of fields is often called a *generation* of quarks and leptons. The consistency of the theory requires that quarks and leptons appear in Nature in equal numbers, organizing themselves into successive generations in this way.

Experimental Consequences of the GWS Theory

Now that we have a fundamental theory for the coupling of W and Z bosons to fermions, we can work out the consequences of this theory for observable processes mediated by weak bosons. This analysis should reproduce the effective Lagrangian description of the weak interactions used in Chapters 17

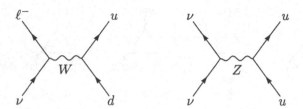

Figure 20.3. Some processes with virtual W and Z boson exchange.

and 18, and also predict additional observable effects of weak boson exchange. In our discussion here, we will derive only the most basic relations in this subject; we do not have space for a systematic survey of the phenomenology of weak interactions. However, we encourage the reader to study the experimental foundations of the weak interactions, which contain many beautiful illustrations of the principles of quantum field theory.[‡]

At energies low compared to the vector boson masses, the couplings of the weak bosons have their major effects through processes that involve virtual weak boson exchange. These processes are shown in Fig. 20.3. We will derive the Feynman rules for massive gauge bosons in Chapter 21. Meanwhile, it is reasonable to guess that the W and Z boson propagators are given by

$$\langle W^{\mu +}(p) W^{\nu -}(-p)\rangle = \frac{-ig^{\mu\nu}}{p^2 - m_W^2}, \qquad \langle Z^\mu(p) Z^\nu(-p)\rangle = \frac{-ig^{\mu\nu}}{p^2 - m_Z^2}. \quad (20.89)$$

We will see in Section 21.1 that these propagators give correct expressions for diagrams with W and Z exchange up to terms of order (m_f/m_W), where m_f is a fermion mass.

First consider the W exchange diagram in Fig. 20.3, in the limit of energies low compared to the W mass. We can then neglect the p^2 term in the denominator of the W propagator (20.89). Taking the W coupling from Eq. (20.79), we find that the diagram can be described by the effective Lagrangian

$$\Delta \mathcal{L}_W = \frac{g^2}{m_W^2} J_W^{\mu -} J_{\mu W}^+ $$

$$= \frac{g^2}{2m_W^2} \left(\bar{e}_L \gamma^\mu \nu_L + \bar{d}_L \gamma^\mu u_L \right) \left(\bar{\nu}_L \gamma_\mu e_L + \bar{u}_L \gamma_\mu d_L \right). \quad (20.90)$$

The coefficient is often written in terms of the *Fermi constant*,

$$\frac{G_F}{\sqrt{2}} = \frac{g^2}{8m_W^2}. \quad (20.91)$$

The various terms in this effective Lagrangian reproduce the expressions we have already written in Eqs. (17.31), (18.28), and (18.29). Since these interactions among leptons and quarks are mediated by the exchange of an

---

[‡]The experimental successes of the theory of weak interactions are reviewed in the book of Commins and Bucksbaum (1983).

electrically charged vector boson, they are called collectively *charged-current* interactions. The effective Lagrangian (20.90) turns out to provide an impressively successful description of the phenomenology of charged-current weak interactions. We have described its use in high-energy neutrino scattering, but it has comparable successes in nuclear β-decay, muon decay, and a variety of other processes.

In a similar way, we can work out the effective Lagrangian resulting from virtual Z^0 exchange. We find

$$
\Delta\mathcal{L}_Z = \frac{g^2}{2m_Z^2} J_Z^\mu J_{\mu Z}
$$
$$
= \frac{4G_F}{\sqrt{2}} \left(\sum_f \bar{f}\gamma^\mu (T^3 - \sin^2\theta_w Q) f \right)^2, \tag{20.92}
$$

where the sum in the second line runs over all left-handed and right-handed flavors, and we have used relation (20.73) to simplify the prefactor. We say that the effective Lagrangian (20.92) mediates *neutral current* weak interaction processes. Notice that, if we define $SU(2)$ gauge currents as

$$
J^{\mu a} = \sum_f \bar{f}\gamma^\mu T^a f, \tag{20.93}
$$

then the effective Lagrangians of W and Z exchange can be written together in the simple form

$$
\Delta\mathcal{L}_W + \Delta\mathcal{L}_Z = \frac{4G_F}{\sqrt{2}} \left[(J^{\mu 1})^2 + (J^{\mu 2})^2 + (J^{\mu 3} - \sin^2\theta_w J_{EM}^\mu)^2 \right]. \tag{20.94}
$$

This expression becomes manifestly invariant under an unbroken global $SU(2)$ symmetry in the limit $g' \to 0$ or $\sin^2\theta_w \to 0$. We will discuss this observation further at the end of this section.

The neutral current effective Lagrangian (20.92) contains terms that couple together all of the various species of quarks and leptons. These terms violate parity, and so distinguish themselves from the effects of strong and electromagnetic interactions. For example, Eq. (20.92) predicts the existence of neutral current deep inelastic neutrino scattering events, in which a high-energy neutrino shatters a nucleon but does not convert to a final-state muon or electron. This process is analyzed in Problem 20.4. Similarly, the neutral-current interaction predicts the presence of parity-violating effects in electron deep inelastic scattering. It also predicts a parity-violating electron-nucleon interaction that should mix atomic energy levels, and a similar parity-violating nucleon-nucleon interaction. Within the GWS theory, the strengths of all of these various effects are predicted in terms of the Fermi constant and one additional parameter, the value of $\sin^2\theta_w$. Thus, the GWS theory can be tested by observing each of these effects and asking whether a single value of this parameter can account for the strengths of all of these disparate processes.

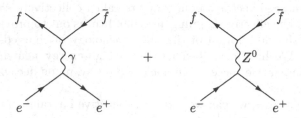

Figure 20.4. Diagrams contributing to the process $e^+e^- \to f\bar{f}$ in the Glashow-Weinberg-Salam theory.

Further tests of the GWS theory are available at higher energies. The process $e^+e^- \to f\bar{f}$ is affected in an essential way, since the theory contains a new diagram with s-channel Z^0 exchange, which interferes with the standard photon exchange diagram, as shown in Fig. 20.4. It is straightforward to work out the effects of this interference using the methods of Section 5.2, so we have left this analysis as Problem 20.3.

As the center-of-mass energy approaches m_Z, the Z^0 appears directly as a resonance in the e^+e^- annihilation cross section. Similarly, both the W and the Z can be observed as resonances in quark-antiquark annihilation, viewed as a parton subprocess in proton-antiproton scattering. The positions of these resonances are predicted from G_F, $\sin^2\theta_w$, and the value of e or α, according to Eqs. (20.72) and (20.91). Using these relations, we find

$$m_W^2 = \frac{\pi\alpha}{\sqrt{2}G_F \sin^2\theta_w}, \qquad m_Z^2 = \frac{\pi\alpha}{\sqrt{2}G_F \sin^2\theta_w \cos^2\theta_w}. \qquad (20.95)$$

The detailed shape of the Z^0 resonance is shown in Fig. 20.5. The experimental measurements shown are compared to a theoretical curve with the resonance position adjusted for the best fit. The height and width of the resonance are then predicted by the GWS theory. The resonance is broadened to higher energies by processes in which the electron and positron radiate collinear photons before annihilation; this correction was discussed in Problem 5.5.

Because the Lagrangian of the GWS theory treats left- and right-handed fermions as distinct species with completely different quantum numbers, the couplings of the Z^0 to left- and right-handed fermions differ signficantly. One manifestation of this is the presence of a *polarization asymmetry*, a net polarization of fermions produced in the decay $Z^0 \to f\bar{f}$, or an asymmetry in the inverse process of Z^0 production. This asymmetry can be read directly from the form of the Z^0 current given in (20.80):

$$
\begin{aligned}
A_{LR}^f &= \frac{\Gamma(Z^0 \to f_L \bar{f}_R) - \Gamma(Z^0 \to f_R \bar{f}_L)}{\Gamma(Z^0 \to f_L \bar{f}_R) + \Gamma(Z^0 \to f_R \bar{f}_L)} \\
&= \frac{(\frac{1}{2} - |Q_f|\sin^2\theta_w)^2 - (Q_f \sin^2\theta_w)^2}{(\frac{1}{2} - |Q_f|\sin^2\theta_w)^2 + (Q_f \sin^2\theta_w)^2}.
\end{aligned}
\qquad (20.96)
$$

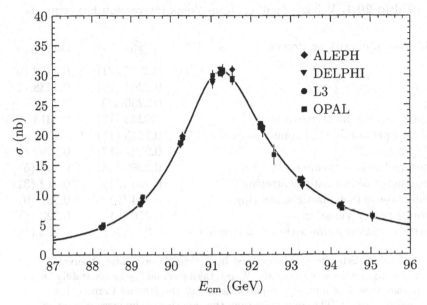

Figure 20.5. The total cross section for e^+e^- annihilation to hadrons for E_{cm} close to the Z^0 boson mass, as measured by the ALEPH, DELPHI, L3, and OPAL experiments and compiled by the Particle Data Group, *Phys. Rev.* **D50**, (1994), Fig. 32.14. References to the original articles are given there. The solid curve is the prediction of the GWS theory.

For a realistic value $\sin^2 \theta_w = 0.23$, this expression gives a 15% asymmetry for charged leptons and a 95% asymmetry for d, s, and b quarks. The asymmetry can be checked experimentally for leptons by measuring the polarization of τ leptons at the Z^0 resonance, or by measuring the relative cross sections for producing the resonance using left- versus right-handed electrons. For quarks, the asymmetry can be determined indirectly from the forward-backward production asymmetry on the resonance, as explained in Problem 20.3.

Because the weak neutral current has so many different manifestations, the GWS theory of weak interactions can be subjected to a stringent test by comparing the values of the parameter $\sin^2 \theta_w$ needed to account for each of its predicted effects. Table 20.1 presents the values of $\sin^2 \theta_w$ extracted from a wide variety of weak interaction neutral current effects and asymmetries. In all cases, one-loop radiative corrections must be included to analyze the experiment at the required level of accuracy. These radiative corrections involve some subtlety. First, one must adopt a specific renormalization convention that defines $\sin^2 \theta_w$ and use it consistently in all calculations. The table shows results for two different choices of this convention. In both conventions, the values of weak-interaction observables are taken to be functions of α, G_F, and a third independent parameter. In the first column this parameter is the mass ratio m_W/m_Z, and, following the tree-level expression (20.73), we consider

Table 20.1. Values of $\sin^2\theta_w$ from Weak Interaction Experiments

Observed Quantity or Process	s_W^2	$\sin^2\theta_{w\,\overline{MS}}$
m_Z	0.2247 (21)	0.2320 (6)
m_W	0.2264 (25)	0.2338 (22)
Γ_Z	0.2250(18)	0.2322 (6)
Lepton f-b asymmetries at the Z^0	0.2243 (17)	0.2315 (11)
All pair-production asymmetries at the Z^0	0.2245 (17)	0.2317 (8)
A_{LR}^e at the Z^0	0.2221 (17)	0.2292 (10)
Deep inelastic neutrino scattering	0.2260 (48)	0.233 (5)
Neutrino-proton elastic scattering	0.205 (31)	0.212 (32)
Neutrino-electron elastic scattering	0.224 (9)	0.231 (9)
Atomic parity violation	0.216 (8)	0.223 (8)
Parity violation in inelastic e^- scattering	0.216 (17)	0.223 (18)

The values listed here are obtained by fitting experimental observations by adjusting the value of s_W^2 or $\sin^2\theta_{w\,\overline{MS}}$, taking α and G_F as accurately known parameters. The numbers in parentheses are the standard errors in the last displayed digits. The conversion from the experimentally measured quantities to s_W^2 or $\sin^2\theta_{w\,\overline{MS}}$ depends on the value of the top quark mass and the mass of the Higgs boson. These values assume a top quark mass of 169 GeV and a Higgs mass of 300 GeV; the quoted errors include an uncertainty of 17 GeV in the top quark mass and a range from 60 GeV to 1000 GeV for the Higgs mass. The differences in the relative errors between the two columns reflect the importance of this theoretical uncertainty. Some observables depend weakly on α_s; these values assume $\alpha_s(m_Z) = 0.120 \pm .007$. This table is taken from the article of P. Langacker and J. Erler for the Particle Data Group, *Phys. Rev.* **D50**, 1304 (1994). That article contains a full set of references and a discussion of the sources of uncertainty in these determinations.

this ratio to define a renormalized value of $\sin^2\theta_w$:

$$s_W^2 \equiv 1 - \frac{m_W^2}{m_Z^2}. \tag{20.97}$$

In the second column, the third parameter is $\sin^2\theta_w$ computed from the weak interaction coupling constants defined by minimal subtraction (Eq. (11.77)). The differences between different definitions of $\sin^2\theta_w$ appear at the level of one-loop computations and can reveal interesting physics; this subject is discussed in Section 21.3.

A second subtlety is that the one-loop corrections to weak neutral current processes depend on the value of the t quark mass, which has only recently been determined and is still somewhat poorly known. The dependence on the t quark mass is relatively strong, for interesting reasons that we will discuss in Section 21.3. The one-loop corrections also depend weakly on properties of the particles responsible for the spontaneous symmetry breaking.

We can see from Table 20.1 that a wide variety of effects due to the weak neutral current have been observed, with magnitudes accounted for by a single, consistent value of $\sin^2 \theta_w$. This remarkable concordance of theory and experiment gives us confidence that the Glashow-Weinberg-Salam theory is indeed the correct description of weak and electromagnetic interactions.

Fermion Mass Terms

We now return to the problem of writing mass terms for the quarks and leptons. Recall that one cannot put ordinary mass terms into the Lagrangian, because the left- and right-handed components of the various fermion fields have different gauge quantum numbers and so simple mass terms violate gauge invariance. To give masses to the quarks and leptons, we must again invoke the mechanism of spontaneous symmetry breaking.

We began this section by assuming that a scalar field ϕ acquires a vacuum expectation value (20.58), in order to give mass to the W and Z bosons. This scalar field needed to be a spinor under $SU(2)$ and to have $Y = 1/2$ in order to produce the correct pattern of gauge boson masses. With these quantum numbers, we can also write a gauge-invariant coupling linking e_L, e_R, and ϕ, as follows:

$$\Delta \mathcal{L}_e = -\lambda_e \overline{E}_L \cdot \phi \, e_R + \text{h.c.} \tag{20.98}$$

Here the $SU(2)$ indices of the doublets E_L and ϕ are contracted; notice also that the charges Y of the various fields sum to zero. The parameter λ_e is a new dimensionless coupling constant. If we replace ϕ in this expression by its vacuum expectation value (20.58), we obtain

$$\Delta \mathcal{L}_e = -\frac{1}{\sqrt{2}} \lambda_e v \, \bar{e}_L e_R + \text{h.c.} + \cdots. \tag{20.99}$$

This is a mass term for the electron. The size of the mass is set by the vacuum expectation value of ϕ, rescaled by the new dimensionless coupling:

$$m_e = \frac{1}{\sqrt{2}} \lambda_e v. \tag{20.100}$$

Since the electron mass is proportional to v, one might expect that the masses of the electron and the W boson should be of the same order. In fact, taking the observed values, $m_e/m_W \sim 6 \times 10^{-6}$. Since λ_e is a renormalizable coupling, it must be treated as an input to the theory. Thus the GWS theory allows the electron to be very light, but it cannot explain why the electron is so light compared to the W boson.

We can write mass terms for the quark fields in the same way. Notice that, in the following expression, both terms are invariant under $SU(2)$ and have zero net Y:

$$\Delta \mathcal{L}_q = -\lambda_d \overline{Q}_L \cdot \phi \, d_R - \lambda_u \epsilon^{ab} \overline{Q}_{La} \phi_b^\dagger u_R + \text{h.c.} \tag{20.101}$$

Substituting the vacuum expectation value of ϕ from Eq. (20.58), these terms become

$$\Delta \mathcal{L}_q = -\frac{1}{\sqrt{2}} \lambda_d v \, \bar{d}_L d_R - \frac{1}{\sqrt{2}} \lambda_u v \, \bar{u}_L u_R + \text{h.c.} + \cdots, \tag{20.102}$$

standard mass terms for the d and u quarks. The GWS theory thus gives the relations

$$m_d = \frac{1}{\sqrt{2}} \lambda_d v, \qquad m_u = \frac{1}{\sqrt{2}} \lambda_u v. \tag{20.103}$$

As with the electron, the theory parametrizes but does not explain the small values of the d and u quark masses observed experimentally.

When additional generations of quarks are introduced into the theory, there can be additional coupling terms that mix generations. Alternatively, we can diagonalize the Higgs couplings by choosing a new basis for the quark fields. We will show that this is always possible in Section 20.3. However, this simplification of the Higgs couplings causes a complication in the gauge couplings. Let

$$u_L^i = (u_L, c_L, t_L) \,, \qquad d_L^i = (d_L, s_L, b_L) \tag{20.104}$$

denote the up- and down-type quarks in their original basis, and let $u_L^{\prime i}$ and $d_L^{\prime i}$ denote the quarks in the basis that diagonalizes their Higgs couplings. This latter basis is the physical one, since it is the basis that diagonalizes the mass matrix. The two bases are related by unitary transformations:

$$u_L^i = U_u^{ij} u_L^{\prime j}, \qquad d_L^i = U_d^{ij} d_L^{\prime j}. \tag{20.105}$$

In this new basis, the W boson current takes the form

$$J_W^{\mu+} = \frac{1}{\sqrt{2}} \bar{u}_L^i \gamma^\mu d_L^i = \frac{1}{\sqrt{2}} \bar{u}_L^{\prime i} \gamma^\mu (U_u^\dagger U_d)_{ij} d_L^{\prime j}. \tag{20.106}$$

This expression is conventionally written

$$J_W^{\mu+} = \frac{1}{\sqrt{2}} \bar{u}_L^{\prime i} \gamma^\mu V_{ij} d_L^{\prime j}, \tag{20.107}$$

where V_{ij} is a unitary matrix called the Cabibbo-Kobayashi-Maskawa (CKM) matrix. The off-diagonal terms in V_{ij} allow weak-interaction transitions between quark generations. For example, restricting to two generations for simplicity and writing

$$V_{1j} d_L^{\prime j} = \cos\theta_c d_L^\prime + \sin\theta_c s_L^\prime, \tag{20.108}$$

the term proportional to $\sin\theta_c$ allows an s quark to decay weakly to a u quark. We have made use of this structure in our discussion of the effective Lagrangian for K meson decays in Section 18.2. We will discuss CKM flavor mixing and its symmetry properties in more detail in Section 20.3.

It is interesting to note that there is no term within the structure we have described that gives a mass to the neutrino. If we wanted to generalize Eq. (20.98) to allow a neutrino mass term, we would have to introduce a new fermion field ν_R that is completely neutral under $SU(2) \times U(1)$. Then we could write the Higgs coupling

$$\Delta \mathcal{L}_\nu = -\lambda_\nu \epsilon^{ab} \overline{E}_{La} \phi_b^\dagger \nu_R + \text{h.c.}, \tag{20.109}$$

which would give the ν_e a mass, presumably comparable to that of the electron. But we know from experiment that neutrino masses are extremely small; the mass of the ν_e is known to be less than 10 eV. This extreme suppression of the neutrino masses would be naturally explained if the states ν_R do not exist. We will show in Section 20.3 that this assumption also implies that there are no transitions between leptons of different generations; this result is also in accord with very strong experimental bounds.

The Higgs Boson

This discussion of fermion mass generation emphasizes that the scalar field that causes spontaneous breaking of the gauge symmetry is an important ingredient in the structure of the Glashow-Weinberg-Salam theory. We should therefore ask whether it has any more direct manifestations.

To investigate this question, we will work in the *unitarity gauge*, analogous to that used for the Abelian model in Eq. (20.12). Let us parametrize the scalar field ϕ by writing

$$\phi(x) = U(x) \frac{1}{\sqrt{2}} \begin{pmatrix} 0 \\ v + h(x) \end{pmatrix}. \tag{20.110}$$

The two-component spinor on the right has an arbitrary real-valued lower component, given by the vacuum expectation value of ϕ plus a fluctuating real-valued field $h(x)$ with $\langle h(x) \rangle = 0$. This spinor is acted on by a general $SU(2)$ gauge transformation $U(x)$ to produce the most general complex-valued two-component spinor. We can now make a gauge transformation to eliminate $U(x)$ from the Lagrangian. This reduces ϕ to a field with one physical degree of freedom.

An explicit renormalizable Lagrangian that leads to a vacuum expectation value for ϕ is

$$\mathcal{L} = |D_\mu \phi|^2 + \mu^2 \phi^\dagger \phi - \lambda (\phi^\dagger \phi)^2. \tag{20.111}$$

The minimum of the potential energy occurs at

$$v = \left(\frac{\mu^2}{\lambda} \right)^{1/2}. \tag{20.112}$$

In the unitarity gauge, the potential energy terms in (20.111) take the form

$$\mathcal{L}_V = -\mu^2 h^2 - \lambda v h^3 - \frac{1}{4} \lambda h^4$$
$$= -\frac{1}{2} m_h^2 h^2 - \sqrt{\frac{\lambda}{2}} m_h h^3 - \frac{1}{4} \lambda h^4. \tag{20.113}$$

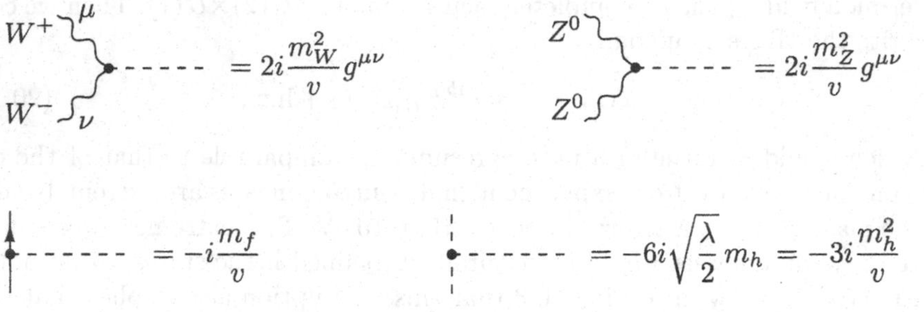

Figure 20.6. Feynman rules for the couplings of the Higgs boson to vector bosons, to fermions, and to itself.

The quantum of the field $h(x)$ is a scalar particle with mass

$$m_h = \sqrt{2}\mu = \sqrt{2\lambda}\, v. \qquad (20.114)$$

This particle is known as the *Higgs boson*. As for the fermions in the GWS theory, the Higgs boson has a mass whose general magnitude is controlled by the vacuum expectation value v, but whose precise value is determined by a new, unspecified, renormalizable coupling constant.

The expansion of the kinetic energy term of (20.111) in unitarity gauge yields the gauge boson mass term (20.62), plus additional terms involving the Higgs boson field. Explicitly,

$$\mathcal{L}_K = \frac{1}{2}(\partial_\mu h)^2 + \left[m_W^2 W^{\mu+} W_\mu^- + \frac{1}{2} m_Z^2 Z^\mu Z_\mu \right] \cdot \left(1 + \frac{h}{v} \right)^2, \qquad (20.115)$$

where m_W and m_Z are given by Eqs. (20.63).

Finally, the fermion mass terms in Eqs. (20.98) and (20.101) lead to direct couplings of the Higgs boson to fermions. Evaluating these terms in unitarity gauge, we find that, for any quark or lepton flavor, the Higgs boson couples according to

$$\mathcal{L}_f = -m_f \bar{f} f \left(1 + \frac{h}{v} \right). \qquad (20.116)$$

The Higgs boson couplings in Eqs. (20.113), (20.115), and (20.116) lead to the Feynman rules shown in Fig. 20.6.

In general, the couplings of the Higgs boson to other particles of the weak interaction theory are proportional to the masses of those particles. Thus, the particles that are most easily made in the laboratory have very weak couplings to the Higgs boson, which makes it difficult to observe this particle. In any event, the Higgs boson has not yet been found. As of this writing, the Higgs boson that we have just described has been searched for and excluded for values of m_h below 60 GeV. If the self-coupling λ is large, however, the Higgs boson could have a mass as large as 1000 GeV; thus, a large dynamic range remains unexplored.

The phenomenological properties of the Higgs boson are worked out in more detail in the Final Project of Part III.

A Higgs Sector?

Since there is no experimental evidence for the existence of the simple Higgs boson contained in the GWS model, it is worth asking whether the W and Z bosons might acquire mass by a more complicated mechanism. There are two aspects to this question.

First, is it certain that the W and Z bosons are gauge bosons of a spontaneously broken $SU(2) \times U(1)$ symmetry? The evidence for this idea comes from the universality of the couplings of the various quarks and leptons to the weak interactions. This universality is tested in the fact that the same value of the Fermi constant describes all charged-current weak-interaction processes, and that this same strength of coupling combined with a single value of $\sin^2 \theta_w$ describes the whole range of weak neutral current phenomena. We have seen, especially in the discussion of Chapter 16, that the principle of local gauge invariance leads naturally to the prediction of universal, flavor-independent, coupling constants. No other principle is known that would explain this striking regularity. Thus there is compelling evidence that the underlying theory of the weak interactions is a spontaneously broken gauge theory.

However, it is quite possible that the mechanism of the spontaneous breaking of $SU(2) \times U(1)$ is more complicated than the simple model of a single scalar field that we have written in Eq. (20.111). In principle, the breaking of $SU(2) \times U(1)$ might be the result of the dynamics of a complicated new set of particles and interactions, which we will refer to as the *Higgs sector*. Experiment gives us only three properties of this new sector: First, it must generate the masses of the quarks and leptons. Second, it must generate the masses of the W and Z bosons. The third piece of information, which is the only nontrivial one, comes from the relation (20.73) between weak boson masses in the GWS theory,

$$m_W = m_Z \cos\theta_w. \tag{20.117}$$

This relation is satisfied experimentally to better than 1% accuracy, that is, to the level of one-loop radiative corrections. Whatever complicated mechanism we invoke to generate the spontaneous breaking of $SU(2) \times U(1)$, it should reproduce this relation in a natural way.

To understand the implications of relation (20.117), we must analyze the gauge boson mass matrix without assuming that $SU(2) \times U(1)$ is broken by the expectation value of a scalar field. Actually, it is possible to compute the gauge boson mass matrix under much less restrictive assumptions, using the argument given at the very end of Section 20.1. There we constructed the gauge boson mass matrix from the matrix elements for gauge currents to create or destroy Goldstone bosons. We will now show that relation (20.117) follows for a large class of models of $SU(2) \times U(1)$ breaking for which these matrix elements satisfy certain simple properties.

Any model of weak-interaction symmetry breaking must contain some set of fields that is responsible for the spontaneous breaking of $SU(2) \times U(1)$. Think of this sector of the theory as a field theory with a global $SU(2) \times U(1)$ symmetry, which is promoted to a local symmetry through its coupling to gauge bosons. In the theory with global symmetry, this symmetry is spontaneously broken to $U(1)$. Since three continuous symmetries are spontaneously broken, this sector must supply three Goldstone bosons, which will eventually be eaten by W^+, W^-, and Z^0. Call these three bosons π_a, where $a = 1, 2, 3$. Let $J^{\mu a}$ be the $SU(2)$ symmetry currents of the new sector, and let $J^{\mu Y}$ be the $U(1)$ current. The gauge boson mass matrix will then be constructed from the matrix elements (20.46), which here take the form

$$\langle 0 | \, J^{\mu A} \, | \pi_b(p) \rangle = -ip^\mu F^A{}_b, \tag{20.118}$$

with $A = 1, 2, 3, Y$ and $b = 1, 2, 3$. Using the method of Eq. (20.55), we find that the gauge boson vacuum polarization contains the pole term

$$-\frac{i}{k^2} (g_A F^A{}_c)(g_B F^B{}_c), \tag{20.119}$$

summed over c, where $g_A = g$ for $A = 1, 2, 3$ and $g_A = g'$ for $A = Y$. Then we can identify the gauge boson mass matrix as

$$m^2_{AB} = g_A g_B F^A{}_c F^B{}_c. \tag{20.120}$$

To reproduce the known form of the weak gauge boson mass matrix, we must now place constraints on the $F^A{}_b$. First we must insure that the photon remains massless. This follows if the linear combination of charges (20.69) annihilates the vacuum. In the language of Eq. (20.118), we must insist that the corresponding linear combination of currents cannot excite a Goldstone boson:

$$\langle 0 | \left(J^{\mu 3} + J^{\mu Y} \right) | \pi^a(p) \rangle = 0. \tag{20.121}$$

We can also achieve relation (20.117), using the following additional assumption: The symmetry-breaking sector has an $SU(2)$ global symmetry, under which the three Goldstone bosons and the three $SU(2)$ gauge currents transform as triplets, which remains exact when the $SU(2)$ gauge symmetry is spontaneously broken. This global $SU(2)$ symmetry implies that, if $A = a = 1, 2, 3$ in Eq. (20.118),

$$\langle 0 | \, J^{\mu a} \, | \pi^b(p) \rangle = -iF p^\mu \delta^{ab}, \tag{20.122}$$

where F is a parameter with the dimensions of mass. Combining (20.122) and (20.121), we have

$$\langle 0 | \, J^{\mu Y} \, | \pi^3(p) \rangle = +iF p^\mu. \tag{20.123}$$

Inserting this form for $F^A{}_b$ into (20.120), we find the gauge boson mass matrix

$$m^2 = F^2 \begin{pmatrix} g^2 & & & \\ & g^2 & & \\ & & g^2 & -gg' \\ & & -gg' & g'^2 \end{pmatrix}, \tag{20.124}$$

where the matrix acts on the gauge boson $(A_\mu^1, A_\mu^2, A_\mu^3, B_\mu)$. The eigenvectors of this matrix are precisely (20.63) and (20.64). To reproduce the eigenvalues, we need only define

$$v = 2F. \tag{20.125}$$

We have now shown that the GWS relation between the W and Z boson masses is not special to the situation in which the gauge symmetry is broken by a single scalar field. This relation follows from the much more general assumption of an unbroken global $SU(2)$ symmetry of the Higgs sector. This symmetry is often called *custodial SU(2) symmetry*.* We have seen this symmetry already as the global $SU(2)$ symmetry of the weak-interaction effective Lagrangian (20.94).

For the case of a single scalar field, the custodial symmetry arises in the following way: If we write the field ϕ in terms of its four real components, the Lagrangian (20.111) (ignoring the gauge couplings) has $O(4)$ global symmetry. The vacuum expectation value of ϕ breaks this symmetry down to $O(3)$, that is, $SU(2)$.

However, there are many other quantum field theories that break $SU(2)$ spontaneously while leaving another global $SU(2)$ symmetry unbroken. One rather complex example is given by QCD with two massless flavors, if we identify the gauged $SU(2)$ with the symmetry generated by U_L in (19.82) and identify the custodial $SU(2)$ with vectorial isospin symmetry. A copy of the familiar strong interactions with a mass scale large enough to give $F = 125$ GeV would be a perfectly acceptable model for the Higgs sector. (Unfortunately, it is not easy in this model to generate masses for the quarks and leptons.)

The question of the nature of the Higgs sector and the explicit mechanism of $SU(2) \times U(1)$ breaking is probably the most pressing open problem in the theory of elementary particles. We will discuss this question further in the Epilogue.

20.3 Symmetries of the Theory of Quarks and Leptons

Putting together the theory of strong interactions described in Chapter 17 and the theory of weak and electromagnetic interactions described in the previous section, we have now constructed a complete description of elementary particle interactions. It is interesting to investigate the symmetries of this theory, to ask what might be the fundamental symmetries of the underlying description of Nature.

We have already seen, in the arguments leading up to Eq. (15.17), that the Lagrangian of a gauge theory is highly restricted by the conditions of renormalizability and gauge invariance. In this section, we will construct the

---

*P. Sikivie, L. Susskind, M. Voloshin, and V. Zakharov, *Nucl. Phys.* **B173**, 189 (1980).

most general renormalizable Lagrangian consistent with the $SU(3) \times SU(2) \times U(1)$ gauge symmetries of the strong, weak and electromagnetic interactions. We can then ask what further global symmetries we must impose on this theory in order to give it the global symmetries that we see in Nature.

As a first step, we will ignore the Higgs scalar field and the mass terms of quarks, leptons, and gauge bosons. Then the Lagrangian of the theory of quarks and leptons is entirely specified by gauge invariance and renormalizability. We have

$$\mathcal{L}_K = -\frac{1}{4} \sum_i (F_{i\mu\nu}^a)^2 + \sum_J \bar{\psi}_J (i\slashed{D}) \psi_J, \qquad (20.126)$$

where the index i runs over the three factors of the gauge group and the index J runs over the various multiplets of chiral fermions.

In principle, we could add to (20.126) the following pseudoscalar pure gauge operators:

$$\Delta \mathcal{L}_\theta = \sum_i \frac{\theta_i g_i^2}{64\pi^2} \epsilon^{\mu\nu\lambda\sigma} F_{i\mu\nu}^a F_{i\lambda\sigma}^a. \qquad (20.127)$$

These terms are apparently odd under both P and T. However, we saw at the end of Section 19.2 that terms of this form can be generated or canceled by making a change of variables in the effective action. For example, the change of variables on the right-handed electron field

$$e_R \to e^{i\alpha} e_R \qquad (20.128)$$

produces, according to (19.78) or (19.79), a correction to the Lagrangian involving the P- and T-odd combination of field strengths for the $U(1)$ gauge field

$$\Delta \mathcal{L} = \alpha \cdot \frac{g'^2}{32\pi^2} \epsilon^{\mu\nu\lambda\sigma} F_{\mu\nu} F_{\lambda\sigma}. \qquad (20.129)$$

The coefficient of (20.129) differs from the corresponding coefficient in (19.79) because we transform only the right-handed chiral component of the electron field. If we were to transform another fermion field, of hypercharge Y, we would find a similar shift, with the coefficient proportional to Y^2. If this new field coupled to the $SU(2)$ or $SU(3)$ gauge fields, we would also find terms proportional to those field strengths. Thus, we can eliminate the term in (20.127) involving the $U(1)$ field strengths by making the change of variables (20.128) with $\alpha = -\frac{1}{2}\theta_1$. We can eliminate all three terms in (20.127) by making appropriate chiral rotations on three fermion multiplets, say, e_R, E_L, and Q_L. The change of variables (20.128), which rotates the right-handed electron field, is not symmetric under parity and, in fact, changes the definition of the parity operation. By making this change of variables, we are choosing new coordinates in which the P and T transformation properties of the whole theory are as simple as possible.

Let us now investigate the properties of the Lagrangian (20.126) under P, C, and T. The couplings of the QCD gauge bosons are invariant to each of these symmetries separately. However, the couplings of the $SU(2)$ gauge bosons violate P and C as much as possible. Recall from Section 3.6 that P converts a left-handed electron to a right-handed electron, and that C converts a left-handed electron to a left-handed positron. Each of these operations converts a particle that couples to $SU(2)$ gauge bosons to one that does not. However, the combination of these two operations interchanges left-handed particles with right-handed antiparticles. Thus the combined operation CP is a symmetry of (20.126). This Lagrangian is also invariant under time reversal.

Thus, we see that the discrete symmetries of C and P, on the one hand, and CP and T, on the other, stand on a very different footing in gauge field theories. Any chiral gauge theory will naturally violate C and P. At this level in our analysis, it is a mystery why C and P should be observed to be approximate symmetries of Nature. On the other hand, every theory of gauge bosons and massless fermions respects CP and T. It is known experimentally that Nature contains some interaction that violates CP, since the CP selection rules are weakly violated in the decays of the K^0 meson. But to find a source for this violation, we must add terms to our basic gauge theory (20.126).

We must, first of all, add dynamics to (20.126) that will cause the spontaneous breaking of $SU(2) \times U(1)$. We will begin by working with the simplest model with one Higgs scalar field ϕ. The most general renormalizable Lagrangian for ϕ is

$$\mathcal{L}_\phi = \left| D_\mu \phi \right|^2 + \mu^2 \phi^\dagger \phi - \lambda (\phi^\dagger \phi)^2. \tag{20.130}$$

The Hermiticity of $\mathcal{L}_\phi$ implies that the parameters μ^2 and λ are real. Thus this Lagrangian respects P, C, and T. As discussed at the end of the previous section, this Lagrangian also automatically has the custodial $SU(2)$ symmetry required to produce the mass relation (20.117).

Finally, we must add the terms that couple the Higgs field to the quarks and leptons. Here, renormalizability and gauge invariance provide the weakest constraints, and there are many allowed interactions. We will first analyze the coupling of ϕ to the quark fields, and then generalize this discussion to the leptons.

In writing the Higgs field couplings to the quarks, we should recall that there are known to be three generations of quarks and leptons. Thus there are three doublets of left-handed quarks:

$$Q_L^i = \begin{pmatrix} u^i \\ d^i \end{pmatrix}_L = \left(\begin{pmatrix} u \\ d \end{pmatrix}_L, \begin{pmatrix} c \\ s \end{pmatrix}_L, \begin{pmatrix} t \\ b \end{pmatrix}_L \right). \tag{20.131}$$

There are six right-handed quarks, three with $Y = \frac{2}{3}$ and three with $Y = -\frac{1}{3}$:

$$u_R^i = (u_R, c_R, t_R), \qquad d_R^i = (d_R, s_R, b_R). \tag{20.132}$$

When we couple gauge fields to these quarks, we replace the ordinary derivatives with covariant derivatives. This automatically gives all of the quarks the same coupling to QCD and all quarks of the same type the same coupling to

the weak interactions. It does not allow mixing between the various quark flavors. However, the coupling of the Higgs field to the quarks does not follow
from a gauge principle and so need not have any of these restrictions. Unless
we require quark flavor conservation by postulating a new discrete symmetry
of the theory, the Higgs couplings will, in general, mix the various flavors.

If we do not impose any additional symmetries on the theory, we must
write the most general renormalizable gauge-invariant coupling with the structure of Eq. (20.101):

$$\mathcal{L}_m = -\lambda_d^{ij}\overline{Q}_L^i \cdot \phi\, d_R^j - \lambda_u^{ij}\epsilon^{ab}\overline{Q}_{La}^i \phi_b^\dagger u_R^j + \text{h.c.}, \tag{20.133}$$

where λ_d^{ij} and λ_u^{ij} are general, not necessarily symmetric or Hermitian,
complex-valued matrices. The operation of CP interchanges the operators
written in (20.133) with their Hermitian conjugates without changing the coefficients; thus, CP is equivalent to the substitutions

$$\lambda_d^{ij} \to (\lambda_d^{ij})^*, \qquad \lambda_u^{ij} \to (\lambda_u^{ij})^*. \tag{20.134}$$

CP would be a symmetry of (20.133) if the matrices λ^{ij} were real-valued;
however, there is no principle that requires this. Without the imposition of
further symmetry requirements, it seems that (20.133) does maximum violence
to all discrete and flavor conservation symmetries.

However, just as we were able to eliminate the T-violating terms (20.127)
by making a chiral rotation, we can simplify the form of (20.133) by appropriate chiral transformations. To find the required transformations, diagonalize
the Hermitian matrices obtained by squaring λ_d and λ_u. Define unitary matrices U_u and W_u by

$$\lambda_u\lambda_u^\dagger = U_u D_u^2 U_u^\dagger, \qquad \lambda_u^\dagger\lambda_u = W_u D_u^2 W_u^\dagger, \tag{20.135}$$

where D_u^2 is a diagonal matrix with positive eigenvalues. Then

$$\lambda_u = U_u D_u W_u^\dagger, \tag{20.136}$$

where D_u is the diagonal matrix whose diagonal elements are the positive
square roots of the eigenvalues of (20.135). We can define unitary matrices U_d
and W_d in a similar way and decompose λ_d as

$$\lambda_d = U_d D_d W_d^\dagger. \tag{20.137}$$

Now make the change of variables

$$u_R^i \to W_u^{ij} u_R^j, \qquad d_R^i \to W_d^{ij} d_R^j. \tag{20.138}$$

This eliminates the unitary matrices W_u and W_d from the Higgs coupling
(20.133). Since each of the three u_R^i and each of the three d_R^i have the same
coupling to the gauge fields, W_u and W_d commute with the corresponding
covariant derivatives. Thus, under (20.138),

$$\sum_i (\bar{u}_R^i(i\slashed{D})u_R^i + \bar{d}_R^i(i\slashed{D})d_R^i) \to \sum_i (\bar{u}_R^i(i\slashed{D})u_R^i + \bar{d}_R^i(i\slashed{D})d_R^i), \tag{20.139}$$

and so W_u and W_d disappear from the theory.

The analogous transformation on the left-handed fields also makes a dramatic simplification. Make the change of variables

$$u_L^i \to U_u^{ij} u_L^j, \qquad d_L^i \to U_d^{ij} d_L^j. \tag{20.140}$$

This transformation eliminates U_u, U_d from the terms in (20.133) that involve the lower component of the Higgs field. In unitarity gauge, only these terms survive. By combining the diagonal elements of D_u and D_d with the vacuum expectation value of the Higgs field, we can relate these elements to quark masses:

$$m_u^i = \frac{1}{\sqrt{2}} D_u^{ii} v, \qquad m_d^i = \frac{1}{\sqrt{2}} D_d^{ii} v. \tag{20.141}$$

With this replacement, (20.133) takes the form

$$\mathcal{L}_m = -m_d^i \bar{d}_L^i d_R^i \left(1 + \frac{h}{v} \right) - m_u^i \bar{u}_L^i u_R^i \left(1 + \frac{h}{v} \right) + \text{h.c.} \tag{20.142}$$

This has the standard form of quark mass terms and Higgs boson couplings. The transformations (20.138) and (20.140) thus convert the quark fields to the basis of mass eigenstates. In this basis, the mass terms and Higgs couplings are diagonal in flavor and conserve P, C, and T.

Since left-handed u and d quarks have identical couplings to QCD, the matrices U_u and U_d commute with the QCD couplings in the covariant derivative. However, u_L and d_L are mixed by the weak interactions, and so we must investigate the effect of (20.140) on the $SU(2) \times U(1)$ couplings more carefully. This is most easily done by referring to the Lagrangian (20.79). The matrices U_u and U_d cancel out of the pure kinetic terms in the first line of (20.79). They also cancel out of the electromagnetic current J_{EM}^μ; for example,

$$\bar{u}_L^i \gamma^\mu u_L^i \to \bar{u}_L^i U_u^{\dagger ij} \gamma^\mu U_u^{jk} u_L^k = \bar{u}_L^i \gamma^\mu u_L^i. \tag{20.143}$$

By the same logic, U_u and U_d cancel out of the Z^0 boson current.

However, in the current that couples to the W boson field, we find

$$J^{\mu+} = \frac{1}{\sqrt{2}} \bar{u}_L^i \gamma^\mu d_L^i \to \frac{1}{\sqrt{2}} \bar{u}_L^i \gamma^\mu \left(U_u^\dagger U_d \right)^{ij} d_L^j. \tag{20.144}$$

That is, the charge-changing weak interactions link the three u_L^i quarks with a unitary rotation of the triplet of d_L^i quarks, with this rotation given by the unitary matrix

$$V = U_u^\dagger U_d. \tag{20.145}$$

The matrix V is known as the *Cabibbo-Kobayashi-Maskawa* (CKM) mixing matrix.

The matrix V can have complex elements, but we can remove phases from V by performing phase rotations of the various quark fields. Before analyzing the case of three generations, it is useful to consider the case of two generations—u, d, c, and s. In this case, V is a 2×2 unitary matrix. Such a matrix has 4 parameters; we can write its most general form as

$$V = \begin{pmatrix} \cos\theta_c e^{i\alpha} & \sin\theta_c e^{i\beta} \\ -\sin\theta_c e^{i(\alpha+\gamma)} & \cos\theta_c e^{i(\beta+\gamma)} \end{pmatrix}. \tag{20.146}$$

One parameter of V is a rotation angle, and the other three are phases. We can remove these phases by performing the change of variables on the quark fields

$$q_L^i \to \exp[i\alpha^i]q_L^i. \tag{20.147}$$

This global phase rotation has no effect on any term of the Lagrangian except for the weak charged current (20.144). A phase rotation that is equal for all four quark flavors cancels out of (20.144). However, the other three possible phase transformations are just what we need to eliminate α, β, and γ.

When we have chosen the phases of the quark fields in this way, V takes the form

$$V = \begin{pmatrix} \cos\theta_c & \sin\theta_c \\ -\sin\theta_c & \cos\theta_c \end{pmatrix}. \tag{20.148}$$

Then the quark terms in the weak charged current can be written

$$J^{\mu+} = \frac{1}{\sqrt{2}}\left(\cos\theta_c \bar{u}_L\gamma^\mu d_L + \sin\theta_c \bar{u}_L\gamma^\mu s_L - \sin\theta_c \bar{c}_L\gamma^\mu d_L + \cos\theta_c \bar{c}_L\gamma^\mu s_L\right). \tag{20.149}$$

We have already seen, in Eqs. (18.31) and (18.32), that this is the way the s quark enters the weak interactions. The angle θ_c is the Cabibbo angle, as defined in Eq. (18.30).

The same set of arguments can be made for the theory with three generations. Here V is a general unitary 3×3 matrix. Such a matrix has 9 parameters. Of these, 3 are rotation angles; this is the number of parameters of an $O(3)$ rotation. The remaining 6 parameters are phases. We can remove these phases by making phase rotations of quark fields as in (20.147), but the overall phase is redundant, so we can remove only 5 of these phases. The final form of V contains 3 angles, of which one is the Cabibbo angle, and one phase. After all the transformations we have made, this one phase that makes some couplings of the W^+ to quarks complex is the only remaining parameter that violates CP.

We began this argument from a Lagrangian for the quark-Higgs boson coupling that seemed to violate all possible flavor symmetries and all discrete spacetime symmetries. However, by making changes of variables on the fermion fields, we have been able to dramatically simplify the form of the Lagrangian. If we keep only those terms involving the massless gauge bosons, the photon and the gluons, plus the mass terms and interactions written in (20.142), we see that this set of terms conserves P, C, T, and all flavor symmetries. This dramatic simplification occurs because the unbroken gauge symmetry of Nature, the gauge symmetry of QCD and QED, is nonchiral and can be written as acting on Dirac fermions. Since we have omitted only the

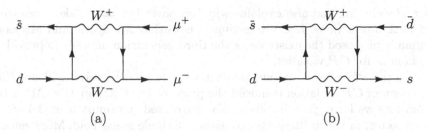

Figure 20.7. Higher-order diagrams that seem to give the leading contributions to flavor-changing weak neutral current processes: (a) $K^0 \to \mu^+\mu^-$; (b) $K^0 \leftrightarrow \overline{K}^0$.

effects mediated by the massive W and Z bosons, this much of the analysis already guarantees that Nature will appear, to a high degree of approximation, to respect the three separate discrete symmetries and all quark flavor conservation laws. Notice that we did not assume any fundamental global symmetries, but depended only on the assignment of gauge quantum numbers in the $SU(3) \times SU(2) \times U(1)$ gauge theory.

If we include the Z boson and the weak neutral current, we have a theory that violates P and C through Z exchange but that respects CP. In addition, this theory respects all flavor conservation laws. We describe this situation by saying that there is no flavor-changing weak neutral current. The experimental evidence for this statement is quite impressive. The best tests come from the study of the neutral K^0 meson, which is an $s\bar{d}$ bound state and so could decay by Z^0 exchange if this boson coupled to a flavor-changing current. In fact, the decay $K^0 \to \mu^+\mu^-$ is highly suppressed, to the level of the one-loop weak interaction correction shown in Fig. 20.7(a). Similarly, the interconversion of K^0 and $\overline{K}^0$, which could proceed directly if the Z^0 could change flavor, is suppressed to the level of the contribution shown in Fig. 20.7(b).

On the other hand, W bosons couple to currents that can change quark flavor, in a pattern parametrized by the Cabibbo angle and the other angles in the CKM matrix. Thus, heavy quark flavors decay by W boson exchange processes. Since the W couples to a current that contains only left-handed quarks, it mediates an interaction that violates P and C maximally. This violation of discrete symmetries is concealed from our ordinary experience because the amplitude for W exchange is small. However, this P and C violation is a dramatic qualitative feature of weak decays.

Since the coupling of the W to quarks contains an irreducible phase, these couplings in principle can violate CP. However, we have seen that this phase can be removed in a theory with only two generations. This means that the phase of the CKM matrix can have physical consequences only in a process that involves all three generations. Typically, this means that the CKM phase can contribute only to weak interaction loop corrections or to complicated exclusive decay processes. Thus the $SU(3) \times SU(2) \times U(1)$ theory can account for CP violation, and also explains why this effect is much weaker even than the weak interactions. It is interesting to note that Kobayashi and Maskawa originally proposed the existence of the third generation in order to provide a mechanism for CP violation.[†]

On the other hand, at this moment there is no conclusive evidence that the origin of CP violation is indeed the phase of the CKM matrix. All of the arguments we have given in this section have used the simplest model of the Higgs sector, in which this sector consists of a single scalar field. More general models of the Higgs sector may leave behind a more complicated set of quark-Higgs couplings than appear in (20.142), and some of these may violate CP. In addition, there may be terms in the Higgs sector itself that lead to CP violation. The origin of the observed CP violation is still an open problem that needs both theoretical and experimental exploration.

Before leaving this subject, we must discuss one more aspect of this argument that is still mysterious. To simplify the Lagrangian of the gauge theory of quarks to its final form, we needed to make chiral changes of variables in the functional integral. We saw in Section 19.2, and we reviewed at the beginning of this section, that such changes of variables produce the new P- and T-violating terms written in Eq. (20.127). It can be shown, using the fact that these terms are total derivatives, that the terms involving $SU(2)$ and $U(1)$ field strengths have no observable effects. However, the term involving QCD field strengths can induce an electric dipole moment for the neutron, a T-violating effect that has been searched for and excluded at an impressive level of accuracy. Thus the P- and T-violating combination of QCD field strengths cannnot be allowed to appear in the Lagrangian. On the other hand, if the original up and down quark Higgs coupling matrices were of the most general possible form, it seems that this cannot be avoided. This problem is known as the *strong CP problem*. To solve this problem, one must either constrain the Higgs coupling matrices, violating the spirit of the argument we have just concluded, or one must add additional structure to the Higgs sector.[‡]

Finally, let us discuss the general form and simplification of the Higgs boson couplings to leptons. When we wrote the Glashow-Weinberg-Salam Lagrangian in the previous section, we noted that no gauge field coupled to the right-handed neutrino. Thus, we chose to eliminate this particle from the theory. We might need right-handed components of the neutrinos to construct neutrino mass terms, but at the moment there is no evidence for nonzero neutrino masses. Thus, in the remainder of this section, we will assume that there are no right-handed neutrinos and work out the consequences of this assumption.[*]

[†]M. Kobayashi and T. Maskawa, *Prog. Theor. Phys.* **49**, 652 (1973).

[‡]The strong CP problem, its proposed solutions, and their unexpected implications are reviewed by R. D. Peccei in *CP Violation*, C. Jarlskog, ed. (World Scientific, 1989).

[*]In generalizations of the $SU(2) \times U(1)$ model, neutrinos can acquire Majorana

Generalizing Eq. (20.133), we can write the most general coupling of a Higgs boson to three generations of leptons. Since there are no right-handed neutrinos, the only possible coupling is

$$\mathcal{L}_m = -\lambda_\ell^{ij} \overline{E}_L^i \cdot \phi \, e_R^j + \text{h.c.} \tag{20.150}$$

To diagonalize this coupling, represent λ_ℓ in the form

$$\lambda_\ell = U_\ell D_\ell W_\ell^\dagger, \tag{20.151}$$

and eliminate the matrices U_ℓ and W_ℓ by the changes of variables

$$e_L^i \to U_\ell^{ij} e_L^j, \qquad \nu_L^i \to U_\ell^{ij} \nu_L^j, \qquad e_R^i \to W_\ell^{ij} e_R^j. \tag{20.152}$$

Since we are now making the same change of variables on the two components of the weak doublet E_L^i, this change of variables commutes with the $SU(2)$ interactions in the covariant derivative. Thus the unitary matrices U_ℓ and W_ℓ completely disappear from the theory. The result is a theory of leptons that conserves CP exactly and also conserves the lepton number of each generation. This last result is very accurately tested experimentally. For example, there is no evidence for the generation-changing muon decay processes $\mu^- \to e^- \gamma$ or $\mu^- \to e^- e^- e^+$; the branching ratios for these processes are known to be below 10^{-10}.

We have seen, then, that the $SU(3) \times SU(2) \times U(1)$ gauge theory of quarks and leptons does an excellent job of accounting for the symmetries and conservation laws that are observed in elementary particle phenomena. It predicts which symmetries should be exact in Nature and which should be approximate. For approximate symmetries, it gives an accurate estimate of the level of symmetry violation. Most remarkably (except for the one issue of the strong CP problem), none of these predictions depend on any underlying global discrete or flavor symmetries in the fundamental equations. The global symmetries that we observe in Nature follow only from gauge invariance and the specific representation assignments that we made in constructing our gauge theory description.

---

mass terms that are naturally very small. These models also respect the constraints on lepton flavor mixing described in the next paragraph. For an introduction to these ideas on neutrino mass, see P. Ramond, in *Perspectives in the Standard Model*, R. K. Ellis, C. T. Hill, and J. D. Lykken, eds. (World Scientific, 1992).

Problems

20.1 Spontaneous breaking of $SU(5)$. Consider a gauge theory with the gauge group $SU(5)$, coupled to a scalar field Φ in the adjoint representation. Assume that the potential for this scalar field forces it to acquire a nonzero vacuum expectation value. Two possible choices for this expectation value are

$$
\langle\Phi\rangle = A \begin{pmatrix} 1 & & & & \\ & 1 & & & \\ & & 1 & & \\ & & & 1 & \\ & & & & -4 \end{pmatrix} \quad \text{and} \quad \langle\Phi\rangle = B \begin{pmatrix} 2 & & & & \\ & 2 & & & \\ & & 2 & & \\ & & & -3 & \\ & & & & -3 \end{pmatrix}.
$$

For each case, work out the spectrum of gauge bosons and the unbroken symmetry group.

20.2 Decay modes of the W and Z bosons.

(a) Compute the partial decay widths of the W boson into pairs of quarks and leptons. Assume that the top quark mass m_t is larger than m_W, and ignore the other quark masses. The decay widths to quarks are enhanced by QCD corrections. Show that the correction is given, to order α_s, by Eq. (17.9). Using $\sin^2\theta_w = 0.23$, find a numerical value for the total width of the W^+.

(b) Compute the partial decay widths of the Z boson into pairs of quarks and leptons, treating the quarks in the same way as in part (a). Determine the total width of the Z boson and the fractions of the decays that give hadrons, charged leptons, and invisible modes $\nu\bar{\nu}$.

20.3 $e^+e^- \to$ hadrons with photon-Z^0 interference

(a) Consider a fermion species f with electric charge Q_f and weak isospin I^3_L for its left-handed component. Ignore the mass of the f. Compute the differential cross section for the process $e^+e^- \to f\bar{f}$ in the standard electroweak model. Include the effect of the Z^0 width using the Breit-Wigner formula, Eq. (7.60). Plot the behavior of the total cross section as a function of CM energy through the Z^0 resonance, for u, d, and μ.

(b) Compute the forward-backward asymmetry for $e^+e^- \to f\bar{f}$, defined as

$$
A^f_{FB} = \frac{(\int_0^1 - \int_{-1}^0)d\cos\theta(d\sigma/d\cos\theta)}{(\int_0^1 + \int_{-1}^0)d\cos\theta(d\sigma/d\cos\theta)},
$$

as a function of center of mass energy.

(c) Show that, just on the Z^0 resonance, the forward-backward asymmetry is given by

$$
A^f_{FB} = \frac{3}{4}A^e_{LR}A^f_{LR}.
$$

(d) Show that the cross section at the peak of the Z^0 resonance is given by

$$
\sigma_{\text{peak}} = \frac{12\pi}{m_Z^2}\frac{\Gamma(Z^0 \to e^+e^-)\Gamma(Z^0 \to f\bar{f})}{\Gamma_Z^2},
$$

where Γ_Z is the total width of the Z^0. Notice that both the total width of the Z^0 and the peak height are affected by the presence of extra invisible decay modes. Compute the shifts in Γ_Z and σ_{peak} that would be produced by a hypothetical fourth neutrino species, and compare these shifts to the cross section measurements shown in Fig. 20.5.

20.4 Neutral-current deep inelastic scattering.

(a) In Eq. (17.35), we wrote formulae for neutrino and antineutrino deep inelastic scattering with $W^\pm$ exchange. Neutrinos and antineutrinos can also scatter by exchanging a Z^0. This process, which leads to a hadronic jet but no observable outgoing lepton, is called the *neutral current* reaction. Compute $d\sigma/dxdy$ for neutral current deep inelastic scattering of neutrinos and antineutrinos from protons, accounting for scattering from u and d quarks and antiquarks.

(b) Next, consider deep inelastic scattering from a nucleus A with equal numbers of protons and neutrons. For such a target, $f_u(x) = f_d(x)$, and similarly for antiquarks. Show that the formulae in part (a) simplify in such a situation. In particular, let R^ν, $R^{\bar{\nu}}$ be defined as

$$R^\nu = \frac{d\sigma/dxdy(\nu A \to \nu X)}{d\sigma/dxdy(\nu A \to \mu^- X)}, \qquad R^{\bar{\nu}} = \frac{d\sigma/dxdy(\bar{\nu} A \to \bar{\nu} X)}{d\sigma/dxdy(\bar{\nu} A \to \mu^+ X)}.$$

Show that R^ν and $R^{\bar{\nu}}$ are given by the following simple formulae:

$$R^\nu = \frac{1}{2} - \sin^2\theta_w + \frac{5}{9}\sin^4\theta_w(1 + r),$$

$$R^{\bar{\nu}} = \frac{1}{2} - \sin^2\theta_w + \frac{5}{9}\sin^4\theta_w(1 + \frac{1}{r}),$$

where

$$r = \frac{d\sigma/dxdy(\bar{\nu} A \to \mu^+ X)}{d\sigma/dxdy(\nu A \to \mu^- X)}.$$

These formulae remain true when R^ν and $R^{\bar{\nu}}$ are redefined to be the ratios of neutral- to charged-current cross sections integrated over the region of x and y that is observed in a given experiment.

(c) By setting r equal to the observed value—say, $r = 0.4$—and varying $\sin^2\theta_w$, the relations of part (b) generate a curve in the plane of R^ν versus $R^{\bar{\nu}}$ that is known as *Weinberg's nose*. Sketch this curve. The observed values of R^ν, $R^{\bar{\nu}}$ lie close to this curve, near the point corresponding to $\sin^2\theta_w = 0.23$.

20.5 A model with two Higgs fields.

(a) Consider a model with two scalar fields ϕ_1 and ϕ_2, which transform as $SU(2)$ doublets with $Y = 1/2$. Assume that the two fields acquire parallel vacuum expectation values of the form (20.23) with vacuum expectation values v_1, v_2. Show that these vacuum expectation values produce the same gauge boson mass matrix that we found in Section 20.2, with the replacement

$$v^2 \to (v_1^2 + v_2^2).$$

(b) The most general potential function for a model with two Higgs doublets is quite complex. However, if we impose the discrete symmetry $\phi_1 \to -\phi_1$, $\phi_2 \to \phi_2$, the most general potential is

$$V(\phi_1, \phi_2) = -\mu_1^2 \phi_1^\dagger \phi_1 - \mu_2^2 \phi_2^\dagger \phi_2 + \lambda_1 (\phi_1^\dagger \phi_1)^2 + \lambda_2 (\phi_2^\dagger \phi_2)^2$$
$$+ \lambda_3 (\phi_1^\dagger \phi_1)(\phi_2^\dagger \phi_2) + \lambda_4 (\phi_1^\dagger \phi_2)(\phi_2^\dagger \phi_1) + \lambda_5 ((\phi_1^\dagger \phi_2)^2 + \text{h.c.}).$$

Find conditions on the parameters μ_i and λ_i so that the configuration of vacuum expectation values required in part (a) is a locally stable minimum of this potential.

(c) In the unitarity gauge, one linear combination of the upper components of ϕ_1 and ϕ_2 is eliminated, while the other remains as a physical field. Show that the physical charged Higgs field has the form

$$\phi^+ = \sin\beta\, \phi_1^+ - \cos\beta\, \phi_2^+ ,$$

where β is defined by the relation

$$\tan\beta = \frac{v_2}{v_1}.$$

(d) Assume that the two Higgs fields couple to quarks by the set of fundamental couplings

$$\mathcal{L}_m = -\lambda_d^{ij} \overline{Q}_L^i \cdot \phi_1 d_R^j - \lambda_u^{ij} \epsilon^{ab} \overline{Q}_{La}^i \phi_{2b}^\dagger u_R^j + \text{h.c.}$$

Find the couplings of the physical charged Higgs boson of part (c) to the mass eigenstates of quarks. These couplings depend only on the values of the quark masses and $\tan\beta$ and on the elements of the CKM matrix.

Chapter 21

Quantization of Spontaneously Broken Gauge Theories

In Chapter 20 we saw that when a gauge symmetry is spontaneously broken, the gauge bosons acquire mass. This phenomenon allowed us to construct a realistic theory of the weak interactions. Up to this point, however, we have discussed spontaneously broken gauge theories only in a simplistic way. To isolate the physical degrees of freedom, we have used the device of going to the unitarity gauge. However, it is not at all clear what the rules of perturbation theory are in this gauge, or how the unitarity gauge constraint is maintained when we compute Feynman diagrams. We have also seen that the Goldstone bosons that are absorbed into the massive gauge bosons play an important role in formal arguments about these theories, so we would like to quantize these theories in a gauge that does not eliminate these particles from the beginning.

In this chapter we will address these problems, by carrying out the formal gauge-fixing of theories with spontaneously broken gauge symmetry using the Faddeev-Popov method. We will define a class of gauges, called the R_ξ gauges, almost all of which contain the Goldstone bosons of the original spontaneous symmetry breaking. These particles cancel the effects of other unphysical particles in the formalism to maintain the unitarity of the theory. These cancellations are a more intricate version of the cancellations between gauge and ghost degrees of freedom that we saw in Chapter 16. However, we will see in Section 21.2 that a theory does not forget that it contains Goldstone bosons and that, under some circumstances, the properties of the Goldstone bosons in the theory without gauge couplings can carry over to the theory with massive gauge bosons.

Finally, having defined the perturbation theory and clarified the role of the Goldstone bosons in spontaneously broken gauge theories, we will carry out some explicit loop calculations of interest in the theory of weak interactions. Here we will see applications of the ideas of Chapter 11, that a theory with spontaneously broken symmetry can be renormalized with the counterterms of the symmetric Lagrangian. In Section 21.3 we will show through some examples that this result applies with equal force to gauge theories, and that it endows the weak-interaction gauge theory with substantial predictive power.

21.1 The R_ξ Gauges

In our discussion of the low-energy effective Lagrangian for weak interactions, we proposed in Eq. (20.89) the following expression for the propagator of a massive gauge boson:

$$\langle A^\mu(p) A^\nu(-p) \rangle \overset{?}{=} \frac{-ig^{\mu\nu}}{p^2 - m^2}. \tag{21.1}$$

This expression is a natural first guess, generalizing the Feynman-'t Hooft gauge. However, it is unsatisfactory in a number of ways.

The most important of these defects concerns the treatment of gauge boson polarization states. The propagator (21.1) contains four components, corresponding to the transverse, longitudinal, and timelike polarizations. We saw in Chapters 5 and 16 that, for massless gauge bosons, the unphysical longitudinal and timelike components cancel in computations. For a massive gauge boson, however, the longitudinal polarization state corresponds to a real physical particle; we do not want it to cancel. Expression (21.1) does not take this change into account.

An Abelian Example

To understand this and other formal problems that arise for gauge theories with spontaneously broken symmetry, we need to carefully redo the Faddeev-Popov quantization of these theories. To begin, we will quantize the spontaneously broken Abelian gauge theory introduced in Eq. (20.1):

$$\mathcal{L} = -\tfrac{1}{4}(F_{\mu\nu})^2 + |D_\mu \phi|^2 - V(\phi), \tag{21.2}$$

with $D_\mu = \partial_\mu + ieA_\mu$. Here $\phi(x)$ is a complex scalar field. However, it will be most convenient to analyze the model by writing ϕ in terms of its real components,

$$\phi = \frac{1}{\sqrt{2}} (\phi^1 + i\phi^2). \tag{21.3}$$

Then the infinitesimal local symmetry transformation is

$$\delta\phi^1 = -\alpha(x)\phi^2, \qquad \delta\phi^2 = \alpha(x)\phi^1, \qquad \delta A_\mu = -\frac{1}{e}\partial_\mu\alpha. \tag{21.4}$$

Let us assume that $V(\phi)$ forces the scalar field to acquire a vacuum expectation value: $\langle \phi^1 \rangle = v$. Then we should change variables by a shift:

$$\phi^1(x) = v + h(x); \qquad \phi^2 = \varphi. \tag{21.5}$$

The field ϕ^2 or φ is the Goldstone boson. The Lagrangian (21.2) now takes the form

$$\mathcal{L} = -\tfrac{1}{4}(F_{\mu\nu})^2 + \tfrac{1}{2}(\partial_\mu h - eA_\mu\varphi)^2 + \tfrac{1}{2}(\partial_\mu\varphi + eA_\mu(v+h))^2 - V(\phi). \tag{21.6}$$

This Lagrangian is still invariant under an exact local symmetry,

$$\delta h = -\alpha(x)\varphi, \qquad \delta\varphi = \alpha(x)(v+h), \qquad \delta A_\mu = -\frac{1}{e}\partial_\mu\alpha. \qquad (21.7)$$

Thus, in order to define the functional integral over the variables (h,φ,A_μ), we must introduce Faddeev-Popov gauge fixing.

Starting from the functional integral

$$Z = \int \mathcal{D}A\mathcal{D}h\mathcal{D}\varphi \; e^{i\int\mathcal{L}[A,h,\varphi]}, \qquad (21.8)$$

we can introduce a gauge-fixing constraint as we did in Section 9.4. Following the steps leading from Eq. (9.50) to Eq. (9.54), we find

$$Z = C \cdot \int \mathcal{D}A\mathcal{D}h\mathcal{D}\varphi \; e^{i\int\mathcal{L}[A,h,\varphi]} \, \delta(G(A,h,\varphi)) \det\left(\frac{\delta G}{\delta\alpha}\right), \qquad (21.9)$$

where C is a constant proportional to the volume of the gauge group and $G(A,h,\varphi)$ is a gauge-fixing condition. Alternatively, we can introduce the gauge-fixing constraint as $\delta(G(x) - \omega(x))$ and integrate over $\omega(x)$ with a Gaussian weight, as in the derivation of Eq. (9.56). This gives

$$Z = C' \cdot \int \mathcal{D}A\mathcal{D}h\mathcal{D}\varphi \; \exp\left[i\int d^4x\left(\mathcal{L}[A,h,\varphi] - \tfrac{1}{2}(G)^2\right)\right] \det\left(\frac{\delta G}{\delta\alpha}\right). \quad (21.10)$$

The gauge-fixing function G is arbitrary, but we can simplify our formalism by choosing it appropriately.

An especially convenient choice of the gauge-fixing function is

$$G = \frac{1}{\sqrt{\xi}}(\partial_\mu A^\mu - \xi ev\varphi). \qquad (21.11)$$

When we form G^2, the term quadratic in A_μ will provide the same gauge-dependent addition to the gauge field action that we saw in the derivation of Eqs. (9.58) and (16.29). In addition, the cross term between A_μ and φ is engineered to cancel the quadratic term of the form $\partial_\mu\varphi A^\mu$ coming from the third term of (21.6). With this choice, the quadratic terms of the gauge-fixed Lagrangian $(\mathcal{L} - \tfrac{1}{2}G^2)$ are

$$\mathcal{L}_2 = -\frac{1}{2}A_\mu\left(-g^{\mu\nu}\partial^2 + \left(1 - \frac{1}{\xi}\right)\partial^\mu\partial^\nu - (ev)^2 g^{\mu\nu}\right)A_\nu$$
$$+ \frac{1}{2}(\partial_\mu h)^2 - \frac{1}{2}m_h^2 h^2 + \frac{1}{2}(\partial_\mu\varphi)^2 - \frac{\xi}{2}(ev)^2\varphi^2. \qquad (21.12)$$

The mass term for the h field comes from the expansion of $V(\phi)$, as in (20.6). The mass term for the gauge field comes from the Higgs mechanism, that is, from the third term of (21.6). Notice that the formalism also produces a mass for the Goldstone boson φ:

$$m_\varphi^2 = \xi(ev)^2 = \xi m_A^2. \qquad (21.13)$$

The fact that this mass is gauge-dependent is a signal that the Goldstone boson is a fictitious field, which will not be produced in physical processes.

To complete the Faddeev-Popov quantization procedure, we must derive the Lagrangian of the ghosts. This Lagrangian depends on the gauge variation of G, which can be computed by inserting (21.7) into (21.11). We find

$$\frac{\delta G}{\delta \alpha} = \frac{1}{\sqrt{\xi}}\left(-\frac{1}{e}\partial^2 - \xi e v(v+h)\right). \tag{21.14}$$

The determinant of this operator can be accounted for by including a set of Faddeev-Popov ghosts with the Lagrangian,

$$\mathcal{L}_{\text{ghost}} = \bar{c}\left[-\partial^2 - \xi m_A^2\left(1 + \frac{h}{v}\right)\right]c, \tag{21.15}$$

where $m_A = ev$ as in Eq. (21.13). Since this is an Abelian gauge theory, the ghost field does not couple directly to the gauge field. It does, however, couple to the physical Higgs field, so it cannot be completely ignored as in QED.

From the quadratic terms in the Lagrangians for A_μ, h, φ, and the ghosts, we can readily find the propagators for these fields. All four propagators are shown in Fig. 21.1. The only complicated case is that of the gauge field. The term in (21.12) involving A_μ involves an operator whose Fourier transform is

$$g^{\mu\nu}k^2 - \left(1 - \frac{1}{\xi}\right)k^\mu k^\nu - m_A^2 g^{\mu\nu}$$
$$= \left(g^{\mu\nu} - \frac{k^\mu k^\nu}{k^2}\right)(k^2 - m_A^2) + \left(\frac{k^\mu k^\nu}{k^2}\right)\frac{1}{\xi}(k^2 - \xi m_A^2). \tag{21.16}$$

The inverse of this matrix gives the A_μ field propagator:

$$\langle A^\mu(k)A^\nu(-k)\rangle = \frac{-i}{k^2 - m_A^2}\left(g^{\mu\nu} - \frac{k^\mu k^\nu}{k^2}\right) + \frac{-i\xi}{k^2 - \xi m_A^2}\left(\frac{k^\mu k^\nu}{k^2}\right)$$
$$= \frac{-i}{k^2 - m_A^2}\left(g^{\mu\nu} - \frac{k^\mu k^\nu}{k^2 - \xi m_A^2}(1 - \xi)\right). \tag{21.17}$$

Notice that the transverse components of the A field and the component h of the Higgs field acquire the masses m_A, m_h that we found in Section 20.1. The unphysical components of A, the Goldstone bosons, and the ghosts all acquire the same gauge-dependent mass $\sqrt{\xi}m_A$.

ξ Dependence in Perturbation Theory

Because the parameter ξ was introduced only in the gauge fixing, we expect it to cancel out of all computations of expectation values of gauge-invariant operators and of S-matrix elements. This cancellation can be proved to all orders in perturbation theory by using the BRST symmetry of the gauge-fixed Lagrangian.* Here, however, we will simply illustrate the cancellation of ξ in a simple example.

---

*See, for example, Taylor (1976).

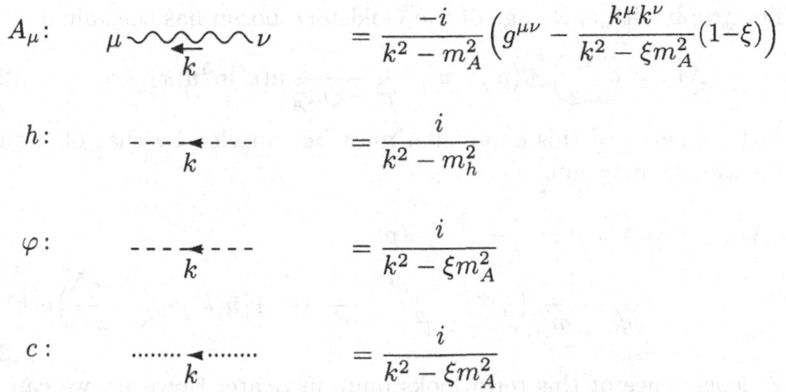

$A_\mu:$ $\mu \sim\!\!\!\sim\!\!\!\sim\!\!\!\sim \nu$ $\quad k$ $\qquad = \dfrac{-i}{k^2 - m_A^2}\left(g^{\mu\nu} - \dfrac{k^\mu k^\nu}{k^2 - \xi m_A^2}(1-\xi)\right)$

$h:$ $\quad ---\!\!\blacktriangleleft\!---$ $\quad k$ $\qquad = \dfrac{i}{k^2 - m_h^2}$

$\varphi:$ $\quad ---\!\!\blacktriangleleft\!---$ $\quad k$ $\qquad = \dfrac{i}{k^2 - \xi m_A^2}$

$c:$ $\quad \cdots\cdots\!\blacktriangleleft\!\cdots\cdots$ $\quad k$ $\qquad = \dfrac{i}{k^2 - \xi m_A^2}$

Figure 21.1. Propagators of the gauge field, Higgs fields, and ghosts in the Abelian model with spontaneously broken symmetry.

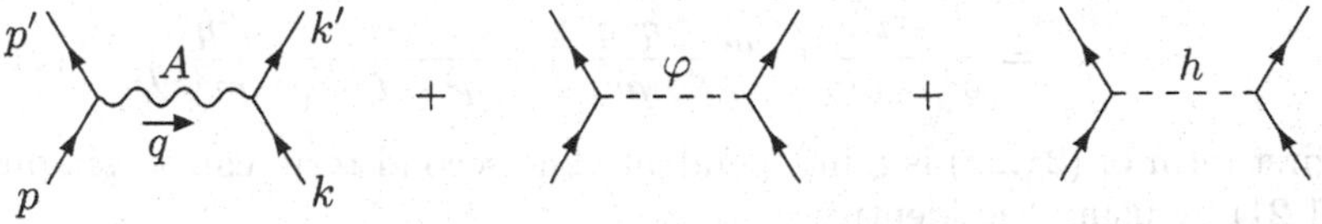

Figure 21.2. Diagrams contributing to fermion-fermion scattering at leading order in the Abelian model with spontaneous symmetry breaking.

Consider coupling a fermion to the spontaneously broken gauge theory through a chiral interaction:

$$\mathcal{L}_f = \bar{\psi}_L(i\slashed{D})\psi_L + \bar{\psi}_R(i\slashed{\partial})\psi_R - \lambda_f(\bar{\psi}_L \phi \psi_R + \bar{\psi}_R \phi^* \psi_L), \qquad (21.18)$$

with $D_\mu = \partial_\mu + ieA_\mu$ as before. This is a stripped-down, Abelian version of the coupling of fermions to the weak interaction gauge theory. The fermion ψ receives a mass

$$m_f = \lambda_f \frac{v}{\sqrt{2}} \qquad (21.19)$$

from the spontaneous symmetry breaking. (This theory has an axial vector anomaly that would render loop calculations inconsistent, but we will analyze it only at the level of tree diagrams.)

In this theory, the leading-order diagrams contributing to fermion-fermion scattering are those shown in Fig. 21.2. Notice that the contribution from the exchange of the unphysical particle φ must be included, since this particle appears in the Feynman rules. The ghosts do not appear in this process until the one-loop level. Since the propagator of the physical Higgs particle h is independent of ξ, the cancellation of the ξ dependence must take place between the transverse and longitudinal components of A_μ and the Goldstone boson φ.

The graph with exchange of the Goldstone boson has the value

$$i\mathcal{M}_\varphi = \left(\frac{\lambda_f}{\sqrt{2}}\right)^2 \bar{u}(p')\gamma^5 u(p)\frac{i}{q^2 - \xi m_A^2}\bar{u}(k')\gamma^5 u(k). \tag{21.20}$$

The ξ dependence of this expression must be canceled by that of the gauge boson exchange diagram,

$$i\mathcal{M}_A = (-ie)^2 \bar{u}(p')\gamma_\mu\left(\frac{1-\gamma^5}{2}\right)u(p)$$

$$\times \frac{-i}{q^2 - m_A^2}\left(g^{\mu\nu} - \frac{q^\mu q^\nu}{q^2 - \xi m_A^2}(1-\xi)\right)\bar{u}(k')\gamma_\nu\left(\frac{1-\gamma^5}{2}\right)u(k). \tag{21.21}$$

The ξ dependence of this term looks quite intricate. However, we can make some simplifications by rewriting the gauge boson propagator as

$$\frac{-i}{q^2 - m_A^2}\left(g^{\mu\nu} - \frac{q^\mu q^\nu}{m_A^2} + q^\mu q^\nu\left[\frac{1}{m_A^2} - \frac{1}{q^2 - \xi m_A^2}(1-\xi)\right]\right)$$

$$= \frac{-i}{q^2 - m_A^2}\left(g^{\mu\nu} - \frac{q^\mu q^\nu}{m_A^2}\right) + \frac{-i}{q^2 - \xi m_A^2}\left(\frac{q^\mu q^\nu}{m_A^2}\right). \tag{21.22}$$

The first term of (21.22) is ξ-independent. The second term can be simplified in (21.21) by using the identity

$$q^\mu \bar{u}(p')\gamma_\mu\left(\frac{1-\gamma^5}{2}\right)u(p) = \frac{1}{2}\bar{u}(p')\left[(\not{p} - \not{p}') - (\not{p} - \not{p}')\gamma^5\right]u(p)$$

$$= \frac{1}{2}\bar{u}(p')\left[\not{p}'\gamma^5 + \gamma^5\not{p}\right]u(p) \tag{21.23}$$

$$= m_f \bar{u}(p')\gamma^5 u(p),$$

and the analogous identity on the other fermion line. After making these rearrangements and inserting the explicit values $m_f = \lambda_f v/\sqrt{2}$ and $m_A = ev$, the gauge boson exchange amplitude (21.21) takes the form

$$i\mathcal{M}_A = (-ie)^2 \bar{u}(p')\gamma_\mu\left(\frac{1-\gamma^5}{2}\right)u(p)\frac{i}{q^2 - m_A^2}\left(g^{\mu\nu} - \frac{q^\mu q^\nu}{m_A^2}\right)\bar{u}(k')\gamma_\nu\left(\frac{1-\gamma^5}{2}\right)u(k)$$

$$+ \left(\frac{\lambda_f}{\sqrt{2}}\right)^2 \bar{u}(p')\gamma^5 u(p)\frac{-i}{q^2 - \xi m_A^2}\bar{u}(k')\gamma^5 u(k). \tag{21.24}$$

The second term of (21.24) precisely cancels the Goldstone boson exchange diagram (21.20). The terms that remain in the fermion-fermion scattering amplitude are independent of ξ.

This demonstration merits two additional comments. First, throughout this book, we have become accustomed to dotting the gauge boson momentum into a gauge boson vertex and finding zero or contact terms. However, in spontaneously broken gauge theories, we typically find a different result. The fermionic current $\bar{\psi}\gamma^\mu(1-\gamma^5)\psi$ is not conserved, with the nonconservation being proportional to the fermion mass. This allows the manipulation (21.23) to contribute terms proportional to the Higgs boson vacuum expectation value, which interplay with the Goldstone boson contributions. We will discuss this point further, and find a physical application of it, in Section 21.2.

The second point concerns the final form of the gauge-invariant sum of the gauge boson and Goldstone boson exchange diagrams. These give just the result we would have found by neglecting the Goldstone boson and computing the gauge boson exchange using the first term of (21.22) as the propagator:

$$\langle A_\mu(q) A_\nu(-q) \rangle = \frac{-i}{q^2 - m_A^2} \left(g^{\mu\nu} - \frac{q^\mu q^\nu}{m_A^2} \right). \tag{21.25}$$

The tensor structure represents a gauge boson polarization sum. To identify what vectors are summed over, notice that, if the vector boson is on-shell, and if we boost to its rest frame, this structure becomes precisely the projection onto the three purely spatial directions. These are the three polarization states of an on-shell massive vector particle. In a general frame, still for q^μ on-shell, the tensor in (21.25) remains the projection onto physical polarization states:

$$\sum_{\epsilon^\mu q_\mu = 0} \epsilon^\mu \epsilon^{\nu*} = -\left(g^{\mu\nu} - \frac{q^\mu q^\nu}{m_A^2} \right). \tag{21.26}$$

Thus, in the cancellation of the ξ-dependent parts of the gauge boson propagator, we also find that the Goldstone boson diagram cancels the contribution of the unphysical timelike polarization state of the gauge boson, leaving over the required three physical polarizations.

The perturbation theory rules that we have developed have a very different character for different values of ξ. Thus, it is even more true in the case of spontaneously broken symmetry that we can find different special simplifications by choosing different values of this gauge parameter. For $\xi = 0$, Lorentz gauge, the Goldstone boson is massless and has exactly the couplings it has in the ungauged model of symmetry breaking, while the gauge boson propagator is purely transverse:

$$\mu \overset{}{\underset{k}{\sim\!\!\sim\!\!\sim\!\!\sim}} \nu = \frac{-i}{k^2 - m_A^2} \left(g^{\mu\nu} - \frac{k^\mu k^\nu}{k^2} \right); \qquad - - - \overset{\blacktriangleleft}{\underset{k}{}} - - - = \frac{i}{k^2}. \tag{21.27}$$

This gauge is especially useful for analyzing models of symmetry breaking. Both propagators have poles at $k^2 = 0$. However, we know that there are no corresponding physical particles, because these poles move away from $k^2 = 0$ as we change ξ, while the S-matrix must be ξ-independent.

For $\xi = 1$, we recover the simple form of the gauge boson propagator given in (21.1). This choice of the gauge boson propagator is not consistent, however, unless we also include Goldstone boson exchanges in which the Goldstone boson is also assigned the mass m_A:

$$\mu \overset{}{\underset{k}{\sim\!\!\sim\!\!\sim\!\!\sim}} \nu = \frac{-ig^{\mu\nu}}{k^2 - m_A^2}; \qquad - - - \overset{\blacktriangleleft}{\underset{k}{}} - - - = \frac{i}{k^2 - m_A^2}. \tag{21.28}$$

This gauge, still called the Feynman-'t Hooft gauge, is the most convenient one for general higher-order computations.

For any finite value of ξ, the gauge boson and Goldstone boson propagators fall off as $1/k^2$ and thus obey the general power-counting analysis of Section 10.1. It follows that, in any one of these gauges, the perturbation theory will be renormalizable, in the sense that the divergences are removed by a finite set of counterterms. Furthermore, the analysis of Section 11.6 tells us that the only counterterms required are those that are symmetric under the original global symmetry of the theory. However, we should require one further condition of our renormalization procedure: We should insist that the counterterms preserve local gauge invariance, and, in particular, preserve the property that S-matrix elements and the matrix elements of gauge-invariant operators are independent of ξ. This result was proved to all orders in perturbation theory by 't Hooft and Veltman and by Lee and Zinn-Justin.[†] Thus, in the gauge defined by any finite value of ξ, we can, in principle, straightforwardly compute a physical quantity to any order. The gauges defined by the possible values of ξ are known as the *renormalizability*, or R_ξ, gauges.

By taking the limit $\xi \to \infty$ of the R_ξ gauges, we find a gauge with very different simplifying features. In this limit, the unphysical degrees of freedom, which have masses proportional to $\sqrt{\xi}$, disappear from the theory. The gauge boson and Goldstone boson propagators become:

$$\underset{\mu \;\; \overset{\longleftarrow}{k} \;\; \nu}{\sim\!\!\sim\!\!\sim\!\!\sim} \;=\; \frac{-i}{k^2 - m_A^2}\left(g^{\mu\nu} - \frac{k^\mu k^\nu}{m_A^2}\right); \qquad \underset{k}{\text{---}\!\blacktriangleleft\!\text{---}} \;=\; 0. \qquad (21.29)$$

The gauge boson propagator contains exactly the three spacelike polarization states. In this gauge, the only singularities of Feynman diagrams correspond to the propagation of physical intermediate states. Thus, the unitarity of the S-matrix follows from the Cutkosky rules, as in the globally symmetric theories considered in Section 7.3, without the need to worry about the cancellation of unphysical states.[‡] The $\xi \to \infty$ limit of the R_ξ gauges thus gives the quantum-mechanical realization of the *unitarity* (or U) gauge, introduced in Eq. (20.12).

It is not straightforward to prove renormalizability directly in the U gauge. In this gauge, the gauge boson propagator falls off more slowly than $1/k^2$ at large k. This signals trouble for the evaluation of loop diagrams. Typically, in fact, individual loop diagrams will diverge as $\log \xi$ or worse as $\xi \to \infty$. Still, the gauge invariance of the S-matrix implies that these divergences must cancel in the sum of all diagrams contributing to a given process, so that this sum has a smooth limit as $\xi \to \infty$. There is no difficulty of principle with the fact that we use one gauge to prove the renormalizability of spontaneously

---

[†] G. 't Hooft and M. J. G. Veltman, *Nucl. Phys.* **B50**, 318 (1972), B. W. Lee and J. Zinn-Justin, *Phys. Rev.* **D5**, 3121, 3137, 3155 (1972), **D7**, 1049 (1973).

[‡] In the more sophisticated language of Section 16.4, the crucial identity (16.54), which is required for the unitarity of the S-matrix, is true manifestly.

broken gauge theories and another gauge to prove their unitarity. In fact, this method of argumentation makes natural use of the underlying symmetries of the theory.

Non-Abelian Analysis

Now that we have thoroughly examined the R_ξ gauges for an Abelian gauge theory, we are ready to generalize to the non-Abelian case. There is no difficulty in being completely general, so let us consider a Yang-Mills gauge theory with gauge group G, spontaneously broken by the vacuum expectation value of a scalar field.

We will build on our classical analysis of this system following Eq. (20.13). As in that analysis, it will be most convenient to write the scalars as a multiplet ϕ_i of real-valued fields. Then the gauge transformation of the ϕ_i takes the form

$$\delta\phi_i = -\alpha^a(x)T^a_{ij}\phi_j, \tag{21.30}$$

where the T^a_{ij} are real, antisymmetric representation matrices of G. Similarly, the transformation of the gauge fields is

$$\delta A^a_\mu = \frac{1}{g}\partial_\mu\alpha^a - f^{abc}\alpha^b A^c_\mu = \frac{1}{g}(D_\mu\alpha)^a. \tag{21.31}$$

(If the gauge group is not simple, the coupling g need not be the same for every a.) The Lagrangian invariant under these gauge transformations is

$$\mathcal{L} = -\tfrac{1}{4}(F^a_{\mu\nu})^2 + \tfrac{1}{2}(D_\mu\phi)^2 - V(\phi), \tag{21.32}$$

with

$$D_\mu\phi_i = \partial_\mu\phi_i + gA^a_\mu T^a_{ij}\phi_j. \tag{21.33}$$

Assume that the potential $V(\phi)$ is minimized at a point where some of the components of ϕ acquire vacuum expectation values. As in (20.16), define

$$\langle\phi_i\rangle = (\phi_0)_i. \tag{21.34}$$

We will expand ϕ_i about this value:

$$\phi_i(x) = \phi_{0i} + \chi_i(x). \tag{21.35}$$

It will be convenient to divide the space of values χ_i into two subspaces. The vectors $T^a\phi_0$ correspond to symmetry transformations of the vacuum expectation value of ϕ. The field fluctuations along these directions are the Goldstone bosons. Let $\{n_i\}$ be an orthonormal basis for this subspace; then the unit vectors n_i are in 1-to-1 correspondence with the Goldstone bosons. The field fluctuations orthogonal to all of the vectors $T^a\phi_0$ correspond to the (massive) physical scalar fields of the spontaneously broken gauge theory.

In the discussion to follow, the vectors $T^a\phi_0$ will play an important role. We should then recall the notation for these vectors that we introduced in Eq. (20.51):

$$F^a_i = T^a_{ij}\phi_{0j}. \tag{21.36}$$

The matrix $F^a{}_i$ is not generally square; it has one row for each gauge generator, and one column for each component of ϕ. However, many of its elements are zero. Its nonzero elements connect the spontaneously broken gauge generators and the Goldstone bosons. In Eq. (20.56), we showed that the gauge boson masses generated through the Higgs mechanism can be written

$$m^2_{ab} = g^2 F^a{}_j F^b{}_j. \tag{21.37}$$

To give a concrete example of a matrix $F^a{}_j$, let us compute it in the GWS electroweak theory. Following the conventions introduced in Eq. (20.14), we should rewrite the Higgs field of the GWS model in terms of four real scalar fields. A convenient parametrization is

$$\phi = \frac{1}{\sqrt{2}} \begin{pmatrix} -i(\phi^1 - i\phi^2) \\ v + (h + i\phi^3) \end{pmatrix}. \tag{21.38}$$

The fields ϕ^i are the Goldstone bosons, and h is the massive Higgs boson. The vacuum state is simply

$$\phi_0 = \frac{1}{\sqrt{2}} \begin{pmatrix} 0 \\ v \end{pmatrix}.$$

The real representation matrices are

$$T^a = -i\tau^a = -i\frac{\sigma^a}{2}, \qquad T^Y = -iY = -i\frac{1}{2}.$$

A simple computation then shows, for instance, that $T^1\phi_0$ equals $v/2$ times a unit vector in the ϕ^1 direction. Filling in the remaining components of $F^a{}_i$, with $a = 1, 2, 3, Y$ and $i = 1, 2, 3$, we find

$$gF^a{}_i = \frac{v}{2} \begin{pmatrix} g & 0 & 0 \\ 0 & g & 0 \\ 0 & 0 & g \\ 0 & 0 & -g' \end{pmatrix}. \tag{21.39}$$

We do not need to include the components of $F^a{}_i$ along the direction of the physical Higgs field h; the vectors $T^a\phi_0$ are all orthogonal to this direction.

If we insert (21.35) into (21.32) as a change of variables, we find, for the quadratic terms in the Lagrangian,

$$\mathcal{L}_2 = -\tfrac{1}{2}A^a_\mu\left(-g^{\mu\nu}\partial^2 + \partial^\mu\partial^\nu\right)A^a_\nu + \tfrac{1}{2}(\partial_\mu\chi)^2$$
$$+ g\partial^\mu\chi_i A^a_\mu F^a{}_i + \tfrac{1}{2}(m^2_A)^{ab}A^a_\mu A^{\mu b} - \tfrac{1}{2}M_{ij}\chi_i\chi_j, \tag{21.40}$$

where $(m^2_A)^{ab}$ is the gauge boson mass matrix (21.37) and

$$M_{ij} = \frac{\partial^2}{\partial\phi_i\partial\phi_j}V(\phi)\bigg|_{\phi_0}. \tag{21.41}$$

We proved in Eq. (11.13) that

$$n_i M_{ij} = 0 \tag{21.42}$$

for all possible directions n_i in the subspace spanned by the $T^a\phi_0$, so the Goldstone bosons are massless.

To study the quantum theory of this system we start with the functional integral

$$Z = \int DAD\chi\, e^{i\int \mathcal{L}[A,\chi]}. \tag{21.43}$$

Using the Faddeev-Popov gauge-fixing procedure, we define this integral, analogously to (21.10), as

$$Z = C' \cdot \int DAD\chi\, \exp\left[i\int d^4x\left(\mathcal{L}[A,\chi] - \tfrac{1}{2}(G)^2\right)\right]\det\left(\frac{\delta G}{\delta \alpha}\right), \tag{21.44}$$

for an arbitrary gauge-fixing function $G(A,\chi)$. The R_ξ gauges are defined by the choice

$$G^a = \frac{1}{\sqrt{\xi}}\left(\partial_\mu A^{a\mu} - \xi g F^a{}_i\chi_i\right). \tag{21.45}$$

Note that G involves only the components of χ that lie in the subspace of the Goldstone bosons.

The gauge-fixing term adds to the Lagrangian the following set of quadratic terms:

$$(-\tfrac{1}{2}G^2)_2 = \tfrac{1}{2}A^a_\mu\left(\tfrac{1}{\xi}\partial^\mu\partial^\nu\right)A^a_\nu + g\partial_\mu A^{a\mu}F^a{}_i\chi_i - \tfrac{1}{2}\xi g^2\left[F^a{}_i\chi_i\right]^2. \tag{21.46}$$

The term that mixes A_μ and χ_i is arranged to cancel between (21.40) and (21.46). The final quadratic Lagrangian for the gauge and Goldstone boson fields is

$$\mathcal{L}_2 = -\frac{1}{2}A^a_\mu\left(\left[-g^{\mu\nu}\partial^2 + \left(1 - \frac{1}{\xi}\right)\partial^\mu\partial^\nu\right]\delta^{ab} - g^2 F^a{}_i F^b{}_i g^{\mu\nu}\right)A^b_\nu$$
$$+ \frac{1}{2}(\partial_\mu\chi)^2 - \frac{1}{2}\xi g^2 F^a{}_i F^a{}_j\chi_i\chi_j. \tag{21.47}$$

The mass matrices of gauge bosons and Goldstone bosons in this Lagrangian are closely related to one another. The gauge boson mass matrix is

$$(m_A^2)^{ab} = g^2 F^a{}_i F^b{}_i = g^2(FF^T)^{ab}. \tag{21.48}$$

In an R_ξ gauge, the timelike components of the gauge bosons acquire the mass matrix

$$\xi m_A^2 = \xi g^2(FF^T)^{ab}. \tag{21.49}$$

At the same time, the Goldstone bosons acquire the mass matrix

$$(m_G^2)_{ij} = \xi g^2 F^a{}_i F^a{}_j = \xi g^2(F^T F)_{ij}. \tag{21.50}$$

The two matrices (21.49) and (21.50) have different numbers of zero eigenvalues, but their nonzero eigenvalues are in 1-to-1 correspondence. This is precisely the correspondence induced by the Higgs mechanism between the massive gauge bosons and the Goldstone bosons that they absorbed to gain mass.

Finally, we must construct the ghost Lagrangian. This is found from the gauge variation of the gauge-fixing term G^a. Inserting (21.30) and (21.31) into (21.45), we find

$$\frac{\delta G^a}{\delta \alpha^b} = \frac{1}{\sqrt{\xi}}\left(\frac{1}{g}(\partial_\mu D^\mu)^{ab} + \xi g(T^a \phi_0) \cdot T^b(\phi_0 + \chi)\right). \tag{21.51}$$

Thus, the ghost Lagrangian is

$$\mathcal{L}_{\text{ghost}} = \bar{c}^a\left[-(\partial_\mu D^\mu)^{ab} - \xi g^2 (T^a \phi_0) \cdot T^b(\phi_0 + \chi)\right]c^b. \tag{21.52}$$

Notice that the ghosts have exactly the same mass matrix (21.49) as the unphysical components of the gauge bosons. This Lagrangian also contains both the familiar coupling of the ghosts to the gauge fields and the coupling to the physical Higgs fields that we found in the Abelian case (21.15).

We have now computed the kinetic energy terms for gauge fields, scalar fields, and ghosts in an R_ξ gauge. It is straightforward to convert these results to the calculation of propagators for these fields; the computations are exactly the same as in the Abelian case. We find for the three propagators

$$\begin{array}{c}\mu \nu \\ a \!\!\!\!\!\sim\!\!\!\!\sim\!\!\!\!\sim\!\!\!\!\sim\!\! b \\ k\end{array} = \left(\frac{-i}{k^2 - g^2 F F^T}\left[g^{\mu\nu} - \frac{k^\mu k^\nu}{k^2 - \xi g^2 F F^T}(1-\xi)\right]\right)^{ab},$$

$$\begin{array}{c}i \text{-}\text{-}\text{-}\!\!\blacktriangleleft\!\!\text{-}\text{-}\text{-} j \\ k\end{array} = \left(\frac{i}{k^2 - \xi g^2 F^T F - M^2}\right)_{ij},$$

$$\begin{array}{c}a \cdots\cdots\!\!\blacktriangleleft\!\!\cdots\cdots b \\ k\end{array} = \left(\frac{i}{k^2 - \xi g^2 F F^T}\right)^{ab}. \tag{21.53}$$

All of these equations involve the matrix F defined in Eq. (21.36); the appearance of a matrix in the denominator should be interpreted as a matrix inverse. The scalar field propagator also includes the mass matrix (21.41) of the physical Higgs bosons. There is no conflict between this matrix and the mass matrix of the Goldstone bosons, since they project onto orthogonal subspaces.

Although the preceding discussion has been extremely abstract, it is not hard to specialize to a particular example. So consider, once again, the GWS electroweak theory, for which the matrix $F^a{}_i$ is given by Eq. (21.39).

The gauge boson mass matrix in the GWS theory is

$$g^2 F F^T = \frac{v^2}{4}\begin{pmatrix} g^2 & 0 & 0 & 0 \\ 0 & g^2 & 0 & 0 \\ 0 & 0 & g^2 & -gg' \\ 0 & 0 & -gg' & g'^2 \end{pmatrix},$$

in agreement with Eq. (20.124). (The g on the left-hand side should be interpreted as g' for the fourth component of F.) Diagonalizing this matrix gives the familiar relations (20.62). Thus, in the basis of mass eigenstates, the four gauge-boson propagators decouple to give simply

$$\mu \,\overset{\longleftarrow}{\underset{k}{\sim\!\sim\!\sim\!\sim}}\, \nu \;\; = \frac{-i}{k^2 - m^2}\left[g^{\mu\nu} - \frac{k^\mu k^\nu}{k^2 - \xi m^2}(1-\xi) \right], \qquad (21.54)$$

where m^2 is m_W^2, m_Z^2, or, for the photon, zero. Notice that, for the photon, this expression precisely reproduces Eq. (9.58).

The mass matrix of the Goldstone bosons in the GWS theory is

$$\xi g^2 F^T F = \xi \frac{v^2}{4}\begin{pmatrix} g^2 & 0 & 0 \\ 0 & g^2 & 0 \\ 0 & 0 & g^2 + g'^2 \end{pmatrix}.$$

These fields therefore have the propagator

$$\overset{\longleftarrow}{\underset{k}{-\!-\!-\!-\!-}} \;\; = \frac{i}{k^2 - \xi m^2}, \qquad (21.55)$$

with $m^2 = m_W^2$ for ϕ^1 and ϕ^2 (the bosons eaten by the $W^\pm$) and $m^2 = m_Z^2$ for ϕ^3 (the boson eaten by the Z). The field $h(x)$, which is the physical Higgs field, propagates independently with a mass determined by the Higgs potential (and no factor of ξ in the propagator).

Finally, there are four ghost fields. According to Eq. (21.53), these have the propagator

$$\overset{\longleftarrow}{\underset{k}{\cdots\!\cdots\!\cdots}} \;\; = \frac{i}{k^2 - \xi m^2}, \qquad (21.56)$$

with the same values of m^2 as the four gauge bosons.

The Feynman rules for the interaction vertices of these particles are complicated to write out, due to the large number of possible combinations. However, it is quite straightforward to generate these rules by expanding the weak interaction Lagrangian and reading off the vertices term by term. We will work out a few examples in the following section.*

21.2 The Goldstone Boson Equivalence Theorem

From the results of the previous section, we see that perturbative calculations in the R_ξ gauges involve intricate cancellations among unphysical particles. Sometimes, however, these unphysical particles can still leave their footprints in physical observables. In this section we will see that, in the high-energy limit, the unphysical Goldstone boson that is eaten by a massive gauge boson still controls the amplitude for emission or absorption of the gauge boson in its longitudinal polarization state.

---

*The complete Feynman rules for the weak-interaction gauge theory are given in Appendix B of Cheng and Li (1984).

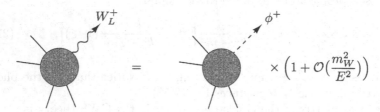

Figure 21.3. The Goldstone boson equivalence theorem. At high energy, the amplitude for emission or absorption of a longitudinally polarized massive gauge boson becomes equal to the amplitude for emission or absorption of the Goldstone boson that was eaten by the gauge boson.

When we introduced the Higgs mechanism for vector boson mass generation, we pointed out that it involves a certain conservation of degrees of freedom. A massless gauge boson, which has two transverse polarization states, combines with a scalar Goldstone boson to produce a massive vector particle, which has three polarization states. When the massive vector particle is at rest, its three polarization states are completely equivalent, but when it is moving relativistically, there is a clear distinction between the transverse and longitudinal polarization directions. This suggests that a rapidly moving, longitudinally polarized massive gauge boson might betray its origin as a Goldstone boson. The strongest version of this idea is expressed in Fig. 21.3: The amplitude for emission or absorption of a longitudinally polarized gauge boson becomes equal, at high energy, to the amplitude for emission or absorption of the Goldstone boson that was eaten. Remarkably, this statement is precisely correct, as a consequence of the underlying local gauge invariance. This *Goldstone boson equivalence theorem* was first proved by Cornwall, Levin, Tiktopoulos, and Vayonakis.[†]

Formal Aspects of Goldstone Boson Equivalence

The proof of the Goldstone boson equivalence theorem is based on the Ward identities of the spontaneously broken gauge theory. To give a complete proof of the theorem, we would have to construct and analyze these Ward identities in some detail. However, it is possible to understand the idea of the proof by examining the special case of the theorem in which a single massive vector boson is emitted or absorbed in a scattering process. The analysis of this special case requires only the relatively simple Ward identity satisfied by a current between on-shell states.[‡]

[†]J. M. Cornwall, D. N. Levin, and G. Tiktopoulos, *Phys. Rev.* **D10**, 1145 (1974); C. E. Vayonakis, *Lett. Nuov. Cim.* **17**, 383 (1976). For an illuminating discussion of the equivalence theorem, see B. W. Lee, C. Quigg, and H. Thacker, *Phys. Rev.* **D16**, 1519 (1977).

[‡]For a careful derivation of the equivalence theorem, including processes involving

To prepare for a discussion of longitudinal vector bosons, we need some simple kinematics. A vector boson at rest has momentum $k^\mu = (m, 0, 0, 0)$ and a polarization vector that is a linear combination of the three orthogonal unit vectors

$$(0, 1, 0, 0), \qquad (0, 0, 1, 0), \qquad (0, 0, 0, 1). \qquad (21.57)$$

If we boost this particle along the $\hat{3}$ axis, its momentum boosts to $k^\mu = (E_{\mathbf{k}}, 0, 0, k)$. The three possible polarization vectors are now the three unit vectors satisfying

$$\epsilon^\mu k_\mu = 0, \qquad \epsilon^2 = -1. \qquad (21.58)$$

Two of these are the first two vectors in (21.57); these give the transverse polarizations. The third vector satisfying (21.58) is the longitudinal polarization vector

$$\epsilon_L^\mu(k) = \left(\frac{k}{m}, 0, 0, \frac{E_{\mathbf{k}}}{m} \right), \qquad (21.59)$$

which is the boost of the third vector in (21.57). An important and somewhat counterintuitive feature of (21.59) is that it becomes increasingly parallel to k^μ as k becomes large. In fact, component by component,

$$\epsilon_L^\mu(k) = \frac{k^\mu}{m} + \mathcal{O}(m/E_{\mathbf{k}}) \qquad (21.60)$$

as $k \to \infty$. Since the components of k^μ are growing as k, this statement is consistent with the requirement that $\epsilon_L \cdot k = 0$ while $k \cdot k = m^2$.

With this kinematic situation in mind, let us analyze the Ward identity satisfied by a gauge current matrix element between on-shell states. It is simplest to work in Lorentz gauge ($\xi = 0$), where the gauge-fixing term (21.45) does not involve the Goldstone boson fields. The Ward identity can then be written as follows:

$$0 = k^\mu \left(\quad \right) = k^\mu \left(\quad + \quad \right). \qquad (21.61)$$

In the last expression we have written the matrix element as the sum of two pieces. First, the current can couple directly into a one-particle-irreducible vertex function $\Gamma^\mu(k)$. This gives the class of diagrams that contribute to the scattering of a gauge boson from the external states. However, for a spontaneously broken gauge theory, there is an additional term, which is not one-particle-irreducible, in which the current creates a Goldstone boson and it is this particle that couples to the external states through a 1PI vertex $\Gamma(k)$.

Let us write the relation linking the gauge current and the Goldstone boson state as

$$\langle 0 | J^\mu | \pi(k) \rangle = -iF k^\mu, \qquad (21.62)$$

multiple absorptions and emissions of massive vector bosons, see M. S. Chanowitz and M. K. Gaillard, *Nucl. Phys.* **B261**, 379 (1985).

as in Eq. (20.46). Then the argument leading to Eq. (20.56) tells us that the gauge boson mass is given by

$$m = gF, \tag{21.63}$$

where g is the gauge boson coupling constant.

With these identifications, we can write the Ward identity that follows from the conservation of the gauge current:

$$k_\mu \langle J^\mu \rangle = 0, \tag{21.64}$$

between on-shell states. Writing each term shown in (21.61) in terms of the appropriate one-particle-irreducible vertex function, we find

$$k_\mu \Gamma^\mu(k) + k_\mu \left(ig F k^\mu \right) \frac{i}{k^2} \Gamma(k) = 0. \tag{21.65}$$

Thus,

$$k_\mu \Gamma^\mu(k) = m\Gamma(k). \tag{21.66}$$

Now use this equation in the limit of large gauge boson momentum. Since the gauge boson vertex is one-particle-irreducible, the momenta of propagators inside the vertex are not, in general, collinear with k^μ. Then, according to (21.60), we may replace k^μ/m by the longitudinal polarization vector. Notice that this would not be permissible (but, also, is not necessary) in the second term of (21.65). Our final result is

$$\epsilon_{L\mu}(k)\Gamma^\mu(k) = \Gamma(k), \tag{21.67}$$

as $k \to \infty$, with an error of order m^2/k^2. That is, in the high-energy limit, the couplings of longitudinal gauge bosons become precisely those of their associated Goldstone bosons.

The equivalence theorem can be derived in another way, using the counting of physical states in spontaneously broken gauge theories, which we discussed below Eq. (21.26). In the previous section, we saw that, at least at the tree level, unitarity is maintained in spontaneously broken gauge theories by the cancellation of diagrams that produce timelike-polarized gauge bosons against diagrams that produce Goldstone bosons.

The situation is most clear in Feynman-'t Hooft gauge. There, the numerator of the gauge boson propagator is $-g^{\mu\nu}$. We can write this in terms of polarization vectors as

$$-g^{\mu\nu} = \sum_{i=1,2,3} \epsilon_i^\mu(k)\epsilon_i^{\nu*}(k) - \frac{k^\mu k^\nu}{m^2}. \tag{21.68}$$

The last term is the contribution from unphysical timelike polarization states. The unitarity of the S-matrix requires that, when a Cutkosky cut through a diagram puts a gauge boson propagator on-shell, the contribution of this piece

Figure 21.4. Decay of a t quark into $W^+ + b$.

must be canceled by a Cutkosky cut that runs through a Goldstone boson line. The required cancellation is

$$-\left|\frac{k_\mu}{m}\Gamma^\mu(k)\right|^2 + |\Gamma(k)|^2 = 0, \tag{21.69}$$

or, diagrammatically,

$$(-1)\left|\ \ \ \overset{W_t^+}{\nearrow}\ \ \right|^2 + \left|\ \ \ \overset{\phi^+}{\nearrow}\ \ \right|^2 = 0 .$$

Once again, since $\Gamma^\mu(k)$ is a one-particle-irreducible vertex, we can use (21.60) to replace (k^μ/m) by the longitudinal polarization vector $\epsilon_L^\mu(k)$ for a high-energy gauge boson. Then (21.69) becomes just the square of (21.67).

Through these formal arguments, we can see, at least to the tree level in processes with single gauge boson emission, that the equivalence theorem must be valid. However, it is much more illuminating to see the equivalence theorem at work in explicit calculations for interesting physical processes. We will now illustrate its influence in two examples.

Top Quark Decay

The first example is the weak decay of the top quark. This charge $+2/3$ quark is sufficiently heavy that it can decay to a real W^+ through $t \to W^+ + b$. The diagram for this decay is given by the simple gauge vertex shown in Fig. 21.4.

Let us first try to guess the magnitude of the top quark width. The squared matrix element will contain a factor of g^2, times some expression with dimensions of mass. Since the width should be large if the top quark mass is heavy, a first guess might be

$$\Gamma \sim \frac{g^2}{4\pi} m_t. \tag{21.70}$$

The correct expression, however, turns out to be enhanced by a factor of $(m_t/m_W)^2$.

The amplitude for this decay can be read from Eq. (20.80):

$$i\mathcal{M} = \frac{ig}{\sqrt{2}}\bar{u}(q)\gamma^\mu\left(\frac{1-\gamma^5}{2}\right)u(p)\epsilon_\mu^*(k). \tag{21.71}$$

(We set the relevant CKM factor equal to 1.) We will now turn this amplitude into an expression for the decay rate of the top quark. For simplicity, we will ignore the mass of the b quark in this computation.

Squaring the amplitude in (21.71) according to our standard methods, and then averaging over initial and summing over final spins, we find

$$\frac{1}{2} \sum_{\text{spins}} |\mathcal{M}|^2 = \frac{g^2}{2} \left[q^\mu p^\nu + q^\nu p^\mu - g^{\mu\nu} q \cdot p \right] \sum_{\text{polarizations}} \epsilon_\mu^*(k) \epsilon_\nu(k). \quad (21.72)$$

We can sum explicitly over physical gauge boson polarizations by inserting the expression (21.26) for the polarization sum. This gives

$$\frac{1}{2} \sum_{\text{spins}} |\mathcal{M}|^2 = \frac{g^2}{2} \left[q^\mu p^\nu + q^\nu p^\mu - g^{\mu\nu} q \cdot p \right] \left[-g_{\mu\nu} + \frac{k_\mu k_\nu}{m_W^2} \right]$$
$$= \frac{g^2}{2} \left[q \cdot p + 2 \frac{(k \cdot q)(k \cdot p)}{m_W^2} \right]. \quad (21.73)$$

For $m_b = 0$,

$$2q \cdot p = 2q \cdot k = m_t^2 - m_W^2, \qquad 2k \cdot p = m_t^2 + m_W^2. \quad (21.74)$$

Then

$$\frac{1}{2} \sum_{\text{spins}} |\mathcal{M}|^2 = \frac{g^2}{4} \frac{m_t^4}{m_W^2} \left(1 - \frac{m_W^2}{m_t^2} \right) \left(1 + 2 \frac{m_W^2}{m_t^2} \right). \quad (21.75)$$

After multiplying by phase space, we find

$$\Gamma = \frac{g^2}{64\pi} \frac{m_t^3}{m_W^2} \left(1 - \frac{m_W^2}{m_t^2} \right)^2 \left(1 + 2 \frac{m_W^2}{m_t^2} \right). \quad (21.76)$$

This is larger than our initial estimate (21.70) by a factor $(m_t/m_W)^2$.

It is not difficult to find the origin of this enhancement, by using the Goldstone boson equivalence theorem. In the gauge theory of weak interactions, the top quark obtains its mass from its coupling to the Higgs sector. The relation between the top-Higgs coupling λ_t and the top quark mass is written in Eq. (20.103). The top quark can be heavy only if λ_t is large. But then the amplitude for the top quark to decay to a Goldstone boson will be enhanced above (21.70) by the factor

$$\frac{\lambda_t^2}{g^2} = \frac{m_t^2}{2m_W^2}, \quad (21.77)$$

which is in fact the enhancement we found in (21.76).

To make the comparison more precise, we will now compute the prediction of the equivalence theorem for the top quark decay rate into a longitudinally polarized W^+ boson. Recall from (20.101) that the term in the weak interaction Lagrangian that couples t and b to the Higgs field is

$$\Delta \mathcal{L} = -\lambda_t \epsilon^{ab} \overline{Q}_{La} \phi_b^\dagger t_R + \text{h.c.} \quad (21.78)$$

Figure 21.5. Decay of a t quark into a Goldstone boson and a b quark.

Decompose the Higgs field as in (21.38), and write

$$\phi^{\pm} = \frac{1}{\sqrt{2}}(\phi^1 \pm i\phi^2). \tag{21.79}$$

These are the fields of the charged Goldstone bosons that are eaten by the $W^{\pm}$. Including the Goldstone boson in the theory adds a process $t \to \phi^+ + b$, shown in Fig. 21.5. This process is mediated by the Lagrangian term

$$\Delta\mathcal{L} = \lambda_t \bar{b}_L \phi^+ t_R, \tag{21.80}$$

which leads to the decay amplitude

$$i\mathcal{M} = i\lambda_t \bar{u}(q)\left(\frac{1+\gamma^5}{2}\right)u(p). \tag{21.81}$$

From this expression, we easily find

$$\tfrac{1}{2}\sum_{\text{spins}} |\mathcal{M}|^2 = \lambda_t^2 \, q \cdot p. \tag{21.82}$$

If we now ignore the mass of the Goldstone boson, or, equivalently, consider the limit $m_t \gg m_W$, we find for the top quark decay rate

$$\Gamma = \frac{\lambda_t^2}{32\pi} m_t = \frac{g^2}{64\pi} \frac{m_t^3}{m_W^2}, \tag{21.83}$$

in agreement with the leading term of (21.76) in this limit. Our results imply that only the production of the longitudinal polarization state of the W^+ is enhanced; this is easily checked directly by substituting explicit polarization vectors into (21.72).

In our derivation of (21.76), we summed over the physical polarization states of the emitted W^+; one might say that we used the prescription of the U gauge to sum over polarizations. We could equally well have used the prescription of Feynman-'t Hooft gauge, replacing

$$\sum_i \epsilon_\mu^*(k)\epsilon_\nu(k) \to -g_{\mu\nu}, \tag{21.84}$$

and also adding the contribution of the Goldstone boson emission diagram, treating the Goldstone boson as a massive particle with mass m_W. With these prescriptions, the gauge boson matrix element gives

$$\frac{1}{2}\sum_{\text{spins}}|\mathcal{M}|^2 = \frac{g^2}{2}(2q\cdot p) = \frac{g^2}{2}(m_t^2 - m_W^2). \tag{21.85}$$

The Goldstone boson emission diagram gives

$$\frac{1}{2}\sum_{\text{spins}}|\mathcal{M}|^2 = \lambda_t^2\, q\cdot p = \frac{g^2}{4}\frac{m_t^2}{m_W^2}(m_t^2 - m_W^2). \tag{21.86}$$

The sum of these contributions indeed reproduces (21.75) and thus gives the same result (21.76) for the total decay rate. In Feynman-'t Hooft gauge, the enhancement due to the large coupling of the top quark to the Higgs sector shows up explicitly in the Goldstone boson emission contributions to the total rate of W^+ production.

$e^+e^- \to W^+W^-$

Our second example is more complicated, but also contains more interesting physics. This is the reaction $e^+e^- \to W^+W^-$. In this reaction, the equivalence theorem does not lead to an enhancement of the cross section, but, rather, directs a cancellation between Feynman diagrams. As we will see, this cancellation is essential for the internal consistency of the theory.

In Problem 9.1, we computed the cross section for e^+e^- annihilation into a pair of charged scalar particles, as in Fig. 21.6(a), and found the result

$$\frac{d\sigma}{d\cos\theta}(e^+e^- \to \phi^+\phi^-) = \frac{\pi\alpha^2}{4s}\sin^2\theta \tag{21.87}$$

at energies much larger than the scalar mass. Just as for e^+e^- annihilation to fermion pairs, this cross section falls as $1/s$ at high energy. It can be shown that this behavior is required by unitarity: Since the electron and positron annihilate through a pointlike vertex, the annihilation takes place in only one partial wave. Unitarity puts a limit on the amplitude in this partial wave, requiring that $\mathcal{M}$ be bounded by a constant, and thus that σ be bounded by $1/s$ at high energy.*

The same unitarity argument applies to e^+e^- annihilation to vector bosons. Here, however, it is much less obvious that Feynman diagrams actually produce a cross section consistent with unitarity. Consider the contribution of Fig. 21.6(b). We would expect that the square of this diagram should contain a contribution to the cross section of the form of the scalar contribution (21.87) multiplied by the dot product of polarization vectors:

$$\frac{d\sigma}{d\cos\theta}(e^+e^- \to W^+W^-) \sim \frac{\pi\alpha^2}{4s}\cdot|\epsilon(k_+)\cdot\epsilon(k_-)|^2, \tag{21.88}$$

---

*Partial-wave analysis for relativistic collisions is discussed in Perkins (1987), Chapter 4.

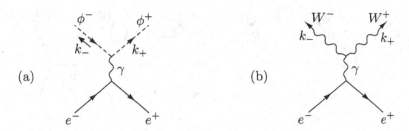

Figure 21.6. Electron-positron annihilation through a virtual photon (a) to charged scalar bosons, (b) to W bosons.

where k_+ and k_- are the momenta of the outgoing W bosons. For transversely polarized W bosons, this term is well behaved, but for longitudinally polarized W's it leads to problems. Using the approximation (21.60) for the longitudinal polarization vectors, we find

$$\epsilon(k_+) \cdot \epsilon(k_-) \to \frac{k_+ \cdot k_-}{m_W^2} \to \frac{s}{4m_W^2} \tag{21.89}$$

for $s \gg m_W^2$. This leads to a cross section that grows much faster than is allowed by unitarity. In principle, the cross section could be brought back down to a proper behavior by the addition of contributions from higher orders in perturbation theory, but this would be a most unpleasant resolution. It would imply that the theory of W bosons becomes strongly coupled at energies such that

$$\frac{s}{4m_W^2} \sim \left(\frac{g^2}{4\pi}\right)^{-1}, \tag{21.90}$$

corresponding to center-of-mass energies of order 1000 GeV. But if the theory of W bosons is strongly coupled at short distances, it is hard to understand why, at large distances, it should become the simple, weak-coupling theory that we observe.

Fortunately, there is another possible resolution of this problem. In the weak interaction gauge theory, there are three Feynman diagrams that contribute to the amplitude for $e^+e^- \to W^+W^-$ at the tree level; these are shown in Fig. 21.7. Each diagram separately produces a cross section that grows in the same manner as (21.88). However, it is possible that the badly behaved terms might cancel among the three diagrams, leaving a more proper high-energy behavior. If this miraculous cancellation were to occur, it would allow the theory of W bosons to be consistently weakly coupled up to very high energies.

Although such a cancellation seems unlikely at first sight, it is actually required by the Goldstone boson equivalence theorem. The theorem states that, at high energy, the cross section for producing longitudinal W bosons should be equal to the cross section for producing the corresponding scalar Goldstone bosons. But we know that scalar cross sections behave as $1/s$, as

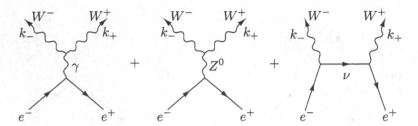

Figure 21.7. Diagrams contributing to $e^+e^- \to W^+W^-$ in the weak interaction gauge theory.

indicated in (21.87). Thus, somehow, the gauge boson cross section must also conspire to produce this result. We will now show this explicitly. We will see that the required cancellations are directed by the Ward identities of the gauge theory.

To prepare for this calculation, we need the Feynman rules for the vertices shown in Fig. 21.8. The Feynman rules for the couplings of the electron to W, Z, and γ can be read directly from (20.80). The relative strengths of these couplings are determined by the $SU(2) \times U(1)$ quantum numbers of the left- and right-handed components of the electron. It is equally straightforward to construct the couplings of the Goldstone bosons to Z and γ. Since the boson ϕ^+ has electric charge 1, the photon coupling is just that found in Problem 9.1. The Z coupling is determined with the additional information that the ϕ^+ has $I^3 = +1/2$. All of these expressions are shown in Fig. 21.8.

The three-gauge-boson vertices that appear in Fig. 21.7 arise from the cubic terms in the gauge field action. Since the $U(1)$ field strength is linear in gauge fields, these come only from the kinetic term of the $SU(2)$ gauge field. To identify the specific pieces we need, we must rewrite this cubic term in the basis of mass eigenstates given by (20.63) and (20.64). This can be done as follows:

$$
\begin{aligned}
-\tfrac{1}{4}(F^a_{\mu\nu})^2 &\to -\tfrac{1}{2}(\partial_\mu A^a_\nu - \partial_\nu A^a_\mu)g\epsilon^{abc}A^{\mu b}A^{\nu c} \\
&= -g(\partial_\mu A^1_\nu - \partial_\nu A^1_\mu)A^{\mu 2}A^{\nu 3} + g(\partial_\mu A^2_\nu - \partial_\nu A^2_\mu)A^{\mu 1}A^{\nu 3} \\
&\quad - g(\partial_\mu A^3_\nu - \partial_\nu A^3_\mu)A^{\mu 1}A^{\nu 2} \\
&= ig\big[(\partial_\mu W^+_\nu - \partial_\nu W^+_\mu)W^{\mu-}A^{\nu 3} - (\partial_\mu W^-_\nu - \partial_\nu W^-_\mu)W^{\mu+}A^{\nu 3} \\
&\quad + \tfrac{1}{2}(\partial_\mu A^3_\nu - \partial_\nu A^3_\mu)(W^{\mu+}W^{\nu-} - W^{\mu-}W^{\nu+})\big]. \quad (21.91)
\end{aligned}
$$

Finally, inserting $A^3_\mu = \cos\theta_w Z_\mu + \sin\theta_w A_\mu$ and $g = e/\sin\theta_w$, we find the Feynman rules shown in Fig. 21.9.

Before examining the amplitude for e^+e^- annihilation to vector boson

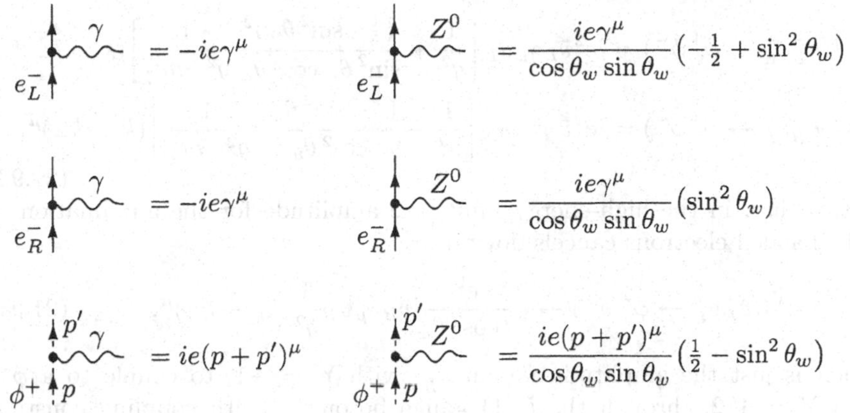

$$\text{(diagram: } e_L^- , \gamma \text{)} = -ie\gamma^\mu$$

$$\text{(diagram: } e_L^- , Z^0 \text{)} = \frac{ie\gamma^\mu}{\cos\theta_w \sin\theta_w}\left(-\tfrac{1}{2} + \sin^2\theta_w\right)$$

$$\text{(diagram: } e_R^- , \gamma \text{)} = -ie\gamma^\mu$$

$$\text{(diagram: } e_R^- , Z^0 \text{)} = \frac{ie\gamma^\mu}{\cos\theta_w \sin\theta_w}\left(\sin^2\theta_w\right)$$

$$\text{(diagram: } \phi^+ , \gamma \text{)} = ie(p+p')^\mu$$

$$\text{(diagram: } \phi^+ , Z^0 \text{)} = \frac{ie(p+p')^\mu}{\cos\theta_w \sin\theta_w}\left(\tfrac{1}{2} - \sin^2\theta_w\right)$$

Figure 21.8. Feynman rules of the weak-interaction gauge theory for electrons and scalars coupling to photons and Z bosons.

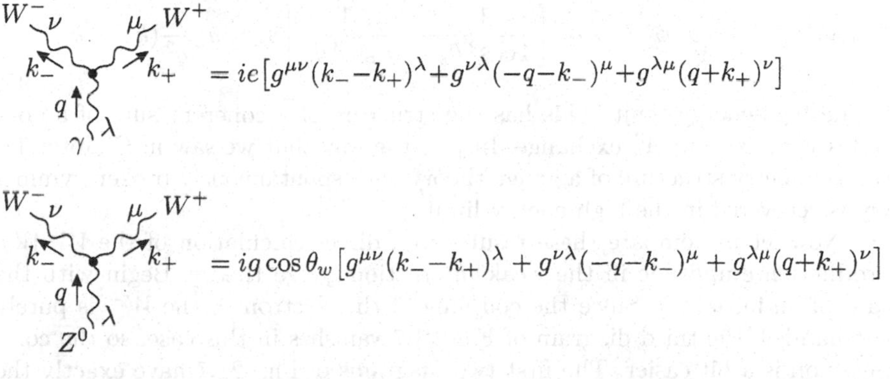

$$\text{(diagram: } W^-_\nu , W^+_\mu , \gamma_\lambda ; k_-, k_+, q\text{)} = ie\left[g^{\mu\nu}(k_- - k_+)^\lambda + g^{\nu\lambda}(-q-k_-)^\mu + g^{\lambda\mu}(q+k_+)^\nu\right]$$

$$\text{(diagram: } W^-_\nu , W^+_\mu , Z^0_\lambda ; k_-, k_+, q\text{)} = ig\cos\theta_w\left[g^{\mu\nu}(k_- - k_+)^\lambda + g^{\nu\lambda}(-q-k_-)^\mu + g^{\lambda\mu}(q+k_+)^\nu\right]$$

Figure 21.9. Feynman rules of the weak-interaction gauge theory for $WW\gamma$ and WWZ vertices.

pairs, we will first work out the amplitude for production of a pair of charged scalars. The equivalence theorem predicts that the amplitude for production of two longitudinal W bosons should become equal to this amplitude at high energy. Assembling vertices from Fig. 21.8, we find that, for an electron of either helicity, the amplitude to annihilate to scalars through a virtual photon is

$$i\mathcal{M}(ee \to \gamma^* \to \phi^+\phi^-) = ie^2\bar{v}\gamma_\mu u \frac{1}{q^2}(k_+ - k_-)^\mu, \qquad (21.92)$$

where k_+, k_- are the momenta of the scalars and $q = k_+ + k_-$. The corresponding amplitude for annihilation through a virtual Z^0 depends on the e^+e^- helicities. Adding these contributions to the preceding expression, we find

$$
i\mathcal{M}(e_L^- e_R^+ \to \phi^+\phi^-) = ie^2 \bar{v}_L \gamma_\mu u_L \left[\frac{1}{q^2} + \frac{(\frac{1}{2} - \sin^2\theta_w)^2}{\sin^2\theta_w \cos^2\theta_w} \frac{1}{q^2 - m_Z^2} \right](k_+ - k_-)^\mu,
$$

$$
i\mathcal{M}(e_R^- e_L^+ \to \phi^+\phi^-) = ie^2 \bar{v}_R \gamma_\mu u_R \left[\frac{1}{q^2} - \frac{(\frac{1}{2} - \sin^2\theta_w)}{\cos^2\theta_w} \frac{1}{q^2 - m_Z^2} \right](k_+ - k_-)^\mu.
$$

$$(21.93)$$

Notice that, in the high-energy limit, the amplitude for the annihilation of right-handed electrons cancels down to

$$
i\mathcal{M}(e_R^- e_L^+ \to \phi^+\phi^-) \to i\frac{e^2}{2\cos^2\theta_w} \bar{v}_R \gamma_\mu u_R \frac{1}{q^2}(k_+ - k_-)^\mu, \qquad (21.94)
$$

which is just the amplitude for an e_R^-, with $Y = -1$, to couple to a ϕ^+, with $Y = 1/2$, through the $U(1)$ gauge boson B_μ with coupling constant $g' = e/\cos\theta_w$. This expression reflects the fact that the e_R^- has no direct coupling to the $SU(2)$ gauge bosons. Similarly, the amplitude for left-handed electrons tends to

$$
i\mathcal{M}(e_L^- e_R^+ \to \phi^+\phi^-) \to ie^2 \left[\frac{1}{4\cos^2\theta_w} + \frac{1}{4\sin^2\theta_w} \right] \bar{v}_L \gamma_\mu u_L \frac{1}{q^2}(k_+ - k_-)^\mu
$$

$$(21.95)$$

in the high-energy limit. This has the structure of a coherent sum of amplitudes with B_μ and A_μ^3 exchange. In just the way that we saw in Chapter 11, the symmetry structure of a gauge theory with spontaneously broken symmetry is recovered in the high-energy limit.

Now let us compare these results to a direct calculation of the W^+W^- production amplitude in the weak interaction gauge theory. Begin with the case of an initial e_R^-. Since the coupling of the electron to the W^- is purely left-handed, the third diagram of Fig. 21.7 vanishes in this case, so the computation is a bit easier. The first two diagrams of Fig. 21.7 have exactly the same structure and sum to

$$
i\mathcal{M}(e_R^- e_L^+ \to W^+W^-) = \bar{v}_R \gamma_\lambda u_R \left[(-ie)\frac{-i}{q^2}(ie) + \frac{ie\sin\theta_w}{\cos\theta_w} \frac{-i}{q^2 - m_Z^2} \frac{ie\cos\theta_w}{\sin\theta_w} \right]
$$

$$
\cdot \left[g^{\mu\nu}(k_- - k_+)^\lambda + g^{\lambda\nu}(-q - k_-)^\mu + g^{\lambda\mu}(k_+ + q)^\nu \right] \epsilon_\mu^*(k_+)\epsilon_\nu^*(k_-).
$$

$$(21.96)$$

This equation is valid in any of the R_ξ gauges, since, if we ignore the electron mass,

$$
q^\lambda \bar{v}_R \gamma_\lambda u_R = 0. \qquad (21.97)
$$

The second line of Eq. (21.96) contains the enhancement for longitudinal W bosons mentioned above. If we approximate the longitudinal polarization vectors by (21.60) and drop terms that do not grow as $s \to \infty$, this line becomes

$$
\left[g^{\mu\nu}(k_- - k_+)^\lambda + g^{\lambda\nu}(-q - k_-)^\mu + g^{\lambda\mu}(k_+ + q)^\nu \right] \frac{k_{+\mu}}{m_W} \frac{k_{-\nu}}{m_W}
$$

$$= \frac{1}{m_W^2} \left[k_+ \cdot k_- (k_- - k_+)^\lambda - 2k_- \cdot k_+ k_-^\lambda + 2k_+ \cdot k_- k_+^\lambda \right] + \mathcal{O}(1) \cdot (k_- - k_+)^\lambda$$

$$= \frac{s}{2m_W^2} (k_+ - k_-)^\lambda + \cdots. \tag{21.98}$$

On the other hand, the expression in brackets in the first line of (21.96) cancels almost completely, to

$$-ie^2 \left(\frac{1}{q^2} - \frac{1}{q^2 - m_Z^2} \right) = +ie^2 \frac{m_Z^2}{q^2(q^2 - m_Z^2)}.$$

Using both of these simplifications, we find

$$i\mathcal{M}(e_R^- e_L^+ \to W_L^+ W_L^-) = \bar{v}_R \gamma_\lambda u_R \left[(ie^2) \frac{m_Z^2}{s^2} \right] \frac{s}{2m_W^2} (k_+ - k_-)^\lambda. \tag{21.99}$$

By inserting the relation $m_W = m_Z \cos\theta_w$, we see that this amplitude is identical to (21.94), as required by the equivalence theorem.

For the amplitude with an initial e_L^-, the computation is somewhat more involved. Now all three diagrams of Fig. 21.7 contribute, and since the last diagram has a different kinematic structure, it will be less clear how the diagrams combine together. In what follows, we will demonstrate the cancellation of the unitarity-violating enhanced terms, and we will indicate how the terms one order smaller in m_W^2/s assemble into the correct structure. However, we will not account rigorously for all of these smaller terms. The full calculation of these diagrams is the subject of Problem 21.2.

For the case of an initial e_L^-, the first two diagrams of Fig. 21.7 sum to the expression

$$= \bar{v}_L \gamma_\lambda u_L \left[(-ie) \frac{-i}{q^2} (ie) + \frac{ie(-\frac{1}{2} + \sin^2\theta_w)}{\sin\theta_w \cos\theta_w} \frac{-i}{q^2 - m_Z^2} \frac{ie \cos\theta_w}{\sin\theta_w} \right]$$

$$\cdot \left[g^{\mu\nu}(k_- - k_+)^\lambda + g^{\lambda\nu}(-q - k_-)^\mu + g^{\lambda\mu}(k_+ + q)^\nu \right] \epsilon_\mu^*(k_+) \epsilon_\nu^*(k_-), \tag{21.100}$$

which differs from (21.96) only in the coupling of the electron to the virtual Z^0. For longitudinal W bosons, we can simplify this expression as we did (21.96), obtaining

$$= \bar{v}_L \gamma_\lambda u_L (ie^2) \left[\frac{m_Z^2}{s(s - m_Z^2)} - \frac{1}{2\sin^2\theta_w} \frac{1}{s - m_Z^2} \right] \frac{s}{2m_W^2} (k_+ - k_-)^\lambda. \tag{21.101}$$

The second term in brackets is a potentially dangerous contribution, which must be canceled by the diagram with t-channel neutrino exchange. This diagram has the value

$$W_L^- \quad\quad W_L^+ \qquad = \left(\frac{ig}{\sqrt{2}}\right)^2 \bar{v}_L \gamma^\mu \frac{i(\not{\ell} - \not{k}_-)}{(\ell - k_-)^2} \gamma^\nu u_L(\ell) \epsilon_\mu^*(k_+) \epsilon_\nu^*(k_-), \qquad (21.102)$$

where ℓ is the initial electron momentum. Approximating the longitudinal polarization vectors as before, we have

$$W_L^- \quad\quad W_L^+ \qquad = -i\frac{g^2}{2} \bar{v}_L \frac{\not{k}_+}{m_W} \frac{(\not{\ell} - \not{k}_-)}{(\ell - k_-)^2} \frac{\not{k}_-}{m_W} u_L(\ell). \qquad (21.103)$$

Now we manipulate this expression as if we were proving a Ward identity. Using the fact that $u_L(\ell)$ satisfies the Dirac equation,

$$(\not{\ell} - \not{k}_-)\not{k}_- u_L(\ell) = -(\not{\ell} - \not{k}_-)^2 u_L(\ell) = -(\ell - k_-)^2 u_L(\ell), \qquad (21.104)$$

expression (21.103) reduces to

$$W_L^- \quad\quad W_L^+ \qquad = i\frac{g^2}{2} \bar{v}_L \frac{\not{k}_+}{m_W^2} u_L(\ell). \qquad (21.105)$$

Finally, using Eq. (21.97), we can rewrite this expression as

$$W_L^- \quad\quad W_L^+ \qquad = ie^2 \frac{1}{2\sin^2\theta_w} \frac{1}{2m_W^2} \bar{v}_L \gamma_\lambda u_L(\ell)(k_+ - k_-)^\lambda. \qquad (21.106)$$

This term cancels the dangerous high-energy behavior of (21.101). To see that the sum of diagrams has the correct high-energy limit, however, the approximations that we have used are not quite adequate. In particular, the correction to relation (21.60) for the polarization vectors is of order m_W^2/s and must be taken into account. When all of the corrections of order m_W^2/s are included, it turns out that the sum of the s-channel diagrams (21.101) is unchanged, while the expression for the neutrino exchange diagram (21.106) is multiplied by the factor $(1 + 2m_W^2/s)$. Then the sum of all three diagrams gives

$$i\mathcal{M}(e_L^- e_R^+ \to W_L^+ W_L^-) = ie^2 \bar{v}_L \gamma_\lambda u_L(\ell)(k_+ - k_-)^\lambda \frac{1}{s}$$
$$\cdot \left[\frac{1}{2\cos^2\theta_w} - \frac{1}{4\cos^2\theta_w \sin^2\theta_w} + \frac{1}{2\sin^2\theta_w} \right]. \qquad (21.107)$$

The middle term in brackets cancels half of each of the other two terms, to give an expression that agrees precisely with Eq. (21.95).

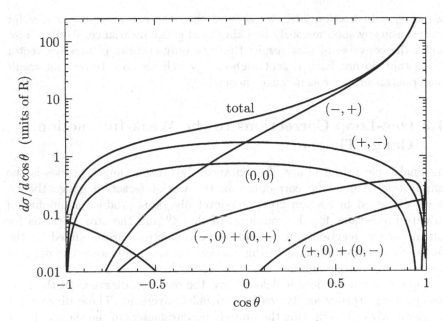

Figure 21.10. The differential cross section for $e_L^- e_R^+ \to W^+ W^-$, in units of R (Eq. (5.15)), at $E_{\mathrm{cm}} = 1000$ GeV. The various curves show the contributions to the total from individual helicity states of W^- and W^+; these are denoted (h_-, h_+), where each helicity takes the values $(+, -, 0)$. The contributions from the $(+, +)$ and $(-, -)$ states are too small to be visible. Notice that both the $W_L^- W_L^+$ cross section, denoted $(0,0)$, and the $(+, -)$ cross section become proportional to $\sin^2 \theta$ at very high energy.

The calculation of Problem 21.2 gives for the complete annihilation amplitude

$$i\mathcal{M}(e_L^- e_R^+ \to W_L^+ W_L^-) = ie^2 \bar{v}_L \gamma_\lambda u_L (k_+ - k_-)^\lambda \frac{1}{s}$$

$$\cdot \left[\frac{1}{2\sin^2\theta_w} \left\{ -\frac{s}{s - m_Z^2} \left(\frac{m_Z^2}{2m_W^2} + 1 \right) + \frac{2}{\beta^2} \right. \right. \tag{21.108}$$

$$\left. \left. -\frac{8m_W^2}{s\beta^2(1 + \beta^2 - 2\beta\cos\theta)} \right\} + \frac{m_Z^2}{m_W^2} \left(\frac{\frac{1}{2}s + m_W^2}{s - m_Z^2} \right) \right],$$

where $\beta = (1 - 4m_W^2/s)^{1/2}$ is the W boson velocity. The high-energy limit of this expression indeed reproduces (21.107). The contributions to the differential cross section for $e_L^- e_R^+ \to W^+ W^-$ from this and the other possible helicity states are plotted in Fig. 21.10.

These cancellations among the diagrams of Fig. 21.7 occur by virtue of the Ward identities of the gauge theory. That is, they occur only because the theory has an underlying local gauge invariance. At the beginning of our discussion, we argued that these cancellations are necessary to insure that the theory remains, in a consistent way, weakly coupled up to arbitrarily high energy. In Section 20.1, we showed that one can generate masses for vector bosons by spontaneously breaking local gauge invariance. We have now argued the converse of that result: that the only theories of massive vector bosons that do not have violent high-energy behavior are those that result from spontaneously broken gauge theories.[†]

21.3 One-Loop Corrections to the Weak-Interaction Gauge Theory

The final topic in our study of spontaneously broken gauge theories is the computation of one-loop corrections in the weak-interaction gauge theory. As we discussed in Section 20.2, tree-level diagrams produce a number of intricate predictions for the couplings of the Z^0 and the cross sections for neutral current reactions. In general, these predictions are modified by the effects of one-loop diagrams. In this section we will study some examples of these one-loop corrections.

As in any renormalizable field theory, the one-loop diagrams of the electroweak gauge theory are typically ultraviolet divergent. These divergences can be absorbed by adjusting the underlying parameters of the theory. These adjustments define a set of counterterms which, by renormalizability, render the full set of one-loop diagrams of the theory finite. Those amplitudes that are not adjusted by hand then become predictions of the theory.

In Chapter 11, we saw that this general procedure, which applies to any renormalizable field theory, gives especially rich information when applied to a theory with spontaneous symmetry breaking. In a theory with spontaneously broken symmetry, the amplitudes of the theory vary markedly for different particles in the same multiplet of the original symmetry. However, the counterterms of the theory respect the symmetry relations. Thus, the adjustment of an amplitude for one particle leads to definite predictions for other particles that are not related by any manifest symmetry.

Theoretical Orientation, and a Specific Problem

At the end of Section 11.6, we presented a useful framework for organizing calculations of the predictions of renormalizable theories with spontaneous symmetry breaking. We defined a *zeroth-order natural relation* to be a relation among observable quantities in the theory that is true for any values of the parameters in the Lagrangian. Since the counterterms of the theory shift the values of the underlying parameters without adding new terms, a zeroth-order natural relation will not be corrected by these counterterms. Thus, if the theory is renormalizable, the one-loop corrections to a zeroth-order natural

---

[†]This statement is proved systematically in the paper of Cornwall, Levin, and Tiktopoulos cited at the beginning of this section.

relation will be finite, and will in fact be definite predictions from the quantum structure of the field theory. Though we discussed this idea originally in theories with spontaneously broken global symmetry, it applies equally well to theories with spontaneously broken gauge symmetry. In this section, we will apply this idea to derive finite one-loop corrections to relations in the weak-interaction gauge theory.

It is easy to find zeroth-order natural relations in the electroweak theory. The leading-order predictions given in Section 20.2 involve a relatively small number of free parameters. Many of these predictions are made for energies at which the quark and lepton masses can be ignored; then they depend only on the coupling constants g and g' and the vacuum expectation value v, which sets the scale of spontaneous symmetry breaking. The remaining ingredients of the weak-interaction theory are given in terms of these parameters; for example,

$$m_W = g\frac{v}{2}, \qquad m_Z = \sqrt{g^2 + g'^2}\,\frac{v}{2},$$

$$e = \frac{gg'}{\sqrt{g^2 + g'^2}}, \qquad \frac{G_F}{\sqrt{2}} = \frac{g^2}{8m_W^2} = \frac{1}{2v^2}. \tag{21.109}$$

Even in this set of quantities, we have four relations that depend on three underlying parameters, so there is one relation of observable quantities that is independent of the parameters of the Lagrangian.

Since many of the predictions of the weak interaction gauge theory are determined by the parameter $\sin^2\theta_w$, it is useful to define $\sin^2\theta_w$ in terms of observables and then use this definition as a basis for constructing natural relations. In our discussion of the precision tests of electroweak theory in Section 20.2, we used the definition

$$s_W^2 \equiv 1 - \frac{m_W^2}{m_Z^2} \tag{21.110}$$

as a standard for comparison of different experiments. But since the three most accurately known weak-interaction observables are α, G_F, and m_Z, it is useful to construct another physical definition of $\sin^2\theta_w$ based on these three quantities. Define θ_0 such that

$$\sin 2\theta_0 \equiv \left(\frac{4\pi\alpha_*}{\sqrt{2}G_F m_Z^2}\right)^{1/2}, \tag{21.111}$$

where α_* is the running coupling constant of QED evaluated at the scale $Q^2 = m_Z^2$. The renormalization group insists that it is the value of the electric charge at the weak-interaction scale that enters precision electroweak predictions, and this observation is confirmed by summing radiative correction diagrams involving light quarks and leptons. The current best values of the quantities in Eq. (21.111) give

$$s_0^2 \equiv \sin^2\theta_0 = 0.2307 \pm 0.0005. \tag{21.112}$$

Thus, this quantity provides a very accurate standard of reference.

Once Eq. (21.111) is taken to define a reference value of $\sin^2 \theta_w$, the equations of Section 20.2 that connect $\sin^2 \theta_w$ to other observables become zeroth-order natural relations. For example, the tree-level equations

$$\frac{m_W^2}{m_Z^2} = \cos^2 \theta_w, \qquad A_{LR}^e = \frac{(\frac{1}{2} - \sin^2 \theta_w)^2 - (\sin^2 \theta_w)^2}{(\frac{1}{2} - \sin^2 \theta_w)^2 + (\sin^2 \theta_w)^2} \qquad (21.113)$$

are natural relations linking four observables of the weak interactions. The corrections to these relations will be well-defined predictions of the theory.

In principle, we could now compute all of the one-loop diagrams that correct the parameters m_W, m_Z, G_F, α, and A_{LR}^e. However, this is a very complicated exercise, requiring an extensive technical apparatus.[‡] In this section we will focus on radiative corrections from one simple source that can be considered independently. Aside from the question of anomalies, the electroweak theory does not restrict the number of quark or lepton generations. Thus, it is sensible, and gauge invariant, to compute the one-loop corrections due to one quark or lepton doublet. For definiteness, we consider the effects of the (t, b) quark doublet.

By focusing on the radiative corrections due to heavy quarks, we dramatically simplify the calculational task before us. The various observables of the weak-interaction gauge theory are extracted from the measurement of scattering amplitudes with light fermions, leptons or quarks, in the initial and final states. For example, G_F is measured from the strength of a low-energy weak-interaction process, usually chosen to be the rate of muon decay: $\mu \to \nu_\mu e^- \bar{\nu}_e$. For any such process, there are one-loop corrections of many kinds, as shown in Fig. 21.11. In addition to corrections to the vector boson propagator, there are vertex corrections, box diagrams, and diagrams with real photon emission. In general, the contributions of the various classes of diagrams are not gauge invariant; rather, gauge invariance results from cancellations between the classes of diagrams in Fig. 21.11(b), (c), and (d). However, since heavy quarks do not couple directly to the light leptons, the (t, b) doublet contributes only the single diagram shown in Fig. 21.11(f), which must be gauge invariant by itself. This same conclusion applies to the (t, b) correction to other leptonic weak interaction processes. If we ignore the CKM angles that mix the t and b with other species, the conclusion extends also to weak-interaction processes involving light quarks.

A similar situation occurs with other species of particles, such as those of the Higgs sector. The coupling of Higgs sector particles to a light quark or lepton is proportional to the fermion's mass, which we can often ignore. Thus the most important contributions from Higgs-sector particles are propagator corrections. The case in which the spontaneous symmetry breaking is produced by a single scalar field ϕ is particularly straightforward to analyze;

[‡]A detailed theoretical discussion of one-loop corrections to the electroweak theory can be found in W. Hollik, *Fortscr. d. Physik* **38**, 165 (1990).

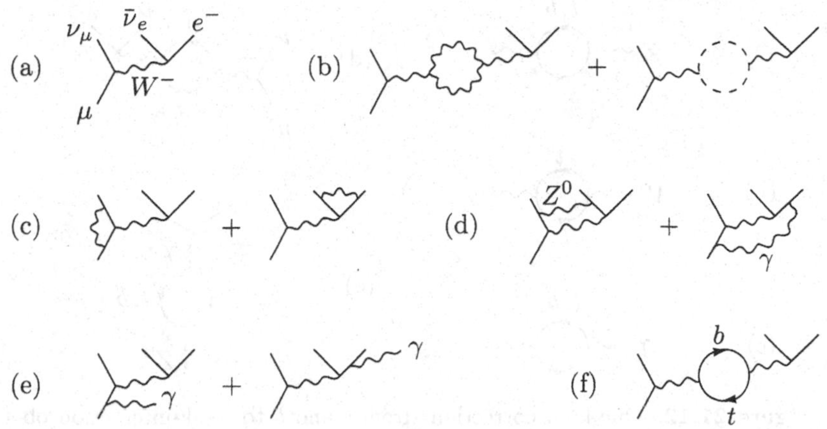

Figure 21.11. Examples of radiative corrections to μ decay in the weak interaction gauge theory: (a) lowest-order diagram; (b) propagator corrections; (c) vertex diagrams; (d) box diagrams; (e) real photon corrections; (f) the contribution of the (t, b) doublet.

this is done in Problem 21.4. Loop corrections from particles that do not couple directly to the external fermions are often termed *oblique*, since they enter the low-energy weak interactions only indirectly.

Influence of Heavy Quark Corrections

Our task, then, is to compute the corrections to relations (21.113) due to the (t, b) doublet. These two relations depend on five observable quantities— m_Z, m_W, A^e_{LR}, α, and G_F—with the last two parameters entering through θ_w and Eq. (21.111). We will express these five quantities as functions of the bare parameters g, g', and v, with corrections proportional to combinations of t and b vacuum polarization diagrams. The zeroth-order terms will naturally cancel out when we compute the corrections to the relations (21.113).

The loop amplitudes that we require are shown in Fig. 21.12. To deal with these contributions most straightforwardly, we introduce a uniform notation for vacuum polarization amplitudes. Denote the vacuum polarization amplitude involving the gauge bosons I and J as

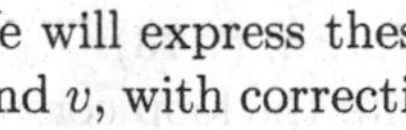

$$\underset{I}{\overset{\mu}{\sim}}\!\!\bigcirc\!\!\underset{J}{\overset{\nu}{\sim}} \;=\; i\Pi^{\mu\nu}_{IJ}(q), \qquad (21.114)$$

where I and J may be γ, W, or Z. When the gauge bosons are massive, the vacuum polarization amplitudes need not be transverse by themselves, so $\Pi^{\mu\nu}_{IJ}(q)$ need not vanish at $q^2 = 0$. Thus, we will change our notation from the case of QED and write the decomposition of $\Pi^{\mu\nu}_{IJ}(q)$ into tensor structures as

$$\Pi^{\mu\nu}_{IJ}(q) = \Pi_{IJ}(q^2)g^{\mu\nu} - \Delta(q^2)q^\mu q^\nu. \qquad (21.115)$$

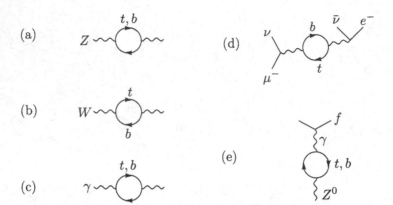

Figure 21.12. One-loop corrections from t and b to weak-interaction observables: (a) m_Z; (b) m_W; (c) α; (d) G_F; (e) A^e_{LR}.

In all of the examples to follow, the factors q^μ will dot into currents of light leptons, to give zero as in Eq. (21.97). Thus the form factor $\Delta(q^2)$ will drop out of our calculations. Our previous result that $\Pi^{\mu\nu}(q)$ vanishes in QED at $q^2 = 0$ appears in this formalism as the set of constraints

$$\Pi_{\gamma\gamma}(0) = \Pi_{\gamma Z}(0) = 0. \qquad (21.116)$$

For the other amplitudes, our sign conventions are chosen so that a positive value of $\Pi_{IJ}(m^2)$ gives a positive mass shift to the gauge boson. Let us also define

$$\Pi'_{\gamma\gamma}(0) = \left.\frac{d\Pi_{\gamma\gamma}}{dq^2}\right|_{q^2=0} ; \qquad (21.117)$$

this is the quantity we called $\Pi(0)$ in Eq. (7.73).

Now we use this notation to write the loop corrections to each of the five observables. The first two diagrams in Fig. 21.12 are simply mass corrections, and so, straightforwardly,

$$m_Z^2 = (g^2 + g'^2)\frac{v^2}{4} + \Pi_{ZZ}(m_Z^2),$$
$$m_W^2 = g^2\frac{v^2}{4} + \Pi_{WW}(m_W^2). \qquad (21.118)$$

Note that both vacuum polarization amplitudes are evaluated at the poles in the respective propagators. To evaluate the shift of α by one-loop corrections, we consider the effect of Fig. 21.12(c) on the low-energy Coulomb potential. The values of the leading-order propagator and the one-loop correction combine to give the factors

$$\frac{-ie^2}{q^2}\left(1 + i\Pi_{\gamma\gamma}(q^2) \cdot \frac{-i}{q^2}\right), \qquad (21.119)$$

where, in this equation, e^2 is given in terms of bare variables as in (21.109). Thus, the observed value of α, in the limit $q^2 \to 0$, is modified according to the relation

$$4\pi\alpha = \frac{g^2 g'^2}{g^2 + g'^2}\left(1 + \Pi'_{\gamma\gamma}(0)\right). \tag{21.120}$$

In a similar way, the diagrams of Fig. 21.12(d) give a modified strength of the 4-fermion weak interaction process that leads to μ decay. The leading and one-loop diagrams sum to

$$\frac{g^2}{q^2 - m_W^2}\left(1 + i\Pi_{WW}(q^2)\frac{-i}{q^2 - m_W^2}\right). \tag{21.121}$$

Then the effective strength of the weak interaction vertex at $q^2 = 0$ is shifted as follows:

$$\frac{G_F}{\sqrt{2}} = \frac{1}{2v^2}\left(1 - \frac{\Pi_{WW}(0)}{m_W^2}\right). \tag{21.122}$$

Notice that, in the approximation of keeping only oblique corrections, the strength of every low-energy weak interaction amplitude is corrected by this same factor.

Finally, the polarization asymmetry A^e_{LR} is corrected by a (t, b) loop diagram according to Fig. 21.12(e). The analogous diagram with an intermediate Z^0 is summed into the Z^0 propagator and does not affect the form of the vertex. At zeroth order, the coupling of the Z^0 to any left- or right-handed light fermion is given, according to Eq. (20.71), by

$$\sqrt{g^2 + g'^2}\left(T^3 - \frac{g'^2}{g^2 + g'^2}Q\right). \tag{21.123}$$

The coefficient of Q is the bare value of $\sin^2\theta_w$. The loop diagram in Fig. 21.12(e) adds to this a contribution

$$i\Pi_{Z\gamma}(q^2)\frac{-i}{q^2} \cdot (ieQ). \tag{21.124}$$

To discuss asymmetries at the Z^0 resonance, we set $q^2 = m_Z^2$. The term (21.124) adds to the piece of (21.123) proportional to Q; thus it shifts the bare value of $\sin^2\theta_w$. When we include this correction, the Z^0 coupling takes the form

$$\sqrt{g^2 + g'^2}\left(T^3 - s_*^2 Q\right), \tag{21.125}$$

where

$$s_*^2 = \frac{g'^2}{g^2 + g'^2} - \frac{e}{\sqrt{g^2 + g'^2}}\frac{\Pi_{\gamma Z}(m_Z^2)}{m_Z^2}. \tag{21.126}$$

The asymmetries at the Z^0 resonance discussed in Section 20.2 are computed as ratios of these couplings. Thus, to include the oblique radiative correction to A^f_{LR}, for any light fermion species f, we reevaluate formula (20.96), using s_*^2 in place of the zeroth-order $\sin^2\theta_w$.

We might, in fact, say that s_*^2 gives an additional way to define $\sin^2 \theta_w$ from observable quantities, to be compared to the definitions s_W^2 given in (21.110) and s_0^2 given in (21.111). Speaking strictly, the value of $\sin^2 \theta_w$ determined by the asymmetries at the Z^0 depends on the quark or lepton quantum numbers through vertex corrections that are not included in the analysis above. However, these species-dependent corrections are small and can be systematically subtracted to define a universal s_*^2 that determines the weak interaction asymmetries of all fermion species.*

The three definitions of $\sin^2 \theta_w$ all agree at zeroth order but receive different radiative corrections. If we include only the oblique corrections, it is easy to produce compact formulae for the three quantities. From (21.126), we have

$$s_*^2 = \frac{g'^2}{g^2 + g'^2} - \sin \theta_w \cos \theta_w \frac{\Pi_{\gamma Z}}{m_Z^2}. \tag{21.127}$$

In the prefactor of the one-loop correction, we can ignore the distinction between the bare and renormalized values of $\sin^2 \theta_w$. We can obtain a similar expression for s_W^2 by taking the ratio of the two formulae in (21.118):

$$s_W^2 = \frac{g'^2}{g^2 + g'^2} - \frac{1}{m_Z^2} \left(\Pi_{WW}(m_W^2) - \frac{m_W^2}{m_Z^2} \Pi_{ZZ}(m_Z^2) \right). \tag{21.128}$$

Finally, we can evaluate the oblique corrections to $\sin^2 \theta_0$ defined by (21.111). This is most readily done by writing $\delta \theta_0$ for the difference between the true and the bare value of θ_0, and then expanding (21.111) as follows:

$$2 \cos 2\theta_0 \, \delta \theta_0 = \frac{1}{2} \sin 2\theta_0 \left[\frac{\delta \alpha}{\alpha} - \frac{\delta G_F}{G_F} - \frac{\delta m_Z^2}{m_Z^2} \right]. \tag{21.129}$$

The shifts of α, G_F, and m_Z^2 can be read from (21.120), (21.122), and (21.118). Then we can reconstruct

$$\sin^2 \theta_0 = \frac{g'^2}{g^2 + g'^2} + 2 \sin \theta_0 \cos \theta_0 \delta \, \theta_0. \tag{21.130}$$

Assembling the pieces and evaluating the coefficients of the vacuum polarization diagrams to zeroth order, we obtain

$$\sin^2 \theta_0 = \frac{g'^2}{g^2 + g'^2}$$
$$+ \frac{\sin^2 \theta_w \cos^2 \theta_w}{\cos^2 \theta_w - \sin^2 \theta_w} \left[\Pi_{\gamma\gamma}'(0) + \frac{1}{m_W^2} \Pi_{WW}(0) - \frac{1}{m_Z^2} \Pi_{ZZ}(m_Z^2) \right]. \tag{21.131}$$

It is not difficult to discover that each of the equations (21.127), (21.128), and (21.131) contains ultraviolet divergences. However, if the weak interaction gauge theory is renormalizable, these divergences should cancel when we

---

*This is explained clearly in D. Kennedy and B. W. Lynn, *Nucl. Phys.* **B322**, 1 (1989).

compute the corrections to any zeroth-order natural relation. In the situation that we consider, renormalizability implies that the various definitions of $\sin^2 \theta_w$ should differ only by expressions that are ultraviolet-finite.

We are now almost prepared to check this prediction explicitly. We can clarify the structure of the ultraviolet divergences in our relations for the various quantities $\sin^2 \theta_w$ by recasting the vacuum polarization amplitudes to make more explicit the quantum numbers to which the gauge bosons couple. Recall from Eq. (20.71) that the Z boson couples to the combination of $SU(2)$ and electromagnetic quantum numbers $(T^3 - \sin^2 \theta_w Q)$. Similarly, the W bosons couple to $T^{\pm}$, or, equivalently, to T^1, T^2. It is useful to break up the vacuum polarization amplitudes into terms that depend on these specific quantum numbers. We will also extract the coupling constants indicated in (20.71). Thus we replace

$$\Pi_{\gamma\gamma} = e^2 \Pi_{QQ},$$

$$\Pi_{\gamma Z} = \left(\frac{e^2}{\sin\theta_w \cos\theta_w}\right)\left[\Pi_{3Q} - \sin^2\theta_w \Pi_{QQ}\right],$$

$$\Pi_{ZZ} = \left(\frac{e}{\sin\theta_w \cos\theta_w}\right)^2 \left[\Pi_{33} - 2\sin^2\theta_w \Pi_{3Q} + \sin^4\theta_w \Pi_{QQ}\right],$$

$$\Pi_{WW} = \left(\frac{e}{\sin\theta_w}\right)^2 \Pi_{11},$$

(21.132)

where Q denotes the electric charge and $1, 2, 3$ denote the components of weak-interaction $SU(2)$.

A vacuum polarization amplitude can always be viewed as an expectation value of a pair of currents. From this viewpoint, the quantities on the right-hand side of (21.132) are expectation values of currents with definite quantum numbers. For example, Π_{33} is an expectation value of a pair of $SU(2)$ currents $J^{\mu 3}$. Acting on the standard fermions, J^{μ}_a is a left-handed current and J^{μ}_Q is a vector current.

The ultraviolet divergences in the expectation values of currents in (21.132) have the form

$$\Pi_{33} \sim (A + Bq^2)\log\Lambda^2,$$

$$\Pi_{11} \sim (A + Bq^2)\log\Lambda^2,$$

$$\Pi_{3Q} \sim (Bq^2)\log\Lambda^2,$$

$$\Pi_{QQ} \sim (Cq^2)\log\Lambda^2.$$

(21.133)

We will demonstrate this explicitly later in this section. However, we can understand this structure from the following rough argument: Since the symmetry of the theory should be recovered at large momentum, the amplitudes Π_{33} and Π_{11}, which differ only by their orientation in the symmetry space, should have the same ultraviolet divergences. The divergence in the slope of Π_{3Q} should be related to that in the slope of Π_{33} because $Q = T^3 + Y$ and Π_{3Y} is unimportant asymptotically since $\text{tr}[T^3 Y] = 0$. We pointed out in Eq. (21.116) that Π_{3Q} and Π_{QQ} vanish at $q^2 = 0$; thus they have no q^2-independent divergences.

Now we will rewrite the two zeroth-order natural relations in (21.113) in such a way that we can apply (21.133). To do this, we take the differences of Eqs. (21.127), (21.128), and (21.131) to obtain

$$
s_*^2 - \sin^2 \theta_0 = \frac{\sin^2 \theta_w \cos^2 \theta_w}{\cos^2 \theta_w - \sin^2 \theta_w} \left\{ \frac{\Pi_{ZZ}(m_Z^2)}{m_Z^2} - \frac{\Pi_{WW}(0)}{m_W^2} - \Pi_{\gamma\gamma}'(0) \right.
$$

$$
\left. - \frac{\cos^2 \theta_w - \sin^2 \theta_w}{\sin \theta_w \cos \theta_w} \frac{\Pi_{\gamma Z}(m_Z^2)}{m_Z^2} \right\},
$$

$$
s_W^2 - s_*^2 = -\frac{\Pi_{WW}(m_W^2)}{m_Z^2} + \frac{m_W^2}{m_Z^2} \frac{\Pi_{ZZ}(m_Z^2)}{m_Z^2} + \sin \theta_w \cos \theta_w \frac{\Pi_{\gamma Z}(m_Z^2)}{m_Z^2}.
$$

$$(21.134)$$

Inserting (21.132), and also using the relation $m_W = m_Z \cos \theta_w$ in the coefficients of terms already of one-loop order, we find after some algebra

$$
s_*^2 - \sin^2 \theta_0 = \frac{e^2}{(\cos^2 \theta_w - \sin^2 \theta_w) m_Z^2} \left\{ [\Pi_{33}(m_Z^2) - \Pi_{11}(0) - \Pi_{3Q}(m_Z^2)] \right.
$$

$$
\left. + \sin^2 \theta_w \cos^2 \theta_w [\Pi_{QQ}(m_Z^2) - m_Z^2 \Pi_{QQ}'(0)] \right\},
$$

$$
s_W^2 - s_*^2 = \frac{e^2}{\sin^2 \theta_w m_Z^2} [\Pi_{33}(m_Z^2) - \Pi_{11}(m_W^2) - \sin^2 \theta_w \Pi_{3Q}(m_Z^2)].
$$

$$(21.135)$$

If indeed the ultraviolet divergences of the vacuum polarization integrals have the structure of (21.133), then the divergent part of each expression in brackets in (21.135) vanishes, and the weak interaction gives definite, finite predictions for the differences of s_*^2, s_W^2, and $\sin^2 \theta_0$.

Computation of Vacuum Polarization Amplitudes

We can verify the divergence structure (21.133) by computing the vacuum polarization diagrams for t and b quarks explicitly. Rather than computing these one by one, it is easiest to compute, once and for all, the most general fermionic vacuum polarization amplitudes, and then to recover the amplitudes required in the previous paragraph as special cases of these.

Consider, then, the two vacuum polarization amplitudes shown in Fig. 12.13. The diagrams are built from two fermion propagators with different masses m_1 and m_2, linked by left- or right-handed currents. We call the vacuum polarization amplitude with two left-handed currents $\Pi_{LL}^{\mu\nu}(q^2)$, and that with one left and one right-handed current $\Pi_{LR}^{\mu\nu}(q^2)$. Since the vacuum polarizations depend on only one momentum and two vector indices, there is no way that they can contain an invariant involving $\epsilon^{\mu\nu\rho\sigma}$. Thus, the amplitudes with other combinations of currents are related to these by

$$
\Pi_{RR}^{\mu\nu}(q^2) = \Pi_{LL}^{\mu\nu}(q^2), \qquad \Pi_{RL}^{\mu\nu}(q^2) = \Pi_{LR}^{\mu\nu}(q^2). \qquad (21.136)
$$

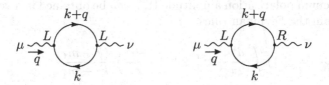

Figure 21.13. Elementary vacuum polarization amplitudes of fermionic currents.

In addition, there is no difficulty in regularizing these diagrams using dimensional regularization with an anticommuting γ^5, the regularization prescription we endorsed at the end of Section 19.4. The vacuum polarization of a vector current is reconstructed as

$$\Pi^{\mu\nu}_{VL}(q^2) = \Pi^{\mu\nu}_{LL}(q^2) + \Pi^{\mu\nu}_{RL}(q^2). \tag{21.137}$$

The vacuum polarization of purely left-handed currents is given by

$$\begin{aligned}
&= (-1) \int \frac{d^4k}{(2\pi)^4} \, \text{tr}\left[(i\gamma^\mu)\left(\frac{1-\gamma^5}{2}\right) \frac{i(\slashed{k}+m_1)}{k^2-m_1^2} \right. \\
&\qquad\qquad\qquad \left. \cdot (i\gamma^\nu)\left(\frac{1-\gamma^5}{2}\right) \frac{i(\slashed{k}+\slashed{q}+m_2)}{(k+q)^2-m_2^2} \right] \\
&= -\int \frac{d^4k}{(2\pi)^4} \, \text{tr}\left[\gamma^\mu \slashed{k} \gamma^\nu (\slashed{k}+\slashed{q})\left(\frac{1+\gamma^5}{2}\right)\right] \cdot \frac{1}{(k^2-m_1^2)((k+q)^2-m_2^2)}.
\end{aligned} \tag{21.138}$$

The prefactor (-1) comes from the fermion loop. There is no possible tensor structure antisymmetric in μ and ν, so we can now drop the γ^5 term. From here, the calculation proceeds as in Section 7.5. We combine denominators using

$$\frac{1}{(k^2-m_1^2)((k+q)^2-m_2^2)} = \int\limits_0^1 dx \, \frac{1}{(\ell^2-\Delta)^2}, \tag{21.139}$$

where

$$\ell = k + xq, \qquad \Delta = xm_2^2 + (1-x)m_1^2 - x(1-x)q^2. \tag{21.140}$$

Then, integrating with dimensional regularization and following the steps leading to Eq. (7.90), we find

$$\begin{aligned}
&= -\frac{4i}{(4\pi)^{d/2}} \int\limits_0^1 dx \, \frac{\Gamma(2-\frac{d}{2})}{\Delta^{2-d/2}} \left[g^{\mu\nu}\left(x(1-x)q^2 \right.\right. \\
&\qquad\qquad \left.\left. -\tfrac{1}{2}(xm_2^2+(1-x)m_1^2)\right) - x(1-x)q^\mu q^\nu \right].
\end{aligned} \tag{21.141}$$

Notice that both $\Pi^{\mu\nu}_{LL}$ and its first derivative with respect to q^2 are logarithmically divergent.

The vacuum polarization amplitude $\Pi^{\mu\nu}_{LR}$ can be obtained in a very similar fashion. From the Feynman rules,

$$
\begin{aligned}
\sim\!\!\!\bigcirc\!\!\!\sim \;{}_{L\quad R} &= (-1)\int \frac{d^4k}{(2\pi)^4}\, \mathrm{tr}\left[(i\gamma^\mu)\left(\frac{1-\gamma^5}{2}\right)\frac{i(\slashed{k}+m_1)}{k^2-m_1^2}\right.\\
&\qquad\qquad \left.\cdot(i\gamma^\nu)\left(\frac{1+\gamma^5}{2}\right)\frac{i(\slashed{k}+\slashed{q}+m_2)}{(k+q)^2-m_2^2}\right]\\
&= -\int \frac{d^4k}{(2\pi)^4}\,\mathrm{tr}\left[\gamma^\mu m_1\gamma^\nu m_2\left(\frac{1+\gamma^5}{2}\right)\right]\frac{1}{(k^2-m_1^2)((k+q)^2-m_2^2)}.
\end{aligned}
$$
(21.142)

From here, the same manipulations as in the previous paragraph lead to

$$
\sim\!\!\!\bigcirc\!\!\!\bullet\!\!\!\sim \;{}_{L\quad R} = -\frac{2i}{(4\pi)^{d/2}}\int_0^1 dx\,\frac{\Gamma(2-\frac{d}{2})}{\Delta^{2-d/2}}\left[g^{\mu\nu}m_1 m_2\right].
$$
(21.143)

As a check, we can use (21.141), (21.143), and (21.136), setting $m_1 = m_2 = m$, to assemble the QED vacuum polarization of vector currents. We find

$$
\begin{aligned}
\Pi^{\mu\nu}_{VV}(q^2) &= e^2\left[\Pi^{\mu\nu}_{LL} + \Pi^{\mu\nu}_{LR} + \Pi^{\mu\nu}_{RL} + \Pi^{\mu\nu}_{RR}\right]\\
&= \frac{-8e^2}{(4\pi)^{d/2}}\int_0^1 dx\,\frac{\Gamma(2-\frac{d}{2})}{\Delta^{2-d/2}}\left[x(1-x)q^2 g^{\mu\nu} - x(1-x)q^\mu q^\nu\right],
\end{aligned}
$$
(21.144)

where now $\Delta = m^2 - x(1-x)q^2$. This coincides precisely with our result from Section 7.5.

As we argued below (21.115), only the terms in the vacuum polarization amplitudes proportional to $g^{\mu\nu}$ will enter our expressions for weak-interaction radiative corrections. Thus, we can summarize the calculation of the basic vacuum polarization amplitudes by quoting the results for this leading form factor:

$$
\begin{aligned}
\Pi_{LL}(q^2) = \Pi_{RR}(q^2) &= -\frac{4}{(4\pi)^{d/2}}\int_0^1 dx\,\frac{\Gamma(2-\frac{d}{2})}{\Delta^{2-d/2}}\Big[x(1-x)q^2\\
&\qquad\qquad -\tfrac{1}{2}(xm_2^2 + (1-x)m_1^2)\Big];\\
\Pi_{LR}(q^2) = \Pi_{RL}(q^2) &= -\frac{2}{(4\pi)^{d/2}}\int_0^1 dx\,\frac{\Gamma(2-\frac{d}{2})}{\Delta^{2-d/2}}\left[m_1 m_2\right].
\end{aligned}
$$
(21.145)

From these terms, we can assemble any desired vacuum polarization of t and b quarks in the weak-interaction gauge theory. To make use of these expressions more easily, we will expand the quantities (21.145) in the limit $d \to 4$. If we set $\epsilon = 4 - d$, the integrands of the expressions above simplify according to

$$\frac{1}{(4\pi)^{d/2}} \frac{\Gamma(2-\frac{d}{2})}{\Delta^{2-d/2}} \to \frac{1}{(4\pi)^2} \left[\frac{2}{\epsilon} - \gamma + \log(4\pi) - \log \Delta \right]. \qquad (21.146)$$

Let

$$E = \frac{2}{\epsilon} - \gamma + \log(4\pi) - \log(M^2), \qquad (21.147)$$

where M is an arbitrary subtraction scale. It is useful to define

$$b_0(12X) \equiv b_0(m_1^2, m_2^2, q_X^2) = \int_0^1 dx \, \log\left(\Delta(m_1^2, m_2^2, q_X^2)/M^2\right),$$

$$b_1(12X) \equiv b_1(m_1^2, m_2^2, q_X^2) = \int_0^1 dx \, x \log\left(\Delta(m_1^2, m_2^2, q_X^2)/M^2\right),$$

$$b_2(12X) \equiv b_2(m_1^2, m_2^2, q_X^2) = \int_0^1 dx \, x(1-x) \log\left(\Delta(m_1^2, m_2^2, q_X^2)/M^2\right).$$

$$(21.148)$$

The abbreviated notation will prove useful below. In (21.148), X labels a momentum scale; we will need $q_X = 0, m_W, m_Z$. Note that for equal masses,

$$b_1(11X) = \tfrac{1}{2}b_0(11X). \qquad (21.149)$$

With this notation,

$$\Pi_{LL}(q_X^2) = -\frac{4}{(4\pi)^2} \left[\left(\tfrac{1}{6}q_X^2 - \tfrac{1}{4}(m_1^2 + m_2^2)\right)E - q_X^2 b_2(12X) \right.$$
$$\left. + \tfrac{1}{2}\left(m_2^2 b_1(12X) + m_1^2 b_1(21X)\right) \right] \qquad (21.150)$$

and

$$\Pi_{LR}(q_X^2) = -\frac{2}{(4\pi)^2} \left[m_1 m_2 E - m_1 m_2 b_0(12X) \right]. \qquad (21.151)$$

We can now reconstruct all of the specific vacuum polarization amplitudes that appear in Eq. (21.135) in terms of divergences proportional to E and finite parts proportional to the b_i. The simplest is the expectation value of electromagnetic currents, which is given in our present notation by

$$
\Pi_{QQ}(q_X^2) = -3 \cdot \frac{8}{(4\pi)^2} \Big[\big(\tfrac{2}{3}\big)^2 \big(\tfrac{1}{6} q_X^2 E - q_X^2 b_2(ttX)\big)
$$
$$
+ \big(\tfrac{1}{3}\big)^2 \big(\tfrac{1}{6} q_X^2 E - q_X^2 b_2(bbX)\big) \Big]. \tag{21.152}
$$

The prefactor 3 is the trace over colors. As we expect from QED, (21.152) contains a divergence only in a term proportional to q_X^2. The divergent parts of the other amplitudes are

$$
\Pi_{33}(q_X^2) = -\frac{12}{(4\pi)^2} \cdot \tfrac{1}{2} \big[\tfrac{1}{6} q_X^2 - \tfrac{1}{4}(m_t^2 + m_b^2) \big] E + \cdots,
$$
$$
\Pi_{11}(q_X^2) = -\frac{12}{(4\pi)^2} \cdot \tfrac{1}{2} \big[\tfrac{1}{6} q_X^2 - \tfrac{1}{4}(m_t^2 + m_b^2) \big] E + \cdots, \tag{21.153}
$$
$$
\Pi_{3Q}(q_X^2) = -\frac{12}{(4\pi)^2} \cdot \tfrac{1}{2} \big[\tfrac{1}{6} q_X^2 \big] E + \cdots.
$$

These divergences indeed follow the pattern claimed in Eq. (21.133), and thus the predictions of the weak interaction gauge theory given in (21.135) are free of ultraviolet divergences.

The Effect of m_t

Using the notation we have developed, we can write the finite parts of the relations (21.135) in a compact form. The first relation becomes

$$
s_*^2 - \sin^2 \theta_0 = \frac{3\alpha}{\pi(\cos^2 \theta_w - \sin^2 \theta_w)} \Big\{ \big(\tfrac{1}{4} - \tfrac{1}{3}\big) b_2(ttZ) + \big(\tfrac{1}{4} - \tfrac{1}{6}\big) b_2(bbZ)
$$
$$
- \tfrac{1}{4} \Big(\frac{m_t^2}{m_Z^2} [b_1(ttZ) - b_1(bt0)] + \frac{m_b^2}{m_Z^2} [b_1(bbZ) - b_1(tb0)] \Big)
$$
$$
+ 2 \sin^2 \theta_w \cos^2 \theta_w \big(\tfrac{4}{9} [b_2(ttZ) - b_2(tt0) - m_Z^2 b_2'(tt0)]
$$
$$
+ \tfrac{1}{9} [b_2(bbZ) - b_2(bb0) - m_Z^2 b_2'(bb0)] \big) \Big\}. \tag{21.154}
$$

Similarly, the second relation becomes

$$
s_W^2 - s_*^2 = \frac{3\alpha}{\pi \sin^2 \theta_w} \Big\{ \big(\tfrac{1}{4} - \tfrac{1}{3} \sin^2 \theta_w\big) b_2(ttZ) + \big(\tfrac{1}{4} - \tfrac{1}{6} \sin^2 \theta_w\big) b_2(bbZ)
$$
$$
- \tfrac{1}{4} \cos^2 \theta_w b_2(tbW)
$$
$$
- \tfrac{1}{4} \Big(\frac{m_t^2}{m_Z^2} [b_1(ttZ) - b_1(btW)] + \frac{m_b^2}{m_Z^2} [b_1(bbZ) - b_1(tbW)] \Big) \Big\}. \tag{21.155}
$$

Though it is now straightforward to work out the complete expressions for the relations (21.154) and (21.155), we will content ourselves here with identifying the most important term in the limit in which the t quark mass becomes large. Notice that, in each of these expressions, there are terms with coefficients proportional to m_t^2/m_Z^2. These are easiest to understand within the simpler combination of vacuum polarization amplitudes

$$\Pi_{11}(0) - \Pi_{33}(0) = \frac{12}{(4\pi)^2}\frac{1}{4}\left[m_t^2\big(b_1(tt0) - b_1(bt0)\big) + m_b^2\big(b_1(bb0) - b_1(tb0)\big)\right]$$

$$= \frac{3}{16\pi^2}\int\limits_0^1 dx\left\{xm_t^2\log\frac{m_t^2}{M^2} + (1-x)m_b^2\log\frac{m_b^2}{M^2}\right.$$

$$\left. - (xm_t^2 + (1-x)m_b^2)\log\frac{xm_t^2 + (1-x)m_b^2}{M^2}\right\}$$

$$= -\frac{3}{16\pi^2}\int\limits_0^1 dx\left\{xm_t^2\log\frac{xm_t^2 + (1-x)m_b^2}{m_t^2} + \mathcal{O}(m_b^2)\right\}$$

$$= \frac{3}{16\pi^2}\cdot\frac{1}{4}m_t^2 + \mathcal{O}(m_b^2) \tag{21.156}$$

for $m_t \gg m_b$. If m_t is also much greater than m_Z, one can find a contribution proportional to m_t^2/m_Z^2 in each of the relations (21.154), (21.155) by replacing the argument $q_X^2 = m_Z^2$ with $q_X^2 = 0$, using (21.156), and ignoring all other contributions. One can show, by detailed examination of (21.154) and (21.155), that this procedure gives the complete leading term in m_t. The result is

$$s_*^2 - \sin^2\theta_0 = -\frac{3\alpha}{16\pi(\cos^2\theta_w - \sin^2\theta_w)}\frac{m_t^2}{m_Z^2} + \cdots,$$

$$\tag{21.157}$$

$$s_W^2 - s_*^2 = -\frac{3\alpha}{16\pi\sin^2\theta_w}\frac{m_t^2}{m_Z^2} + \cdots,$$

where the omitted terms are of order α with no enhancement.

The enhancement factor m_t^2/m_Z^2 is exactly the one that we found in our study of top quark decay in Section 21.2. It reflects the fact that some electroweak couplings of the top quark are effectively proportional to λ_t, the top quark coupling to the Higgs sector, instead of simply to the weak interaction coupling g.

The complete numerical evaluation of the formulae for s_*^2 and s_W^2 is shown in Fig. 21.14. To compare the results of this section to experiment, we have included, in addition to the top quark effect, the m_t-independent one-loop corrections from loops containing W and Z bosons and light quarks and leptons. In the figure, the predictions are compared to the value of s_*^2 obtained from the measurement of the Z^0 polarization and forward-backward asymmetries and the value of s_W^2 obtained from measurement of the W boson mass.

According to the figure, the weak interaction gauge theory requires the top quark mass radiative correction (or a similar radiative correction from some other heavy particle) for its consistency with experiment. The top quark is predicted to have a mass approximately equal to 170 GeV. A recent analysis

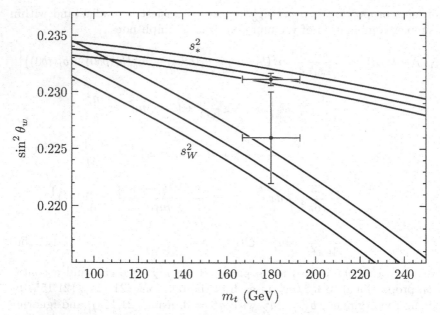

Figure 21.14. Dependence of s_*^2 and s_W^2 on the top quark mass, for fixed α, G_F, m_Z. The three curves in each group correspond to three different values of the Higgs boson mass: 100, 300, 1000 GeV from bottom to top. The curves are compared to values of s_*^2 and s_W^2, taken from the article of Langacker and Erler quoted in Table 20.1, and the CDF/D0 value of the top quark mass given in Eq. (21.159).

of all neutral current weak interaction data has given the prediction[†]

$$m_t = 169 \pm 24 \text{ GeV}. \tag{21.158}$$

Just as this book was being completed, the CDF and D0 experiments at Fermilab announced the observation of the production of top quark pairs in proton-antiproton scattering. From kinematic fits to events believed to contain top quarks, these experiments reported[‡]

$$m_t = 180 \pm 13 \text{ GeV}. \tag{21.159}$$

The discovery of the top quark in just the range required by precision electroweak measurements is quite remarkable. We can only conclude that, in the domain of weak interactions as well as those of electromagnetic, strong, and scalar interactions that we have studied earlier, the fluctuations predicted by quantum field theory make their imprint on the phenomena of Nature.

[†]P. Langacker and J. Erler, in *Review of Particle Properties*, *Phys. Rev.* **D50**, 1304 (1994).

[‡]F. Abe, et. al., *Phys. Rev. Lett.* **74**, 2626 (1995); S. Abachi, et. al., *Phys. Rev. Lett.* **74**, 2632 (1995).

Problems

21.1 Weak-interaction contributions to the muon $g - 2$. The GWS model of
the weak interactions leads to two new contributions to the anomalous magnetic moments of the leptons. Because these contributions are proportional to $G_F m_\ell^2$, they are
extremely small for the electron, but for the muon they might possibly be observable.
Both contributions are larger than the contribution of the Higgs boson discussed in
Problem 6.3.

(a) Consider first the contribution to the muon electromagnetic vertex function that
involves a W-neutrino loop diagram. In the R_ξ gauges, this diagram is accompanied by diagrams in which W propagators are replaced by propagators for
Goldstone bosons. Compute the sum of these diagrams in the Feynman-'t Hooft
gauge and show that, in the limit $m_W \gg m_\mu$, they contribute the following
term to the anomalous magnetic moment of the muon:

$$a_\mu(\nu) = \frac{G_F m_\mu^2}{8\pi^2 \sqrt{2}} \cdot \frac{10}{3}.$$

(b) Repeat the calculation of part (a) in a general R_ξ gauge. Show explicitly that
the result of part (a) is independent of ξ.

(c) A second new contribution is that from a Z-muon loop diagram and the corresponding diagram with the Z replaced by a Goldstone boson. Show that these
diagrams contribute

$$a_\mu(Z) = -\frac{G_F m_\mu^2}{8\pi^2 \sqrt{2}} \cdot \left(\frac{4}{3} + \frac{8}{3} \sin^2 \theta_w - \frac{16}{3} \sin^4 \theta_w \right).$$

21.2 Complete analysis of $e^+ e^- \to W^+ W^-$.

(a) Using explicit polarization vectors, work out the amplitudes for $e^+ e^- \to$
$W^+ W^-$ from left- and right-handed electrons to states in which the W^+ and
W^- have definite helicity. For the cases in which both W bosons have longitudinal polarization, verify that Eq. (21.99) gives the correct high-energy limit
for right-handed electrons, and verify the complete expression (21.108) for left-
handed electrons. For the cases in which one W is longitudinally polarized and
the second is transversely polarized, show that the individual diagrams give contributions to the amplitudes that grow like $\sqrt{s}$, but that the complete amplitudes
fall as $1/\sqrt{s}$.

(b) Show that the contributions to $e_L^- e_R^+ \to W^- W^+$ found in part (a) reproduce
Fig. 21.10, and that the differential cross section for $e_R^- e_L^+ \to W^- W^+$ is about
30 times smaller. How many of the qualitative features of the figure can you
understand physically?

21.3 Cross section for $d\bar{u} \to W^- \gamma$. Compute the amplitudes for $d\bar{u} \to W^- \gamma$
for the various possible initial and final helicities. Ignore the quark masses. In this
approximation, only the annihilation amplitude from $d_L \bar{u}_R$ is nonzero. Show that the
scattering amplitudes for all final helicity combinations vanish at $\cos \theta = -1/3$, where
θ is the scattering angle in the center-of-mass system. Compute the differential cross
section as a function of $\cos \theta$.

21.4 Dependence of radiative corrections on the Higgs boson mass.

(a) Consider the contributions to weak-interaction radiative corrections involving the physical Higgs boson h^0 of the GWS model. The couplings of the h^0 were discussed near the end of Section 20.2. Show that, if we ignore terms proportional to the masses of light fermions, the Higgs boson contributes one-loop corrections to the processes considered in Section 21.3 only through vacuum polarization diagrams. It follows that the contributions to vacuum polarization amplitudes that depend on the Higgs boson mass are gauge invariant.

(b) Draw the vacuum polarization diagrams in Feynman-'t Hooft gauge that involve the Higgs boson, and compute the dependence of the various vacuum polarization amplitudes on the Higgs boson mass m_h.

(c) Show that, for $m_h \gg m_W$, the natural relations discussed in Section 21.3 receive corrections

$$s_*^2 - s_0^2 = \frac{\alpha}{\cos^2\theta_w - \sin^2\theta_w} \frac{(1 + 9\sin^2\theta_w)}{48\pi} \log \frac{m_h^2}{m_W^2},$$

$$s_W^2 - s_*^2 = \alpha \frac{5}{24\pi} \log \frac{m_h^2}{m_W^2}.$$

The effect of varying m_h is displayed in Fig. 21.14 and is included as a theoretical uncertainty in the prediction (21.158). More accurate experiments might allow one to predict m_h from its effect on electroweak radiative corrections.

Final Project

Decays of the Higgs Boson

At the end of Section 20.2, we discussed the mystery of the origin of spontaneous symmetry breaking in the weak interactions. The simplest hypothesis is that the $SU(2) \times U(1)$ gauge symmetry of the weak interactions is broken by the expectation value of a two-component scalar field ϕ. However, since we have almost no experimental information about the mechanism of this symmetry breaking, many other possibilities can be suggested.

Eventually, this problem should be resolved by experimental observation of the particles associated with the symmetry breaking. To form incisive experimental tests, we should compute the properties expected for these particles. We saw in Section 20.2 that, if the symmetry is indeed broken by a single scalar field ϕ, the symmetry-breaking sector contributes only one new particle, a scalar h^0 called the Higgs boson. The mass m_h of this particle is unknown. However, the couplings of the h^0 to known fermions and bosons are completely determined by the masses of those particles and the weak interaction coupling constants. Thus, it is possible to compute the amplitudes for production and decay of the h^0 in some detail. More complicated models of $SU(2) \times U(1)$ symmetry breaking typically contain one or more particles that share some properties with the h^0. Thus, this study is a useful starting point for the more general analysis of experimental tests of these models.

In this Final Project you will compute the amplitudes for the Higgs boson h^0 to decay to pairs of quarks, leptons, and gauge bosons. The computations beautifully illustrate the working of perturbation theory for non-Abelian gauge fields. Those decays of the Higgs boson that involve quarks and gluons bring in aspects of QCD. Thus, this exercise reviews all of the important technical methods of Part III. Except for a question raised at the end of part (a), the problem relies only on material from unstarred sections of Part III. The material in Chapter 20 plays an essential role. Material from Chapter 21 enters the analysis only in parts (b) and (f), and the other parts of the problem (except for the final summary in part (h)) do not rely on these. If you have studied Section 19.5, you will have some additional insight into the results of parts (c) and (f), but this insight is not necessary to work the problem.

Consider, then, the minimal form of the Glashow-Weinberg-Salam electroweak theory with one Higgs scalar field ϕ. The physical Higgs boson h^0 of this theory was discussed in Section 20.2, and we listed there the couplings of this particle to quarks, leptons, and gauge bosons. You can now use that information to compute the amplitudes for the various possible decays of the h^0 as a function of its mass m_h. You will discover that the decay pattern has a complicated dependence on the mass of the Higgs boson, with different decay modes dominating in different mass ranges. The dependences of the various decay rates on m_h illustrate many aspects of the physics of gauge theories that we have discussed in Part III.

In working this exercise, you should consider m_h as a free parameter. For the other parameters of weak-interaction theory, you might use the following values: $m_b = 5$ GeV, $m_t = 175$ GeV, $m_W = 80$ GeV, $m_Z = 91$ GeV, $\sin^2 \theta_w = 0.23$, $\alpha_s(m_Z) = 0.12$.

(a) Compute first the rate for $h^0 \to f\bar{f}$, where f is a quark or lepton of the standard model. After a completely trivial computation, you should find

$$\Gamma(h^0 \to f\bar{f}) = \left(\frac{\alpha m_h}{8 \sin^2 \theta_w}\right) \cdot \frac{m_f^2}{m_W^2}\left(1 - \frac{4m_f^2}{m_h^2}\right)^{3/2} \cdot N_c(f),$$

where $N_c(f) = 1$ for leptons, 3 for quarks. If you have studied Chapter 18, you might improve this result for the case in which the fermion f is a quark, by computing the leading-log QCD corrections for the case $m_h \gg m_q$. Remember that the quark mass m_q is determined at the quark threshold $M^2 \sim m_q^2$.

(b) If $m_h > 2m_W$, the Higgs boson can decay to W^+W^-; if it is just a bit heavier, it can also decay to $Z^0 Z^0$. Compute the decay widths to these final states. You can check your result in the following way: If $m_h \gg m_W$, the dominant contribution to the decay comes from production of longitudinally polarized W or Z bosons, and this contribution can be estimated as follows:

$$\Gamma(h^0 \to W^+W^-) \approx \Gamma(h^0 \to \phi^+\phi^-), \quad \Gamma(h^0 \to Z^0 Z^0) \approx \Gamma(h^0 \to \phi^3\phi^3),$$

where $\phi^\pm$, ϕ^3 are the Goldstone bosons of the Higgs sector and the quantities on the right-hand sides of these relations are computed in the *ungauged* Higgs theory. Explain why this statement should be true, and verify it explicitly.

(c) The third important decay mode of the Higgs is the decay to 2 gluons. The amplitude for this decay is generated by diagrams involving quark loops. Compute these diagrams, using dimensional regularization. The diagrams will be finite, but nevertheless there is a subtle contribution which apparently depends on the regulator. Check that you have computed the amplitude correctly by verifying that it is gauge invariant. Then square the amplitude and construct the decay rate. You should find

$$\Gamma(h^0 \to 2g) = \left(\frac{\alpha m_h}{8 \sin^2 \theta_w}\right) \cdot \frac{m_h^2}{m_W^2} \cdot \frac{\alpha_s^2}{9\pi^2} \cdot \left|\sum_q I\left(\frac{m_h^2}{m_q^2}\right)\right|^2,$$

where the sum runs over all quark species and $I(m_h^2/m_q^2)$ is a form factor with the property that $I(x) \to 1$ as $x \to 0$ and $I(x) \to 0$ as $x \to \infty$. This property implies that the dominant contribution to the decay rate comes from very heavy quarks. You need not evaluate $I(x)$ explicitly at this stage; just leave it in the form of a Feynman parameter integral.

(d) The existence of the process $h^0 \to 2g$ implies the existence of the inverse process $g + g \to h^0$, which is a mechanism for production of Higgs bosons in proton-proton collisions. Using the parton model, derive a relation between the partial width $\Gamma(h^0 \to 2g)$ and the total cross section for $pp \to h^0 + X$. Compute this cross section numerically (in nanobarns) for a 30 GeV Higgs for pp collisions of center of mass energy 1–40 TeV. (1 TeV $= 10^3$ GeV.) For the purposes of this problem set (though this is not actually a good approximation) you may ignore the Q^2 dependence of the gluon distribution function and take simply

$$f_g(x) = 8 \cdot \frac{1}{x}(1-x)^7.$$

You may also set $I(m_h^2/m_t^2) = 1$; this is correct to about 10%.

(e) The final decay mode that you should consider is $h^0 \to 2\gamma$. Consider first the contribution from the loop diagrams involving quarks and leptons. Show that the result is simply expressed in terms of the form factor $I(m_h^2/m^2)$ that you derived in part (c).

(f) Next, compute the contribution to this decay amplitude from the loop diagram involving W bosons, and the various diagrams one must add to this to obtain a gauge-invariant result. It is easiest to work in Feynman-'t Hooft gauge. Add this contribution to that of very heavy quarks and leptons, each with electric charge Q_f. Your result should reduce to the following expression in the limit $m_h \ll m_W$:

$$\Gamma(h^0 \to 2\gamma) = \left(\frac{\alpha m_h}{8 \sin^2 \theta_w}\right) \cdot \frac{m_h^2}{m_W^2} \cdot \frac{\alpha^2}{18\pi^2} \cdot \left| \sum_f Q_f^2 \, N_c(f) - \frac{21}{4} \right|^2.$$

(g) Now work out the detailed behavior of the form factor $I(x)$ defined in part (c). Reduce your expression from part (c) to a one-parameter integral, then evaluate this integral numerically. Plot $I(m_h^2/m_t^2)$ over the range 50 GeV $< m_h <$ 500 GeV, and compute the decay width $\Gamma(h^0 \to 2g)$ numerically (in keV) over this range. The computation of part (f) introduces an addition form factor; compute this function in the same way.

(h) Finally, put together all the pieces. Find the branching fraction of the h^0 into each of its five major decay modes $b\bar{b}$, $t\bar{t}$, gg, W^+W^-, Z^0Z^0, for Higgs bosons of mass 50 GeV – 500 GeV.

Epilogue

Chapter 22

Quantum Field Theory at the Frontier

In this textbook we have surveyed the most important ideas of quantum field theory. Working from the basic concepts that come from fusing relativity, quantum mechanics, and fields, we have built an elaborate structure, which includes such remarkable elements as coupling constant renormalization and non-Abelian gauge fields. We have seen at many points that the strange and abstract elements of this structure actually intersect with observation and even produce explanations for unexpected aspects of the behavior of elementary particles.

In the course of our study, we have arrived at a complete theory of the strong, weak, and electromagnetic interactions of elementary particles. Each element of this theory has been described as a quantum field theory, and these quantum field theories have turned out to have very similar structure as gauge theories coupled to fermions. At various points in our discussion, we have noted that these theories have passed stringent quantitative experimental tests. We have not had space to describe the wide variety of experiments that contribute to our faith in these theories, but today almost all particle physicists consider this $SU(3) \times SU(2) \times U(1)$ gauge theory as established. In fact, most of these people refer to this theory condescendingly as 'the standard model'.

Though the best data to support the standard model have come from experiments of the past five years, the ideas behind it are much older. Most of the theoretical developments described in this book were concluded in the 1970s, a generation removed from the current frontier of physics. But this does not mean that quantum field theory is irrelevant to that frontier, any more than quantum mechanics and electrodynamics have lost their relevance after many years of exploration. On the contrary, the theory of elementary particles—like other areas of physics that depend on quantum fluctuations in continua—still holds deep mysteries, and quantum field theory remains the principal tool for exploration of these questions.

In this final chapter, we will flash forward to the present day and discuss the relevance of quantum field theory to current questions in the physics of the fundamental interactions. We will present what are, in our view, the outstanding unsolved problems of elementary particle physics and describe how quantum field theory is being used to confront these problems. Many of these applications involve aspects of quantum field theory that are beyond the scope of this book. These include the use of quantum field theory in the regime beyond the reach of perturbation theory and the use of quantum field theory to explore the special properties of theories with higher spin and local symmetry. In these areas our discussion will be mainly qualitative, but we will give references that provide points of entry into each of these subjects.

It should be obvious that our discussion in this final chapter will express our personal opinions and by no means represents the consensus of experts in quantum field theory. In addition, any collection of 'current problems' in a rapidly developing area of research should quickly become dated. In fact, we hope the readers of this book will quickly make this chapter obsolete by solving the problems that we highlight here.

22.1 Strong Strong Interactions

One paradoxical aspect of our discussion of the strong interactions is that all of our concrete results were obtained by assuming that these interactions are weak. At large momentum transfer, we argued, this assumption is actually valid due to asymptotic freedom. Still, it is uncomfortable that we have left the most obvious questions about strongly interacting particles—for example, their masses and low-energy interactions—in a mysterious regime excluded from our analysis.

To work with QCD in the region where the strong interactions are strong, we need to answer three questions: First, how do we describe the forces that bind quarks together into hadrons? Second, what is an appropriate description of the quark-antiquark and three-quark systems bound by those forces? And finally, how do we compute scattering amplitudes and matrix elements of currents using these bound states?

At this moment, there is no derivation of the complete force between quarks from the QCD Lagrangian. Explicit calculations can be done only in the limits of weak and strong coupling. In the weak-coupling limit, the result is a Coulomb potential with an asymptotically free coupling constant. The strong coupling limit, on the other hand, gives a linear potential which confines color in the way that we described, but did not derive, at the end of Section 17.1. The derivation of this result is quite unusual and brings in a new set of mathematical methods.

So far in this book, we have not discussed a strong coupling approximation to a quantum field theory. There is a simple reason for this: In a quantum field theory in which the coupling g^2 is very large, the elementary particles or their bound states typically acquire masses that grow with g^2. For $g^2 \to \infty$, these masses become comparable to the cutoff Λ and the field theory ceases to have a local continuum description.

Wilson proposed to solve this problem in a radical way, by replacing spacetime with a lattice of discretely spaced points. Such a lattice is easiest to visualize in Euclidean spacetime, and so we can use a functional integral over fields on a lattice to approximate Euclidean Green's functions. Such a theory can have a well-defined strong coupling limit. A theory of this type is very similar to a lattice model of a magnetic system.

In fact, we can understand this construction of a quantum field theory by using the concepts of Chapter 13. A lattice theory with fluctuating spin variables at each lattice site is described in the large by a quantum field theory of scalar fields with the symmetry of the underlying spin variables. Typically, the strong-coupling limit of the quantum field theory corresponds to the high-temperature limit of the magnet, in which the correlation length is much smaller than the lattice spacing. Decreasing the coupling constant corresponds to decreasing the temperature. Eventually, the coupling constant comes close to a fixed point of the renormalization group, and one can use this fixed point to define a limit of the lattice functional integral in which the lattice spacing is taken to zero.

To build a lattice model of the strong interactions, one needs to find a set of variables on the discrete lattice that correspond in the large to non-Abelian gauge fields. Wilson proposed that the fundamental variables for such a theory should be the line elements from one lattice vertex v_1 to a neighboring vertex v_2,

$$U(v_2, v_1) = P \exp\left[ig \int dx^\mu \, A_\mu^a t^a\right]. \tag{22.1}$$

To construct the lattice gauge theory with gauge group G, one should integrate over a finite group transformation U for each link of the lattice. Taking a product of these U matrices around a closed path, one can construct gauge-invariant observables, just as we did in Section 15.3. An appropriate Lagrangian can also be constructed as a sum of gauge-invariant products of the U matrices about elementary closed loops of the lattice.*

In a spin system, the defining property of the high-temperature phase is the exponential decay of correlations

$$\langle \vec{s}(0) \cdot \vec{s}(\mathbf{x}) \rangle \sim \exp\left[-|\mathbf{x}|/\xi\right] \tag{22.2}$$

as $|\mathbf{x}| \to \infty$. The analogous property of the gauge-invariant correlation function of U matrices around a closed path P is

$$\left\langle \prod_P U \right\rangle \sim \exp\left[-A/\xi^2\right], \tag{22.3}$$

where A is the area spanned by the path. This behavior is in fact seen explicitly in the expansion of Wilson's lattice gauge theory for strong coupling. A pair of color sources that stand a distance R apart for a Euclidean time T are represented by a large rectangular loop of width R and length T. For such a

---

*This construction was introduced by K. Wilson, *Phys. Rev.* **D10**, 2445 (1974). The construction is described pedagogically in M. Creutz, *Quarks, Gluons, and Lattices* (Cambridge University Press, Cambridge, 1983).

loop, we can compare the result (22.3) to the expression for the energy of an excited state in Euclidean time,

$$\langle \exp[-H_E T] \rangle \sim \exp[-RT/\xi^2].$$ (22.4)

Then we see that static sources of gauge charge, in the strong-coupling limit, are attracted to one another by a potential energy

$$V(R) \sim R/\xi^2$$ (22.5)

at sufficiently large R. Similarly, when one introduces dynamical quarks into a lattice gauge theory and studies their properties in the strong-coupling limit, configurations with large separation of color sources are suppressed in the Euclidean functional integral by factors of the form of (22.3). The strong-coupling limit then predicts the permanent confinement of quarks into color-singlet bound states.

The argument we have just given applies equally well to gauge theories based on Abelian or non-Abelian symmetry groups. But non-Abelian gauge theories have the important additional property of asymptotic freedom. In this context, that implies that a theory with weak coupling at short distances flows to a theory with strong coupling at large distances. If we imagine integrating out short-distance degrees of freedom as we described in Section 12.1, and if there is no zero of the β function or other barrier to the renormalization group flow, we should eventually arrive at an effective theory for which the strong-coupling expansion is a good approximation. Thus, in the particular case of non-Abelian gauge theories, asymptotic freedom allows us to connect a short-distance picture based on free quarks and gluons to a large-distance picture based on color confinement.

It would be wonderful if the strong-coupling picture that we have described led to mathematical equations in continuum spacetime describing the motion of permanently confined quarks and antiquarks. Many authors have tried to write such equations by imagining the area suppression of the Wilson loop correlation function (22.3) to result from a physical surface that spans the loop. For the large rectangular loop associated with color sources, this surface can be interpreted physically as the lines of color electric flux that run from one source to the other (as in Fig. 17.1), swept out through Euclidean time. At one moment of Euclidean time, this surface can be idealized as an abstract one-dimensional excitation, often called a *string*. Unfortunately, the quantum properties of an idealized string turn out to be very complicated, since each small element of the string must be considered as an independent quantum degree of freedom. The only systems of string equations that have actually been solved have bizarre features, including unwanted massless particles. Up to now, no one has succeeded in writing an equation for the quark-confining string that is useful for quantitative calculations of quark bound states.[†]

[†] For one approach to color confinement from a picture involving Wilson loops and strings, see A. A. Migdal, *Phys. Repts.* **102**, 199 (1983).

However, the lattice regularization of a non-Abelian gauge theory suggests another approach to quantitative calculations in strong-interaction theory. By approximating QCD by a lattice gauge theory with a nonzero lattice spacing and a finite spacetime volume, we reduce the functional integral to a finite number of bounded integrations, that is, an integral over $SU(3)$ group matrices for each of the finite number of links in the lattice. A lattice of size, for example, 20^4 allows the lattice spacing to be smaller than the size of a hadron while the full size of the lattice is much larger than a hadronic radius. Then one can compute correlation functions by evaluating the integrals numerically, by the Monte Carlo method. Since the functional integral with a finite lattice spacing is related to the original functional integral with zero lattice spacing by integrating out short-distance degrees of freedom, the lattice approximation can be systematically improved by computing the short-distance effects perturbatively, using asymptotic freedom to justify a weak-coupling analysis.[‡]

This numerical method has now become the principal theoretical tool for quantitative calculations in hadron physics. This method currently gives the masses of the low-lying mesons and baryons to accuracies of 10–20%; it also allows the calculation of weak interaction matrix elements of hadrons at the 25% level. As computers become more powerful, this numerical method can be pushed to higher accuracy.

Eventually, it will be interesting to ask whether these nonperturbative numerical calculations are consistent with our precision knowledge of the perturbative region of QCD. At the time of this writing, the first such comparison has been made: We have listed in Table 17.1 a value of α_s from ψ and Υ spectroscopy. In this calculation, the experimentally determined masses of $\bar{c}c$ and $\bar{b}b$ bound states are compared to values computed numerically with lattice regularization. The comparison of these values gives the required bare coupling constant of the lattice theory, which can be converted to a value of $\alpha_s(m_Z)$ in the convention of the table. The resulting estimate for $\alpha_s(m_Z)$ does agree reasonably well with purely perturbative determinations.

What is the future of nonperturbative calculations in hadron physics? On the one hand, we expect to see further development of numerical lattice methods. These methods have hardly begun to address problems of hadron-hadron scattering and multiparticle matrix elements, and this seems an important direction for the future. In addition, these methods should eventually supply an engineering understanding of hadrons at the percent level or better. On the other hand, we hope also to see a quantitative continuum approach to hadron structure, in which dynamical quarks interact with some appropriate type of string degrees of freedom.

[‡]For an introduction to numerical lattice gauge theory, see *From Actions to Answers*, T. DeGrand and D. Toussaint, eds. (World Scientific, Singapore, 1990).

22.2 Grand Unification and its Paradoxes

If we put aside our questions about the low-energy, nonperturbative behavior of QCD, the $SU(3) \times SU(2) \times U(1)$ gauge theory gives an apparently complete description of elementary particle interactions at those energies that we have probed experimentally. But what happens beyond our current reach? Does this theory need modification, or could it continue to be valid at much higher energies?

The $SU(3) \times SU(2) \times U(1)$ gauge theory contains three independent gauge coupling constants, and the observed values of these couplings are larger for the larger components of the gauge group. This pattern can be explained by a bold hypothesis about the behavior of the gauge couplings at very high energy. If at some very large energy scale, these three couplings were equal, the values of the $SU(3)$ and $SU(2)$ couplings would increase at smaller momentum scales due to their asymptotically free renormalization group equations, while the value of the $U(1)$ coupling would decrease, resulting in the observed pattern of couplings at low energies. An even bolder hypothesis would be that the three gauge symmetries are subgroups of a single large symmetry group, which is spontaneously broken at very high energy scales. The simplest choice for this larger symmetry is $SU(5)$. In that theory, the coupling constants of $SU(3) \times SU(2) \times U(1)$ have the following relation to the underlying $SU(5)$ coupling at the scale of $SU(5)$ breaking:

$$g_5 = g_3 = g = \sqrt{\frac{5}{3}}\, g'. \tag{22.6}$$

The idea that the $SU(3) \times SU(2) \times U(1)$ gauge group is embedded in a larger simple group is known as *grand unification*; the particular choice of $SU(5)$ as the unifying group is due to Georgi and Glashow.[*] The observed quarks and leptons can be seen to fit neatly into an anomaly-free chiral representation of $SU(5)$; this embedding leads to a natural explanation of the fractional charges of quarks.[†]

Within this framework, we can extrapolate the values of the three coupling constants from the energy scale of m_Z upward. The result of this extrapolation is shown as the solid lines in Fig. 22.1. The coupling constants do come close together at very high energies, though they do not actually meet. The dashed lines in the figure show the evolution with a modified set of renormalization group equations, to be explained in Section 22.4; with this choice, the three couplings meet accurately within their current uncertainties. In any event, the evolution of coupling constants occurs on a logarithmic scale in energy, so grand unification cannot be achieved without assuming an enormous value—of order 10^{16} GeV—for the symmetry-breaking scale.

---

[*]H. Georgi and S. L. Glashow, *Phys. Rev. Lett.*, **32**, 438 (1974). The remarkable hubris of this paper makes it required reading for every student.

[†]For a pedagogical introduction to grand unification, see Ross (1984).

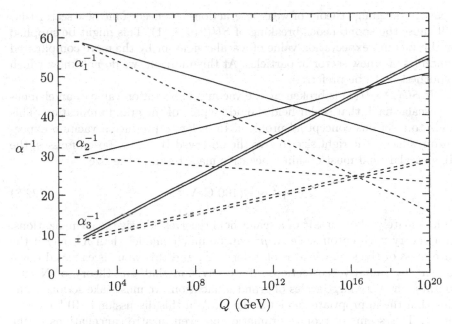

Figure 22.1. Extrapolation in energy of the coupling constants of the $SU(3) \times SU(2) \times U(1)$ gauge model, g_3, g, and $\sqrt{5/3}g'$. The solid lines are plotted using the β functions corresponding to the known set of elementary particles; the dashed lines are plotted using the β functions corresponding to a supersymmetric multiplet of particles.

The idea of a grand unification at such enormous energies raises many difficult questions, but it also suggests a wonderful opportunity. There is another enormous energy scale in quantum field theory, the scale at which the gravitational attraction of elementary particles becomes comparable to their strong, weak, and electromagnetic interactions. Conventionally, one defines the *Planck scale* as the energy for which the gravitational interaction of particles becomes of order 1:

$$m_{\text{Planck}} = (G_N/\hbar c)^{-1/2} \sim 10^{19} \text{ GeV}. \qquad (22.7)$$

However, already at energies of order 10^{18} GeV, the gravitational attraction of particles is comparable to the gauge force due to the vector bosons of a grand unified theory. Though this scale is still slightly higher than the scale at which the standard model coupling constants meet, it is not unreasonable to hope that grand unification is somehow related to the unification of gravity with the forces of elementary particle physics.

On the other hand, the introduction of this large scale into physics leads to a number of conceptual problems. The first of these problems, which one meets immediately upon suggesting this extension of the standard model, is the Higgs boson mass. In our discussion at the end of Section 20.2, we came to a somewhat ambiguous conclusion about the nature of the Higgs boson. As a part of the gauge theory of weak interactions, we need some new sector that will cause the spontaneous breaking of $SU(2) \times U(1)$. This might be supplied by the vacuum expectation value of a scalar field, or by the more complicated dynamics of a new sector of particles. At this moment, we do not know which hypothesis is to be preferred.

If $SU(2) \times U(1)$ is broken by the vacuum expectation value of an elementary scalar field, that scalar field should be part of the grand unification. This leads to a serious conceptual problem. In order to produce a vacuum expectation value of the right size to give the observed W and Z boson masses, the Higgs scalar field must obtain a negative mass term, of the size

$$-\mu^2 \sim -(100 \text{ GeV})^2. \tag{22.8}$$

Unfortunately, the (mass)^2 of a scalar field receives additive renormalizations. In a theory with cutoff scale Λ, μ^2 can be much smaller than Λ^2 only if the bare mass of the scalar field is of order $-\Lambda^2$, and this value is canceled down to $-\mu^2$ by radiative corrections. If we envision that our theory of Nature contains the very large scales of grand unification, we must take seriously the idea that the appropriate value to take for Λ in this discussion is 10^{16} GeV or larger. This seems to require dramatic and even bizarre cancellations in the renormalized value of μ^2.

We met a situation of this type in the theory of phase transitions. At zero temperature, a ferromagnet typically has a spin expectation value of the order of the underlying atomic parameters. As the temperature is raised, or as some other variable in the system is changed, the magnetization decreases. Finally, by fine adjustment of the temperature, we can arrive at a situation where the system approaches a critical point. In the very near vicinity of this point, the expectation value of the spin field is much smaller than the value predicted from atomic parameters, and the system is described by an approximately massless continuum scalar field theory.

In statistical mechanics, this picture of the light scalar field makes sense because there is an experimenter sensitively adjusting a dial. In the theory of weak interactions, there is no one obviously making a fine adjustment that gives the (mass)^2 of the Higgs boson a value 28 orders of magnitude or more below its natural value. Thus, it is a mystery why the Higgs boson mass should be so small compared to the grand unification scale. Particle physicists refer to this question as the *gauge hierarchy problem*.

How can one naturally arrange a Higgs field mass term to be so much smaller than the underlying mass scale of the fundamental interactions? One possible strategy would be to arrange for a symmetry of the fundamental Lagrangian that forbids the Higgs boson mass term and that is very weakly broken. This idea turns out to be very difficult to implement. To build a theory of this type, one would need to create a scalar field theory in which additive radiative corrections to the Higgs boson mass must cancel to any foreseeable order in perturbation theory. But the Higgs mass term is very simple in form,

$$\Delta \mathcal{L} = \mu^2 |\phi|^2, \tag{22.9}$$

and it is hard to imagine any principle that would keep this term from being generated by radiative corrections. There is one proposal for a symmetry with this property, but it requires the introduction of a profound principle called *supersymmetry* that entails deep modifications of fundamental physics. In particular, it requires a large number of new elementary particles, some of which should have masses below 1000 GeV, within the reach of the next generation of accelerators. We will discuss this possibility further in Section 22.4.

In this discussion, the problem of the Higgs mass stemmed from the hypothesis that the Higgs boson was an elementary particle. An alternative viewpoint, already suggested at the end of Section 20.2, is that the Higgs boson is a composite state bound by a new set of interactions. This idea also leads to observable experimental consequences, since the mass scale of these new interactions must be close to the weak interaction mass scale. In the simplest theories of this type, the symmetry breaking of the Higgs sector is modeled on the dynamical chiral symmetry breaking of the strong interactions, which we discussed in Section 19.3. The new strong interactions required by the theory lead to a spectrum of new particles with masses of about 1000 GeV.[‡] Thus, the two conflicting hypotheses on the nature of the sector that breaks $SU(2) \times U(1)$ both lead to new phenomena observable at future accelerators, and possibly even at present ones.

Just as these two different theories of the Higgs sector present completely different answers to the question of why the weak-interaction symmetry $SU(2) \times U(1)$ should be spontaneously broken, they also imply completely different answers to the question of the origin of the quark and lepton masses. In a model in which the Higgs field is elementary, the quark and lepton masses are derived from the renormalizable couplings of fermions to the Higgs field. These couplings would presumably be part of the grand unification and could be predicted only by theories that made explicit reference to the grand unification scale. In principle, the knowledge of these couplings could give us clues as to the details of the grand unification. Even if the Higgs field is composite, we cannot avoid the fact that the generation of masses for the quarks and leptons requires the breaking of $SU(2) \times U(1)$. Thus, these mass terms must arise from couplings of the quarks and leptons to the Higgs sector of interactions. In this class of models, the interactions leading to the quark and lepton masses must arise at energies close to the scale of the Higgs sector strong interaction and may eventually be observable experimentally.

From either viewpoint, it is still mysterious why the spectrum of quarks

---

[‡]The properties of these models of the Higgs sector, known to specialists as *technicolor* models, are described in R. Kaul, *Rev. Mod. Phys.* **55**, 449 (1983) and K. D. Lane, in *The Building Blocks of Creation*, S. Raby, ed. (World Scientific, 1993).

and leptons covers 5 orders of magnitude, from the electron at 0.5 MeV to the top quark at 175 GeV. It is also not understood what gives rise to the pattern of quark mixings encoded in the CKM matrix and the magnitude of CP violation. Even with detailed confirmation of the standard model, these questions seem today very far from solution.

The enormous mass scale of grand unification can also enter one more physical quantity, one that poses an even greater paradox than that of the Higgs boson mass. When we first quantized a field in Section 2.3, we discovered that the energy density of the vacuum in free scalar field theory received an infinite positive contribution from the zero-point energies of the various modes of oscillation. With a cutoff scale Λ, this zero-point energy is given roughly by

$$\langle 0| H |0\rangle \sim \Lambda^4. \tag{22.10}$$

At many other points in our discussion, we found similarly large contributions to the vacuum energy. The filling of the Dirac sea in the quantization of the free fermion theory led to a downward shift in the vacuum energy with a similar ultraviolet divergence. Spontaneous symmetry breaking gives a finite but still possibly large shift in the vacuum energy density,

$$\Delta \langle 0| H |0\rangle \sim -cv^4, \tag{22.11}$$

with dimensionless c, for a field vacuum expectation value v. The spontaneous breaking of the weak interaction $SU(2) \times U(1)$ symmetry and of the strong interaction chiral symmetry both would be expected to shift the vacuum energy density in this way.

In elementary particle physics experiments, this shift of the vacuum energy is unobservable. Experimentally measured particle masses, for example, are energy differences between the vacuum and certain excited states of H, and the absolute vacuum energy cancels out in the calculation of these differences. However, there is a way that the absolute vacuum energy could potentially be observed, through the coupling of the vacuum energy to gravity. According to Einstein, the energy-momentum tensor of matter $\Theta^{\mu\nu}$ is the source of the gravitational field. A vacuum energy density $\langle 0| H |0\rangle = \lambda$ contributes to this source a term

$$\Theta^{\mu\nu} = N\big(\Theta^{\mu\nu}\big) + \lambda g^{\mu\nu}, \tag{22.12}$$

where the first term on the right is subtracted to have zero vacuum expectation value. The vacuum energy term has the form of Einstein's cosmological constant and thus potentially affects the expansion of the universe.

In fact, measurements of the cosmological expansion exclude a large cosmological constant. The current limit is

$$\lambda < 10^{-29} \text{ g/cm}^3 \sim (10^{-11} \text{ GeV})^4. \tag{22.13}$$

We have no understanding of why λ is so much smaller than the vacuum energy shifts generated in the known phase transitions of particle physics, and so much again smaller than the underlying field zero-point energies. The discrepancy in λ between the experimental bound (22.13) and naive intuition is 120 orders of magnitude! The solution to this problem may come from one of many sources. It may be that the formalism of the quantum field theory of gravity requires that the vacuum energy be subtracted from the energy-momentum tensor that appears in Einstein's equations of gravity. It may be that there is a new physical mechanism coming from particle physics or from gravity itself that sets the total vacuum energy to zero. Or it may be that the overall scale of energy-momentum is genuinely ambiguous and is set by a cosmological boundary condition. At this moment, all of these possibilities are just guesses. All we know for certain is that the unification of quantum field theory and gravity cannot be straightforward, that there is some important concept still missing from our current understanding.*

22.3 Exact Solutions in Quantum Field Theory

From the idea of grand unification, with its great promise and mystery, we turn to the study of model quantum field theories that are so simple that they can be solved exactly. Throughout this book, we have stressed the intrinsic complexity of quantum field theory and the importance of using perturbation theory as a replacement for exact knowledge. But there are a variety of quantum field theories for which exact solutions are known. In this section, we will describe some of these and review the insights we have gained from them.

In searching for exact solutions to quantum field theory models, there is no reason to restrict our attention to four-dimensional spacetime. In fact, we have seen examples of two-dimensional theories with similar complexity of renormalization and short-distance behavior. At the same time, these theories occupy a one-dimensional space, and their degrees of freedom can be visualized as links in a chain. This allows some powerful simplifications.

In our discussion of the axial anomaly in two dimensions in Section 19.1, we showed that the photon of two-dimensional massless QED becomes a massive boson. More detailed examination of this theory shows that this boson is a noninteracting particle. The theory is originally formulated in terms of fermions, interacting through Coulomb forces. However, it is possible to exactly rewrite the theory as a theory of a scalar field that creates and destroys fermion-antifermion pairs. Heuristically, a particle and an antiparticle moving down the light-cone in one-dimensional space do not separate and therefore comprise one bosonic degree of freedom. In a wide class of models, the bosonic theories rewritten in this way are free-field theories. A remarkable model of this type is the *Thirring model*,

$$\mathcal{L} = \bar{\psi} i \partial \!\!\!/ \psi - \frac{g}{2} \bar{\psi} \gamma^\mu \psi \bar{\psi} \gamma_\mu \psi, \tag{22.14}$$

---

*The cosmological constant problem and a variety of unsuccessful solutions are reviewed in S. Weinberg, *Rev. Mod. Phys.* **61**, 1 (1989).

in two dimensions. In this model, the replacement of the fermion field by a boson field leads to a free field theory. Using this field theory, one can compute correlation functions of fermion bilinears explicitly and show directly that these operators have anomalous dimensions. In renormalization-group language, the model contains a line of fixed points parametrized by the coupling constant g.[†]

A more general class of two-dimensional models can be solved by visualizing them in a Hamiltonian picture as a one-dimensional chain of coupled field operators. The prototype of this method is a problem in the statistical mechanics of magnets, the one-dimensional chain of coupled spins. Consider a long chain of N discrete sites, with a spin-1/2 system at each site. The Pauli sigma matrices $\boldsymbol{\sigma}_i$ act on the two-dimensional Hilbert space at the site i. The Hamiltonian for the spin chain is then

$$H = \sum_i \left(-J \boldsymbol{\sigma}_i \cdot \boldsymbol{\sigma}_{i+1} \right). \tag{22.15}$$

Since

$$\boldsymbol{\sigma}_i \cdot \boldsymbol{\sigma}_{i+1} = 2(\sigma_i^+ \sigma_{i+1}^- + \sigma_i^- \sigma_{i+1}^+) + \sigma_i^3 \sigma_{i+1}^3, \tag{22.16}$$

this Hamiltonian conserves the number of up spins. The state with all spins down is an eigenstate of the Hamiltonian, and the states with one spin up in a state of definite momentum are also eigenstates. In 1934, Bethe analyzed the problem of two spins up and computed their S-matrix. He then discovered an amazing fact, that by multiplying the S-matrices for the two-spin problem, he could find the exact eigenstates of the Hamiltonian for any number of spins up. By considering $N/2$ spins up, he found the ground state of the system. This technique, now known as *Bethe's ansatz*, has been used to solve a wide variety of one-dimensional problems in condensed matter physics and quantum field theory. For example, this technique has been used by Andrei and Lowenstein to solve the Gross-Neveu model presented in Problem 11.3 and to demonstrate that the spectrum of this model has the properties expected from asymptotic freedom.[‡]

Even where it is not possible to solve a model for all values of its parameters, it is sometimes possible to find exact information about two-dimensional models at points where they contain massless fields. It is well known that a variety of classical two-dimensional partial differential equations can be solved by exploiting techniques of complex variables. For example, the two-dimensional Laplace equation $\nabla^2 \phi = 0$ is invariant to conformal mappings $z \to w(z)$,

---

[†]For an introduction to these models, see S. Coleman, *Phys. Rev.* **D11**, 2088 (1975), *Ann. Phys.* **101**, 239 (1976).

[‡]For an introduction to Bethe's ansatz and its generalizations, see N. Andrei, K. Furuya, and J. H. Lowenstein, *Rev. Mod. Phys.* **55**, 331 (1983), L. D. Faddeev, in *Recent Advances in Field Theory and Statistical Mechanics*, J. B. Zuber and R. Stora, eds. (North-Holland, Amsterdam, 1984), and R. J. Baxter, *Exactly Solved Models in Statistical Mechanics* (Academic Press, London, 1982).

where $z = x + iy$. Two-dimensional quantum field theories with massless particles often have this conformal symmetry at the classical level, though generically it is anomalous. In special systems, however, these anomalies vanish and the quantum theory is invariant to conformal mapping. These theories typically contain operators with anomalous dimensions, indicating that each such theory is a new, nontrivial fixed point of the renormalization group. The conformal symmetry of the theory can be used to compute these anomalous dimensions.

As an example of this class of theories, consider the two-dimensional nonlinear sigma model in which the basic field is not a unit vector, as we discussed in Section 13.3, but rather a unitary matrix of a Lie group G. The Lagrangian of this theory is

$$\mathcal{L} = \frac{1}{4g^2} \int d^2x \ \text{tr}\left[\partial_\mu U^\dagger \partial^\mu U\right]. \tag{22.17}$$

Like the theory of Section 13.3, this model is asymptotically free. However, Witten has shown that, by adding to this Lagrangian a particular perturbation of a rather complicated form first written by Wess and Zumino, one can find a fixed point of the renormalization group with manifest $G \times G$ global symmetry. This theory is conformally invariant, and all operator correlation functions can be computed using the conformal symmetry.*

One result of the nonperturbative exploration of quantum field theory was the realization that field theories can contain particle states that are not simply related to the quanta of the original fields. In the weak-coupling limit of a quantum field theory, such new states can appear as new solutions of the classical field equations. Consider, for example, ϕ^4 theory in two dimensions in the broken-symmetry phase. The equation of motion is

$$\frac{\partial^2}{\partial t^2}\phi - \frac{\partial^2}{\partial x^2}\phi - \mu^2\phi + \lambda\phi^3 = 0. \tag{22.18}$$

Treating this equation as a classical partial differential equation, we can find the time-independent solution

$$\phi(x) = \frac{\mu}{\sqrt{\lambda}} \tanh \frac{x\mu}{\sqrt{2}}. \tag{22.19}$$

This is a field configuration that begins in one well of the potential at $x = -\infty$ and crosses over to the other well as $x \to +\infty$. This solution has an energy of order μ/λ, larger by a factor of $1/\lambda$ than the mass of a ϕ quantum. Since the original equation (22.18) was Lorentz-covariant, the boosts of this solution must also be solutions to the classical partial differential equation. It is natural to suggest that, in the ϕ^4 quantum field theory, these solutions form a new set of massive particles. Such solutions, and the particles corresponding to

---

*For an introduction to conformally invariant two-dimensional quantum field theories, see P. Ginsparg, in *Fields, Strings, Critical Phenomena*, E. Brezin and J. Zinn-Justin, eds. (North-Holland, Amsterdam, 1989).

them, are often called *solitons*, borrowing a more specialized term from the literature on two-dimensional partial differential equations.[†]

Many examples are now known of particles that are associated in this way with classical solutions of a quantum field theory. In theories with spontaneously broken symmetry, the appearance of such particles is often related to the topology of the set of vacuum states; the ϕ^4 theory above gives a simple example of this relation. These examples are not limited to two dimensions but can also occur in theories that are potentially realistic. Such solutions can have magical properties. One interesting example is found in the $SU(2)$ gauge theory with a Higgs scalar field in the vector representation, the Georgi-Glashow model considered in Section 20.1. 't Hooft and Polyakov showed that this theory has a classical solution in which the Higgs field ϕ_a has the form

$$\phi_a(\mathbf{x}) = f(|\mathbf{x}|)x_a. \tag{22.20}$$

They showed that, when the gauge theory is interpreted as a unified model of weak and electromagnetic interactions, this solution is a magnetic monopole! In addition, particles that arise as heavy classical states in the weak coupling limit can have a more intricate relation to the dynamics of the theory when the coupling is increased. For example, in theories of the type of two-dimensional QED or the Thirring model in which fermions can be replaced by bosons, a weak-coupling limit is obtained by adding to the theory a large fermion mass. Then the original fermions are recovered from the bosonic representation of the theory as classical solutions very similar to that given in (22.19).

In some theories, one can find classical solutions of the Euclidean field equations. These solutions, called *instantons*, are localized in Euclidean time as well as in space. Thus, they are interpreted as quantum processes that modify the effective Hamiltonian of a quantum field theory. The most famous example of an instanton is found in four-dimensional non-Abelian gauge theories. It was shown by 't Hooft that this field configuration leads to a quantum process that violates the conservation of the $U(1)$ axial current in QCD. We have explained in Section 19.3 that this violation of current conservation is exactly what is needed to explain the spectrum of light mesons in QCD.

There is probably much more to be learned, especially about the strong-coupling behavior of gauge theories, by deeper analysis of the classical solutions to the field equations, and of the interrelations of the many exactly or partially solvable two-dimensional field theories.

[†]For an introduction to the use of solutions of the classical field equations in the analysis of problems in field theory, see S. Coleman (1985), Chaps. 6 and 7, and R. Rajaraman, *Solitons and Instantons* (North-Holland, Amsterdam, 1982).

22.4 Supersymmetry

Among the properties that a quantum field theory might possess to make it more beautiful or more mathematically tractable, there is one higher symmetry with particularly far-reaching implications. This is a symmetry that relates fermions and bosons, known (without hyperbole) as *supersymmetry*. In this section, we will introduce some of the purely mathematical consequences of supersymmetry, and then discuss the question of whether the true field equations of Nature could be supersymmetric.

A generator of supersymmetry is an operator that commutes with the Hamiltonian and converts bosonic into fermionic states. Such an operator must carry half-integer spin, in the simplest case spin $1/2$. Let Q_α, with $\alpha = 1, 2$, be the left-handed spinor components of this operator. Their Hermitian conjugates, $Q_\beta^\dagger$, form a right-handed spinor. The anticommutator $\{Q_\alpha, Q_\beta^\dagger\}$ is a 2×2 matrix with positive diagonal elements; thus it cannot vanish. This matrix commutes with H but transforms nontrivially under Lorentz transformations. A Lorentz-covariant expression for this anticommutator is

$$\{Q_\alpha, Q_\beta^\dagger\} = 2\sigma_{\alpha\beta}^\mu P^\mu, \tag{22.21}$$

where P^μ is a conserved vector quantity. Such quantities are severely restricted; a theorem of Coleman and Mandula states that, if a quantum field theory in more than two dimensions has a second conserved vector quantity in addition to the energy-momentum 4-vector, the S-matrix equals 1 and no scattering is allowed. Thus the only possible choice for P^μ in Eq. (22.21) is the total energy-momentum. The Coleman-Mandula theorem also rules out any higher-spin conservation laws. This eliminates the possibility that a supersymmetry generator could have spin $3/2$ or higher. The most general possibility is a collection of spin-$1/2$ operators with the anticommutation relations

$$\{Q_\alpha^i, Q_\beta^{j\dagger}\} = 2\delta^{ij}\sigma_{\alpha\beta}^\mu P^\mu, \tag{22.22}$$

with $i, j = 1, \ldots, N$. In the following discussion, we will mainly consider only the simplest case, $N = 1$.[‡]

The algebra (22.22) of conserved quantities has profound conseqences for the theory. Since the right-hand side of (22.22) is the total energy-momentum, it involves every field in the theory. To reproduce this algebra, the left-hand side must also involve every field. The representations of this algebra pair every bosonic state with a fermionic state at the same energy, and vice versa. If supersymmetry is an exact symmetry of the quantum field theory, it must act on every sector of the theory. In a realistic model, even the gravitational field must have a fermionic partner. This means that Einstein's equations of gravity must be generalized to a new set of geometrical equations that involve a fermionic (spin-$3/2$) field.

[‡]An excellent introduction to the formalism of supersymmetry is J. Wess and J. Bagger, *Supersymmetry and Supergravity* (Princeton University Press, 1983).

The first consequences of making a quantum field theory supersymmetric are easy to understand. For every (complex) scalar field, one must introduce a chiral fermion field. The self-interactions of the bosonic fields are related to the interactions of these fields with the fermions; for example, a possible interaction Lagrangian with coupling constant λ is

$$\Delta\mathcal{L} = -\lambda^2 |\phi|^2 - \tfrac{1}{2}\lambda\psi^T \sigma^2 \psi. \tag{22.23}$$

We have written a more general supersymmetric Lagrangian in Problem 3.5. Similarly, for every gauge field, one must introduce a chiral fermion in the adjoint representation of the gauge group. This fermion, called the *gaugino*, mediates interactions of the scalar fields with their fermionic partners whose strength is given by the gauge coupling g.

The special relation between the bosonic and fermionic interactions leads to great simplifications in the renormalization of supersymmetric theories. Some of these simplifications can be anticipated. Since supersymmetry requires that each scalar particle have a fermionic partner of the same mass, these particles must have the same mass renormalization. But we have seen that the fermion mass is multiplicatively renormalized and thus is only logarithmically divergent, while a scalar mass term is additively renormalized and thus can be quadratically divergent. Supersymmetry must imply that the quadratic divergences of scalar mass terms automatically vanish. In fact, these cancellations occur in every order of perturbation theory, with loop diagrams involving bosons canceling against diagrams with virtual fermions. To see another simplification required by supersymmetry, take the vacuum expectation value of the anticommutation relation (22.21). The vacuum state has zero momentum: $P^i |0\rangle = 0$. If the vacuum state is supersymmetric, $Q_\alpha |0\rangle = Q_\beta^\dagger |0\rangle = 0$. Then Eq. (22.21) implies

$$\langle 0| H |0\rangle = 0. \tag{22.24}$$

We have noted already that bosonic fields give positive contributions to the vacuum energy through their zero-point energy, and fermionic fields give negative contributions. We now see that, in a supersymmetric model, these contributions cancel exactly, not only at the leading order but to all orders in perturbation theory.

Deeper examination of supersymmetric theories leads to additional, and quite unexpected, cancellations in renormalization theory. For example, one can show that the coupling constants in scalar-fermion self-interactions, such as λ in (22.23), are renormalized only through field strength renormalizations. Thus the relative size of two different scalar interactions remains unchanged. If a particular type of renormalizable interaction is omitted, it cannot be generated by renormalization, in contrast to the case in ordinary field theory. The simplest supersymmetry does not constrain the renormalization of gauge couplings, but higher supersymmetries can have a profound effect: In $N = 2$ supersymmetric models, the β function vanishes if the leading-order term is arranged to be zero. In $N = 4$ supersymmetric models, this cancellation is automatic and $\beta(g) = 0$ exactly. These models give examples of four-dimensional quantum field theories with no ultraviolet divergences.*

Supersymmetry thus endows a quantum field theory with remarkable, even magical properties. But is it possible that the true equations of Nature could possess such a high degree of symmetry? Since we are certain that there is no charged boson with the same mass as the electron, we know that supersymmetry cannot be an exact symmetry of Nature. But it is tempting to guess that it might be a spontaneously broken symmetry of the underlying equations.

In fact, this conjecture has fruitful consequences for the grand unified theories that we discussed in Section 22.2. The problem of the Higgs boson mass that we highlighted in that section has an elegant solution in supersymmetry models. In a supersymmetric version of the standard model, the Higgs field is one of a large number of scalar fields with various $SU(3) \times SU(2) \times U(1)$ quantum numbers. For all of these scalar fields, the mass terms get only a logarithmic multiplicative renormalization. If supersymmetry were broken in such a way as to give mass differences of a few hundred GeV between the observed fermionic quarks and leptons and their scalar partners, one would also find a Higgs boson (mass)^2 of the correct size. There are good reasons, which follow from more detailed properties of the theory, why it is the Higgs field, rather than some other scalar field, that obtains a vacuum expectation value.†

If this set of ideas is correct, the scalar partners of quarks and leptons would be light enough to be discovered experimentally in the near future. In that case, these scalar particles and the fermionic partners of gauge bosons would affect the renormalization of coupling constants between present energies and the grand unification scale. This might potentially disturb the prospects for grand unification, but, instead, it improves them: the dashed lines of Fig. 22.1, with a more impressive meeting of the three coupling constants, were generated by replacing the conventional β functions with ones including the supersymmetric partners.

The last of the problems discussed in Section 22.2 is also ameliorated by the introduction of supersymmetry. In a grand unified theory with broken supersymmetry, those momentum scales that are much larger than the mass differences of supersymmetry partners give no contribution to the vacuum energy. Thus the natural size of the cosmological constant in these theories is $\lambda \sim (100 \text{ GeV})^4$. This reduces the cosmological constant problem to a discrepancy of 50 orders of magnitude—but this is not nearly enough.

---

*Supersymmetric models with vanishing β function are reviewed by P. West, in *Shelter Island II*, R. Jackiw, N. N. Khuri, S. Weinberg, and E. Witten, eds. (MIT Press, Cambridge, 1985).

†Supersymmetric models of quarks and leptons, and their observable consequences, are reviewed in H. P. Nilles, *Phys. Repts.* **110**, 1 (1984), and in H. E. Haber and G. L. Kane, *Phys. Repts.* **117**, 75 (1985).

It is an exciting prospect that supersymmetric partners of the particles of the standard model might soon be seen in experiments. What we anticipate, in any event, is that the experiments of the next generation will make a definite choice between this hypothesis for the nature of the Higgs sector and the other possibilities discussed in Section 22.2. Either way, we will have advanced our knowledge one step toward the truly fundamental equations.

22.5 Toward an Ultimate Theory of Nature

What are these fundamental equations? Do they involve quantum field theory, or some very different organizing principle? Any answer to this question must be completely speculative. Nevertheless, there are some principles, and an example, that can guide this search.

For all the attention we have given in this book to the basic interactions of particle physics, we have given very little attention to gravity. In part, this is because the quantum theory of gravity has no known observational consequences. But it is also true that the quantum theory of gravity is still ill-formed and uncertain. If gravity is treated as a weak-coupling field theory with Feynman diagrams, one quickly finds that the divergences of these diagrams render the theory nonrenormalizable. This is no surprise, because gravity is a theory in which the coupling constant has inverse mass dimensions, with the mass scale m_{Planck} given by (22.7). In our general philosophy of renormalization, all of the complexity of this theory should be concentrated at the scale m_{Planck}.

At the scale where quantum fluctuations of the gravitational field are important, we must expect profound changes in physics. If these changes occur within the context of quantum field theory, they will at the least entail fluctuating spacetime geometry and topology. But it seems equally probable that quantum field theory will actually break down at this scale, with continuous spacetime replaced by a new discrete or nonlocal geometry.

One particular model for the behavior of spacetime at very small distances is *string theory*, the dynamics of abstract one-dimensional extended objects. In Section 22.1, we mentioned that such objects seemed to occur naturally in attempts to describe quark confinement in QCD, but that the detailed properties of these objects made them unsuitable for strong interaction phenomenology. Among the disappointing properties of these systems were the appearance of massless spin-2 states of the string, and a constraint that the dimension of spacetime must be increased unless the spectrum of the theory contained many massless spin-1 states. In 1974, Scherk and Schwarz made the remarkable suggestion that string theory was a correct mathematical description of a different problem, the unification of elementary particle interactions with gravity. They interpreted the spin-2 quantum as the graviton and the spin-1 quanta as gauge bosons of a gauge theory.[‡] A decade later,

---

[‡]J. Scherk and J. H. Schwarz, *Nucl. Phys.* **B81**, 118 (1974).

Green and Schwarz put this conjecture on a firmer footing by showing that a particular string theory could be interpreted as a grand unified theory in ten spacetime dimensions, with all gravitational and gauge Ward identities automatically satisfied and all anomalies automatically canceling. Since that time, many other solutions to the constraint equations of string theory have been found, some of which correspond to unified models of gauge interactions and gravity in four dimensions. These models can naturally incorporate supersymmetry and, under that condition, give ultraviolet-finite results for all scattering amplitudes, including those of gravitons.*

String theories solve the ultraviolet divergence problems of quantum field theory by rejecting the idea that elementary particles are pointlike objects with contact interactions. Rather, in string theory, quarks, leptons, gauge bosons, and gravitons are extended loops of string excitation which thus interact nonlocally. Since particles cannot be definitely localized, spacetime itself takes on a nonlocal character. In some sense, distances much less than the Planck length $1/m_{\text{Planck}}$ do not exist in the string description of gravity. As yet, it is not clear how to understand intuitively the sort of geometry that string theory requires. This mathematical problem is now being actively investigated.

If indeed the truly fundamental geometry of Nature is nonlocal, discrete, or discontinuous in some other way, then the grand program for the fundamental interactions that we have set forth in this book must be altered in an essential way. The most elementary equations of Nature will not be gauge-invariant quantum field theories, but instead theories built from very different elements. Even the principles of model construction will be different from those based on gauge and Lorentz invariance that we have discussed here.

On the other hand, quantum field theory will still play an essential role in the interpretation of this structure. All of the processes we now observe, even the elementary particle processes at the highest energies currently accessible, occur over distances 15 orders of magnitude larger than the sizes of the strings or other fluctuating structures that appear in the underlying equations. The relation of experimental observations to these fundamental structures is thus very similar to the relation of macroscopic observations to the underlying atomic structure of matter. In the study of matter, we use a classical, Newtonian description of atoms to bridge this gap and to relate the properties of gases, liquids, and solids to underlying atomic properties. We might say that the quantum theory of atoms gives rise to a set of effective Newtonian equations that is extremely powerful in the macroscopic domain. Especially in the theory of gases, this Newtonian description was also used as a tool to realize the existence of atoms and to derive their properties.

---

*A technical introduction to string theory and its use in building unified models has been given by M. B. Green, J. H. Schwarz, and E. Witten, *Superstring Theory*, 2 vols. (Cambridge University Press, 1987).

Similarly, whatever the nature of Planck-scale physics, it leads to some effective continuum quantum field theory. This quantum field theory might well be an accurate approximation to the underlying physics already at distances of 100 Planck lengths, corresponding to momenta of 10^{17} GeV. From here to the scale of weak interactions, and from there up to the wavelength of light, and from there to the size of the universe, quantum field theory can be treated as the basic framework for the equations of physics. By recognizing the symmetries of the particular set of field equations that Nature has provided us, we can learn to compute all of the details of physical processes over this whole enormous domain. And, by contemplating the origin of these symmetries, perhaps we will also be able to see through to the next level and unlock the true structure of spacetime.

Reference Formulae

This Appendix collects together some of the formulae that are most commonly used in Feynman diagram calculations.

A.1 Feynman Rules

In all theories it is understood that momentum is conserved at each vertex, and that undetermined loop momenta are integrated over: $\int d^4p/(2\pi)^4$. Fermion (including ghost) loops receive an additional factor of (-1), as explained on page 120. Finally, each diagram can potentially have a symmetry factor, as explained on page 93.

ϕ^4 **theory:** $\mathcal{L} = \frac{1}{2}(\partial_\mu\phi)^2 - \frac{1}{2}m^2\phi^2 - \frac{\lambda}{4!}\phi^4$

$$\text{Scalar propagator:} \qquad \frac{i}{p^2 - m^2 + i\epsilon} \qquad (A.1)$$

$$\phi^4 \text{ vertex:} \qquad = -i\lambda \qquad (A.2)$$

$$\text{External scalar:} \qquad = 1 \qquad (A.3)$$

(Counterterm vertices for loop calculations are given on page 325.)

Quantum Electrodynamics: $\mathcal{L} = \bar{\psi}(i\partial\!\!\!/ - m)\psi - \frac{1}{4}(F_{\mu\nu})^2 - e\bar{\psi}\gamma^\mu\psi A_\mu$

$$\text{Dirac propagator:} \qquad \frac{i(p\!\!\!/ + m)}{p^2 - m^2 + i\epsilon} \qquad (A.4)$$

$$\text{Photon propagator:} \qquad \frac{-ig_{\mu\nu}}{p^2 + i\epsilon} \qquad (A.5)$$

(Feynman gauge; see page 297 for generalized Lorentz gauge.)

QED vertex: $= iQe\gamma^\mu$ (A.6)

$(Q = -1$ for an electron)

External fermions: $= u^s(p)$ (initial)   (A.7)

$= \bar{u}^s(p)$ (final)

External antifermions: $= \bar{v}^s(p)$ (initial)   (A.8)

$= v^s(p)$ (final)

External photons: $= \epsilon_\mu(p)$ (initial)   (A.9)

$= \epsilon_\mu^*(p)$ (final)

(Counterterm vertices for loop calculations are given on page 332.)

Non-Abelian Gauge Theory:

$$\mathcal{L} = \bar{\psi}(i\not{\partial} - m)\psi - \tfrac{1}{4}(\partial_\mu A_\nu^a - \partial_\nu A_\mu^a)^2 + gA_\mu^a\bar{\psi}\gamma^\mu t^a\psi$$
$$- gf^{abc}(\partial_\mu A_\nu^a)A^{\mu b}A^{\nu c} - \tfrac{1}{4}g^2(f^{eab}A_\mu^a A_\nu^b)(f^{ecd}A^{\mu c}A^{\nu d})$$

The fermion and gauge boson propagators are the same as in QED, times an identity matrix in the gauge group space. Similarly, the polarization of external particles is treated the same as in QED, but each external particle also has an orientation in the group space.

Fermion vertex: $= ig\gamma^\mu t^a$ (A.10)

3-boson vertex: $=$ $gf^{abc}\big[g^{\mu\nu}(k-p)^\rho$
$+ g^{\nu\rho}(p-q)^\mu$ (A.11)
$+ g^{\rho\mu}(q-k)^\nu\big]$

4-boson vertex: $=$ $-ig^2\big[f^{abe}f^{cde}(g^{\mu\rho}g^{\nu\sigma} - g^{\mu\sigma}g^{\nu\rho})$
$+ f^{ace}f^{bde}(g^{\mu\nu}g^{\rho\sigma} - g^{\mu\sigma}g^{\nu\rho})$ (A.12)
$+ f^{ade}f^{bce}(g^{\mu\nu}g^{\rho\sigma} - g^{\mu\rho}g^{\nu\sigma})\big]$

Ghost vertex: $= -g f^{abc} p^\mu$ (A.13)

$$b, \mu$$
$$a \quad p \quad c$$

Ghost propagator: $a \cdots\!\blacktriangleleft\!\cdots b = \dfrac{i\delta^{ab}}{p^2 + i\epsilon}$ (A.14)

(Counterterm vertices for loop calculations are given on pages 528 and 532.)

Other theories. Feynman rules for other theories can be found on the following pages:

Yukawa theory	page 118
Scalar QED	page 312
Linear sigma model	page 353
Electroweak theory	pages 716, 743, 753

A.2 Polarizations of External Particles

The spinors $u^s(p)$ and $v^s(p)$ obey the Dirac equation in the form

$$
\begin{aligned}
0 &= (\slashed{p} - m)u^s(p) = \bar{u}^s(p)(\slashed{p} - m) \\
&= (\slashed{p} + m)v^s(p) = \bar{v}^s(p)(\slashed{p} + m),
\end{aligned}
\tag{A.15}
$$

where $\slashed{p} = \gamma^\mu p_\mu$. The Dirac matrices obey the anticommutation relations

$$\{\gamma^\mu, \gamma^\nu\} = 2g^{\mu\nu}. \tag{A.16}$$

We use a chiral basis,

$$\gamma^\mu = \begin{pmatrix} 0 & \sigma^\mu \\ \bar{\sigma}^\mu & 0 \end{pmatrix}, \qquad \gamma^5 = \begin{pmatrix} -1 & 0 \\ 0 & 1 \end{pmatrix}, \tag{A.17}$$

where

$$\sigma^\mu = (1, \boldsymbol{\sigma}), \qquad \bar{\sigma}^\mu = (1, -\boldsymbol{\sigma}). \tag{A.18}$$

In this basis the normalized Dirac spinors can be written

$$u^s(p) = \begin{pmatrix} \sqrt{p \cdot \sigma}\, \xi^s \\ \sqrt{p \cdot \bar{\sigma}}\, \xi^s \end{pmatrix}, \qquad v^s(p) = \begin{pmatrix} \sqrt{p \cdot \sigma}\, \eta^s \\ -\sqrt{p \cdot \bar{\sigma}}\, \eta^s \end{pmatrix}, \tag{A.19}$$

where ξ and η are two-component spinors normalized to unity. In the high-energy limit these expressions simplify to

$$u(p) \approx \sqrt{2E} \begin{pmatrix} \frac{1}{2}(1 - \hat{p} \cdot \boldsymbol{\sigma})\xi^s \\ \frac{1}{2}(1 + \hat{p} \cdot \boldsymbol{\sigma})\xi^s \end{pmatrix}, \qquad v(p) \approx \sqrt{2E} \begin{pmatrix} \frac{1}{2}(1 - \hat{p} \cdot \boldsymbol{\sigma})\eta^s \\ -\frac{1}{2}(1 + \hat{p} \cdot \boldsymbol{\sigma})\eta^s \end{pmatrix}. \tag{A.20}$$

Using the standard basis for the Pauli matrices,

$$\sigma^1 = \begin{pmatrix} 0 & 1 \\ 1 & 0 \end{pmatrix}, \qquad \sigma^2 = \begin{pmatrix} 0 & -i \\ i & 0 \end{pmatrix}, \qquad \sigma^3 = \begin{pmatrix} 1 & 0 \\ 0 & -1 \end{pmatrix}, \qquad \text{(A.21)}$$

we have, for example, $\xi^s = \begin{pmatrix} 1 \\ 0 \end{pmatrix}$ for spin up in the z direction, and $\xi^s = \begin{pmatrix} 0 \\ 1 \end{pmatrix}$ for spin down in the z direction. For antifermions the physical spin is opposite to that of the spinor: $\eta^s = \begin{pmatrix} 1 \\ 0 \end{pmatrix}$ corresponds to spin *down* in the z direction, and so on.

In computing unpolarized cross sections one encounters the polarization sums

$$\sum_s u^s(p)\bar{u}^s(p) = \not{p} + m, \qquad \sum_s v^s(p)\bar{v}^s(p) = \not{p} - m. \qquad \text{(A.22)}$$

For polarized cross sections one can either resort to the explicit formulae (A.19) or insert the projection matrices

$$\left(\frac{1+\gamma^5}{2} \right), \qquad \left(\frac{1-\gamma^5}{2} \right), \qquad \text{(A.23)}$$

which project onto right- and left-handed spinors, respectively. Again, for antifermions, the helicity of the spinor is opposite to the physical helicity of the particle.

Many other identities involving Dirac spinors and matrices can be found in Chapter 3.

Polarization vectors for photons and other gauge bosons are conventionally normalized to unity. For massless bosons the polarization must be transverse:

$$\epsilon^\mu = (0, \boldsymbol{\epsilon}), \qquad \text{where } \mathbf{p} \cdot \boldsymbol{\epsilon} = 0. \qquad \text{(A.24)}$$

If $\mathbf{p}$ is in the $+z$ direction, the polarization vectors are

$$\epsilon^\mu = \frac{1}{\sqrt{2}}(0, 1, i, 0), \qquad \epsilon^\mu = \frac{1}{\sqrt{2}}(0, 1, -i, 0), \qquad \text{(A.25)}$$

for right- and left-handed helicities, respectively.

In computing unpolarized cross sections involving photons, one can replace

$$\sum_{\text{polarizations}} \epsilon_\mu^* \epsilon_\nu \longrightarrow -g_{\mu\nu}, \qquad \text{(A.26)}$$

by virtue of the Ward identity. In the case of massless non-Abelian gauge bosons, one must also sum over the emission of ghosts, as discussed in Section 16.3. In the massive case, one must in addition include the emission of Goldstone bosons, as discussed in Section 21.1.

A.3 Numerator Algebra

Traces of γ matrices can be evaluated as follows:

$$\text{tr}(\mathbf{1}) = 4$$
$$\text{tr}(\text{any odd} \# \text{ of } \gamma\text{'s}) = 0$$
$$\text{tr}(\gamma^\mu \gamma^\nu) = 4g^{\mu\nu}$$
$$\text{tr}(\gamma^\mu \gamma^\nu \gamma^\rho \gamma^\sigma) = 4(g^{\mu\nu}g^{\rho\sigma} - g^{\mu\rho}g^{\nu\sigma} + g^{\mu\sigma}g^{\nu\rho}) \qquad \text{(A.27)}$$
$$\text{tr}(\gamma^5) = 0$$
$$\text{tr}(\gamma^\mu \gamma^\nu \gamma^5) = 0$$
$$\text{tr}(\gamma^\mu \gamma^\nu \gamma^\rho \gamma^\sigma \gamma^5) = -4i\epsilon^{\mu\nu\rho\sigma}$$

Another identity allows one to reverse the order of γ matrices inside a trace:

$$\text{tr}(\gamma^\mu \gamma^\nu \gamma^\rho \gamma^\sigma \cdots) = \text{tr}(\cdots \gamma^\sigma \gamma^\rho \gamma^\nu \gamma^\mu). \qquad \text{(A.28)}$$

Contractions of γ matrices with each other simplify to:

$$\gamma^\mu \gamma_\mu = 4$$
$$\gamma^\mu \gamma^\nu \gamma_\mu = -2\gamma^\nu$$
$$\gamma^\mu \gamma^\nu \gamma^\rho \gamma_\mu = 4g^{\nu\rho} \qquad \text{(A.29)}$$
$$\gamma^\mu \gamma^\nu \gamma^\rho \gamma^\sigma \gamma_\mu = -2\gamma^\sigma \gamma^\rho \gamma^\nu$$

(These identities apply in four dimensions only; see the following section.) Contractions of the ϵ symbol can also be simplified:

$$\epsilon^{\alpha\beta\gamma\delta}\epsilon_{\alpha\beta\gamma\delta} = -24$$
$$\epsilon^{\alpha\beta\gamma\mu}\epsilon_{\alpha\beta\gamma\nu} = -6\delta^\mu_\nu \qquad \text{(A.30)}$$
$$\epsilon^{\alpha\beta\mu\nu}\epsilon_{\alpha\beta\rho\sigma} = -2(\delta^\mu_\rho \delta^\nu_\sigma - \delta^\mu_\sigma \delta^\nu_\rho)$$

In some calculations, it is useful to rearrange products of fermion bilinears by means of *Fierz identities*. Let $u_1, \ldots, u_4$ be Dirac spinors, and let $u_{iL} = \frac{1}{2}(1 - \gamma^5)u_i$ be the left-handed projection. Then the most important Fierz rearrangement formula is

$$(\bar{u}_{1L}\gamma^\mu u_{2L})(\bar{u}_{3L}\gamma_\mu u_{4L}) = -(\bar{u}_{1L}\gamma^\mu u_{4L})(\bar{u}_{3L}\gamma_\mu u_{2L}). \qquad \text{(A.31)}$$

Additional formulae can be generated by the use of the following identities for the 2×2 blocks of Dirac matrices:

$$(\sigma^\mu)_{\alpha\beta}(\sigma_\mu)_{\gamma\delta} = 2\epsilon_{\alpha\gamma}\epsilon_{\beta\delta}; \qquad (\bar{\sigma}^\mu)_{\alpha\beta}(\bar{\sigma}_\mu)_{\gamma\delta} = 2\epsilon_{\alpha\gamma}\epsilon_{\beta\delta}. \qquad \text{(A.32)}$$

In non-Abelian gauge theories, the Feynman rules involve gauge group matrices t^a that form a representation r of a Lie algebra G. The symbol G also denotes the adjoint representation of the algebra. The matrices t^a obey

$$[t^a, t^b] = if^{abc}t^c, \qquad \text{(A.33)}$$

where the structure constants f^{abc} are totally antisymmetric. The invariants $C(r)$ and $C_2(r)$ of the representation r are defined by

$$\text{tr}[t^a t^b] = C(r)\delta^{ab}, \qquad t^a t^a = C_2(r)\cdot\mathbf{1}. \tag{A.34}$$

These are related by

$$C(r) = \frac{d(r)}{d(G)} C_2(r), \tag{A.35}$$

where $d(r)$ is the dimension of the representation. Traces and contractions of the t^a can be evaluated using the above identities and their consequences:

$$t^a t^b t^a = [C_2(r) - \tfrac{1}{2}C_2(G)]t^b$$
$$f^{acd}f^{bcd} = C_2(G)\delta^{ab} \tag{A.36}$$
$$f^{abc}t^b t^c = \tfrac{1}{2}iC_2(G)t^a$$

For $SU(N)$ groups, the fundamental representation is denoted by N, and we have

$$C(N) = \frac{1}{2}, \qquad C_2(N) = \frac{N^2-1}{2N}, \qquad C(G) = C_2(G) = N. \tag{A.37}$$

The following relation, satisfied by the matrices of the fundamental representation of $SU(N)$, is also very helpful:

$$(t^a)_{ij}(t^a)_{k\ell} = \frac{1}{2}\left(\delta_{i\ell}\delta_{kj} - \frac{1}{N}\delta_{ij}\delta_{k\ell}\right). \tag{A.38}$$

A.4 Loop Integrals and Dimensional Regularization

To combine propagator denominators, introduce integrals over Feynman parameters:

$$\frac{1}{A_1 A_2 \cdots A_n} = \int\limits_0^1 dx_1 \cdots dx_n\, \delta(\textstyle\sum x_i - 1) \frac{(n-1)!}{[x_1 A_1 + x_2 A_2 + \cdots x_n A_n]^n} \tag{A.39}$$

In the case of only two denominator factors, this formula reduces to

$$\frac{1}{AB} = \int\limits_0^1 dx\, \frac{1}{[xA + (1-x)B]^2}. \tag{A.40}$$

A more general formula in which the A_i are raised to arbitrary powers is given in Eq. (6.42).

Once this is done, the bracketed quantity in the denominator will be a quadratic function of the integration variables p_i^μ. Next, complete the square and shift the integration variables to absorb the terms linear in p_i^μ. For a one-loop integral, there is a single integration momentum p^μ, which is shifted to a momentum variable ℓ^μ. After this shift, the denominator takes the form

$(\ell^2 - \Delta)^n$. In the numerator, terms with an odd number of powers of ℓ vanish by symmetric integration. Symmetry also allows one to replace

$$\ell^\mu \ell^\nu \to \frac{1}{d} \ell^2 g^{\mu\nu}, \tag{A.41}$$

$$\ell^\mu \ell^\nu \ell^\rho \ell^\sigma \to \frac{1}{d(d+2)} (\ell^2)^2 \left(g^{\mu\nu} g^{\rho\sigma} + g^{\mu\rho} g^{\nu\sigma} + g^{\mu\sigma} g^{\nu\rho} \right). \tag{A.42}$$

(Here d is the spacetime dimension.) The integral is most conveniently evaluated after Wick-rotating to Euclidean space, with the substitution

$$\ell^0 = i\ell_E^0, \qquad \ell^2 = -\ell_E^2. \tag{A.43}$$

Alternatively, one can use the following table of d-dimensional integrals in Minkowski space:

$$\int \frac{d^d\ell}{(2\pi)^d} \frac{1}{(\ell^2 - \Delta)^n} = \frac{(-1)^n i}{(4\pi)^{d/2}} \frac{\Gamma(n - \frac{d}{2})}{\Gamma(n)} \left(\frac{1}{\Delta} \right)^{n - \frac{d}{2}} \tag{A.44}$$

$$\int \frac{d^d\ell}{(2\pi)^d} \frac{\ell^2}{(\ell^2 - \Delta)^n} = \frac{(-1)^{n-1} i}{(4\pi)^{d/2}} \frac{d}{2} \frac{\Gamma(n - \frac{d}{2} - 1)}{\Gamma(n)} \left(\frac{1}{\Delta} \right)^{n - \frac{d}{2} - 1} \tag{A.45}$$

$$\int \frac{d^d\ell}{(2\pi)^d} \frac{\ell^\mu \ell^\nu}{(\ell^2 - \Delta)^n} = \frac{(-1)^{n-1} i}{(4\pi)^{d/2}} \frac{g^{\mu\nu}}{2} \frac{\Gamma(n - \frac{d}{2} - 1)}{\Gamma(n)} \left(\frac{1}{\Delta} \right)^{n - \frac{d}{2} - 1} \tag{A.46}$$

$$\int \frac{d^d\ell}{(2\pi)^d} \frac{(\ell^2)^2}{(\ell^2 - \Delta)^n} = \frac{(-1)^n i}{(4\pi)^{d/2}} \frac{d(d+2)}{4} \frac{\Gamma(n - \frac{d}{2} - 2)}{\Gamma(n)} \left(\frac{1}{\Delta} \right)^{n - \frac{d}{2} - 2} \tag{A.47}$$

$$\int \frac{d^d\ell}{(2\pi)^d} \frac{\ell^\mu \ell^\nu \ell^\rho \ell^\sigma}{(\ell^2 - \Delta)^n} = \frac{(-1)^n i}{(4\pi)^{d/2}} \frac{\Gamma(n - \frac{d}{2} - 2)}{\Gamma(n)} \left(\frac{1}{\Delta} \right)^{n - \frac{d}{2} - 2}$$
$$\times \frac{1}{4} \left(g^{\mu\nu} g^{\rho\sigma} + g^{\mu\rho} g^{\nu\sigma} + g^{\mu\sigma} g^{\nu\rho} \right) \tag{A.48}$$

If the integral converges, one can set $d = 4$ from the start. If the integral diverges, the behavior near $d = 4$ can be extracted by expanding

$$\left(\frac{1}{\Delta} \right)^{2 - \frac{d}{2}} = 1 - (2 - \frac{d}{2}) \log \Delta + \cdots. \tag{A.49}$$

One also needs the expansion of $\Gamma(x)$ near its poles:

$$\Gamma(x) = \frac{1}{x} - \gamma + \mathcal{O}(x) \tag{A.50}$$

near $x = 0$, and

$$\Gamma(x) = \frac{(-1)^n}{n!} \left(\frac{1}{x+n} - \gamma + 1 + \cdots + \frac{1}{n} + \mathcal{O}(x+n) \right) \tag{A.51}$$

near $x = -n$. Here γ is the Euler-Mascheroni constant, $\gamma \approx 0.5772$. The following combination of terms often appears in calculations:

$$\frac{\Gamma(2-\frac{d}{2})}{(4\pi)^{d/2}} \left(\frac{1}{\Delta}\right)^{2-\frac{d}{2}} = \frac{1}{(4\pi)^2}\left(\frac{2}{\epsilon} - \log\Delta - \gamma + \log(4\pi) + \mathcal{O}(\epsilon)\right), \quad (A.52)$$

with $\epsilon = 4 - d$.

Notice that Δ is positive if it is a combination of masses and *spacelike* momentum invariants. If Δ contains timelike momenta, it may become negative. Then these integrals acquire imaginary parts, which give the discontinuities of S-matrix elements. To compute the S-matrix in a physical region, choose the correct branch of the function by the prescription

$$\left(\frac{1}{\Delta}\right)^{n-\frac{d}{2}} \rightarrow \left(\frac{1}{\Delta - i\epsilon}\right)^{n-\frac{d}{2}}, \quad (A.53)$$

where $-i\epsilon$ (not to be confused with ϵ in the previous paragraph!) gives a tiny negative imaginary part.

Traces in Eq. (A.27) that do not involve γ^5 are independent of dimensionality. However, since

$$g^{\mu\nu} g_{\mu\nu} = \delta^\mu_{\;\mu} = d \quad (A.54)$$

in d dimensions, the contraction identities (A.29) are modified:

$$\begin{aligned}
\gamma^\mu \gamma_\mu &= d \\
\gamma^\mu \gamma^\nu \gamma_\mu &= -(d-2)\gamma^\nu \\
\gamma^\mu \gamma^\nu \gamma^\rho \gamma_\mu &= 4g^{\nu\rho} - (4-d)\gamma^\nu \gamma^\rho \\
\gamma^\mu \gamma^\nu \gamma^\rho \gamma^\sigma \gamma_\mu &= -2\gamma^\sigma \gamma^\rho \gamma^\nu + (4-d)\gamma^\nu \gamma^\rho \gamma^\sigma
\end{aligned} \quad (A.55)$$

A.5 Cross Sections and Decay Rates

Once the squared matrix element for a scattering process is known, the differential cross section is given by

$$d\sigma = \frac{1}{2E_A 2E_B |v_A - v_B|} \left(\prod_f \frac{d^3 p_f}{(2\pi)^3} \frac{1}{2E_f}\right)$$

$$\times \left|\mathcal{M}(p_A, p_B \rightarrow \{p_f\})\right|^2 (2\pi)^4 \delta^{(4)}(p_A + p_B - \textstyle\sum p_f). \quad (A.56)$$

The differential decay rate of an unstable particle to a given final state is

$$d\Gamma = \frac{1}{2m_A}\left(\prod_f \frac{d^3 p_f}{(2\pi)^3}\frac{1}{2E_f}\right)\left|\mathcal{M}(m_A \rightarrow \{p_f\})\right|^2 (2\pi)^4 \delta^{(4)}(p_A - \textstyle\sum p_f). \quad (A.57)$$

For the special case of a two-particle final state, the Lorentz-invariant phase space takes the simple form

$$\left(\prod_f \int \frac{d^3 p_f}{(2\pi)^3}\frac{1}{2E_f}\right)(2\pi)^4 \delta^{(4)}(\textstyle\sum p_i - \textstyle\sum p_f) = \int \frac{d\Omega_{\rm cm}}{4\pi}\frac{1}{8\pi}\left(\frac{2|\mathbf{p}|}{E_{\rm cm}}\right), \quad (A.58)$$

where $|\mathbf{p}|$ is the magnitude of the 3-momentum of either particle in the center-of-mass frame.

A.6 Physical Constants and Conversion Factors

Precisely known physical constants:

$$c = 2.998 \times 10^{10} \text{ cm/s}$$
$$\hbar = 6.582 \times 10^{-22} \text{ MeV s}$$
$$e = -1.602 \times 10^{-19} \text{ C}$$
$$\alpha = \frac{e^2}{4\pi\hbar c} = \frac{1}{137.04} = 0.00730$$
$$\frac{G_F}{(\hbar c)^3} = 1.166 \times 10^{-5} \text{ GeV}^{-2}$$

The values of the strong and weak interaction coupling constants depend on the conventions used to define them, as explained in Sections 17.6 and 21.3. For the purpose of estimation, however, one can use the following values:

$$\alpha_s(10 \text{ GeV}) = 0.18$$
$$\alpha_s(m_Z) = 0.12$$
$$\sin^2\theta_w = 0.23$$

Particle masses (times c^2):

e :	0.5110 MeV		p :	938.3 MeV
μ :	105.6 MeV		n :	939.6 MeV
τ :	1777 MeV		$\pi^\pm$:	139.6 MeV
$W^\pm$:	80.2 GeV		π^0 :	135.0 MeV
Z^0 :	91.19 GeV			

Useful combinations:

Bohr radius: $\quad a_0 = \dfrac{\hbar}{\alpha m_e c} = 5.292 \times 10^{-9} \text{ cm}$

electron Compton wavelength: $\quad \lambdabar = \dfrac{\hbar}{m_e c} = 3.862 \times 10^{-11} \text{ cm}$

classical electron radius: $\quad r_e = \dfrac{\alpha\hbar}{m_e c} = 2.818 \times 10^{-13} \text{ cm}$

Thomson cross section: $\quad \sigma_T = \dfrac{8\pi r_e^2}{3} = 0.6652 \text{ barn}$

annihilation cross section: $\quad 1 \text{ R} = \dfrac{4\pi\alpha^2}{3E_{\text{cm}}^2} = \dfrac{86.8 \text{ nbarn}}{(E_{\text{cm}} \text{ in GeV})^2}$

Conversion factors:

$$(1 \text{ GeV})/c^2 = 1.783 \times 10^{-24} \text{ g}$$
$$(1 \text{ GeV})^{-1}(\hbar c) = 0.1973 \times 10^{-13} \text{ cm} = 0.1973 \text{ fm};$$
$$(1 \text{ GeV})^{-2}(\hbar c)^2 = 0.3894 \times 10^{-27} \text{ cm}^2 = 0.3894 \text{ mbarn}$$
$$1 \text{ barn} = 10^{-24} \text{ cm}^2$$
$$(1 \text{ volt/meter})(e\hbar c) = 1.973 \times 10^{-25} \text{ GeV}^2$$
$$(1 \text{ tesla})(e\hbar c^2) = 5.916 \times 10^{-17} \text{ GeV}^2$$

A complete, up-to-date tabulation of the fundamental constants and the properties of elementary particles is given in the *Review of Particle Properties*, which can be found in a recent issue of either *Physical Review* **D** or *Physics Letters* **B**. The most recent *Review* as of this writing is published in *Physical Review* **D50**, 1173 (1994).

Bibliography

Mathematical Background and Reference

Cahn, Robert N., *Semi-Simple Lie Algebras and Their Representations*, Benjamin/Cummings, Menlo Park, California, 1984. Written by a physicist, for physicists.

Carrier, George F., Krook, Max, and Pearson, Carl E., *Functions of a Complex Variable*, McGraw-Hill, New York, 1966. An excellent practical introduction to methods of complex variables and contour integration.

Gradshteyn, I. S. and Ryzhik, I. M., *Table of Integrals, Series, and Products* (trans. and ed. by Alan Jeffrey), Academic Press, Orlando, Florida, 1980.

Physics Background

Baym, Gordon, *Lectures on Quantum Mechanics*, Benjamin/Cummings, Menlo Park, California, 1969. A concise, informal text that is especially rich in nontrivial applications.

Fetter, Alexander L. and Walecka, John Dirk, *Theoretical Mechanics of Particles and Continua*, McGraw-Hill, New York, 1980. Includes several chapters on continuum mechanics and useful appendices of mathematical reference material.

Feynman, Richard P., *QED: The Strange Theory of Light and Matter*, Princeton University Press, Princeton, New Jersey, 1985. A transcription of four lectures given to a general audience, presenting Feynman's approach to quantum mechanics, including Feynman diagrams. Highly recommended.

Feynman, Richard P. and Hibbs, A. R., *Quantum Mechanics and Path Integrals*, McGraw-Hill, New York, 1965. An introduction to the use of path integrals in nonrelativistic quantum mechanics.

Goldstein, Herbert, *Classical Mechanics* (second edition), Addison-Wesley, Reading, Massachusetts, 1980. Chapter 12 introduces classical relativistic field theory.

Jackson, J. D., *Classical Electrodynamics* (second edition), Wiley, New York, 1975.

Landau, L. D. and Lifshitz, E. M., *The Classical Theory of Fields* (fourth revised English edition, trans. Morton Hamermesh), Pergamon, Oxford, 1975. Contains a succinct development of electromagnetic theory from the Lagrangian viewpoint.

Landau, L. D. and Lifshitz, E. M., *Statistical Physics* (third edition, Part 1, trans. J. B. Sykes and M. J. Kearsley), Pergamon Press, 1980. An insightful if concise textbook of statistical mechanics, containing the original pedagogical exposition of Landau's theory of phase transitions.

Reichl, L. E., *A Modern Course in Statistical Physics*, University of Texas Press, Austin, 1980. A complete textbook of statistical mechanics.

Schiff, Leonard I., *Quantum Mechanics* (third edition), McGraw-Hill, New York, 1968.

Shankar, Ramamurti, *Principles of Quantum Mechanics*, Plenum, New York, 1980. A very clear presentation of the basic theory.

Taylor, Edwin F. and Wheeler, John Archibald, *Spacetime Physics* (second edition), Freeman, New York, 1992. An elementary but insightful introduction to special relativity.

Taylor, John R., *Scattering Theory*, Robert E. Krieger, Malabar, Florida, 1983 (reprint of original edition published by Wiley, New York, 1972). A very clear development of scattering theory for nonrelativistic quantum mechanics.

Relativistic Quantum Mechanics and Field Theory

Bailin, D. and Love, A., *Introduction to Gauge Field Theory* (revised edition), Institute of Physics Publishing, Bristol, 1993. Develops the theory entirely from the path integral viewpoint.

Balian, Roger and Zinn-Justin, Jean (eds.), *Methods in Field Theory*, North-Holland, Amsterdam, 1976. Proceedings of the 1975 Les Houches Summer School in Theoretical Physics, including lectures on functional methods, renormalization, and gauge theories.

Berestetskii, V. B., Lifshitz, E. M., and Pitaevskii, L. P., *Quantum Electrodynamics* (second edition, trans. J. B. Sykes and J. S. Bell), Pergamon, Oxford, 1982. An excellent reference for QED applications.

Bjorken, James D. and Drell, Sidney D., *Relativistic Quantum Mechanics*, McGraw-Hill, New York, 1964. Develops Feynman diagrams using intuitive arguments, without using fields.

Bjorken, James D. and Drell, Sidney D., *Relativistic Quantum Fields*, McGraw-Hill, New York, 1965. Redevelops Feynman diagrams from the field viewpoint, using canonical quantization.

Brown, Lowell S., *Quantum Field Theory*, Cambridge University Press, New York, 1992. A careful treatment of the foundations of quantum field theory and its application to scattering processes.

Coleman, Sidney, *Aspects of Symmetry*, Cambridge University Press, Cambridge, 1985. Informal lectures on a number of topics involving gauge theories and symmetry, given between 1966 and 1979.

Collins, John, *Renormalization*, Cambridge University Press, Cambridge, 1984. A careful development of the technical machinery needed for all-orders proofs of renormalizability, operator product expansion, and factorization theorems.

Deser, Stanley, Grisaru, Marc, and Pendleton, Hugh, *Lectures on Elementary Particles and Quantum Field Theory*, MIT Press, Cambridge, 1970, vol. 1. Four extremely useful summer school lectures.

Gross, Franz, *Relativistic Quantum Mechanics and Field Theory*, Wiley, New York, 1993. Includes a number of topics in "advanced quantum mechanics", and an introductory chapter on bound states.

Itzykson, Claude and Zuber, Jean-Bernard, *Quantum Field Theory*, McGraw-Hill, New York, 1980. A comprehensive textbook.

Jauch, J. M. and Rohrlich, F., *The Theory of Photons and Electrons* (second edition), Springer-Verlag, Berlin, 1976. An authoritative monograph on QED.

Kaku, Michio, *Quantum Field Theory: A Modern Introduction*, Oxford University Press, New York, 1993. Contains brief introductory chapters on a number of advanced topics.

Kinoshita, T., ed., *Quantum Electrodynamics*, World Scientific, Singapore, 1990. A collection of review articles on precision tests of QED.

Mandl, F. and Shaw, G., *Quantum Field Theory* (revised edition), Wiley, New York, 1993. The easiest book on field theory; introduces QED and some electroweak theory using canonical quantization.

Ramond, Pierre, *Field Theory: A Modern Primer* (second edition), Addison-Wesley, Redwood City, California, 1989. Contains very nice treatments of the Lorentz group, path integrals, ϕ^4 theory, and quantization of gauge theories.

Ryder, Lewis H., *Quantum Field Theory*, Cambridge University Press, Cambridge, 1985. A concise treatment of the more formal aspects of the subject.

Sakurai, J. J., *Advanced Quantum Mechanics*, Addison-Wesley, Reading, 1967. Develops Feynman diagrams without using fields.

Schweber, Silvan S., *QED and the Men Who Made It: Dyson, Feynman, Schwinger, and Tomonaga*, Princeton University Press, Princeton, 1994. An excellent history of the subject up to about 1950.

Schwinger, Julian (ed.), *Selected Papers on Quantum Electrodynamics*, Dover, New York, 1958. Reprints of important papers written between 1927 and 1953.

Sterman, George, *Introduction to Quantum Field Theory*, Cambridge University Press, Cambridge, 1993. An introductory textbook with special emphasis on the essentials of perturbative QCD.

Elementary Particle Physics

Aitchison, Ian J. R. and Hey, Anthony J. G., *Gauge Theories in Particle Physics* (second edition), Adam Hilger, Bristol, 1989. An elementary introduction to gauge theories, concentrating mostly on tree-level processes.

Barger, Vernon and Phillips, Roger J. N., *Collider Physics*, Addison-Wesley, Menlo Park, California, 1987. Basic discussion of the application of QCD to high-energy collider phenomenology.

Cahn, Robert N. and Goldhaber, Gerson, *The Experimental Foundations of Particle Physics*, Cambridge University Press, Cambridge, 1989. Reprints of many original papers, supplemented by introductory overviews, additional references, and exercises. Highly recommended.

Cheng, Ta-Pei and Li, Ling-Fong, *Gauge Theory of Elementary Particle Physics*, Oxford University Press, New York, 1984. An advanced, authoritative monograph.

Commins, Eugene D. and Bucksbaum, Philip H., *Weak Interactions of Leptons and Quarks*, Cambridge University Press, Cambridge, 1983. A thorough review of both theory and experiment.

Field, Richard D., *Applications of Perturbative QCD*, Benjamin/Cummings, Menlo Park, 1989. A useful description of the techniques needed for QCD calculations beyond the leading order.

Georgi, Howard, *Weak Interactions and Modern Particle Theory*, Benjamin/Cummings, Menlo Park, California, 1984. Advanced, insightful treatment of selected topics.

Griffiths, David, *Introduction to Elementary Particles*, Wiley, New York, 1987. A good undergraduate-level survey.

Halzen, Francis and Martin, Alan D., *Quarks and Leptons: An Introductory Course in Modern Particle Physics*, Wiley, New York, 1984. Uses Feynman diagrams throughout, and concentrates on gauge theories.

Perkins, Donald H., *Introduction to High Energy Physics* (third edition), Addison-Wesley, Menlo Park, California, 1987. A good introduction to phenomena, with relatively little emphasis on Feynman diagrams and gauge theories.

Quigg, Chris, *Gauge Theories of the Strong, Weak, and Electromagnetic Interactions*, Benjamin/Cummings, Menlo Park, California, 1983. A very nice overview of gauge theories and their experimental tests.

Ross, Graham G., *Grand Unified Theories*, Benjamin/Cummings, Menlo Park, California, 1984. A clear introduction to gauge theories that unify the interactions of particle physics.

Taylor, J. C., *Gauge Theories of Weak Interactions*, Cambridge University Press, Cambridge, 1976. A concise treatment of the standard model and related theoretical issues.

Condensed Matter Physics

Abrikosov, A. A., Gorkov, L. P., and Dzyaloshinskii, I. E., *Quantum Field Theoretical Methods in Statistical Physics* (second edition), Pergamon, Oxford, 1965. A classic, but very terse, exposition of the application of Feynman diagrams to condensed matter problems.

Anderson, P. W., *Basic Notions of Condensed Matter Physics*, Benjamin/ Cummings, Menlo Park, California, 1984. An informal overview of the concepts of broken symmetry and renormalization as applied to condensed matter systems.

Fetter, Alexander L. and Walecka, John Dirk, *Quantum Theory of Many-Particle Systems*, McGraw-Hill, New York, 1971. A straightfoward introduction to the use of Feynman diagrams in nuclear and condensed matter physics.

Ma, Shang-Keng, *Modern Theory of Critical Phenomena*, Benjamin/Cummings, 1976. An introduction to the use of renormalization group methods in the theory of critical phenomena.

Mattuck, Richard D., *A Guide to Feynman Diagrams in the Many-Body Problem*, McGraw-Hill, New York, 1967. A clear and easy introduction to the use of Feynman diagrams in solid state physics.

Parisi, Giorgio, *Statistical Field Theory*, Benjamin/Cummings, 1988. Far-ranging applications of ideas from quantum field theory to problems in statistical mechanics.

Stanley, H. Eugene, *Introduction to Phase Transitions and Critical Phenomena*, Oxford University Press, Oxford, 1971.

Zinn-Justin, Jean, *Quantum Field Theory and Critical Phenomena* (third edition), Oxford University Press, Oxford, 1996. A treatise on the application of renormalization theory to the study of critical phenomena.

Corrections to This Book

A list of misprints and corrections to this book is posted on the World-Wide Web at the URL 'http://www.slac.stanford.edu/~mpeskin/QFT.html', or can be obtained by writing to the authors. We would be grateful if you would report additional errors in the book, or send other comments, to mpeskin@slac.stanford.edu or to dschroeder@cc.weber.edu.

Index

A_{LR}^f (polarization asymmetry at Z^0), 710–712, 728, 760–763, 771–772

Abelian group, 487, 496

Abelian Higgs model, 690–692, 732–739

Accelerated track, xiv

Acknowledgments, xvii

Action, 15, 277

Active transformations, 35–37

Ad hoc subtraction, 195, 221–222, 230

Adjoint representation, 499, 501–502

Adler-Bell-Jackiw anomaly, 661–667, 672–676, 686, 688, 705–707
operator equation, 661

ALEPH experiment, 711

Alloys, binary, 267–268, 272, 437

α (critical exponent), 441, 446–447

α (fine-structure constant), xxi, 125, 809
measurements of, 197–198
radiative corrections to, 252–257, 335, 424, 762–764

α_* ($\alpha(m_Z)$), 759

α_s, 260–262, 550, 573, 712
measurements of, 593–595
running of, 551–553, 594–595

α^i (Dirac matrices), 52

Altarelli-Parisi equations, xvi, 590–593, 597–598, 644–647, 649
splitting functions for QED, 587
splitting functions for QCD, 590
initial conditions, 587, 591

Ambiguity in choice of momentum scale, 552, 573

Amplitude, in quantum mechanics, 276

Amplitude ($\mathcal{M}$), see Scattering amplitude

Amputated diagrams, 109, 113–114, 175, 227–229, 331

Analogies to statistical mechanics, 266–275, 292–294, 364, 367 (table), 395, 440–442, 788

Analytic continuation to Euclidean space, 293

Analytic properties of amplitudes, 211–237, 252–253, 619–621, 633, 808

Angular momentum, 6–7, 38–39, 60, 144, 147

Annihilation, see e^+e^- annihilation, Cross section

Annihilation operators, 26, 58, 123

Anomalous dimensions, 427, 447, 613–615, see also γ

Anomalous magnetic moment, 188, 210
of electron, 196–198
of muon, 197–198, 773

Anomaly, see Axial vector anomaly, Adler-Bell-Jackiw anomaly, Trace anomaly

Anomaly cancellation, 705–707, 799

Anomaly coefficient, 680–681, 687–688

Anomaly-free theories, 679–681, 786

Anticommutation relations, Dirac, 40, 56

Anticommuting numbers, 73, 298–302, 518

Antifermions, 59, 61, 804

Antiferromagnets, 449–450

Antilinear operator, 67

Antiparticles, 4, 14, 29

Antiquarks, see Quarks, Parton distribution functions

Antiscreening, 542–543

Printed in the United States
by Baker & Taylor Publisher Services

Printed in the United States
by Baker & Taylor Publisher Services